CLINICAL EXERCISE PHYSIOLOGY

SECOND EDITION

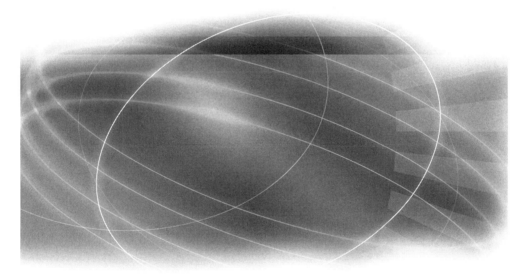

Jonathan K. Ehrman, PhD, FACSM
Henry Ford Hospital

Paul M. Gordon, PhD, MPH, FACSM
University of Michigan

Paul S. Visich, PhD, MPH
Central Michigan University

Steven J. Keteyian, PhD, FACSM
Henry Ford Hospital

Editors

Human Kinetics

Library of Congress Cataloging-in-Publication Data

Clinical exercise physiology / Jonathan K. Ehrman . . . [et al.], editors. -- 2nd ed.
   p. ; cm.
  Includes bibliographical references and index.
  ISBN-13: 978-0-7360-6565-8 (hard cover)
  ISBN-10: 0-7360-6565-2 (hard cover)
  1. Exercise therapy.   2. Clinical exercise physiology.   I. Ehrman, Jonathan K., 1962-
  [DNLM: 1. Exercise Therapy--Case Reports.   2. Exercise--physiology--Case Reports.
  WB 541 C641 2009]
  RM725.C582 2009
  615.8'2--dc22

                   2008022748

ISBN-10: 0-7360-6565-2
ISBN-13: 978-0-7360-6565-8

The Web addresses cited in this text were current as of April 30, 2008, unless otherwise noted.

Acquisitions Editor: Michael S. Bahrke, PhD; **Developmental Editors:** Maureen Eckstein and Amanda Ewing; **Assistant Editors:** Christine Bryant Cohen and Kyle G. Fritz; **Copyeditor:** Bob Replinger; **Proofreader:** Kathy Bennett; **Indexer:** Craig Brown; **Permission Manager:** Dalene Reeder; **Graphic Designer:** Joe Buck; **Graphic Artist:** Kathleen Boudreau-Fuoss; **Cover Designer:** Bob Reuther; **Photo Asset Manager:** Laura Fitch; **Photo Office Assistant:** Jason Allen; **Art Manager:** Kelly Hendren; **Associate Art Manager:** Alan L. Wilborn; **Illustrators:** Jennifer Gibas and Tammy Page; **Printer:** Sheridan Books

Printed in the United States of America   10   9

The paper in this book is certified under a sustainable forestry program.

Human Kinetics
Web site: www.HumanKinetics.com

United States: Human Kinetics, P.O. Box 5076, Champaign, IL 61825-5076
800-747-4457
email: humank@hkusa.com

Canada: Human Kinetics, 475 Devonshire Road Unit 100, Windsor, ON N8Y 2L5
800-465-7301 (in Canada only)
email: info@hkcanada.com

Europe: Human Kinetics, 107 Bradford Road, Stanningley, Leeds LS28 6 AT, United Kingdom
+44 (0) 113 255 5665
email: hk@hkeurope.com

Australia: Human Kinetics, 57A Price Avenue, Lower Mitcham, South Australia 5062
08 8372 0999
e-mail: info@hkaustralia.com

New Zealand: Human Kinetics, P.O. Box 80, Torrens Park, South Australia 5062
0800 222 062
e-mail: info@hknewzealand.com

CONTENTS

PART I   Introduction to Clinical Exercise Physiology   1

Chapter 1   Introduction   3

Jonathan K. Ehrman, PhD, FACSM, Paul M. Gordon, PhD, MPH, FACSM,
  Paul S. Visich, PhD, MPH, and Steven J. Keteyian, PhD, FACSM

Chapter 2   Behavioral Approaches to Physical Activity Promotion   17

Gregory W. Heath, DHSc, MPH

Chapter 3   Pharmacotherapy   31

Diana Kostoff, PharmD, BCPS

Chapter 4   General Interview and Examination Skills   61

Steven J. Keteyian, PhD, FACSM

Chapter 5   Graded Exercise Testing and Exercise Prescription   77

Paul S. Visich, PhD, MPH, and Jonathan K. Ehrman, PhD, FACSM

Chapter 32 Stroke   583

Christopher J. Womack, PhD, FACSM

FOREWORD

The idea that exercise and physical activity improve health and well-being is well appreciated by public health officials, clinicians, and the public. What may be less well understood are its far-reaching benefits. Specifically, exercise not only aids in primordial prevention (preventing the development of disease-specific risk factors) but also promotes primary prevention (preventing the first occurrence of a disease-specific event) and secondary prevention (preventing the reoccurrence of a disease-related event). It is this latter form of prevention, secondary prevention, to which much of *Clinical Exercise Physiology* is devoted. Although we often appreciate the important role that exercise serves in the prevention and treatment of cardiovascular diseases, the favorable influence that it imparts across a broad spectrum of chronic diseases is made obvious in this book.

The foundational chapters in part I provide information on which subsequent chapters draw. The distinguished group of editors—Ehrman, Gordon, Keteyian, and Visich—and their expert contributors provide 27 additional population- or disease-specific chapters, each with up-to-date information regarding the role, methods, and expectations of how to assess, train, and prescribe exercise in patients with clinically manifest disease. Additionally, for both the student and the clinical exercise professional new to the field, *Clinical Exercise Physiology* provides a well-balanced, uniform, and evidence-based summary of the epidemiology, pathology, and treatment strategies associated with each disease.

No longer can we overlook the important role that exercise testing and training play in determining future risk and improving outcomes. Whether we care for patients with cancer, diabetes, cystic fibrosis, or arthritis, the symptoms or chief complaints of loss of physical function, the early onset of fatigue, and progressive exercise intolerance are common to essentially all patients who have a chronic disease. These concerns position exercise, with its long-standing history of improving fitness and physical function, as a key adjunctive treatment able to improve not only exercise tolerance but in many instances quality of life and clinical outcomes as well.

Another important but potentially less recognizable contribution to come about from this second edition of *Clinical Exercise Physiology* is its role in further linking and advancing clinical exercise physiology as a profession. Following the standardization of the graduate curriculum for clinical exercise physiology in 2004, the creation in 2008 of a professional member association designated solely for clinical exercise physiologists (i.e., Clinical Exercise Physiology Association), and the ongoing offering of clinical certifications through the American College of Sports Medicine (ACSM), this book represents the definitive body of work and a repository of up-to-date and comprehensive clinical exercise information from which students and practitioners can draw. Clearly, with this second edition of *Clinical Exercise Physiology*, editors Ehrman, Gordon, Keteyian, and Visich have again done much to advance clinical exercise physiology as an allied health profession.

Whether you are studying in graduate school in preparation to enter the field, preparing to take an ACSM certification examination, or are now working in the field and desire to maintain your skills and learn new ones, you will likely find that this book meets your needs. And if you are like me, you will quickly come to appreciate the far-reaching and beneficial features and effects that exercise offers to a variety of patients who have a chronic disease or disability.

Philip A. Ades, MD
Professor of Medicine, Division of Cardiology
Director, Cardiac Rehabilitation and Preventive
    Cardiology
University of Vermont College of Medicine
Burlington, VT

CONTRIBUTORS

Ann L. Albright, PhD, RD
Director, Division of Diabetes Translation
    Centers for Disease Control
    Atlanta, GA

Krista A. Barbour, PhD
    Department of Psychiatry and Behavioral Sciences
    Duke University Medical Center

Michael J. Berry, PhD
    Department of Health and Exercise Science
    Wake Forest University
    Winston-Salem, NC

Kyle D. Biggerstaff, PhD
Associate Professor, Department of Kinesiology
    Texas Woman's University
    Denton, TX

Julie Biller, MD
    Children's Hospital of Wisconsin
    Milwaukee, WI

Brian W. Carlin, MD
Assistant Professor of Medicine
    Drexel University College of Medicine
    Philadelphia, PA
    Allegheny General Hospital
    Pittsburgh, PA

Chad C. Carroll, PhD
    Human Performance Laboratory
    Ball State University
    Muncie, IN

A.S. Contractor, MD
Head, Dept. of Preventive Cardiology and
    Rehabilitation
    Asian Heart Institute
    Bombay, India

Michael J. Danduran, MS
    Children's Hospital of Wisconsin
    Milwaukee, WI

David Donley, MS
    Div. of Exercise Physiology
    West Virginia University
    Morgantown, WV

Jonathan K. Ehrman, PhD, FACSM
    Preventive Cardiology
    Henry Ford Hospital
    Detroit, MI

Emma Fletcher, MA
    School of Health Sciences
    Central Michigan University
    Mount Pleasant, MI

David R. Gater, Jr., MD, PhD
Chief, Spinal Cord Injury
    Hunter Holmes McGuire Veterans Affairs Medical
        Center
Professor, Physical Medicine and Rehabilitation
    Virginia Commonwealth University
    Richmond, VA

Neil F. Gordon, MD, PhD
Medical Director, INTERVENT USA, Inc.
    Savannah, GA

Paul M. Gordon, PhD, MPH, FACSM
Associate Professor and Director
    Laboratory for Physical Activity and Exercise
        Intervention Research
    University of Michigan
    Ann Arbor, MI

Gregory W. Heath, DHSc, MPH
    Department of Health and Human Performance
    University of Tennessee at Chattanooga
    Department of Medicine
    Chattanooga, TN

Kim M. Huffman, MD, PhD
    Physical Medicine and Rehabilitation
    Veterans Affairs Medical Center
    Division of Rheumatology, Allergy, and Clinical
        Immunology
    Department of Medicine
    Duke University Medical Center
    Durham, NC

Steven J. Keteyian, PhD, FACSM
    Division of Cardiovascular Medicine
    Henry Ford Hospital
    Detroit, MI

Diana Kostoff, PharmD, BCPS
Clinical Pharmacist
  Henry Ford Health System
  Detroit, MI

Virginia B. Kraus, MD, PhD
Division of Rheumatology, Allergy, and Clinical
  Immunology
Associate Professor, Department of Medicine
  Duke University Medical Center
  Durham, NC

Charles P. Lambert, PhD
Research Assistant Professor, Division of Geriatrics
  and Nutritional Science
  Department of Internal Medicine
  Washington University School of Medicine
  St. Louis, MO

Richard M. Lampman, PhD
Director of Research, Department of Surgery
Associate Director of Research
  St. Joseph Mercy Hospital
  Ann Arbor, MI
Adjunct Associate Professor
  Department of Physical Medicine and
    Rehabilitation
  University of Michigan Medical School
  Ann Arbor, MI

Nicole Y.J.M. Leenders, PhD
  The Ohio State University
  Columbus, OH

Timothy R. McConnell, PhD
  Department of Exercise Science
  Bloomsburg University
  Bloomsburg, PA

David C. Murdy, MD
  Internal Medicine, Dean Medical Center
Clinical Instructor, University of Wisconsin
  Janesville, WI

David L. Nichols, PhD
Associate Professor
  Department of Kinesiology and Institute for
    Women's Health
  Texas Woman's University
  Denton, TX

Patricia Painter, PhD
Associate Professor
  Department of Medicine–Division of Renal
    Diseases/Hypertension
  University of Minnesota
  Minneapolis, MN

Jan Perkins, PhD
  School of Rehabilitation and Medical Sciences
  Central Michigan University
  Mt. Pleasant, MI

Farah A. Ramírez-Marrero, PhD, MSc
Professor, Department of Physical Education and
  Recreation
  University of Puerto Rico, Río Piedras Campus
  San Juan, Puerto Rico
KL2 Scholar, Mayo Clinic
  Rochester, Minnesota

James Raper, CFNP, DNS
  School of Nursing and School of Medicine
  University of Alabama at Birmingham
  Birmingham, AL

Amy E. Rauworth, MS
Associate Director, National Center on Physical
  Activity and Disability (www.ncpad.org)
  Department of Disability and Human Development
  University of Illinois at Chicago

James H. Rimmer, PhD
Director, National Center on Physical Activity and
  Disability (www.ncpad.org)
  Department of Disability and Human Development
Professor, University of Illinois at Chicago
  Chicago, IL

Michael Saag, MD
  School of Medicine
  University of Alabama at Birmingham
  Birmingham, AL

William Saltarelli, PhD
  School of Health Sciences
  Central Michigan University
  Mt. Pleasant, MI

John R. Schairer, DO
  Division of Cardiovascular Medicine
  Henry Ford Hospital
  Detroit, MI

Anil Singh, MD
Pulmonary and Critical Care Medicine Fellow
  Allegheny General Hospital
  Pittsburgh, PA

Karandeep Singh, MD
  Department of Cardiology
  Geisinger Medical Center
  Danville, PA

Barbara Smith, RN, PhD, FACSM
School of Nursing
University of Maryland, Baltimore
Baltimore, MD

Kerry J. Stewart, EdD
Professor of Medicine
Division of Cardiology
Johns Hopkins School of Medicine
Baltimore, MD

Ann M. Swank, PhD
Department of Health and Sport Sciences
Exercise Physiology Laboratory
University of Louisville
Louisville, KY

Chad D. Swank, MS
Physical Therapist
Baylor Institute for Rehabilitation
Dallas, TX

Paul S. Visich, PhD, MPH
School of Health Sciences
Central Michigan University
Mt. Pleasant, MI

Seth W. Wolk, MD
Vascular Surgeon
Department of Surgery
St. Joseph Mercy Hospital, Ann Arbor, MI
Adjunct Clinical Associate Professor, Department
of Surgery
University of Michigan Medical School
Ann Arbor, MI

Christopher J. Womack, PhD, FACSM
Associate Professor
James Madison University
Harrisonburg, VA

C. Mark Woodard, MS
Department of Health and Exercise Science
Wake Forest University
Winston-Salem, NC

Joshua S. Wooten, MS
Research Associate
Institute for Women's Health
Texas Woman's University
Denton, TX

J. Tim Zipple, DScPT, OCS
School of Rehabilitation and Medical Sciences
Central Michigan University
Mt. Pleasant, MI

PREFACE

Five years have passed since the first edition of *Clinical Exercise Physiology* was published, and much has occurred in both the field of practice and the science behind it. This text has quickly become the primary textbook for students preparing to become clinical exercise physiologists as well as an excellent resource for people preparing to sit for the American College of Sports Medicine Registered Clinical Exercise Physiologist (RCEP) examination. In 2003 clinical exercise physiology as a profession was just taking shape. Now it is blossoming into a profession that works within the healthcare system to deliver evidence-based care to a broad range of patients with chronic diseases. Although much important work remains, if two of the original purposes for this book were to disseminate the research associated with clinical exercise physiology and to provide a comprehensive resource for people working in the field, we are largely accomplishing what we set out to do.

Continued evidence of the expanding role of clinical exercise physiology in healthcare includes the conduct of an increasing amount of clinical exercise physiology–related research and the incorporation of the results from such research into evidence-based guidelines for the treatment of patients with a variety of diseases. In 2008 a professional membership organization (the Clinical Exercise Physiology Association) was established to serve practitioners in the field through advocacy, the offering of continuing education, and membership networking. Additionally, organizations such as the American College of Sports Medicine (ACSM), the American Council on Exercise (ACE), and the American Society of Exercise Physiologists (ASEP) continue to offer several relevant certification and registry examinations, helping to provide uniformity to the level of preparedness of those working in the field or aspiring to do so. Finally, between the late 1960s and early 1980s, numerous undergraduate- and graduate-level exercise science or physiology curriculums were developed in U.S. universities. The subspecialty of clinical exercise physiology emerged over time from such programs, such that in 2004 clinical exercise physiology was formally recognized by the Commission on the Accreditation of Allied Health Education Programs (CAAHEP) as a field of study that required a formal and uniform graduate-level curriculum.

Although the day-to-day duties of people working today as clinical exercise physiologists mostly involve patients with cardiovascular disease, clinical exercise physiologists are now more than ever helping to care for patients with cancer, musculoskeletal disorders, and metabolic diseases such as chronic kidney disease and diabetes. Therefore, the timing is right for this revised version of what we believe has become a staple in the preparation of students interested in clinical exercise physiology. This second edition of *Clinical Exercise Physiology* is, we believe, better organized than its predecessor was. The initial part of the book presents five foundational chapters, including an excellent review of the history of clinical exercise physiology, a description of the essentials of the physical examination, and a comprehensive review of the pharmacology that confronts the practicing clinical exercise physiologist. Part II focuses on the general populations of children, the elderly, females, and those with depression. The remaining chapters, the core of the book, cover specific diseases and conditions. The chapters are organized into six parts, each relating to a physiology system: metabolic, cardiovascular, pulmonary, immunologic and hematologic, orthopedic and musculoskeletal, and neuromuscular.

Based on reviews and feedback from students and professors, we kept the general construction of the disease-related chapters in parts III through VIII of this second edition of *Clinical Exercise Physiology* the same. Each of the chapters in these sections begins with an introduction to the specific disease that includes the definition and scope of the disease and a discussion of the relevant pathophysiology. A focus on the medical and clinical considerations follows, including signs and symptoms, diagnosis, exercise testing, and evidence-based treatment. Each chapter concludes with an overview of the exercise prescription for the disorder being discussed, with special emphasis placed on any unique disease-specific issues that might alter the exercise prescription.

Each chapter also contains several practical application boxes that provide additional information summarizing unique chapter-specific information. In each of the disease-specific chapters two of these practical application boxes focus on the exercise prescription and on practical information to consider when interacting with the patient. A third practical application box reviews the relevant exercise-training literature and discusses the physiological adaptations to exercise training and the ways that exercise can influence primary and secondary disease prevention. Finally, each chapter contains an actual patient case, progressing from initial presentation and diagnosis to therapy and exercise treatment. Each case study concludes with several questions aimed at facilitating group discussion in the classroom or for the individual learner to consider when preparing for the RCEP examination.

To keep abreast of trends in the field, four new chapters were added to cover metabolic syndrome, depression, stroke, and cerebral palsy. Additionally, each chapter has undergone a thorough revision to ensure that the material is consistent with current science and practice guidelines. The text also features a test package and presentation package for instructors—new to this edition and available online at www.HumanKinetics.com/ClinicalExercisePhysiology.

• **Test package.** Created with Respondus 2.0, the test package includes approximately 457 true-or-false and multiple-choice questions. With Respondus, instructors can create versions of their own tests by selecting from the question pool; select their own test forms and save them for later editing or printing; and export the tests into a word-processing program.

• **Image bank.** Includes approximately 380 PowerPoint slides containing tables, illustrations, and photos from the text that instructors can use for class discussion and illustration. The slides can be used directly within PowerPoint, or they can be printed to make transparencies or handouts for distribution to students. Instructors can easily add, modify, and rearrange the order of the slides. A blank PowerPoint template is also provided so instructors can create a personalized PowerPoint presentation.

Few, if any, graduate clinical exercise physiology programs in the United States currently provide students with the breadth of information required to sit for the ACSM RCEP examination. Those who plan to study to take this or any similar certification examination should understand that no single text provides in-depth coverage of all the clinical populations that benefit from physical activity and exercise. But this text may be as close as one can come. In all, *Clinical Exercise Physiology* addresses 27 different diseases and populations.

Besides serving as textbook for students studying in the field, *Clinical Exercise Physiology* is an excellent resource guide that the professional will want to have on her or his desk or office bookshelf. The features of consistent organization, case studies, discussion questions, up-to-date references, and feature boxes are designed to provide information required for effective study. In fact, the content was developed based on the KSAs (i.e., knowledge, skills, abilities) of the ACSM RCEP examination. We hope that this text serves as a valuable textbook for the student and as a useful desk reference for the practicing clinical exercise physiologist.

ACKNOWLEDGMENTS

My sincere thanks for this second edition of *Clinical Exercise Physiology* go to the many contributors from both the first and second editions. Thanks for your time and talent in pulling this together. And thanks to Drs. Gordon, Keteyian, and Visich for their collaboration. Finally, my family always has my deep appreciation for putting up with me and supporting me in my professional and "play" endeavors. Janel, Joshua, Jacob, Jared, and Johanna, I love you all very much.

Jonathan K. Ehrman

As we continue to see the value of exercise with various chronic diseases, I am excited to see how this textbook has become accepted in our field. Only through the expertise of many contributors has this book been possible. I thank each of them. As a professor for the past 15 years I am excited that this comprehensive book in clinical exercise physiology can be offered to colleagues and, most important, to students in the field of exercise physiology who are interested in improving the health of others. I continue to thank my wonderful wife, Diane, who is always supportive of my endeavors; my two sons, Matt and Tim; and my parents, Frank and Mary, who have always encouraged their children to give their best effort in whatever they choose to do (success = drive + intelligence).

Paul Visich

Many thanks to everyone at Human Kinetics who helped us prepare and bring life to this second edition of a textbook that we again believe will do much to advance and support the rapidly expanding practice and profession of clinical exercise physiology. Thanks as well to my coeditors, contributing authors, and reviewers for the tireless efforts put forth on behalf of this book. My never-ending thanks to W. Douglas Weaver, MD, chief, division of cardiovascular medicine at Henry Ford Hospital, for his continued support of my effort to contribute to my chosen profession. As always, many blessings to Lynette, Stephanie, Courtland, Jacob, and Aram for their support of my professional interests. Finally, in loving memory of Albert Z. and Virginia Keteyian.

Steven J. Keteyian

The clinical exercise physiology specialty is now taking shape, and knowledge in the area is rapidly expanding. We are blessed to have a distinguished group of authors who have worked diligently to impart state-of-the-art knowledge about their respective subspecialties. Many thanks go to them for their tireless efforts as well as to the reviewers and staff who often go unnoticed behind the scenes to assure an exceptional final product.

To my wonderful wife, Ina, and my children, Joshua, Natalie, and Liam—thanks for putting up with me and providing much love, support, and understanding. Finally, in loving memory of my father, Edwin W. Gordon III, who always believed in me.

Paul M. Gordon

Introduction to Clinical Exercise Physiology

As mentioned in the preface, although the day-to-day duties of people working in our field today mostly involve patients with cardiovascular disease, more than ever before clinical exercise physiologists now contribute to the evidence-based care provided to patients with cancer, musculoskeletal disorders, and metabolic diseases such as chronic kidney disease and diabetes. As a result, the chapters in this first part of Clinical Exercise Physiology not only review the rich history and expanding scope of our profession and the foundational knowledge that we use each day in the areas of exercise testing and exercise prescription but also address other key components that we must become proficient in if we are to contribute to the care of patients across a broad range of chronic diseases, a great number of which are addressed in this book. The other topics deserving of attention are behavioral approaches, pharmacology, and general evaluation and examination skills. A thorough reading of these chapters will provide the necessary background for the important work that lies ahead.

Chapter 1 is an introduction to clinical exercise physiology. With each passing year clinical exercise physiology as a profession deepens its roots, along with a variety of other allied health professions, into the delivery of evidence-based healthcare. This chapter reviews the history, current practice, and future directions for exercise physiology in disease prevention and management, as well as the various professional organizations that are helping to shape the field. Now leading the way is the Clinical Exercise Physiology Association. This organization has done much to ensure that our profession thrives and plays an important part in the care of patients with a variety of chronic diseases.

Chapter 2 centers on behavioral approaches. Although most of the material in this textbook and in the classroom pertains to the knowledge and physical skills needed to write safe exercise prescriptions, interpret exercise responses, and lead or supervise exercise, the fact remains that at end of the day, after we have written our prescriptions and measured and tested our patients, we are in the business of changing human behavior. This challenge differs little from the one that confronts the practicing physician or nurse. They, too, play an important role and should be relied on to help educate and motivate patients to take an active part in improving their own health. To that end, this chapter reviews and applies the various behavioral approaches known to help patients adopt long-term habits aimed at improving their health. The information presented here can be applied across demographic categories and disease conditions.

Chapter 3 focuses on pharmacotherapy. Those who want to help care for and understand the underlying pathophysiology of a particular condition should take the time to understand how and why a particular medication is being used to help treat it. Adopting this approach will help the clinical exercise physiologist understand the underlying disease process and improve his or her ability to develop a proper exercise prescription and conduct safe exercise evaluations and training sessions. Learning the essential clinical pharmacology principles taught in this chapter, along with the several common groups of medications that are covered, will go a long way toward preparing a person to work effectively in the field. Because new medications are often being evaluated for most chronic conditions, the same mind-set should be applied to learn about each new drug along the way.

Chapter 4 discusses general evaluation and examination skills. The purpose of this chapter is to identify the evaluation and examination skills needed to determine, on any given day, whether a patient can safely exercise. Rarely is a chronic disorder stagnant. Instead, diseases are dynamic. As a result, the clinical exercise physiologist must regularly inquire about or recognize signs, symptoms, and other evidence of possible disease progression. This approach applies whether a patient is being prepped for surgery or about to undergo a graded exercise stress test. Safety remains paramount. Although this chapter will help the clinical exercise physiologist learn about what questions to ask, what signs to look for, and how to interpret the evaluations performed by others, there is no substitute for taking additional time to practice newly acquired observational, history-taking, and examination skills on patients.

Chapter 5 looks at graded exercise testing and prescription. Like the other chapters in part I, this chapter reviews and provides foundational information needed to work through later chapters of the book. Fortunately, graded exercise testing and prescription are topics that most have likely had prior coursework in, freeing them up to begin to think about conducting an exercise test or writing an exercise prescription in a manner that integrates exercise physiology knowledge with the unique clinical issues germane to the chronic condition of the current patient. This approach will serve students well when they are someday working an 8 h shift in an exercise-testing lab, allowing them to integrate information about patients presenting with a variety of clinical conditions, using an approach that considers all aspects of a patient's condition and a physician-directed care plan.

Introduction

Jonathan K. Ehrman, PhD, FACSM

Paul M. Gordon, PhD, MPH, FACSM

Paul S. Visich, PhD, MPH

Steven J. Keteyian, PhD, FACSM

Since the previous edition of this textbook was published in 2002, much has changed in relation to professionalization of the field of clinical exercise physiology. This introduction briefly reviews issues relevant to this topic.

THE PAST, PRESENT, AND FUTURE OF CLINICAL EXERCISE PHYSIOLOGY

Clinical exercise physiology is a subspecialty of exercise physiology that investigates the relationship of exercise and chronic disease; the mechanisms and adaptation by which exercise influences the disease process; and the importance of exercise testing and training in the prevention, evaluation and treatment of these diseases. Someone who practices clinical exercise physiology (i.e., the clinical exercise physiologist) must be knowledgeable about the broad range of exercise responses that occur both within a disease class and across different chronic diseases. Although the response of various organ systems is usually the focus, important behavioral, psychosocial, and spiritual issues are usually present as well. A proficient clinical exercise physiologist must have a knowledge base in several fields:

- Anatomy
- Physiology (organ systems and exercise)
- Chemistry (organic and biochemistry)
- Biology (cellular and molecular)
- Psychology (behavioral medicine, counseling)

In addition, a clinical exercise physiologist must complete an undergraduate or graduate degree in the field, perform a clinical internship of up to 600 h (diagnostic and functional exercise testing; exercise assessment, prescription, and supervision; counseling; and education), and pass a certification examination that is accredited using standards set by an accrediting agency that is independent and outside of the organization offering the certification. See practical application 1.1 on p. 4 for more information.

The Past

The formal use of exercise in the assessment and treatment of chronic disease has existed for more than 40 yr, but exercise physiology can trace its roots back to the late 1800s (Fernand LaGrange's textbook *Physiology of Bodily Exercise* published in 1889) and the early 1900s (e.g., the Harvard Fatigue Laboratory). Although at that time much of the focus was on the physiological response to exercise in healthy and athletic populations, in the late 1930s Sid Robinson and colleagues

Scope of Practice

The following are definitions of the (clinical) exercise physiologist or specialist from several organizations. Careful reading of these definitions reveals that no clear consensus is in place regarding the title of the person who works with patients in an exercise or rehabilitative setting. Titles include exercise physiologist, exercise specialist, clinical exercise specialist, certified exercise physiologist, and registered exercise physiologist.

American College of Sports Medicine (ACSM)

The exercise specialist (ES) is a healthcare professional certified by the ACSM to deliver a variety of exercise assessment, training, rehabilitation, risk factor identification, and lifestyle management services to people with or at risk for cardiovascular, pulmonary, and metabolic diseases. These services are typically delivered in cardiovascular or pulmonary rehabilitation programs, physicians' offices, or medical fitness centers. The ACSM clinical exercise specialist is competent to provide exercise-related consulting for research, public health, and other clinical and nonclinical services and programs.

The ACSM registered clinical exercise physiologist (RCEP) provides exercise management for patients with a broad spectrum of chronic diseases or disabilities. This national registry for clinical exercise physiologists catalogues allied health professionals who work in the preventive or rehabilitative application of exercise and physical activity for populations in which exercise has been shown to provide a therapeutic or functional benefit.

American Society of Exercise Physiologists (ASEP)

ASEP answers the scope of practice in the following way:

Who is an exercise physiologist? An **exercise physiologist** *is a person who has an academic degree in exercise physiology or who is certified by ASEP to practice exercise physiology (through the exercise physiologist certified exam [EPC]), or who has a doctorate degree with an academic degree or emphasis in exercise physiology from an accredited college or university.*

American Council on Exercise (ACE)

As defined by ACE, the ACE–certified advanced health + fitness specialist demonstrates the ability to provide in-depth preventative and post-rehabilitative fitness programming that addresses common diseases and disorders seen on a daily basis.

Canadian Society of Exercise Physiology (CSEP)

The CSEP–certified exercise physiologist performs assessments/evaluations; prescribes conditioning exercise; and provides exercise supervision/monitoring, counseling, healthy lifestyle education, and outcome evaluation for "apparently healthy" individuals or populations with medical conditions, functional limitations, or disabilities through the application of physical activity and exercise, for the purpose of improving health, function, and work or sport performance.

Although some may consider that differences in the above definitions for the clinical exercise professional is simply a matter of semantics, such differences often lead to confusion among the public relative to which title or type of exercise professional they should seek when referred to or considering participation in a clinical exercise program. Additionally, among other health care professionals there may be confusion about the job titles and the duties of those who hold them. For instance, in many institutions someone who performs the technical duties in a noninvasive cardiology laboratory is often titled a cardiovascular technician (an occupation defined by the Department of Labor). But since the duties performed by the cardiovascular technician can overlap with the above presented descriptions of the (clinical) exercise physiologist, it is not uncommon for cardiovascular technicians to be hired into exercise physiologist-type positions.

Recently, a request for a new occupational definition was submitted to the Department of Labor because the previous definition of exercise physiologist, available as late as 1991, is no longer listed among the occupation titles cited by the Department of Labor (http://online.onetcenter.org). The term *exercise physiology* is used only in the description of the occupation of recreation and fitness studies teacher. That job description states that the occupation is responsible for teaching exercise physiology courses. The following information is a portion of a proposal for a new detailed occupation for the Standard Occupational Classification (SOC):

Occupational Title: Exercise Physiologist

Occupational definition: Exercise physiologists administer exercise and fitness evaluations and provide exercise programming, counseling, and education for the purpose of improving or maintaining physical function and fitness and general health.

Nature of the work: Exercise physiologists assess, plan, and implement evidence-based fitness programs that include exercise and physical activities designed to improve cardiorespiratory (aerobic) function, body composition, muscular strength, muscular endurance, and flexibility. Exercise physiologists provide exercise and physical activity services to healthy adults and children and to persons who have or are at risk of chronic diseases or medical conditions.

Work environment: The work environment of the exercise physiologist can include, but is not limited to health and fitness facilities, community fitness centers, corporate wellness programs, hospitals (inpatient and outpatient), medical clinics, medical fitness centers, university-based fitness programs, and/or university or professional athletic training facilities. Exercise physiologists with advanced training conduct physical activity and exercise research and publish peer-reviewed articles in scientific literature. Exercise physiologists with advanced training may also be employed by colleges and universities to teach undergraduate and graduate courses for students in health sciences, allied health, medicine, kinesiology, athletic training, and physical education.

studied the effects of the aging process on exercise performance. Outside of the U.S., other countries have also contributed greatly to the knowledge base of exercise physiology, notably the Scandinavian and other European countries in the 1950s and 1960s.

The development of the modern-day clinical exercise physiologist dates back to the 1960s, around the time when the term *aerobics* was popularized by Dr. Ken Cooper . . . a time when regular exercise was considered important to maintaining optimal health (6) and prolonged bed rest was found to be associated with marked loss of exercise tolerance (7). In addition, pioneers such as Herman Hellerstein demonstrated that bed rest was detrimental to people with heart disease as well. These findings gave way to the development of inpatient and, subsequently, outpatient cardiac rehabilitation programs throughout the 1970s and into the 1990s. An excellent, in-depth historical perspective of exercise physiology is provided by Wilmore, Costill, and Kenney (9).

The Present

Although published in 1996, *Physical Activity and Health: A Report of the Surgeon General* remains a landmark report

Diseases and Conditions Related to a Lack of Exercise

Cancer (colon, breast, reproductive)

Cardiovascular disease (cardiac, peripheral vascular)

Falling

Health-related quality of life

Non-insulin-dependent diabetes

Mood and mental health

Obesity

Osteoarthritis

Osteoporosis

Overall mortality

Adapted from the Surgeon General's Report on Physical Activity and Health, 1996.

for the field of clinical exercise physiology (8). Using an evidence-based approach, it identified numerous chronic diseases and conditions in which a lack of exercise or physical activity places people at greater risk (see "Diseases and Conditions Related to a Lack of Exercise" on p. 5).

Exercise limits disability and improves outcome for other diseases and conditions as well, including multiple sclerosis, spinal cord injury, chronic obstructive pulmonary disease, and many others. For these, exercise training should be part of a comprehensive treatment plan. Several population groups are at increased risk for developing a chronic disease or disability because of physical inactivity, as compared to other certain groups, including women, children, and people of certain races and ethnicities. Today, more than ever before, clinical exercise physiology is at the forefront of advances in clinical care and public policy directed at improving health and lowering future disease risk through regular exercise training.

Because of the surgeon general's report, published practice guidelines, position statements, and ongoing research, cardiac rehabilitation programs have evolved into sophisticated exercise training and behavioral management programs administered by multidisciplinary teams composed of clinical exercise physiologists, nurses, physicians, dietitians, physical therapists, psychologists, and others. In addition, many rehabilitation programs have expanded the focus of their "maintenance," or Phase 3 and 4 programs, to include patients with other chronic diseases known to respond well to regular exercise.

The Future

Expansion of employment opportunities for those with clinical exercise physiology credentials is currently evolving from working primarily in cardiac rehabilitation programs and teaching to now include other disease management programs in which exercise plays an important role. These conditions and diseases include obesity, peripheral arterial disease, diabetes, and cancer. Additionally, more professionals are gaining employment in clinical research trials at both the sponsor and site investigation locations. Noninvasive testing laboratories, including those specializing in cardiopulmonary exercise testing, often find the clinical exercise professional to be an asset. Finally, the emerging importance of physical activity and health and an increase in both professional and public awareness of the skills of clinical exercise physiologists have led to increasing employment in nonclinical settings. These settings include, but are not limited to, personal training, corporate fitness programs, medically affiliated fitness centers, professional and amateur sports consulting, and weight management programs.

Recent reports from the American College of Sports Medicine, American Physiological Society, American Association of Cardiovascular and Pulmonary Rehabilitation, and the Medical Fitness Association suggest that approximately 5,200 professionals in the United States consider themselves exercise physiologists.

The profession of clinical exercise physiology has a unique body of knowledge that includes exercise prescription development and implementation of both primary and secondary prevention services. This information has been published through the years in a growing number of biomedical journals that contribute to the clinical exercise physiology body of knowledge (see "Selected Biomedical Journals").

This body of knowledge has led to the emergence of evidence-based recommendations for both the public and patients with a chronic disease. In addition, the clinical exercise physiologist is uniquely trained to identify

Selected Biomedical Journals

ACSM's Health and Fitness Journal
American Journal of Cardiology
American Journal of Clinical Nutrition
American Journal of Physiology
American Journal of Sports Medicine
Annals of Internal Medicine
Canadian Journal of Applied Physiology
Circulation
European Journal of Applied Physiology
European Journal of Sport Science
International Journal of Obesity
Journal of Aging and Physical Activity
Journal of Applied Physiology
Journal of Cardiopulmonary Rehabilitation
JAMA, Journal of the American Medical Association
Medicine and Science in Sports and Exercise
New England Journal of Medicine
Pediatric Exercise Science
Sports Medicine
The Physician and Sportsmedicine
Research Quarterly for Exercise and Sport

individual lifestyle-related issues that promote poor health and to design and implement a behavior-based treatment plan aimed at modifying lifestyle factors. Thus, the profession of clinical exercise physiology and the clinical exercise physiologist fill a void in healthcare that is becoming increasingly important, especially in the United States, as the average age of the population rapidly increases (the first of the baby boom generation began retiring in 2007), resulting in an increase in the number of people living with chronic diseases throughout developed countries.

The involvement of exercise physiologists in a variety of settings has grown dramatically over the last 30 yr. Historically, most exercise physiologists were engaged in human performance–related research or academic instruction. But many now provide professional advice and service in clinical, preventive, and recreational fitness programs located in health clubs, corporate facilities, and hospital-based complexes. Exercise physiologists provide a number of important services, including fitness assessments or screenings, exercise testing and outcome assessments, exercise prescriptions or recommendations, exercise leadership, and exercise supervision. To date in the United States, only Louisiana has opted to license clinical exercise physiologists to sanction their delivery of specified and defined services to consumers. However, efforts are underway in other states such as Massachusetts, Minnesota, California, and Kentucky. Practical application 1.2 presents information about licensure of the exercise physiologist.

Practical Application 1.2

The Challenge of Licensure

Efforts to enact legislation that requires the licensure of exercise physiologists will likely be driven by evidence concerning consumer safety. Evidence justifying licensure must first be gathered and published and then be appreciated by some reasonable segments of both the public and state legislators. In large part, this effort will likely focus on whether such legislation is necessary to protect the public from harm because of services provided by exercise physiologists. Some healthcare provider groups (i.e., physical therapists) may well oppose licensing efforts for exercise physiologists, as they seek to protect what they believe to be their scope of practice. Although such opposition is expected and occurs, one must appreciate the fact that exercise physiologists can deliver relevant services and often do so in a more cost effective manner, as well as provide unique services (e.g., prescription of exercise in patients with known disease, evaluation of program effectiveness) that are outside the scope of other health care providers.

The lack of applicable state licensing requirements for exercise physiologists has, in some respects, muddied the waters for those who attempt to assess and define the practice roles for such professionals in the delivery of services to consumers. This confusion may be particularly true from the perspective of the legal system, where questions arise relative to a variety of concerns related to service delivery. Important legal questions include the following:

- What services may exercise physiologists lawfully provide given the absence or lack of licensure in almost every state?
- What practices performed by exercise physiologists may be prohibited as a matter of law because of state statutes regarding unauthorized practice of medicine?
- What practices performed by exercise physiologists may be prohibited as a matter of law because of unauthorized practice of other licensed healthcare services (e.g., physical therapy, nursing, dietetics)?
- What potential liabilities may exercise physiologists face when their delivery of service results in harm, injury, or death attributable to alleged negligence or malpractice?
- What recognition may be given to exercise physiologists and their opinions in a variety of legal settings (such as evaluating disability or working capacity in matters involving insurance, personal injury, or workers' compensation)?

PROFESSIONAL ORGANIZATIONS AND CERTIFICATIONS

Literally hundreds of exercise- and fitness-related certifications are available, but only a few exist for the purpose of certifying exercise professionals to work with people with chronic diseases. The primary professional organizations that provide clinically oriented exercise physiology certifications are the American Council on Exercise (ACE), the American College of Sports Medicine (ACSM), and the Canadian Society of Exercise Physiology (CSEP). The American Society for Exercise Physiologists (ASEP) offers a certification that combines assessment of both clinical and nonclinical content areas. Table 1.1 lists the aspects of each of these organizations' certifications, which are also reviewed in the next several paragraphs.

American Council on Exercise

The ACE has a long history of certifying exercise and fitness professionals, predominantly in the area of personal training (2). Nevertheless, with the foreseeable growth in the aging population and the rise in the prevalence of chronic diseases, particularly obesity, ACE has developed four certifying levels for the general public. Besides the personalized trainer certification, other available certifications include advanced health and fitness specialist, group fitness instructor, and lifestyle and weight management consultant. The ACE clinical advanced health and fitness specialist is most closely aligned with the clinical certifications offered by other credentialing bodies. This certification is designed to test individual competencies to work with chronic diseased populations, using a written examination either in a live or computer-based setting. To sit for the examination, one must have proof of current cardiopulmonary resuscitation (CPR) certification, have a bachelor's degree in an exercise-related field, and a current ACE personal trainer certification or NCCA-accredited certification. In addition, 300 h of practical experience designing and implementing exercise programs for apparently healthy or high-risk people is required.

American Society of Exercise Physiologists

The American Society of Exercise Physiologists was organized to provide professional visibility for the increasing numbers of academically trained exercise physiologists (4). With that, the ASEP Board of Directors sought to develop a certification that would assess the candidate's academic and technical competence in the area of exercise physiology. The purpose was to help college-prepared exercise physiologists attain key positions in the health, fitness, rehabilitative, and research fields of professional work.

The ASEP exercise physiology certification (EPC) has both a written and practical examination. The EPC certification measures both academic knowledge and practical skills, as a means to help ensure professional competence for organizations that are hiring exercise physiologists. Requirements to sit for the examination include an academic degree in exercise physiology or a concentration in exercise science and proof of academic competency (C or 2.0 GPA or better) in specific topic areas. In addition, candidates must be current members of ASEP and must have completed 250 h of documented, hands-on laboratory or internship experience.

American College of Sports Medicine

The American College of Sports Medicine has been internationally recognized as a leader in exercise and sports medicine for over 50 years (3). Besides clinical exercise physiologists, the ACSM is comprised of professionals with expertise in a variety of sports medicine fields, including physicians, general exercise physiologists and those from academia, physical therapists, athletic trainers, dietitians, nurses, and other allied health professionals.

Since 1974 ACSM has been recognized throughout the industry for its certification program. Although ACSM does provide certifications for the exercise professional interested in preventive exercise programming (certified personal trainer, health fitness instructor), the ACSM rehabilitative track is most appropriate for the clinical exercise physiologist. For example, the ACSM clinical exercise specialist certification is designed for the professional who plans to administer and supervise an exercise program for patients with cardiovascular or pulmonary disease, or the patient with metabolic syndrome. The general requirements for taking the exercise specialist certification examination include a minimum of a bachelor's degree in a relevant health field or the equivalent (i.e., a 2 yr degree and 2 yr of experience in a cardiac rehabilitation program), 600 h of internship in a cardiopulmonary rehabilitation setting, and current basic cardiac life support certification (BCLS). The test format is a computer-based examination designed to test

Table 1.1   Comparison of Certifications for Clinical Exercise Physiologists

	Advanced Health + Fitness Specialist (ACE)*	Exercise physiologist certified (ASEP)*	Clinical exercise specialist (ACSM)*	Registered clinical exercise physiologist (ACSM)*	Certified exercise physiologist (CSEP)*
Requirements	BS in exercise science or related field, current ACE Personal Trainer Certification or another National Commission for Certifying Agencies (NCCA)-accredited certification Current CPR certification 300 h of practical experience	Academic degree with major or emphasis in exercise science, exercise physiology, or physiology (type of degree not specified) with C grade or better in five of the nine content areas Current ASEP membership >250 h laboratory or internship experience	BS in any health field >600 h of practical, clinical experience Current CPR or BCLS certification	MS in exercise science, exercise physiology, or physiology >600 h clinical experience, preferably across all domains Current CPR or BCLS certification	4 yr degree in physical activity, exercise sciences, kinesiology, human kinetics, or related field, with a specialization in health-related fitness and work and sport-related fitness applications for asymptomatic and symptomatic populations >300 h of related experience
Written examination	150 questions Specific exam content available on ACE Web site; exam study materials available for purchase Content domains: Program assessment Program design Program implementation Professional responsibility	200 questions Approximate weightings and content areas: Exercise physiology (cardiorespiratory training) (36%) Cardiac rehabilitation (including ECGs and health fitness) (18.5%) Exercise metabolism and regulation (11.5%) Kinesiology (including neuromuscular) (10.5%) Research (6.5%) Sports biomechanics (6%) Environmental exercise physiology (6%) Sports nutrition (5%)	Computer-based exam at Pearson Vue authorized testing centers 115–170 questions over 3.5 h time limit Approximate weightings and content areas: Health appraisal and fitness exercise testing (26%) Exercise prescription (training) and programming (19%) Exercise physiology and related exercise science (10%) Pathophysiology and risk factors (10%) Human behavior (5%) Safety, injury prevention, and emergency procedures (5%	Computer-based exam at Pearson Vue authorized testing centers 140–175 questions over 3.5 h time limit Approximate weightings and content areas: Health appraisal, fitness and clinical exercise testing (25%) Exercise prescription and programming (21%) Exercise physiology and related exercise science (19%) Medical and surgical management (13%) Pathophysiology and risk factors (9%) Human behavior (5%)	Written examination requiring 75% passing score

(continued)

Table 1.1 *(continued)*

	Advanced Health + Fitness Specialist (ACE)*	Exercise physiologist certified (ASEP)*	Clinical exercise specialist (ACSM)*	Registered clinical exercise physiologist (ACSM)*	Certified exercise physiologist (CSEP)*
Written examination *(continued)*			Nutrition and weight management (2%) Patient management and medications (2%) Program administration, quality assurance, and outcome assessment (2%) Medical and surgical management (2%)	Safety, injury prevention, and emergency procedures (4%) Program administration, quality assurance, and outcome assessment (4%)	
Practical examination	No practical testing	9 content areas	By computer-based exam	By computer-based exam	"Objective standard practical examination"
Recertification requirements	Every 2 yr Requires 20 h continuing education credits (CEC); CPR and first aid certification not required but can receive CECs for those certifications	Every 5 yr Requires 25 continuing education credits	Every 3 yr Requires 60 continuing education credits and CPR certification	Every 3 yr Requires 60 continuing education credits and CPR certification	Every 2 yr Requires CPR certification, 12 fitness appraisals, and 30 professional development credits

*Information gathered from the ACE (2), ACSM (3), ASEP (4), and CSEP (5) Web sites and promotional materials.

the candidate's knowledge and practical competence in graded exercise testing, exercise prescription, exercise leadership, and more.

In response to the growing body of evidence outlining the benefits of exercise for both disease prevention and rehabilitation across a variety of diseases and populations (e.g., cardiovascular, pulmonary, metabolic, musculoskeletal, neuromuscular, immunology, oncology, geriatric, obstetric, and pediatric), in 1998 the ACSM developed a registry for clinical exercise physiologists. This registry is designed to credential the clinical exercise physiologist who cares for patients with a chronic disease. To become an ACSM registered clinical exercise physiologist (RCEP) a minimum of a master's degree in exercise physiology, exercise science or physiology is required along with 600 h of documented practical experience. Practical hours should be accumulated from experiences across various diseases and conditions to help prepare candidates for both examination and employment. Goals of the RCEP certification are (1)

to provide the consumer and employer with the assurance that successful candidates are adequately prepared to work with patients with a variety of chronic diseases and special populations and (2) to improve the visibility and acceptance of the clinical exercise physiologist among the public and other health professionals. The testing format is a computer-based examination that is the same as that used for the exercise specialist certification. The use of computer-based testing ensures that the the examination is delivered and scored without variation worldwide.

Canadian Society of Exercise Physiology (CSEP)

The CSEP–certified exercise physiologist title was developed and approved as a certification in 2006 (5). It combines the previous CSEP personal fitness and lifestyle consultant (PFLC) and exercise therapist (ET) certification scopes with an emphasis on populations with disabilities or chronic diseases. Certification requires a university degree in physical activity, exercise science, kinesiology, human kinetics, or related field and passing written and practical examinations. The CSEP offers regular workshops to assist candidates in preparing for the national examination process, although attendance is not a requirement to sit for the examination.

Clinical Exercise Physiology Association

Established originally in 2008 as an affiliate society of ACSM, the Clinical Exercise Physiology Association (CEPA) is an autonomous professional member organization with the sole purpose of advancing the scientific and practical application of clinical exercise physiology for the betterment of health, fitness, and quality of life among patients at high risk or living with a chronic disease. Although no clinical exercise certifications are offered through CEPA per se, the focus of this organization is to serve the profession of clinical exercise physiology and practicing clinical exercise physiologists through advocacy, education, and career development.

PROFESSIONALIZATION OF CLINICAL EXERCISE PHYSIOLOGY

The biomedical literature continues to expand on the knowledge that physical activity plays a direct role preventing the development of many different chronic diseases, as well as playing an important role in inhibiting disease progression. Given this and the increasing number of people 65 yr of age and older in the population, employment opportunities for the clinical exercise physiologist continue to expand.

The previously mentioned submission to the federal government for the establishment of an exercise physiologist occupation under the Standard Occupational Classification system recognizes the initial development and subsequent landmark work completed by the Committee on the Accreditation for the Exercise Sciences (CoAES). Starting in April 2004, operating under the auspices of the Commission on Accreditation of Allied Health Education Programs (CAAHEP), CoAES accomplished its goal of establishing guidelines and standards for postsecondary academic institutions for personal fitness trainer and academic programs at both the bachelor's and master's degree levels. Academic programs throughout the U.S. are now applying and receiving accreditation at the undergraduate (exercise science) and graduate (applied exercise physiology and clinical exercise physiology) levels (see "CoAES Sponsors"). Using an internal process, the American Society of Exercise Physiologists also accredits academic programs. No other organization currently accredits academic exercise science or exercise physiology programs in North America.

Besides planning to attend a school that offers an exercise physiology curriculum accredited through the CoAES, students interested in becoming clinical exercise physiologists should seek the guidance of professionals in the field to select a program that also provides a well-rounded set of practical, laboratory, and research experiences aimed at best preparing them to successfully

CoAES Sponsors

American Alliance for Health, Physical Education, Recreation and Dance	www.aahperd.org
American Association of Cardiovascular and Pulmonary Rehabilitation	www.aacvpr.org
American College of Sports Medicine	www.acsm.org
American Council on Exercise	www.acefitness.org
American Kinesiotherapy Association	www.akta.org
Cooper Institute	www.cooperinst.org
Medical Fitness Association	www.medicalfitness.org
National Strength and Conditioning Association	www.nsca.com
National Academy of Sports Medicine	www.nasm.org

work in the field. Currently, although most clinical exercise physiologists are employed in cardiopulmonary rehabilitation programs, more programs designed to help care for patients with other chronic diseases and disabilities are being launched each week. These other programs are usually targeted at patients with chronic kidney disease, cancer, chronic fatigue, arthritis and metabolic syndrome.

Those interested in cardiac or pulmonary rehabilitation should become familiar with the American Association of Cardiovascular and Pulmonary Rehabilitation (AACVPR). The AACVPR publishes a directory of programs in the United States and in other countries, which is an ideal resource for identifying potential employers. The AACVPR also offers a career hotline to members and nonmembers that lists open positions and provides a site to post resumes (4). The ACSM's monthly journal *Medicine and Science in Sports and Exercise* (*MSSE*) lists exercise physiology–related employment and educational positions. In addition, posted positions can be accessed through the ACSM Web site (3). Finally, the Web site www.exercisejobs.com lists positions nationwide and provides the opportunity to post a resume online.

Another area of the field experiencing growth involves programs that use exercise in primary prevention. Many hospitals and corporations recognize that regular exercise training can reduce future medical expenses and increase productivity, with many of these programs implemented in a medical fitness center setting. In fact, most medically based fitness facilities now exclusively hire people with exercise-related degrees, as a means to ensure the safety for an increasingly diverse clientele that includes both healthy people and individuals with clinically manifest disease.

Those seeking employment or attempting to maintain professional certification would also be wise to join one or more professional organizations. Besides joining one or more of the national-level organizations previously mentioned, many clinical exercise physiologists decide to also join a regional chapter of the ACSM or a state chapter of the AACVPR, ASEP, or CSEP. These state or regional chapters all hold local educational programs that provide professionals an avenue to gain continuing education credits and an opportunity to network with those who make decisions to hire clinical exercise physiologists.

Several other professional organizations publish or present exercise-related research or information that focuses on specific diseases or disorders. These include the American Heart Association (AHA), the American College of Cardiology (ACC), the American Diabetes Association (ADA), the American Academy of Orthopedic Surgeons, the American Cancer Society, the Multiple Sclerosis Society, and the Arthritis Foundation. Table 1.2 lists the Web sites for these and other organizations, most of which serve as excellent information resources. Finally, clinical exercise physiologists should stay abreast of current research in general exercise physiology, as well as research specific to their disease area of interest. The body of scientific information specific to clinical exercise physiology continues to grow rapidly. Regular journal reading is an excellent way to keep yourself current in the field.

After a person has begun practicing as a clinical exercise physiologist, he or she should consider using several safeguards. Only one state, Louisiana, now requires licensing of exercise physiologists, which means that almost anyone else in the other 49 states can claim to be an exercise physiologist. Fortunately, several other states are now considering legislation that will limit the practice of clinical exercise physiology to those trained to do so. The aforementioned national and local professional societies are useful for obtaining updates of ongoing licensure developments within each state. A major issue related to licensure and professionalization is the risk that practicing professionals assume when working with clients, as discussed in practical application 1.2. Professionals must protect themselves against potential litigation related to their practice. Practical application 1.3 describes an interesting case that highlights both the potential risk and the important role that a clinical exercise physiologist can play in litigation. Clinical exercise physiologists should consider malpractice insurance for such incidents. In a hospital setting the institution often covers malpractice insurance, but if working in fitness, small clinical, ambulatory care, or personal-training settings, the exercise professional should inquire about and obtain if necessary his or her own insurance coverage. Organizations such as the American College of Sports Medicine offer discounted group rates for malpractice insurance for the exercise professional.

Limiting legal risk is important to the practicing exercise physiologist. Within the confines of the civil justice system, under which personal injury and wrongful death cases are determined, clinical exercise physiologists could be held accountable to a variety of professional standards and guidelines. These standards and guidelines, sometimes referred to as practice guidelines, can apply to some practices carried out by clinical exercise physiologists and others who provide services within health and fitness facilities. A variety of professional organizations have developed and published guidelines or standards dealing with the provision of service by fitness professionals, including clinical exercise physiologists.

Table 1.2 Selected Internet Sites of Chronic Disease Organizations and Institutes

Category	Organization or institute	Internet site
Endocrinology and metabolic disorders	American Diabetes Association	www.diabetes.org
	National Institute of Diabetes and Digestive and Kidney Diseases	www.niddk.nih.gov
	National Kidney Foundation	www.kidney.org
	North American Association for the Study of Obesity	www.naaso.org
Cardiovascular diseases	American Heart Association	www.americanheart.org
	American College of Cardiology	www.acc.org
	Heart Failure Society of America	www.hfsa.org
	National Heart, Lung, and Blood Institute	www.nhlbi.nih.gov
	Vascular Disease Foundation	www.vdf.org
Respiratory diseases	American Lung Association	www.lungusa.org
	National Heart, Lung, and Blood Institute	www.nhlbi.nih.gov
Oncology and immune diseases	American Cancer Society	www.cancer.org
	National Cancer Institute	www.nci.nih.gov
Bone and joint diseases and disorders	American Academy of Orthopedic Surgeons	www.aaos.org
	American College of Rheumatology and Association of Rheumatology Health Professionals	www.rheumatology.org
	Arthritis Foundation	www.arthritis.org
	International Osteoporosis Foundation	www.osteofound.org
	National Institute of Arthritis and Musculoskeletal and Skin Diseases	www.niams.nih.gov
	Spondylitis Association of America	www.spondylitis.org
Neuromuscular disorders	National Multiple Sclerosis Society	www.nmss.org
	National Institute of Neurological Disorders and Stroke	www.ninds.nih.gov
	National Spinal Cord Injury Association	www.spinalcord.org
Special populations	American Geriatrics Society	www.americangeriatrics.org
	National Institute on Aging	www.nia.nih.gov
	National Institute of Child Health and Human Development	www.nichd.nih.gov
	The National Women's Health Information Center	www.4woman.gov

Legal Case

To appreciate what can be at stake for health professionals who become involved in courtroom litigation, consider this case of an exercise physiologist working in the clinical setting. She wrote of her litigation experiences in the first person. Excerpts follow.

"The hospital where I work is a tertiary care facility.... The staff members in Cardiac Rehabilitation are required to have a master's degree in exercise physiology, American Heart Association basic cardiac life support certification, and American College of Sports Medicine certification as an exercise specialist....

"In May of 1991, Cardiac Rehabilitation was consulted to see a 78 yr old woman 4 d post–aortic valve surgery. Her medical history included severe aortic stenosis, coronary artery disease, left ventricular dysfunction, carotid artery disease, atrial fibrillation, hypertension, and noninsulin dependent diabetes.... She was in atrial fibrillation but had stable hemodynamics, and was ambulating without problems in her room.

"The patient was first seen by Cardiac Rehabilitation 7 d postsurgery. Chart review ... revealed no contraindications to activity.... Initial assessment revealed normal supine, sitting, standing, and ambulating hemodynamics. She walked independently in her room, and approximately 100 ft (30 m) in 5 min ... using the handrails for support.... General instructions for independent walking later that day were given....

"On postop day 8, there was documentation in the chart regarding the patient's need for an assist device while ambulating.... The patient ambulated approximately 200 ft (60 m) with a quad cane.... She appeared stable....

"On postop day 9, I was informed by the patient's nurse, as well as by documentation in the chart, that the patient had fallen in her room that morning, apparently hitting her head on the floor.... A computed tomography scan of the head was negative.

"On postop day 10, I observed the patient sitting with her daughter in the hall.... Her mental status had noticeably changed. She did not recognize her daughter, where she was, or why she was in the hospital.... No further formal exercise was performed.... The patient was transferred to intensive care and later died of a massive brain hemorrhage.

"Two years after the incident, a hospital lawyer contacted me and informed me that the family had brought suit against the hospital contending that the patient fell because she was not steady enough to walk independently.... I was called to testify and explain my participation in her care....

"I met with the hospital's defense lawyer on two separate occasions. The first meeting, the lawyer inquired about my education, certification, years of employment, and what specifically my job entailed.... Two weeks before trial, the hospital's defense lawyer again questioned me.... He also played the role of the plaintiff's lawyer and reworded similar questions in a slightly different manner. The defense lawyer encouraged me to maintain consistency in my answers, and remain calm and assured.... To prove our case, we had to establish, by a preponderance of the evidence (i.e., more likely than not) that the patient was indeed ambulating independently prior to her fall. My documentation was crucial....

"Answering the plaintiffs' lawyer's questions during cross-examination was like a mental chess game.... My answers were crucial, as was my composure and the belief that all my training and certification *did* qualify me as a professional healthcare provider.... Although the only hard evidence I had to work with in the courtroom was my chart notes, the judge did allow me to supplement them with oral testimony....

"Whether you document in the patient's chart or on a separate summary sheet that is filed in your office, your documentation is a permanent record that provides powerful legal evidence....

"Later I was told that my testimony helped the hospital prove the patient's fall was not a result of negligence. The court ruled in favor of our hospital."

Practically, professionals use standards and guidelines statements to identify probable and evidence-based benchmarks of expected service delivery owed to patients and clients, which helps ensure appropriate and uniform delivery of service.

In the course of litigation, these standards statements are used through expert witnesses to establish the standard of care to which providers will be held accountable in the event of patient injury or death. In years past, expert witnesses often used their subjective and personal opinions to establish the standard of care and then provided an opinion as to whether that standard was violated in the actual delivery of service. Today, most professional organizations use an evidence-based approach to establish standards of care, to achieve the following purposes:

- Achieve a consensus of standards of care among professionals that practitioners could aspire to meet in their delivery of services in accordance with known and established benchmarks of expected behavior
- Reduce cases of negligence and findings of negligence or malpractice
- Minimize the significance of individually and subjectively expressed opinions in court proceedings
- Reduce potential inconsistent verdicts and judgments that arose from the expression of individual opinions

Practice guidelines are a reference readily available to compare the actual delivery of service to what the profession as a whole has determined to be established benchmarks. Consequently, clinical exercise physiologists should review relevant published practice guidelines and then consider in what situations the standards might act as a shield to protect against negligent actions. Others may engage in the same process to determine whether such standards should be used as a sword to attack the care provided to clients in particular cases. Clinical exercise physiologists can use the following strategies to limit exposure and risk.

- In programs that include patients with known disease, and in which diagnosis, prescription, or treatment is the probable or actual reason for performing the procedures, a physician should be involved in a significant way. This involvement must be meaningful and real if it is to be legally effective.

Examples include referring patients, receiving and acting on results provided by rehabilitation staff, and discussing findings with the rehabilitation or therapy staff.

- In programs that have patients who demonstrate a high risk of suffering an adverse event or injury during exercise, the physician should be within immediate proximity to the participant during initial testing. The physician need not be watching the patient or the electrocardiogram monitor, but she or he should be controlling the staff during the procedure. Thereafter, the physician's proximity of supervision in further exercise testing should be dictated by the medical interpretation of each patient's initial test result.
- Evaluate and screen participants before recommending or prescribing activity, especially in rehabilitative or preventive settings.
- Secure a medically mandated consent-type document when involved in procedures such as a graded exercise test.
- Develop and assist individuals with implementing a safe and evidence-based exercise prescription that is aimed at a patient's goals and addresses disease-related disability.
- Recognize and refer people who have conditions that need evaluation before commencing or continuing with various activity programs.
- Provide feedback to referring professionals relative to the progress of participants in activity.
- Provide appropriate, timely, and effective emergency care, as needed.

CONCLUSION

Although the profession of clinical exercise physiology has evolved tremendously over the past 15 yr, in many respects it remains in its infancy compared with other allied health professions. Aspiring and practicing clinical exercise physiologists require up-to-date information to prepare for or to maintain certification and to continue to provide contemporary and evidence-based patient care. This text provides a comprehensive and practical review for the most common chronic diseases. Each chapter serves as a guide for important issues regarding client–clinician interaction through the development of a comprehensive exercise prescription.

Behavioral Approaches to Physical Activity Promotion

Gregory W. Heath, DHSc, MPH

The clinical exercise physiologist can use a number of behavioral strategies to assess and counsel individual patients about changing their physical activity behavior. The behavioral strategies discussed are intended to be used in the context of supportive social and physical environments. A nonsupportive environment is considered a barrier to participation in regular physical activity. Consequently, if this barrier is not altered, change is unlikely to occur. Thus, one of the clinician's goals as part of the counseling strategy is to identify environmental barriers with the client, include steps on how to overcome these barriers, and build supportive social and physical environments. Although no guarantees can be made, the literature suggests that if clinicians take a behavior-based approach to physical activity counseling, within the context of a supportive environment, they may experience greater success in getting their clients moving. Therefore, information is presented about the role of social and contextual settings in promoting health- and fitness-related levels of physical activity. The most important task of the clinical exercise physiologist is to guide a client into a lifelong pattern of regular, safe, and effective physical activity.

BENEFITS OF PHYSICAL ACTIVITY

The exercise physiologist who understands both the physiological basis for activity and the effect of pathology on human performance is well positioned for physical activity counseling. To be an effective counselor, however, the clinical exercise physiologist also needs to understand human behavior in the context of the individual client's social and physical milieu. This chapter seeks to underscore some of the important theories and models of behavior that have been shown to be important adjuncts for clinicians seeking to help people make positive changes in their physical activity behavior.

Persons who engage in regular physical activity have an increased physical working capacity (1); decreased body fat (2); increased lean body tissue (2); increased bone density (3); and lower rates of coronary heart disease (CHD) (4), diabetes mellitus, hypertension (5), and cancer (6). Increased physical activity is also associated with greater longevity (7). Regular physical activity and exercise can also help people improve their mood and motivational climate (8), enhancing their quality of life, improving their capacity for work and recreation, and altering their rate of decline in functional status (9).

When promoting planned exercise and physical activity, the counselor must pay attention to specifically designed outcomes. Notably, the health and fitness outcomes of a well-designed exercise prescription need to be accounted for. Finally, a number of physiological, anatomical, and behavioral characteristics should also be considered to ensure a safe, effective, and enjoyable exercise experience for the participant.

Health Benefits of Physical Activity

Physical activity has been defined as any bodily movement produced by skeletal muscles that result in caloric expenditure (10). Because caloric expenditure uses energy and energy utilization enhances weight loss or weight maintenance, caloric expenditure is important in the prevention and management of obesity, CHD, and diabetes mellitus. The *Healthy People 2010* Physical Activity and Fitness Objective 22.2 (11) highlights the need for every person to engage in regular, moderate physical activity for at least 30 min per day, preferably daily. Current research suggests that engaging in moderate physical activity for at least 30 min per day will help ensure that calories are expended with the conferring of specific health benefits. For example, daily physical activity equivalent to a sustained walk for 30 min per day would result in an energy expenditure of about 1,050 kcal per week. Epidemiologic studies suggest that a weekly expenditure of 1,000 kcal could have significant individual and public health benefits for CHD prevention, especially among those who are initially inactive. More recently, the American College of Sports Medicine and the Centers for Disease Control and Prevention concluded that the scientific evidence clearly demonstrates that regular, moderate-intensity physical activity provides substantial health benefits (12). Therefore, after an extensive review of the physiological, epidemiological, and clinical evidence, an expert panel formulated the following recommendation:

Every U.S. adult should accumulate 30 min or more of moderate-intensity physical activity on most, preferably all, days of the week.

This recommendation has recently been restated to provide more clarity (13). Moderate levels of physical activity on 5 d per week, or vigorous levels on 3 d per week, is now the stated minimum recommendation. Notice also that the recommendation emphasizes the benefits of moderate physical activity that can be accumulated

in bouts of 10 min of exercise. Intermittent activity has been shown to confer substantial benefits. Therefore, the recommended 30 min of activity can be accumulated in shorter bouts of 10 min spaced throughout the day. Although the accumulation of 30 min of moderate-intensity physical activity has been shown to confer important health benefits, this recommendation is not intended to represent the optimal amount of physical activity for health but instead a minimum standard or base on which to build to obtain more specific outcomes related to physical activity and exercise. Specifically, selected fitness-related outcomes may be a desired result for the physical activity participant, who may seek the additional benefits of improved **cardiorespiratory fitness, muscle endurance, muscle strength, flexibility,** and **body composition**. Additionally, specific health outcomes related to physical activity may also be desirable. For instance, the recommendations provided by the Institute of Medicine (IOM) for the prevention of weight gain and weight regain set the duration of physical activity at 60 min and 90 min of moderate-intensity activity on 5 or more d per week, respectively (IOM Report 2004).

Fitness Benefits of Physical Activity

Regular vigorous physical activity helps achieve and maintain higher levels of cardiorespiratory fitness than moderate physical activity does. Health-related fitness comprises five components: cardiorespiratory fitness, muscle strength, muscle endurance, flexibility, and enhanced body composition (see "Examples of Health and Fitness Benefits of Physical Activity").

Cardiorespiratory fitness, or aerobic capacity, describes the ability of the body to perform high-intensity activity for a prolonged period without undue physical stress or fatigue. Having higher levels of cardiorespiratory fitness enables people to carry out their daily occupational tasks and leisure pursuits more easily and with greater efficiency. Vigorous physical activities such as the following help achieve and maintain cardiorespiratory fitness and can contribute substantially to caloric expenditure:

Very brisk walking	Skating
Jogging or running	Rope jumping
Lap swimming	Soccer
Cycling	Basketball
Fast dancing	Volleyball

These activities may also provide additional protection against CHD over moderate forms of physical

Examples of Health and Fitness Benefits of Physical Activity

Health benefits	Fitness benefits
Reduction in premature mortality	Cardiorespiratory fitness
Reduction in cardiovascular disease risk	Muscle strength and endurance
Reduction in colon cancer	Flexibility
Reduction in type 2 diabetes mellitus	Enhanced body composition
Improved mental health	

activity. Higher levels of cardiorespiratory fitness can be achieved by increasing the frequency, duration, or intensity of an activity beyond the minimum recommendation of 20 min per occasion, three occasions per week, at more than 50% of aerobic capacity (14).

Muscular strength and endurance are the ability of skeletal muscles to perform hard or prolonged work (15). Regular use of skeletal muscles helps improve and maintain strength and endurance, which greatly affects the ability to perform the tasks of daily living without undue physical stress and fatigue. Examples of tasks of daily living include home maintenance and household activities such as sweeping, gardening, and raking. Engaging in regular physical activity such as weight training or the regular lifting and carrying of heavy objects appears to maintain essential muscle strength and endurance for the efficient and effective completion of most activities of daily living throughout the life cycle (16). The prevalence of such physical activity behavior is still low with recent prevalence estimates indicating that only 19.6% of the adult population engages in strength training at least twice per week (*MMWR*, July 21, 2006, 55(28): 769-772, "Trends in Strength Training—United States 1998-2004").

Musculoskeletal flexibility describes the range of motion in a joint or sequence of joints. Joint movement throughout the full range of motion helps improve and maintain flexibility. Those with greater total-body flexibility may have a lower risk of back injury (17). Older adults with better joint flexibility may be able to drive an automobile more safely (18). Engaging regularly in stretching exercises and a variety of physical activities that require a person to stoop, bend, crouch, and reach may help maintain a level of flexibility compatible with quality activities of daily living.

Excess body weight occurs when too few calories are expended and too many consumed for individual metabolic requirements (19). The maintenance of an acceptable ratio of fat to lean body weight is another desired component of health-related fitness. The results of weight-loss programs focused on dietary restrictions alone have not been encouraging. Physical activity burns calories, increases the proportion of lean to fat body mass, and raises the metabolic rate (20). Therefore, a combination of both caloric control and increased physical activity is important for attaining a healthy body weight. The guidelines from the IOM and the *Dietary Guidelines for Americans* have highlighted the importance of increasing the duration of moderate-intensity physical activity to 60 or 90 min per day as the necessary dose to sustain weight loss (IOM of the National Academies of Science. Dietary reference intakes for energy, carbohydrate, fiber, fat, fatty acids, cholesterol, protein, and amino acids macronutrients. Washington, DC: National Academy Press, 2002; *Dietary Guidelines for Americans 2005* U.S. Department of Health and Human Services and U.S. Department of Agriculture).

ISSUES RELATED TO REGULAR PHYSICAL ACTIVITY PARTICIPATION

In designing any exercise prescription, the professional needs to consider various physiological, behavioral, psychosocial, and environmental physical and social (57) variables that are related to participation in physical activity (21). Two commonly identified determinants of physical activity participation are self-efficacy and social support.

Self-efficacy, a construct from **social cognitive theory**, is most characterized by the person's confidence to exercise under a number of circumstances. It appears to be positively associated with greater participation in physical activity. **Social support** from family and friends has consistently been shown to be associated with greater levels of physical activity participation. Incorporating some mechanism of social support within the exercise prescription appears to be an important strategy for enhancing compliance with a physical activity plan (22). Common barriers to participation in physical activity are time constraints and injury. These barriers can be taken into account by encouraging participants to include physical activity as part of their lifestyle, not only engaging in planned exercise but also incorporating transportation, occupational, and household physical activity into their daily routine. A quiz of common barriers that prevent participation in physical activity is provided (table 2.1) to help steer the clinician's and participant's awareness and target strategies to improve compliance.

Participants can also be counseled to help them prevent injury. A program of low- to moderate-intensity physical activities is more likely to be adhered to than high-intensity activities during the early phases of an exercise program. Moreover, moderate activity is less likely to cause injury or undue discomfort (23). A number of physical and social environmental factors can also affect physical activity behavior (24). Family and friends can be role models, provide encouragement, or be companions during physical activity. The physical environment often presents important barriers to participation in physical activity, including lack of bicycle trails and walking paths away from vehicular traffic, inclement weather, and unsafe neighborhoods (25). Sedentary behaviors such as excessive television viewing or computer use may also deter people from being physically active.

Table 2.1 Barriers to Being Active Quiz: What Keeps You From Being More Active?

Directions: Listed here are reasons that people give to describe why they do not get as much physical activity as they should. Please read each statement and indicate how likely you are to say each of them:				
How likely are you to say?	**Very likely**	**Somewhat likely**	**Somewhat unlikely**	**Very unlikely**
1. My day is so busy now that I just don't think I can make the time to include physical activity in my regular schedule.	3	2	1	0
2. None of my family members or friends like to do anything active, so I don't have a chance to exercise.	3	2	1	0
3. I'm just too tired after work to get any exercise.	3	2	1	0
4. I've been thinking about getting more exercise, but I just can't seem to get started.	3	2	1	0
5. I'm getting older so exercise can be risky.	3	2	1	0
6. I don't get enough exercise because I have never learned the skills for any sport.	3	2	1	0
7. I don't have access to jogging trails, swimming pools, bike paths, and so forth.	3	2	1	0
8. Physical activity takes too much time away from other commitments—work, family, and so on.	3	2	1	0
9. I'm embarrassed about how I will look when I exercise with others.	3	2	1	0
10. I don't get enough sleep as it is. I just couldn't get up early or stay up late to get some exercise.	3	2	1	0
11. It's easier for me to find excuses not to exercise than to go out to do something.	3	2	1	0
12. I know of too many people who have hurt themselves by overdoing it with exercise.	3	2	1	0
13. I really can't see learning a new sport at my age.	3	2	1	0
14. It's just too expensive. You have to take a class or join a club or buy the right equipment.	3	2	1	0

How likely are you to say?	Very likely	Somewhat likely	Somewhat unlikely	Very unlikely
15. My free times during the day are too short to include exercise.	3	2	1	0
16. My usual social activities with family or friends do not include physical activity.	3	2	1	0
17. I'm too tired during the week and I need the weekend to catch up on my rest.	3	2	1	0
18. I want to get more exercise, but I just can't seem to make myself stick to anything.	3	2	1	0
19. I'm afraid I might injure myself or have a heart attack.	3	2	1	0
20. I'm not good enough at any physical activity to make it fun.	3	2	1	0
21. If we had exercise facilities and showers at work, then I would be more likely to exercise.	3	2	1	0

Follow these instructions to score yourself:

- Enter the circled number in the spaces provided, putting the number for statement 1 in space 1, statement 2 in space 2, and so on.
- Add the three sources on each line. Your barriers to physical activity fall into one or more of seven categories: lack of time, social influences, lack of energy, lack of willpower, fear of injury, lack of skill, and lack of resources. A score of 5 or above in any category shows that the barrier is an important one for you to overcome.

1 =	8 =	15 =	Row sum =	Lack of time
2 =	9 =	16 =	Row sum =	Social influence
3 =	10 =	17 =	Row sum =	Lack of energy
4 =	11 =	18 =	Row sum =	Lack of willpower
5 =	12 =	19 =	Row sum =	Fear of injury
6 =	13 =	20 =	Row sum =	Lack of skill
7 =	14 =	21 =	Row sum =	Lack of resources

Centers for Disease Control and Prevention, www.cdc.gov/.

Risk Assessment

Exercise prescription may be fulfilled in at least three different ways: (a) a program-based level that consists primarily of supervised exercise training (26); (b) exercise counseling and exercise prescription followed by a self-monitored exercise program (27); and (c) community-based exercise programming that is self-directed and self-monitored (28).

Within supervised exercise programs and self-monitoring programs offering exercise counseling and prescription, participants should complete a brief medical history, risk factor questionnaire, and preprogram evaluation (29). More information on the medical history and risk factor questionnaire and preprogram evaluation is discussed in chapter 4.

In a community-based, self-directed program, medical clearance is left to the judgment of the individual partici-pant. An active physical activity promotion campaign in the community seeks to educate the population regarding precautions and recommendations for moderate and vigorous physical activity (30). These messages should provide information that participants must know before beginning a regular program of moderate to vigorous physical activity. This information should include the following:

1. Awareness of preexisting medical problems (i.e., CHD, arthritis, osteoporosis, or diabetes mellitus)

2. Consultation before starting a program, with a physician or other appropriate health professional, if any of the previously mentioned problems are suspected

3. Appropriate mode of activity and tips on different types of activities

4. Principles of training intensity and general guidelines about rating of perceived exertion and training heart rate

5. Progression of activity and principles of starting slowly and gradually increasing activity time and intensity

6. Principles of monitoring symptoms of excessive fatigue

7. Making exercise fun and enjoyable

Theories and Models of Physical Activity Promotion

Historically, the most common approach to exercise prescription taken by health professionals has been direct information. In the past, the counseling sequence often consisted of the following:

1. Exercise assessment, usually cardiorespiratory fitness measures

2. Formulation of the exercise prescription

3. Counseling the patient regarding

   mode (usually large-muscle activity),

   frequency (three to five sessions per week),

   duration (20-30 min per session), and

   intensity (assigned target heart rate based on the exercise assessment) (31)

4. Review of the exercise prescription by the health professional and participant

5. Follow-up

6. Visits (reassessments and revising of the exercise prescription)

7. Phone contact

Most of the research evaluating this traditional approach to exercise prescription has not been too favorable in terms of its results in long-term compliance and benefits (32). That is, most people who begin an exercise program drop out during the first 6 mo. Clinicians use this traditional information-sharing approach because it is easiest for them, requires less time, and is prescriptive in nature. But it is not interactive with the client. More recently, contemporary theories and models of human behavior have been examined and developed for use in exercise counseling and interventions (33). These theories, referred to as cognitive–behavioral techniques, represent the most salient theories and models that have

been used to promote the initiation of and adherence to physical activity. These approaches vary in their applicability to physical activity promotion. Some models and theories were designed primarily as guides to understanding behavior, not as guides for designing intervention protocols. Others were specifically constructed with a view toward developing cognitive–behavioral techniques for physical activity behavior initiation and maintenance. Consequently, the clinical exercise physiologist may find that the majority of the theories summarized in table 2.2 will assist in understanding physical activity behavior change. A more detailed description of each theory is presented elsewhere. Moreover, other reviewed theories have evolved sufficiently to provide specific intervention techniques to assist in behavior change.

The Patient-Centered Assessment and Counseling for Exercise and Nutrition (PACE and PACE +) materials were developed for use by the primary care provider in the clinical setting targeting apparently clinically healthy adults (44). The materials have been evaluated for both acceptability and effectiveness in a number of different clinical settings (43). Sample materials taken from PACE are included in practical application 2.1 on p. 24. These materials are intended to provide a quick look at the steps in assessing and counseling an individual for physical activity. The materials incorporate many of the principles from the theoretical constructs previously reviewed in table 2.2. For further explanation of PACE materials, visit their Web site at www.paceproject.org. Wankel et al. (46) demonstrated the effectiveness of cognitive–behavioral techniques in enhancing physical activity promotion efforts, in which the use of increased social support and decisional strategies improved adherence to exercise classes among participants. Martin et al. (47) demonstrated through a series of studies the positive effects of personalized praise and feedback and the use of flexible goal setting among participants on exercise class adherence. Participants in the enhanced intervention group demonstrated an 80% attendance rate during the intervention compared with the control group's 50% attendance rate (47). McAuley et al. (48) successfully emphasized strategies to increase self-efficacy and thereby increase physical activity levels among adult participants in a community-based physical activity promotion program. These successful strategies included social modeling and social persuasion to improve compliance and exercise adherence. Promoting physical activity through home-based strategies holds much promise and might prove to be cost effective (49). Through tailored mail and telephone interventions,

Table 2.2 Summary of Theories and Models Used in Physical Activity Promotion

Theory or model	Level	Key concepts
Health belief model (33)	Individual	Perceived susceptibility
		Perceived severity
		Perceived benefits
		Perceived barriers
		Cues to action
		Self-efficacy
Relapse prevention (34,35)	Individual	Skills training
		Cognitive reframing
		Lifestyle rebalancing
Theory of planned behavior (36,37)	Individual	Attitude toward behavior
		Outcome expectations
		Value of outcome expectations
		Subjective norm
		Beliefs of others
		Motive to comply with others
		Perceived behavioral control
Social cognitive theory (38,39)	Interpersonal	Reciprocal determinism
		Behavioral capability
		Self-efficacy
		Outcome expectations
		Observational learning
		Reinforcement
Social support (40)	Interpersonal	Instrumental support
		Informational support
		Emotional support
		Appraisal support
Ecological perspective (57)	Environmental	Multiple levels of influence:
		Intrapersonal
		Interpersonal
		Institutional
		Community
		Public policy
Transtheoretical model (41,42,43,44)	Individual	Precontemplation
		Contemplation
		Preparation
		Action
		Maintenance

Adapted from K. Glanz and B.K. Rimer, 1995, *Theory-at-a-glance: A guide for health promotion practice* (U.S. Department of Health and Human Services).

significant levels of social support and reinforcement have been shown to enhance participants' self-efficacy in complying with exercise prescription, thus significantly improving levels of physical activity (50).

Finally, lifestyle-based physical activity promotion has been demonstrated to increase levels of moderate physical activity among adults. Lifestyle-based physical activity focuses on home- or community-based participation in forms of activity that include much of a person's daily routine (e.g., transport, home repair and maintenance, yard maintenance) (51). This approach evolved from the idea that physical activity health benefits may accrue from an accumulation of physical activity minutes over the course of the day (1). Because lack of time is a common barrier to regular physical activity, some researchers recommend promoting lifestyle changes whereby a person can enjoy physical activity throughout the day as part of his or her lifestyle. Taking the stairs at work, taking a walk during lunch, and walking or biking for transportation are all effective forms of lifestyle physical activity.

Ecological Perspective

A criticism of most theories and models of behavior change is that they emphasize individual behavior change and pay little attention to sociocultural and physical environmental influences on behavior (52). Recently, interest has developed in ecological approaches to increasing participation in physical activity (53). These approaches place the creation of supportive environments on par with the development of personal skills and the reorientation of health services. Creation of supportive physical environments is as important as intrapersonal factors when behavior change is the defined outcome. Stokols (55) illustrated this concept of a health-promoting environment by describing how physical activity could be promoted by establishing environmental supports, such as bike paths, parks, and incentives to encourage walking or bicycling to work. An underlying theme of ecological perspectives is that the most effective interventions occur on multiple levels. Interventions that simultaneously influence multiple levels and multiple settings (e.g.,

Practical Application 2.1

The PACE + Model

Telling patients what to do does not work, especially over the long term. An effective behavioral model helps facilitate long-term changes by telling patients how to change. PACE (Patient-Centered Assessment and Counseling for Exercise and Nutrition) is a comprehensive approach to physical activity and nutrition counseling that uses materials developed by a team of researchers at San Diego State University. The curriculum draws heavily on the stages of change model, which suggests that individuals change their habits in stages. Taking into account each person's readiness to make changes, PACE provides tailored recommendations for patients in each stage. PACE offers three different counseling protocols. Empirically derived behavioral strategies are applied in each protocol. The development of the PACE model began in 1990, funded originally by the Centers for Disease Control and Prevention, the Association of Teachers of Preventive Medicine, and San Diego State University. The original PACE materials, first released in 1994, dealt only with physical activity. The program was originally developed to overcome barriers to physician counseling for physical activity—especially lack of time for counseling, lack of standardized counseling protocols, and lack of training in behavioral counseling. Counseling was designed to be delivered in 2 to 5 min during a general patient checkup.

The PACE + materials were thoroughly tested and found to be acceptable and usable by healthcare providers and patients across the United States. Physicians also found PACE to be practical, improving their confidence in counseling patients about physical activity (52). In a controlled efficacy study of 212 sedentary adults, patients who received PACE counseling increased their minutes of weekly walking by 38.1 compared with 7.5 among the control group. Additionally, 52% of the patients who received PACE counseling adopted some physical activity compared with 12% of the control group (53).

Since these earlier studies, the current PACE + materials have been revised to include the recommendations from *Physical Activity and Health: A Report of the Surgeon General* (1) and also address nutrition behaviors such as decreasing dietary fat consumption; increasing fruit, vegetable, and fiber consumption; and balancing caloric intake and expenditure for weight control (54).

schools, work sites) may be expected to lead to greater and longer lasting changes and maintenance of existing health-promoting habits. In addition, investigators have recently demonstrated that behavioral interventions primarily work by means of the mediating variables of intrapersonal and environmental factors (56). Mediating variables are those that facilitate and shape behaviors—we all have a set of intrapersonal factors (e.g., personality type, motivation, genetic predispositions) and environmental factors (e.g., social networks such as family, cultural influences, and the built or physical environment). But few researchers have attempted to delineate the role of these mediating factors in facilitating health behavior change. Sallis et al. (57) recently described how difficult it is to assess the effectiveness of environmental and policy interventions because of the relatively few evaluation studies available. However, based on the experience of the New South Wales (Australia) Physical Activity Task Force, a model has been proposed to help understand the steps necessary to implement these interventions (58). Figure 2.1 presents an adaptation of this model as prepared by Sallis (57) and outlines the necessary interaction between planning, advocacy, organizations, policies, and environments to make such interventions a reality. Another pragmatic model that appears to have relevance for the promotion of physical activity has been proposed by McLeroy and colleagues (59). This model specifies five levels of determinants for health behavior:

1. Intrapersonal factors, including psychological and biological variables, as well as developmental history
2. Interpersonal processes and primary social groups, including family, friends, and coworkers
3. Institutional factors, including organizations such as companies, schools, health agencies, or healthcare facilities
4. Community factors, which include relationships among organizations, institutions, and social networks in a defined area

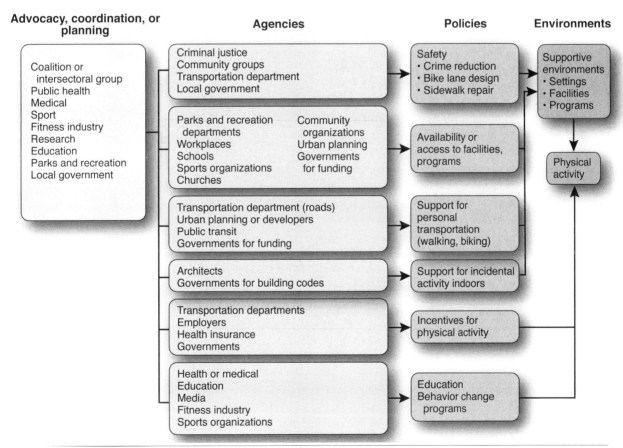

Figure 2.1   Conceptualization of the development of policy and environmental interventions to promote physical activity.

5. Public policy, which consists of laws and policies at the local, state, national, and supranational levels

Important in implementing this concept of behavioral determinants is realizing the key role of behavioral settings, which are the physical and social situations in which behavior occurs. Simply stated, human behavior such as physical activity is shaped by its surroundings; a person in a supportive social environment with access to space and facilities is more likely to be active. The determinant role of selected behavioral settings is important: Some are designed to encourage healthy behavior (e.g., sports fields, gymnasiums, health clubs, and bicycle paths), whereas others encourage unhealthy (or less healthy) behaviors (e.g., fast-food restaurants, vending machines with high-fat and high-sugar foods, movie theaters). Exercise physiologists need to understand the environment in which their clients live. These structures (e.g., fields, gymnasiums, community centers) are part of each person's living environments; people who disregard them are less likely to be active. For physical activity providers, an important adjunct in assessing and prescribing physical activity interventions for participants is understanding the physical and social contexts in which their patients live. This information can be obtained from various sources and can be at the level of the individual or at the general community level. When individual physical activity behavior information is coupled with sociodemographic, physical, and social context information, physical activity interventions can be further tailored to maximize the participant's physical activity behavior change and maintenance plan. Exercise physiologists cannot alter the client's physical environment, but they should be able to address environmental barriers and provide insights into how to overcome those barriers. In the long run, we all should be a part of changing our environments for the better.

An example of such tailoring for physical activity promotion that alters physical activity behaviors includes the work of Linenger et al. (59), in which efforts to increase physical activity among naval personnel were accomplished by a multifactorial environmental and policy approach to physical activity promotion. These investigators compared an "enhanced base" with a "control base." The enhancements included increasing the number of bike trails on base, acquiring new exercise equipment for the local gym, opening a women's fitness center, instituting activity clubs, and providing release time for physical activity and exercise (60). The changes were positive for those living on the enhanced base—that is, they increased their physical activity levels.

Another example, this time emphasizing an incentive-based approach to promoting physical activity, is the work of Epstein and Wing (61). Although this work was undertaken some time ago, the lessons are relevant in the inactive culture prevalent today. In this study, contracts and a lottery were used to boost exercise attendance. Participants increased their overall physical activity levels. Compared with a "usual care" comparison group, adherence and activity levels were significantly improved and sustained (58). But some researchers urge caution in using an incentive-based approach because they believe that over the long term participants never internalize the health behavior, meaning that they are likely to revert to sedentary habits after the incentive is removed or loses its appeal. Nevertheless, incentives have been proved effective in the short term.

Additional community-based environmental efforts to influence physical activity behavior have included the use of signs in public settings to increase use of stairs and walkways (62,63). These latter studies are examples of single intervention efforts that can be carried out in concert with systematic exercise prescription efforts among individuals. Thus, the increase in stair usage as a result of a promotional campaign can help people meet their prescribed energy expenditure requirements. Table 2.3 outlines some of the barriers that hinder people in becoming more physically active. Also listed are some suggested solutions, although these can vary from client to client.

Useful resources for clinical exercise physiologists are the recent evidence-based recommendations for physical activity promotion in communities (www.thecommunityguide.org/pa/default.htm/) (see table 2.4 on p. 28). The evidence base for these recommendations provides insights into how exercise practitioners can integrate their clinical efforts to assess and counsel patients into supportive and reinforcing environments. The recommendations are summarized with respect to informational, behavioral, social, and environmental approaches to promoting physical activity (64, 65).

CONCLUSION

Within the past decade, physical activity has emerged as a key factor in the prevention and management of chronic conditions. Although the role of exercise in health promotion has been appreciated and practiced for decades, recent findings regarding the mode, frequency, duration, and intensity of physical activity have modified exercise prescription practices. Included in these modifications has been the delineation between health and fitness

Table 2.3 Tips on Overcoming Potential Barriers to Regular Physical Activity

Barriers	Suggestions for overcoming physical activity barriers
Lack of time	Identify available time slots. Monitor your daily activities for 1 wk. Identify at least three 30 min time slots you could use for physical activity.
	Add physical activity to your daily routine. For example, walk or ride your bike to work or shopping, organize school activities around physical activity, walk the dog, exercise while you watch TV, park farther from your destination, and so on.
	Make time for physical activity. For example, walk, jog, or swim during your lunch hour, or take fitness breaks instead of coffee breaks.
	Select activities that require minimal time, such as walking, jogging, or stair climbing.
Social influence	Explain your interest in physical activity to friends and family. Ask them to support your efforts.
	Invite friends and family members to exercise with you. Plan social activities that involve exercise.
	Develop new friendships with physically active people. Join a group, such as the YMCA or a hiking club.
Lack of energy	Schedule physical activity for times in the day or week when you feel energetic.
	Convince yourself that if you give it a chance, physical activity will increase your energy level; then, try it.
Lack of motivation	Plan ahead. Make physical activity a regular part of your daily or weekly schedule and write it on your calendar.
	Invite a friend to exercise with you on a regular basis. Then both of you write it on your calendars.
	Join an exercise group or class.
Fear of injury	Learn how to warm up and cool down to prevent injury.
	Learn how to exercise appropriately considering your age, fitness level, skill level, and health status.
	Choose activities that involve minimal risk.
Lack of skill	Select activities that require no new skills, such as walking, climbing stairs, or jogging.
	Exercise with friends who are at the same skill level as you are.
	Find a friend who is willing to teach you some new skills.
	Take a class to develop new skills.
Lack of resources	Select activities that require minimal facilities or equipment, such as walking, jogging, jumping rope, or calisthenics.
	Identify inexpensive, convenient resources available in your community (community education programs, park and recreation programs, worksite programs, etc.).
Weather conditions	Develop a set of regular activities that are always available regardless of weather (indoor cycling, aerobic dance, indoor swimming, calisthenics, stair climbing, rope skipping, mall walking, dancing, gymnasium games, etc.)
	Look on outdoor activities that depend on weather conditions (cross-country skiing, outdoor swimming, outdoor tennis, etc.) as bonuses—extra activities possible when weather and circumstances permit.
Travel	Put a jump rope in your suitcase and jump rope.
	Walk the halls and climb the stairs in hotels.
	Stay in places with swimming pools or exercise facilities.
	Join the YMCA or YWCA (ask about reciprocal membership agreement).
	Visit the local shopping mall and walk for 30 min or more.
	Take a portable audio player and listen to your favorite upbeat music as you exercise.

(continued)

Table 2.3 *(continued)*

Barriers	Suggestions for overcoming physical activity barriers
Family obligations	Trade babysitting time with a friend, neighbor, or family member who also has small children.
	Exercise with the kids—go for a walk together, play tag or other running games, or get aerobic dance or exercise music for kids (several are on the market) and exercise together. You can spend time together and still get your exercise.
	Hire a babysitter and look at the cost as a worthwhile investment in your physical and mental health.
	Jump rope, do calisthenics, ride a stationary bicycle, or use other home gymnasium equipment while the kids are busy playing or sleeping.
	Try to exercise when the kids are not around (e.g., during school hours or their nap time).
	Encourage exercise facilities to provide child care services.
Retirement years	Look on your retirement as an opportunity to become more active instead of less. Spend more time gardening, walking the dog, and playing with your grandchildren. Children with short legs and grandparents with slower gaits are often great walking partners.
	Learn a new skill that you've always been interested in, such as ballroom dancing, square dancing, or swimming.
	Now that you have the time, make regular physical activity a part of every day. Go for a walk every morning or every evening before dinner. Treat yourself to an exercise bicycle and ride every day while reading a favorite book or magazine.

Content in the "Personal Barriers" taken from *Promoting physical activity: A guide for community action*, 1999 (USDHHS).

Table 2.4 Summary of Recommended Physical Activity Interventions—Guide to Community Preventive Services

Intervention	Recommendation
Informational approaches to increasing physical activity	
Community-wide campaigns	Recommended (strong evidence)
Point-of-decision prompts	Recommended (sufficient evidence)
Mass media campaigns	Insufficient evidence
Behavioral and social approaches to increasing physical activity	
Individually adapted health behavior change	Recommended (strong evidence)
Health education with TV and video turnoff	Insufficient evidence
College-age physical and health education	Insufficient evidence
Family-based social support	Insufficient evidence
School-based physical education	Recommended (strong evidence)
Community social support	Recommended (strong evidence)
Environmental and policy approaches to increasing physical activity	
Creation or enhanced access to places for physical activity	Recommended (strong evidence)
Community-scale urban design and land use	Recommended (sufficient evidence)
Street-scale urban design and land use	Recommended (sufficient evidence)
Transport policy and practices	Insufficient evidence

Reprinted from Guide to Community Preventive Services. Available: www.thecommunityguide.org/pa/pa.pdf.

outcomes relative to the physical activity prescription. Most important, new approaches to physical activity prescription and promotion that emphasize a behavioral approach with documented improvements in compliance have now become available to health professionals. Behavioral science has contributed greatly to the understanding of health behaviors such as physical activity. Behavioral theories and models of health behavior have been reexamined in light of physical activity and exercise. Although more research is needed to develop successful, well-defined applications that are easily adaptable for intervention purposes, behavioral principles and guidelines have evolved that are designed to help the health professional understand health behavior change and guide people into lifelong patterns of increased physical activity and improved exercise compliance.

New frontiers in the application of exercise prescription to specific populations as well as efforts to define the specific dose (frequency, intensity, duration) of physical activity for specific health and fitness outcomes are now being explored. As this information becomes available, it must be introduced to the participant through the most effective behavioral paradigms, such as the models discussed in this chapter. Moreover, positive changes in the participant's physical and social environments must occur to enhance compliance with exercise prescriptions. In turn, increased levels of physical activity among all people will improve health and function.

Case Study

Medical History

Mrs. KY is a 45 yr old Caucasian female. She is married and has two teenage sons. She is employed as a senior manager at a large bank and reports experiencing an above-average level of tension and stress. She presents at the referral of her primary care physician, who has observed that the client has elevated blood pressure and cholesterol levels that may be attributed to her stressful and highly sedentary job. In addition, the client admitted that she would like to lose 30 to 40 lb (14 to 18 kg) and improve her fitness so that she can ride her bike with her husband on a new community rail trail recently installed in her neighborhood.

Diagnosis

Mrs. KY is a sedentary but otherwise healthy middle-aged female with significant risk factors for CVD including obesity, dyslipidemia, and psychosocial stress. At the request of her physician she seeks to start exercising as part of a disease prevention program.

Exercise Test Results

The client is 5 ft 6 in. (168 cm) tall and weighs 196 lb (89 kg). Her body mass index is 31.7. She has a resting heart rate of 85 beats $\cdot$ min^{-1} and a resting blood pressure of 136/89 mmHg. Her total cholesterol is 198 mg $\cdot$ dl^{-1} untreated, and her high-density lipoproteins are 34 mg $\cdot$ dl^{-1}. Her graded treadmill stress test reveals that she has a $\dot{V}O_2$ max of 20.5 ml $\cdot$ kg^{-1} $\cdot$ min^{-1}, which is normal for an unfit woman of her age range. Her electrocardiogram was unremarkable at rest, as well as during and following her test. In addition, she reported smoking from age 17 to 40. She complains of occasional joint stiffness in her hands and ankles.

The client describes herself as nonathletic and admits to never participating in an organized sport or exercise setting. She is aware of the benefits of exercise but did not feel the incentive to begin a formal program until her doctor's recommendation. The client jokes that although her workday is highly organized and structured, the rest of her life is chaotic and that only with the support of her husband and kids does anything get done at home. She laments that her eating habits are atrocious and that she is so tired when she gets home from work that she only has energy to make dinner before crashing in front of the television. She presents to the clinician to start a workout program that will help her achieve her goals.

Exercise Prescription

The exercise plan including the traditional components—frequency, intensity, duration, and modality (discussed in detail in later chapters)—may be tailored to address specific risk factors. The subject's medical history,

(continued)

however, clearly indicates that this person had not prioritized exercise participation until she received her doctor's recommendation. Moreover, she presents with numerous potential barriers to engaging in a physically active lifestyle as well as behaviors that contribute to her overweight status. The clinician should assist the participant in establishing awareness and developing strategies to address those barriers. Furthermore, the clinician should consider tailoring strategies for motivating the participant toward the adoption of a healthy physically active lifestyle.

Discussion Questions

1. Applying the transtheoretical model, at what stage of exercise adoption is this client?

2. Based on your response to question 1, what types of interventions are most appropriate for this stage of change and why?

3. If you used the health belief model, what factors would you emphasize to achieve optimal exercise adherence?

4. How would Bandura's social cognitive theory be relevant to fostering exercise adherence for this client?

3

Pharmacotherapy

Diana Kostoff, PharmD, BCPS

The study of the action of a drug on the body is referred to as *pharmacology*, whereas the art and science of individualizing drug therapy based on patient-specific characteristics are referred to as *therapeutics* or *pharmacotherapy*. This chapter outlines basic information about the pharmacology and therapeutics of commonly used drug therapies. It is not meant to serve as a comprehensive reference, but instead provides general information in an easy-to-use format. Much of the textual information is summarized in tabular form. This chapter is organized into eight general categories based on systems or disease states or situations (drugs of abuse) that the clinical exercise physiologist may commonly encounter when working with patients. The effects of the agents on exercise capacity, if any, are noted in the text. Agents banned by either the United States Olympic Committee (USOC) or United States Anti-Doping Agency (USADA) (1) or the National Collegiate Athletic Association (NCAA) (2) are noted as well.

METABOLIC DISEASES AND AGENTS

Several patient populations may be prescribed various metabolic agents for treatment (table 3.1). Cancer and renal patients may be anemic because of treatments or their diseases and may need agents to improve oxygen-carrying capacity (3). Insulin and hypoglycemic agents are necessary for controlling blood glucose levels among people with diabetes. The clinician should thoroughly understand these forms of treatments.

Anemias

Anemia is a decrease in the red blood cells, which carry hemoglobin, a substance that transports oxygen. If the concentration of hemoglobin falls, the oxygen-transporting capacity of the body diminishes. Persons with anemia have decreased exercise tolerance and are more prone to fatigue. The primary causes of anemia include reduced red blood cell and hemoglobin production, **hemolysis** of red blood cells, and loss of blood. Although an inadequate dietary intake of several nutrients may reduce the production of red blood cells and hemoglobin, the most common cause of anemia throughout the world is iron deficiency (3).

Sports anemia is a term often used to describe a low hemoglobin condition that is relatively common at the beginning of rigorous training or when training requirements increase. This condition is characterized by exhaustion and fatigue. After adaptation to training, sports anemia seems to subside. The severity and exact causes of this condition have not yet been determined. Possible explanations for this condition are inadequate dietary iron intakes by athletes or the use of protein for tasks other than red blood cell production during the early training stages. As a result, athletes may have higher daily iron requirements than average (4). Iron requirements are determined by physiological losses and the needs determined by growth. The daily recommended

Table 3.1   Metabolic Agents and Their Exercise Effects

Medications	Heart rate	Blood pressure	ECG	Exercise capacity
Iron supplementation	↔	↔	↔	May increase in those with iron deficiency but no effect in those with normal iron levels
Erythropoietin	↔	↑	↔	May increase in those with anemia; questionable effect in persons with normal hemoglobin
NSAIDs and selective COX-2 inhibitors	↔	↔	↔	Do not affect exercise directly but by reducing pain may allow the person to perform

Note. ECG = electrocardiogram; NSAIDs = nonsteroidal anti-inflammatory drugs; COX-2 = cyclooxygenase-2; ↔ = no change; ↑ = increase.

dietary intake for iron is 8 mg in adult males and post-menopausal females but 18 mg in menstruating females (5). Children require more iron because of growth, as do pregnant women (5).

Iron Supplementation

Iron-deficiency anemia is treated with orally administered iron supplements. Ferrous sulfate, the least expensive of iron preparations, is the treatment of choice. Iron supplementation carries several side effects (3):

Heartburn

Nausea

Upper-gastrointestinal discomfort

Constipation

Diarrhea

Iron poisoning in adults is rarely fatal, but children between 12 and 24 mo are susceptible to death from acute iron toxicity. Multivitamins and prenatal vitamins should be kept out of the reach of children for these reasons, and a child-resistant cap should be used if children are present in the household.

Iron has no effect on exercise capacity in persons with normal serum iron levels. In athletes who are iron deficient, however, supplementation can increase hemoglobin levels with subsequent increases in oxygen-carrying capacity and performance enhancement (6) (figure 3.1).

Erythropoietin

Erythropoietin (EPO) is a naturally occurring hormone secreted by the kidneys when hemoglobin concentrations are low. EPO stimulates the bone marrow stem cells to make red blood cells (3). Although iron is required as a cofactor for red blood cell production, the bone marrow is essential for providing the cells. A synthetic form of the hormone is available as epoetin alfa (Epogen, Procrit)

and darpepoetin alfa (Aranesp) and used clinically in persons who have lost the ability to make the hormone on their own (patients with cancer, human immunodeficiency virus, or kidney failure) (3,5). EPO is generally well tolerated, but side effects can include hypertension, seizures, and thrombosis (7,8).

EPO has been used by endurance athletes to enhance performance by artificially increasing the oxygen-carrying capacity of the blood. By increasing production of red blood cells, however, EPO increases the viscosity of the blood, which can lead to thrombosis. Use of EPO to increase hemoglobin is different from blood doping. In doping, an athlete typically receives a blood transfusion (his or her own blood previously donated) before competition to increase hemoglobin concentrations (9). EPO products and blood doping are banned by the USOC and the NCAA (1,2).

Diabetes

Several types of medications are used to treat high blood glucose levels found in the patient with diabetes. Patients with type 1 diabetes will be on supplemental or replacement insulin, which promotes glucose uptake. Patients with type 2 diabetes may be treated with oral hypoglycemic agents (sulfonylureas, biguanides, thiazolidinediones, and short-acting insulin secretagogues), which lower blood glucose by aiding insulin production in the pancreas or allowing the cells to use insulin more effectively. These patients may also take α-glucosidase inhibitors, which reduce the blood glucose concentration after a meal (10). In severe cases, patients with type 2 diabetes will be prescribed oral hypoglycemic agents in conjunction with insulin. None of the agents used in the treatment of diabetes mellitus will directly affect exercise capacity. But when glucose availability is disrupted and a patient experiences **hypoglycemia**, the ability to continue exercise can be greatly diminished.

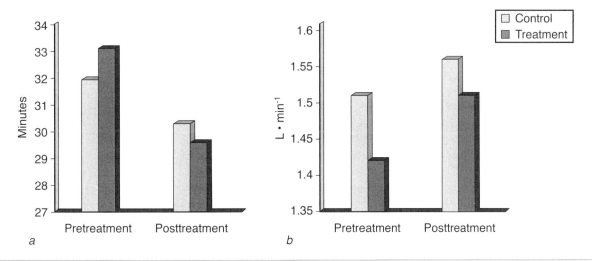

Figure 3.1   Changes in (a) 15K time trial and (b) oxygen consumption following iron supplementation in iron-depleted nonanemic women. *Note.* In (a), a group before versus after difference was observed ($p < .05$), and a difference between the iron-supplemented group and control group was observed after treatment ($p < .05$). In (b), a before versus after increase was observed within both groups ($p < .05$).

Reprinted, by permission, from P.S. Hinton et al., 2000, "Iron supplementation improves endurance after training in iron-depleted, nonanemic women," *Journal of Applied Physiology* 88: 1103-1111. Used with permission from American Physiology Society.

Insulin

Insulin, secreted by the β-cells of the pancreas, is the principal hormone required for proper glucose use in normal metabolic processes. Currently, most patients take a recombinant human form of insulin that is manufactured in different preparations to increase its duration of action. Insulin is available in short-acting form such as regular insulin, which can be administered intravenously or subcutaneously. It is also formulated in long-acting form such as insulin detemir and glargine, which are administered as subcutaneous injections. The most common adverse effects of insulin are hypoglycemia and weight gain (10). Athletes have used insulin to increase muscle stores of glycogen and reduce protein breakdown (1). The USADA permits the use of insulin only to treat insulin-dependent (type 1) diabetes mellitus (1).

Sulfonylureas

These agents are used to treat type 2 diabetes mellitus, also known as non-insulin-dependent diabetes mellitus. They bind to the plasma membrane of functional β-cells in the pancreas, leading to calcium influx and insulin release from secretory granules. Therefore, sulfonylureas enhance insulin secretion. Common agents in this class include chlorpropamide, tolbutamide, glipizide, glyburide, and glimepiride. Patients should be advised to monitor their blood glucose regularly and to eat at regular intervals to avoid wide fluctuations in blood glucose. The most common adverse effect is hypoglycemia.

Biguanides

Metformin is an oral hypoglycemic agent not related to the sulfonylureas. It decreases hepatic glucose production, decreases intestinal absorption of glucose, and improves insulin sensitivity. Insulin secretion remains unchanged, but plasma blood glucose decreases. Patients with impaired renal function (serum creatinine $>1.5 \text{ mg} \cdot \text{dl}^{-1}$) should not take metformin because it can lead to metabolic (lactic) acidosis and death. Other side effects include nausea and gastrointestinal upset and weight loss.

Thiazolidinediones

Thiazolidinediones (TZDs or glitazones) are antidiabetic agents that increase insulin sensitivity through a variety of mechanisms. They also decrease hepatic gluconeogenesis and increase insulin-dependent muscle glucose uptake. Agents in this class include pioglitazone and rosiglitazone. In early 2000 the agent troglitazone was removed from the market by the Food and Drug Administration (FDA) because of several reports of hepatictoxicity with resulting liver failure. Other agents in this class have the propensity to cause liver toxicity and therefore should be closely monitored. In addition, hypoglycemia can occur when thiazolidinediones are used in combination with insulin or sulfonylureas. Other adverse effects include weight gain, fluid retention, and edema. Thus, thiazolidinediones are contraindicated in patients with New York Heart Association class III and IV heart failure, and caution should be used when given to patients with class I and II heart failure or any other underlying cardiac diseases (11,12).

Short-Acting Insulin Secretagogues

Nateglinide and repaglinide are two drugs in this class of antidiabetic agents. Both drugs are rapidly absorbed, have a short half-life, and have a short duration of action (up to 4 h) (10). They stimulate insulin secretion from the β-cells of the pancreas like sulfonylureas (10). In addition, they both require the presence of glucose to stimulate this insulin secretion (10). Therefore, as glucose levels normalize, less insulin secretion occurs, which decreases the risk of hypoglycemia, the most common side effect with these agents (10).

α-Glucosidase Inhibitors

The α-glucosidase inhibitors block enzymes in the small intestine, slowing the breakdown of sucrose and complex carbohydrates (13). Acarbose and miglitol are currently the available medications in use. Because these drugs work in the small intestine, gastrointestinal side effects are common and limit the use of these agents (10).

DPP-4 Inhibitors

Incretins, such as glucagon-like peptide (GLP)-1, improve pancreatic function, but these incretins are also inactivated by the enzyme dipeptidyl peptidase (DPP)-4 (14). Vildagliptin is a novel, potent, orally active, highly selective DPP-4 inhibitor that enhances the antidiabetic actions of incretins (14). Treatment with vildagliptin is well tolerated with minimal side effects such as dizziness, headache, pruritis, and nasopharyngitis (15). Hypoglycemia can be experienced as with the other antidiabetic agents (15,16).

Other Injectable Agents

Two available medications are given by subcutaneous injection. Pramlintide acetate (Symlin) injection is used in diabetic patients treated with insulin. It is a synthetic analog of human amylin, which is a naturally occurring hormone produced by pancreatic β-cells that contributes to glucose control in response to food intake (17). Pramlintide acetate is coadministered with insulin, and therefore hypoglycemia is the major adverse effect. Other side effects include nausea, vomiting, decreased appetite, stomach pain, tiredness, dizziness, and injection site reactions (17).

Unlike, pramlintide acetate, exenatide (Byetta) injection is approved only for patients with type 2 diabetes (18). It differs in chemical structure and pharmacological action from all other diabetic agents. Exenatide is an incretin mimetic agent that works to enhance the antihyperglycemic effects of incretins (18). Its major side effects are nausea, vomiting, diarrhea, jitters, dizziness, headache, decreased appetite, injection site reactions, and gastrointestinal effects (18). Hypoglycemia is seen mostly in combination with oral antiglycemic agents (18).

Obesity and Metabolic Syndrome

Obese patients often take multiple drugs for the components of metabolic syndrome and symptomatic treatments such as analgesics for arthritis and bronchodilators to assist in breathing. Medications for weight loss are typically not prescribed unless individuals have a body mass index (BMI) $\geq 30\,\mathrm{kg\cdot m^{-2}}$ or $\geq 27\,\mathrm{kg\cdot m^{-2}}$ in the presence of cardiovascular heart disease (CHD) risk factors, such as diabetes and hypertension.

An effective antiobesity drug should reduce energy intake from food, stimulate energy expenditure, or cause both effects for a net result of weight loss (19). Current drugs on the market act mainly on energy intake and require patients to make lifestyle changes in diet and exercise to achieve maximal effectiveness (19). These current agents have been shown to produce a weight loss of 8 to 13 lb (3.6 to 5.8 kg) beyond what can be achieved by diet alone, to maintain weight below a patient's baseline weight, and to improve most cardiovascular risks in relation to weight loss (19). In addition, antiobesity medications can indirectly improve exercise tolerance through weight loss.

Medications used for metabolic syndrome are mentioned throughout this chapter. Angiotensin II receptor blockers, other antihypertensive agents, and lipid-lowering medications are in the cardiovascular section of this chapter. Metformin and other antidiabetic drugs were discussed earlier. Rimonabant and the other antiobesity medications will be discussed in the following sections.

Orlistat

Orlistat (Xenical, Alli) is an intestinal lipase inhibitor, taken three times a day with meals, that causes malabsorption of 30% of dietary fat (19). It leads to 5 to 10% weight loss in 50 to 60% of patients (19). Orlistat can be used long term. Patients should be advised to eat a low-fat diet to prevent having have steatorrhea (19). The side effect profile includes mainly gastrointestinal effects such as abdominal pain, flatulence, fecal urgency, or incontinence (20). These adverse events improve with continued use of orlistat (20). Orlistat, available under the tradename Alli, was approved by the FDA in February 2007 to be sold over the counter and thus is readily accessible to the public.

Sibutramine

Another weight-loss agent available by prescription for long-term treatment is sibutramine (Meridia). Sibutramine promotes and prolongs satiety by inhibiting the reuptake of noradrenaline and serotonin (19). The expected weight loss is 5 to 10% in 60 to 70% of patients (19). Heart rate can increase because of the inhibition of noradrenaline reuptake (19). Some patients, in particular those who do not lose weight, may see a rise in blood pressure (19). Blood pressure must be monitored during the first 3 mo of treatment (19). Sibutramine should not be used in patients with history of coronary heart disease, stroke, congestive heart failure, or arrhythmias (20). Dry mouth, anorexia, insomnia, constipation, headache, dizziness, and nausea are additional side effects (19).

Rimonabant

Rimonabant has yet to get FDA approval in the United States for a primary indication for metabolic syndrome. It is the first cannabinoid-1 receptor antagonist and the first medication available for metabolic syndrome as its primary indication. Stimulation of the cannabinoid-1 receptors in the brain promotes eating (19). Therefore, blockade of the cannabinoid-1 receptors by rimonabant produces weight loss by reducing the urge to eat. Stimulation of these same receptors in peripheral tissues also results in low concentrations of high-density lipoprotein (HDL) cholesterol and decreased insulin resistance and inflammation, which all increase cardiovascular risk. Thus, by antagonizing the cannabinoid-1 receptors, rimonabant may produce improvements in cardiovascular risk profiles independent of weight loss (19). Clinical trial results suggest that rimonabant produces 5 to 10% weight loss in 60 to 70% of patients (19). Patients need to be assessed for anxiety, mood, and depression during treatment (19). Other adverse side effects include dizziness, nausea, and diarrhea (19).

Noradrenergic Agents

Phentermine, phendimetrazine, and diethylpropion are to be used only short term for weight-loss treatment (20). Their mechanism of action primarily relies on enhanced noriephinephrine or dopamine concentrations that stimulate the hypothalamus and result in decreased appetite (20). Therefore, these drugs are known as stimulants. The stimulants can produce weight loss, but many patients regain the weight during therapy or after discontinuation of therapy (21). These agents can also cause significant increases in blood pressure, heart rate, palpitations, and arrhythmias. They should not be used in hypertensive patients or in those with unstable cardiovascular diseases. Diabetic patients may see changes in their insulin or oral hypoglycemic doses after initiation of these medications (20). Insomnia is another major complaint. Because all three agents are stimulants, they are banned by the USOC, USADA, and NCAA (1,2).

CARDIOVASCULAR DISEASES AND AGENTS

Cardiovascular disease (CVD) is the number one killer of men and women in developed countries (22). Cardiovascular disease can encompass angina, arrhythmias, hypertension, myocardial infarction, congestive heart failure (CHF), and hypercholesteremia. Patients with CVD often receive multiple medical therapies that can affect exercise capacity. The clinical exercise physiologist is likely to encounter patients with CVD as they enter cardiac rehabilitation programs or other lifestyle modification programs designed for weight management and reduction of CVD risk factors (see table 3.2 on p. 36).

Angina

The symptoms of angina are caused by a mismatch between oxygen supply (decreased because of blockages in the coronary arteries) and oxygen demand (23). Patients with chronic stable angina will have enough blood flow at rest to meet metabolic needs because of internal processes that cause the coronary arteries to dilate. With exercise, however, the internal mechanisms are not sufficient to dilate the arteries further. A mismatch between myocardial oxygen supply and demand thus occurs, causing anginal symptoms. The only way to increase oxygen supply is through surgery, but medication can help decrease oxygen demand and thereby decrease anginal symptoms (23).

There are four types of antianginal agents: β-adrenergic antagonists, calcium channel antagonists, long-acting nitrate formulations for the prevention of chronic stable angina, and fast-acting nitrates used to relieve acute angina (23). All antianginals improve exercise capacity in the patient with chronic stable angina by decreasing anginal symptoms with exertion. They function as antianginal agents by lowering heart rate and blood pressure

Table 3.2 Cardiovascular Agents and Their Exercise Effects

Treatment choice	Medications	Heart rate	Blood pressure	ECG	Exercise capacity
Diuretic agents	Hycrochlorothiazide Furosemide spironolactone	↔ (R and E)	↔ or ↓ (R and E)	↔ or PVCs (R). May cause PVCs and false-positive test results if hypokalemia occurs; may cause PVCs if hypomagnesemia occurs (E)	↔, except possibly in patients with CHF
Antianginal agents	Nitrates	↑ (R) ↑ or ↔ (E)	↓ (R) ↓ or ↔ (E)	↑ HR (R) ↑ or ↔ HR (E) ↓ ischemia (E)	↑ in patients with angina; ↔ in patients without angina; ↑ or ↔ in patients with CHF
Antiarrhythmic agents[a]	Quinidine Disopyramide	↑ or ↔ (R and E)	↓ or ↔ (E)	↑ or ↔ HR (R). May prolong QRS complex and Q-T interval (R). Quinidine may result in false-negative test results (E)	↔
	Procainamide	↔ (R and E)	↔ (R and E)	May prolong QRS complex and Q-T interval (R); may result in false-positive test results (E)	↔
	Phenytoin Tocainide Mexiletine	↔ (R and E)	↔ (R and E)	↔ (R and E)	↔
	Flecainide Moricizine	↔ (R and E)	↔ (R and E)	May prolong QRS complex and Q-T interval (R) ↔ (E)	↔
	Propafenone	↓ (R) ↓ or ↔ (E)	↔ (R and E)	↓ HR (R) ↓ or ↔ HR (E) Prolongs P-R interval	↔
	Amiodarone Bretylium Sotalol Ibutilide Dofetilide	↓ (R and E)	↔ (R and E)	↓ HR (R) ↔ (E) Prolong Q-T interval and P-R interval to a lesser extent	↔
Calcium channel-blocking agents					
Antihypertensive agents	Verapamil Diltiazem	↓	↓	Prolong P-R interval; may cause heart block	No effect
	Amiodipine Isradipine Nifedipine Felodipine	No effect	↓	No effect	No effect

36

Treatment choice	Medications	Heart rate	Blood pressure	ECG	Exercise capacity
β-adrenergic blocking agents					
	Propranolol Metoprolol Timolol	↓	↓	Prolong P-R interval; may cause heart block	↓
Other antiadrenergic agents					
	Prazosin Terazosin Doxazosin	No effect	↓	No effect	No effect
	Carvedilol Labetalol	↓	↓	Prolong P-R interval; may cause heart block	May decrease
Centrally acting agents					
	Methyldopa Cloindine	May decrease	↓	May prolong P-R interval	May decrease
Vasodilators					
	Hydralazine Minoxidil	No effect	No effect	No effect	No effect
Angiotensin-converting enzyme inhibitors					
	Captopril Enalapril Lisinopril	No effect	No effect	No effect	No effect but may increase in patients with heart failure
Angiotensin receptor blockers					
	Losartan Valsartan Irbesartan	No effect	No effect	No effect	No effect
Cardiac glycoside agents	Digoxin	↓ (R)	No effect	May prolong P-R interval or cause heart block or various arrhythmias at toxic doses	No effect
Antilipemic agents	Lovastatin Simvastatin Fluvastatin Atorvastatin Cholestyramine Gemfibrozil	No effect	No effect	No effect	No effect
Anticoagulant agents	Warfarin	↔ (R and E)	↔ (R and E)	↔ (R and E)	↔

Note. [a] = All antiarrhythmic agents may cause new or worsened arrhythmias (proarrhythmic effect); ECG = electrocardiogram; R = rest; E = exercise; PVC = premature ventricular contraction; CHF = congestive heart failure; HR = heart rate; ↔ = no change; ↓ = decrease; ↑ = increase.

to decrease the work of the heart (decrease myocardial oxygen demand) by reducing preload on the heart.

β-Adrenergic Antagonists

Because β-blockers reduce the resting heart rate and blood pressure, they therefore reduce the myocardial demand for oxygen (24). See table 3.2 and the section "Hypertension, Congestive Heart Failure, Myocardial Infarction, and Acute Coronary Syndromes" for more detailed information on β-blockers.

Calcium Channel-Blocking Agents

Calcium channel antagonists prevent the blood vessels from constricting and thus prevent coronary artery spasm (25). They also dilate arteries in the peripheral vascular system to decrease the force that the heart has to pump against (decreasing oxygen demand). Certain calcium antagonists, such as verapamil and diltiazem, also slow the heart rate (26). Table 3.2 and the section "Hypertension, Congestive Heart Failure, Myocardial Infarction, and Acute Coronary Syndromes" have additional details about calcium channel blockers.

Nitrates

Nitrates such as nitroglycerin cause dilation of the blood vessels. Nitrates can be short acting or long acting. Short-acting nitroglycerin is available as a sublingual tablet or an oral spray. A tablet of nitroglycerin placed under the tongue or inhalation of the oral spray usually relieves an episode of angina in 1 to 3 min; the effect of these short-acting nitrates lasts 30 min. Anyone with chronic stable angina must carry nitroglycerin tablets or spray at all times (23). Long-acting nitrates are available as tablets, skin patches, or paste. Tablets are taken one to four times daily. Nitro paste and skin patches, in which the drug is absorbed through the skin over many hours, are also effective. Side effects of nitrates include the following (23):

Headache

Low blood pressure

Dizziness

Fainting

Arrhythmias

Cardiac arrhythmias (irregular heartbeat) can range from life threatening to asymptomatic (27). Drug treatment of cardiac arrhythmias is sometimes required either to prevent fatal arrhythmias or to decrease symptoms or other adverse effects that could be caused by the arrhythmia. Optimal therapy of cardiac arrhythmias requires accurate diagnosis and assessment of risk, optimal management of any precipitating factors, and careful selection of antiarrhythmic drug therapy if required.

Antiarrhythmic Agents

Antiarrhythmic drugs work by blocking calcium, sodium, and potassium channels in the electrical conduction system of the heart (27). The most serious adverse effect of antiarrhythmic agents is proarrhythmia, or the precipitation of a life-threatening arrhythmia from the antiarrhythmic drug itself. For that reason, judicious use of antiarrhythmic drugs is prudent. All antiarrhythmic agents can decrease exercise tolerance by decreasing contractility of the heart, which lowers cardiac output. In addition, exercise can sometimes precipitate an arrhythmia in susceptible individuals. See table 3.2 on p. 36 and 37 for a list of antiarrhythmic agents and their effects on heart rate, blood pressure, electrocardiogram, and exercise.

Hypertension, Congestive Heart Failure, Myocardial Infarction, and Acute Coronary Syndromes

Several classes of cardiovascular drugs are used to treat high blood pressure (28). They work by different mechanisms, and it is unclear if any agent is more effective at preventing the clinical sequela of hypertension. All are effective at lowering blood pressure.

Many of the agents used for the treatment of hypertension are also used for other cardiovascular indications. Diuretics can be used to lower blood pressure but are also used in patients with CHF to decrease the edema that they may experience. Though β-blockers are used to treat hypertension, they are the most prescribed drugs for patients during and after a myocardial infarction.

Anticoagulation is important in certain patients with cardiovascular diseases to prevent the formation of clots that can lead to stroke, myocardial infarction, and pulmonary embolism. Antiplatelet drugs are used for prevention and treatment in patients with myocardial infarction or in patients who receive stents during angioplasty. All these agents will be discussed in the following sections.

Diuretics

Thiazide diuretics are mostly used for the treatment of hypertension but may be used in patients who have mild edema from congestive heart failure or vascular insufficiency. The mechanism for their antihypertensive effects is not well defined but may be attributable, in part, to altered sodium handling by the kidney. Decreased blood volume is not likely to contribute to the antihypertensive effects because a decrease in blood pressure is noticed at doses far below those required to elicit net fluid losses (29). Thiazide diuretics have the following effects:

- Increase the urinary excretion of sodium and chloride in equivalent amounts
- Inhibit reabsorption of sodium and chloride in the ascending limb of the loop of Henle and the distal tubules
- Increase potassium and bicarbonate excretion
- Decrease calcium and uric acid excretion (29)

The thiazide diuretics include bendroflumethiazide, benzthiazide, chlorothiazide, chlorthalidone, hydrochlorothiazide, indapamide, methyclothiazide, metolazone, polythiazide, quinethazone, and trichlormethiazide. These agents are generally well tolerated, but some side effects occur:

- Exacerbations of diabetes or gout in certain people
- Phototoxicity (regular use of sunscreen advised)
- Electrolyte depletion, which may manifest as muscle cramping or weakness

Loop diuretics inhibit the reabsorption of sodium and chloride in the loop of Henle. These extremely potent diuretics can cause profound diuresis with subsequent electrolyte depletion (30). Therefore, they are commonly used to decrease congestion and fluid overload in patients with CHF. Potassium supplementation is often required. The loop diuretics include bumetanide (Bumex), ethacrynic acid (Edecrin), furosemide (Lasix), and torsemide (Demedex). The side effects of loop diuretics are the same as listed previously, and the potential to cause electrolyte disturbances increases.

Potassium-sparing diuretics interfere with sodium reabsorption in the distal tubule, thus inhibiting water and sodium reabsorption without affecting potassium (29). Because potassium is filtered at the glomerulus and then absorbed parallel to sodium throughout the proximal tubule and the ascending limb of the loop of Henle, only minor amounts reach the distal tubule. Potassium-sparing diuretics include amiloride, spironolactone, and triamterene. **Hyperkalemia** is the most serious side effect of the potassium-sparing diuretics, and concomitant potassium supplementation should be avoided (29). Coadministration of angiotensin-converting enzyme (ACE) inhibitors can precipitate hyperkalemia with these drugs. In addition, spironolactone (Aldactone)

may sometimes cause gynecomastia (enlargement of the breasts). Diuretics do not affect exercise response but are banned by both the USOC and the NCAA. They can be used to dilute the urine, mask the presence of prohibited substances, and reduce weight quickly (1,2).

Calcium Channel-Blocking Agents

Besides the calcium channel blockers (CCB) verapamil and diltiazem, which are also class IV antiarrhythmic agents, several other CCBs can be used to treat hypertension (31). These agents are collectively known as the dihydropyridine CCBs. Whereas the nondihydropyridine CCBs, verapamil and diltiazem, lower heart rate by inhibiting calcium influx in conduction tissue, the dihydropyridine CCBs cause vasodilation by inhibiting calcium influx in vascular smooth muscle (26). In fact, a common effect of short-acting dihydropyridines (nifedipine, isradipine) is that they may actually increase heart rate through a reflex mechanism resulting from rapid vasodilation (26). These agents are generally well tolerated but may cause lower extremity edema in some patients. Sustained-release products or long-acting agents such as amlodipine (Norvasc) are preferred over short-acting agents (28). Long-acting agents aid compliance by offering once daily dosing and cause less activation of the sympathetic nervous system because onset of vasodilation is more gradual. Dihydropyridine CCBs exhibit antiplatelet effects and can cause bruising. They also lower blood pressure and systemic vascular resistance. CCBs can increase exercise tolerance in patients with chronic stable angina by decreasing angina on exertion.

β-Adrenergic Blocking Agents

Catecholamines and β-adrenergic blocking agents compete with each other for available β-receptor sites. Propranolol, nadolol, timolol, penbutolol, carteolol, sotalol, and pindolol inhibit both the β_1-receptor (cardiac muscle) and β_2-receptors (bronchial and vascular smooth muscle) (32). Acebutolol, atenolol, betaxolol, bisoprolol, esmolol, and metoprolol are specific to the β_1-receptors at usual doses (32). At higher doses, specificity weakens. Nonspecific agents can precipitate wheezing in patients with reactive airway disease. In addition, unopposed β-adrenergic activity in the peripheral vasculature can cause vasoconstriction and may impair blood flow in patients with peripheral vascular disease (32). Metoprolol and bisoprolol decrease mortality rates in patients with New York Heart Association class II or III CHF (33). In patients experiencing an acute myocardial infarction, β-blockers must be initiated within 24 h and patients must be discharged home on β-blocker therapy if they have no contraindications to therapy (34). Patients with diabetes mellitus receiving insulin or oral hypoglycemic agents should be cautioned that β-blockers blunt the tachycardia associated with hypoglycemia (32). They can also cause sexual dysfunction. All β-blockers depress myocardial contractility, decrease heart rate and blood pressure, and can decrease exercise tolerance (32). In addition, β-blockers suppress tremors and can be abused by athletes in shooting sports. For this reason, they are banned by the NCAA for riflery and shooting only (2) and by the USOC in bobsled, ski jumping, freestyle skiing, archery, gymnastics, modern pentathlon, riflery and shooting, and sailing (1).

α-Adrenergic Blocking Agents

Prazosin, terazosin, and doxazosin selectively block postsynaptic α_1-adrenergic receptors (35). These drugs dilate both arterioles and veins. Orthostatic or postural hypotension can occur and is more common when these agents are initiated (35). Caution should be advised when using α-blockers and other vasodilators with erectile dysfunction medications, such as Viagra, because hypotensive episodes may occur. Otherwise, α_1-adrenergic blockers are generally well tolerated. The α-blockers do not affect exercise tolerance.

α/β-Adrenergic Blocking Agents

Labetalol and carvedilol are combination α/β-blockers. They inhibit β_1- and β_2-receptors as well as peripheral α_1-adrenergic receptors. Carvedilol has been shown to decrease mortality rates in patients with New York Heart Association class II or III CHF (36). It can cause dizziness and worsening of chronic heart failure when first initiated, and therefore it should be started at the lowest dose possible and titrated upward over several weeks. Labetalol and carvedilol can decrease exercise tolerance because of their β-adrenergic-blocking effects (33).

Centrally Acting Agents

Methyldopa, methyldopate, clonidine, reserpine, guanfacine, and guanabenz lower blood pressure by stimulating central inhibitory α-adrenergic receptors. Reduction in sympathetic outflow from the central nervous system decreases peripheral vascular resistance, renal vascular resistance, heart rate, and blood pressure (35). Sedation is a common side effect with these agents. Tolerance to centrally acting agents may occur and requires the addition of diuretics to the antihypertensive regimen to maintain efficacy. Although methyldopa is used less frequently than most antihypertensives, it is the drug of choice for pregnant patients with hypertension because of extensive experience in managing pregnant hyperten-

sive women with this agent (37). Clonidine (Catapres) is sometimes also used to aid in withdrawal from narcotic and nicotine addiction. These agents do not affect exercise response (35).

Vasodilators

Hydralazine and minoxidil exert a peripheral vasodilating effect through a direct relaxation of vascular smooth muscle. This results in a decrease in peripheral vascular resistance and a subsequent increase in cardiac output. Minoxidil can cause serious adverse effects including **pericardial effusion,** which is a collection of fluid in the lining of the heart that can lead to cardiac failure. Minoxidil can also result in excessive growth of hair on the palms of the hands and face, which has led to the development of topical formulations to treat male pattern baldness. These agents do not affect exercise response (38).

Angiotensin-Converting Enzyme Inhibitors

Angiotensin-converting enzyme (ACE) inhibitors prevent the formation of the potent vasoconstrictor angiotensin II (39). Reduction in circulating angiotensin II concentrations results in decreased peripheral vascular resistance with a subsequent increase in cardiac output and no change in heart rate. Most patients with diabetes mellitus are on an ACE inhibitor to protect their kidneys (28). In addition, ACE inhibitors are first-line treatment for chronic heart failure (40). Patients who have left ventricular systolic dysfunction and have a history of myocardial infarction should also be on an ACE inhibitor (41). ACE inhibitors are generally well tolerated, but a small percentage of patients may develop a cough. In addition, angioedema, a life-threatening allergic reaction, can rarely occur with all ACE inhibitors (39). ACE inhibitors increase exercise duration and tolerance in patients with congestive heart failure but otherwise have no effect on exercise capacity.

Angiotensin Receptor Blockers

Angiotensin receptor blockers (ARBs) inhibit the binding of angiotensin II to angiotensin II type 1 receptors, which block the vasoconstriction and aldosterone-secreting effects of angiotensin II (42). ARBs are generally used when patients cannot tolerate ACE inhibitors. They have a role in hypertension, CHF, and myocardial infarction in patients with left ventricular systolic dysfunction, and in patients with diabetes (28,40,41). Some available agents include losartan (Cozaar), valsartan (Diovan), candesartan (Atacand), and irbesartan (Avapro). ARBs do not affect exercise tolerance.

Aldosterone Blockers

Aldosterone blockers work at the final step of the renin–angiotensin–aldosterone system by preventing the binding of aldosterone to its receptors (43). The use of aldosterone blockers should be considered in patients with hypertension and heart failure (28). The two commonly used aldosterone blockers are the nonselective blocker spironolactone (Adactone) and the selective blocker eplerenone (Inspra). Side effects of spironolactone include antiandrogenic effects (gynecomastia, breast pain, impotence, decreased libido, menstrual irregularities) because of its activity at androgen and progesterone receptors and hyperkalemia because it has potassium-sparing effects (43). Hyperkalemia is also a side effect of eplerenone, but eplerenone is not associated with antiandrogenic effects because it is highly selective for the aldosterone receptor (43). These drugs do not affect exercise response.

Cardiac Glycoside Agents

The cardiac (or digitalis) glycosides include digitoxin and digoxin (44). Cardiac glycosides have a direct action on cardiac muscle that increases force and velocity of cardiac muscle contractions (inotropic actions). These inotropic effects result from inhibition of the sodium–potassium–adenosine triphosphatase pump in the sarcolemmal membrane that leads to increases in intracellular calcium (44). In addition, cardiac glycosides increase vagal tone to the atrioventricular node, which decreases heart rate and prolongs the P-R interval on the electrocardiogram (44). Serum drug concentrations of digoxin are often used to adjust dosing and check for toxicity. Cardiac glycosides are used in chronic heart failure and for heart rate control in atrial fibrillation or flutter (44). Digoxin toxicity may manifest as severe gastrointestinal complaints, visual disturbances, heart block, or any number of ventricular arrhythmias (44). Although digoxin slows the heart rate and increases the P-R interval on the electrocardiogram, these effects are obliterated by exercise when vagal tone to the atrioventricular node is overcome by the sympathetic activation produced by exercise (54). Digoxin has no effect on exercise capacity.

Anticoagulant Agents

Warfarin is used to prevent clot formation and subsequent stroke that can result from atrial fibrillation or prosthetic heart valves or to prevent recurrence of venous clots that cause deep venous thrombosis or pulmonary embolism (45). Warfarin interferes with the hepatic synthesis of vitamin K–dependent clotting factors, which results in depletion of factors VII, IX, X, and II (prothrombin)

(46). Because these factors are involved in stimulating coagulation, inhibition results in decreased propensity to form a blood clot. Warfarin has a narrow therapeutic index and extreme interindividual variability. For these reasons, it must be intensely monitored to ensure that the level of anticoagulation remains within acceptable parameters (45). Monitoring of warfarin therapy is achieved with the international normalized ratio (INR), a globally standardized measurement of anticoagulation (46). Common side effects of warfarin include bleeding, and patients should be counseled about signs and symptoms of acute blood loss (46). In addition, many significant drug interactions occur with warfarin, and patients should not stop or start any new drugs without notifying their healthcare providers. Patients should be advised to avoid activity that may lead to trauma, such as rock climbing, contact sports (basketball and football), and motor sports. Otherwise, warfarin has no effect on exercise capacity.

Antiplatelet Agents

Aspirin produces its antiplatelet effects by inhibiting the synthesis of thromboxane A_2 by irreversibly inhibiting the synthesis of cyclooxygenase-1. Aspirin also has anti-inflammatory properties. Its role in acute coronary syndromes such as unstable angina and non-ST and ST myocardial infarctions is well established (34,47). Most cardiac patients will be on a daily dose of aspirin (75–162 mg) if they can tolerate the side effects of bleeding and gastrointestinal upset and do not have an aspirin allergy (34).

Clopidogrel and ticlodipine are thienopyridines. They work by blocking the adenosine diphoshate (ADP) receptors on platelets. Clopidogrel has become the preferred medication because ticlodipine is associated with neutropenia and its use in patients requires frequent monitoring of complete blood count (48). Clopidrogrel has a role in patients with ST-segment elevation myocardial infarction or acute coronary syndrome (ACS) with an aspirin allergy, in patients with non-ST-segment elevation ACS managed medically, and in patients receiving percutaneous coronary intervention (PCI) (41,47,49). The most serious side effect with clopidogrel is bleeding.

Hypercholesteremia

Cholesterol is produced by the liver and can also enter the bloodstream through dietary absorption. Although cholesterol is needed for normal cell function, excess amounts in the bloodstream increase the risk of atherosclerotic diseases (50). Medications used to treat dyslipidemia are classified in the following categories: statins, bile acid sequestrants, nicotinic acid, fibric acids, probucol, and cholesterol absorption inhibitors. Estrogen replacement therapy (ERT) and hormone replacement therapy (HRT) are no longer used in the treatment of hypercholesteremia because of findings that they may increase the risk of cardiovascular disease in women (51,52).

The statin agents are currently the most effective and commonly prescribed lipid-lowering medications. They competitively inhibit 3-hydroxy-3-methyl-glutaryl-coenzyme A (HMG-CoA) reductase, the rate-limiting enzyme in cholesterol synthesis (50,53). The statins potently decrease low-density lipoprotein (LDL) cholesterol and modestly increase high-density lipoprotein (HDL) cholesterol (53). Common side effects include bloating and abdominal cramping. Liver function tests can be elevated, but hepatic failure is extremely rare. **Rhabdomyolysis** with renal dysfunction or acute renal failure secondary to **myoglobinuria** has occurred. Patients should be advised to report myalgias or muscle weakness as soon as possible. In addition, muscle soreness without other complications has been reported with atorvastatin and fluvastatin (50). Risk of systemic muscle damage and breakdown is increased when these agents are taken with cyclosporine, erythromycin, gemfibrozil, fibric acid derivatives, or lipid-lowering doses of niacin (53). Table 3.3 lists additional medications used to treat dysplipidemia (54–57).

PULMONARY DISEASES AND AGENTS

The major pulmonary diseases that will be covered are asthma, chronic obstructive pulmonary disease (COPD), and cystic fibrosis. These pulmonary diseases are treated with numerous classes of medications. The classes of drugs are not disease specific but are used to treat the symptoms, causes, or consequences of the disease. Pulmonary diseases, especially asthma and COPD, are common and have a direct effect on exercise tolerance.

Asthma and COPD

Pharmacologic treatment is used to prevent and control asthma symptoms, reduce the frequency and severity of exacerbations, and reverse airflow obstruction (58). Asthma is a chronic disorder, as is COPD. Asthma medications are classified as controller (taken daily to achieve and maintain control of persistent asthma) and rescue (quick relief to provide prompt reversal of airflow

obstruction and bronchoconstriction) (58). COPD has many similarities to asthma, and many of the medications used in asthma are also used in the treatment of COPD.

Cystic Fibrosis

Cystic fibrosis (CF) is a common genetic disease (59). The disease affects a number of organ systems, but the effects on the pulmonary system are important to an exercise physiologist. A number of drugs are used to help alleviate and treat the pulmonary symptoms that a patient with CF may experience. Many of these agents affect exercise tolerance. Table 3.4, on p. 44, includes medications used in lung disorders such as asthma, COPD, and CF (58,59).

IMMUNOLOGIC DISEASES AND AGENTS

The two major immunologic diseases discussed in this section are cancer and acquired immune deficiency syndrome (AIDS)–human immunodeficiency virus (HIV). Several medications are used to treat these two diseases. Most of these agents do not influence exercise capacity.

Cancer

Chemotherapy may be given as primary, palliative, adjuvant, or neoadjuvant therapy (60). Chemotherapy is the primary treatment for only a few cancers (leukemias, lymphomas, choriocarcinomas, and testicular cancer) (60). Chemotherapy is used palliatively to decrease tumor size and improve patient symptoms and quality of life (60,61). Adjuvant chemotherapy is defined as the use of systemic agents to eradicate any stray cancer cells remaining in the body following local treatment such as surgery or radiation (60,61). Neoadjuvant chemotherapy is used to decrease tumor burden so that other treatment modalities are more effective (60,61). The goals of treatment with chemotherapy are cure, control of the disease by stopping cancer growth and metastasis to extend life and provide quality of life, or palliation (61).

The era of modern cancer chemotherapy began in 1941, when nitrogen mustard was administered to lymphoma patients (60,62). Currently, a number of antineoplastic drugs, biologically directed therapies, and immune therapies are available. Cancer chemotherapy agents are commonly categorized by their mechanism of action or by their origin (60).

Table 3.3 Medications for the Treatment of Dyslipidemia

Medications	Class and primary effects	Exercise effects	Other effects	Special considerations
Atorvastatin Fluvastatin Lovastatin Pravastatin Rosuvastatin Simvastatin	Statins: competitive inhibitors of the enzyme HMG-CoA reductase, reducing the hepatic synthesis of cholesterol	May experience myopathy, myalgia, muscle weakness, muscle cramps during exercise	↓ LDL ↑ HDL ↓ TG	Major and rare side effects include hepatotoxicity and myopathy. Common side effects include myalgia and constipation. Other potential side effects include muscle weakness and muscle cramps. Liver function tests need to be monitored
Cholestyramine Colestipol Colesevelam	Bile acid sequesterants: bind bile acid in the gut, promoting increased cholesterol to bile acid synthesis	None	↓ LDL ↑ HDL ↑ TG	Contraindicated in patients with hypertriglyceridemia. Side effects include constipation, bloating, epigastric fullness, flatulence, and nausea. Can interfere with absorption of nutrients and medications. Must separate administration times from other medications
Niacin (immediate release) Niaspan (extended release)	Nicotinic acid: reduces hepatic synthesis of VLDL, which reduces the synthesis of LDL	None	↓ LDL ↑ HDL ↓ TG ↓ BP	Side effects include skin irritation and flushing; may worsen glucose tolerance and activate gout and hepatotoxicity. Start patients at low doses and slowly increase
Clofibrate Gemfibrozil Fenofibrate	Fibrates: mainly TG-lowering by TG hydrolysis	When combined with statins, may cause muscle weakness or pain during exercise	↓ LDL ↑ HDL ↓ TG	Common side effects are upper-gastrointestinal disturbances. May increase liver function tests. Risk of myopathy in combination with statin
Probucol	Inhibits LDL oxidation and increases LDL clearance	None	↓ LDL ↓ HDL	Can prolong QT interval on ECG
Ezetimibe	Cholsteral absorption inhibitor: inhibits intestinal absorption of cholesterol and phytosterols	May cause fatigue, or when combined with statins may cause muscle weakness or pain during exercise	↓ LDL ↑ HDL ↓ TG	Side effects include stomach pain and feelings of tiredness. Myopathy is more common when combined with a statin

LDL = low-density lipoprotein; VLDL = very low-density lipoprotein; HDL = high-density lipoprotein; TG = triglycerides; BP = blood pressure; ECG = electrocardiogram.

Antineoplastic or Cytotoxic Agents

The most known chemotherapeutic agents are the antineoplastic, or cytotoxic, drugs. These agents work at specific phases of the cell cycle, such as the synthesis of DNA or of the mitotic spindle (62). Cytotoxic drugs have many serious adverse effects. Because many of these agents affect DNA synthesis, rapidly dividing cells such as hair follicles, bone marrow, and the intestinal mucosa are more sensitive to the toxic effects of chemotherapy (60). Common side effects include hair loss, myelosuppression such as anemia, neutropenia, thrombocytopenia, mucositis, nausea and vomiting, and infertility (60). Drug-specific toxicities include doxorubicin-induced or daunorubicin-induced cardiotoxicity and bleomycin-related pulmonary toxicity (60).

Alkylating agents

The alkylating chemotherapy drugs are amongst the oldest and most useful antineoplastic agents (60). They can damage DNA during any phase of the cell cycle by inhibiting DNA replication. They work by bonding highly reactive

Table 3.4 Medications or Therapies for the Treatment of Various Pulmonary Disorders

Medications or therapies	Class and primary effects	Exercise effects	Special considerations
Albuterol (short-acting β-agonist) Bitolterol (short-acting β-agonist) Pirbuterol (short-acting β-agonist) Terbutaline (short-acting β-agonist) Salmeterol (long-acting β-agonist) Ipratropium (anticholinergic) Theophylline (methylxanthine)	Bronchodilators—decrease bronchospasm by opening up the airways	Improved ventilation during exercise, especially in patients with bronchospasm. Short-acting and long-acting β-agonists may cause cardiovascular stimulation. Salmeterol can cause skeletal muscle tremor and hypokalemia. Ipratropium may increase or have no effect on heart rate. Theophylline may produce premature ventricular contractions on the EKG at rest and during exercise	Timing of administration before exercise will affect gains. Short-acting β-agonists and ipratropium are used for rescue relief. Salmeterol and theophylline are controller medications. Salmeterol and albuterol are used in prevention of exercise-induced bronchospasm. Salmeterol is not used to treat acute symptoms or exacerbations. Ipratropium is treatment of choice for bronchospasm because of β-blockers. Side effect of ipratropium is dry mouth. Theophylline has gastrointestinal side effects
Prednisone (systemic steroid) Prednisolone (systemic steroid) Methylprednisone (systemic steroid) Beclomethasone (inhaled steroid) Budesonide (inhaled steroid) Flunisolide (inhaled steroid) Fluticasone (inhaled steroid) Triamcinolone (inhaled steroid) Zafirlukast (leukotriene modifier) Zileuton (leukotriene modifier) Cromolyn sodium (cromone) Nedocromil (cromone) Omalizumab (anti-IgE) Ibuprofen (NSAID)	Anti-inflammatories—decrease inflammatory component of the disease	None	Steroids may increase risk of osteoporosis, hypertension, diabetes, adrenal suppression, obesity, cataracts, muscle weakness, bruising, skin thinning, growth suppression. Inhaled steroids can additionally cause oral thrush, cough, and dysphonia. Inhaled steroids are used as controller medications. Systemic steroids can be used as controller and rescue therapy. Leukotriene modifiers are used as controller therapy. Drug interactions and liver toxicity are possible. Cromolyn sodium and nedocromil are controller medications and can be used as preventative treatment before exposure to exercise or known allergen. Side effects are cough and sore throat. Omalizumab is used in patients with elevated IgE level who are uncontrolled on inhaled steroids
A variety of oral or intravenous antibiotics (i.e., amoxicillin)	Antibiotics—prevent, treat, or decrease bacterial load	None	
Tobramycin Gentimycin Colistin Polymixin B	Inhaled antibiotics—prevent, treat, or decrease bacterial load	None	May induce bronchospasm, vocal alteration, or hoarseness

Medications or therapies	Class and primary effects	Exercise effects	Special considerations
Nebulized hypertonic saline Dnase N-acetylcysteine	Mucolytics—reduce viscosity of pulmonary secretions or sputum	Improved mucociliary clearance; enhanced cough; some improvement in pulmonary function	May induce bronchospasm or cough. N-acetylcysteine has unpleasant taste and odor
Loratadine Cetirizine Fexofenadine	Antihistamines	Improved ventilation with exercise	Timing of administration before exercise will affect gains
Airway clearance techniques Mucous clearance techniques			Lack of appropriate techniques could result in decreased exercise ability

alkyl groups with nucleophilic groups of proteins and nucleic acids (60). Some of the most commonly used alkylating agents are cyclophosphamide, carmustine, cisplatin, dacarbazine, ifosfamide, oxaliplatin, and temozolomide.

Antimetabolite agents

These chemotherapeutic medications resemble naturally occurring nuclear structural components, such as nucleotide bases, or inhibit enzymes involved in the synthesis of DNA and proteins (60). Common medications include fluorouracil (5-FU), capecitabine, cytarabine (ara-C), gemcitabine, hydroxyurea, 6-mercaptopurine, and methotrexate.

Antimicrotubule agents

The antimicrotubule agents (docetaxel, paclitaxel, estramustine, vinblastine, vincristine, vinorelbine) disrupt microtubule function needed for the formation of the mitotic spindle and therefore interfere with mitosis during cell division (60).

Topoisomerase-active agents

Irinotecan, topotecan, daunorubicin, doxyrubicin, epirubicin, etoposide, idarubicin, and mitoxantrone are topoisomerase inhibitors. They inhibit the topoisomerase I or II enzymes, which are necessary to maintain DNA structure during replication and transcription (60).

Biologically Directed Agents

Biologically directed therapies, such as endocrine therapies, monoclonal antibodies, tyrosine kinase inhibitors, proteosome inhibitors, and tumor vaccines, target malignant cells more specifically or the biochemical processes that control neoplastic cell growth (60). Endocrine therapies are mostly used in breast (anastrazole, exemestane, letrozole, tamoxifen), prostate (bicalutamide, flutamide, goserelin, leuprolide), and endometrial (megestrol acetate) cancers because these cancers rely on hormones for growth. The monoclonal antibiodies are bevacizumab, cetuximab, rituximab and trastuzumab. Gefitinib (Iressa) and imatinib (Gleevac) are tyrosine kinase inhibitors, and bortezomib (Velcade) is a proteosome inhibitor.

Immune-Directed Agents

Immune therapy involves stimulating the patient's immune system to fight the cancer (60). Interferon alfa and aldesleukin (interleukin-2, IL-2) are used in immunotherapy. A common side effect of both therapies is flulike syndrome: fever, malaise, chills, headache, and fatigue. Because interferon alfa may cause excessive daytime sedation, it should be administered in the evening (60). An adverse effect of aldesleukin is vascular or capillary leak syndrome, which can manifest as hypotension, pulmonary, and peripheral edema (60). In addition, patients with a history of cardiac arrhythmia may require cardiac monitoring during aldesleukin therapy (60).

AIDS–HIV

The major cause of AIDS is infection with the retrovirus human immunodeficiency virus type 1 (HIV-1) (63). The goals of therapy for the treatment of HIV-1 infected patients are to reduce HIV-related morbidity and mortality, improve quality of life, restore and preserve immunologic function, and maximally and durably inhibit the replication of the HIV-1 virus to less than detection levels by using antiretroviral therapy (64).

There are four classes of antiretroviral therapy: the nucleoside–nucleotide reverse transcriptase inhibitors (NRTIs), the nonnucleoside reverse transcriptase inhibitors (NNRTIs), the protease inhibitors (PIs), and the fusion inhibitors (FIs) (64). Their mechanisms of action target different points of the HIV viral replication cycle. The

HIV virus is a retrovirus because it works backward by transcribing RNA into DNA to replicate in human cells using an enzyme called reverse transcriptase (63). NRTIs and NNRTIs are inhibitors of reverse transcriptase. PIs block the protease enzyme, which is responsible for viral maturation, resulting in the production of immature, non-infectious virions (63–65). FI or entry inhibitors work by preventing the fusion of HIV with the target cell (63). Currently only one FI is available on the market.

Table 3.5 summarizes the antiretroviral agents available for HIV-1 infected people (63,64,66). Clinicians may use these agents in multiple combinations. Most have many adverse effects and drug interactions. The NRTIs all have a risk for lactic acidosis, hepatic steatosis, and fat redistribution or accumulation (66). Also, long-term use of NRTIs may result in a decrease in $\dot{V}O_2$ maximum. Protease inhibitors can affect glucose metabolism and cause fat redistribution, hypertriglyceridemia, and hypercholesterolemia (66). In addition, protease inhibitors have several drug interactions, as do efavirenz, nevirapine, and delavirdine (63). None of the antiretroviral agents has a direct effect on exercise capacity.

ORTHOPEDIC AND MUSCULOSKELETAL DISEASES AND AGENTS

The exercise physiologist may encounter a number of patients with orthopedic and musculoskeletal diseases.

Table 3.5   Medications for the Treatment of HIV-1 Infection

Medications	Exercise effects	Special considerations
Nucleoside and nucleotide reverse transcriptase inhibitors (NRTIs)		
Abacavir (Ziagen)	None	Side effects: headache, nausea, diarrhea, malaise, hypersensitivity reactions
Abacavir, lamivudine, and zidovudine (Trizivir)—combination product	None	See individual components for side effects
Didanosine (Videx; Videx EC)	None	Side effects: diarrhea, nausea, rash, headache, fever, hyperuricemia, pancreatitis, peripheral neuropathy. Rarely used
Emtricitabine (Emtriva)	None	Side effects: headache, diarrhea, nausea, rash, skin hyperpigmentation, exacerbation of hepatitis B on stopping use of drug
Emtricitabine and tenofovir (Truvada)—combination product	None	See individual components for side effects
Lamivudine (Epivir)	None	Well tolerated; side effects seen when given with zidovudine. Exacerbation of hepatitis B can occur on stopping use of drug
Lamivudine and abacavir (Epzicom)—combination product	None	See individual components for side effects
Lamivudine and zidovudine (Combivir)—combination product	None	Side effects: headache, nausea, diarrhea, abdominal pain, insomnia. See individual components for side effects
Stavudine (Zerit)	None	Side effects: diarrhea, nausea, vomiting, headache, peripheral neuropathy, pancreatitis, hyperlipidemia. Highest incidence of lipoatrophy, which can be exacerbated by exercise, hyperlipidemia, and lactic acidosis
Tenofovir (Viread)	None	Side effects: headache, nausea, vomiting, exacerbation of hepatitis B can occur on stopping use of drug, cases of renal dysfunction
Zalcitabine (Hivid)	None	Side effects: oral ulcers, rash, pancreatitis, peripheral neuropathy, stomatitis, hepatic failure in patients with hepatitis B
Zidovudine (Retrovir; AZT)	None	Side effects: nausea, anorexia, vomiting, headache, asthenia, insomnia, myalgias, malaise, nail pigmentation, macrocytosis, bone marrow suppression

Medications	Exercise effects	Special considerations
Nonnucleoside reverse transcriptase inhibitors (NNRTIs)		
Delavirdine (Rescriptor)	None	Side effects: nausea, diarrhea, vomiting, headache, skin rash, increase in liver enzymes
Efavirenz (Sustiva)	None	Side effects: dizziness, insomnia, dreams, somnolence, confusion, rash, severe depression, suicidal ideation, elevation of liver enzymes, false-positive urine test for cannabinoid, pregnancy category D. Stop use of efavirenz 1–2 wk before taking other drugs to avoid development of resistance
Nevirapine (Viramune)	None	Side effects: rash, severe life-threatening skin reactions, hepatotoxicity. Stop use of nevirapine 1–2 wk before taking other drugs to avoid development of resistance
Protease inhibitors (PIs)		
Atazanavir (Reyataz)	None	Side effects: asymptomatic unconjugated hyperbilirubinemia, jaundice, diarrhea, nausea, abdominal pain, headache, rash. Prolongation of PR interval (first-degree AV block) reported; rarely second-degree block
Fosamprenavir (Lexiva)	None	Side effects: skin rash, nausea, vomiting, diarrhea, Steven-Johnson syndrome, hemolytic anemia
Indinavir (Crixivan)	None	Side effects: ↑ indirect bilirubin, nausea, vomiting, diarrhea, kidney stones, nephrolithiasis. Patients should be well hydrated
Lopinavir and ritonavir (Kaletra)— combination product	None	Side effects: nausea, vomiting, diarrhea, lipid abnormalities, pancreatitis, increase in liver enzymes, oral solution product 42% alcohol
Nelfinavir (Viracept)	None	Side effects: nausea, vomiting, diarrhea
Ritonavir (Norvir)	None	Side effects: bitter aftertaste, nausea, vomiting, diarrhea, paresthesias, lipid abnormalities, hepatitis, pancreatitis, ↑ CPK and uric acid. Used primarily to enhance the action of other PIs. Many potentially fatal drug interactions
Saquinavir (Invirase) and ritonavir	None	Side effects: diarrhea, abdominal discomfort, nausea, headache, increased liver enzymes
Tipranavir (Aptivus)	None	Side effects: nausea, vomiting, diarrhea, abdominal pain, rash, lipid effects, hepatitis, hepatic failure
Fuzeon inhibitor (FI)		
Enfuvirtide (Fuzeon; T20)	None	Side effects: local injection site reactions because injection is subcutaneous, diarrhea, nausea, fatigue, peripheral neuropathy, insomnia, ↓ appetite, myalgia, eosinophilia, ↑ bacterial pneumonia, hypersensitivity reactions

The major orthopedic disease that will be covered in this chapter is osteoporosis. Arthritis and lower-back pain are the musculoskeletal diseases that will be discussed that may affect a patient's exercise capacity.

Osteoporosis

Osteoporosis is characterized by low bone mass, micro-architectural deterioration of bone, compromised bone strength, and an increase in fracture risk (67). The major complication of osteoporosis is fracture, which may lead to bone pain, disability, and surgery (68). The consequences of osteoporosis will affect the patient's exercise tolerance.

Most of the current drugs approved by the Food and Drug Administration (FDA) for osteoporosis are considered antiresorptive therapy. They halt the loss of bone or even increase bone mass by inhibiting bone resorption, while having no effect on bone formation. See table 3.6 on p. 49 for a summary of medications used in the prevention and treatment of osteoporosis.

Calcium Supplementation

Adequate calcium intake is considered standard for the prevention and treatment of osteoporosis (68). Calcium carbonate is the most commonly used dietary calcium supplement. Most calcium supplements should be taken with food because absorption is increased. Doses vary from 200 to 1,500 mg per day of elemental calcium. The National Osteoporosis Foundation (NOF) recommends at least 1,200 mg of dietary calcium daily, including supplements if necessary (67). Many drug interactions deal with absorption, so caution should be used. Common side effects include constipation, gas, and upset stomach.

Vitamin D Supplementation

Vitamin D is also responsible for maintaining calcium homeostasis. Normal doses of vitamin D are 200 to 1,000 units per day. NOF states that people at risk for a deficiency should take 400 to 800 units daily of vitamin D (67). If malabsorption, drug interaction, or deficiency issues are present, then higher doses may be necessary. The patient's calcium levels must be monitored closely because hypercalcemia is a major side effect.

Bisphosphonates

Bisphosphonates are the most powerful of the antiresorptive drugs being used to prevent and treat osteoporosis. Bisphosphonates increase bone mass by drastically reducing bone resorption (69), similar to the actions of estrogen or calcitonin. Bisphosphonates have been shown effective in reducing bone loss and decreasing fracture risk in postmenopausal women and men, as well as increasing bone strength (70–72). Bisphosphonates may also be useful in preventing the loss of bone that may occur from long-term corticosteroid treatment (73,74). Oral bisphosphonates are poorly absorbed, and absorption can decrease if the recommended mode of administration is not followed. Side effects are the following (68):

Nausea

Abdominal pain

Dyspepsia

Active ulcer

Patients who begin bisphosphonate therapy should be monitored for any of these side effects. Recently approved once-weekly and once-monthly doses of bisphosphonates appear to have equal efficacy and side effects (75,76), and the extended dosing interval may improve compliance (77).

Selective Estrogen Receptor Modulators (SERMs)

Raloxifene is the first SERM with FDA approval for use in prevention and treatment of osteoporosis. The increases in BMD that are shown when taking raloxifene or other SERMs are generally less than those seen with other antiresorptive medication (78,79), but reduction of fracture risk has been demonstrated. The major side effects of raloxifene use are increased risk of venous blood clots and hot flashes, but favorable changes in serum lipids have been demonstrated (80). Raloxifene has also been shown to reduce the risk of breast cancer (81). SERMs are prohibited by the USADA because they can change the response to estrogen in the human body (1).

Calcitonin

Calcitonin is a hormone used for calcium regulation by the body. Salmon calcitonin, in either an injectable or a nasal spray preparation, has been proved effective in both increasing low bone mass and decreasing fracture risk in postmenopausal women (82–84). It is approved only for treatment of osteoporosis, not prevention.

Anabolic Agents

Anabolic agents increase bone formation and usually result in increases in bone mineral density (BMD) much greater than those seen with antiresorptive therapy. But the increase in bone formation also results in an increase in bone turnover, and it is thought that a reduction in overall bone turnover is important in reducing fracture risk (85). Nevertheless, the one anabolic agent that has received FDA approval for treatment of osteoporosis, teriparatide, has been found to increase BMD to twice the degree that aldendronate does (86) as well as to reduce fracture risk (87). Other potential anabolic agents under investigation include sodium fluoride, growth hormone, insulin-like growth factor I, and statins, but results are not yet conclusive and long-term trials that evaluate fracture risk have not be conducted.

Estrogen and Hormone Replacement Therapy

Estrogen replacement therapy (ERT, estrogen alone) and hormone replacement therapy (HRT, estrogen in combination with progestin) have been used in the treatment and prevention of osteoporosis in postmenopausal women. Studies have shown that ERT or HRT can halt the loss of bone mass and often increase it (88–90). The results of the Women's Health Initiative (WHI) also established the efficacy of ERT and HRT in the reduction of fracture risk. The major finding of the WHI, however, was that both HRT and ERT resulted in increased risk of cardiovascular disease and that HRT also increased the risk for certain cancers (51,52). These results from the WHI suggest that ERT and HRT should probably no longer be recommended as a treatment for osteoporosis because other equally effective therapies are available (51,52,91). If ERT or HRT is used, the lowest effective dose should be administered.

Arthritis and Lower-Back Pain

Arthritis is an inflammatory condition that involves the joints. Some clinical symptoms include fatigue, weakness, joint pain and swelling, muscle aches, and stiff muscles (92). Lower-back pain includes pain in the lumbosacral area that may have a musculoskeletal cause. Patients may experience acute or chronic lower-back pain.

A number of disease states require the use of medications for pain. Patients with lower-back pain use analgesics to relieve their symptoms. Patients with osteoarthritis and rheumatoid arthritis can take a combination of different classes of drugs to help manage their diseases. Pharmacological therapies can be administered orally, topically, or directly into joints. Topical agents include capsaicin cream. Two commonly used oral drug groups are the nonsteroidal anti-inflammatory drugs (NSAIDs) and the selective COX-2 inhibitors. Additional disease-modifying agents and classes of medications are summarized in table 3.7.

Table 3.6 Medications for the Prevention and Treatment of Osteoporosis

Medications	Class and primary effects	Exercise effects	Other effects	Special considerations
Calcium	Supplement—helps maintain adequate calcium levels in body	None	Used for prevention and treatment	May decrease absorption of many drugs. Food helps increase absorption so calcium should be taken with food. Side effects: constipation, gas, upset stomach, kidney stones (rare)
Vitamin D	Supplement—helps maintain calcium homeostasis	Helps maintain muscle function and decreases pain, which can help with exercise tolerance	Used for prevention and treatment of osteoporosis and falls	Cholestryamine, colestipol, orlistat, mineral oil can decrease absorption of vitamin D. Some drugs can increase metabolism of vitamin D
Alendronate Risedronate Ibandraonate	Biphosphonates—reduce bone resorption	Causes GI distress that may interfere with exercise	Used for prevention and treatment of osteoporosis	Do not coadminister with any medications including calcium. Must advise patient to remain upright for at least 30 min after consumption of drug. Major side effects: nausea, GI irritation, perforation, ulceration, and bleeding
Raloxifene	SERM—reduces bone resorption	None	Used for prevention and treatment of osteoporosis; has positive effects on lipid profile and breast cancer prevention	Side effects: hot flashes, leg cramps, and venous thromboembolism
Calcitonin	Reduces bone resorption	None	Used for treatment of osteoporosis	Intranasal product can cause rhinitis and nose bleeds

(continued)

Table 3.6 *(continued)*

Medications	Class and primary effects	Exercise effects	Other effects	Special considerations
Teriparatide	Parathyroid hormone— anabolic agent that stimulates bone formation	None	Use for patients at high risk of osteoporosis-related fracture who cannot or will not take or have failed bisphosphonate therapy	Pain at injection site because injection is subcutaneous, dizziness, leg cramps. Contraindicated in Paget's disease of the bone, unexplained elevations in alkaline phosphatase, or history of skeletal radiation therapy

Table 3.7 Medications for the Treatment of Arthritis and Lower-Back Pain

Medication	Primary effects	Exercise effects	Special considerations
Nonnarcotic analgesic			
Acetaminophen Paracetomol	CNS analgesic and antipyretic effects (no peripheral anti-inflammatory activity)	May decrease rating of perceived exertion, may inhibit posteccentric exercise skeletal muscle protein synthesis (93)	Used in OA, RA, AS, lower-back pain; caution required in liver and renal disease
Nonsteroidal anti-inflammatory drugs (NSAID) Aspirin Ibuprofen Naproxen Ketoprofen Indomethocin Sulindac Diclofenac Tolmetin Nabumetone Meloxicam Piroxicam Etodolac Ketorolac	Anti-inflammatory analgesic, antipyretic, and antiplatelet effects: inhibit cyclooxygenases 1 and 2 to decrease prostaglandin synthesis	Improve postexercise muscle soreness, may decrease postexercise skeletal muscle protein synthesis (of unknown clinical significance) (108), may decrease postexercise renal function (glomerular filtration rate) and increase risk of renal failure (acute tubular necrosis) in sodium- and volume-depleted states (94,95)	Used in OA, RA, AS, lower-back pain; caution required in renal and liver disease; increases risk of gastrointestinal bleeding and ulceration; produces reversible platelet inhibition (except aspirin, which is irreversible); decreases renal blood flow and worsens hypertension; may increase cardiovascular risk including thrombosis, myocardial infarction, and stroke
Cyclooxygenas II specific inhibitors: Celecoxib	Anti-inflammatory, analgesic, and antipyretic effects: specifically inhibit cyclooxygenase 2 to decrease prostaglandin (PGI2) synthesis	Probably similar to NSAIDs	Used in OA, RA, AS, lower-back pain; caution required in renal and liver disease; less gastrointestinal side effects than traditional NSAIDs; does not alter platelet aggregation; potential increase in cardiovascular events because of COX-2 specific reduction of PGI_2, producing a relative increase in thromboxane A2, which favors platelet aggregation and thrombosis

Medication	Primary effects	Exercise effects	Special considerations
Narcotic analgesics			
Tramadol	Analgesic effects, weak inhibition of norepinephrine and serotonin reuptake, weak opioid activity	None noted	Used in OA, RA, AS, lower-back pain; caution required in renal and liver disease; less addictive and less potential than other opioids
Hydroxycodone Oxycodone Morphine Hydromorphone Methadone Codeine Fentanyl	Analgesic effects, stimulate central opiate receptors	Appear not to alter blood pressure or heart rate during isometric exercise (96,97), proportionally reducing ventilation and metabolic rate during exercise (98)	Used in OA, RA, AS, lower-back pain; caution required in respiratory, renal, and liver disease; can cause hypotension and bradycardia
Nutraceuticals			
Glucosamine Chondroitin S-adenosyl-L-methionine (SAM-e) Methylsulfonyl-methane (MSM)	May enhance cartilage repair and inhibit cartilage degradation, but little data exists	None noted	Used in OA. Because nutraceuticals are not FDA regulated, scientific data validating efficacy are generally less comprehensive and abundant than for FDA-regulated agents and regulations related to standardized production are less stringent
Corticosteroids			
Prednisone Prednisolone Methylprednisolone Dexamethasone	Potent anti-inflammatory effects. Hormones bind intracellular receptors with numerous effects on immune system and inflammatory components	Associated with reduction of approximately three beats per minute in resting and maximal heart rates without effect on peak $\dot{V}O_2$, ventilatory threshold, maximal lactate, or blood pressure (99). Prolonged use associated with muscle weakness wasting, and myopathy. Intra-articular injection into arthritic joint can enhance tolerance of activity and exercise and is associated with minimal side effects	Used in OA (intra-articular injections), RA (intra-articular injections, oral), AS (intra-articular injections, oral), lower-back pain. Prolonged use has many side effects including the following: possible tuberculosis reactivation and increased risk of other infections; precipitate or exacerbate diabetes mellitus; promote muscle wasting (upregulation of ubiquitin-proteosome pathway, primarily in type 2 fibers) and myopathy (100); reduce bone density. Edema and hypertension. Associated with increased cardiovascular disease and events including atherosclerosis, myocardial infarctions, stroke, and congestive heart failure
Disease-modifying agents			
Methotrexate Leflunomide Hydroxy-chloroquine Gold Azathiprine Sulfasazine	Anti-inflammatory effects, multiple mechanisms	None noted beyond increased tolerance of exercise as expected because of improvement of disease with effective disease modification	Used in RA, AS. Caution required in immunosuppressed states, renal and hepatic disease. Most associated with serious toxicities (hepatic, renal, pulmonary, hematologic), requiring monitoring by experienced clinicians. Not used in treatment of osteoarthritis (except rarely hydroxychloroquine). Hydroxychloroquine can rarely cause muscle weakness

(continued)

Table 3.7 *(continued)*

Medication	Primary effects	Exercise effects	Special considerations
Disease-modifying agents—biologics			
Infliximab Etanercept Adalimumab Anakinra Rituximab	Anti-inflammatory effects, binds inflammatory molecules or cells	None noted beyond increased tolerance of exercise as expected because of improvement of disease with effective disease modification	Used in RA, AS. Caution required in immunosuppressed states and congestive heart failure. May induce tuberculosis (infliximab and adalimumab) and hepatitis B reactivation (rituximab); may increase risk of nontuberculosis infections (primarily pulmonary) and malignancy (lymphoma) development; may worsen heart failure (infliximab, etanercept, adalimumab). Associated with hematologic toxicities (anemia, leukopenia, lymphopenia, thrombocytopenia)

OA = osteoarthritis; RA = rheumatoid arthritis; AS = ankylosing spondylitis.

The USADA prohibits all glucocorticosteroids when administered orally, rectally, intravenously, or intramuscularly (1). In addition, USADA also bans a list of narcotic medications, including fentanyl, hydromorphone (Dilaudid) methadone, meperidine (Demerol), morphine, oxycodone (Oxycontin, Percocet, Percodan), and pentazocine (Talwin) (1).

Nonsteroidal Anti-Inflammatory Drugs

Common nonsteroidal anti-inflammatories (NSAIDs) include ibuprofen, naproxen sodium, ketorolac, sulindac, and etodolac (many are available without a prescription). These agents have both anti-inflammatory and fever-reducing activities. Although the exact mechanism of action is unknown, the main mechanism of action appears to be inhibition of cyclooxygenase (COX) activity (both COX-1 and COX-2) and prostaglandin synthesis (production of certain types of prostaglandins is important for inflammation) (101). They are also effective analgesics and can be used to relieve mild to moderate pain. Chronic use of these agents can erode the gastric mucosa with subsequent gastrointestinal bleeding (102). In addition, renal damage also has been reported with repeated use of higher than recommended daily doses of NSAIDs (103). Patients may also experience some gastric upset with short-term use, but this effect can be minimized by taking these agents with food or milk. Age appears to increase the risk of adverse drug reactions, so patients over the age of 65 yr should be cautioned not to exceed the recommended daily dose. NSAIDs do not affect exercise capacity directly but may facilitate continued exercise by decreasing pain and inflammation resulting from exercise-induced injuries.

Selective COX-2 Inhibitors

Celecoxib (Celebrex) is the only NSAID available that selectively inhibits COX-2. Selective inhibition of COX-2 decreases inflammatory prostaglandins without a subsequent decrease in other prostaglandins that protect the gastric mucosa. For this reason, it is believed that COX-2 inhibitors may be less likely to cause gastric ulceration. Common side effects include the following (104):

Abdominal pain

Diarrhea

Dyspepsia

Flatulence

Nausea

But these side effects occur less often than they do with other NSAIDs. Concern has been raised about the possible prothrombotic potential of COX-2 inhibitors and potential increase in cardiovascular event rates (105). Rofecoxib (Vioxx) and Valdecoxib (Bextra) were withdrawn from the market because of increased risk of cardiovascular events. Like other NSAIDs, COX-2 inhibitors do not affect exercise capacity directly but may facilitate continued exercise by decreasing pain and inflammation resulting from exercise-induced injuries.

NEUROMUSCULAR DISEASES AND AGENTS

A patient with cerebral palsy or spinal cord injury will be challenging to the physiologist. These diseases have a neuromuscular component, and patients may experience spasticity. A number of medications with different mechanisms of action can be used to help these patients.

Cerebral Palsy

Cerebral palsy is a condition of posture and motor impairment that results from an injury to the developing central nervous system (106). The common characteristics of the syndrome are spasticity, movement disorders, muscle weakness, ataxia, and rigidity (106). Cerebral palsy is a disease of children.

Spinal Cord Injury

Spinal cord injury most often affects young adults or adults. The patient's condition depends on where the injury to the spinal cord occurs. Because the spinal cord is a collection of nerves, injury to this area can cause neuromuscular symptoms and many other sensory and motor losses.

Antispasticity Agents

Oral, intramuscular, and intravenous medications are available for neuromuscular spasticity. Baclofen, dantrolene sodium, and diazepam can be administered orally. Baclofen serum levels peak 2 to 3 h after oral administration, which necessitates multiple daily doses (107). Other medications that are used are anticonvulsants or seizure drugs.

Intravenous medications used to manage spasticity in children with cerebral palsy include intrathecal baclofen, botulinum toxin A, alcohol, and phenol (106). Continuous intrathecal baclofen (CIB) is administered through a subcutaneous pump and has been found to be an effective treatment in reducing tone (106). An overdose of CIB can cause unresponsiveness and profound hypotonia requiring ventilation (107). Neuromuscular blocking agents such as botulinium toxin A, phenol, and alcohol are injected into the spastic muscle to reduce tone (117). See table 3.8 for additional information (106).

PSYCHIATRIC DISEASES AND AGENTS

A number of psychiatric disorders are present in our society. This chapter focuses on anxiety, depression, and substance abuse. Anxiety and depression can have a direct effect on exercise tolerance and ability. These disorders are relatively common, occur more frequently in women than in men, and are influenced by other factors, such as genetics and social issues (108,109). Many times, patients have a combination of anxiety and depression.

Substance abuse can occur because many medications have abuse potential. Some of the drugs used in the treatment of anxiety can be considered drugs of abuse. Many of the pain medications, such as the narcotics, have additive properties. Anabolic steroids and human growth hormone are other agents that people abuse. Many of the medications considered drugs of abuse can affect exercise capacity. These are discussed in detail in the following sections.

Anxiety

Anxiety can produce psychological (e.g., worry) and physiological arousal (e.g., tachycardia or shortness of breath) if it becomes excessive (108). Therefore, as mentioned earlier, anxiety can affect exercise capacity directly.

Antianxiety Agents

The benzodiazepines are the major class of medications used to treat acute anxiety symptoms. Alprazolam (Xanax), diazepam (Valium), and lorazepam (Ativan) are members of this class. All benzodiazepines are equally effective when used at equipotent doses. They work by binding to GABA receptors and inhibiting the activity of GABA, which may play a role in anxiety (108). The most common side effects include drowsiness, sedation, and other central nervous system (CNS) symptoms. A concern with the use of benzodiazepines is physical dependence and abuse (108). Therefore, their use cannot be stopped abruptly because rebound and withdrawal symptoms can occur. Because benzodiazepines have a tendency for dependency, antidepressants, in particular the selective serotonin reuptake inhibitors (SSRIs), have become the drugs of choice for chronic treatment of anxiety disorders, especially when depression is also present (108). Antianxiety drugs may decrease heart rate and blood pressure by controlling anxiety. Their affects on EKG are variable during rest or may result in false positive or false negative test results during exercise.

Depression

Depression can manifest as a number of emotional and physical symptoms, such as a loss of interest and

Table 3.8　Medications for the Treatment of Spasticity

Medications	Class and primary effects	Exercise effects	Special considerations
Baclofen Clonidine Diazepam Dantrolene Tizanidine	Antispasticity—inhibit the reflexes that cause increased tone	Clonidine, tizanidine: Monitor BP carefully. Baclofen: Check pulse regularly	Side effects: sedation, nausea, hypotonicity, dizziness, loss of seizure control
Phenobarbital Phenytoin Valproic acid Gabapentin Toprimate	Anticonvulsants—work through many pathways to decrease transmission of brain activity	Phenytoin: Monitor BP carefully Phenobarbital, phenytoin, valproic acid: Check pulse regularly and if possible watch for abnormalities on EKG Phenobarbital, valproic acid gabapentin: Watch for signs of dyspnea and apnea	Side effects: hepatic toxicity, fatigue, ataxia, dizziness, decreased cognitive function Toprimate: Maintain proper hydration to avoid kidney stones
Alcohol Phenol Botulinum toxin A	Neuromuscular blockers—denervate muscles and nerves	None	Side effects: Alcohol and phenol: pain on injection, possible permanent muscle fibrosis, burning, tingling, and numbness sensations Botulinum A toxin: Pain on injection, muscle soreness, rash, fatigue, weakness, flulike symptoms, and allergic reaction

pleasure in usual activities, sad or pessimistic mood, fatigue, sleep and appetite disturbances, gastrointestinal complaints, and palpitations (109). The cause of depression is unknown. One theory of many is that depression occurs because of decreased levels of neurotransmitters (noriepinephrine, serotonin, and dopamine) in the brain (109). The goals of antidepressant therapy are to reduce the symptoms of acute depression, to facilitate the patient to a level of functioning present before the onset of illness, and to prevent further episodes of depression (109). Antidepressants have equivalent efficacy when administered in comparable doses (109). They are classified by pharmacologic action.

Selective Serotonin Reuptake Inhibitors (SSRIs)

The most commonly prescribed class of antidepressants, selective serotonin reuptake inhibitors (SSRIs) increase serotonin levels. Agents include fluoxetine (Prozac), sertaline (Zoloft), paroxetine (Paxil), citalopram (Celexa), escitalopram (Lexapro), and fluvoxamine (Luvox). The most common adverse effects are nausea, vomiting, diarrhea, sexual dysfunction in males and females, headache, insomnia, and fatigue (109).

Mixed Serotonin and Norepinephrine Reuptake Inhibitors

These medications potentiate the activity of norepinephrine and serotonin. Venlafaxine (Effexor) is an agent in this class of medication. Its side effect profile is similar to the SSRIs, but it can also cause a dose-related increase in diastolic blood pressure. Nefazodone (Serzone) and trazodone (Desyrel) can cause orthostatic hypotension among other adverse effects. Additional agents include bupropion (Wellbutrin), maprotiline (Ludiomil), mirtazapine (Remeron), and amoxapine (Asendin).

Tricyclic Antidepressants

Amitriptyline (Elavil), imipramine, nortriptyline, and desipramine are the major members of this class. These medications affect additional neurotransmitters and produce many pharmacologic effects and unwanted but expected side effects. These side effects include dry mouth, constipation, blurred vision, urinary retention, dizziness, tachycardia, memory impairment, orthostatic hypotension, cardiac conduction delays, heart block, arrhythmias, weight gain, sexual dysfunction, and excessive perspiration. Patients with significant cardiac disease should exercise caution when taking tricyclic antidepressants (109).

Monoamine Oxidase Inhibitors (MAOIs)

MAOIs increase the concentrations of norepinephrine, serotonin, and dopamine. Phenelzine (Nardil) and tranylcypromine (Parnate) are currently available in the United States. Postural hypotension is the most common side effect. MAOIs should not be taken concurrently with certain foods (high tyramine content) or drugs because hypertensive crisis can occur.

St. John's Wort

St. John's Wort is an herbal medication available over the counter. The side effects are mild (109). Many drug interactions can occur with St. John's Wort and other medications. Patients need to inform health professionals if they are taking any herbal medications.

Substance Abuse

Abuse of prescription, legal, and illegal drugs is a major problem facing society today (110) (table 3.9). Commonly abused legal substances include nicotine, alcohol, and caffeine (111). Commonly abused prescription drugs include central nervous system stimulants such as amphetamines and diet pills and central nervous system depressants such as benzodiazepines, barbiturates, and narcotic analgesics (opiates). Illegal drugs of abuse include stimulants such as cocaine, depressants like heroin, and other agents such as LSD and ecstasy (110). The abuse potential of anabolic steroids and human growth hormone will be discussed in detail in the following sections.

The essential feature of substance dependence is the continued use of the substance despite adverse substance-related problems. Repeated use of the drug is often associated with the development of tolerance, withdrawal,

Table 3.9 Drugs of Abuse and Narcotic Analgesics and Their Exercise Effects

Medications	Heart rate	Blood pressure	ECG	Exercise capacity
Nicotine	↑ or ↔ (R and E)	↑ (R and E)	↑ or ↔ HR May provoke ischemia, arrhythmias (R and E)	↔ , except ↓ or ↔ in patients with angina
Alcohol	↔ (R and E)	Chronic use may have role in ↑ BP (R and E)	May provoke arrhythmias (R and E)	↔
Caffeine	Variable effects depending on previous use		May provoke arrhythmias	Variable effects on exercise capacity
CNS stimulants				
Amphetamines Diet pills Cocaine	↑ (R and E)	↑ (R and E)	↑ or ↔	No effect
CNS depressants				
Benzodiazepines Barbiturates Opiates	May decrease HR by controlling anxiety. Withdrawal may precipitate anxiety and increased HR	No effect	No effect	No effect

Note. ECG = electrocardiogram; R = rest; E = exercise; BP = blood pressure; CNS = central nervous system; ↔ = no change; ↑ = increase; ↓ = decrease.

Table 3.10 Hormones and Steroids and Their Exercise Effects

Medications	Heart rate	Blood pressure	ECG	Exercise capacity
Anabolic steroids	↔	May increase	↔	Performance enhancing
Androgenic steroids	↔	May increase	↔	Performance enhancing
Human growth hormone	↔	↔	↔	Performance enhancing

Note. ECG = electrocardiogram; ↔ = no change.

and compulsive use, but it is possible to meet criteria for dependence in the absence of physical dependence (110).

Anabolic Steroids

These agents are of particular importance to the exercise physiologist. See table 3.10 for the affect of anabolic steroids and human growth hormone on exercise. Although these agents are often used to treat various illnesses, they are often abused, particularly by athletes and bodybuilders because of their reputed performance-enhancing effects. The purpose of this section is to alert the clinician to the myriad dangers associated with abuse of these substances.

Anabolic steroids are a family of hormones that include the natural male hormone, testosterone, together with a large number of synthetic compounds (112). All drugs in this family possess both anabolic and **androgenic** effects. Although various steroids differ in the relative amounts of anabolic and androgenic effects they produce, none even approaches being purely anabolic or purely androgenic. Examples of anabolic steroids include oxymetholone (Anadrol), oxandrolone (Oxandrin), stanozolol (Winstrol), methandrostenolone, nandrolone, testosterone, and tetrahydrogestrinone (THG). The anabolic effects include increased protein synthesis, decreased nitrogen excretion, and consequent gains in muscle size and strength (113). The androgenic effects are the "masculinizing" effects of the drugs, such as growth of male hair patterns and male sexual characteristics. Anabolic steroids promote tissue-building processes and reverse catabolic, or tissue-depleting, processes. They may be used clinically for various illnesses, such as patients with severe burn injuries (114,115). These products are controlled substances because of their abuse potential.

Anabolic steroids are taken as pills or are injected. Steroid abusers may take doses 10 to 100 times greater than the medically recommended doses (116). Users often combine several types of steroids to boost their effectiveness—a method called stacking (116). In another method, called cycling, users take steroids for 6 to 12 wk or more, stop for several weeks, and then start again (116).

Athletes, as well as some coaches, trainers, and physicians, report significant increases in muscle mass,

strength, and endurance from steroid use. In acknowledgment of these effects, the International Olympic Committee has placed more than 20 anabolic steroids and related compounds on its list of banned drugs (1). No well-controlled studies have documented that the drugs improve agility, skill, cardiovascular capacity, or overall athletic performance.

Use of agents in amounts of up to 40 times the recommended therapeutic dose has been reported. An abuse or addiction syndrome occurs with the chronic use of these drugs (117). Long-term use can lead to preoccupation with drug use, difficulty stopping despite side effects, and drug craving (118). A type of withdrawal syndrome may be noted as well when drug concentrations fluctuate; symptoms are similar to those seen with alcohol, cocaine, and narcotic withdrawal. To detect steroid use or abuse, the exercise professional should be aware of physical, psychological, and behavioral changes.

Males who take large doses of anabolic steroids typically experience changes in sexual characteristics. Although derived from a male sex hormone, the drug can trigger a mechanism in the body that can shut down the healthy functioning of the male reproductive system. Several side effects are possible (116):

Shrinking of the testicles

Reduced sperm count

Impotence

Baldness

Difficulty or pain in urinating

Development of breasts

Enlarged prostate

Females may experience "masculinization" as well as other problems (116):

Growth of facial hair

Changes in or cessation of the menstrual cycle

Enlargement of the clitoris

Deepened voice

Breast reduction

For both sexes, chronic use of anabolic steroids may cause the following side effects (116):

Acne

Jaundice

Trembling

Swelling of feet or ankles

Bad breath

Increase in low-density lipoprotein cholesterol

Reduction in high-density lipoprotein cholesterol

High blood pressure

Heart attacks

Enlargement of the left ventricle of the heart

Liver damage and cancers

Aching joints

Increased chance of injury to tendons, ligaments, and muscles

Increased risk of blood clots

Psychological effects of anabolic steroid abuse are significant. Wide mood swings ranging from periods of violent, even homicidal, episodes known as roid rages to bouts of depression when drug use stops have been shown to occur. In addition, anabolic steroid users may suffer from paranoid jealousy, extreme irritability, delusions, and impaired judgment stemming from feelings of invincibility (116).

When used in combination with exercise training and a high-protein diet, anabolic steroids can increase the size and strength of muscles, improve endurance, and decrease recovery time between workouts. All anabolic steroids are banned by the USOC, USADA, and the NCAA (1,2).

Human Growth Hormone

Human growth hormone (GH), which consists of 191 amino acids linked in a specific sequence, is secreted from the anterior pituitary gland (119). After being secreted by the pituitary gland, circulating levels of GH stimulate production of insulin-like growth factor-1 (IGF-1) from the liver. Most of the positive effects of GH are mediated by the IGF-1 system. Both growth hormone (GH) and IGF-1 are anabolic, nonandrogenic compounds that decrease body fat and increase lean muscle mass.

Practical Application 3.1

Client–Clinician Interaction

The clinical exercise physiologist must obtain information about all medications and their dosages from the patient during the medical evaluation. The mechanism of action for each medication should be clearly understood, and the effect of each on cardiorespiratory parameters (heart rate, blood pressure, electrocardiogram) should be noted before testing or prescribing exercise. The clinician should bear in mind that some medications will influence the patient's exercise capacity. For instance, a β-blocker can improve the exercise capacity in a patient with ischemic disease but hampers the exercise capacity of a healthy person. In some cases, medications can be added or removed while the patient is exercising in the program. If graded exercise test data is unavailable with the new medications, it may be necessary to use several precautions until appropriate exercise responses can be established. These may include using electrocardiogram monitoring for signs of ischemia or decreased cardiac output; administering prophylactic nitroglycerin to high-risk patients with coronary artery disease before beginning exercise; and avoiding exercise in the presence of angina, dyspnea, or extreme fatigue. Another critical issue that a clinician often confronts is the timing between administering the medication and the exercise session. Taking a medication just before exercise (i.e., breakfast and morning exercise) may not be optimal because the elapsed time may be insufficient to allow the medication to take full effect, particularly with longer-acting medications that influence heart rate or blood pressure.

On occasion, the clinician may become aware of drug abuse by a client. Whether drug abuse is intentional or not (e.g., the client becomes overreliant on painkillers), the clinician should counsel the patient about drug abuse and refer the patient to his or her primary care provider for appropriate treatment. Positive encouragement and support can encourage the patient to seek the appropriate treatment. As such, the clinician must maintain professional conduct by using sensitivity and discretion in handling the matter and maintaining patient confidentiality.

Selected populations have lower than normal circulating GH, notably the aging population, those with hypothalamic–pituitary disorders, and patients with acute diseases associated with heightened tissue breakdown, including AIDS, malnutrition, postoperative wounds, infections, bony fractures, and burns. Administration of synthetic GH increases circulating levels of IGF-1, skin thickness, and bone mineral content and reduces fat mass (120).

Many people tout GH supplementation as a veritable fountain of youth. Among athletes, doping with GH has become an increasing problem during the last 10 yr despite lack of evidence of its benefits (1). GH has a reputation among GH users of being fairly effective, but the few controlled studies that have been performed in athletes with normal GH levels showed that GH produced no significant performance-enhancing effects.

The most common side effect of excess GH is a condition known as acromegaly (119). When GH is given after growth is completed, thickening of the bones and skull can occur, leading to bony changes that alter the patient's facial features: The brow and lower jaw protrude, the nasal bone enlarges, and spacing of the teeth increases (119). Other symptoms of acromegaly include thick, coarse, oily skin; enlarged lips, nose, and tongue; deepening of the voice attributable to enlarged sinuses and vocal cords; snoring attributable to upper-airway obstruction; excessive sweating and skin odor; fatigue and weakness; headaches; impaired vision; abnormalities of the menstrual cycle and sometimes breast discharge in women; and impotence in men (119). Enlargement of body organs, including the liver, spleen, kidneys, and heart, may also occur.

CONCLUSION

The widespread use of pharmacological agents in our society requires the clinician to be familiar with the effect of these agents on exercise and exercise capacity. Whether involved in the training of elite athletes or in cardiac rehabilitation, the exercise physiologist is likely to encounter patients who are taking various pharmacological agents. This chapter is not a comprehensive overview of all therapeutic agents available. Instead, it has targeted specific types of therapies likely to be encountered by exercise physiologists in daily practice.

Case Study

Medical History

Mr. MT is a 46 yr old white male with a family history of cardiovascular disease, high blood pressure (150/94), and hypercholesterolemia (total cholesterol = 284, high-density lipoprotein = 35). Because of his extremely high risk of cardiac disease, his primary care physician has started him on the following medications:

Metoprolol, 100 mg per day (a β-blocker)

Atorvastatin, 10 mg per day (a cholesterol-lowering agent)

Niacin, 1,000 mg per day (a cholesterol-lowering agent)

Aspirin, 81 mg per day

Diagnosis

He is referred to you for an exercise prescription as part of a comprehensive lifestyle modification program to deal with his cardiovascular risk factors.

Exercise Test Results

No previous exercise test results are available.

Exercise Prescription

The goal is to determine an appropriate exercise regimen given this patient's current medical therapy. Given the lack of a preliminary exercise test and the potential influence of his medical therapy (β-blocker) on heart rate responses, the rating of perceived exertion (RPE) is used to determine the appropriate exercise intensity.

In addition, the subject's target heart rate is determined by plotting the heart rate responses against a graded workload. Given this assessment, the following exercise prescription is determined:

Frequency = 5 d per week

Intensity = 78 to 95 beats · min^{-1}

RPE = 11 to 15

Duration = 30 min

Mode = aerobic

At a scheduled follow-up visit, the patient complains of increased sluggishness throughout the day and increased fatigue when performing the prescribed exercise.

Discussion Questions

1. What are some of the reasons that the patient could be experiencing these symptoms?
2. What changes could you make in the patient's current exercise and medication regimen to decrease his symptoms and increase his quality of life?
3. Given the change in medical therapy, how might you update the exercise prescription?

4

General Interview and Examination Skills

Steven J. Keteyian, PhD, FACSM

Clinical exercise physiologists work in settings that require them to assess patients with various health problems. This chapter focuses on the elements of the clinical evaluation conducted by a physician or physician extender, as well as the measurements that the clinical exercise physiologist may need to obtain to determine whether the patient can exercise safely.

The clinical evaluation of any patient usually involves two steps. First, a general interview is conducted to obtain historical and current information. A physical assessment or examination follows, the extent of which may vary based on who is conducting the examination and the nature of the patient's symptoms or illness. After these are completed, a brief numerical list is generated to summarize the assessment, relative to both prior and current findings and diagnoses. Finally, a numerical plan is generated to indicate the one, two, or three key actions that are to be taken in the care of the patient. This chapter describes in detail the general interview and physical examination components of the clinical evaluation.

GENERAL INTERVIEW

The general interview is a key step in establishing the patient database, which is the working body of knowledge that the patient and the clinical exercise physiologist will share throughout the course of treatment. This database is primarily built from information obtained from the patient's hospital or clinical records. But as the clinical exercise physiologist, you will need to interview the patient to obtain information that is missing, as well as update data to address any changes in the patient's clinical status since his or her last clinic visit of record. With experience you will learn which information is incomplete or requires questioning the patient. "Essentials of Clinical Evaluation for the New Patient Referred" on p. 62 lists the relevant components, some of which you will need to enter into the patient file or database.

Reason for Referral

The reason for referral for exercise therapy is generally self-explanatory and may include one or more of the following: to improve exercise tolerance, improve muscle strength, increase range of motion, or provide relevant intervention and behavioral strategies to reduce future risk. But the clinical exercise physiologist may need to reconcile the physician's reason for referral and the patient's understanding of the need for therapy. Differences can exist between the two. For example, consider the 48 yr old sedentary patient employed as an automotive

Acknowledgment: Much of the writing in this chapter was adapted from the first edition of *Clinical Exercise Physiology.* As a result, we wish to gratefully acknowledge the previous efforts of and thank Peter A. McCullough, MD, MPH, FACC.

Essentials of Clinical Evaluation for the New Patient

- General interview
  - Reasons for referral
  - Demographics (age, gender, ethnicity)
  - History of present illness (HPI)
  - Current medications
  - Allergies
  - Past medical history
  - Family history
  - Social history
- Physical examination
- Assessment
- Plan

Reprinted, by permission, from P.A. McCullough, 1999, *Clinical Exercise Physiology* 1(1): 33-41.

company executive who undergoes single-vessel coronary artery angioplasty with stent deployment and returns to work in just a few days without physical limitations. Unfortunately, sometimes these patients perceive that they are cured after their coronary revascularization and surmise that they do not need to engage in rehabilitation or make lifestyle adjustments. When such a patient is referred for cardiac rehabilitation, he or she must understand that coronary artery disease is a dynamic disorder that is influenced, just like his or her original problem, by lifestyle habits and medications. The clinical exercise physiologist plays an important role in enforcing long-term compliance to physical activity, hypertension and diabetes management, proper nutrition, and medical compliance—all key components of secondary prevention.

Demographics

Patient demographics such as age, sex, and ethnicity are the basic building blocks of clinical knowledge. A great deal of medical literature describes the relationship between this type of demographic information and health problems. For example, nearly 40% of the adult Americans with at least one type of heart disease are age 65 yr or older (13). Age is also the most powerful risk factor for osteoarthritis, with 68% of those over the age of 65 having some clinical or radiographic evidence of the disorder. Finally, age is an independent predictor of

survival in virtually every cardiopulmonary condition, including acute myocardial infarction, stroke, peripheral arterial disease, and chronic obstructive pulmonary disease (11). Because these and other age-related diseases can influence a patient's ability to exercise, age becomes a key piece of information to consider when developing an exercise prescription.

Sex or gender also relate to outcomes such as behavioral compliance or disease management in patients with chronic diseases. For example, **rheumatoid arthritis** (RA) is more common in women than men (3:1 ratio) and seems to have an earlier onset in women. Also, although the onset of cardiovascular disease is, in general, 10 yr later in women than in men, the morbidity and mortality after revascularization procedures (i.e., coronary bypass or angioplasty) are higher in women (9). Finally, keep in mind that exercise capacity, as measured by peak oxygen consumption, decreases progressively in men and women from the third through the eighth decade. A recent paper by Ades and colleagues (1) showed that the rate of decline in men with age (-0.242 ml $\cdot$ kg^{-1} $\cdot$ min^{-1} per year) is greater than that rate of decline for women (-0.116 ml $\cdot$ kg^{-1} $\cdot$ min^{-1} per year). These and other sex-based difference remain an area of intense investigation as clinical scientists strive to determine which biological or socioeconomic factors account for the poorer outcomes sometimes observed in women. Additionally, a few data describe the positive or negative effects of exercise as an intervention for many chronic diseases in women, especially those with cardiopulmonary disease (3). When an exercise prescription is developed for women, unique compliance- and disease-related barriers and confounders need to be solved to improve exercise-related outcomes.

A great deal of information is available about differences in health status between various ethnic groups. Most of these differences are attributable to socioeconomic status and access to care, but a few ethnic-related differences are worth mentioning. For example, obesity, hypertension, renal insufficiency, and left ventricular hypertrophy are all more common in African American patients with cardiovascular disease than in their age- and sex-matched Caucasian counterparts. Likewise, diabetes and insulin resistance are more common in Hispanic Americans and some tribes of Native Americans (4).

The clinical exercise physiologist should consider these issues when developing, implementing, and evaluating an exercise treatment plan. Also, this information may influence the clinician's decision about which risk factors to address first. Finally, program expectations and outcomes may be influenced as well.

History of Present Illness

The purpose of this element is to record and convey the primary information related to the condition that led to the patient to be referred to a clinical exercise physiologist. The history usually begins with the "chief complaint," which is usually one sentence that sums up the patient's comments. The body of the history of present illness is a paragraph that summarizes the manifestations of the illness as they pertain to pain, mobility, nervous system dysfunction, or alterations in various other organ system functions (e.g., circulatory, pulmonary, musculoskeletal, skin, and gastrointestinal). Important elements of the illness are reviewed such as the date of onset, chronicity of symptoms, types of symptoms, exacerbating or alleviating factors, major interventions, and current disease status. Traditionally, this is a paragraph that describes events in the patient's own words. A practical approach for the clinical exercise physiologist is to incorporate reported symptoms with information from the patient's medical record.

For patients with cardiovascular disease, the features of chest pain should be described (table 4.1). Such a description can help in the future application of diagnostic testing, when it comes to assessing pretest probability of underlying obstructive coronary artery disease (5). Standard classifications should be used, if possible, such as the Canadian Cardiovascular Society functional class for angina or the New York Heart Association functional class (table 4.2) (2). For patients with pain attributable to muscular, orthopedic, or abdominal problems, the important elements of the illness such as chronicity, type of symptoms, and exacerbating or alleviating factors need to be addressed.

Table 4.1 Features of Chest Pain

Type of discomfort	Quality	Radiation	Exacerbating and alleviating factors
Typical*	Heaviness, pressure—like with squeezing stress	To neck, jaw, back, left arm, less commonly the right arm	Worsened with exertion or relieved with rest or nitroglycerin
Atypical	Sharp, stabbing, pricking, tingling	None	None clearly present; can happen anytime
Noncardiac	Discomfort clearly attributable to another cause	Not applicable	Not applicable

*Note the difference between typical stable and typical unstable angina. The difference between these two type of angina pertains to no change in (or stable pattern for) intensity, duration, or frequency of pain in the past 60 d, as well as no change in precipitating factors.

Reprinted, by permission, from P.A. McCullough, 1999, *Clinical Exercise Physiology* 1(1): 33-41.

Table 4.2 Classification of Cardiovascular Disability

Class I	Class II	Class III	Class IV
New York Heart Association functional classification			
Patients with cardiac disease but no physical marked limitations (e.g., undue fatigue, palpitation, dyspnea, or anginal pain)	Patients with cardiac disease resulting in slight limitation (fatigue, dyspnea, angina) of physical activity	Patients with cardiac disease with marked limitation of physical activity	Patients with cardiac disease who are unable to carry out physical activity without discomfort
	Comfortable at rest	Symptoms such as fatigue, palpitations, and dyspnea occur with less than ordinary physical activity	Symptoms may even be present at rest
			Physical activity worsens symptoms

(continued)

Table 4.2 *(continued)*

Class I	Class II	Class III	Class IV
Specific activity scale			
Can perform activities requiring >7 METS such as doing outdoor work (shovel snow), play basketball, jog or walk 5 miles · hr^{-1}, and carry objects	Can perform activities requiring >5 METS but <7 METS; sexual intercourse without stopping, rake leaves, weed garden, in-line skate, walk at 4 miles · hr^{-1} on level ground	Can perform activity requiring between >2 and <5 METS; shower without stopping, make bed, clean windows, sweep garage, walk at 4 miles · hr^{-1} on level ground, bowl, golf	Cannot perform activities requiring >2 METS
			Cannot carry out activities listed in Class III
Canadian Cardiovascular Society functional classification			
Routine daily activity does not cause angina	Slight limitation of ordinary physical activities such as climbing stairs rapidly, doing exertion after meals, in cold or windy conditions, under emotional stress, or within a few hours after awakening	Marked limitation of routine daily activities such as walking one or two blocks on level ground or climbing more than one flight of stairs in normal conditions	Unable to carry out physical activity without discomfort
Angina occurs with rapid pace or prolonged exertion	No symptoms when walking more than two blocks on level ground or climbing more than one flight of ordinary stairs at a normal pace and in normal conditions		Angina may be present at rest

Reprinted, by permission, from American Alliance of Health, Physical Education, Recreation and Dance, 2004, *Physical education for lifelong fitness: The physical best teacher's guide* (Champaign, IL: Human Kinetics), 127.

Medications and Allergies

A current medication list is an essential part of the clinical evaluation, especially for the practicing clinical exercise physiologist, because certain medications can alter physiological responses during exercise. In fact, asking about current medications is an excellent segue into obtaining relevant past medical history. Compare the medications that patients state they are taking against what you think they should be taking given the medical information they report. For example, medical therapy for patients with chronic heart failure usually includes a diuretic agent, a β-adrenergic blocking agent (i.e., β-blocker), and a vasodilator such as an angiotensin-converting enzyme inhibitor. If during your first evaluation of a new patient with chronic heart failure you learn that the patient is not taking one of these agents, you should ask if his or her doctor has prescribed it in the past. In doing so, you may learn whether the patient has been found to be intolerant to an agent, or you may simply refresh his or her memory.

When describing the current medical regimen, be sure to include frequency, dose of administration, and time taken during the day. The latter may be especially important if a medication affects heart rate response at rest or during exercise, because you must allow sufficient time (usually 2 to 3 h minimum) between when the medication is taken and when the patient begins to exercise. Specifically, the medication must have time to be absorbed and exert its therapeutic effect.

A medical allergy history, with a comment on the type of reaction, is also a necessary part of your patient database. If a patient is unaware of any drug allergies, note this as "no known allergies (nka)" in your database. Note as well medicines that the patient does not tolerate (e.g., "Patient is intolerant to nitrates, which cause severe headache").

Medical History

This section should contain a concise, relevant list of past medical problems with attention to dates. Be sure that your list is complete because orthopedic, muscular, neurological, gastrointestinal, immunological, respiratory, and cardiovascular problems all have the potential to influence the exercise response and the type, progression, duration, and intensity of exercise.

For example, for patients with coronary heart disease, the record must include the severity of coronary lesions, types of conduits used if bypass was performed, target vessels, and most current assessment of left ventricular function (i.e., ejection fraction). For patients with cerebral and peripheral disease, the same degree of detail is needed with respect to the arterial beds treated.

Among patients with intrinsic lung disease, attempt to clarify asthma versus chronic obstructive pulmonary disease attributable to cigarette smoking. Such information may help explain why certain medications are used when the patient is symptomatic and why others are part of a patient's long-term, chronic medical regimen. Inquire about and investigate other organ systems as well. For example, if a diagnostic exercise test is ordered for a patient with intermittent claudication, using a dual-action stationary cycle may be better than using a treadmill. Additionally, knowing about any previous low-back pain is important when developing an exercise prescription.

Family History

This element should be restricted to known, relevant heritable disorders in first-degree family members (parents, siblings, and offspring). Relevant heritable disorders include certain cancers (e.g., breast), adult-onset diabetes mellitus, familial hypercholesterolemia, sudden death, and premature coronary artery disease defined as new-onset disease before the age of 45 in men or 55 in women. While assessing family history, you may wish to discuss with the patient that first-degree family members should be screened for pertinent risk factors and possible early disease detection, if indicated.

Social History

This section collects information about marital or significant partner status, employment status, relevant transportation, housing information, and routine and leisure activities. Because long-term compliance to healthy behaviors is influenced, in part, by conflicts with transportation, work hours, and childcare and family responsibilities, inquire about and discuss these issues. Conclude by making reasonable suggestions to improve long-term compliance.

Also, inquire about diet, nicotine or alcohol use, illegal substance abuse, and exercise. Because prior physical activity and dietary habits often predict a patient's ability to comply with behavioral change, give extra attention to obtaining an accurate physical activity and dietary history. Therefore, assess both prior and current habits and ask specific questions about type and frequency of the habit over a fixed period.

For example, if a patient states that she does a lot of walking, ask how many times in the past month she walked for 20 min or more. Does she walk by herself? What time of the day does she walk? How do her current exercise habits differ from 6 mo ago? Collecting this information is an important element of the evaluation, because it identifies those behaviors that may need to be reinforced or modified to reduce future risk. After all, many of the job duties that clinical exercise physiologists perform boil down to helping patients initiate and maintain healthy behaviors.

Assessing a patient's risk factors or risk profile helps the clinician not only quantify future risk but also focus future primary and secondary preventive strategies. But for the patient with established disease (e.g., coronary artery disease), listing non-modifiable risk factors (age, family history, menopausal status) is not useful.

PHYSICAL EXAMINATION

Physicians and physician extenders are taught to take a complete head-to-toe approach to the physical examination. For every part of the body examined, an orderly process of inspection, palpation, and, if applicable, **auscultation** and percussion is followed. The clinical exercise physiologist, however, can take a more focused approach and concentrate on abnormal findings, based on patient complaints or symptoms and information from prior examinations performed by others.

Specifically, you must develop the skills needed to determine whether, on any given day, it is safe to allow exercise by a patient who presents with signs or symptoms that may or may not be related to a current illness. For example, consider the patient with a history of dilated cardiomyopathy who complains of being short of breath and having to sleep on three pillows the previous two nights, just so that he could breathe more comfortably. This complaint should raise a red flag for you, because it may indicate that the patient is experiencing pulmonary edema attributable to heart failure. Besides asking other questions, you should assess body weight, peripheral edema, and lung sounds before allowing the person to exercise. A telephone conversation with the patient's

Red Flag Indicators of a Change in Clinical Status

The following are red flag indicators that, if detected by the clinical exercise physiologist, should be discussed with a physician before allowing a patient to exercise:

- New onset or change in pattern of shortness of breath or chest pain
- Complaint of recent syncope (loss of consciousness) or near syncope
- Neurologic symptoms suggestive of transient ischemic attack (vision or speech disturbance)
- Recent fall
- Lower-leg pain at rest (also called critical leg ischemia)
- Severe headache
- Pain in bone area (i.e., in patients with history of cancer)
- Unexplained tachycardia (>100 beats · min^{-1}) or bradycardia (<40 beats · min^{-1})
- Systolic blood pressure >200 mmHg or <86 mmHg; diastolic blood pressure >110 mmHg
- Pulmonary rales or active wheezing

doctor concerning your findings, if meaningful, might also be warranted.

At no time should your examination be represented as being performed in lieu of the evaluation conducted by a professional licensed to do so. Still, you are responsible for ensuring patient safety. Therefore, the information and data that you gather are important. They should be communicated to the referring physician and become part of the patient's permanent medical record. Identifying the relevant aspects of the baseline examination will provide a reference point for future comparison, particularly if complications occur. Important red flags that should be identified and evaluated by the clinical exercise physiologist and, if needed, a physician are shown in "Red Flag Indicators of a Change in Clinical Status."

General State

An initial comment about the patient's general appearance is warranted. This quick, subjective evaluation is usually made soon after meeting the patient. Assess whether the patient appears as follows:

Comfortable or distressed or anxious

Healthy or frail

Well-developed or undernourished

Blood Pressure, Heart Rate, and Respiratory Rate

Expected competencies of the clinical exercise physiologist are the accurate determination of blood pressure in each arm, pulse (heart) rate, **body mass index (BMI)**, and respiratory rate. A difference of more than 20 mmHg between the right and left brachial systolic blood pressures, taken in close time proximity, may indicate significant subclavian atherosclerosis and deserves additional evaluation by a physician. Additionally, the identification of either **hypertension** (i.e., systolic consistently >140 mmHg or diastolic consistently >90 mmHg) or **hypotension** is important.

Hypotension can be either symptomatic or asymptomatic, with the latter sometimes attributable to the blood pressure–lowering effect of medications used to treat a problem other than **hypertension**. For example, patients with chronic heart failure are often given medications designed to decrease peripheral vascular resistance. As a result, recording a blood pressure of 90/70 mmHg in these patients is not uncommon.

Tachycardia (heart rate >100 beats · min^{-1}) after 15 min of sitting rest is always abnormal and indicates extremely impaired left ventricular systolic function, severely impaired pulmonary function, hyperthyroidism, anemia, volume depletion, or the rare effect of medications such as over-the-counter decongestants or appetite suppressants. **Bradycardia** (heart rate <60 beats · min^{-1}) can indicate either abnormal or normal physiological function. In a sedentary patient with a history of prior illnesses, a heart rate of less than 40 beats · min^{-1} may represent an underlying problem and should be brought to the attention of the patient's physician.

Obesity

An extremely important chronic "vital sign" is an assessment of obesity. The best measure of obesity is the BMI. The BMI is mass in kilograms divided by the height in meters squared (kg · m^{-2}). You should record in your database the BMI of every patient whom you are evaluating for the first time. Practically, you accomplish this by weighing the patient in kilograms and then asking for

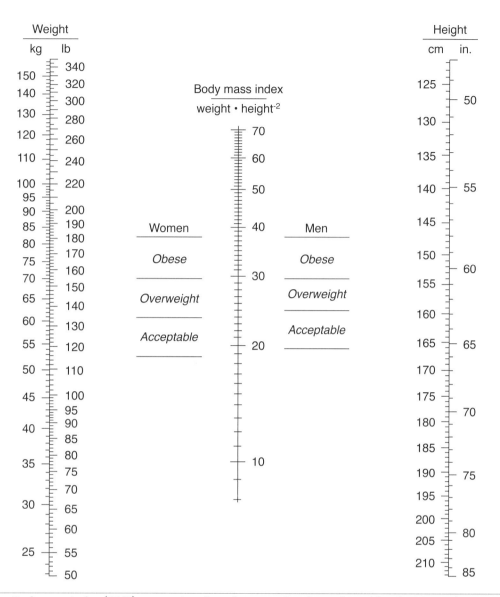

Figure 4.1   Body mass index (BMI) conversion chart that is used with weight and height. To use this nomogram, place a straight edge between the body weight (without clothes) in kilograms or pounds located on the left-hand line and the height (without shoes) in cm or in. located on the right-hand line. The point where the straight edge intersects the middle line indicates the BMI.

Reprinted from the CDC.

self-reported height in feet and inches. An easy-to-use conversion chart such as the one in figure 4.1 should be accessible for the quick calculation of BMI.

Obesity distribution is obtained by asking the patient to measure his or her own waist circumference with a measuring tape. In general, a waist circumference greater than 40 in. (100 cm) in a male or more than 35 in. (90 cm) in a female is indicative of central adiposity. Android or central obesity is one component of the "deadly quartet," which also includes insulin resistance, hypertension, and dyslipidemia (7). This simple measure of waist circumfer-

ence is usually preferred over the waist-to-hip ratio, which involves two measures and, as a result, is more subject to measurement error.

Pulmonary System

The thorax should be inspected and deformities of the chest wall and thoracic spine noted (see figure 4.2 on p. 68). Thoracic surgical incisions and implantable pacemaker or defibrillator sites should also be inspected and palpated. Redness or tenderness of any incision is

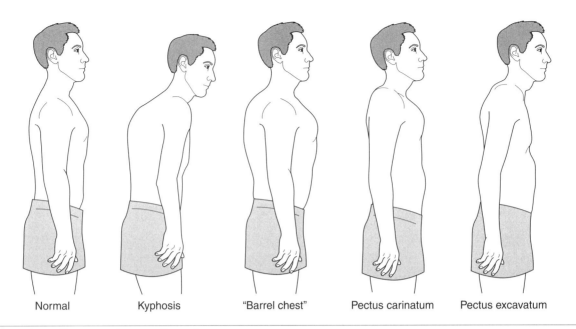

Normal | Kyphosis | "Barrel chest" | Pectus carinatum | Pectus excavatum

Figure 4.2 Common chest configurations.

Reprinted from *Physical Diagnosis: History and Examination*, M.H. Swartz, pg. 261, © 1999, with permission from Elsevier Science.

always abnormal and often signifies a wound infection, which should prompt a physician evaluation. With the patient sitting, the lungs should be auscultated with the diaphragm of the stethoscope in both anterior and posterior positions, and breath sounds should be characterized as normal, decreased, absent, coarse, wheezing, or crackling (i.e., rales). Decreased or absent breath sounds should prompt percussion of the chest wall for dullness. Dullness signifies a pleural effusion, which is an abnormal collection of fluid in the pleural space that does not readily transmit sound. This finding on physical examination would prompt withholding exercise and notifying the patient's physician. Coarse breath sounds can signify pulmonary congestion or chronic bronchitis. Crackles or rales can be caused by atelectasis (inadequate alveolar expansion after thoracic surgery), pulmonary edema attributable to congestive heart failure, or intrinsic lung disease such as pulmonary fibrosis.

Cardiovascular System

With the patient supine, the cardiac examination should start with inspection of the anterior chest wall. A cardiac pulse that can be visualized on the chest wall is often abnormal and represents a left ventricular hyperdynamic state. Standing at the patient's right side and while the patient lies comfortably, palpate the heart with the right hand. Make an effort to characterize where you feel or palpate the cardiac **point of maximal cardiac impulse** (PMI), or cardiac apex, as shown in figure 4.3. A normal

PMI is the size of a dime and is located in the fourth or fifth intercostal space at the midclavicular line. The PMI should be characterized as normal, diffuse (enlarged), hyperdynamic, or sustained. The location of the PMI should be identified as normal or laterally displaced. A diffuse and laterally displaced PMI indicates left ventricular systolic impairment, often with enlargement of that chamber. If you feel two cardiac impulses, this finding often indicates a right and left ventricular heave

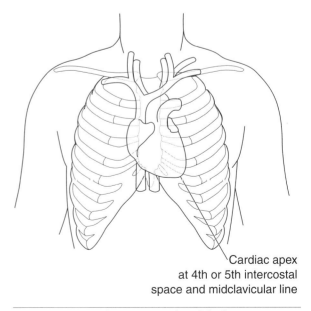

Cardiac apex at 4th or 5th intercostal space and midclavicular line

Figure 4.3 Surface topography of the heart.

Auscultation of the Human Heart

Auscultation of, or listening to the sounds made by, the heart is but one part of a comprehensive cardiac examination. Initiate a habit of auscultating the heart in a systematic fashion that is the same for every patient whom you evaluate. Establishing a uniform approach will more quickly familiarize you with normal heart sounds and help you identify abnormal sounds. To aid your concentration, try to auscultate the heart in a quiet room. Approach the patient from the right side and do all that you can to minimize anxiety. Attempt to warm your stethoscope before using it and communicate with the patient as you progress through this part of the physical examination. Remember that maintaining patient modesty is always a priority.

Auscultation is usually done with the patient lying on his or her back; before an exercise test, however, auscultation can be performed with the patient in the sitting position. As mentioned earlier, when auscultating the heart for cardiac sounds, do so systematically. Begin by placing the diaphragm of the stethoscope firmly on the chest wall in the lower-left parasternal region. First, characterize the rhythm as regular, occasionally irregular, or irregularly irregular. An irregularly irregular rhythm is usually attributable to atrial fibrillation. The diaphragm of the stethoscope is best used to hear high-pitched sounds, whereas the bell portion is used for low-pitched sounds.

Next, move the stethoscope to the point on the chest where the first heart sound (S_1, the sounds of mitral and tricuspid valves closing) is best characterized. For most people, this location is found at the apex of the heart at the left midclavicular line and near the fourth and fifth intercostal spaces. You will hear two sounds. The first sound (S_1) is the louder and more distinct sound of the two. Then, move your stethoscope upward and to the right side of the sternal border at the second intercostal space. This location is generally the best place to characterize the second heart sound (S_2, the sounds of aortic and pulmonic valves closing) because it is louder and more pronounced here.

Soft heart sounds occur with low cardiac output states, obesity, and significant pulmonary disease (e.g., diseased or hyperinflated lung tissue between the chest wall and the heart). Loud heart tones occur in thin people and in hyperdynamic states such as pregnancy. The second heart sound normally splits with inspiration as right ventricular ejection is delayed with the increased volume it receives from the augmented venous return of inspiration. This delays or splits the pulmonic component of S_2 (sometimes referred to as P_2) from the aortic component of S_2 (referred to as A_2, figure 4.4).

With practice and as you begin to care for patients with various cardiac problems, you will become exposed to and appreciate third (S_3) and fourth (S_4) heart sounds. S_3 and S_4 are low-pitched sounds that are heard during diastole and best appreciated by using the bell portion of the stethoscope. The presence of either of these two heart sounds is most often associated with a heart problem and should be brought to the attention of a physician. S_3 is best heard at the apex and occurs right after S_2. S_4 is also well heard at the apex and occurs just before S_1. An S_3 commonly indicates severe left ventricular systolic impairment with volume overload and dilation. An S_4 commonly indicates chronic stiffness or poor compliance of the left ventricle, usually attributable to long-term hypertension. As you can appreciate, learning and identifying heart sounds will require a great deal of practice.

The clinical exercise physiologist should be able not only to appreciate systole and diastole but also, with advanced training, to listen for murmurs in

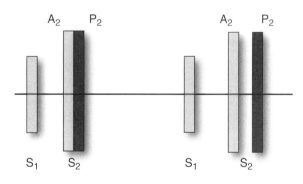

Figure 4.4 Normal physiologic splitting of the second heart sound because of augmented venous return to the right heart during inspiration.

Reprinted, by permission, from P.A. McCullough, 1999, *Clinical Exercise Physiology* 1(1): 33-41.

(continued)

▲ Practical Application 4.1 *(continued)*

the mitral, tricuspid, pulmonic, and aortic areas (figure 4.5). Murmurs are characterized by the timing in the cardiac cycle (systolic, diastolic, or both), location where best heard, radiation, duration (short or long), intensity, pitch (low, high), quality (musical, rumbling, blowing), and change with respiration (8). A central concept to keep in mind while listening with the stethoscope is that the sounds heard are attributable to changes in blood velocity and the movement of cardiac valve leaflets, both of which are driven by pressure gradients and result in flow. Systolic murmurs are more common and are often characterized as ejection type (e.g., diamond-shaped or holosystolic). Diastolic murmurs are distinctly less common.

Exercise physiologists working in the clinical setting must acquire the basic skills needed to auscultate the heart.

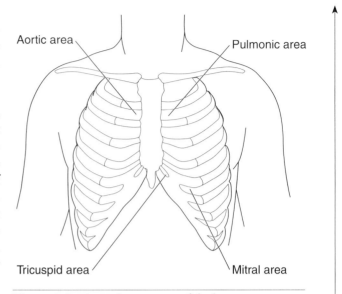

Figure 4.5 Ausculatory areas of the heart.

Reprinted, by permission, from P.A. McCullough, 1999, *Clinical Exercise Physiology* 1(1): 33-41.

---

suggestive of biventricular impairment in patients with heart failure.

An essential component in the examination of the heart is auscultation. An introduction to the assessment of heart sounds is reviewed in practical application 4.1 on p. 69. Practice listening to heart and lung sounds in apparently healthy people and patients whom you help care for who are known to have cardiac and pulmonary disorders as described in their medical records.

The cardiovascular physical examination should also include some evaluation of the status of peripheral vascular circulation. Extremities that are well perfused with blood are warm and dry. Poorly perfused extremities are often cold and clammy. Measuring and grading the characteristics of the arterial pulse in a region assesses adequacy of blood flow in arteries. Arterial pulses are graded (6) as follows: 3 = bounding, 2 = normal, 1 = reduced or diminished, and 0 = absent or nonpalpable. Using a stethoscope, the clinical exercise physiologist can also listen for **bruits**, which are high-velocity swooshing sounds created as blood becomes turbulent when it flows past a narrowing artery or through a tortuous artery. Volume should be assessed, and bruits should be characterized as soft or loud. Bruits detected in the carotid arteries that were not previously mentioned in the medical record should be brought to the attention of a physician, because they may indicate severe carotid atherosclerosis. A bruit heard in the abdomen during diastole (diastolic bruit), near the level of the umbilicus,

may indicate stenosis of the renal artery and should also be brought to the attention of a physician. Bruits in the common femoral arteries are suggestive of peripheral arterial disease but by themselves do not call for immediate physician evaluation.

Peripheral **edema** (e.g., swelling of the lower legs, ankles, or feet) is a cardinal sign of congestive chronic heart failure. Because of elevated left ventricular end-diastolic pressure and consequently the backward cascade of increased pressure to the left atrium, pulmonary capillaries, pulmonary artery, and right-sided cardiac chambers, increased hydrostatic forces move extracellular fluid from within the blood vessels into the tissue spaces of the lower extremities. Edema is graded on a 1 to 3 scale, with 1 being mild, 2 being moderate, and 3 being severe. Additionally, "pitting edema" can be present, which is easily identified by pressing a thumb into an edematous area (e.g., distal anterior tibia) and observing that an indentation remains. A patient with 3+ pitting edema of the lower legs and ankles obviously has a great deal of fluid that has left the vascular compartment and moved into the surrounding tissue.

Not all edema, however, result from congestive heart failure. Minor edema can be a side effect of many medications such as slow-channel calcium entry blockers (e.g., nifedipine). In addition, chronic venous incompetence associated with prior vascular surgery, obesity, or lymphatic obstruction can cause edema in the setting of normal cardiac hemodynamics and heart function. A practical point for the clinical exercise physiologist is that an increase in edema or body mass (>1.5 kg) over a 2 or 3 d period is often the first sign of worsening congestive heart failure and warrants a call to a physician.

Musculoskeletal System

Approximately one person in seven suffers from some sort of musculoskeletal disorder. The history of present illness or past medical history should note the major areas of discomfort and self-reported limitation of motion. In addition, prior major orthopedic surgeries, such as a hip or knee joint replacement, should be noted.

The approach to the musculoskeletal physical examination should be grounded in observation. For example, observe the patient as he or she gets up from a chair and walks into a rehabilitation area, gets on to an examination table, or handles personal belongings. Observe gait and characterize it as normal (narrow based, steady, deliberate), antalgic (limping because of pain), slow, hemiplegic (attributable to weakness or paralysis), shuffling (parkinsonian), wide based (cerebellar ataxia or loss of position information), foot drop, or slapping (sensory ataxia or loss of position information) (figure 4.6). An antalgic gait is a limp, which reflects unilateral pain and compensation for that pain. A slow gait is often a tipoff for back disease, hip arthritis, or underlying neurological problems. Hemiplegic, shuffling, wide-based, foot-drop, and slapping gaits all represent compensation for underlying neurologic disease such as a spinal cord injury, cerebellar dysfunction (e.g., attributable to alcohol), or midbrain dysfunction (e.g., Parkinson's disease). These "neurological gaits" are all unsteady and leave the patient prone to falling. For safety reasons, the clinical exercise

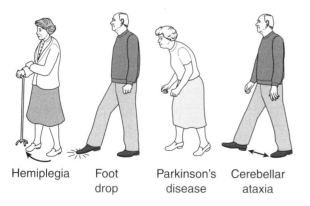

Hemiplegia   Foot   Parkinson's   Cerebellar
              drop   disease   ataxia

Figure 4.6 Description of common gait abnormalities.

Reprinted from *Physical Diagnosis: History and Examination*, M.H. Swartz, pg. 452, © 1999, with permission from Elsevier Science.

Table 4.3 Terminology for Describing Joint Motion

Motion	Definition
Flexion	Motion away from zero position
Extension	Motion toward zero position
Dorsiflextion	Movement toward dorsal surface
Plantar (or palmar) flexion	Movement toward plantar (or palmar) surface
Adduction	Movement toward midline
Abduction	Movement away from midline
Inversion	Turning of plantar foot surface inward
Eversion	Rotation of plantar foot surface outward
Internal rotation	Rotation of anterior limb surface inward
External rotation	Turning of anterior limb surface outward
Pronation	Rotation of palm downward
Supination	Rotation of palm upward

Reprinted from *Physical Diagnosis: History and Examination*, M.H. Swartz, pg. 261, © 1999, with permission from Elsevier Science.

physiologist must pay special attention to gait and modify the exercise prescription as necessary.

The core of the musculoskeletal physical examination is an assessment of range of motion of the moveable joints. Important terminology for describing limb motion is given in table 4.3. As needed, palpation of the major joints (elbows, wrists, hips, knees, and ankles) can be performed to note thickening of the joint capsules, swelling or effusion, and tenderness of ligaments or tendons. Redness, warmth, swelling, and fever are all signs of active inflammation and require evaluation by a physician before proceeding with exercise testing or therapy. If these signs are found in conjunction with a prosthetic joint, such as a total knee replacement, they may indicate an infection and require immediate contact with the patient's surgeon.

In addition to joint health, muscle strength can be examined and graded on a scale of 0 to 5, with 0 indicating flaccid paralysis and 5 indicating sufficient power to overcome the resistance of the examiner. Muscle stiffness and soreness (not related to exercise) should be noted because they are often the sign of a chronic underlying inflammatory condition that requires medical evaluation.

Low-Back Pain

Because low-back pain (see also chapter 28) is one of the most common physical complaints in human medicine, it is worth mentioning that the etiology of this problem can range from a mild muscle strain to a life-threatening ruptured abdominal aortic aneurysm. The clinical exercise physiologist must have a rational approach to this problem and tailor aspects of care based on etiology, severity, and prognosis. In the young, stable patient with no evidence of neurological compromise (e.g., radiating pain, numbness), a physician evaluation is likely not necessary before exercise testing or therapy. In the geriatric patient (i.e., >65 yr old), however, new-onset low-back pain that has not been evaluated by a physician deserves referral to the primary care physician or a back specialist before exercise testing or training because a spontaneous compression fracture attributable to osteoporosis may have occurred.

Arthritis

A final musculoskeletal condition worth mentioning, one that is especially common among the elderly, is arthritis (see also chapter 26). The two most common types of arthritis are rheumatoid arthritis (RA) and **osteoarthritis** (OA). Rheumatoid arthritis (RA) is a chronic inflammatory condition manifested by functional disability and early morning stiffness (>1 h) followed by improvement through the rest of the day. Most damage experienced in RA occurs in the first 5 yr, and approximately 40% of patients who suffer from RA become completely disabled in 15 yr. Life expectancy falls by 3 to 18 yr.

RA begins insidiously with fatigue, anorexia, generalized weakness, and vague musculoskeletal symptoms until the appearance of synovitis (inflammation of joint lining) becomes apparent. It has a predilection for the small joints in the hand, especially the metacarpophalangeal joints (knuckles). RA results in pain, inflammation, thickening of the joint capsule, lateral deviation of the fingers, and significant disability. RA can involve large joints and the spine, especially the cervical vertebrae. RA is usually apparent to the clinical exercise physiologist by history and the presence of disease-modifying antiarthritic drugs (DMARDs) (see chapter 3).

Patients with RA can generally exercise safely unless neck pain or lancinating pains in the shoulders or arms are reported. These symptoms are indicative of cervical spine involvement and require a physician's attention. In some cases, bracing or surgery is needed to prevent cervical spine subluxation and paralysis.

Osteoarthritis (OA) is the most common skeletal disorder in adults. It can result from longstanding wear and tear on large and small joints including shoulders, elbows, wrists, hands, hips, knees, and back. The correlation between the pathological severity of OA and patient symptoms is poor. Patients with OA may show signs of synovitis and secondary muscle spasm.

Exercise rehabilitation may improve the symptoms of OA in one location (e.g., the back), only to worsen pain in another location (e.g., hips and knees). The clinical exercise physiologist should take a pragmatic approach and work in a coordinated fashion with physical and occupational therapy colleagues to find activities that minimize joint loading at involved sites.

Nervous System

Like the examination of the musculoskeletal system, a neurological examination performed by the clinical exercise physiologist mainly involves observation. In general, the clinician should make a comment in the patient file regarding level of understanding, orientation, and cognition. Obvious disabilities of speech, balance, and muscle tremor and disabilities of the eyes, ears, mouth, face, and swallowing should be cross-referenced for confirmation with the patient's medical record.

The practicing clinical exercise physiologist rarely performs a detailed neurological examination, so this chapter will not describe the conduct for such an examination. Nevertheless, a notation regarding gross **hemiparesis** or complete paralysis of a limb should be made. As mentioned previously (figure 4.6), "neurological gaits" should be identified as well. For obvious reasons, patients with any of these problems may require an appropriately modified exercise prescription.

Although the clinical exercise physiologist may work with the spinal cord–injured patient (see chapter 29) or patients with multiple sclerosis (see chapter 30), the most common neurological problem that most exercise physiologists encounter will be patients who have suffered a previous cerebrovascular event or stroke. Stroke is defined as an ischemic insult to the brain resulting in neurological deficits that last for more than 24 h (see chapter 32). The etiology of stroke includes local arterial thrombosis, cardiac and carotid thromboembolism, and intracranial hemorrhage. Risk factors for stroke include the following:

> History of stroke
> Atrial fibrillation
> Left ventricular dysfunction
> Aneurysms
> Carotid artery stenoses
> Uncontrolled hypertension

If the clinical exercise physiologist determines that any of these risk factors are new, he or she should notify the physician before the patient starts exercise. Common abnormalities after a stroke are a loss or diminished limb function, drooping of one side of the face, drooling, and garbled speech (dysarthria).

Metabolic and Other Organ Systems

Signs and symptoms of abnormalities of metabolism and of other organ systems of the body (e.g., endocrine, immune, and hematological systems) are generally less common and less obvious on physical examination. Therefore, they are typically beyond the scope of practice for most clinical exercise physiologists. Nevertheless, the clinical exercise physiologist should consider some important points in these areas.

Diabetes

Before starting an exercise program, patients with diabetes should have a foot examination to look for blisters, cuts, and abrasions. Any new signs of infection should be reported to a doctor. Also, diabetes is associated with a higher probability of silent ischemia, multivessel coronary disease, and decreased left ventricular function. Therefore, the clinical exercise physiologist should determine whether symptomatic versus asymptomatic ischemia is present, based on previous graded exercise tests, functional studies, and angiographic information.

Evaluation of the diabetic must include notation of the doses of diabetic medications and the presence of any diabetic complications such as peripheral or autonomic **neuropathies**. The clinical exercise physiologist should be aware of the signs and symptoms of hypoglycemia, such as sudden weakness, pallor, sweatiness, and confusion, and should be prepared to treat with orange juice (with sugar added) or a carbohydrate source.

Obesity

We previously described how obesity is identified and quantified by BMI. Obesity (BMI >30) is the most rapidly growing metabolic disorder in the United States. Identifying during examination exactly where excess fat is stored is important, because the site of excess fat deposition is related to overall **mortality** rate and increased risk of various diseases. In general, women tend to store fat in the lower half of the body around the hip and thigh area (gynoid obesity), whereas men tend to store fat on the upper part of the body around the abdomen (android obesity). The latter of the two types is associated with the greater risk for developing heart disease, diabetes, and hypertension.

Given the unlikely occurrence of rapid shifts in the U.S. human gene pool over the past 30 to 50 yr, lack of physical activity and absolute increases in caloric intake are likely responsible for the secular increase in mean population body weight. Patients with obesity are likely to have a reporting bias. During your evaluation, be aware that some obese patients commonly self-report a caloric intake that is less than needed even to maintain a body weight that has not changed or increased over the past several months or year.

Anemia

A history of anemia should be noted in the general evaluation. Anemia is generally defined as a hemoglobin less than 12 g · dl^{-1} in women and less than 13 g · dl^{-1} in men. Complaints of new-onset excessive fatigue or weakness, excessive paleness, excessive vaginal bleeding, or gastrointestinal bleeding should be reported to the physician before starting exercise therapy. In addition, many patients with cancer, renal disease, and human immunodeficiency virus (HIV) suffer from chronic anemia that is attributable to either the disease itself or some of the drugs used to treat the disease. Anemia is commonly observed in patients age 80 yr and older and patients who undergo cardiac or pulmonary surgery. For example, after coronary artery bypass surgery, a patient's hemoglobin may fall to 10 g · dl^{-1}. Less common is the incidence of anemia among highly active, usually younger, female athletes with poor dietary habits.

Inquire about any history or evidence of easy bruising or excessive gum or nasal bleeding. Skin pallor, especially pallor of the conjunctiva (inner eyelid) and the oral mucosa, is an examination finding supportive of anemia. Unusual bruising on the trunk and extremities should be noted as well. Again, if these are new, a physician should be notified before the patient begins a particularly strenuous exercise program.

Human Immunodeficiency Virus

You will be alerted to HIV infection by the clinical history and multiple drugs used to treat HIV (see chapter 25). If tuberculosis has been identified in a patient who is HIV-positive, records should indicate a completed treatment course with medical therapy and the patient should be on maintenance therapy before admission to exercise sessions with other patients.

Principles of exercise therapy should include careful attention to changes in exercise tolerance and dyspnea. Patients with fever and marked changes in general strength or dyspnea on exertion should be referred to a physician for evaluation of opportunistic infection.

Widespread use of antiretroviral therapy has brought out an unexpected side effect profile including dyslipidemia, muscle atrophy, abnormal adipose deposition on the back of the neck ("buffalo hump"), and early atherosclerosis (10). Be on the lookout for the buffalo hump on the posterior neck, indicators of poor muscle strength or endurance, and abnormal lipoprotein values in patients with HIV.

Cancer

The clinical exercise physiologist increasingly encounters patients with various malignancies. Because cancer can involve more than one organ or organ system, the clinical evaluation should identify the organ systems or tissues involved, the date of first discovery, and the treatment received. The clinical exercise physiologist should also assess the patient's understanding of the cancer status: cured, in remission, or active. In general, most cancers with no objective evidence of recurrence 5 yr after treatment are considered cured. Pain in the bone area, often a manifestation of metastases to bone, is a red flag for the clinical exercise physiologist and should prompt post-

ponement of exercise therapy until the issue is discussed with the patient's physician. For patients in remission or with active neoplastic processes, anemia with fatigue and decreased appetite are common.

CONCLUSION

This chapter provides a platform for the clinical evaluation of the new patient by the clinical exercise physiologist. Special emphasis is given to those day-to-day interview and examination skills that you may need to help you decide whether a patient can exercise safely. Also, we have emphasized that the information you gather through a clinical evaluation, when viewed in conjunction with existing medical record information and physical examination findings reported by others, provides a point of reference should complications occur during exercise treatment. If the evaluation is broadened to include a comprehensive assessment of future health risk and consideration of disease, your clinical evaluation also becomes a useful decision-making tool and guide for patient education.

Case Study

Ms. WY

Medical record number: 123-45-678

Service date: October 17, 2007

DOB: September 12, 1965

A 42 yr old Caucasian female seen for a graded exercise test before starting a cardiorespiratory and resistance-training exercise program.

Medical History

An 8 yr history of HIV disease with no opportunistic infections. She now complains of chronic fatigue and lack of energy. She has shortness of breath on moderate exertion such as carrying a full laundry basket up one flight of stairs. Patient denies any chest pain or discomfort at rest or with exertion.

Medications: efavirenz, zidovudine, didanosine.

Medical history: HIV disease for 8 yr, no complications.

Family history: negative for premature coronary artery disease.

Social history: unemployed, unmarried, lives alone.

Lifestyle history: unrestricted diet, sedentary with very rare physical exertion.

Risk factors: nonsmoker; fasting lipoprotein profile from medical record of March 12, 2007:
total cholesterol = 266 mg $\cdot$ dl^{-1}, low-density lipoprotein cholesterol = 185 mg $\cdot$ dl^{-1}, high-density lipoprotein cholesterol = 47 mg $\cdot$ dl^{-1}, triglycerides = 150 mg $\cdot$ dl^{-1}; no diabetes in the medical record.

Fasting blood glucose (taken in office this morning): 88 mg $\cdot$ dl^{-1}.

Physical Examination

General appearance: appears chronically ill, pale.

Vital signs: blood pressure = 110/50 mmHg in both arms, pulse = 108 beats · min^{-1}, respiration = 22 breaths · min^{-1}, mass = 66 kg, height = 5 ft 4 in. (163 cm), BMI = 25, waist circumference = 30 in.

Head and neck: no oral lesions, pale conjunctival (inner eyelid) and oral mucosal membranes, "buffalo hump" noted.

Cardiovascular: no carotid bruits, no jugular venous distension, clear lungs, normal point of maximal impulse, regular rate and rhythm, normal S_1 and S_2, 2/6 short early peaking systolic ejection murmur best heard in the pulmonic area without radiation, no extra heart sounds, no edema, normal peripheral vascular exam.

Skin: no rashes.

Musculoskeletal: normal gait, joints are normal, good range of motion throughout.

Neurologic: normal cranial nerves, normal motor exam, sensory exam reveals decrease in light touch sensation from toes to the knees bilaterally.

Resting electrocardiogram before exercise test: sinus rhythm, rate = 110 beats · min^{-1}, PR interval = 0.14 s, QRS duration = 0.09 s, no Q waves and no ST or T wave abnormalities. Computer interpretation is normal electrocardiogram.

Diagnosis

(a) HIV-positive with associated exercise intolerance and fatigue; (b) likely medical therapy–induced hyperlipidemia and deposition of fat (buffalo hump) in posterior neck region.

Plan

(a) Complete graded exercise test; (b) initiate cardiorespiratory training program and resistance or weight training program aimed at increasing muscle mass, muscle endurance, and aerobic capacity; (c) initiate counseling about low-fat diet and refer to registered dietitian for specific meal planning and counseling.

Discussion Questions

1. What is your assessment of the functional class? Explain your answer.
2. Are any red flags present on the clinical evaluation that require contacting, at least by telephone, her physician or another supervising physician before exercise testing?
3. What is the likely cause of the resting tachycardia?
4. Do any preventive cardiology issues need to be addressed in the near future?
5. Does exercise present special risks to physical injury?

5

Graded Exercise Testing and Exercise Prescription

Paul S. Visich, PhD, MPH

Jonathan K. Ehrman, PhD, FACSM

This chapter discusses the importance or value of graded exercise testing (GXT) and the general development of an exercise prescription. Current recommendations for supervision of GXT are reviewed. In addition, absolute and relative contraindications for performing and terminating a GXT are discussed, along with commonly used protocols and modes for testing. And, an in-depth assessment of resting and exercise ECGs is reviewed, so that the reader has a clear understanding of the importance of interpreting the patient's ECG before, during, and after completing a GXT. The theory and practical application of an exercise prescription is then presented and is to be used as the basis of all exercise training discussions in the population and disease specific chapters that follow.

EXERCISE TESTING

Graded exercise testing (GXT) was first used in approximately 1846, when Edward Smith began to evaluate the response of different physiological parameters (heart rate, respiratory rate, inspired air) during exertion (15). Since that time, graded exercise testing has become a valuable tool to evaluate functional capacity in many settings. Even with the addition of the **radionuclide agents**, **stress echocardiogram**, and pharmacological stress testing, GXT with a 12-lead ECG remains the first choice in evaluating for cardiovascular ischemia in most individuals who have the ability to exert themselves physically. Additionally, recent analysis of GXT data has resulted in the development of several valuable prognostic equations. This chapter provides a general overview of graded exercise testing commonly used in today's clinical setting. Other chapters in this textbook address unique features of population-specific exercise testing. This chapter focuses on the information that a clinical exercise physiologist, along with other allied health professionals, should acquire to perform a graded exercise test (GXT) in physician-supervised and nonsupervised settings.

Personnel

Before 1980 GXT in a clinical setting was primarily (90%) supervised by a cardiologist (63). Since that time, GXTs have been performed by many healthcare professionals (internists, family practitioners, physician assistants, clinical exercise physiologists and exercise specialists, nurses, and physical therapists). The use of healthcare professionals rather than cardiologists has come about because of better understanding of graded exercise testing over the years, cost-containment initiatives, time constraints on physicians, and more sophisticated ECG analysis (computerized exercise ST-segment interpretation). The American Heart Association guidelines for exercise testing laboratories state that the paramedical personnel listed previously, when appropriately trained and possessing specific performance skills (e.g., American College of Sports Medicine certification), can safely supervise clinical GXTs (53). In addition, paramedical personnel should be trained in basic life support, and advanced cardiac life support is highly suggested. See "Graded Exercise Testing Skills for Paramedical Professionals."

For graded exercise testing of patients with high-risk medical conditions, such as heart failure or high-grade dysrhythmias, direct physician supervision is suggested.

Graded Exercise Testing Skills for Paramedical Professionals

- Knowledge of absolute and relative contraindications for exercise testing
- Ability to communicate properly with the client to complete a medical history and informed consent
- Ability to explain to the client the purpose of completing the graded exercise test (GXT), procedures of the test, and the responsibilities of the client during the test
- Competence in cardiopulmonary resuscitation certified by American Heart Association basic cardiac life support and preferably advanced cardiac life support
- Knowledge of specificity, sensitivity, and predictive value of a positive and negative test and diagnostic accuracy of exercise testing in different patient populations
- Understanding of causes that produce false-positive and false-negative test results
- Knowledge of appropriate mode of activity and exercise protocol (e.g., Bruce, Naughton, Balke-Ware) for each individual, based on his or her medical history
- Knowledge of normal and abnormal hemodynamic responses to graded exercise (blood pressure and heart rate response) in different age groups and with various cardiovascular conditions
- Knowledge of normal and abnormal acute cardiovascular physiological responses to exercise for different age groups and cardiovascular conditions
- Knowledge of metabolic data collected during a GXT and knowledge of how to interpret the data (e.g., maximal oxygen uptake, metabolic equivalent) for different medical conditions
- Knowledge of 12-lead electrocardiography and changes in the electrocardiogram that may result from exercise, especially ischemia, arrhythmias, and conduction abnormalities
- Knowledge of proper lead placement and skin preparation for a 12-lead electrocardiogram
- Knowledge and skills for accurately taking blood pressure under resting and exercise conditions
- Knowledge of how to run and troubleshoot the medical equipment used for graded exercise testing (treadmill, electrocardiogram, bicycle ergometer, metabolic cart)
- Ability to communicate with the client to assess signs and symptoms of cardiovascular disease before, during, and after the GXT
- Knowledge of how to assess signs and symptoms by using appropriate scales (e.g., chest pain, shortness of breath, rating of perceived exertion)
- Knowledge of the appropriate time to end the GXT and absolute and relative indications for test termination
- Knowledge of knowing when to ask for physician support when the physician is not directly involved in the GXT
- Ability to communicate results of the GXT to the supervising physician.

Reprinted, by permission, from G.P. Rodgers et al., 2000, "Clinical competence statement on stress testing: A report of the American College of Cardiology/American Heart Association/American College of Physicians—American Society of Internal Medicine Task Force," Circulation 102: 1726-1738.

Otherwise, having a supervising physician in the immediate area and readily available to respond to emergencies and questionable interpretations is acceptable. Four studies found that the average morbidity and mortality rates during a GXT with physician supervision (>85% of the tests directly supervised by a physician) were 3.6 and 0.44 per 10,000 tests, respectively (8,56,57,63). In three studies involving nonphysician supervision, average morbidity and mortality rates of 2.4 and 0.77 per 10,000 tests were observed, respectively (24,39,42). These data suggest that no differences occur in morbidity and mortality rates related to graded exercise testing when direct physician and paramedical staff supervision are compared. In addition, when a symptom-limited GXT is performed on a high-risk population (**left ventricular dysfunction** with **ejection fraction** <35%), nonphysician supervision has been observed to be safe, when a physician is immediately available (63). Although new Medicare guidelines require direct physician supervision when testing high-risk patients, current findings suggest that paramedical professions can safely complete these tests, at a significant cost savings.

Indications

Although the overall risk of death during a GXT is small (0 to 5 sudden cardiac deaths per 10,000 tests), before a GXT is performed it is important to understand the purpose or reason for completing the test and, most important, to be sure that the benefits outweigh the risks (30,60). If a question arises concerning the purpose for completing the GXT, the referring physician should be contacted for clarity.

With respect to indications for GXT, the American College of Cardiology (www.acc.org) and the American Heart Association (www.americanheart.org) provide guidelines for exercise testing for various clinical situations. For each clinical situation, three classes exist. Class I represents conditions in which there is agreement that a GXT is justified. In Class IIa the evidence is conflicting, but testing is considered useful. In class IIb the evidence is conflicting, but the justification for testing is less well established. In class III there is general agreement that testing is not useful or effective and is potentially harmful. In addition, the weight of the evidence for testing is defined as the highest ranking (A) if the data were derived from multiple randomized clinical trials involving large numbers of patients. An intermediate ranking (B) is given if the data were derived from a limited number of randomized clinical trials that involved a small number of patients or from careful analyses of nonrandomized studies or observational registries. The lowest rank (C)

is given if expert consensus was the primary basis for the recommendation (29).

In most cases, a GXT is the test of choice because of its noninvasive nature and lower cost. The diagnostic value of determining the presence of coronary artery disease (CAD) is greatest when evaluating individuals with an intermediate to high probability of CAD, which is based on the person's age, sex, and the presence of symptoms (typical and atypical). Typical angina symptoms in 30 yr old males and females represent an intermediate likelihood of CAD, whereas when males reach age 40 and females reach age 50, typical angina symptoms represent a high likelihood of CAD. In respect to atypical (probable) angina, intermediate likelihood of CAD occurs at age 30 or older in males and at age 50 or older in females (39). Conversely, there appears to be little value in improving patient outcomes when exercise testing individuals who are asymptomatic.

In individuals with documented CAD, graded exercise testing is used to evaluate the severity of the disease and potential progression of disease. This information can be beneficial in evaluating the need for further intervention. In individuals who have an abnormal response such as an ST depression of 1 mm or greater at low work levels (e.g., unable to complete stage 1 of the Bruce protocol), a higher mortality rate (5%) is observed compared with those who were able to complete three stages without any ST changes (<1% mortality rate) (70).

In respect to minimizing healthcare costs, decreasing the length of stay in the hospital is looked on favorably. Submaximal GXT can be completed prior to discharge for prognostic assessment, and maximal GXT can be completed 14 to 21 d after discharge in uncomplicated patients (29). If the GXT is interpreted as normal (nonischemic, normal hemodynamic response with exertion, and no significant dysrhythmias, and the patient achieves an acceptable **metabolic equivalent** [MET] level), this information aids the physician in deciding to discharge the patient, whereas an abnormal response will potentially lead to further diagnostic testing.

Following coronary revascularization, a GXT may be used to assess the outcome of the procedure and help determine a safe and effective exercise prescription. To evaluate the effectiveness of medical therapy, a GXT is used as an objective measure to evaluate changes in the onset of dysrhythmias, chest pain, and hypertension. In individuals with documented CAD, a GXT can be used yearly to evaluate disease progression. The onset of chest pain, ST depression, ischemia, or shortness of breath that occurs at a lower **rate pressure product** or workload can suggest disease progression. Although a change in a GXT may suggest disease progression, other factors may promote unfavorable changes without necessarily increasing disease progression. Factors that can have a negative effect include the following:

Substantial weight gain

Significant decrease or stoppage of aerobic exercise training

Decrease or discontinuation in medications that negatively influence aerobic capacity

When a person has a substantial weight gain, the cardiovascular system must work harder for a given workload, which can enhance the onset of chest pain, ST depression, and ischemia. If a person stops aerobic exercise or decreases activity substantially to the point of a significant decline in **functional capacity** (**FC**), abnormal signs or symptoms could result prematurely, if underlying CAD is present. Finally, alterations in a patient's medical regimen could alter the onset of abnormal signs or symptoms. **Calcium channel blockers**, **β blockers**, **angiotension converting enzyme inhibitors**, and **nitrates** are common medications given to cardiac patients, because they decrease myocardial oxygen consumption (per given workload) by decreasing HR, contractility, and vascular resistance. Ultimately, these medications delay the onset of ischemia and increase a person's work tolerance in the presence of ischemic heart disease. Decreasing the dosage or frequency of these drugs can potentially promote abnormal signs or symptoms at an earlier workload. Therefore, keeping accurate records of the patient's drug regimen is important, especially before a GXT.

Based on the severity of disease, a GXT can be a useful tool to **stratify** patients. In people at high risk (i.e., abnormal GXT response at low workloads; <5 METs), further medical intervention may be required. But if an abnormal ECG response does not occur until the person reaches 10 METs, the prognosis is much more favorable and additional medical or surgical intervention may not be needed. Graded exercise testing can also be used for the primary purpose of evaluating FC. Assessment of a patient's FC can aid in determining the level of risk and determine the need for further medical intervention. The American Association of Cardiovascular and Pulmonary Rehabilitation suggest that a patient who achieves a FC >7 METS be considered at low risk (along with absence of complex ventricular dysrhythmias, angina, or other symptoms and a normal hemodynamic response with exercise). Conversely, a high-risk patient is someone who has angina or other unusual symptoms (e.g., unusual shortness of breath, dizzy, lightheaded, etc.) before achieving 5 METS (73).

After assessing a patient's FC following a symptom-limited GXT, along with assessing ST changes and angina

symptoms, the prognosis can be estimated by using the **Duke nomogram** to determine the 5 yr survival and average annual mortality rate, as shown in figure 5.1 (43).

In addition, heart rate (HR) in recovery has an independent influence on mortality rate. It appears that HR recovery at 2 min best predicts mortality rate, and that a HR decrease of less than 22 beats per minute seated produced a **hazard ratio** of 2.6, irrespective of b-blocker use. However, HR recovery is not able to discriminate those with coronary disease, and the authors suggest that HR recovery should supplement other predictors of mortality rate, such as the Duke nomogram (61).

The assessment of FC can be completed by measurement (open circuit spirometry) or prediction, based on speed and grade on a treadmill or work rate (kilogrammeters per minute) on a bicycle ergometer. Predicted measurements, however, tend to overestimate values, which can potentially lead to misclassifying a patient's level of risk. Because of the inconvenience and the expense of directly measuring a patient's FC, many clinical facilities rely on predicted metabolic equations.

The standard ACSM metabolic equations (2) are commonly used, but they rely on achieving a steady state, which does not occur at maximal effort and leads to overestimation, as previously mentioned (see "Equations to Predict Energy Expenditure" on p. 82). Other metabolic prediction equations exist for predicting FC but are specific to the Bruce GXT protocol, use of handrail support, ramping on a treadmill, and watt increment per minute on a bicycle ergometer (table 5.1). Because numerous maximal GXT protocols are used in the field, additional equations are needed for predicting FC.

Just before the patient is discharged from the hospital, a GXT may be used. This test is commonly a submaximal effort (following myocardial infarction) to assess future risk and verify that the patient can safely resume activities of daily living following discharge. In addition, this GXT can be used to determine a safe exercise prescription for home-based and initial cardiac rehabilitation programs. The need for a GXT before starting an exercise program depends on age, sex, number of risk factors, signs or symptoms related to cardiopulmonary disease, and documented cardiopulmonary disease. The ACSM provides specific recommendations detailing when a physical examination and GXT should be completed and whether a physician is needed to provide supervision before starting an exercise program (2).

Contraindications

Although the risk of death during a GXT is very small (1 out of every 2,500 tests), GXT is not recommended in

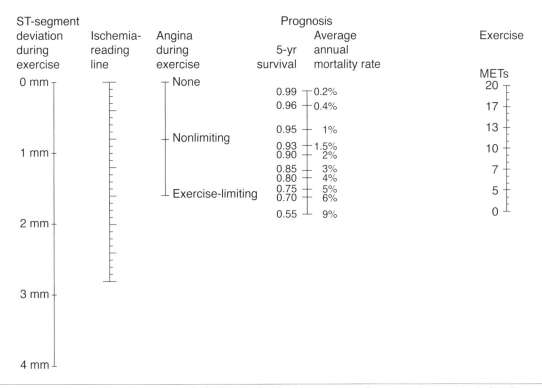

Figure 5.1 Duke nomogram. Step 1: Mark the largest level of ST-segment deviation. Step 2: Mark the observed degree of angina. Step 3: Connect the marks from steps 1 and 2 with a straight line and mark where this line crosses the ischemia line. Step 4: Mark the total METs achieved during the testing. Step 5: Connect the marks from steps 3 and 4 with a straight line. The point where this line crosses the prognosis line indicates the 5 yr survival rate and annual mortality rate for the subject.

Table 5.1 Estimating Functional Capacity at Maximal Effort

Standard treadmill test	
Using the Bruce Protocol without holding the handrail	$\dot{V}O_2max$ (ml · kg^{-1} · min^{-1}) = 14.8 – 1.379 (time in min) + 0.451 (time^2) – 0.012 (time^3)
Using the Bruce Protocol while holding the handrail	$\dot{V}O_2max$ (ml · kg^{-1} · min^{-1}) = 2.282 (time in min) + 8.545
Ramping treadmill protocols	
Individualized ramp protocol	$\dot{V}O_2max$ (ml · kg^{-1} · min^{-1}) = 0.72x + 3.67 x = predicted $\dot{V}O_2$ based on peak speed and grade using ACSM walking equation
Standardized Bruce ramping protocol	$\dot{V}O_2max$ (ml · kg^{-1} · min^{-1}) = 3.9 (time in min) – 7.0
Cycle ergometry (based on the final power completed in a 15 W · min^{-1} protocol	
Males	$\dot{V}O_2max$ (ml · min^{-1}) = 10.51 (power in W) + 6.35 (body mass in kg) – 10.49 (age in yr) + 519.3
Females	$\dot{V}O_2max$ (ml · min^{-1}) = 9.39 (power in W) + 7.7 (body mass in kg) – 5.88 (age in yr) + 136.7

Note. ACSM = American College of Sports Medicine.

Equations to Predict Energy Expenditure

Walking

$ml \cdot kg^{-1} \cdot min^{-1} = (0.1 \cdot S) + (1.8 \cdot S \cdot G) + 3.5$

Most appropriate between 50 and 100 m · min^{-1}, or between 1.9 and 3.7 mph

OK above this range (up to 5 mph [134 m · min^{-1}]) if the patient is truly walking

S = speed in m · min^{-1}, G = percent grade expressed as a fraction

Running

$ml \cdot kg^{-1} \cdot min^{-1} = (S \cdot 0.2) + (0.9 \cdot S \cdot G) + 3.5$

Most accurate for speeds >134 m · min^{-1}, or 5 mph, or down to 80 m · min^{-1}, or 3 mph, if truly jogging

S = speed in m · min^{-1}, G = percent grade expressed as a fraction

Leg ergometer

$ml \cdot kg^{-1} \cdot min^{-1} = (1.8 \, W \cdot M^{-1}) + 7.0$

Most appropriate for power outputs between 50 and 200 W

W = ergometer work rate in kg (m · min^{-1}), M = body mass in kg

Arm ergometer

$ml \cdot kg^{-1} \cdot min^{-1} = (3 \, W \cdot M^{-1}) + 3.5$

Most appropriate for power outputs between 25 and 125 W

W = ergometer work rate in kg · m · min^{-1}, M = body mass in kg

Stepping

$ml \cdot kg^{-1} \cdot min^{-1} = (0.2 \cdot F) + (1.33 \cdot 1.8 \cdot H \cdot F) + 3.5$

Most appropriate for stepping rates between 12 and 30 steps · min^{-1}, and step height is in meters (1 in. = 0.0254 m)

F = stepping frequency per minute, H= step height in meters

Reprinted, by permission, from American College of Sports Medicine, 2005, *ACSM's guidelines of exercise testing and prescription*, 7th ed. (Philadelphia, PA: Lippincott, Williams, and Wilkins).

some situations because the risks of completing the test outweigh the benefits (29). Current guidelines separate **absolute** versus **relative contraindications for test termination** (see "Contraindications to Graded Exercise Testing" on page 83). Absolute contraindications represent potential serious medical consequences if a GXT is completed, whereas relative contraindications suggest that a potential medical concern is present and that further inquiry (benefits versus risk) should be considered before the GXT is conducted (29).

In addition to contraindications of GXT, there are conditions in which GXT is not recommended for diagnostic purposes because of the decreased ability to identify CAD (i.e., ST-T changes). Resting ECG abnormalities that limit the diagnostic abilities of a GXT are:

Complete left bundle branch block

Preexcitation syndrome (Wolff-Parkinson-White)

Left ventricular hypertrophy

Digoxin therapy

Greater than 1 mm of ST-segment depression

Electronically paced ventricular rhythm (29)

To increase sensitivity to detect **exertional ischemia**, a GXT with **nuclear perfusion** or **echocardiography** is a better diagnostic procedure to consider when an individual has these resting abnormalities. If a GXT is being performed for reasons other than evaluating exertional myocardial ischemia (i.e., functional capacity

Contraindications to Graded Exercise Testing

Absolute Contraindications

1. Acute myocardial infarction (within 2 d) or other acute cardiac event
2. Significant change on the electrocardiogram suggesting ischemia, myocardial infarction, or other acute cardiac event
3. Unstable angina not stabilized by medical therapy
4. Uncontrolled cardiac dysrhythmias causing symptoms or hemodynamic compromise
5. Symptomatic severe aortic stenosis
6. Uncontrolled symptomatic heart failure
7. Acute pulmonary embolus or pulmonary infarction
8. Acute myocarditis or pericarditis
9. Suspected or known dissecting ventricular or aortic aneurysm
10. Acute infections (influenza, rhinovirus)

Relative Contraindications

1. Left main coronary stenosis
2. Moderate stenotic valvular heart disease
3. Severe arterial hypertension (systolic blood pressure >200 mmHg or diastolic blood pressure >110 mmHg)
4. Tachycardic or bradycardic dysrhythmias
5. Hypertrophic cardiomyopathy and other forms of outflow tract obstruction
6. Mental or physical impairment that leads to inability to exercise adequately or is exacerbated with exercise
7. High-degree atrioventricular block
8. Ventricular aneurysm
9. Chronic infectious disease (AIDS, mononucleosis, hepatitis)
10. Electrolyte abnormalities (hypokalemia, hypomagnesemia)
11. Uncontrolled metabolic diseases (diabetes, myxedema, thyrotoxicosis)

Performing a graded exercise test on a person with a relative contraindication may be acceptable if the benefits outweigh the risks.

Adapted from G. M. Adams, 1998, *Exercise physiology: Laboratory manual*, 3rd ed. (Boston, WCB McGraw-Hill); adapted from American College of Sports Medicine, 1998, "Position stand. The recommended quantity and quality of exercise for developing and maintaining cardiorespiratory and muscular fitness, and flexibility in healthy adults," *Medicine and Science in Sports and Exercise* 30(6): 975-991.

or efficacy of medical therapy to control dysrhythmias or hypertension), the GXT may still be appropriate.

Graded Exercise Testing Protocols

There are numerous protocols used to perform graded exercise testing (see table 5.2 on p. 84). The supervisor of the GXT must use a protocol that enables the client to achieve maximal effort without premature fatigue. The client should exercise for a minimum of 6 min but not more than 12 min. This will allow enough time for significant physiologic adaptations to exercise to occur, but will reduce the likelihood of ending the GXT due to skeletal muscle fatigue. The GXT supervisor is responsible for choosing the most appropriate protocol. When reviewing the medical history of a client, evaluation of exercise and activity habits can help determine an appropriate exercise protocol. Ideally, each clinical site should have specific protocols for testing clients with a variety of functional capacities and should avoid changing protocols if possible from one test to the next on a client (unless the exercise time is <6 min or >12 min). Such consistency facilitates making comparisons, such as defining the workload that corresponds to the onset of chest pain or ST changes.

Table 5.2 Commonly Used Treadmill and Bicycle Protocols

Protocol	Stage	Time (min)	Speed in mph (kph)	Grade (%)	VO$_2$ (ml · kg^{-1} · min^{-1})	METs
Bruce[a]	1	3	1.7 (2.7)	0.0	8.1	2.3
	2	3	1.7 (2.7)	5.0	12.2	3.5
	3	3	1.7 (2.7)	10.0	16.3	4.6
	4	3	2.5 (4.0)	12.0	24.7	7.0
	5	3	3.4 (5.5)	14.0	35.6	10.2
	6	3	4.2 (6.8)	16.0	47.2	13.5
	7	3	5.0 (8.0)	18.0	52.0	14.9
	8	3	5.5 (8.9)	20.0	59.5	17.0
Naughton	1	2	1.0 (1.6)	0.0	8.9	2.5
	2	2	2.0 (3.2)	0.0	14.2	4.1
	3	2	2.0 (3.2)	3.5	15.9	4.5
	4	2	2.0 (3.2)	7.0	17.6	5.0
	5	2	2.0 (3.2)	10.5	19.3	5.5
	6	2	2.0 (3.2)	14.0	21.0	6.0
	7	2	2.0 (3.2)	17.5	22.7	6.5
Balke-Ware	1	1	3.3 (5.3)	2.0	15.5	4.4
	2	1	3.3 (5.3)	3.0	17.1	4.9
	3	1	3.3 (5.3)	4.0	18.7	5.3
	4	1	3.3 (5.3)	5.0	20.3	5.8
	5	1	3.3 (5.3)	6.0	21.9	6.3
	6	1	3.3 (5.3)	7.0	23.5	6.7
	7	1	3.3 (5.3)	8.0	25.1	7.2
Ellestead	1	3	1.7 (2.7)	10.0	16.3	4.6
	2	2	3.0 (4.8)	10.0	26.0	7.4
	3	2	4.0 (6.4)	10.0	33.5	9.6
	4	3	5.0 (8.0)	10.0	41.0	11.7
	5	2	5.0 (8.0)	15.0	53.1	15.2
	6	3	6.0 (9.7)	15.0	63.0	18.0
Substandard Balke	1	2	2.0 (3.2)	0.0	8.9	2.5
	2	2	2.0 (3.2)	2.5	11.3	3.2
	3	2	2.0 (3.2)	5.0	13.7	3.9
	4	2	2.0 (3.2)	7.5	16.1	4.6
	5	2	2.0 (3.2)	10.0	18.5	5.3
	6	2	2.0 (3.2)	12.5	20.9	6.0
	7	2	2.0 (3.2)	15.0	23.3	6.7

Protocol	Stage	Time (min)	Speed in mph (kph)	Grade (%)	VO$_2$ (ml · kg^{-1} · min^{-1})	METs
Standard Balke	1	2	3.0 (4.8)	2.5	15.2	4.3
	2	2	3.0 (4.8)	5.0	18.8	5.4
	3	2	3.0 (4.8)	7.5	22.4	6.4
	4	2	3.0 (4.8)	10.0	26.0	7.4
	5	2	3.0 (4.8)	12.5	29.6	8.5
	6	2	3.0 (4.8)	15.0	33.2	9.5
	7	2	3.0 (4.8)	17.5	36.9	10.5
Superstandard Balke	1	2	3.4 (5.5)	2.0	15.9	4.5
	2	2	3.4 (5.5)	4.0	19.2	5.5
	3	2	3.4 (5.5)	6.0	22.5	6.4
	4	2	3.4 (5.5)	8.0	25.7	7.4
	5	2	3.4 (5.5)	10.0	29.0	8.3
	6	2	3.4 (5.5)	12.0	32.3	9.2
	7	2	3.4 (5.5)	14.0	35.6	10.2

	Stage	Duration (min)	RPM	kg · m · min^{-1}: watts	VO$_2$ (ml · kg^{-1} · min^{-1})[b]	METs
Standard bicycle test	1	2 or 3	50.0	150: 25	10.9	3.1
	2	2 or 3	50.0	300: 50	14.7	4.2
	3	2 or 3	50.0	450: 75	18.6	5.3
	4	2 or 3	50.0	600: 100	22.4	6.4
	5	2 or 3	50.0	750: 125	26.3	7.5
	6	2 or 3	50.0	900: 150	30.1	8.6

Note. When trying to determine the appropriate protocol, consider the client's age, current functional capacity, activity level, and medical history and disease status. MET = metabolic equivalent.

[a]VO$_2$ is calculated while the client is walking through stage 6. If the client is running at stage 6, the VO$_2$ would be 42.4 ml · kg^{-1} · min^{-1} and 12.1 METs. The conventional Bruce protocol begins at 1.7 mph (2.7 kph) and a 10% grade; modified versions start at 1.7 mph (2.7 kph) at 0% or 1.7 mph (2.7 kph) and 5%. [b]VO$_2$ and METs determined for a 70 kg individual. Values need to be adjusted based on body weight.

Protocol Considerations

For diagnostic purposes, the Bruce treadmill protocol is the most widely used test because of its long-standing history, training familiarity, frequent citation in the medical literature, and normative data for many populations. In people who are not frail, do not have extremely low functional capacity, and are free from orthopedic problems that are irritated by walking uphill, the Bruce protocol is acceptable for diagnostic purposes. But using the Bruce protocol does have negative consequences; these include large increments in workload (2 to 3 METs) that make it difficult to determine a person's functional capacity, 3 min stages that for some are too fast to walk and too slow to run, and a large vertical component, which can lead to premature leg fatigue. When trying

to determine whether the Bruce protocol is appropriate, the GXT supervisor should ask the client whether he or she can walk several flights of stairs without stopping. Because of these concerns, other treadmill protocols should be considered. Commonly used protocols with smaller MET increments between stages include the Balke-Ware and Naughton protocols. These protocols advance in 1 and 2 min stages, respectively, with increments of 1 MET or less. These protocols are commonly used with the elderly and people who are deconditioned because of a chronic medical problem such as heart or pulmonary disease. One of the more common problems with these protocols is that the supervisor underestimates the client's functional capacity and the person exercises for longer than 12 min. More recently, **ramping treadmill protocols** along with the more automated ECG treadmill or bicycle ergometer systems are available. The advantage of ramping tests is that there are no large incremental changes, ramp rates can be adjusted, and the client's functional capacity can be more accurately determined. Companies that include a ramping option allow conversion of standardized protocols (i.e., Bruce) to a ramping format, in which treadmill speed and grade can change in small time increments (i.e., 6 to 15 s). When the Bruce protocol was compared between the standard 3 min stage and ramping in 15 s intervals, similar hemodynamic changes were observed but the ramping test produced a significantly greater duration in time and a higher peak MET level. In addition, subjects perceived the ramping protocol to be significantly easier than the standard Bruce protocol (72).

Equipment Considerations

If a treadmill cannot be used, the bicycle ergometer is the usual second choice. But when comparing the treadmill to the bicycle ergometer, a person will achieve a 5% to 20% lower functional capacity on the bicycle ergometer, based on leg strength and conditioning level (2). Bicycle ergometer testing is helpful for diagnostic reasons and to assess functional capacity in clients with weight-bearing and gait problems. In electronically braked bicycles, ramping protocols are commonly used, during which a slow increase in workload occurs during each minute. A typical bicycle ramping protocol consists of an increase between 15 and 25 W · min^{-1}. In more standardized bicycle ergometers, a work rate increase of 150 kpm · min^{-1} or 25 W every 2 to 3 min stage is commonly used for the general population. In frail clients or in individuals with a low functional capacity, smaller work rate increments should be considered. In contrast, in heavier clients, the standard incremental increase of 25 W may be insufficient to reach maximal effort within 12 min. With respect to cadence, 50 to 60 rev · min^{-1} are suggested with standard bicycle ergometers. In the electronically braked bicycles, cadence is not a major concern, because resistance automatically changes, based on cadence, to maintain a specific work rate (e.g., 100 W). Among heavier and more active clients who need to accomplish a high workload to reach maximum effort, increasing work rate by 300 kgm · min^{-1} or 50 W every 2 or 3 min may be more appropriate. After the client achieves 75% of his or her predicted maximal HR or a **rating of perceived exertion** (**RPE**) of 3 (0 to 10 scale) or 13 (6 to 20 scale), the workload would increase by 150 kgm · min^{-1} every stage or 25 W, so that the ending workload could be more clearly defined. This protocol may more accurately determine the client's functional capacity. In the past, arm ergometry testing was used for diagnostic purposes in clients with severe orthopedic problems, **peripheral vascular disease**, lower-extremity amputation, and neurological conditions that inhibited the ability to exercise on a treadmill or bicycle ergometer. In a clinical setting, however, the arm ergometer is now replaced with **pharmacological stress testing**. Clients for whom an arm ergometry GXT may still be beneficial include those with symptoms of myocardial ischemia during dynamic upper-body activity or those who suffered a myocardial infarction and plan to return to an occupation that requires upper-body activity. Because of a smaller muscular mass in the arms versus the legs, a typical arm protocol involves 10 to 15 W increments for every 2 to 3 min stage at a cranking rate of 50 to 60 rev · min^{-1}.

Graded Exercise Testing Procedures

Before the start of the GXT, a number of procedures must be completed.

1. Reading and signing of the informed consent
2. Medical history and physical assessment
3. Assessment of resting 12-lead ECG, BP, and heart sounds
4. Instructions on completing the GXT and the responsibility of the patient

If resting measurements are acceptable, the GXT can be started. During the GXT, the client needs close monitoring for abnormalities and the supervisor needs to be responsible for discontinuing the test at the appropriate time.

Patient Instructions

The patient must receive clear instructions on how to prepare for the GXT along with an appropriate explanation of the test. A lack of proper instruction can delay the test, increase patient anxiety, and potentially increase the health risk to the patient. Instructions for patient preparation should include information on clothing; footwear; food consumption; avoidance of alcohol, cigarettes, caffeine, and over-the-counter medicines that could influence the GXT results; and whether the patient should take his or her prescribed drug regimen. A prescribed medication may be discontinued because certain agents can inhibit the ability to observe ischemic responses during exercise (e.g., β-blockers blunt HR and BP, which can inhibit the occurrence of ST-segment changes associated with ischemia). However, this is not a requirement. Note that if a physician wants the patient to discontinue medications before GXT, the patient should receive those instructions from the physician. The process of weaning a person off various medications can vary, and the rebound effect can be a concern. This area is generally not the responsibility of the paramedical professional, unless instructed by the physician. Recognize that the client may be anxious before completing the GXT, and the test supervisor should answer all questions in a professional and caring manner. If the test supervisor is prepared and confident, the client will be more at ease and potentially more willing to give his or her best effort.

Medical Evaluation

Before any testing, the test supervisor should understand the individual's medical history. A medical history will help determine whether any contraindications to the GXT are present, will aid in choosing an appropriate testing protocol, and will help identify areas that may require more supervision during the test (e.g., history of hypertension, previous knee injury). A comprehensive medical history should evaluate height and weight; assess risk factors and any recent signs or symptoms of cardiovascular disease; document previous cardiovascular disease; determine other chronic diseases that may influence the outcome (e.g., arthritis, gout); and obtain information about current medications (including dose and frequency), drug allergies, recent hospitalizations or illnesses, and exercise and work history. A plethora of medical history forms are available, and the test supervisor should be comfortable with the form that she or he is using and be sure that the questions asked are appropriate to the subject's sex and age. In addition to medical history, a brief physical examination of heart and lung sounds and peripheral edema is usually required. Specific assessment skills are presented in chapter 4. Finally, all clients should complete an informed consent. An informed consent properly informs clients of what they will be asked to do with respect to exercise, complications associated with completing the test (risks and discomforts), test supervisor responsibilities, and client responsibilities. Informed consent provides the client both an opportunity to ask questions about the procedure and freedom to refuse participation. If the client agrees to participate, he or she must sign the form, and clients under 18 yr of age must have a parent or guardian sign the consent. Examples of both consent and medical history forms are available in Pina et al. (53).

ECG Preparation and Resting Measurements

The clarity of the ECG tracing during exercise is of utmost importance, especially for diagnostic purposes. Therefore, the client needs to be properly prepared. For males, hair from all electrode positions should be removed with a razor or battery-operated shaver. Conductance through the skin surface is critical for achieving a clear ECG recording and shaving allows the electrodes to lay flat on the skin. Removal of body oils, lotions and skin abrasion are also required for optinal conductance. Silver chloride electrodes are ideal because they offer the lowest offset voltage and are the most dependable in minimizing **motion artifact** (53). After electrodes are in place and cables are attached, each electrode placement site should be assessed for excessive resistance. On the newer ECG stress systems, a built-in assessment determines whether the electrode sites have excessive resistance that would require reprepping of a specific site. Another good test is to lightly tap on the electrode and observe the ECG. If the ECG produces a great deal of motion artifact, check the electrode and lead wire for potential interference. Adequate electrode preparation can decrease the chances of motion artifact, whereas muscle artifact cannot be reduced unless upper-body skeletal muscle activity is decreased. Because skin moves while the patient walks on the treadmill, keeping electrode and wire movement to a minimum is important, especially in large-breasted and obese individuals. Using some sort of wrap or mesh tubing over the electrodes and lead wires may help decrease movement.

After the patient has rested for a several minutes, 12-lead ECG and BP should be recorded in both the supine and standing positions. Evaluating both conditions allows the test supervisor to assess the effects of body position on hemodynamics. The resting ECG should be compared with a prior resting ECG for any changes. If any significant differences contraindicate starting the GXT (see section on contraindications), the paramedical supervisor should contact the supervising physician or the referring physician with the updated information, so that a decision can be made about completing the GXT.

Test Procedures

After the tester prepares the patient, chooses the appropriate protocol, and reviews the resting ECG for any abnormalities that may inhibit the evaluation, the GXT can start after the patient receives instruction on how to use the related mode of testing (treadmill, bicycle ergometer). In elderly individuals asked to walk on a treadmill for the first time, the test supervisor should take time to make sure that they are comfortable walking on the unit with little or no rail support assistance. Clients who are not comfortable may stop prematurely simply because of apprehension. In addition to the supervisor of the GXT (physician or paramedical professional), a technician is needed to monitor the patient's BP and potential signs and symptoms of exercise intolerance. ECG, HR, RPE, and BP should be recorded at the end of each stage. Approximately 30 to 45 s should be given to measure BP manually. Automated BP units are commonly used for resting measurements, but their accuracy under exercise conditions continues to be of question.

As the exercise test progresses from one stage to the next, testing staff must observe the client and communicate constantly. Before the test begins, the patient must understand that he or she is responsible for communicating at the onset of any discomfort. At minimum, at the end of each stage of testing, staff should ask clients how they are feeling and if they are having any discomfort. At each stage the supervisor should record any discomfort that the client has, including the absence of any stated discomfort, because doing so implies that the question was asked. During the GXT, the ECG must be continually monitored. A 12-lead ECG should be reviewed at the end of each stage, because it is possible that ECG changes can take place in any of the lead combinations. Between stages, the supervisor should monitor leads that reflect the different walls of the myocardium. If gas exchange is analyzed during the GXT, communication becomes more difficult because of the use of a mouthpiece or mask. Establishing specific hand signals is important so that symptoms (e.g., chest pain, shortness of breath) along with RPE can be accurately determined. A client can be apprehensive during her or his first experience with the measurement of gas exchange because breathing and speaking abilities are inhibited. The test supervisor must properly instruct the client on the use of the gas exchange apparatus and allow the client to wear the gas exchange apparatus while resting, so that she or he is comfortable before starting the test.

Exercise Data

The use of RPE with exercise testing is now common and provides a monitor of how hard the client perceives his or her work and how much longer he or she will be able to continue. The client's RPE is generally recorded during the last 15 s of each stage. Two RPE scales are commonly used: the 6- to 20-point category scale and the 0- to 10-point category-ratio scale.

To improve the accuracy of these scales, the following instructions (42) should be given to the client:

> *During the exercise test we want you to pay close attention to how hard you feel the exercise work rate is. This feeling should reflect your total amount of exertion and fatigue, combining all sensations and feelings of physical stress, effort, and fatigue. Don't*

concern yourself with any one factor such as leg pain, shortness of breath, or exercise intensity, but try to concentrate on your total, inner feeling of exertion. Try not to underestimate or overestimate your feelings of exertion; be as accurate as you can.

Other scales that are helpful in evaluating clients' specific symptoms include angina, **dyspnea**, and peripheral vascular disease scales (table 5.3). The angina scale is beneficial because it evaluates the intensity of chest pain and helps determine whether the test should be terminated based on standard criteria (see "Indications for Termination of Graded Exercise Test" on p. 91). Another important point is to document the onset of angina with the corresponding MET level and **rate pressure product** (**RPP**). The progression of angina can be evaluated with the scale, can be beneficial when comparing previous test results (i.e., whether the intensity of angina is occurring at a lower or higher workload), and may provide information about the progression of CAD. The dyspnea scale, which concerns rating the intensity and progression of dyspnea, is commonly used when testing patients with pulmonary disease and serves a similar role as the scale used with clients who have angina. The peripheral vascular disease scale is beneficial when evaluating clients with documented or questionable **claudication**, based on their medical history.

In respect to prescribing exercise, the information gathered from the scales mentioned here during the GXT can be beneficial. By comparing a client's RPE with a prescribed training HR range, the clinician can determine whether the client can achieve and sustain the intensity level. Be aware, however, that differences may exist when mode of testing differs from mode of training. Cardiac clients perceive exercise to be more difficult during exercise training compared with exercise testing (13). Besides using RPE, comparing prescribed training HR ranges to the onset of angina, dyspnea, and claudication symptoms can be beneficial in determining a safe and feasible training range. Finally, although these scales can be helpful, they need to be interpreted and considered in conjunction with other medical findings.

As clients approach their maximum effort, they may become anxious and want to stop the test, so the supervisor and technician must be prepared to record peak values (ECG, BP, RPE). Following is a suggested order of events at maximal effort:

1. Record BP 30 to 45 s before the test stops.

2. Record ECG 10 s before the test stops.

3. Record symptoms and RPE immediately before test termination.

Table 5.3   Angina, Dyspnea, and Peripheral Vascular Disease Scales

	Angina scale
1+	Light, barely noticeable
2+	Moderate, bothersome
3+	Moderately severe, very uncomfortable
4+	Most severe or intense pain ever experienced
	Dyspnea scale
1+	Mild, noticeable to patient but not observer
2+	Mild, some difficulty, noticeable to observer
3+	Moderate difficulty but patient can continue
4+	Severe difficulty, patient cannot continue
	Peripheral vascular disease scale
1+	Definite discomfort or pain but only of initial or modest levels (established but minimal)
2+	Moderate discomfort or pain from which the patient's attention can be diverted by a number of common stimuli (e.g., conversation, interesting TV show)
3+	Intense pain from which the patient's attention cannot be diverted except by catastrophic events (e.g., fire, explosion)
4+	Excruciating and unbearable pain

4. Reduce workload to a comfortable level. Consider short active recovery.

5. Stop treadmill, seat patient and monitor for up to 8 minutes or until near physiologic baseline is achieved.

6. Remove the gas exchange instrument after seating patient, if applicable.

Submaximal GXTs are often terminated when the client achieves a certain MET level or a percentage of maximal HR (i.e., 85% of predicted maximum HR). If HR is affected by medical therapy such as a β-blocking agent, an RPE of 13 may be used to terminate a submaximal test. Submaximal tests are commonly used in patients being discharged from a hospital, to make sure that future risk of a cardiovascular event is sufficiently low and that returning to activities of daily living is safe. Symptom-limited or maximal GXTs are commonly used for diagnostic purposes and assessment of functional capacity. A symptom-limited test refers to a test that is terminated because of the onset of symptoms that put the client at increased risk for further medical problems. A maximal test terminates when an individual reaches his or her maximal level of exertion, without being limited by any abnormal signs or symptoms as listed in "Indications for Termination of Graded Exercise Test." When a client gives a maximal effort, the test is normally terminated because of volitional or voluntary fatigue. Several criteria are often used to determine whether a person has reached maximal effort:

1. A plateau in $\dot{V}O_2$ (<2.1 ml · kg^{-1} · min^{-1} increase)

2. A respiratory exchange ratio value greater than 1.1

3. Venous blood lactate exceeding 8 to 10 mM (23)

These criteria rely on having a metabolic cart to measure oxygen consumption and a lactate analyzer, which is not feasible at many clinical facilities. Therefore, other criteria are commonly used in conjunction with voluntary fatigue; these include an RPE greater than 17, a plateau in HR despite an increasing workload or attainment of 85% of predicted maximal HR in non-β-blocked clients, and attainment of a 240,000 value for the rate pressure product.

When to Stop a GXT

The supervisor of the GXT must be aware of and understand potential complications that may occur and know when to terminate the GXT. This knowledge is especially crucial if a physician is not immediately present. Knowledge of the normal and abnormal physiological responses that take place with GXT is imperative. Reasons for test termination are presented as absolute and relative contraindications (see "Indications for Termination of Graded Exercise Test"). Absolute contraindications are considered high-risk indications (except a subject's desire to stop and technical difficulties) that have the potential to result in serious complications if the test is continued. Therefore, the test should be terminated immediately. If the physician is not immediately present, he or she should be contacted and informed of the outcome as soon as possible. With respect to relative contraindications, these findings might represent reasons for test termination, but generally the test is not stopped unless other abnormal signs or symptoms occur simultaneously. The reason for performing the GXT may also influence relative reasons for test termination. For example, when evaluating a patient for suspected CAD, if the test supervisor observes left bundle branch block with exertion which precludes interpretation of ST changes, this is a likely reason for test termination. Each facility must have policies and procedures that describe absolute and relative indications for test termination. Doing so eliminates any confusion about why a test was terminated and helps protect the facility from potential negligence if complications arise. If a question ever arises about whether a test should be stopped based on what is observed, the tester should always err on the conservative side and consider safety of the patient the highest priority (i.e., the test can always be repeated if necessary).

ECG Analysis

This section discusses analysis of the 12-lead ECG, which is based on recognizing normal and abnormal responses during a GXT. This section should not replace an in-depth study of ECG analysis, but it serves as a review for GXT-specific issues. When reviewing the resting ECG before the GXT, the clinician should also assess the client's medical history to see whether any discrepancies are present (e.g., the client states no previous myocardial infarction, but the current ECG shows significant Q waves throughout the lateral leads). Also important is a comparison of the client's current resting 12-lead ECG with a previous ECG, especially when an abnormality is detected (e.g., the client's current resting ECG rhythm shows atrial fibrillation; is this dysrhythmia new or old?). When clinically significant differences are detected on the resting ECG and appear to be a new finding, the supervising or referring physician should be informed before the test. When determining the safety of performing the GXT, refer to "Contraindications to Graded Exercise Testing."

Resting ECG Abnormalities

The most common reason to complete a GXT with 12-lead ECG is to assess potential CAD. As a result, accurate interpretation of ST changes is important. Specific ECG abnormalities that prevent using ST changes to assess ischemia are left bundle branch block, right bundle branch block (unable to interpret ST changes

in anterior leads, V1–V3, but lateral and inferior leads are interpretable), and Wolff-Parkinson-White (WPW) syndrome (figures 5.2, 5.3, and 5.4 on p. 92-93). With respect to ST depression on the resting ECG attributable to left ventricular hypertrophy or digoxin use, a difference of opinion exists about the usefulness of ST changes during exercise. If less than 1 mm of resting ST depression is present, it is generally accepted that GXT represents a reasonable test to assess exercise-induced ischemia; specificity, however, is reduced. In conditions in which a client has greater than 1 mm ST depression at rest, GXT with 12-lead ECG alone offers little diagnostic information. Use of cardiac imaging by echocardiography or radionuclide testing is preferred (24).

Exercise ECG Changes

During the GXT, one person should consistently observe the ECG recording to identify the onset of any ECG change. The evaluation of the ST segment is of great importance because of its ability to suggest the onset of ischemia. In addition, the onset of dysrhythmias should be identified, especially those that are related to absolute and relative contraindications for test termination (see "Indications for Termination of Graded Exercise Test" on p. 92). Although the onset and progression of ST depression is in most cases subtle, dysrhythmias can occur suddenly and can be brief, intermittent, or sustained.

All 12 ECG leads should be monitored; however, V5 is most likely to demonstrate ST segment depression, whereas the inferior leads (II, III, and aVF) are associated with a relatively higher incidence of false-positive findings (i.e., ST depression occurs, but it truly is not related to ischemia). V5 represents the most diagnostic lead because when true ischemia occurs with exertion, it is most commonly observed in this lead which reflects a large area of the left ventricle. When the test supervisor recognizes ST-segment changes, he or she should note when (time and work rate) it begins, the **morphology**, the magnitude, and any additional changes that take place at each stage. Most ECG stress systems provide summaries of ST segments throughout the testing period, which is beneficial but needs to be verified. In addition, documenting any symptoms associated with the ST changes (e.g., chest pain, shortness of breath, dizziness) is important.

• **ST-Segment Depression:** Of the potential ST-segment changes, ST-segment depression is the most frequent response and is suggestive of **subendocardial ischemia**. One or more millimeters of horizontal or downsloping ST-segment depression that occurs 0.08 s

Indications for Termination of Graded Exercise Test

Absolute Indications

1. Decrease in systolic blood pressure of >10 mmHg from baseline blood pressure despite an increase in workload, when accompanied by other evidence of ischemia

2. Moderate to severe angina (2–3 or more)

3. Increasing nervous system symptoms (e.g., ataxia, dizziness, or near syncope)

4. Signs of poor perfusion (cyanosis or pallor)

5. Technical difficulties monitoring the electrocardiogram or systolic blood pressure

6. Subject's desire to stop

7. Sustained ventricular tachycardia

8. ST elevation (>1.0 mm) in leads without diagnostic Q waves (other than V1 or a VR)

Relative Indications

1. Decrease in systolic blood pressure of >10 mmHg from baseline blood pressure despite an increase in workload, in the absence of other evidence of ischemia

2. ST or QRS changes such as excessive ST depression (>2 mm horizontal or downsloping ST-segment depression) or marked axis shift

3. Arrhythmias other than sustained ventricular tachycardia, including multifocal PVCs, triplets of PVCs, supraventricular tachycardia, heart block, or bradyarrhythmias

4. Fatigue, shortness of breath, wheezing, leg cramps, or claudication

5. Development of bundle branch block or intraventricular conduction delay that cannot be distinguished from ventricular tachycardia

6. Increasing chest pain

7. Hypertensive response

8. A systolic blood pressure of >250 mmHg or a diastolic blood pressure of >115 mmHg

Note. PVC = premature ventricular contractions.

Adapted from *Journal of American College of Cardiology,* Vol. 27, American College of Cardiology/American Heart Association, "ACC/AHA guidelines for the perioperative cardiovascular evaluation for noncardiac surgery: Report of the American College of Cardiology/American Heart Association Task Force on Practice Guidelines (Committee on Perioperative Cardiovascular Evaluation for Noncardiac Surgery)," pgs. 910-48. Copyright 1996, with permission from Elsevier.

(i.e., 80 msec) past the J point is recognized as a positive test for myocardial ischemia. When ST-segment changes of this type occur along with typical angina symptoms, the likelihood of CAD is extremely high (approximately 95%). In addition, the greater the ST depression, the more leads with ST depression, and the more time it takes for the ST depression to resolve in recovery, the more likely it is that significant CAD is present. In addition, in some cases ST-segment depression is observed only in recovery and should be treated as an abnormal response. Additionally, J-point depression with upsloping ST segments that are greater than 1.5 mm depressed at 0.08 s past the J-point are suggestive of exercise-induced ischemia. But if the rate of upsloping ST depression is gradual (<1 mV $\cdot$ s^{-1}), the probability of CAD is increased (18), although the rate of false-positive tests will be higher. As the rate of false-positive tests increases, specificity decreases and sensitivity increases (i.e., the test is more likely to suggest that CAD is present when it truly is not, but is also more likely to detect people who truly have CAD). The negative consequence to decreased specificity is that client anxiety will increase because of the need for additional diagnostic testing (i.e., radionuclide or echocardiography GXT).

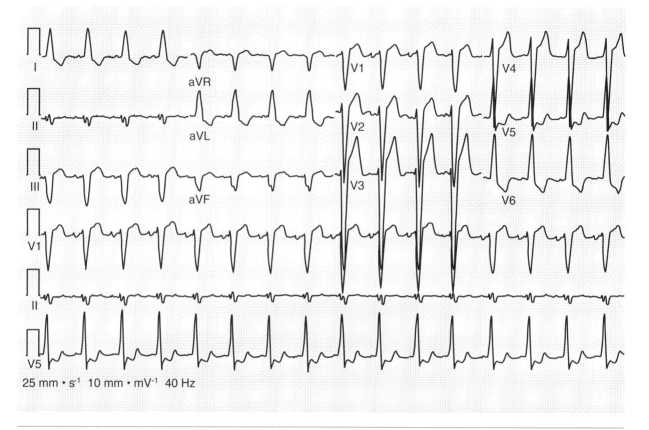

Figure 5.2 Electrocardiogram showing left bundle branch block.

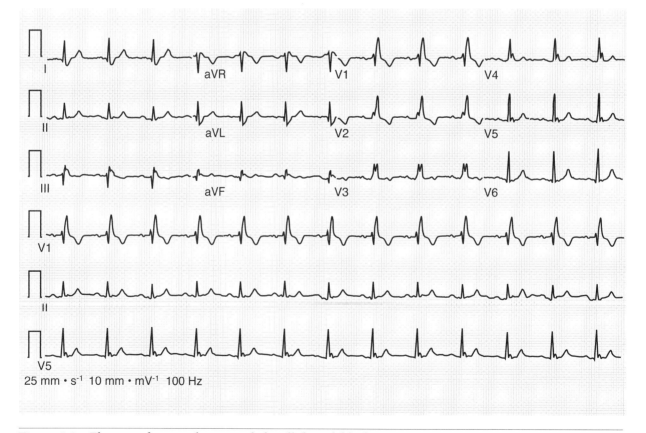

Figure 5.3 Electrocardiogram showing right bundle branch block.

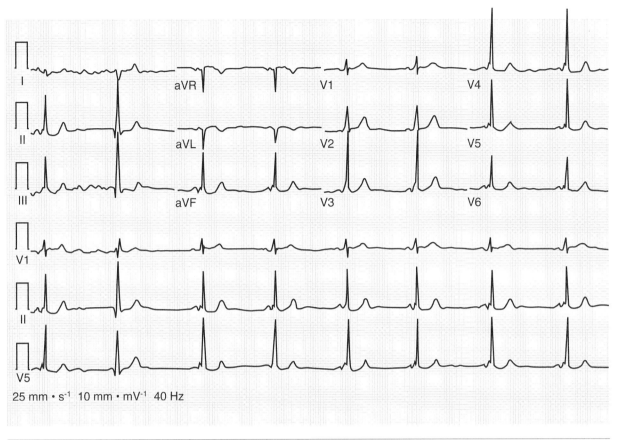

25 mm · s⁻¹ 10 mm · mV⁻¹ 40 Hz

Figure 5.4 Electrocardiogram showing Wolff-Parkinson-White preexcitation abnormality.

• **ST-Segment Elevation:** ST-segment elevation or J-point elevation observed on a resting ECG attributable to early repolarization is not abnormal in healthy people, but this should be documented on the GXT report. With exertion, this type of ST elevation normally returns to the isoelectric line. ST-segment elevation with exertion (assuming the resting ECG is normal) may suggest **transmural ischemia** or a coronary artery spasm. This type of ST-segment elevation is associated with serious arrhythmias and is an absolute reason for stopping the test (see "Indications for Termination of Graded Exercise Test"). When Q waves are present on the resting ECG from a previous infarction, ST elevation with exertion may suggest a left ventricular **aneurysm** or wall motion abnormality, which is a potential reason for discontinuing the GXT and referral for further testing. ST-segment elevation can localize the ischemic area and the arteries involved, whereas this is not always the case with ST-segment depression (50,65).

• **T-Wave Changes:** When exercise is started in healthy individuals, initially T-wave amplitude gradually decreases. Later, at maximal exercise, T-wave amplitude increases. These T-wave changes are believed to be asso-

ciated with an increase in stroke volume. An increase in serum potassium following exercise is postulated to be responsible for increased T-wave amplitude. In the past, T-wave inversion with exertion was thought to reflect an ischemic response. But it is now believed that flattening or inversion of T waves is not associated with ischemia. T-wave inversion is common in the presence of left ventricular hypertrophy. Inverted T waves present on the ECG at rest that normalize with exertion may reflect ischemia. Normalization of T waves, however, is commonly observed in subjects with and without exercise-induced ischemia. Normalization of T waves is also present during ischemic responses associated with coronary spasms. This finding has the greatest significance under resting conditions. Overall, T-wave changes with exertion are not specific to exercise-induced ischemia but should be correlated with other signs and symptoms. Note that T-wave changes frequently follow ST changes and can be difficult to isolate with exertion.

• **R-Wave Changes:** The normal response of R waves during exercise is an increase in amplitude to a submaximal HR; then as the individual approaches maximum, R-wave amplitude decreases below resting values.

Previously, it was suggested that increased R-wave amplitude at peak exercise is suggestive of CAD. But note that if the effort was submaximal, R-wave amplitude may not have had the opportunity to decrease. If an individual was able to give a maximal effort and R waves increased in amplitude with no other signs or symptoms of CAD, the predictive power of R-wave changes alone and CAD is poor. In general, it appears that abnormal R-wave changes provide no substantial additional information, unless combined with other abnormal signs or symptoms.

• **Dysrhythmias:** In relation to ST changes with exertion, dysrhythmias are of equal importance and potentially more life threatening, based on the suddenness in which dysrhythmias may appear. Although a GXT is most commonly used to diagnosis potential CAD, this test is also commonly used to evaluate symptoms (e.g., near syncope) attributable to dysrhythmias. In addition, GXT may be used to evaluate the effectiveness of medical therapy in controlling dysrhythmias. The supervisor of the GXT must have strong knowledge of dysrhythmia detection and be able to respond appropriately and quickly. When dysrhythmias appear during the GXT, the onset of the dysrhythmias and any signs or symptoms associated with the dysrhythmias should be documented, along with any other ECG changes (e.g., ST depression). Concerning the potential dysrhythmias, the supervisor should be knowledgeable about three major areas, each requiring premature termination of the test:

1. Supraventricular dysrhythmias that compromise cardiac function
2. Ventricular dysrhythmias that have the potential to progress to life-threatening dysrhythmias
3. The onset of high-grade conduction abnormalities

• **Supraventricular Dysrhythmias:** Isolated premature atrial and junctional contractions and short runs of supraventricular tachycardia are generally considered benign and provide little diagnostic value relative to CAD. Supraventricular tachycardia may be related simply to an increased adrenergic response that occurs with physical exertion, although sustained supraventricular tachycardia is generally considered an absolute indication for test termination, if the client begins to experience symptoms that may be related to the fast rate of supraventricular tachycardia (e.g., dizziness, syncope). The onset of supraventricular tachycardia is most noticeable at low levels of exertion, during which a large change in HR can be easily observed. If the dysrhythmias persist following test termination, the supervising physician should be informed immediately if he or she is not in the testing room. Additionally, exercise-induced atrial flutter or fibrillation is generally considered a reason for terminating the GXT.

• **Ventricular Dysrhythmias:** In general, ventricular dysrhythmias are considered to be of greater importance than supraventricular dysrhythmias, because they can more easily progress to more life-threatening dysrhythmias. Occasional (fewer than six per minute) ventricular premature beats, however, are benign and do not warrant stopping the test. Estimates are that 20% to 30% of the healthy population and 50% to 60% of the cardiac population experience ventricular premature beats under resting conditions. Generally, ventricular premature beats at rest that disappear with exertion are considered benign, but when ventricular premature beats increase in frequency with exertion, there is more reason to suspect an underlying cardiac problem (i.e., cardiomyopathy, CAD). Ventricular premature beats can be either unifocal or multiform in nature. Unifocal beats originate from the same place within the ventricle, whereas multifocal beats suggest that the abnormality is originating from multiple places in the ventricles. Multiform ventricular premature beats are considered of greater significance, because of a greater risk to progress to more dangerous ventricular dysrhythmias. The time in which the ventricular premature beat occurs during the cardiac cycle is important. Ventricular premature beats that occur on the T wave are referred to as the R-on-T phenomenon and, though rare, are more likely to induce ventricular tachycardia. Their presence is a reason for test termination, especially in the presence of ischemia. Ventricular couplets or pairs should be noted but generally are not considered an absolute reason for test termination. If the rate of ventricular couplets increases or if they become multiform, this circumstance is considered a relative contraindication for test termination. Ventricular premature beats that occur as a triplet (ventricular tachycardia) at a rate between 100 to 150 beats · min^{-1} are also considered a relative contraindication. Ventricular tachycardia that reaches 150 to 250 beats · min^{-1} is an absolute contraindication for test termination, because of the increased likelihood of progressing to a faster rate and eventually ventricular flutter and ventricular fibrillation. The test supervisor must recognize ventricular tachycardia and record the onset. Generally, ventricular tachycardia is easily recognized by the wide QRS complex; the onset of left bundle branch block, however, can resemble ventricular tachycardia. If an individual converts to bundle branch block, P waves still precede the QRS complex, whereas this is not true in ventricular tachycardia. Because heart rate increases during exercise, however, identifying

P waves is often difficult. If the tester has difficulty trying to distinguish between these two dysrhythmias, the test should be terminated.

EXERCISE PRESCRIPTION

Optimal adaptation to exercise training requires the development of, and adherence to, an exercise prescription. An exercise prescription is a specific guide provided to an individual for the performance of an exercise-training program. Despite the term *prescription* within, the development of an exercise prescription does not necessarily require approval of a physician. In some situations, however, especially with clinical patients, a physician's signature may be required (e.g., possible requirement for Medicare reimbursement for cardiac rehabilitation) or desired (i.e., to limit liability). Individuals practicing clinical exercise physiology must possess the knowledge, skills, and practical experience to combine both art and science when developing an exercise prescription (see practical application 5.1).

The primary purposes of the exercise prescription are to provide a reliable and valid guide for optimal health and physical fitness improvements and to provide a safe environment in which to achieve these improvements. Specificity of the exercise prescription is relevant to the clinical population. The American College of Sports Medicine (ACSM) has published several position statements on special populations including healthy people; those with coronary artery disease, osteoporosis, hypertension, diabetes; and older people (2,4–7). The exercise prescription can also be specific to the five health-related components of physical fitness:

1. Cardiorespiratory endurance (aerobic fitness): ability of the cardiorespiratory system to supply oxygen to active skeletal muscles during prolonged submaximal exercise and the ability of the skeletal muscles to perform aerobic metabolism.

Practical Application 5.1

The Art of Exercise Prescription

Definition: The art of exercise prescription is the successful integration of exercise science with behavioral techniques that results in long-term program compliance and attainment of the individual's goals (2). Unlike chemistry or physics, physiology and psychology are not exact sciences. We cannot always predict physiological or psychological responses because numerous factors have influence. These include, but are not limited to, age, physical and environmental condition, sex, previous experiences, and diet. When developing an exercise prescription, we should follow the basic guidelines provided in this chapter. By doing so, we can elicit the desired response both during a single exercise-training session and over the course of an extended training period. When we consider effectors of the exercise response, not all people respond as expected, especially those with chronic disease. For example, people with coronary artery disease often require modification of exercise intensity because of myocardial ischemia. Those who are extremely deconditioned may not be able to handle a typical intensity level for their age. Others may have a variable training response based on their type and severity of disease. *ACSM's Guidelines for Exercise Testing and Prescription* (2) lists several reasons for altering an exercise prescription in selected individuals:

- Variance in objective (physiological) and subjective (perceptual) responses to an exercise-training bout
- Variance in the amount and rate of exercise-training responses
- Differences in goals between individuals
- Variance in behavioral changes relative to the exercise prescription

Each of these should be considered for each person during both the initial development and subsequent review of the exercise prescription. A modified exercise prescription should not be considered adequate until follow-up is performed. As a rule, a person's exercise prescription should be reevaluated weekly until the parameters of the exercise prescription appear to be adequate to improve health-related behaviors and selected physiological indexes while maintaining safety.

2. Skeletal muscle strength: peak ability to produce force. Force may be developed by **isometric**, **dynamic**, or **isokinetic** contraction.

3. Skeletal muscle endurance: ability to produce a submaximal force for an extended period.

4. Flexibility: the ability of a joint to move through its full, capable **range of motion**.

5. Body composition: the relative percentage of fat or nonfat mass of the body weight. Chapter 12 provides details regarding body composition assessment and exercise prescription for fat or weight loss.

Each of these components of physical fitness is related to at least one aspect of health, and each can be improved by regular, specific exercise training. Each component of health-related physical fitness is positively influenced by exercise training, likely reducing the risk of a primary or secondary chronic disease (10,51,68). Thus, the importance of regular exercise and physical activity for overall health is well established. The general benefits of exercise training are presented in table 5.4 (68). Several principles of exercise prescription must be considered during its development:

1. Specificity of training

2. Progressive overload

3. Reversibility

The exercise physiologist must also consider several aspects of a person's psychological state that are relevant to beginning and adhering to an exercise-training program (see chapter 2). Parts II through VII of this text use this chapter as a baseline to provide specific guidelines for exercise prescription development with respect to specific clinical diseases and conditions.

Exercise-Training Sequence

A comprehensive training program should include flexibility, resistance, and cardiorespiratory (aerobic) exercises. The order of the exercise-training routine is important for both safety and effectiveness. Information on this topic, however, is lacking. Generally, it is recommended that flexibility training take place following a warm-up period or following an aerobic or resistance-training routine to reduce the risk of muscular injury and soreness. In a clinical population, if an aerobic bout and a resistance-training bout take place on the same day, the best approach is to perform the activity that is the primary focus of that day's training first. Specific information on sequence is provided in the chapters about chronic diseases and conditions.

Goal Setting

A comprehensive exercise prescription should consider goals of each person. Common goals include these:

Feeling better

Looking better

Losing weight

Preparing for competition

Improving general health and reducing primary or secondary risk of disease

Reducing the negative effects of a chronic disease or condition on the ability to perform physical activity

People with specific diseases often have goals that relate directly to reversing or reducing the progression of their disease and its side effects. Because of these and other goals, a clinical exercise physiologist must have a comprehensive understanding of how to alter a general exercise prescription to provide the best chance of success to achieve a desired goal. Also, the exercise physiologist should assess whether goals are realistic and discuss them with patients when they are not.

Principles of Exercise Prescription

To gain the optimal benefits of exercise training, regardless of the area of emphasis (i.e., cardiorespiratory, strength, muscular endurance, body composition, range of motion), the following principles must be followed.

Specificity of Training

The principle of specificity of training states that the body will adapt in specific ways to specific types of exercise. Changes in physiological function occur in response to either an acute bout of exercise (short-term exercise) or a chronic series of exercise (long-term exercise). In more simplistic terms, specificity of training states that what a person does for exercise relates directly to the improvements that occur. For example, several studies report a range of 5% to 15% higher $\dot{V}O_2$peak difference in normal, healthy people tested using a treadmill versus a cycle ergometer (33,44,45,47,69,74). Although part of this difference is related to a smaller total muscle mass used during cycling, much of it is related to specific training adaptations. To illustrate this point further, note that trained cyclists are able to achieve a similar $\dot{V}O_2$peak on the cycle compared with a treadmill, but the opposite is not true for a trained runner. Crossover effects from

Table 5.4　Clinical Benefits of Exercise Training

Physiological, health and disease	Psychological, quality of life
Improved cardiorespiratory and musculoskeletal fitness	Improved sleep patterns
Improved metabolic, endocrine, and immune function	Reduced depression and anxiety
Reduced overall mortality rate	Improved health behavior
Reduced cardiovascular disease risk	Improved mood levels
Reduced cancer risk (colon, breast)	Overall improved health-related quality of life
Reduced risk of osteoporosis and osteoarthritis	Reduced risk of falling
Reduced risk of non-insulin-dependent diabetes mellitus	Reduced risk of obesity

Adapted from the Surgeon General's Report on Physical Activity and Health (68).

Table 5.5　Selected Modes and Types of Exercise Training and General Skeletal Muscle Adaptations

Mode	Adaptation	Comment
I. Cardiovascular: general cardiorespiratory endurance and improved body composition (aerobic)		
Walking	Leg endurance	Easy to perform for most
Running	Leg endurance	Increased injury risk over walking
Cycling	Leg endurance	Good for those with walking difficulty
Rowing	Arm and leg endurance	Often difficult to perform
Stair stepping	Leg endurance	Seated is better option for many clinical patients
Swimming	Arm and leg endurance	Poor technique will require a lot of exertion to perform; easy on joints
Rope skipping	Primarily leg endurance	Likely difficult for most clinical patients to perform
Elliptical training	Leg and arm endurance	Easy on joints
Cross-country skiing	Leg and arm endurance	Stationary machines require good coordination
II. Power: submaximal and peak exercise power increases (anaerobic)		
Sprint running	Leg power	Any of the sprinting modes are potentially dangerous for most clinical populations
Sprint cycling	Leg power	
Sprint swimming	Arm and leg power	
Resistance training	Muscle-specific power	Excellent for most clinical patients but requires proper instruction

one mode to another can also occur and are likely the result of a combination of central cardiac improvements (e.g., increased stroke volume and cardiac output) and involvement of specifically trained skeletal muscles that are then used in an alternative exercise mode (e.g., train the legs by running and then use the legs for kicking while swimming), which is the basis for the **cross-training** concept.

Based on the specificity concept, a general recommendation for exercise prescription development is that a training regimen should use modes of exercise that provide adaptations as close as possible to those desired (table 5.5). The sidebar on p. 98 provides questions that can be asked of a person for whom an exercise professional is developing an exercise prescription. This approach

will help ensure that the exercise prescription is specific to that person.

Progressive Overload

The progressive overload principle refers to the relationship noted between the dose of exercise and the benefits gained. There appears to be a physical activity or exercise threshold at which physiological benefits and disease and mortality benefits are gained and even a threshold beyond which benefits are lost. For instance, in the Harvard Alumni study, a dose response was reported for all-cause mortality from a caloric expenditure of less than $500\,kcal \cdot wk^{-1}$ to 3,000 to 3,499 $kcal \cdot wk^{-1}$ (51). The all-cause mortality rate rose slightly, however, when more than 3,500 kcal was expended per week.

Overload refers to the increase in total work performed above and beyond that normally performed on a day-to-day basis. For example, when a person performs walking as part of an exercise-training regimen, the pace and duration should be above those typically encountered on a daily basis to gain fitness benefits. Progressive overload is the gradual increase in the amount of work performed in response to the continual adaptation of the body to the work. Applying this principle would relate to walking more often, farther, or at a faster pace. Overload is often applied by using the FITT principle. FITT is an acronym for frequency, intensity, time (i.e., duration), and type of the exercise that is performed.

Frequency

Frequency is the number of times that an exercise routine or physical activity is performed (per week or per day).

Intensity and Duration

Intensity and duration are presented together because they are of equal importance when determining the total volume or amount of exercise or physical activity.

The intensity of exercise or physical activity refers to either the objectively measured work or the subjectively determined level of effort performed by an individual. Typical objective measures of work that are important to the clinical exercise professional include oxygen consumption ($\dot{V}O_2$, or **metabolic equivalent**), caloric expenditure (**kilocalories [kcal]** or **joules [J]**), and power output (kilograms per minute [kg/min] or watts [W]). The anaerobic or lactate threshold may also be used to determine exercise intensity; but it is impractical to use as a guide during exercise training. The subjective level of effort can be evaluated by either a verbalized statement from a person performing

Questions to Ask a Person When Developing an Exercise Prescription

Specificity

- What are your specific goals when performing exercise?
- Do you want to walk farther?
- Do you want to do more activities of daily living?
- Do you want to perform something that you currently cannot?
- Do you want to feel better?
- Is there something else?

Mode

- What types of exercise or activity do you like the best?
- What types of exercise do you like the least?

Frequency

- Do you know how many days per week of exercise or physical activity are required for you to reach your goals?
- How often during a week do you have 30 to 60 min of continuous free time?

Intensity

- Do your goals include optimal improvement of your fitness level?
- Or are your goals primarily related to your health?

Time

- How much time per day do you have to perform an exercise routine?
- What is the best time for you to exercise?
- Can you get up early or take 30 to 40 min at lunch time?

exercise (e.g., "I'm tired" or "This is easy") or by using a standardized scale (e.g., Borg rating of perceived exertion). The patient must be taught the proper use of this assessment to obtain accurate indications of perceived effort (49).

Duration (or time in the FITT acronym) refers to the amount of time that is spent performing exercise or physical activity. During exercise training, the duration of training is typically accumulated without interruption or with very short rest periods to gain fitness benefits. To the contrary, the Centers for Disease Control and Prevention (CDC)–ACSM statement on physical activity states that every U.S. adult should accumulate 30 min or more of physical activity each day of the week, suggesting that people can gain health-related benefits from the accumulation of duration throughout a day (52).

Reversibility

The reversibility principle refers to the loss of exercise-training adaptations because of inactivity. Positive adaptations accrue at a rate specific to the overloaded physiological processes. Typically, most untrained people can expect a 10% to 30% improvement in $\dot{V}O_2$peak and work capacity following an 8 to 12 wk period of training (7). Alterations in other physiological variables, such as body weight and blood pressure, may take a variable amount of time. As a rule, less fit people can expect to achieve gains at a faster rate and to a greater relative degree than more fit people. Maintaining improvements in fitness is an important issue. If a minimal training volume is not maintained, training effects will be lost. This reversal of fitness is often called deconditioning or detraining.

Cardiorespiratory Endurance

Cardiovascular conditioning to improve aerobic endurance requires that individuals perform modes of training that use large muscle groups and are continuous and repetitive or rhythmic (3). Common leg exercise modes include walking, running, cycling, skating, stair stepping, rope skipping, and group aerobics (e.g., dance, step, tae-bo, spinning, water, seated). Exercise using strictly the arms is limited but includes upper-body crank ergometry, dual-action cycles using only the arms, and wheelchair ambulation. Several other popular modes of exercise use both the arms and legs: rowing, swimming, and some types of stationary equipment (e.g., dual-action [arm and leg] cycles, cross-country skiing, elliptical trainers, seated dual-action steppers).

Specificity of Training

General benefits gained from aerobic exercise training appear to be independent of any specific type or mode of training. For instance, the Harvard Alumni study reported that men who were physically active and had a high weekly caloric expenditure, regardless of mode, had a lower incidence of all-cause mortality than those who were less active (51). Further interpretation of the Harvard Alumni database demonstrates that all-cause mortality rate is improved in those who perform higher intensity exercises than those who perform less vigorous activity (41). These reports suggest that the specific type of physical activity is less important for general mortality benefits than the amount and intensity of the activity.

Progressive Overload

To gain benefits from cardiorespiratory training, people must follow the principle of progressive overload (FITT principle).

Frequency

Recommendations for aerobic activity frequency vary between sources. The ACSM position stand titled "The Recommended Quantity and Quality of Exercise for Developing and Maintaining Cardiorespiratory and Muscular Fitness, and Flexibility in Healthy Adults" states a desired frequency of 3 to 5 d per week of cardiorespiratory training (7). A slightly different recommendation was published in a joint recommendation paper from the CDC and the ACSM (52) which recommends "physical activity on most, preferably all, days of the week." The difference in these recommendations is twofold. First, the ACSM position stand recommends exercise to improve cardiorespiratory fitness level, whereas the CDC–ACSM recommendation focuses on improving overall health. Cardiorespiratory fitness and health benefits related to the performance of regular exercise training or physical activities, respectively, are not mutually exclusive. The difference in these recommendations appears to lie in the desired effects of exercise or physical activity. The second difference in these recommendations is that the CDC–ACSM joint statement recommends a reduced intensity of activity. Therefore, to expend a sufficient amount of calories per week (e.g., 2,000 as recommended by Paffenbarger), the person must increase number of days (i.e., >3–5) that physical activity is performed (51).

Research suggests that an increased time commitment to exercise training of more than 3 d per week may not be an efficient use of time for the person with little time to spare (9,35,54). In fact, in a study that held total exercise volume (i.e., intensity, duration, frequency) constant, no difference in $\dot{V}O_2$peak was reported for those who exercised 3 versus 5 d per week (62). Still, exercising more than 5 d per week may play a positive role by increasing total caloric expenditure, optimizing health improvements, and reducing all-cause mortality rates.

Intensity and Duration

Intensity and duration of training are interdependent with respect to the overall training load. Generally, the higher the intensity of an exercise-training bout, the shorter the duration, and vice versa. The selection of the intensity and duration when developing an individual's exercise program should be made by considering several factors, including the current cardiovascular conditioning level of the individual, the existence of underlying chronic diseases such as coronary artery disease or obesity, the possibility of an adverse event, and the individual's goals.

Practical Application 5.2

Determining the Appropriate Heart Rate Range

It is impossible for an individual to guide his or her exercise intensity using $\dot{V}O_2$. But because the $\dot{V}O_2$ relationship with heart rate is linear in most healthy and diseased individuals between about 50% and 90% of $\dot{V}O_2$peak, heart rate is an excellent guide for exercise intensity. Several methods exist to determine an exercise-training heart rate range. The following reviews how to develop an exercise prescription based on heart rate.

When the **true maximal heart rate** (HRmax) is unknown, the maximal heart rate can be estimated using one of the following equations:

Equation 1: 220 – age = estimated HRmax

Equation 2: 210 – (0.5 × age) = estimated HRmax

Note: Equation 2 produces a higher value than the former for adults (i.e., >20 yr old).
Example: to predict HRmax in a 32 yr old person

Using equation 1: 220 – 32 = 188

Using equation 2: 210 – (0.5 × 32) = 194

The disadvantage of estimating HRmax is that the standard deviation (SD) is ± 10 to 12 beats · min^{-1}, and thus this method has a high variability from person to person.

Another disadvantage of estimating HRmax is the inaccuracy of equations 1 and 2 in people on β-blocker medications. Equation 3 is a relatively recently developed equation for use in these persons (11).

Equation 3: 162 – (0.7 × age) = estimated HRmax

Given the drawback of estimated HRmax, using the HRmax from a stress test is preferable as long as

- the subject attained a true peak exercise capacity on the exercise stress evaluation (e.g., did not stop because of intermittent claudication, arrhythmias, poor effort, severe dyspnea, etc.) and
- the subject took his or her prescribed chronotropic medication (e.g., β-blocker) before the exercise stress and no change has occurred in the type or dose of this medication

Each of these must be considered when using the peak HR from an exercise evaluation to develop a training heart rate range (THRR). Three methods can be used to determine the THRR. Presented here are the heart rate reserve (HRR) and %HRmax methods. The direct method is not used often and thus not presented. See ACSM's *Guidelines for Graded Exercise Testing and Prescription* (2) for more information.

HRR method (a.k.a. Karvonen method):

Equation 4: Step I. HRmax – HRrest = HRR
             Step II. (HRR × desired $\dot{V}O_2$ percentages) + HRrest = THRR

Example: HRmax (estimated or actual) = 170, HRrest = 68, and desired $\dot{V}O_2$ percentages are 50% and 85%

Step I. 170 − 68 = 102 (HRR)

Step II. 102 × .5 + 68 = 119 (lower end)

102 × .85 + 68 = 155 (upper end)

Note: Recently several investigators have noted that a specific %HRR is equivalent to the same % of $\dot{V}O_2$ reserve (10,47,48). $\dot{V}O_2$ reserve is defined as the difference between resting and peak exercise $\dot{V}O_2$ ($\dot{V}O_2$max − $\dot{V}O_2$ rest).

A note about $\dot{V}O_2$ reserve: Although a better relationship exists between HRR and $\dot{V}O_2$ R than between HRR and absolute $\dot{V}O_2$, no alteration in the method for determining the training heart rate range (THRR) using the HRR method should occur based on these findings. We simply now know that the %HRR has closer to a 1:1 relationship with % $\dot{V}O_2$ R than with % peak $\dot{V}O_2$. This means that 50% HRR = 50% $\dot{V}O_2$ R and 75% HRR = 75% $\dot{V}O_2$ R. As recommended by the ACSM (4), the range used for intensity with the HRR method should remain from 40% or 50% to 85% of peak $\dot{V}O_2$ R for cardiac and most other populations.

%HRmax method:

Equation 5: HRmax × desired HR percentages = THR

Example: HRmax (estimated or actual) = 161 and desired HR percentage range is 60% to 90%

161 × .6 = 97 (lower end)

161 × .9 = 145 (upper end)

Note: Because the relationship between $\dot{V}O_2$ and HR is not a straight line, the % peak exercise HR will not match the exact same % $\dot{V}O_2$peak. At lower intensity levels the % peak HR is approximately 10% higher than the % $\dot{V}O_2$peak (i.e., 60% of peak HR is equivalent to about 50% of $\dot{V}O_2$peak). This difference in percentage is reduced at higher intensity levels (i.e., 90% of peak HR is roughly equivalent to 85% of $\dot{V}O_2$peak).

To achieve an adequate training response, most people must exercise at an intensity between 50% and 85% of their $\dot{V}O_2$peak for the cardiorespiratory system to adapt and for aerobic capacity to increase (37). In some clinical populations with extreme deconditioning (e.g., those with heart failure, the elderly, obese people), improvements in $\dot{V}O_2$peak may be noted at intensities as low as 40% of their maximal ability (2). Generally, the lower a person's initial fitness level, the lower the required intensity level to produce adaptations. The upper level for healthy individuals to improve $\dot{V}O_2$peak is typically set at 85%, as recommended by the ACSM (7). This upper level is generally regarded as the threshold between optimal fitness gains and increased risks of orthopedic injury or adverse cardiovascular event. Training at a high intensity may also be difficult for healthy and clinically diseased individuals who have an anaerobic (lactate) threshold that is less than 85% of their $\dot{V}O_2$peak. Practical application 5.2 provides additional information about determining appropriate training heart rate.

ACSM's position stand (7) recommends 20 to 60 min of continuous or intermittent (no less than 10 min per bout) aerobic activity, whereas the CDC–ACSM statement suggests that accumulating a minimum of 30 min of exercise per day is sufficient (7,52). As stated previously, these differing recommendations reflect different goal-related focuses. Both regimens will reduce the primary and secondary risks of chronic disease. Some evidence, however, indicates that exercise training is more effective than accumulating physical activity at positively influencing a variety of cardiovascular disease risk factors such as plasma cholesterol and hypertension in a healthy population (46).

Reversibility

Several studies have investigated the physiological effects of detraining before or after a period of conditioning. The classic study of Saltin et al. (59) reported on the effects of bed rest over a 3 wk period in five subjects. All subjects had reductions in cardiac output resulting from reduced stroke volume, and this result was related to reductions in total heart volume. This trend was reversed when training was implemented. Another study of seven subjects who exercise trained for 10 to 12 mo followed by 3 mo of detraining evaluated the time-course effect of inactivity (17). $\dot{V}O_2$peak decreased by 7% from 0 to 12 d of detraining and by another 7% from 21 to 56 d. The early reduction in $\dot{V}O_2$peak was the result of a near equal reduction in stroke volume. The later reduction was the result of a decline in arteriovenous oxygen difference. Note that the absolute reduction in $\dot{V}O_2$ was greater in the complete bed rest study than in the study that allowed limited daily activity (17,59). This finding suggests that a minimal amount of activity can be effective at maintaining fitness levels or can attenuate the effects of deconditioning.

The take-home points of the reversibility concept are the following:

- Cardiorespiratory conditioning can be maintained with a reduced level of exercise training.

- The more sedentary a person becomes, the greater the loss of fitness is.

- All people, no matter what their conditioning level or disease status, are prone to deconditioning.

- A sedentary lifestyle results in an additive effect on the loss of fitness that naturally occurs with aging.

Skeletal Muscle Strength and Endurance

Resistance training improves muscular strength and power and reduces levels of muscular fatigue. The definition of muscular strength is the maximum ability to develop force or tension by a muscle. The definition of muscular power is the maximal ability to apply a force or tension at a given velocity. See "Benefits of Resistance Training."

For general skeletal muscle conditioning, the focus should be on the primary muscle groups. Figure 5.5 on p. 103 demonstrates common resistance-training maneuvers that focus on these muscle groups. Proper technique is important during resistance training to reduce the risk of injury and increase the effectiveness of an exercise. General recommendations include the following:

Lift throughout the range of motion unless otherwise specified.

Breathe out (exhale) during the lifting phase and in (inhale) during the recovery phase.

Do not arch the back.

Do not recover the weight passively by allowing weights to crash down before beginning the next lift (i.e., always control the recovery phase of the lift).

Benefits of Resistance Training

- Improved muscular strength and power
- Improved muscular endurance
- Improvements in cardiorespiratory fitness
- Reduced effort for activities of daily living as well as leisure and vocational activities
- Improved flexibility
- Reduced skeletal muscle fatigue
- Elevated skeletal density and improved connective tissue integrity
- Reduced risk of falling
- Improved body composition
- Possible reduction of blood pressure values
- Improved glucose tolerance
- Possible improvement in blood lipid profile

In certain clinical populations the following may be prudent:

Ensure that a fully equipped crash cart and staff trained in advanced cardiac life support are available.

Monitor blood pressure before and after a resistance-training session and periodically during a session.

Ensure that a clinical exercise professional familiar with resistance training in the specific clinical population conducted the initial patient orientation and regularly reevaluates lifting technique.

Constantly assess for signs and symptoms of exercise intolerance that may occur during resistance training.

Instruct participants always to train with a partner.

Resistance exercises should be sequenced so that large-muscle groups are worked first and smaller groups thereafter (31). If smaller muscle groups are trained first (i.e., those associated with fine movement), the large-muscle groups may become fatigued earlier when they are used.

Resistance training, like other types of training, can be general or specialized to result in specific adaptations. Most of the specialized training routines have little relation to improvement of skeletal muscle fitness for healthy and chronically diseased individuals. These

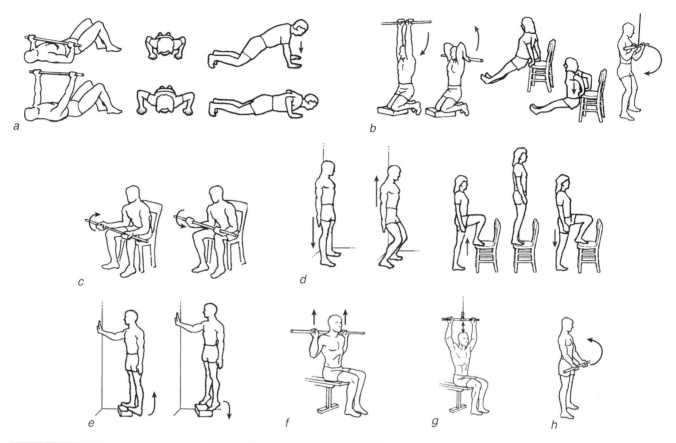

Figure 5.5 Examples of specific resistance-training exercises. (*a*) bench press: keep knees bent, do not arch back, keep hands slightly wider than shoulders, consider alternative chair version of exercise; (*b*) elbow extensor: do not arch back, tighten buttocks or wear a weight belt to support back, consider alternative triceps exercise; (*c*) double wrist curl: support forearms on thighs; (*d*) half squat: support back on wall, place feet away from wall so that knees do not move past the ankles, slide down wall slightly and back up, consider alternative chair version of exercise; (*e*) toe raises: can perform on stairs, use hands for balance; (*f*) shoulder press: keep back straight, keep hands wider than shoulders; (*g*) pull-downs: keep back straight, keep hands wider than shoulders; (*h*) arm curls: do not arch back, consider variations supporting elbows.

populations will improve sufficiently with a standard resistance-training program as suggested by the ACSM (2). Circuit programs (aerobic and resistance training) are a popular method to incorporate a cardiovascular stimulus during a resistance-training program (27,48). But although impressive strength gains have been reported, only modest cardiorespiratory benefits result, or even none at all (26,34).

Specificity of Training

Resistance training can be performed to provide general benefits or specific adaptations. For specificity, the lifting routine should closely resemble the movements in which gains in muscular fitness are desired. But because the components of skeletal muscle fitness are related, any type of resistance-training program will provide benefits in each area of muscular fitness (i.e., strength, power,

endurance). Thus, the resistance-training program for most people should be designed in a general fashion to promote overall muscular fitness improvement.

Progressive Overload

The ACSM recommends performing resistance exercise training for 8 to 12 repetitions per set to produce maximum improvement in both skeletal muscle strength and endurance (7). In general, the greater the overload, the greater the improvement. Excessive and prolonged overload, however, can lead to **maladaptation**, which increases the risk of overtraining effects and skeletal muscle injury.

Frequency

Most studies report optimal gains when subjects perform resistance training from 1 to 3 d per week (18,25,32).

The ACSM recommends performing a general or circuit resistance-training program on at least 2 or 3 d per week (7). There is little evidence that substantial gains can be realized from performing resistance exercise on more than 3 d per week.

Intensity and duration

The intensity of resistance training is also important in determining the load or overload placed on the skeletal muscle system to produce adaptation. For maximal strength and endurance improvement, the resistance should be at the person's 8- to 12-**repetition maximum** (RM) so that the person is at or near maximal exertion at the end of the repetitions (although this level of resistance may not be appropriate for some clinical conditions). The RM can be determined using either a direct or an indirect method (14,28,64,71).

The total training load placed on the skeletal muscle system is a combination of the number of repetitions performed per set and the number of sets per resistance exercise. The ACSM recommends between one and three sets per exercise (7). Several well-designed studies indicate little benefit of performing resistance training for more than one set per resistance exercise (21). If more than one set is performed per exercise, it may be prudent to keep the between-set period to a minimum to reduce the total exercise time. Generally, no more than 1 min of rest is required between sets.

Reversibility

Maximal adaptations in skeletal muscle strength and endurance can be expected within 8 to 12 wk (55). The anticipated mean improvement in strength is about 25% to 30% (22). As with aerobic training, reductions in total resistance-training volume without complete cessation of training allow maintenance of much of the gained resistance-training effects (31). Complete loss of training adaptations will likely occur after several weeks to months of inactivity.

Flexibility Training

Flexibility is the ability to move a joint throughout a full capable range of motion (ROM) (see figure 5.6). Poor flexibility is associated with the following:

Increased risk of lower-back pain (acute and chronic)

Increased risk of muscle or tendon injury (38,40,76)

Reduced ability to perform activities of daily living and tasks requiring high-intensity muscular exertion (74,75)

Reduced postural stability, which increases the risk of falling (36)

Several devices can be used to assess ROM. These include a goniometer, which is a protractor device; the Leighton flexometer, which is strapped to a limb and reveals the ROM in degrees as the limb moves around its joint; and the sit-and-reach box, which assesses the ability to forward flex the torso while in a seated position and measures torso forward flexion as attributed to lower-back, hamstring, and calf flexibility. An excellent review of the methodology of the sit-and-reach test is provided by Adams (1). In some clinical disease states, flexibility may decrease as the course of the disease progresses (e.g., multiple sclerosis, osteoporosis, obesity). These and any other populations would benefit from regular ROM assessment and an exercise-training program designed to enhance flexibility. A well-rounded flexibility program focuses on the

neck,

shoulders,

upper trunk,

lower trunk and back,

hips,

knees, and

ankles.

The following are brief descriptions of the three primary stretching modes of flexibility training.

1. Static: A stretch of the muscles surrounding a joint that is held without movement for a period of time (e.g., 10–30 s) and may be repeated several times.

2. Ballistic: A method of rapidly moving a muscle to stretch and relax quickly for several repetitions.

3. Proprioceptive neuromuscular facilitation (PNF): A method whereby a muscle is isometrically contracted, relaxed, and subsequently stretched. The theory is that the contraction activates the **muscle spindle receptors** or **Golgi tendon organs**, which results in a reflex relaxation (i.e., inhibition of contraction) in either the **agonist** or **antagonist** muscle. There are two types of PNF stretching (16,19):

• Contract–relax occurs when a muscle is contracted and then relaxed and passively stretched. Enhanced relaxation is theoretically produced through the muscle spindle reflex.

- Contract–relax with agonist contraction occurs initially in the same manner as contract–relax, but during the static stretch the opposing muscle is contracted. This action theoretically induces more relaxation in the stretched muscle through a reflex of the Golgi tendon organs.

Static and ballistic types of stretching are simple to perform and require only basic instruction. PNF stretching is somewhat complex and may require close supervision by a clinical exercise professional. Static stretching is typically believed to be the safest method for enhancing the range of motion of a joint. Both ballistic and PNF

stretching may increase the risk of delayed-onset muscle soreness and muscle fiber injury. PNF is possibly the most effective of the methods of stretching at improving joint ROM (7,20,58).

Specificity of Training

Flexibility of a joint and the increase in range of motion from a flexibility-training program are specific to a joint. The flexibility of a joint depends on the joint structure, the surrounding muscles and tendons, and the use of that joint for activities. A joint used during physical activity, especially if it requires a good ROM, will typically demonstrate good flexibility.

a

b

(continued)

Figure 5.6 Examples of specific range-of-motion exercises: (*a*) thighs; (*b*) lower back; (*c*) abdominal muscles; (*d*) arms, shoulders, and neck; (*e*) hamstrings, calves, and ankles.

Figure 5.6 *(continued)*

Progressive Overload

Flexibility routines should be performed as often as possible. As ROM increases, an individual should enhance the stretch to a comfortable degree. This practice will produce optimal increases in ROM.

Frequency

A stretching routine can be performed a minimum 2 to 3 d per week (7). As stated previously, however, daily stretching is advised for optimal improvement in ROM.

Intensity and Duration

As previously stated, all types of stretching routines should be held for 10 to 30 s. For PNF this should follow a 6 s contraction period. Each stretch should also be performed for three to five repetitions and to a point of only mild discomfort or a feeling of stretch. Figure 5.6 provides many of the common maneuvers used in a comprehensive flexibility-training program.

Reversibility

Few data exist on the rate of loss of ROM. Many factors are involved in the rate of loss of ROM: injury, specific individual physiology, degree of overall inactivity, and posture. The reintroduction of a flexibility-training routine should result in rapid improvements in ROM.

CONCLUSION

The GXT continues to be a useful tool and is generally the first diagnostic tool used to assess the presence of significant coronary artery disease with or without nuclear perfusion or echocardiography imaging. In past years, a cardiologist normally supervised the GXT. But as knowledge of testing has increased in regard to identifying clients who are at increased risk for further complications, as ECG assessment technology has advanced, as cost-containment issues have become more important, and as cardiologists' time has become more constrained, properly trained clinical exercise physiologists now often perform graded exercise testing. Data from the test can be used to design an exercise-training program.

Any type of exercise-training routine, whether it is cardiorespiratory conditioning, resistance training, or ROM training, should follow the FITT principle to ensure an optimal rate of improvement and safety during training. When an exercise physiologist is working with specific clinical populations, modifications of these general principles may be necessary to ensure a safe and effective training routine. The chapters in this text about specific populations and diseases contain information specific to adapting the general exercise testing and training principles to these individuals.

Case Study

Medical History

Ms. WB was a high school and college cross country runner. She is Caucasian, 40 years old, and the mother of two children, ages 8 and 6. Her children are in school and she works full-time. She wishes to begin a regular exercise-training routine. She has an interest in improving all five areas of her health-related physical fitness. She does not smoke and has no known serious medical condition. Her body mass index is 31.4 and she has a goal of 50 lb weight loss (currently 64 in. tall and 180 lb). She has not run regularly in 15 years and only walks occasionally.

Diagnosis

Ms. WB is evaluated by her physician and cleared for participation in an exercise program. She is apparently healthy. With her doctor's permission she decides to take advantage of her local hopsital's performance assessment program to establish a baseline fitness assessment from which to gain specific exercise-training recommendations.

Exercise Test Results

A symptom-limited exercise text is performed and the following reported:

Protocol: ramp running

Rest HR: 76

Rest blood pressure: 136/88

Rest electrocardiogram: normal

Peak HR: 187

Peak blood pressure: 210/90

Exercise electrocardiogram: rare premature atrial contractions and premature ventricular contractions; 0.5 mm J-point depression with quickly upsloping ST segments at peak exercise.

$\dot{V}O_2$peak: 32.7 ml $\cdot$ kg^{-1} $\cdot$ min^{-1}

$\dot{V}O_2$ at anaerobic threshold: 25.6 ml $\cdot$ kg^{-1} $\cdot$ min^{-1}

Peak exercise pace: 8.2 mph

Pace at anaerobic threshold: 6.3 mph

(continued)

Other Procedures

Body fat: 32%

Bench press 1-repetition maximum (1RM): 40 lb

Sit-and-reach: lowest quintile (poor)

Leg press 1 RM: 150 lb

Discussion Questions

1. What would be an appropriate comprehensive exercise prescription for Ms. WB?

2. Ms. SJ is 56 years old and healthy. She wants to begin her own exercise program. What are your general recommendations and concerns?

3. Mr. BW performed a stress test on a treadmill with the following results: stage 4 of Bruce protocol (4.2 mph, 16% grade), HRmax = 190, maximum blood pressure = 200/70, $\dot{V}O_2$peak = 43.2 ml $\cdot$ kg^{-1} $\cdot$ min^{-1}, no symptoms, stopped because of leg fatigue. His resting HR and blood pressure were normal. What specific intensity recommendations can you make for him with regard to beginning a jogging program? Use each of the HR based methods discussed in this chapter.

PART II

Special Populations

Part II discusses various populations that vary in regard to the value of exercise and appropriate exercise prescriptions for the purpose of preventing health problems and making favorable changes in health. The clinical exercise physiologist needs to be knowledgeable in many areas related to specific populations and be aware of the benefits and risks for each population.

Chapter 6 focuses on children. Our children have a great opportunity to avoid chronic diseases if they start living a healthy lifestyle at an early age. But with the rise in obesity observed in our children and the relationship of obesity with other CVD risk factors, concern has increased about the ramifications that may occur over time. To have a favorable influence on our children's health as it pertains to chronic diseases, intervention programs need to begin at an early age. This is supported by the fact that early blood vessel changes are observed in children who have multiple CVD risk factors. The clinical exercise physiologist should be knowledgeable about developing safe and effective lifestyle modification programs that address major factors that are negatively influencing the health of our children.

Chapter 7 deals with the elderly. As we are all aware, our society is becoming older. To meet the needs of an aging society, we need to address many chronic health concerns. We must encourage and educate our elderly population to stay physically active so that they can maintain activities of daily living and live a more independent life for a longer period. When an elderly person plans to start an exercise program, physician clearance is important, because most older people will have some risk factors that should be assessed. Medical clearance is a greater concern when an older person is interested in exercising above moderate intensity. The clinical exercise physiologist should be aware of health issues that affect the elderly and the benefits of physical activity in this population.

Chapter 8 examines female-specific issues. Following the passage of Title IX, women became much more physically active. With the increase in sports participation by women, certain health concerns (psychological and physiological) surfaced. Initially the health concerns for women were addressed as the female athlete triad. More recently the American College of Sports Medicine has described this health concern as it relates to energy availability. The health concerns associated with this abnormality potentially have long-term ramifications and are typically associated with sports in which weight is associated with an athlete's performance. The clinical exercise physiologist should be aware of specific health issues that challenge female athletes and be knowledgeable in assessing these health concerns.

Chapter 9 studies depression and exercise. Depression is the most prevalent psychiatric disease or disorder in the United States. Patients with depression often require more attention by the clinical exercise physiologist than do nondepressed people. Having the knowledge to recognize symptoms is critical for the exercise physiologist, especially when working with patients that have recently been through a life-threatening medical event. Exercise has been shown to be an effective treatment for depression and may be an alternative to medication.

6

Children

William Saltarelli, PhD

Pediatric health specialists generally agree that the habitual physical activity behavior of children and adolescents is a major health issue (72). High levels of physical activity may play an important role in determining cardiorespiratory, skeletal, and psychological health of children (79,83). The positive influence of habitual physical activity on various conditions such as coronary artery disease, hypertension, obesity, stroke, hypercholesterolemia, colon cancer, diabetes, osteoporosis, and low functional capacity in adults is well documented (18). But the benefits of physical activity and its influence on those diseases and risk factors in children are not as clear. To address this issue in children, an advisory committee (International Consensus Conference on Physical Activity Guidelines for Adolescents) composed of scientists interested in the influence of physical activity and health in children was formed in 1992 to develop a consensus on the effects of physical activity on youth and adolescents. The results of this effort were published in the November 1994 issue of *Pediatric Exercise Science*. The following brief synopsis of the committee's findings represents the most recent and comprehensive scientific review of the topic. The focus of this chapter is to review what is known about physical activity and children, including basic exercise physiology, exercise testing protocols, control of cardiovascular disease risk factors, and suggestions for prescribing exercise for this special population. These topics are extremely important in light of the fact that our children are becoming less active and more obese than previous generations. These trends have increased the health risks of our children now and in their future.

SCOPE

The most recent data on the physical activity patterns of children come from the Youth Risk Behavior Surveillance System (YRBSS) completed under the auspices of the Centers for Disease Control and Prevention in 1999. The YRBSS includes national, state, and local school-based surveys of children and adolescents age 9 to 13 yr (27). These surveys were large-scale cross-sectional studies that consisted of self-reported 7 d recall questions. The best way to answer the question of whether today's children are physically active is to compare the YRBSS's data with Healthy People 2000 objectives (72). *Healthy People 2000* is a set of disease prevention and health promotion objectives that the United States was to achieve by the year 2000. The objectives were created by scientists associated with the Centers for Disease Control and Prevention, and the document identified a wide range of public health priorities and specific, measurable objectives.

Light to Moderate Physical Activity

The goal for Healthy People 2010 with respect to moderate physical activity was to have children and adolescents (ages 6–17) engage in moderate physical activity for at least 30 min each day, at least 5 d per week. The YRBSS data found that only 26.7% of students nationwide had walked or bicycled for at least 30 min on 5 or more days per week preceding the survey. African American and Hispanic students were more likely than white students to have participated in moderate physical activity (27). Also, females were less active than males, and activity decreased with age in both sexes.

Vigorous Physical Activity

The goal of Healthy People 2000 was for at least 75% of children and adolescents (ages 6–17) to engage in vigorous physical activity that promotes cardiorespiratory fitness 3 or more days per week for at least 20 min. The YRBSS survey data revealed that 64.7% of children and adolescents engage in vigorous physical activity to the intensity of sweating and breathing hard for 20 min per session, 3 or more days per week. Males (72.3%) were more likely to engage in vigorous activity than females (57.1%). White children were more vigorously active than African Americans and Hispanics, and vigorous activity declined for all groups as they got older. In addition, for both boys and girls, the level of vigorous physical activity declined markedly from age 12 to 18. For example, 70% of 12 yr old boys reported that they were active, whereas 48% of 18 yr olds were active.

Flexibility and Strength Development Activities

The goal of Healthy People 2000 was to increase to at least 40% the proportion of children and adolescents who perform physical activities that enhance and maintain muscular strength, muscular endurance, and flexibility. This objective was addressed by two survey questions in the YRBSS. The first question asked if students participated in stretching exercises (e.g., toe touching, knee bending, and leg stretching) on 3 or more days preceding the survey. Results indicated that 51.3% of the students had participated in these activities. In addition, ninth-grade girls (59.8%) were significantly more likely than older girls were to do stretching exercise (41%).

The second question focused on strengthening exercise (e.g., push ups, sit ups, and weightlifting) performed on 3 or more days of the 7 d preceding the survey. Overall, boys (58.1%) were more likely to participate in this activity than girls (43.2%). As with stretching, ninth-grade girls (52%) were more likely to participate in strengthening exercises than older girls were (34%). Overall, more than half of students nationwide (51.4%) had engaged in strengthening exercises on 3 of the 7 d preceding the survey.

PATHOPHYSIOLOGY

The following is a summary of the physical activity guidelines for adolescents as developed by the International Consensus Conference on Physical Activities Guidelines for Adolescents and published in the November 1994 issue of *Pediatric Exercise Science*. The following specific topics address the influence of physical activity on aerobic fitness, blood pressure, obesity, blood lipids, skeletal health, and psychological health (83).

Aerobic Fitness

In adults, it is generally accepted that habitual physical activity is related to aerobic capacity (maximal oxygen uptake or $\dot{V}O_2$max) and that fit individuals have lower mortality and morbidity rates (19). But these relationships are not as clear in children. Lack of physical activity is known to be an independent risk factor for cardiovascular disease in adults, but again little is known about the influence of physical activity on these risk factors in children. Recent evidence strongly links cardiovascular disease risk factors with atherosclerotic lesions in children as young as 2 yr of age (90,92,94).

The correlation between physical activity and aerobic capacity in children is small to moderate, with correlation coefficients (R) around .16 (63). Morrow and Freedson also concluded from their review of the literature that neither walking nor daily physical activity provides the intensity to increase aerobic capacity in children. They suggested that the lack of relationship between physical activity and aerobic fitness could be attributable to one of the following:

- Available measurement techniques for physical activity are poor.
- Children have a generally acceptable level of fitness unrelated to physical activity.
- Aerobic systems are not trainable in children.

Analysis of data from the National Children and Youth Fitness Survey by Dotson and Ross (32) concluded that children who scored lowest in the 1 mi (1.6 km) run–walk (estimating $\dot{V}O_2$max) reported participating in fewer high-intensity cardiorespiratory activities. Other conclusions concerning activity and aerobic capacity from the review by Morrow and Freedson (63) included these:

1. Walking does not appear to provide a significant stimulus to increase aerobic power in children.
2. Daily physical activity probably has a weak association with aerobic capacity ($\dot{V}O_2$max) in children.

3. Specific intensity of activity necessary for aerobic benefit is not known precisely but may be at a higher intensity than for adults.

Blood Pressure

Estimates are that 2.8 million children and adolescents in the United States have hypertension (11). Hypertension in children is defined as an average systolic or diastolic blood pressure above the 95th percentile for age and sex (68) (see table 6.1 for specific normal and hypertensive values). Studies on adults and children have consistently shown that physical activity does not lower resting blood pressure in normotensive individuals. But this is not the case with hypertensive subjects, as shown by numerous controlled studies (40,92). A more extensive review of the subject can be found in an article by Alpert and Wilmore (3), who recommended that hypertensive children engage in regular physical activity to increase basic fitness and reduce levels of fatness.

Adiposity and Obesity

The third National Health and Nutrition Examination Survey (NHANES III) reported that approximately one

child in five in the United States is overweight (49,65). The NHANES definition of overweight is a body mass index (BMI) greater than the 95th percentile. Analysis of past NHANES data revealed that over the past 30 yr, the number of obese children in the United States has more than doubled. Tables 6.2 and 6.3 on p. 114 provide data on different body composition measurements including overweight and overfat definitions and normal values.

Recent studies have reviewed the cardiovascular risk of adolescent obesity (35,60,94). Overweight children were at least 2.4 times more likely to have elevated total cholesterol, low-density lipoprotein cholesterol, triglycerides, low high-density lipoprotein cholesterol, low fasting insulin, and elevated blood pressure. In addition, Must and colleagues (65) found that adults who were overweight as children have increased mortality and morbidity rates irrespective of adult weight. The causes of childhood obesity are complex and were reviewed in depth by Schlicker et al. (87). Low physical activity (sedentary lifestyle) and unhealthy diet are important contributing factors.

Blood Lipids

The atherosclerotic process has been shown to begin in early childhood, and this process is influenced by

Table 6.1 Abnormal Blood Pressure Values in Children of Various Ages

	Percentile	Age	Boys	Girls
Systolic blood pressure (mmHg)	90th, borderline hypertension	10	115	115
		12	119	119
		14	125	125
		16	130	130
	95th, hypertension	10	119	119
		12	125	123
		14	128	128
		16	134	134
Diastolic blood pressure (mmHg)	90th, borderline hypertension	10	75	75
		12	77	77
		14	78	78
		16	81	81
	95th, hypertension	10	80	80
		12	83	81
		14	83	82
		16	85	85

Note. Blood pressure values are reported in the literature by percentiles of height. This table reflects values at the 50th percentile for height. Normal blood pressure <90th percentile; hypertension >95th percentile.

Data from Bartosh and Aronson, 2004, "American Academy of Pediatrics The Fourth Report on the diagnosis, evaluation, and treatment of high blood pressure in children and adolescents," *Pediatrics* 114(2): 555-576.

Table 6.2 Obesity Risk—BMI

Age	Boys BMI (85%) (at risk for obesity)	Boys BMI (95%) (obese)	Girls BMI (85%) (at risk for obesity)	Girls BMI (95%) (obese)
10	19.5	22.5	20.0	23.0
11	20.2	23.2	20.5	23.5
12	20.2	23.2	21.5	25.5
13	23.0	25.2	22.5	26.2
14	22.5	26.0	26.2	27.0
15	23.5	26.5	24.0	28.0
16	24.0	27.5	24.5	29.0
17	25.0	28.0	25.0	29.5
18	25.5	29.0	25.5	30.5

Note. >85th percentile is considered at risk for overfat; >95th percentile is overfat. BMI = body mass index.

Data from National Center for Health Statistics, *Prevalence of overweight among children and adolescents.*

Table 6.3 Obesity Risk: Percent Fat and Sum of Skinfold Measurements

Risk of obesity	Sum of skinfolds (mm)		Body fat (%)	
	Boys	Girls	Boys	Girls
Optimal	21–25	16–34	10–20	15–27
Moderate risk	25–33	34–42	20–25	17–32
High	>33	>42	>25	>32

Note. For ages 10–16 years. Percent fat and sum of skinfolds from calf and triceps.

Data from T.G. Lohman, 1992, *Advances in body composition assessment* (Champaign, IL: Human Kinetics), 94.

hyperlipidemia (17). In addition, it has been shown that abnormal blood lipids and lipoproteins track from childhood to adulthood (50,69).

The importance of hyperlipidemia in children can also be found in the conclusions of the Report of the Expert Panel on Blood and Cholesterol (67):

- Atherosclerosis or its precursors begin in young people.
- Elevated cholesterol levels early in life play a role in the development of adult atherosclerosis.
- Eating patterns and genetics affect blood cholesterol levels and the risk of coronary heart disease.
- Lowering blood lipid levels in children and adolescents may be beneficial.

Table 6.4 presents normal lipid levels for children 5 to 19 yr of age.

Atherosclerosis

Most people state that the number one reason that children need to be more physically active is to reverse the alarming trend of our children becoming fatter. Although this reason is valid, a more global reason is to help change the modifiable cardiovascular disease risk factors. Cardiovascular disease (CVD) is the leading cause of death in the United States, responsible for almost one million deaths each year. This figure represents 37% of all deaths, which is more than the next five leading causes of death combined (11). In 2006 Americans are expected to pay $40.1 billion in medical costs and disability as a result of the complications of CVD, which include myocardial infarction, stroke, aneurysm, and peripheral vascular disease.

Because the clinical manifestations of atherosclerosis are usually not visible until later in life, the morbidity and mortality associated with CVD is often considered an adult problem. Most researchers now agree, however, that atherosclerosis is a continuous process that begins early in life. Recent autopsy data from the Pathobiological Determinants of Atherosclerosis in Youth study (PDAY) and the Bogalusa Heart Study have found young children have what are referred to as fatty streaks associated with certain blood vessels (17,90). These fatty streaks found in aortas and coronary blood vessels are not raised above the intimal surface but contain foam cells of mononuclear or smooth muscle origin. In addition, these benign fatty streaks contain minimal extracellular lipid or connective tissue. Bogalusa researchers have found children with three or more risk factors for CVD had 35.0% of their aortas and 11.5% of their coronary arteries covered with fatty streaks (17). These innocuous or benign fatty streaks appear not to cause immediate problems but can progress to more clinically relevant fibrous plaques or raised fatty streaks. Observed in adolescents and young adults, these fatty plaques are now elevated above the intimal surface and contain lipid-filled foam cells in addition to aggregates of extracellular lipid. These raised lesions are sometimes called transitional or intermediate lesions because evidence shows that they can become calicified significantly changing vessel histology that can contribute to adult CVD-related death and disability (59). Overwhelming evidence has suggested that the prevalence and severity of these histological blood vessel changes are strongly associated with certain CVD risk factors (17, 59,60), including obesity, hypertension, abnormal blood lipids, abnormal blood glucose metabolism, cigarette smoking, and physical inactivity, and are common in childhood (17,14,99). Further data show that a large percentage of American children have more than one of these risk factors (12). The incidence of metabolic syndrome also provides evidence of the extent of CVD risk factors in children. Metabolic syndrome is defined as having three or more of the following abnormalities: high triglycerides, low HDL-C, abdominal obesity, high fasting glucose, and hypertension. Recent work completed in Rochester, New York, found that 40.9% of children between the ages of 12 and 19 have at least one risk factor and that 4.2% had thee or more risk factors (30). Autopsy data from the Bogalusa Heart Study has shown a direct relationship between the number of risk factors that a child has and the extent of fatty streaks within blood vessels (17). According to Berenson and coworkers, subjects in the Bogalusa Heart Study with zero, one, two, and three or four risk factors had, respectively, 19.1%, 30.3%, 37.9%, and 35.0% of the intimal surface covered with fatty streaks in the aorta (17). Research has indicated that early intervention can improve the CVD risk profiles of children with multiple risk factors (41). Physical activity is known to modify many of the CVD risk factors and could possibly slow the progression of fatty streak development by modifying risk factors in childhood (12).

Table 6.4 Optimal and At-Risk Levels for Blood Lipids for Children

Lipid fraction	Optimal level (mg · dl⁻¹)	Risk level (mg · dl⁻¹)	Reference
HDL-C	>60	<35	McGill (59)
TC	<170	>200	NCEP (67)
Non-HDL	<130	>190	McGill (59)
LDL	<114	>130	NCEP (67)

Note. Suggested values for optimal and at-risk levels have been determined by normative data and autopsy studies. The literature is inconsistent regarding exact numbers for children.

The following section contains information about the influence of physical activity on many of these CVD risk factors and other health outcomes.

Skeletal Health

The problem of bone loss, osteoporosis, and bone fracture risk in our society is well documented (26,35,61). Although this disease process is complex, the pathophysiology includes the following:

- Failure to attain a sufficiently high peak bone mass (bone mineral density) during the growing years
- Failure to maintain peak bone mass for a sufficient period during the adult years
- Accelerated bone loss in later years (13)

Estimates are that 90% of the total adult bone mass is deposited by the end of adolescence (57). The relationship of bone mineral content to physical activity in children appears to be similar to what is observed in adults; high-impact weight-bearing activities increase bone mineral content.

The following recommendations were put forth by Bailey and Martin (13), from their review of physical activity and skeletal health in children.

1. A lifelong commitment to physical activity and exercise must be made.
2. Weight-bearing activities are better than weight-supported activities such as swimming and cycling.
3. Short, intense daily activity is better than prolonged activity done infrequently.
4. Activities that increase muscle strength should be promoted because these will enhance bone density.
5. Activities should work all large-muscle groups.
6. Immobilization and periods of immobility should be avoided. When this is not possible (as in bed

rest during sickness), even brief daily weight-bearing movements can help reduce bone loss.

Psychological, Social, and Emotional Health

Psychological and mental health is extremely important in children and adolescents and includes the following: depression, anxiety, stress, self-esteem, self-concept, hostility, anger, and intellectual function (25,91). In a more recent review of the literature, Tortolero, Tayler, and Murray (93) concluded the following:

1. A strong relationship exists between physical activity and improved self-efficacy, greater perceived physical competence, greater perceived health and well-being, and decreased depression and stress.

2. Moderate positive relationships have been found between physical activity and self-concept, self-esteem, and greater alcohol use (a negative outcome).

3. An inconclusive relationship exists between physical activity and body image, academic functioning, social skills, anxiety, hostility, aggression, suicide, sexual activity, and tobacco use (93).

ASSESSMENT OF PHYSICAL ACTIVITY IN CHILDREN

Physical activity is a complex behavior and therefore a challenge to assess, especially in children and adolescents. When considering all the movements and activities of children and adolescents, it is easy to conclude that there is no best method of assessment. An excellent general review of the most popular methods of physical activity assessment was presented by Harro et al. (42). For large-scale field assessments such as in physical education classes, methods such as heart-rate monitors, activity monitors, direct observation, and self-report questionnaires seem most appropriate. Welk and Wood (96) provided an excellent review of the attributes of various assessment techniques.

Because of their simplicity, paper-and-pencil recall assessments are popular. The Seven-Day Physical Activity Recall developed by Sallis et al. (84), Physical Activity Questions for Children (44), and the Previous Day Physical Activity Recall (PDPAR) are examples of these easily administered assessments. The PDPAR is used in Fitnessgram, which provides computer feedback to children. Children enter their physical activity levels for each 30 min block of time during the day. Fitnessgram recommends that children record one weekend and two weekdays of physical activity. To help prompt the responses, the children are provided with a sample list of activities divided into categories based on the activity pyramid. The activity pyramid categories include lifestyle, aerobic activity, aerobic sports, muscular activity, flexibility, and rest. For each activity, students are asked to rate the intensity of the activity with categories of light, moderate, and vigorous activity (37). Practical application 6.1 describes client–clinician interaction between the exercise physiologist and children.

This section briefly reviews each area of fitness (aerobic, anaerobic, muscular strength and endurance, body composition, and flexibility) with respect to age- and sex-related differences. Refer to table 6.5 for information on metabolic, anaerobic, thermoregulatory, and cardiorespiratory factors of children and their relationship to adult responses.

Aerobic Capacity

Laboratory evaluation of cardiorespiratory capacity (aerobic capacity) is indicated for children for a variety of reasons including diagnosis of medical conditions, assessment of exercise-induced symptoms, measurement of exercise response following surgery, or use in exercise prescription. With the interest in youth fitness, aerobic tests have taken on more importance. The American College of Cardiology and the American Heart Association have published selection criteria for identifying children to be tested for cardiorespiratory endurance in a clinical setting (38). Refer to this comprehensive reference for clinical indicators and criteria.

The most commonly used indicator of cardiorespiratory capacity is maximal oxygen uptake, which is defined as the highest volume of oxygen that can be consumed during an exercise bout per unit of time. $\dot{V}O_2$max is commonly expressed in absolute terms $(L \cdot min^{-1})$ and in relative terms $(ml \cdot kg^{-1} \cdot min^{-1})$. Before puberty, absolute $\dot{V}O_2$max is slightly higher in boys compared with girls, and it improves with age and growth in both sexes. Following puberty, however, this relationship changes as boys increase $\dot{V}O_2$max with age and with growth of their cardiovascular, pulmonary, and skeletal systems.

$\dot{V}O_2$max (absolute) in girls in this same age group remains relatively constant (15). This sex difference is possibly attributable to the increase in nonmetabolic weight (fat) in females, a decrease in habitual physical activity, or both. Many exercise physiologists prefer to

Client–Clinician Interaction

The pediatric exercise physiologist Oded Bar-Or (15) often states in his presentations that "children are not small adults" when referring to responses and adaptations to exercise. This advice is important to keep in mind when evaluating the exercise performance of children. The following is a brief review of special considerations that must be addressed when testing children.

Laboratory Environment

- The lab must be safe and well illuminated.
- Staff should be trained and have warm, friendly personalities to establish a positive relationship with children as soon as they enter the facility.

Unique Safety Issues

- Two testers are essential to ensure constant visual and verbal contact with the child.
- During treadmill testing, two spotters should be used—one at the subject's side and one behind the subject.

Pretest Protocols

1. Establish a relationship with the children as they enter the facility.
2. Completely explain the test to both parents and children. Be sure that both know exactly what will take place and what the child will be asked to do.
3. Following the explanation of the procedure, parental consent and the child's assent documents should be completed. A child's assent document (for children age 6 and older) should be written in age-appropriate language and be short and to the point.
4. Although offering an incentive to the child is controversial, rewards can be extremely helpful in eliciting a maximal effort.

Laboratory Equipment

When appropriate, testing equipment should be modified for the size and maturity of the subject.

Table 6.5 Physiological Characteristics of the Exercising Child

Function	Comparison with adults	Implications for exercise prescription
Aerobic		
$\dot{V}O_2$max ($L \cdot min^{-1}$)	Lower function of body mass	
$\dot{V}O_2$max ($ml \cdot kg^{-1} \cdot min^{-1}$)	Similar	Can perform endurance tasks reasonably well
Submaximal oxygen demand (economy)	Cycling: similar (18% to 30% mechanical efficiency); walking and running: higher metabolic cost	Greater fatigability in prolonged high-intensity tasks (running and walking); greater heat production in children at a given speed of running

(continued)

Table 6.5 *(continued)*

Function	Comparison with adults	Implications for exercise prescription
Anaerobic		
Glycogen stores	Lower concentration and rate of utilization of muscle glycogen	
Phosphofructokinase (PFK)	Glycolysis limited because of low level of PFK	Ability of children to perform intense anaerobic tasks that last 10 to 90 s is distinctly lower than that of adults
Phosphagen stores	Stores and breakdown of ATP and CrP are the same	Same ability to deal metabolically with brief intense exercise
Oxygen transient	Children reach steady state faster than adults do. Shorter half-time of oxygen increase in children	Children reach metabolic steady rate faster; children contract a lower oxygen deficit and recover faster. Children, therefore, are well suited to intermittent activities
LAmax	Lower maximal blood lactate levels Lower at a given percent of $\dot{V}O_2$max	May be reason why children perceive a given workload as easier
Heart rate at lactate threshold	Higher	
Cardiovascular		
Maximal cardiac output (Qmax) Q at a given $\dot{V}O_2$	Lower because of size difference.	Immature cardiovascular system means children are limited in bringing internal heat to surface for dissipation when exercising intensely in the heat
Maximal stroke volume (SVmax)	Lower because of size and heart volume difference	
Stroke volume at a given $\dot{V}O_2$	Lower	
Maximal heart rate (HRmax)	Higher	Up to maturity HRmax is between 195 and 215 beats $\cdot$ min^{-1}
Heart rate at submax work	At a given power output and at relative metabolic load, child has a higher heart rate	Higher heart rate compensates for lower stroke volume
Oxygen-carrying capacity	Blood volume, hemoglobin concentration, and total hemoglobin are lower in children	
Oxygen content in arterial and venous blood ($CaO_2 - CvO_2$)	Somewhat higher	Potential deficiency of peripheral blood supply during maximal exertion in hot climates
Blood flow to active muscle	Higher	
Systolic and diastolic pressures	Lower maximal and submaximal	No known beneficial or detrimental effects on working capacity of children
Cardiopulmonary response		
Maximal minute ventilation $\dot{V}$Emax (L $\cdot$ min^{-1})	Smaller	Early fatigability in tasks that require large respiratory minute volumes
$\dot{V}$Emax (ml $\cdot$ kg^{-1} $\cdot$ min^{-1})	Same as adolescents and young adults	

Function	Comparison with adults	Implications for exercise prescription
Cardiopulmonary response *(continued)*		
$\dot{V}E$submax; ventilatory requivalent	$\dot{V}E$ at any given $\dot{V}O_2$ is higher in children	Less efficient ventilation means a greater oxygen cost of ventilation; may explain the relatively higher metabolic cost of submaximal exercise
Respiratory frequency and tidal volume	Marked by higher rate (tachypnea) and shallow breathing response	Children's physiologic dead space is smaller than that of adults; therefore, alveolar ventilation is still adequate for gas exchange
Perception (RPE)	Exercising at a given physiological strain is perceived to be easier by children	Implications for initial phase of heat acclimatization
Thermoregulatory		
Surface area	Per unit mass is approximately 36% greater in children (percentage is variable, depends on size of child, i.e., surface area per mass may be higher in younger children and lower in older ones)	Greater rates of heat exchange between skin and environment. In climatic extremes, children are at increased risk of stress
Sweating rate	Lower absolute and per unit of surface area. Greater increase in core temperature required to start sweating	Great risk of heat-related illness on hot, humid days because of reduced capacity to evaporate sweat; lower tolerance time in extreme heat
Acclimatization to heat	Slower physiologically, faster subjectively	Children require longer and more gradual program of acclimatization; special attention required during early stages of acclimatization
Body cooling in water	Faster because of higher surface per heat-producing unit mass; lower thickness of subcutaneous fat	Potential hypothermia
Body core heating during dehydration	Greater	For prolonged activity, children must hydrate well before and take in fluid during activity

Reprinted with permission of the American College of Sports Medicine, *ACSM's Certified News,* 8: 1, p. 2, © American College of Sports Medicine 1997. 1-800-486-5643.

express aerobic capacity in relative terms using mass in kilograms as the relative factor.

Maximum aerobic power in relative terms remains relatively constant across a wide range of ages (6–16) in boys, reflecting an increase in body mass and an equal increase in the ability to use oxygen. In girls, however, $\dot{V}O_2$max in relative terms declines continuously, especially at puberty, and the decline continues through life (15).

The best measure of aerobic capacity in adults is a test of maximum oxygen uptake. When discussing the concept of maximum oxygen uptake ($\dot{V}O_2$max) in children, it is tempting to compare the responses and values of children to those of adults. Keep in mind that children are not small adults. In adults, exercise physiologists usually equate high relative $\dot{V}O_2$max values with superior endurance performance. This concept does not hold true for children, especially very young children. For example, compare the $\dot{V}O_2$max of an average healthy young adult from a treadmill protocol of about 40 to 45 ml · kg^{-1} · min^{-1} to the average $\dot{V}O_2$max of a 10 yr old child of about 50 to 55 ml · kg^{-1} · min^{-1}. The average 1 mi (1.6 km) run time (performance) of these adults will be about 7 to 8 min, whereas the average time of the children will be much slower, about 10 to 12 min. Therefore, high $\dot{V}O_2$max in children does not necessarily produce good endurance performance times. Many other variables influence running performance including body composition, skill, running economy, practice, and pacing. In addition, studies comparing endurance performance with laboratory $\dot{V}O_2$max determination indicate lower correlations ($R = .6–.7$) (55) in children than the much stronger correlation of about $R = .9$ in adult studies (31). A final note is that $\dot{V}O_2$max (relative) remains relatively constant through the growing years in children while their endurance performance steadily improves.

Medical Evaluation

Before aerobic capacity is tested in children and adolescents, preliminary information should be gathered (as with adults) including a complete medical history, parental consent, child's assent, and resting physiological measures. Acquiring the child's assent and ensuring that the subject and parents fully understand the test and what will be expected of them are extremely important. Establishing a positive relationship between the subject and all laboratory personnel is critical to the success of the test. A positive and safe atmosphere is essential. Resting physiological data of heart rate (HR), blood pressure, and electrocardiogram should be measured to ensure that the child is healthy.

Contraindications

Contraindications to testing are similar to American College of Sports Medicine (ACSM) guidelines for adults and are reviewed for children by Rowland (80). Contraindications include the following:

- Acute inflammatory cardiac disease (e.g., pericarditis, myocarditis, acute rheumatic heart disease)
- Uncontrolled congestive heart failure
- Acute myocardial infarction
- Acute pulmonary disease (e.g., acute asthma, pneumonia)
- Severe systemic hypertension (e.g., blood pressure >240/120 mmHg)
- Acute renal disease (e.g., acute glomerulonephritis)
- Acute hepatitis (within 3 mo after onset)

- Drug overdose affecting cardiorespiratory response to exercise (e.g., digitalis toxicity, quinidine toxicity) (80)

Careful observation of the subject for indications to terminate the test is critical to overall test safety. If an electrocardiograph is administered, adult termination criteria should be applied (ACSM guidelines). Other termination criteria for children include the following:

- Excessive increase in systolic blood pressure over 240 mmHg or diastolic pressure over 120 mmHg
- Progressive decrease in systolic blood pressure
- Pallor or clamminess of the skin
- Pain, headache, dyspnea, or nausea (8)

There is much confusion about the definition of a $\dot{V}O_2$max test in children. Attaining clear adult criteria of a plateau of oxygen uptake in children is extremely difficult (15), probably because children stop because of fatigue more often than adults do or because their cardiovascular systems are different. Many researchers prefer to use the term *peak oxygen uptake* rather than *maximal oxygen uptake* to measure endurance capacity. Rowland even suggested the term *exhaustive* to indicate the best effort that the child is willing to provide (80). Our laboratory uses a three-criterion format. Children must attain two of the first three criteria listed subsequently for a peak test to be achieved. Also listed are other criteria found in the literature.

1. Attainment of 95% of age-predicted maximum HR
2. Respiratory exchange ratio (RER) greater than 1.05
3. Rating of perceived exertion (Borg RPE) of 18 or greater (75)

Additional criteria:
4. Subjective signs of exhaustion
5. Stopping because of fatigue
6. Lactate levels of about 6 to 7 mmol · L^{-1} for 7 to 8 yr olds or 4 to 5 mmol · L^{-1} for 5 to 6 yr olds (69)

Success in eliciting a maximal effort takes planning and specific strategies. Children must understand that a maximal effort is expected, and this concept must be reinforced throughout the test by constant encouragement. Statements such as "You're looking great," "Keep going," and "You can go more stages" really help. Although controversial on ethical grounds, rewards or incentives for each stage have been used to motivate children.

The physiological differences between cycle and treadmill protocols must also be taken into account when interpreting maximal testing data. Their differences follow those seen in adults and include the following:

1. Peak heart rate and $\dot{V}O_2$max are greater during treadmill protocols.
2. RER at maximal effort is usually higher in cycle ergometry.

RECOMMENDATIONS AND ANTICIPATED RESPONSES

Most laboratory testing is performed with progressive workloads applied on cycle ergometers or motorized treadmills. Rowland (78) presented a complete treatment of this topic including mode and protocol selection. As with adults, cycle tests hold some advantage over those performed on a treadmill. Cycle ergometers are safer and more appropriate, especially with obese subjects, and are more conducive to data collection (blood pressure, HR, RPE, and lactate). But there are some concerns with cycle tests:

- Some children may end the test prematurely because of leg fatigue.
- Some children have trouble keeping to a standard cadence or pedal speed.
- Cycle data have been shown to be less reliable than treadmill data.
- RER and blood lactate have been found to be higher on cycles.

One advantage of motorized treadmills is that children are familiar with walking and running. But the use of this mode requires some practice. Children need 3 to 5 min to become accustomed to and feel comfortable with the machine. Two spotters should be used, one in back and one to the side, especially when metabolic data are being collected.

- Treadmill ergometry: Most protocols used to assess the aerobic capacity of children, whether maximal or submaximal in nature, are modifications of adult protocols. In clinical settings, the Balke protocol (see table 6.6) has been modified for children by altering stage duration and accommodating different fitness levels.

When $\dot{V}O_2$max is assessed in healthy children, a good approach is first to establish a comfortable running speed for each child, usually between 3 and 5 mph (5 and 8 kph). This initial trial also lets children become accustomed to the equipment and laboratory personnel. Speed is kept constant at this predetermined rate throughout the test. Subsequent stages are 2 min in duration, and the grade is increased 2% in each stage until volitional fatigue or a maximum criterion is achieved.

- Cycle ergometry: Similar to treadmill protocols, cycle ergometry protocols for children are usually modifications of those used with adults. Most cycle protocols require a pedal cadence of 50 to 60 rpm. The Adams submaximal progression continuous cycle protocol, also called a PWC170, can be used to evaluate relative aerobic power in children (1).This protocol consists of three stages that are usually submaximal in intensity. Heart rate is monitored during each of the 6 min stages. Performance is evaluated according to the mechanical power that the child produces at the HR of 170 beats/min (PWC170). The resistance of the initial stage depends on the child's weight in kilograms, as shown in table 6.7 on p. 122. Modification of the PWC170 has included shorter (3 min) stages and a higher heart rate of 190 (PWC190) (79).

- Field tests of aerobic capacity: Many field tests have been used to estimate aerobic capacity in children (47,81). The 1 mi (1.6 km) run–walk is one of the cardiorespiratory endurance measures included in both the Physical Best (5,6) and Fitnessgram (37) health fitness test batteries. Numerous studies have addressed the

Table 6.6 Modified Balke Treadmill Protocol

Subject	Speed in mph (kph)	Initial grade (%)	Increment (%)	Stage duration (min)
Poorly fit	3.00 (4.8)	6	2	2
Sedentary	3.25 (5.2)	6	2	2
Active	5.00 (8.0)	0	2.5	2
Athlete	5.25 (8.4)	0	2.5	2

Reprinted, by permission, from T.W. Rowland, 1993, *Pediatric laboratory exercise testing: Clinical guidelines* (Champaign, IL: Human Kinetics), 6.

reliability and validity of these tests to estimate $\dot{V}O_2$max (23,46). Regression equations for both the 1 mi (1.6 km) run–walk and pacer tests (multistage shuttle run) have been developed to estimate $\dot{V}O_2$max. These equations are presented in "Oxygen Uptake Equations for Run–Walk and Pacer Tests."

• The multistage shuttle run test can be used to estimate $\dot{V}O_2$max according to the procedure described by Leger et al. (51). Each subject runs back and forth on a 20 m course starting at a speed of 8.5 kph (2.36 m · s^{-1}). The running speed is increased by 0.5 kph (0.14 m · s^{-1}) every minute. The running pace is regulated by a prerecorded audiotape, which signals when the subject needs to be at one or the other end of the 20 m course. Subjects keep completing subsequent stages until they cannot keep up with the progression of speed. Their speed at the last stage completed is used to calculate $\dot{V}O_2$max. Table 6.8 gives healthy zone standards for the 1 mi (1.6 km) run–walk, pacer laps, and $\dot{V}O_2$max.

Oxygen Uptake Equations for Run–Walk and Pacer Tests

Mile (1.6 km) run–walk

$\dot{V}O_2$max (ml · kg^{-1} · min^{-1})
= .21 (age × sex) – 0.84 (BMI)
= 8.41 (MRW) + 0.34 (MRW^2) + 108.94

Where: age = age in years; sex = 0 for females and 1 for males; BMI = body mass index (kg · m^{-2}); MRW = run–walk time in minutes.

Pacer: multistage shuttle

$\dot{V}O_2$max (ml · kg^{-1} · min^{-1})
= [3.238 × (max speed)] – [3.248 × (age)]
+ [0.1536 × (max speed × age)] + 31.025

Where: max speed (kph) = 8.5 + 0.5 (number of stages completed + age in years)

Data from K.J. Cureton et al., 1995, "A generalized equation for prediction of VO2 peak from one-mile run/walk performance," *Medicine and Science in Sports and Exercise* 27(3):445-45 and Cooper Institute, 2001, *FITNESSGRAM text administrator's manual*, 2nd ed. (Champaign, IL: Human Kinetics).

Table 6.7 Adams Submaximal Progressive Continuous Cycling Protocol, Initial Power Settings by Body-Weight Groups

Body weight (kg)	First-stage power (W)	Second-stage power (W)	Third-stage power (W)
<30	16.5	33	50
30–39.9	16.5	50	83
40–59.9	16.5	50	100
≥60	16.5	83	133

Adapted, by permission, from F.H. Adams, L.M. Linde, and H. Miyake, 1961, "The physical working capacity of normal children," *Pediatrics* 28: 55; G.M. Adams, 1988, *Exercise physiology laboratory manual* (New York: McGraw-Hill).

Anaerobic Capacity

Children produce less anaerobic power and capacity than adults do, and anaerobic power increases with age (15). The lower anaerobic power and capacity of children has been attributed to lower lactate production, possibly attributable to lower levels of glycolytic enzymes such as phosphofructokinase, lower sympathetic activity, lower glycogen storage capabilities, or higher ventilatory thresholds (expressed as a percentage of $\dot{V}O_2$max) (15). In addition, recovery $\dot{V}O_2$ (oxygen debt) and maximal exercise blood pH have been found to be less in children. All or some of these factors may contribute to the lower anaerobic capability in children. When males and females are compared with respect to anaerobic performance, males generally outperform females at all ages, and the rate of improvement with age is greater in males. Medical evaluation concerns and contraindications for anaerobic capacity testing are similar to those described for aerobic capacity testing.

The evaluation of anaerobic capacity in pediatric populations has been used to follow the progress of neuromuscular diseases (e.g., cerebral palsy and muscular dystrophy) (15). In healthy children, the Wingate test is used along with various field tests to evaluate athletic performance.

• Wingate anaerobic cycle power test: The Wingate anaerobic power test was developed to assess both short-term and moderate-term anaerobic capacities in adults. This test lasts 30 s and requires a subject to pedal a cycle ergometer as fast as possible. The number of revolutions is counted in each 5 s period for a total of 30 s. From the number of revolutions completed, peak power (most revolutions in any 5 s interval) and 30 s anaerobic capacity can be determined. The test can be performed by the legs or by arms. Bar-Or (15) stated that the Wingate anaerobic test is feasible for use with healthy and disabled children

Table 6.8 Healthy Zone Standards for Aerobic Capacity

	Mile (1.6 km) run–walk (min)		Pacer (laps)		$\dot{V}O_2max$ $(ml \cdot kg^{-1} \cdot min^{-1})$	
	Boys	Girls	Boys	Girls	Boys	Girls
10	9:00–9:30	9:30–12:30	17–55	35–37	42–52	39–47
11	8:30–11:00	9:00–12:00	23–61	37–49	42–52	38–46
12	8:00–10:30	9:00–12:00	29–68	13–40	42–52	37–45
13	7:30–10:00	9:00–11:30	35–74	15–42	42–52	36–44
14	7:00–9:30	8:30–11:00	41–80	18–44	42–52	35–43
15	7:00–9:00	8:00–10:30	46–85	23–50	42–52	35–43
16	7:00–8:30	8:00–10:30	52–90	28–56	42–52	35–43
17	7:00–8:30	8:00–10:00	57–64	34–61	42–52	35–43
≥18	7:00–8:30	8:00–10:00	57–94	34–61	42–52	35–43

Data from Cooper Institute, 2001, *FITNESSGRAM text administrator's manual*, 2nd ed. (Champaign, IL: Human Kinetics).

as young as 6 yr old. Practical application 6.2 provides the specific protocol.

• Field tests of anaerobic capacity: A simple test of anaerobic performance is the timed 50 yd (45.7 m) run. Included in many fitness-testing batteries, this test provides information on children's anaerobic capacity. Note, however, that performance on a 50 yd (45.7 m) run is influenced by running experience, genetics, and motor performance, especially in young children.

• Standing vertical jump: Sargent (86) developed the vertical jump test to measure leg power in adults. The application of this test in children is the same as in adults. Subjects are required to jump vertically as high as they can. Subjects' ability to exert leg power is determined from the height of the jump compared with the height reached by fingers when the children are standing erect with the arms extended. This field test is easily learned and perfected by children and adolescents. Two numerical values can be used to represent results of the vertical jump test. The simpler number is the difference between the initial reach and jump performance. But peak power in watts can also be calculated by the following formula:

$$Peak\ power\ (watts) = [78.5 \times VJ\ (cm)] + [60.6 \times mass\ (kg)] - [15.3 \times height\ (cm)] + 431$$

where: VJ = vertical jump (jump height minus reach height), mass = body mass in kg, and height = body height in cm. An easier method that uses the preceding equation is the Lewis nomogram, which uses jump height and weight (2).

Muscular Strength and Endurance

As recently as the late 1970s, pediatric exercise experts and medical doctors believed that prepubescent children could not benefit from strength or resistance training because this developmental group lacked the prerequisite circulatory hormones (95). Furthermore, many believed that the stress imposed by this training was not safe and could injure bones, especially at the growth plates. Since that time, numerous controlled studies have provided compelling evidence that strength or resistance training produces strength gains in both prepubescent girls and boys (70,88,97). As with adults, the effectiveness of training appears to depend on intensity, volume, and duration. But these specific factors have not been established with certainty in children. An excellent reference is a document sponsored by American Orthopedic Society for Sports Medicine (24), which concludes that strength training for the prepubescent

• improves muscular strength and endurance,
• improves motor skills,
• protects against injury (sports),
• has positive psychological benefits, and
• provides a forum for the introduction of safe and proper training.

Medical Evaluation

One of the more serious injury concerns is the potential for strength or resistance training to cause skeletal damage to the epiphysis or growth plates of children.

Wingate Protocol

Equipment

Monarch mechanically braked cycle or arm ergometer

Protocol

1. The subject performs a 3 to 4 min warm-up.

2. The subject observes a 3 min rest period.

3. Standard braking force should be based on subject's weight (in kilograms) from the Wingate protocol table (table 6.9).

4. The subject begins pedaling at 0 kp resistance, and then the proper load calculated from the Wingate protocol table is applied.

5. Count the maximum revolution in 30 s as well as for each 5 s segment. Be sure to give encouragement.

6. Provide a 3 min cool-down at low resistance. Cool-down is important to help the cardiovascular system recover.

7. Peak power is calculated from the number of revolutions in the best 5 s interval. The kilopond (kp) is a unit of force (gravational) and a kilogram (kg) is a unit of mass. While they are not technically the same they are used interchangeably when calculating work or power at normal acceleration of gravity.

   Peak anaerobic power ($kgm \cdot 5 s^{-1}$) = revolutions in 5 s × force (kp) x flywheel circumference

   Conversion to watts = peak = $2 \times kgm \cdot 5 s^{-1}$.

8. Anaerobic capacity is calculated from the total revolutions per 30 s.

   Anaerobic capacity in 30 s ($kgm \cdot 30 s^{-1}$) = revolutions $\cdot 30 s^{-1}$ × flywheel circumference × force (kp)

   Conversion to watts = anaerobic capacity = $kgm \cdot 30 s^{-1}$.

Table 6.9 Wingate Protocol

	Leg (kp)	Arm (kp)	Leg (kp)	Arm (kp)
	Girls		Boys	
20–24.9	1.3–1.7	0.8–1.0	1.4–1.8	0.8–1.1
25–29.9	1.7–2.0	1.0–1.2	1.8–2.0	1.1–1.3
30–34.9	2.0–2.3	1.2–1.4	2.1–2.5	1.3–1.5
35–39.9	2.3–2.7	1.4–1.6	2.5–2.7	1.5–1.6
40–44.9	2.7–3.0	1.6–1.8	2.8–3.2	1.7–1.9
45–49.9	3.0–3.3	1.8–2.0	3.2–3.5	1.9–2.1
50–54.9	3.3–3.7	2.0–2.2	3.5–3.9	2.1–2.3
55–59.9	3.7–4.0	2.2–2.4	3.9–4.2	2.3–2.5
60–64.9	4.0–4.3	2.4–2.6	4.2–4.6	2.5–2.8
65–69.9	4.3–4.7	2.6–2.8	4.6–4.9	2.8–3.0

Adapted, by permission, from T.W. Rowland, 1993, *Pediatric laboratory exercise testing: Clinical guidelines* (Champaign, IL: Human Kinetics), 169.

Although there are some reports of epiphyses fractures during late puberty, weight training has been reported as the cause in only one case. Most injuries caused by training are of the muscle strain nature, which result from improper lifting techniques or attempts at maximal lifts. No evidence indicates that weight training is more risky with respect to musculoskeletal injury than other youth sports (20,21,73,74).

Contraindications

Children must be emotionally mature enough to begin a strength program. Adults usually suggest resistance training to children, but the child must be enthusiastic about participation. Otherwise, contraindications are absent, unless the child has preexisting musculoskeletal problems.

The following are specific guidelines for resistance training for the preadolescent put forth by Blimke (21), including safety and program information.

- Encourage resistance training as only one of a variety of normal recreational and sport activities.

- Encourage using a variety of training modalities, such as free weights, springs, machines, and body weight.

- Discourage interindividual competition and stress the importance of personal improvement.

- Discourage extremely high-intensity (loading) efforts, such as maximal or near-maximal lifts with free weights or weight machines.

- Avoid isolated eccentric training.

- Encourage a circuit system approach to capitalize on possible cardiorespiratory benefits.

- If using weight-training machines, select either those that have been designed for children or those for which the loads and levers can be easily adjusted to accommodate the reduced strength capacity and size of children.

- Provide experienced supervision, preferably by an adult, when free weights or training machines are used in training.

- Preclude physical and medical contraindications.

- Provide instruction in proper technique and demand that children use that technique.

- Have children warm up with calisthenics and stretches.

- Begin with exercises that use body weight as resistance before progressing to free weights or weight-training machines.

- Individualize training loads when using free weights and training machines.

- Train all major muscle groups and both flexors and extensors.

- Exercise muscles through their entire range of motion.

- Alternate days of training with rest days, and do not allow children to train more than three times per week.

- When children use free weights or machines, they should progress gradually from light loads, high repetitions (>15), and few sets (two to three) to heavier loads, fewer repetitions (6 to 8), and moderate numbers of sets (three to four).

- Instruct children to cool down after training with stretching exercises for major joints and muscle groups.

- When selecting equipment, check for durability, stability, sturdiness, and safety.

- Instruct children to heed sharp or persistent pain as a warning and seek medical advice.

Recommendations and Anticipated Responses

The assessment of muscular strength and endurance in children is needed for normal evaluation and for proper training prescriptions. Although many types of machines (isokinetic, variable resistance, Nautilus) are used, this chapter presents assessment with free weights and the protocols of popular fitness test batteries that all feature low-cost and simple equipment. In children, repetition maximum (RM) can easily be determined. Kramer and Fleck (48) suggested that RM for six or fewer repetitions is sufficient to measure strength (see "Estimating Children's Repetition Maximum" on p. 126).

Many programs have as their primary goal to increase muscular endurance. Kramer and Fleck (48) stated that having the child do as many repetitions as possible at a specified percentage of his or her body weight or 6RM resistance is sufficient to evaluate muscular endurance. They also suggested using 60% to 80% of 6RM to test relative local muscle endurance. Tests are terminated when repetitions lack proper technique or safety is a factor (48).

For evaluating the strength and endurance of abdominal muscles, the most popular field tests include the sit-up and curl-up. Fitnessgram uses the curl-up, as described in detail in *Fitnessgram Test Administration Manual* (37). The advantage over the traditional sit-up

Estimating Children's Repetition Maximum

1. The child warms up with 5 to 10 repetitions at 50% of the estimated 6RM.

2. After 1 min of rest and some stretching, the child performs six repetitions at 70% of the estimated 6RM.

3. The child repeats step 2 at 90% of the estimated 6RM.

4. After about 2 min of rest, depending on the effort needed to perform the 90% set, the child performs six repetitions with 100% or 105% of the estimated 6RM.

5. If the child successfully completes six repetitions in step 4, add 2.5% to 5% of the resistance used in step 4 and have the child attempt six repetitions after 2 min of rest. If the child does not complete six repetitions in step 4, subtract 2.5% to 5% of the resistance used in step 4 and have the child attempt six repetitions after 2 min of rest.

6. If the first part of step 5 is successful (the child lifts 2.5 to 5% more resistance than used in step 4), retest the child starting with a higher resistance after at least 24 h of rest, because fatigue will greatly affect performance of more sets. If the second part of step 5 is successful (the child lifts 2.5 to 5% less than the resistance used in step 4), this is the child's 6RM. If the second part of step 5 is not successful (the child does not lift 2.5 to 5% less resistance used in step 4), retest the child after at least 24 h of rest, starting with less resistance.

Adapted, by permission, from W.J. Kramer and S.J. Fleck, 1993, *Strength training for young athletes* (Champaign, IL: Human Kinetics).

is that the curl-up minimizes the use of the hip flexors used in a sit-up and is safer with respect to low-back and spine compression. The student being tested lies on a mat with knees bent at 140°, arms straight and parallel to the trunk, and palms resting on the mat. Curl-ups are completed in a slow and controlled manner at a cadence of 20 per minute (one curl every 3 s). Students are stopped after they complete a maximum of 75 curl-ups or lose proper form or cadence. Used more often, however, is the traditional sit-up (6). After a signal is given, the student sits up (touching forearms to the thighs) as many times as possible in 1 min. Traditionally, the pull-up or chin-up has been used to assess arm and shoulder strength. A major problem with this test is that most children cannot complete one repetition.

These zero scores are not helpful in assessment and can be demoralizing to children. Many believe that the pull-up is not appropriate for children who cannot accomplish as least one repetition. With this problem in mind, Fitnessgram includes the 90° push-up, modified pull-up, and flexed-arm hang.

The 90° push-up begins with the student being tested assuming a prone position on the mat with the hands placed under the shoulders; the fingers stretched out; the legs straight, parallel, and slightly apart; and the toes tucked under. The student pushes up off the mat with the arms until the arms are straight. The back should be kept in a straight line during the movement. A push-up is then recorded when the student lowers the body using the arms until the elbows bend to a 90° angle and the upper arm is parallel to the floor. Push-ups continue at a rhythm of 20 per minute, or 1 push-up every 3 s. Students are stopped when their form or cadence falters (37).

Body Composition

Body composition assessment has recently become important in the field of pediatric exercise and medicine. Body composition has an important influence on both field and laboratory performance in children (78). No specific medical issues need to be addressed before assessing body composition. In assessing body composition, the tester should conduct the test in a place that respects the child's privacy, and the test must be sensitive to the child's self-image. The way in which individual results are presented to the child should include parental involvement.

Methods commonly available for measuring body composition in children include the following:

- Hydrostatic weighing
- Skinfold measurement

- BMI
- Bioelectrical impedance
- Dual-energy X-ray absorptiometry (DEXA)

Choosing a method of body composition assessment for children is difficult because of many developmental factors that influence the accuracy of various prediction equations. Hydrostatic weighing is considered the gold standard for the estimation of percent body fat in adults. As outlined by Lohman (53), this method is problematic for use in children not only because the process is difficult for children (i.e., being submerged and blowing air out of lungs) but also because children undergo changes in their chemical composition of fat-free mass (FFM), especially during puberty (53). For example, protein and mineral content increase about 5% and the water content decreases about 9% from birth to adulthood (75). In fact, the chemical maturity of FFM is not reached until late adolescence (55). Therefore, the use of adult equations for relating body density to percent fat in children has been found to overestimate fat between 7% and 13% (22). When performing hydrostatic weighing on children, Lohman proposed using the Siri equation for estimating body fat from density. This method incorporates age-specific values for FFM.

A common method to estimate body fatness is the anthropometric measurement of skinfold thickness. Although it has drawbacks, this simple and relatively inexpensive approach estimates body fat with an accuracy of 3% to 4%. Testers should practice the method to develop accuracy before reporting values to children. Slaughter et al. (89) provided accurate equations using two skinfold sites (either the calf and triceps or calf and subscapularis). These sites are relatively easy to locate and measure in children. Notice that Slaughter's regression equations account for racial and developmental differences (see "Prediction Equations of Percent Fat From Triceps and Calf or From Triceps and Subscapularis Skinfolds in Children"). The simple sum of skinfolds (calf and triceps or triceps and subscapularis) can be useful not only when normal values are compared but also when weight loss (or fat loss) is considered. When the sum of skinfolds is used as a criterion measure, the problem of accurate regression equations is eliminated.

BMI $(kg \cdot m^{-2})$ is the easiest and least invasive body composition measurement. Many large epidemiological studies use this method because research has shown that the risk of chronic disease increases with BMI. A weakness of BMI is the inability to distinguish between muscle tissue and fat; therefore, some subjects with extensive muscle mass may have a high BMI. This method should

be considered when skinfold measurements are not feasible. Table 6.2 presents healthy BMI values.

Bioelectrical impedance analysis (BIA) is a simple noninvasive technique used to estimate body composition. This method is based on passing a small electrical current through the body that is affected by differences in tissue conductivity. The result is then used to estimate fat-free mass and percent fat. This technique is considered valid and reliable in certain adult populations, but not in

Prediction Equations of Percent Fat From Triceps and Calf or From Triceps and Subscapularis Skinfolds in Children

Triceps and Calf Skinfolds
% fat = 0.735 × SF + 1.0 males, all ages
% fat = 0.610 × SF + 5.0 females, all ages

Triceps and Subscapularis Skinfolds (>35 mm)
% fat = 0.783 × SF + I males
% fat = 0.546 × SF + 9.7 females

Triceps and Subscapularis Skinfold (<35 mm)
% fat = 1.21 (× SF) – 0.008 (× SF)2 + I males
% fat = 1.33 (× SF) – 0.013 (× SF)2 + 2.5 females
(2.0 blacks, 3.0 whites)

Table 6.10 Age, Race, Sex, and Adjustments to Percent Fat Predications

I = Intercept which varies with maturation level and racial group for males as follows:

Age	Black	White
Prepubescent	–3.5	–1.7
Pubescent	–5.2	–3.4
Postpubescent	–6.8	–5.5
Adult	–6.8	–5.5

Note. SF = skinfold. Calculations were derived by using the equation in Slaughter et al. (84).

Reprinted, by permission, from T.G. Lohman, 1992, *Advances in body composition assessment* (Champaign, IL: Human Kinetics), 94.

children. Problems exist with children because their rapid growth rates and subsequent body chemistry changes (43). For these reasons, BIA is not recommended for use in estimating body composition in children.

Research studies on children are increasingly reporting body composition measured by dual-energy X-ray absorptiometry (DEXA) scans. This new technology passes a low-level x-ray source through the body and is capable of estimating whole-body and regional bone density as well as lean tissue and fat. Recent studies have confirmed the safety, reliability, and validity of using DEXA to estimate body fat in children and adults. This technique may well become the gold standard of body composition assessment in the near future.

Flexibility

Flexibility is defined as the ability to move joints through a full range of motion. A widely accepted concept is that children are extremely flexible and that therefore flexibility should not be a priority in activities or training. Exceptions seem to be children in sports such as gymnastics and dancing, in which flexibility is required and its importance is appreciated. Nevertheless, flexibility training is recommended at all ages to ensure safe activity. Most research studies show a decline in flexibility as children become older. Clark (29) concluded that boys tend to lose flexibility after age 10 and girls after age 12. In fact, Milne et al. (62) found that flexibility in both boys and girls declined between kindergarten and second grade. Girls as a group are usually more flexible than boys, which may reflect the activities in which girls participate. In addition, the YRBSS data showed that girls were more apt to engage in flexibility training than boys (28). Contraindications and specific medical concerns are absent, unless the child has preexisting musculoskeletal problems.

- Trunk extension test (trunk lift): This test is included in the Fitnessgram test battery (37). The objective is to lift the upper body off the floor using the large muscles of the back and hold this position to allow for measurement. The subject lies on a mat in a prone position with toes pointed and hands placed under the thighs. The subject lifts the upper body off the floor in a slow, controlled manner to a maximum height of 12 in (30 cm). The subject holds this position long enough for the tester to measure the distance from the floor to the chin.

- Traditional sit-and-reach flexibility test: The Physical Best health-related test battery includes the traditional sit-and-reach test as a measure of low-back and hamstring flexibility. This test is completed with a sit-and-reach box. With shoes removed and knees straight, students are asked to bend forward as far as possible. Although the sit-and-reach test is a component of many health-related fitness test batteries, it has been criticized because it may test flexibility of the hamstrings only, not the lower back (29).

- Back-saver sit-and-reach: Another version of the sit-and-reach is an optional test included in Fitnessgram called the back-saver sit-and-reach. This test is similar to the traditional sit-and-reach except that each leg is tested separately. This test is reported to cause less pressure on the anterior portion of the lumbar vertebra. Stretching one hamstring at a time avoids excessive flexing of the lumbar spine and hyperextension of both knees (37).

Practical Application 6.3

Literature Review

Aerobic Fitness

The plasticity of aerobic fitness in children is currently being debated and studied. Plasticity refers to the extent that normal growth-related changes (improvements) in maximal aerobic capacity can be altered by changes in physical activity. More succinctly stated, plasticity refers to the question, can increases in the level of physical activity improve aerobic fitness ($\dot{V}O_2$max)? Plasticity, therefore, can be thought of as trainability. Numerous studies have established that improvement in $\dot{V}O_2$max after a period of aerobic training is less substantial in children than it is in adults. Rowland (75) reviewed 13 studies that attempted to correlate habitual activity in children with their level of aerobic capacity ($\dot{V}O_2$max). He found that only five of the studies concluded that a significant correlation exists between levels of physical activity and aerobic capacity (78). All these studies measured habitual physical activity of children and were not training studies.

A possible explanation for the lack of support for a significant training effect for habitual physical activity of children can be found in the literature. For example, studies assessing the intensity of physical activity in children have consistently shown that only a small percentage of children meet the guidelines that call for at least 20 min of sustained activity eliciting between 60% and 90% of maximal heart rate. Specifically, Armstrong et al. (12) found that only 13% of boys and 6.5% of girls attained heart rates over 160 beats · min^{-1} for a 20 min period during a 3 d assessment period.

If physical activity is not a major contributor to $\dot{V}O_2$max, then what about formal endurance training? Adult training using ACSM guidelines for intensity, duration, mode, and frequency (large-muscle groups, rhythmic activities, 20 to 60 min, 3 to 5 d per week equivalent to 65 to 90% maximum heart rate) usually produces between a 5% and 35% improvement in 12 wk (8). Mahon (50) reviewed three controlled training studies on children less than 8 yr of age. In these studies, the experimental group showed a 12.5% increase in $\dot{V}O_2$max, whereas the control group increased only 7.5%. Studies with children 8 to 13 yr old showed an average 13.8% increase in $\dot{V}O_2$max, and the controls increased only 0.7% (59). For adolescents 13 yr of age or older, an increase of 6.8% in $\dot{V}O_2$max was found, and the control group had no change in $\dot{V}O_2$max. These data show that children and adolescents can adapt to training by increasing $\dot{V}O_2$max but at a much lower rate than adults. When considering $\dot{V}O_2$max changes in children, recognize that initial fitness and genetic endowment can also influence responses.

In the previously mentioned training studies, the average frequency was 3 d per week, the average duration was 30 min, and the intensities were greater than 160 beats · min^{-1} and as high as 85% HRmax. The modes were continuous running, weight training, aerobics, and jumping rope. Clearly, children as young as 8 yr old can increase $\dot{V}O_2$max with training, but their increase will not be as great as that of adults.

Most children are not active enough to improve aerobic capacity or $\dot{V}O_2$max. Most studies that have shown improvement in $\dot{V}O_2$max have been studies that evaluated high-intensity training. Studies that have evaluated children not in formal training programs have found little correlation between physical activity and aerobic capacity. The goals, therefore, of promoting physical activity in children should be as follows:

1. To promote physical activity habits that can be carried to adulthood
2. To modify cardiovascular disease factors including blood pressure, blood lipids, and body composition

Anaerobic Fitness

Little information is available concerning exercise prescription and trainability of anaerobic systems in children by employing short-burst activities. Rowland (75) indicated that the major reason for this lack of information is that an accurate, noninvasive method of assessing anaerobic metabolism similar to $\dot{V}O_2$max in aerobic systems does not exist (80). Consequently, short-burst activities, which are common in the habitual activity of children, are poorly understood. The limited available research indicates that children can increase their anaerobic power by training. For example, studies by Grodjinovski et al. (39), Rotstein et al. (77), and Sargent et al. (86) have shown that children improve anaerobic performance from 4% to 14% following interval-type training. Grodjinovski's 6 wk training study consisted of one group riding a cycle ergometer for five 10 s all-out bouts followed by three 30 s all-out bouts (39). The other group ran three all-out 40 m runs followed by three all-out 150 m runs during each training session. Both groups trained 3 d per week. Improvement in anaerobic performance was about 4% in each group. Although this study shows that anaerobic improvement can be improved by training, it also gives some information on duration, intensity, and frequency of training. This study indicates a training frequency of 3 wk of short-duration and high-intensity efforts.

EXERCISE PRESCRIPTION

As stated in the most recent edition of ACSM's *Guidelines for Exercise Testing and Prescription* (9), the art of exercise prescription is the successful integration of exercise science with behavioral techniques that result in long-term program compliance and attainment of the individual's goals. This task is extremely challenging with adults and even more difficult with children. Compounding the problem is the lack of specific definitive information on the appropriate duration and intensity of exercise for children.

The International Consensus Conference on Physical Activity Guidelines for Adolescents concluded that all youth should be physically active daily as part of play, games, sports, work, transportation, recreation, physical education, or planned exercise. These guidelines also stated that additional aerobic benefits may be achieved by engaging in moderate to vigorous physical activities that use large-muscle groups, such as running, cycling, and swimming. The general intensity, frequency, and duration of these activities should be a minimum of 3 d per week for a minimum of 30 min at an intensity of 75% heart rate reserve (83). For younger children, the National Association for Sports and Physical Education (32) issued physical activity guidelines for elementary school–aged children and recommended the following:

- Elementary school–aged children should accumulate at least 30 to 60 min of age-appropriate and developmentally appropriate physical activity from a variety of activities on all, or most, days of the week.

- An accumulation of more than 60 min, and up to several hours per day, of age-appropriate and developmentally appropriate activity is encouraged.

- Some of the child's activity each day should be in periods lasting 10 to 15 min or more and should include moderate to vigorous activity. This activity will typically be intermittent in nature, involving alternating moderate to vigorous activity with brief periods of rest and recovery.

- Children should not have extended periods of inactivity.

Cardiovascular Training

The ACSM (9) set forth general recommendations in 1988 for aerobic exercise prescription in children and adolescents:

- Although children are generally quite active, they usually choose to participate in activities that consist of short-burst, high-energy exercise. Children should be encouraged to participate in sustained activities that use large-muscle groups.

- The type, intensity, and duration of exercise activities need to be based on the maturity of the child, medical status, and previous experiences with exercise.

- Regardless of age, the exercise intensity should start low and progress gradually.

- Because of the difficulty in monitoring heart rates with children, a modified Borg scale is a practical method of monitoring exercise intensity in children.

- Children are involved in a variety of activities throughout the day. A specific time should be dedicated to sustained aerobic activities.

- The duration of the exercise session will vary depending on the age of the children, their previous exercise experience, and the intensity of the exercise session.

- Because inducing children to respond to sustained periods of exercise is often difficult, the session periods need to be creatively designed.

An important omission in the ACSM guidelines is proper intensity. Rowland (73) and others have suggested the use of adult guidelines to increase aerobic fitness, and studies have shown higher intensity to be effective. But these guidelines seem proper for trained athletes. Sallis and Patrick (78) suggested a more moderate approach.

- Guideline 1: All adolescents should be physically active daily, or nearly every day, as part of play, games, sports, work, transportation, recreation, physical education, or planned exercise, in the context of family, school, and community activities (consistent with objective 1.3 from Healthy People 2000) (72). Adolescents should do a variety of physical activities as part of their daily lives. These activities should be enjoyable, involve a variety of muscle groups, and include some weight-bearing activities. The intensity or duration of the activities is probably less important than the fact that energy is expended and a habit of daily activity is established. Adolescents are encouraged to incorporate physical activity into their lifestyles by doing such things as walking up stairs, walking or riding a bicycle for errands, having conversations while walking with friends, parking at the far end of parking lots, and doing household chores.

- Guideline 2: Adolescents should engage in three or more sessions per week of activities that last 20 min or more and that require moderate to vigorous levels of exertion (consistent with objective 1.4 from Healthy People 2000) (72). Moderate to vigorous activities are those that require at least as much effort as brisk or fast walking. A diversity of activities that use large-muscle groups are recommended as part of sports, recreation, chores, transportation, work, school physical education, or planned exercise. Examples include brisk walking, jogging, stair climbing, basketball, racket sports, soccer, dance, swimming laps, skating, strength (resistance) training, lawn mowing, strenuous chores, cross-country skiing, and cycling.

Note. Vigorous physical activities are rhythmic, repetitive activities that require the use of large-muscle groups to elicit a heart response of 60% or more of a subject's maximum heart rate adjusted for age. An exercise heart rate of 60% of maximum heart rate for age is sufficient for cardiorespiratory conditioning. Maximum heart rate equals roughly 220 beats · min^{-1} minus age in years.

Specific aerobic fitness prescription guidelines for children and adolescents are reviewed in *The Physical Best Teacher's Guide* (7). Practical application 6.4 contains these guidelines, which are divided into three objectives: basic health-related fitness, intermediate health-related fitness, and athletic performance. This approach allows for differences in abilities, interests, and fitness objectives of children.

Alternative methods of calculating optimal heart rate in children are available. Most methods of calculating optimal aerobic training zones for children require an estimate of maximum heart rate. Adult formulas such as 220 minus age are usually used for children. As stated

by Rowland, however, maximal heart rate determined by treadmill and cycle studies remains constant across the pediatric years (79). Therefore, adult formulas strongly dependent on age may not be appropriate. Direct maximal heart rate is best for determining exercise prescriptions. An alternative formula suggested by Tanaka et al. is max HR = 208 − .7 × age in years. This formula is slightly less dependent on age. Although this formula was developed from adult data, it required further study in children (86).

Resistance Training

Training prescriptions for children show great variability but seem generally to follow adult prescriptions with the exception of lower resistance and high repetitions. Practical application 6.5 on p. 132 reviews training principles for children with respect to improving muscular strength and endurance for basic health and athletic performance.

Practical Application 6.4

Cardiovascular Training Recommendations

	Basic health-related fitness	Intermediate health-related fitness	Athletic performance fitness
Frequency	3 times per wk	3-5 times per wk	5-6 times per wk
Intensity	50%-60% HRmax	60%-75% HRmax	65%-90% HRmax
Time	30 min total, accumulated	40-60 min total, accumulated	60-120 min total, accumulated
Type	Walking, jogging, dancing, games, and activities that require minimal equipment demands	Jogging, running, fitness-based games and activities, intramural and local league sports	Training programs, running, aerobics, interscholastic, and community sports programs
Overload	Not necessary to bring child to overload during base level	Be creative with activity to increase tempo or decrease rest period; 1-3 times per wk	Program design should stress variable intensities and durations to bring student into overload; 2-3 times per wk
Progression and specificity	Let student "get the idea" of movement. Progression is minimal	Introduce program design and incorporate variation	Specific sets, repetitions, and exercises to meet desired outcome

Note. HR = heart rate.

Reprinted, by permission, from AAHPERD, 1999, *Physical education for lifelong fitness: The physical best teacher's guide* (Champaign, IL: Human Kinetics), 88.

Practical Application 6.5

Training Principles Applied to Muscular Strength and Muscular Endurance, Based on Fitness Goals

	Basic health-related fitness	Intermediate health-related fitness	Athletic performance fitness
Frequency	Two or three times per week; allow for minimum 1 d rest between training sessions	Three or four times per week; alternating upper- and lower-body segments allows for consecutive training days	Four or five times per week; training activities specific to sport participation
Intensity	Very light, <40% of "projected" maximal effort*	Light to moderate, 50–70% of "projected" maximal effort*	Specific load requirements for sport participation
Time	One or two sets of 6–12 repetitions	One to three sets of 6–15 repetitions	Three to five sets of 5–20 repetitions
Type	Body weight, single-joint, and multijoint activities involving major muscle groups*	Resistance exercises such as leg press, bench press, pull-ups, and additional presses and pulls*	Advanced sport-specific, multijoint lifts (clean pulls, power presses, Olympic-style lifts)
Overload	Not necessary to bring child to overload during base level	Introduce one of the components of overload; one or two times per week	Program design should stress variable intensity and duration to take student into overload; two or three times per week
Progression and specificity	Let student get the idea of correct movement; progression is minimal	Introduce program design and incorporate variation	Specific sets, repetitions, and exercises to meet desired outcome

*Projected maximal effort (one-repetition max, or 1RM) can be calculated from submaximal testing. For example, if a child bench-presses 125 lb (57 kg), 10 times the calculated 4RM = (number of reps × 0.03) + 1 × 125 = 0.3 + 1 × 125 = 162.5 1RM.

Reprinted, by permission, from AAHPERD, 1999, *Physical education for lifelong fitness: The physical best teacher's guide* (Champaign, IL: Human Kinetics), 100.

Supervision to ensure proper technique is most important when prescribing resistance training to children (82).

Range of Motion

Flexibility, or adequate range of motion, has been identified as one of the components of health-related fitness. Many questions exist concerning flexibility training:

- Should low-intensity aerobic activity be performed before stretching?
- Which general method of stretching—static or ballistic—is better or safer?
- How much time should be dedicated to flexibility training in children and adolescents?

Although a review of the literature provides no definitive answers to these questions, general principals of flexibility training can be stated. *The Physical Best Teacher's*

Guide states that children should engage in at least 5 min of low-level aerobic activity before stretching. With respect to the static versus ballistic question, *The Physical Best Teacher's Guide* suggests static flexibility training for the child whose objective is basic or intermediate health-related fitness, and a more ballistic approach when the objective is increasing athletic performance. The final question concerning time and emphasis on flexibility training depends on the specific objective and amount of time available. Although no clear recommendations can be found in the literature, a suggestion of 5% of the available time seems reasonable. Practical application 6.6 summarizes the FITT principle for flexibility training (7).

Practical Application 6.6

The FITT Principle Applied to Flexibility Training, Based on Fitness Goals

	Basic health-related fitness	Intermediate health-related fitness	Athletic performance fitness
Frequency	Before and after each activity or exercise session (minimum of three times per week)	Before and after each activity or exercise session (daily)	Before and after each training session
Intensity	To mild tension, or slight muscular discomfort	To mild tension, or slight muscular discomfort	To mild tension, or slight muscular discomfort, at a level appropriate for sport participation
Time	10–15 s, two times per stretch	10–15 s, three times per stretch	Dependent on static, dynamic, or ballistic (usually conducted by qualified trainer or coach)
Type	Static; major muscle groups	Static; major muscle groups, introduction of dynamic stretching	Usually dynamic or ballistic; major muscle groups and sport-specific stretches
Overload	Not necessary at base level	Ask student to identify level of stretch intensity; if appropriate for activity, have student stretch slightly farther than previous same stretch	Because dynamic and ballistic stretches dominate advanced level, overload is not appropriate to ballistic stretching
Progression and specificity	Start easy into stretch; slow movements with minimal applied resistance to muscle involved	Stretch major core muscles first and then move to extremities; introduce dynamic flexibility	Start with easy multijoint dynamic movements. Progress to more resistive dynamic movements, followed by moderate static or proprioceptive neuromuscular facilitation stretching

Reprinted, by permission, from AAHPERD, 1999, *Physical education for lifelong fitness: The physical best teacher's guide* (Champaign, IL: Human Kinetics), 127.

CONCLUSION

This chapter focuses on physical activity and the basic exercise physiology of children and adolescents. Of primary interest is the question, are the physical activity patterns of children adequate for optimal health? The answer to this question is not entirely clear, but recent survey data indicate that only 64% of children and adolescents engage in vigorous physical activity and only 26% of the same group engaged in light to moderate physical activity within 7 d of the survey. Physical activity seems to have a positive effect on obesity, blood lipids, skeletal health, and psychological factors. More research is needed to quantify and qualify the amount of activity that will maximally influence these factors.

Another largely unanswered question in this age group involves proper exercise prescription guidelines. Most published guidelines for the enhancement of aerobic capacity, anaerobic capacity, muscular strength, and flexibility are simply modified from adult research. This chapter summarizes the most recent research with respect to the proper intensity, duration, and frequency of activity to enhance these physiological attributes. Guidelines are also presented to design safe and effective training programs for children and adolescents.

Case Study

JT is a normal 10 yr old white male. He is sedentary most of the time and would rather play computer games than play outdoors or participate in sports. JT's mother is concerned about his long-term health. She asked a staff member at a university human performance laboratory to assess his health and suggest healthy lifestyle changes. The tests and results follow:

Body Composition Results

Height: 157.5 cm

Weight: 70.5 kg

Skinfold thickness: calf 23 mm and triceps 22 mm

Cardiorespiratory endurance:
$\dot{V}O_2$max (treadmill) = 41 ml $\cdot$ kg^{-1} $\cdot$ min^{-1} and 2.9 L $\cdot$ min^{-1}

Mile run = 9 min 30 sec

Pacer test = 19 laps

Flexibility: Sit-and-reach = 22 cm

Muscular strength:
Sit-ups per minute = 20
Pull-ups = 0

Blood lipids and glucose:
Total cholesterol = 200 mg $\cdot$ dl^{-1}
High-density lipoprotein = 32 mg $\cdot$ dl^{-1}
Triglycerides = 190 mg $\cdot$ dl^{-1}
Low-density lipoprotein = 220 mg $\cdot$ dl^{-1}
Glucose = 119 mg $\cdot$ dl^{-1}
Blood pressure = 130/86 mmHg

Discussion Questions

1. Is JT at risk with respect to body composition, body mass index, percent fat, or sum of skinfolds?

2. Is JT at risk with respect to cardiorespiratory endurance in treadmill testing, mile run, or pacer results?

3. Is JT at risk with respect to blood lipids, glucose, or blood pressure?

4. As you sit down with JT and his mother following testing, how would you approach the subject with the results?

5. What suggestions would you give JT to improve his results? If you suggest increasing physical activity, how would you justify this to him? Explain why physical activity is or is not beneficial for JT.

6. Write an exercise prescription for JT to improve the following:

   Aerobic endurance

   Muscular strength and endurance

   Flexibility

   Blood pressure

   Blood test results (high-density lipoprotein and cholesterol)

7

The Elderly

Nicole Y.J.M. Leenders, PhD

Geriatrics is a branch of medicine that deals with the problems and diseases associated with elderly people (>65 yr) and the aging process. As people age, physiological processes change. These alterations are natural but are affected by certain conditions such as genetics, inactivity, and environmental factors. For the purpose of description, the elderly are divided into the **old age** (65–74 years of age), the **very old age** (75–84 years of age), and the **oldest old** (older than 85 yrs of age) (59,91). Although these groups chronologically categorize the elderly, the elderly are a heterogeneous group of people in terms of physiological status, and overlap is present between groups. As discussed in the next section, the elderly are continually increasing in absolute numbers of the population. This chapter emphasizes the importance of recognizing the elderly, their potential limitations, and how they can improve their health and fitness status by exercise training.

SCOPE

Since 1900 the over 65 yr old population in the United States has grown by 200% and now constitutes at least 12% of the population. By 2030 approximately 72 million people in the United States will be over the age of 65, representing 20% of the population (1). Today, a person reaching the age of 65 has an average life expectancy of an additional 18.5 yr (1).

A child born in 2003 in the United States has a life expectancy of 78 yr, compared with only 48 yr in 1900.

Older women outnumber older men (21.1 million to 15.2 million, respectively). Minority populations are predicted to represent 24% of the elderly population in 2020, up from 16% in 2000. Currently, on average, whites outlive blacks by 6 yr (44).

The prevalence of chronic disease and functional impairment increases with age. In 2004 greater than 50% of the elderly in the United States had at least one type of **disability** (physical or nonphysical), and multiple chronic diseases and impairments are common (1).

The high prevalence of these chronic diseases results in increased health and medical care costs and reliance on medical services and assisted living. Money spent in 2002 for care of the elderly was 43% of total healthcare expenditure (98). The five most expensive health conditions are heart disease, cancer, trauma, mental disorders, and pulmonary disorders. Falls are a leading cause of injury, disability, and death of the elderly (73,100).

A decline in general health status and reduced mobility negatively affects an older person's ability to carry out activities of daily living (ADLs). For instance, many have difficulty with bathing, dressing, and preparing food (13), and about one-third perceive their health as fair to poor. Not surprisingly, the rate of physical inactivity decreases with age (figure 7.1). Only 25% of the elderly report regular participation in sustained **physical activity** five or more times per week for 30 min or more per occasion (106). Blacks and males are less active than whites and females, respectively, across every age group (47).

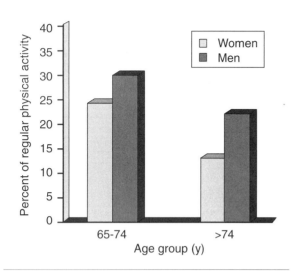

Figure 7.1 Percentage of older individuals who report engaging in regular physical activity (in moderate-intensity physical activity 5 d a week for 30 min each time or in vigorous activity at least three times a week for 20 min each time).

Adapted from NHIS 2002–2004.

PHYSIOLOGY AND PATHOPHYSIOLOGY

Anatomical and physiological changes of several organ systems can lead to functional disability and increased risk of premature death (100). Table 7.1 reviews the effects of aging on the organ systems. Sudden decline of organ function is often attributable to chronic disease, whereas aging effects result in gradual impairment (86). There are two major theories of aging:

1. The genetic theory proposes that aging is programmed by cellular signals from birth to death.
2. The damage theory proposes that aging is caused by progressive translation and transcription errors in cells.

Diet, environmental factors, and genetics influence changes in organ system function. Each of these may affect gene expression. The aging process may be attenuated by modifying factors that enhance the aging process (5,97). Table 7.2 reviews the major organ systems and specific effects associated with aging. Many of these changes are augmented by physical inactivity and are attenuated when exercise volume is maintained as aging occurs.

Table 7.1 Aging Effects on Organ Systems

Organ system	Effects
Skeletal muscle	Sarcopenia: loss of mass, strength, contractile speed, and power. Loss of mass at rate of 1.2 kg per decade from the fifth to the ninth decade attributable to hypoplasia and atrophy (33,40,42,62,84)
	Hypertrophy attributable to compensatory loss of motor units (4). Motor unit remodeling: reduced axonal sprouting rate, nerve fiber contacts per muscle fiber, and type 2 fiber density (attributable to cross-reinnervation) (28,109); increase in motor unit size (41)
	Myosin phenotype: increased slow isoform (54,86). Increased risk of osteoporosis, fraility, fractures, and arthritis. Decreased flexibility (96)
Body composition	Decreased lean mass and total water; increased fat; reduced bone mass starting in the third decade and accelerating in the fifth, increasing osteoporosis risk (14,38,40,57,88,95,103)
Cardiovascular	Arterial stiffening and reduced dilatory capacity; increased peripheral resistance and blood pressure; ventricular thickening; increased left ventricular end-diastolic volume; reduced baroreceptor sensitivity and orthostatic tolerance; 5–15% per decade decline in peak $\dot{V}O_2$ attributable to lower peak cardiac output (heart rate and stroke volume) and augmented by an increase in body fat; increased cardiovascular disease risk (25,37,38,58,62,77,88,98,104,106)
Metabolic	Insulin insensitivity and glucose intolerance, and elevated plasma insulin levels leading to increased type 2 diabetes risk; increased obesity prevalence (109)
Respiratory	Increased chest wall stiffening and elastic recoil; pulmonary artery stiffening and increased pressure; reduced inspiratory and expiratory capacity; reduced pulmonary function (VC, FEV1); increased dead space volume; reduced peak ventilation by up to 35% (25,100).

Organ system	Effects
Nervous	Reduced stimulus for visceral arterial constriction for blood redistribution during exercise; reduced cardiac β-adrenergic stimulation; central nervous deficits leading to decreased cognition, memory, learning ability, reaction time, and sleep; altered gait and balance; increased risk of dementia and Alzheimer's; impaired hearing and sight (9,10,19,76,95)
Energy expenditure and energy intake	Decrease in daily energy expenditure attributable to decrease in resting metabolic rate related to a decrease in fat-free mass and increase in fat mass; low caloric and protein intake (80,81,82,92,109)
Thermoregulation	Decreased ability to regulate body temperature when homeostatis is challenged; decreased amount of sweat per active sweat gland; reduced response to increased blood flow during exercise attributable to structure and response of cutaneous blood vessels; inadequate ability to reduce splanchnic blood flow during exercise (50,51,91)

Note. VC = vital capacity; FEV_1 = forced expiratory volume in 1 s.

Table 7.2 Physiological Changes Associated With Aging

System	Change
Cardiovascular	Rest • Heart rate decreases • Systolic and diastolic blood pressure increases Maximum exercise • Heart rate decreases • Cardiac output decreases • Arterial-venous oxygen difference decreases • Stroke volume, no change • Oxygen consumption decreases • Cardiac and vascular responses to β-adrenergic stimulation decreases
Respiratory	Maximum exercise • Maximum ventilation decreases • Tidal volume decreases • Breathing frequency increases • Vital capacity decreases • Residual volume increases
Musculoskeletal	Muscle mass and strength decrease Elasticity in connective tissue decreases Balance decreases Coordination decreases Bone density decreases
Metabolic	Glucose tolerance decreases Insulin action decreases Metabolic rate decreases
Thermoregulation	Thirst sensation decreases Skin blood flow decreases Sweat production per sweat gland decreases

Adapted from H.C. Barry and S.W. Eathorne, 1994, "Exercise and aging. Issues for the practitioner," *Medical Clinics of North America* 78: 357-376; E.F. Binder et al., 1999, "Peak aerobic power is an important component of physical performance in older women," *Journal of Gerontology* 54A: M353-356; A.C. King and J.E. Martin, 1988, Physical activity promotion: Adoption and maintenance. In *American College of Sports Medicine's resource manual for guidelines for exercise testing and prescription,* 3rd ed. (Baltimore: Williams & Wilkins), 564-569; E.G. Lakatta, 1999, "Cardiovascular aging research: The next horizons," *Journal of American Gerontology Society* 47: 613-625; R.J. Spina et al., 2004, "Absence of left ventricular and arterial adaptations to exercise in octogenarians," *Journal of Applied Physiology* 497: 1654-1659; U.S. Department of Health and Human Services, 1996, *Physical activity and health: A report of the surgeon general* (Atlanta, GA: U.S. Department of Health and Human Services, Centers of Disease Control and Prevention, National Center for Chronic Disease Prevention and Health Promotion).

MEDICAL AND CLINICAL CONSIDERATIONS

As people age they become more engaged in the medical system for both prevention and treatment. The following section addresses these issues.

Chronic Medical Conditions

The most common chronic diseases present in the elderly are coronary artery disease, arthritis, hypertension, diabetes mellitus, and obesity. Each of these is covered in detail in other chapters of this textbook. Of these, hypertension is the most prevalent, coronary artery disease is the leading cause of death (24), and Alzheimer's disease is the ninth-leading cause of death (44).

History and Physical Exam

When necessary (i.e., for high-risk individuals), a qualified physician should perform a thorough medical history and physical examination. The American College of Sports Medicine recommends that men older than 45 yr of age and women older than 55 yr of age have a physical exam before beginning a program of **vigorous-intensity physical activity** (i.e., >70% $\dot{V}O_2$max) (3). The general methods of a medical evaluation are presented in chapter 4. For the elderly, emphasis should be placed on assessing specific areas of risk, including the cardiovascular and musculoskeletal systems (3,6,42). For an office evaluation of functional capacity for performing ADLs, questions about bathing, dressing, and getting in and out of bed should be asked. A standardized questionnaire, such as the Yale Physical Activity Scale, which is specifically designed for the elderly population, may be used to assess activity level (25).

Diagnostic Testing

Many older individuals have a history of falling, are **frail**, and have limited ability to perform ADLs. This circumstance does not necessarily preclude a person from exercise participation, but it does direct the focus of a preexercise examination. Agility, coordination, gait, and balance should be assessed to identify those at risk of falling. For example, difficulty standing up from a chair, visible difficulty with a slow walking speed, and poor ability to maintain balance during a push on the shoulder are predictors of falling (22,73). A goniometer can be used to obtain measurements of range of motion (ROM) in joints such as the hip, ankle, knee, shoulder, and elbow.

Limited ROM in these joints is related to limitations in performing ADLs. Table 7.3 reviews preexercise evaluation tests. Evaluation of mental (depression) and intellectual impairment (dementia, Alzheimer's disease) may also be made, because these may affect the ability to perform appropriate exercise training and maintain exercise compliance.

Exercise Testing

Low- to moderate-risk elderly people can participate in **moderate-intensity exercise**, defined as an intensity of 3 to 6 **metabolic equivalents (METs)**, or 40% to 60% peak $\dot{V}O_2$, without an exercise test (3). For screening, individuals should complete a questionnaire to identify any preexisting conditions that may preclude exercise (table 7.4). Based on responses, consultation with a physician or an exercise test may be recommended before the person begins an exercise program (31). If the person has one or more risk factors or any signs or symptoms of coronary artery disease, or if he or she wishes to begin a high-intensity exercise program, an exercise evaluation (physical exam and exercise test) should be performed (3). Specific exercise-testing methods are presented in chapter 5.

The exercise evaluation includes a review of the absolute and relative contraindications to exercise testing. Common contraindications in the elderly include elevated resting blood pressure (diastolic >115 mmHg or systolic >200 mmHg), moderate valvular heart disease; electrolyte abnormalities; complex ventricular ectopy; ventricular aneurysm; uncontrolled metabolic diseases such as diabetes; and neuromuscular, musculoskeletal, and rheumatoid disorders (3). Table 7.5 presents a summary of exercise-testing specifics.

The elderly are at increased risk of obesity, osteoporosis, hypertension, low aerobic fitness level, poor balance, ambulatory instability, neuromuscular incoordination, and vision and hearing impairment. The exercise professional must consider these when selecting an exercise-testing protocol (92,93). Walking is appropriate for most. Handrail support to provide balance assistance is important when elderly people use a treadmill. But using a handrail will artificially elevate the maximal attained MET level, and this effect should be taken into consideration. Cycling should be considered for those with balance or gait problems. Performing an exercise test during the morning hours may be best, because many people are less fatigued then than they are later in the day (86). The maximal attainable work rate during a graded exercise test for an older individual is likely less than 7 METs. A low-intensity testing protocol with small increments in work rate (1.0–2.0 METs per stage) is generally recom-

Table 7.3 Preexercise Training Evaluations

Test	Measurement	Outcome	Risk
Chair stand	Stand from a chair of standard height, unaided and without using arms	Ability Time required	Unable >2.0 s
Step-ups	Step-ups onto a single 23 cm step in 10 s	Ability Number of times	Unable <3 in 10 s
Walking speed	6 m walk	Time Number of steps RPE Heart rate Blood pressure Gait abnormalities such as asymmetry	<0.6 m · s^{-1}
Tandem walk	Walking along a 2 m line, 5 cm wide	Number of errors (off line, touching examiner or another object)	>eight errors
One-leg stand	Stand on one leg	Ability Time	<2 s
Functional reach	Maximal distance a person can reach forward beyond arm's length while maintaining a fixed base of support in the standing position	Inches	
Timed "up and go"	Stand up from standard chair, walk distance of 3 m, turn, walk back to chair, and sit down again	Time	>10 s
Range of motion	Using a goniometer, assess the following: shoulder abduction (SA), flexion (SF), extension (SE); elbow flexion (EF), extension (EE); hip flexion (HF), extension (HE); knee flexion (KF), extension (KE); ankle dorsiflexion (DF), plantar flexion (PF)	Degrees	<90° (SA); <150° (SF); <20° (SE); <140° (EF); <20° (EE); <90° (HF); within 10° (HE); <90° (KF); not within <10° full KE; unable to perform DF and PF

Note. RPE = rate of perceived exertion.

Adapted from E.F. Binder et al., 1999, "Peak aerobic power is an important component of physical performance in older women," *Journal of Gerontology* 54A: M353-356; W.J. Chodzko-Zajko and K.A. Moore, 1994, "Physical fitness and cognitive function in aging," *Exercise in Sport and Science Review* 22: 195-220; R.J. Kuczmarski et al., 1994, "Increasing prevalence of overweight among U.S. adults," *Journal of the American Medical Association* 272: 205-211; M.C. Nevitt et al., 1989, "Risk factors for recurrent non-syncopal falls," *JAMA* 261: 2663-2668; M.L. Pollock et al., 2000, "Resistance exercise in individuals with and without cardiovascular disease: Benefits, rationale, safety, and prescription. An advisory from the committee on exercise, rehabilitation, and prevention, council on clinical cardiology, and the American Heart Association," *Circulation* 101: 828-833.

mended. Popular protocols that fit these parameters include the Naughton and Balke protocols, as well as cycling increases of 25 to 30 W per stage. Ramping-type protocols can also be considered. Chapter 5 provides details about exercise-testing protocols.

Peak heart rate, stroke volume, cardiac output, and $\dot{V}O_2$ are lower in the elderly than they are in younger people. Potential mechanisms include cardiac **β-recep-** **tor** down regulation (75), physical inactivity, and skeletal muscle morphological changes. And because maximal power output is lower in the elderly, at any absolute submaximal work rate an elderly person is closer to anaerobic threshold compared with someone who is younger (83,92). Cardiac output at maximal exercise is generally 20% to 30% lower compared with younger individuals (75).

Table 7.4 Exercise Questionnaire

Questions	Yes	No
A. Do you get chest pains while at rest or during exertion?		
B. If the answer to question A is yes, is it true that you have not had a physician diagnose those pains yet?		
C. Have you ever had a heart attack?		
D. If the answer to question C is yes, was your heart attack within the last year?		
E. Do you have high blood pressure?		
F. If you don't know the answer to question E, was your blood pressure reading more than 150/100?		
G. Are you short of breath after extremely mild exertion and sometimes even at rest or at night in bed?		
H. Do you have any ulcerated wounds or cuts on your feet that do not seem to heal?		
I. Have you lost 10 lb (4.5 kg) or more in the past 6 months?		
J. Do you get pain in your buttocks or the back of your legs—thighs and calves—when you walk?		
K. While at rest, do you frequently experience fast irregular heartbeats or, at the other extreme, very slow beats? (Although a low heart rate can be a sign of an efficient and well-conditioned heart, a very low rate can indicate a complete heart block.)		
L. Are you currently being treated for any heart or circulatory condition such as vascular disease, stroke, angina, hypertension, congestive heart failure, poor circulation in the legs, valvular heart disease, blood clots, or pulmonary disease?		
M. As an adult, have you ever had a fracture of the hip, spine, or wrist?		
N. Did you fall more than twice in the past year (no matter what the reason)?		
O. Do you have diabetes?		

Key: If the answer to any of these questions is yes, an individual should be advised to undergo an evaluation by a physician.

Reprinted, by permission, from W.J. Evans, 1999, "Exercise training guidelines for the elderly," *Medicine and Science in Sports and Exercise* 21: 12-17.

Table 7.5 Exercise Testing Specifics for the Elderly

Test type	Mode	Protocol specifics	Clinical measures	Clinical implications	Special consideration
Cardiovascular	Treadmill or ergometer	Low intensity with small increments in work rate, steady state or ramp	Estimated METs or peak $\dot{V}O_2$, heart rate, ECG, blood pressure	Describing or determining exercise prescription	Mornings may be better, high incidence of undiagnosed heart disease
Strength	Weight machines	Modified 1RM with focus on muscles of ADL	Weight lifted, repetitions		Agility, balance, coordination, and gait deficits may affect safety and ability to perform testing
Range of motion	Goniometer	Hip, ankle, knee, shoulder and elbow, lower back and hamstrings	Degrees of motion	Limited ROM in these joints is related to limitations in performing ADL	Evaluation of mental and intellectual impairment as it may affect ability to perform testing

Mental deficiencies such as dementia or Alzheimer's disease may limit the ability of an individual to comprehend directions during a graded exercise test. These individuals should be tested in a comfortable environment, preferably with a familiar person present. Similar recommendations can be made for those who have hearing or vision impairments. For those people, a clear description by video or audio is recommended (2,3).

With regard to strength and ROM testing, general testing procedures, such as the one-repetition maximum (1RM) test, can be used. But because a 1RM test may increase the risk of injury, the exercise professional may administer the modified 1RM evaluation or simply provide general recommendations for low-resistance strength training without prior strength assessment in most elderly individuals. Using weight machines is best, because they reduce balance requirements and the risk of injury. The focus of evaluation should include those muscles used for typical ADLs, such as the hip extensors, flexors, and abductors; knee extensors and flexors; ankle dorsiflexors and plantar flexors; and shoulder extensors, flexors, and abductors (53).

Treatment of Chronic Medical Conditions

Most elderly people have some type of chronic disease such as hypertension and heart disease. They are limited in their ability to perform ADLs. The older the individual is, the more prevalent these conditions tend to be. On the other hand 37% of noninstitutionalized older people rate their health as excellent or very good (1), but this may be a reflection of their increasingly sedentary lifestyle to counteract their debilitation. Although it is known that the elderly, regardless of age or physical condition, can benefit from regular physical activity, few of them are regularly active (figure 7.1). Changing to an active lifestyle can be a challenge but doing so has the potential to preserve skeletal muscle strength and endurance, enhance body composition, improve the ability to perform ADLs, and enhance the quality of life. Additionally, activity can potentially reduce the risk of complications from established chronic disease. Other chapters in this text provide specific exercise recommendations and clinical treatment modalities for individuals with chronic diseases such as heart disease, diabetes, arthritis, and cancer.

EXERCISE PRESCRIPTION

Until recently, for most elderly individuals the 1995 joint recommendation of the American College of Sports Medicine (ACSM) and the Centers for Disease Prevention was appropriate (78,106). But in 2007 the ACSM and the American Heart Association (AHA) published an updated recommendation on physical activity that is specific to older adults (72). Specifically, this special report states the following:

Regular physical activity, including aerobic activity and muscle-strengthening activity, is essential for healthy aging. This preventive recommendation specifies how older adults, by engaging in each

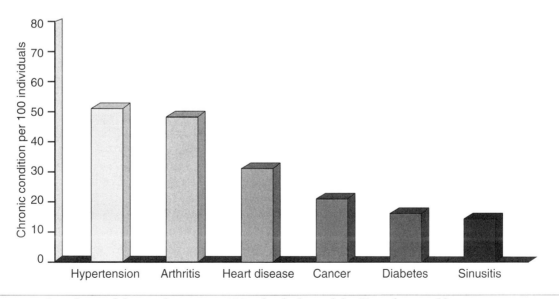

Figure 7.2   Prevalence of chronic diseases per 100 individuals in adults 65 yr of age or older.
Adapted from Administration on Aging, *Profile of older Americans,* 2005.

recommended type of physical activity, can reduce the risk of chronic disease, premature mortality, functional limitations, and disability. (72)

Given this preamble the following sections present specific recommendations for aerobic and resistance

training. The report also specifically states that these exercise recommendations are in addition to general recommendations to live a more active life.

Practical application 7.1 discusses client–clinician interaction for elderly clients who are participating in supervised exercise training.

Client–Clinician Interaction

Many older individuals may not have a spouse, close children, or friends to rely on for socialization, assistance, and support (30). Although social relationships may change with age from family to more formalized organizations or nonfamily members, many elderly live in social isolation and are lonely. This circumstance is important because epidemiological studies demonstrate a relationship between social support and both mental and physical health (18,38,108). Furthermore, several studies demonstrate lack of social support to be a major risk factor for depression, morbidity, and mortality (11,43,76,77,90,99,102).

Participation in an organized physical activity program provides an excellent opportunity for interaction and forming social networks, which in turn can contribute to feelings of well-being and improve quality of life and physical performance (74,82). Furthermore, a favorable attitude on the part of significant others (i.e., spouse, friends) toward an exercise program is frequently associated with better compliance with the program. The clinical exercise professional can help by identifying persons who require social support. In a structured exercise setting, the exercise professional should also be prepared to provide social support to these individuals by making conversation.

Exercise programs that offer either supervised or community-based sessions can increase social interaction for elderly persons (43,51). Examples of methods to improve social interaction for the elderly who are participating in an exercise program include a "buddy" exercise system, in which individuals are matched up with those of similar ability, age, and interest when performing exercise; social occasions including picnics, holiday parties, and birthday celebrations; and placement of exercise equipment that allows people to face each other. Each of these opportunities can be facilitated by an exercise professional responsible for programming in the structured exercise setting (e.g., cardiac rehabilitation programs and senior community centers).

Because many older individuals have at least one chronic disease and are sedentary or only minimally active, beginning an exercise program with intermittent bouts of moderate-intensity activity may be best. This approach will allow most people to achieve early success with an exercise regimen and may improve compliance by reducing pain, fatigue, and injuries. A goal should be to work toward 30 min of continuous exercise. Additionally, persons beginning an exercise program should be instructed on proper exercise technique and training load increases to limit their susceptibility to injury. The clinical exercise professional should also stress that regular physical activity can enhance quality of life. The elderly person may find it difficult to perform physical activity or exercise at home because of lack of space or equipment. Alternatives include senior citizen and community centers, health clubs and church facilities, shopping malls, and community swimming pools. These locations may also provide fellowship that is important for the single person. Social activities that may be of interest to the elderly include an organized walk or bicycle tour or competitive events such as the Senior Olympics or masters swimming (2,6). Whatever facility is used or event is performed, it must be safe and free from barriers. Additionally, many older people, especially those with a low fitness level, have a low tolerance for heat and cold. Thus, environmental conditions, fluid intake, and clothing are important considerations (5,50,92).

Exercise Recommendations

In general, an active lifestyle preserves and enhances skeletal muscle strength and endurance, flexibility, cardiorespiratory fitness, and body composition, which otherwise would diminish and increase chronic disease risk with advancing age. The exercise professional should therefore develop exercise interventions for the elderly that increase daily physical activity levels (i.e., energy expenditure) and thereby delay or counteract the effects of risk factors on chronic disease. Practical application 7.2 provides specific exercise recommendations for the elderly.

Cardiovascular Training

The ACSM–AHA special report on physical activity for older adults recommends that they participate at least 5 d per week for 30 min at moderate intensity (<70% heart rate reserve or 5 to 6 on a 10-point exertion scale) or at vigorous intensity on 3 d per week for 20 min. Standard exercise training guided by heart rate or work rate may also be employed by the elderly (72).

Low cardiorespiratory fitness is a risk factor for cardiovascular disease and all-cause mortality (10). Low $\dot{V}O_2$peak is associated with reduced ability to perform ADLs including climbing stairs and brisk walking (7). A small improvement in cardiovascular fitness is associated with a lower risk of death and in low-fit people can substantially increase their functional abilities (29). Healthy, sedentary older men and women can increase their cardiorespiratory fitness by performing aerobic exercise training (27,32,54,64,75,95,101).

Common physical activities that the elderly population should engage in are walking (outdoors, indoors, or treadmill), cycling, gardening, swimming (laps or water aerobics), and golfing while walking with a pull cart (106). Those who are unable to perform ambulatory activities may be candidates to perform seated chair activities, stationary cycling, and water activities. Tai chi, Pilates, yoga, and similar activities may also be beneficial to improve strength and balance (57,87). High-intensity activities such as running, rowing, using aerobic or gravity riders, or using stair steppers may not be appropriate. Low- to moderate-intensity exercise programs can be performed daily. Higher intensity or vigorous exercise sessions (>70% heart rate reserve or 7 to 8 on a 10-point exertion scale) should be performed only 3 to 5 d per week (3) to allow for recovery days, which is important for the elderly because they recover more slowly from exercise than younger people do. Older individuals with low exercise capacity may benefit from multiple daily sessions of short duration, whereas the more capable person can benefit from three

sessions per week with exercise bouts performed once per day (3,5). The appropriate intensity can be prescribed using the heart rate reserve, heart rate percent, or rating of perceived exertion methods (see chapter 5).

The healthy older individual should perform exercise for at least 30 min, but exercising longer than 60 min should likely be discouraged because of possible diminishing returns for general health benefits. The optimal dose of exercise and physical activity has not been completely clarified and likely varies in subgroups of individuals based on genetic endowment, age, sex, health status, body composition, and other factors. If a person beginning an exercise program is predominantly sedentary, has severe chronic disease, or has an extremely low fitness level, a minimum of 30 min of continuous exercise may not be possible. Sessions of as little as 10 min, performed two or three times a day are appropriate in this situation (72). This approach still provides general health benefits (3). During the initial stages of an exercise program, the intensity should be low (40–60% heart rate or $\dot{V}O_2$ reserve). This period is important for promoting adherence to physical activity. Intensity and duration can be increased every 2 to 3 wk, as tolerated, until the desired total exercise volume is reached. The exercise professional working with elderly clients must recognize indicators of musculoskeletal or orthopedic injury, boredom, and decreased exercise tolerance, which may lead to cessation of exercise training (5,42,50).

Resistance Training

Strength training has the potential to improve functional capacity and quality of life of the elderly person (34). In general, most elderly individuals can participate in a resistance-training program that is individually designed. Before initiating a resistance-training program, those with hypertension or arthritis or at risk of osteoporotic fracture should be evaluated by a physician (12,31). The ACSM–AHA recommendation is to perform 8 to 10 different resistance exercises, for 10 to 15 repetitions on at least 2 nonconsecutive days per week (72). The recommended intensity level, using the 10-point exertion scale, should range from 5 to 6 for moderate efforts and 7 to 8 for vigorous efforts (72). Interestingly, the ACSM–AHA recommendation does not advise the number of sets to be performed. A reasonable suggestion is to keep the number of sets per session to no more than two or three. Also, some evidence indicates that increasing the number of repetitions to 20 and decreasing the amount of resistance lifted may promote muscle endurance (31).

Free weights can be used if individual supervision is provided. Exercise with free weights, however, has an

Exercise Prescription for the Elderly

Mode	Frequency	Intensity	Duration	Special considerations
Aerobic training				
More often: walking, cycling, pool activity, seated aerobics, ADLs. Less often: jogging, swimming laps, rowing, aerobic dance	Moderate-intensity exercise three times per week and vigorous-intensity exercise five times per week. General physical activity should be performed daily	ADLs at comfortable pace. Low intensity at 40% HRR or <5 on the 10-point exertion scale. Moderate intensity at 50–70% HRR or 5–6 on the 10-point exertion scale. Vigorous intensity at >70% HRR or 7–8 on the 10-point exertion scale	Goal is 30 min continuous and up to 60 min for low and moderate intensities, but only 20 min for vigorous-intensity exercise. Intervals may be as short as 10 min	May need to start with short bouts initially and build to 30 min continuously Higher intensity may be difficult, especially for those who are sedentary Consider beginning at low to moderate intensity Progress duration and frequency initially, because this approach may improve compliance Comorbidities such as arthritis, osteoporosis, and heart disease need to be considered See relevant chapters for further special considerations
Resistance training				
More often: multistation machine type (e.g., Universal), elastic bands, hand weights Less often: free weights	Two or more times per week	5–6 (moderate) or 7–8 (vigorous) on the 10-point exertion scale	10–15 repetitions for strength gains 20–30 min per session Consider up to 20 repetitions to improve endurance	Free weights may be difficult for a person to balance, so assistance should always be available Focus on major muscle groups that are used to perform ADLs (shoulders, legs). See chapters that review heart diseases, osteoporosis, and arthritis for further special considerations
Range-of-motion training				
Static stretching: See chapter 5 for examples Balance training to reduce falling risk	Minimally two times per week Maximally daily, but especially following an aerobic or resistance-training session	Subject should feel a mild stretch without inducing pain. Progress range of stretch based on lack of discomfort experienced No specific balance recommendations	5–30 min total, with two 30 s bouts on each muscle group. Involve all major muscle groups (neck, shoulders, arms, lower back, quadriceps, hamstrings, calves, ankles) No specific balance routines are recommended, but yoga or tai chi may be beneficial	Avoid ballistic movements and the Valsalva maneuver during the stretching routine. A brief routine performed before or after aerobic or resistance exercise may only focus on the muscle groups used and last as little as 5 min. Consider using alternative methods such as seated movements and use of a towel or elastic band to assist those with difficulties performing standard flexibility training. An overall routine can last up to 20–30 min. See chapters that review osteoporosis and arthritis for further special considerations

Note. ADL = activities of daily living; HRR = heart rate reserve; RM = repetition maximum.

inherent stability component that may make exercising in this way difficult for some elderly persons. If no weight equipment is available, milk jugs filled with sand, cans of food, or elastic bands can be used for resistance (31,104). Exercise professionals should be able to explain the use of these items to elderly individuals. Proper technique is important to limit the risk of injury. Correct breathing patterns should be taught so that participants avoid the **Valsalva** maneuver (i.e., breath holding), which can reduce venous return and increase blood pressure. At least 48 h of rest should occur between sessions for optimal recovery to reduce the risk of injury (31,60,82,106).

A standard program (see chapter 5) should be followed: two to three sets of a selected weight that can be lifted 8 to 12 times (31,60). The resistance-training session should last between 20 and 30 min. Progress should be monitored, and progression of intensity and resistance can occur every 2 to 3 wk.

Range of Motion

Both flexibility (ROM) and balance training are recommended in the ACSM–AHA report (72). The recommendation is to perform these activities at least 2 d per week for at least 10 min each time. Certainly, these exercises can be performed as often as daily. Performing them in conjunction with aerobic or resistance training may be useful. The objective of ROM exercise is to increase or maintain joint ROM and decrease stiffness, and balance training can reduce the risk of falling in community-dwelling individuals. These goals are important for the elderly, because improved ROM and balance can promote prolonged independent living. Besides traditional ROM exercises, activities such as yoga, dancing, and tai chi can be performed (15). Before performing an ROM routine, the participant should begin with a warm-up period of light-intensity aerobic-type exercise. ROM exercises should then focus on all the major joints of the body (i.e., hip, back, shoulder, knee, trunk, and neck region). The degree of stretch can increase as the person can tolerate. Chapter 5 reviews general recommendations.

EXERCISE TRAINING

In healthy, sedentary elderly men and women, $\dot{V}O_2$peak can increase 10% to 15% if moderate-intensity (40–50% $\dot{V}O_2$peak) training is performed. Greater improvements of up to 30% are observed when exercise training is performed 3 or more days per week, for 30 to 45 min at an intensity greater than 50% of $\dot{V}O_2$peak (27,54,94,95). In

elderly men, increases in $\dot{V}O_2$peak resulting from endurance training are attributable to increased cardiac output and a widening of the arteriovenous oxygen difference. Increases in cardiac output are attributable entirely to an increased stroke volume. Maximal heart rate does not change, or decreases slightly, with endurance training (92,101). Increases in stroke volume are mediated by left ventricular volume overload **hypertrophy** (27) and increased blood volume, which enhances the Frank-Starling response. The increase in $\dot{V}O_2$peak in older women observed after a 9 mo training program was attributable to a widened arteriovenous oxygen difference without hemodynamic improvement (95). Smaller improvements in $\dot{V}O_2$peak (12%), without improvements in LV funtion and hypertrophy, are observed in octogenarians after exercising for 13 mo at 83% peak heart rate (96).

Aerobic training also improves insulin sensitivity and glucose homeostasis in the elderly (19,20,46). Insulin action and skeletal muscle **GLUT 4** concentrations improved in elderly people who used a cycle ergometer at 70% of $\dot{V}O_2$peak for 60 min on 7 consecutive days (20). These changes may reduce the risk of type 2 diabetes or help control blood glucose levels. Improvement in mitochondrial function and content of mitochondria with exercise training (63) may also enhance insulin action.

Although many studies demonstrate improved cardiorespiratory function following moderate- to high-intensity exercise training, this type of training may not always be the best choice. For instance, when an elderly group participated in a structured physical activity program, spontaneous physical activity during the remainder of the day decreased (41,65). Combining strength and endurance training is also beneficial for elderly people. One study showed that after 6 mo of combined resistance and endurance training, older healthy individuals increased their $\dot{V}O_2$peak (11%) and their upper- and lower-body strength (21). The ability to carry out normal daily tasks such as carrying groceries, transferring laundry, vacuuming, making a bed, climbing stairs, and floor sweeping improved and translated to carrying 14% more weight and moving 10% faster. The amount of improvement depended on the initial conditioning level, with the least fit demonstrating the most improvement (21).

Elderly individuals, including the oldest old and very frail elderly, demonstrate physiological adaptations to strength training (7,34,35,38,45,105). The extent of adaptation depends on the frequency, volume, mode, type of training, and initial training state (31). Resistance-training programs lasting from 8 wk to 1 yr can increase muscle strength and mass in the elderly, regardless of age and sex.

Increases in strength are the result of both muscle hypertrophy and neuromuscular adaptation (7,105). Several studies reported large increases in skeletal muscle strength in the elderly that are proportionally greater than increases in cross-sectional area (34,35,37). Besides promoting strength gains, regular resistance training improves gait, balance, and overall functional capacity (14,34,45). Strength training also increases bone mineral density and content (71), increases metabolic rate (16), assists with maintenance of body weight by decreasing fat mass and increasing lean mass, and improves insulin action and plasma levels (66). The gains observed during resistance training will persist for several weeks after training ends (58).

Several studies demonstrated that regular ROM training improves the flexibility of the spine, hips, ankles, knees, and shoulders in older individuals (61,62,67–70,85), whereas few have shown no effect (28). Improvements in flexibility can increase the effective range of strength gains and improve ambulatory ability.

CONCLUSION

More than half of elderly people have at least one disability or chronic condition. Participation in a regular physical activity or an exercise program has many physiological health benefits including reducing the risk and lessening the effect of many chronic diseases. Physical activity of light to moderate intensity helps improve health, whereas moderate- to high-intensity physical activity with an emphasis on aerobic endurance improves cardiorespiratory function ($\dot{V}O_2$) as well as health. Elderly individuals also demonstrate improvements during resistance training by increasing muscle mass and strength. This type of exercise improves gait, balance, overall functional capacity, and bone health. Regular physical activity and exercise also produces psychological benefits. In general, elderly people can improve physical and mental health by performing regular physical activity, and all medical and exercise professionals should encourage exercise and physical activity.

Case Study

Medical History

Mrs. KA is a 70 yr old white female who has a history of falling and osteoarthritis in her hands. She has been a homemaker for most of her life and has never performed any type of regular physical activity. She is widowed and lives by herself. She has smoked half a pack of cigarettes a day for the past 40 yr. Her body mass index is 30, and her blood pressure is 140/76 mmHg.

Diagnosis

Mrs. KA went to see her doctor with complaints about her inability to complete normal daily living tasks such as carrying bags, walking around the block without worry of falling, and going up and down her basement stairs. She was unable to stand from a chair of standard height without using her arms and was unable to step up onto a 23 cm step. She also said that she felt lonely during most days of the week. A resting electrocardiogram showed no abnormalities. No significant diagnostic findings were made. The physician recommended that she begin an exercise routine and ordered an exercise test.

Exercise Test Results

The patient underwent an exercise test. Her electrocardiogram demonstrated isolated premature ventricular contractions and normal ST segments at peak exercise. The following values were noted:

- Resting heart rate = 84
- Resting blood pressure = 138/80
- Peak heart rate = 145
- Peak blood pressure = 220/90
- Estimated $\dot{V}O_2$ peak = 13 ml · kg^{-1} · min^{-1}

Development of Exercise Prescription

Mrs. KA wishes to become stronger so that she can perform general ADLs and remain independent. Her physician referred her to the local YMCA to join the elderly adult fitness program. She went somewhat reluctantly but after consulting with the exercise professional on staff, she set a goal of becoming more active on a regular basis to maintain her independent living status.

Discussion Questions

1. What is an appropriate exercise activity and exercise intensity for Mrs. KA?
2. What are the major concerns for Mrs. KA during exercise activity?
3. Based on Mrs. KA's medical history and living situation, what are important barriers to her continued participation in the physical activity program?
4. What special precautions were likely taken during her exercise test?
5. What motivational strategies can be used to ensure her exercise adherence?
6. What are the major risks of Mrs. KA's participation in an exercise program?
7. What other tests might you perform on Mrs. KA (diagnostic or functional)?
8. What might be appropriate recommendations to progress Mrs. KA to walking three times a week for 30 min or on a daily basis?

Female-Specific Issues

Farah A. Ramírez-Marrero, PhD, MSc

Around the world, women's involvement in physical activity, regular exercise, and sports has increased tremendously during the past several decades. An emphasis on the health benefits of physical activity and exercise is partly responsible for this change. Another reason is the influence of media attention to women's sports, especially in the United States with the passage of the Title IX legislation in 1972, which helped reduce societal prejudices and obstacles for females to participate in sports. Young girls now find sports-related female role models who are physically fit, strong, and competitive. Additionally, research consistently confirms the multiple physiological and psychological benefits of increased physical activity, exercise, and sports participation among women of all ages.

As the number of women involved in vigorous exercise training and competition increases, so has the awareness and concerns about potential menstrual cycle, reproductive function, metabolic, thermoregulatory, and orthopedic disturbances increased among women who exercise train. **Disordered eating** behaviors, **amenorrhea**, and **osteoporosis** are clinical conditions linked to what has been known for years as the female athlete triad. In 2007 the American College of Sports Medicine (ACSM) updated their position stand on the triad (1), now described as three interrelated spectrums: optimal energy availability to **low energy availability** (with or without eating disorders), eumenorrhea to amenorrhea, and optimal bone health to osteoporosis. This chapter emphasizes the clinical considerations of the female athlete triad, although it also discusses other clinical aspects pertaining to females.

FEMALE-SPECIFIC ISSUES

Menstrual cycle changes during intense training and competition were reported by elite female athletes as early as the 1950s and 1960s (36). Many female athletes welcomed the reduced frequency or cessation of their menses and justified it as a natural, harmless result of their intense exercise training. Health professionals believed that these changes were temporary and, therefore, were not concerned about long-term health consequences. But young female athletes and nonathletes who experience the cessation of their menstrual cycle have an increased risk of low **bone mineral density** (BMD) (17). In addition, many female athletes and coaches believe that low body weight is necessary for success in certain sports. The attempt to achieve unrealistically low body weight may explain the disordered eating behaviors sometimes observed in female competitive athletes.

Disordered eating behaviors and the consequential low energy availability increase the risk of developing menstrual cycle disturbances and eventual bone demineralization. Reductions in BMD at a young age may never be regained and may result in premature risk of osteoporotic fractures. In 1992 the ACSM Women's Task Force addressed the three interrelated medical disorders known as the female athlete triad. The triad include disordered eating (e.g., anorexia nervosa, bulimia nervosa, eating disorder not otherwise specified, or other subthreshold disordered eating behavior), amenorrhea (i.e., absence or suppression of menstruation), and osteoporosis (i.e., reduction in bone mass density). The ACSM published a position stand in 1997 to establish that the female athlete triad was a serious syndrome that affected sports performance and caused medical and psychological morbidity and mortality, and that the triad could affect girls and women who participate in a variety of physical activities (2). The Medical Commission of the International Olympic Committee also published a position stand on the female athlete triad (50) that recognized the physical and psychological health risks of the triad, and highlighted the importance of managing, treating, and preventing its potential consequences. More recently the ACSM published an update of their position stand of the triad (1). Their recommendations for diagnosis, prevention, and treatment of the female athlete triad are included in this chapter.

Note that the female athlete triad is not exclusively a problem of female athletes. Although the triad spectrum of low energy availability leading to amenorrhea and reduced BMD is more likely to occur in athletic and nonathletic females obsessed with the attainment of unrealistically low body weight, the spectrum of low energy availability leading to potential risk of low BMD is also observed among male athletes and nonathletes who engage in disordered eating behaviors to purposely reduce body weight (9). Male jockeys who train and compete in horse-racing events are probably the most extreme example of male athletes who pursue unrealistically low body weight.

Low energy availability could be the result of an imbalance between energy intake and energy expenditure, probably resulting from high levels of exercise training. Some endurance athletes may have difficulties consuming enough calories to balance the energy expenditure associated with sports training. Therefore, they unintentionally experience transient periods of low energy availability that are not necessarily associated with the triad syndrome and its health consequences (30). Most of the available data on the female athlete triad are cross-sectional and includes the limitation that potentially confounding preexisting factors are usually not taken into consideration (30). There is a need for prospective data that link amenorrhea or disordered eating with osteoporotic fractures, infertility, or cardiovascular mortality in former female athletes later in life. The components of the triad will be discussed here, considering all these perspectives and the awareness that the vast majority of female and male athletes have multiple health benefits because of their engagement in sports training and competition.

Low Energy Availability With or Without Disordered Eating

Transient states of negative energy balance are required for weight-loss programs. A negative energy balance could be achieved by increasing energy expenditure through exercise or by reducing energy intake through dietary modifications. After the desired body weight is reached, dietary intake is normalized to achieve energy balance and maintenance of body weight. Ideally, trained professionals supervise weight-loss programs to prevent excessive caloric imbalance or nutrient deficiencies that lead to clinical manifestations. But negative energy balance and low energy availability could be present when athletes intentionally or unintentionally engage in high levels of energy expenditure through exercise training and insufficient dietary energy consumption. These practices usually escape diagnosis and professional supervision and, when sustained for relatively long periods, can increase the risk of developing amenorrhea and low BMD.

Losing body weight can become an ongoing, frustrating struggle for women who constantly strive to achieve what they believe is the "right" body shape for sport or life. Abnormal eating behaviors could be the result of multiple attempts to reach unrealistically low body weight or body fat, and are defined as a subclass of multiple abnormal eating practices (106). These practices include fasting, diet pills, laxatives, diuretics, vomiting, and binge eating, and are related to the disorders of anorexia nervosa and bulimia nervosa (2).

Anorexia nervosa is a psychological disorder characterized by an exaggerated pursuit of thinness or an obsessive fear of becoming obese. The disorder is characterized by a constant energy deficit that results in a considerable loss of body weight (i.e., at least 15% below ideal body weight) and fat, and a refusal to maintain a minimally normal body weight for age and height (25,53,93). Females who are anorexic are at risk of developing amenorrhea. Amenorrhea is one of the diagnostic criteria cited by the American Psychiatric Association (DSM IV) for anorexia nervosa (53).

Bulimia nervosa is characterized by recurrent episodes of binge eating, typically followed by purging, which may be self-induced vomiting, use of laxatives, or use of diuretics (25,91,97). People with bulimia experience a fear of becoming obese. But unlike anorexia nervosa, bulimia may not be accompanied by weight loss. Even if purging does not occur, fasting or excessive exercise is usually present as a compensatory behavior (60,100). Some experts classify bulimia as a variant of anorexia nervosa.

Another classification of disordered eating behavior is known as eating disorder not otherwise specified (EDNOS). This classification is used when a disordered eating behavior is present but criteria for a specific eating disorder are not (50). For example, EDNOS is diagnosed in a female who has the criteria for anorexia nervosa but has regular menses and body weight in the normal range. The American College of Physicians has estimated that 10% to 15% of adolescent girls and young women have clinically evident anorexia or bulimia nervosa (25). Others have estimated that in adolescent and young adult women, the prevalence of anorexia nervosa and the prevalence of bulimia are approximately 1% and 2% to 4%, respectively (53).

Published reports suggest that female athletes have a similar or higher prevalence of disordered eating behaviors compared with the general female population, with a range of 15% to 62% for selected sports (2,25,65,85). In contrast, an assessment of the prevalence of disordered eating among 1,445 females competing in Division II of the National Collegiate Athletics Association (NCAA) revealed that no athlete met the established clinical criteria for anorexia nervosa and only 1% met the criteria for bulimia (52). This study suggested, however, that some female student-athletes have subclinical indicators associated with disordered eating behaviors (i.e., 9% bulimia and 3% anorexia nervosa), recently referred to as anorexia athletica.

Bachner-Melman et al. (11) tested the risk of developing anorexia athletica by comparing anorexic symptoms and personality profiles of four groups of women: anorexic, aesthetic athletes, nonaesthetic athletes, and nonathletic controls. The aesthetic athletes did not show evidence of anorexic symptoms or negative attitudes toward eating or weight, and their personality profiles were not different from the nonaesthetic athletes and controls. The authors suggested that sport participation more likely influences the general well-being and psychological health of female athletes.

Moreover, Reinking and Alexander (83) found that female athletes competing in NCAA Division I were not at a higher risk of disordered eating symptoms compared with women not competing in collegiate sports (7% versus 13%, respectively). Also, a lower prevalence of disordered eating behavior among elite female athletes (20.2%) compared with nonathletic controls (37%) was reported by Klungland Torstveit and Sundgot-Borgen (54). But these authors also reported a higher proportion of elite female athletes (27%), particularly those competing in leanness sports, having both disordered eating and menstrual dysfunction compared with nonathletic controls (14%). This result could be associated with a higher energy drain caused by the combination of low energy intake and high energy expenditure in the female athlete group.

Some recognized sociocultural, biological, and psychological factors associated with the development of disordered eating behavior are presented in table 8.1 on p. 152. Ethnic factors may play a role, because Caucasians and Hispanics appear to be at higher risk of disordered eating behaviors than African American adolescent females are (85). Strong societal and cultural influence on human behavior may cause many young females to connect thinness with success, beauty, and power. A successful and confident woman, as observed on television and in magazines, is represented by a tall and thin, sometimes skeletal-looking, female. In addition, the need to be accepted and liked by family members and friends combined with low self-esteem may be important triggering factors for disordered eating behaviors.

Female athletes with disordered eating behaviors are more likely to be highly competitive and have a perfectionist attitude. The particular sport in which they train and compete might also influence the development of disordered eating behaviors. Female athletes who compete in aesthetic sports (e.g., dancing, acrobatics, gymnastics, cheerleading, diving, artistic swimming, and figure skating) and are diagnosed with anorexia athletica often have abnormal eating and body image symptoms, although their symptomatology is not associated with eating disorders (11). Sports that require weight classifications and sports in which the athlete's appearance is evaluated by a panel of judges are more likely to include athletes at risk of developing eating disorders (53). Potentially high-risk sports, not arranged in any particular order, include the following (106):

Dance	Volleyball
Figure skating	Swimming
Gymnastics	Diving
Distance running	Horse racing
Cycling	Martial arts
Cross-country skiing	Rowing

Amenorrhea

Amenorrhea is classified as primary or secondary depending on the conditions in which it develops. **Primary amenorrhea** was previously defined as no menarche by age 16 or no sexual development by age 14 (1,2,70,106).

Table 8.1 Factors Associated With Disordered Eating Behavior

Sociocultural	Biological	Psychological	Sports related
Thinness equals success	Abnormal neurochemistry controlling sense of hunger	Dysfunctional family	Pressure to reach a specific weight or low body fat to optimize performance
Thinness equals beauty		Physical or sexual abuse	Lack of nutritional knowledge
Thinness equals power		Low self-esteem	Emphasis on lean appearance that is evaluated by judges
		Lack of identity	Weight classification
		Competitiveness	High self-expectations
		Perfectionism	Repetitive exercise routines
		Compulsiveness	
		Distorted body image	
		Tendency for depression	
		Extreme preoccupation for weight or diet	

The revised definition of primary amenorrhea includes the absence of menstrual cycles by age 15 in previously nonmenstruating girls, even when other normal postpubertal development is present (82). Secondary amenorrhea is generally diagnosed when menstrual bleeding has not occurred for at least three to six consecutive menstrual cycles in women who already had at least one previous menstruation (1,2,106). Practical application 8.1 reviews the literature pertaining to mechanisms of low energy availability, amenorrhea, and exercise.

Scope

Women who perform vigorous exercise training or restrict energy intake resulting in low energy availability could be at risk of developing menstrual cycle disturbances. Table 8.2 lists various types of menstrual disturbances. Females who train and compete in sports that require strenuous physical training and who are pressured to maintain low body weight or a lean body physique are at increased risk of developing menstrual cycle disturbances. The prevalence of amenorrhea in the general population is estimated between 2% and 5%, whereas in women engaged in vigorous exercise training the prevalence is estimated between 5% and 46% (44). Among elite runners and professional ballet dancers, the prevalence is estimated at 40% and 66%, respectively (70,80).

Taking into consideration that athletic amenorrhea is more likely to occur in the presence of low energy availability, a look into the prevalence of both conditions is warranted. In a unique assessment of all Norwegian female athletes compared with nonathlete controls, the prevalence of menstrual cycle disturbances was determined to be 31% and 25%, respectively (54). The same authors reported a 27% prevalence of both disordered eating and menstrual dysfunction among female athletes, particularly those competing in leanness sports, compared with 14% prevalence in nonathletes (55). But when all the components of the triad were considered, no differences were observed between the elite athletes and nonathletes (4.3% versus 3.4%, respectively).

Note that the prevalence of menstrual cycle disturbances in athletic and nonathletic females is likely underreported because abnormal luteal phases and anovulatory cycles are not always obvious, because these can occur even when menstrual bleeding is present.

Physiology and Pathophysiology

A normal menstrual cycle occurs when there is precise synchronization of hormonal events in the hypothala-

Table 8.2 Types of Menstrual Disturbances

Type	Definition
Delayed menarche	Menstruation not started by age 16
Shortened luteal phase	Duration less than normal 10 to 16 d
Anovulatory cycles	Menstrual cycle without egg release
Oligomenorrhea	Irregular or inconsistent menstrual cycles
Amenorrhea	Complete cessation of menstrual cycle

Literature Review of Low Energy Availability, Amenorrhea, and Exercise

In a series of studies, Loucks and coworkers (39,40,42,43) evaluated the independent and combined effect of exercise training and dietary restriction on the menstrual cycle. Low blood levels of T_3 (i.e., 3,5,3'-triiodothyronine) were used to identify energy deficiency. Although a low T_3 level is not associated with sickness, it accurately reflects thyroid function and is commonly used to detect hypothyroidism and starvation. Low T_3 levels are consistently observed among amenorrheic athletes compared with regularly menstruating athletes (38,51).

In one of the studies (39), six groups of women were compared: sedentary and energy balanced, sedentary and energy deprived, light exercise and energy balanced, light exercise and energy deprived, heavy exercise and energy balanced, and heavy exercise and energy deprived. A low T_3 level was found in each of the energy-deprived groups but not in the energy-balanced groups, regardless of exercise-training status. The researchers suggested that dietary modifications to prevent energy deficiencies might prevent the low T_3 levels and athletic amenorrhea. These results suggest that it may not be necessary to modify the exercise-training routine (i.e., reduce training volume) to restore a normal menstrual cycle.

A threshold of energy intake that abruptly induces T_3 suppression was identified to occur at energy intakes of 20 to 25 kcal $\cdot$ kg^{-1} of lean body mass per day (40). In a separate observation, when a low T_3 level was induced by exercise alone, without dietary restriction, the suppression of luteinizing hormone (LH) pulsatility was smaller than previous investigations instituting energy deficits by combining both methods. This finding suggests a possible protective effect of exercise against menstrual cycle dysfunction resulting from low energy availability (42). Aggressive 24 h refeeding protocols in energy-deficient subjects increased T_3 levels without restoring LH pulsatility (42).

In summary, exercise by itself does not appear to trigger menstrual cycle disturbances, but it is part of the equation. Energy deficiency caused by increased exercise energy expenditure or reduced energy intake will compromise the energy available for reproductive functions and thyroid metabolism, alter LH secretion, and likely result in amenorrhea. Aggressive refeeding for a relatively long period (i.e., >24 h) might be needed to completely reverse LH pulse suppression.

mus, anterior pituitary gland, and ovaries (table 8.3 on p. 154 and figure 8.1 on p. 155). The cascade of events that regulates the menstrual cycle is initiated by the hypothalamus with the release of gonadotropin-releasing hormone (GnRH). As females mature, the hypothalamus begins consistent hourly bursts of GnRH secretion. At about age 10, the pituitary begins to secrete more follicle-stimulating hormone and **luteinizing hormone**, which culminates with the initiation of monthly menstrual cycles beginning between ages 11 to 16 (i.e., **menarche**). The control and timing of these pulses determine the normal events of the menstrual cycle (43).

Athletic amenorrhea is diagnosed when the suppression of the menstrual cycle is attributed to intense exercise training and low energy availability. The correct diagnosis requires that pregnancy, abnormalities of the reproductive system, failure of the ovaries, pituitary abnormalities, and tumors be ruled out as the cause

(70). One proposed mechanism for athletic amenorrhea is an imbalance of the hypothalamic–pituitary–gonadal axis (60). Infertility is a consequence of menstrual cycle suppression. A more serious negative effect is reduced estrogen production, also known as hypoestrogenism (64). Low estrogen levels in premenopausal women are related to premature bone mineral loss, increased risk of stress fractures and osteoporosis (31), negative lipid profiles, and increased cardiovascular risk factors (71).

Because the psychological and physiological profile of amenorrheic females is highly variable, identifying the precise causative factors of athletic amenorrhea is difficult. From all the possible causes listed in table 8.4 on p. 154 (6,37,60,64,65,70,94,110), the research evidence suggest that low energy availability is more likely the causative factor for athletic amenorrhea.

The low body fat hypothesis is based on early observations of amenorrheic runners who were more likely

Table 8.3 Normal Menstrual Cycle Events

Gland	Function
Hypothalamus	Starts secreting gonadotropin-releasing hormone (GnRH) in a pulsatile fashion every 60-90 minutes
Pituitary gland	Secretes follicle-stimulating hormone (FSH) and luteinizing hormone (LH) in a pulsatile fashion
Ovaries	Follicular phase Approximately 14 days Secretes estrogens (primarily estradiol) to stimulate the growth of the endometrial lining A marked increase in estradiol at the end of the follicular phase may serve as a signal for the LH surge from the pituitary gland Ovulation: The LH surge from the pituitary gland stimulates the release of an egg (ovum) from the follicle Luteal phase Approximately 10-16 days The follicle transforms into a corpus luteum Starts the secretion of progesterone to prepare the endometrium for the fertilized egg If no egg is implanted, estrogen and progesterone levels drop and the endometrial lining degenerates The drop in estrogen and progesterone levels may stimulate the hypothalamus to release GnRH and start the cycle again

Table 8.4 Potential Causes of Athletic Amenorrhea

Potential cause	Effect
Low body weight or fat	Delayed menarche and amenorrhea
Intense training before menarche	Delayed menarche
Increased psychological stress	Increased endogenous opioids = depressed GnRH production
Low energy availability	Abnormal luteal phase

to have lower initial body weight for height, greater weight loss with onset of exercise training, and a lower percent body fat than regularly menstruating runners (27,37,95,100,108). The suggestion was that those females with a body fat level below a critical value were at high risk of becoming amenorrheic (37). A critical value of body fat has not been clearly established. But a hormone secreted by adipose cells, leptin, has been associated with females' reproductive function (109). Leptin secretion is influenced by fat deposits and energy balance, and low leptin levels have been observed in amenorrheic compared with eumenorrheic women athletes (109). Because leptin receptors are in hypothalamic nerves controlling GnRH pulse generation (21), leptin could be a factor in the signal-

ing to the reproductive axis when energy stores are dangerously low (109).

The physical and psychological stress hypothesis suggests that the demands of sports training and competition stimulate the secretion of stress-related neurohormones including endogenous opioids (i.e., endorphins), dopamine, and cortisol. These neurohormones are believed to have a direct and indirect effect on GnRH production and secretion from the hypothalamus (6,70,95,106), resulting in disruption of the menstrual cycle.

The low energy availability hypothesis suggests that a dietary energy deficit (i.e., consumption of fewer calories than required for a given basal metabolic rate and physical activity energy expenditure level) generates a signal that somehow disrupts the GnRH **pulsatile** secretion

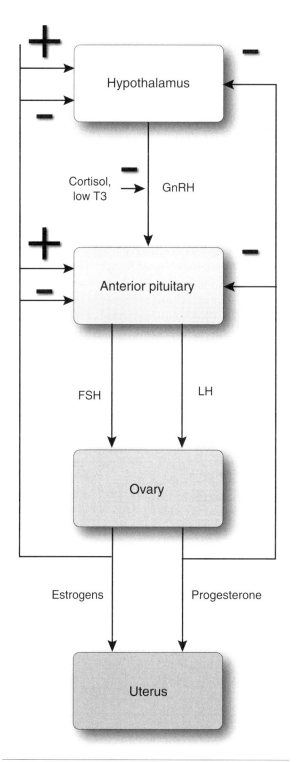

Figure 8.1 Hormonal feedback: events that regulate the menstrual cycle. GnRH = gonadotropin-releasing hormone; FSH = follicle-stimulating hormone; LH = luteinizing hormone.

Reprinted, by permission, from A.J. Pearl, 1993, *Athletic female* (Champaign, IL: Human Kinetics), 47.

and results in an abnormal luteal phase of the menstrual cycle. As previously suggested, this signal could be associated with leptin secretion. If the energy deficit persists, the body's natural reaction will be to conserve essential physiological functions (e.g., cardiovascular, thermoregulation) and compromise less critical functions (e.g., reproduction, growth) (53,65,66).

Hyperandrogenism is another kind of menstrual irregularity that is even less common among female athletes and will be mentioned only briefly in this section. This type of irregularity is more likely to be observed in female athletes who compete in sports or activities that do not require leanness for success, and might result in follicular impairment, amenorrhea, and anovulatory cycles (108). Also, if the hyperandrogenism persists for an extended period in young females, increased cardiovascular risk might develop because of the accompanying hypoestrogenism condition.

Osteoporosis

Osteoporosis is defined as a decrease in bone density that enhances bone fragility and increases the risk of fractures (58,106). Athletes who perform weight-bearing and weight-training activities have up to a 40% higher BMD compared with sedentary individuals (28). Female athletes with disordered eating behaviors leading to low energy balance and menstrual cycle disturbances are at high risk of suffering serious consequences on bone health. Despite the mechanical loading effect of weight-bearing exercise training, female athletes who are amenorrheic tend to have a lower BMD, lose bone at an accelerated rate over time (e.g., 2–6% per year), and have a higher prevalence of stress fractures than both menstruating athletes and sedentary women (2,106). The spinal BMD of young amenorrheic female athletes is similar to the spinal BMD of a 70 to 80 yr old postmenopausal women (33). Skeletal tissue responds to mechanical loading and hormone replacement therapy. Therefore, both regular physical activity and normal plasma estrogen levels are crucial for the maintenance of normal BMD (38). For more information on the scope and pathophysiology of the disease, please see chapter 27.

Energy availability is another important factor recently exposed in the study of bone health. Depressed T_3 and IGF-1 levels observed during chronic energy deficit conditions could affect bone mineralization because both hormones have bone formation properties (109). Low energy availability induced by prolonged exercise could trigger metabolic hormone actions to conserve energy (i.e., increased cortisol, catecholamine, glucagon, and growth hormone and reduced insulin, T_3, and IGF-

1) and maintain body homeostasis (114). Some of these hormones reduce bone formation and promote bone resorption, and therefore increase the risk of developing low BMD. Although most of these studies have been conducted among women athletes, some evidence suggests that low energy availability leading to low BMD can also affect male athletes (114).

There are many reports of low BMD in female athletes with menstrual cycle disturbances, but the actual prevalence of osteoporosis in young female athletes is not known. A review of published studies on the prevalence of low BMD in female athletes with menstrual disturbances is presented in table 8.5. Reduced BMD in amenorrheic athletes has been associated with increased risk of stress fractures (33,59,108). An interesting case report of a nontraumatic femoral shaft fracture was reported in a 32 yr old white female with **oligomenorrhea** during the 13th mi of a half marathon (33). This case report highlights the potential health implications of menstrual cycle disturbances for the competitive as well as the recreational female athlete, although no information was available on that particular female runner's eating behavior or any other potential risk factors for low BMD.

Data on the prevalence of menstrual dysfunction plus low BMD, disordered eating plus low BMD, and disordered eating plus menstrual dysfunction in female athletes and nonathletes show a higher prevalence of eating disorders plus menstrual dysfunction among athletes (27%) compared with nonathletes (14%), but a lower prevalence of eating disorders plus low BMD in the athletes (10%) compared with nonathletes (15%) (54). Nonathletes also had a higher prevalence of menstrual dysfunction plus low BMD (12%) compared with athletes (5%). In another report among 112 U.S. collegiate athletes, Beals and Hill (10) reported a low prevalence (0.9%) of the combined risk of disordered eating, menstrual dysfunction, and low BMD. Only one of the female athletes evaluated had disordered eating and low BMD. Although disordered eating plus menstrual dysfunction can increase the risk of progressive bone loss and possible **osteoporosis** development, weight-bearing exercise training in which impact activities are included provides protection against bone demineralization.

Table 8.5 Bone Mineral Density (BMD) in Female Athletes With Menstrual Cycle Disturbances

Subjects	Site	Observations	Authors
38 amenorrheic athletes	Spinal bone	22–29% less bone mass than eumenorrheic athletes	Cann et al., 1984 (16)
14 amenorrheic athletes	Lower lumbar vertebrae	Lower BMD compared with 14 eumenorrheic athletes matched for age, height, weight, sport, dietary intake, and training program	Drinkwater et al., 1984 (30)
22 amenorrheic professional ballet dancers	Spine, wrist, metatarsals	BMD lower compared with eumenorrheic ballet dancers	Warren et al., 1991 (105)
29 amenorrheic athletes	Lumbar spine, femoral neck, trochanter, Ward triangle, intertrochanteric region, femoral shaft, tibia	BMD lower compared with 20 eumenorrheic athletes	Rencken et al., 1996 (83)
186 elite athletes, 145 nonathlete controls	Total body, lumbar spine, femoral neck	5.4% athletes and 12.4% controls with menstrual dysfunction and low bone mass density	Klungland Torstvert and Sundgot-Borgen, 2005 (53,54)
91 oligomenorrheic runners	Total body, spine, hip	Lower bone mass density in oligomenorrheic compared with eumenorrheic runners	Conn et al., 2003 (25)

Low BMD observed during an extended period of menstrual cycle disturbance may never return to normal, even with hormone replacement therapy or the resumption of normal menses. The use of exogenous estrogens and calcium supplementation in amenorrheic athletes may halt the loss of bone mass but has little effect on regaining lost bone density (32). Therefore, hormone replacement therapy is useful only for acute, short-term interventions because young female athletes are less likely to adhere to prolonged medical treatments. Females who perform vigorous exercise training must be educated about the importance of regaining a normal menstrual cycle by increasing their caloric intake, reducing training volume, or both. Formerly amenorrheic female athletes who decreased training volume by 10% to 20% and increased their body weight by 2% to 3% had a 6% increase of BMD during the year immediately following menstrual cycle resumption (32).

The mechanical loading effect of exercise training on BMD is considered a protective factor. This effect, however, is sport specific. The anatomical site that receives the direct impact caused by the sport-specific bodily movement appears to receive most of the osteogenic benefit (88,98,106). Young female gymnasts compared with young distance runners with similar menstrual irregularities have a higher whole-body lumbar spine and femoral neck BMD (88). In another study that compared young female gymnasts with long-distance runners, it was reported that gymnasts, in spite of having higher self-reported menstrual irregularities, also had higher BMD compared with the runners (14). Although menstrual disorders are not an issue in male endurance runners, a recent study revealed that they have the same threat of low spine bone density as their female counterparts do (47). The osteogenic advantage that may be associated with impact-loading sports appears to persist from childhood to adulthood (113). But interindividual variability (e.g., genetic predisposition for resisting or developing one or more of the female triad disturbances) and the challenge of identifying high-risk sports make it difficult to draw conclusions regarding the influence of genetics and exercise training on bone health status.

MEDICAL AND CLINICAL CONSIDERATIONS

Females with signs and symptoms of low energy availability should be closely monitored for menstrual dysfunction and bone health and, if necessary, appropriately treated to achieve energy balance, normal menstrual function, and maintenance of BMD. Once identified, the behavioral or neuroendocrine mechanisms can be treated and serious health consequences can be prevented. In the case of female or male athletes with low energy availability, ideal treatment should preferably be under the guidance of a multidisciplinary team that includes a physician, dietitian, psychologist, exercise professional, coaches, and close family members. The team goal is to educate its members about harmful eating and training practices and their results, to help restore a normal energy-balanced diet and, if appropriate or applicable, to help resume a normal menstrual cycle.

Female athletes also face other clinical issues that might predispose them to different types of injuries and health problems when evaluation and education are not provided. Some of these issues are painful menses or **dysmenorrhea**, hydration and regulation of body temperature, exercise training during pregnancy, use of **anabolic steroids**, and orthopedic considerations.

Low Energy Availability With or Without Disordered Eating

Female athletes and nonathletes with disordered eating behaviors are often in denial. They typically try to disguise their body image by wearing baggy or loose clothing, and they avoid eating in front of other people. Some characteristic signs and symptoms of females who have anorexia and bulimia are presented in table 8.6 on p. 158. These characteristics are divided into two categories: starvation in females who have anorexia and purging in females who have bulimia. But women who suffer from anorexia or **bulimia** share many signs and symptoms, including self-criticism of their bodies, frequent mood swings, menstrual irregularities, and extreme preoccupation with food, calories, and body weight.

Serious medical complications can result from a **disordered eating** behavior. Some of the most damaging health problems are caused by the abuse of laxatives and diuretics. These drugs will not help reduce body fat but will cause excessive loss of body water, leading to dehydration and electrolyte imbalance. A low potassium level (hypokalemia) can induce cardiac arrhythmias resulting in inadequate cardiac output and possibly death. Laxative abuse can also result in gastrointestinal complications like bleeding, loss of normal colonic peristalsis, and gastric dilation (81). Some recommended laboratory evaluations that aid in the diagnosis of disordered eating include the following (53):

Table 8.6 Disordered Eating in Female Athletes: Signs and Symptoms of Anorexia Nervosa (Starvation) and Bulimia (Purging)

Signs	Symptoms
Anorexia nervosa	
Bradycardia	Chronic fatigue
Hypotension	Lethargy
Hypothermia	Cold intolerance
Lanugo	Cold and discolored extremities
Dehydration	Dry hair
Electrolyte imbalance	Dry skin
Compulsive, excessive exercise	Constipation
Loss of fat and muscle	Abdominal pain
Extreme restriction of food intake	Diarrhea
Weight fluctuations over short periods	Light-headedness
Anemia	Headaches
Frequent consumption of water and diet sodas	Lack of concentration
Amenorrhea	Stress fractures
Osteoporosis	Depression
Bulimia	
Enlarged parotid glands	Chest or throat pain
Edema at face and extremities	Fatigue, headaches
Tooth enamel erosion	Abdominal pain, constipation, or diarrhea
Excessive use of laxatives	Bloating
Metacarpal calluses	Depression
Weight fluctuations	Hysterical behavior
Disappearing or making trips to the bathroom after eating large meals	

Adapted from L.A. Marshall, 1994, Clinical evaluation of amenorrhea. In *Medical and orthopedic issues of active and athletic women*, edited by R. Agostini (Philadelphia: Hanley & Belfus), 152-163; C. Pomeroy and J.E. Mitchell, 1992, Medical issues in the eating disorders. In *Eating, body weight and performance in athletes*, edited by K.D. Brownell, J. Rodin, and J.H. Wilmore (Philadelphia: Lea & Febiger), 202-221; American College of Sports Medicine, 2000, *ACSM's guidelines for exercise testing and prescription*, 6th ed. (Philadelphia: Lippincott, Williams & Wilkins); C.L. Otis, 1992, "Exercise-associated amenorrhea," *Clinical Sports Medicine* 11(2): 351-362.

Urinalysis to assess changes in pH (normal pH = 4.6–8.0)

Complete blood count (e.g., screen for infection, malignancy, anemia, and any inflammatory process)

Blood chemistry evaluation (e.g., electrolytes, liver, kidney, and thyroid functions)

Electrocardiography if bradycardia (i.e., pulse <50 beats · min^{-1}) or electrolyte abnormalities are present. Look for the following electrocardiographic signs:

1. Low voltage (QRS amplitude <0.5 mV in all limb leads)
2. Low (<3 mm) or inverted T-waves (usually represents ischemia)
3. Prolonged QT intervals (normal = 0.34 –0.42 s)

A team physician or a family physician should be initially consulted to identify disordered eating behaviors and arrange for the appropriate treatment course, including referrals to a registered dietitian and, if needed, to an eating disorder specialist. Questions regarding food (i.e., meals per day, yesterday's diet, eating pattern), weight (i.e., highest and lowest weight last year, ideal competition weight, satisfaction with present weight, weight-loss practices), menstrual cycle (i.e., menarche, regularity, last menstruation, oral contraceptive use), and training (i.e., changes in training regime, outside training activities, injuries) could be introduced during the initial assessment (50). Treatment for females with disordered eating is intended to help normalize eating patterns and behavior, increase self-esteem, and improve communication about issues related to their condition. Providing a comfortable, nonjudgmental environment where those with disordered eating can get helpful information, direction, and support is important (25). If the patient does not have normal eating patterns and has low energy availability but is otherwise healthy, nutritional counseling might be the only intervention needed. But if risks of serious eating disorder are present, a stronger approach that provides both nutritional counseling and psychological counseling is recommended. Supervision of every meal might initially be required, and in extreme cases hospitalization might be the best option. See "Eating Disorders Information and Screening Available on the Internet."

Nutritional counseling should include evaluation of dietary intake (e.g., dietary recall questionnaires), assessment of exercise training and exercise energy expenditure (e.g., physical activity questionnaires, pedometers or accelerometers), and body composition assessment (e.g., skinfolds or circumferences) on a monthly or bimonthly basis. Female athletes or young girls at risk of disordered eating behaviors need assistance to determine a normal healthy body weight and percent body fat range and to perceive these as normal. One method commonly used is to have the individual sign a written contract to consume a prescribed amount of calories per day or gain a goal amount of body weight per week. Failure to comply with the agreement would mean that sports training and competition is put on hold or completely rescinded. Intervention from a mental health professional is crucial to evaluate life stressors, develop strategies to cope with daily situations, and assess issues of self-esteem, all of which may be related to the underlying problem (53).

Exercise should be limited when electrolyte and electrocardiogram abnormalities are present. Physicians might consider hospitalization when the following signs are present: body weight 30% below normal for age and height, hypotension, dehydration, electrolyte abnormalities, severe psychological depression, and failure to respond to treatment. Approximately 20% of nonathletes treated for disordered eating behavior are unable to overcome their struggle with body image and weight control. A 19% mortality rate of **anorexia nervosa** patients is reported (81). The mortality rate of people with bulimia is not known. Suicide, cardiovascular failure, gastrointestinal perforations, or bloodstream infections are typically the immediate cause of death (53). But most patients with disordered eating behavior can improve their condition with proper treatment and supervision (90).

Eating Disorders Information and Screening Available on the Internet

The information and questionnaires presented in the following Web sites are intended for educational purposes only. They do not constitute a professional evaluation. If a disordered eating behavior is suspected, you must consult a qualified healthcare professional for a thorough clinical evaluation.

Eating Disorder Test (http://caringonline.com/eatdis/misc/edtest.htm)

National Eating Disorder Screening Program (www.nmisp.org/eat.htm)

What's Going On With Me? Evaluating Eating and Exercise Habits (www.NationalEatingDisorders.org)

University of Iowa Weight Management Questionnaire (http://www.uihealthcare.com/depts/obesity/weight-historyquestionnaire.pdf)

Eating Disorder Screening Quiz (www.psychcentral.com/eatingquiz.htm)

Eating Disorder Screening Quiz (http://quiz.ivillage.co.uk/uk_diet/tests/eatingdisorders.htm)

Eating Attitudes Test (http://psychcentral.com/quizzes/eat.htm)

Anorexia-Bulimia Questionnaire (http://caringonline.com/eatdis/misc/smedtest.htm)

Amenorrhea

Women who exercise vigorously and experience cessation of their menstrual cycle should have a clinical examination. The primary factors to be considered in the exam are presented in "Evaluation of Athletic Amenorrhea." Amenorrhea is likely the result of many factors that affect the normal hypothalamic function. Body weight and body fat reduction, as well as physical and emotional stress, are the most recognized factors of amenorrhea associated with increasingly high physical activity levels, high energy expenditure, and low energy intake.

Recommended laboratory tests for the diagnosis of amenorrhea are included in table 8.7 (70). It is important to remember that before amenorrhea can be diagnosed, pregnancy and any abnormalities of the pituitary gland and reproductive tract must be ruled out.

Females with amenorrhea who are able to match total daily caloric intake with total daily energy expenditure, either by increasing caloric intake (e.g., 3–5%) or decreasing exercise-training dose (e.g., 10–20%), usually regain a normal menstrual cycle within 2 or 3 mo (32). If an active female is not willing to make these changes, then the recommended treatment could be (70,80) as follows:

1. Hormone replacement therapy (e.g., 0.625 mg of conjugated estrogen on days 1–25 and progesterone on days 14–25) and increased calcium intake (e.g., 1,200–1,500 mg per day) to prevent low BMD and possibly coronary artery disease risk factors. If the woman is sexually active, the use of low-dose oral contraceptives to prevent bone loss is acceptable (e.g., <50 g estrogen daily) (96).

2. Bromocriptine administration to reduce prolactin levels if hyperprolactinemia is diagnosed.

3. Nutritional and psychological counseling.

4. Reassessment every 3 to 6 mo.

Osteoporosis

Osteoporosis is a silent disease with little or no possibility of being detected in young women unless medical counseling is sought for related conditions such as menstrual cycle dysfunction, eating disorders, or stress fractures. The **hypoestrogenic** status of amenorrheic, energy-deficient females stimulates the process of bone demineralization. These weakened, porous bones are less resistant to the forces of intense and continuous skeletal muscle contractions. Therefore, energy-deficient and amenorrheic females are also at a higher risk of experiencing stress fractures, which could be an indicator of premature osteoporosis.

Evaluation of Athletic Amenorrhea

Menstrual cycle history
Age at menarche
Previous menstrual irregularities
Use of birth control pills
Pregnancy

Nutritional habits
Vegetarian diet
Low body weight
Low body fat
Eating disorders

Physical activity
Sudden increases in frequency, duration, and intensity of exercise
More than one exercise session a day

Social and emotional support
Family relationships
Relationships with friends
Coping skills

Estrogen deficiency
Vaginal dryness
Dyspareunia
Hot flashes

Androgen excess
Polycystic ovarian syndrome
Baldness
Acne
Facial, chest, and abdominal hair

Others
Galactorrhea
Stress fractures
Low bone density

Adapted from L.A. Marshall, 1994, Clinical evaluation of amenorrhea. In *Medical and orthopedic issues of active and athletic women*, edited by R. Agostini (Philadelphia, PA: Hanley & Belfus), 152-163.

Early detection of bone demineralization is difficult because not all young females with menstrual cycle disturbances will develop low bone mass density and its subsequent signs and symptoms. Female athletes in particular should have their menstrual status and dietary habits regularly checked. They need to be educated about bone health status and should consult a healthcare professional when needed. If signs or symptoms of a dis-

Table 8.7 Laboratory Tests for the Diagnosis of Amenorrhea

Test for	Normal values
Pregnancy	+ or −
Elevated thyroid-stimulating hormone	$2–210\,mU \cdot L^{-1}$
Reduced T3 (3,5,3'-triiodothyronine)	$1.2–1.5\,mmol \cdot L^{-1}$ or $110–230\,mg \cdot dl^{-1}$
Elevated prolactin (midmorning before breakfast)	$3–19\,mg \cdot m^{-1}$
Progestin challenge	Vaginal bleeding within 10 d of progesterone injection or oral medroxyprogesterone acetate
Follicle-stimulating hormone (FSH)	$>40\,IU \cdot L^{-1}$ or $<5\,IU \cdot L^{-1}$ with no withdrawal bleeding after progestin challenge
Luteinizing hormone (LH)	$6–30\,IU \cdot L^{-1}$
Elevated LH/FSH	>2:1
Elevated testosterone	$24–47\,mg \cdot dl^{-1}$
Elevated dehydroepiandrosterone sulfate	$50–400\,mg \cdot dl^{-1}$
Bone mineral density	$1.0–1.1\,g \cdot cm^{-2}$

Adapted from S.A. Marsh and D.G. Jenkins, 2002, "Physiological responses to the menstrual cycle: Implications for the development of heat illness in female athletes," *Sports Medicine* 32: 601-614.

ordered eating behavior or menstrual irregularities are observed or suspected, bone density evaluations might be considered.

Measurements of BMD should be compared with same-age, same-sex, and same-race controls. Fractures are more likely to occur when BMD is two standard deviations below the normal value or below the 90th percentile of values from a like group with osteoporotic fractures. After diagnosis, proper treatment should be designed to help recover normal menstrual cycles and estrogen levels. As previously mentioned, females at risk must increase dietary energy intake to balance energy expenditure and consider increasing daily calcium intake (e.g., $1,200–1,500\,mg \cdot day^{-1}$). Estrogen replacement therapy and oral contraceptives are also alternatives for females at risk who are unwilling to change diet or exercise patterns. There is no guarantee that bone mass will recover to normal levels. A more realistic goal would be to attenuate or stop the demineralization process. Follow-up evaluations are desirable to assess the effect of intervention in sustaining bone health in young females with menstrual dysfunction and low energy availability. Prepaticipation screening for high school and university student-athletes has been suggested as an appropriate means to achieve early identification of those at risk of low bone density or menstrual dysfunction (92).

Dysmenorrhea

Dysmenorrhea is defined as painful menstruation without pelvic abnormalities in which symptoms like vomiting, headaches, back pain, diarrhea, and dizziness are usually present. The prevalence of dysmenorrhea is reported to be between 17% and 81%. Dysmenorrhea is thought to be the leading cause of menstrual disorder among young females and the leading cause of short-term high-school absenteeism in the United States (7,57), affecting performance in academics, sports, and social activities. Nonsteroidal anti-inflammatory drugs that help to inhibit prostaglandin secretion are used effectively to treat dysmenorrheal symptoms. A small number of females also report using exercise to alleviate these symptoms (7). Some studies report a 35% to 40% prevalence rate of dysmenorrhea among female athletes (79,87). Others report a much lower prevalence rate in athletes compared with nonathletes (35,51) and much less intense symptoms when females engage in sports activities (51,72).

Thermoregulation

The ability to control and regulate body temperature at rest and during exercise is influenced by many factors such as body surface area, body fat, menstrual cycle, physical fitness level, and thermoregulatory effectiveness (i.e., the ability to dissipate body heat). The average woman has more body fat, a lower physical fitness level, a higher skin and core temperature, a higher heart rate, and a lower sweat rate compared with the average man (8,49,103). Many of these characteristics could be beneficial during exercise in the cold (e.g., more body fat, higher body temperature). During exercise in hot environments, however, many of these characteristics could reduce the ability to regulate body temperature. For example, having a lower sweat rate might reduce the ability of the body to cool through sweat evaporation, although having a smaller body size and thus a larger ratio of surface area to mass allows women to have faster heat exchange between the skin and the environment (8). In addition, the hormonal changes during the menstrual cycle are known to cause body temperature fluctuations.

The menstrual cycle is characterized by two distinctive phases: follicular (preovulatory) and luteal (postovulatory). During the luteal phase, progesterone levels increase and estrogen levels start to increase before ovulation and remain elevated. Body core temperature during the luteal phase is approximately 0.3 to 0.6 °C higher than it is during the follicular phase (20,104). This difference is attributed to progesterone secretion. Influences of FSH on resting body temperature appear to be part of the overall mechanism of thermoregulation control. For example, estrogen is associated with cutaneous vasodilation and the onset of sweating at a lower body temperature, whereas progesterone appears to shift the threshold for the onset of cutaneous vasodilation to a higher body temperature (18,19). Changes in water and sodium distribution are also apparent during the luteal phase because of the combined effect of progesterone and aldosterone (101). Estrogen also affects fluid balance by indirectly stimulating thirst through increases in arginine vasopressin levels that induce fluid retention (100). Although the increased core temperature and threshold for the onset of sweating caused by the elevated progesterone levels might indicate high risk for heat-related illness during the luteal phase of the menstrual cycle, the menstrual cycle phase does not appear to influence a woman's ability to tolerate exercise in hot and humid conditions (18,49,69). Women appear to compensate for the increased threshold for the onset of sweating and the lower sweat rate by increasing cutaneous vasodilation and therefore increasing conductive heat dissipation (49).

The physiological interrelationship between the various hormonal changes during the menstrual cycle and their influence on thermoregulatory and fluid balance responses is not completely understood. Clearly, however, women exposed to exercise in hot environments are able to improve thermoregulatory mechanisms. Both heat acclimation and physical training decrease the threshold of skin vasodilation and sweating response and increase sweat production in trained females (8,38,56). These changes help improve the rate of heat dissipation during exercise. Exercise training also improves time to exhaustion and results in lower core body temperature

and heart rates at submaximal work rates when hydration levels are kept normal.

Dehydration during exercise causes body temperature to increase more rapidly because of reduced sweat production. If sweat is not available at the skin surface, the cooling effect of sweat evaporation is absent, and the body's ability to dissipate heat diminishes. Dehydration also affects the cardiovascular system (i.e., reduced stroke volume and blood pressure, reduced maximal oxygen consumption), alters muscle metabolism, and reduces the intensity at which anaerobic threshold occurs (73,75). Therefore, proper hydration before, during, and after exercise is extremely important for female athletes and athletes in general. Table 8.8 presents current recommendations for fluid, calcium, and iron intake for female athletes.

Menstrual cycle disturbances do not appear to affect the thermoregulatory effectiveness of athletic females during exercise in hot environments. Women with menstrual dysfunction, however, appear to have lower core body temperature at rest and delayed heat production during exercise in cold environments compared with regularly menstruating women (4,103). During exercise, amenorrheic women also have a reduced heart rate response and a higher skin blood flow than eumenorrheic women, suggesting a reduction in heat storage by improved thermoregulatory responses. Low progesterone level is postulated as a reason for the reduced cardiovascular strain and better heat dissipation observed in amenorrheic females at rest and during exercise. However, Rikenlund et al. (86) reported an impairment in flow-mediated dilation that was correlated with high levels of cholesterol and low-density lipoproteins in amenorrheic athletes compared with regularly menstruating athletes, athletes with oligomenorrhea, and nonathlete controls. The authors suggest that hypoestrogenism is likely the cause of the observed endothelial dysfunction and unfavorable lipid profile among the group of amenorrheic athletes. The long-term clinical implications of these observations in the female athlete are yet to be elucidated.

Pregnancy

Exercise training during pregnancy is a common practice. Scientific evidence suggests that physical activity of moderate intensity improves maternal and fetal well-being. Women who continue to exercise during pregnancy usually experience the following (3,15,24,45):

- Improved cardiovascular function
- Limited weight gain and body fat retention
- Improved digestion

Table 8.8 Recommendations: Fluid and Other Nutrients

	Amount	Timing
Fluids Should be cooler than ambient temperature (15–22 °C). Flavored drinks might enhance palatability and promote voluntary drinking	**Before exercise**	
	16 oz (480 ml)	2 h before
	8–16 oz (240–480 ml)	10–15 min before
	During exercise	
	3–10 oz (90–300 ml)	Every 10–20 min
	After exercise	
	At least 1 pint for every pound of body weight lost (1 L for every kg lost)	Unrestricted
	Amount	Notes
Calcium intake	800–1,200 mg · d^{-1} if normally menstruating 1,200–1,500 mg · d^{-1} if amenorrheic	Calcium carbonate is the recommended supplement
Iron intake	From animal and vegetable sources Vitamin C helps absorption	No supplements recommended Could become toxic
Monitor diet to enhance adequate intake of vitamins, minerals, proteins, and calories		

Adapted from A. Moquin and R.S. Mazzeo, 2000, "Effect of mild dehydration on the lactate threshold in women," *Medicine and Science in Sports and Exercise* 32(2): 396-402.

Reduced constipation

Reduced back pain

Improved attitude and mental state

Easier labor or reduction in possible complications during labor

Reduced odds of cesarean delivery

Faster recovery

Better fitness level

Moderate exercise appears to reduce diastolic blood pressure in pregnant women at risk of hypertension (111). The offspring of exercising pregnant women may have a reduced fat cell growth rate without compromising other body cell growth, a high stress tolerance, and an advanced neurological developmental rate (13,24). In addition, vigorous exercise during pregnancy in previously recreational exercisers had no negative effects on gestational age or birth weight (33).

Regular physical training and pregnancy often influence the same physiological variables, resulting in an additive effect. For example, both exercise training and pregnancy improve cardiovascular fitness level (i.e., increased maximal oxygen consumption), cardiac output, heart rate response to exercise, total blood volume, red blood cell volume, and ventilatory response to exercise (23,24,98). Many of these adaptations are associated with increased total body weight, amount of metabolically active tissue, and hormonal changes that occur during pregnancy. The conservative point of view emphasizes that exercise during pregnancy is potentially harmful. Medical and safety concerns regarding pregnancy and exercise are presented in table 8.9 on p. 164. But scientific evidence is lacking in support of a cause-and-effect relationship between exercise and pregnancy-related problems such as ectopic pregnancies, spontaneous abortion, fetal or placenta abnormalities, premature membrane rupture, or premature labor. Moreover, regular aerobic exercise during pregnancy or postpartum does not alter the quality and quantity of breast milk.

Light to moderate physical activity (i.e., intensity <60% $\dot{V}O_2$max, duration 20–30 min) is recommended for pregnant women who have not previously engaged in vigorous exercise training. No evidence indicates that running, jumping, or increasing body temperature causes abnormal or difficult pregnancies. But the consensus is that pregnant women should avoid starting an intense exercise-training program because of risk of injury and compromised cardiovascular function. In addition, exercise should be modified or ceased if a pregnant woman experiences any of the following (3,23):

Calf pain or swelling of ankles, hands, or face

Acute illness

Decreased fetal movement

Vaginal bleeding in early pregnancy

Table 8.9 Exercise and Pregnancy: Medical and Safety Concerns for the Mother and Fetus

Mother		Fetus	
Concern	Solution	Concern	Solution or effect
Poor balance while running or jogging because of shifts in weight distribution and center of gravity	Slow down, run cautiously, and never run alone	Direct fetal trauma. Tissue and fluid surrounding fetus provides protection	No scientific data
Overheating and dehydration. Pregnancy elevates body core temperature by approximately 0.5 °C, elevating resting metabolic rate by 15–20%. Excessive sweating might reduce blood volume	Drink plenty of fluids before, during, and after exercise. Use appropriate exercise clothing or avoid exercise during extremely hot and humid weather	Hyperthermia and reduced fetal blood flow	Might cause neural tube defects, growth retardation, reduced birth weight, or fetal abnormalities
Leg, hip, and abdominal pain. Reduced circulation to lower extremities during late pregnancy, extra weight to carry	Never forget to stretch and warm up before any exercise session. Wear cushioned and comfortable shoes	Reduced fetal blood flow	No scientific data
Nutrient availability. Pregnancy increases energy requirements by approximately 300 kcal/d	Pregnant women are expected to increase from 25 to 40 lb (11 to 18 kg)	Substrate availability and hypoxia. Reduced fetal glucose and oxygen availability	Might cause growth retardation, reduced birth weight, or fetal abnormalities
Reduced oxygen availability for aerobic exercise. Cardiovascular drift: added blood circulation to placenta	Modify exercise intensity. Never exercise to the point of fatigue or exhaustion. Avoid intense and prolonged exercise. Monitor heart rate and rates of perceived exertion	Reduced fetal blood flow. Intense exercise redistributes blood flow—more to muscles and less to other areas, including the placenta	Light to moderate physical activities are considered safe for mother and fetus
Musculoskeletal injury. Ballistic movements and sudden postural changes can increase the risk of injury. The risk of injury for fit pregnant females should be lower	Continuous aerobic exercises are more acceptable than intermittent anaerobic exercises	Umbilical cord entanglement. Can cause reduced blood flow to important fetal organs	No scientific data

Note. A physician with background in exercise physiology should be consulted before any exercise program is considered during pregnancy. Ask about contraindications to exercise and a list of high-risk sports to avoid during pregnancy.

Persistent nausea or vomiting, dizziness, and head-aches

Chest pain or excessive palpitations

Sudden onset of abdominal or pelvic pain

Regular physical activity during pregnancy is safe when necessary precautions are taken to ensure the safety of both the pregnant woman and the fetus. Comfort level, rest periods, nutrition, body position, and hydration should be continuously monitored during exercise training in pregnant woman. Exercise in the supine position after the first semester and dehydration can reduce fetal cardiac output, blood flow, and adequacy of nutrients to the baby. Each of these has the potential to have an adverse effect on the normal process of growth and development. Pregnant women should avoid any indicators of dehydration (e.g., thirst, weight loss of 0.8–1.0 kg) and consume enough calories (e.g., approximately 300 kcal · d^{-1} above normal) to support the metabolic demands of both pregnancy and exercise (23,98,108). To ensure proper hydration, pregnant women should follow the recommendations presented in table 8.8.

Maximal exercise testing during pregnancy is usually avoided unless it is a medical necessity (3). A detailed exercise prescription is not necessary to provide a safety measure for most healthy, physically active pregnant women (23). But some prescription recommendations are presented here. For general health and cardiovascular fitness, three to four exercise sessions per week of 30 min duration are considered safe for pregnant women (23,108). The use of heart rate to monitor exercise intensity could present a problem because individual heart rates vary widely during pregnancy. Nonetheless, heart rates exceeding 150 beats · min^{-1} have been associated with reduced fetal blood flow (109). The use of rating of perceived exertion (e.g., Borg scale = 13–14) is considered more reliable than heart rate for monitoring exercise intensity during pregnancy. Pregnant women should never reach the point of fatigue or exhaustion during exercise because the metabolic demand might shift blood away from the placenta, thus compromising fetal blood flow. Resistance training is usually considered safe as long as hydration, exercise intensity, and perceived exertion are carefully monitored. Range-of-motion or flexibility routines can be used as long as movements pressuring the abdominal area or placing pregnant women on their backs are avoided. These body positions may reduce fetal blood flow.

Anabolic Steroids

The use of anabolic steroids to improve strength, sports performance, and physical appearance has become an issue among athletic females who have strong incentives to seek any competitive edge. Many young female athletes competing for athletic scholarships perceive the use of these drugs as an "investment" in their future (39,112). A prevalence of 0.5% to 2% of anabolic steroid use among female adolescents has been reported (5,39). Despite many legal and educational efforts, steroid use and abuse, particularly among the athletic population, has not declined in the United States (112).

A general perception among many athletic females is that anabolic steroids improve muscle strength and size, training intensity, and performance (68). The scientific evidence has shown that anabolic steroids potentially increase protein synthesis rate, muscle size, and muscle strength when used in combination with heavy resistance or strength training. Anabolic steroids can also benefit female athletes by reversing the catabolic effect of stress-induced cortisol levels known to increase during intense exercise training and competition, particularly in those who are amenorrheic.

Many negative side effects have been reported among female anabolic steroid users. The continuous self-administration of multiple anabolic androgenous steroids increases appetite, aggressiveness, irritability, sex drive, acne, body hair, voice deepening, and clitoral size among females (68). Continuous use of anabolic steroids also tends to cause altered dietary practices and chronic preoccupation with body physique (42). Other side effects include menstrual abnormalities, changes in lipoprotein profiles, and mental depression symptoms associated with anabolic steroid withdrawal (42,68). Anabolic steroids induce menstrual cycle irregularities by interfering with the amplitude or daily total of GnRH pulses, leading to **anovulatory** cycles.

Although these events can cause hypoestrogenism, they are more likely to cause hyperandrogenism, with relatively low negative effects on bone mass health. But the abuse of anabolic steroids could predispose athletic women to premature **atherosclerosis** increasing the risk of cardiovascular diseases.

Orthopedic Issues

Each year, an estimated seven million Americans receive medical attention for sports-related injuries (26). Although sports injuries and related emergency visits to the hospital are higher in males compared with females (16,99), the risk of sustaining a sports-related injury, such as serious knee injury, is four to six times more likely in female athletes than in male athletes who participate in sports that require frequent quick stops, jumps, and cutting movements (46,48,8). For instance, in sports like handball, soccer, basketball, and alpine skiing, women experience proportionately more injuries to the anterior cruciate ligament and lateral ankle sprains compared with men (29,40,78,89). Most of these injuries occur in noncontact situations when the athlete lands from a jump, makes a lateral pivot while running, or performs a high-speed plant-and-cut movement (46,78).

More than 30,000 knee injuries occur each year among intercollegiate and high school female athletes in the United States (46). The higher incidence of knee injuries observed in women's sports compared with men's is attributed to anatomical factors such as greater pelvic width, poor alignment of the lower extremity, knee instability, increased tibial rotation, and an increased Q angle (i.e., male = 13°, female = 18°) (22,48). Other factors including weaker quadriceps musculature and hormonal changes may alter ligament compliance and bone strength (40,48,67).

Studies have looked at the relationship between menstrual cycle and risk of sports-related injuries among athletic females. Some have observed an increased risk of injury in female athletes who have irregular menstrual periods (12,59). Others have found a higher risk of injury during the premenstrual phase compared with the menstrual flow phase of the cycle (74,78). The suggested mechanism is a potential increase in ligamentous laxity and decreased neuromuscular performance associated with the fluctuation of female sex hormones throughout the menstrual cycle (46). Because oral contraceptives stabilize hormonal levels during the menstrual cycle, they have been associated with reduced risk of sports-related injury (46,74). Moreover, the physiological and psychological symptoms associated with the menstrual cycle (e.g., discomfort, swelling, irritability) can affect fitness components such as balance and coordination, which increases the risk of injury in the female sports participant. These observations remain speculative because no supporting evidence from controlled research studies has been published to date.

EXERCISE PRESCRIPTION

Most female athletes are not underweight, under fat, or overstressed. Most athletic women do not experience abnormal reproductive function. But some women athletes might be obsessed with their weight and body image. These women are at risk of engaging in behaviors leading to one or more of the triad disorders previously described, and their health could be at risk. In addition, other female athletes could unintentionally experience low energy availability and might be at risk of menstrual disturbances and eventual low BMD if not appropriately identified and treated.

All female athletes should have a complete physical examination prior to sports training and competition. This examination should include information about eating behavior, history of menstrual irregularities, history of stress fractures, mental depressive symptoms, pressure to lose weight, and history of exercise intensity, frequency, and duration. If any sign indicates that the athlete is at risk of one or more of the triad disorders, a multidisciplinary team should be consulted (i.e., sport nutritionist, sport psychologist, team physician) for follow-up visits and more clinical evaluations before continued exercise training. Physical fitness testing is also appropriate to help the athletic female achieve specific goals and allow for comparisons with normative data. Fitness testing should include measurement and evaluation of body composition, cardiorespiratory endurance, muscular strength and endurance, flexibility, balance, and coordination. An individualized exercise prescription for each of these areas should be developed for each female athlete. Chapter 5 reviews the general concepts of exercise prescription that also apply to these women. Practical application 8.2 gives tips for working with triad disorder patients.

Practical Application 8.2

Client–Clinician Interaction

Strict confidentiality must be assured at all times for the female athlete–clinician interaction to be successful. The interdisciplinary team must convey the message that they will respect the athlete's desire to remain lean and will help her remain healthy and strong for her sport's training and competition. Emotional and behavioral issues like fear of eating and gaining weight and poor dietary selection must be addressed early in the interaction. If low energy availability is detected, but the athlete is unaware of her behavior, nutritional counseling may be the only recommendation and no additional clinical interactions may be necessary. Other cases might require negotiation between the clinician and the female athlete to establish minimal criteria to continue training in the sport. Suspension from training must be the last option considered because this extreme action might result in more damage to the athlete's self-esteem. But female athletes must be aware that they will be considered injured athletes and that lack of progress in their treatment might imply no further sport competition.

Sundgot-Borgen and Torstveit (104) suggested that female athletes agree with these minimal criteria in a written contract to allow continued training: (1) comply with all treatment strategies as well as possible, (2) be closely monitored by health professionals in an ongoing basis, (3) understand that treatment precedes training and competition, and (4) participate in no competition while in treatment. The clinical interaction may include discussions between the clinician and the female athlete regarding the recommended decision tree for disordered eating and amenorrhea published by the IOC Medical Commission Working Group Women in Sport (49).

Exercise prescription for women with signs and symptoms of the female triad must include a determination of the proper amount of exercise training versus the amount of dietary intake so that energy balance is achieved. Athletic women should include resistance and flexibility exercises in their training programs.

CONCLUSION

Participation in physical activities, exercise, and sports is important for the health and well-being of all girls and women. Those instances in which females use excessive exercise and abnormal eating patterns to achieve an unrealistically low body weight and slim image need to be assessed and addressed. Eating disorders, amenor-rhea, and osteoporosis are the components of the athletic female triad and represent a real health threat for some girls and women. Sports training during pregnancy, thermoregulation, and the use of performance-enhancing drugs are other issues that need to be discussed with female athletes. Clinical exercise professionals have the responsibility of educating, evaluating, and treating athletic girls and women so that they can achieve success in sport and life.

Case Study

Medical History

Ms. CS is a 16 yr old Hispanic female athlete referred to the Sports Health and Exercise Science Center located at the Olympic Center in Salinas, Puerto Rico. The team handling the case consisted of a family physician, a sports nutritionist, a sports psychologist, and a psychiatrist. The athlete had no record of previous medical problems. She was 62 in. (157 cm) tall and weighed 110 lb (50 kg) (BMI = 20 kg · m^{-2}). She trained and competed in long jump and triple jump. Her typical training schedule consisted of two sessions of approximately 2 hr · d^{-1}, 5 d · wk^{-1} (strength, endurance, and specific event practice sessions). Her estimated exercise caloric expenditure was 500–1,000 kcal · d^{-1}.

Diagnosis

After Ms. CS complained of feeling tired most of the time during practice, her coach referred her to the sports medicine clinic. During the evaluation with the family physician Ms. CS explained that she had been experiencing fatigue and dizziness for about a year. She was also sleeping more than usual. Because she was concerned about keeping a certain body weight for her sport, she started inducing vomiting, but now the vomiting "comes naturally." She also expressed feeling extremely sad sometimes because she felt alone with no family interested or involved in her life.

Test Results

A dietary recall analysis revealed that her caloric intake was about 850 kcal · d^{-1}. Her blood tests showed high hematocrit and serum sodium levels. The psychologist and psychiatrist diagnosed clinical depression.

Treatment

The athlete met with an exercise physiologist and sports nutritionist for educational sessions regarding sport performance, risk of injuries, and energy availability. The agreement with the nutritionists was to start a dietary plan of 1,500 kcal · d^{-1} for 2 wk. She would then be reevaluated. An antidepressant was prescribed.

(continued)

Discussion Questions

1. What other clinical problems might place this athlete at risk?

2. What other screening and clinical tests must be performed for a complete diagnosis and treatment plan?

3. Is this a case of inadequate eating or disordered eating? Why?

4. In what sports are athletes more susceptible for disordered eating behaviors?

5. What exercise or training recommendations should be given to this athlete?

6. Do other nutritional evaluations or recommendations need to be addressed in this case?

9

Depression and Exercise

Krista A. Barbour, PhD

Depression is the most prevalent psychiatric disease or disorder in the United States. The term *depression* is often used to describe varying levels of psychological distress, ranging from a dysphoric mood state to diagnosis of a clinical disorder, such as **major depressive disorder** (MDD). This chapter focuses on the clinical syndrome of MDD and includes discussion of subthreshold depressive symptoms where appropriate. Because depression often develops out of or accompanies many of the chronic diseases listed in this book, we introduce it relatively early so that the student interested in the clinical exercise physiology profession can appreciate the confounding factors that clinicians face when treating someone who is recovering from a chronic disease. For example, the clinical exercise physiologist must feel comfortable and familiar working not only with patients with diabetes but also with patients with diabetes who are experiencing depression.

In the United States, psychiatric disorders are diagnosed using criteria outlined in the *Diagnostic and Statistical Manual of Mental Disorders, Fourth Edition* (DSM-IV; 1). The DSM-IV diagnostic criteria for MDD are presented on p. 170. In summary, a diagnosis of MDD requires endorsement of at least five symptoms, one of which must be either depressed mood or diminished interest or pleasure, that have been present during the same 2 wk period.

SCOPE

A recent estimate of the lifetime prevalence of MDD in community-based adults was found to be 16.6%, and a 12 mo prevalence estimate was 6.7% (27). Kessler and colleagues (28) found that of those individuals meeting criteria for MDD, 80% were classified as having moderate or severe episodes. Thus, MDD is a prevalent disorder that people often experience at a significant level of severity. If not treated, most episodes of depression will last for several months before remitting (31). Furthermore, MDD tends to be episodic; many individuals who recover from an episode of MDD will experience a recurrence of the disorder (17).

MDD is consistently found to be more common in women than in men, and this gender difference holds across cultures (50). Studies examining differences in prevalence of MDD between ethnic minorities and Caucasians have yielded mixed results, but both African Americans and Latinos appear to report less depression compared with Caucasians (27). But subgroups of ethnic minorities may be at increased risk for MDD because of difficulties in meeting basic needs, such as food and shelter (40). The clinical exercise physiologist should consider this possibility when working with patients with low socioeconomic backgrounds.

Besides the great personal cost of MDD, the public health burden is enormous. For example, MDD has been linked to greater healthcare utilization, decreased quality of life, missed time at work, and increased rates of attempted suicide (25). When compared with patients with chronic medical disease (e.g., diabetes, hypertension), people with depression report similar or worse functioning (51). For the year 1990 the economic burden of depression in the United States (including direct treatment costs, mortality costs of depression-related

Summary of DSM-IV Diagnostic Criteria for Major Depressive Episode

Five (or more) of the following symptoms are present during the same 2 wk period and represent a change from previous functioning (Note: At least one of the symptoms is either number 1 or number 2).

1. Depressed mood most of the day, nearly every day
2. Diminished interest or pleasure in all or most activities
3. Significant change in weight or appetite
4. Insomnia or hypersomnia
5. Psychomotor agitation or retardation
6. Fatigue or loss of energy
7. Feelings of worthlessness or guilt
8. Diminished ability to think or concentrate
9. Recurrent thoughts of death or suicide

Note. The symptoms cause clinically significant distress or impairment, are not caused by the effects of a substance or general medical condition, and are not better accounted for by bereavement or another psychiatric disorder.

suicide, and indirect costs associated with depression in the workplace) was estimated at approximately $44 billion (22). Depression ranked fourth in the list of conditions with the greatest global disease burden for the year 2000 (48). Clearly, MDD is a prevalent, recurrent disorder associated with significant morbidity and economic costs.

PATHOPHYSIOLOGY

Depression is widely viewed as an interaction between genetics and the environment. As a result, both neurochemical and psychosocial variables have been implicated to contribute to the pathophysiology of depression. Specifically, three neurotransmitters systems—norepinephrine, serotonin, and dopamine—have been theorized to play a significant role in the onset and maintenance of depression. These theories of neurotransmitter dysregu-

lation are complex but generally suggest neurotransmitter deficiencies in individuals with depression. In addition, hyperactivity in the hypothalamus–pituitary–adrenal axis (a marker of reaction to stressors) has been implicated in the pathophysiology of depression. Also, a widely studied psychosocial contribution to the etiology of depression involves the role that negative cognitions play in causing depressed mood. Exercise training may serve to improve depressive symptomatology through alterations in neurotransmitter functioning or improvement in negative evaluations of self (e.g., enhanced self-efficacy and self-worth).

MEDICAL AND CLINICAL CONSIDERATIONS

The gold standard for the assessment of depression is a structured diagnostic interview administered by a trained mental health professional, but an extensive assessment of that kind is impractical in most exercise settings. Fortunately, a number of tools or measures exist that allow exercise physiologists to screen for depressive symptoms and to refer patients to the appropriate mental health professional for further assessment when necessary.

Signs and Symptoms

Besides helping to administer these pen-and-paper screening tests, exercise physiologists should be cognizant of the cardinal signs and symptoms that might alert them to the fact that intervention by a trained mental health professional is urgent. Examples of such include self-reported symptoms of hopelessness and suicidal ideations. Patients who report these symptoms should be carefully monitored and the appropriate healthcare professional should be consulted (e.g., the patient's primary care provider).

History and Physical Exam (or Diagnostic Testing)

A number of self-report measures are available to assess the presence and severity of depressive symptoms. These questionnaires generally consist of a series of written questions with multiple-choice responses that patients complete on their own. Widely used examples include the Beck Depression Inventory-II (BDI-II; 6), the Center for Epidemiological Studies Depression Scale (CES-D; 41), and the Hospital Anxiety and Depression Scale (HADS;

54). These measures are used to obtain information about the type and severity of depressive symptoms. For additional depression management strategies, see practical application 9.1.

Self-report questionnaires are not designed to diagnose MDD and should not be substituted for a diagnostic clinical interview conducted by a trained mental health professional. Instead, these screening questionnaires enable clinical exercise physiologists and other clinicians to triage each case and develop a plan for action. For example, one patient's depressive symptoms may be mild enough to manage within the exercise setting, whereas another patient may benefit from referral to a mental health provider for further evaluation of mood. Often, it is not practical for a mental health specialist to assist in the interpretation of these questionnaires, so specific screening questions have been developed for use by primary care physicians. For example, the Primary Care Evaluation of Mental Disorders (PRIME-MD) is an assessment procedure (available in English and Spanish) intended to detect psychiatric disorders in primary care patients (45). As a brief screen, evidence suggests that two simple questions from the PRIME-MD may be used to identify individuals who may need further evaluation for depression and thus would be appropriate for use in the exercise setting (47,52). The two PRIME-MD depression-screening questions are as follows:

1. During the past month, have you been bothered by having little interest or pleasure in doing things?

2. During the past month, have you been bothered by feeling down, depressed, or hopeless?

A positive screen is indicated if the answer to either question is yes.

Practical Application 9.1

Client–Clinician Interaction

Exercise professionals who work with patients with depression may find the experience both challenging and rewarding. As noted elsewhere in the chapter, symptoms of depression such as fatigue and loss of interest in people and activities may adversely affect a patient's ability to adhere to an exercise program.

Patients with depression often require more attention by the exercise professional compared with nondepressed individuals. The exercise professional's consideration may be especially important during the initiation phase of an exercise program. For example, many patients with depression feel a lack of self-worth and may not believe in their ability to participate successfully in an exercise program. The loss of energy that occurs with depression may further contribute to a patient's reluctance to begin exercising. Thus, depressed patients may need more reassurance and positive reinforcement (even for just showing up to an exercise-training session) when they begin the program. In addition, the cognitive symptoms of depression such as indecisiveness and difficulty in concentrating may lead to problems in recalling the exercise prescription or remembering to set exercise goals. Occasional reminders of such exercise information can be helpful (presenting information in both oral and written forms is a good idea).

Depression often is accompanied by social problems such as family conflict and unemployment. The exercise professional should be aware that such stressful life situations may present barriers to participation in an exercise program. For example, a depressed patient who is going through a divorce and is between jobs may be less likely to consider exercise a priority.

When clinicians encounter a patient with depressive symptoms that interfere significantly with exercise participation, the appropriate approach may be to refer him or her to a mental health provider for treatment to improve quality of life and, potentially, to achieve motivation to exercise. Several treatments for depression have shown success, including antidepressant medication and cognitive-behavioral therapy. Patients with depression should also be informed that exercise training has been shown to reduce depressive symptoms in both healthy and medically ill populations.

In summary, depressed patients will most likely require greater attention to ensure continued adherence to the exercise prescription compared with patients who are not depressed. Because depression is associated with drop out from exercise programs, staff should closely monitor depressed clients and refer them for treatment when appropriate.

Finally, in the assessment of depression, recognize that several of the symptoms of depression (especially those within the somatic realm, such as weight loss and fatigue) are also symptoms of a variety of medical diseases. Thus, the diagnosis of depression can be complicated in patients who also are medically ill (12).

Treatment of Depression

Many people with depression do not seek treatment. In a recent household survey of 9,282 adults aged 18 yr and older (49), 6.7% of respondents met criteria for MDD. Of these individuals, 56.8% had sought treatment for MDD within the previous 12 mo period. Unfortunately, only 37.5% of those who sought treatment received treatment that was considered even minimally adequate (i.e., based on current, evidence-based treatment guidelines). These results suggest that MDD is untreated or inadequately treated for most individuals. Indeed, the undertreatment of MDD has been recognized as a significant public health problem in the United States (24), particularly among ethnic minority groups (37).

Except in cases of severe depression (for which care is usually rendered by a psychiatrist, psychologist, or other mental health professional), most depressed patients who seek treatment are treated by primary care physicians (53). Patients who present with depression in the primary care setting are more likely to be offered medication than psychotherapy or a combination of medication and psychotherapy (42).

Antidepressant Medication

Antidepressant medication is the most widely used treatment for depression in the United States (37; see practical application 9.2 for a list of widely used antidepressant medications). Results from numerous randomized clinical trials have shown the efficacy of antidepressant medications, particularly for patients with a severe level of depression (46). **Selective serotonin reuptake inhibitors** (SSRIs), a class of widely used antidepressant medications, has been shown to be safe in a variety of patient populations, including patients with diabetes, cardiovascular disease, and cancer (21).

Unfortunately, evidence suggests that a significant minority of people do not respond to treatment with antidepressant medication (4). In addition, adherence to antidepressant medication regimens often is inadequate relative to treatment guidelines (14), and early discontinuation of treatment is associated with recurrence of MDD (36). These findings underscore the importance of identifying alternatives to pharmacotherapy in the treatment of depression. Two such treatments that have received empirical support are **cognitive-behavioral therapy** (CBT) and regular exercise.

Practical Application 9.2

Pharmacology— Commonly Prescribed Antidepressant Medication

Medication	Class	Exercise effect	Other effects
Bupropion	Miscellaneous	Rapid heart beat, dizziness	Restlessness, insomnia, headache, tremor, dry mouth, confusion, nausea, constipation, menstrual complaints
Citalopram	SSRI	None identified	Nausea, nervousness, insomnia, diarrhea, rash, agitation, sexual side effects
Duloxetine	SNRI		
Escitalpram	SSRI		
Fluoxetine	SSRI		
Paroxetine	SSRI		
Sertraline	SSRI		
Venlafaxine	SNRI		

Note. Buproprion does not share a chemical structure with other types of antidepressant medications and as a result is considered a miscellaneous or "other" antidepressant medication. SSRI = selective serotonin reuptake inhibitors; SNRI = serotonin and norepinephrine reuptake inhibitors.

Cognitive-Behavioral Therapy

Cognitive-behavioral therapy emphasizes the important influence that negative thoughts have on emotion and behavior (5). For people with MDD, treatment focuses on modifying maladaptive thoughts as well as addressing deficits in behavior (e.g., unassertiveness, isolating oneself from others) that lead to and maintain depression. For example, a depressed patient undergoing CBT might be taught to identify cognitive distortions associated with her or his depression, to challenge those distortions, and to replace them with more realistic thoughts. Because patients with depression often lack motivation, another therapeutic exercise might involve planning a daily schedule of activities in which to engage. Following the schedule may result in decreased boredom and loneliness as well as increased motivation.

In summary, the goal of CBT is to reduce symptoms of depression by changing maladaptive thoughts and behaviors. CBT is widely used and has been demonstrated to be as effective as pharmacotherapy in treating MDD (18). Table 9.1 provides a brief list of CBT resources that offer more information about this treatment.

Exercise Testing

Intuitively, several of the key symptoms of depression might be expected to affect performance during exercise testing. For example, a patient with psychomotor retardation and fatigue may demonstrate poorer performance relative to nondepressed patients. Using the PRIME-MD and the BDI-II, Lavoie and colleagues (29) assessed depression in 1,367 patients referred for exercise stress testing. Thirteen percent of the sample was found to meet criteria for MDD. Compared with nondepressed patients, individuals with MDD exhibited a lower peak heart rate when expressed as a percentage of age-predicted maximum, a lower peak MET level, and shorter total exercise duration. Similar results were found in a sample of outpatient veterans with stable coronary artery disease (43). Thus, some researchers (29) recommend routine depression

screening of patients who are undergoing exercise testing. For patients with significant depression, pharmacological stress testing may be preferred as a way to minimize the effect of motivational factors on test performance.

Exercise as a Treatment for Depression

Several studies have found exercise to be an effective treatment for depression (34,35,44). In a recent meta-analysis (30), the researchers concluded that exercise was superior to no treatment and was as effective as cognitive therapy in treating depression. To date, only one study has examined the use of exercise compared with antidepressant medication (sertraline hydrochloride) as a treatment for MDD in older adults. In this study (8), 156 middle-aged and older adults diagnosed with MDD were randomized to supervised exercise, sertraline, or a combination of exercise and sertraline. The 16 wk exercise treatment consisted of three weekly sessions of aerobic activity. By the end of the treatment period, each of the three treatment groups experienced a significant reduction in their levels of depression. The treatments did not differ significantly from one another in efficacy, suggesting that exercise may be a viable alternative to medication in the treatment of MDD.

A follow-up assessment completed 6 mo after treatment was initiated (3) showed that those individuals assigned to exercise alone experienced lower rates of depression than did those who received medication or a combination of exercise and medication. In addition, only 9% of remitted participants in the exercise group relapsed, compared with more than 30% of participants in the medication and combination groups. Another finding was that 64% of participants who received the exercise treatment continued to exercise following completion of the program. Self-reported exercise among all participants was associated with a 50% reduction in risk of depression 6 mo after the study ended. In sum, exercise may be as effective as antidepressant medication in the reduction of depressive symptoms.

Table 9.1 Cognitive-Behavioral Therapy (CBT) Resources

Resource	Description
Feeling Good: The New Mood Therapy Revised and Updated	Self-help book written by David Burns that uses the basic principles of CBT (11)
Association for Behavioral and Cognitive Therapies Web site (www.aabt.org)	Web site that can be used to locate psychotherapists by geographical location
Cognitive Therapy: Basics and Beyond	Book written by Judith Beck that describes the cognitive model and therapy in detail (7)

A number of potential mechanisms may be responsible for the reduction in depressive symptoms associated with exercise (10). For example, the central monoamine theory suggests that exercise corrects dysregulation of the central monoamines believed to lead to depression. Psychological factors also may be responsible for exercise-related improvements in mood. One hypothesis is that exercise reduces depression through increases in self-esteem and self-efficacy. Other potential psychological mechanisms include the distraction from negative emotion provided by exercising as well as the behavioral activation occurring with exercise, which is also an important component of CBT for MDD. The issue of how much exercise is needed to achieve an anti-depressant effect is also an important topic, one that is discussed in practical application 9.3. An exercise prescription for patients with depression appears in practical application 9.4 on p. 176.

EXERCISE PRESCRIPTION

Developing an exercise prescription for people with depression will likely differ little from the prescription used for healthy individuals. Clinicians should be aware, however, that several symptoms of depression (e.g., loss of interest, fatigue) may interfere with participation in exercise, and that **anxiety** disorders can further complicate matters. Concerning anxiety, although its relationship with exercise is not as well studied as that between depression and exercise, evidence suggests that anxiety does affect exercise participation and compliance rates (see "Anxiety and Exercise").

For individuals whose depression is considered stable enough to begin an exercise program (e.g., they are not suicidal, their depression is not so severe that it prevents active participation), the question arises about how to manage depressive symptoms that may affect the course of participation in exercise. First, recognize that significant comorbidity exists between depression and other chronic diseases; people with depression are unlikely to present with depression alone. Furthermore, depression can affect the course of chronic disease. For example, the presence of depression is associated with more complications and increased mortality in patients with diabetes (13,26). Evidence also identifies depression

Practical Application 9.3

Literature Review

In general, studies have demonstrated the effectiveness of exercise in treating depression. One question that remains unanswered is the dose of exercise required to obtain an antidepressant effect. Specifically, what frequency, intensity, and duration are most beneficial in treating patients with depression? Dunn and colleagues (16) recently completed a trial that was designed to address this issue. In this trial, 80 sedentary adults diagnosed with major depressive disorder (MDD) were randomized to undergo 12 wk of one of five aerobic exercise-training treatment conditions: low energy expenditure and 3 d of exercise training per week, high energy expenditure and 3 d of exercise training per week, low energy expenditure and 5 d of exercise training per week, high energy expenditure and 5 d of exercise training per week, or stretching and flexibility control.

Results of the trial showed that exercise conducted at the high energy expenditure dose (consistent with public health recommendations) was effective in reducing depressive symptoms (47% from baseline measurement) over the 12 wk treatment period. In contrast, although participants randomized to the low energy expenditure dose did experience some reduction in depression over the treatment period (30% from baseline measurement), those participants did not respond significantly better than participants in the control condition did (29% from baseline measurement). Regarding frequency of training, no significant difference in treatment response was found between those participants who exercised 3 d per week relative to those who exercised 5 d per week. This result identifies total energy expenditure as the key aspect of exercise dose related to remission of MDD, regardless of days per week exercised.

In summary, this trial demonstrated that an exercise-training dose consistent with public health recommendations for energy expenditure is effective for the treatment of MDD in previously sedentary adults. Importantly, an exercise prescription that includes lower energy expenditure does not appear to be beneficial in reducing depressive symptoms. Thus, exercise professionals should encourage depressed patients to achieve at least the minimum recommended levels of energy expenditure.

as a powerful and independent risk factor for cardiac outcomes in patients with coronary heart disease (such as recurrent myocardial infarction and mortality; 32). For cancer patients, untreated depression has been linked to poorer treatment adherence, increased hospital stays, and mortality (39).

Besides often being present as a comorbidity in patients with another chronic disease, depression is associated with unhealthy lifestyle behaviors such as tobacco and alcohol use (2), poor diet, and physical inactivity (9).

Adherence and Exercise

Depression has been found to be associated with decreased adherence to treatment regimens across a wide variety of medical illnesses. In a quantitative review of 12 studies that included patients with renal failure, cardiovascular disease, arthritis, and cancer, patients with co-existing depression were three times more likely to be nonadherent to treatment recommendations as were nondepressed patients (15). This finding has been replicated in more recent studies of patients with diabetes (33) and coronary heart disease (19), as well as with individuals referred to exercise programs. For example, results of studies of depressed patients enrolled in cardiac rehabilitation have demonstrated that depression status at program entry is predictive of number of sessions attended as well as dropping out (20). Thus, patients with depression may find it more difficult to stay engaged in an exercise program compared with patients who are not depressed, and specific symptoms of depression such as fatigue and a loss of interest in people and activities may

Anxiety and Exercise

In patients with depression, anxiety may be the single best predictor of both treatment dropout and failure to benefit from treatment. For example, participants enrolled in an exercise treatment program who also rated themselves as high in anxiety during the screening period were at greater risk of dropping out of the treatment program prematurely. In addition, if patients (with comorbid anxiety and depression) completed the treatment program, they were less likely to experience remission of depression (23).

Exercise professionals may encounter several types of anxiety:

Social phobia: This disorder is characterized by an intense fear of making mistakes or looking foolish in public. This fear often leads to avoidance of certain people, places, or social events.

Posttraumatic stress disorder (PTSD): PTSD arises in response to experiencing or witnessing a traumatic event. Although many people exposed to trauma temporarily experience stress-related symptoms, those with PTSD continue to struggle with intrusive thoughts and nightmares, as well as increased arousal (e.g., anger) and avoidance of reminders of the trauma.

Panic disorder: People with panic disorder experience sudden, unexpected periods of extreme fear known as panic attacks. Some symptoms of panic attacks include sweating, heart palpitations, a feeling of choking, dizziness, and a fear of dying.

Obsessive-compulsive disorder (OCD): OCD is characterized by disturbing, uncontrollable thoughts (obsessions). To alleviate these thoughts, individuals engage in repetitive behaviors (compulsions) in an effort to prevent some feared situation (e.g., excessive hand washing to prevent contamination).

Generalized anxiety disorder (GAD): People with GAD experience excessive worry across a number of life domains (e.g., family, work) that is difficult to control. The worry is associated with insomnia, muscle tension, and restlessness.

Exercise professionals should familiarize themselves with the symptoms of anxiety disorders. Remember that occasional, minimal symptoms of anxiety are a normal part of life. To meet criteria for a disorder, symptoms must result in impaired functioning in an aspect of a person's life. If a patient appears to be experiencing significant anxiety, the exercise professional may wish to discuss this with him or her and make a referral for treatment as needed. Both CBT and anxiolytic medication have been shown to be effective in treating anxiety disorders.

Practical Application 9.4

Summary of Exercise Prescription

Mode	Frequency	Intensity	Duration	Special considerations
Gross motor activities such as walking or biking	Five times per week	60%–75% of heart rate reserve	40–60 min	Untreated or undertreated depression can negatively affect adherence to exercise. Facilitate care to prevent depression symptoms from interfering with exercise participation

interfere with adherence to an exercise regimen (see "Strategies for Improving Exercise Adherence in Patients With Depression").

O'Neal and colleagues (38) have offered recommendations for working with depressed people in the supervised exercise setting. First, they emphasize that when working with depressed individuals, nonadherence should be expected. Exercise professionals should avoid judging or blaming the patient for his or her depression, because doing so will likely lead to guilt and a sense of failure that may cause the person to drop out from the exercise program. Instead, when nonadherence occurs, it should be viewed as a learning opportunity. That is, lapses in exercise participation can be used to identify an individual's unique barriers to adherence. The exercise professional can then assist him or her in finding ways around these obstacles. Finally, when working in exercise settings, the exercise professional should be familiar with the symptoms of depression and have some knowledge of treatment options. When depression is identified, the exercise professional should express warmth and empathy toward the patient while taking care to keep an appropriate clinician–client boundary. That is, the exercise professional should not attempt to be the patient's psychotherapist but should instead have referral sources available.

CONCLUSION

In summary, depression is a prevalent condition that can affect exercise testing as well as level of participation in exercise programs. Screening for depressive symptoms that may interfere with exercise and subsequent referral for treatment is essential. The practicing clinical exercise physiologist should appreciate that the depressed person who exercises is at risk for noncompliance because of the disorder itself, that depression is common in patients with other chronic diseases, and that associated depressive symptoms (e.g., loss of interest, fatigue) may interfere with exercise habits if such therapy is being used to help treat the disorder. People with depression will likely require increased attention from exercise professionals to ensure adequate adherence to the exercise prescription.

Fortunately, several treatments have shown success in the treatment of depression, including antidepressant medication, cognitive-behavioral therapy, and exercise training. Patients who exhibit significant depressive symptoms should be approached in an empathetic manner and encouraged to seek treatment to improve quality of life and gain maximal benefit from exercise.

Strategies for Improving Exercise Adherence in Patients With Depression

- Most important, work to establish good rapport with patients. Positive feedback and empathy from exercise staff can go a long way toward promoting adherence.

- At the initiation of an exercise program, review with patients their unique barriers to participation (e.g., work responsibilities, family issues). These barriers should then be discussed with patients to find ways to overcome or minimize these obstacles.

- Educate patients about the benefits of exercise for physical health and depression. Elicit from patients other benefits of exercise specific to them and periodically remind them of those benefits.

- Patients are more likely to adhere to exercise training if the experience is enjoyable. Work with patients to increase their satisfaction with the program (e.g., switching equipment used, varying the time of day).

- Help patients develop realistic exercise goals (e.g., gradual increase in number of exercise sessions per week).

- Encourage patients to reward themselves for participation in exercise. Emphasize the importance of positive reinforcement for accomplishments. Even simple rewards (e.g., a new book, a pleasant dinner out) can be powerful motivators.

- Recommend to patients that they talk to family members and friends about their exercise program and goals. Such people are often a valuable resource in offering encouragement for patients' participation in exercise.

- Remember that untreated or undertreated depression is likely to have a negative effect on adherence to exercise. Encourage patients to seek treatment for depression if symptoms appear to interfere with exercise participation.

Case Study

History and Diagnosis

Ms. TH is a 62 yr old female who joined your medical-fitness center with the intention of reinitiating an exercise program. She is motivated to exercise because her primary care provider recommended that she increase her activity level because of recent weight gain. Ms. TH revealed that she has struggled with her mood for the past year and that her primary care physician recently diagnosed her with major depression. Her physician started her on sertraline to manage her symptoms. She noted that before becoming depressed she enjoyed walking around her neighborhood several times per week but has been sedentary since becoming depressed. Her medical history is significant for hypothyroidism, for which she takes daily levothyroxine.

Exercise Test Results

Ms. TH recently underwent exercise testing, results of which showed that she exercised for 6 min (Bruce protocol, approximately 7 METs) and achieved a maximum heart rate of 136 beats per minute (resting = 73 beats per minute) and blood pressure of 165/76 mmHg (resting = 118/76). The test was free of ischemia, chest pain, and arrhythmia. It was noted that, although she was not experiencing symptoms, she required much prompting throughout to continue the test.

(continued)

Exercise Prescription

Ms. TH's exercise prescription would not differ significantly from that of other healthy but sedentary adults. Specifically, over several weeks and as tolerated, the prescription should progressively increase her activity levels from being inactive to exercising 3 to 5 d per week. During walking or cycling, she should try to achieve an intensity that corresponds to 60 to 70% of heart rate reserve, which is 110 to 117 beats per minute.

Discussion Questions

1. Based on her exercise test results, how might you interact with Ms. TH as she begins her exercise program?

2. What do you note in Ms. TH's history that is promising in terms of starting an exercise program?

3. What can you tell Ms. TH about the effects of exercise on depression?

4. What symptoms should you monitor to decide whether Ms. TH's depression is worsening while she is participating at your center? How might you handle this situation?

5. Given her age, we would predict a peak heart rate around 158 beats per minute. Assuming that she is not on a β-adrenergic blocking agent, what might account for a peak heart rate that appears submaximal in nature?

PART III

Endocrinology and Metabolic Disorders

Issues related to abnormal metabolic function of the human body are complex and extremely important with respect to the potential detrimental health effects that may occur. Some of these conditions result in their own specific effects or hazards to health, such as diabetes, high blood pressure, and renal failure. But these and the others listed in this part (metabolic syndrome, obesity, dyslipidemia) increase the risk for a multitude of other chronic diseases including, but not limited to, cardiovascular disease and cancer.

Chapter 10 delves into how conditions such as diabetes, obesity, hypertension, and dyslipidemia, each of which is the topic of a subsequent chapter, tend to cluster to make up what is termed *metabolic syndrome*. Although specific definitions of metabolic syndrome are subject to debate, it is clear that when these conditions are combined they generate greatly increased risk for cardiac and other diseases.

Diabetes has risen to epidemic proportions in the United States and throughout the world. Certainly, a strong behavioral component influences the risk and development of type 2 diabetes. Chapter 11 presents information about the rise of diabetes prevalence and the role that the clinical exercise physiologist can play in the prevention and treatment of diabetes with exercise. Given that the prevalence is only expected to increase throughout the next several decades, this topic has become more important than ever before.

As with diabetes, the rate of rise of obesity, especially over the past decade or two, has been unprecedented. The link of obesity to diabetes and heart disease is undeniable, and obesity is linked as well to increased risks of cancer, arthritis, disability, hypertension, and many other chronic diseases. As shown in chapter 12, weight-loss strategies can be effective, and exercise physiologists can play an active role in implementing these plans. Additionally, strong evidence suggests that exercise is one of the most important parts of weight-loss maintenance. The clinical exercise physiologist will continue to be looked upon to develop and implement sustainable exercise programs for those actively involved in a weight-loss program.

Although hypertension awareness, diagnosis, and treatment have been enhanced tremendously since the 1960s, hypertension remains a leading contributor to as many as 10% of all deaths in the United States. Exercise training can undoubtedly enhance both the prevention and treatment of hypertension. Chapter 13 provides specific exercise recommendations for both at-risk and hypertensive disease populations.

Although the role that dyslipidemia plays in the development of atherosclerosis is debatable, the association is irrefutable. Therefore, therapies such as statins and nutritional counseling have been developed to treat abnormal blood lipid values. Exercise can play a role in this treatment regimen with respect to controlling weight, which can positively affect lipids, and raising HDL levels. Chapter 14 presents this information in detail.

The extent of kidney failure is on the rise throughout the United States. Many people with failing kidneys live with other chronic diseases including heart failure, diabetes, and hypertension. These patients often are treated with hemodialysis. As discussed in chapter 15, exercise is an important treatment modality for the comorbid conditions associated with renal failure. Additionally, because hemodialysis requires up to 3 to 4 h several times per week, exercise has increasingly been incorporated into these settings over the past decade.

Metabolic Syndrome

David Donley, MS

The metabolic syndrome is a term for a constellation of endogenous risk factors that increase the risk of developing atherosclerotic **cardiovascular disease** (ASCVD), **type 2 diabetes mellitus**, and recently reported, kidney disease (16). Although metabolic syndrome is a widely recognized condition, the diagnosis and classification of this disease is less than clear. Recently, the number of metabolic disturbances associated with this syndrome has been shown to include **microalbuminuria**, **hyperuricemia**, **fatty liver disease**, deficient fibrinolysis, and increased **C-reactive protein** (CRP) levels indicating chronic inflammation (7,9). As a result of these recent discoveries, several health agencies have revised their criteria for diagnosing metabolic syndrome. Currently, four health agencies provide similar yet unique criteria for the diagnosis of the metabolic syndrome. These four agencies and their criteria for metabolic syndrome are listed in table 10.1. Notable risk factors that are common in all four definitions are **abdominal obesity**, elevated **blood pressure** and glucose levels, and **dyslipidemia** characterized by low levels of high-density lipoprotein (**HDL**) **cholesterol** and elevated **triglycerides** (30).

SCOPE

The prevalence of metabolic syndrome varies depending on the defining criteria that are used to diagnosis the syndrome. Most of the epidemiological studies that examine the prevalence of metabolic syndrome use either the definitions set forth by the World Health Organization (WHO) or the ATP III guidelines. Using the WHO criteria, 25.1% of the U.S. population meets the diagnostic criteria for metabolic syndrome, whereas using the ATP II guidelines, 23.9% of the population meets the diagnostic criteria for metabolic syndrome. On closer examination, the prevalence rates of metabolic syndrome increase dramatically with age. Ford and colleagues reported the prevalence of the metabolic syndrome is <10% in individuals aged 20–29 yr, 20% in individuals aged 40–49 yr, and 45% in individuals aged 60–69 yr. These findings suggest that the estimated prevalence of metabolic syndrome is a person's age minus 20.

Within the United States, the prevalence of metabolic syndrome also varies significantly by race and ethnicity. The highest prevalence rates of metabolic syndrome have been reported in Mexican Americans at 31.9% compared with 23.8% among non-Hispanic whites and 21.6% among African Americans. As one would expect, metabolic syndrome is primarily seen in **overweight** and obese people. The prevalence of metabolic syndrome in normal weight (BMI <25) men and women (4.6% and 6.2% respectively) is much lower than in overweight (BMI 25.0–29.9) men and women (22.4% and 28.1% respectively) and obese (BMI ≥30) men and women (59.6% and 50.0% respectively). With the number of overweight and obese people reaching epidemic levels within the United States and worldwide, the prevalence of metabolic syndrome will also reach epidemic proportions and place enormous burdens on world economies. Current predictions are that by the year 2020 as much

Table 10.1 Criteria for Diagnosing Metabolic Syndrome

World Health Organization (4,5): Diabetes or impaired fasting glucose, or impaired glucose tolerance or insulin resistance plus two or more of the factors listed below
American College of Endocrinology (10): Fasting serum glucose 110–125 mg · dl^{-1} or 2 h post glucose challenge (75 mg) of 140–200 mg · dl^{-1}
NCEP ATP III (1): Three or more of the factors listed below
International Diabetic Federation (3): Central obesity (waist circumference ≥94 cm for Europid women and ≥80 cm for Europid men, with ethnic-specific cut points for other groups) plus two or more of the factors listed below

	Body mass index	Waist-to-hip ratio	Triglycerides (mg · dl^{-1})	High density lipoprotein (mg · dl^{-1})	Blood pressure (mmHg)	Microal-buminuria	Fasting serum glucose (mg · dl^{-1})
World Health Organization (4,5)	>30	>0.85 (female) >0.90 (male)	≥150	<39 (female) <35 (male)	≥140/90	≥ 20 mcg · min^{-1} or ≥ 30 mcg · mg^{-1} albumin-to-creatinine ratio	See definition above
ACE	Obesity is included in a list of factors that increase the likelihood of IRS		≥150	<50 (female) <40 (male)	≥130/85		See definition above
NCEP		Abdominal circumference >88 cm (female) and >102 cm (male)	≥150	<50 (female) <40 (male)	≥130/85		≥110
IDP			≥150 or specific treatment	<50 (female) or specific treatment <40 (male) or specific treatment	≥130/85 or specific treatment		≥100 or previously diagnosed diabetes

as 40% of the population will have metabolic syndrome. As a result metabolic syndrome has often been referred to as "the disease of the new millennium" (13).

PATHOPHYSIOLOGY

Although the etiology of metabolic syndrome is not well understood, many theories or factors have been proposed to explain this clustering of multiple vascular risk factors (i.e., abdominal obesity, atherogenic dyslipidemia, **hypertension, insulin resistance**, and **glucose intolerance**). Although metabolic syndrome is likely a product of the interaction of multiple factors, evidence

suggests that the development of insulin resistance and the presence of central or visceral adiposity are central components of metabolic syndrome and can potentially account for many of the metabolic disturbances associated with metabolic syndrome (e.g., endothelial dysfunction, dyslipidemia, hypertension, chronic inflammation). See figure 10.1 for an explanation.

Excessive adipose tissue is associated with insulin resistance, dyslipidemia, and hypertension, all of which are key components to metabolic syndrome. Additionally, visceral adipose tissue has been shown to be an even better predictor of increased risk of CVD in mildly obese and obese subjects.

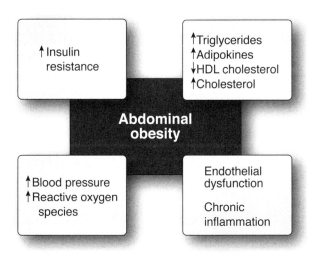

Figure 10.1 Components of metabolic syndrome.

Recent research demonstrates that white adipose tissue is not just a storage site for triglycerides but also an endocrine tissue capable of secreting numerous bioactive molecules that affect metabolism. This finding may help explain how obesity increases a person's risk for insulin resistance and CVD. Leptin, adiponectin, tumor necrosis factor alpha (TNF-α), resistin, angiotensinogen, interlukin-6 (IL-6), plasminogen activator inhibitor (PAI-1), and C-reactive protein (CRP) are all adipokines that have been shown to be secreted from adipose cells. Disruption in the release or receiving of these adipokines has been linked to insulin resistance, type 2 diabetes mellitus, hyperlipidemia, and vascular diseases. These findings have led some to proclaim adipokines as the missing link between insulin resistance and cardiovascular disease.

Note that not everyone identified as having metabolic syndrome is at high cardiovascular risk based on traditional CAD risk factor analysis. But metabolic syndrome is a precursor to the development of diabetes unless appropriate interventions are initiated at an early stage of the syndrome. The identification of metabolic syndrome allows for initiation of early interventions in individuals who are destined to develop diabetes and thereby place them at high risk for CVD.

MEDICAL AND CLINICAL CONSIDERATIONS

Specific diagnostic criteria for assessing metabolic syndrome has not been uniformly accepted, although there is consensus over the primary symptoms. Following diagnosis, the major objectives for clinical management are to reduce the underlying causes of metabolic syndrome such as weight reduction and increasing physical activity

levels. Moreover, risk factors associated with the disease also need to be treated.

Signs and Symptoms

The signs and symptoms associated with metabolic syndrome can vary depending on which diagnostic criteria are used. Several signs, however, are extremely common in metabolic syndrome, and these have been referred to as the core components of metabolic syndrome. These signs include hypertension, abnormal glucose metabolism, dyslipidemia, and obesity (specifically, abdominal obesity). In addition, several other abnormalities have been linked to metabolic syndrome and may present in patients with metabolic syndrome. These include microalbuminuria, hyperuricemia, fatty liver disease, deficient fibrinolysis, inactivity or sedentary lifestyle, cigarette smoking, and chronic inflammation as assessed through elevated levels of C-reactive protein (CRP).

History and Physical Exam

The diagnosis of metabolic syndrome can vary depending on the defining guidelines that are used. Nevertheless, the identification of the core components of metabolic syndrome (hypertension, abnormal glucose metabolism, dyslipidemia, and obesity) should be made during a routine physical examination followed by a fasting lipid panel. A careful review of the patient's medical history will also provide beneficial insights in starting a comprehensive lifestyle approach for the patient.

Diagnostic Testing

Additional testing such as a glucose tolerance test may be warranted to rule out diabetes. The criteria for that diagnosis are reviewed in chapter 11. People with metabolic syndrome can benefit greatly from a comprehensive exercise program. Because these individuals are at a threefold increase of risk for coronary artery disease and ischemic cerebrovascular disease, they must be more closely monitored during the initial phases of exercise. In particular, cardiovascular (heart rate and blood pressure) and metabolic (blood glucose) responses to exercise should be evaluated because hypertension and impaired glucose metabolism are core components of metabolic syndrome.

Exercise Testing

According to the American College of Sports Medicine, people diagnosed with metabolic syndrome should be categorized as either moderate or high risk (depending on the diagnosing criteria employed) for the likelihood that untoward events will occur during exercise participation (2). Therefore, exercise testing should be performed with caution and with the intent of determining functional capacity to derive a safe individual training intensity (see table 10.2). Standardized treadmill protocols that use relatively small workload increases are recommended and are often well tolerated by those with metabolic syndrome. But protocols can, and should, be tailored to the specific elements of the syndrome (e.g., type 2 diabetes mellitus, obesity, hypertension).

Table 10.2 Components of Initial Fitness Assessment for Individuals With Metabolic Syndrome

Test type	Mode	Protocol specifics	Clinical measures	Clinical implications	Special considerations
Cardiovascular	• Treadmill (most common) • Cycle ergometer	Small (≤1 MET) staged or ramp protocol in case of treadmill. 25 W staged or ramp protocol in case of cycle ergometer	• Blood pressure • Heart rate • ECG • Pre and post blood glucose levels	Because of the increased likelihood of CVD and type 2 diabetes mellitus, abnormal cardiovascular and metabolic responses to exercise are more likely to occur and therefore need to be monitored closely	Treadmill protocols are favorable because the exercise program is likely to consist of walking and because walking is an activity that subjects are familiar with. Cycle ergometers are common when the level of obesity exceeds the safe operating capacity of some treadmills. Individuals with diabetes and severe peripheral neuropathy may also be more comfortable on a cycle ergometer
Strength	Variable resistance machines (e.g., Nautilus) are often less intimidating than free weights	1RM or estimation of 1RM from 8RM tests of major muscle groups (8–12 stations)	• Blood pressure • Heart rate • Blood glucose levels	Because resistance training normally results in an increase in diastolic blood pressure and because individuals with metabolic syndrome are likely to be hypertensive, blood pressure should be monitored closely. Instructions should be given on proper breathing techniques (avoiding Valsalva) during initial strength testing	
Range of motion					If the individual is obese, he or she may have reduced range of motion because of increased fat mass around joints

Treatment

The primary goals in the management of metabolic syndrome are to reduce the risk for clinical atherosclerotic disease and to prevent type 2 diabetes mellitus. Initial treatment strategies center on treating the underlying risk factors such as obesity, hypertension, and dyslipidemia. Because obesity is strongly associated with metabolic syndrome, treatments designed to promote weight loss (e.g., caloric restriction, increased physical activity, pharmacological agents, and surgical procedures when necessary) should be a central component of the treatment of metabolic syndrome in obese individuals. For obese people the goal for weight loss should be to achieve a decline of about 7% to 10% from baseline total body weight during a period of 6 to 12 mo (17). Reducing total body weight along with developing the ability to maintain weight loss has been shown to have a favorable effect on most, if not all, metabolic risk factors. Table 10.3 identifies the recommendations for the management of metabolic syndrome.

Pharmacological Management

Weight loss can be attained through a combination of restricting caloric intake, initiating an exercise program, and using weight-loss drugs in those with class II or class III obesity. A complete description of weight-loss drugs is provided in chapter 3. Weight-loss medications currently approved by the Food and Drug Administration (FDA) include sibutramine (Meridia) and orlistat (Xenical) for long-term use and phentermine for short-term use. Rimonabant represents a new class of drugs (selective endocannabinoid CB1 receptor antagonist) that has shown to produce weight loss and improvements in metabolic syndrome components independent of weight loss (32). Observers speculate that rimonabant may obtain FDA approval in the United States with a primary indication for metabolic syndrome, a first to obtain this primary indication. However, this medication has yet to receive approval from the FDA. Table 10.4 on p. 186 lists common pharmacological treatments.

Besides promoting weight loss, pharmacological management is useful in the management of hyperlipidemia, insulin resistance, and hypertension that may be associated with metabolic syndrome. Statin medications have been shown to lower **LDL cholesterol** and reduce the risk for atherosclerotic cardiovascular disease events in patients with metabolic syndrome (6). Triglyceride-lowering classes of drugs such as fibrates and nicotinic acid have also been reported to decrease the risk for ASCVD in patients with metabolic syndrome and type 2 diabetes mellitus (31). Besides lowering plasma levels of triglycerides, fibrate types of medications have also been shown to lower plasma levels of LDL cholesterol and increase plasma levels of HDL cholesterol (31). This effect is particularly beneficial in a metabolic syndrome in which plasma HDL cholesterol levels are often below acceptable levels. See table 10.4 for commonly used medications in individuals with metabolic syndrome.

Physical Activity

Regular physical activity has long been known to provide a strong protective effect against ASCVD and all-cause mortality (22). This protective effect may be through reductions in plasma LDL cholesterol, **VLDL cholesterol**, increased plasma HDL cholesterol, reductions in blood pressure, or improvements in insulin sensitivity. All these favorable effects have been reported following regular physical activity (1).

Table 10.3 Management Strategies for Treatment of Metabolic Syndrome

Lifestyle modifications	Athrogenic dyslipidemia	Hypertension	Insulin resistance	High risk of cardiovascular disease
1. Dietary modification (decreased total calories, increased soluble fiber) 2. Weight reduction 3. Increased physical activity (60 min activity daily) 4. Smoking cessation	1. Statins 2. Fibrates 3. Nicotinic acid 4. Omega-3 fish oils 5. High soluble fiber diet	1. ACE inhibitor or ARB class of drugs 2. Decreased sodium intake 3. Increased physical activity 4. Decreased psychological stress	1. Increased physical activity 2. Thiazolidinediones 3. Metformin 4. Meglitinides 5. Acarbose	1. Supervised physical activity 2. Low-fat diet 3. Low-dose aspirin

Table 10.4 Medications Commonly Prescribed for Individuals With Metabolic Syndrome

Medications	Class and primary effects	Exercise effects	Other effects	Special considerations
Telmisartan, irbesartan, valsartan	Angiotension II receptor blocker Blood pressure	↓ exercising blood pressure	↓ resting blood pressure	↓ new-onset type 2 diabetes mellitus in at-risk patients
Lovastatin, simvastatin, pravastatin, rosuvastatin	Statins Lipid lowering	No effect	↓ LDL cholesterol and some ↑ HDL cholesterol	Many statins are metabolized by cytochrome p450 and therefore have higher risks of drug–drug interactions
Gemfibrozil, fenofibrate	Fibrates Lipid lowering	Clofibrate may ↑ arrhythmias and angina in patients with prior MI	↓ triglycerides, ↑ HDL cholesterol	Often used with markedly elevated triglycerides
Niacin	Nicotinic acid Increase in HDL	Possibly ↓ blood pressure	↓ total cholesterol, LDL cholesterol, triglycerides	Most effective at ↑ HDL cholesterol. Can worsen glycemic control in patients with metabolic syndrome or diabetes
Omega-3 fatty acids	Fish oil	No effect	↓ triglycerides and improve insulin resistance	No hyperglycemic medication currently licensed for use in metabolic syndrome without type 2 diabetes mellitus
Metformin	Oral antihyperglycemic	No effect		
*Rimonabant	Selective CB1 endocannabinoid receptor antagonist Weight reduction			Preliminary indications show improvements in cardiovascular risk profiles independent of weight loss

*Expected to be FDA approved, but is not at present, with a primary indication for the metabolic syndrome (the first medication with a primary indication for metabolic syndrome). However, at present the FDA is not recommending it as an antiobesity treatment.

EXERCISE PRESCRIPTION

The exercise prescription for individuals with metabolic syndrome will depend on the presence and severity of the underlying risks (e.g., obesity, type 2 diabetes mellitus, hypertension, and established CHD). Because obesity is present in most individuals with metabolic syndrome, the exercise prescription guidelines (see table 10.5) should be based on those for obese patients. See chapter 12 (obesity). But care should be taken not to overlook other CVD risks such as hypertension, type 2 diabetes mellitus, and dyslipidemia when designing the exercise prescription. For the specific exercise prescription guidelines for each of these conditions, see chapters 11 (diabetes), 13 (hypertension), and 14 (dyslipidemia and hyperlipidemia).

Table 10.5 Exercise Presciption

Mode	Frequency	Intensity	Duration
Aerobic: walking, cycling, swimming	5–7 d per week	50-75% $\dot{V}O_2R$	45–60 minutes
Resistance exercise: 8–10 exercises for major muscle groups	2 d per week	12–15RM	One set
Flexibility: static stretching	Postexercise		10–30 s per exercise of each major muscle group

In general, because the emphasis of the exercise prescription is on obesity and weight loss, continuous or intermittent, low (40% to 60% $\dot{V}O_2R$ or HRR) to moderate (50% to 75% $\dot{V}O_2R$ or HRR) intensity aerobic exercises should be performed initially with goals of improving cardiorespiratory fitness and attaining energy expenditure goals (200–400 kcal per session) within reasonable time limits (45 to 60 min per session). Activities such as brisk walking, swimming, and cycling are usually well tolerated by those with metabolic syndrome. Resistance and flexibility exercises should be prescribed based on the presence of underlying CVD risks. Most individuals with metabolic syndrome will be overweight, and therefore the exercise prescription should be designed to maximize caloric expenditure. But because people with metabolic syndrome are at high risk of developing cardiovascular disease, additional physiological measures (e.g., blood pressure, heart rate, ECG) should be recorded during the initial fitness assessment.

Educating people with metabolic syndrome about their particular CVD risks and the ways in which physical activity affects those risks should be emphasized. Including physical activity in other parts of the person's lifestyle is also beneficial for furthering weight loss and lowering ASCVD risk. Practical application 10.1 provides information about client–clinician interaction. Individuals should be encouraged to minimize sedentary activities (e.g., television viewing) during leisure time and to replace those activities with activities such as walking, gardening, household chores, resistance training, or any activity that requires continual movement.

EXERCISE TRAINING

Regular physical activity has been reported to improve several cardiovascular risk factors associated with metabolic syndrome (20,27). For example, regular physical activity has been reported to reduce plasma LDL and

Practical Application 10.1

Client–Clinician Interaction

Motivating people with metabolic syndrome can be a challenge because many of these individuals may feel fine physically and therefore do not understand the seriousness of their condition. With the exception of increased body weight, the symptoms of metabolic syndrome are visually unrecognizable. Therefore, patient education on the seriousness of each of their conditions should be emphasized during the first visit. Although a large number of the people with metabolic syndrome are overweight and will benefit from even modest weight loss, patients need to be made aware of benefits other than weight loss that are associated with regular physical activity. In fact, weight loss attributed to physical activity without dietary manipulation is usually small. But regular physical activity has been shown to increase insulin sensitivity and HDL cholesterol and lower LDL cholesterol and resting systolic and diastolic blood pressure independent of weight loss. The benefits associated with regular physical activity may be sufficient in some people to the extent that pharmacological intervention may not be warranted. In the absence of weight loss, improvements in cardiovascular risks such as resting blood pressure, lipid profiles, and glucose tolerance may provide the psychological motivation needed to continue to adhere to a physically active lifestyle.

VLDL cholesterol levels, raise plasma levels of HDL cholesterol, lower blood pressure, and improve insulin sensitivity (8). Additionally, physical activity is associated with successful weight reduction and maintenance (21). Perhaps the most supportive evidence for regular physical activity is that moderate and vigorous physical activities are associated with a decreased risk of metabolic syndrome, independent of age, smoking habits, and alcohol use (29). Additionally, increased levels of cardiovascular fitness levels have been shown to elicit a strong protective effect against all-cause and cardiovascular mortality in men with metabolic syndrome (19).

Case Study

Medical History

Mr. GW is a 47 yr old white male computer technician who has been referred by his primary care physician to an exercise program for CVD risk reduction. He has a positive family history of heart disease; his father had a nonfatal myocardial infarction at age 53. Currently he is 5 ft 10 in. (178 cm) and weighs 240 lb (109 kg). He has been taking Avapro (150 mg qd) for the last 5 yr to control his hypertension. He reports being sedentary at work and home.

Diagnosis

Aside from occasional low-back pain resulting from prolonged sitting, his medical history is unremarkable. The results from his exercise test, body composition assessment, and fasting blood lipid profile were obtained before he started the exercise program. The results are as follows.

Exercise Test Results

Resting heart rate was 84 beats $\cdot$ min^{-1}, and resting blood pressure was 120/86. The subject performed a modified Bruce treadmill protocol and achieved a peak MET level of 8.0 at a heart rate of 164 beats $\cdot$ min^{-1} (98% of predicted max), blood pressure of 194/78, and an RPE of 18. An occasional premature ventricular contraction (PVC) was noted at rest and during recovery, but this resolved with exercise. Body composition testing using the BodPod system revealed a body fat percentage of 38%. Waist circumference was 130 cm.

Total cholesterol = 230 mg $\cdot$ dl^{-1}
HDL = 33 mg $\cdot$ dl^{-1}
LDL = 175 mg $\cdot$ dl^{-1}
Triglycerides = 240 mg $\cdot$ dl^{-1}

Exercise Prescription
Initial exercise program

Frequency = 5 d per week
Intensity = 98 to 115 beats $\cdot$ min^{-1} (60–70% max heart rate)
Resistance training, 8 to 12 reps, one set
Duration = 30 to 45 min
Mode = 20 to 30 min aerobic and 10 to 15 min resistance on opposite days

Exercise progression

Frequency = 5 d per week

Intensity = Progress toward upper end of heart rate intensity range as tolerated. Continue to add weight to resistance exercises and possibly add an additional set if time permits.

Duration = Gradually increase to 60 min aerobic plus time needed to complete two to three sets of resistance training.

The subject should be educated about the signs and symptoms of exercise intolerance as well as the significance of his health conditions (i.e., hypertension and metabolic syndrome). Blood pressure should be monitored initially during resistance training. Proper breathing while performing resistance training needs to be demonstrated to prevent Valsalva. Additionally, the subject should be taught how to self-monitor exercise intensity by checking his heart rate and using the Borg rating of perceived exertion scale. Weight loss will significantly help with the metabolic risk factors present. Therefore, a decrease in total caloric consumption and a low-fat diet combined with physical activity should be used to achieve a reduction in total body weight of 7 to 10% from baseline during the first 12 mo. The addition of statin pharmacotherapy may also be beneficial in lowering LDL cholesterol. Increased physical activity in daily activities such as taking the stairs and parking farther from the door when possible should also be emphasized.

Discussion Question

How can an exercise program help treat this person's metabolic syndrome? Increased physical activity will help normalize blood pressure, reduce LDL and total cholesterol levels, possibly raise HDL cholesterol, improve insulin sensitivity, and help with weight reduction. Independent of these risk reductions, regular physical activity has been shown to elicit a strong protective effect against cardiovascular disease and all-cause mortality.

11

Diabetes

Ann L. Albright, PhD, RD

Diabetes mellitus (diabetes) is a group of metabolic diseases characterized by an inability to produce sufficient insulin or use it properly, resulting in hyperglycemia (10). Insulin, a hormone produced by the β-cells of the **pancreas,** is needed by muscle, fat, and the liver to utilize glucose. The hyperglycemia resulting from diabetes places people with this disease at risk for developing microvascular diseases including retinopathy and nephropathy, macrovascular disease, and various neuropathies (both autonomic and peripheral).

SCOPE

Approximately 24 million people in the United States have diabetes. One-quarter of these are undiagnosed, in large part because symptoms may develop gradually and years can pass before severe symptoms appear (28,34). Even before symptom development, however, these individuals are at increased risk for developing complications (15,46,69,82,108).

Diabetes continues to become increasingly common in the United States. From 1980 through 2004, the number of Americans with diagnosed diabetes more than doubled (from 5.8 million to 14.6 million). Currently, 7.8% of the United States population has diabetes, and 1.6 million new cases of diabetes among people 20 years and older were diagnosed in 2007 (28). Estimates are that the number of Americans with diagnosed diabetes will double in the next 15 to 20 years. Diabetes is a worldwide

problem. The Centers for Disease Control and Prevention consider diabetes to be at epidemic proportions in the United States. The reasons for the epidemic are likely threefold:

1. An increasingly sedentary lifestyle and poor eating practices, resulting in a rise in overweight and obesity
2. The increase in high-risk ethnic populations (see below) in the United States
3. Aging of the population

Diabetes is currently the seventh-leading cause of death in the United States (28). African Americans, Hispanics, American Indians, Alaskan Natives, Native Hawaiians, other Pacific Islanders, and Asians have higher rates of diabetes than the non-Hispanic white population does (28). Death rates of middle-aged people with diabetes are twice those of people without diabetes. As serious as these mortality statistics are, they underestimate the effect of diabetes. Because people with diabetes usually die from the complications of this disease, diabetes is underreported as the underlying or contributing cause of death and is listed on only half of the death certificates of those with the disease (16,26). The economic effect of diabetes is staggering. The estimated direct costs (for medical treatment and services) and indirect costs (in time lost from work) of diabetes are estimated to be $174 billion per year (8). A large portion of the economic burden of diabetes is attributable to long-term complications and hospitalizations.

PATHOPHYSIOLOGY

The various forms of diabetes affect the options for treatment. All forms share the risk for developing complications. This section reviews the types of diabetes and associated complications.

Diabetes Categories

The American Diabetes Association recognizes four categories of diabetes, as listed in "Etiologic Classification of Diabetes Mellitus."

Type 1 Diabetes

Type 1 diabetes is caused by β-cell destruction resulting in an absolute deficiency of insulin. Consequently, insulin must be supplied by regular injections or an insulin pump. Those with type 1 diabetes are prone to develop diabetic ketoacidosis when marked hyperglycemia occurs. Approximately 5% to 10% of those with diabetes have type 1 (10).

Type 1 diabetes comprises two subgroups: immune-mediated and **idiopathic.** Type 1 immune-mediated diabetes was formerly known as juvenile-onset or insulin-dependent diabetes. This form of the disease usually occurs before age 30. Most cases occur in childhood or adolescence, and symptoms appear to develop abruptly. Type 1 immune-mediated diabetes is considered an **autoimmune** disease in which the immune system attacks the body's own tissues. Type 1 idiopathic diabetes is a new subgroup and represents only a small number of people with β-cell destruction. Estimates are that 10% to 20% of Caucasians who develop diabetes in adulthood may have immune-mediated β-cell destruction in which the disease develops over several months to years (10,121). These patients have variable insulin deficiency and only intermittently require insulin treatment.

Type 2 Diabetes

Type 2 diabetes was formerly called adult-onset or non-insulin-dependent diabetes. This is the most common form of the disease and affects approximately 90% to 95% of all those with diabetes (10). The onset of type 2 diabetes usually occurs after age 40, although it is seen at increasing frequency in adolescents (29,100).

The pathophysiology of type 2 diabetes is complex and multifactorial. Insulin resistance of the peripheral tissues and defective insulin secretion are common features. With insulin resistance, the body cannot effectively use insulin in the muscle or liver even though sufficient insulin is being produced early in the course of the disease (31,84,90,106). Type 2 diabetes is progressive.

Etiologic Classification of Diabetes Mellitus

Type 1 Diabetes
- β-cell destruction, usually leading to absolute insulin deficiency
- Immune mediated
- Idiopathic

Type 2 Diabetes
May range from predominant insulin resistance with relative insulin deficiency to predominant secretory defect with insulin resistance

Other Specific Types
- Genetic defects of β-cell function
- Genetic defects in insulin action
- Diseases of the exocrine pancreas
- Endocrinopathies
- Drug or chemical induced
- Infections
- Uncommon forms of immune-mediated diabetes
- Other genetic syndromes sometimes associated with diabetes

Gestational Diabetes Mellitus
- Any degree of glucose intolerance with onset or first recognition during pregnancy
- Patients with any form of diabetes may require insulin treatment at some stage of their disease. Such use of insulin does not, of itself, classify the patient.

Over time, the pancreas cannot increase insulin secretion enough to compensate for the insulin resistance, and hyperglycemia occurs (figure 11.1). The treatment options are medical nutrition therapy and exercise and, if medication is needed, oral agents or insulin. Ketoacidosis rarely occurs.

A clear genetic influence is present for type 2 diabetes. The risk in children of parents with type 2 diabetes for developing diabetes is about two times greater than

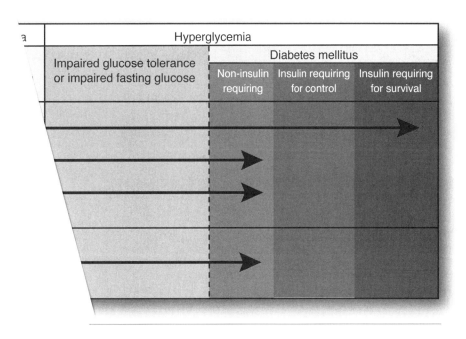

	Hyperglycemia			
	Impaired glucose tolerance or impaired fasting glucose	Diabetes mellitus		
		Non-insulin requiring	Insulin requiring for control	Insulin requiring for survival

...s Association. From Diabetes Care®, Vol. 31, Suppl. 1, 2008; S55-S60. Reprinted with permission ...es Association.

normal (64). Along with genetic influences, other risk factors are present. Obesity contributes significantly to insulin resistance, and most (80%) people with type 2 diabetes are overweight or obese at disease onset (25,67). An abdominal distribution of body fat (e.g., male belt size >40 in. [>102 cm] and female >35 in. [>89 cm]) is associated with type 2 diabetes (61). The risk of developing type 2 diabetes also increases with age, lack of physical activity, history of **gestational diabetes**, and presence of hypertension or dyslipidemia (49,120). The combination of hypertension, dyslipidemia, obesity, and diabetes is often termed *metabolic syndrome* or *cardiometabolic risk*. See chapter 10 for more information on metabolic syndrome.

Other Specific Types

The third category of diabetes, termed *other specific types*, accounts for only 1% to 2% of all diagnosed cases of diabetes (10). In these cases, certain diseases, injuries, infections, medications, or genetic syndromes cause the diabetes. This form may or may not require insulin treatment.

Gestational Diabetes

Gestational diabetes occurs during 7% of pregnancies (10). It is usually diagnosed during pregnancy by an oral glucose tolerance test, often performed routinely during the second trimester. Risk factors for developing this form of diabetes include family history of gestational

diabetes, previous delivery of a large birth weight (>9 lb [4 kg]) baby, and obesity. Although glucose tolerance usually returns to normal after delivery, approximately 50% of the women who develop gestational diabetes will go on to develop type 2 diabetes within 10 yr and should receive frequent long-term follow-up (79).

Complications of Diabetes

Associated complications of diabetes are categorized as acute and chronic. This section reviews those complications.

Acute Complications

The acute complications of diabetes are hyperglycemia (high blood sugar) and **hypoglycemia** (low blood sugar). Each of these acute complications must be quickly identified to ensure proper treatment and reduce the risk of serious consequences.

Hyperglycemia

The manifestations of hyperglycemia are as follows:

1. Diabetes out of control
2. Diabetic ketoacidosis
3. Hyperosmolar nonketotic syndrome (30)

Diabetes out of control is a term used to describe blood glucose levels that are frequently above the patient's **glycemic goals** (see following discussion and table 11.1).

Table 11.1 Suggested Treatment Goals for Blood Glucose, Blood Pressure, and Lipids

Glycemic control		Key concepts in setting glycemic goals:
A1C	<7.0%*	• A1C is the primary target for glycemic control.
Preprandial capillary plasma glucose	70–130 mg · dl^{-1} (3.9–7.2 mmol · L^{-1})	• Goals should be individualized based on: duration of diabetes, pregnancy status, age, comorbid conditions, hypoglycemia unawareness, individual patient considerations
Peak postprandial capillary plasma glucose^	<180 mg · dl^{-1} (<10.0 mmol · L^{-1})	
Blood pressure	<130/80 mmHg	• More stringent glycemic goals (i.e., a normal A1C, <6%) may further reduce complications at the cost of increased risk of hypoglycemia.
Lipids**		
LDL	<100 mg · dl^{-1} (<1.6 mmol · L^{-1})	
Triglycerides	<150 mg · dl^{-1} (<1.7 mmol · L^{-1})	• Postprandial glucose may be targeted if A1C goals are not met despite reaching preprandial glucose goals
HDL	>40 mg · dl^{-1} (>1.0 mmol · L^{-1})***	

*Referenced to a nondiabetes range of 4.0–6.0% using a DCCT-based assay.

^Postprandial glucose measurements should be made 1–2 h after the beginning of the meal, generally peak levels in patients with diabetes.

**Current NCEP/ATP III guidelines suggest that in patients with triglycerides ≥200 mg · dl^{-1}, the "non-HDL" cholesterol (total cholesterol minus HDL) be utilized. The goals is ≤130 mg · dl^{-1}.

***For women, it has been suggested that the HDL goal be increased by 10 mg · dl^{-1} (14).

High blood glucose levels cause the kidneys to excrete glucose and water, which causes increased urine production and dehydration. Symptoms of high blood glucose levels and dehydration are

- headache,
- weakness, and
- fatigue (30).

The best treatment for a patient with diabetes out of control includes drinking plenty of non-carbohydrate-containing beverages, regular self-monitoring of blood glucose, and, when instructed by a healthcare professional, an increase in diabetes medication. Frequent high blood glucose levels damage target organs or tissues, which increases the risk of chronic complications.

Diabetic ketoacidosis occurs in patients whose diabetes is in poor control and in whom the amount of **effective insulin** is very low or absent. This result is much more likely to occur in those with type 1 diabetes. **Ketones** form because without insulin, the body cannot use glucose effectively and a high amount of fat metabolism occurs to provide necessary energy. A by-product of fat metabolism in the absence of adequate carbohydrate is ketone body formation by the liver, causing an increased risk of coma and death. Ketone levels in the blood are approximately 0.1 mmol · L^{-1} in a person without diabetes and can be as high as 25 mmol · L^{-1} in a person with diabetic ketoacidosis. This level of ketosis can be evaluated with a urine dipstick test. Other symptoms of ketoacidosis include abdominal pain, nausea, vomiting, rapid or deep breathing, and sweet- or fruity-smelling breath. Exercise is contraindicated in anyone experiencing diabetic ketoacidosis.

Hyperglycemic hyperosmolar nonketotic syndrome occurs in patients with type 2 diabetes when **hyperglycemia** is profound and prolonged. This circumstance is most likely to happen during periods of illness or stress, in the elderly, or in those who are undiagnosed (30). The syndrome results in severe dehydration attributable to rising blood glucose levels, resulting in excessive urination. Extreme dehydration eventually leads to decreased mentation and possible coma. Exercise is contraindicated during periods of hyperglycemic hyperosmolar nonketotic syndrome.

Hypoglycemia

Hypoglycemia (also called insulin shock and insulin reaction) is a potential side effect of diabetes treatment and usually occurs when blood glucose levels drop below 60 to 70 mg · dl^{-1}. Hypoglycemia may occur in the presence of the following factors:

- Too much insulin or selected antidiabetic oral agents
- Too little carbohydrate intake
- Missed meals
- Excessive or poorly planned exercise (30)

Hypoglycemia can occur either during exercise or several hours later. Postexercise, late-onset hypoglycemia generally occurs following moderate- to high-intensity exercise that lasts longer than 30 min. This kind of hypoglycemia results from increased insulin sensitivity, ongoing glucose utilization, and physiological replacement of glycogen stores through gluconeogenesis (3). Patients should be instructed to monitor blood glucose before and periodically after exercise to assess glucose response. This approach is also recommended in clinical exercise programs, such as cardiac rehabilitation, especially in patients new to exercise. Suggested recommendations for pre- and postexercise blood glucose assessment are provided later in this chapter.

The two categories of symptoms of hypoglycemia are autonomic and neuroglycopenic. As blood glucose decreases, **glucagon,** epinephrine, growth hormone, and cortisol are released to help increase circulating glucose. Autonomic symptoms such as shakiness, weakness, sweating, nervousness, anxiety, tingling of the mouth and fingers, and hunger result from epinephrine release. As the blood glucose delivery to the brain decreases, neuroglycopenic symptoms such as headache, visual disturbances, mental dullness, confusion, amnesia, seizures, or coma may occur (30).

Some people with diabetes lose their ability to sense hypoglycemic symptoms (termed hypoglycemia unawareness). By instituting tight control of blood glucose, the threshold may be lowered so that symptoms do not occur until blood glucose drops quite low. Intensity of control may need to be slightly reduced to alleviate hypoglycemia unawareness. Hypoglycemic unawareness may also result from autonomic neuropathy. To the contrary, patients who have been in poor control may sense low blood glucose symptoms at levels with much higher values than 60 to 70 mg · dl^{-1}. Treatment of hypoglycemia consists of testing blood glucose to confirm hypoglycemia and, if the person is conscious, consumption of approximately 15 g of carbohydrate (e.g., glucose, sucrose, or lactose) that does not contain fat. Commercial products (glucose tablets) are available that allow a person to eat a precise amount of carbohydrate. Other sources include 1 C (240 ml) of nonfat milk, 1/2 C (120 ml) of orange juice, one-half can (180 ml) of regular soda, six or seven Life Savers, 2 tbsp (30 ml) of raisins, or 1 tbsp (15 ml) of sugar, honey, or corn syrup. The person with diabetes should wait about 15 or 20 min to allow the symptoms to resolve. If necessary, she or he should consume another 15 g of carbohydrate (43). If the patient becomes unconscious because of hypoglycemia, an injection of glucagon should be administered. If glucagon is not available, 911 should be called immediately.

Chronic Complications

Diabetes is the leading cause of adult-onset blindness, nontraumatic lower-limb amputation, and end-stage renal failure (51). In addition, those with diabetes are at two to four times the normal risk of heart disease and stroke (28). The hyperglycemia of diabetes is considered of primary importance in developing the chronic complications, along with hypertension and hyperlipidemia. Tight blood glucose control can reduce the risk of developing diabetic complications in patients with either type 1 or 2 diabetes (32,107). The clinical exercise professional who is involved in the exercise training of people with diabetes must obtain information about the presence and stage of complications. The clinician should then use this information when developing an exercise prescription and behavior modification plan designed to help those with diabetes reduce their risk of developing or amplifying the complications of the disease. Cardiac rehabilitation programs may be suitable for patients, including those at high risk, who wish to incorporate an exercise program into their lifestyle. The chronic complications are more clearly described when considered in three categories:

1. Macrovascular (large vessel or atherosclerotic) disease, which includes coronary artery disease with or without angina, myocardial infarction, cerebrovascular accident, and peripheral arterial disease

2. Microvascular (small vessel) disease, which includes diabetic retinopathy (eye disease) and diabetic nephropathy (kidney disease)

3. Neuropathy that involves both the peripheral and autonomic nervous systems (30)

Macrovascular disease

Diabetes is a risk factor for macrovascular disease. The vessels to the heart, brain, and lower extremities can be affected. Figure 11.2 on p. 196 illustrates the relationship of insulin resistance (including hyperglycemia) to coronary artery disease. Blockage of the blood vessels in the legs results in peripheral artery disease, intermittent claudication (see chapter 19), and exercise intolerance (7). Reduction and control of vascular risk factors are especially important in those with diabetes. The methods used for this purpose are similar to those used for coronary heart disease. The chapter about myocardial infarction (chapter 16) in this text reviews the vascular risk factor control methods in detail. The symptoms of peripheral arterial disease can be improved with exercise training, as reviewed in chapter 19. Note that the

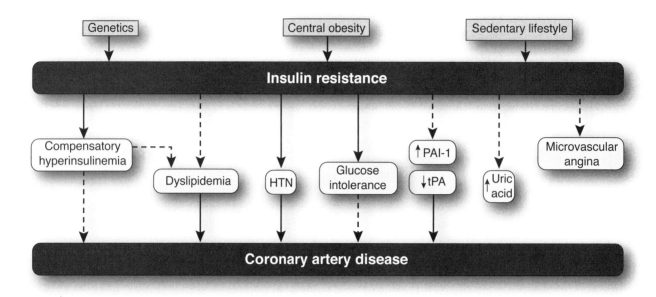

Figure 11.2 Components of insulin resistance syndrome.

This article was published in *Diabetes mellitus diagnosis and treatment*, 4th ed., M.B. Davidson, pgs. 267-298, Copyright Elsevier 1998.

National Cholesterol Education Program (NCEP) Adult Treatment Panel III (ATP III) guidelines recommend that lipid and other risk factor treatment for those with diabetes be to the level of those with coronary artery disease (see chapter 14).

Microvascular disease

Microvascular disease causes retinopathy and nephropathy, which cause abnormal function and damage to the small vessels of the eyes and kidneys, respectively. The ultimate result of retinopathy can be blindness, whereas end-stage renal failure is the most serious complication of nephropathy (see chapter 15). Prevention or appropriate management requires periodic (often yearly) dilated eye examinations and renal function tests, along with optimal blood glucose and blood pressure control. The exercise professional must give careful attention to the stage of complications when prescribing exercise for those with microvascular involvement; this topic is discussed in detail in the exercise prescription section of this chapter.

Peripheral and autonomic neuropathy

Both peripheral and autonomic neuropathy have implications for exercise. **Peripheral neuropathy** typically affects the legs before the hands. Patients initially experience sensory symptoms (paresthesia, burning sensations, and hyperesthesia) and loss of tendon reflexes. As the complication progresses, the feet become numb and patients are at high risk for foot injuries because they have difficulty realizing when they are injured. Muscle weakness and atrophy can also occur. Foot deformities can

result, causing areas to receive increased pressure from shoe wear or foot strike, placing them at risk for injury (30). The large number of lower-limb amputations from diabetes is the result of loss of sensation that places the patient at risk for injury and from diminished circulation attributable to peripheral artery disease. This circumstance impairs healing and can lead to severe reductions in blood flow, potential gangrene, and amputation. Persons with diabetes must be given instruction on how to examine their feet and practice good foot care. Foot care is especially important when someone with peripheral neuropathy begins an exercise program, because increased walking and cycle pedaling increase the risk of foot injury. Diabetic autonomic neuropathy may occur in any system of the body (e.g., cardiovascular, respiratory, neuroendocrine, gastrointestinal). Many of these systems are integral to the ability to perform exercise (1). **Cardiovascular autonomic neuropathy** is manifested by high resting heart rate, attenuated exercise heart rate response, abnormal blood pressure, and redistribution of blood flow response during exercise. This combination can severely limit exercise capacity and physical functioning.

MEDICAL AND CLINICAL CONSIDERATIONS

In the clinical setting, laboratory tests and examinations are used to diagnose diabetes or to facilitate ongoing monitoring. The following sections review these purposes.

Signs and Symptoms

The signs and symptoms of diabetes include excessive thirst (**polydipsia**), frequent urination (**polyuria**), unexplained weight loss, infections and cuts that are slow to heal, blurry vision, and fatigue. Many who develop type 1 diabetes have some or all of these symptoms, but those with type 2 diabetes may remain asymptomatic. One-fourth of the population with diabetes do not know that they have the disease (28).

History and Physical Examination

Patients who present for a clinic visit should have a thorough medical history review (see chapter 4). Those presenting with the risk factors or signs and symptoms for diabetes who have not been previously diagnosed with having diabetes should be evaluated appropriately. The evaluation includes performing the associated diagnostic testing presented in the next section.

The physical examination should focus on potential indicators of diabetes complications. These may include elevated resting pulse rate, loss of sensation or reflexes especially in the lower extremities, foot sores or ulcers that heal poorly, excessive bruising, and retinal vascular abnormalities.

The American College of Sports Medicine recommends that those with known metabolic disease have a medical exam and clinical exercise test before participation in moderate (40–60% $\dot{V}O_2$max) and vigorous intensity (>60% $\dot{V}O_2$max) exercise training (5). When reviewing the medical history of a patient with diabetes for exercise-training clearance, the exercise professional should consider the following:

1. The presence or absence of acute and chronic complications and, if chronic complications exist, the stage of complications
2. Laboratory values for hemoglobin A1C, plasma glucose, lipids, and **proteinuria**
3. Blood pressure
4. Self-monitoring blood glucose results
5. Body weight and body mass index
6. Medication use and timing
7. Exercise history
8. Nutrition plan, particularly timing, amount, and type of most recent food intake
9. Other non-diabetes-related health issues

Diagnostic Testing

The American Diabetes Association recommends that all people over age 45 be tested for diabetes, which should be repeated at 3 yr intervals (10). Testing should occur earlier or more frequently if someone has any of the following risk factors:

1. Is overweight or obese (>120% desirable body weight or a body mass index >27 $kg \cdot m^{-2}$)
2. Has a first-degree relative with diabetes
3. Is a member of a high-risk ethnic population
4. Delivered a baby weighing more than 9 lb (4 kg) or was diagnosed with gestational diabetes
5. Is hypertensive (>140/90 mmHg)
6. Has a high-density cholesterol less than or equal to 35 $mg \cdot dl^{-1}$ or a triglyceride of greater than or equal to 250 $mg \cdot dl^{-1}$
7. Had an impaired fasting glucose or glucose tolerance test

Three criteria are used to diagnose diabetes (see "Criteria for the Diagnosis of Diabetes Mellitus" on p. 198). In the absence of hyperglycemia with acute metabolic decompensation (i.e., ketoacidosis), these criteria should be confirmed on a subsequent day. The goal of lowering the value considered diagnostic for diabetes (i.e., 126 $mg \cdot dl^{-1}$) is to treat those earlier who are at risk for microvascular complications. Those found to meet the criteria for the diagnosis of diabetes should be told they have diabetes, not "borderline" diabetes. The latter explanation may give the patient the impression that the disease is not serious. The method of therapy used to treat diabetes should not be taken to determine the seriousness of the disease. Regardless of treatment, diabetes is a serious disease that requires diligent self-care and appropriate medical intervention. A fasting blood glucose ranging from 100 to 125 $mg \cdot dl^{-1}$ is considered a risk factor for developing diabetes and is termed impaired fasting glucose. When an oral glucose tolerance test is used, a 2 h postload glucose between 140 $mg \cdot dl^{-1}$ and 199 $mg \cdot dl^{-1}$ is termed impaired glucose tolerance. Impaired fasting glucose and impaired glucose tolerance are now called prediabetes. This group of patients should receive instruction and encouragement to lower their risk of developing diabetes, including beginning an exercise-training program.

Exercise Testing

Most people with diabetes can benefit from participating in regular exercise. Participation in exercise is not without risk, however, and each individual should be assessed for safety (table 11.2). Priority must be given to minimizing the potential adverse effects of exercise through appropriate screening, program design, monitoring, and patient education (7). Exercise testing may be viewed as a barrier or unnecessary for some patients. Discretion must be used to determine the need for exercise testing. The clinical exercise professional must be prepared to provide input to the physician to assist in this decision-making process. Practical application 11.1 provides information about client–clinician interaction. Table 11.3 summarizes exercise-testing specifics.

Criteria for the Diagnosis of Diabetes Mellitus

1. Symptoms of diabetes plus casual plasma glucose concentration ≥ 200 mg $\cdot$ dl^{-1} (11.1 mmol $\cdot$ L^{-1}). Casual is defined as any time of day without regard to time since the last meal. The classic symptoms of diabetes include polyuria, polydipsia, and unexplained weight loss.
   or
2. Fasting plasma glucose ≥ 126 mg $\cdot$ dl^{-1} (7.0 mmol $\cdot$ L^{-1}). Fasting is defined as no caloric intake for at least 8 h.
   or
3. Two-hour plasma glucose ≥ 200 mg $\cdot$ dl^{-1} (11.1 mmol $\cdot$ L^{-1}) during an oral glucose tolerance test. The test should be performed as described by the World Health Organization (2), using a glucose load containing the equivalent of 75 g of anhydrous glucose dissolved in water.

Table 11.2 Potential Adverse Effects of Exercise in Patients With Diabetes

Cardiovascular	Microvascular	Metabolic	Musculoskeletal and traumatic
Cardiac dysfunction and arrhythmias attributable to ischemic heart disease (often silent) Excessive increments in blood pressure during exercise Postexercise orthostatic hypotension	Retinal hemorrhage Increased proteinuria Acceleration of microvascular lesions	Worsening of hyperglycemia and ketosis Hypoglycemia in patients on insulin or insulin secretagogue oral therapy	Foot ulcers (especially in presence of neuropathy) Orthopedic injury related to neuropathy Accelerated degenerative joint disease Eye injuries and retinal hemorrhage

Practical Application 11.1

Client–Clinician Interaction

The interaction between the client and the clinician at the time of exercise evaluation, especially during ongoing exercise-training visits, is important. Living with diabetes poses many challenges and fears for the patient and his or her family. The exercise professional must be aware of the psychosocial components of living with a chronic disease and must be able to apply strategies to help the patient maintain participation in exercise.

The clinician should consider the following guidelines. Treat the patient as an individual who is much more than his or her diagnosis. Be cautious about referring to the patient as a diabetic, because this terminology labels the patient by the disease. Remember that the person usually needs to apply a great deal of effort and discipline to live with diabetes. Acknowledge that diabetes is challenging and listen to the patient's particular challenges. In general, do not use terms like *noncompliance* when discussing an exercise program. Inherent in the definition of noncompliance is the concept that a person is not following rules or regulations enforced by someone else. This concept is incongruent with self-management and patient empowerment, which consider the patient to be the key member of the healthcare team. The healthcare professional should not make decisions for the patient. Instead, the clinical exercise professional should equip the patient with information so that the patient can make his or her own decisions.

The following are some strategies for exercise maintenance. Ask the patient to consider the following questions: How easily can I engage in my activity of choice where I live? How suitable is the activity in terms of my physical attributes and lifestyle (1,77)? Have the patient identify exercise benefits that she or he finds personally motivating. Be sure that exercise goals are not too vague, ambitious, or distant (3). Establish a routine to help exercise become more habitual. Have the patient identify any social support systems that he or she may have. Provide positive feedback to the patient.

Table 11.3 Exercise Testing Summary

Test type	Mode	Protocol specifics	Clinical measures	Clinical implications	Special considerations
Cardiovascular	Treadmill Ergometer (leg or arm)	Low level for many ($\leq$2 MET per stage or $20\,W \cdot min^{-1}$ increases in work rate)	Peak $\dot{V}O_2$ or estimated METs; Heart rate and blood pressure responses; 12-lead ECG	Watch for ischemia and arrhythmias because it is often undiagnosed and patients are at high risk for heart disease	Chest pain due to myocardial ischemia may not be perceived in those with neuropathy (also may blunt peak HR achieved); Peripheral vascular diseased patients likely should use the cycle ergometer mode; Consider testing blood glucose before test to reduce the risk of hypoglycemia
Strength	Machine weights; Isokinetic dynamometer	1RM or indirect 1RM method	Strength and power		1RM may not be recommended in those with severe disease and who are sedentary; Those with retinopathy often should not perform resistance training
Range of motion	Sit-and-reach; Goniometry		Major muscle groups range of motion		Do not hold breath; any exercise may result in excessive blood pressure response

Contraindications for exercise testing are listed in chapter 5. Table 11.3 reviews exercise testing for cardiovascular, strength, and range-of-motion assessment. A physician should be consulted before exercise testing if any absolute contraindication exists. Exercise testing may need to be postponed until it can be safely conducted.

Age, duration of diabetes, and presence of complications should be considered before a patient begins an exercise test. In most cases, standard methods should be used (7). Physician supervision is recommended (5). Chapter 5 provides details on protocol selection and summary information on exercise testing is shown in table 11.3.

Because of the high risk for cardiovascular disease in those with diabetes, low-level treadmill protocols or cycle or arm ergometer modes are recommended for the following groups:

1. Type 1 diabetes and over 30 yr old or had diabetes longer than 15 yr

2. Type 2 diabetes and over 35 yr old

3. Type 1 or type 2 diabetes plus one or more other coronary artery disease risk factors

4. Suspected or known coronary artery disease

5. Any microvascular or neurological diabetic complications (7,12)

6. Peripheral arterial disease or peripheral neuropathy

Treatment

There is currently no cure for diabetes. The disease must be managed with a program of exercise, **medical nutrition therapy**, self-monitoring of blood glucose, diabetes self-management education, and, when needed, medication. The patient and his or her healthcare team must work together to develop a program to achieve individual treatment goals. Few diseases require the same level of ongoing daily patient involvement as that required by diabetes. Because so much patient involvement is required in diabetes, patients must receive information and training on disease management. Other members of the healthcare team may include the patient's primary care physician or an endocrinologist, a nurse practitioner, a physician assistant, a diabetes educator, a registered dietitian, a clinical exercise professional, a behavioral or psychosocial counselor, and a pharmacist. In many instances, these healthcare professionals work together in a diabetes education program. The American Diabetes Association, along with many other contributing organizations, has developed standards for diabetes education programs (13). The patient must understand and be involved in developing appropriate treatment goals, which are developed with consideration of the patient's desires, abilities, willingness, cultural background, and comorbidities. Suggested treatment goals for blood glucose, blood pressure, and plasma lipids from the American Diabetes Association are provided in table 11.1. **Evidence-based** care guidelines include regular hemoglobin A1C testing, dilated eye exam, foot exam, blood pressure monitoring, lipid panel, renal function tests, smoking cessation counseling, flu or pneumococcal immunizations, and diabetes education (see "Basic Guidelines for Diabetic Care") that should be followed to help ensure appropriate care. The patient should be educated about the purpose and importance of the medical tests and feedback on her or his results.

Basic Guidelines for Diabetes Care

Physical and Emotional Assessment

Blood Pressure, Weight/BMI—Every visit.
- For adults: Blood pressure target goal <130/80 mmHg.
- For children: Blood pressure target goal <90th percentile adjusted for age, height, and gender; BMI (body mass index)-for-age <85th percentile.

Foot Exam (for adults)—Thorough visual inspection *every diabetes care visit:* pedal pulses, neurological exam yearly.

Dilated Eye Exam (by trained expert)—
- Type 1: Five years post diagnosis, then every year.
- Type 2: Shortly after diagnosis, then every year. *Note:* Internal quality assurance data may be used to support less frequent testing.

Depression—Probe for emotional/physical factors linked to depression *yearly;* treat aggressively with counseling, medication, and/or referral.

Dental—Oral exam—comprehensive periodontal evaluation at least *twice yearly;* periodontal maintenance *two to four times a year.*

Lab Exam

A1C (HbA1c)—
- Quarterly, if treatment changes or if not meeting goals; one–two times/year if stable.
- Target goal <7.0% or <1% above lab norms.
- For children: Modify as necessary to prevent significant hypoglycemia.

Microalbuminuria (Albumin/Creatinine Ratio)—
- Type 1: Begin with puberty once the duration of diabetes is *more than five years* unless proteinuria has been documented.
- Type 2: Begin at diagnosis, then *every year* unless proteinuria has been documented.

Glomerular Filtration Rate (GFR)—Estimate whenever yearly chemistries are checked.

Blood Lipids (for adults)—On *initial visit,* then *yearly* for adults. Target goals (mg · dl^{-1}): cholesterol, triglycerides <150; LDL <100; HDL >40 for men; HDL >50 for women.

Self-Management Training

Management Principles and Prevention of Complications—
- Initially and ongoing: Focus on helping the patient achieve the AADE 7 self-care behaviors: healthy eating, being active, monitoring, taking medications, problem solving, healthy coping, and reducing risks. Screen for problems with and barriers to self-care; assist patient to identify achievable self-care goals.
- For children: *As appropriate* for developmental stage.

Self-Glucose Monitoring—
- Type 1: Typically test *four times a day.*
- Type 2 and others: *As needed* to meet treatment goals.

Medical Nutrition Therapy (by trained experts)—
- Initially: Assess needs/condition, assist patient in setting nutrition goals.
- Ongoing: Assess progress toward goals, identify problem areas.

Physical Activity—Initially and ongoing: Assess and prescribe physical activity based on patient's needs/condition.

Weight Management—Initially and ongoing: Must be individualized for patient.

Interventions

Preconception, Pregnancy, and Postpartum Counseling and Management—
- Consult with high-risk, multidisciplinary perinatal/neonatal programs, and providers where available (e.g., California Diabetes and Pregnancy Program "Sweet Success").
- For adolescents: Age appropriate counseling advisable, beginning with puberty.

Aspirin Therapy (for adults)—*75-162 mg · day^{-1}* as primary and secondary prevention of cardiovascular disease unless contraindicated.

Smoking Cessation—Ask, advise, assess readiness to quit, and assist at every visit, adjusting the frequency as appropriate to the patient's response. Refer to California Smokers' Helpline 1-800-NO-BUTTS (662-8887).

Immunizations—Influenza and pneumococcal, *per CDC recommendations.*

Developed by the Diabetes Coalition of California and the California Diabetes Program, revised July 2007. For further information: www.calidiabetes.org or (916) 552-9888.

Medical nutrition therapy, often the most challenging aspect of therapy, is essential to the management of diabetes. Nutrition recommendations were developed by the American Diabetes Association (11). These guidelines promote individually developed dietary plans based on metabolic, nutrition, and lifestyle requirements in place of a calculated caloric prescription. This approach is appropriate because a single diet cannot adequately treat all types of diabetes or individuals. Consideration must be given to each macronutrient (i.e., protein, fat, carbohydrate) when developing a nutrition plan for the person with diabetes. Protein intake should be approximately 10% to 20% of daily caloric intake, because no evidence indicates that lower or higher intake is of value. Based on the risk of atherosclerosis, fat intake should be limited, with less than 10% from saturated fats and up to 10% from polyunsaturated fats. Cholesterol intake should be limited to 300 mg daily. Carbohydrate and monosaturated fat make up the remaining calories and need to be individualized based on glucose, lipid, and weight goals. The most common nutritional assumption about diabetes is that sugars should be avoided and replaced with starches. Little evidence supports this assumption (18,74). Priority should first be given to the total amount of carbohydrate consumed rather than the source, because all carbohy-

drates can raise blood glucose. Nutritional value must also be considered.

Self-monitoring of blood glucose is also an important part of managing diabetes. No standard frequency for self-monitoring has been established, but it should be performed frequently enough to help the patient meet treatment goals. Increased frequency of testing is often required when initiating an exercise program to assess blood glucose before and after exercise and to allow safe exercise participation. Patients must be given guidance about how to use the information to make exercise, food, and medication adjustments. Those who require medication (insulin or oral agents) must understand how their medications work with food and exercise to ensure the greatest success and safety. The clinical exercise professional must understand diabetes medications so that she or he can safely prescribe exercise and provide guidance on exercise training to patients with diabetes. Refer to

Prevention and Treatment of Abnormal Blood Glucose Before and After Exercise

Preexercise Hypoglycemia

Blood glucose levels should be monitored before an exercise session to determine whether the person can safely begin exercising. The preexercise assessment of blood glucose and carbohydrate consumption is conducted to prevent exercise-induced hypoglycemia. Consideration must be given to how long and intense the exercise session will be and whether the patient is managed with medication (insulin or oral agents). The following guidelines can be used to determine whether additional carbohydrate intake is necessary (44).

If blood glucose is less than 100 mg · dl^{-1} and the exercise will be of low intensity and short duration (e.g., bike riding or walking for <30 min), 10 to 15 g of carbohydrate should be consumed. If blood glucose is greater than 100 mg · dl^{-1}, no extra carbohydrate is needed. If blood glucose is less than 100 mg · dl^{-1} and exercise is of moderate intensity and moderate duration (e.g., jogging for 30–60 min), 30 to 45 g of carbohydrate should be consumed. If blood glucose is 100 to 180 mg · dl^{-1}, then 15 g of carbohydrate is needed. If blood glucose is less than 100 mg · dl^{-1} and exercise is of moderate intensity and long duration (e.g., 1 h of bicycling), then 45 g of carbohydrate should be consumed. If blood glucose is 100 to 180 mg · dl^{-1}, then 35 to 45 g of carbohydrate is needed. Remember that these guidelines may need to be modified in certain cases. For instance, someone trying to lose weight might benefit from a medication adjustment rather than increased food intake.

Preexercise Hyperglycemia

If the preexercise blood glucose is greater than 250 mg · dl^{-1}, urine should be checked for ketones. If ketones are present (moderate to high) or if blood glucose is greater than 300 mg · dl^{-1} irrespective of whether ketones are present, exercise should usually be postponed until glucose control is improved. The blood glucose values given previously are guidelines and actions should be verified with the patient's physician. Patients who use medication as part of diabetes treatment should be assessed to determine whether the timing and dosage of medication will allow exercise to have a positive effect on blood glucose. For example, a patient who uses insulin and had blood glucose of 270 mg · dl^{-1}, had no ketones, and had taken regular insulin within 30 min will see a reduction in blood glucose from both the insulin and exercise. If this patient has not just administered fast-acting insulin and the previous insulin injection has run its duration, the patient is underinsulinized and additional insulin is needed to help reduce the blood glucose before he or she exercises. In this case, exercise would likely increase blood glucose level. In all cases, adding additional medication must be cleared by a clinician with prescriptive authority. Those with type 2 diabetes who are appropriately managed by diet and exercise alone will usually experience a reduction in blood glucose with low to moderate exercise. Timing of exercise after meals can often help many patients with type 2 diabetes reduce **postprandial** hyperglycemia. Blood glucose should be monitored after an exercise session to determine the patient's response to exercise.

Postexercise Hypoglycemia

A patient will more likely experience hypoglycemia (usually <70 mg · dl^{-1}) after exercise than during exercise because of the replacement of muscle glycogen, which uses blood glucose (115). Periodic monitoring of blood glucose is necessary in the hours following exercise to determine whether blood glucose is dropping. More frequent monitoring is especially important when initiating exercise. If the patient is hypoglycemic, he or she needs to take appropriate steps to treat this medical emergency as previously discussed.

Postexercise Hyperglycemia

In poorly controlled diabetes, insulin levels are often too low, resulting in an increase in counterregulatory hormones with exercise. This circumstance causes glucose production by the liver, enhanced free fatty acid release by adipose tissue, and reduced muscle uptake of glucose. The result is an increased blood glucose level during and after exercise. High-intensity exercise can also result in hyperglycemia. In this case, the intensity and duration of exercise should be reduced as needed.

chapter 3 for specific information about both oral diabetes medications and insulin.

EXERCISE PRESCRIPTION

Exercise is a vital component of diabetes management. Exercise is considered a method of treatment for type 2 diabetes because it can improve insulin resistance. Although exercise alone is not considered a method of treating type 1 diabetes because of the absolute requirement for insulin, it is still an important part of a healthy lifestyle for those individuals.

Special Exercise Considerations

When developing an exercise prescription for persons with diabetes, the exercise professional should consider the topic of fitness versus the health benefits of exercise. Methods to enhance maximal oxygen uptake are often extrapolated to the exercise prescription for disease prevention and management (7). Changes in health status, however, do not necessarily parallel increases in maximal oxygen uptake. In fact, evidence strongly suggests that

regular participation in light- to moderate-intensity exercise may help prevent diseases such as coronary artery disease, hypertension, and type 2 diabetes but will not have an optimal effect on maximal oxygen uptake (4,36,50). Therefore, when frequency and duration are sufficient, exercise performed at an intensity below the threshold for an increase in maximal oxygen uptake can be beneficial to health (47). Exercise must be prescribed with careful consideration given to risks and benefits. The consequences of disuse combined with the complications of diabetes are likely to lead to more disability than the complications alone (1). Exercises that can be readily maintained at a constant intensity, and in which there is little interindividual variation in energy expenditure, are preferred for those with complications where more precise control of intensity is needed (7).

Macrovascular Disease

Macrovascular disease is a complication that often affects patients with diabetes. The primary macrovascular diseases are coronary artery disease and peripheral artery disease. Chapters 17 and 19 review specifics regarding preexercise evaluation and exercise prescription for coronary and peripheral artery disease. These approaches should be incorporated for patients with diabetes and coronary and peripheral artery disease.

Peripheral Neuropathy

The major consideration in patients with peripheral neuropathy is the loss of protective sensation in the feet and legs that can lead to musculoskeletal injury and infection. Non-weight-bearing activities are recommended to minimize the risk of injury (table 11.4). Proper footwear and examination of the feet are especially important for these patients. The clinical exercise professional should reinforce instruction given to the patient on self-examining the feet, learn how to recognize related injuries, and encourage the patient to have her or his feet examined regularly.

Autonomic Neuropathy

Cardiovascular autonomic neuropathy can affect the patient with diabetes. This disease is manifested by abnormal heart rate, abnormal blood pressure, and redistribution of blood flow. Patients with cardiovascular autonomic neuropathy have a higher resting heart rate and lower maximal exercise heart rate than those without this condition (57). Thus, estimating peak heart rate in this population may lead to an overestimation of the training heart rate range if heart rate–based methods are used (see chapter 5). Early warning signs of ischemia may be absent in these patients. The risk of exercise hypotension and sudden death increases (39,58). An active cool-down reduces the possibility of a postexercise hypotensive response. Exercise in this patient population should focus on lower-intensity activities in which mild changes in heart rate and blood pressure can be accommodated (48). Because of difficulty with thermoregulation, these patients should be advised to stay hydrated and not to exercise in hot or cold environments (12).

Retinopathy

The exercise recommendations for those with diabetic retinopathy are contingent on the stage of the complication and should focus on limiting systolic blood pressure and jarring activities. Table 11.5 provides information on

Table 11.4 Exercises for Patients With Diabetes and Loss of Protective Sensation

Contraindicated exercise	Recommended exercise
Treadmill	Swimming
Prolonged walking	Bicycling
Jogging	Rowing
Step exercises	Chair exercises
Arm exercises	Other non-weight-bearing exercises

Copyright © 2004 American Diabetes Association. From Diabetes Care®, Vol. 27, Suppl. 1, 2004; S58-S62. Reprinted with permission from *The American Diabetes Association.*

Table 11.5 Considerations for Activity Limitation in Those With Diabetic Retinopathy

Level of DR	Acceptable activities	Discouraged activities	Ocular reevaluation
No DR	Dictated by medical status	Dictated by medical status	12 mo
Mild NPDR	Dictated by medical status	Dictated by medical status	6–12 mo
Moderate NPDR	Dictated by medical status	Activities that dramatically elevate blood pressure (e.g., power lifting, heavy lifting, Valsalva maneuver)	4–6 mo
Severe NPDR	Dictated by medical status	Activities that substantially increase systolic blood pressure (e.g., Valsalva maneuver, active jarring, surgery, boxing, heavy competitive sports)	2–4 mo
PDR	Low-impact, cardiovascular conditioning (e.g., swimming, walking, low-impact aerobics)	Strenuous activities (e.g., Valsava maneuver, pounding or jarring)	1–2 mo

Note. DR = diabetic retinopathy; NPDR = nonproliferative diabetic retinopathy; PDR = proliferative DR.

Copyright © 2004 American Diabetes Association. From Diabetes Care®, Vol. 27, Suppl. 1, 2004; S58-S62. Reprinted with permission from *The American Diabetes Association.*

selection of appropriate activities based on severity of retinopathy. For optimal improvement and safety, exercise should be conducted in a supervised environment when retinopathy is significant (68,110,119).

Nephropathy

Elevated blood pressure is related to the onset and progression of diabetic nephropathy. Placing limits on low- to moderate-intensity activity is not necessary, but strenuous exercises should likely be discouraged in those with diabetic nephropathy because of the elevation in blood pressure (12). Patients on renal dialysis or who have received a kidney transplant can also benefit from exercise (6). See chapter 15 for details about renal failure.

Exercise Recommendations

Endurance, resistance, and range-of-motion exercise training are all appropriate modes for most patients with diabetes. Patients who are trying to lose weight (especially those with type 2 diabetes) should expend a minimum cumulative total of 1,000 kcal per week in aerobic activity and participate in a well-rounded resistance-training program (22,41). Patient interests, goals of therapy, type of diabetes, medication use (if applicable), and presence and severity of complications must be carefully evaluated in developing the exercise prescription. The following exercise prescription recommendations are guidelines, and individual patient circumstances must always determine the specific prescription (7). Practical application 11.3 on p. 206 presents a summary of the exercise prescription recommendations.

Mode

Personal interest and the desired goals of the exercise program should drive the type of physical activity that is selected. Caloric expenditure is often a key goal for those with diabetes. Walking is the most commonly performed mode of activity (38). Walking is a convenient, low-impact activity that can be used safely and effectively to maximize caloric expenditure. Non-weight-bearing modes should be used if necessary (e.g., if the patient has peripheral arterial disease or peripheral neuropathy). For a given level of energy expenditure, the health-related benefits of exercise appear to be independent of the mode.

Intensity

Programs of moderate intensity are preferable for most people with diabetes because the cardiovascular risk and chance for musculoskeletal injury is lower and the likelihood of maintaining the exercise program is greater. Some low-fit individuals, however, may increase $\dot{V}O_2$peak at an intensity level as low as 40% of $\dot{V}O_2$peak. Exercise should generally be prescribed at an intensity of 60% to 80% of maximal heart rate, 50% to 75% of $\dot{V}O_2$max, or a rating of perceived exertion of 12 to 13 (7). Chapter 5 provides specifics for determining and calculating proper exercise intensity.

Frequency

The frequency of exercise should be 3 to 7 d per week. Exercise duration, intensity, weight-loss goals, and personal interests determine the specific frequency. Additionally, the blood glucose improvements with exercise in those with diabetes are seen for greater than 12 but less than 72 h (111). These data indicate that exercise done on 3 nonconsecutive days each week, and ideally 5 or more days per week, is recommended. Those who take insulin and have difficulty balancing caloric needs with insulin dosage may prefer to exercise daily. This schedule will result in less daily adjustment of insulin dosage and caloric intake than a schedule in which exercise is performed every other day or sporadically, and it will reduce the likelihood of a hypoglycemic or hyperglycemic response. In addition, patients who are trying to lose weight will maximize caloric expenditure by participating in daily physical activity (7).

Duration and Rate of Progression

Exercise duration and the rate of progression can be at the standard levels for a chronic diseased population. Chapter 5 reviews the specifics of determining appropriate duration and progression of exercise.

Resistance Training

Some evidence supports the inclusion of resistance training in a patient's program. In nondiabetic subjects, resistance training improves glucose tolerance and insulin sensitivity (56,80,97). The limited data available for diabetes suggest that resistance training is safe and effective (37,42). All patients should be screened for contraindications before they begin resistance training. Proper instruction and monitoring are also needed. A recommended resistance-training program consists of a minimum of 8 to 10 exercises involving major muscle groups performed with a minimum of one set of 10 to 15 repetitions to near fatigue. Resistance-training exercises should be done at least 2 d per week (7). Modifications such as lowering the intensity of lifting, preventing exercise to the point of exhaustion, and eliminating the amount of sustained gripping or isometric contractions should be considered to ensure safety. Chapter 5 provides

Summary of Exercise Prescription

Training method	Mode	Intensity	Frequency	Duration	Progression	Goals	Special considerations
Aerobic	Walking, cycling, swimming	50–75% of maximal aerobic capacity	Three to five times per week or most days of the week	20–60 min	Rate of progression depends on many factors including baseline fitness, age, weight, health status, and individual goals. Usually, the best approach is to alter duration rather than intensity. Exercise should begin at a comfortable intensity, be no more than 10–15 min in duration, and then gradually increase as tolerated. After the desired duration is reached, intensity may likewise be gradually increased	Patient dependent Energy expenditure of 700 to 2,000 kcal per week	Note. All special considerations listed in the chapter apply to all these training methods: Avoid exercise during peak insulin action time Search for vascular and neurological complications, including silent ischemia Warm-up and cool-down are important Promote patient education Assess for proper footwear and inspect feet daily Avoid extreme environmental temperatures Avoid exercise when blood glucose control is poor Adequate hydration should be maintained Instruct patient on blood glucose monitoring and to follow guidelines to prevent hyper- and hypoglycemic events
Resistance	Free weights, machines, elastic bands	About 60% of 1RM	At least two times per week, but never on consecutive days	10–15 repetitions per set, one to two sets per type of specific exercise	As tolerated When perceived exertion is below "fairly light" (i.e., 11 on the Borg 6–20 scale) progression of weight/resistance by 5–10% may be attempted		
Range of motion	Static stretching		Postaerobic exercise session	10 to 30 s per exercise of each muscle group	As tolerated May increase range of stretch as long as patient does not complain of pain (acute or chronic)		

specific information about general resistance training that should be considered for these patients.

Timing

Exercise should be performed at the time of day most convenient for the participant. Because of the risk of hypoglycemia, those taking insulin should give careful consideration to the time of day that they perform exercise. They should not exercise when insulin action is peaking. Because exercise acts like insulin in that it promotes peripheral glucose uptake, the combination of exercise and peak insulin action increases the risk of hypoglycemia. Because of this effect and the need to replace muscle glycogen, hypoglycemia is more likely to occur after exercise than during exercise (101). Exercising late in the evening for those on insulin and some oral medications is not recommended because of the possible occurrence of hypoglycemia when sleeping.

EXERCISE TRAINING

Exercise has long been recognized as an important component of diabetes care (71). Benefits for those with diabetes are seen with both acute and chronic exercise. Acute bouts of exercise can improve blood glucose, particularly in those with type 2 diabetes (24,72). The response of blood glucose to exercise is related to pre-exercise blood glucose level as well as the duration and intensity of exercise. Several studies about type 2 diabetes have demonstrated a reduction in blood glucose levels that is sustained into the postexercise period following mild to moderate exercise (55,65,81,105). The reduction in blood glucose is attributed to an attenuation of hepatic glucose production with a normal increase of muscle glucose utilization (23,62,81). The effect of acute exercise

on blood glucose levels in those with type 1 diabetes and lean patients with type 2 diabetes is more variable and unpredictable.

Physiological Adaptations

A rise in blood glucose with exercise can be seen in patients who are extremely insulin deficient (usually type 1) and with short-term, high-intensity exercise (20,54,62). Also, an elevation in blood glucose has been shown in patients with type 2 diabetes, but this result occurred with short-term, high-intensity exercise (82). Most of the benefits of exercise for those with diabetes come from regular, long-term exercise. These benefits can include improvements in metabolic control (glucose control and insulin resistance), hypertension, lipids, body composition and weight loss, and psychological well-being. Epidemiological evidence supports the role of exercise in the primary prevention or delay of type 2 diabetes (85). Early studies showing an increase in type 2 diabetes in societies that had abandoned traditional active lifestyles suggest a relationship between physical activity and diabetes (114). Several cross-sectional studies have shown that blood glucose and insulin values after an oral glucose tolerance test were significantly higher in less active, compared with more active, individuals (35,40,73,87,91,112). Prospective studies of several groups have also demonstrated that a sedentary lifestyle may play a role in the development of type 2 diabetes (45,50,75,76,88,96). Some of the early data in support of exercise in the prevention of type 2 diabetes come from a 6 yr clinical trial in which subjects with impaired glucose tolerance were randomized into one of four groups: exercise only, diet only, diet plus exercise, or control. The exercise group was encouraged to increase daily

Practical Application 11.4

Literature Review

Evidence from a randomized multicenter trial demonstrated that type 2 diabetes can be prevented or postponed in adults (>25 yr old) with impaired glucose tolerance. The Diabetes Prevention Program (DPP) evaluated the effectiveness of (1) intensive lifestyle modification (low-calorie diet, moderate physical activity of 30 min a day 5 d per week; (2) standard care plus the drug metformin; (3) standard care plus placebo. Participants who made lifestyle changes reduced their risk of getting type 2 diabetes by 58%. The lifestyle intervention was effective for all participants of all ages (>25 yr old) and ethnic groups studied. Participants with standard care plus metformin reduced their risk for getting type 2 diabetes by 31%. The exercise professional has a critical role to play in helping implement the lifestyle changes necessary for achieving the DPP results in the population at large.

physical activity to a level that was comparable to a brisk 20 min walk. The incidence of diabetes in the exercise intervention groups was significantly lower than in the control group (85). A randomized, multicenter clinical trial of type 2 diabetes prevention in those with impaired glucose tolerance at numerous sites around the United States found a 58% reduction in the incidence of type 2 diabetes with lifestyle intervention that had goals of 7% weight loss and 150 min of physical activity per week (33).

Like acute exercise, exercise training can improve blood glucose. Exercise training has been shown to improve glucose control as measured by hemoglobin A1C or glucose tolerance, primarily in those with type 2 diabetes (52,70,92,105). These studies used training programs ranging from 6 wk to 12 mo, and improved glucose tolerance was seen in as little as 7 consecutive days of training in subjects with early type 2 diabetes (94). Improvements in blood glucose deteriorate within 72 h of the last bout of exercise, emphasizing the need for consistent exercise (99). Following exercise training, insulin-mediated glucose disposal is improved. Insulin sensitivity of both skeletal muscle and adipose tissue can improve with or without a change in body composition (53,66,78). Exercise may improve insulin sensitivity through several mechanisms, including changes in body composition, muscle mass, capillary density, and glucose transporters in muscle (GLUT 4) (2). The effect of exercise on insulin action is lost within a few days, again emphasizing the importance of consistent exercise participation (7). The data supporting a positive effect of exercise on blood pressure come primarily from studies done in subjects without diabetes. Two studies showed a reduction in blood pressure with exercise training in those with type 2 diabetes (69,98). Additional research is needed to improve understanding of the effect of exercise training on blood pressure in diabetes. This information is necessary because essential hypertension is present in more than 60% of those with type 2 diabetes (9). Information about the effect of exercise on lipids in diabetes shows primarily positive results. Most data are available on type 2 diabetes. Increased aerobic capacity in people with type 2 diabetes is related to a less atherogenic lipid profile. Improvements have been demonstrated in triglycerides, total cholesterol, and the ratio between high-density lipoprotein cholesterol and total cholesterol following physical training in patients with type 2 diabetes (17,19,94,95). Most of these studies included dietary modifications, so isolating the effect of exercise is difficult. Weight loss is often a therapeutic goal for those with type 2 diabetes, because 80% of those with type 2 are obese. Moderate weight loss has been shown to improve glucose control (113,115) and decrease insulin resistance. Medical nutrition therapy and exercise combined are more effective than either alone in achieving moderate weight loss (93,116,117). Visceral or abdominal body fat is negatively associated with insulin sensitivity so that increased abdominal body fat decreases peripheral insulin sensitivity. This body fat is a significant source of free fatty acids and may be preferentially oxidized over glucose, contributing to hyperglycemia (21,86). Exercise results in preferential mobilization of visceral body fat, likely contributing to the metabolic improvements (83,118). Exercise is one of the strongest predictors of success of long-term weight control (63). This feature of exercise is extremely important because weight lost is often regained.

Psychological Benefits

Psychological benefits of regular exercise have been demonstrated in those without diabetes including reduced stress, reduction in depression, and improved self-esteem (59,102,104). These benefits are equally applicable to those with diabetes, and because of the burden of diabetes, exercise professionals should encourage clients with

Practical Application 11.5

Amputation

Diabetes is the leading cause of nontraumatic lower-limb amputations. Amputations occur because of impaired sensation and circulation in the extremity, resulting in wounds that are not able to heal properly. Because the energy cost of walking increases markedly for someone with an amputation and a prosthetic limb, walking exercises are more difficult. In addition, prolonged walking may cause trauma and ulceration of the stump. Various upper-body exercises including chair exercises, weights, arm ergometry, or non-weight-bearing exercise such as swimming may be better choices. The clinical exercise staff should closely monitor the patient, and the patient should have regular visits with her or his health care professional.

diabetes to exercise to maximize their psychological well-being (109).

CONCLUSION

Living with a chronic illness poses special issues. Diabetes management requires ongoing dedication, and the patient must cope with complications if they develop and, at the very least, deal with the threat of their development. Exercise training should be an essential component of the treatment plan for patients with diabetes because it can improve blood glucose control, lipid levels, blood pressure, and body weight; reduces stress; and has the potential to reduce the burden of this metabolic disease.

Case Study

Medical History

Medication: Glucophage taken two times per day, Captopril (for blood pressure control and protection of kidneys), and Lipitor for control of hyperlipidemia. Laboratory values: Last Hb A1C = 8.8% (normal 4–6%); cholesterol 200 mg · dl^{-1}, low-density lipoprotein cholesterol 130 mg · dl^{-1}; high-density lipoprotein cholesterol 35 mg · dl^{-1}; triglycerides 160 mg · dl^{-1}; microproteinuria. Physical exam: blood pressure 130/80 mmHg; resting heart rate 70 beats · min^{-1}; height 5 ft 11 in. (190 cm); weight 230 lb (104 kg) with 27% body fat (skinfold). Complications history: Acute periodic episodes of diabetes out of control but has never experienced hyperosmolar nonketotic syndrome or **diabetic ketoacidosis**. Chronic two-vessel bypass surgery 5 yr ago, moderate peripheral neuropathy, and early (stage 3) diabetic nephropathy.

Diagnosis

Mr. SR is 63 yr old and was diagnosed with type 2 diabetes 5 yr ago.

Exercise Test Result

No abnormal electrocardiogram changes, maximum blood pressure 180/83 mmHg; maximum heart rate 150 beats · min^{-1}; $\dot{V}O_2$max 25.5 ml · kg^{-1} · min^{-1}; random blood glucose before test 180 mg · dl^{-1}.

Exercise Prescription

The goals of the exercise program, mutually agreed on by the patient and the clinical exercise professional, are to lose weight and improve body composition, improve blood glucose levels, and reduce risk for another cardiac event. When asked about interests and hobbies, the patient indicates that he enjoys traveling, wine tasting, playing with his dog, and classic movies. Participation in a supervised exercise program and frequent contact with an exercise professional is advised. A warm-up and cool-down of static stretches and low-intensity aerobic activity are prescribed.

Mode: Stationary cycling or water exercise. Low- to non-weight-bearing activities are selected because of the peripheral neuropathy. Walking his dog and walking while traveling are discussed. The patient must take care of his feet and do these activities as safely as possible.

Frequency: 3 to 5 d per week with a goal of increasing to daily.

Intensity: This patient is taught to monitor heart rate and to use the rating of perceived exertion (RPE) scale. The intensity is prescribed at 60% of maximum heart rate (150 beats · min^{-1}), or 90 beats · min^{-1}. An RPE rating of 12 to 13 on a 6- to 20-point Borg scale is advised.

Duration: An initial duration of 15–30 min is suggested and should eventually be increased to 60 min per session to facilitate weight loss.

Rate of progression: Attention is first given to frequency of exercise. After he has reached 5 d per week or more, duration will be increased.

Other information: Mr. SR is instructed to increase his blood glucose monitoring frequency to assess the effect of exercise on his blood glucose control.

(continued)

Case Study *(continued)*

Discussion Questions

1. Is this patient likely to have problems with hypoglycemia during or following exercise? Why or why not?

2. What are potential risks of exercise for this patient? What cautions should he be given?

3. What strategies can be given to help the patient stay motivated and maintain a regular exercise program?

4. What general nutrition suggestions might be helpful for this patient?

5. What other healthcare team members should this patient work with as he begins his exercise program? Why?

6. What else should be added to his program to help him attain his goals?

12

Obesity

David C. Murdy, MD

Jonathan K. Ehrman, PhD, FACSM

Obesity is a chronic disease associated with, if not the principal cause of, much disease, disability, and discrimination. Many providers still see obesity as a social or moral problem and stigmatize its treatment, adding to the considerable barriers to understanding and treating this expanding epidemic. Short-term intervention has limited effectiveness, and long-term success is rare without ongoing medical efforts or weight-loss surgery. Clinicians should approach obesity as a disease in itself and manage its comorbidities as well. Few medical treatments have such far-ranging positive effects as assisting an overweight or obese patient achieve and sustain a medically significant weight loss.

Exercise is an essential component in the management of obesity, along with diet and lifestyle change. These three elements of therapeutic lifestyle change are critical first steps in the prevention and treatment of many medical conditions but are overlooked because of the complexity of their practical application. Exercise combined with calorie reduction, lifestyle change, and in some cases weight-loss medication and surgery can be best provided by medical clinicians who work in a team environment with a long time horizon. Using this approach, obesity can be reduced in a cost- and care-efficient manner and its comorbidities can be controlled or eliminated. Within this approach, clinical exercise physiologists are playing an increasingly important role. This trend is expected to strengthen over the next several decades.

The term **obesity** is defined as a severe excess of fat in proportion to lean body mass. The term **overweight** is defined as a body weight that exceeds a reference threshold value. In most cases excess weight is due to being overfat. Reference values are derived from population data and are specific to certain characteristics including sex and height. Distributions of this data are determined, and criterions are then based on threshold values associated with increased risk or morbidity and mortality. Because many assessments are used to determine threshold values, the definitions of overweight and obesity vary (60).

Many methods of assessing fatness (skinfolds, bioelectrical impedance, underwater weighing, and dual-energy x-ray absorptiometry) are available. Using these expensive and time-consuming techniques requires technical skill, patient cooperation, and space, all of which may preclude their use. A more practical approach for the clinical setting is to use **body mass index** (**BMI**). BMI is recommended by the National Institutes of Health to classify overweight and obesity and to estimate relative risk of associated disease (76). BMI indicates overweight for height but does not discriminate between fat mass and lean tissue. This is demonstrated in figure 12.1 on page 212. The BMI does, however, significantly correlate with total body fat (42). Therefore, BMI is an acceptable measure of overweight and obesity in the clinical setting. BMI is calculated as weight in kilograms divided by height in meters squared. BMI can also be determined using pounds and inches. The following are the BMI formulae:

$$BMI = kg \cdot m^{-2} \text{ or } BMI = lb \cdot in.^{-1} \times 704.5$$

The classification of overweight and obesity by BMI is based on the 1998 "Clinical Guidelines on the Identification, Evaluation, and Treatment of Overweight and Obesity in Adults" and is shown in table 12.1 (76).

Medically significant obesity refers to adults who have gained 30 lb (14 kg) or more as adults and as a result have increased their waist circumference >6 in. (15 cm) (12). The NIH guidelines defines this population as adults with a BMI >30 kg $\cdot$ m^{-2} or >25 kg $\cdot$ m^{-2} with a family history of obesity or an obesity related comorbidity (76).

Although BMI is recommended to evaluate obesity, patients may request to determine their healthy weight range for their height. *Dietary Guidelines for Americans, Fifth Edition* recommends healthy weights for Americans (30). These guidelines are similar to the oft-cited Metropolitan Life Insurance ideal weight tables (table 12.2) and provide a range of weight that vary by frame size (68). Frame size can be specifically determined using elbow widths, but usually individuals subjectively self-select themselves into a frame category (i.e., small, medium, large). In addition, table 12.1 provides ranges of body weight–related categories based on ideal body weight and body fat percentage.

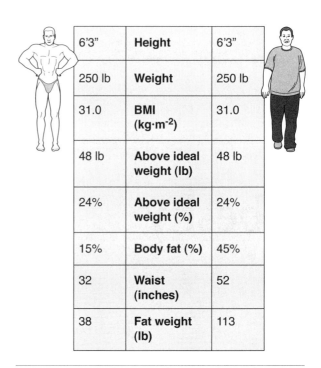

6'3"	Height	6'3"
250 lb	Weight	250 lb
31.0	BMI (kg·m⁻²)	31.0
48 lb	Above ideal weight (lb)	48 lb
24%	Above ideal weight (%)	24%
15%	Body fat (%)	45%
32	Waist (inches)	52
38	Fat weight (lb)	113

Figure 12.1 Both men in this example are the same height and weight and thus have the same BMI and weight relative to ideal weight. But it is apparent by physical appearance that the person on the left has a lower body fat percentage than the man on the right. The body fat, waist, and fat weight determination clarify this difference.

SCOPE

Table 12.2 shows the prevalence of normal weight, overweight, and obesity in the population using the 2006 report from the U.S. Department of Health and Human Services (74). Of note is an increase in the extreme obesity category from 1988 (2.9%) to 2002 (4.9%). In the same period the prevalence of overweight and obesity also rose in every age, gender, and ethnic category. And throughout the United States, each state demonstrated an increase in the prevalence of obesity with a continual rise throughout the 1990s and into the early 21st century (figure 12.2).

Table 12.1 Body-Weight-Related Classifications (65,73)

	Underweight	Normal	Overweight	Mildly obese (class I)	Moderately obese (class II)	Morbidly obese (class III)
BMI**	<18.5	18.5–24	25–29	30–34	35–39	≥40
Weight*		0–10%	10–20%	20–40%	40–100%	>100%
% fat***	<20	20–25	26–31	32–37	38–45	>45

*Weight: percent over standard height–weight tables (see table 12.2).

**BMI: body mass index (kg $\cdot$ m^{-2}), adapted from NIH's Clinical Guidelines on the Identification, Evaluation, and Treatment of Overweight and Obesity in Adults (75).

***Fat (%): calculated body fat expressed as percent of total weight.

Adapted from National Institutes of Health, 1998, *Clinical guidelines on the identification, evaluation, and treatment of overweight and obesity in adults* (Bethesda, MD: National Institutes of Health).

Table 12.2 Current (2001–2004) U.S. Population and Age-Adjusted Body Weight Demographics (74)

Population	Overweight and obese	Normal weight or underweight	Overweight	Obese
Males	71%	29%	37%	34%
Females	61%	39%	28%	33%
All	66%	34%	34%	32%
White male	71%			
Black male	67%			
Hispanic male	76%			
White female	57%			
Black female	80%			
Hispanic female	73%			

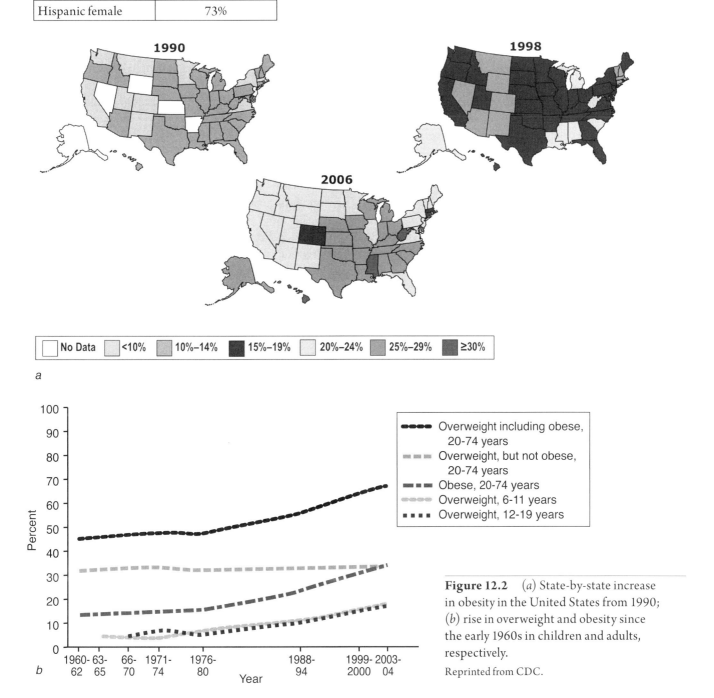

Figure 12.2 (*a*) State-by-state increase in obesity in the United States from 1990; (*b*) rise in overweight and obesity since the early 1960s in children and adults, respectively.

Reprinted from CDC.

213

Although obesity often begins in childhood and early adolescence, 70% of all obesity begins in adulthood. Approximately 17 to 19% of children are overweight, and many will develop obesity and carry their obesity into adulthood (74,78). The risks of childhood obesity persisting into adult obesity depend on the severity of obesity, age of onset, and parental obesity (111). After age 25 the average person gains 1.5 lb (0.7 kg) of fat each year. This weight gain is attributed to an environment that discourages physical activity and promotes excessive food intake including large-portion sizes and high fat content.

Weight that is 20% above desirable carries increased health risk. The patterns of fat distribution also affect risk (75). Central or android obesity (upper body, with a waist size of ≥35 in. [88 cm] for women or ≥40 in. [100 cm] for men) carries a higher risk for diabetes and coronary heart disease than lower-body obesity (40,87,88). The enzyme lipoprotein lipase that regulates the storage of fats as triglyceride is more active in abdominal obesity and therefore increases fat storage. Upper-body obesity is measured by the waist-to-hip ratio (W/H). Increased health risk is present in women when the W/H ratio exceed 0.8 and in men when the W/H ratio exceeds 1.0 (93). A more important factor in fat distribution is distinguishing abdominal visceral fat from subcutaneous fat. Visceral fat lies deep within the body cavities and is associated with a higher risk than subcutaneous fat because of the metabolic characteristics, which include insulin resistance and glucose intolerance. Visceral fat is measured by magnetic resonance imaging (MRI), which is expensive and unavailable to most practitioners.

A more practical measurement of visceral fat uses sagittal diameters (91). This technique requires the patient to lie on his or her back. The sagittal diameter is obtained by measuring the distance from the examination table to a horizontal level placed over the abdomen at the site of the iliac crest. This promising technique is currently the best practical predictor of visceral fat. The assessment of visceral fat provides additional information regarding health risk and should be used in counseling patients about realistic weight loss. Realistic weight loss should be based on total weight loss and the resulting redistribution of fat.

For those who are overweight and obese, direct medical costs (including preventive, diagnostic, and treatment services) in 1998 accounted for 9% of total U.S. medical expenditures (36). Andreyeva et al. suggested that each increment of 5 in BMI (e.g., 30 to 35, 35 to 40) results in a 25 to 100% increase in the lifetime healthcare expenditures for an individual (8). In 2007 dollars, this figure may be as much as $100 billion (99). As noted in table 12.3, Medicaid and Medicare paid up to half of this expenditure, which exemplifies the growing burden of overweight and obesity on federal and state spending. Indirect costs relate to morbidity and mortality costs and are difficult to capture. Morbidity costs result from decreased productivity and absenteeism. Mortality costs are the value of future income lost by premature death.

Obesity appears to reduce life expectancy, and this effect is most powerful in those who develop obesity earlier in life (41). Flegal et al. (using NHANES III data) reported that obesity, but not overweight, is related to excess premature death (37). They estimated 112,000 excess deaths per year because of obesity and reported that as obesity levels became more severe (class I to II to III) the mortality rate increased. They also reported no effect on longevity in the cohort defined as overweight (BMI = 25–29.9). This finding is similar to the findings of NHANES I and II. In addition, several studies have found a protective effect on mortality for overweight when other conditions are present such as heart failure and revascularization per percutaneous intervention (27,52).

Table 12.3 Medical Spending, in Billions of Dollars, Attributable to Overweight and Obesity

Insurance category	Overweight and obesity		Obesity	
	MEPS (1998)	**NHA (1998)**	**MEPS (1998)**	**NHA (1998)**
Out-of-pocket	$7.1	$12.8	$3.8	$6.9
Private	$19.8	$28.1	$9.5	$16.1
Medicaid	$3.7	$14.1	$2.7	$10.7
Medicare	$20.9	$23.5	$10.8	$13.8
Total	$51.5	$78.5	$26.8	$47.5

Note. Calculations are based on data from the 1998 Medical Expenditure Panel Survey (MEPS) merged with the 1996 and 1997 National Health Interview Surveys, and healthcare expenditures data from National Health Accounts (NHA). MEPS estimates do not include spending for institutionalized populations, including nursing-home residents (36).

Data from 1998 Medical Expenditure Panel Survey (MEPS); The 1996 and 1997 National health interview surveys; Health care expenditures data from National Health Accounts (NHA).

Americans spend over $46 billion each year in efforts to control their weight. These efforts include participation in commercial and clinical weight-loss programs, and purchases of food and nonfood supplements. In recent years several professional guidelines have been written that have focused on obesity, including *Dietary Guidelines for Americans* (101); *Physical Activity and Health: A Report of the Surgeon General* (103); NIH's *Clinical Guidelines on the Identification, Evaluation, and Treatment of Overweight and Obesity in Adults* (75); the ACSM position stand (2); the American Heart Association scientific statement (84); and *Healthy People 2010: National Health Promotion and Disease Prevention Objectives* (102).

PATHOPHYSIOLOGY

Obesity results from long-standing positive energy balance. The simplicity of this restatement of the first law of thermodynamics can blind us to the complex physiology of food intake and calorie expenditures and tempt us to tell patients only to "eat less and run more." Such an approach is oversimplistic and ineffective. Positive energy balance has a myriad of contributors in current society, ranging from increasing availability of lower-cost foods that some refer to as a toxic food environment to decreasing physical activity at work and in leisure. Average daily calorie intake has increased by over 200 kcal over the last several decades (43) as food costs have fallen dramatically and as more calories are consumed outside the home. At the same time physical activity has fallen because of advances in the workplace.

Beyond daily behavioral and dietary influences, recent research has identified an ever-increasing number of genetic and physiologic factors that point to a large array of neurological and peripheral endocrine messengers that influence food intake and nutrient utilization and that regulate weight in a way that often frustrates patients' weight-loss efforts (57).

A revolution in the understanding of obesity began with the identification of **leptin**, particularly when it was first produced in a laboratory in the 1990s. Leptin is secreted by fat cells and was found to regulate body weight in mice. Administration of leptin to mice genetically altered to be deficient in leptin reduced their extreme obesity and poor growth, resulting in marked reduction in food intake and weight, and return of normal growth and metabolic function. This research proved that a molecular defect could be the basis of some forms of obesity and provide a new approach for the treatment of obesity (11). Humans produce leptin from their fat cells in proportion to their weight and, particularly, their girth.

Humans are not leptin deficient, however, and exogenous leptin administration has limited benefit for weight reduction except in the rarest of cases of human obesity because of genetic absence of this fat cell hormone. Nevertheless, leptin research changed the perception of many clinicians and researchers. Instead of shunning obesity or considering it an untreatable moral failing, they now see it as a complex behavioral and neuroendocrine disorder that may be unlocked with additional study (18).

Genetic causes of obesity are a feature of rare disorders such as Prader-Willi syndrome (21) but are thought to be a factor in at least half of human obesity based on studies of twins raised separately or children who were adopted (97). Genetics plays a significant part in explaining responses to overfeeding or weight loss in metabolic lab settings, which points to variation in inheritable control of food intake, fat storage, and energy expenditure (15). These patients do not conveniently follow equations of calorie intake or energy utilization.

Body fat distribution is also genetically determined and gender specific, and predictable changes throughout life may confound patients' efforts at weight control. Increasing age reduces growth hormone and gonadal hormone secretion, which may predispose to greater visceral fat storage with its links to metabolic and cardiovascular abnormalities linked to hypertension, coronary artery disease, and diabetes mellitus (14).

Energy expenditure has complex genetic and environmental variation that can promote weight gain in some and frustrate weight loss in others. The **set-point theory** of weight regulation can be seen in metabolic lab studies of overfeeding compared with calorie restriction. Weight loss decreased total and resting energy expenditure, which slowed further weight loss. Weight gain through overfeeding was associated with an increase in energy expenditure, which slowed further weight gain. These studies show that after weight loss, individuals require 15% fewer calories to maintain their lower weight. This tendency causes weight-loss patients to return to higher weights unless they restrict calories long term or expend calories through greater physical activity (61).

Variation in energy regulation is seen in genetic studies of the Pima Indians, who are at significantly increased risk of obesity and its comorbid conditions (98). Efforts to augment energy expenditure through medications have been complicated by adverse effects of increased blood pressure and heart rate. This area will remain a target for ongoing pharmaceutical research because resting metabolic activity accounts for about 70% of energy expenditure. The metabolic cost of food digestion accounts for about 10% of energy expenditure. The remainder is physical activity, which is influenced by sedentary activities such as television and computer viewing compared with intentional exercise or dedicated increases in lifestyle activities.

Of the millions of calories eaten or expended over a lifetime, a mere fraction of 1% can result in clinically significant obesity. The neurological and peripheral control system for this process is biased toward preserving weight in times of famine and receives signals from fat cells and the gastrointestinal tract. An interesting observation in patients following gastric bypass weight-loss surgery is the dramatic reduction in **ghrelin**, a potent appetite-increasing gut hormone, which may explain, in part, the success of this surgical procedure (26). Ghrelin, produced during stomach distention, triggers the central nervous system appetite stimulant **neuropeptide Y (NPY)**, growth hormone, and norepinephrine. NPY and norepinephrine predominantly stimulate carbohydrate intake. Weight loss can increase serum ghrelin levels, which may explain part of the process that makes sustaining weight loss difficult.

Cholecystokinin, serotonin, and peptide YY are among a group of CNS and gut peptides involved in satiety and reduced food intake (62). Serotonin has been targeted in the treatment of depression with selective serotonin reuptake inhibitors such as fluoxetine, but these medications have had limited effect in weight-loss treatment. Nutrient selection during weight-loss treatment may play a factor because appetite-reducing peptide YY is increased during high-protein diets. This may partially explain the appetite-reducing effect of high-protein diets. Table 12.4 lists the monoamines and peptides that affect appetite.

The pathophysiology of obesity has been known for millennia. The degree of obesity predicts the morbidity and mortality for a long list of common afflictions in humans (20). Obesity and extreme obesity are associated with an increased rate of death from all causes, particularly from cardiovascular disease. In the United States estimates of annual excess deaths due to obesity range from 111,000 to 365,000 (3,37,71). Obesity and sedentary lifestyle is considered the second-leading cause of preventable death in America and may overtake tobacco abuse within the next decade. During the last 40 yr other

cardiovascular risk factors have decreased; high total cholesterol levels (≥240) decreased from 34% to 17%, high blood pressure (BP ≥140/≥90) decreased from 31% to 15%, and smoking dropped from 39% to 26%. Weight-related cardiometabolic risks have increased at an accelerated rate; diagnosed diabetes mellitus rose from 1.8% to 6.0%, and metabolic syndrome increased from 27% to 56% for men (49). The effects of obesity are so dramatic that projections of life expectancy, which have risen with declining cardiovascular risks over 40 yr, are predicted to decline as the full weight of increasing obesity is felt in society (79). For obese individuals, life expectancy decreased by about 7 yr compared with normal-weight individuals in the Framingham Study (82).

Type 2 diabetes mellitus is strongly associated with obesity; up to 80% of cases are weight related. Compared with people with a BMI of 22, those with a BMI of 35 have a 61-fold increased risk of developing diabetes (25). The Diabetes Prevention Project showed that even a modest 7% weight loss reduced the progression of prediabetes to diabetes by 58% over 4 yr, exceeding medication treatment (i.e., metformin) by 25% (59). Obesity, particularly abdominal (or visceral) obesity is also part of the current diagnostic criteria for **metabolic syndrome**, as explained in chapter 10 (50). Although the definition and meaning of metabolic syndrome has sparked controversy (56), several studies have shown that this clustering of risk factors is associated with a doubling of risk for diabetes mellitus and cardiovascular morbidity and mortality.

Intensive lifestyle intervention is currently being studied in the NIH-sponsored, multicenter, Look AHEAD trial (65). This decade-long study is evaluating the benefit of modest weight loss through low-calorie diets, meal replacement programs, and weight-loss medications coupled with thorough behavioral education, support, and exercise over 5 yr of initial treatment and up to 6 yr of follow-up in over 5,000 individuals. The 1 yr results of the Look AHEAD trial showed an average 8.6% weight loss in the group treated with intensive lifestyle interventions compared with a 0.7% loss in their diabetes support and education group.

Hypertension often increases in obese people and may be present in up to half of those seeking weight-loss treatment. In the Framingham Heart Study excess body weight explained over 25% of the incidence of hypertension (113). A 10 kg weight loss can reduce blood pressure by 10 mmHg, which is comparable to the effect required to achieve FDA approval for a new antihypertensive drug.

Cardiovascular morbidity and mortality is greater in obese individuals, and the role of adipose tissue, particularly visceral fat mass, is becoming more clearly established as a key coronary disease risk factor. Women

Table 12.4 Monoamines and Peptides That Affect Feeding

Stimulatory	Inhibitory
Norepinephrine	Leptin
Neuropeptide Y	Cholecystokinin
Opioids	Serotonin
Melanin-concentrating hormone	CRH or urocortin

in the Nurse Health Study showed a fourfold increased risk of death from cardiovascular disease if their BMI was over 32 compared with those with a BMI under 19 (112). The INTERHEART study showed that central obesity, particularly elevated waist circumference, explained 20% of the risk of first myocardial infarction (118). Obesity may also increase cardiovascular disease through elevated LDL-cholesterol levels and reduced HDL-cholesterol levels. Heart failure, mostly diastolic dysfunction, is increased twofold in obese people. This consequence is often largely reversible with significant weight loss. Atrial fibrillation and flutter are increased about 50% in obese individuals. Stroke risk is also doubled for obese women, based on the findings of the Nurses Health Study. In addition, venous thromboembolic disease, including DVT and pulmonary embolus, is increased in obese individuals, particularly those who are sedentary (108).

Respiratory illness, with adverse effects on exercise capacity, is also common in obese people (86). Obesity is the greatest predictor of obstructive sleep apnea, a syndrome of interrupted sleep with snoring and apneic periods that lowers oxygen saturation levels to levels associated with potentially lethal ventricular arrhythmias. Dyspnea and asthma are increased with obesity through adverse effects on respiratory mechanics and through GERD-associated bronchospasm.

Obesity is associated with increased gastroesophageal disease and hepatobiliary illness. GERD is increased through reflux of stomach acid into the esophagus because of increased intra-abdominal pressure, which may cause esophagitis and lead to esophageal cancer. Gallstones, cholecystitis, and biliary dyskinesia (biliary colic without cholelithiasis) are increased in obese individuals because of increased biliary excretion of cholesterol (96). Active weight loss may increase the risk of cholelithiasis and cholecystitis because cholesterol is removed from reducing fat stores and secreted, where it may crystallize in the gallbladder. Weight-loss diets with daily modest amounts of fat may empty the gallbladder and reduce this risk. Nonalcoholic fatty liver disease is linearly related to obesity and can lead to cirrhosis and liver failure but is often reversible with weight loss (9).

Osteoarthritis, particular weight-bearing joint disease, is increased with increasing obesity. Elevated weight multiplies the effects on the hips, knees, and ankles and increases the risk of foot pain and plantar fasciitis, which may further increase weight through secondary reductions in physical activity. Significant weight loss is associated with a 50% reduction in arthritic joint pain and may postpone or obviate the need for joint replacement in some individuals.

Depression and eating disorders, particularly binge-eating disorder, are increased in obese women, especially those seeking professional help with weight loss. Evaluation and management of these issues are important to eliminate them as barriers to weight loss. Psychosocial function is decreased by obesity, and outright prejudice is common toward obese people, particularly women, who may be significantly less likely to marry or to complete their education, and thus are likely to face increased poverty and lower annual incomes (48).

MEDICAL AND CLINICAL CONSIDERATIONS

Although medically significant obesity exists in most patients who seek medical care, only a minority present requesting medical help with weight reduction. Increasing the awareness of those not aware of the relationship between their weight and medical problems is an effective way of helping them become determined to address their weight. Compassion and understanding coupled with flexible and practical weight-loss recommendations are essential to building rapport with obese patients. Most individuals seeking professional help with weight loss have a BMI of 38 or more, and invariably they have attempted to lose weight several times in the past. Proper assessment of the barriers to and benefits of weight loss should be the initial step in the clinical evaluation of the obese patient. Comprehensive assessment of obesity can lead to appropriate treatment and effective long-term control of obesity and its related comorbidities, as is shown in "Health Consequences of Obesity."

Signs and Symptoms

Although obesity would seem the most obvious medical condition, its key determinants, weight and height, are often omitted in medical records or ignored in patient problem lists. Ideally, patients should be measured at each visit in lightweight clothing without shoes. Comparable weights would ideally be at the same time of day (intraday variation of up to 2% is common) with empty bladders. Appropriate scales that can measure extreme weights (up to 600 lb [270 kg]) are essential, or wheelchair scales should be used, because many patients who weigh more than 600 lb (270 kg) cannot stand to be weighed. A wall-mounted statiometer can be used for heights, although patients typically report accurate height, but not weight. Waist circumference should be measured at the umbilicus while the patient is standing, because measurement at that location is the best predictor of central, or visceral, obesity and its significance is not related to a patient's height. Patients should not suspend their breathing

Health Consequences of Obesity (53)

Greatly increased risk (relative risk >3)

Diabetes

Hypertension

Dyslipidemia

Breathlessness

Sleep apnea

Gall bladder disease

Moderately increased risk (relative risk about 2–3)

Coronary heart disease or heart failure

Osteoarthritis of the knees

Hyperuricemia and gout

Complications of pregnancy

Increased risk (relative risk about 1–2)

Cancer

Impaired fertility or polycyctic ovary syndrome

Low-back pain

Increased risk of anesthesia

Fetal defects arising from maternal obesity

Based on D. Haslam, N. Sattar, and M. Lean, 2006, "ABC of obesity. Obesity time to wake up," *British Medical Journal* 333: 640-642.

during waist measurements, and a measuring tape should not be overtightened.

Body fat measurements, typically by bioimpedance but also through underwater weighing, skinfold measurements, or DXA scans, can be performed, but they add little to therapeutic management and have significant methodological errors and costs. The U.S. surgeon general recommends that BMI be added to blood pressure and pulse as a routinely recorded vital sign because further evaluation and management of obesity and its related comorbidities depend on it.

History and Physical Exam

Most clinicians and patients are aware that a blood pressure of 150/90 mmHg or a cholesterol of 240 mg/dl or more is associated with increased cardiovascular risk, and immediate aggressive therapy would be started. Few are aware, however, that a BMI of 33 conveys the same risk of cardiovascular death (17). Additionally, only 25% of physician visits with patients who are obese address issues of weight reduction (95).

Therefore, many overweight and obese patients who begin to work with an exercise professional have not been approached about their weight by a medical professional. The exercise professional should address this issue by simply asking the patient about the subject. Assessing the patient's readiness to address his or her weight can be achieved with the simple question "Have you been trying to lose weight?" Potential responses are the following:

No, and I do not intend to in the next 6 mo (precontemplation).

No, but I intend to in the next 6 mo (contemplation).

No, but I intend to in the next 30 d (preparation).

Yes, but for less than 6 mo (action).

Yes, for more than 6 mo (maintenance).

The response of the exercise professional should be appropriate to the patient's readiness to lose weight. For instance, patients who are not intending to lose weight might simply be educated about the health risks of being overweight and the benefits of losing weight. Those who are preparing to lose weight soon might be best served by directing them into a clinically based weight-loss program.

The evaluation of the obese person, whether she or he is seeking medical help for weight control or for routine care of unrelated medical concerns, should include relevant factors of history and physical exam as well as laboratory testing that can better characterize the person's risks from obesity, its often silent comorbid conditions, and barriers to effective treatment. The medical approach to obesity should consider, in compassionate cooperation with the patient, the risks of obesity, appropriate treatment, and the most effective form of treatment.

Identification of medical causes of obesity including illnesses, medications, and lifestyle changes known to increase weight must be done. Examples include diabetes mellitus, polycystic ovary syndrome, Cushing's syndrome (although rare), many medications (some neuroleptics, psychotropics, anticonvulsants, antidiabetics, β-blockers, ACE inhibitors, and hormone therapies), and behavioral changes such as smoking cessation, job changes, injuries, sleep habits, and life stressors. History of obesity should be obtained whenever possible including the age of onset of obesity, weight gain since adolescence, changes in weight distribution, peak and lowest maintained adult weight, any history of weight loss (intentional or unintentional), and method used to lose weight. Assessment of readiness for weight loss, social support, weight-control skills, and history of any past or present eating disorders (binge eating, bulimia nervosa, or, rarely, anorexia nervosa) should be considered. Family history of obesity predisposes toward weight problems, and social history can identify key issues such as divorce, job changes, and upcoming weddings or class reunions that can affect weight-loss efforts. An early family history or cardiovascular disease may motivate an overweight patient to control his or her cardiovascular risk factors. A thorough review of systems can help identify medical conditions such as joint pain or asthma that may be overlooked barriers to weight control.

An exercise history should identify current intentional exercise or opportunities for increased lifestyle activities, evaluate any exercise barriers and patient-specific benefits to help in the assessment of any contraindications to moderate or vigorous exercise, and develop a comprehensive exercise plan. Exercise testing may be appropriate for the assessment of cardiovascular disease or to explain dyspnea on exertion, which is common in obese patients.

Diagnostic Testing

Laboratory testing should screen for diabetes (fasting blood sugar), elevated cholesterol and triglycerides as well as LDL and HDL cholesterol levels (lipid profile), and clinical or subclinical hypothyroidism (TSH and free T4). Comprehensive chemistry profiles can assess for nonalcoholic fatty liver and renal disease. Electrocardiograms are rarely necessary except to evaluate specific cardiovascular problems such as elevated blood pressure or palpitations.

Exercise Testing

Routine exercise testing in the overweight and obese population is not indicated. When performed, it is primarily to assess for the presence of coronary artery disease (38). Testing may also be performed to determine functional capacity and develop an exercise prescription based on heart rate. Several studies demonstrate that obese patients can exercise to maximal exertion on a variety of treadmill protocols (38).

Although walking is the preferred mode of exercise for testing, it is not always practical in those who are obese. These patients, especially those with BMI values greater than $40 \text{ kg} \cdot \text{m}^{-2}$, often have concomitant gait abnormalities and joint-specific pain during weight-bearing

exercise. Seated devices such as upper-body ergometers, stationary cycles, or seated stepping machines offer excellent alternatives that allow patients to achieve maximal exercise effort in a non-weight-bearing mode. Despite this, McCullough et al. reported that of a group of 43 patients referred for bariatric surgery (mean BMI = $48 \pm 5 \text{ kg} \cdot \text{m}^{-2}$), only one could not perform a walking protocol (67).

Testing should be performed with the normal routine (see chapter 5). Prediction equations for MET from the work rate achieved on an exercise device are typically inaccurate in people who are obese. Assessment of cardiopulmonary gas exchange provides an accurate measurement of exercise ability. Gallagher et al. reported a peak oxygen consumption (peak $\dot{V}O_2$) level of $17.8 \pm 3.6 \text{ mL} \cdot \text{kg}^{-1} \cdot \text{min}^{-1}$ (equivalent to 5.1 METs) in a morbidly obese group of patients who achieved peak RER values greater than 1.10 (46). No complications related to exercise testing were reported in this cohort, suggesting that exercise testing is safe in this extremely obese population. See table 12.5 on p. 220 for a review of exercise-testing methods.

Treatment

Although all weight-loss treatment should include specific diet, behavioral, and exercise prescriptions, and in some cases pharmacotherapy or weight-loss surgery, the specifics of treatment must be matched to the patient's circumstances, including his or her BMI and related considerations. Patients at lower BMI and without comorbid conditions should be offered less intensive treatment than patients with extreme obesity (BMI ≥40) who have significant obesity-related comorbidities such as diabetes, hypertension, and obstructive sleep apnea. Figure 12.3 on p. 221 provides the strategy used for weight management as discussed in this chapter.

Treatment goals need to consider both the medical benefits of modest (10%) weight loss and the patient's expectations. The NIH has recommended a 10% weight loss within 4 to 6 mo and weight-loss maintenance as an initial weight-loss goal because this amount is related to several health-related benefits. See "Health Benefits of a 10% Weight Loss" on p. 220. Improvements in obesity-related functional limits and medical comorbidities should be identified (24). Patients, on the other hand, commonly want to lose 35% of their weight to attain their dream weight, seek to lose 25% to reach a satisfactory weight, and would consider a weight loss of 17% to be disappointing (44). Few commonly prescribed weight-loss programs achieve average weight losses that match patient's expectations. Physician expectations are not much lower; they consider a weight-loss level of 13% to

Table 12.5 Review of Exercise-Testing Methods

Test type	Mode	Protocol specifics	Clinical measures	Clinical implications	Special considerations
Cardiovascular	• Treadmill • Cycle ergo-meter	Typically low level becasue of deconditioned state (e.g., ≤2 MET increments)	Standard ECG, blood pressure, rating of perceived exertion	• Typically only necessary to perform if at risk for ischemic heart disease • Also may be useful for setting an exercise prescription based on heart rate	• Must use equipment that can handle high body weight • Non-weight-bearing equipment useful if joint problems
Strength	Resistance machines	• Standard • Assess major muscle groups (e.g., chest, arms, legs)	Strength of various muscle groups	Serves as a baseline for developing a resistance training program for use in a weight-loss or weight maintenance program	May consider performing a 2RM to 10RM assessment if patient is limited
Range of motion	• Sit-and-reach • Goniometry	Standard	Range of motion of major joints	• Used to identify joints with limited range to focus on • Also used to assess progress	• May have difficulty with standard sit-and-reach because of requirement to sit on floor • Excess fat tissue may limit range in affected joints

Health Benefits of a 10% Weight Loss (53)

Blood Pressure
Decline of about 10 mmHg in systolic and diastolic blood pressure in patients with hypertension (equivalent to most BP medications)

Diabetes
Decline of up to 50% in fasting glucose for newly diagnosed patients

Prediabetes
- >30% decline in fasting or 2 h insulins
- >30% increase in insulin sensitivity
- 40 to 60% decline in the incidence of diabetes

Lipids
- 10% decline in total cholesterol
- 15% decline in LDL cholesterol
- 30% decline in triglycerides
- 8% increase in HDL cholesterol

Mortality
- >20% decline in all-cause mortality
- >30% decline in deaths related to diabetes
- >40% decline in deaths related to obesity

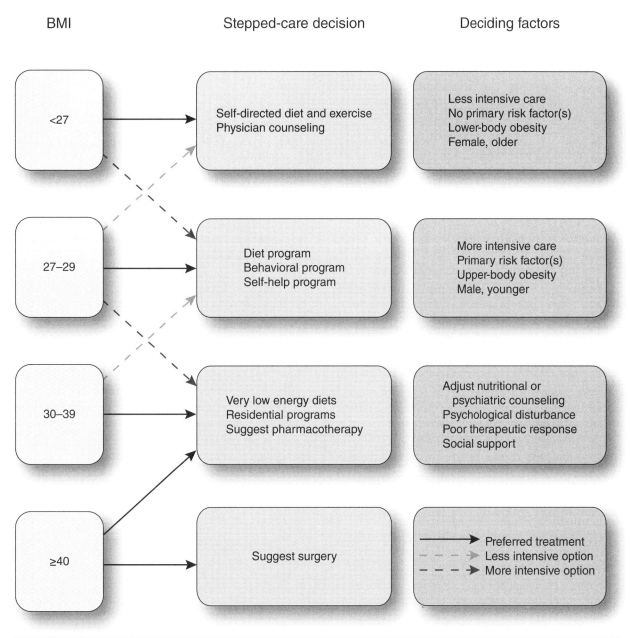

BMI | Stepped-care decision | Deciding factors

<27 — Self-directed diet and exercise / Physician counseling — Less intensive care / No primary risk factor(s) / Lower-body obesity / Female, older

27–29 — Diet program / Behavioral program / Self-help program — More intensive care / Primary risk factor(s) / Upper-body obesity / Male, younger

30–39 — Very low energy diets / Residential programs / Suggest pharmacotherapy — Adjust nutritional or psychiatric counseling / Psychological disturbance / Poor therapeutic response / Social support

≥40 — Suggest surgery — Preferred treatment / Less intensive option / More intensive option

Figure 12.3 A systematic approach to management based on BMI and other risk factors (117).

Reprinted, by permission, from B. Gumbiner, 2001, *Obesity* (Philadelphia, PA: American College of Physicians).

be disappointing. Patient expectations have been remarkably resistant to change, which may contribute to treatment dissatisfaction, treatment discontinuation, and treatment recidivism.

Diet Therapy

To lower weight, energy balance must be negative. Calorie reduction is the essential first step. For normal adults, 10 kcal per pound (22 kcal per kilogram) are required to maintain weight. A bell shape describes the variation in average energy expenditure of about 20%. Hence, some individuals will require 12 kcal per pound (26 kcal per kilogram) and others only 8 kcal per pound (18 kcal per kilogram) per day to maintain weight. The lowest calorie level for weight maintenance is about 1,200 kcal daily, even for those at bed rest. The minimum calorie intake to assure adequate intake of essential micro- and macronutrients is 500 kcal per day, commonly provided under medical supervision in very low calorie diets. Exercise and activity levels affect maintenance calorie levels by 25% or more, depending on the degree of physical activity, and physical activity needs to be considered in determining maintenance calorie levels.

Hypocaloric diets for weight loss typically set intake at 500 to 750 kcal less than predicted maintenance requirements. Typically, a deficit of 3,500 kcal is needed to lose 1 lb (7,700 kcal to lose 1 kg), and such diets can average about a 1 lb (0.45 kg) per week weight loss. If losses are slower, more aggressive diet therapy should be considered. Many popular variations in hypocaloric diet composition have been developed. Because of the high calorie content of fat (9 kcal per gram) compared with carbohydrates or proteins (4–5 kcal per gram) and the heart health benefits for cholesterol lowering, most national guidelines recommend low-fat diets.

Recently, higher-protein and lower-carbohydrate diets have become favored by patients because of greater weight losses and better satiety. Higher-protein diets tend to promote satiety, and lower carbohydrate levels can promote greater fat utilization (ketosis), which can boost the rate of weight loss. Overall reviews of popular commercial diet programs have shown about 5% weight loss at 1 yr and a 3% weight loss for standard hypocaloric diets at 2 yr (100). Although published data is supportive of higher-protein and lower-carbohydrate diets, no specific low-calorie diet composition is clearly superior to any other, and all fall short of patient and provider expectations.

Recent research has focused on more structured lower-calorie diets using meal replacement supplements. These diets are combined with exercise and lifestyle efforts to increase weight loss and achieve greater long-term results (45). Better portion control is achieved in these programs, which often use higher protein content to promote satiety and reduce snacking. Most use 1,000 kcal diets initially and combine weight-loss medications to optimize results. Wadden and colleagues showed 18% weight-loss results at 6 mo and 17% at 12 mo when a structured meal replacement diet was followed by a low-calorie diet in conjunction with behavioral and lifestyle changes while taking the weight-loss medication sibutramine (104).

Because most patients who present for medical help to lose weight will have BMIs of 38 or more, even structured meal replacement programs may not be sufficient. When rapid weight loss is critical, medically supervised very low-calorie diets (<800 kcal per day; a.k.a. VLCD) can be used. These diets routinely consist of highly engineered powdered supplements rich in protein. Weight losses can begin at 1 lb (.45 kg) per day but average 3 to 5 lb (1.4 to 2.3 kg) weekly for a typical 16 wk period (73). First developed in the 1920s, these diets resurfaced in the 1960s in surgical research centers. Popularized in the 1970s, such "last-chance" diets fell from favor when excess protein losses led to deaths from ventricular arrhythmias. Reformulated in the 1980s with better-absorbed nutrients, these diets, often called liquid protein diets, peaked in popularity when celebrities used them for rapid success, although most people later regained significant weight after resuming their previous eating and activity patterns. In studies comparing very low-calorie diets to low-calorie diets, outcomes were similar at 1 yr (106). Now, most very low-calorie diets are combined with long-term combination treatment. Among completers, weight loss ranges from 10 to 35%, averaging over 15% at 2 and 4 yr (6).

Practical Application 12.1

Literature Review

The VLCD diet plan, using high biological grade protein-based supplement, is often criticized for supposedly increasing the risk of putting excess weight back on or for hampering future success with weight-loss attempts. Li et al. (63) present data on over 480 individuals who used a 700 to 800 kcal per day VLCD diet (a.k.a. complete meal replacement) from two to four times. Initial attempt weight loss was 21.3 kg for women and 28.8 kg for men. For those who restarted the VLCD, there was no difference in total weight loss at restart attempt 2, 3, or 4 as compared with the initial attempt when assessed for rate of weight lost per week. The authors' conclusion was that their data refute the notion that repeated VLCD dieting reduces the effect of subsequent weight-loss diet attempts. This finding supports a modern approach to clinical weight management that suggests that those desiring to lose large amounts of weight and maintain this weight loss should consider the periodic or continuous use of high biological grade protein-based supplement.

Behavioral Therapy

Providing any level of diet advice without behavior change is typically futile. Most medically significant weight-loss efforts require frequent contact and support to adopt healthier weight behaviors necessary for weight loss and maintenance of weight loss. Regular accountability, problem solving, and skill building are necessary over a 20 wk period to establish long-term success. Such behavioral change can be supported by individual or group therapy and augmented by phone and Internet follow-up. Weight-loss efforts typically move from precontemplation to contemplation to determination and then to action phases. Maintenance efforts must follow action steps in weight loss; otherwise, relapse is common. Motivation and realistic goal setting must be supported in a compassionate environment focused on measurable progress (70).

Record keeping and review predicts success because most people will make better choices when made aware of the significance of those choices. Stimulus control helps patients identify stress and emotional eating cues and make other choices. Unhealthy eating behaviors like eating while driving or eating in front of a television or computer screen should be discouraged. Increased intentional exercise or lifestyle activity such be planned and monitored. Cognitive restructuring is used to detect black-and-white thinking and to help patients avoid an all-or-nothing pattern. Addressing emotional issues such as depression and shame with supportive therapy is essential, and referral for significant mental health or eating disorders may be necessary. Nutrition education and planning for maintenance, including relapse, can help reduce recidivism and the need for retreatment (19).

Exercise Therapy

Certainly, exercise and physical activity are important to avoid becoming overweight or obese. But for the treatment of overweight and obesity, exercise alone has not shown long-term weight-loss success (115). Exercise in conjunction with diet therapy or other treatment modalities, however, is effective in slightly accelerating weight loss (115). In the National Weight Control Registry (NWCR) over 90% of the successful subjects combined exercise with diet therapy (58). And evidence from the NWCR suggests that regular exercise of 60 to 90 min on most days of the week, expending 2,500 to 2,800 kcal per week, may be required to maintain large amounts of weight loss for the long term (30,58). Regardless of the timing of exercise and its effectiveness in causing or maintaining weight loss, all overweight and obese patients will likely demonstrate improvements in cardiovascular function and physical fitness as a result. Additionally, exercise can improve self-esteem, which may improve adherence to weight-loss-based treatments. the exercise prescription for the overweight and obese persons is reviewed later in this chapter.

Pharmacotherapy

Medications have been used to reduce weight for over 100 yr. The first medication advertised for weight loss was thyroid extract. Now known to cause more muscle loss than fat losses, thyroid replacement, necessary for hypothyroidism, is no longer used for weight loss. Weight-loss drugs have faced significant prejudice. Many people expect long-term benefits from short-term treatment, something not expected with antihypertensives, for example, and are reluctant to use them at all. Although not the "cure," some weight-loss medications are an important tool for achieving and maintaining medically significant weight loss.

Weight-loss drugs are now appropriately recommended for individuals with a BMI ≥30 or BMI ≥27 if they have obesity-related comorbidities. All studies designed to approve recent weight-loss medications incorporated standard behavioral, diet, and exercise changes and showed average weight losses of 8 to 10% over 6 mo with continued success when treatment was continued for 2 to 4 yr. Most produce a 4 lb (1.8 kg) weight loss within the first month. All weight-loss medications are associated with clinically significant improvements, based on the degree of weight loss, in blood sugar, blood pressure, and cholesterol profiles. Many of these drugs rival disease-specific medications for outcomes for these commonly treated conditions (64,94).

Phentermine (Adipex-P, Ionamin) is the most commonly prescribed weight-loss drug in the United States. Approved in 1958 and currently recommended for short-term use, it acts as an appetite suppressant but can cause dry mouth, palpitations, and anxiety. It was part of the famous phen–fen combination therapy effort pioneered in the mid-1990s by a study published by Weintraub until the fenfluramines (Pondimin and Redux) where withdrawn from the market after fears of cardiac valve abnormalities were discovered. Cleared by the FDA, phentermine remains the least expensive common weight-loss medication.

Sibutramine (Meridia), approved in 1997, is a serotonin and noradrenergic reuptake inhibitor, which decreases appetite, promotes satiety, and has modest beneficial metabolic effects. It produces 5 to 10% weight loss in two-thirds of treated patients taken for up to 2 yr (55). Sibutramine can cause dry mouth and in rare cases temporary increases in blood pressure and pulse that require early initial monitoring. It is not contraindicated in patients with controlled high blood pressure. This drug may become generically available after 2007.

Orlistat (Zenical), approved in 1999, is an intestinal lipase inhibitor that causes a 30% malabsorption of dietary fat. It is not absorbed into the bloodstream, and patients who eat high-fat meals may have oily diarrhea. Patients who follow a typical fat diet do not have significant GI symptoms from fat malabsorption. Orlistat produces an energy deficit of about 180 kcal daily (28). Orlistat has been approved by the FDA for over-the-counter sale.

Approved in parts of Europe, rimonabant (Accomplia) was not approved by the FDA for the U.S. market. The first of several cannabinoid-1 receptor blockers, it works centrally to block cravings. Rimonabant produces a 5 to 10% weight loss in two-thirds of patients treated and improves lipid and glucose risk factors in a weight-loss-related manner. Increased depression seen in some people who take rimonabant is a concern of the FDA (31,83).

Other medications have been recommended "off label" for weight loss. Metformin has been shown to be weight neutral in trials for weight loss, but it is an insulin sensitizer and improves fertility in women with polycystic ovarian syndrome (42). Fluoxitine and sertraline have been used for eating disorders, including night-eating syndrome, but have not been shown to reduce weight alone. Some SSRIs and many psychiatric medications are associated with weight gain and increased risk of diabetes (47).

Topiramate (Topamax), a drug approved for epilepsy that was shown to have a side effect of weight loss, has been studied for weight loss, but the trial was withdrawn because of cognitive side effects (confusion, word-finding problems) (16). Some studies, however, have shown that it has unique potential for reducing binge-eating syndrome, a problem in up to 40% of patients presenting for medical assistance for weight loss. Topiramate has recently become generically available, but its manufacturer was investigating stereo-isomers that may have fewer side effects.

Buproprion (Wellbutrin), an antidepressant, has been shown to promote modest weight loss (5). It is an activating drug related chemically to an older weight-loss drug, diethyproprion, and has not been pursued by its manufacturer for special FDA approval for weight loss. It may be the most weight-loss friendly choice of antidepressant in depressed weight-loss patients.

Combinations of medications have become a promising arena for weight-loss medications, and several are likely to become available in the next few years. When combined with group lifestyle modification programs, medications have clearly have increased weight loss compared with medication alone. Wadden and colleagues showed that patients who received sibutramine in addition to comprehensive lifestyle modification lost an average of 12.1 kg compared to only 5 kg for those treated with sibutramine alone at 1 yr. Individuals treated with lifestyle modification alone or sibutramine combined with brief therapy lost about 7 kg. Patients in this trial who kept food intake records among the top third lost 18.1 kg compared with only 7.7 kg for those in the bottom third of record keepers. This study underscores the importance of comprehensive lifestyle change for weight-loss success (104).

Surgical Therapy

The fastest growing area of obesity treatment is surgical procedures to restrict the stomach or cause malabsorption of food or both. Frustration with less invasive medical therapies now leads over 250,000 people to have one of these types of procedures each year. Surgery for weight loss has been performed for over 50 yr, and newer laparoscopic techniques can cause patients to lose a third of their weight (>50% of their excess weight) within 18 mo. Such weight loss results in marked reduction in obesity-related medical conditions and improved life expectancy in these seriously obese patients (22). Recent research has confirmed reduced mortality with weight-loss surgery (3).

Although not without significant risks (up to 1% for death and 15% for morbidity), surgery can lead to long-term weight loss. Additionally, patients undergoing bariatric surgery must commit to a lifelong program of restricted diet, lifestyle changes, vitamin supplementation, and follow-up testing to assure safety. Weight regain after 2 yr can mitigate some of the initial benefits such that, in the Swedish Obesity Study, longer-term weight loss at 10 yr averaged between 15 and 24% of initial weight depending on the procedure chosen (90).

Surgery is typically restricted to those with a BMI ≥40 or those with a BMI ≥35 if they have obesity-related comorbid conditions. Many insurers require a 6 to 12 mo trial of comprehensive medical therapy before surgery, and many surgeons require that patients attempt to lose 5 to 10% of initial weight before surgery to reduce complications, aid exposure interoperatively, and allow a laparoscopic approach (94).

The initial intestinal bypass procedures (jejunoileal bypass) of the 1950s were abandoned because of excessive nutritional deficiencies and were replaced by gastric banding procedures (gastric stapling) and vertical banded gastroplasty from the 1960s through the early 1990s (72). Roux-en-Y gastric bypass, combining the restrictive effect of gastric stapling to the malabsorptive effect of intestinal bypass, became the gold standard in the mid-1990s (66). Now done laparoscopically, this procedure is still recommended for those with BMIs of ≥50. Laparoscopic adjustable gastric banding has been adopted as a lower-risk procedure that has fewer postoperative complications and almost no mortality for those at lower weights. This procedure places an adjustable restriction on the stomach, which is adjusted to promote early satiety and lower calorie intake. Outside the United States this device has become the procedure of choice (32).

Comprehensive Long-Term Therapy

The imperative for health providers interested in obesity and for those who face the many comorbid conditions directly related to obesity is to match patients to treatment options that will achieve meaningful medical and personal benefits and not merely tell patients to "eat less and run more." The NIH-sponsored Look AHEAD trial is designed around intensive biweekly comprehensive treatment over 5 yr in hopes of reducing obesity-related morbidity and mortality from diabetes in obese subjects (65). This level of treatment also requires combinations of diet, exercise, behavioral, and pharmacologic therapy, and in many cases weight-loss surgery for those who fail less invasive approaches.

The maintenance of weight loss also requires diligent follow-up to offset the metabolic penalty of the reduced obesity (61). The Study to Prevent Regain (STOP) looked at face-to-face and Internet options for long-term weight-loss management and found that both have a role in preventing the regain of weight so commonly seen with termination of treatment (114). The chronic disease model applies to obesity as much as it does to the comorbidities of obesity such as diabetes mellitus. Comprehensive approaches that depend on long-term lifestyle training will be those that can tame obesity and reduce its effect on mortality in this century

EXERCISE PRESCRIPTION

The American College of Sports Medicine guidelines recommend that an exercise program focus on physical activities that result in an energy expenditure of more than 2,000 kcal per week to promote and maintain weight loss (1). These recommendations are beyond the general recommendation of 1,000 to 2,000 kcal expenditure per week for general health benefits. Expenditure of 2,000 kcal per week is equal to about 150 to 300 min

per week or 50 to 60 min per day of physical activity and exercise. But this figure is less than the 2,500 to 2,800 kcal per week expenditure recommended by the National Weight Control Registry, as mentioned previously in this chapter (115). Therefore, the following exercise prescription recommendations are based on a weekly caloric expenditure of 2,000 to 2,800 kcal per week. This goal range is appropriate for all obese individuals, although some obesity class II patients and most class III patients will have to progress gradually to these higher levels of energy expenditure. Counseling about physical activity provided by a clinical exercise physiologist will help people develop realistic goals, establish appropriate exercise progression schedules, and gain control of their exercise programs (81).

Cardiovascular Exercise

Initially, exercise and physical activity should focus on cardiovascular (i.e., aerobic) modes. The primary reason for this approach is to maximize the amount of caloric expenditure per unit of time spent exercising. To achieve the 2,000 to 2,800 kcal per day expenditure, the predominance of exercise must be aerobic. Although resistance training may provide added benefits, the caloric expenditure of resistance training is less than that of aerobic exercise because (1) it is performed discontinuously, (2) a single training session incorporates less exercise time than an aerobic session does, and (3) resistance training should be performed on only 2 or 3 d per week. Resistance training should be considered only after a regular aerobic program that meets the weekly caloric expenditure goal is in place for 1 to 2 mo.

Mode selection is important for enhancing adherence. Some patients may have preexisting musculoskeletal issues that may prevent some types of cardiovascular exercise. These issues often concern pain in the lower back, hip, knee, and ankle joints that may be chronic in nature and may improve as weight is lost. The clinical exercise physiologist should assess these issues and make recommendations to avoid this type of pain.

In general, aerobic exercise should be categorized as either weight bearing or nonweight bearing. When possible, walking is the best form of exercise to perform for several reasons. Walking has few disadvantages; all patients have experience with the activity. There are few reports of muscular soreness and injuries from walking. Walking is available to most patients and does not require special facilities. Neighborhoods, parks, walking trails, shopping malls, fitness centers, and so on offer walking opportunities. A minimum amount of attention is necessary, so socializing is easy and convenient. If a patient wishes to walk on a treadmill, care should be taken to assess the weight limits of the treadmill. Many are rated to only 350 lb (160 kg). Treadmills especially designed for obese individuals are available. Some patients may wish to jog, an activity that has the advantages of increased caloric expenditure per unit of time performed compared with walking and may result in greater improvements in fitness levels. But jogging should be avoided when possible, especially in patients with no previous jogging history or who have a preexisting musculoskeletal issue that may be aggravated by jogging. Some class I patients may be appropriate candidates for jogging.

Some non-weight-bearing exercise options include stationary cycling, recumbent cycling, seated stepping, upper-body ergometry, seated aerobics, and water activities. The clinical exercise physiologist should adapt these modes of exercise by providing larger seats and stable equipment. People who are obese often have difficulty getting on or off these types of equipment or moving through the range of motion required by a given piece of equipment. For some individuals, seated aerobics may be an excellent option to reduce the typical orthopedic limitations that some people experience, including back, hip, knee, and ankle pain. Another advantage is that seated or chair aerobics can be performed in the comfort of a person's home. Water provides a less weight-bearing option to walking or aerobic dance activities performed on land. The buoyancy of water takes much of the body weight off the joints. Additionally, patients who experience heat intolerance with other activities are often more comfortable performing water-based exercise. Most patients are not efficient swimmers, so swimming laps should be avoided. An experienced exercise leader can make a workout session fun and effective. For example, the resistance of the water can be used creatively to increase intensity. Many patients will not consider water activities because of the effort necessary to get into and out of the pool and because of their concern about their appearance in a bathing suit. The exercise physiologist should work to overcome these issues by using zero-entry pools and locations where the public does not have a direct view of the facility.

Frequency

Behavioral changes in activity must be consistent and long lasting if the patient is to lose weight and maintain weight loss over the long term. Daily exercise and physical activity at the recommended levels of duration and intensity are required to achieve and sustain long-term, significant weight loss. All people who are obese can exercise daily, typically from the beginning of a program. Key factors are to minimize the duration and intensity

initially to avoid excessive fatigue or muscle soreness that may sabotage the patient's willingness to exercise the next day.

Intensity

The intensity of exercise must be adjusted so that the patient can endure up to 1 h of activity each day. For those who have never exercised previously, intensity in the range of 50% to 60% of peak $\dot{V}O_2$ (50–60% of heart rate reserve) is typically low enough for sustained exercise. As an individual progresses, a goal of 60 to 80% of heart rate reserve is adequate. People without significant comorbid conditions can perform at these intensities in either a supervised or a nonsupervised setting. Many individuals who are obese are hypertensive and may be taking a β-blocker. This possibility must be considered when prescribing intensity using heart rate. Typical rating of perceived exertion values of 11 to 15 (6 to 20 scale) may be substituted.

Duration

For those with little or no previous recent exercise history, beginning with 20 to 30 min each day is appropriate. Breaking this exercise time into two or three sessions per day of shorter duration (5 to 15 min) may be required for highly deconditioned people. Progression of approximately 5 min every 1 to 2 wk until the person can perform at least 60 min of exercise is typically appropriate. This progression scheme is intended to increase compliance to the duration of each session as well as to daily exercise. An accumulation of time over several sessions in a day is as beneficial as one continuous work bout for calorie expenditure. Besides performing this intentional exercise duration, all obese people should be continuously encouraged to maximize daily physical activity by considering all options. For instance, they could park at the far end of the parking lot when visiting a store or get off one or two stops early when taking public transportation. The duration of daily physical activity should typically not be restricted unless the patient appears to be suffering from effects of excessive activity. But the contribution of incidental exercise to total caloric expenditure is significant and may be as beneficial as the planned exercise (5).

Intensity and duration must be manipulated so that the intensity is low enough to allow suitable duration to expend the recommended caloric energy. For many obese patients, the intensity will not be great enough to improve cardiovascular fitness. The initial focus, however, should be on weight loss and therefore caloric expenditure. As the exercise progresses and the patient improves, higher-intensity activities should be encouraged. Patients should be encouraged to increase the duration from 20 min per day to 60 min per day on every day of the week so that they expend >2,000 kcal per week. An exercise program for obese patients should include both the supervised and nonsupervised phases with adaptations in modes, intensity, duration, and frequency to provide adequate calorie expenditure while preventing soreness and injury. Patients with existing comorbidities should preferably participate in a supervised exercise program 3 to 5 d per week with a prescribed intensity and duration to treat their comorbidities. Patients should be physically active 60 min each day including the days of supervised exercise; therefore, they may have to supplement the time with walking to accumulate 60 min. The remaining days of the week (2–4 d) can be nonsupervised with self-reported exercise to accumulate 60 min of physical activity each day.

Resistance Exercise

If resistance training is incorporated, careful attention must be given to beginning this type of program. Strength equipment may not be an option for some morbidly obese individuals. In general, the exercise prescription for obese people should include resistance intensity in the range of 60% to 80% of an individual's 1RM, performed for 8 to 15 repetitions for two sets each, with 2 to 3 min of rest after each bout. This plan will allow the person to perform 6 to 10 exercises in a 20 to 30 min session. Resistance exercises can be performed maximally on 2 to 3 d per week. These exercises should focus on the major muscle groups of the chest, shoulders, upper and lower back, abdomen, hips, and legs. The primary acute benefit of the prescribed resistance program is to improve muscle endurance and, secondarily, to increase muscle strength. For obese individuals the long-term benefit may be related to a higher resting metabolic rate (RMR) and protection of lean mass loss during rapid weight-loss attempts.

Range of Motion

Obese patients may have a reduced range of motion as a result of increased fat mass surrounding joints of the body, in conjunction with a lack of stretching. As a result, these patients often respond slowly to changes in body position and have poor balance. Therefore, range of motion may improve spontaneously with weight loss. Still, to the degree possible, patients should perform a brief flexibility routine focused on the legs, lower back, and arm and chest regions. Normal flexibility routines (see chapter 5) are recommended as tolerated. See practical application 12.3 on p. 228 for a summary of this exercise prescription.

EXERCISE TRAINING

Low fitness levels add to the risk of mortality in overweight and obese people (109). In fact, low fitness is similar in risk to diabetes, hypertension, elevated cholesterol, and smoking. Ample evidence indicates that regular aerobic exercise training improves physical functioning,

Practical Application 12.2

Client–Clinician Interaction

Obese and overweight individuals are often simply told by their doctors to lose weight. The physician may even provide them with diet literature or send them to a registered dietitian. But the recommendation for exercise is often to "go walk a mile a day" or something similar. The physician is likely comforted by the patients' apparent acceptance of this exercise strategy, but far more often than not the patient does not begin, or adhere to, regular exercise. Therefore, the role of a clinical exercise physiologist in a weight management setting can be extremely important. The exercise physiologist must be able to discuss the realistic expectations of exercise for weight loss, design a program that begins at a person's given level of readiness and comfort; is properly designed for progression, and provides an avenue for regular follow-up for program adjustments to enhance adherence. The clinical exercise physiologist must be comfortable with the obese clientele, especially when discussing exercise options that may not be appropriate. For instance, some patients may wish to jog, and the exercise physiologist must be prepared to discuss the issues that may confront a morbidly obese individual when jogging. Others may be reluctant to consider using a swimming pool because of issues of appearance or ease of entering and exiting the pool. Time may not be available to build rapport, so the exercise physiologist must be able to discuss these issues almost immediately with any patient.

Summary of Exercise Prescription

Training method	Mode	Intensity	Frequency	Duration	Progression	Goals	Special considerations and comments
Aerobic	• Walking • Consider any non-weight-bearing mode where appropriate	50 to 80% of heart rate reserve	Progress to daily	Progress to 60 to 90 min	• Initial bouts are typically 10 to 20 min if person has no recent exercise history • Progress duration and frequency initially and intensity later • Consider beginning at lower end of intensity for very deconditioned people	• Achieve regular exercise pattern • Achieve 60 to 90 min per day by the time weight maintenance program begins	• Non-weight-bearing modes should be considered if joint pain or injury exists • Watch for indications of hyperthermia • Provide guidelines on water consumption during exercise
Resistance	• Machines • Free weights • Elastic bands • Calisthenics	• 10 to 15RM (i.e., 10–15 repetitions per set) • RPE of 11–15 (6–20 scale)	2 to 3 d per week	30 min involving two sets per major muscle group with minimum 1 min rest between sets	As tolerated to maintain 10 to 15 reps per set at an RPE of 11 to 15	• Regular resistance training • Improved skeletal muscle strength and endurance • Maintenance of lean mass during rapid weight-loss phase	• Because of range-of-motion limitations, some equipment may be difficult to use (e.g., machines) • Because of high incidence of hypertension, consider reducing breathhold or Valsalva maneuver
Range of motion	Static and proprioceptive or passive stretching	Within comfortable ranges	Daily	10 to 30 s per major joint	Increased range of motion as tolerated	Enhanced range of motion	Consider certain stretching techniques that may be difficult for some obese or overweight patients (e.g., poor balance, coordination, inability to sit on floor, etc.)

independent of weight loss, in those who are obese or overweight (23). Although not specific to the obese and overweight individual, Blair et al. have reported that men who increase their cardiovascular fitness level by one quartile have significant reductions in their long-term morbidity and mortality profile. Gulati et al. report similar benefits for women (13,51).

The focus of exercise training in the overweight and obese person should initially be on caloric expenditure, which is best achieved with aerobic-based training. Although dietary changes appear to be more effective than structured exercise alone at reducing body weight, data suggest that exercise alone can result in weight loss similar to caloric restriction (89). The amount of exercise required to be performed, however, is likely beyond the capacity of most obese individuals with no recent exercise history (e.g., 700 kcal per day for 3 mo). Some data support a synergistic effect of combined caloric restriction and exercise for weight loss in women (34). This finding, however, has not been universally replicated (105). Jakicic et al. reported that exercise intensity (moderate or vigorous) and duration (moderate or high) adjusted to expend between 1,000 and 2,000 kcal per week, combined with calorie intake restriction, did not have an effect on the amount of weight lost over 12 mo in a group of obese women (54). Others have demonstrated that diet combined with increased daily lifestyle physical activity may be as effective as a program of diet and intentional exercise (6).

Exercise training is likely most important in the weight maintenance process of weight control (69). Data from the National Weight Control Registry project indicate that regular exercise training that expends more than 2,000 kcal per week is a strong predictor of long-term weight-loss maintenance (58). Walking has shown to be effective for long-term (2 yr) weight-loss maintenance in women (39).

The expectations for cardiovascular exercise training on hemodynamic and other physiologic system adaptations are similar to that of the nonobese or overweight population. Increases in relative peak $\dot{V}O_2$ of 28% have been reported following a cardiovascular training program (110). Blood pressure declines acutely following exercise, and accumulation of exercise over time may chronically reduce blood pressure (80). These blood pressure effects appear to be independent of weight loss (35). An important adaptation to exercise in the overweight or obese population is an increase in insulin sensitivity and improved glucose metabolism (33,85). But reports give conflicting evidence about whether exercise alone, in the absence of weight loss, results in enhanced insulin sensitivity in women (85). The Diabetes Prevention Project demonstrated that lifestyle intervention of weight loss and exercise training was superior to medical therapy for preventing or delaying the development of diabetes in obese individuals with prediabetes (59). Improvements in lipoprotein profiles, vascular function, inflammatory biomarkers, and the risk of blood clot development may also be positive effects of exercise training, independent of weight loss, in obese people (29,77,107). But the long-term effects of these exercise-related responses on the development of disease has not been studied.

An important issue with weight loss is loss of skeletal muscle tissue. Recently Weiss et al. reported on a group of overweight individuals who lost approximately 10% of their initial body weight through either caloric restriction or cardiorespiratory training (6 d per week for 60 min, about 320 kcal expended per day). Thigh skeletal muscle loss and knee extensor strength declined in both groups following the intervention, but to a significantly greater degree in the caloric restriction group, suggesting a beneficial effect of exercise training to ameliorate loss of lean mass during weight loss. The use of resistance training to stimulate increases in lean mass may also be beneficial for reducing lean mass losses (10). No evidence suggests that resistance training during caloric restriction for weight loss results in additive weight loss (2). Table 12.6 on p. 230 briefly reviews the anticipated responses to exercise training in obese individuals.

Table 12.6 Review of Exercise-Training Responses

Cardiorespiratory endurance	Skeletal muscle strength	Skeletal muscle endurance	Flexibility	Body composition
Improved peak $\dot{V}O_2$ similar to that of normal-weight individuals (about 15–30% on average)	Evidence that aerobic and resistance training maintains or enhances skeletal muscle strength	Likely improvements similar to that expected in normal-weight individuals	• No randomized studies available that assess flexibility training in the obese • Expect normal increases in flexibility with ROM training but may be limited by excessive fat tissue	• Standard exercise training alone does not result in a significant reduction in weight • Up to 700 kcal per day expenditure from exercise will reduce weight • Evidence that aerobic and resistance training may preserve lean mass during weight-loss attempts

CONCLUSION

Clinical exercise physiologists, especially those who have a strong background in behavioral or lifestyle counseling, can play a critical role in the primary prevention, treatment phase, and secondary prevention of overweight and obesity. Patients will benefit from seeking out exercise physiologists who can accurately put into perspective the proper exercise program for an overweight or obese person and offer realistic expectations for exercise.

Case Study

Medical History

Mr. JL (age 52) came to an orientation session for a weight management program weighing 368 lb (167 kg). He was 69 in. (175 cm) tall, and his waist measurement was 64 in. (163 cm). He is married and works a full-time, sedentary job. He had a goal of attaining a weight of 190 to 200 lb (86 to 91 kg). He indicated that he would be disappointed if he achieved a weight of 260 lb (118 kg). He had diabetes, high blood pressure, and heel spurs. Medications include glyburide, metformin, and protonix. His baseline HR was 68 beats · min^{-1}, and his BP was 130/82 mmHg. Mr. JL reports success in weight loss previously with a variety of methods but gains it back over time. He is currently at his highest adult weight. He has difficulty exercising, primarily because of lack of time.

Diagnosis

Morbid obesity with secondary comorbid conditions of diabetes and high blood pressure. Increased risk of weight cycling because of previous successful weight-loss attempts with subsequent regain.

Exercise Test Results

A modified Bruce protocol test was performed on the patient within the past several months because of a complaint of atypical chest pain to his primary care physician. Results indicated a peak work capacity estimated at 10 METs (12 min test), normal HR, excessive BP increases (peak BP = 246/116 mmHg), and no indication of myocardial ischemia.

Exercise Prescription

The patient self-selected a diet based on reducing caloric intake by 500 kcal per day using foods purchased at grocery stores. The diet is based on the exchange system and focuses on low-fat eating and portion control. The patient is not interested in exercising regularly.

Discussion Questions

1. Considering the diet and exercise prescription information, what are your thoughts on the likelihood of success for this patient?

2. Discuss his goal and the weight goal that he would consider disappointing.

3. What diet plan might have been suggested for this patient?

4. Given the patient's apathy for exercise, how might you go about developing and implementing an exercise plan for him?

Ultimately, this patient did do well on his initial diet and exercise plan. He was persuaded through counseling to switch to a VLCD diet plan. Over the succeeding 9 mo he lost 168 lb (76 kg) and then switched to a weight maintenance diet. He also worked up to stationary cycling 5 d per week for 50 to 60 min.

5. Given the previous information, discuss why this patient was more successful after the changes in diet and exercise plan.

6. What do you think might have occurred with respect to his comorbid medical conditions?

7. What behavioral issues and plan would be prudent to discuss and develop for this patient?

8. Discuss the patient's exercise routine and its importance to weight loss and maintenance.

9. What other plans might be put in place to help this patient achieve long-term weight-loss maintenance?

Hypertension

A.S. Contractor, MD

Neil F. Gordon, MD, PhD

Hypertension is defined as a transitory or sustained elevation of systemic arterial blood pressure (BP) to a level likely to induce cardiovascular damage or result in other adverse consequences (37). The seventh report of the Joint National Committee on Prevention, Detection, Evaluation, and Treatment of High Blood Pressure (JNC VII) defines hypertension as having a **systolic** BP of 140 mmHg or greater, having a **diastolic** BP of 90 mmHg or greater, or taking antihypertensive medication (9). The report also defines an additional class of patients with systolic BP ranging from 120 to 139 mmHg or diastolic BP ranging from 80 to 89 mmHg as prehypertensive, or at heightened risk of developing hypertension in the future (9). The classification of BP for adults (age 18 or older) is shown in table 13.1.

In more than 95% of cases, the **etiology** of hypertension is unknown, and it is called essential, **idiopathic**, or primary hypertension. Secondary hypertension is **systemic** hypertension with a known cause (43). Table 13.2 lists the causes of secondary hypertension and their relative frequency. Although essential hypertension and secondary hypertension are the major classifications of hypertension, several other descriptive terms are used to define various types of hypertension. Isolated systolic hypertension is defined as systolic BP of 140 mmHg or more and diastolic BP of less than 90 mmHg. Malignant hypertension is the syndrome of markedly elevated BP associated with **papilledema**. In these cases, the diastolic BP is usually far greater than 140 mmHg. **White-coat hypertension** is the situation in which a person's BP is elevated when measured by a physician or other health-care personnel but is normal when measured outside a healthcare setting.

SCOPE

Hypertension affects about 65 million Americans in the age group of 20 and older, and its prevalence increases with age (16). In 2003 hypertension was listed as a primary or contributing cause of death in about 277,000 of the more than 2.4 million U.S. deaths. From 1993 to 2003 the death rate from hypertension increased 29.3%, and the actual number of deaths rose 56.1% (39).

Nearly one in three Americans has hypertension (16). Estimates are that among adults older than age 50, the lifetime risk of developing hypertension approaches 90% (40). According to the National Health and Nutrition Examination Survey III (NHANES III 1999–2000), the prevalence of hypertension among the different ethnic populations and sexes living in the United States is as follows (39):

- For non-Hispanic whites, 30.6% of men and 31.0% of women
- For non-Hispanic blacks, 41.8% of men and 45.4% of women
- For Mexican Americans, 27.8% of men and 28.7% of women

  (Estimates are age adjusted.)

Table 13.1 Classification and Management of Blood Pressure for Adults

BP classification	SBP mmHg*	DPB mmHg*	Lifestyle modification	Initial drug therapy	
				Without compelling indication*	**With compelling indications†**
Normal	<120	And <80	Encourage	No antihypertensive drug indicated	Drugs for compelling indications‡
Prehypertension	120–139	Or 80–89	Yes		
Stage 1 hypertension	140–159	Or 90–99	Yes	Antihypertensive drugs indicated	Drugs for compelling indications.‡ Other antihypertensive drugs, as needed
Stage 2 hypertension	≥160	Or ≥100	Yes	Antihypertensive drugs indicated. Two-drug combination for most†	

Note. DBP = diastolic blood pressure; SBP = systolic blood pressure.

*Treatment determined by highest BP category.

†Initial combined therapy should be used cautiously in those at risk for orthostatic hypotension.

‡Compelling indications include heart failure, postmyocardial infarction, high coronary artery disease risk, diabetes, chronic kidney disease, and recurrent stroke prevention. Treat patients with chronic kidney disease or diabetes to BP goal of <130/80 mmHg (10).

Data from A.V. Chobanian et al., 2003, "The seventh report of the Joint National Committee on Prevention, Detection, Evaluation, and Treatment of High Blood Pressure: The JNC 7th report," *Journal of the American Medical Association* 289: 2560-2572.

Table 13.2 Prevalence of Various Forms of Hypertension in the General Population and in Specialized Referral Clinics

Diagnosis	General population, %	Specialty clinic, %
Essential hypertension	92–94	65–85
Renal hypertension:		
Parenchymal	2–3	4–5
Renovascular	1–2	4–16
Endocrine hypertension:		
Primary aldosteronism	0.3	0.5–12
Cushing's syndrome	<0.1	0.2
Pheochromocytoma	<0.1	0.2
Oral contraceptive induced	0.5–1	1–2
Miscellaneous	0.2	1

Reprinted, by permission, from G.H. Williams, 1998, Approach to the patient with hypertension. In *Harrison's principles of internal medicine*, edited by E. Braunwald et al. (New York: McGraw-Hill Companies). With permission of The McGraw-Hill Companies.

On average, blacks have higher BP than nonblacks, as well as increased risk of BP-related complications, particularly stroke and kidney failure (9). Awareness of hypertension increased from the NHANES II (1976–1980) survey to the phase 1 NHANES III (1988–1991) survey, but it has reached a plateau since then and did not show an increase in the phase 2 NHANES III (1991–1994) survey. The NHANES survey in 1999–2000 showed similar levels of awareness and marginally improved treatment and control rates (table 13.3). In 2006, the direct and indirect costs due to hypertension were $63.5 billion. This figure includes health expenditures (direct costs, which include the cost of physicians and other professionals, hospital and nursing-home services, medications, home health, and other medical durables) and lost productivity resulting from **morbidity** and mortality (indirect costs) (39).

Table 13.3 Trends in the Awareness, Treatment, and Control of High Blood Pressure in Adults, United States, 1976–2000

	National Health and Nutrition Examination Survey, %			
	1976–80	**1988–91**	**1991–94**	**1999–2000**
Awareness	51	73	68	70
Treatment	31	55	54	59
Control*	10	29	27	34

*Systolic blood pressure below 140 mmHg and diastolic blood pressure below 90 mmHg and on antihypertensive medication.

Data are percentage of adults aged 18 to 74 yr with systolic blood pressure of 140 mmHg or greater, diastolic blood pressure of 90 mmHg or greater, or taking antihypertensive medication.

Based on A.V. Chobanian et al., 2003, "The seventh report of the Joint National Committee on prevention, detection, evaluation, and treatment of high blood pressure: The JNC 7 report," *Journal of American Medical Association* 289: 2560-2572.

PATHOPHYSIOLOGY

A variety of systems are involved in the regulation of BP: renal, hormonal, vascular, peripheral, and central adrenergic systems. BP is the product of cardiac output (CO) and total peripheral resistance (TPR): BP = CO × TPR. The pathogenic mechanisms leading to hypertension must lead to increased TPR by inducing vasoconstriction, to increased CO, or to both. Hypertension is frequently associated with a normal CO and elevated TPR.

Essential hypertension tends to cluster in families and represents a collection of genetically based diseases and syndromes with a number of underlying inherited biochemical abnormalities. Factors considered important in the genesis of essential hypertension include genetic factors, salt sensitivity, inappropriate renin secretion by the kidneys, and the environment. Environmental factors that have been implicated in the development of hypertension include obesity, physical inactivity, and alcohol and salt intake.

Although secondary hypertension forms a small percentage of cases of hypertension, recognizing these cases is important because they can often be improved or cured by surgery or specific medical therapy. Nearly all the secondary forms of hypertension are renal or **endocrine** hypertension. Renal hypertension is usually attributable to a derangement in the renal handling of sodium and fluids, leading to volume expansion or an alteration in renal secretion of vasoactive materials resulting in a systemic or local change in arteriolar tone. Endocrine hypertension is usually attributable to an abnormality of the adrenal glands.

Untreated hypertension leads to premature death, the most common cause being heart disease, followed by stroke and renal failure. Hypertension damages the **endothelium**, which predisposes the individual to **atherosclerosis** and other **vascular pathologies**. In the presence of **hyperlipidemia** and a damaged endothelium, atherosclerotic plaque develops, whereas in its absence, the **intima** thickens. Increased **afterload** on the heart caused by hypertension may lead to left **ventricular hypertrophy** and is an important cause of **congestive heart failure**. Hypertension-induced vascular damage can lead to stroke and transient ischemic attacks as well as end-stage renal disease. A meta-analysis of nine studies, involving 420,000 people, revealed that prolonged increases in usual diastolic BP of 5 and 10 mmHg were associated with at least 34% and 56% increases in stroke risk and with at least 21% and 37% increases in coronary heart disease (CHD) risk, respectively (24). Moreover, although both systolic and diastolic BP are important, in persons older than age 50 systolic BP is a much more important cardiovascular disease risk factor than diastolic BP (9).

MEDICAL AND CLINICAL CONSIDERATIONS

The clinical evaluation of a person with hypertension should be aimed at assessing secondary forms of hypertension, assessing factors that may influence therapy, determining whether target organ damage is present, and identifying other risk factors for cardiovascular disease. Establishing an accurate pretreatment baseline BP is also vital.

Signs and Symptoms

Hypertension is often referred to as the silent killer because most patients do not have specific symptoms related to their high BP. Headache is popularly considered a symptom of hypertension, although it occurs only in severe hypertension; most commonly, such headaches are localized to the **occipital** region and are present on awakening in the morning. Dizziness, palpitations, and easy fatigability are other complaints related to elevated BP. Some symptoms, such as **epistaxis**, **hematuria**, and blurring of vision, are attributable to underlying vascular disease, which may have contributed to the hypertension.

History and Physical Exam

A thorough medical history should assess the duration and severity of hypertension and symptoms and signs, if any. The history should include questions concerning the individual's risk factors for CHD and stroke, and symptoms and signs of CHD, heart failure, renal disease, and endocrine disorders. Information should be obtained about the past and present use of medications and about lifestyle habits. A comprehensive medical history will help in the treatment of primary hypertension and in the diagnosis and treatment of causes of secondary hypertension.

Diagnostic Testing

An accurate BP reading is the most important part of the diagnostic evaluation. A person is classified as hypertensive based on the average of two or more BP readings, at each of two or more visits after an initial screening visit.

The auscultatory method of BP measurement with a properly calibrated and validated instrument should be used. Persons should be seated quietly for at least 5 min in a chair (rather than on an exam table), with both feet placed on the floor, and the arm supported at heart level. Caffeine, exercise, and smoking should be avoided for at least 30 min before measurement. Measurement of BP in the standing position is indicated periodically, especially in those at risk for postural hypotension. An appropriate-sized cuff (cuff bladder encircling at least 80% of the arm) should be used to ensure accuracy. At least two measurements should be made (9). Systolic BP is the point at which the first of two or more Korotkoff sounds is heard (phase 1), and diastolic BP is the point before the disappearance of Korotkoff sounds (phase 5).

Measurement of BP at home by the patient or family or automated ambulatory monitoring helps verify the diagnosis of hypertension and response to treatment. Advantages of self-measurement of BP include distinguishing sustained hypertension from white-coat hypertension. Table 13.4 provides follow-up recommendations based on the initial set of BP measurements.

Laboratory tests for hypertension should include urinalysis, **hematocrit**, blood chemistry (sodium, potassium, creatinine, and lipid profile), and an electrocardiogram (9). These routine tests will help determine the presence of target organ damage and other risk factors. From an exercise prescription point of view, the electrocardiogram is an important test, because it may reveal the presence of arrhythmias and baseline ST-segment changes. Other tests that can be of value, depending on indications, include **creatinine clearance**; microscopic urinalysis; chest X-ray; echocardiogram; and serum calcium, phosphate, and uric acid.

Exercise Testing

Hypertension is one of the major risk factors for CHD, and a graded exercise test is a useful screening tool for

Table 13.4 Recommendations for Follow-Up Based on Initial Blood Pressure Measurements for Adults

Initial blood pressure, mmHg*	Follow-up recommended[†]
Normal	Recheck in 2 yr
Prehypertension	Recheck in 1 yr**
Stage 1 hypertension	Confirm within 2 mo**
Stage 2 hypertension	Evaluate or refer to source of care within 1 mo. For those with higher pressures (e.g., >180/110 mmHg), evaluate and treat immediately or within 1 wk depending on clinical situation and complications

*If systolic and diastolic categories are different, follow recommendations for shorter time follow-up (e.g., 160/86 mmHg should be evaluated or referred to source of care within 1 mo).

[†]Modify the scheduling of follow-up according to reliable information about past BP measurements, other cardiovascular risk factors, or target organ disease.

**Provide advice about lifestyle modifications.

Based on A.V. Chobanian et al., 2003, "The seventh report of the Joint National Committee on prevention, detection, evaluation, and treatment of high blood pressure: The JNC 7 report," *Journal of American Medical Association* 289: 2560-2572.

hypertensive patients. The American College of Sports Medicine (ACSM), however, does not recommend mass exercise testing to identify those at high risk for developing hypertension because of an exaggerated BP response (42). Hypertensive individuals with an additional CHD risk factor, males older than 45 yr of age, and females older than 55 yr of age should perform an exercise test with electrocardiogram monitoring before starting a vigorous exercise program (42).

Contraindications

The American College of Cardiology, American Heart Association, and ACSM guidelines on exercise testing state that severe arterial hypertension, defined as systolic BP greater than 200 mmHg or diastolic BP greater than 110 mmHg at rest, is a relative contraindication to exercise testing (19,42).

Recommendations and Anticipated Responses

Standard exercise-testing methods and protocols may be used for persons with hypertension (42). Before the hypertensive person undergoes graded exercise testing, obtaining a detailed health history and baseline BP in both the supine and standing positions is important. The person should be taking his or her usual antihypertensive medications when exercise testing is performed for the purpose of exercise prescription. Certain medications, especially β-blockers, affect BP at rest and during exercise and may affect the heart rate response to exercise.

Abnormal BP Response

Blood pressure is the product of CO and TPR. Normally during exercise, the TPR decreases, but the increase in CO more than compensates for the decrease in TPR and systolic BP increases. Diastolic BP usually remains the same or may decrease slightly because of the decrease in TPR. But hypertensive patients often experience an increase in diastolic BP both during and after exercise. They are often unable to reduce TPR to the same extent as normotensive people (those with normal BP). Impaired endothelial function in the early stage and, later, a reduced lumen-to-wall thickness could be responsible for the increased resistance during exercise (18).

Indications for Terminating Graded Exercise Test

A significant decrease in systolic BP (>10 mmHg) from baseline systolic BP despite an increase in workload is an indication for terminating an exercise test. An excessive increase in BP, defined as a systolic BP greater than 250 mmHg or diastolic BP greater than 115 mmHg, is an indication for terminating an exercise test (42).

Predictive Value of BP Response

An exaggerated BP response to graded exercise testing in normotensive people has been associated with an accentuated future risk of developing hypertension. An exaggerated response can be arbitrarily defined as a level of BP higher than that expected for the individual being tested. Data from the Framingham Heart Study (36) showed that an exaggerated diastolic BP response to exercise is predictive of risk for new-onset hypertension in normotensive men and women and that an elevated recovery systolic BP is predictive of the future development of hypertension in men. The Framingham Heart Study is one of the few studies on the predictive value of exercise BP that included large numbers of men and women. After multivariate adjustment, an exaggerated diastolic BP response during stage 2 of the Bruce treadmill protocol was observed to have the strongest association with new-onset hypertension in both men (odds ratio = 4.16) and women (odds ratio = 2.17). The study defined an exaggerated exercise BP response as either a systolic or diastolic BP above the 95th percentile of sex-specific, age-predicted values, during stage 2 of the Bruce protocol. These values are shown in table 13.5 on p. 238 (36).

Studies at the Cooper Clinic (28) and Mayo Clinic (1) revealed that an exaggerated exercise BP response had an odds ratio of 2.4 for predicting future hypertension. The subjects were mostly men, and the average follow-up time was about 8 yr. The Mayo Clinic study also found that exercise hypertension was a significant predictor for total cardiovascular events but not for death or any individual cardiovascular event.

In a review article, Benbassat and Froom (5) found that the prevalence of hypertension on follow-up among normotensive subjects with a hypertensive response to exercise testing was 2.06 to 3.39 times higher than that among subjects with a normotensive response. But the predictive value was limited because 38.1% to 89.3% of those with a hypertensive response to exercise did not have hypertension on follow-up, and a normotensive response only marginally reduced the risk of future hypertension.

After 17 yr of follow-up in 4,907 men, Filipovsky et al. (17) found that the exercise-induced increase of systolic BP was a risk factor for death from cardiovascular as well as noncardiovascular causes independent of resting BP. Similar findings have been shown in other studies (30). Some studies, however, showed no additional prognostic information regarding total mortality rates and

Table 13.5 Sex-Specific Predicted 95th Percentile Values for Systolic and Diastolic BP at Stage 2 of Exercise (on the Bruce Protocol) for Different Age Groups

Age	Exercise SBP	Exercise DBP
Men		
20–24	190	93
25–29	193	97
30–34	196	101
35–39	198	103
40–44	201	105
45–49	204	106
50–54	208	107
55–59	211	107
60–64	214	107
65–69	218	106
Women		
20–24	165	92
25–29	169	95
30–34	173	98
35–39	177	100
40–44	181	102
45–49	186	103
50–54	190	104
55–59	195	104
60–64	199	103
65–69	204	102

Adapted, by permission, from J.P. Singh et al., 1999, "Blood pressure response during treadmill testing as a risk factor for new-onset hypertension. The Framingham heart study," *Circulation* 99(14): 1831-6.

cardiovascular events from exercise BP readings (15). According to the ACSM, mass exercise testing is not advocated to identify those at high risk for developing hypertension in the future because of an exaggerated exercise BP response.

Treatment

The goal of prevention and management of hypertension is to reduce morbidity and mortality rates by the least intrusive means possible. This goal may be achieved through lifestyle modification alone or in combination with pharmacological treatment. People with a systolic BP of 120 to 139 mmHg or a diastolic BP of 80 to 89 mmHg should be considered prehypertensive and also require lifestyle intervention.

The BP goal is <130/80 mmHg in patients with hypertension with diabetes or chronic kidney disease and <140/90 mmHg in other patients with hypertension. Because most patients with hypertension, especially those older than age 50, will reach the diastolic BP goal after systolic BP is at the goal level, JNC VII recommends that the primary focus be on achieving the systolic BP goal. Figure 13.1 depicts the algorithm recommended by JNC VII for the treatment of hypertension.

Lifestyle modifications (table 13.6 on p. 240) help in controlling BP as well as other risk factors for cardiovascular disease. A review of randomized controlled trials of over 6 mo duration analyzing the effect of weight reduction in reducing BP found a decrease of 5.2/5.2 mmHg and 2.8/2.3 mmHg in hypertensive and normotensive participants, respectively (14). Weight reduction enhances the effects of antihypertensive medications and positively affects other cardiovascular risk factors, such as diabetes and **dyslipidemia**.

A recent meta-analysis of lifestyle interventions in BP management showed statistically significant reductions in systolic BP/diastolic BP of 5.0/3.7 mmHg by following an improved diet (13). Most of these diets included a target of weight reduction. The Dietary Approaches to Stop Hypertension (DASH) eating plan is a diet rich in fruits, vegetables, and low-fat dairy products with reduced content of dietary cholesterol and saturated and total fat. The plan has been shown to reduce BP by 8 to 14 mmHg (35,41).

Patients should be questioned in detail about their current alcohol consumption, because excessive alcohol consumption is a risk factor for high BP. The JNC VII report recommends that men who drink alcohol should be counseled to limit their daily intake to no more than 1 oz (30 ml) of ethanol and women and lighter-weight individuals should be told not to exceed 0.5 oz (15 ml). These quantities are equivalent to two drinks per day in most men and one drink per day in women. A drink is 12 oz (360 ml) of beer, 5 oz (150 ml) of wine, or 1.5 oz (45 ml) of 80-proof liquor (44).

Epidemiological data demonstrate a positive association between sodium intake and level of BP. Patients with essential hypertension may be classified as salt sensitive and salt resistant, based on the absolute changes in BP that originate from dietary salt intake (34). African Americans, older people, and patients with hypertension or diabetes are more sensitive to changes in dietary sodium than are others in the general population. A recent review of randomized controlled trials of 6 mo or longer duration in adults over the age of 45 found a small but statistically significant effect of lowered BP through

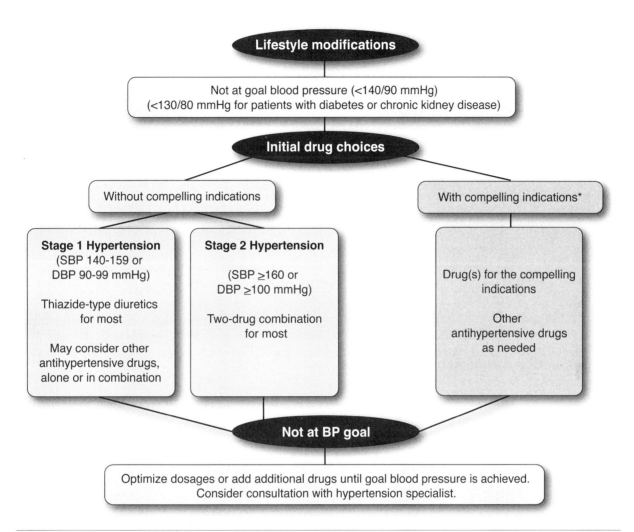

Figure 13.1 The algorithm recommended by JNC VII for the treatment of hypertension. DBP = diastolic blood pressure; SBP = systolic blood pressure. *Compelling indications include heart failure, postmyocardial infarction, high coronary heart disease risk, diabetes, chronic kidney disease, and recurrent stroke prevention. Treat patients with chronic kidney disease or diabetes to BP goal of <130/80 mmHg.

Reprinted, by permission, from A.V. Chobanian et al., 2003, "The seventh report of the Joint National Committee on prevention, detection, evaluation, and treatment of high blood pressure: The JNC 7th report," *Journal of American Medical Association* 289: 2560-2572.

salt reduction. Salt reduction resulted in pooled net systolic/diastolic BP changes of 2.9/2.1 mmHg in hypertensive individuals and 1.3/0.8 mmHg in normotensive individuals (14). The level of BP reduction was related to the level of salt reduction. A moderate sodium restriction to no more than 100 mmol per day (2,400 mg of sodium) is recommended in the JNC VII report. The average American sodium consumption is more than 4,100 mg a day in men and 2,750 mg per day in women (9).

The decision to initiate pharmacological therapy should be guided by the degree of BP elevation, the presence of target organ damage, and the presence of clinical cardiovascular disease or other cardiovascular risk factors (see table 13.1 and figure 13.1). The presence of cardiovascular risk factors is assessed during the initial

evaluation of the patient with hypertension. Their presence independently modifies the risk for future cardiovascular disease. After the clinician has determined the person's BP and the presence of risk factors, target organ damage, and clinical cardiovascular disease, the person's risk group can be determined. This classification into a risk group helps guide therapeutic decisions.

Most patients with hypertension who require drug therapy in addition to lifestyle modification will require two or more antihypertensive medications to achieve goal BP. If BP is >20/10 mmHg above the goal, consideration should be given to initiating antihypertensive therapy with two agents, one of which should usually be a thiazide-type diuretic. Thiazide-type diuretics should be used in drug treatment for most patients with

Table 13.6 Lifestyle Modifications to Prevent and Manage Hypertension*

Modification	Recommendation	Approximate SBP production (range)[†]
Weight reduction	Maintain normal body weight (body mass index 18.5–24.9 kg · m^{-2}).	5–20 mmHg · 10 kg^{-1}
Adoption of DASH eating plan	Consume a diet rich in fruits, vegetables, and low-fat dairy products with reduced content of saturated and total fat.	8–14 mmHg
Dietary sodium reduction	Reduce dietary sodium intake to no more than 100 mmol per day (2.4 g sodium or 6 g sodium chloride).	2–8 mmHg
Physical activity	Engage in regular aerobic physical activity such as brisk walking (at least 30 min per day, most days of the week).	4–9 mmHg
Moderation of alcohol consumption	Limit consumption to no more than two drinks (e.g., 24 oz [720 ml] beer, 10 oz [300 ml] wine, or 3 oz [45 ml] of 80-proof whiskey) per day in most men and to no more than one drink per day in women and lighter-weight persons.	2–4 mmHg

DASH indicates dietary approaches to stop hypertension.

*For overall cardiovascular risk reduction, stop smoking.

[†]The effects of implementing these modifications are dose and time dependent and could be greater for some individuals.

From A.V. Chobanian et al., 2003, "The seventh report of the Joint National Committee on prevention, detection, evaluation, and treatment of high blood pressure: The JNC 7th report," *Journal of American Medical Association* 289: 2560-2572.

uncomplicated hypertension, either alone or combined with drugs from other classes. Certain high-risk conditions are compelling indications for the initial use of other antihypertensive drug classes. Compelling indications include heart failure (diuretics, β-blockers, ACE inhibitors or ARBs, and aldosterone antagonists), postmyocardial infarction (β-blockers, ACE inhibitors or ARBs, and aldosterone antagonists), patients at high risk for CHD (diuretics, β-blockers, ACE inhibitors or ARBs, and calcium channel blockers), diabetes (diuretics, β-blockers, ACE inhibitors or ARBs, and calcium channel blockers), chronic kidney disease (ACE inhibitors or ARBs), and recurrent stroke prevention (diuretics and ACE inhibitors or ARBs). A complete listing of hypertensive medications is reviewed in chapter 3.

After initiation of drug therapy, most patients should return for follow-up and adjustment of medications at approximately monthly intervals until the BP goal is reached. More frequent follow-up may be needed for patients with stage 2 hypertension or with complicating comorbid conditions. Serum potassium and creatinine should be monitored at least once or twice per year. Follow-up visits can usually be at 3 to 6 mo intervals after BP is at goal and stable (9).

EXERCISE PRESCRIPTION

Exercise training has been recommended as one of the important lifestyle modifications for the prevention and management of hypertension (2). When compared with active and fit individuals, those who are sedentary have a 20% to 50% increased risk of developing hypertension. Endurance exercise training by individuals who are at high risk for developing hypertension will reduce the increase in BP that occurs with age (42).

Although regular aerobic exercise has been shown to reduce BP, the mechanisms responsible for this remain largely unknown. Some evidence shows that exercise training is associated with a decrease in plasma norepinephrine levels, which may be responsible for the decrease in BP (2). The kidneys play an important role in BP regulation. In this respect, exercise training may decrease BP by improving renal function in patients with essential hypertension. Another postulated mechanism is that regular physical activity causes favorable changes in arterial structure, which would presumably reduce peripheral vascular resistance (38). Hyperinsulinemia has been postulated to raise BP by renal sodium retention, sympathetic nervous activation, and induction of vascular smooth muscle hypertrophy (25,33). Hypertension and hyperinsulinemia, along with insulin resistance, abdominal obesity, increased triglycerides, and decreased high-density lipoprotein, often are clustered together to form what has been called the metabolic syndrome. Even a single bout of exercise has a well-known insulin-like effect and dramatically increases skeletal muscle glucose transport. Exercise training increases insulin sensitivity, which can decrease serum insulin and BP (4).

ACSM recommendations (20,42) for aerobic exercise programming for patients with hypertension include the following:

- Frequency: 3 to 7 d per week
- Intensity: 40% to 70% $\dot{V}O_2R$ or HRR (typically, this intensity corresponds to a rating of perceived exertion of 11–14)
- Time: 30 to 60 min of continuous or intermittent (minimum of 10 min bouts) aerobic activity
- Type: Primarily aerobic exercise supplemented by resistance exercise

The frequency, duration, and intensity of aerobic exercise should be modulated to achieve a weekly energy expenditure of 700 (initial goal) to 2,000 (long-term goal) kcal per week. Patients should also be advised to increase the amount of leisure-time activity.

The clinician should be aware that β-blockers attenuate the heart rate response to exercise; β-blockers, calcium channel blockers, and vasodilators may cause postexertion hypotension (an adequate cool-down may be especially important for patients taking these medications); and certain diuretics may cause a decrease in serum potassium levels, thereby predisposing the patient to arrhythmias.

Moderate-intensity resistance training is an important component of a well-rounded exercise program for the prevention, treatment, and control of hypertension. A recent meta-analysis which pooled data from studies published between 1996 and 2003 showed that the overall effect of resistance training was a decrease of 3.2 mmHg in systolic BP and a decrease of 3.5 mmHg in diastolic BP (7).

When performing resistance training, hypertensive patients generally should adhere to the American Heart Association's guidelines (31) for patients with cardiovascular disease (which has been endorsed by the ACSM). These include the following:

- Frequency: 2 to 3 d per week
- Sets: one
- Reps: 10 to 15
- Stations or devices: 8 to 10 exercises that condition the major muscle groups

Special Exercise Considerations

Those with more marked elevations in BP (>160/100) should add endurance training to their treatment only after initiating drug therapy. Individuals should not be allowed to exercise on a given day if their resting systolic BP is more than 200 mmHg or diastolic BP is more than 110 mmHg. Although BP termination criteria for exercise testing are established at >250/115 mmHg, lower BP thresholds for termination of an exercise-training session may be prudent (i.e., >220/105 mmHg) (42). Diuretics and β-blockers may impair thermoregulation during exercise in hot or humid environments. Those taking these medications should be well informed about signs and symptoms of heat intolerance, and know how to make prudent modifications in the exercise routine to prevent heat illness.

Strength or resistance training is not recommended as the only form of exercise training for people with hypertension. With the exception of circuit weight training, resistance training has not consistently been shown to lower BP. Resistance training is recommended as a component of a well-rounded fitness program but not when done independently.

Practical Application 13.1

Summary of Exercise Prescription

Mode	Frequency	Intensity	Duration	Special considerations
Aerobic: large-muscle-group activities such as walking, jogging, cycling	3–7 d per week	40–70% $\dot{V}O_2R$	30–60 min	No exercise if resting systolic BP exceeds 200 mmHg or diastolic BP exceeds 110 mmHg
Resistance exercise: 8–10 exercises for major muscle groups	2–3 d per week	40–60% 1RM	One set of 10–15 reps	

Exercise Recommendations

The exercise recommendations for individuals with hypertension should take into account their medical history, current BP levels, presence of cardiovascular disease, and its risk factors. Comorbid conditions such as diabetes, CHD, and heart failure should be adequately controlled before the start of exercise training. The program developed should include cardiovascular endurance training, resistance training, and flexibility exercises. Additional information relating to the exercise prescription for patients with hypertension can be found in the client–clinician interaction.

Cardiovascular Training

The ACSM, in its position statement on physical activity, physical fitness, and hypertension, states that moderate-intensity cardiovascular exercise training appears effective in lowering BP acutely and chronically (2). For those with hypertension, endurance exercise, corresponding to 40 to 70% of $\dot{V}O_2R$, is recommended to maximize the benefits and minimize possible adverse effects of more vigorous exercise (2). This intensity range corresponds to approximately 11 to 14 on the Borg rating of perceived exertion (RPE) 6-to-20 scale. The mode (large-muscle activities), frequency (3–7 d per week), and duration (30–60 min of continuous or intermittent aerobic activity) are similar to those recommended for healthy adults (42). The ACSM now views exercise and physical activity for health and fitness in the context of an exercise dose continuum. That is, there is a dose response to exercise by which benefits are derived through varying quantities of physical activity ranging from approximately 700 to 2,000 or more kilocalories of effort per week (3). A more detailed review of exercise and hypertension is presented in the literature review.

Resistance Training

In the past, hypertensive patients have been discouraged from participating in resistance training because of fear of overloading an already compromised **myocardium** (27). These fears were increased by a study performed by McDougall et al., who recorded pressures in excess of 400/200 mmHg in weight lifters during high-intensity resistance exercise (27). Similarly, there have been concerns that resistance training could give rise to increased arterial stiffness with a subsequent deleterious effect on TPR. Studies by Cortez-Cooper et al. (11) and Miyachi et al. (29) that used protocols consisting of high-intensity supersets showed a significant increase in the carotid augmentation index along with a decrease in the central arterial compliance in men and women. But a study by

Practical Application 13.2

Client–Clinician Interaction

Patients with hypertension are at heightened risk for atherosclerotic cardiovascular disease. Atherosclerotic cardiovascular disease is by far the leading cause of death in hypertensive patients. Therefore, besides assisting the patient with an appropriate exercise prescription, the clinician should attempt to educate the patient about atherosclerotic cardiovascular disease and its risk factors. In view of this, education also should be provided about factors that may help minimize the risk for exercise-related cardiac complications. These points include the importance of an adequate warm-up and cool-down and the warning symptoms and signs of an impending cardiac event. Drug therapy is often needed to optimize hypertension management and to facilitate cardiovascular disease risk reduction. Consequently, hypertensive patients are often receiving treatment with one or more medications. The clinician should educate the patient about the effect, if any, of specific medications on exercise performance and training. The clinician also should emphasize to the patient the importance of taking medications as prescribed and not discontinuing drug therapy without notifying his or her personal physician. If the clinician believes that the patient may be experiencing medication-related adverse effects, the clinician should refer the patient to his or her personal physician. The clinician is likely to interact with the patient on many occasions throughout the course of a year. Therefore, the clinician will probably have an opportunity to measure the patient's BP on many occasions. When discussing the patient's BP recordings, the clinician must strike a balance between not alarming the patient about minor day-to-day fluctuations in BP and expressing appropriate concern about excessive elevations in systolic or diastolic BP. But if the clinician believes that the patient's BP is not under adequate control, the clinician should consult with the patient's physician.

Literature Review

Reviews of studies on the BP-lowering effects of exercise training and other lifestyle modifications in hypertensive people have shown vastly differing results, perhaps because of the different study inclusion criteria and study designs. All meta-analyses, however, concluded that BP decreases significantly in response to exercise training. Recently, Dickinson and colleagues conducted a meta-analysis of trials from 1998 to 2006 (13). Only randomized controlled trials with at least 8 wk of follow-up were included. A total of 105 studies that randomized 6,805 participants were identified. These trials showed that BP reductions occurred because of a host of lifestyle interventions. Robust statistically significant effects were found for improved diet, aerobic exercise, alcohol and sodium restriction, and fish oil supplements. With aerobic exercise a mean reduction in systolic BP of 4.6 mmHg and a mean reduction in diastolic BP of 2.4 mmHg occurred.

The ACSM, in its position stand on exercise and hypertension, found that endurance training significantly reduced systolic BP and diastolic BP, and those reductions were greater in the hypertensive population as compared with the normotensive population. Fagard et al. found significantly greater absolute reductions in systolic BP/diastolic BP in hypertensive patients (7.4/5.8 mmHg) versus normotensive patients (2.6/1.8 mmHg) (2). There are, however, concerns over long-term BP-lowering effects of exercise being attenuated, mostly because of high levels of exercise dropout.

Research shows that 24 hr ambulatory BP monitoring may be more predictive of target organ damage than casual resting measures, but nighttime BP is less influenced by exercise training than daytime BP. Previous trials have led to conflicting results, and even in trials showing a significant reduction in ambulatory BP, the magnitude of reduction was generally less than that observed in casual BP (2).

Recently, it has been hypothesized that improvements in BP and plasma lipid induced by exercise training in hypertensive individuals may be genotype dependent. Hagberg et al. (22) found evidence to support the possibility that ACE, apoE, LPL PvuII, and Hind III **genotypes** may identify hypertensive individuals likely to reduce BP the most with exercise training.

The prevalence of modifiable CHD risk factors is higher in hypertensive individuals. Exercise training has been shown to have a beneficial effect on obesity, lipid profiles, and insulin sensitivity. Dengel et al. (12) showed that a 6 mo intervention of aerobic exercise and weight loss substantially reduced BP, improved insulin sensitivity by 39%, and resulted in a 50% reduction in the number of metabolic abnormalities associated with the metabolic syndrome in obese, hypertensive, middle-aged men. Brown et al. (8) found that an aerobic exercise program of 7 d duration improved insulin sensitivity in African American hypertensive women independent of changes in fitness levels, body composition, or body weight. Subjects performed aerobic exercise on the treadmill or stationary bicycle for 50 min on 7 consecutive days. The exercise intensity corresponded to 65% of their maximum heart rate reserve.

Exercise training does not, however, necessarily have to be performed for extended periods to achieve significant reductions in systolic and diastolic BP. A recent report by Guidry and coworkers showed that even short bouts of exercise at low intensity (15 min in duration at 40% $\dot{V}O_2$max) had comparable effects on BP reduction to longer and more intense bouts of exercise (20 min at 60% $\dot{V}O_2$max) (21).

Blair et al. (6) found that hypertensive men who were more fit had lower death rates compared with less-fit men. Between 1970 and 1981, these authors tested 1,832 men who reported a history of hypertension but were otherwise healthy. Mortality surveillance was conducted on the group through 1985. The inverse relation between fitness and all-cause mortality held even after investigators adjusted for the influence of age, serum cholesterol, resting systolic BP, body mass index, current smoking, and length of follow-up.

Most studies show a beneficial effect of exercise on CHD risk factors, even if the exercise is not enough to increase fitness level or decrease body weight.

The rationale for resistance training as an adjunct to aerobic exercise for controlling BP stems from multiple studies. In a recent meta-analysis, Cornelissen and Fagard pooled data from studies published between 1996 and 2003 that included nine randomized controlled trials involving 341 participants (10). The overall effect of resistance training was a decrease of 3.2 mmHg in systolic BP and a decrease of 3.5 mmHg in diastolic BP. At present insufficient data are available on the effects of resistance training on ambulatory BP.

Guidelines for a Circuit Weight-Training Program

- Select 8 to 12 exercises to create a well-balanced program.
- Establish a conservative 1RM in each exercise or a 10RM on three to four key exercises.
- Use 40 to 60% of 1RM.
- Do 10 to 15 reps in 30 to 60 s.
- Do one set of each exercise in a circuit pattern, alternating between upper and lower body and moving from large-muscle-group exercises to small-muscle-group exercises.
- Begin with a 45 s rest between sets and gradually reduce to a 15 to 30 s rest.
- Train 2 or 3 d a week.
- Machine weights are preferable for hypertensive patients.
- Emphasize full range of motion and proper posture during all exercises.
- Warm up with stretching and 10 to 15 min of moderate aerobic exercise (11–13 on the RPE scale).
- Perceived exertion during the circuit should be 12 to 14.
- Avoid straining or heavy lifting; emphasize aerobic circuit training.
- Use proper breathing technique to avoid the Valsalva maneuver.
- Maintain a firm handgrip, but not too tight, to avoid a pressor response that may cause excessive rises in BP.
- When you can comfortably complete two to three circuits with a given load, increase the load by 2.5 to 10 lb (1.1 to 4.5 kg).
- Be sure to adhere to the medical regimen prescribed by your physician.
- Exercise caution if you have diabetic retinopathy or any other condition resulting in raised intraophthalmic pressures.

Reprinted, by permission, from T. LaFontaine, 1997, "Resistance training for patients with hypertension," *Strength Condition* 19: 5-9.

Rakobowchuk and coworkers that used a circuit-training protocol showed no changes in arterial compliance after 3 mo of resistance training in young men. The authors hypothesized that circuit weight training would be more appropriate for a hypertensive population (32).

Kelly et al. (26) in a meta-analysis of progressive resistance exercise found a 2% and 4% reduction in systolic BP and diastolic BP, respectively, and concluded that progressive resistive exercise is efficacious in reducing resting systolic BP and diastolic BP. Harris and Holly (23) evaluated a circuit weight-training program in male subjects with BP between 140/90 and 160/99 mmHg. Subjects exercised at approximately 79% of their maximum heart rate, 3 d a week, for 9 wk. They improved their muscular strength and cardiovascular endurance and lowered their diastolic BP from 96 to 91 mmHg. Resting heart rate and systolic BP did not change. See "Guidelines for a Circuit Weight-Training Program." Resistance training for hypertensive people should ideally involve lower resistance with higher repetitions. The recommendations are to do one set of 8 to 10 different exercises that condition the major muscle groups, 2 to 3 d a week. Each set should consist of 10 to 15 repetitions (31). Circuit weight training, defined as lifting a weight equal to 40% to 60% of 1RM for 10 to 20 reps in a 30 to 60 s period, has also been found to be beneficial for hypertensive people. After a rest of 15 to 45 s, the person moves to the next exercise (27). BP should be monitored frequently before, during, and after resistance training during the initial few weeks of participation. The improvement in strength from a resistance-training program will help hypertensive persons better perform both occupational and leisure tasks and will enhance their quality of life (27).

Range of Motion

Flexibility exercises should be included in the exercise routine. They should include a variety of upper- and lower-body range-of-motion activities, which should be performed on at least 2 to 3 d of the week. At least four repetitions per muscle group should be done, and each static stretch should be held for 10 to 30 s (3). The goal of these exercises is to reduce the risk of musculoskeletal injury and improve the individual's flexibility.

CONCLUSION

Hypertension affects more than 65 million Americans and is one of the leading causes of death. In more than 95% of cases, the etiology of hypertension is unknown. Hypertension is one of the major risk factors for CHD and is often found clustered with other CHD risk factors.

Lifestyle modification helps control BP and reduces other CHD risk factors. Exercise training helps prevent the development of hypertension and reduces the BP of those with hypertension. Chronic endurance training has been shown to reduce both systolic and diastolic BP by about 5 to 10 mmHg. Studies have shown that cardiovascular exercise training at somewhat lower intensity (40–70% of $\dot{V}O_2R$) appears to lower BP as much as, or more than, exercise at higher intensity. The mode (large-muscle activities), frequency (3–7 d per week), and duration (30–60 min of continuous or intermittent aerobic activity) are similar to those recommended for healthy adults. Although limited data suggest that resistance training has a favorable effect on resting BP, the magnitude of the acute and chronic BP reductions are less than those reported for endurance exercise. Resistance training for hypertensive people should involve lower resistance and more repetitions.

Case Study

Medical History

Mr. AB, a 60 yr old Caucasian male, comes to your fitness center for an exercise program. He is apparently healthy but has a 15 yr history of hypertension. His goal is to improve his health and fitness, and reduce his risk for heart disease. In the past he did heavy resistance training and power lifting. At present he has a body mass index of 31 kg $\cdot$ m^{-2}. His medications include aspirin (325 mg daily) and atenolol (50 mg daily). He has a family history of premature CHD, his LDL cholesterol is elevated (165 mg/dl), his resting BP is 148/88 mmHg, and his resting heart rate is 60 beats per minute. He drinks three to four cans of beer with dinner each evening.

Diagnosis

Mr. AB has a lengthy history of hypertension. Additional risk factors include obesity, age, family history, dyslipidemia, sedentary lifestyle, and elevated alcohol consumption.

Exercise Test Results

During a recent maximal exercise test, he exercised for 6 min using the Bruce treadmill protocol and achieved a maximum heart rate of 130 beats per minute and BP of 240/90 mmHg. He did not develop any significant arrhythmias or ST-segment changes.

Exercise Prescription

Mr. AB achieved 81% of his age-predicted HR max. The test was negative for underlying cardiovascular ischemia. The initial prescription should be set at a modest intensity (40% $\dot{V}O_2R$), and emphasis should be placed on building exercise duration to increase energy expenditure. Non-weight-bearing activities should be emphasized initially to minimize risk of musculoskeletal injury.

Discussion Questions

1. Is it necessary for Mr. AB to exercise at high intensity to help manage his hypertension?
2. How will atenolol (a β-blocker) affect the prescription of a target heart rate range for Mr. AB?
3. What is an appropriate aerobic exercise prescription for Mr. AB?
4. What advice would you give Mr. AB regarding going back to heavy resistance training and power lifting?
5. What additional lifestyle modifications should be discussed with Mr. AB to optimize his BP control?

Hyperlipidemia and Dyslipidemia

Kyle D. Biggerstaff, PhD

Joshua S. Wooten, MS

A linear association between elevated blood cholesterol concentrations and coronary artery disease (CAD), and atherosclerotic disease in general, is well established. A variety of abnormal lipid patterns have been identified that may contribute to the development of atherosclerotic diseases. This chapter discusses the medical and clinical relevance of the primary lipid abnormalities seen in the general population. Furthermore, the potential for exercise to improve lipid profiles is reviewed, and specific recommendations are provided for prescribing exercise in the hyperlipidemic patient.

Unfortunately, the term *hyperlipidemia* has been incorrectly used by much of the medical community to mean any form of lipid abnormality. More appropriately, hyperlipidemia is any condition that elevates fasting blood triglyceride or cholesterol concentrations. But when genetic, environmental, and pathological factors combine to alter blood lipid and lipoprotein concentrations abnormally, the condition is better termed *dyslipidemia*. Although severe forms of dyslipidemia are linked to genetic defects in cholesterol metabolism, less severe forms may result secondarily either because of other diseases (e.g., diabetes) or as a result of combining a specific genetic pattern with various environmental exposures such as diet, exercise, body composition, and smoking. Secondary dyslipidemia may result from diabetes mellitus, hypothyroidism, renal insufficiency, nephrotic kidney disease, or biliary obstruction.

SCOPE

The prevalence of high blood cholesterol in the United States has been on the decline. The Third National Health and Nutrition Examination Survey (NHANES III; 1988–1994) demonstrated that the public's intake of saturated fat and total fat declined. NHANES III also showed that blood cholesterol concentrations dropped. Between 1978 and 1994, average total cholesterol concentrations among U.S. adults fell from 213 to 203 mg/dl. In NHANES 1999–2002, no further decrease in total cholesterol occurred, and the percentage of adults with total cholesterol greater than 240 mg/dl was only 17% (10). The fact that less than 20% of the adult population has a total cholesterol concentration less than 240 mg/dl, achieves one of the Healthy People 2010 objectives. In addition, these findings may have contributed to the 30.2% decline in coronary heart disease (CHD) mortality rates from 1993 to 2003 (62). The decline in blood cholesterol concentrations is likely related to several factors. Cholesterol awareness among the public has improved. The latest Cholesterol Awareness Survey of physicians and the public noted that from 1983 to 1995, the percentage of the public who had ever had their blood cholesterol checked increased from 35% to 75%. Alternatively, the CDC has reported that the percentage of adults who have had their blood cholesterol checked in the previous 5 yr has increased from 67.6% in 1991 to 73.1% in 2003 (11). Because of greater awareness of total cholesterol concentrations, more Americans will receive treatment than in the past. The targeted efforts to treat lower blood cholesterol concentrations are probably attributable to the increased availability of effective medications that have relatively minor side effects and the evidence suggesting that use of the medications appears to lower CHD mortality rates.

PATHOPHYSIOLOGY

Lipids are not water soluble and consequently are transported through the plasma compartment as a lipoprotein. A lipoprotein is spherical in shape, made up of a lipid core that contains triglyceride, free and esterified cholesterol, and phospholipid and surrounded by an apolipoprotein (apo) to provide water solubility (figure 14.1). Lipoproteins differ slightly by apo, cholesterol (free and esterified), and triglyceride content, resulting in differing density ranges as determined by ultracentrifugation and electrophoretic mobility.

The primary lipoprotein classifications (see table 14.1) include chylomicron, very low-density lipoprotein (VLDL), low-density lipoprotein (LDL), and high-density lipoprotein (HDL). The chylomicron is synthesized

from the intestinal absorption of dietary fat and is responsible for the transport of dietary fat to extrahepatic tissue. Chylomicrons are catabolized in the liver via the apo E receptor-mediated pathway. VLDL (figure 14.2) is synthesized in the liver and is the primary mechanism for transporting endogenous triglycerides. The VLDL is converted into an intermediate density lipoprotein (IDL) following the removal of triglyceride at extrahepatic tissues. Lipoprotein lipase (LPL) hydrolyzes and removes triglycerides transported by VLDL, thereby promoting the conversion of VLDL into IDL. The IDL, which is higher in cholesteryl ester content, is rapidly converted into LDL. The conversion of IDL to LDL is promoted by further LPL hydrolysis and being acted on by the intravascular enzyme cholesterol ester transfer protein (CETP).

The primary carrier of cholesterol in the bloodstream is LDL, usually the most prevalent lipoprotein form in humans. Elevated LDL cholesterol (LDL-C) is linearly associated with an increase in CAD risk (7). Lipoprotein (a), or Lp(a), is a subclassification of LDL and is also highly atherogenic. Lp(a) contains apo (a), a protein that is linked to apo B and has a similar structure to plasminogen. Consequently, Lp(a) may inhibit normal fibrinolytic activity, increasing the likelihood of the development of a thrombus (42). The cardioprotective function of HDL is at least in part performed through a reverse cholesterol transport mechanism. Free cholesterol from circulating lipoproteins and possibly extrahepatic tissue is taken up and transported by circulating HDL to the liver, where it is catabolized (see figure 14.3 on p. 250). The exchange of lipids and biotransformation of lipoproteins can occur in the intravascular space and are also catalyzed by another enzyme, cholesterol ester transfer protein (CETP). CETP exchanges LDL or HDL cholesteryl ester for triglyceride from triglyceride-rich lipoproteins (TRL). This exchange leads to more triglyceride-rich LDL or HDL particles. Chylomicrons and VLDL are the primary TRL.

The association between elevated plasma cholesterol and increased incidence of CAD is well established. A clear and direct correlation has been observed in individuals who have plasma cholesterol concentrations of $200 \text{ mg} \cdot \text{dl}^{-1}$ or higher. An even stronger association has been observed between elevated LDL-C concentration and atherosclerotic diseases. Even within the LDL subclass, smaller and denser LDL particles are linked to the atherogenic process (7). In addition, a greater number of smaller and denser LDL particles is associated with increased risk for CAD (57). In combination with traditional lipid risk factors for CAD, LDL size is a strong predictor for CAD (57). Furthermore, in addition to LDL, it appears that any other lipid containing apo B

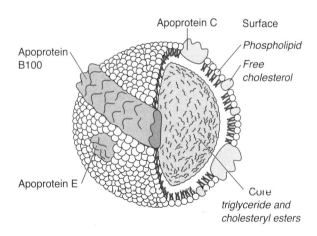

Figure 14.1 General structure of lipoproteins (a schematic representation of very low-density lipoprotein).

Reprinted from *William's textbook of endocrinology*, 9th ed., J. Wilson et al., copyright 1998, with permission from Elsevier Science.

Table 14.1 Lipoprotein Classifications and Corresponding Forms of Dyslipidemia

Lipoprotein class	Major lipids	Apolipoproteins	Density (g · ml⁻¹)	Diameter	Conditions where elevated
Chylomicron	Dietary triglyceride	Major: A-IV, B-48, B-100, H Minor: C-I, C-II, C-III, A-I, A-II, E	<0.95	800–5,000	LPL deficiency, apo C-II deficiency, apo E-2
VLDL	Endogenous triglycerides	Major: B-100, C-III, E Minor: A-I, A-II, B-48, C-II, D, G	0.96–1.006	300–800	Familial combined hyperlipidemia; familial hypertriglyceridemia; apo C-II deficiency; type III hyperlipoproteinemia
IDL		Major: B-100, C-III, apo A Minor: C-I, C-II, E	1.006–1.019	250–350	Type III hyperlipoproteinemia
LDL		Major: B-100, apo A Minor: C-I, C-II	1.055–1.120	180–280	Familial hypercholesterolemia; familial combined hypercholesterolemia
HDL		Major: A-I, A-II, F Minor: A-IV, C-I, C-II, C-III, E	1.063–1.120	50–120	Hyperalphalipoproteinemia, CETP deficiency
HDL₂			1.063–1.125	90–120	
HDL₃			1.125–1.210	50–90	

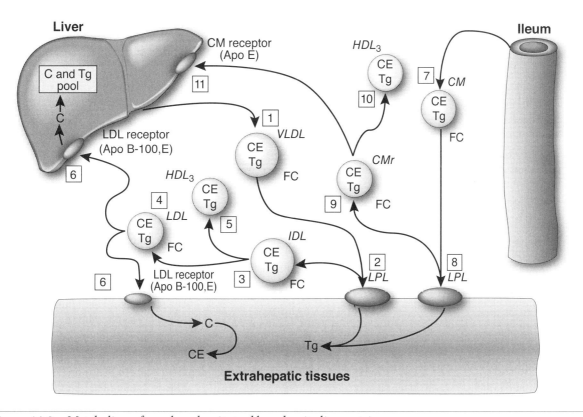

Figure 14.2 Metabolism of very low-density and low-density lipoproteins.

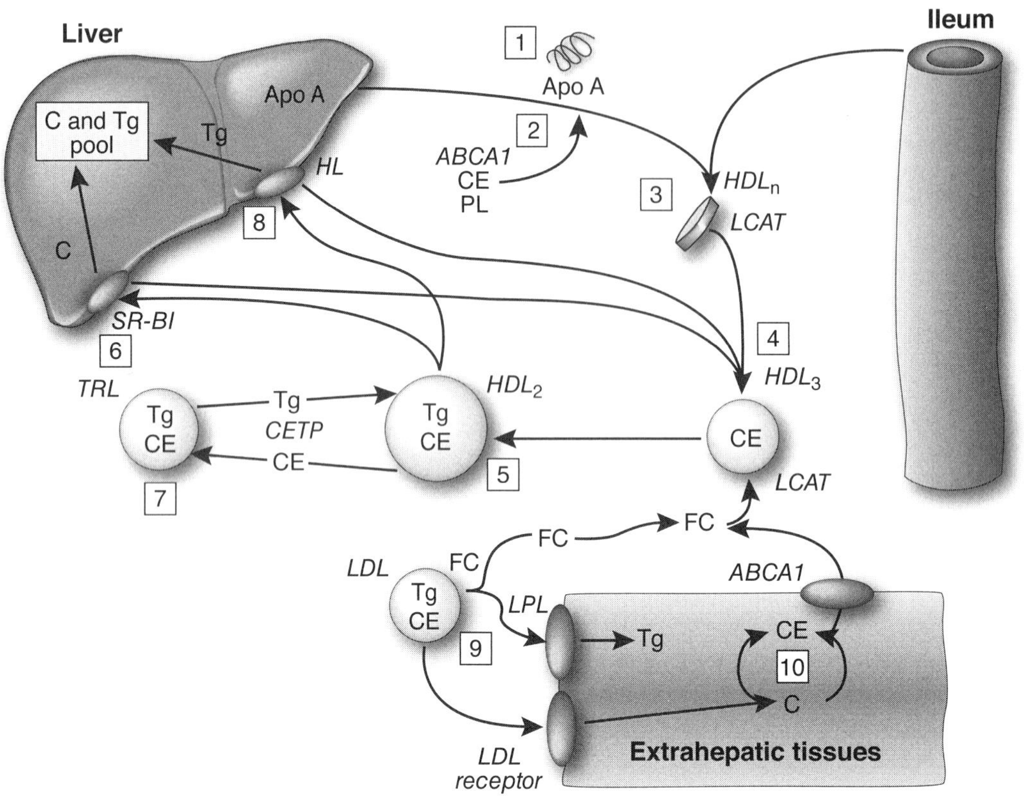

Figure 14.3 Metabolism of high-density lipoproteins.

can be atherogenic. Thus, patients with elevated VLDL containing apo B are at risk for premature development of atherosclerosis.

The atherosclerotic process appears to occur mainly in large to medium-sized elastic and muscular arteries of the heart, brain, or extremities. Numerous studies have suggested that the first step in atherogenesis occurs in response to an injury causing endothelial dysfunction in the arterial wall (60). Several possible factors that can create endothelial dysfunction have been observed:

Elevated and modified LDL

Free radicals caused by cigarette smoking

Hypertension

Diabetes mellitus

Elevated plasma homocysteine concentrations

Genetic alterations

Infectious organisms such as herpes viruses and *C. pneumoniae* (53)

As a result, the endothelial wall is rendered more permeable and develops an adhesiveness to platelets, leukocytes, and other procoagulant factors. Highly susceptible areas to progressive atherogenesis include

places in the arterial tree where bifurcations, curvatures, and branching occur, resulting in alterations in blood flow and increased turbulence (53). Turbulent blood flow against the endothelium appears to induce a release of procoagulant factors such as vascular cell adhesion molecule-1 (VCAM-1), leaving these sites prone to progressive lesion development.

The initiation of lesion development appears to begin with the attachment of circulating monocytes, T-cells, and other atherogenic lipoproteins (e.g., adhesive molecules, oxidized LDL) to the endothelial surface. These particles modify the endothelial surface, which begins a cascade of events resulting in the proliferation of plaque within the vascular wall (13). In addition, circulating LDL enters this space, where it undergoes oxidation. Nearby macrophages take up the oxidized LDL and begin to take the appearance of a foam cell. As lipid continues to accumulate in the foam cells, smooth muscle cell proliferation occurs through the release of growth factors. At this point, the first visible signs, fatty streaks, begin to appear. The smooth muscle cells along with monocytes begin to migrate to the intimal space within the artery wall. As this cycle of interactions continues, the fatty streaks mature into a fibrous lesion, which begins to extend into the lumen of the artery. The foam cells continue to accumulate LDL, which begins to crystallize. At this point, the foam cells can rupture, spilling the cholesterol contents, which leads to further proliferation of lipid deposition throughout the lesion. Smooth muscle cells, accompanied by collagen synthesis, continue the progression of the lesion. Advanced stages of lesion development include calcification, which hardens the plaque.

A high level of HDL-C lowers the risk of CHD development (30,38). Circulating HDL is involved in a reverse cholesterol transport process, where cholesterol from circulating lipoproteins and possibly lesions is gathered up by the HDL in the intravascular space and transported to the liver for catabolism. Consequently, low plasma HDL-C (<40 mg · dl^{-1}) is considered a risk factor for CHD. Promising work is currently being performed on ATP-binding cassette and scavenger-receptor proteins to elucidate the mechanisms regulating cholesterol efflux from extrahepatic tissues to HDL during reverse cholesterol transport (49,51,61).

MEDICAL AND CLINICAL CONSIDERATIONS

Several dyslipidemic and hyperlipidemic conditions can present themselves clinically. Although many of the lipid abnormalities can be genetically influenced and most conditions can place the individual at risk for developing premature CHD, some may actually lower risk (e.g., hyper-α-lipoproteinemia).

Signs and Symptoms

The dyslipidemic condition can be classified based on which lipid or lipoprotein is abnormal. Although classifying conditions for treatment purposes is helpful, some of the following classifications can be identified only through extensive testing that may not be available in every clinical setting. Nevertheless, the following are dyslipidemia conditions.

Hypercholesterolemia

Blood cholesterol concentrations greater than 240 mg/dl are classified as hypercholesterolemia. This concentration of total cholesterol or higher greatly elevates risk of CHD and corresponds to the 80th percentile of the adult U.S. population. Severe forms of hypercholesterolemia conditions include the following conditions.

Familial hypercholesterolemia

Characterized by a hyper-β-lipoproteinemia, this genetic condition results from defective LDL receptors or a lack of LDL receptors, which reduce LDL clearance rates. Patients with this disease present with an elevated total cholesterol and LDL concentrations often over 260 mg · dl^{-1}, and these concentrations may reach as high as 500 mg · dl^{-1}. Xanthomas and atheromas may be present as a result of excessive lipid deposition. These patients are at increased risk for premature atherosclerosis. Familial hypercholesterolemia occurs in approximately 1 in 500 people in the United States.

Familial polygenic hypercholesterolemia

A more common genetic condition, familial polygenic hypercholesterolemia manifests as high total cholesterol and LDL-C concentrations (>220 mg · dl^{-1}) and occurs in approximately 1 in 100 people. Premature atherosclerotic disease is commonly observed in this population, but to a lesser extent than in the familial hypercholesterolemia patient.

Familial combined hypercholesterolemia

Characterized by elevated total cholesterol and triglyceride concentrations, familial combined hypercholesterolemia is seen in multiple individuals within the same family. This condition is common in the United States, found in approximately 1% of the general population. Patients affected with this condition are at increased risk for premature atherosclerotic disease.

Hypertriglyceridemia

Elevated fasting triglyceride concentrations may be classified as hypertriglyceridemia. Triglycerides may be either consumed in the diet (chylomicron) or synthesized (VLDL) in the liver. In addition to familial combined hypercholesterolemia, additional forms of hypertriglyceridemia may result from the following conditions.

Apolipoprotein E2 genotype

The three primary apo E isoforms are E2, E3, and E4. The apo E2 genotype does not allow efficient binding to the apo E receptor, which is needed for triglyceride hydrolysis. Consequently, patients having the apo E2 genotype are unable to hydrolyze triglycerides efficiently in the liver. The condition results in an increase in both chylomicron and VLDL-triglyceride (VLDL-Tg) concentrations.

Familial lipoprotein lipase (LPL) deficiency

Caused by either a deficiency in apo C-II, which is responsible for LPL activation, or an insufficient production of LPL, this condition limits the clearance of chylomicron and VLDL. Besides having elevated triglyceride concentrations, patients typically have low values of HDL-C and LDL-C.

Familial hypertriglyceridemia

This genetic condition causes abnormally high concentrations of endogenously produced triglycerides. An extremely high VLDL-Tg concentration is present in the patient's lipid profile. Cholesterol concentrations also rise in proportion to the concentration of triglycerides, resulting in an elevated CAD risk.

Hyperlipoproteinemia

Hyperlipoproteinemia is a generic term used to imply an elevated lipoprotein class. Virtually all dyslipoproteinemias result from a defect in the stage of lipoprotein development, transportation, or destruction. A hyperlipoproteinemia may result from many causes including a familial trait, as a secondary characteristic of another disease (e.g., thyroid disease, diabetes), or possibly because of excess body weight, lack of exercise, and poor nutrition.

Hyper-β-lipoproteinemia

A condition characterized by excess LDL-C concentrations usually as a result of too much β-band protein (apo B). Hyper-β-lipoproteinemia may result from excess production but is most likely the result of a reduced binding capacity at the liver for LDL hydrolysis. As an example, people with a familial polygenic hypercholesterolemia have a hyper-β-lipoproteinemia pattern and are at increased risk for developing atherosclerosis.

Hyper-α-lipoproteinemia

A rare condition characterized by an increased concentration of HDL-C. The etiology of this genetic disease is likely linked to both increased synthesis of the apo A (α-band) and a reduction in HDL clearance. Nevertheless, this dyslipidemia is apparently beneficial to health, because elevated HDL is inversely related to atherosclerotic diseases.

Hypolipoproteinemia

Hypolipoproteinemia is a generic term used to identify a low lipoprotein concentration. Individuals suffering from hypolipoproteinemia, as in a hyperlipoproteinemia, may have developed the condition from a familial trait, as a secondary characteristic of another disease (e.g., thyroid disease, diabetes), or possibly because of excess body weight, lack of exercise, and poor nutrition. Specific forms of hypolipoproteinemia include the following conditions.

A-β-lipoproteinemia

A rare genetic disorder characterized by absence of β-band lipoprotein (LDL) in the plasma, A-β-lipoproteinemia is caused by a defect in apo B synthesis. Note that no chylomicrons or VLDL are formed.

Familial hypo-β-lipoproteinemia

This genetic condition manifests a lower than normal LDL concentration (10–50%), but chylomicrons are developed. The condition probably stems from a lack of synthesis of apo B-100, which is associated with the LDL.

Hypo-α-lipoproteinemia

A term used for low concentrations of HDL-C (<40 mg · dl^{-1}), hypo-α-lipoproteinemia can result from a lack of synthesis of apo A or insufficient LPL activity. Low HDL-C concentrations may also result from behavioral influences such as excess body weight or smoking.

History and Physical Exam

A patient entering an exercise program with no history of CHD may not even know his or her lipid and lipoprotein risk classification, even though much effort has recently been made to improve awareness and management of CHD risk factors among the public. As such, patient education should be provided, and a complete lipid and lipoprotein profile should be periodically obtained in accordance with current National Cholesterol Education Program (NCEP) screening guidelines. Although a medical evaluation is not required for the hyperlipidemic or dyslipidemic patient to begin exercise, the clinician should be careful to probe for any signs or symptoms of underlying disease, particularly among those who present with multiple CHD risk factors. A complete medical evaluation is usually provided to patients with known CHD and will likely include a current lipoprotein profile.

Diagnostic Testing

General dyslipidemia conditions can be identified through a complete blood lipid profile obtained through forearm venipuncture following a 12 h fast (water only). Most clinical laboratories provide a measure of total cholesterol, LDL-C, HDL-C, and triglycerides. More advanced testing may be needed to identify a specific dyslipidemia classification such as those mentioned earlier. Enzymatic measurements, lipoprotein subtractions, and apo analyses are usually obtained only when the patient is under the care of a lipid specialist.

General guidelines for the evaluation of blood lipids and lipoproteins have been established. The NCEP (26) first published a systematic approach to the management of high cholesterol in 1988. The program establishes

Table 14.2 National Cholesterol Education Program Adult Treatment Panel III Classification

LDL cholesterol (mg · dl⁻¹)	Classification
<100	Optimal
100–129	Near optimal
130–159	Borderline high
160–189	High
≥190	Very high
Total cholesterol (mg · dl⁻¹)	**Classification**
<200	Desirable
200–239	Borderline high
≥240	High
HDL cholesterol (mg · dl⁻¹)	**Classification**
<40	Low
≥60	High
Triglyceride (mg · dl⁻¹)	**Classification**
<150	Normal
150–199	Borderline high
200–499	High
≥500	Very high

Note. LDL = low-density lipoprotein; HDL = high-density lipoprotein.

goals for both individual management and the nation as a whole. The NCEP Adult Treatment Panel III (ATP III) lipoprotein classifications (48) are listed in table 14.2. In general, the program calls for all adults age 20 and older to have total cholesterol measured at least once every 5 yr.

Exercise Testing

The American Heart Association (AHA) and the American College of Sports Medicine (ACSM) have developed exercise-testing procedures and protocols that can guide the clinician (see chapter 5). Although no specific testing modifications are needed for this population, a cautious approach to the evaluation process is prudent given the probabilities for underlying disease. Clinicians should be familiar with the contraindications for exercise and exercise testing before treating this population, and certain individuals may need a medical evaluation. In addition,

risk stratification of patients should be completed before graded exercise testing. *ACSM's Guidelines for Exercise Testing and Prescription, Seventh Edition* (3) defines low-, moderate-, and high-risk stratification categories based on risk factors for CAD. People at moderate or high risk will need a medical examination before a graded exercise test and possibly medical supervision during the test (3).

Contraindications

The ACSM has developed contraindications for exercise (3), which are reported in chapter 5. By itself, the atherogenic blood lipid profile is not a contraindication. Patients with dyslipidemia, however, may present with existing disease or anomalies that may preclude them from exercise. Consequently, a thorough review of the patient's medical history and additional physical assessments should be conducted to rule out any established risk during exercise training.

Recommendations and Anticipated Responses

People with elevated plasma cholesterol, LDL-C in particular, are considered at risk for CAD; consequently, care must be taken to ensure that there are no signs or symptoms of underlying disease. This point is particularly important among middle-aged patients with a familial hyperlipidemia that may accelerate the risk for premature CAD. If signs or symptoms are present, or if the patient has known disease, the exercise test should follow the protocol established for patients with known disease. In contrast, patients who have no known disease may undergo normal exercise-testing protocols for patients considered at risk. The exercise evaluation is an effective tool for ruling out underlying disease and determining the patient's physical fitness, and it can be used to guide the formulation of the exercise prescription. Table 14.3 lists the recommendations for exercise testing in the dyslipidemic patient.

Treatment

The intensity of an individual's treatment depends on his or her overall risk status. Typically, people at higher risk for CHD should be given a more aggressive treatment. Patients with high blood cholesterol may be categorized into three general levels of risk (27):

CHD and CHD risk equivalent: persons having CHD, other atherosclerotic disease (e.g., peripheral vascular disease or carotid artery disease), or diabetes

Table 14.3 Exercise Testing in the Dyslipidemia Patient

Test type	Mode	Protocol specifics	Clinical measures	Clinical implications	Special considerations
Cardiovascular	Cycle ergometer Treadmill	YMCA Bruce (see chapter 5 for additional graded exercise-testing protocols)	HR, BP, ECG (YMCA) HR, BP, ECG, $\dot{V}O_2$, $\dot{V}CO_2$ (Bruce)	Determine response to graded exercise of HR, BP, ECG, $\dot{V}O_2$	Need medical supervision for males ≥45 or females ≥55 yr of age, or more than one CAD risk factor
Strength	Free weights Machine weights	1RM bench press and leg press	1RM		
Range of motion (ROM)	Static stretching	Sit-and-reach (hip flexion)	ROM at hip joint		

Note. 1RM = one-repetition maximum; BP = blood pressure; ECG = electrocardiogram; HR = heart rate; $\dot{V}CO_2$ = volume of carbon dioxide expired; $\dot{V}O_2$ = volume of oxygen consumed.

Multiple risk factors: patients with no known CHD but at high risk because of high blood cholesterol in combination with other CHD risk factors

0-1 risk factors: patients with no known CHD who have no other CHD risk factors

Over the past two decades, several pharmacological agents (reviewed in chapter 3) have been introduced that can effectively control many hyperlipidemia or dyslipidemia conditions. In accordance with national cholesterol guidelines, however, physicians frequently allow patients to attempt to control influential behavioral factors such as weight loss, dietary alterations, and exercise before choosing to use a medication. Several treatments are reviewed next. Recommendations for exercise prescription for prevention and improvement of dyslipidemia are addressed later in the chapter.

Diet Therapy

The NCEP ATP III clinical guidelines recommend a first-line intervention to improve cholesterol concentrations before administering medications through therapeutic lifestyle changes (TLC) that include diet and exercise. The recommended nutrient composition of the TLC diet includes reducing total fat intake to 25% to 35% of total calories. Allowing for this, saturated fat should be less than 7%, polyunsaturated fat 10% or less, and monounsaturated fats less than 20% of total calories. If LDL-C concentrations have not reached targeted goals after 6 wk, increases in dietary fiber and plant stanols or sterols are recommended. As part of the TLC approach, weight reduction may be recommended to achieve optimal weight goals. Success in adhering to the

nutrient recommendations may require use of a qualified nutritional professional or registered dietician.

Patients switching from a standard Western diet (35–40% fat calories, with 15–20% saturated fat) to the TLC-recommended diet can expect an average decrease in their total blood cholesterol of 5 to 10% (14,50). Unfortunately, this form of dieting has poor adherence rates. Nevertheless, aggressive dietary changes are an important component in controlling the patient's total cholesterol concentration and should not be overlooked. Allison et al. (2) reported that 35% of cardiac patients were able to reduce total cholesterol concentrations to NCEP-recommended concentrations using diet therapy, thus eliminating the need for costly medications.

The effect that various types of dietary fat have on serum total cholesterol concentrations has been reviewed extensively. The recommendation to restrict dietary saturated fats, in particular, stems from several studies that have identified a clear and direct association between the amount of dietary saturated fat consumed and total cholesterol concentrations (28). High amounts of saturated fat intake appear to decrease LDL clearance rates and, thus, prolong the exposure of LDL within the circulation. In a similar manner, a high amount of dietary cholesterol also reduces LDL clearance from the circulation and increases LDL synthesis. A reduced LDL catabolism and subsequent longer lifecycle within the circulation may promote the formation of the smaller, dense LDL subfraction, which has greater atherogenic properties (7). Polyunsaturated fats appear to have little effect on plasma cholesterol concentrations; however, high polyunsaturated fat diets may reduce plasma HDL-C cholesterol concentration. Moreover, the long-term consequences of this type of diet are unknown. Consequently, the AHA currently recommends that polyunsaturated fats account for no more than 10% of total calories. In addition, polyunsaturated fats are often hydrogenated to harden their appearance (e.g., margarine). This process alters the chemical structure of the double bonds in the fatty acid from a cis to a trans configuration. Trans fatty acids appear to act in a similar manner as saturated fats, raising plasma cholesterol (43) and LDL-C and lowering HDL-C concentrations (4).

Recently, research has begun to indicate that carbohydrates should not be used as a substitute for saturated fats. A high-carbohydrate diet may reduce HDL-C concentrations and raise triglycerides. Rather, saturated and trans fatty acids should be replaced with monounsaturated fats. Monounsaturated fats, such as oleic acid, have a single double-bond configuration and appear to lower plasma cholesterol concentrations while maintaining and possibly elevating HDL-C concentrations (4).

Omega-3 fatty acids, which are long-chain fatty acids that contain a double bond in the n-3 position, may lower triglycerides; however, few data support the effectiveness of omega-3 fatty acids in the prevention CVD. Nevertheless, cross-sectional data from populations whose diets are rich in omega-3 fats clearly demonstrate that these people have lower CVD rates (37).

Plasma lipids and lipoproteins may also be modified by other dietary sources. Diets high in soluble fiber can sequester bile and consequently reduce LDL-C concentrations by approximately 5% to 10% (33), but HDL-C concentrations may also be reduced. Other dietary components such as garlic (68) and walnuts (54) may help to lower plasma cholesterol concentrations. Additional research is needed to support these preliminary findings. A recent investigation determined the independent and combined effects of plant stanol esters, which are known to improve cholesterol, and regular aerobic exercise on blood lipid profiles (see practical application 14.1). The data provide compelling insights on the usefulness of combining therapeutic approaches.

In general, patients should be encouraged to increase the consumption of fruits, vegetables, and high-fiber foods.

Practical Application 14.1

Literature Review

Therapeutic lifestyle changes like improved dietary practices and regular exercise are encouraged by the NCEP ATP III. Increasing dietary consumption of fatty acids derived from plants is associated with improved total cholesterol and LDL-C concentrations. Specifically, plant stanol esters (PSE) may improve total cholesterol and LDL-C by altering absorption of cholesterol by the intestine, decreasing hepatic secretion of VLDL-C, and potentially modifying peripheral expression of LDL receptors. Similarly, aerobic exercise can stimulate elevated HDL-C and reduced triglyceride concentrations if a sufficient caloric expenditure is achieved each week through exercise. The mechanisms regulating lipoprotein metabolism influenced by frequent exercise are thought to be intravascular enzyme transfer proteins like lipoprotein lipase activity (LPLa) and cholesterol ester transfer protein activity (CETPa). The purpose of this investigation was to determine the independent and combined effects of PSE supplementation and regular aerobic exercise on blood lipid profiles. This investigation was a double-blind, randomized, placebo-control study.

Participants for this investigation were men (n = 14) and postmenopausal women (n = 12) between 40 and 65 yr of age. These participants were nonsmoking and sedentary, were not currently consuming PSE margarine or blood lipid medications except hormone replacement, and were not hyperlipidemic. Furthermore, they had not been diagnosed with cardiovascular, metabolic, or pulmonary disease. Because this investigation was designed to study the effects of exercise and PSE specifically, and not the effects of weight change, participants whose body weight changed by more than 2 kg or whose body fat percentage changed more than 3% were excluded from the final analysis.

Each participant recorded 3 d dietary records that were analyzed for composition and used to provide dietary guidelines to assist each participant in remaining weight stable. Additionally, physical activity records were kept to determine whether additional physical activity might be contributing to any lipid alterations that occurred during the study period. Participants were assigned to either a PSE supplementation (PSES) or control (CON) group based on a gender-stratified randomized system for a total of 9 wk and 3 d. The PSES

(continued)

group consumed 3 tbsp (45 ml) of margarine (Benecol Light) per day containing a total of 42 g of PSE. The CON group also consumed 3 tbsp (45 ml) of margarine (I Can't Believe It's Not Butter Light—Soft) per day that did not contain PSE. Both margarines contain 5 g of fat. An exercise-training program was added to the PSES and CON protocols after 4 wk. The exercise program consisted of treadmill walking at an intensity estimated by heart rate that elicited 65% of $\dot{V}O_2R$ until 400 kcal were expended during each exercise session. This exercise protocol was performed 5 d per week until the end of wk 9. Thus, the exercise protocol provided an exercise energy expenditure of 2,000 kcal $\cdot$ wk^{-1}.

All blood samples were collected in a 10 to 12 h fasted state at the same time of day. Blood samples were collected at three different points across the study: baseline, after 4 wk of PSES or CON, and after the completion of the exercise-training protocol combined with MS and CON. Two blood samples were taken at each point separated by 24 h, which were then pooled to provide an average value. The blood samples collected at the end of the exercise-training period were collected 72 h and 96 h following the final exercise session to eliminate the possibility of the measurements representing a potential acute effect of the final exercise session rather than a training response.

Compliance with the dietary intervention (PSE = 41.4 g $\cdot$ d^{-1} versus CON = 41.3 g $\cdot$ d^{-1}) and exercise training protocol (98% completion; 415 kcal $\cdot$ d^{-1}) were excellent. Average heart rates were between 130 and 134 beats $\cdot$ min^{-1}. By design, body weight and body fat percentage did not change across the study, although the exercise protocol produced a significant increase (4%) in $\dot{V}O_2$max. With the exception of dietary fat consumption, no dietary variables changed during the 9 wk period. Dietary fat did increase after the first 4 wk of margarine consumption in both the PSES and CON groups. At the end of the 9 wk study, dietary fat consumption returned to a percentage similar to baseline.

All blood lipid and intravascular enzyme measurements were corrected for changes in plasma volume that occurred during the study. Plasma volume fell 2% during margarine supplementation, but then rose 1% after exercise training. Total cholesterol was significantly reduced after 4 wk of PSE consumption and remained at this concentration when exercise was added. Total cholesterol was significantly elevated above baseline concentration when exercise was added to CON. Furthermore, total cholesterol was not different from baseline or the final concentration when on the CON-only intervention. Similar to total cholesterol, LDL-C and triglyceride concentrations were significantly reduced by PSE consumption and were not altered by CON. In those who consumed PSE, HDL-C concentration was not altered by PSE or by exercise training. Conversely, CON did not alter HDL-C, but the addition of exercise to CON stimulated a significant elevation of HDL-C concentration. As expected given the response of total cholesterol and HDL-C, the total cholesterol-to-HDL-C ratio was significantly reduced by PSE consumption and maintained with the addition of exercise. The total cholesterol-to-HDL-C ratio was elevated during CON and returned to baseline when exercise was added. No significant changes were detected in LPLa or CETPa.

These data suggest that HDL-C is independently increased by exercise and that PSE independently promotes reduced total cholesterol, HDL-C, and triglyceride concentrations. Furthermore, PSE appeared to minimize the beneficial effect of exercise on HDL-C concentration. Thus, the combination of exercise training and PSE supplementation does not appear to provide a synergistic benefit, and may in fact attenuate the positive effect of exercise training on HDL-C. Although it is unfortunate that the synergistic effect did not exist, several important positive effects did occur upon PSE supplementation that was maintained with the addition of exercise. It might be important to consider the potential outcome of PSE supplementation and exercise in patients with high total cholesterol, triglyceride, and high HDL-C concentrations. Another important aspect is that weight change was an exclusion criterion during the study. In many clinical settings with dyslipidemic patients, weight loss will be a primary goal. The present investigation does not address the role of PSE supplementation and exercise training in this situation, but this may be an extremely beneficial situation. Additionally, this investigation did not attempt to determine the acute response to exercise. This acute response of lipoproteins could be beneficial when combined with PSE supplementation.

Based on S. Alhassan, 2006, "Blood lipid responses to plant stanol ester supplementation and aerobic exercise training," *Metabolism* 55: 541-549.

Moreover, saturated and trans fatty acids should be minimized and replaced with monounsaturated and omega-3 fatty acids.

Pharmacological Management

If diet therapy is insufficient to restore appropriate LDL-C concentrations, drug therapy may be initiated. Pharmacological treatments should be considered as an adjunct to diet therapy and not as a replacement. Usually a 6 mo diet therapy trial is sufficient to determine its effectiveness, but individuals at higher risk, such as those with existing CAD, may be given a shorter diet-alone trial before resorting to drug therapy.

The clinician and patient should carefully weigh the benefits versus the costs (health insurance, adverse effects) of drug therapy. Moreover, any improvement in cholesterol attributed to the medication may be reversed if drug therapy is discontinued. Medications for treatment of dyslipidemia are reviewed in chapter 3. Common medications can influence lipid and lipoprotein profiles (table 14.4).

Smoking Cessation

Cigarette smoking is known to negatively influence blood lipid profiles. Specifically, both male and female cigarette smokers have lower HDL-C (32,39,58). Additionally, lifetime nonsmoking individuals exposed to frequent large doses of environmental, or secondhand, tobacco smoke possess HDL-C concentrations that are similar to those of smokers, both of which are significantly lower than HDL-C concentrations of nonsmokers (47). Fortunately, upon cessation of cigarette smoking, HDL-C concentra-

tion rapidly (30–60 d) returns to a concentration similar to those of nonsmokers (45,46,52).

Prevention

Because hyperlipidemia is an independent risk factor for atherosclerotic diseases including cardiac, peripheral, and cerebral vascular disease, steps should be taken to control the lipid profile. As a method of prevention, people should be educated and encouraged to follow the NCEP guidelines for cholesterol testing. Healthcare providers have not been consistent in adhering to NCEP guidelines with patients (22), and consequently, more attention to disease prevention guidelines is warranted. Among families in which premature disease is suspected, efforts should be made to screen children early (>2 yr) to intervene therapeutically and prevent the aggressive development of disease. In addition, people need to understand the potentially modulating effects that other risk factors place on blood lipids. For example, efforts should be made to prevent weight gain because added weight has a negative influence on the lipid profile (high LDL-C and low HDL-C). In addition to weight maintenance, improved body composition, glucose tolerance, and smoking cessation may positively affect the lipid profile. Thus, efforts to minimize other risk factors are critical for preventing hyperlipidemia. Although exercise has a direct, positive influence on the lipid profile, it can also indirectly improve lipids through weight loss, reduced adiposity, and improved glucose tolerance.

Based on the general risk classifications mentioned previously, cholesterol management guidelines have

Table 14.4 Effect of Common Medications on Lipid and Lipoprotein Profiles

Medication	Triglycerides	VLDL	LDL	HDL	Comments
β-blockers		Increase		Decrease	Those with intrinsic sympathomimetic activity have little effect
Thiazides		Increase	Increase		Dose dependent
Oral contraceptives			Increase or decrease	Increase or decrease	Depends on dose of progestogen and estrogen used
Hormone replacement therapy			Decrease	Increase	Depends on dosage
Corticosteroids, cyclosporine	Increase	Increase	Increase		
Dilantin				Increase	
Alcohol	Increase[a]	Increase[a]		Increase[b]	

Note. VLDL = very low-density lipoprotein; LDL = low-density lipoprotein; HDL = high-density lipoprotein. [a]Alcohol intake >3% of total calories; [b]not more than 12 oz (360 ml) per day.

been developed to target LDL-C concentrations. The guidelines are listed in table 14.5. Emphasis is placed on LDL-C concentrations to help guide the appropriate treatment recommendations. As such, a full fasting lipoprotein analysis should be conducted at the point of screening. The current recommendations for cholesterol management using a full lipoprotein profile emphasize the concentration of LDL-C, because it is the principal constituent behind the atherosclerotic process.

EXERCISE PRESCRIPTION

In general, regular exercise along with a balanced diet is recommended as part of the ATP III therapeutic lifestyle changes for improving the lipid profile. Special considerations for developing the exercise prescription along with currently supported exercise recommendations are discussed next. See practical application 14.2 on p. 260 for a summary of the exercise prescription for the dyslipidemia patient.

Special Exercise Considerations

The mechanisms behind the alterations in blood lipid profiles following exercise training are not fully understood, although many of the improvements may be attributed to enzymatic changes. Hydrolysis of TRL particles, such as VLDL and chylomicrons, is performed by LPL. This action is increased by both acute exercise (23) and exercise training (56). The elevated LPL activity is directly related to the lowering of plasma triglyceride concentrations. Following lipid hydrolysis, the remnant particles may be transformed into nascent HDL or accepted by circulating HDL for reverse cholesterol transport. Besides having an increase in LPL activity, exercise-trained individuals have an elevated lecithin cholesterol acyltransferase (LCAT) activity (29), which is responsible for maturation of HDL through cholesterol esterification. Another enzyme, CETP, apparently decreases following exercise training (55) and may also play a role in the maturation of HDL. Esterified cholesterol is removed by CETP from HDL and LDL to larger triglyceride-rich particles such as VLDL and chylomicron. People with CETP deficiency typically have very high HDL-C concentrations. Nevertheless, the role of CETP following exercise training has not been fully elucidated. The amount of circulating small dense LDL may also be reduced by increased CETP activity (18). Hepatic triglyceride lipase generally decreases following exercise training (29). The decline in hepatic triglyceride lipase activity prolongs the circulating half-life of HDL because of a decrease in HDL catabolism by the liver. The increased metabolic half-life of the HDL likely aids in the increased large cholesteryl ester–rich HDL$_2$ subfraction observed following exercise training.

Individuals with familial dyslipidemia may not experience the same enzymatic responses, nor expect the same alterations in lipid profiles as healthy individuals following exercise training. Exercise alone has not been shown to be effective at increasing HDL-C concentrations among subjects with hypo-α-lipoproteinemia or low HDL-C. Moreover, LPL-deficient individuals with familial hypertriglyceridemia may not experience the same level of improvement in the lipid profiles as healthy individuals do. Nevertheless, exercise training among individuals with a familial dyslipidemia is important in reducing other risk factors (e.g., body composition, glucose tolerance, hypertension) that can influence the lipid profile as well as independently affect CVD mortality and morbidity rates.

Table 14.5 National Cholesterol Education Program: Clinical Guidelines for Cholesterol Management

Risk category	LDL goal	LDL level at which to initiate therapeutic lifestyle changes	LDL level at which to consider drug therapy
CHD or CHD equivalent	<100 mg · dl^{-1}	≥100 mg · dl^{-1}	≥130 mg · dl^{-1} (100–129 mg · dl^{-1}: drug optional)
Multiple risk factors[a]	<130 mg · dl^{-1}	≥130 mg · dl^{-1}	≥130mg · dl^{-1} with 10–20% risk[b] ≥160 mg · dl^{-1}l with <10% risk[b]
Zero or one risk factor[a]	<160 mg · dl^{-1}	≥160 mg · dl^{-1}	≥190 mg · dl^{-1} (160–189 mg · dl^{-1}: drug optional)

Note. LDL = low-density lipoprotein; CHD = coronary heart disease.

[a]Risk factors include cigarette smoking, hypertension, low HDL, family history of premature CHD (CHD in males <55 yr, females <65 yr), age (men ≥45 yr, women ≥55 yr). [b]Risk of developing CHD over a 10 yr period as determined by Framingham scoring procedures.

The goals for exercise prescription may vary slightly depending on the form of dyslipidemia. People with congenital hyperlipidemia need to adhere to prescribed medications and use exercise to target other risk factors such as weight loss and reduced adiposity, because the direct effect of exercise on blood lipids may be limited in this population. Furthermore, excess body weight and adiposity should be reduced as a primary objective of any hyperlipidemia patient because body mass index (BMI) is directly associated with total cholesterol concentrations. Thus, an exercise prescription that targets weight loss can profoundly affect the lipid profile. Reductions in total cholesterol, LDL-C, and triglycerides along with increases in HDL-C have been observed following weight loss (59).

The exercise prescription should first address any chronic disease (i.e., CAD, diabetes) issues. Consequently, if a patient presents with CAD, the immediate exercise prescription goals may not allow an optimal training level for lipid control. Greater strides may take place in controlling lipid levels after the subject has achieved sufficient physical fitness and attended to other immediate needs. Furthermore, among individuals with metabolic disorders in whom lipid abnormalities, such as diabetes, often develop, efforts need to be placed on weight control and a regimented diet to control blood glucose concentrations. In doing so, lipid concentrations, at least in part, may be controlled.

Among people who are otherwise healthy, the exercise prescription may be written to address hyperlipidemia more specifically. Although limited data are available to tailor the exercise prescription toward optimal improvements in blood lipids, the preliminary guidelines presented here are based on previous investigations, and further research is needed to specifically tailor the exercise prescription to modify blood lipids. Although findings of several investigations have supported the benefits of exercise training on the lipid profile, the majority of studies were performed in subjects with a normal lipid status. Very few data exist regarding the benefits of exercise training among the dyslipidemia populations.

Several months of exercise training may be required to provide any significant or lasting improvements to the blood lipid profile. Consequently, added encouragement along with an evaluation of the patient's behavioral and environmental milieu may be needed to help the patient stay on course. Further encouragement should be taken from several recent studies that have identified favorable improvements in the blood lipid profile after a single exercise session (12,18,23,24). Moreover, these improvements have been identified in both trained and untrained individuals and among those with elevated total cholesterol. These improvements appear to be only transient, lasting only 48 to 72 h after exercise, which should encourage patients to exercise at least every other day.

Exercise Recommendations

The minimum amount of exercise training required to improve blood lipids is not fully known, but the results of previous studies indicate that the quantity of exercise is likely the most important factor. A dose–response relationship has been identified between the volume of physical activity and HDL-C cholesterol (70). Furthermore, this relationship has been identified in both men (71) and women (69). Although the specific combination of exercise frequency and duration for lipid improvements has not been fully elucidated, a minimum of 1,000 kcal/wk has been recommended (66). *ACSM's Guidelines for Exercise Testing and Prescription, Seventh Edition* (3) suggests that dyslipidemic individuals should follow an exercise prescription similar to that of overweight or obese individuals (see table 14.6). Thus, the goal of this exercise prescription should be to maximize energy expenditure (3). People who are able to achieve higher amounts of weekly activity will likely benefit more.

Accordingly, aerobic exercise becomes the foundation of the exercise prescription. Specifically, aerobic modes of exercise should be performed at 40 to 70% of heart rate reserve or $\dot{V}O_2$ reserve. To emphasize weight control

Table 14.6 Exercise-Training Benefits in the Dyslipidemia Patient

Cardiorespiratory endurance	Skeletal muscle strength	Skeletal muscle endurance	Flexibility	Body composition
Increase $\dot{V}O_2$max, decrease resting and submaximal HR, decrease resting and submaximal BP, increase HDL-C, decrease triglyceride	Maintain or increase muscular strength	Maintain or increase muscular endurance	Maintain or increase flexibility	If weight loss is a goal, decrease percent body fat, decrease BMI, reduce waist circumference

Note. BMI = body mass index; BP = blood pressure; HDL-C = high-density lipoprotein cholesterol; HR = heart rate; $\dot{V}O_2$max = maximal oxygen consumption.

through caloric expenditure ($\geq$2,000 kcal $\cdot$ wk^{-1}), aerobic exercise should be performed frequently (5 or more days per week) and for a fairly long duration (40–60 min $\cdot$ d^{-1}; 200–300 min $\cdot$ wk^{-1}). Note that the prescribed exercise intensity is relatively moderate in support of exercise performed frequently and for relatively long duration. Furthermore, if the dyslipidemic person is currently overweight or obese, low-impact or non-weight-bearing exercise modes like cycling, walking, and water exercise should be considered to reduce the risk of developing orthopedic complications in the legs, ankles, knees, and hips. Exercise performed to improve or maintain muscular strength, muscular endurance, and flexibility are important to include for dyslipidemic patients to provide

Practical Application 14.2

Exercise Prescription for the Dyslipidemia Patient

Training method	Mode	Intensity	Frequency	Duration	Progression	Goals	Special considerations and comments
Aerobic	Walk, jog, cycle, swim	40–70% $\dot{V}O_2R$ or HRR	3–5 d $\cdot$ wk^{-1} or more	40–60 min	Weeks 1–4: 40–50% HRR or $\dot{V}O_2R$; 3 d $\cdot$ wk^{-1} for 10–20 min; 60–140 kcal $\cdot$ d^{-1} Weeks 5–12: 50–60% HRR or $\dot{V}O_2R$; 3–5 d $\cdot$ wk^{-1} for 20–40 min; 140–325 kcal $\cdot$ d^{-1} After 3 mo: 55–70% HRR or $\dot{V}O_2R$; 5 d $\cdot$ wk^{-1} or more for 40–60 min; 300–550 kcal $\cdot$ d^{-1}	$\geq$2,000 kcal $\cdot$ wk^{-1}	Maximize caloric expenditure if weight loss is appropriate
Resistance	Free weights, machine weights, elastic bands	One set of 8–15 reps to fatigue	2–3 d $\cdot$ wk^{-1}	8–10 different exercises for major muscle groups	Weeks 1–4: 1 d $\cdot$ wk^{-1}, 1 set of 12–15 reps, 8–10 exercises Weeks 5–12: 2 d $\cdot$ wk^{-1}, one or two sets of 8–12 reps, 8–10 exercises After 3 mo: 2–3 d $\cdot$ wk^{-1}, one to three sets of 8 to 12 reps to fatigue, 8–10 exercises	Maintain or increase lean body mass	Be aware of additional CAD risk factors, especially hypertension
Range of motion	Static	Stretch to end of range of motion without pain	4–7 d $\cdot$ wk^{-1}	15–30 s $\cdot$ stretch^{-1}, two to four times per stretch	Weeks 1–4: 2–3 d $\cdot$ wk^{-1} Weeks 5–12: 3–5 d $\cdot$ wk^{-1} After 3 mo: 4–7 d $\cdot$ wk^{-1}	Increase range of motion of hip, thigh, and lower back	

Note. HRR = heart rate reserve; $\dot{V}O_2R$ = $\dot{V}O_2$ reserve.

a well-rounded approach to physical fitness. Resistance and flexibility exercises are not at the core of the exercise-based treatment for dyslipidemia primarily because these modes of exercise do not substantially contribute to the overall caloric expenditure goals that appear to be beneficial for improvements in blood lipid concentrations. Additionally, some data suggest that resistance exercise improves lipid profiles. Resistance and flexibility exercise prescriptions for dyslipidemic patients are not different from those for healthy adults.

Because dyslipidemia is a common component of the metabolic syndrome, dyslipidemic patients frequently present with multiple CAD risk factors, especially obesity, impaired fasting glucose or type 2 diabetes, and hypertension. When multiple risk factors are present, the exercise prescription guidelines listed earlier should be modified so that all risk factors are considered (3). Typically, this guideline would mean making the exercise prescription appropriate for the most serious complication present.

Mode

Aerobic exercise has proven to be effective at improving the blood lipid profile. Both reductions in triglycerides and increases in HDL-C have been observed following endurance exercise training. In contrast, resistance training may not be as effective. Cross-sectional data indicate that elite resistance-trained athletes have blood lipid profiles similar to their sedentary counterparts. Resistance-training studies that have identified improvements in blood lipids have either used a circuit-training protocol (which attempts to resemble aerobic exercise) or involved substantial reductions in body weight. Enhanced fat utilization during aerobic activity may be the metabolic stimulus needed to improve blood lipid concentrations (18). Another important consideration is that weight loss, considered an important goal for improving the lipid profile, may be better achieved through aerobic exercise training, during which higher caloric expenditure may be maintained. Nevertheless, resistance training should be considered an important adjunct to any successful fitness program.

Frequency

The frequency of exercise needed to improve the lipid profile also needs further examination. Studies evaluating the effects of a single exercise session on HDL-C indicate that elevations in HDL-C can occur 18 to 24 h after exercise and remain elevated for 48 to 72 h (12,18,23–25,65). Furthermore, over a 24 wk training period, these acute increases in HDL-C continued in response to the acute exercise session (12). This informa-

tion gives rise to speculation that the increase in HDL-C following exercise training is attributed, in part, to the transient increase observed following the acute exercise session. Furthermore, the transient rise in HDL-C in response to a single exercise session may have important day-to-day health benefits. Consequently, exercise sessions should occur at least three or four times per week, and no more than 1 d should separate exercise sessions. To maximize energy expenditure, aerobic exercise should be performed 5 d per week or more.

Intensity

The cross-sectional differences in HDL-C concentrations between endurance-trained athletes and sedentary counterparts led many to believe that the lipid improvements were related to the intensity of exercise. Although endurance-trained athletes typically engage in vigorous exercise, recent studies using sedentary individuals have been able to observe blood lipid improvements following moderate physical activity (50–60% maximum heart rate) (15). Consequently, moderate-intensity exercise is sufficient to develop and sustain an improved lipid profile and will likely ensure a better exercise adherence rate. Among healthy individuals, though, maintaining progressive and optimal improvements in HDL-C may require an eventual increase in training intensity. An acute increase in HDL-C following a single exercise session in moderately trained males appears to depend, in part, on training intensity (23). An aerobic exercise prescription that promotes weight loss and potentially improves lipid and lipoprotein profiles may be through exercise at 40 to 70% of $\dot{V}O_2$ reserve or heart rate reserve (3). The potential risks and benefits of raising the patient's exercise-training volume need to be carefully reviewed beforehand, and a prudent plan for progression should be developed.

Duration

The exercise bout should last at least 30 min with an optimal goal of 40 to 60 min (3). Improvements in blood lipids are evident, however, even when exercise time is broken up throughout the day (19). Because weight loss is also an effective modulator of blood lipids, extending the caloric expenditure rate to 2,000 kcal·wk^{-1} will likely help with both goals (3).

EXERCISE TRAINING

The effect of exercise on lipoproteins has been studied for approximately 50 yr (41). More recently, several reviews and meta-analyses described the response of blood lipids and lipoproteins following exercise in general (16,17) and in specific populations including women (35), postmenopausal women (5), children and adolescents (20), and overweight and obese adults (36). Another review compared acute versus training responses (63). The interaction between the effects of exercise and dietary interventions on blood lipids are beginning to be studied, as well (1). A summary of exercise-training benefits for a dyslipidemic patient is provided in table 14.6.

Exercise has become a recommended and valuable therapeutic treatment for improving the blood lipid profile. Caution must be used when interpreting its therapeutic value, at least with regard to the blood lipid profile. Most exercise studies have been conducted in participants with normal lipid status, and consequently few data exist regarding the effects of exercise training among people with various dyslipidemia conditions. Thus, despite the commonly held notion that exercise may improve the lipid profile in those with abnormal lipids, this may simply not be the case. For instance, some individuals appear to have intractable lipid profiles. One hypothesis behind the lack of a favorable blood lipid response to exercise training may be related to apo E genotype. Individuals with a specific apo E genotype may not be able to hydrolyze triglycerides adequately and may have limited improvements in HDL-C. Keep in mind that physical activity is recommended even among patients whose lipid profile appears to be resistant to exercise in order to minimize other risk factors (e.g., obesity and glucose intolerance).

In many people, exercise has a powerful modulating effect on many components of the lipid profile. Cross-sectional studies have long identified a clear association between physical activity or physical fitness and improved lipid profile (18). Aerobically trained athletes typically have lipid profiles that are lower in triglycerides and higher in HDL-C than their sedentary counterparts. Furthermore, few if any improvements in total cholesterol and LDL-C have been observed. In contrast, prospective studies, while showing blood lipid improvements, have not established a strong relationship between physical activity and blood lipid concentrations. Although some of this discrepancy can be related to the lack of sufficient length of the training study (<6 mo), other potential reasons remain. For example, a selection bias may have existed in many cross-sectional studies, implying that highly fit individuals or elite athletes may have other characteristics, some of which may be genetic, that give rise to the large differences observed. Furthermore, the large improvements reported from cross-sectional studies may also be related to the volume of exercise training, changes in body composition, failure to control for recent exercise, dietary intake, or weight loss.

Although many of the improvements in the lipid profile can require several months, benefits may accrue from a single exercise bout (12). Both a decrease in plasma triglycerides and an increase in HDL-C have been observed 18 to 48 h following exercise (18). Furthermore, the transient changes observed following an acute exercise bout still occur after 24 wk of training, suggesting that exercise training and a single session of exercise exert distinct and interactive effects (12). These effects on HDL-C and triglyceride concentrations also appear to be associated with the total energy expenditure during exercise (63).

Exercise training has been shown to lower plasma triglycerides (65), especially when baseline concentrations

Practical Application 14.3

Client–Clinician Interaction

The clinician should discuss with the patient in clear and understandable terms that aerobic exercise can substantially improve blood lipid concentrations. When performed in conjunction with a gradual weight control program, aerobic exercise can be a powerful tool in the patient's treatment. The clinician should inform the patient that beneficial responses, particularly in HDL-C and triglyceride concentrations, could be derived from both a single exercise session and from aerobic exercise training. In general, both these acute and chronic responses to exercise appear to be a function of caloric expenditure resulting from aerobic exercise. Thus, the ultimate goal of the clinical exercise physiologist when working with dyslipidemic patients is to promote frequent bouts of relatively long-duration aerobic exercise resulting in substantial energy expenditure each week. One measure of the success of the patient and clinical exercise physiologist might be a high rate of exercise adherence by the dyslipidemic patient, not only across several months of training but also during each week of training because of the beneficial acute responses following a single exercise session of sufficient energy expenditure.

are elevated (18,72). Although acute decreases have been observed following a single prolonged (>60 min) exercise bout (56,64) and following shorter exercise bouts (12), habitually active individuals have low triglycerides even when they have not recently exercised (18). Furthermore, a greater reduction in triglyceride concentrations is more likely following exercise-training programs when weight loss has occurred (72).

Perhaps the most prominent effect that exercise training has on the lipid profile is on HDL-C. Exercise-training studies indicate that HDL-C may increase following exercise training by 5% to 15% (18). But little or no increase in HDL-C may occur following exercise training among people with initially low concentrations (75) or unless weight loss occurs (18). The increase in HDL-C following training appears to be specific to the HDL_2 subfraction, which may provide stronger atherogenic protection.

Alterations in total cholesterol and LDL-C as a result of exercise training have been less convincing. Because total cholesterol is made up of all cholesterol-carrying lipoproteins, changes within these various components can prevent any observable change in total cholesterol. Because LDL-C and HDL-C are usually the principle carriers of cholesterol, the ratio of total cholesterol to HDL-C may provide a stronger indication of relative movements within

the lipid profile and an indication of atherogenic risk. When the ratio of total cholesterol to HDL-C is calculated, studies have indicated that exercise training can lower this ratio if weight loss has occurred (59). Furthermore, the decrease in this ratio cannot be solely attributed to an increase in HDL-C. Although exercise training is not very effective at lowering LDL-C unless weight loss or decreased adiposity (59,2) occurs, changes in LDL subfractions have been observed. Exercise-trained individuals have less small dense LDL subfraction (LDL3) (67,74), which is considered more atherogenic.

The consumption of a meal rapidly alters blood lipid profiles. Most well-controlled laboratory studies do not represent realistic dietary patterns, because blood lipid concentrations are measured in the fasting state. Given that humans routinely eat throughout the day, the timing of exercise following a meal may be important with respect to changes in blood lipid profiles. Exercise performed 1 to 12 h before a high-fat meal may reduce postprandial hypertriglyceridemia (34,73). Benefits may result from performing intermittent exercise as well as continuous exercise. Postprandial triglyceride concentration was reduced following intermittent exercise to concentrations similar to those observed following continuous exercise when a high-fat meal was consumed 1 d following exercise (44).

Case Study

Medical History

Ms. VB, a 42 yr old white accountant, has been referred to an exercise program for CVD risk reduction following a recent medical examination. She has a family history of heart disease. Her father had a nonfatal myocardial infarction at age 52. She is 5 ft 6 in. (168 cm) tall and weighs 190 lb (86 kg). Her medical history revealed that she has been prehypertensive for the past 5 yr. She currently does not exercise regularly. She has smoked one pack of cigarettes a day since she was age 17. A fasting blood lipid profile obtained during her medical examination revealed the following:

Total cholesterol = 224 mg · dl^{-1}

HDL-C = 35 mg · dl^{-1}

LDL-C = 166 mg · dl^{-1}

Triglycerides = 230 mg · dl^{-1}

Diagnosis

Ms. VB has numerous CVD risk factors including dyslipidemia, which places her at moderate risk according to ACSM risk stratification criteria.

Exercise Test Results

She has a resting heart rate of 78 beats · min^{-1} and resting blood pressure of 134/86 mmHg. Her graded exercise test determined that her $\dot{V}O_2$max was 29 ml · kg^{-1} · min^{-1}.

(continued)

Exercise Prescription

Initial exercise program

Frequency = 3 d · wk^{-1}

Intensity = 126 to 138 beats · min^{-1} or 13.7 to 16.2 ml · kg^{-1} · min^{-1} (40–50% heart rate reserve or $\dot{V}O_2$ reserve)

Rating of perceived exertion = 11 to 13

Duration = 10 to 20 min · d^{-1}, end goal 20 min · d^{-1} or 30 to 60 min · wk^{-1}

Energy expenditure = 60 to 140 kcal · d^{-1} or 180 to 520 kcal · wk^{-1}

Mode = aerobic

Exercise progression (over first 6–12 wk)

Frequency = 3 to 5 d · wk^{-1}

Intensity = 138 to 150 beats · min^{-1} or 16.2 to 18.8 ml · kg^{-1} · min^{-1} (50–60% heart rate reserve or $\dot{V}O_2$ reserve)

Rating of perceived exertion = 11 to 13

Duration = 20 to 40 min · d^{-1}, end goal 40 min · d^{-1} or 60 to 120 min · wk^{-1}

Energy expenditure = 140 to 325 kcal · d^{-1} or 700 to 1,600 kcal · wk^{-1}

Exercise progression (after 3 mo)

Frequency = 5 or more d · wk^{-1}, alternating

Intensity = 144 to 162 beats · min^{-1} or 17.5 to 21.3 ml · kg^{-1} · min^{-1} (55–70% heart rate reserve or $\dot{V}O_2$ reserve)

Rating of perceived exertion = 12 to 15

Duration = 40 to 60 min · d^{-1} or 200 to 300 min · wk^{-1}

Energy expenditure = 300 to 550 kcal · d^{-1} or ≥2,000 kcal · wk^{-1}

The patient should be educated on the signs and symptoms of exercise intolerance. Supplemental weight training should be introduced using modest amounts of weight (12–15 reps) and basic exercises for general conditioning purposes. Her lipid profile would likely benefit from gradual weight loss and by eliminating smoking. She should also be placed on a step 2 diet to control saturated fat intake. Total fat intake should be 25 to 35% of total calories. In addition, saturated fat should be less than 7%, polyunsaturated fat should be less than or equal to 10%, and monounsaturated fat should be less than 20% of total calories. The patient's negative caloric balance from diet and exercise be between 500 and 1,000 kcal · d^{-1}. After 3 to 6 mo, a repeat blood lipid profile should be conducted to reassess the patient's status. Following reassessment, the patient may need to be placed on drug therapy, but performing a 3 to 6 mo follow-up will determine the effect of a behavioral approach.

Discussion Questions

1. What objectives should be addressed in hopes of improving the patient's lipid profile?

2. For a patient who has low HDL cholesterol (<40 mg · dl^{-1}), what specific recommendations would you make?

3. What is the importance of an exercise program for patients who have a familial hyperlipidemia?

15

End-Stage Renal Disease

Patricia Painter, PhD

Chronic kidney disease (CKD) results from structural renal damage and progressively diminished renal function. After it begins, the disease progresses to end-stage renal disease (ESRD), requiring some form of renal replacement therapy such as dialysis or transplantation.

SCOPE

Approximately 308,000 patients with kidney failure live in the United States (1). Although CKD affects people of every age, race, and walk of life, 42% of the ESRD population is over the age of 60 years. Several studies have indicated that dialysis patients have lower educational and income levels than those of the general population (2).

In the United States, before 1972, access to dialysis treatment was limited, and selection of patients for treatment was made by committees of medical professionals, clergy, and lay people. Essentially, these committees decided who would receive the lifesaving therapy of dialysis. In 1972 Congress passed landmark legislation that extended Medicare coverage to patients with ESRD. This legislation hinged on the expectation of successful vocational rehabilitation of these patients (an expectation that has not been realized). Renal replacement therapy is expensive. The estimated cost of dialysis is $51,000 per patient per year; kidney transplant costs less over time ($18,000 per year) (1). Although the cost of renal replacement therapy has remained relatively constant since 1972, the population of patients with ESRD is

increasing annually (100 people are diagnosed with ESRD each day). Thus, the cost to the Medicare program is substantial for the number of patients involved. Additionally, ESRD patients are qualified for disability payments, another cost to government programs (2). Access to, payment for, and preferences of treatment vary in other countries.

Although the overall outcomes and well-being of patients with renal failure have significantly improved because of advances in technology and pharmacology that improve these patients' potential for rehabilitation, it is generally acknowledged that rehabilitation has not been addressed nationally in this patient group in a sustained, consistent, and integrated fashion (2). Low levels of physical functioning contribute significantly to the low levels of rehabilitation, thus indicating the need for physical rehabilitation as a part of the routine medical therapy of these patients.

PATHOPHYSIOLOGY

Damage to the kidney can result from long-standing diabetes mellitus, hypertension, autoimmune diseases (e.g., lupus), glomerulonephritis, pyelonephritis, some inherited diseases (i.e., polycystic kidney disease, Alport's syndrome), and congenital abnormalities. The damaged kidney initially responds with higher filtration and excretion rates per nephron, which masks symptoms until only 10% to 15% of renal function remains. Progressive renal failure results in loss of both excretory and regulatory functions, resulting in uremic syndrome. Uremia is characterized by fatigue, nausea, malaise, anorexia, and subtle neurological symptoms. Patients present with these symptoms and often with peripheral edema, pulmonary edema, or congestive heart failure. Diagnosis is made from elevated serum creatinine, blood urea nitrogen, and reduced glomerular filtration rate (3).

The loss of the excretory function of the kidney results in the buildup of toxins in the blood, any of which can negatively affect enzyme activities and inhibit systems such as the sodium pump, resulting in altered active transport across cell membranes and altered membrane potentials. The loss of regulatory function of the kidneys results in the inability to regulate extracellular volume and electrolyte concentrations, which adversely affects cardiovascular and cellular functions. Most patients are volume overloaded, resulting in hypertension and often congestive heart failure. Other malfunctions in regulation include impaired generation of ammonia and hydrogen ion excretion, resulting in metabolic acidosis and decreased production of erythropoietin, which is the primary cause of the anemia of ESRD (4,5).

Normal substances may be excessively produced or inappropriately regulated in response to renal failure. Parathyroid hormone may be the most important of these. Parathyroid hormone is produced in excess secondary to hyperphosphatemia, reduced conversion of vitamin D to its most active forms, and malabsorption and impaired release of calcium ions from bone. The attempt to maintain adequate circulating calcium ion concentrations in the face of hypocalcemia results in hyperparathyroidism and renal osteodystrophy (4,5).

Several metabolic abnormalities are associated with uremia, including insulin resistance and hyperglycemia. Hyperlipidemia is characterized in patients treated with dialysis by hypertriglyceridemia with normal (or low) total cholesterol concentrations. Several interventions associated with the dialysis treatment or immunosuppression therapy (following transplant) can contribute to these metabolic abnormalities (4,5).

MEDICAL AND CLINICAL CONSIDERATIONS

Renal failure produces specific signs and symptoms, and the diagnosis is strongly related to the evaluation of serum creatinine and blood urea nitrogen. The treatment of renal failure depends on creatinine clearance. Dietary adjustment for protein, sodium, and fluid intake plays an important role in the initial management of renal failure. If dietary intervention is not successful, renal replacement therapy is required. Currently, the three alternatives for

renal replacement therapy are hemodialysis, peritoneal dialysis, or transplantation. Although transplantation is the preferred method, patients need to be free of other life-threatening illnesses to be considered for transplantation. Hemodialysis is the most common therapy for renal failure, although it requires significant time throughout the week at a renal center. The third alternative for renal therapy replacement, peritoneal dialysis, is the method least used in United States, but it is more frequently used in other countries.

Signs and Symptoms

Deterioration in renal function results in an overall decline in physical well-being. Signs include anemia, fluid buildup in tissues, loss of bone minerals, and hypertension. Patients experience fatigue, shortness of breath, loss of appetite, restlessness, change in urination patterns, and overall malaise.

History and Physical Exam

The medical history will usually focus on renal-related issues, including cause of renal failure, current renal function (if any), and current treatment. The comorbidities will be listed, and treatments for each. There will rarely be any information on physical functioning or recommendations for activity. There is often much more information that may be more helpful in the dialysis chart (for those on dialysis). The dialysis chart includes evaluations by the social worker and dietitian, both of which more often address limitations in physical functioning (in terms of activities of daily living or need for assistance in the home). In the dialysis chart there is also typically a problem list that is more concise, and includes what has been implemented to address the medical concern. Also of interest may be the short- and long-term patient care plan which is a multidisciplinary plan that is included in the dialysis chart. This information will give the physiologist a better 'view' of the overall plan for the patient in terms of renal replacement therapy, and social considerations. The clinical exercise physiologist should pay attention to any cardiac history and the type and frequency of dialysis treatment in order to develop the best strategy for exercise that considers the treatment burden experienced by the patient.

Diagnostic Testing

Diagnosis of renal failure is typically made by determination of levels of serum creatinine and blood urea nitrogen through a blood test. Renal biopsy can be done to determine the etiology of disease, and a renal scan or intravenous pyelogram can be performed to rule out obstruction or congenital abnormalities that may contribute to increased creatinine levels in the blood.

Treatment

Treatment of chronic renal failure consists of medical management until the creatinine clearance is less than 5 ml · min^{-1}, at which time more aggressive therapy is required. Management is directed at minimizing the consequences of accumulated nitrogenous waste products normally excreted by the kidneys. Dietary measures play a primary role in the initial management, with very low-protein diets being prescribed to decrease the symptoms of uremia and possibly to delay the progression of the disease. In addition to protein restriction, dietary sodium and fluid restrictions are critical (4), because the fluid regulation mechanisms of the kidney are deteriorating in function. Any excess fluid taken remains in the system and, with progressing deterioration in renal function, will ultimately result in peripheral edema, congestive heart failure, and pulmonary congestion.

Progressive deterioration of renal function will ultimately require the initiation of some form of renal replacement therapy to maintain life. Treatment options include hemodialysis (performed in center or at home),

peritoneal dialysis, or transplantation. The decision to initiate dialysis is determined by many factors, including cardiovascular status, electrolyte levels (specifically potassium), chronic fluid overload, severe and irreversible oliguria or anuria, significant uremic symptoms, and excessively abnormal laboratory values (usually creatinine >8–12 mg · dl^{-1}, blood urea nitrogen >100–120 mg · dl^{-1}) and creatinine clearance (<5 ml · min^{-1}). Renal replacement therapy does not correct all signs and symptoms of uremia and often presents the patient with other concerns and side effects to deal with. Table 15.1 lists laboratory values for healthy patients versus those undergoing dialysis.

Hemodialysis

Hemodialysis is the most common form of renal replacement therapy in the United States. Approximately 63% of all patients undergo hemodialysis in a center or at home. In other countries, some patients prefer more home-based treatments such as peritoneal dialysis (discussed later). Hemodialysis is a process of ultrafiltration (fluid removal) and clearance of toxic solutes from the blood. It necessitates vascular access by way of an arteriovenous connection that uses either a prosthetic conduit or native vessels. Two needles are placed in the fistula; one directs blood out of the body to the artificial kidney (dialyzer), and the other directs blood back into the body. The

Table 15.1 Normal Laboratory Values Compared With Typical Values for Dialysis Patients

Laboratory value	Normal range	Typical range for dialysis patients[a]
Hemoglobin (g · dl^{-1})	12.0–16.0	[b]
Hematocrit (%)	37.0–47.0	[b]
Sodium (mEq · L^{-1})	136.0–145.0	135.0–142.0
Potassium (mEq · L^{-1})	3.5–5.3	4.0–6.0
Chloride (mEq · L^{-1})	95.0–110.0	95.0–100.0
HCO$_3$ (mEq · L^{-1})	22.0–26.0	18.6–23.4
Albumin	3.7–5.2	2.7–3.3
Calcium	9.0–10.6	9.0–12.0
Phosphorous	2.5–4.7	2.5–6.0
BUN (mg · dl^{-1})	5.0–25.0	60.0–110.0
Creatinine (mg · dl^{-1})	0.5–1.4	12.0–15.0
PH	7.35–7.45	7.38–7.39
Creatinine clearance (ml · min^{-1})	85.0–150.0	0 (or minimal residual)
Glomerular filtration rate (ml · min^{-1})	90.0–125.0	0 (or minimal)

Note. BUN = blood urea nitrogen.

[a]Assuming well-dialyzed, stable patient. [b]Hematocrit and hemoglobin levels depend on the level of erythropoietin treatment. With no treatment, hematocrit may be as low as 17%. With treatment, it can be normalized, but most patients are treated to a range of 33% to 37%.

dialyzer has a semipermeable membrane that separates the blood from a dialysis solution, which creates an osmotic and concentration gradient to clear substances from the blood. Factors such as the characteristics of the membrane, transmembrane pressures, blood flow, and dialysate flow rate determine removal of substances from the blood. Manipulation of the blood flow rate, dialysate flow rate, dialysate concentrations, and time of the treatment can be used to remove more or less substances and fluids (4).

The duration of the dialysis treatment is determined by the degree of residual renal function, body size, dietary intake, and clinical status. A typical dialysis prescription is 3 to 4 h three times per week. Complications of the dialysis treatment include hypotension, cramping, problems with bleeding, and fatigue. Significant fluid shifts can occur between treatments if the patient is not careful with dietary and fluid restrictions. Table 15.2 and "Long-Term Complications of Dialysis" list the complications.

Peritoneal Dialysis

Approximately 8% of patients in the United States are treated with peritoneal dialysis. Other countries tend to have a higher percentage of patients treated with this form

of dialysis. This form of therapy is accomplished by introducing a dialysis fluid into the peritoneal cavity through a permanent catheter placed in the lower abdominal wall. The peritoneal membranes are effective for ultrafiltration of fluids and clearance of toxic substances in the blood of uremic individuals. The dialysis fluid is of a given osmotic and concentration to provide gradients to remove fluid and substances. The fluid is introduced either by a machine (cycler), which cycles fluid in and out over an 8 to 12 h period at night, or manually by 2 to 2.5 L bags that are attached to tubing and emptied by gravity into and out of the peritoneum. The latter process, known as continuous ambulatory peritoneal dialysis, allows the patient to dialyze continuously throughout the day. Continuous ambulatory peritoneal dialysis requires exchange of fluid every 4 h using sterile technique (4). Table 15.3 lists complications associated with peritoneal dialysis.

Patients may choose peritoneal dialysis so that they can experience more freedom and less dependency on a center for use of a machine. This method of treatment allows patients to travel and dialyze on their own schedules. Patients who have cardiac instability may also be placed on peritoneal dialysis because the method does not involve the major fluid shifts experienced with

Table 15.2 Complications Associated With Hemodialysis Treatment

Complication	Pathophysiology
Hypotension	Decreased plasma volume with slow refilling
	Impaired vasoactive or cardiac responses
	Vasodilation
	Autonomic dysfunction
Cramping	Contraction of intravascular volume
	Reduced muscle perfusion
Anaphylactic reactions	Reaction to dialysis membrane (usually at first use)
Pyrogen or infection-induced fever	Bacterial contamination of water system
	Systemic infection (often at the access site)
Cardiopulmonary arrest	Dialysis line disconnection
	Air embolism
	Aberrant dialysate composition
	Anaphylactic membrane reaction
	Electrolyte abnormalities
	Intrinsic cardiac disease
Itching	Unknown etiology
"Restless legs"	Unknown etiology

Based on K.L. Johansen, 1999, "Physical functioning and exercise capacity in patients on dialysis," *Advances in Renal Replacement Therapy* 6(2): 142.

Long-Term Complications of Dialysis

Metabolic abnormalities
- Metabolic acidosis
- Hyperlipidemia (type 4)
  - Increased triglycerides
  - Increased very low-density lipoprotein cholesterol
  - Decreased high-density lipoprotein cholesterol
  - Normal total cholesterol
- Hyperglycemia
- Other hormonal dysfunction

Malnutrition

Cardiovascular disease
- Hypertension
- Ischemic heart disease
- Congestive heart failure
- Pericarditis

Uremic osteodystrophy (secondary hyperparathyroidism)

Peripheral neuropathy

Amyloidosis

Severe physical deconditioning

Frequent hospitalizations

Continuation of progressive complications in diabetic patients

Table 15.3 Complications Associated With the Peritoneal Dialysis Treatment

Complication	Comments
Infections	Possible at exit site or along catheter—"tunnel infection"
	May be result of a break in sterile procedures during exchange
Peritonitis	Most frequent complication of peritoneal dialysis
Hypotension	Excessive ultrafiltration and sodium removal
Hernia, leaks	Associated with the increased intra-abdominal pressure

hemodialysis. Peritoneal dialysis may be preferable for diabetic patients, because they can inject insulin into their dialysate and achieve better glucose control.

Complications of peritoneal dialysis include problems with the catheter or catheter site, infection, hernias, low-back pain, and obesity. Hypertriglyceridemia is a problem caused by the exposure and absorption of glucose from the dialysate. Patients may absorb as many as 1,200 kcal from the dialysate per day, contributing to the development of obesity and hypertriglyceridemia (4).

Renal Transplantation

Transplantation of kidneys is the preferred treatment of ESRD. In the United States, 12,000 kidney transplants are performed each year (28% of ESRD patients). The source of the kidneys available for transplant can be a living relative, an unrelated individual, or a cadaver. Because of the shortage of organs available for transplantation and improvements in immunosuppression medications, living nonrelated transplants are becoming more frequent. Patients considered for transplant are generally healthier and younger than the general dialysis population, although there are no age limits to transplantation. Patients with severe cardiac, cerebrovascular, or pulmonary disease and neoplasia are not considered candidates. Table 15.4 on p. 270 lists long-term complications of transplantation.

Following transplantation, patients are placed on immunosuppression medication, which includes combinations of glucocorticosteroids (prednisone), cyclosporine derivative, and monoclonal antibody therapy. New immunosuppression medications are constantly being developed, allowing for minimization of side effects by altering therapies or combinations of therapies. Patients may experience rejection early (acute) or later (chronic), which is detected by elevation of creatinine. Rejection is treated immediately with increased dosing of immunosuppression (mostly prednisone), with a tapering back to maintenance dose. Patients must remain on immunosuppression for the lifetime of the transplanted organ. Nationwide 1 yr graft survival is 85%, and patient survival is 90%. Five-year rates are 67% graft survival and 85% patient survival. Causes of graft loss include chronic rejection (25%), cardiovascular deaths (20.3%), infectious deaths (8.7%), acute rejection (10.2%), technical complications (4.7%), and other deaths (10.2%) (see table 15.4). Short-term transplant survival has been improved with new immunosuppression medications, leaving the major challenges of long-term survival of graft and patients to be investigated. Loss of kidney results in the need to return to dialysis (6).

Table 15.4 Long-Term Complications Associated With Transplantation

Complication	Comments
Rejection	Can be acute or chronic; in most cases treated with increased immunosuppression dosages
Cardiovascular disease	Most frequent cause of death posttransplant
	All known risk factors are prevalent, and immunosuppression medications may exacerbate risk
Infections	Immunosuppression may increase infection risk
Musculoskeletal disorders	Glucocorticoid therapy (prednisone) reduces bone density and causes muscle protein breakdown
Obesity	Very prevalent, often associated with prednisone therapy, but more likely attributable to imbalance between calorie intake and expenditure (i.e., lifestyle issues)

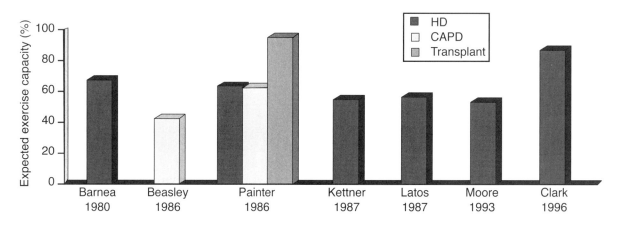

Figure 15.1 Exercise capacity in patients with end stage renal disease. HD = hemodialysis; CAPD = continuous ambulatory peritoneal dialysis.

This article was published in *Advances in Renal Replacement Therapy,* Vol. 5(2), K.L. Johansen, "Physical functioning and exercise capacity in patients on dialysis," pg. 142, Copyright Elsevier 1999.

Complications of kidney transplantation are primarily related to immunosuppression therapy and include infection, hyperlipidemia, hypertension, obesity, steroid-induced diabetes, and osteonecrosis. The incidence of atherosclerotic cardiovascular disease is four times higher in kidney transplant recipients than the general population, and cardiovascular risk factors are prevalent in most patients.

Exercise Testing

Most patients on dialysis (see treatment section) are severely limited in exercise capacity, primarily by leg fatigue. Peak oxygen uptake is reported to be only 60% to 70% of normal age-expected levels, and it can be as low as 39% (7,18) (figure 15.1). The degree to which exercise is limited in these patients is difficult to determine because of the complex nature of uremia, which affects nearly every organ system of the body. Reduced exercise

capacity is almost certainly a multifactorial problem that is influenced by anemia, muscle blood flow, muscle oxidative capacity, myocardial function, and the person's activity levels. Muscle function may be affected by nutritional status, dialysis adequacy, hyperparathyroidism, and other clinical variables (19).

Most studies that have measured oxygen uptake have included only the healthiest patients; thus, most patients have an even lower exercise capacity and may be unable to perform exercise testing. Information obtained from exercise testing in this patient group is not diagnostically useful, because most patients stop exercise because of leg fatigue and do not achieve age-predicted maximal heart rates. Although many patients have abnormal left ventricular function (32), most patients have conditions that make interpretation of stress electrocardiogram difficult, including left ventricular hypertrophy (LVH) with strain patterns, electrolyte abnormalities, and digoxin effects on the electrocardiogram. Thus, stress testing is not necessarily recommended before initiation of exercise training, and requiring stress testing may prevent some patients from becoming more physically active (20). Because exercise capacity is so low, most patients will not train at levels that are much above the energy requirements of their daily activities. Therefore, risk associated with such training is minimal. Heart rate is not recommended for determining training intensity because of the effects of antihypertensive medications and fluid shifts on heart rates. Thus, exercise testing is not needed to develop a training heart rate prescription. Practical application 15.1 describes client–clinician interaction for patients with end-stage renal disease.

Physical performance may best be tested in dialysis patients by using tests such as stair climbing, 6 min walk test, sit-to-stand-to-sit test, or gait speed testing. These tests, which have been standardized and used in many studies of elderly people, have been shown to predict outcomes such as hospitalization, discharge to nursing home, and mortality rate (21). A walking–stair-climbing test was validated in hemodialysis patients by Mercer et al. (33). In dialysis patients, these tests are effective in demonstrating improvements from exercise counseling interventions (22). Additionally, self-reported physical-functioning scales such as those on the SF-36 Health Status Questionnaire are highly predictive of outcomes in dialysis patients, specifically hospitalization and death (23). Exercise training improves scores on these self-reported scales in hemodialysis patients (22).

Exercise capacity is similarly low in peritoneal dialysis patients (8,13,34). Following successful renal transplant, exercise capacity increases significantly, to near sedentary normal predicted values (8,13,24). Renal transplant recipients who were active and who participated in the 1996 U.S. Transplant Games had exercise capacity that averaged 115% of normal age-predicted values (25). Exercise testing with standard protocols is more appropriate for transplant recipients who are able to push themselves in their training programs above their daily levels of activity. Exercise heart rate responses are normalized after transplant. The major abnormality noted in transplant recipients is excessive blood pressure response to exercise.

Practical Application 15.1

Client–Clinician Interaction

Exercise professionals can best serve the needs of dialysis patients if they understand the patients' treatment regimens and the setting in which they receive treatment. Most hemodialysis patients receive treatments 3 times per week for 3 to 4 h. They are transported either by friends and family or by medical transportation services, so they have minimal flexibility.

The staffing at outpatient dialysis clinics typically consists of a charge nurse and patient care technicians who are not medically trained except for administration of dialysis. The schedules for dialysis are quite tight, and the opportunity for patient education is minimal in terms of both staff and time. Therefore, attending exercise class at a supervised program may be a significant barrier to exercise participation. Thus, to optimize adherence, every effort should be made to implement home exercise programs or programs at the dialysis clinic.

For a program to be successful, the exercise professional should first interact with and educate the dialysis staff, which takes the support of the administration in coordinating in-service training. The dialysis staff is close to the patients and can be influential in their participation (or nonparticipation) in exercise. This support is critical to the efforts of the patient and the exercise personnel. Educational and motivational programs at the dialysis clinic not only increase staff and patients' awareness of the importance of exercise but also change the environment of the clinic from one of illness to one of wellness.

Following kidney transplant, many patients are afraid to exert themselves vigorously. That fear comes from lack of information provided by the transplant service, weakness despite a significant improvement in overall health, fears or concerns on the part of the patient's family, and lack of experience with exercise (because of health concerns before transplant). Exercise is not routinely addressed following kidney transplant—either at the time of transplant or in the routine follow-up care in clinic. The exercise professional should attempt to educate the transplant team about the importance of exercise and, as much as possible, become part of the routine care team so that exercise counseling is incorporated into posttransplant care.

EXERCISE PRESCRIPTION

The exercise prescription for patients on dialysis should include flexibility and range of motion, strengthening, and cardiovascular exercises. Weight management considerations may be needed for many transplant recipients. For the dialysis patient, the key to prescription is understanding the multiple barriers to exercise that may exist. These include general feelings of malaise, time requirements of treatment, lack of encouragement and information provided by nephrology healthcare workers, fear, and accustomization or adaptation of life-styles to low levels of functioning. Thus, any prescription should start slowly and progress gradually to prevent discouragement and additional feelings of fatigue or muscle soreness (29). Practical application 15.2 reviews the literature about exercise training and end-stage renal disease.

Special Exercise Considerations

When patients are diagnosed with end-stage renal failure, most are never given information on exercise and physi-

Practical Application 15.2

Literature Review

Several studies have reported the effects of cardiovascular exercise training in dialysis patients. The type and duration of these studies were variable (range 8 wk to 12 mo), but all studies compared exercise tests before and after training in the same individuals. Most studies were performed before recombinant human erythropoietin became available, most included a small number of patients, and only a few provided adequate control or comparison data.

Most studies showed an improvement in $\dot{V}O_2$max after exercise training, with an average increase of 16.4% across all studies (range 0–52%) (9,11,12,17,18,22,35). Although the improvement was similar to that seen in healthy people, training did not normalize $\dot{V}O_2$max, and the posttraining values remain well below age-predicted values. Besides the improvements in oxygen uptake, improvements were reported in hematocrit (before the availability of erythropoietin), blood pressure control, and lipid profiles. Many of these improvements have not been duplicated in other studies, and most studies had too few subjects to be conclusive (19). None of these studies included measures of quality of life or self-reported functioning. Impressive improvements have been observed in muscle fiber size and in atrophy following a program of cardiovascular exercise training plus sports activity and strengthening exercises (34). Also, positive changes in cardiac function following physical training have been observed (32,36).

One recent study showed significant improvements in gait speed and the sit-to-stand-to-sit test following exercise counseling interventions for independent home exercise and in-center cycling. This study also reported significant improvements in the physical scales of the SF-36 Health Status Questionnaire (22).

Most exercise-training studies incorporated exercise on nondialysis days, three times per week.

The studies reported major problems in adherence to exercise that required patients to attend on their nondialysis days at a supervised center. The dialysis treatment provides an ideal opportunity to capture an audience. A high rate of adherence to bicycle exercise training during the dialysis treatment was observed (12). This programming was well tolerated within the first hour of the treatment and did not interfere with the dialysis treatment. $\dot{V}O_2$peak changes were comparable with those seen in studies in which cardiovascular exercise was performed on nondialysis days.

One exercise-training study that included patients who were treated with peritoneal dialysis showed significant improvements in $\dot{V}O_2$peak (16.2%) and physical-functioning dimensions of quality of life (34). Two studies reported significant improvements in exercise capacity in renal transplant recipients (27,28). In a randomized controlled trial of independent home exercise over the first year of transplant, we found significant improvements in $\dot{V}O_2$peak with home exercise (increase in $\dot{V}O_2$peak to 30 ml $\cdot$ kg^{-1} $\cdot$ min^{-1} at 1 yr) compared with those in the usual care group (24 ml $\cdot$ kg^{-1} $\cdot$ min^{-1} at 1 yr; unpublished data). The values for the usual care group were not much different than values reported for high-functioning dialysis patients.

cal activity. If they ask, typically they are told to take it easy or not to overdo it. This advice poses questions and plants doubt in the minds of patients and their families, who will be extremely protective. Patients do not know how much activity is too much activity, and because they do not feel well and are fatigued, they opt for no activity. The dialysis staff, who see the patients regularly for their treatments, often reinforce an inactive lifestyle. Not surprisingly, then, many patients are skeptical about becoming physically active. Patients interact primarily with their dialysis providers and thus receive little information from others. Therefore, the exercise professional must take the time to learn about dialysis and transplant to understand what patients must deal with daily or three times per week. This learning experience could entail watching a patient being put on the dialysis machine and visiting with a few patients during their treatment. Patient support groups are also a good source of information. Patients often talk freely about their experiences. By listening carefully, the exercise professional will understand more about patient responses to major changes in lifestyle such as initiating exercise, and they may be able to devise more effective ways to motivate patients. The exercise professional should reach out to dialysis staff about how exercise can benefit their patients and assure them that the programs initiated will be safe and will not interfere with the treatments (31). This education should also include ideas about how the dialysis staff can encourage patients to be physically active. Additional encouragement and reinforcement can greatly facilitate patient efforts in rehabilitation. Likewise, lack of support and understanding on the part of the dialysis staff can sabotage efforts to increase patients' activity.

Most exercise professionals practice in their own laboratory and depend on referrals of patients to their exercise facility. Many patients may be unable to participate in exercise at the designated facility but may benefit significantly from counseling on home independent exercise. Thus, exercise professionals should reach out to other healthcare providers (other than cardiologists and pulmonologists), educate them on the services that they can offer their patients, and discuss the benefits of exercise for their patient groups. Although the nephrology community is becoming more interested in improving physical functioning of patients, most nephrologists and kidney transplant staff are not familiar with how exercise may benefit their patients or how to evaluate physical functioning or prescribe exercise. A trained professional who knows about the problems associated with dialysis and transplant may be a welcome addition to the patient care team.

Exercise Recommendations

The timing of exercise in relation to the dialysis treatment should be considered. Hemodialysis patients can exercise anytime, although they may feel best on their nondialysis days and may be able to tolerate higher intensity or duration. Most feel extremely fatigued following their dialysis treatment, and problems with hypotension may develop following the treatment when vasodilation induced by physical activity occurs. Immediately before the dialysis treatment, some patients may have excessive fluid in their systems, because they are unable to rid their bodies of fluid taken in between treatments. Thus, they may not tolerate as much exercise before dialysis, because of increased volume overload on the left ventricle; increased blood pressure at rest (and during exercise), which may increase ventricular preload and afterload; and, in extreme cases, pulmonary congestion. Exercise should be deferred if the patient is experiencing shortness of breath related to excess fluid status. No specific guidelines regarding the upper limit of fluid weight gain contraindicate exercise, although the guidelines for blood pressure established by American College of Sports Medicine (37) should be followed. Practical applications 15.3 on p. 274 and 15.4 on p. 275 discuss exercise prescription for patients who are undergoing dialysis and those with transplants.

The ideal mode of exercise and the ideal time for the patient to exercise, in terms of adherence and convenience, may be recumbent stationary cycling during the hemodialysis treatment. Although this form of exercise does not interfere with the dialysis treatment, most facilities are unwilling to have cycles in the clinic for their patients. Thus, independent home exercise may be the best approach for these patients.

Cycling during dialysis is best tolerated during the first 1 to 1.5 h of the treatment because after that time the patient has a greater risk of being hypotensive when in the chair, making cycling difficult. This response is caused by the continuous removal of fluid throughout the treatment, which decreases cardiac output, stroke volume, and mean arterial pressure at rest (38). Therefore, after 2 h of dialysis, cardiovascular decompensation may preclude exercise (38).

For patients treated with continuous ambulatory peritoneal dialysis, the exercise may be best tolerated at a time when the abdomen is drained of fluid, which allows for greater diaphragmatic excursion and less pressure against the catheter during exertion, reducing the risk of hernias or leaks around the catheter site (30). Patients may choose to exercise in the middle of a dialysis exchange—after draining fluid and before introducing the new dialysis

Exercise Prescription for Patients Treated With Dialysis

Mode	Frequency	Intensity	Duration	Progression
Cardiovascular exercise: walking, cycling, swimming, low-level aerobics, stepping	4 to 5 d per week	RPE of 12 to 15 (on 6- to 20-point scale)	Work up to 30 min of continuous exercise	Start with intervals of intermittent exercise and gradually increase the work intervals
Strength exercise: resistance bands, isometric, very low-level hand and ankle weights, body-weight resistance	2 to 3 d per week	Sets: three sets of exercises for major muscle groups	Repetitions: 12 to 15 of each exercise	Start with one set of 12 repetitions with 1 to 2 lb (0.5 to 1 kg) weights; increase gradually

Special Considerations for Dialysis Patients

Patients will have extremely low fitness levels.

Timing of exercise sessions should be coordinated with dialysis sessions.

Patients will experience frequent hospitalizations and setbacks.

Gradual progression is critical.

Heart rate prescriptions are typically invalid—use of RPE is recommended.

Maximal exercise testing is typically not tolerated well by most patients and is not diagnostically useful for those with coronary artery disease because of peripheral muscle fatigue.

Performance-based testing is more feasible and useful.

One-repetition maximum testing for strength is not recommended because of secondary hyperparathyroidism-related bone and joint problems.

Prevalence of orthopedic problems will be significant.

Motivating patients is often a challenge.

Every attempt should be made to educate dialysis staff about the benefits of exercise so that they can help motivate patients to participate.

fluid. This option requires capping off the catheter for exercise, a technique that must be discussed with the dialysis nurse.

Mode

There is no restriction on the type of activity that can be prescribed for dialysis or transplant patients. Range-of-motion and strengthening exercises are critical for most patients because they are stiff and weak after long periods of inactivity. Because many patients have weak muscles and joint discomfort, non-weight-bearing cardiovascular activity may be best tolerated. As for anyone else, if jarring activity causes joint discomfort, then a change in mode of exercise is indicated. The access site for the hemodialysis may be in the arm or upper leg. This circumstance should not inhibit activity at all, although many patients are told not to use the arm with their fistula in it. This restriction is typically given by the vascular surgeon at the time of placement and pertains only to the time of healing (i.e., 6–8 wk). The only precaution for the fistula is to avoid any activity that would close off the flow of blood (e.g., having weights lying directly over the top of the vessels). Although the patient should be protective of the access site, use of the extremity will increase flow through it and actually help develop muscles around the access site, which should make placement of needles easier.

Patients with a peritoneal catheter should avoid full sit-ups and activities that involve full flexion at the hip. They can accomplish abdominal strengthening by using isometric contractions and crunches. Swimming may be a challenge

Practical Application 15.4

Exercise Prescription for ESRD Treated With Transplantation

Mode	Frequency	Intensity	Duration	Progression
Cardiovascular exercise: walking, jogging, cycling, swimming, aerobics, stepping, sports	4 to 5 d per week	65% to 80% of peak heart rate; RPE of 12 to 15 (on 6- to 20-point scale)	Work up to 30 min of continuous exercise (longer duration for weight management)	Start with intervals of intermittent exercise and gradually increase the work intervals until continuous exercise is tolerated
Strength exercise: resistance band, weight machines, hand weights, free weights, calisthenics	3 d per week	Sets: three sets of exercises for major muscle groups	Repetitions: 12 to 15 of each exercise	Start with one set of 12 repetitions with low weights; increase gradually

Special Considerations: Transplant

Patients are initially weak, so gradual progression is recommended.

Patients may experience a lot of orthopedic and musculoskeletal discomfort with strenuous exercise.

Weight management often becomes an issue following transplant.

Patients and their families are often fearful of "overexertion"; thus, gradual progression should be stressed.

Prednisone may delay adaptations to resistance training.

Exercise should be decreased in intensity and duration during episodes of rejection, not curtailed completely.

Patients may experience frequent hospitalizations during the first year posttransplant. Because patients are immunosuppressed, every effort must be made to avoid infectious situations (e.g., strict sterilization procedures for testing and training equipment).

for those with peritoneal catheters because of the possibility of infection. Patients must be advised to cover the catheter with some protective tape and to clean around the catheter exit site after swimming. Freshwater lake swimming is not recommended, whereas swimming in chlorinated pools and in the ocean involves less risk of infection.

Although transplant recipients are often told not to participate in vigorous activities, the main concern is any contact sport that may involve a direct hit to the area of the transplanted kidney (e.g., football). Vigorous activities and noncontact sports are well tolerated by transplant recipients who have worked to build adequate muscle strength and cardiovascular endurance through a comprehensive general conditioning program.

Frequency

Range-of-motion exercises should be encouraged daily. Hemodialysis patients will feel especially stiff after their dialysis session because of 3 to 4 h of sitting as well as removal of fluid and often cramping. Stretching during and after the dialysis treatment may relieve this stiffness. Muscle strengthening should be done 3 d per week. Cardiovascular exercise should be prescribed for at least 3 d per week, although a prescription of 4 to 6 d per week may be most beneficial.

Intensity

Cardiovascular exercise intensity should be prescribed by using a rating of perceived exertion (RPE), because heart rates are highly variable in dialysis patients as a result of fluid shifts and vascular adaptations to fluid loss during the dialysis treatment. Many patients initially may tolerate only a few minutes of very low-level exercise, which means that warm-up and cool-down intensities are irrelevant. These individuals should just be encouraged to increase duration gradually at whatever level they can tolerate. After they achieve 20 min of continuous exercise, warm-up, conditioning, and cool-down phases can be incorporated, with an RPE of 9 to 10 for warm-up and cool-down and 12 to 15 for the conditioning time (on a 6- to 20-point scale).

Duration and Progression

The exercise professional must start patients slowly and progress them gradually. In practice, this means that the patient should determine the duration of activity that he or she can comfortably tolerate during the initial sessions. This duration will be the starting duration of activity. If the patient tolerates only 2 to 3 min of exercise, the prescription may be for several intervals of 2 to 3 min, with a gradual decrease in rest times to progress the patient to continuous activity. A progression in duration of 2 to 3 min per session or per week is recommended, depending on individual tolerance. Extremely weak patients may need to start with a strengthening program of low weights and high repetitions and range-of-motion exercise before initiating any cardiovascular activity. The progression should gradually work up to 20 to 30 (or more) min of continuous activity at an RPE of 12 to 15 (on a 6- to 20-point scale).

When a patient begins cycling during dialysis, the initial session is usually limited to 10 min, even if the patient is able to tolerate a longer duration. This precaution assures the dialysis staff and the patient that cycling does not have any adverse effects on the dialysis treatment. The patient can then progress duration in subsequent sessions according to tolerance, as described previously. RPE is also used for intensity prescription during the dialysis treatment, because removal of fluid from the beginning to the end of dialysis can cause resting and exercise heart rates (standard submaximal level) to vary by 15 to 20 beats (38).

EXERCISE TRAINING

The results of an exercise-training program in renal patients will be variable. Most often there is an improvement in physical functioning (as measured by exercise testing or physical performance testing or self-reported functioning on questionnaires). In some dialysis patients (and possibly diabetic transplant recipients), particularly those who are very low functioning and have multiple comorbidities, there may not be dramatic improvements. However, the natural course of the condition is for a deterioration in functioning, thus, if the outcome of the exercise intervention is to maintain functioning, that is a positive result. Most patients will experience improvement in muscle strength, and often an increase in lean muscle mass. This increase in lean mass may have implications for hemodialysis patients, since the amount of fluid removed during dialysis is gauged by body weight. Thus, if their body weight (lean mass) increases, their target weight for dialysis may need to be adjusted. Many dialysis patients will also experience improvements in blood pressure, many requiring reduction in antihypertensive agents. Diabetic patients may experience improved glucose control, thus requiring adjustments in insulin requirements.

Diaysis and transplant recipients respond to exercise training with similar magnitude of changes in strength and exercise capacity as normal healthy individuals. However, they may not achieve similar maximal levels of functioning. Likewise, the time course for improvement may be longer in patients with renal disease. For transplant recipients on prednisone, progression in muscle strengthening may be slower, and absolute gains may take longer than healthy individuals, however they can achieve normal levels of muscle strength, counteracting the negative effects of prednisone on the muscles.

Most patients will experience significant improvements in energy level and ability to perform their activities of daily living, and may experience fewer symptoms associated with dialysis such as muscle stiffness, cramping, and hypotensive events during dialysis. If exercise is performed during the hemodialysis treatment, there may be improved clearance of toxins. Overall, quality of life improves—particularly in the physical domains.

CONCLUSION

Exercise prescription for patients with renal failure depends on their treatment. The prescription must be individualized to the patient's limitations. It should include the type of exercise (cardiovascular, range of motion, strengthening), frequency of exercise, timing of exercise in relation to treatment, duration, intensity (prescribed primarily based on RPE), and progression. The progression should be gradual in those who are extremely debilitated. The starting levels and progression must be according to tolerance, because fluctuations in well-being, clinical status, and overall ability frequently change with changes in medical status. Hospitalization or a medical event (e.g., clotting of the fistula or placement of a new fistula) may set a patient back in the progression of his or her program, requiring frequent evaluation of the prescription. The goal is for patients to become more active in general and, if possible, for them to work toward a regular program of 4 to 6 d per week of cardiovascular exercise, 30 min or more per session at an intensity of 12 to 14 on the RPE scale. Strengthening exercise should be recommended three times per week.

Medical History

Mrs. HN is a 68 yr old Hispanic female with known ESRD. She has been on hemodialysis for 28 mo, and her treatment prescription is for 3 d per week with 3 h treatment sessions each day. She has a graft in her right upper arm as her access site. She presents with the complaints of lack of energy, weakness, and decreased endurance. Her nephrologist refers her to the staff exercise physiologist.

Diagnosis

The exercise physiologist reviews Mrs. HN's chart to find that her ESRD is secondary to long-term non-insulin-dependent diabetes (15 yr). Mrs. HN has also developed severe peripheral neuropathy as a result of her diabetes.

Exercise Test Results

The exercise physiologist then conducts a battery of physical function tests to assess Mrs. HN's physical ability. These tests consist of the sit-to-stand test, the 6 min walk, and the 20 ft (6 m) gait speed test at both a comfortable and a fast pace. The results of these tests are as follows: sit-to-stand test, 33.01 s, which is 28% of normal age-predicted values (39); 6 min walk, 350 ft (107 m); 20 ft normal gait speed, 55.01 cm/s, which is 42% of normal age-predicted values (40). Her self-reported physical function scale on the SF-36 Health Status Questionnaire is 55 (average age value is 84). During her walking tests, Mrs. HN exhibits poor balance and endurance as a result of her peripheral neuropathy and general weakness, respectively. Physical activity questionnaires are administered to assess her current activity as well as degree of difficulty of those activities.

Exercise Prescription

With the assessment complete, an exercise prescription is developed. Mrs. HN is first counseled on exercise as it relates to her diabetes and glycemic control. Written information is also provided. The exercise prescription continues only when it is certain that Mrs. HN fully understands the balance between exercise and glycemic control. Because of her poor balance and endurance, a stationary bicycle is the preferred mode for cardiovascular exercise. She is asked to begin with a frequency of 3 to 4 d per week, on nondialysis days, because she generally feels better on those days. The duration of the exercise is 10 min with two bouts each exercise day, totaling 20 min of exercise each exercise day. The initial prescribed intensity should be light to moderate, or enjoyable. She is asked to progress gradually each week with a goal of 30 min of continuous cycling 3 to 4 d per week, minimum. Her initial exercise prescription also includes various flexibility exercises for the upper and lower body as well as for the back. She is asked to do these exercises daily. Strengthening exercises are also prescribed. Again, both upper- and lower-body exercises that use the major muscle groups are encouraged. These exercises are prescribed for 3 d per week on nonconsecutive days. She will initially perform the exercises without resistance weight and gradually progress to performing them with weight. Her initial prescription consists of one set of 10 repetitions of each exercise. The exercise physiologist reviews Mrs. HN's progress weekly at her dialysis treatments. Progression and exercise participation are noted in the patient's chart. At the end of 8 wk of exercise, Mrs. HN's physical functioning is again assessed with the battery of physical function tests and the activity questionnaires.

Discussion Questions

1. Why is performance-based testing preferred for this patient?
2. What causes Mrs. HN to have poor balance during walking? How does this condition affect her exercise prescription?
3. Why is no heart rate prescription given to Mrs. HN for her cycling program?
4. How would the cycling program differ if she chose to cycle during the dialysis treatment instead of at home?
5. Does anything in her history suggest that cycling during dialysis might be advantageous for Mrs. HN?

PART IV

Cardiovascular Diseases

A reduction in cardiovascular disease (CVD) mortality has been observed over the last 20 years, but the incidence of CVD has changed minimally. At the same time, many advances have been made in the technology of assessing and correcting significant disease. But with the negative societal trend toward being overweight and obese, and physically inactive, we can expect many healthcare dollars to be directed toward CVD. The clinical exercise physiologist should be familiar with the areas addressed in the chapters in part IV because all these forms of CVD can be positively altered by an appropriate rehabilitation program that addresses specific lifestyle interventions.

Chapter 16 centers on myocardial infarction. The prevalence of myocardial infarctions continues to be very high in our society. The value of secondary prevention in this population has been shown to reduce the risk of mortality. These patients can strongly benefit from positive lifestyle alterations by participating in a cardiac rehabilitation program. The clinical exercise physiologist plays an important role in assessing the patient's cardiovascular limitations through graded exercise testing and developing an appropriate exercise prescription. This chapter provides a clear understanding of the disease process and preventive measures that can be used to ward off further events.

Chapter 17 discusses revascularization of the heart. Revascularization procedures have become a common way to address significant coronary artery disease. Emergency percutaneous coronary intervention is becoming the standard way of decreasing the risk of myocardial damage. The challenges that the clinical exercise physiologist must address vary according to the clinical procedure completed. He or she must have

a good understanding of the procedures and potential issues that may arise when performing a graded exercise test and prescribing exercise.

Chapter 18 deals with chronic heart failure. With the prevalence of CVD and the growing older population, the incidence of chronic heart failure in our society is increasing. Estimates are that heart failure (HF) will increase two- to threefold over the next 10 years. Based on the large increase in HF, the clinical exercise physiologist will be required to have a strong knowledge base to work with HF patients in a rehabilitation setting, especially because few guidelines currently exist for prescribing exercise in these patients.

Chapter 19 explores peripheral artery disease. Peripheral artery disease (PAD) is common in our society, especially in those with existing CVD, because the major risk factors for CVD are similar to those of PAD. Significant PAD can severely limit a person's exercise tolerance, depending on when the subject develops intermittent claudication. The degree of a person's PAD will have a large influence on testing procedures offered. The clinical exercise physiologist must learn specific strategies to develop an appropriate exercise program for reducing a patient's symptoms of PAD.

Chapter 20 considers pacemakers and implantable cardioverter defibrillators (ICDs). As our population becomes older, the need for the use of pacemakers and ICDs becomes greater. Because of the complexity of these devices, the clinical exercise physiologist needs to be knowledgeable in how they work, how they can be influenced when a graded exercise test is performed, and what issues may arise under normal exercise training (e.g., what exercise heart rate should be avoided to prevent the risk of premature firing of an ICD).

Myocardial Infarction

Paul S. Visich, PhD, MPH

Emma Fletcher, MA

A myocardial infarction (MI) refers to the death of myocardial muscle cells that occurs when a substantial decrease or complete disruption of blood flow through a coronary artery deprives the downstream tissue of oxygen for an extended period. Typically, an MI is caused by thrombus formation in a coronary artery that has a degree of established coronary artery disease or atherosclerosis. The term **acute MI** refers to the sudden occurrence of ischemia that leads to myocardial damage and subsequent infarction (3,52).

SCOPE

Having an MI is one of the most prevalent events resulting in hospitalization in the United States and other Western countries. Approximately 565,000 new attacks and 300,000 recurrent MIs occur annually in the United States. The risk of a subsequent MI, sudden death, heart failure, and stroke within 6 yr post MI is substantial (161). The American Heart Association (AHA) estimates that approximately 40% of those with an MI will die within 1 yr following their event (161).

About half of patients experiencing an acute MI never reach a hospital emergency department. Mortality rates increase for every 30 min that elapse after the onset of symptoms (18). The delay occurs for a variety of reasons, including the following:

- Failure to recognize symptoms

- Failure to act on the perception of symptoms
- Failure of timely transport to an acute care facility (18,58,67,97,122,161)

The majority of MIs occur in persons older than 65 yr of age. Incidence rates differ between males and females. In 2006 the AHA estimated that males experienced an MI about 1.4 times more than females. Females, however, are more likely to die within 1 yr post MI, primarily because females tend to have heart attacks later in life and mortality risk increases with age of MI onset. The annual rate of first MI is similar in black and white people until age 45 for females and age 65 for males. From there on, black people have a higher annual rate of first MI (12).

MI results in a major expenditure for diagnosis and treatment. The estimated individual cost of a fatal acute MI is $17,532 per event and the cost of a nonfatal acute MI is $15,540 per event in the first year following the event (135,185). This amounts to approximately $5.54 billion annually for new MI cases, a large burden on the U.S. healthcare system.

PATHOPHYSIOLOGY

The pathophysiology of CAD or atherosclerosis is a complex process that can result in an MI. Atherosclerosis is characterized by an accumulation of lipids, inflammatory cells, and connective tissue within the arterial intimal layer. The atherosclerotic plaque usually develops at irregular intervals and occurs in most of the major arteries in the body. However, atherosclerosis most often occurs in the area of a **bifurcation** within the main coronary artery branches (i.e., left main, left anterior descending, circumflex, right coronary artery). A single artery may develop multiple discrete lesions along its length.

A recent investigation reported finding coronary artery atherosclerotic lesions in 29.4% of male and female victims of noncardiac trauma (age 16–50). Severe disease was noted in 5.4% of these victims, the majority of which were observed in the left anterior descending coronary artery. Among the samples, the incidence of atherosclerotic plaque was greater in males (31%) than in females (14%) and the relative frequency of cases increased with age (57).

The genesis of atherosclerosis is slow and progressive and has a long asymptomatic phase (36,176). Factors that likely play a direct or indirect role in atherosclerotic development include diabetes (i.e., metabolic injury), physical inactivity, obesity, smoking, hypertension, blood concentrations of homocysteine, C-reactive protein (CRP) and interleuken-6 (IL6), the aging process, genetic makeup, and gender-related factors (15,27,32,77,79,85, 95,111,139,140,174,177). Other potential or emerging risk factors include but are not limited to mental stress, blood concentration of inflammatory proteins such as lipoprotein-a and serum amyloid A, tissue plasminogen activator (tPA) and plasminogen activator inhibitor type 1 (PAI-1) activity, and blood concentration of certain adhesion molecules such as intercellular adhesion molecule-1 (ICAM-1) and vascular adhesion molecule-1 (VCAM-1) (29,169,170,182). Factors unknown at this time probably contribute as well.

Endothelial dysfunction (EDF) is the earliest detectable physiological manifestation of atherosclerosis (176,186). Under normal conditions the endothelium is involved in the regulation of vascular tone and inflammatory and thrombogenic processes through the secretion of substances, such as prostacyclin and nitric oxide (NO) (36,46). Both prostacyclin and NO inhibit platelet activation and promote vasodilation (60). Nitric oxide also inhibits the expression of adhesion molecules responsible for the recruitment of inflammatory cells. Atherosclerosis disrupts the normal functioning of the endothelium, which leads to an increased permeability to plasma lipids and inflammatory cells (particularly monocytes and T lymphocytes). The combination of EDF and increased levels of atherogenic lipoproteins leads to the development of the early atherosclerotic lesion manifested as **foam cells**, which form pale yellow fatty streaks in the arterial intimal layer (36). These initial changes are likely

to occur at a very early age. Risk factors for the initiation of atherosclerosis have been identified in children as young as 2 yr of age (110). However, the progression from a foam cell to an **atheroma** occurs significantly more often and at younger ages in people living in developed, affluent countries.

As **atherosclerosis** progresses, vascular smooth muscle cells (VSMC) migrate from the medial layer of the vessel to form a fibrous cap over the lipid core through the production of extracellular matrix proteins such as collagen and elastin (36,176). The progression of atherosclerosis is hypothesized to depend on the existence and interaction of several factors:

1. Hemodynamic forces (e.g., elevated blood pressure, turbulent blood flow)
2. Levels of plasma atherogenic lipoproteins (e.g., low-density lipoprotein [LDL], lipoprotein-a [LPa], high-density lipoprotein [HDL])
3. Disruption of the intimal layer by free radical formation (i.e., oxidation), smoking (i.e., chemical injury), or high blood pressure (i.e., mechanical stress injury)

Atherosclerosis remains clinically silent until either of two mechanisms occurs (36). If the lesion size is increased to such an extent that blood flow becomes restricted and nutrient supply cannot meet demand, then symptoms of ischemia (such as angina) will occur. However, **plaque** growth does not always lead to lumen **stenosis** (176). As the plaque grows, the vessel remodels or expands, thereby preserving the lumen diameter and therefore blood flow. Significant narrowing of the lumen due to plaque buildup only occurs when the artery can no longer expand any further (36,176).

Alternatively, the lesion can fissure, rupture or ulcerate causing **platelet aggregation** and ultimately an intravascular thrombosis, leading to angina and/or MI (36,48,176). This process is most likely to occur in smaller lesions (i.e., <50% narrowing) that have not yet formed a fibrous cap (i.e., unstable lesions). In fact, 85% of all infarct related lesions are less than 75% occluded (Stary, 1995). This is in contrast to previous beliefs that thrombosis development most likely occurred in atheromas significantly narrowing the arterial lumen (e.g., >75% narrowing). Typically, these lesions are more stable and less likely to rupture (47,48,151,141).

The amount and severity of myocardial damage resulting from thrombotic occlusion depends on the location of the occlusion, whether blood flow is completely or partially disrupted, the duration of the blood flow disruption, the amount of collateral blood vessel circulation, and the state of myocardial oxygen demand. The more severe and prolonged the blood flow limitation the more likely infarction will occur, and this results in cellular death or necrosis. These cells lose wall integrity, resulting in the release of several proteins into circulation, such as troponin (I and T), creatinine kinase and its isoenzyme creatine kinase-MB, and lactate dehydrogenase (refer to diagnostic and laboratory evaluations) that are used in the diagnosis of MI. The loss of cell integrity results in the initial thinning of the myocardial wall, which is susceptible to rupture (i.e., aneurysm). A dense fibrous network of connective tissue replaces these cells over time, resulting in permanent scarring that improves the integrity of the infarct tissue. This area of the myocardium can no longer contribute to the contractile process and becomes either **akinetic** (i.e., without movement) or **dyskinetic** (i.e., moving in the opposite direction, or bulging). Four to six weeks after MI, the integrity of the myocardial wall in the infarcted tissue will have reached its strongest point. Other potential hazards during the MI recovery period include dysarrhythmias, **cardiogenic shock,** myocardial rupture, thrombosis, and heart failure (161).

MEDICAL AND CLINICAL CONSIDERATIONS

Often, the initial symptoms of coronary artery disease are exposed during a bout of physical exertion because of a mismatch of myocardial oxygen demand and supply. Investigations have reported that the relative risk of experiencing a MI is 10 times greater during vigorous exertion (i.e., >6 METS) than at other times, with the greatest risk among individuals that normally have a low level of physical activity (56).

Patients experiencing an MI can be identified by specific signs and symptoms exhibited during a physical examination. Most often these are determined in the emergency room. However, silent MIs are often discovered at the physician's office during a routine examination when an electrocardiogram (ECG) reveals **significant Q waves.**

Signs and Symptoms

Pain is the most common complaint in the patient experiencing a MI. Patients often describe their pain as a squeezing or burning sensation or a heavy or crushing discomfort in the chest that may radiate to the arm, shoulder, or neck regions (15). This symptom is the result of severe myocardial ischemia and is termed **angina pectoris.** If the patient has had previous episodes of angina, the dis-

comfort felt during the MI is typically similar in nature (i.e., location, description) but more intense. Other possible symptoms associated with an MI include:

Unusual shortness of breath (i.e., dyspnea)

Profound weakness

Profuse sweating (i.e., diaphoresis)

Loss of consciousness (i.e., syncope)

Confusion

Dizziness

Nausea

Often, during an evolving MI, patients become anxious and restless. Many, especially males, deny symptoms and do not seek immediate attention. They may do this, in part, because they do not consider the discomfort experienced before an MI painful, and therefore do not think that the discomfort is related to the heart. The AHA states that almost 50% of patients delay seeking medical assistance from the onset of symptoms by more than 4 h (161).

About a quarter of patients with an evolving anterior MI are tachycardic and hypertensive, likely the result of an elevation in sympathetic activity in response to a reduction in cardiac output. Elevated sympathetic activity increases heart rate and contractility to counteract a reduction in cardiac output.

About 50% of patients with an evolving inferior MI have an elevated parasympathetic nervous tone resulting in bradycardia and hypotension. Chapter 4 provides specifics regarding the cardiovascular physical examination.

Diagnostic Testing

When MI is suspected during the physical examination, the next step in the differential diagnosis involves diagnostic and laboratory testing. These tests are divided into three groups:

1. Laboratory indicators of tissue damage (biochemical markers)
2. ECG
3. Cardiac imaging

Blood analysis can be performed to determine whether tissue damage has occurred to the myocardium. Polymorphonuclear leukocytosis and **erythrocyte sedimentation** rates are both indicative of myocardial damage. The time course of polymorphonuclear leukocytosis is from immediately after the onset of pain until 3 to 7 d after resolution of the MI. The erythrocyte sedimentation rate also increases after an MI and remains elevated for 1 to 2 wk.

Myocardial injury can also be recognized by the appearance of various proteins (cardiac biomarkers) released into circulation when the myocardial walls lose their integrity (9). Generally in an unperfused heart the relationship between the concentration of biomarkers released into circulation and the degree of muscle tissue lost is directly proportional (the higher the concentration of the cardiac biomarker, the greater the loss of muscle tissue).

The most common biochemical markers used for MI diagnosis include troponin (I and T), creatinine kinase (CK) and its isoenzyme CK-MB, and lactate dehydrogenase (LDH) (9,78,184). Normal ranges for these biomarkers vary among laboratories, but typical ranges include 105–333 IU · L^{-1}, 24–170 IU · L^{-1}, 0–10.4 ng · ml^{-1}, and 0–00.4 ng · ml^{-1} for LDH, CK, CK-MB, and troponin, respectively.

The preferred biomarker for myocardial damage is troponin (I or T) because it is highly specific to myocardial tissue (8,9,132). After myocardial injury, troponin levels will rise within 4 to 6 h and peak at 12 to 24 h. Troponin I will remain elevated for 6 to 8 d, and troponin T will remain elevated for as long as 7 to 10 d following initial injury.

Both CK and CK-MB levels will rise within 4 to 6 h after myocardial injury. CK will peak at 12 to 36 h and remain elevated for 3 to 4 d, whereas CK-MB will peak at 12 to 24 h and remain elevated for only 2 to 3 d. An elevated CK can also be indicative of overall skeletal muscle damage and is not specific for myocardial injury.

LDH begins to increase later, at about 8 to 12 h, peaks around 2 to 3 d, and remains elevated for 7 to 10 d (78). Table 16.1 summarizes the typical cardiac biomarker response to acute myocardial injury.

Electrocardiographic changes associated with an MI are categorized as follows:

1. Acute
2. Evolving or resolving
3. Chronic

Infarctions are commonly described in two ways depending on the amount of myocardial tissue involved. Patients presenting with an acute **transmural** MI have ECG changes that include hyperpolarization of the T waves and ST-segment elevation (figure 16.1). Patients experiencing a subendocardial MI typically demonstrate ST-segment depression. A subendocardial MI can be differentiated from transient ischemia if the ST-segment depression persists despite a reduction in myocardial oxygen demand (e.g., stopping exercise or physical activity) and blood pressure (i.e., reduced afterload). Myocardial oxygen demand is estimated by the **double product** (heart rate and systolic blood pressure).

Infarctions are further described according to which of the coronary arteries are involved and the location in the ventricle. Anterior MIs, for example, result from lesions in the left anterior descending coronary artery and involve the anterior wall of the left ventricle. In contrast, inferior MIs usually result from lesions in the right coronary artery. Lateral, posterior, apical, and septal regions of the heart are other common infarct sites. Note that extensive MIs involving multiple sites can also occur (53).

During the **evolving MI** and **resolving MI** phases, Q waves develop. In addition, ST-segment changes return toward baseline, and the T waves may invert. The ECG during the **chronic MI** phase is characterized by fully developed Q waves and either inverted or resolved T waves (figure 16.2). This ECG is typically observed when a patient enters a supervised cardiac rehabilitation program. A portion of people (almost 30% of all transmural MIs) will not demonstrate significant Q waves following an MI (105). These MIs are termed non-Q-wave MIs. Note that the ECG is more variable and less specific than the other methods of MI diagnosis, including plasma enzyme elevation and cardiac imaging.

Cardiac imaging provides information regarding wall motion and blood perfusion. Because wall motion abnormalities are apparent in nearly all MIs, their detection is useful for diagnosis. Two-dimensional echocardiography is used to visualize the left ventricular walls and assess the

Table 16.1 Cardiac Biomarkers

Biomarker	Normal range	Initial rise (h)	Peak (h)	Elevation (d)
Troponin (I and T)	0–00.4 ng · ml^{-1}	4–6	12–24	6–8 and 7–10
CK	14–170 IU · L^{-1}	4–6	12–36	3–4
CK-MB	0–10.4 ng · ml^{-1}	4–6	12–24	2–3
LDH	105–333 IU · L^{-1}	8–12	48–72	7–10

Data from M. Kemp et al., 2004, "Biochemical markers of myocardial injury," *British Journal of Anesthesia* 93(1): 63-73.

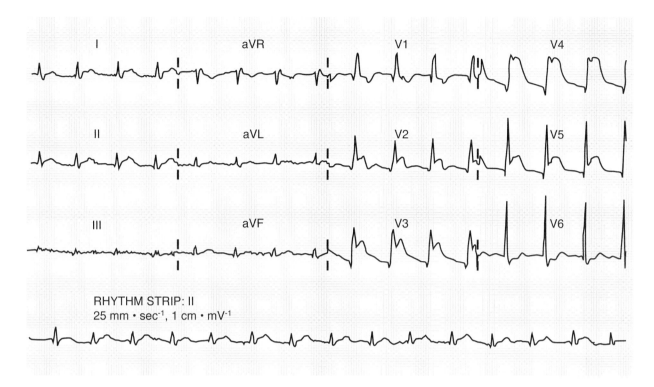

Figure 16.1 12-lead ECG.

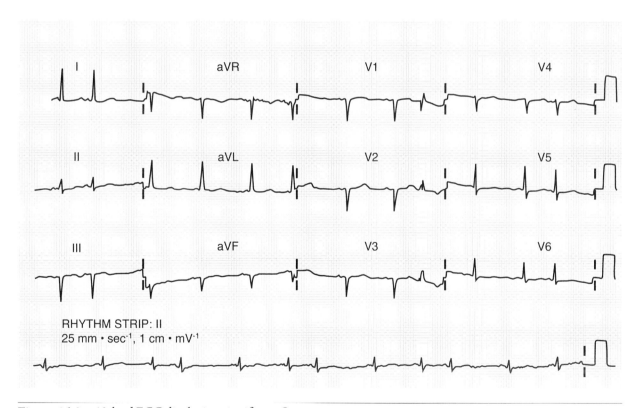

Figure 16.2 12-lead ECG displaying significant Q-waves.

amount of movement. Note that the echocardiographic analysis cannot distinguish between wall motion abnormalities caused by an old, healed MI and an acute MI. Radionuclide ventriculography, often termed multiple-gated analyses (MUGA), uses technetium-99m stannous pyrophosphate, which tags to red blood cells. By imaging tagged red blood cells, healthcare workers can assess wall motion and ejection fraction (EF). Both two-dimensional echocardiography and MUGA can be performed during exercise or pharmacological (e.g., with dobutamine or diprydimole) stress evaluation. Other procedures, such as computed tomography and magnetic resonance imaging, may also be used to assess ventricular wall motion.

Radionuclide imaging procedures are used to assess either blood flow defects or wall motion abnormalities. Myocardial perfusion is assessed with isotopes such as **thallium 201** or technetium-99m sestamibi. Once injected, these radioisotopes are taken up by the myocardium proportionally to blood flow. On the X-ray image (immediately following exercise or pharmacological stress), areas of limited isotope uptake are darkened and reflect areas in which blood flow is reduced or absent. Reimaging several hours later that continues to show limited uptake can reveal permanent tissue damage (MI). If areas that initially reflected limited isotope uptake are now showing good **reperfusion**, the presence of reversible ischemia is suggested.

Following an acute MI, the risk of future morbidity and mortality is determined by the existence of residual myocardial ischemia (evident as exertional angina or ST-segment changes on the ECG) and the extent of left ventricular damage or dysfunction. The patient's EF provides an indication of ventricular dysfunction (53). A normal EF is between 50 and 60% at rest (123). The relative risk of sudden death or cardiac arrest with resuscitation is two to three times higher in patients with an EF <30% as compared with patients with an EF >40%. In addition, each 5% reduction in EF increases the risk of these events by 21% within the first 30 d post MI. This risk has been shown to increase despite pharmacological intervention (figure 16.3) (148).

Prognostic Evaluations Post MI

An algorithm developed jointly by the American College of Cardiology (ACC) and AHA provides strategies to follow when evaluating a post-MI patient (55). The strategy chosen is based on whether the patient is considered at high risk of a future event. High-risk patients typically should have a cardiac catheterization performed before

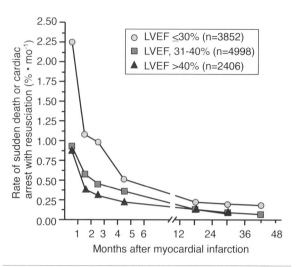

Figure 16.3 Rate of sudden death or cardiac arrest with resuscitation over the course of the trial in the three categories of left ventricular ejection fraction (LVEF). The analysis was restricted to patients for whom data on LVEF were available. The average rate (percentage per month) is shown at the midpoint of each period.

Adapted, by permission, from S.D. Solomon, 2005, "Sudden death in patients with myocardial infarction and left ventricular dysfunction, heart failure, or both," *New England Journal of Medicine* 352(25): 2581-2588. Copyright © 2005 Massachusetts Medical Society. All rights reserved.

hospital discharge to determine whether revascularization is warranted. Stable, lower-risk post-MI patients who have not had a cardiac catheterization should undergo a submaximal exercise evaluation 4 to 7 d post MI and before hospital discharge or a symptom-limited evaluation 14 to 21 d after discharge (55). Further diagnostic evaluation is then performed if the exercise evaluation is considered abnormal.

Exercise Testing

Graded exercise testing is used in the post-MI patient to evaluate prognosis and functional status. The clinical exercise physiologist must consider the status of a patient before testing and be knowledgeable of the exercise response in these patients. Practical application 16.1 reviews the process of interaction between the clinical exercise professional and the post-MI patient. The determination of whether to perform an exercise evaluation lies with whether the specific risks of the evaluation, based on an individual's status, outweigh the potential benefits. Given the array of possible scenarios with which each patient may present, the best approach is to assess each patient individually before making a decision about whether to perform an exercise test. To establish the level of risk, patients are stratified into low-, intermediate-, or high-risk groups based on medical history, physical assessment, and initial 12-lead ECG and cardiac markers (55). Guidelines for patient risk stratification developed separately by the AHA and the American Association of Cardiovascular and Pulmonary Rehabilitation (AACVPR) are reviewed in *ACSM's Guidelines for Exercise Testing and Prescription* (15).

Contraindications

The primary purpose of performing an exercise evaluation soon after an MI is to allow the detection of exertional myocardial ischemia. In patients who have recently experienced an acute MI, the AHA, the American College of Cardiology (ACC), and American College of Sports Medicine (ACSM) have published separate

recommendations with respect to the minimal amount of time that should elapse before an exercise test is performed. According to the ACC and AHA, exercise testing can be performed after the patient has been stabilized or is symptom free (with respect to ischemia and heart failure) for a minimum of 8 to 12 h or 2 to 3 d in low- and medium-risk patients, respectively. Exercise testing is contraindicated in patients with high risk for further ischemic events (55,137). The AHA, ACC, and ACSM also review absolute and relative contraindications to exercise (see chapter 5 on exercise testing).

Exercise test protocols

Exercise test protocols can be either submaximal or symptom limited. Submaximal testing is demonstrated to be safe in stable patients (63), but it is now used less

Practical Application 16.1

Client–Clinician Interaction

The post-MI patient, like most patients with a chronic disease, will come into contact with healthcare system personnel on numerous occasions. The patient referred to a cardiac rehabilitation program will, depending on length of stay in the program, make contact with the staff on 10 to 36 occasions over a 1 to 3 month period. This number rises to hundreds of contacts over a period of years in those who continue in a phase 3 or phase 4 cardiac rehabilitation program. Therefore, the potential to affect behavioral change may be greater for cardiac rehabilitation clinicians than the patient's physician because of the greater total amount of contact hours.

Cardiac rehabilitation staff must develop skills to use when approaching the post-MI patient to discuss behavior change. The early post-MI patient is often overwhelmed when dealing with issues related to the MI. These issues include fear of death or disability, depression, family support, and medication side effects. The cardiac rehabilitation professional must develop the ability to assess patient concerns. The clinician with this skill will be more likely to approach the post-MI patient at the right time, and in the right setting, to deal with issues that are both important and within the patient's ability to handle. A patient commonly must make many changes to reduce his or her risk of a mortal or morbid event. Doing this is likely to be an overwhelming task. The exercise professional must be able to address these issues with the post-MI patient in a manner that motivates the patient to make behavioral changes.

The initial patient contact is important. The clinician must first establish a relationship before discussing behavioral change. The exercise professional is more likely to be successful if he or she has an approach that is friendly, courteous, patient, kind, and empathetic. With this approach, a patient is more likely to be honest and open. In selected patients, several informal discussions may be necessary before serious issues can be discussed. Some patients may have difficulty discussing issues with particular exercise professionals, such as members of the opposite sex and those with large age discrepancies (i.e., a young exercise professional and an older patient).

Motivation to make behavioral changes will vary among post-MI patients. Some patients will be in a state of denial when entering a cardiac rehabilitation program, whereas others may have already committed themselves to change. Others will be somewhere between these extremes on the readiness-to-change continuum. An exercise professional may at times have to search to find what motivates a person to make a change. Examples include remaining healthy and active to play with grandchildren, improving fitness levels to return to full employment and leisure activities, and maintaining the will to remain alive. People are motivated in different ways, and the exercise professional must strive to cultivate an individual's motivational cues.

frequently than in the past (15). Treadmills or cycle ergometers are the most common modes of testing. The protocol is low level, with work rate increments of 1 to 2 metabolic equivalents (METs) every 2 to 3 min. Examples of commonly used low-level treadmill protocols are the modified Naughton, the modified Bruce, the Balke, and the Stanford (15,49). Blood pressure, ECG, and symptom monitoring should be performed throughout the exercise test. These tests are discontinued at a predetermined endpoint. Common criteria for test termination include any one of the following:

- A peak heart rate between 120 and 130 bpm
- Achieving 70% of predicted maximal heart rate
- A peak work rate of 5 METS (15,55)

A symptom-limited evaluation is different from a submaximal test because no endpoint is predetermined based on heart rate or work rate. These tests are designed to continue until the patient demonstrates signs or symptoms such as the following:

- Angina
- ST-segment depression ≥ 2 mm
- Hypotension (i.e., a decrease in systolic blood pressure ≥ 10 mmHg from rest)
- Ventricular arrhythmias
- Fatigue (55)

Symptom-limited testing has been found to induce an ischemic response nearly twice as often as submaximal tests, is a better estimation of functional capacity, and therefore is more useful for exercise prescription (55,69,70,73,168). But there are concerns regarding the prognostic value of ST-segment displacement seen in early post-MI patients, who are typically deconditioned, because evidence indicates that this may lead to an unnecessary cardiac catheterization (136). For example, an investigation of 236 post-MI patients who performed an exercise evaluation 3 wk after an acute MI reported that exercise-related parameters, such as ST-segment displacement and exercise duration, were poor predictors of cardiac prognosis (1,84). Therefore, these predischarge tests are often used only for functional assessment and hemodynamic evaluation purposes.

An early post-hospitalization exercise test should be performed 14 to 21 d postdischarge to assess prognosis and functional capacity if no predischarge evaluation was undertaken. If the patient did undergo a submaximal test before discharge, then a post-hospitalization test should be performed 3 to 6 wk following discharge (137). Post-hospitalization exercise testing is similar to the symptom-limited predischarge test with respect to test endpoints. The protocol used, however, is commonly set at a higher level. Stage increments in work rate range from 1 to 3 METs. The most common treadmill protocols used are the modified or standard Bruce, the Ellestad, and the Naughton (12). During symptom-limited testing, the ECG and blood pressure are assessed before each stage increment throughout the evaluation. Gas exchange assessment is useful in accurately quantifying exercise capacity and is important in the post-MI patient who develops heart failure (see chapter 5). Ratings of perceived exertion (RPE) provides an evaluation of the patient's subjective level of exertion and can be used later to guide exercise training.

The post-hospitalization test is better for predicting prognosis than the predischarge test. Indicators of prognosis including functional capacity, ST-segment changes, and blood pressure response can be used to determine prognosis and guide medical management (15,55). For instance, patients without indications of ischemia who achieve at least 5 METs are candidates for medical management. Those who become ischemic, show decreased systolic blood pressure, or cannot achieve 5 METs should be referred for cardiac catheterization, because their short-term prognosis, without intervention, is poor (55).

Besides standard ECG exercise testing, other procedures that may enhance the sensitivity of the test can be used. These procedures are typically used when the resting ECG demonstrates ST-wave abnormalities, as often seen in post-MI patients, especially those with ventricular conduction abnormalities associated with bundle branch block, left ventricular hypertrophy, Wolff-Parkinson-White, and digitalis therapy. Myocardial perfusion imaging (i.e., thallium, sestamibi) provides information regarding both infarct size and peri-infarct ischemia (i.e., ischemia in the tissue surrounding infarcted myocardium). Evidence suggests that infarct size is related to prognosis (90). Exercise echocardiographic evaluation improves both the sensitivity and **specificity** of standard exercise testing and is performed either immediately following exercise on a treadmill or cycle ergometer or following infusion of a drug (e.g., dobutamine or adenosine). A negative echocardiographic evaluation indicates reduced risk of future MI and overall death (81).

The response of the MI patient to exercise testing depends on factors such as the time course of the post-MI period, the amount of myocardial damage, the degree of left ventricular dysfunction, the pre-MI level of exercise or physical activity, and medications. The post-MI patient is commonly deconditioned. However, the ability of the skeletal muscles to extract oxygen and thus produce a large arteriovenous oxygen difference ($\dot{V}O_2$ =

cardiac output × arteriovenous oxygen difference) is not limited to a great extent in post-MI patients compared with healthy normal individuals (136). Therefore, the poor exercise ability of the post-MI patient is related to myocardial function. For example, a group of male patients evaluated at 3 wk post MI had a mean $\dot{V}O_2$peak of 20.5 ml · kg^{-1} · min^{-1} compared with 37.5 ml · kg^{-1} · min^{-1} in a group of healthy male subjects (65). Reduced peak exercise cardiac output, related to a comparatively low peak heart rate (137 ± 19 bpm versus 170 ± 13 bpm, respectively) and possible left ventricular dysfunction, were probably responsible for part of this difference. Similar $\dot{V}O_2$peak (19.4 ml · kg^{-1} · min^{-1}) and heart rate values (144 ± 22 bpm) were reported in another study of post-MI patients (109). In this study, peak cardiac output was 12 L · min^{-1}. By using this cardiac output value and the peak heart rate, researchers determined that peak stroke volume was 83 ml · beat^{-1}. These cardiac output and stroke volume values are lower than reported values in normally active individuals (22 L · min^{-1} and 112 ml · beat^{-1}, respectively). Thus, the reduced $\dot{V}O_2$peak noted in post-MI patients is directly related to depressed cardiac function (i.e., decreased cardiac output).

Treatment

In the setting of an acute MI, the primary focus of treatment is pain relief and reperfusion to salvage myocardium that would otherwise infarct. Chronic treatment focuses on revascularization procedures, medical management, and risk factor reduction.

Acute Treatment

Treatment of ischemic or anginal pain associated with MI includes sublingual nitroglycerin and morphine. Nitroglycerin relieves pain by reducing myocardial oxygen demand (reduced venous return) and increasing oxygen supply (coronary artery dilation). Common side effects include hypotension, **tachycardia**, and dizziness. Nitroglycerin should not be administered if the systolic blood pressure is below 100 mmHg. Morphine is effective at pain reduction through its narcotic effects. Several thrombolytic agents (e.g., aspirin, tissue plasminogen activator, and streptokinase) are available to lyse clots responsible for coronary artery occlusion. The goal is to reperfuse the affected tissue to salvage myocardium that otherwise would infarct. Supplemental oxygen should be used to increase oxygen supply in patients with an oxygen saturation <90%. A flow rate of 4 to 5 L · min^{-1} is commonly administered by nasal cannula (14,172).

Percutaneous coronary intervention (PCI) and coronary artery bypass surgery are also used as primary treatment strategies for acute MI. From a review of the literature, Kristensen et al. (80), Antman et al. (14) and Thun et al. (166) conclude that primary angioplasty substantially decreases both short- and long-term re-infarction, and to a lesser degree, decreases stroke and mortality in younger and older patients. However, this conclusion does not negate the importance of pre-hospitalization fibrinolysis. When administered within 2 h of symptom onset, this therapy is associated with a lower mortality rate than primary angioplasty predominantly because of increased myocardial salvage and a decrease in cardiogenic shock (13,153). Some suggest that revascularization procedures be used only after MI to prolong life or relieve unacceptable symptoms despite optimal medical management (16). Chapter 17 is devoted to revascularization.

Admittance to a coronary care unit should occur when the patient is clinically stable. Bed rest is typically used in the initial stages of an acute MI to reduce myocardial oxygen demand and the risk of a large infarct size. By the second to fourth day, however, patients should be regularly sitting up several times a day and beginning to ambulate because functional capacity decreases by 1% during each day of bed rest. Early ambulation will increase orthostatic and hemodynamic stress and reduce the amount of negative consequences (e.g., orthostatic hypotension, plasma volume loss) after an MI.

Chronic Treatment

Chronic (long-term) treatment of coronary artery disease can be subdivided into primary and secondary prevention. Primary prevention of coronary artery disease is an important measure put in place to reduce the risk of a first MI. Secondary prevention refers to activities that aim to prevent successive events from occurring.

Primary Prevention

The incidence of coronary artery disease is influenced by the individual's age, gender, and number of risk factors and signs or symptoms of cardiovascular disease. Individuals at an increased risk are: male, older (i.e., >45 yr for males and >55 yr for females), have two or more cardiovascular risk factors, and are symptomatic (i.e., have angina-like chest discomfort, dyspnea or orthopnea, exertional fatigue, dizziness or syncope, palpitations) (15). Positive risk factors for coronary artery disease include the following:

- Family history of MI, revascularization or sudden death in a first-degree relative (occurring before 55 or 65 yr of age in male and female first-degree relatives, respectively)

- Current cigarette smoking (including those who quit within the previous 6 mo)
- Hypertension (i.e., systolic blood pressure ≥140 mmHg, or diastolic blood pressure ≥90 mmHg, or taking antihypertensive medication)
- Dyslipidemia (i.e., LDL >130 mg · dl^{-1}, or HDL <40 mg · dl^{-1}, or total cholesterol >200 mg · dl^{-1}, or on lipid-lowering medication)
- Impaired fasting glucose (i.e., fasting glucose ≥100 mg · dl^{-1})
- Obesity (i.e., BMI >30kg · m^{-2})
- Inactive lifestyle (i.e., individuals not accumulating ≥30 min of moderate physical activity on most days of the week) (15)

Primary prevention revolves around risk factor modification. Regular exercise training or increased daily physical activity levels are especially useful for reducing the risk of coronary artery disease and MI (20,86,118). Exercise training has been shown to have multiple cardioprotective mechanisms. These include improved endothelial function, reduced plasma CRP levels, favorable modification of the positive risk factors for coronary artery disease, and potentially anti-ischemic and antithrombolytic effects (86). Regular exercise training may also improve thrombolytic effects during nonexercise periods.

Exercise, however, also slightly elevates the risk of an acute MI (179). Approximately 5 to 10% of MIs are associated with vigorous physical activity (162), and the relative risk of MI is 10 times greater during exertion than at other times (56). Each year approximately 75,000 people in the United States suffer an MI soon after performing vigorous exercise (104). This number, however, is fewer than 10 per 100,000 persons (164,171). Note that the relative risk of exercise-induced events is highest among people who are the least physically active and those who are unaccustomed to vigorous physical activity (56).

The mechanism of action of MI during exercise may be related to the **prothrombotic** effect of vigorous exercise. Exertion acutely alters the structure of the coronary arteries and increases the hemodynamic forces acting on the coronary arteries. These mechanical forces increase stress on the less flexible atherosclerotic areas and may contribute toward plaque rupture during vigorous exercise (56,130). Therefore, the person with existing coronary artery disease is at a higher risk of **thrombus** development during exercise than a person without coronary disease. Prophylactic aspirin therapy is commonly recommended for males over 40 yr of age to reduce the risk of coronary thrombosis, especially during exercise.

Secondary Prevention

A comprehensive secondary preventive approach focuses on reducing the risk factors of coronary artery disease in conjunction with optimized medical therapy. Secondary measures are effective at preventing MI in those with established coronary artery disease or previous MI (31). Long-term medical therapy typically includes β-blockers, angiotensin-converting enzyme inhibitors (ACE inhibitors), aspirin, anticoagulation therapy, and lipid management.

Extensive investigations have studied β-blockers in placebo-controlled trials in post-MI patients and unequivocally demonstrate reductions in the development of heart failure, cardiogenic shock, ventricular fibrillation, and all-cause mortality and sudden death (19,66,82,143,160). Evidence also shows that both fatal and nonfatal reinfarction rates are reduced (34,112,149). The use of β-blocker therapy should begin as soon as possible and continue indefinitely in those patients where its use is not contraindicated (147). Calcium channel blockers (i.e., calcium antagonists) may be useful for controlling ischemia. This result is achieved through a decrease in systolic blood pressure, because of a reduction in afterload through reduced vasoconstriction. Calcium channel blockers, however, have not been shown experimentally to reduce mortality rate and may in fact increase mortality rate in some post-MI patients (159). Therefore, they should only be used in patients who cannot tolerate β-blockade.

ACE inhibitors have proven effective at reducing the risk of heart failure development, reducing all-cause and cardiovascular-related re-hospitalizations and lowering all-cause mortality in post-MI patients with reduced ejection fractions (e.g., <40%) (2,24,87,89). The beneficial effect on mortality and hospitalization rates has been found to be maintained for at least 10 to 12 yr post MI (24).

Prolonged antiplatelet therapy reduces the risk of recurrent MI, stroke, or vascular death (64). Aspirin is the most superior antiplatelet agent in this population (14). Aspirin should be administered in the emergency room setting, and the patient should take a daily dose of 75 to 162 mg/d thereafter. Plavix (clopidogrel) or coumadin (Warfarin) anticoagulation therapy is effective in patients who cannot use aspirin (147). In addition, coumadin use is indicated in patients at risk of developing an atrial or ventricular thrombus.

Statin therapy has proven highly effective at lipid management and concomitantly reducing cardiovascular morbidity and mortality (50,129,138,154). From an extensive review of the literature, Wilt et al. (181) conclude that statin therapy reduced nonfatal MIs, all-cause mortality, and CHD mortality by 25%, 16%, and 23%, respectively. A comprehensive outline of the medications listed is discussed in chapter 3.

Table 16.2 lists the risk factors and generalized recommendations for risk factor alteration as outlined by the ACC and AHA (147).

Cardiac Rehabilitation

Secondary prevention in the post-MI patient should include a referral to a cardiac rehabilitation program (CRP) (14). A well-designed CRP provides the post-MI patient with a risk factor assessment, goal setting, education, and exercise therapy (59,86). The scientific evidence in favor of CRPs is strong and points toward the following:

1. Improved overall survival
2. Enhanced quality of life and self-image
3. Reductions in revascularization procedures
4. Reductions in subsequent nonfatal MI
5. Improved exercise tolerance and functional capacity
6. Beneficial effects on lipid levels, hypertension, symptoms of angina and dyspnea, weight loss, smoking behavior, and stress level
7. Return to a normal lifestyle (31,59,86,121,152)

In particular, CRPs can reduce total mortality by 20% and cardiac related mortality by 26% (158). Cardiac rehabilitation is implemented in three phases. Phase 1 begins in the hospital before discharge. The primary goal is to deter the negative effects of bed rest and introduce lifestyle changes with the aim of minimizing the risk of a future cardiac event (15). Phase 2 commences shortly after discharge and continues for 4 to 12 wk depending on the rehabilitation facility and insurance reimbursement (145,165). Currently, Medicare covers exercise programs for patients who have been referred by a doctor who have had

- an MI within the last 12 mo,
- coronary bypass surgery,
- stable angina pectoris,
- angioplasty,
- valvular surgery, or
- cardiac transplant (98).

During phase 2, patients are encouraged to adopt a healthier lifestyle by altering behaviors that may predispose them to further events. Phase 3 of cardiac rehabilitation is a maintenance program usually carried out at home or in the community and designed to continue indefinitely. The main goals are continued focus on physical fitness and further risk factor reduction (145).

Table 16.2 Secondary Prevention for Patients With Coronary Artery Disease

Action item	Recommendation
Smoking	Complete cessation
Lipid management	LDL <100 mg · dl^{-1} (optional goal <70 mg · dl^{-1}), HDL >40 mg · dl^{-1}, triglycerides <200 mg · dl^{-1}, non-HDL <130 mg · dl^{-1}
Physical activity	Minimum goal is 30 to 60 min 5 d per week (optimal goal is daily activity)
Blood pressure	<140/90 mmHg or <130/80 mmHg if patient has chronic kidney disease or diabetes
Weight control	BMI between 18.5 and 24.9 kg · m^{-2}, waist circumference: women <35 in. (89 cm), men <40 in. (102 cm)
Diabetes management	Initiate lifestyle and pharmacotherapy to achieve near-normal fasting plasma glucose (i.e., Hb$_{A1C}$ <7%)
Antiplatelet or anticoagulants	Start and continue aspirin 75 to 162 mg · d^{-1} indefinitely. If aspirin is contraindicated, consider clopidogrel 75 mg · d^{-1} or warfarin (INR of 2.0–3.0)
Renin-angiotensin-aldosterone system blockers	In patients with left ventricular ejection fraction (LVEF) ≤40%, hypertension, diabetes, or chronic kidney disease, start and continue ACE inhibitors indefinitely. Consider angiotensin receptor blockers (ARBs) in patients who are intolerant to ACE inhibitors and have heart failure or LVEF ≤40%. Aldosterone blockade is indicated in patients without significant renal dysfunction or hyperkalemia, who are already receiving treatment with an ACE inhibitor and β-blocker, have a LVEF ≤40%, and have either diabetes or heart failure
β-blockers	Start in all patients and continue indefinitely, observing the usual contraindications

Adapted from S.C. Smith et al., 2006, "AHA/ACC guidelines for secondary prevention for patients with coronary and other atherosclerotic disease: 2006 update," *Circulation* 113: 2363-2372.

Smoking cessation should focus on complete discontinuation of all types of smoking, including exposure to environmental tobacco smoke (147). Smoking can cause coronary artery spasm, adversely affect β-antagonist drug effects, and increase mortality rate in post-MI patients (79,111). The rates of reinfarction and mortality are significantly reduced following smoking cessation (68,88,131,150,175). Results from a large meta-analysis suggest that long-term smoking cessation following MI reduces the relative risk of death by 15 to 61% (180). Continued behavioral management is important because of the high rate of relapse of smokers who quit (up to 50% within 1 yr). Most patients who are successful at quitting smoking do so only after several attempts. Typically, the willingness to discontinue cigarette smoking is greatest during and immediately after the acute MI. But this period may be extremely difficult for the patient, who commonly loses focus on the behavior modification necessary for lifestyle changes.

Lipid management should be initiated in post-MI patients presenting with a LDL level greater than 100 mg · dl^{-1} or total cholesterol level greater than 200 mg · dl^{-1}, as outlined by the National Cholesterol Education Program (62). Both LDL and total cholesterol levels are predictive of future MI in patients with coronary artery disease and are related to the progression of atherosclerosis (28,62,120,134,142,183). Post-MI patients who exercise train demonstrate slight reduction in LDL cholesterol, significant reduction in triglycerides, and elevation of HDL cholesterol (17,33,167). Chapter 14 provides a comprehensive review of methods for treating lipid disorders.

To reduce cardiovascular risk, blood pressure values are recommended to remain below 140/90 mmHg (or below 130/80 mmHg if the patient also has chronic kidney disease or diabetes) (147). Hypertension is an independent predictor of morbidity and mortality rates in post-MI patients, and aggressive blood pressure control can significantly reduce cardiac-related death in this population (30,75,107). Chapter 13 reviews treatment strategies for hypertension.

According to the AHA (2006), obesity is becoming a global epidemic in both children and adults. Obesity, defined as a BMI value greater than or equal to 30 kg · m^{-2}, is associated with increased risk of cardiovascular disease, morbidity, and mortality (124). Chapter 12 reviews obesity and weight-loss methods.

Physical inactivity is generally believed to be related to increased risk of future MI in patients who have had a previous infarction (174). The Corpus Christi Heart Project showed that patients who remained active or increased their activity after their initial MI decreased all-cause mortality rates by 79% and 89% respectively compared with patients who remained sedentary (152).

Additionally, regular physical activity is useful in treating several risk factors related to reinfarction including hypertension, obesity, smoking, diabetes, and dyslipidemia (30,124,174). Patients who are more physically active are also more likely to be successful at modifying their behavior as it relates to the cardiovascular risk factors. Finally, research shows a trend toward a reduction in nonfatal reinfarction with exercise training. For example, patients who increased their activity following an MI decreased their risk of nonfatal reinfarction by 78% (152). But this evidence is equivocal because others have reported that people with a previous MI are not protected from the recurrence of a nonfatal MI following cardiac rehabilitation (94,114,116). Even so, the benefits of exercise for reducing the risk of cardiac and all-cause death and improving quality of life make exercise training an important part of the therapeutic regimen for the post-MI patient (86,94,158,165).

Compared with other treatment strategies, cardiac rehabilitation services are considered highly cost-effective for post-MI patients (117,121,128,165,185). The reduction in all-cause mortality rate as a result of CRPs is similar to the reduction provided by β-blockade therapy in post-MI patients (25).

Unfortunately, despite these benefits, few patients are referred to CRPs, particularly exercised-based CRPs (86,165). Of the eligible patients, the estimated participation rate is only 10 to 20% per year (7). The factors contributing to the poor referral rate include

- poor patient motivation,
- inadequate reimbursement,
- poor accessibility, and
- limited public awareness of CRPs and their benefits (7,165).

To rectify underutilization, home-based programs have been suggested as an alternative setting for CRPs (86). Evidence suggests that home-based aerobic exercise is beneficial in the post-MI population, particularly in the elderly (38,39,72,83,91,144). Studies suggest that post-MI home-based cardiac rehabilitation is more cost effective, has more prolonged positive effects, and may be preferable in low-risk and older (>75 yr of age) low-risk patients. The assignment of lower-risk patients to home-based CRPs would allow for increased access to hospital-based CRPs for medium- and high-risk patients (91,144).

EXERCISE PRESCRIPTION

The exercise prescription for the post-MI patient should follow standard procedures presented in chapter 5. Specific modifications and considerations should be made when indicated. Open communication is an effective tool for assessing and educating the post-MI patient.

Phase 1 of the CRP exercise prescription should focus on providing an orthostatic workload to counteract the negative effects of bed rest. For example, functional capacity ($\dot{V}O_2$max) is reduced by 1% for every day of bed rest (37). Orthostatic stress is achieved by having the patient spend time in an upright position, either seated or standing. Stable patients should be progressed to this as early as possible. Orthostatic activity loads the cardiovascular system by activating the sympathetic nervous system. The result is an increase in heart rate and blood pressure (afterload), which increases myocardial oxygen consumption. Patients with residual ischemia or other acute MI complications must be carefully and slowly progressed to lengthening bouts of upright posture. Ambulation should begin on the second or third day after the patient becomes clinically stable.

While they are hospitalized, all patients should begin a range-of-motion program. This program is designed to keep the joints and skeletal muscles from losing compliance and to maintain flexibility, which is often lost during prolonged bed rest. Range-of-motion exercise can be started passively by an exercise professional while the patient is in bed or sitting in a chair. Specifically, improved range of motion will allow greater ease of movement and reduce the risk of skeletal muscle injury. An exercise professional (e.g., clinical exercise physiologist, exercise specialist, physical therapist) should meet with the patient to discuss the importance of early ambulation and flexibility with regard to preparation for the patient's return home. See chapter 5 for specifics of proper range-of-motion exercise.

The optimal amount of exercise for these patients depends on their medical history, clinical status, and symptoms. During the hospitalization course, the patient should be progressed from walking several feet to several hundred feet (a hundred meters) two to four times a day. Initially, the patient may not be physically capable of carrying out continuous ambulation, so the duration usually begins with intermittent bouts lasting 3 to 5 min with rest periods interspersed. These rest periods can include walking at a slower pace or complete rest, as tolerated (15).

Structured exercise-training sessions (i.e., phase 2 of cardiac rehabilitation) should start as soon as possible following hospital discharge (121,165). In developing an exercise prescription for the post-MI patient, the exercise professional should consider the patient's

- current functional capacity and level of risk;
- exercise response (ischemia, blood pressure, heart rate, arrhythmias, RPE, work rate);
- exertional symptoms (angina, dyspnea, claudication, dizziness); and
- individual goals.

Most of these factors can be determined through an exercise stress test, which is recommended prior to starting the CRP (see prognostic evaluations post MI). But depending on the particular program philosophy, a post-MI patient may begin and even complete a 4 to 12 wk phase 2 CRP without ever having a maximal exercise evaluation.

The post-MI patient whose conditions are uncomplicated can begin moderate exertion within 1 wk following discharge (165). When determining the type of setting where a patient will exercise, the clinical exercise physiologist must consider the drawbacks of each setting and the goals of the patient. For instance, the hospital-based or community center–based cardiac rehabilitation setting is less convenient, but in cardiac rehabilitation undertaken at home, the patient has no daily interaction with staff and other patients. Regardless of which setting is selected, standard exercise prescription guidelines should guide the exercise training. Practical application 16.2 on p. 294 provides exercise-training recommendations for post-MI patients.

At this time not enough information is available to make recommendations regarding patients at risk. Reports suggest, however, that post-MI patients are most vulnerable to cardiac rupture in the initial period following the acute phase of an inferior MI. The literature contains no reports of myocardial rupture during the outpatient rehabilitation phases. In fact, Mineo et al. performed low-level exercise training (<4–6 METs), without complication, in two patients who suffered previous uncomplicated myocardial rupture (103).

A more controversial risk is whether or not exercise is beneficial or harmful to patients with exertional ischemia. Two theories describe the effects of exertional ischemia:

1. Prolonged exertional ischemia causes transient myocardial stunning. This circumstance can result in loss of cardiovascular function, demonstrated by depressed ejection fraction and elevated end-diastolic and end-systolic volumes attributable to myocardial cell hibernation (10,119).

2. Ischemia during regular exertion results in preconditioning, which is associated with the development of collateral circulation and reduced risk of MI and improved survival post MI (22,76).

Exercise-Training Recommendations

Aerobic and resistance training may be best initiated in a monitored setting (e.g., cardiac rehabilitation). The clinical exercise physiologist should consider comorbid conditions such as hypertension, diabetes, chronic obstructive pulmonary disease, and peripheral arterial disease when developing and implementing an exercise program.

Type	Modes	Intensity	Frequency and duration	Special considerations
Aerobic	Walking, cycling (leg ergometry or outdoors), stepping (seated or upright), elliptical trainer, swimming or other water activities, upper-body ergometry	40 to 60% HRR (can be increased to 85% HRR if tolerated) Use RPE of 11 to 13 (on 6–20 scale) or a HR 20 bpm above rest if no exercise evaluation was performed	One or two bouts per day, 5 d per week, 30–45 min per session	Adjust intensity for ischemia or hemodynamic instability if necessary. Goals are to increase endurance and to reduce risk of a secondary event and mortality. Supplement with lower-intensity physical activity on nonexercising days. Consider timing of exercise test and training time of day in β-blocked patients, because HR response may be altered
Resistance	Machines or light free weights	40 to 60% of direct or indirect 1RM or weight lifted comfortably 8 to 15 times. RPE of 12 to 13 (on 6–20 scale)	2 to 3 d per week, 8 to 10 exercises (working the major muscle groups), 10 to 15 repetitions (10–20 min per day)	Refrain from resistance exercise for 2 to 5 wk post-MI. Avoid Valsalva maneuver. Goals are to increase skeletal muscle strength and endurance. Circuit training may be used by selected patients
Flexibility	Passive	Hold to the point of mild discomfort	Daily for 5 to 10 min	Work throughout the full range of movement for each major joint. Avoid ballistic stretching. Avoid breath holding. Goals are to improve flexibility and reduce injury risk

Note. HRR = heart rate reserve.

Adapted from American Association of Cardiovascular and Pulmonary Rehabilitation, 2004, *Guidelines for cardiac rehabilitation and secondary prevention programs,* 4th ed. (Champaign, IL: Human Kinetics); L. Armstrong et al., 2006, *ACSM's guidelines for exercise testing and prescription,* 7th ed. (Philadelphia: Lippincott, Williams and Wilkins); G.F. Fletcher et al., 2001, "Exercise standards for testing and training: a statement for healthcare professionals from the American Heart Association," *Circulation* 104:1694-1740; M.L. Pollock et al., 2000, "Resistance exercise in individuals with and without cardiovascular disease: Benefits, rationale, safety, and prescription. An advisory from the committee on exercise, rehabilitation, and prevention, council on clinical cardiology, and the American Heart Association," *Circulation* 101: 828-833.

Safety Considerations With Exercise Training

Although CRPs have many benefits, the risks associated with the exercise component should not be ignored. Despite low incidence, the major complications during exercise include cardiac arrest, MI, myocardial rupture, and death (21,49,63,71,73,165).

Because this issue of myocardial stunning versus collateral development and recruitment during exercise training remains controversial, the prudent approach is to remain on the conservative side when training post-MI patients with inducible ischemia. As little ischemia as possible should be allowed with the assumption that repeated bouts of ischemia can be harmful. Refer to the cardiovascular training section for safely setting the upper limits of exercise.

The incidence of a sudden cardiac event during exercise depends on the type and intensity, and the use of continuous ECG monitoring. Low-intensity exercise such as walking or cycling, and continuous monitoring has been found to have the lowest incidence (49). According to the AHA, the average occurrence of cardiac arrest, nonfatal MI, and death is approximately 1 in every 117,000, 220,000, and 750,000 patient hours of exercise, respectively (162).

Therefore, although complications during exercise can arise, the occurrence of these are rare and CRPs can be determined as safe for the majority of post-MI patients.

Cardiovascular Training

Post-MI patients improve their physical work capacity through endurance training. In randomized controlled trials, post-MI patients participating in this type of exercise demonstrated greater improvements in functional capacity than non-endurance-trained post-MI patients (26,39,40,61,68,90–92,101,102,106,113,115, 133,150,157,173). An increase of 10% to 30% in $\dot{V}O_2$peak is common following 8 or more weeks of exercise training (4,26,38,40,42,43,45,51,54,61,68,91–93,108,115, 146,150,156,173,178). The mechanism for enhanced functional capacity is related to increases in stroke volume, cardiac output, maximal oxygen consumption, minute ventilation, exercise time, work rate, lactate threshold, and a widened arteriovenous oxygen difference (4,43,44,108,121). Echocardiographic findings also show that exercise training will increase ejection fraction and reduce end-systolic volume, suggesting improvement in left ventricular contractile function in post-MI patients (43,45).

Following hospital discharge, patients commonly attend cardiac rehabilitation at least 3 d per week and exercise for a minimum of 30 min per session (15). These patients should also begin to increase their amount of physical activity and exercise at home. The upper end of the frequency range for exercise prescription may be increased to 6 or 7 d per week and the duration may be increased to 40 to 60 min in selected patients. But the increased risk of musculoskeletal injury and cardiovascular events should be considered (126,162).

Each exercise-training session should begin with a 5 to 10 min warm-up period that consists of active aerobic exercise at a reduced intensity level compared with the actual training session. The warm-up may reduce the risk of arrhythmia and symptomatic ischemia during exercise training. Active cool-down invokes the skeletal muscle pump of the lower legs, which assists the recovery hemodynamic response and allows for a safer and faster return to preexercise cardiovascular status. Insufficient cool-down can result in lower-limb venous pooling leading to hypotension, dizziness, syncope, arrhythmias, and myocardial ischemia (49,99).

Typical modes of aerobic exercise focus on walking but may also include other forms. To ensure optimal compliance with the program, a needs analysis should be performed to determine the modes of exercise that the patient enjoys or would like to partake in.

Various methods can be employed to monitor exercise intensity. When a symptom-limited exercise evaluation has been performed, the exercise intensity can be guided by using a HR–based method with adjustment for ischemia or other signs and symptoms. Practical application 16.2 provides exercise intensity recommendations for post-MI patients. Reduced exercise intensity may be prescribed in selected patients, such as those who are severely deconditioned and those limited by cardiac complications. In these patients, positive adaptations can occur at exercise intensities as low as 40% of HRR. A goal is to increase exercise intensity slowly to 60% to 85% HRR as the training effect occurs (15,163).

In those CRPs in which patients have not had a maximal exercise evaluation, the exercise intensity should be based on the following ACSM recommendations:

- A rating of perceived exertion (RPE) of 11 to 13 (6–20 scale)
- A heart rate (HR) not exceeding 20 bpm above the resting HR (15)

Many patients who exercise independently find pulse monitoring without a HR monitor difficult, and this circumstance often causes unnecessary anxiety. An alternative method used to guide intensity is the talk test. Patients are instructed to exercise to the onset of dyspnea or to the highest work rate that still permits comfortable conversation. The intensity should then be maintained at or slightly below this level (163). Brawner et al. (23) found that patients with coronary artery disease who exercise at this intensity are typically working at 65% of their **heart rate reserve** (HRR), which is similar to the HR elicited at the ventilatory threshold and is within the recommended training range for cardiorespiratory fitness. However, approximately 12.5% of individuals may be able to exercise above 85% of HRR and continue to talk comfortably. Because of this individual variation in both comfortable talking and RPE, an objective variable should be used (e.g., heart rate or work rate), when available, to guide exercise intensity in post-MI patients, particularly in the immediate postdischarge period.

Another common approach is to monitor the post-MI patient electrocardiographically, by telemetry, for the duration of phase 2 participation. In fact, for those with Medicare insurance, this method is a requirement for reimbursement (98). But the trend now is to move away from continuous ECG monitoring, especially in low-risk and non-exercise-induced **ischemia** patients. Investigations have demonstrated that ECG monitoring during phase 2 cardiac rehabilitation resulted in few significant findings and little physician referral or other intervention (74,100).

Regardless of the methods employed for monitoring exercise intensity, signs and symptoms of exercise intolerance should be closely monitored in post-MI patients while they perform exercise. Safe exercise intensity is a HR 10 bpm below the HR associated with the following signs and symptoms:

- Angina
- Hypotension (i.e., plateau or decrease in systolic blood pressure)
- Hypertension (i.e., systolic blood pressure >250 mmHg or diastolic blood pressure >115 mmHg)
- Ischemia (i.e., horizontal or downsloping ST-segment depression ≥1 mm)
- Increased frequency of ventricular dysrhythmias or other significant ECG disturbances (e.g., high-degree atrioventricular block, atrial fibrillation, supraventricular tachycardia)
- Onset of left ventricular dysfunction or moderate to severe **wall motion** abnormalities
- Other signs or symptoms of exercise intolerance (e.g., nausea, dyspnea, cyanosis, dizziness, pallor skin color, confusion, cold or clammy skin, palpitations) (15)

After several weeks, the post-MI patient should progress to increased intensity and duration of exercise training and physical activity. The ACSM recommends that the rate of progression be based on an individual's functional capacity. Table 16.3 provides an example of exercise progression using intermittent exercise in

those patients with functional capacity above and below 4 METs.

After the initial 4 to 12 wk of exercise training, much of the potential physiological adaptation will have taken place. Nevertheless, lifelong exercise training and increased physical activity should be encouraged. Training at this point can be carried out in a supervised setting such as a phase 3 CRP or in a nonsupervised setting such as at home or at a public fitness facility. Patients who are candidates for a nonsupervised setting include those who have

- an estimated exercise tolerance or functional capacity (FC) of ≥7 METs, or a measured FC ≥5 METs, or a FC two times greater than their occupational demand;
- no indications of myocardial ischemia (i.e., <1 mm ST-segment depression);
- stable or absent cardiac symptoms;
- no severe arrhythmias or signs of exercise intolerance; and
- self-motivation to continue exercise training, knowledge about the disease process, and knowledge about how to exercise train safely (15).

Resistance Training

Resistance training increases skeletal muscle strength and endurance. Additionally, Ades et al. (6) found that an intense resistance-training program in elderly females with coronary artery disease improves the ability to perform activities of daily living, balance, coordination, and flexibility. General resistance-training guidelines and modalities are presented in chapter 5.

The ACSM, AACVPR, and AHA have published separate guidelines that provide indications for resistance training in cardiac diseased populations (11,15,125). According to the AHA, post-MI patients should refrain from resistance training for 2 to 3 wk following their event (125). The ACSM and AACVPR, however, suggest that the patient wait a minimum of 5 wk and should undertake 4 wk of consistent endurance training in a supervised CRP before beginning resistance training (11,15). The general criteria for entry into a resistance-training program include approval from a medical director or personal physician, absence of or controlled hypertension (i.e., systolic blood pressure <160 mmHg or diastolic blood pressure <100 mmHg), peak work capacity of at least 5 METs, stabilized heart failure symptoms and arrhythmias, and no evidence of severe valvular disease (11,15,125). Practical application

Table 16.3   Exercise Progression for Cardiac Patients Using Intermittent Exercise

Functional capacity (FC) >4 METs					
Week	Exercise intensity (%FC)	Total minutes at %FC	Minutes spent exercising	Minutes spent at rest*	Reps
1	50–60	15–20	3–5	3–5	3–4
2	50–60	15–20	7–10	2–3	3
3	60–70	20–30	10–15	Optional	2
4	60–70	30–40	15–20	Optional	2
FC ≤4 METs					
Week	Exercise intensity (%FC)	Total minutes at %FC	Minutes spent exercising	Minutes spent at rest*	Reps
1	40–50	10–15	3–5	3–5	3–4
2	40–50	12–20	5–7	3–5	3
3	50–60	15–25	7–10	3–5	3
4	50–60	20–30	10–15	2–3	2
5	60–70	25–40	12–20	2	2
6	Continue with two reps of continuous exercise with one rest period or progress to a single exercise bout				

Note. The time spent at rest can include complete rest or exercise at a lower intensity.

Adapted from L. Armstrong et al., 2006, *ACSM's guidelines for exercise testing and prescription,* 7th ed. (Philadelphia: Lippincott, Williams and Wilkins).

16.2 provides resistance-training recommendations for post-MI patients.

The type of resistance used will vary from patient to patient. Weight machines, dumbbells or hand weights, barbells, pulley weights, elastic bands, and calisthenics can be adapted for most patients (125). Because of the general level of deconditioning, post-MI patients will likely gain strength and endurance with the initiation of any resistance-training program.

The intensity level should be set at a resistance that can be comfortably handled for 10 to 15 repetitions, the patient's RPE should not exceed 11 to 13 (6 to 20 scale), and breath-holding or the Valsalva maneuver should be avoided (15,125). During resistance exercise the Valsalva maneuver can reduce venous return because of elevated intrathoracic pressure, which results in an initial elevation of mean arterial pressure that increases afterload and myocardial work. Within a few seconds, a secondary reduction in cardiac output occurs, caused by reduced venous return, and mean arterial pressure subsequently decreases. This combination of events can be problematic for the post-MI patient with residual ischemia or reduced left ventricular function. Normal breathing (i.e., exhalation during the exertion phase of the lift) during exercise reduces these risks.

Some investigations have used a range of 40% to 60% of 1RM for the initial intensity range (15,35,41,96,

156). The direct or indirect 1RM method may be used (see chapter 5) (5,41). Lighter resistance training (i.e., <20 lb [9 kg]) can begin early in the rehabilitative process (e.g., in the latter stages of phase 2 CRPs) and progress slowly to greater intensity after 4 to 12 wk of training. The resistance should be increased by approximately 2 to 5 lb · wk^{-1} (0.9 to 2.3 kg · wk^{-1}) and 5 to 10 lb · wk^{-1} (2.3 to 4.5 kg · wk^{-1}) for the arms and legs, respectively (15).

Resistance training should not be performed daily or on consecutive days in post-MI patients, because most are deconditioned and prone to excessive fatigue, injury, and noncompliance when the exercise load becomes too great (125,127). To reduce the risk of noncompliance, the following are suggested:

- Begin with light resistance to reduce the risk of skeletal muscle soreness.
- Progress the training load slowly and within the patient's capability.
- Keep training sessions to 15 to 45 min.

If a circuit-type program is used, cardiac patients should begin with one circuit per session and gradually progress to two or three circuits per session. Focus should be on the major muscle groups (see chapter 5).

A clinical exercise professional should regularly assess each patient for indications of maladaptations and progress the patient when appropriate. As with aerobic exercise, the resistance-training session should be discontinued if any indicator of exercise intolerance occurs. A potential danger of resistance training in the post-MI patient is a markedly elevated systolic or diastolic blood pressure. Systolic blood pressure increases myocardial oxygen demand, and large increases may produce ischemia. However, the rate pressure product during resistance training does not typically exceed values at peak aerobic exercise until the load reaches at least 80% of 1RM. Conversely, elevated diastolic pressure enhances coronary blood perfusion and may satisfy elevations in myocardial oxygen demand by increasing oxygen availability to the myocardial tissue.

Overall, resistance training in the post-MI patient appears to be safe and effective. Stewart et al. (156) reported no complications in a group of post-MI patients who performed a circuit program of 10 to 15 repetitions at 40% of their 1RM. But avoidance of resistance training may be prudent in unstable patients, and patients with impaired left ventricular function should be monitored closely (49,125).

CONCLUSION

Suffering an MI is one of the primary causes of hospitalization and death worldwide, and as such, is a major economic burden. Regular exercise has been shown to mediate the atherosclerotic process and therefore to protect against both initial and subsequent MIs, as well as mortality from all causes. Therefore, habitual exercise training is an accepted and important part of the rehabilitative process for patients who have had an MI.

Case Study

Medical History

Mr. RG is a 48 yr old white male who previously had an unremarkable medical history. He is sedentary and has been for most of his adult life. He works as a foreman in an auto factory. He has smoked two packs of cigarettes a day for 30 yr, and his body mass index is 28.4. His total cholesterol is 274 mg · dl^{-1}, LDL is 145 mg · dl^{-1}, and HDL is 24 mg · dl^{-1}. He does not know his typical blood pressure value.

Diagnosis

Mr. RG is presented to the emergency department with substernal and left arm discomfort and diaphoresis. This condition woke him approximately 1 h before he usually arose and worsened on the drive to work. He saw the nurse at the factory and was then transported by ambulance to the hospital. Sublingual nitroglycerin and aspirin were given on the way to the hospital to treat his symptoms. The patient was immediately given morphine on arrival at the hospital. His blood pressure was 180/110 mmHg. An ECG demonstrated ST-segment elevation and tall T waves in leads V1 through V6 and II, III, and aVf. Blood was drawn and stat analyzed for creatine phosphokinase and lactate dehydrogenase. These values were both elevated above resting values. From this information, the diagnosis of acute myocardial infarction was made. The plan was to evaluate for thrombolytic versus primary angioplasty therapy.

Mr. RG underwent percutaneous coronary intervention (PCI) with stent placement of his right coronary artery and left coronary artery 2 h after arrival at the hospital. The circumflex artery also had an 80% blockage that was anatomically difficult and thus was not revascularized. A subsequent in-hospital dobutamine echocardiography examination revealed a lateral wall motion abnormality at the highest dose of dobutamine.

Exercise Test Results

Following stabilization and release from the hospital, Mr. RG was referred for cardiac rehabilitation. The patient began phase 2 at 2 wk post MI and was subsequently scheduled for an exercise test to evaluate his functional status, prognosis, and ischemic status and to develop an exercise prescription.

He walked on a treadmill using a modified Bruce protocol. His medications were metoprolol, Zocor, aspirin, and captopril. His ECG was interpreted as anterior and inferior MIs with repolarization abnormality. The following are the results:

Protocol: Bruce

Time: 10:46 (stage 4 of standard Bruce protocol)

Resting heart rate = 88

Resting blood pressure = 148/94 mmHg

Peak heart rate = 136

Peak blood pressure = 226/100 mmHg

$\dot{V}O_2$peak = 21.4 ml $\cdot$ min^{-1} $\cdot$ kg^{-1}

Symptoms: angina at peak exercise (+3/4); shortness of breath; right calf "cramping"

ECG: ST-segment depression = 1 mm horizontal at heart rate = 115 and 2.5 mm horizontal at peak exercise in leads V5, V6, and aVL

Development of Exercise Prescription

Mr. RG wishes to progress to jogging and has a goal of running a marathon. He is a candidate for coronary artery bypass surgery but wishes to manage his condition medically. He has retired from his job and appears willing to make the necessary changes to control his risk of a future event.

Discussion Questions

1. What would be general recommendations for a home-based exercise program immediately following hospitalization?

2. What is an appropriate exercise prescription for Mr. RG for the initial stages of phase II cardiac rehabilitation?

3. Based on your knowledge of Mr. RG's medical history, what might be his greatest challenges? How would you handle this in a cardiac rehabilitation setting?

4. Assuming that Mr. RG improves as expected during the phase 2 program, what might you recommend for him to progress to jogging and subsequently running a marathon?

5. What are Mr. RG's risks during exercise?

6. What is the significance of Mr. RG's symptoms during his exercise evaluation?

7. How would you approach Mr. RG for behavior modification of his coronary artery disease risk factors?

17

Revascularization of the Heart

Timothy R. McConnell, PhD

Karandeep Singh, MD

When a person has coronary artery disease, clinical procedures may be elected to restore myocardial blood flow with the specific intent of symptom relief and improved morbidity and mortality. Over the past decade, referral patterns to cardiac rehabilitation for patients having coronary artery revascularization have evolved beyond patients who have undergone surgical revascularization with coronary artery bypass surgery (45). Today, because of advances in coronary invasive technology, cardiac rehabilitation programs must also prepare for a growing number of people who have experienced percutaneous interventions including percutaneous transluminal coronary **angioplasty** alone and, more frequently, in combination with **stent** therapy, which involves the placement of a mesh tube along the artery wall to prevent reocclusion (1,24). Even though minor convalescent differences exist among the different percutaneous interventional procedures, the standards of practice and expected outcomes for cardiac rehabilitation are similar (13,45); thus, this chapter focuses on patients who have undergone coronary artery bypass surgery or percutaneous transluminal coronary angioplasty with or without stent therapy.

CORONARY ARTERY BYPASS SURGERY

Coronary artery bypass surgery (CABS) involves revascularization by using a venous graft from an arm or leg or an arterial graft (both free, not intact such as a section, or from a regional intact native vessel, e.g., **internal mammary, gastroepiploic**) to provide blood flow to the myocardium beyond the site of the occluded or nearly occluded area in a coronary artery. Although CABS has traditionally involved a **sternotomy** and the use of a **heart and lung bypass**, technical advances now permit a growing number of procedures to be performed:

1. Procedures performed through small port incisions (or "minisurgery") using microscopic procedures

2. Procedures performed with robotic technology

3. Surgery performed on the beating heart without the use of cardiopulmonary bypass

Subsequent to these technical advances, postoperative morbidity has significantly decreased. The postsurgical hospital stay for CABS patients without complications is now less than 5 d.

As a result of the evolution of revascularization procedures (particularly percutaneous intervention), the role for CABS has changed, now being reserved for the following patients:

1. Patients who are post–percutaneous transluminal coronary angioplasty (PTCA) with restenosis
2. Patients who are no longer candidates for angioplasty but still have target vessels offering preservation of left ventricular systolic function
3. Those with multivessel disease not amenable to angioplasty
4. Those with technically difficult vessel lesions, for example, on the curve of a vessel or in a distal location not readily amenable to angioplasty

Subsequently, the number of surgical revascularization procedures has declined but still play an essential role for higher risk occlusive disease. Successful CABS results in an increased functional capacity (19 to 29%), improved cycle ergometer or treadmill performance, variable improvements in left ventricular function, increased maximal heart rate, increased rate pressure product, and a reduction in ST-segment depression (14,27). Combined with medical therapy, CABS may more effectively relieve significant residual exercise-induced symptomatic or silent myocardial ischemia. Thus, the symptom relief, improved functional capacity, and improved quality of life may be the most practical and important patient benefits of CABS.

PERCUTANEOUS INTERVENTIONS

Coronary angioplasty, which is also called percutaneous transluminal coronary angioplasty (PTCA), is less invasive than CABS. Several techniques have been developed for use in restoring adequate blood flow in diseased coronary arteries. Often the procedure is combined with stent therapy to reduce the likelihood that the artery will reocclude.

Percutaneous Transluminal Coronary Angioplasty

PTCA is a well-established, safe, and effective revascularization procedure for patients with symptoms attributable to coronary artery disease. The procedure may use one or more techniques alone or in combination to open the vessel:

1. Balloon dilation is most commonly used in conjunction with stent placement (figure 17.1).
2. Rotational atherectomy may be applied to central bulky lesions (figure 17.2) in a minority of cases.
3. Directional atherectomy and laser may be used to debulk large lesions, but the risk of vessel wall perforation or dissection may be greater. The use of these devices is limited to a few centers in contemporary practice.

The complications of angioplasty are acute vessel closure (rebound vasoconstriction) or chronic restenosis, thrombotic distal **embolism,** myocardial infarction (MI), arrhythmias, and bleeding.

Stent Therapy

To reduce the risk of acute closure and restenosis of coronary arteries after PTCA (20), several models of intracoronary stents have been advocated. Stent therapy is frequently used in conjunction with one of the previously described techniques to preserve the patency of the vessel. Stents are stainless steel mesh tube bridges that are

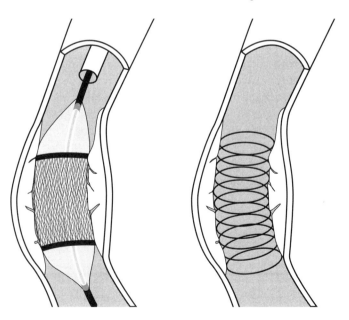

Figure 17.1 Percutaneous transluminal coronary angioplasty balloon catheter and two types of stents: latticed steel (left) and the coiled stent (right).

Reprinted, by permission, from T.E. Feldman and R.M. Gunnar, 1998, "Revascularization after CABG: Atherectomy and stenting," *Hospital Practice* 33(1):43.

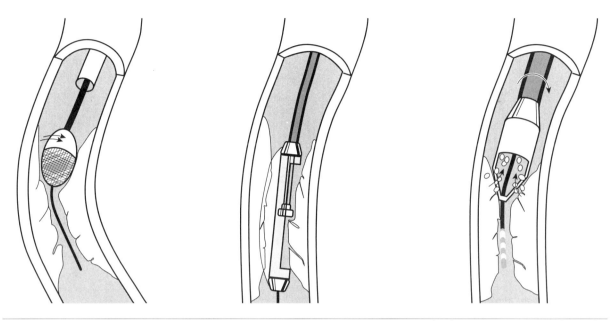

Figure 17.2 Intracoronary atherectomy procedures: rotational device. Catheter (left); directional device (center); and extraction catheter (right).

Reprinted, by permission, from T.E. Feldman and R.M. Gunnar, 1998, "Revascularization after CABG: Atherectomy and stenting," *Hospital Practice* 33(1):43.

advanced on the end of a balloon catheter, passed across the **culprit lesion**, and expanded. The stent, serving as a permanent intravascular prosthesis, compresses the lesion, resulting in an open vessel. This is the final treatment following balloon angioplasty or debulking in over 95% of the cases. After removal of the balloon catheter, the stent remains permanently in the coronary artery and is eventually covered with endothelium, becoming part of the luminal wall structure.

A quantum leap in the last 3 yr has been the availability of drug-eluting stents. The metal scaffolding prevents acute closure and also provides a vehicle for local drug delivery. This innovation has reduced the late problem of instant restenosis to a great extent (35,44,46).

Many stent procedures require a one-night hospital stay or are a same-day procedure. The loss in functional capacity following percutaneous coronary intervention (PCI) is less than that following bypass procedure. Subsequently, PCI patients begin cardiac rehabilitation at higher functional capacity, quality of life, and **self-efficacy** (14,27).

MEDICAL AND CLINICAL CONSIDERATIONS

The success rate of a revascularization procedure may be predicted, in part, by the patient's age, other existing comorbidities, and severity and location of the lesion.

Coronary Artery Bypass Surgery

Elective CABS improves the likelihood of long-term survival in patients who have the following:

- Significant left main coronary artery disease
- Three-vessel disease
- Two-vessel disease with a proximal left anterior descending stenosis
- Two-vessel disease and impaired left ventricular function (1)

For those experiencing failed angioplasty with persistent pain or hemodynamic instability, acute MI with persistent or recurrent ischemia refractory to medical therapy, **cardiogenic shock**, or failed PTCA with an area of myocardium still at risk, revascularization by CABS offers effective relief of angina pectoris and improves the patient's quality of life (1). The 1 yr occlusion rate of the grafts is approximately 15%, and the ensuing years have only a small additional annual occlusion rate (2). The CABS patient's postoperative education should include wound care, appropriate management of recurring symptoms, and risk factor modification.

Percutaneous Transluminal Coronary Angioplasty

In select cases of unstable angina, PTCA has an acute success rate of 84% to 90%. Following successful PTCA, restenosis occurs in approximately 25% of patients, almost always within the first 6 mo (1,3). The several predictors of restenosis are listed in order of importance in table 17.1 (36).

Patients who have had PTCA in the setting of unstable angina should have close surveillance following hospital discharge and should be advised to seek prompt medical attention in the event of a recurrence of the symptoms that were occurring before their PTCA (16).

Stent Therapy

Improved technology currently gives procedural success rates in excess of 95% in most centers. Acute closure and restenosis remain as limitations to short- and long-term success, respectively, although the incidence of both these complications has decreased dramatically in recent years. The incidence of thrombosis and acute closure is in the range of 1 to 2% with use of thienopyridenes (clopidogrel or ticlopidine); therefore, chronic **anticoagulation therapy** is no longer required. Restenosis rate ranges from 25% to 40% with bare metal stents and less than 10% with drug-eluting stents (3,29).

EXERCISE TESTING

The graded exercise test (GXT) is commonly used for continued diagnosis of possible ischemic myocardium, prognostication, and the establishment of functional status for exercise prescription purposes. Although an integral component for exercise prescription, the timing of the GXT is somewhat controversial. Standard administration procedures and contraindications to testing, discussed in chapter 5, should be followed. Practical application 17.1 outlines exercise testing for revascularized patient.

Coronary Artery Bypass Surgery

Requiring all patients to have an exercise test after successful bypass surgery for the purpose of beginning a supervised and monitored exercise program is of questionable clinical benefit and an unnecessary financial burden (31). The patient's coronary anatomy is known, and unless surgical complications or postsurgical symptoms are present, the chance of detecting unknown ischemia is extremely low. In addition, because of the acute convalescent period, the patient may not be able to give a physiological maximal effort, sacrificing test sensitivity.

A more opportune time for testing the patient is after incisional pain has resolved, blood volumes and hemoglobin concentrations have normalized, and skeletal muscular strength and endurance have improved from participating in low-level exercise and activity. At least 3 to 4 wk postsurgery, the patient will be able to give a near-maximal physiological effort resulting in test results with greater diagnostic accuracy for assessing functional capacity, determining return-to-work status, or recommending the resumption of physically vigorous recreational activities. For patients whose surgical revascularization was not successful or who are experiencing symptoms suggestive of ischemia, a clinical exercise test before starting an exercise program is recommended. All testing procedures should follow professional guidelines (4-6) as noted in chapter 5.

Table 17.1 Predictors of Restenosis After PTCA–Stent

PTCA	Stent
Degree of residual stenosis after PTCA	Lesion eccentricity
Diameter of the parent vessel	Diameter of the parent vessel
Number of diseased vessels	Type of vessel stented (artery versus vein)
Degree of reduction of the stenosis	Location of stent in vessel
Presence or type of coronary dissection	Presence of multiple stents
Presence of documented variant angina	Recurrence of unstable angina

Note. All predictors are positively associated with risk for the revascularized vessel to reocclude. PTCA = percutaneous transluminal coronary angioplasty.

Exercise Testing

Test type	Mode	Protocol specifics	Clinical measures	Clinical implications	Special considerations
Cardiovascular	Treadmill Cycle (if treadmill not possible)	Bruce, Ellestad (for younger or more physically fit individuals) Naughton or Balke-Ware (for older, deconditioned, or symptomatic patients) Ramping protocols appear more tolerable for many patients Pharmacologic for those unable to exercise	Heart rate and rhythm Blood pressure 12-lead electrocardiogram Symptoms Rating of perceived exertion Nuclear or echocardiographic imaging as prescribed Gas exchange analysis as prescribed	Rhythm disturbances Hemodynamics Myocardial ischemia Ischemic threshold Perception of work difficulty Ischemic myocardium LV function Dyspnea with exertion	CABG: Chest and leg wounds (4 to 12 wk for complete healing) PTCA: reocclusion—recurrence of previous symptoms
Strength	Isometric Isotonic Isokinetic	Peak force RM procedures as described 4RM or 8RM Peak torque	Maximal strength of the muscle or muscle group tested Blood pressure Heart rate Symptoms	Functional fitness	CABS: Incisional healing No Valsalva
Range of motion	Trunk flexion	Sit-and-reach	Posterior leg and lower-back flexibility	Functional fitness	Orthopedic complication that may preclude testing

Testing procedures outlined in *ACSM Guidelines for Exercise Testing and Prescription, Seventh Edition*.

Valve Surgery

- Disease or condition: For severe regurgitant (leaking) or stenotic (narrowing) valve disease, valve replacement may be essential for symptom relief and improved exercise tolerance. Regarding the open-heart surgical process, precautions similar to those for the bypass surgery patient should be followed with consideration to a few special considerations.

- Special considerations: Symptom resolution may not occur immediately following surgery. Symptoms may gradually resolve because of heart remodeling after valve replacement, such as aortic stenosis. Avoid strength training for severe stenotic or regurgitant valvular disease. Isometric exercises are not recommended. Common exercise-induced symptoms include shortness of breath, fatigue, and dizziness or light-headedness.

- Exercise prescription: Follow procedures similar to revascularization surgery.

Percutaneous Transluminal Coronary Angioplasty

Debate exists regarding the proper timing of stress testing in PTCA patients. Several reports of acute thrombotic occlusion associated with exercise testing shortly after successful PTCA have been reported, although these have not borne out as relevant in clinical practice. Although no chest pain is reported during the test, ischemia within 1 h after testing has been reported (8,40). The mechanisms for the apparently abnormal test responses are unclear but are possibly related to the following:

1. Higher levels of **platelet aggregation** during exercise testing

2. An increase in **thromboxane** A2

3. Platelet activation and hyperreactivity increase during exercise

4. Increased arterial wall stress associated with increased coronary blood flow

5. The higher blood pressure that occurs during exercise, which may traumatize an already disrupted **intima** (8,23)

On the other hand, exercise testing of patients with PTCA has been accepted standard practice, particularly for those with incomplete revascularization. A large body of evidence supports the use of early postprocedure exercise testing (1–2 d) to evaluate the functional status of the PTCA patient (7,42).

Stent Therapy

Controversy with regard to safety of early testing after stent placement has essentially been laid to rest. Most authorities now accept that performing a stress test after coronary stenting is safe. The accuracy of these tests, however, is still debated. In particular, there have been reports of false-positive stress results early on after coronary stenting. As with CABS and PTCA, the need to test all patients after stent therapy before starting cardiac rehabilitation is debatable. For the successfully revascularized patient, an exercise test may be redundant and may not provide any further useful clinical information before the patient starts a supervised exercise program. The primary concerns with PTCA and stent are **reocclusion** and **restenosis**. Subsequent restenosis may not be detected immediately following the procedure. The patient can exercise in a supervised exercise program and be tested at a later date if symptoms recur or for assessment of functional capacity before return to work.

Additionally, a poor response to exercise training, no improvement in functional capacity, may be indicative of restenosis (28).

EXERCISE PRESCRIPTION

Many body composition changes (loss of lean body mass) occur within the first few days of bed rest (43), supporting the need for early exercise intervention during hospital admission. Patients who perform typical ward activities and moderate, supervised ambulation do not suffer the magnitude of loss in lean body tissue as those who remain inactive. Early standing and low-level activities including range of motion and slow ambulation may be all that are required to deter postsurgical lean body mass loss while in the hospital (43).

After hospital discharge, many positive physiological adaptations occur in revascularized patients who participate in a supervised exercise program (45):

- Improved cardiac performance at rest and during exercise
- Improved exercise capacity of 20% to 25%
- Greater total work performed
- Improved angina-free exercise tolerance, much of which is attributable to peripheral muscular adaptations (18)
- Improved neurohumoral tone (25)

Patients in such a program gain the following:

- They more often achieve full working status.
- They have fewer hospital readmissions.
- They are less likely to smoke at 6 mo following completion of exercise therapy (5).

When we compare the physiological and psychosocial outcomes between CABS, PTCA–stent, and MI patients at the beginning and end of 12 wk of cardiac rehabilitation, some group trends are apparent. CABS patients may begin with lower functional capacities and lower ratings of **quality of life** and self-efficacy attributable to the surgical recuperative process, but they obtain greater improvement during the program and obtain similar or greater values than other cardiac patients at program completion, regardless of age (9,14,27). This result may reflect lower rates of **ischemia** than MI patients, greater confidence in their ability, and the potential psychological feeling that "something was done" about their heart disease and that they are "cured." Regardless of age, the CABS patients demonstrate functional improvement but may require a longer training period to obtain the same magnitude of effect (39). The PTCA–stent groups have not suffered the loss in functional capacity because of the more prolonged recuperative process following a myocardial infarction or bypass surgery and have greater functional capacity when starting cardiac rehabilitation (14,27).

Exercise prescription guidelines for **revascularization** patients have been published by the American College of Sports Medicine (6), American Association of Cardiovascular and Pulmonary Rehabilitation (5), and American Heart Association (4).

Special Exercise Considerations

Although revascularized patients are equally knowledgeable about risk factors as post-MI patients, they are less compelled to make changes (15). Post-MI patients initiate considerably greater lifestyle changes than revascularized patients do (34). Patients undergoing revascularization may be less motivated to adhere to risk factor behavior change because of a perception of being less sick or cured, which has a negative effect on compliance with risk factor modification (16,21,33).

Revascularized patients may encounter, or anticipate, restrictions differently than other cardiac patients do. Most PTCA patients are capable of resuming normal activities of daily living following hospital discharge. But patients frequently perceive considerable restrictions after the procedure including all activities of daily living—leisure activities, sexual activity, and early return to work (11,37,41).

Depression remains prevalent in patients with coronary heart disease after major cardiac events (CABS and PTCA included). Cardiac rehabilitation does reduce the prevalence and severity of depression. Therefore, cardiac patients should be routinely screened and offered the benefits of comprehensive cardiac rehabilitation including psychosocial support and pastoral care (32).

Spouses may be more likely to seek information about the patient's psychological reactions and recovery, whereas patients are more likely to seek information about their physical condition and recovery (33). Patients tend to be more positive than spouses, who tend to be more fearful of the future (17). Also, patients and spouses differ in their views on the causes of coronary artery disease and about the responsibility for lifestyle changes and the management of health and stress (38). Therefore, assessing both patient's and spouse's educational needs is important. "Healthcare Considerations for PTCA

Patients" lists important concerns in the rehabilitation of the post-PTCA patient. Exercise may be contraindicated for patients who continue to be symptomatic postevent or postprocedure, particularly at low workloads (<5.0 **METS**).

Coronary Artery Bypass Surgery

Primary concerns for the CABS patient when entering outpatient cardiac rehabilitation is the state of incisional healing and sternal stability, hypovolemia, and low hemoglobin concentrations. During the initial patient interview, the rehabilitation professional needs to ensure that the surgical wound has no signs of infection, significant draining, or instability. Questions should focus on the following:

Excessive or unusual soreness and stiffness

Cracking or motion in the sternal region

Whether the patient is sleeping at night

How the patient's chest and leg incisions are responding to his or her current activities of daily living since discharge

Healthcare Considerations for PTCA Patients

- Awareness of other cardiac risk factors
- Control of hypertension, obesity, and smoking
- Progressive exercise and weight reduction
- Counseling services for weight reduction, stress management, and smoking cessation
- Maintaining close contact between health professionals
- Identification of stressful factors
- Organizing and maintaining long-term follow-up records
- Reinforcing the noncurative nature of PTCA as a cardiac treatment modality
- Encouraging revascularization patients, and PTCA patients in particular, to take a proactive approach to improve health outcomes

Adapted, by permission, from B. Gaw, 1992, "Motivation to change lifestyle following percutaneous transluminal coronary angioplasty," *Dimensions of Critical Care Nursing* 11: 68-74.

Also, knowing how the patient performed during the inpatient program may help determine how soon he or she can begin the outpatient program and at what level the patient can begin exercising. For example, was the patient out of bed, upright, and walking soon after surgery without problems? If not, was the patient's lack of activity attributable to extreme physical discomfort, clinical or orthopedic difficulties, or lack of motivation?

Historically, surgical patients did not begin cardiac rehabilitation for 4 to 6 wk postsurgery or longer and avoided upper-extremity exercise for even longer periods. Today, standard practice is for patients to begin the outpatient program soon after discharge, many times within a week of surgery. For the uncomplicated revascularized patient, light upper-extremity exercises are now prescribed, including range-of-motion exercises, light hand weights progressing to light resistive machinery, and gradually progressive upper-extremity ergometry beginning at zero resistance.

Percutaneous Transluminal Coronary Angioplasty

The primary concern for the PTCA or stent patient is restenosis. At the patient's initial orientation session, questioning should address the presence of signs or symptoms indicative of their anginal equivalent. Education should include knowledge of symptoms, including anginal equivalents, management of angina (e.g., how to use nitroglycerin, going to the emergency department), precipitating factors (exertion or anxiety related), and care of the catheter insertion site. Patients with PTCAs and stents may begin the outpatient program as soon as they are discharged from the hospital or immediately following the procedure if it is performed on an outpatient basis.

Exercise training may alleviate the progression of coronary artery stenosis after PTCA by inhibiting smooth muscle cell proliferation, lowering serum lipids, and causing hemostatic changes (26). Aerobic training for 30 to 40 min, four to six times per week for 12 wk, improves treadmill time and myocardial perfusion and reduces the restenosis rate at 3 mo following PTCA (26,27). Other benefits include an improved sense of well-being, relief of depression, stress reduction, and sleep promotion.

As a result of angioplasty with improved techniques of revascularization, more patients with low risk profiles are being referred to cardiac rehabilitation (i.e., patients with a greater exercise capacity, no evidence of ischemia, normal left ventricular function, and no arrhythmias). Specific examples include patients who are younger, have single-vessel disease, and did not experience an MI before their PTCA. Regarding exercise prescription, these individuals may be treated similarly to apparently healthy individuals with the addition of education concerning the recognition of anginal equivalents, self-monitoring, self-care, and risk factor modification. Optimized medical therapy along with appropriately prescribed exercise training can be an alternative approach to interventional strategies in selected patients who are asymptomatic (19). When PCI is the therapy of choice, it should be combined with daily physical exercise and increased physical activity to optimize success (19). Cardiac rehabilitation results in early and sustained improvement in quality of life and is highly cost effective (47). In addition, angioplasty patients commonly experience restenosis. Supervised exercise training and education will improve recognition of signs and symptoms associated with closure. Most important, angioplasty patients need instruction concerning appropriate exercise training, dietary modifications, medications, and general risk factor reduction to slow or reverse the coronary disease process.

Because the PTCA patient remains on complete bed rest while the sheath is in situ for approximately 18 to 24 h, the immobilization often causes back pain. Appropriate flexibility exercises that enhance ROM often help to resolve low-back pain.

Stent Therapy

Because stents are placed by using the same catheter procedure as in the PTCA, the same considerations exist. But the risk for thrombosis is greater following stent therapy. Consequently, patients are often placed on anticoagulant therapy for preventive purposes. Although no specific contraindications preclude exercise following recent stent placement, proceeding with similar caution is prudent.

Exercise Recommendations

Over the past decade, the average length of hospital stay for cardiovascular patients has decreased and is still decreasing by more than 1/2 d per year. Currently, the hospital stay for uncomplicated cases of CABS is usually 2 to 5 d. For PTCA–stents, the stay is 1 to 2 d, or they are done on an outpatient basis, managed in an acute recovery suite, and discharged on the same day. Although cardiac rehabilitation begins as soon as possible during hospital admission, the shorter length of hospital stay has changed the inpatient program to basic range-of-motion exercises and ambulation, and the educational focus is on discharge planning—teaching about medications, home activities, and follow-up appointments. Educational topics previously covered in the inpatient setting are now the responsibility of the outpatient program. Moreover, cardiac rehabilitation professionals must make every effort to enroll patients in an outpatient program. Practical application 17.2 details exercise prescription for the revascularized patient.

Cardiovascular Training

The initial exercise prescription is based on information gained from the patient's orientation interview to the outpatient cardiac rehabilitation program. Patients are questioned concerning the presence of signs or symptoms, their activity while in the hospital, and their activity level since their return home from the hospital.

Initially, the patients are closely observed and monitored to establish appropriate exercise intensities and durations that are within their tolerance. A starting program may include treadmill walking (5–10 min), cycle ergometry (5–10 min), combined arm and leg ergometry (5–10 min), and upper-body ergometry (5 min). Initial intensities approximate 2 to 3 METs (multiple of resting oxygen uptake of 3.5 ml kg^{-1} · min^{-1}). The patient's heart rate, blood pressure, rating of perceived exertion, and signs and symptoms are monitored and recorded. Programs are

Practical Application 17.2

Exercise Prescription for the Revascularized Patient

Training	Mode	Frequency	Intensity	Duration	Special considerations
Aerobic	Treadmill Walking Cycle Combined arm and leg exercise Rowing Stepper Combination of above or others to insure adequate utilization of major muscle groups and distribution between upper and lower extremities	Daily	Asymptomatic: 70-85% of HRmax RPE: 11–15 Symptomatic: Below ischemic threshold RPE: 11–15	At least 30 min May be intermittent (3 × 10 min, etc.) or continuous depending on patient tolerance	Initially, need to be concerned with incisional discomfort in chest, arm, and leg of surgical patient. May need to restrict upper-extremity exercises until soreness resolves. Also, for those with PCI, there may be some groin soreness at the catheter insertion site that may restrict certain physical movements.
Resistance	Elastic bands Hand weights Free weights Multistation machines Equipment selection is based on patient progress. A rational progression is the equipment in the order listed	Two or three times per week	Select a weight where the last repetitions feel somewhat or moderately hard without inducing significant straining (bearing down and breath holding)	8 to 10 repetitions	For surgical patients the initial upper-extremity exercises may be range of motion without resistance—progressing initially with elastic bands or 1 to 3 lb (0.5 to 1.5 kg) increments. Further progression depends on sternal healing and stability. Exercises should be selected that employ muscle groups involved in lifting and carrying.
Warm-up and cool-down exercises	Range-of-motion and flexibility exercises	Daily	Static stretching	5–10 min	Exercises should emphasize major muscle groups, especially lower-back and posterior leg muscles.

gradually titrated during the initial sessions to a rating of perceived exertion of 11 to 14 in the absence of any abnormal signs or symptoms. In general, exercise intensity is progressed by 0.5 to 1.0 MET increments (i.e., 0.5 mph [0.8 kph] or 2.0% grade on the treadmill or 12.5–25 W on the cycle). The rate of progression is based on the patient's symptoms, signs of overexertion, rating of perceived exertion, indications of any exercise-induced abnormalities, and prudent clinical judgment on the part of the cardiac rehabilitation staff. Those with greater exercise capacities (PTCA–stent patients with no MI) are started according to their exercise capacities and progressed more rapidly. The selection of exercise modality depends on the person's program objectives. For example, those who are employed in a labor-type occupation or perform many upper-extremity activities at home will spend a greater portion of their exercise time doing upper-extremity exercises. If specific limitations preclude certain exercise modalities, program modifications are made that allow more time on tolerable equipment to obtain the greatest cardiovascular and muscular advantage. For patients who have an exercise test, standard recommended procedures for exercise prescriptions are followed (4–6).

Resistance Training

Muscular strength and endurance exercise training should be incorporated equally with cardiovascular endurance and flexibility exercise training during the early outpatient recovery period. Following revascularization, low-risk patients can perform muscular strength and endurance exercise training safely and effectively (22). Depending on the patient's clinical and physical status, successful approaches for upper- and lower-extremity strength enhancement include 10 to 12 repetitions with a variety of types of equipment that may include elastic bands, Velcro strapped wrist and ankle weights, hand weights, and various multistation machines. Usual guidelines include maintenance of regular breathing patterns (avoiding **Valsalva maneuvers**), selection of weights so that the last repetition of a set is moderately or somewhat hard, and progression when the perception of difficulty decreases. CABS patients may start range-of-motion exercises with light weights of 1 to 3 lb (0.5–1.5 kg) within 4 wk of surgery as long as sternal stability is ensured and excessive incisional discomfort is not present. PTCA–stent patients may start resistance training immediately. Exercises should be selected that will strengthen muscle groups used during normal activities of daily living for lifting and carrying and occupational or recreational tasks (1).

Weights are selected that allow the completion of 10 repetitions with the last 3 repetitions feeling moderately hard. Those who cannot securely hold hand weights should use wrist weights with Velcro straps. Exercises are selected that use upper-extremity muscle groups involved in routine lifting and carrying (figure 17.3). The lower-extremity exercises are used for patients with low functional capacity, and the exercises are performed with weights that wrap around the ankle with Velcro straps. For low functional capacity patients who have difficulty with the exercises described, elastic bands of different thicknesses are used. Patients are progressed from hand weights to hydraulic resistance machines, again using resistances that result in a perception of difficulty of moderately hard for two or three sets of 10 to 12 repetitions.

The potential benefits of resistance training in the revascularized population include improving muscular strength and endurance and possibly attenuating the heart rate and blood pressure response to any given workload (less strain on the heart). General resistance-training guidelines for cardiac rehabilitation are presented in "Patient's Guide for Resistance Training." Risk stratifying patients to determine eligibility is important.

Range of Motion

Each exercise sessions begins with a series of range-of-motion and flexibility exercises designed to maintain or improve the range of motion around joints and maintain or improve flexibility of major muscle groups (see figure 17.4). The exercises begin in the standing position (or seated in a chair, if difficulty standing) with

---

Patient's Guide for Resistance Training

- Choose a weight that can be comfortably lifted for 12 to 15 repetitions.
- Avoid tight gripping during pushing, pulling, and lifting exercises.
- Do not hold your breath during the activity. Exhale during the exertion phase and avoid straining.
- Perform two to three sets of each exercise and train three times per week.
- Rest 30 to 45 s between sets.
- Increase weight modestly (1–2 lb, or 0.5 to 1.0 kg) after you can easily perform 15 to 20 repetitions of a given weight.

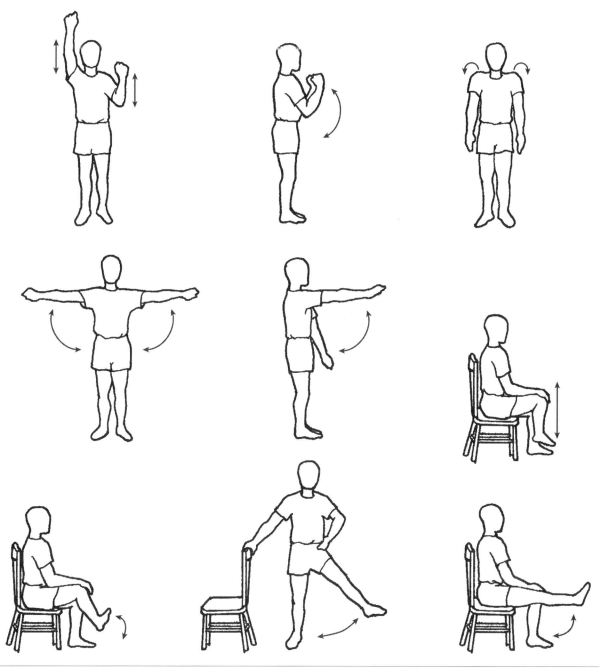

Figure 17.3   Resistance-training exercises.

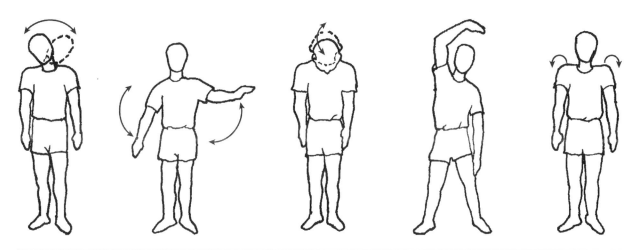

Figure 17.4   An exercise for all seasons: stretching.

the neck progressing downward to the shoulders and trunk and eventually to the lower extremities. The final seated stretches for the improvement of posterior leg muscles and lower-back flexibility may be performed on the floor or in a chair for those with difficulty getting to the floor.

CONCLUSION

Advances in coronary revascularization procedures and an aging population have led to a greater number of patients presenting for rehabilitation following CABS and PCTA. In addition to exercise programming, risk factor modification is essential for prevention of recurrent events (10,30). Furthermore, barring no new symptoms, the GXT may better serve its purpose of assessing functional status if it is postponed until later in the rehabilitation program. Provided that no untoward events occur over the course of rehabilitation, CABS and PTCA patients usually outperform their MI counterparts, achieving greater fitness improvements at a faster rate.

Case Study

Medical History

Mr. TW is a 65 yr old white male; procedures included cardiac catheterization and off-pump two-vessel CABS.

The patient had two previous admissions within the last month for unstable angina. Following PTCA and stent placement at the first admission, he had early recurrence of angina and an anterior-septal infarction. Having declined restudy, he had a trial on medical management. His clinical course was complicated by progressively limiting postinfarct angina.

Cardiac catheterization demonstrated that the left anterior descending coronary artery had developed critical restenosis proximal to the stent. There was a long 60% stenosis attributable to dissection **distal** to the stent. Bypass surgery was advised. Two-vessel off-pump bypass was performed. His postoperative course was uneventful. He was discharged to home convalescence on postoperative day 3.

Diagnosis

Principle diagnosis: post infarct angina

Secondary diagnosis: coronary artery disease, surgical procedures anteroseptal infarct, PTCA to the left anterior descending artery; diabetes mellitus; chronic obstructive pulmonary disease

Medications: Lopressor, Ecotrin, albuterol inhaler, and Micronase with entrance into cardiac rehabilitation in 10 d

Exercise Test Results

An exercise test is not warranted at this stage of recovery.

Exercise Prescription

Outpatient cardiac rehabilitation: Following CABS, the patient returned to cardiac rehabilitation 4 wk postsurgery. The patient had an uncomplicated postsurgical course and had experienced no further symptoms since discharge. The patient was restarted in the program as follows.

Resting heart rate: 72

Resting blood pressure: 142/82

Initial exercise program:

Treadmill walking = 2.0 mph (3.2 kph), 0% grade for 10 min

Combined arm and leg ergometry = 100 W for 10 min

Upper-body ergometry = 30 to 50 W for 5 min

Stepper = 6 METs for 5 min

Hydraulic resistance machine = two sets of three exercises for 10 repetitions

Patient completed 6 wk in the program following surgery at the following workloads. Initially, 6 wk were completed post-PTCA–stent therapy before surgery.

Treadmill walking = 2.5 mph (4.0 kph), 3% grade for 10 min

Combined arm and leg ergometry = 100 to 125 W for 10 min

Upper-body ergometry = 100 W for 5 min

Rower = 50 to 75 W for 10 min

Hydraulic resistance machine = two sets of three exercises for 10 repetitions

Exercise heart rate: 90 to 96 beats · min^{-1}

Exercise rating of perceived exertion: 11

The remainder of program was uneventful. The patient completed a total of 12 wk from his initial start and returned to his home walking program and activities of daily living.

Discussion Questions

1. What differences may exist in the exercise recommendations for the patient and educational guidelines for the first 6 wk (pre-CABS) versus the latter 6 wk (post-CABS)?

2. What kinds of exercise, if any, may need to be avoided after surgery and for how long?

3. At program completion, what type of home exercise would be recommended?

Chronic Heart Failure

Steven J. Keteyian, PhD, FACSM

Heart failure (HF) is "the pathophysiological state in which the heart is unable to pump blood at a rate commensurate with the requirements of metabolizing tissues" (13, p. 503).

The inability of the left ventricle (LV) to pump blood adequately can be due to a failure of systolic or diastolic function. **Systolic dysfunction**, or the inability of cardiac myofibrils to contract or shorten against a load, leads to a reduced **ejection fraction**. Alternately, some individuals with normal systolic contractile function can still present with symptoms of HF. In these patients, the disorder is not associated with an inability of the heart to contract; instead it is attributable to an abnormal increase in resistance to filling of the LV—referred to as **diastolic dysfunction**. Think of diastolic dysfunction as a stiff or noncompliant chamber that is partially unable to expand (reduced distensibility) as blood flows in during diastole.

SCOPE

The public health burden associated with HF is immense, in many ways exceeding other chronic health disorders. Approximately five million people are afflicted with the syndrome, and approximately 550,000 new cases occur each year (64). Because of the aging U.S. population and increased survival of patients with cardiovascular disorders, the prevalence of HF will increase two- to threefold over the next 10 yr.

HF is the leading reason for hospitalizations in people 65 yr of age and older and directly or indirectly con-

tributes to 290,000 deaths annually. The 5 yr mortality rate for a person newly diagnosed with HF is 45%. The economic burden imposed by HF in the United States is enormous, standing at almost $30 billion annually (64).

PATHOPHYSIOLOGY

HF remains a final common denominator for many cardiovascular disorders. Although LV diastolic dysfunction can be an important cause of HF, most exercise research to date has focused on heart failure due to systolic dysfunction. As a result, this chapter primarily addresses this disorder. Please note that in patients with HF attributable to LV systolic dysfunction, some degree of diastolic dysfunction is usually present as well.

Figure 18.1 on p. 316 depicts the complex abnormalities and changes that occur following loss of systolic function. When cardiac cells (myocytes) die because of infarction, chronic alcohol use, long-standing hypertension, disorders of the cardiac valves, viral infections, or yet unknown causes, diminished LV systolic function results. Among all cases of LV systolic dysfunction, approximately 60% are caused by **ischemic heart disease** (i.e., coronary atherosclerosis). For this reason, these patients are commonly referred to as having an ischemic versus a **nonischemic cardiomyopathy**, where a nonischemic cardiomyopathy refers to some other disease process involving heart muscle (e.g., viral cardiomyopathy or alcoholic cardiomyopathy).

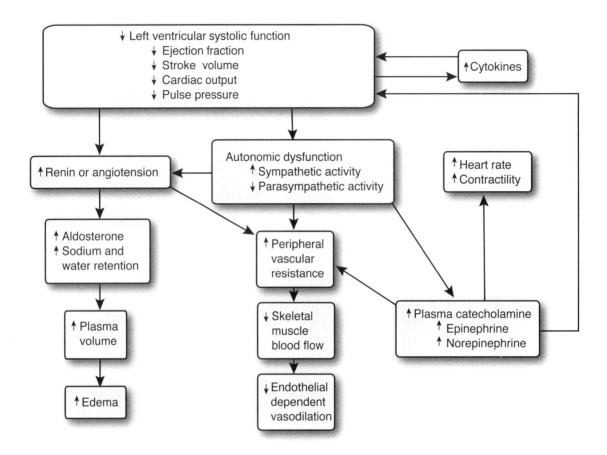

Figure 18.1 Schematic representation describing some of the main physiological and pathophysiological adaptations that occur at rest in patients with heart failure attributable to left ventricular systolic dysfunction.

As figure 18.1 also shows, a variety of physiological adaptations and compensatory changes occur in response to LV systolic dysfunction. Most of the current medical therapies used to treat chronic HF are aimed at modifying one or more of these abnormalities.

Key characteristics unique to the pathophysiology of HF because of systolic and diastolic dysfunction include the following:

1. An ejection fraction that is reduced (systolic) or unchanged or slightly increased (diastolic) at rest

2. An increase in LV mass, with end-diastolic and end-systolic volumes that are increased (systolic failure) or decreased (diastolic failure)

3. Edema or fluid retention because of elevation of diastolic filling pressures or activation of the renin–angiotensin–aldosterone system, causing sodium retention

4. More so with systolic (versus diastolic) heart failure, an imbalance of the autonomic nervous system such that parasympathetic activ- ity is inhibited and sympathetic activity is increased

Additionally, abnormalities of other hormones and chemicals such as an increased release of brain naturetic peptide (BNP), diminished production of nitrous oxide (endothelium-derived relaxing factor), increased endothelin-1, and increased cytokines (e.g., tumor necrosis factor-alpha) all contribute to adverse cardiac and vascular remodeling/function and changes within and around the skeletal muscles.

Substantial clinical evidence now indicates that many of those factors contribute to the remodeling of the LV, reshaping it from an elliptical form to a spherical form. This change in shape itself contributes to a further loss in LV systolic function. Currently, several of the treatment strategies used in patients with HF interrupt this process, referred to as reverse remodeling.

In patients with isolated diastolic dysfunction, LV end-diastolic and systolic volumes are generally reduced, and systolic function (i.e., ejection fraction) may be normal or increased. Table 18.1 describes normal LV characteristics and those associated with LV systolic and diastolic dysfunction.

Table 18.1 Comparison of Typical Left Ventricle Characteristics

	End-diastolic volume (ml)	End-systolic volume (ml)	Stroke volume (ml)[a]	Ejection fraction (%)[b]
Normals	120	55	65	55
Systolic dysfunction	160	110	50	30
Diastolic dysfunction	85	35	50	60

[a]Stroke volume = (end-diastolic volume) − (end-systolic volume). [b]Ejection fraction = (stroke volume)/(end-diastolic volume).

MEDICAL AND CLINICAL CONSIDERATIONS

Before a patient with HF can be cleared for exercise rehabilitation, his or her past and current medical history must be reviewed and functional status evaluated. Exercise testing provides important information about the patient's status, but signs, symptoms, and medications must also be considered.

Signs and Symptoms

Clinically, patients with HF present with several key characteristics or findings, two of which are

1. exercise intolerance as manifested by fatigue or shortness of breath on exertion and

2. fluid retention as evidenced by peripheral edema, recent weight gain, or both.

Those signs and symptoms are often associated with complaints of difficulty sleeping flat or awakening suddenly during the night to "catch my breath." Sudden awakening caused by labored breathing is referred to as **paroxysmal nocturnal dyspnea**.

Labored or difficult breathing while exerting oneself is called **dyspnea on exertion** (DOE), and difficulty breathing while lying supine or flat is referred to as **orthopnea**. Clinically, the severity of orthopnea is rated based on the number of pillows that are needed under a patient's head to prop him or her up sufficiently to relieve dyspnea. For example, so-called three-pillow orthopnea means that a patient needs three pillows under the head and shoulders to breathe comfortably while recumbent.

History and Physical Examination

The signs and symptoms just discussed represent findings that the clinical exercise physiologist may need to evaluate to ensure that on any given day the patient can safely

exercise. For example, a patient's complaint of increased DOE or recent weight gain may or may not be clinically meaningful in HF patients, but such signs warrant further inquiries by the clinical exercise physiologist to ascertain whether they are associated with new or increased ankle edema or fluid accumulation in the lungs. The severity of ankle edema is typically evaluated on a scale of 1 to 3, as discussed in chapter 4. Lung sounds called rales are associated with pulmonary congestion and are best heard using the diaphragm portion of a stethoscope. Rales appear as a crackling noise during inspiration.

Normal first (S_1) and second (S_2) heart sounds are associated with the abrupt closure of the mitral and tricuspid valves, and aortic and pulmonic valves, respectively (see chapter 4). In patients with HF attributable to LV systolic dysfunction, an abnormal third heart sound (S_3) can often be heard when the bell portion of the stethoscope is lightly placed on the chest wall over the apex of the heart. This S_3 sound occurs soon (120–200 ms) after S_2 and is most likely attributable to vibrations caused by the inability of the LV wall and chamber to accept incoming blood during the early, rapid stage of diastolic filling. Listening to the audio materials available in most medical libraries is an effective way to learn both normal and abnormal breath and heart sounds.

Diagnostic Testing

The diagnosis of HF attributable to LV systolic dysfunction, although based on the presence of signs and symptoms, also requires a reduced ejection fraction. Most often, ejection fraction is measured by using an **echocardiogram** machine, although a radionuclide test or cardiac catheterization can assess this parameter as well. Normally, LV ejection fraction is greater than 50%, which means that at least half of the blood in the LV at the end of diastole is ejected into the systemic circulation during systole. In patients with HF attributable to systolic dysfunction, LV ejection fraction is typically reduced below 45%. Severe LV dysfunction may be associated with an ejection fraction of 30% or lower. The decrease in ejection fraction is qualitatively proportional to the amount of myocardium that is no longer functional. Chronic HF attributable to systolic dysfunction is also usually associated with an enlarged LV.

Additionally, in patients with recent onset systolic HF caused by a large anteroseptal myocardial infarction, marked increases in serum troponin and creatine phosphokinase are typically observed in the blood (see chapter 16). In these patients' electrocardiograms (ECGs), changes such as Q waves in leads V1 through V4 are usually evident. In some patients, a cardiac cath-

eterization is performed to determine whether, and to what extent, ischemic heart disease contributed to the problem. Ischemic heart disease is the underlying cause for HF in about 60% of cases.

The diagnosis of HF attributable to LV diastolic dysfunction is somewhat less exact. Although an echocardiogram can be used to evaluate unique characteristics during diastole (i.e., left-sided filling pressures), quite often the diagnosis is made when the clinical syndrome of congestive HF (i.e., fatigue, dyspnea, and peripheral

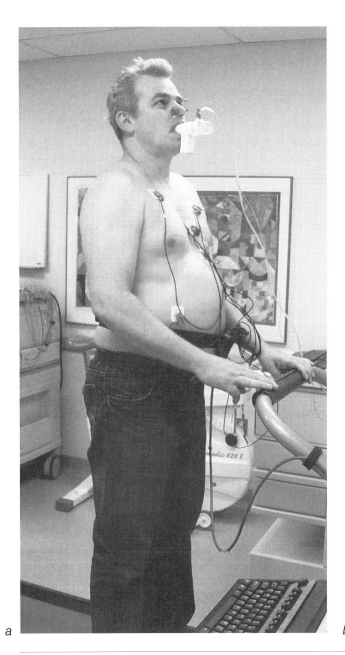 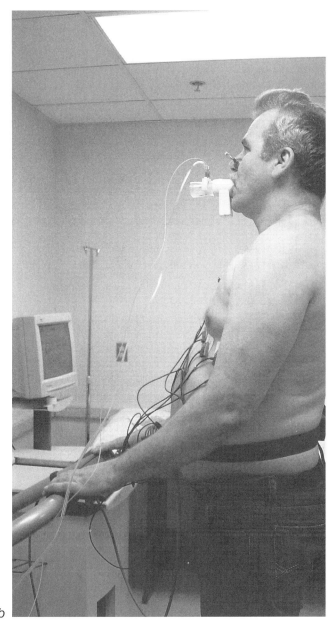

a   b

Figure 18.2   Example of a cardiopulmonary exercise test, often used to estimate prognosis in patients with chronic heart failure: (a) front view, and (b) side view.

and pulmonary edema) requires hospitalization in the presence of a somewhat normal ejection fraction.

Although much of the diagnosis of HF relies on the use of echocardiography and patient symptoms, some laboratory tests such as measuring BNP levels in the plasma may be used to help support or lend weight to the suspected diagnosis of heart failure. Synthesized in excess and released by the myocardium when the ventricles themselves are stretched by mechanical and pressure overload, BNP ($>100\,pg \cdot ml^{-1}$) can be a sensitive index of decompensated HF, even when other clinical signs are nebulous. Serum BNP can, for example, distinguish dyspnea from HF versus pulmonary disease, a particularly useful tool in patients whose physical examination is nonspecific or in those people with poor echocardiographic images.

Despite their utility, neither echocardiography, BNP, nor other assessments of cardiac function quantify the full scope of HF as a disease that has both a cardiac and systemic effect. But assessment of exercise capacity as a means to evaluate integrated physiologic function stands out as an important exception, as measured in terms of maximal oxygen consumption during a cardiopulmonary exercise test (along with other parameters of ventilatory and circulatory efficiency).

Routine evaluation or workup of patients with HF now includes a graded exercise test with measured gas exchange to ventilatory efficiency and determine peak oxygen consumption ($\dot{V}O_2$peak) (figure 18.2). The 3 yr risk of death for patients who achieve a $\dot{V}O_2$peak greater than $17\,ml \cdot kg^{-1} \cdot min^{-1}$ is about 20%, clearly better than 3 yr risk (approximately 50%) in patients who achieve a peak value less than $14\,ml \cdot kg^{-1} \cdot min^{-1}$ (figure 18.3)

(52). Recent and ongoing research suggests that among patients with systolic heart failure being treated with β-adrenergic blocking agents, a peak $\dot{V}O_2$ value below 10 to $12\,ml \cdot kg^{-1} \cdot min^{-1}$ may best indicate marked increase in future 3 yr risk. Ventilatory efficiency is computed as the slope of the relationship of ventilation to carbon dioxide ($\dot{V}_E$-$\dot{V}CO_2$) during exercise, with a value >35, indicating increased future risk of death. Clearly, measured $\dot{V}O_2$peak and $\dot{V}_E$-$\dot{V}CO_2$ can help determine which patients require aggressive medical therapy or possible cardiac transplant.

Although clinicians often use New York Heart Association (NYHA) functional class (see table 18.2 on p. 320) to describe a patient's clinical status, use of that system has limitations in that it does not fully reflect the breadth of the disorder. Table 18.2 shows the various stages in the development of heart failure, as designated by the American College of Cardiology and American Heart Association (33). This staging system covers all patients with left ventricular systolic dysfunction, regardless of whether symptoms are present. Note that for those patients experiencing symptoms (New York Heart Association class II–IV), they fall within stage C and stage D alone.

EXERCISE TESTING

The use of exercise testing in patients with HF provides an enormous amount of useful information. As mentioned previously, information is gathered not only on severity of illness and 3 yr survival but also on response to medications and response to an exercise-training program. Information to guide exercise-training intensity is obtained as well.

Medical Evaluation

The methods for exercise testing patients with HF differ little from the testing used in patients with other types of heart disease. Although most exercise tests conducted in these patients use a steady-state (2–3 min per stage) protocol like the modified Bruce or Naughton (23), a ramp protocol can also be performed with a stationary cycle. With this method, external work rate is increased 10 to 15 W every minute (53). A ramp protocol generally results in less variable data during submaximal exercise because of the more gradual increments in work rate that it provides. $\dot{V}O_2$peak is approximately 10% to 15% lower when measured with cycle ergometry versus a treadmill (53,55).

Because an accurate measure of functional capacity is needed in these patients, the use of prediction equations to estimate $\dot{V}O_2$ is discouraged because they tend

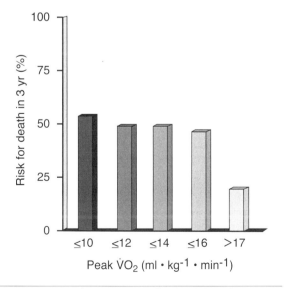

Figure 18.3 Three-year risk for death based on achieved peak $\dot{V}O_2$.

to overpredict functional capacity (23). Measured exercise capacity using a cardiopulmonary cart (figure 18.2) is preferred, and such equipment is usually available in either the cardiac noninvasive or the pulmonary function laboratory of most hospitals. In addition to measuring $\dot{V}_E$-$\dot{V}CO_2$ and $\dot{V}O_2$peak, determining **ventilatory derived lactate threshold** (V-LT) is helpful. An adequate discussion of this parameter is provided elsewhere (23), but an approach often used when determining V-LT is the V-slope method (5).

Contraindications

Twenty years ago, standard teaching in most medical schools was that moderate or harder physical activity should be avoided or withheld in patients with HF. The increased hemodynamic stress that exercise placed on an already weakened heart was thought to worsen heart function further. As a result, most guidelines listed HF as a contraindication to exercise testing. Today, however, patients with stable HF routinely undergo symptom-limited maximum cardiopulmonary exercise testing to evaluate cardiorespiratory function.

All other contraindications to exercise testing still apply to patients with HF, including acute myocardial infarction and malignant arrhythmia. Interestingly, arrhythmias are common in patients with HF; therefore, information detailing a patient's history should be communicated to the person supervising an exercise test. Note that the use of exercise testing to assess myocardial ischemia can be problematic, because many patients with HF present with ECG findings that invalidate or reduce the sensitivity of the test (e.g., left bundle branch block, left ventricular hypertrophy, and nonspecific ST-wave changes attributable to digoxin therapy).

Recommendations and Anticipated Responses

Compared with healthy normal people, patients with HF exhibit differences in their cardiorespiratory and peripheral responses at rest and during exercise (table 18.3) (2). Resting stroke volume and cardiac output are both lower in patients with HF versus controls (stroke volume: approximately 50 versus approximately 75 ml · beat^{-1}; cardiac index: <2.5 versus >2.5 L · min^{-1} · m^{-2}, respectively). Resting heart rate (HR) is increased (HF = 75–105 beats · min^{-1} versus controls = 60–80 beats · min^{-1}) and systolic blood pressure may be reduced, attributable to both the underlying LV systolic dysfunction and the use of afterload-reducing agents such as angiotensin-converting enzyme (ACE) inhibitors.

During submaximal exercise, patients with HF exhibit a higher HR, a lower stroke volume response and attenuated increases in cardiac output and $\dot{V}O_2$ compared with persons without HF (12,60,70). To compensate, the extraction of oxygen in exercising muscles is higher in patients with HF than in persons without HF. Also during exercise, plasma norepinephrine, an endogenous **catecholamine**, is released at increased levels by the sympathetic postganglionic fibers. A disproportionate increase occurs as well in plasma norepinephrine levels, and the magnitude of the increase is generally related to the severity of the illness (42).

Table 18.2 Stages in the Development of Heart Failure: ACC/AHA Guidelines (33)

Stage	Description	Example	New York Heart Association functional class
A (patient at risk)	High risk for heart failure; no anatomic or functional abnormalities; no signs or symptoms	Hypertension, coronary artery disease, diabetes, alcohol abuse, family history	
B (patient at risk)	Structural abnormalities associated with heart failure but no symptoms	Left ventricular hypertrophy, prior myocardial infarction, asymptomatic valvular disease, low ejection fraction	
C (heart failure present)	Current or prior signs and structural abnormalities	Left ventricular systolic dysfunction with or without dyspnea on exertion or fatigue, reduced exercise tolerance	II or III
D (heart failure present)	Advanced structural heart failure with symptoms at rest despite maximal medical therapy	Frequent hospitalizations, awaiting transplant, intravenous support	III or IV

Table 18.3 Resting and Exercise Characteristics of Patients With Heart Failure Attributable to Systolic Dysfunction

	Resting	Submaximal exercise	Peak exercise
Cardiorespiratory			
Ejection fraction	↓	↓	↓
Cardiac output	↓	↓	↓
Stroke volume	↓	↓	↓
Heart rate	↑	↑	↓
Oxygen consumption	↑ ↔	↓	↓
Arterial-mixed venous oxygen difference	↑	↑	↑ ↔
$\dot{V}O_2$ at ventilatory-derived lactate threshold	N/A	↓	N/A
Peripheral			
Arterial blood lactate	↔	↑	↓
Total systemic vascular resistance	↑	↑	↑
Blood flow in active muscle	↔	↓	↓
Skeletal muscle mitochondrial density	↓	N/A	N/A
Skeletal muscle oxidative enzymes	↓	N/A	N/A
Reliance on anaerobic metabolism	N/A	↑	↔

Note. ↑ = increased response compared with healthy subjects; ↓ = decreased response compared with healthy subjects; ↔ = similar responses compared with healthy subjects; NA = not applicable.

Increasing blood flow to metabolically active skeletal muscles is part of a complex interplay between blood pressure and vasoconstriction–vasodilation response to exercise. Patients with HF have a reduced exercise-induced vasodilation attributable to both the increased plasma norepinephrine (42) and impaired endothelial function, the latter because of the lesser release of a local chemical called endothelium-derived relaxing factor (17,44).

Compared with healthy normal persons, at peak exercise patients with HF exhibit a lower power output (30%–40% decrease), lower cardiac output (40% decrease), lower stroke volume (50% decrease), and lower HR (20% decrease) (12,14,32,53,60). Concerning exercise capacity as measured by peak $\dot{V}O_2$, this too is decreased when compared to normal persons, 30% to 35% or more in patients with systolic or diastolic HF (41). Depending on severity of illness, $\dot{V}O_2$peak typically ranges from 8 to 21 ml · kg^{-1} · min^{-1}. For patients not taking a β-adrenergic blocking agent, peak HR often does not exceed 150 beats · min^{-1}. For these patients this blunted or attenuated response of HR to exercise is referred to as **chronotropic incompetence** and represents a characteristic that occurs in 20% to 25% of patients with HF (10). In 1999 Robbins and coworkers (57) showed that, like $\dot{V}O_2$peak, chronotropic incompetence during exercise is a powerful and independent predictor of mortality in patients with HF.

Despite the diminished peak exercise capacity of these patients, the degree of exercise intolerance is not related to the magnitude of LV systolic dysfunction. Among 234 patients tested in our laboratory we observed no relationship (r = .14) between ejection fraction (range = 8–40%) and $\dot{V}O_2$peak (range = 7.5–34.3 ml · kg^{-1} · min^{-1}).

This finding suggests that several other factors besides LV ejection fraction alone contribute to the exercise intolerance that these patients experience. Although this finding may seem a bit perplexing, in that abnormalities in tissues other than the damaged heart are involved, this is in fact the case. Two such noncardiac mechanisms that limit exercise capacity are

1. an inability to dilate peripheral vasculature sufficiently as a means to increase blood flow to the metabolically active muscles (69) and

2. histological and biochemical abnormalities within the skeletal muscle itself (20,47,61).

In HF patients, there is clear evidence of endothelial dysfunction, along with a decrease in the percentage of myosin heavy chain type 1 isoforms, diminished oxidative enzymes, and decreased capillary density. As a result, and when compared with normal people, patients with HF rely more on anaerobic pathways to produce energy earlier during exercise.

Treatment

Over the past 20 yr great strides have been made in the medicines used to treat patients with HF. These advances have led to fewer HF-related deaths, fewer symptoms, and increased exercise tolerance. In patients refractory to optimal medical therapy and who demonstrate a deteriorating clinical state consistent with a 1 yr survival rate less than 50%, cardiac transplant is a consideration.

This section summarizes the guidelines for the medical treatment of patients with HF attributable to LV systolic dysfunction (33). The current medical therapy for these patients includes angiotensin-converting enzyme (ACE) inhibitors, β-adrenergic receptor blockers, diuretics, and, possibly, digoxin or an aldosterone antagonist. ACE inhibitors reduce 5 yr mortality rate by approximately 30% among patients with diagnosed HF. They also improve exercise tolerance, alter the rate of HF progression, influence structural remodeling, and decrease future hospitalizations.

The single most important contribution to the medical management of patients with HF in the past 15 yr has been the use of β-blocker therapy. Initially thought to be contraindicated, β-blockers are now known to be safe, to reduce morbidity and mortality rates by approximately 30%, and to improve resting ejection fraction by approximately 7 percentage points.

Congestion and fluid overload remain important complications in many patients with HF. To address this, diuretic therapy is commonly used. And the role that sodium restriction plays in minimizing fluid congestion cannot be overemphasized. Sodium restriction can help reduce the need for diuretics, and a dietary plan that aggressively restricts sodium intake may actually allow for the discontinuation of diuretic therapy.

The routine use of digoxin for patients with HF, although still unresolved, has been in practice for decades. Although the controversy applies mostly to HF patients with normal sinus rhythm, this agent has been shown to reduce hospitalizations by 27%. The use of digoxin in patients with HF and atrial fibrillation is not challenged. Others agents that may be used to treat patients with HF attributable to systolic dysfunction include antiplatelet and anticoagulation therapies, angiotensin receptor blockers, and an aldosterone antagonist.

For patients with NYHA class II to IV HF who also have electrocardiographic evidence of dyssynchronous contractions of the left and right ventricles (based on a QRS duration greater than 120 ms), a special type of pacemaker therapy (called cardiac resynchronization therapy) is commonly used. As the name implies, cardiac resynchronization therapy involves implanting pacemaker leads in both the right and left ventricles and pacing the ventricles to reestablish the correct firing pattern of one ventricle relative to the other. Resynchronization therapy has been shown to improve cardiac function and exercise tolerance and lowers future hospitalizations and death (33). Additionally, the benefits of implantable defibrillators, either by themselves or combined with cardiac resynchronization pacemakers, are well documented for patients with dilated hearts with low ejection fractions, particularly as a means to reduce mortality that results from the common occurrence of ventricular arrythmias.

In some instances, and despite aggressive attempts to optimize medical therapy, a patient's clinical condition continues to deteriorate such that special medications, cardiac transplantation or mechanical left ventricular assist devices designed to augment circulation and maintain adequate blood flow are required. Without these efforts, many such patients die in months, if not weeks. Cardiac transplantation is a surgical procedure that represents standard therapy for patients with end-stage HF that is refractory to maximal medical therapy. An overview of that procedure is given in practical application 18.1.

An important part of the care for any chronic disease remains secondary prevention. For the exercise professional working with HF patients, this care includes counseling to help manage behavioral habits known to exacerbate the condition. For example, in patients with an ischemic cardiomyopathy, every attempt should be made to help stabilize existing coronary atherosclerosis through aggressive risk factor management and to prevent further loss of cardiac cells (myocytes) attributable to reinfarction.

For the patient with HF attributable to other causes such as illegal substances or alcohol abuse, the exercise physiologist should support healthy behaviors and be alert for signs of relapse that may require referral to a specialist. For all patients with HF, observing a low-sodium diet is an important step toward preventing congestion and fluid overload, thus reducing chances of subsequent hospitalizations. Consistent with this practice, compliance with all prescribed medications, especially diuretics aimed at removing excess fluid, is an important variable that an exercise professional can assess during clinical appointments or before exercise class. See practical application 18.2 on p. 324.

EXERCISE PRESCRIPTION

For patients with HF attributable to LV systolic dysfunction, continuously ECG-monitored exercise is probably

Practical Application 18.1

Cardiac Transplantation

Cardiac transplantation is an effective therapeutic alternative for persons with end-stage HF. Each year, approximately 3,000 to 4,000 such transplants are performed worldwide, with 1 and 3 yr survival rates approximating 84% and 77%, respectively (63). In most patients undergoing cardiac transplant, the atria of the recipient's heart are attached at the level of the atria of the donor's heart.

After surgery, many patients with cardiac transplant continue to experience exercise intolerance. In fact, $\dot{V}O_2$peak is typically between 14 and 22 ml $\cdot$ kg^{-1} $\cdot$ min^{-1} among untrained patients. For this and other reasons, these patients are commonly enrolled in a home-based or supervised cardiac rehabilitation program as soon as 4 wk after surgery. The expected increase in $\dot{V}O_2$peak ranges between 15% and 25% (37,43).

There is an increasing probability of developing accelerated atherosclerosis in the coronary arteries of the donor heart. For this and other reasons, the traditional risk factors for ischemic heart disease such as hypertension, obesity, diabetes, and hyperlipidemia are aggressively treated. Additionally, immune system–mediated rejection of the donor heart remains a constant concern for heart transplant recipients, and therefore most patients receive a variety of immunosuppressive medications. Common agents include cyclosporine, tacrolimus, mycophenolate mofetil, and prednisone.

Patients with cardiac transplant represent a unique physiology, in that the donated heart they received is decentralized. This circumstance means that except for the parasympathetic, postganglionic nerve fibers that are left intact, all other cardiac autonomic fibers are severed. Regeneration of these fibers may occur after 1 yr in some patients, but it is likely best to assume that decentralization, for the most part, is permanent.

Because of the decentralized myocardium, the cardiovascular response of cardiac transplant patients to a single bout of acute exercise differs from the response of normally innervated people. At rest, HR may be elevated at 90 to 100 beats $\cdot$ min^{-1} because parasympathetic (i.e., vagal nerve) influence is no longer present on the sinoatrial node. When exercise begins, HR changes little because (a) no parasympathetic input is present to be withdrawn and (b) no sympathetic fibers are present to stimulate the heart directly. During later exercise, HR slowly increases because of an increase in norepinephrine in the blood. At peak exercise, HR is lower, and both cardiac output and stroke volume are approximately 25% lower than age-matched controls.

Because of the absence of parasympathetic input, the decline of HR in recovery takes longer than normal. This effect is most observable during the first 2 min of recovery, during which HR may stay at or near the value achieved at peak exercise. Systolic blood pressure recovers in a generally normal fashion after exercise (21), which is why this measure can be used to assess adequacy of recovery.

The number of people needing a cardiac transplant each year far exceeds the available number of donors. Consequently, many patients die while awaiting a heart, whereas others receive mechanical left ventricular assist devices or enroll in experimental programs that test new medications or other devices. The magnitude of this donor shortage will likely only increase in the future, as the number of patients with end-stage HF continues to increase.

---

advised for three to nine sessions (1), with the intention that such a practice might help identify abnormal findings and avoid clinical events. After demonstrating for 2 to 3 wk that they can tolerate exercise three times per week, patients can begin a one to two times per week home exercise program. As patients continue to improve and demonstrate no complications to exercise after 12 or more supervised sessions, they can transfer to an all home-based exercise program.

Special Exercise Considerations

Despite the increased attention given to using moderate exercise training in the treatment plan of patients with HF, few HF-specific guidelines exist for prescribing exercise in these patients. Prudent eligibility criteria for exercise training in patients with HF might be as follows:

- Ejection fraction less than 40%
- NYHA class II or III
- Receiving standard drug therapy of ACE inhibitors, β-adrenergic blockade, or diuretics for at least 6 wk
- Absence of any other cardiac or noncardiac problems that would limit participation in exercise

"Contraindications to Exercise in Patients With HF" (2) suggests contraindications to exercise training in patients with HF.

The ejection fraction cutoff of less than 40% was chosen to remain consistent with the majority of the exercise studies conducted in these patients to date. By no means do we intend to infer that patients with an ejection fraction greater than 40% do not benefit from an exercise-training regimen. On the contrary, these patients improve as well. Likewise, besides enrolling stable NYHA class II and III patients in an exercise program, the exercise professional may be able to include carefully selected and motivated class IV patients who are free of pulmonary congestion. We previously reported a case detailing physiological outcomes in an ambulatory class IV patient who underwent 24 wk of exercise training while receiving continuous dobutamine therapy (34). Such patients typically require more supervision during exercise.

Practical Application 18.2

Client–Clinician Interaction

One of the important responsibilities of an exercise professional is to ensure that on any given day, the patient with HF is free of any signs or symptoms that might indicate the need to withhold exercise.

To accomplish this goal, you must not only persuade the patient to verbalize any problems that she or he might be having but also take the initiative to ask the patient key questions. Following are examples of these types of questions:

- How did you sleep over the weekend? Have you had any more bouts of waking up during the night short of breath?
- How has your body weight been over the past four days?
- It's been hot the past couple days. Have you had any increased difficulty breathing?
- Do you ever get that swelling in your ankles anymore?

Although seemingly harmless, each question enables you to assess change in HF-related symptoms. With time and experience, you will develop a sense about when and how to assess these patients.

Also, patients with HF may not tolerate the first few days of their exercise rehabilitation well. Therefore, encouragement and guidance on your part are important. Explain that, over time, the patient can expect to be able to perform routine activities of daily living more comfortably. For some patients, achieving only this goal may be quite fulfilling. Other patients may wish to improve their exercise capacity to the point that they can resume activities that they previously had to avoid. Although such a goal may be realistic, be sure to emphasize that the prudent approach is to advance their training volume in a progressive manner. Trying to do too much too soon may lead to disappointment if their functional improvement does not keep pace with their self-assigned interests.

Finally, when working with patients with HF, emphasize the importance of regular attendance to their exercise regimen. If they need to miss exercise because of personal or medical reasons, let them know that you look forward to seeing them back when they feel better. Because you often see the same patient several times each week over several weeks, you are in a unique position to provide ongoing support, motivational counseling, and monitoring of medical compliance and symptom status over time. The issue of monitoring and intervening upon any lapse in prescribed medications is important because most of the medications that a patient with HF takes have been prescribed because they have been proven to improve survival, lessen symptoms, prevent rehospitalizations, and improve quality of life. All these endpoints are key markers of a successful disease management plan.

Another important exercise consideration for patients with HF is compliance. Using an outcomes data base, we previously conducted a retrospective review of the records of all patients enrolled in the Henry Ford Hospital cardiac rehabilitation program over a 34 mo period. We observed that in patients who suffered a myocardial infarction, nearly 75% of both men and women completed the program. However, among patients with HF and no myocardial infarction the percentages were lower; 71% of men and only 53% of women completed the program.

One reason related to the lower compliance in HF patients might be the fact that they often have comorbidities or other illnesses that interrupt or prevent regular program attendance. For example, arrhythmias, pneumonia, fluctuations in edema, adjustments in medications, and hospitalizations can all affect regular participation. Additionally, some patients with HF do not tolerate their first few exercise sessions well, and they experience fatigue later in the day. To help overcome the poorer compliance in this patient population, the exercise professional should be cognizant of these issues and ask patients to avoid trying to schedule too much on the days that they exercise. Also, patients should be told to expect interruptions in their exercise therapy caused by both HF- and non-HF-related issues. Then, when they feel better, they should make every attempt to return to class.

Exercise Recommendations

Training for cardiorespiratory endurance is an obvious strategy for patients with HF, but muscular strength, muscular endurance, and flexibility training can also improve functional capacity and foster independence.

Contraindications to Exercise in Patients With HF

Relative Contraindications

1. 1.5 to 2.0 kg increase in body mass over previous 3 to 5 d
2. Concurrent continuous or intermittent dobutamine ischemic changes
3. Abnormal increase or decrease of blood pressure with exercise
4. New York Heart Association functional class IV
5. Complex ventricular arrhythmia at rest
6. Recent embolism or thrombophlebitis appearing with exertion
7. Supine resting heart rate $\geq$100 beats $\cdot$ min^{-1}
8. Preexisting comorbidity or behavioral disorder
9. Implantable cardiac defibrillator with heart rate limit set below the target heart rate for training
10. Extensive myocardial infarction resulting in left ventricular dysfunction within previous 4 wk

Absolute Contraindications

1. Signs or symptoms of worsening or unstable heart failure over previous 2 to 4 d
2. As defined by existing guidelines, abnormal blood pressure or early unexpected life-threatening arrhythmia
3. Uncontrolled metabolic disorder (e.g., hypothyroidism, diabetes)
4. Acute systemic illness or fever
5. Recent embolism or thrombophlebitis
6. Active pericarditis or myocarditis
7. Third-degree heart block without pacemaker
8. Significant uncorrected valvular disease except mitral regurgitation because of HF-related LV dilation

Cardiorespiratory Training

Unique issues pertinent to prescribing exercise in patients with HF are described next, including modality, intensity, duration, and frequency of exercise.

- Mode: To improve cardiorespiratory fitness, the exercise professional should select activities that engage large-muscle groups such as stationary cycling or walking. Benefits in exercise capacity or tolerance gained from either of these modalities transfer fairly well to routine activities of daily living. Consistent with this, upper-body ergometry activities can improve function in the upper limbs as well. The exercise prescription for resistance training is described later in this section.

In terms of comparing one modality to another, a review of the exercise-training trials involving patients with HF indicates that similar increases in $\dot{V}O_2$ peak occur with both walking and cycle exercise (6,7,11,18,19,29–31,35,36,39). Many of these studies are shown in table 18.4.

Table 18.4 Review of the Major Exercise-Training Trials

Authors	Training intensity	Training mode	Peak $\dot{V}O_2$ (ml · kg^{-1} · min^{-1}) (before/after)
Coats et al. (11)	60–80% peak HR	Bike ergometer	T: 13.2 ± 0.9/15.6 ± 1.0 C: not given
Kiilavuori et al. (40)	50–60% of peak $\dot{V}O_2$	Bike ergometer	T: 20.7 ± 2.1/23.9 ± 3.0 C: 18.9 ± 1.6/19.1 ± 1.6
Belardinelli et al. (7)	40% peak $\dot{V}O_2$	Bike ergometer	T: +12% C: −5%
Hambrecht et al. (30)	70% peak $\dot{V}O_2$	Bike ergometer	T: 17.5 ± 5.1/23.3 ± 4.2 C: 17.9 ± 5.6/17.9 ± 5.6
Keteyian et al. (35)	60% initial, then 70–80% of peak $\dot{V}O_2$	Bike ergometer, walk, row	T: 16.0 ± 3.7/18.5 ± 4.4 C: 14.7 ± 4.2/15.2 ± 4.8
Kiilavuori et al. (39)	50–60% of peak $\dot{V}O_2$	Bike ergometer, walk, swim, row	T: 19.3 ± 1.6/21.7 ± 2.5 C: 18.3 ± 1.3/18.2 ± 1.5
Dubach et al. (19)	70–80% peak $\dot{V}O_2$	Bike ergometer, walk	T: 19.4 ± 3.0/25.1 ± 4.8 C: 18.8 ± 3.9/19.8 ± 4.3
Keteyian et al. (36)	50–60% initial, then 70–80% of peak $\dot{V}O_2$	Bike ergometer, walk, row	T: 16.1 ± 3.3/18.4 ± 4.2 C: 14.6 ± 3.6/15.3 ± 4.1
Belardinelli et al. (6)	60% peak $\dot{V}O_2$	Bike ergometer	T: 15.7 ± 2.0/19.9 ± 1.0 C: 15.2 ± 2.0/16.0 ± 2.0
Forissier et al. (22)*	HR just below ventilatory threshold	Bike ergometer	T: 20.0 ± 9.0/23.4 ± 11.0 C: 18.1 ± 4.0/21.2 ± 4.0
Curnier et al. (16)*	Within +5 beats of HR at ventilatory threshold	Bike ergometer	T: 20.3 ± 6.1/23.9 ± 6.4 C: 17.9 ± 3.8/20.0 ± 4.4
Linke et al. (46)	70–80% of peak $\dot{V}O_2$	Bike ergometer	T: 16.0 ± 1.2/19.4 ± 1.4 C: 16.9 ± 1.2/16.3 ± 1.3
Giannuzzi et al. (25)	60% of peak $\dot{V}O_2$	Bike ergometer, walk,	T: 13.8 ±3.3/16.2 ± 3.6 C: 13.8 ±2.3/13.7 ± 2.2
Gielen et al. (27)	70% of peak $\dot{V}O_2$	Bike ergometer, walk	T: +29% C: 0%

Note. HR = heart rate; T = treatment, before/after training; C = control, before/after training. *Both groups underwent training. The treatment group received a β-adrenergic blocking agent in addition to usual care medications (the control group received usual care medications only).

• Frequency: Similar to people without HF, those with HF need to engage in a regular exercise regimen of four to five times per week. As mentioned, patients should be counseled to remain regular with their exercise habits and, if their program is interrupted because of personal or medical reasons, to plan on restarting the program as soon as possible. After being in an exercise program for several weeks, patients often comment that they notice less DOE and less fatigue during routine daily activities. Part of exercise prescription involves educating patients about the detraining effect that occurs, which means informing them that many of the benefits that they notice will fade if they stop exercising.

• Intensity: Exercise intensity can be established using variables derived from an exercise test. In patients with HF, different ranges of exercise intensity have been used to improve cardiorespiratory fitness (table 18.4). Relatively similar gains in cardiorespiratory fitness (i.e., $\dot{V}O_2$peak) are evident regardless of whether a light (50–60%), moderate (60–70%), or high (70–80%) intensity regimen was used to guide exercise intensity.

Using the HR reserve method, the first few exercise sessions should be guided at an exercise intensity of about 60%. The exercise professional can have the patient use the rating of perceived exertion scale set at 11 to 14. One reason to start these patients at this lower intensity level is to allow them an opportunity to adjust to the exercise. As mentioned, fatigue later in the day is not uncommon, so restricting exercise intensity at first may help minimize this effect.

As a patient demonstrates improved ability to tolerate therapy, the intensity of effort can increase to 70% to 75% of HR reserve. Again, the patient's rating of perceived exertion should stay between 11 and 14. One study suggests that LV function may decrease further during exercise when patients with LV systolic dysfunction are exercised above V-LT (54). Therefore, a prudent approach might be to keep HR near or below 75% HR reserve, which is likely at or below V-LT for most patients with HF.

• Duration: Most exercise trials involving patients with HF increased training duration up to 30 to 60 min of continuous exercise. On occasion, it may be necessary to start an individual patient with two to three bouts of exercise that are each 4 to 8 min in duration. Interval training such as this has been investigated in patients with HF by Meyer and coworkers (49,50), using a work-to-rest ratio during cycle ergometry of 30 s to 60 s. This model allows greater work rates to be placed on the muscles without sustained higher-intensity stress on the nervous, endocrine, and cardiorespiratory systems.

Resistance Training

The role and benefit of resistance training in patients with HF is increasingly clear. To date, most studies have shown that mild to moderate resistance training is tolerated and improves muscle strength 20 to 45% in stable patients with HF. Given that loss of muscle strength and endurance is common in these patients (51,56,67), it is appropriate to assume that improvements in these measures through resistance training are helpful. But no specific guidelines that address resistance training in patients with HF currently exist.

Instead, resistance-training recommendations in these patients are often drawn from a scientific advisory statement published in 2007 (68). Generally, exercise or lift intensity should start at 40% to 50% of 1RM and then be progressively increased over several weeks to 60% to 70% of 1RM, with the patient starting with one set of 10 to 12 repetitions and progressing to one set of 12 to 15 repetitions. The specific exercises should address the individual needs of the patient but will likely include all the major muscle groups. As patients improve, load should be increased 5% to 10%. Before each training session, however, patients should be evaluated for any signs or symptoms of HF or excessive muscle or joint discomfort.

A summary of the exercise prescriptions used for patients with HF and for patients with cardiac transplant is given in practical applications 18.3 and 18.4 on p. 328-330, respectively.

EXERCISE TRAINING

Improvements with exercise training in patients with HF because of systolic dysfunction can be seen not only in functional and cardiorespiratory measures (practical application 18.5, on p. 330) but also in the skeletal muscle and other organ systems. Conversely, little information is available regarding the physiologic effects and safety of exercise training in patients with HF because of diastolic dysfunction. As a result, the material in this section focuses on exercise-training responses in patients with systolic HF.

Myocardial Function

Exercise training results in a modest increase in peak cardiac output (8%–15%), attributable to increases in both peak HR and peak stroke volume. Interestingly, the 4% to 8% increase in peak HR (i.e., partial reversal of chronotropic incompetence) accounts for up to 40% to 50% of the increase in $\dot{V}O_2$peak (31,35,36). This training

Summary of Exercise Recommendations for Patients With Heart Failure

A limited physical examination should be performed to identify acute signs or symptoms that would prevent participation in exercise. These might include an increase in body mass greater than 1.5 kg during the previous 3 to 5 d, lung sounds (i.e., rales) consistent with pulmonary congestion, complaints of increased difficulty sleeping while lying flat, sudden awakening during the night because of labored breathing, or increased swelling in the ankles or legs. Additionally, a graded exercise test is usually warranted both to identify any important exercise-induced arrhythmias and to develop a target HR range that allows determination of a safe exercise intensity.

Exercise Program

As tolerated, type of exercise, frequency, duration, and intensity of activity should be progressively adjusted to attain an exercise energy expenditure of 700 to 1,000 kcal · week^{-1}. The table below provides a summary of the aerobic, resistance-training, and range-of-motion recommendations.

Training method	Mode	Intensity	Frequency	Duration	Progression	Goals
Aerobic	Cycle or treadmill; arm ergometer as needed	60% to 75% of heart rate reserve or 11 to 14 on rating of perceived exertion scale	Four to five times per week	40 min or more per session; use interval training method as needed	10 min or more per session, up to 40 min or more per session	Improve submaximal and peak endurance
Resistance	Fixed machines, free weights, and bands; six to eight regional exercises	Begin with 40% of 1RM; progress over time to 70% of 1RM	One to two times per week	One set of 12 to 15 repetitions for each of the involved muscle groups	Increase 5% to 10%, as tolerated	Initial focus should be on increasing leg muscle strength and endurance
Range of motion	Static stretching		Before and after each aerobic or resistance-training workout	5 minutes before and 5 to 10 min after each workout, with 10 to 30 s devoted to the major muscle groups and joints		

Special Considerations

- Many patients with HF are inactive and possess a low tolerance for activity. Exercise should be progressively increased in a manner that is individualized for every patient.

- For the first week or so of exercise training, patients may be a bit tired later in the day. To compensate for this, patients should temporarily limit the amount of other home-based activities.

- Exercise-training programs for patients with HF are often interrupted because of both cardiac- and non-cardiac-related reasons. The exercise professional should inform patients that these schedule disruptions are common and encourage them to exercise regularly when they feel well.

Summary of Exercise Recommendations for Patients With Cardiac Transplant

Because patients with cardiac transplant present with a decentralized heart, use of a graded exercise test before they begin an exercise program for the purpose of developing an HR range is not indicated. Such a test, however, will help screen for exercise-induced arrhythmias, quantify exercise tolerance, and serve as a marker to assess exercise outcomes at a future date.

Exercise Program

The exercise professional should modulate type, frequency, duration, and intensity of activity so that the patient attains an exercise energy expenditure of 700 to 1,000 kcal · week^{-1}, as tolerated. A summary of the aerobic, resistance-training, and range-of-motion recommendations is provided below.

Training method	Mode	Intensity	Frequency	Duration	Progression	Goals
Aerobic	Cycle or treadmill; arm ergometer as needed	Use 11 to 14 on rating of perceived exertion scale (37)	Four to five times per week	40 min or more per session	10 min or more per session, up to 40 min or more per session	Improve submaximal and peak endurance
Resistance	Fixed machines, free weights, and bands; six to eight regional exercises	Begin with 40% of 1RM; progress over time to 70% of 1RM	One to two times per week	One set of 12 to 15 repetitions for each of the involved muscle groups	Increase 5% to 10%, as tolerated	Initial focus should be on increasing leg muscle strength and endurance; additional emphasis should then be placed on balanced program aimed at preventing corticosteroid-induced sarcopenia and decrease in bone mineral content
Range of motion	Static stretching		Before and after each aerobic or resistance-training workout	5 minutes before and 5 to 10 min after each work-out, with 10 to 30 s devoted to the major muscle groups and joints		

Special Considerations

- Loss of muscle mass and bone mineral content occurs in patients on long-term corticosteroids such as prednisone. As a result, resistance-training programs play an important role in partially restoring bone health and muscle strength (9). But high-intensity resistance training should be avoided because of increased risk of fracture in bone that has compromised bone mineral content.

- Although isolated cases of chest pain or angina associated with accelerated graft atherosclerosis have been reported, decentralization of the myocardium essentially eliminates angina symptoms in most patients.

(continued)

- A regular exercise program performed within the first few months after surgery may result in an exercise HR during training that is equal to or exceeds the peak HR achieved during an exercise test taken before a patient started training. This response is not uncommon and supports using rating of perceived exertion as the method to guide intensity of effort.

- Marked increases in body fat leading to obesity sometimes occur in cardiac transplant patients. This increase is likely caused by both long-term corticosteroid use and restoration of appetite following illness. Exercise modalities chosen for training should consider any possible limitation caused by excessive body mass.

- Because of the sternotomy, postoperative range of motion in the thorax and upper limbs may be limited for several weeks.

- Although definitive evidence is lacking, approximately 15% of cardiac transplant patients experience calf discomfort during walking. This symptom may be related to the use of cyclosporine.

Practical Application 18.5

Literature Review

Although exercise is an important component in the treatment of HF today, it was not until the late 1980s and early 1990s that sufficient research showed that patients could safely derive benefit. Lee et al. (45), Conn et al. (15), and Squires et al. (59) were the first to describe the structured use of exercise training in patients with HF. Improvements in functional capacity and NYHA functional class were observed. In 1988 two uncontrolled trials (3,62) demonstrated improvements in functional capacity, as measured by $\dot{V}O_2$peak. Since that time, dozens of single-site randomized controlled trials showed that exercise training safely results in a 15% to 25% increase in $\dot{V}O_2$peak (table 18.4). Worth noting is the fact that most of the patients in these trials were taking ACE inhibitors, diuretics, and digoxin, which means that the gain in exercise capacity is in addition to what was derived from standard medical therapy at the time. Two of the trials summarized in table 18.4 were completed with patients taking a β-adrenergic blocking agent, a medication that is now part of the standard treatment regimen for patients with HF (16,22). These two studies suggest that β-blocking agents do not impede the ability of HF patients to experience an increase in peak $\dot{V}O_2$ (approximately 15%–20%).

Likewise, improvements in submaximal exercise tolerance have been observed, as evidenced by 20% to 25% improvements in V-LT. The mechanisms responsible for the increases in peak exercise capacity and V-LT are likely several, given that abnormalities in central transport (i.e., HR and stroke volume), nutritive blood flow to the active skeletal muscles, and skeletal muscle histology and biochemistry all contribute to exercise intolerance.

Although no large-scale trial exists to prove that regular exercise training improves survival or lowers recurrent hospitalizations in patients with HF, such a study is currently being conducted (called HF-ACTION; Heart Failure—A Controlled Trial Investigating Outcomes of Exercise Training) and will report its findings from 2,331 patients in late 2008. A 1999 study using a small sample of 99 patients randomized to either 1 yr of exercise training or no exercise did show an approximate 67% decrease in risk for subsequent deaths and hospitalizations among exercise-trained patients (6). Health-related quality of life also improved in these patients compared with the no-exercise group. Conversely, in another relatively small, single-site trial reported in 2001, McKelvie and associates showed that 12 mo of exercise training had no effect on survival, hospitalization, or quality of life (48).

adaptation of an increase in peak HR differs from that observed in healthy individuals, in whom no increase in peak HR is expected with training (23).

Despite initial concerns to the contrary, these changes in central transport occur without concomitant harmful effects to the myocardium. In fact, numerous trials, some up to 1 yr in duration, show that regular training does not lead to disproportionate worsening of LV ejection fraction or an increase in LV size (18,24). This absence of a training-induced maladaptation within the LV was observed with both moderate-intensity and higher-intensity training protocols. More recently, Gianuzzi and colleagues showed that exercise training not only had no negative effect on myocardial function but actually had a small but favorable effect on reverse remodeling (25).

Skeletal Muscle

In response to exercise training, improvements occur in the skeletal muscle. These include improved ability to dilate the small blood vessels that nourish the metabolically active muscles, which leads to an approximate 25% to 30% increase in local blood flow (30,31). The mechanisms responsible for this include an exercise training–induced decrease in plasma norepinephrine (11,36) and other vasoconstrictor agents (8), as well as increased levels of endothelium-derived relaxing factor (28) and flow-mediated improvement in endothelial function (46).

Although only a few studies investigated the effects of exercise training on intrinsic characteristics of the skeletal muscle, a benefit appears to occur at this level as well. Specifically, the volume density of mitochondria and the enzymes involved with aerobic metabolism (e.g., cytochrome C oxidase) are improved up to 40% (30). In addition, a mild shift in fiber type occurs, involving a small increase in myosin heavy chain type 1 fibers (29,38). Finally, Barnard et al. (4) showed a 31% increase

in single-repetition leg extension strength following 8 wk of combined aerobic and resistance training, compared with no change in a group of HF patients who performed aerobic training only.

Other Organ Systems

As mentioned in the section on physiology and pathophysiology in this chapter, patients with HF also have abnormalities of the autonomic nervous system, such as an increase in sympathetic activity and a decrease in parasympathetic activity. This imbalance likely makes the heart more prone to arrhythmia and less responsive to stressful situations.

Several studies show that exercise training partially restores autonomic, immune, and hormonal function in patients with HF (8,11,40,56,65). Down-regulation of the sympathetic nervous system occurs (58), and parasympathetic activity increases. The levels of circulating norepinephrine in the blood both at rest and during exercise decrease with training (11,36). Finally, Gielen and associates showed the anti-inflammatory effects associated with exercise training (26,27).

CONCLUSION

HF, today's fastest growing cardiac-related diagnosis, represents an immense health burden. Current research establishes that among eligible patients with stable HF, regular exercise training improves exercise tolerance and quality of life. Over the next 10 yr much research will investigate and define to what extent exercise alters both the pathophysiology of the disease and the associated clinical outcomes. Healthcare practitioners involved in the care of such patients should consider regular exercise training when developing their treatment strategies (33).

Case Study

Medical History

Mr. WT is a 55 yr old African American male who was initially hospitalized in 1996. The patient stated he had no history of diabetes, angina, or myocardial infarction. He did have a positive history for cocaine use, as recently as 1 wk before hospitalization. He had a strong family history of ischemic heart disease and denied alcohol and tobacco use. He worked as a computer programmer and was sedentary during leisure.

Diagnosis

At the time of admission, he complained of being unable to lie flat without being short of breath, and the condition had worsened over the previous several days. Cardiac catheterization revealed three-vessel coronary artery

(continued)

disease, which included a 20% lesion in the left main, a 95% stenosis in the left anterior descending artery, total occlusion of the circumflex, and a 70% narrowing in the proximal right coronary artery. An echocardiogram performed during the same admission demonstrated a left ventricular ejection fraction of 30%. Diagnosis was ischemic cardiomyopathy. The patient was discharged from the hospital and scheduled to see a cardiologist in the outpatient clinic later in the week. Medications at the time of hospital discharge were lisinopril, simvastatin, lopressor, digoxin, and furosemide.

The patient was seen in the outpatient clinic 6 d after discharge from the hospital. Blood lipids were obtained and revealed total cholesterol of 227 mg · dl^{-1}, high-density lipoprotein cholesterol of 27 mg · dl^{-1}, and low-density lipoprotein cholesterol of 172 mg · dl^{-1}. A dobutamine echocardiogram showed improved contractility. Based on this test and the cardiac catheterization results, coronary artery bypass surgery was recommended. The patient refused surgery, opting for medical management, lifestyle changes, and cardiac rehabilitation.

Exercise Test Results

Before the patient enrolled in cardiac rehabilitation, a graded exercise test was completed. Resting blood pressure was 138/96 mmHg. Resting ECG showed sinus rhythm with a rate of 72 beats · min^{-1}. Heart rate during seated rest was 74 beats · min^{-1}. Occasional ventricular premature beats were noted at rest, along with left ventricular hypertrophy. An old anterior-lateral infarction pattern was present, age indeterminate.

The patient exercised for 6 min on a treadmill to a peak $\dot{V}O_2$ of 17.2 ml · kg^{-1} · min^{-1}. Chest pain was denied and exercise was stopped because of fatigue and dyspnea. Peak HR was 123 beats · min^{-1}, peak blood pressure was 158/98 mmHg, and 1.0 mm of additional ST-segment depression was observed in lead V6. Isolated ventricular premature beats were again observed.

Exercise Prescription

Mr. WT's goal was to make important and aggressive changes in his lifestyle, including increasing his activity levels from being inactive to exercising 4 to 5 d per week. During rehabilitation he was able to tolerate 30 min of exercise without complication. The training heart rate range in this patient that corresponds to 60% to 70% of heart rate reserve is 103 to 108 beats · min^{-1}.

Discussion Questions

1. Given Mr. WT's history of being inactive during leisure, what is the likelihood that he will be following a regular exercise program 3 yr in the future? What steps can you take now to ensure long-term compliance?

2. Which symptoms did Mr. WT complain of at the time of hospitalization that were consistent with the diagnosis of cardiomyopathy or heart failure?

3. What target levels would you recommend that Mr. WT achieve for his blood lipid levels?

4. Based on data from Mr. WT's exercise test, did he show chronotropic incompetence? Is this a common or uncommon finding in patients with HF?

5. Given the $\dot{V}O_2$peak measured during his exercise test, what magnitude of improvement, if any, would you expect after 12 wk of exercise training? How would this compare with the improvement that would occur in a sedentary, apparently healthy person who also undergoes 12 wk of training?

19

Peripheral Arterial Disease

Richard M. Lampman, PhD

Seth W. Wolk, MD

Peripheral arterial disease (PAD) is a form of arterial vascular disease caused primarily by atherosclerosis. PAD is characterized by chronic luminal narrowing of the arterial beds that supply blood to noncoronary arteries. Of concern, PAD is a marker of systemic atherosclerosis, and as such it is associated with advanced coronary and cerebrovascular disease (67,38). PAD is characterized by diminished functional capacity, limb dysfunction, reduced quality of life, increased cardiovascular ischemic risk, and increased risk of death. For the purpose of this chapter, PAD addresses advanced arterial atherosclerosis of the lower limbs.

SCOPE

The prevalence of PAD is common in the adult population and has been estimated to affect up to 10 million people in the United States (20,61). It is usually diagnosed clinically by either calf pain or an abnormal ankle to brachial systolic blood pressure index (ABI). Estimates are that approximately 9.6% of cardiovascular events are reported in people with PAD, and about 17,400 deaths occur each year (75). Age, gender, and the presence of cardiovascular risk factors are major influences for increasing the prevalence of PAD (67). PAD is also common in other countries; a population study from the Netherlands found the prevalence of advanced PAD for individuals over the age of 55 yr to be 16.9% and 20.5% in women and men, respectively (92). The prevalence rate of PAD

in women lags behind men by 10 yr, with male-to-female ratios varying from 1 to 8 with increasing age (81).

A common symptom of PAD is intermittent **claudication** (IC), which produces symptoms of fatigue, limping, lameness, and pain that occur in people while performing mild physical activity (i.e., walking) and is relieved by rest. Symptoms of IC are caused by a lack of adequate blood flow to the skeletal muscle causing anemia, or **ischemia** of the skeletal muscles. When PAD progresses to symptoms of IC, the prevalence of advanced PAD in those over the age of 50 yr has been reported to be 2% to 7% for men and 1% to 2% for women, and more than 5% to 20% in men over the age of 70 yr (66). In an elderly population of patients with comorbidities, the prevalence of PAD is high in those with cardiovascular risk factors, being 61.4% in diabetics, 21.4% in those with ischemic heart disease, and 13.4% in those with renal failure compared with 4.1% in normal controls (2).

The incidence of PAD in the general population is difficult to assess because many people ignore symptoms and do not seek medical treatment. The incidence of asymptomatic PAD, diagnosed with noninvasive testing, is three to four times more frequent than symptomatic PAD. When PAD has advanced to critical ischemia, the incidence is estimated to be between 0.05% and 0.1% of the population (142). A general increase in the incidence of IC appears to occur up to the age of 60 yr (96).

When IC is present, mortality rates increase two to four times, primarily from concomitant cardiovascular disease (77), and estimates are that 9.6% of cardiovascular events occur in patients with IC (75). Life expectancy is poor. The mortality rate is 3% to 5% per year after IC is present and 20% per year when IC advances to critical ischemia. Coronary artery disease accounts for 50% of deaths, stroke for 15% of deaths, and vascular disease in the stomach for 10% of deaths (28). Because of the frequency of concomitant cardiovascular diseases, the exercise professional must be aware of the status of any comorbidities when exercise testing and prescribing exercise training to people with PAD.

PATHOPHYSIOLOGY

The development of atherosclerosis that affects the intima of the peripheral arterial circulation sets the stage for clinically significant PAD. Atherosclerosis of the arterial wall has been hypothesized to develop in response to specific phospholipids contained in low-density lipoprotein that have become trapped and then oxidized in the endothelium (101). **Endothelial cell** aberrations appear to be associated with abnormal oxidative stresses involving increased endothelial superoxide anion, which leads to endothelium-produced nitric oxide or the accelerated degradation of **endothelium-derived relaxing factors** such as nitric oxide (84,104,137). Vasomotor tone is abnormal in the atherosclerotic vessel because of endothelial dysfunction secondary to increased oxidant degradation of endothelium-derived nitric oxide or prostaglandin endoperoxide, PGH2 (55,138). Reducing a major risk factor, such as hypercholesterolemia, can normalize vascular superoxide anion production and improve endothelium dependent vascular relaxation (105).

The percentage stenosis of the artery, as well as the flow velocity across the arterial lesion, accounts for the degree of the hemodynamic abnormality in the distal extremity (18,65). Flow velocity varies inversely with peripheral resistance. Blood flow at rest may not be influenced by a critical arterial stenosis (i.e., 50–90% stenosis), but the lesion can impair adequate blood flow to muscles to meet metabolic requirements with even mild exercise (148).

Regional blood flow of leg muscles, oxygen consumption rate, and oxygen extraction fraction have been shown to be normal at rest in those with PAD, but of these three variables, muscle regional blood flow and oxygen consumption rates were found higher in the presence of PAD compared with the absence of PAD (26,91). Delayed **hyperemia** appears to occur in the ischemic muscle following exercise, and therefore a lag of postexercise metabolism related to the severity of the disease is present.

Metabolic and neural changes possibly associated with PAD include the following:

1. Impaired energy utilization of claudicating skeletal muscle

2. Impaired resynthesis of high-energy phosphate substrate following exercise

3. Low concentrations of adenosine triphosphate and phosphocreatine (111), leading to severe IC pain

4. Loss of muscle fibers leading to abnormalities of oxidative adenosine triphosphate synthesis

5. Reduced metabolic efficiency of the muscle rather than a direct consequence of inadequate blood flow alone

6. Oxidative stress involving mitochondrial injury

7. Peripheral nerve damage that occurs in the presence of ischemia (33,59), leading to both sensory impairment and motor weakness (30)

More studies are needed to discern whether these metabolic and fiber changes are related to PAD or merely to muscle disuse in those with PAD having IC. Evidence does suggest that muscle dysfunction and poor exercise tolerance are characteristic of those with IC (67). Generalized atrophy in leg muscles may result from a loss of muscle fibers, and weakness may be attributable to denervation of the skeletal muscle. Oxidative metabolism may be inefficient. Unfortunately, these abnormalities impair a patient's ability to ambulate efficiently.

Major risk factors for developing PAD are similar to those for coronary artery and cerebrovascular atherosclerosis. Epidemiological studies show that the incidence and severity of this disease increase with the following factors (15,21,61,96,122,142,91):

- Age
- Smoking
- Non-insulin-dependent diabetes mellitus
- Impaired glucose tolerance
- Hypertension
- Low levels of HDL-cholesterol
- Hypercholesterolemia
- Hypertriglyceridemia
- High blood levels of **apolipoprotein** A-I
- High low-density lipoprotein and low HDL-chol combination
- High levels of **homocysteine** and fibrinogen

- Increased blood viscosity
- Elevated **C-reactive protein** levels

Men, primarily those who smoke cigarettes, appear to have the highest incidence of PAD. Now that women have begun smoking at higher rates, the incidence of PAD in women who smoke is on the increase. Estimates are that approximately 25% of women between the ages of 55 and 74 yr have symptoms of PAD disease; women who are menopausal, smoke cigarettes, and have diabetes mellitus are especially at risk (51). Heavy smokers are at a fourfold risk of developing IC (77).

Diabetes mellitus is a major risk factor associated with the development of PAD (96). Hyperinsulinemia in response to an oral glucose challenge has been shown to be an independent risk factor for PAD in the nondiabetic population (116), and smoking appears to exacerbate the post-glucose hyperinsulinemic response (113). Although those with diabetes have a higher incidence of PAD than those with normal glucose tolerance, combined elevated systolic blood pressure and triglyceride levels in those with diabetes may account for the higher incidence of PAD (88).

PAD has also been closely linked to hypertension and hyperlipidemia (77), to abnormal plasma viscosity (112) and plasma levels of fibrinogen (76), and to hyper-homocysteinemia (74). These metabolic abnormalities should be considered as potentially being influenced by treatment with exercise therapy.

Risk factors related to progression of PAD appear to differ for large and small vessels. Aboyans et al. (1) reported that progression in large vessels was related to smoking, ratio of total to HDL-cholesterol, lipoprotein (a), and C-reactive protein, whereas progression in small vessels was related to diabetes.

MEDICAL AND CLINICAL CONSIDERATIONS

A patient suspected of having advanced vascular disease should be thoroughly evaluated for cardiovascular risk factors and aggressively treated for any underlying advanced vascular or metabolic diseases. Evaluating these clinical conditions is essential to providing appropriate medical treatment. This information is also important for providing appropriate dietary treatment, especially when prescribing an individualized exercise program.

Signs and Symptoms

Impaired exercise capacity resulting from IC in those with advanced PAD can severely a person's ability to meet the occupational, social, and personal demands of daily life. As IC progresses, symptoms occur at progressively shorter periods of physical activity. This progression can threaten or impair functional independence, diminish subjective quality of life, and result in inability to remain gainfully employed (78). Estimates are that 25% of those with IC will need some type of vascular intervention and that less than 5% will require a major amputation. But the atherosclerotic process can progress to obliterate previously nondiseased arteries in both lower extremities (142). When present, IC is associated with a 20% to 40% increase in mortality rates, primarily from cardiovascular disease (77).

PAD is characterized by an initial asymptomatic phase followed by a period of symptomatic IC and finally a phase of critical limb ischemia (ulceration, tissue loss, and **gangrene**). Acute arterial occlusion may occur at any point along the continuum. The prognosis for limb-threatening ischemia is reasonably good. The need for vascular intervention arises in only 25% of men with IC. Only about 5% of patients with IC will require a major amputation of a leg or a foot. The atherosclerotic process can, however, progress to previously nondiseased arteries, and initially unilateral disease may become bilateral (142). Furthermore, because atherosclerosis is a systemic disease, those with PAD have increased cardiovascular morbidity rates (e.g., myocardial infarction, cerebrovascular events, and rupture of aortic aneurysms) and mortality rates when compared with a normal population (67). Walking tests were initially used to evaluate the functional capacity and onset of IC in patients with PAD. More recently, the ability of an older patient with PAD to walk 400 m was reported to be a prognostic factor for mortality and cardiovascular disease, with a reduction in these indices with higher functional capacity (103).

To characterize PAD clinically, a variety of symptom severity scales have been devised. Two popular scales are the Fontaine stages and Rutherford categories (127).

The Fontaine stages are as follows:

- I, asymptomatic
- IIa, mild claudication
- IIb, moderate to severe claudication
- III, ischemic rest pain
- IV, ulceration or gangrene

The Rutherford categories are as follows:

- 0, asymptomatic
- 1, mild claudication
- 2, moderate claudication

- 3, severe claudication
- 4, ischemic rest pain
- 5, minor tissue loss
- 6, major tissue loss

Advanced PAD is manifested by exercise-induced ischemia of the lower limb resulting in symptoms of moderate to severe muscular pain resulting in IC. Rest usually relieves symptoms of pain associated with IC. The pathogenesis of IC is related primarily to restrictive or occlusive atherosclerotic ischemia but may be caused by endothelial dysfunction or thrombosis proximal to the exercising muscle.

History and Physical Examination

After seeing his or her primary care physician, a patient suspected of having advanced PAD should see a vascular surgeon for a more thorough vascular evaluation before embarking on an exercise program. The vascular surgeon's evaluation to confirm advanced PAD will include a medical history, a physical examination, measures of ankle and toe pressures, an ABI for each leg (figure 19.1), a determination of which arteries are involved, an assessment of cardiovascular risk factors, and an assessment of the degree of disability and severity of symptoms. Smoking history, systolic and diastolic blood pressures, lipid and lipoprotein profiles, diabetes status, hemoglobin, serum creatine, height, weight, and body mass index (calculated) should also be included in the initial evaluation. For special research projects, a hypercoagulability screen can be performed, and homocysteine, apolipoprotein A-I [LP(a)], and C-reactive protein levels can be measured.

Exercise training can be a first-line treatment for PAD and is usually the prime reason for referral of a patient with PAD to an exercise physiologist for an individualized exercise prescription. Because those with PAD have many comorbidities, documentation of the following medical problems should be made: IC symptoms, cardiovascular symptoms such as angina, previous myocardial infarction or stroke, degree of muscle weakness, smoking habits, and concurrent diseases such as diabetes or hypertension. Medications (such as insulin) that the patient is taking that may be influenced by routine exercise should also be noted. Because these cardiovascular and metabolic maladies greatly increase the risk for cardiovascular complications that could potentially be exacerbated with vigorous exercise, they should be thoroughly evaluated if suspected, and the patient should receive aggressive medical or surgical treatment before embarking on a vigorous exercise program.

Patients with a maximum metabolic equivalents (METs) level less than 5 METs are probably at higher risk for cardiac complications if they undergo high-risk vascular surgery (12). The risk for those with a low metabolic equivalents level who engage in routine low-intensity exercise therapy has yet to be determined and

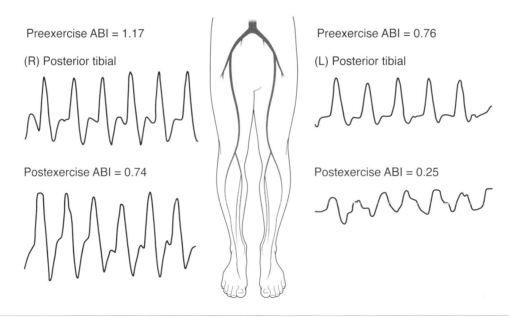

Preexercise ABI = 1.17

(R) Posterior tibial

Postexercise ABI = 0.74

Preexercise ABI = 0.76

(L) Posterior tibial

Postexercise ABI = 0.25

Figure 19.1 Right and left lower extremity Doppler velocity tracings and ankle-brachial index (ABI) at rest and immediately following acute exercise.

may not be a factor in the presence of adequate treatment for comorbidities. Medications to treat hypertension, dyslipidemia, carbohydrate intolerance, left ventricular hypertrophy, platelet aggregation, and coronary artery disease are often used in this patient population because many patients with IC have many concurrent metabolic and cardiovascular diseases. Exercise therapy through a structured individualized exercise program can be of great importance because it can profoundly improve many of these medical conditions. Moreover, after a patient with PAD becomes involved in regular exercise, medications for some or all of these diseases can usually be reduced or even discontinued. Furthermore, exercise treatment can markedly improve functional capacity to the point where a person can perform the activities of daily living.

In general, indications and contraindications to exercise stress testing for those with PAD are similar to those for cardiac patients (4,40). In those who are without known heart disease or are stable with respect to cardiovascular disease and plan to begin a low-level exercise therapy program (walking at their normal walking speed or exercising at equivalent intensity on some mechanical device such as a treadmill), an initial progressive exercise stress test, with multilead electrocardiogram monitoring for detecting cardiac ischemia, may not be necessary. Such a test may be difficult to administer because of IC symptoms that prevent a patient from accomplishing an adequate cardiac workload for diagnostic purposes. In this case, the exercise-training therapy intensity should be kept similar to the person's exercise effort while maintaining a normal walking speed. Guidelines for contraindications to exercise testing and training can be found in *ACSM's Guidelines for Exercise Testing and Prescription* (4).

A peripheral vascular assessment may include resting and postexercise testing on a treadmill for determining resting and exercise ABI, peak heart rate and blood pressure, and time and distances to the onset of claudication and to maximal claudication pain. Peak oxygen consumption measures are usually impractical to obtain because of claudication symptoms that cause a patient to terminate a leg test prematurely before arriving at a peak cardiovascular response level. Attenuated levels of cardiovascular response to exercise testing have been shown in those with PAD (68).

Diagnostic Testing

The diagnosis of advanced PAD can be made clinically by palpation of the patient's peripheral pulses, but lack of a pulse can be attributable to causes other than PAD.

The ankle to brachial index (ABI) is an excellent noninvasive testing method for detecting asymptomatic PAD in office practice, vascular laboratories, and epidemiology studies. The ABI is calculated using the systolic blood pressure obtained from both the dorsalis pedis and the posterior tibial arteries and from both brachial arteries. If the systolic blood pressure in the brachial arteries is unequal, the higher pressure is used, because the lower value may be a result of advanced subclavian or axillary arterial stenosis. A calculated resting ABI should be close to 1.00 and should be recorded to two decimal places. A normal ABI ranges from 1.00 to 1.29. A stenosis of at least 50% is considered important hemodynamically and results in a decreased resting ABI of approximately 0.15 (9). Postexercise ABI may be a useful method for identifying severe ischemia. An ABI can also help detect the severity of PAD and has been shown to be associated with symptoms and cardiovascular mortality, as shown in table 19.1. A resting ABI of less than 0.5 is a significant predictor (relative risk of 2.3) for marked progression of PAD (29).

An ABI at rest has been reported to be 1.08 ± 0.08 for a normal, physically untrained individual and 1.15 ± 0.05 for a physically trained person; these values have been shown to fall 1 min following acute strenuous exercise to 0.70 ± 0.06 and 0.80 ± 0.08, respectively, in untrained and trained subjects (27). Those with PAD who have IC usually present with a resting ABI ranging between 0.50 and 0.90 (66). A resting ABI of less than 0.50 is usually associated with severe limits on activities of daily living. A resting ABI of less than 0.30 is associated with more severe ischemic symptoms and poor prognosis for limb salvage (51).

Noninvasive testing in a population produces a prevalence of PAD about five times higher than would be expected. Hiatt, Hoag, and Hamman (61) demonstrated that to achieve the best clinical diagnosis of PAD, an abnormal dorsalis pedis and posterior tibial ABI in

Table 19.1 Association of Symptoms and Cardiovascular Morbidity Rate (Over 3 Yr) to the Ankle–Brachial Index (ABI) in Patients With Intermittent Claudication (IC)

ABI	Symptoms with walking	Cardiovascular morbidity rate (%)
>0.7–<0.9	None to slight IC	12
0.5–0.7	Debilitating IC	33
<0.3–0.5	Severe IC	60

Table 19.2 Diagnostic Criteria for Determining the Prevalence of Peripheral Arterial Disease (PAD) in Population Studies Using Specific Abnormal Cutoff Points for the Ankle–Brachial Index (ABI)

Criteria used to detect PAD	Detection of PAD and association with cardiovascular risk factors
Abnormal resting ABI in both dorsalis pedis and posterior tibial arteries in the same lower extremity	Higher frequency of PAD and an absent pulse compared with normal subjects or those with one abnormal vessel or an abnormal ABI after exercise Increased risk of PAD with increasing age, non-insulin-dependent diabetes, smoking, hypertension, and hypercholesterolemia
Abnormal resting ABI in one lower extremity	Increased risk of PAD with age and smoking alone
Abnormal ABI in either lower extremity only after exercise	Not associated with any cardiovascular risk factor except male sex

the same leg at rest should be present (see table 19.2). Skin microcirculatory screening and classification have been reported to be clinically useful in detecting critical ischemia that will likely result in amputation of the limb (141).

Murabito et al. (96) have developed an IC risk profile based on the Framingham Heart Study. This profile is an office survey tool that allows for the identification of high-risk patients and can also be used as a basis for educating patients on modifying risk factors for IC. The severity of an ipsilateral limb with advanced PAD can be obtained through the San Diego Claudication Questionnaire (20).

Tests performed in the noninvasive laboratory to evaluate patients with PAD regarding hemodynamic and functional parameters include the following:

- **Doppler ultrasonography**
- Pulse volume recording
- Segmental blood pressure measurement
- Transcutaneous oximetry
- Color-assisted ultrasound imaging
- Exercise testing
- Post occlusive reactive **hyperemia** testing
- Near-infrared spectroscopy

Total body and local metabolic parameters are important to measure in research protocols if appropriate testing equipment is available. Recent evidence suggests that elevated blood levels of C-reactive protein can predict future risk of developing symptomatic PAD (109).

Exercise Testing

An exercise stress test (best performed using a motorized treadmill) can be used to establish the presence of

advanced PAD in a patient at high risk for the disease who has a normal resting ABI. An immediate postexercise fall in the ABI would suggest a high-grade stenotic lesion in the lower limb and could help objectively determine functional impairment. Exercise stress testing also provides baseline hemodynamic information not present at rest and can provide helpful information in the treatment of comorbidities, especially silent myocardial ischemia, that are common in those with advanced PAD (129).

Because the functional capacity of patients with PAD is often low and their physical performance is usually restricted by their symptoms of IC, they present special testing needs when they undergo evaluation of their cardiac status and functional capacity in response to acute exercise. If the major objective is to evaluate the patient's cardiac status (i.e., the presence of myocardial ischemia), the test ideally should be strenuous enough to elicit a pressure rate product (PRP = systolic blood pressure × heart rate) close to 3,000. This double product (PRP) noninvasively estimates myocardial oxygen uptake. A patient with PAD is usually unable to perform an appropriate amount of skeletal muscular work with leg exercise alone to elicit an adequate PRP because of IC symptoms, rather than angina symptoms. Such a patient may need to perform arm work (e.g., arm ergometry) together with leg work (10,24,114) to achieve adequate myocardial stress for detecting myocardial ischemia. The sensitivity for detecting myocardial ischemia in those with IC might be enhanced by including thallium scintigraphy with leg and arm ergometry (10). An alternative method to arm and leg ergometry is to induce a cardiac challenge by using a pharmacological agent such as ergonovine. Although this pharmacological method can be used to help evaluate a patient's cardiac status, it would not provide important information regarding hemodynamic measures and claudication indexes in response to exercise.

Exercise stress testing can be used to evaluate changes in the ABI following acute exercise (see figure 19.1). Functional impairment can be assessed by maximal treadmill walking time, pain-free walking time (initial claudicating distance), and walking time to severe claudication (absolute claudicating time). For research, a run-in phase that consists of two or three treadmill tests can be performed to familiarize patients with treadmill testing (81). The end point of testing is determined by the patient's perception of the severity of pain and may vary from test to test until a patient becomes familiar with the pain sensation. During testing, the patient should not be allowed to use handrail support for his or her body because doing so would influence the reliability of test results (47).

In patients with PAD, either a level walking protocol or a progressive workload treadmill protocol can be used clinically to test for claudication pain and for hemodynamic measurements of the lower limbs in response to exercise (48). Internationally accepted settings for a constant workload are a speed of 2 mph (3.2 kph) and a grade of 12% (best for those with claudication distance between 50 and 150 m) (81).

Although a single-load test results in the workload and walking time following a linear function, a graded test usually follows a progressive incremental workload with respect to time. A graded exercise test starts at a horizontal level, and the slope is incrementally increased at specific time intervals throughout the test. The speed may vary but is usually kept constant because a patient with IC may find it difficult to adjust to speed changes. A typical protocol for a graded treadmill testing protocol is a constant speed of 2 mph (3.2 kph) throughout, 0% grade for 3 min initially, and a 3.5% increase in grade every 3 min thereafter (81).

Other testing protocols may have some utility. Gardner et al. (51) reported that a progressive stair-climbing testing protocol was also reliable for clinically evaluating claudication and hemodynamic responses to exercise in patients with stable IC. Furthermore, similar peak oxygen consumption (O_2peak) values were found in this study between testing protocols when patients with IC underwent level walking, graded walking, or progressive stair climbing. Stair climbing, although it results in similar metabolic, claudication, and peripheral hemodynamic parameters, may not be the test of choice if cardiac parameters are of interest. This guidance is true especially if patients are given an exercise prescription that involves walking, because fewer metabolic demands may be placed on the cardiovascular system with stair climbing (47).

For diagnosing lower-limb arterial insufficiency in an office setting, 30 s of heel raising has been shown to cause changes in ankle pressure 1 min postexercise, expressed as percentage change, similar to those induced by treadmill exercise (5).

Advantages and disadvantages of different testing protocols for patients with PAD are shown in table 19.3. Basically,

Table 19.3 Exercise-Testing Protocols for Patients With Peripheral Arterial Disease

Mode of testing	Advantages	Disadvantages
Treadmill test symptoms	Common form of physiological stress	Cardiac evaluation is limited because of IC
Single-stage test	Good test for those who are extremely deconditioned	May not be strenuous enough for some patients
Multistage continuous	Results in high retest correlation coefficients	Sometimes difficult for patients to adjust to work increments
Ramp protocols	Involves a steady increase in cardio respiratory and hemodynamic responses	Lacks norms for physiological and hemodynamic responses in those with IC
Treadmill	Familiar activity; usually the mode of exercise training	Limited by claudication if a high pressure rate product is required for a cardiac evaluation
Stair stepping	Progressive test is reliable for evaluating symptoms of IC	Lacks norms for physiological responses
Cycle ergometry	Relatively safe testing mode for unstable patients Good test for those with limitations that restrict weight bearing	Not a familiar activity; not best for evaluating calf claudication Unfamiliar method of exercise IC symptoms may differ from those obtained while walking

(continued)

Table 19.3 *(continued)*

Mode of testing	Advantages	Disadvantages
Cycle and arm ergometry	Good test for detecting and evaluating myocardial ischemia in those with IC	Difficult to measure blood pressure responses
Heel raising	Good test for office setting without adequate testing equipment	Unknown workloads for test reproducibility

Note. IC = intermittent claudication.

Practical Application 19.1

Client-Clinician Interaction

Both accelerometers and pedometers have been reported to produce reliable estimates of physical activity on 2 consecutive days (132). Patients need to provide information regarding their quality of life and functional status to their clinicians, because their symptoms of IC usually prevent them from engaging in normal activities of daily living (61). Daily functional status can be assessed by validated questionnaires that a clinician can administer such as the Walking Impairment Questionnaire (which tests ability to walk), the Physical Activity Recall (habitual physical activity level), and the Medical Outcomes Study Short Form 20 or 36 (physical, social, role functioning, well-being, and overall health) (118). Most scores on these questionnaires improve with exercise therapy, more so with treadmill walking than with strength training, and scores continue to improve with exercise therapy over time (61).

The quality-of-life assessment tool, the Short Form 36, is considered a valid survey instrument for evaluating those with IC and many other patient populations (135). The Sickness Impact Profile has been shown to be a sensitive method for evaluating the overall dysfunction in those with IC (6). When administered to patients with IC, the McMaster Health Index Questionnaire revealed significant impairment in general health and physical, social, and emotional function in patients compared with controls (11). Furthermore, because the study found that those measures were not associated with treadmill performance, the clinician needs to consider the many ramifications of the disease itself rather than the functional level of the patient when providing a treatment prescription.

A 6 min walk test, performed outside the laboratory by those with IC, has some clinical utility, because this test has been shown to produce highly reliable measurements that are associated with the functional and hemodynamic severity of IC (94). For more physically functional and otherwise healthy patients, a 6 min walk test can provide the clinician with objective information for evaluating the progress of exercise therapy without expensive laboratory equipment. Furthermore, patients can perform this test on their own and without sophisticated testing equipment.

the mode of testing depends on whether the objective is diagnosis for advanced cardiac disease or IC symptoms, or for prescribing individualized exercise therapy.

$\dot{V}O_2$peak in patients with IC can be predicted by using a multiple regression equation with measures of time to maximal claudication pain and maximal heart rate taken during an incremental graded exercise test (144). Practical application 19.1 discusses methods for evaluating daily physical activity and quality of life in people with PAD.

Treatment

The goals of therapy for patients with IC are to relieve symptoms; prevent progression of atherosclerosis, which may result in severe ischemia (critical limb ischemia), gangrene, and limb loss; prevent other cardiovascular or cerebrovascular events; and improve functional capacity and quality of life (60,67). Treatment options by clinicians for patients with IC include exercise therapy, medical intervention, **angioplasty,** and surgery (table 19.4). Exercise therapy is an efficacious treatment modality for PAD. A minimum of three exercise-training sessions per week must be maintained because people exercising at that frequency showed less decline in their functional capacity compared with those not exercising over time (91). Aggressive risk factor reduction for atherosclerosis should be a high clinical priority (54). Angiotensin-converting enzyme inhibitors and other antihypertensive drugs, antiplatelet therapy (with aspirin or clopidogrel),

Table 19.4 Treatment Options

Medical treatment		
Medication	**Outcomes**	**Comments**
Pentozifylline (an oral for hemorhelogic agent)	May be beneficial to patients with ischemic rest pain	Unclear whether this drug was efficacious for those with only IC (80)
Cilostazol (a drug with vasodilating and antiplatelet properties)	Significant statistical improvement was found in walking distance	Clinical relevance has been questioned because patients experienced side effects of the drug such as headache, abdominal stools, diarrhea, and dizziness (93)
Carnitine or propionyl-L-carnitine	Treatment for patients with IC resulted in increased claudication-limited exercise (60)	
Vasodilators	Treatment for IC has been shown to be ineffective (17)	
Verapamil	An optimal dose has been shown to increase mean pain-free walking distance by 29% and maximal walking distance by 49% in patients with stable IC (8)	
Surgical interventions		
Technique	**Outcomes**	**Comment**
Central reconstructive vascular surgery	May be associated with both short-term and long-term morbidity and mortality rates of the surgical procedure	Usually not performed in most patients with PAD because their major symptom of IC can be successfully improved with daily exercise therapy and, if necessary, medical treatment and weight reduction
Surgical revascularization or PTA	Recommend that patients follow a physical training program after arterial reconstruction surgery for IC to preserve the high metabolic capacity in the peripheral skeletal muscles (70)	Justifiable only if they are relatively safe and effective and will be durable for certain arterial vessels
PTA, laser-assisted angioplasty, and atherectomy for appropriate arteries	PTA may be most effective for an aortoiliac stenosis (143)	May be performed in the presence of severe ischemia (56); <0.5% morbidity
Endoluminal stent placement	Patients with IC had improved hemodynamics	Major complication rate of 11% (98)
Surgical bypass	Normalizing the patient's ABI	Risks of both short- and long-term morbidity (5–10%); mortality (2–3%)
Risk factor reduction programs		
Program	**Outcomes**	**Comment**
Smoking cessation	Smoking is a major risk factor for PAD (a two- to threefold increase)	
Diet and weight reduction	Carrying extra weight of 5 kg or more significantly reduces a PAD patient's claudication distance (147)	Obesity is directly related to claudication distance and is a risk factor for cardiovascular events

(continued)

Table 19.4 *(continued)*

Risk factor reduction programs *(continued)*		
Program	**Outcomes**	**Comment**
Serum lipid and glucose normalizing	Lipid-normalizing therapy has been shown to reduce the risk in the development or worsening of IC (110) Medication, diet, and exercise will help normalize elevated LDL cholesterol (<100 mg · dl⁻¹), hypertriglyceridemia, and low levels of HDL cholesterol (41)	Diabetes is highly associated with PAD and with the progression of vascular disease (41) Because abnormal rheological factors are thought responsible for reduced blood flow in the legs of patients with IC, worsening of their ischemic condition, and promoting progression in clinical symptoms, treatments to normalize hemostatic and rheological factors may prevent further deterioration by PAD (133)
Routine exercise	Exercise therapy has been shown to improve glucose homeostasis, probably by reducing insulin resistance (124)	Exercise therapy has been shown to improve many risk factors for cardiovascular disease
Physical exercise therapy		
Technique	**Outcomes**	**Comment**
	Effectively improves a patient's ability to walk significantly farther before the onset of ischemic pain (137) Improves functional activity and ability to perform activities of daily living	The following were shown to be important for the greatest improvement in distance to walk pain: Frequency: at least three to five sessions per week Intensity: at an exercise level close to maximum pain during training Time: 30 min or more of walking per exercise session Mode of exercise: walking Length of program: greater than 6 mo (45)

Note. IC = intermittent claudication; PAD = peripheral arterial disease; PTA = percutaneous transluminal angioplasty; ABI = ankle–brachial index; HDL = high-density lipoprotein; LDL = low-density lipoprotein.

statins, cilostazol (Pletal), and pentoxifylline (Trental) are recommended medical treatments that the clinician can provide to the client (16,52,54,136).

Exercise therapy, caloric restriction for those who are overweight, increased dietary fiber and a diet rich in monounsaturated fatty acids (olive oil) and n-3 polyunsaturated fatty acids (fish oils) in those who have abnormal lipid blood levels are important lifestyle and dietary changes that can favorably affect comorbidities (53,102,124). Patients with IC are often depressed (97). Therefore, the clinician should consider the patient's psychological status when he or she is being treated for incapacitating IC, especially when the patient with PAD is undergoing smoking cessation. Surgical treatments and angioplasty are costly, can increase morbidity rates, do not address comorbidities, and do not treat associated

mental or physical problems of poor functional capacity and weakness. Criteria for possible surgical intervention when a lower limb is in jeopardy include symptoms of severe ischemia at rest over a 2 wk period, a systolic blood pressure at the ankle less than 50 mmHg, or a systolic blood pressure less than 30 mmHg at the big toe.

EXERCISE PRESCRIPTION

Patients with PAD who desire to improve their medical status and are willing to comply closely with their exercise program should be counseled by their clinician about the many benefits of routine exercise. The clinician needs to discuss expected physiological, hemodynamic, and psychological outcomes so that the patient will understand how he or she will personally benefit from engaging in routine exercise training. An important consideration is that the patient's primary care physician and vascular surgeon support and encourage long-term exercise-training therapy.

Compliance and Adherence Considerations

Cognitive factors such as stress, motivation to participate in exercise training, and the belief that the therapy is beneficial have been shown to relate to improvements in walking distance (125). Rather merely suggesting exercise training to the patient without providing explicit directions, the physician should refer the patient to a specialist (e.g., exercise physiologist or physical therapist) to receive an individualized, progressive, and structured exercise-training program. This individualized approach provides a program of exercise therapy that affords the lowest possible risks of complications and takes into account the patient's comorbidities and medications that may necessitate close supervision by the patient's primary care physician or vascular surgeon. Merely receiving a simple recommendation from a clinician to exercise will not, in most cases, help a patient engage in appropriate exercise therapy, and recidivism will be high. As the patient becomes more physically fit, he or she may proceed to a higher level of exercise-training effort, a decision made by the patient and his or her primary care physician, vascular surgeon, and exercise physiologist or physical therapist.

Patients with IC are often frightened to push themselves physically for fear of damaging their painful claudicating lower-limb skeletal muscles. They need to be assured that walking through their claudication pain while exercise training will not cause lasting harm to their muscles and will ultimately result in better muscle function. In those with IC, no pain discomfort with exercise will result in little physiological gain in improvements in their pain-free walking distance over time with exercise therapy. Patients with PAD, because of the many comorbidities that they have, should clearly understand that if they have chest pain (possibly angina pectoralis) they should stop exercising and, if it persists, go to a local emergency room immediately. But patients should be encouraged to tolerate pain in the legs attributable to claudication as well as they can, and they should be continuously reassured that exercise will not harm the skeletal muscles. Patients need to understand that routine exercise is one of the most beneficial therapies that they can do for their vascular condition as well as for reducing risk factors of their underlying general atherosclerosis.

Proper foot care is important, especially in those with diabetes mellitus, to prevent blisters and possible infections. Properly fitting footwear and socks that do not bunch up in a patient's shoes are crucial to prevent the development of soft-tissue injury or foot ulcers. Daily inspection of the toes and plantar surfaces of the feet is essential to early detection of any abnormality. Patients should be advised to return to their physicians immediately should any changes occur in their feet.

Resistance exercises can supplement walking or other cardiovascular activities. These exercises can be properly selected for strengthening muscles of major muscles groups throughout the body. Selection of muscle groups should be based primarily on those needed to perform activities of daily living but should be extended to muscles that a patient might use for work or recreational activities. The resistance-training effort should be easy to moderate, using free weights and dumbbells or a variety of weight resistance machines, and should include 8 to 10 repetitions performed one to three times with slow concentric and eccentric muscle contractions.

Although patients with IC usually exercise at low intensity, they should begin their aerobic exercise session with a very low walking intensity for 2 to 5 min (warm-up) and end the exercise session the same way (cool-down). These warm-up and cool-down sessions may not be important until a patient can perform at least 10 to 15 min of continuous brisk walking. Please review practical application 19.2 on p. 344 for further information pertaining to exercise prescription in PAD patients.

Recommendations

An aerobic exercise program can be beneficial for PAD patients. Although a patient's overall blood flow in the diseased extremity may not change markedly with exercise therapy, patients will experience physiological adaptations that promote better utilization of oxygen, increase their walking time, and, most important, enhance their quality of life. Walking is the mode of exercise training most effective for reducing IC pain symptoms in patients with advanced PAD. In respect to duration, the goal is to reach 35 to 50 min of activity per session. But with IC symptoms, continuous exercise for this length of time is difficult, so starting with intermittent exercise is strongly suggested. The ability to achieve continuous exercise for the suggested period depends on the severity of disease and the patient's pain threshold. The severity of disease also directly influences the intensity of the walking pace, so the patient should start at or below his or her normal walking pace and progress to tolerable pain. Lastly, in regard to frequency of exercise for PAD patients,

Practical Application 19.2

Exercise Prescription for PAD Patients

Mode of training	Frequency	Intensity	Duration	Special considerations
Intermittent walking (82)	Thee to five times per week	Initially: normal walking pace with goal of effort close to maximum pain while walking	30–50 min	Daily goal: expending 2,000 kcal · wk⁻¹
Resistance training				
Mode of training	Frequency	Repetitions	Sets	Special considerations
Weight training or resistance training	Three to five times per week	8 to 10 repetitions over the full range of muscle function with slower eccentric muscle contractions	One to three sets per session	Muscle group: specifically those of lower extremities and generally major muscle groups throughout the body

Method for aerobic exercise training: intermittent walking bouts of 2 to 5 min followed by a 1 to 2 min recovery period, performed repetitively to accomplish 30 to 50 min of walking per exercise session. Progression accomplished by increasing follow-up: 2 to 6 mo following exercise training. Progression of effort for resistance training: no more than 10% according to readiness.

progressing from two or three exercise sessions per week to daily activity should be the goal for optimal benefits. Enhanced functional performance can be seen soon after the person embarks on a routine exercise program, but it usually takes 6 mo of exercise training to optimize IC pain-relief walking distances. See practical application 19.2 for more information regarding prescribing exercise for patients with PAD.

Mode

Walking is an excellent training method for people with IC. Walking, a rhythmic, dynamic aerobic activity, is the most common weight-bearing activity that is both familiar and, in most cases, enhances the muscles most affected by PAD (95). Walking on a regular basis, with duration and intensity progressed systematically, develops and sustains an individual's physical fitness functional capacity and usually specifically adapts muscle having the most severe claudicating symptoms. Intermittent walking programs, starting at low intensity (at approximately the patient's normal walking speed) and progressing slowly to a more brisk pace over weeks, is an excellent method to improve a patient's fitness level without markedly increasing the risk of adverse events (i.e., musculoskeletal or joint injuries) (82).

Strength training alone or in combination with treadmill training for patients with IC has been compared to determine whether one modality of training was superior (58). These investigators reported that patients with PAD and IC symptoms showed improvements in peak walking times with treadmill exercise or with weight training for skeletal muscles of the legs. The treadmill group, however, showed improvement in $\dot{V}O_2$ peak, but no changes were observed in $\dot{V}O_2$ peak or claudication onset time for those participating in weight training alone. These results further demonstrate the importance of using walking as a major exercise treatment modality for those with IC, but they also show that resistance training can serve as an adjunct treatment for strengthening major muscle groups in the legs.

Frequency

At the onset of exercise therapy, frequency should be a minimum of two to three times per week. As the person adapts to the exercise program and muscle soreness becomes minimal 24 h postexercise, the frequency should gradually increase (i.e., an additional day every 1–2 wk). Ideally, the goal for PAD patients would be to walk daily and, if possible, eventually reach a caloric expenditure of 2,000 kcal · wk⁻¹.

Intensity

The exercise intensity prescribed for patients undergoing therapy has implications when considering time to onset of claudication pain and time to maximal claudication pain but appears not to alter the dissipation of pain during recovery (44). This circumstance is one reason why intermittent exercise-training protocols are ideal for patients with IC. Furthermore, the intensity of effort is not associated with the decrease in ABI with exercise but is related to the recovery time for measures of the ABI (128). These findings are important for prescribing exercise training based on a patient's perception of pain, but patients need to be encouraged to tolerate as much IC discomfort as possible. Because intermittent exercise protocols allow recovery periods interspersed throughout exercise sessions, they are ideal for patients with IC (82). Intermittent exercise therapy can also be effective in increasing total work accomplished and can be followed more closely by patients with advanced PAD and associated IC symptoms.

The intensity of effort can initially be set at a normal walking pace because the patient can readily perform at that level and feels safe doing so. Setting intensity according to a target heart rate often proves impractical because many patients with IC are older and find it difficult to palpate their pulse and monitor it using a wristwatch. The patient interested in monitoring his or her exercise heart rate electronically can use an accurate, commercially available heart rate monitor.

PAD progresses with age. Those who are older (mean age of 75.5 yr), compared with those younger (mean age of 60.4 yr), may show greater impairment in peripheral hemodynamic measurements but do so without exaggerated heart rate or blood pressure responses during a progressive treadmill walking test to maximal leg pain (43). This finding would have implications for training if exercise intensity were based on age-predicted exercise heart rate responses rather than IC symptoms.

An optimal threshold of exercise intensity is not known for those with PAD and most likely varies from one patient to the next depending mostly on the severity of PAD and related symptoms. The intensity need not be high, especially when a patient is embarking on an exercise program. The focus should be on the total time (preferably 35–50 min) of exercise per session. The initial intensity of training chosen for those with PAD should be similar to their normal ambulating pace. This approach to intensity of effort has many advantages, as patients with IC are usually extremely deconditioned because of their IC pain symptoms, which often prevent them from even minimal physical activity. Having patients with IC exer-

cise at their normal walking pace usually allows them to begin and progress in an exercise-training program well within their capabilities. As a patient's physical fitness level improves, an exercise stress test is recommended for evaluating cardiac status and a claudicating index before the intensity of exercise effort is markedly increased to strenuous exercise intensity.

Duration

The duration of physical activity depends on the severity of PAD. The goal is to achieve 35 to 50 min of continuous walking, even at a slow pace. Because of the uniqueness of exertional IC that occurs with PAD, exercise effort needs to be intermittent. Generally, a good starting point is to determine an intensity that the person can maintain for 2 to 5 min. A recovery phase of 1 to 2 min follows. The exercise intensity can begin at the patient's normal walking pace, or slightly slower than normal, and progress to a pace that is equivalent to the person's maximum tolerable pain. Irrespective of the initial walking time, the duration of activity per bout of exercise should ideally be increased by 1 min per week.

EXERCISE TRAINING

Womack et al. (145) showed that exercise rehabilitation for patients with PAD and IC symptoms improved their walking economy as assessed by a decreased slow component of the oxygen cost of ambulation. A meta-analysis of 21 studies showed that exercise training for those with PAD resulted in about a 180% increase in walking distance before the onset of claudication (136). Examination of only randomized controlled studies showed that those with PAD undergoing exercise training showed significant improvements in walking time as compared with those undergoing angioplasty and antiplatelet therapy (85). Pentoxifylline (Trental), a rheologic modifier, has been approved for treating claudication and has been shown to be more efficacious than placebo, but not cilostazol (Pletal). But study results do not support the efficacy of the use of pentoxifylline in general (64,106). Cilostazol has been shown to improve maximal treadmill walking distance as compared with placebo or pentoxifylline (64,108,121). Exercise therapy, however, is superior to either medical treatment in improving walking distance (136).

Patients may be in a supervised rehabilitation setting (45,58,62,120) or an unsupervised home exercise program (7,13,72,82), depending on their needs, especially as it applies to comorbidities that may need close monitoring. Supervised exercise programs have shown about a 150 m walking distance improvement over nonsupervised exercise therapy (13). Home exercise programs, however, are usually conducive to a patient's schedule and are less costly than hospital-based programs. This latter point is important, because many patients are retired and on fixed incomes. In older adults, home-based exercise programs have higher adherence rates as compared with center-based exercise programs (7). Patients may benefit equally by either approach as long as they closely follow an individualized and structured exercise protocol (72,82) rather than merely being told by their physicians to "just go home and exercise" (19,115).

Exercise training can result in improvements in walking distances that exceed those obtained from drug therapies, a marked increase in routine activities of daily living, and usually a reduction in cardiovascular risk (67). If a patient is without symptoms of cardiac disease or has cardiac disease but is otherwise stable regarding cardiovascular status, he or she may be able to exercise safely at home by following a low-intensity exercise program. Low-intensity exercise therapy has been shown to be similar to high-intensity exercise rehabilitation in improving initial IC distances and absolute IC (49). Even if close cardiac monitoring is initially required, most patients can eventually exercise at home on a routine basis. As patients improve their functional status, they may experience some adverse cardiac symptoms, at which time they should be reevaluated by their primary care physician, who may order an exercise stress test, especially because those with PAD often show abnormal cardiovascular response to strenuous exercise (107). See practical application 19.3 for a review of the literature about PAD and exercise.

Muscle function and structural changes occur, mitochondrial bodies increase, oxidative enzymes increase, and free fatty acid oxidation metabolism increases as muscles adapt to exercise training. Most important, the morbidity and mortality rates associated with exercise training are very low in those with IC (35,36,57,63,119). If PAD is present, those who are physically active have been reported to have reduced mortality and cardiovascular events when compared with those who are inactive (50).

Acute Exercise

In those with severe symptoms of IC brought on by exercise, transient ischemia followed by reperfusion occurs following a bout of acute strenuous exercise. This physiological response has led to some controversy about whether exercise is contraindicated for patients with advanced PAD (see table 19.5 on p. 348). Total

antioxidant capacity and renal tubular function have been shown to be abnormal following acute exercise in patients with severe IC (86).

Increased thrombin formation was reported to occur with acute exercise in patients with PAD compared with healthy subjects (99). This finding suggests that acute exercise, also resulting in catecholamine release along with local muscle ischemia, could possibly increase the preexisting prothrombic potential of a diseased arterial wall. Others have reported that major vascular endothelial injury does not occur following acute exercise. This conclusion was reached because concentrations of plasma markers for endothelial damage did not change even though the median ABI changed from 0.96 before exercise to 0.59 following treadmill exercise testing in patients with symptomatic IC (146). If there is a major concern about the possibility of an ischemic-reperfusion injury, drug therapy may provide some protection against systemic vascular endothelial injury following acute strenuous exercise; pentoxifylline has been shown to reduce the post 1 h increase seen in urinary albumin, expressed as a creatinine ratio (139).

Physiological Adaptations and Maladaptations

Some studies have shown that exercise rehabilitation for those with IC does not increase blood flow (32,71,83,89,90,134,149), but others have reported that exercise effectively improved leg blood perfusion (3,34,62,87). A review of studies using exercise-training therapy for those with IC reported improvements in walking tolerance, but blood flow improvements alone could not account for the total improvement in pain-free walking distances (45). Following a 6 mo exercise rehabilitation program, patients showed a 115% increased distance to the onset of claudication pain, and this improvement was independently related to the 27% increase found in blood flow (46).

A possible mechanism by which exercise therapy may improve walking distance in those with IC is by enhancing skeletal muscle oxidative metabolism without altering anaerobic metabolism (87). The benefits of exercise conditioning for patients with IC appear more likely attributable to an improvement in calf muscle oxidative metabolism rather than to changes in skeletal muscle blood flow. Improvement in pain-free walking distance following exercise therapy has been attributed to the following factors:

1. Improved biomechanics of ambulation resulting in decreased metabolic demands (130,132,145)

Literature Review

Williams et al. (144) reported that a structured exercise program along with risk factor reduction for 45 patients with IC resulted in an increase of 122% to 450% in patients' walking distance. Of those patients who smoked, 88% quit smoking after embarking on the exercise program. Important in this report was that exercise therapy was not associated with any morbidity or mortality. Follow-up of these patients after 1 yr showed a significant reduction in hyperlipidemia (both total serum cholesterol and triglycerides levels), and after 2 yr 84% had maintained or improved their walking distances compared with their distances at exit from the structured exercise program. Eighty-three percent of those who quit smoking after starting the exercise-training program remained nonsmokers at follow-up. Others have reported benefits of exercise training and, in general, found routine walking therapy to prolong pain-free walking distance. Feinberg et al. (39) studied the effects of 12 wk of exercise therapy (without an attempt made to modify risk factors for atherosclerosis) in 19 patients with IC. Major endpoints were absolute systolic ankle pressure, ABI, maximum walking time, claudication onset pain time, and the ischemic window (the area under the curve of the exercise-induced decrease of the ankle pressure and its recovery recorded over time). The investigators found 659% and 846% average increases for maximum walking time and claudication pain time, respectively. No change occurred in the absolute ankle pressure or the ABI, and the average ischemic window measures decreased by 58.7%.

Ten published reports with good scientific methodological criteria, although mostly with small sample size, were reviewed by Robeer et al. (123) in an attempt to establish the effect of exercise rehabilitation therapy in older (age 60–76 yr) patients with IC. Scientific criteria examined included the study population, the intervention used, outcome variables examined, and data analysis and presentation. This information was used to develop a weighted scale for the methodological quality of randomized clinical trials employing exercise therapy to evaluate their efficacy. Included in this analysis was the identification of outcome predictors for exercise therapy, but this measure was found in only one of the studies. Of the 10 published studies considered to be of good research design, a 28% to 210% improvement occurred in pain-free maximum walking distance or time. These investigators were unable to determine outcome predictors (i.e., optimal frequency, type, mode, intensity, and duration of exercise or the value of supervision) for exercise training and, therefore, recommended that additional clinical studies be performed to develop optimal exercise rehabilitation programs for patients with IC. Another report from these investigators concluded that these 10 published studies offered no clear answers about a definition of an optimal exercise program, the effect of adherence to an exercise program long term, and how long beneficial effects of exercise last (14).

An earlier review of the literature by Ernst and Fialka (36) resulted in a recommendation that exercise be performed regularly for at least 2 mo and at a fairly high intensity of effort to be beneficial. Continued improvements in functional status, however, have been noted to continue over a 24 wk period of adherence to a structured walking exercise program (117). A model that predicts the outcome of exercise training for those with IC by fitting multistate transition models using autoregressive logistic regression was proposed by de Vries et al. (25). Important covariate parameters that predict outcome of supervised exercise were time, age, ABI, and duration of IC. Patients can markedly improve their functional capacity by participating in either a highly structured home exercise program or a formal supervised exercise program in a clinical setting (72), which especially can improve their perception of their health and well-being (109). An unsupervised home walking program can also improve the ability of patients with IC to walk longer distances without pain (120). This is an important point, because a home program is often more conducive for daily exercise and less expensive for patients compared with traveling to a hospital setting for supervised exercise therapy, even for a short period.

Different modes of training for patients with IC have been demonstrated to show some crossover improvements but are mostly specific to the exercise therapy (e.g., treadmill walking, stair climbing, biking) performed over time in a training program. Thus, if a patient is physically training by exercising on a stair climber or motor-driven treadmill, either mode can improve exercise capacity. But for testing purposes, the training effect will be most apparent on the apparatus used for training (73). This finding has implications for evaluating progress in physical fitness levels, because testing should be performed on an apparatus (e.g., bicycle ergometer, treadmill, and stair climber) similar to that used for exercise therapy.

Table 19.5 Markers Showing Systemic Effects of Ischemia-Reperfusion With a Single Bout of Exercise in Patients With Intermittent Claudication (IC)

Markers	Change in IC with acute exercise
Endothelial damage	
Von Willebrand's factor[a]	NC
Free radicals and neutrophil activation	
Systematic neutrophil count[a]	Increased
Neutrophil elastase	Increased
Neutrophil hydrogen peroxide	NC
Soluble p-selectin	NC
Monocytes	NC platelets
Thrombin formation and fibrin degradation	
Thrombin-antithrombin III complex	Increased
D-dimer[a]	Increased[b]
Tissue plasminogen activator (t-PA)[a]	Increased[b]
Plasminogen activator inhibitor-I antigens[a]	NC
t-PA activity	Increased[b]
Plasmin-α 2-antiplasmin complex	Increased[b]
Plasma catecholamines	Increased[b]
Other	
Urinary microalbumin excretion	Increased
Neutrophil deformability	Increased
Plasma thromboxane	Increased
Interleukin-8	Increased

Note. NC = no change.
[a]Levels at rest are markedly higher in patients with IC compared with controls.
[b]Acute exercise in controls results in a similar increase.

Adapted from A.T. Edwards et al., 1994, "Systemic response in patients with intermittent claudication after treadmill exercise," *British Journal of Surgery* 81(12):1738–1741; U.J. Kirkpatrick et al., 1997, "Repeated exercise induces release of soluble P-selectin in patients with intermittent claudication," *Thrombosis Haemostasis* 78(5):1338-1342; D.R. Lewis et al., 1999, "Vascular surgical society of Great Britain and Ireland: Systemic effects of exercise in claudicants are associated with neurophil activation," *British Journal of Surgery* 86(5): 699-700; E. Mannarino et al., 1989, "Effects of physical training on peripheral vascular disease: A controlled study," *Angiology* 40(1): 5-10; P. Mustonen, M. Lepantalo, and R. Lassila, 1998, "Physical exertion induces thrombin formation and fibrin degradation in patients with peripheral atherosclerosis," *Arteriosclerosis Thrombosis and Vascular Biology* 18(2): 244-249; E.P. Turton et al., 1998, "Exercise-induced neutrophil activation in claudicants: A physiological or pathological response to exhaustive exercise," *European Journal of Vascular Surgery* 16(3): 192-196; K.R. Woodburn et al., 1997, "Acute exercise and markers of endothelial injury in peripheral arterial disease," *European Journal of Vascular and Endovascular Surgery* 14(2): 140-142.

2. Increased collateral circulation resulting in increased peripheral blood flow (3,34,131,133)

3. A reduction in blood viscosity (37,42)

4. An increase in blood cell filterability and a decrease in red cell aggregation (42)

5. Regression of atherosclerosis (62)

6. Increased extraction of oxygen and metabolic substrates resulting from improvements in skeletal muscle oxidative metabolism (22,69,134)

7. Increased pain tolerance (62,149)

Of these reported factors associated with improved walking distance, it appears unlikely that an increased blood flow secondary to increased collateral circulation is the major contributing factor (23,32,100,126,134).

Although some evidence suggests that a low-grade inflammatory response may occur with acute strenuous exercise, regular exercise training for those with IC appears not to cause long-term endothelial inflammation, improves blood **rheology** properties, and may attenuate progression of atherosclerosis. These benefits suggest that exercise therapy is a viable treatment option for those with IC associated with atherosclerosis.

CONCLUSION

This chapter provides a review of information regarding PAD with accompanying IC and the benefits of exercise training for patients with IC symptoms without jeopardy of losing a lower extremity. Limb salvage is usually not the primary goal of treatment because IC is the primary symptom that responds well to conservative measures. Because IC is an expression of systemic atherosclerosis, control of underlying metabolic and cardiovascular comorbidities and maladies should clearly be a priority to reduce cardiac and cerebrovascular morbidity and mortality rates in those with PAD. Intensive risk factor reduction therapy for inhibiting the progression of atherosclerosis, especially abstinence from tobacco products, is a major goal. Exercise therapy can improve the functional capacity of these patients and may also prevent other cardiovascular and metabolic diseases.

Our current knowledge regarding the benefits of exercise comes primarily from studies involving patients who are highly motivated, are fairly young, and have stable comorbidities. Studies still need to be conducted to investigate the frequency, intensity, and time of training necessary for those with IC to gain optimal benefits from exercise therapy. Because our population is aging, efficacy of exercise therapy in those over 65 yr of age, especially in a home-based setting that is less financially demanding, needs to be further investigated. Because the conservative treatments of risk factor reductions and exercise have been shown to be efficacious for patients with IC, third-party payers should be encouraged to be more receptive to reimburse these therapeutic services involving exercise therapy. Insurance companies need to recognize the value of vascular rehabilitation services for medical cost containment, and these entities should strongly support the availability of these services in healthcare facilities. As an important aspect of the overall medical process, healthcare personnel who treat patients with IC need to continue analysis of therapy outcomes to delineate and define appropriate and optimal interventions.

Case Study

Medical History

Ms. RU is a 61 yr old African American female who came to the vascular clinic complaining of right-leg pain symptoms that she had experienced for more than 8 mo. She complained of intermittent tingling sensations in the right foot as well as significant right buttock and thigh pain with ambulation. She is currently able to walk approximately 50 ft (15 m) before the onset of painful leg symptoms. The symptoms make her feel moderately disabled. She denies exertional angina or any major orthopedic problems that would prevent her from engaging in a mild exercise program.

Her history includes coronary artery bypass surgery 10 yr earlier. She underwent a subsequent coronary artery bypass surgery revision and mitral valve placement approximately 7 yr later. A recent cardiac echo showed severely depressed ventricular function with an ejection faction estimated to be about 35%.

Her risk factors for atherosclerosis include a history of hypertension as well as insulin-dependent diabetes mellitus. She smoked cigarettes until 10 yr ago but quit then and has not smoked since. She is also hyperlipidemic, both hypertriglyceridemic and hypercholesterolemic. Her family history is positive for coronary artery disease.

Her physical examination shows that she is a healthy appearing middle-aged female. Her blood pressure was 118/70, and her pulse was 80 and regular. Her femoral, popliteal, and pedal pulses were absent on the right extremity. The right foot was warm with mild delayed capillary refilling. No evidence of ischemic tissue loss or pedal edema was present. On the left extremity, the femoral popliteal and posterior tibial pulses were 2+.

Diagnostic Exercise Testing

She underwent a constant load treadmill test in the vascular laboratory. Results of diagnostic testing showed the resting ABI on the right to be 0.75, which dropped to 0.51 with acute exercise. The right first-digit pressure was 15 mmHg. The left ABI and left first-digit pressures were normal. Doppler velocity waveform analysis was consistent with suprainguinal arterial occlusive disease of the right leg. The arterial Doppler study showed abnormal right lower-extremity waveforms and pressures.

Diagnosis

The medical history and diagnostic exercise test results were consistent with a diagnosis of advanced PAD with IC symptoms of her right leg.

Exercise and Dietary Prescription

Intermittent exercise was recommended, involving walking in the halls of her apartment complex at her normal walking pace. Although she is allowed to exercise on a stationary cycle ergometer for cardiovascular benefits, she was encouraged to choose walking as often as possible to improve metabolism in her gastrocnemius and soleus muscles. The exercise bouts were initially set for 2 to 5 min followed by a 1 to 2 min recovery period (slower

(continued)

walking or sitting). She was instructed not to extend herself beyond the point of mild intermittent claudication pain. She was instructed to repeat the exercise recovery cycle, if possible, 6 to 15 times to ensure 25 to 30 min of exercise per training session initially. She was to progress by increasing the bouts by 1 min per week. Because IC symptoms rather than cardiac dysfunction limited her exercise performance, the patient was instructed to endure calf pain as much as possible but to monitor any chest discomfort closely while exercising.

Her current weight is 160 lb (73 kg). She is approximately 25 to 30 lb (11 to 14 kg) overweight for a woman of 5 ft 5 in. (165 cm). She was encouraged to lose approximately 0.25 to 0.50 lb (0.1 to 0.2 kg) per week to reduce her body weight to a goal of 130 lb (59 kg). The recommended balanced diet included a high-fiber, low-cholesterol, and low-saturated-fat diet that contained approximately 1,700 to 1,800 kcal/d.

The patient markedly increased her ability to ambulate. She initially was able to walk only half the length of the hallway, but eventually she was able to walk at a very slow pace continuously for 20 to 30 min on a daily basis. The patient reported that she felt as if she had more vigor and strength and did not experience such severe claudication symptoms at this level of effort. She denied experiencing any cardiac symptoms throughout her exercise-training program. Because this protocol seemed to be working well and because she had relatively impaired cardiac function (low cardiac ejection fraction), she was maintained on this exercise regime. After 6 mo, she had decreased her body weight to 133 lb (60 kg). Shown in the table below are baseline and follow-up (6 mo) parameters.

Baseline and Follow-Up Parameters

Parameters[a]	Pretraining	Posttraining
ASP	88	87
ABI	0.75	0.73
ICD (m)	16	134
ACD (m)	120	165
Total cholesterol (mg · dl^{-1})	248	205
HDL cholesterol (mg · dl^{-1})	34	40
Triglycerides (mg · dl^{-1})	190	138

Note. ASP = ankle systolic blood pressure (mmHg); ABI = ankle–brachial index; ICD = initial claudication distance; ACD = absolute claudication distance; [a]Hemodynamics of the most severely affected leg.

Discussion Questions

1. Because Ms. RU has poor cardiac function and low functional capacity, how could a daily intermittent exercise protocol be adjusted to optimize her duration of effort?

2. With Ms. RU's low cardiac function, would progression of effort be restricted by claudication symptoms or by cardiac factors?

3. As Ms. RU loses weight, should the intensity of effort be significantly increased?

4. Will physiological adaptations to exercise training be most pronounced in her cardiopulmonary or peripheral vascular system?

5. How often should Ms. RU's exercise prescription be revised, what components (frequency, intensity, duration) could be altered, and what other safe modes of exercise might be recommended?

Pacemakers and Internal Cardiac Defibrillators

Kerry J. Stewart, EdD

The heart of the average person beats about 100,000 times per day. Each contraction results from an electric impulse that is initiated in the sinoatrial (SA) node, passes through the atrioventricular (AV) node, and is spread through the ventricles. An artificial pacemaker maintains a normal heart rate when the intrinsic electrical circuitry of the heart fails. The most common indications for pacemaker implantation are a heart rate that is too slow (symptomatic bradycardia) or fails to increase appropriately with exercise (**chronotropic incompetence**) or an electrical pathway that is blocked (conduction system disease). More recently, pacing systems have been developed to normalize conduction and "resynchronize" the ventricles in patients with heart failure with myocardial conduction slowing, which is usually seen clinically as left bundle branch block (23). This mode of pacing, known as biventricular pacing, uses an additional pacing lead that is programmed to restore cardiac synchrony and mechanical activation that can lead to improvement in hemodynamics, ventricular remodeling, mitral regurgitation, exercise capacity, and quality of life. This mode of pacing will be discussed later in this chapter.

SCOPE AND PATHOPHYSIOLOGY

Rhythm disorders that involve the SA node are classified under the broad term **sick sinus syndrome**. This condition includes the inability to generate a heartbeat or increase the heart rate in response to the body's changing circulation demands. Heart block may cause a loss of AV synchrony, a term that refers to the sequence and timing of the atria and ventricles. Normally, the ventricles contract a fraction of a second after they have been filled with blood following an atrial contraction. Asynchrony may not allow the ventricles to fill with enough blood before contracting. Depending on the patient's specific condition, the artificial pacemaker may replace SA node signals that are delayed or blocked along the pathway between the upper and lower heart, maintain a normal timing sequence between the upper and lower heart chambers, and ensure that the ventricles contract at a sufficient rate. The "2002 Guideline Update for Implantation of Cardiac Pacemakers and Antiarrhythmia Devices" issued by the American College of Cardiology, American Heart

Association, and North American Society for Pacing and Electrophysiology provides a review of the scientific literature and recommendations for which bradyarrhythmias and tachyarrhythmias may optimally be treated with a pacemaker (10).Table 20.1 shows pacemaker types that are used as therapy for different medical conditions (17). The need for pacing can occur at any age. Although some infants require a pacemaker from birth, about 85% of those who need a pacemaker are over the age of 65 yr, with an equal distribution among men and women. About 115,000 pacemakers are implanted in the United States each year. This number is likely to increase because of the growing number of elderly people in the population. In the 1950s external pacemakers were used to treat symptomatic bradycardia. The 1960s produced AV pacemakers that provided a more physiological method of pacing. Today, with miniaturized electronic circuitry, pacemakers improve quality of life by optimizing the hemodynamic state at rest and can produce an appropriate heart response to meet the physiological demands of exercise. Because of newer technology, many patients can maintain or even begin an exercise program after pacemaker implantation. Therefore, the clinical exercise physiologist should understand how pacemakers work to emulate normal cardiac rate, conduction, and rhythm in response to physiological and metabolic needs. She or he should also know about pacemaker programmed settings, how these settings can affect exercise capacity and the exercise prescription, and how to determine whether the patient's response to exercise is appropriate. The clinical exercise physiologist should communicate observations of the patient's responses to exercise to the pacemaker physician, who can reprogram the pacemaker to optimal settings.

PACING SYSTEM

A pacing system consists of separate but closely integrated components that stimulate the heart to contract with precisely timed electrical impulses. The pacemaker (also known as the pulse generator) is a small metal case that contains the circuitry that controls the electrical impulses, along with a battery. Modern pacemakers use lithium batteries that can last for many years, depending on the extent to which the pacemaker is used. Some patients depend on the pacemaker to provide cardiac rhythm and conduction at all times. In others, the pacemaker acts as a backup that fires intermittently when the sinus node fails to produce an appropriate rate or when the conduction system fails to transmit the impulses. The pacing leads are insulated wires connected to a pacemaker. The leads carry the electrical impulse to the heart and the response of the heart back to the pacemaker. These leads are extremely flexible to accommodate both the moving heart and the body. Depending on the type of pacemaker that is implanted, one or two leads may be present. Each lead has at least one electrode that can deliver energy from the pacemaker to the heart and sense information about the electrical activity of the heart.

Pacemaker Implantation

This surgery, performed by surgeons or cardiologists, typically takes an hour or less and is often done as an outpatient procedure. Most patients receive a local anesthetic and remain awake during the surgery. In some cases, general anesthesia and a brief hospital stay are required. The pulse generator is usually implanted below the col-

Table 20.1 Pacemaker Therapies for Different Medical Conditions

Pacemaker varieties	Treatment conditions
Rate responsive for chronotropic incompetence	Patients who need to sustain a heart rate that matches their metabolic needs to their daily lifestyle or condition
Mode switch for managing atrial arrhythmias in patients with bradyarrhythmia	Many patients with sinus node dysfunction and atrioventricular (AV) block experience atrial fibrillation. Mode switch therapy reduces symptoms of atrial fibrillation during dual-chamber pacing
Rate drop response for neurocardiogenic syncope	Patients with carotid sinus syndrome and vasovagal syncope are being treated for their symptoms by preventing their heart rate from falling below a prescribed level
Ablate and pace for atrial fibrillation	Patients with drug-refractory atrial fibrillation have been shown to benefit from ablation of the atrioventricular junction and implantation of a pacing system to maintain an appropriate heart rate
Search AV for patients with intermittent or intact AV conduction	Intrinsic AV activation is generally preferred to a ventricular-paced contraction because it provides improved hemodynamics and extended pacemaker longevity

Adapted from Medtronics. Available: http://www.medtronic.com/brady/clinician/therapies/clinther.html.

larbone just beneath the skin. The leads are threaded into the heart through a vein located near the collarbone. The tip of each lead is then positioned inside the heart. In some cases, the pulse generator is positioned in the abdomen and the pacemaker leads are attached to the outside of the heart. After implantation, the pacemaker can be adjusted with an external programming device. The device works by sending radio frequency signals to the pacemaker via a transmitter placed on the chest.

Temporary External Pacemakers

An external pacemaker pulse generator is a temporary device that is commonly used in emergency and critical care settings, after open heart surgery, or until a permanent pacemaker can be implanted. This device is used outside the body as a temporary substitute for the heart's intrinsic pacing. Many of these devices have adjustments for impulse strength, duration, R-wave sensitivity, and other pacing variables.

Permanent Pacemaker Types

The two basic types of permanent pacemakers are single chamber and dual chamber. Both monitor the heart and send out pacing signals as needed to meet physiological demands.

1. A **single-chamber pacemaker** usually has one lead to carry signals to and from either the right atrium or, more commonly, the right ventricle. This type of pacemaker can be used for a patient whose SA node sends out signals too slowly but whose conduction pathway to the lower heart is intact. A single-chamber pacemaker is also used if there is a slow ventricular rate in the setting of permanent atrial fibrillation. In this case, the tip of the lead is usually placed in the right ventricle.

2. A **dual-chamber pacemaker** has two leads. The tip of one lead is positioned in the right atrium, and the tip of the other lead is located in the right ventricle. This type of pacemaker can monitor and deliver impulses to either or both of these heart chambers. A dual-chamber pacemaker is used when the SA node is unreliable and the conduction pathway to the lower chamber is partly or completely blocked. When pacing does occur, the contraction of the atria is followed closely by a contraction in the ventricles, resulting in timing that mimics the heart's natural way of working.

Pacemakers are categorized by a standardized coding system developed by the North American Society of Pacing and Electrophysiology and the British Pacing and Electrophysiology Group.

Figure 20.1 on p. 354 shows the coding system for pacemaker functions. The letters refer to the chamber paced, the chamber sensed, what the pacemaker does when it senses an event, and other programmable features. This example is a ventricular demand pacemaker that is also rate responsive. The first V indicates that the ventricle is paced. The second V indicates that the pacemaker is programmed to sense for an impulse in the ventricle. The I indicates that when the pacemaker senses the patient's ventricular impulse, the pacemaker is inhibited. The R indicates that the pacemaker is rate responsive or rate adaptive. A sensor in the pacemaker senses physical activity and adjusts the patient's pacing rate according to the level of activity.

Figure 20.2 on p. 355 shows the code for a dual-chamber pacemaker that is also rate responsive. The D stands for dual. Because of two leads, a dual-chamber pacemaker can pace both the atria and the ventricles. Likewise, the pacemaker can sense in the atria and ventricles. The third D indicates that the pacemaker can be either inhibited or triggered. The pacemaker will watch for atrial activity, and if it detects none, it will pace the atrium. After an appropriate AV time interval, the pacemaker will watch for a ventricular depolarization. If this is sensed, the pacemaker will be inhibited. If no ventricular activity is present, the pacemaker will pace the ventricle.

The first V indicates that the ventricle is paced. The second V indicates that the pacemaker senses in the ventricle. The I indicates that the pacemaker will be inhibited. The R indicates that the pacemaker is rate responsive. Ventricular pacing (V) occurs at the programmed **lower rate limit** of 60 beats · min^{-1} when the intrinsic ventricular activity falls below that level. Intrinsic ventricular activity at a faster rate inhibits ventricular pacing.

EXERCISE PHYSIOLOGY

The physiology of exercise for patients with pacemakers is generally the same as for other patients. The difference is how their physiology interacts with the device. For patients who cannot provide an appropriate cardiac output response to exercise, modern pacemakers attempt to increase the cardiac output to meet changing physiological demands. The increase in oxygen uptake from rest to maximal exercise follows this formula:

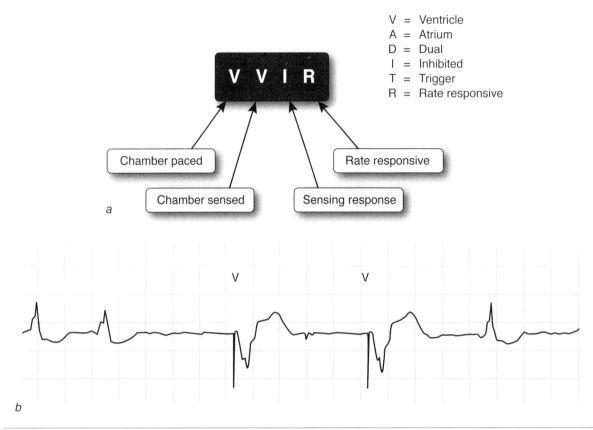

Figure 20.1 (*a*) Coding for a ventricular demand pacemaker that is also rate responsive. The first V indicates that the ventricle is paced. The second V indicates that the pacemaker senses in the ventricle. The I indicates that the pacemaker will be inhibited. The R indicates that the pacemaker is rate responsive. (*b*) VVI operation during atrial fibrillation. Ventricular pacing (V) occurs at the programmed lower rate limit of 60 beats · min^{-1} when the intrinsic ventricular activity falls below that level. Intrinsic ventricular activity at a faster rate inhibits ventricular pacing.

Oxygen uptake is equal to cardiac output multiplied by arteriovenous oxygen difference. From rest to maximal exercise, oxygen uptake can increase 700% to 1,200%, arteriovenous oxygen difference by 200% to 400%, and cardiac output by 200% to 400%. Cardiac output is equal to heart rate multiplied by stroke volume. With exercise, stroke volume can increase by 15% to 20%, whereas heart rate can increase by 200% to 300%. Thus, heart rate is the most important component for increasing cardiac output and is most closely related to metabolic demands. Although AV synchrony contributes to cardiac output, this factor is more important at rest and less important with exercise.

Physiological Pacing

The term *physiological pacing* refers to the maintenance of the normal sequence and timing of the contractions of the upper and lower chambers of the heart. AV synchrony provides higher cardiac output without increasing myocardial oxygen uptake. Dual-chamber pacemakers attempt to provide this physiological beneficial function. The pacemaker senses the patient's sinus node and, in complete heart block, sends an impulse to the ventricle following an appropriate AV timing interval. Although the specific change in cardiac output depends on many factors, the optimal AV delay to produce the maximum cardiac output in normal people is about 150 ms from the beginning of atrial depolarization. The efficiency of cardiac work decreases with a shorter or longer AV interval. In normal subjects, the AV interval shortens with increased heart rate. The pacemaker can also set the AV interval based on heart rate. A dual-chamber pacer can also initiate an atrial impulse in sick sinus syndrome. AV synchrony augments ventricular filling and cardiac output, improves venous return, and assists in valve closure. The loss of atrial function increases atrial pressure and pulmonary congestion. The benefit of AV synchrony is independent of any measure of left ventricular function. The maintenance of normal AV synchrony allows for

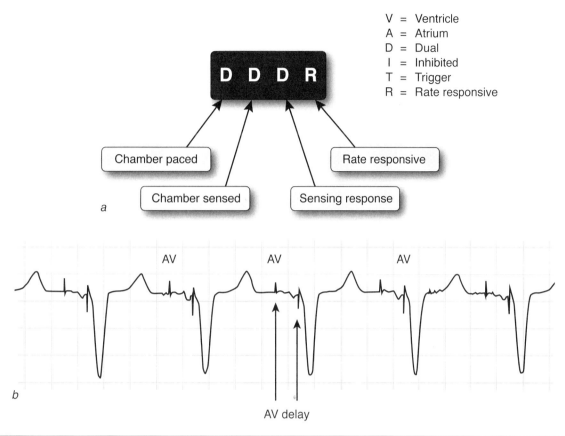

V = Ventricle
A = Atrium
D = Dual
I = Inhibited
T = Trigger
R = Rate responsive

Figure 20.2 (*a*) Coding for a ventricular demand pacemaker that is also rate responsive. The first D stands for dual, indicating pacing in both the atria and the ventricles. The second D indicates sensing capability in both the atria and ventricles. The third D indicates inhibited or triggered, and the R stands for rate responsive. (*b*) DDD operation. Atrial pacing (A) occurs at the programmed lower rate limit of 75 beats · min^{-1}. Because of complete heart block, the pacemaker tracks the atrial rate to pace the ventricle (V) at the same rate after a programmed AV delay.

improved hemodynamic responses with a more normal increase in cardiac output (13). Because of higher cardiac output at any given level of work with synchronous pacing, the arteriovenous oxygen difference is narrower and the serum lactate is lower. Thus, synchronous pacing results (figure 20.3 on p. 356) in less anaerobic metabolism at the same level of work (15). AV synchrony and stroke volume provide their most important contributions to cardiac output at rest, whereas an increase in heart rate is the predominant factor contributing to cardiac output during exercise (figure 20.4 on p. 356).

Mode Switching in Dual-Chamber Pacemakers

Many patients with sinus node dysfunction and AV block develop atrial arrhythmias. The most common arrhythmia is atrial fibrillation. To prevent the dual-chamber pacemaker from tracking or matching every atrial impulse with a ventricular pacing pulse, **mode switching** controls the ventricular rate. The highest rate at which the pacemaker will respond in terms of matching the sinus rate with a ventricular response is known as the maximal tracking rate. Mode switching will temporarily revert to a nontracking mode so that irregular or excessive atrial activity does not drive the ventricles to an extremely high rate. The mode-switching feature is programmable and, depending on the specific pacemaker model, can be adjusted for optimal performance in any given patient.

Rate-Responsive Pacing

The development of rate-responsive pacing (also called rate-adaptive rate or rate-modulated) has dramatically changed the application of pacing with regard to physical activity. The rate-responsive function is used when the native sinus node cannot increase heart rate to meet metabolic demands. Increasing heart rate in response to exercise is probably the single most important factor for

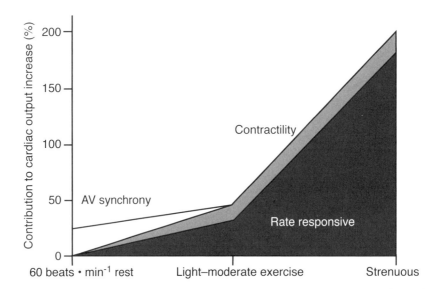

Figure 20.3 The relative contributions of atrioventricular (AV) synchrony, stroke volume (contractility), and heart rate to cardiac output at rest and exercise. Heart rate is the most important contributor to cardiac output during exercise.

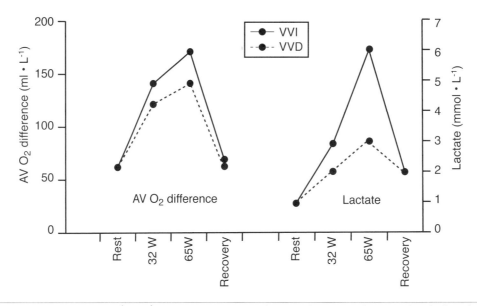

Figure 20.4 Synchronous pacing (VDD) results in less anaerobic metabolism at the same level of work compared with nonsynchronous pacing (VVI).

increasing cardiac output and oxygen uptake. A sudden increase in exercise requires the heart rate to adjust quickly to the workload. Rate-responsive pacemakers can sense the body's physical need for increased cardiac output and produce an appropriate cardiac rate in patients with chronotropic incompetence. The highest rate at which the pacemaker will pace the ventricle in response to a sensor driven rate is known as the maximal sensor rate. When rate-modulated pacing is compared with non-rate-modulated pacing (figure 20.5), we see that exercise

capacity is extremely limited without an appropriate increase in heart rate (6).

Several physiological and metabolic changes occur during exercise as the demand for energy increases:

- Movement that produces vibration and acceleration
- Respiration
- Heat that raises body temperature
- Electricity activity that produces electrocardiographic and electromyographic changes

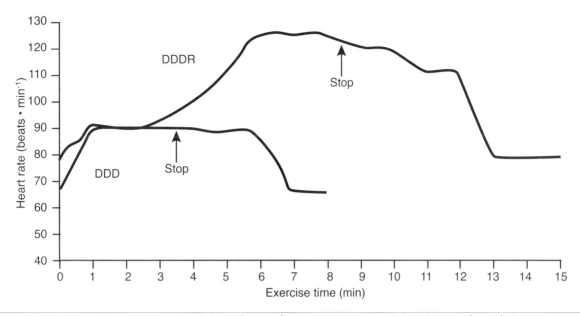

Figure 20.5 A comparison of rate-modulated (DDDR) versus non-rate-modulated pacing (DDD) during treadmill exercise. Without an appropriate increase in heart rate, exercise capacity is extremely limited in a patient with chronotropic incompetence.

Adapted from R.M. Bodenhamer and R.N. Grantham, 1993, "Mode selection: The therapeutic challenge: Adaptive-rate pacing," *St. Paul: Cardiac Pacemakers* 19: 52.

- Carbon dioxide
- Lactic acid, which reduces blood pH
- Changes in intracardiac pressure

Various sensors have been developed to detect these changes and, based on computer algorithms, generate the electrical impulses that are used to pace the heart. The development of optimal sensors and algorithms for rate-modulated pacing systems must meet several requirements:

- The sensor should rapidly detect acceleration and deceleration of physiological changes.
- The response should be proportionate to the exercise workload and metabolic demands.
- The response should be sensitive to both exercise and nonexercise requirements such as posture, anxiety and stress, vagal maneuvers, circadian variations, and fever.
- The response should be specific and not be falsely triggered.

The most common rate-responsive pacemakers detect motion in response to physical activity. Vibration sensors use a piezoelectric crystal located in the pulse generator to detect forces generated during movement. These forces are transmitted to the sensor through connective tissue, fat, and muscle. Acceleration sensors detect body movement in anterior and posterior directions (8). The circuitry is also located in the pulse generator. Because the sensor is not in direct contact with the pacemaker case, no reaction to vibration or pressure occurs.

Producing an appropriate heart rate in response to certain work tasks poses a technological challenge. A simple task of walking up and down stairs can produce different heart rate responses, based on the type of sensor used to sense motion. Compared with a normal heart rate response, an accelerometer produces similar results going up stairs but overestimates the metabolic demand going down stairs. The vibration sensors produce a heart rate that is too low for stepping up and a rate that is too high for stepping down. This result occurs because the vibration of walking down stairs is greater than the vibration of walking up stairs, although the metabolic demand is greater when walking up stairs (4). In contrast to single sensors, dual-sensor rate response provided by activity and minute ventilation may help overcome these types of problems, as observed in appropriate heart rate responses while patients are ascending and descending stairs (2). Advances in pacemaker technology may allow the use of combined or blended sensors and advance algorithms to improve rate performance over a single-sensor system (16). Blended sensors are designed to measure patient workload through respiration and motion, providing optimal rate response during changing levels of activity.

Nevertheless, definitive clinical benefits from blended sensors have yet to be fully established (22).

EXERCISE TESTING

Patients with pacemakers capable of rate modulation should undergo exercise testing to ensure appropriate rate responses (12). Exercise capacity and quality of life are improved by appropriately programmed rate-responsive pacemakers compared with fixed-rate units (25). These devices can be programmed to match the needs of the patient more closely. The primary pacemaker settings can be adjusted to optimize responses to physical activity:

- Sensitivity of the sensor
- Responsiveness to a physiological change
- Rate at which the cardiac rate changes
- Minimum rate at rest and maximal rate at peak activity

Exercise testing is used to guide the adjustment of these settings to improve exercise capacity and reduce symptoms. Exercise testing helps establish upper rate limits and adjust the sensitivity and responsiveness of the sensor. Exercise testing is also used to determine the anginal threshold, if any. Pacing the heart rate beyond the point at which ischemia would occur would not be prudent. Several approaches to exercise testing can be used. These include using informal or formal protocols with or without real-time electrocardiogram (ECG) monitoring and with or without determination of optimal rate-responsive parameters. The patient's health status and lifestyle, the type of pacemaker, and the facilities and experience of personnel will also help determine the specific approach to exercise testing. For many patients, informal exercise testing is a reasonable and less expensive alternative to formal treadmill testing (12). Empiric adjustment of the rate response parameters is common. With informal testing, the patient walks at a self-determined casual pace and at a brisk pace, usually for about 3 min each. The sensor-driven cardiac rate can be determined by examining the ECG. Because pacemakers are capable of storing an electronic record of pacemaker activity, the physician, using a special computer, can also interrogate the pacemaker to examine a histogram display of the heart rate response during the walk. The optimal pacemaker rate is determined empirically. For casual exercise, the target is often 10 to 20 beats $\cdot$ min^{-1} above the lower rate limit. For brisk exercise, the target can be 20 to 50 beats $\cdot$ min^{-1} above the lower rate limit.

This approach to exercise testing is best suited for less active patients who are unlikely to reach their upper rate limit. By examining a display of the sensed atrial rate as measured by an event counter in the pacemaker, or by measuring the heart rate by ECG and asking the patient about symptoms, the physician makes a clinical judgment whether the patient is chronotropically competent. If not, the pacemaker will need to be programmed to elicit an appropriate response. Formal exercise testing allows a chronotropic evaluation that seeks to match the pacemaker-augmented response of the chronotropically incompetent patient to the metabolic requirements of the body (21). Formal exercise testing is typically best for active patients likely to reach the programmed **maximum sensor rate.** Programming the upper rate of rate-adaptive pacing improves exercise performance and exertional symptoms during both low and high exercise workloads compared with a standard nominal value of 120 beats $\cdot$ min^{-1} (7). With formal exercise testing, the protocol selected requires careful consideration. Many protocols are used for exercise testing. Nevertheless, many of the traditional protocols such as the Bruce and Naughton protocols are designed to test for coronary artery disease. Their usefulness in defining optimal programming for rate-responsive pacemakers may be limited. A widely used protocol for assessing patients with pacemakers is the chronotropic assessment exercise protocol (26,27) shown in table 20.2. The advantage of this protocol is that the workload gradually increases to mimic the range of activities of daily living. This protocol allows a more complete assessment of how the pacemaker responds at the lower metabolic equivalent (MET) ranges where patients typically spend most of their time. The chronotropic assessment exercise protocol has five stages at a lesser exercise intensity than the Bruce produces in the second stage (28). Because the Bruce protocol increases by 2 to 3 METs during each 3 min stage, assessing the patient's work capacity and the ability of the pacemaker sensor to provide an adequate hemodynamic response would be difficult.

EXERCISE PRESCRIPTION

Dual-chamber pacemakers are in greater use today. Clinical exercise physiologists need to be familiar with the normal behavior of these devices during exercise. Figure 20.6 shows DDDR operation. The rate at which the sensor-driven heart rate increases follows algorithms that are programmed into the pacemaker. Among the key parameters are the slope of the heart rate increase and decline and the sensitivity of the sensor. With increased

Table 20.2 Chronotropic Assessment Exercise Protocol

Stage	Speed	Grade	Cumulative time	Metabolic equivalents
1	1.4	2	2	2.0
2	1.5	3	4	2.8
3	2.0	4	6	3.6
4	2.5	5	8	4.6
5	3.0	6	10	5.8
6	3.5	8	12	7.5
7	4.0	10	14	9.6
8	5.0	10	16	12.1
9	6.0	10	18	14.3
10	7.0	10	20	16.5
11	7.0	15	22	19.0

physical activity, the pacemaker will follow the sinus rate up to a maximal tracking rate. The activity sensor can be programmed to allow a further increase in the paced rate to the maximal sensor rate in response to physical activity. If the patient continues to exercise, the pacemaker may reach its maximal tracking rate or maximum sensor rate. When this occurs, the pacemaker will not further increase the heart rate. If the patient's native sinus rate continues to increase beyond this point, the pacemaker will switch to an AV block mode because the sinus rate now exceeds the rate at which the pacemaker will permit tracked ventricular pacing. The pacemaker will first switch to a Wenckebach-type block to cause a gradual slowing of the ventricles, and 2:1 AV block ensues if the sinus rate continues to rise. This feature protects against nonexercise sinus tachycardia that might otherwise force the pacemaker to produce ventricular tachycardia. At higher levels of exercise, the metabolic demands will be high, but 2:1 block may occur and slow the ventricular rate. In this situation, the development of 2:1 block is a normal feature of the pacemaker but can cause symptoms because of the sudden drop in heart rate. If this occurs, the patient is exercising too hard or the maximal rate is set too low. If the pacemaker is also rate adaptive, the sensor setting may be too low. The exercise physiologist should record and communicate episodes of abrupt

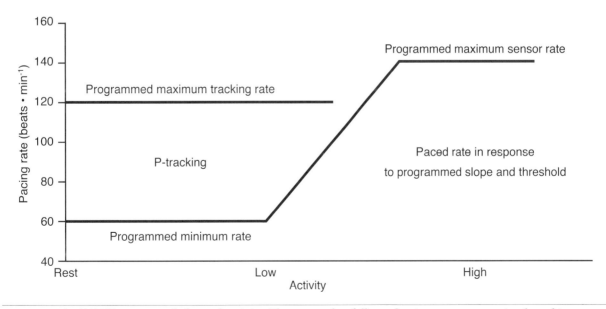

Figure 20.6 DDDR pacing and physical activity. The pacemaker follows the sinus rate to a maximal-tracking rate. In response to physical activity, the sensor-driven response can drive the rate to the maximal sensor rate. The paced rate increases in response to programmed slope and threshold settings.

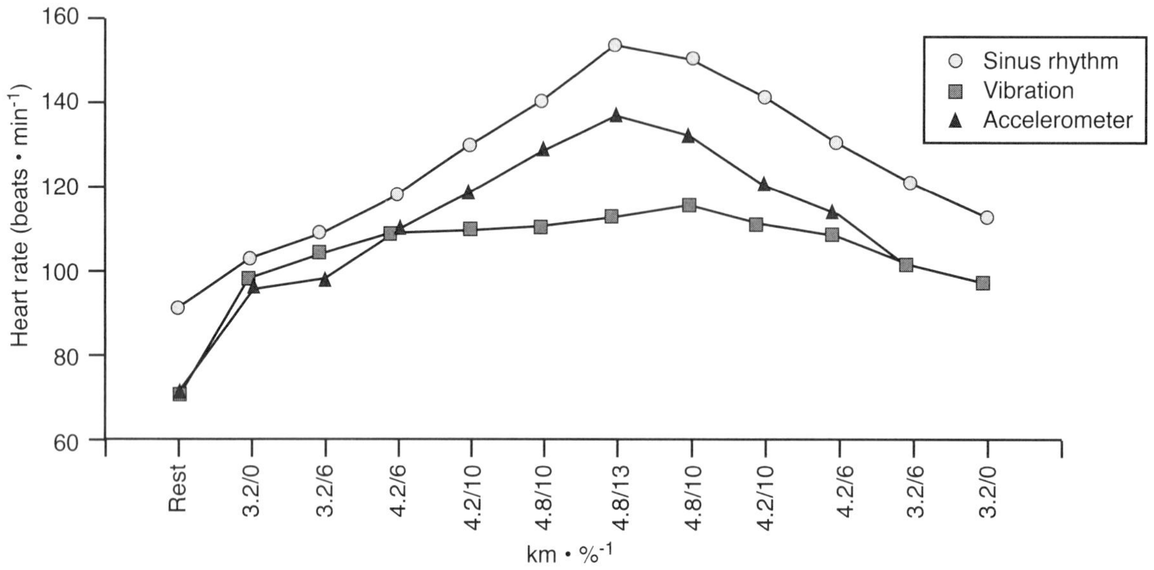

Figure 20.7 Generated forces. The circles show a normal sinus response at rest and with increased treadmill work. Work is increased primarily by raising the slope rather than speed. At rest, sinus rate is about 20 beats above pacemaker rate. This difference is maintained throughout the test. The accelerometer sensor (triangles) is able to produce a heart rate that better matches with the workload compared with the vibration sensor (squares).

Adapted from E. Alt and M. Matula, 1992, "Comparison of two activity-controlled rate-adaptive pacing principles: Acceleration versus vibration," *Cardiology Clinics* 10: 635-638.

decreases in heart rate to the patient's pacemaker physician so that programmed settings can be evaluated for possible change. Several of the exercise modalities that are commonly prescribed in cardiac rehabilitation and adult fitness programs may pose a particular challenge in some patients with activity sensors (3). In the example shown in figure 20.7, because most of the increase in work is accounted for by raising the slope rather than speed on a treadmill, little change in the generated forces would be detected by the vibration sensor. Thus, the heart rate determined by the vibration sensor is too slow for the metabolic demand of the increased work. In this case, the accelerometer sensor is better able to provide a heart rate that more closely matches an appropriate rate response. The clinical exercise physiologist should also be aware of how a vibration sensor responds to outdoor and stationary cycling (figure 20.8). The response of this type of sensor is particularly relevant to cardiac rehabilitation because stationary cycling is the dominant form of exercise in many programs. This sensor response may explain why some patients complain of unusual fatigue and shortness of breath during stationary cycling but not other types of exercise such as treadmill walking. Patients with an artificial pacemaker will require long-term surveillance by their physicians to ensure optimal adjustment of the programming for their individual needs, to maximize the life expectancy of the pacemaker through adjustment

of pacemaker output settings, and to identify and treat complications. Pacemaker follow-up relies on clinical, electrocardiographic, and device assessment. Other tests may include exercise testing, Holter monitoring, and echocardiography. The device assessment requires a specialized programmer to verify pacemaker functions. In some cases, remote interrogation of the pacemaker over telephone lines is done periodically to provide useful information about selected functions of the pacemaker when a more complete test is not deemed necessary. The clinical exercise physiologist can play an important role in the overall evaluation of the patient by providing feedback to the physician about heart rate, blood pressure, and symptomatic responses to exercise. The case study at the end of the chapter illustrates the role of the clinical exercise physiologist in the management of the patient with a pacemaker.

Special Considerations

Exercise prescription requires special attention to the type of pacemaker that is implanted. With fixed-rate pacemakers, cardiac output and arterial pressure are increased by stroke volume (1). Target heart rate cannot be used to guide exercise intensity. Instead, the patient follows ratings of perceived exertion (RPE). It is also important to monitor blood pressure to ensure an

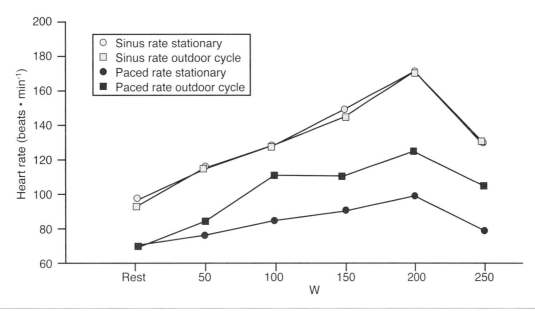

Figure 20.8 Vibration sensor. The upper lines represent a normal sinus rate with outdoor cycling (open squares) and stationary cycling (open circles) at increasing workloads and recovery. The lower lines represent the sensor. The difference of 20 beats at rest is maintained throughout the test during outdoor cycling (filled squares). During stationary cycling (filled circles), the paced cardiac rate is considerably slower than both the sinus rate and the rate during outdoor cycling. This occurs because stationary cycling produces less body motion and vibration.

Adapted from E. Alt and M. Matula, 1992, "Comparison of two activity-controlled rate-adaptive pacing principles: Acceleration versus vibration," *Cardiology Clinics* 10: 635-638.

appropriate intensity. When the sinus node is normal, it is desirable to have the pacemaker "track" native sinus activity by pacing the ventricle after an appropriate AV delay (11). In all cases, the target heart rate must be lower than the anginal threshold in a patient with ischemia (5). Tailoring the exercise prescription and modifying the response rate of the pacemaker based on cardiopulmonary stress testing that determines the anaerobic threshold have been shown to provide functional advantages for patients in cardiac rehabilitation (9). Because rate-responsive pacemakers mediate the heart rate response to exercise, the type of sensor must be taken into account when exercise is prescribed (21). Sensors that detect movement may respond slowly to stationary cycling and increased treadmill slope. Again, the RPE and MET equivalents are extremely useful in establishing the exercise prescription. With modern pacemakers, pacing occurs only when needed. In many patients, such as those with normal sinus function with intermittent heart block, the exercise prescription can be written the same way as those for most other patients. Regarding heart rate monitoring, one study found that the use of dry-electrode heart rate monitors that transmit a signal to a monitor such as those worn on the wrist had no adverse effect on pacemaker function (14).

Exercise Recommendations

Patients with pacemakers can derive benefits from an exercise program similar to those gained by other people. The area of greatest consideration when prescribing exercise is the issue of exercise intensity. Because of the variety of pacemakers, sensors used to detect activity, and mode of exercise prescribed, the appropriate heart rate response can vary considerably. Therefore, when patients with pacemakers start exercising, they should be monitored to make sure that the pacemaker is responding appropriately. See practical application 20.1 on p. 362 for a summary of the relevant exercise prescription.

Mode

Generally all forms of exercise are acceptable in patients with pacemakers, except activities that can cause direct contact with the pacemaker. Therefore, contact sports such as football, soccer, and hockey are generally not recommended. All forms of aerobic exercise are generally acceptable, and most likely carry the greatest benefit in regard to improving overall health and decreasing risk factors for cardiovascular disease. When a patient with a pacemaker performs any form of aerobic exercise, rate-responsive pacemakers should be evaluated to see that they are increasing the heart rate appropriately relative to the intensity of exercise. Pacemakers that rely on vibration or accelerometer sensors to detect body motion during exercise may not produce an adequate response for activities such as stationary cycling and increased treadmill slope. Unusual shortness of breath and fatigue may indicate a lack of **rate-responsive pacing** and need to be monitored during different forms of activity. Weight training may also be acceptable, although weights or bars must not come in contact with the pacemaker.

Frequency

The frequency of activity is based on the goals of the program. If someone is interested in improving his or her health and is exercising at an intensity less than 60% of maximal aerobic capacity, daily activity is recommended. If the person is able and willing to exercise at a higher intensity (60-85% of maximal aerobic capacity), activity on 3 to 5 d a week is recommended.

Intensity

The intensity should be in the recommended range of 40% to 85% of maximal aerobic capacity but is primarily dependent on comorbid conditions (e.g., angina, chronic heart failure). Also, upper limits of the pacemaker (tracking and sensing) can influence the upper limit of exercise intensity. Because heart rate will not increase in a patient with a fixed-rate pacemaker, RPE and MET equivalents need to be used to evaluate exercise intensity. Additionally, blood pressure should be monitored to show appropriate increases with increasing workload. A patient with a fixed-rate pacemaker should not exceed exercise intensity above the point where blood pressure begins to plateau with increasing workload. With dual-chamber and rate-responsive pacemakers, heart rate can be used to determine exercise intensity and should be used along with RPE and METs. Knowledge of maximal tracking

or sensing rates will determine the upper intensity level. Patients should be monitored closely, and activities should be chosen based on the ability of the pacemaker to adjust heart rate with increasing metabolic demands.

Duration

The duration of activity is similar to the general recommended guidelines to promote health and fitness (20-60 min). Duration depends on goals. Ideally, the duration should be adjusted so that the individual achieves an energy expenditure of at least 1,000 kcal per week.

Special Considerations

If a patient goes into second-degree type 1 block (Wenckebach) while exercising, the patient's native sinus rate likely exceeded the pacemaker's maximal tracking or sensor rate. If this occurs, the intensity of exercise should be reduced. If the exercise professional notices a decrease in heart rate well below the patient's tolerable limits, this information should be forwarded to the pacemaker physician so that programmed settings can be evaluated. Practical application 20.2 provides more information about living with a pacemaker.

Automatic Internal Cardioverter Defibrillators

People who survive ventricular fibrillation or other life-threatening ventricular tachycardias are at high risk for recurrence. Although medications that stabilize heart rhythm are available, they are not entirely effective and often produce serious side effects. Increasingly, automatic internal cardioverter defibrillators (ICDs) are being used to control life-threatening ventricular arrhythmias. An ICD is a battery-driven implanted device, similar to a pacemaker, that is programmed to detect and then stop

---

Practical Application 20.1

Summary of Exercise Prescription

The exercise prescription for those who use pacemakers is generally the same as that for others, and many of the same cautions apply. The prescription must consider comorbidities such as angina and chronic heart failure, for example. The area of greatest consideration when prescribing exercise is the issue of exercise intensity, which is determined by the underlying reason for the pacemaker and the type of pacemaker implanted. For patients with an internal cardioverter defibrillator, the exercise heart rate should be kept at least 20 beats below the firing threshold. Activities that might result in contact with an implanted device should be avoided.

Patient Education About Living With a Pacemaker

The clinical exercise physiologist is often a primary source of patient education. Pacemaker patients will ask about what they should be aware of in their day-to-day lives. Advances in pacemaker technology have resulted in continuing improvements. Pacemakers are smaller and better shielded from external interference and magnetic fields than ever before. Recent research has shown that with appropriate protocols, magnetic resonance imaging (MRI) can be performed safely in patients with certain pacemakers and ICD systems (18). Nevertheless, the clinical exercise physiologist should communicate some basic precautions to the patient. This section addresses some common questions and issues.

Sports and Recreational Activities

Many active patients, after appropriate medical clearance, can travel, drive, bathe, shower, swim, resume sexual activities, return to work, walk, hike, garden, golf, fish, and participate in other similar activities. But contact sports that includes jarring, banging, or falling such as football, baseball, and soccer should be avoided. Also, patients should avoid hunting if a rifle butt is braced against the implant site.

Work Activities

Most office equipment is unlikely to generate the type of electromagnet interference that can affect a pacemaker, but equipment with large magnets should not be carried if they are held near the pacemaker. Patients who work with heavy industrial or electrical equipment need to consult with their physicians about resuming work because this equipment may produce high levels of electromagnetic interference that could affect pacemaker function.

Home Activities

People with pacemakers can participate in most activities of daily living and can be reassured that most home electrical devices will not interfere with pacemaker operation. But some precautions are recommended. Cellular phones should be kept at least 6 in. (15 cm) away from the pacemaker site, and phones transmitting above 3 W should be kept at least 12 in. (30 cm) away. The patient should hold the cell phone to the ear opposite the pacemaker site and should not carry a phone in a pocket or on a belt within 6 in. (15 cm) of the pacemaker. Ordinary cordless, desk, and wall telephones are considered safe. The patient should not lift or move large speakers because their large magnets may interfere with the pacemaker. Most general household electrical appliances like televisions and blenders and outdoor tools such as electric hedge clippers, leaf blowers, and lawn mowers do not usually interfere with pacemakers. But the patient should avoid using tools such as chainsaws that require the body to come into close contact with electric spark-generating components. In addition, caution is advised when working near the coil, distributor, or sparkplug cables of a running engine. The safe approach is to turn off the engine before making any adjustments.

Travel

Most people with pacemakers can travel but should tell airport security personnel that they have a pacemaker or other implanted medical devices before going through security systems. Although airport security systems will not affect the pacemaker, the pacemaker's metal case could trigger the metal detection alarm. Home, retail, or library security systems are unlikely to be set off by the pacemaker.

a life-threatening ventricular arrhythmia by delivering an electrical shock directly to the heart. Some models provide tiered therapy by including the capability of providing antitachycardia pacing, cardioversion, and defibrillation, as needed. Modern ICDs are implanted beneath the skin and muscle of the chest or abdomen, and electrodes that sense the heart rhythm and deliver the shock are inserted into the heart through veins. Nevertheless, the site and placement of the electrode wires vary, depending on the patient and model of ICD used. In some cases, electrode patches are sewn to the surface of the heart. Other patients may receive electrodes that are

Client–Clinician Interaction

In a supervised exercise program, the clinician needs to observe whether the client can achieve the desired level of exercise without undue fatigue. Because the increase in heart rate during exercise is the largest contributor to cardiac output, limited exercise capacity may be because of inappropriate heart rate response. Depending on the type of pacemaker, adjustment of the pacemaker settings may allow the heart rate to respond more appropriately to the exercise demand. Carefully observing the client's exercise performance, asking the client about her or his fatigue level, and reporting those findings to the client's physician are key responsibilities of the exercise physiologist.

placed under the skin of the chest near the heart. Rapid technological advances have produced devices that serve as both a pacemaker and an ICD and can be programmed to the patient's individual needs. Microchips inside the device record rhythms and shocks to be used to determine optimal therapy.

Special Considerations

In many cases, the failure of the heart's intrinsic pacing and conduction system is associated with comorbid conditions such as myocardial infarction and chronic heart failure. Many patients with artificial pacemakers or ICDs are elderly and often have limited exercise capacity. The exercise prescription must consider not only the indications for the type of pacemaker implanted but also the limitations to exercise associated with comorbidities. Besides monitoring the patient for appropriate heart rate responses, the exercise physiologist must pay close attention to signs and symptoms that might occur with increased heart rate such as exercise-induced angina, failure of blood pressure to increase or decrease, and marked shortness of breath.

Exercise Recommendations

The American College of Cardiology–American Heart Association–National Association for Sport and Physical Education "Guidelines for Implantation of Cardiac Pacemakers and Antiarrhythmia Devices" (10) provide recommendations for ICD therapy. These guidelines emphasize the need for the physician to establish limitations on the patient's specific physical activities. The guidelines also refer to policies on driving that advise the patient with an ICD to avoid operating a motor vehicle for a minimum of 3 mo and preferably 6 mo after the last symptomatic arrhythmic event to determine the pattern of recurrent ventricular fibrillation or tachycardia. After appropriate evaluation and observation, many patients

with an ICD can participate in exercise programs. In most cases, the guidelines for exercise prescription are similar to those for any other patient with cardiovascular disease and should consider the patient's underlying diagnoses, medications, and symptoms. An exercise stress test is essential for establishing an appropriate exercise prescription. The prescribed target heart rate should be at least 20 beats below the heart rate cutoff point at which the device will shock. The exercise prescription must also take into account the existence of an **angina threshold** or exercise-induced hypotension, because many of these patients have severe coronary artery disease and poor left ventricular function. Furthermore, many patients with ICDs take β-blockers to limit heart rate to control symptoms and to prevent firing of the device. The benefits of pacing are

- alleviation or prevention of symptoms,
- restoration or preservation of cardiovascular function,
- restoration of function capacity,
- improved quality of life,
- enhanced survival, and
- participation in exercise training programs with many forms of physical activity.

Cardiac Resynchronization Therapy

More recently, cardiac resynchronization therapy (**CRT**) or biventricular pacing has been used as a novel adjunctive therapy for patients with advanced heart failure (19). Many of these patients have left bundle branch block or an intraventricular conduction delay, resulting in left ventricular dyssynchrony and a high mortality rate. The efficacy of CRT is based on the reduction in the conduction delay between the two ventricles. CRT is designed to keep the right and left ventricles pumping together by regulating how the electrical impulses are sent through the leads. This therapy contributes to the optimization of the ejection fraction, decrement in mitral regurgitation, and left ventricular remodeling, thus resulting in symptom improvement and enhanced quality of life. Several observational studies and randomized, controlled trials have shown the benefit of CRT in a subgroup of patients with heart failure, with conduction delays (20). Improvements were found in the mean distance walked in 6 min, quality of life, NYHA functional class, peak oxygen uptake total exercise time, reduction of hospitalization, LV function, and reduction of the LV end-diastolic diameter. Recent data suggest that hemodynamic improvements with CRT, previously demonstrated in acute invasive studies, are maintained chronically. In addition, ventricular–arterial coupling, mechanical efficiency, and chronotropic responses are improved after 6 mo of CRT. These findings may explain the improved functional status and exercise tolerance in patients treated with CRT (24). Exercise training after CRT may help to improve exercise tolerance and quality of life further, although no studies have examined this important issue. Note that patients with these devices have serious heart disease, so the usual precautions regarding exercise participation should be applied.

CONCLUSION

Because of the increased prevalence of pacemakers and ICDs, clinical exercise physiologists and cardiac rehabilitation professionals need to know how these devices function and what their limitations are. The two types of pacemakers are single-chamber units and dual-chamber units. Knowledge of the universal coding system is required to understand appropriate pacemaker function (figures 20.1 and 20.2). Initial pacemakers operated at fixed rates and were primarily used for patients who were symptomatic because of bradycardia or high-degree AV blocks. Because of the inability to increase heart rate with exertion in fixed-rate pacemakers, rate-responsive pacemakers have been developed, so that cardiac output can appropriately increase under physical activity. Motion sensors (vibration and accelerometers) are used in rate-responsive pacemakers. Each has advantages and disadvantages. In addition, dual-sensor (activity and ventilation) pacemakers have been developed to enhance normal heart rate response with exertion. In patients with rate-responsive pacemakers, exercise testing should be used to ensure an appropriate increase in heart rate and to allow for adjustment if the unit is not properly functioning. Exercise testing allows optimal programming of the pacemaker to provide maximal hemodynamic benefit and quality of life. When prescribing exercise training with rate-responsive pacemakers, the clinical exercise physiologist must make sure that heart rate increases appropriately with exertion. In activities that do not involve a great deal of change in body movement (cycling, uphill walking), vibration sensors or accelerometers are not able to detect the real difference in activity level. Therefore, if possible, the physician should determine the type of activity that a person is planning to do before implanting a pacemaker. In addition, when prescribing different modes of activity, the clinical exercise physiologist must consider the type of pacemaker. Ideally, patients with rate-responsive pacemakers should be monitored to determine whether the physiological response is acceptable with exertion (i.e., heart rate, ECG, blood pressure, RPE, and METs). In addition, patients should avoid contact sports that carry a risk of direct contact with the pacemaker. Overall, there is not a great deal of limitation in prescribing exercise in pacemaker patients, other than making sure that an appropriate physiological response (increase in heart

Practical Application 20.4

Literature Review

Although it has been shown that cardiac resynchronization therapy has beneficial effects on clinical outcomes and cardiac remodeling, little is known about longer-term effects on myocardial function and exercise tolerance. Steendijk and colleagues (24) studied CRT in 22 patients with chronic heart failure. After 6 mo of this device therapy, marked improvements occurred in NYHA class, quality-of-life scores, 6 min walk distance, left ventricular ejection fraction, stroke work at rest and with increased heart rates, and there was evidence of reverse remodeling. These results demonstrate that hemodynamic improvements with CRT can be maintained with chronic therapy and were associated with improved functional status and exercise tolerance.

rate or blood pressure) occurs with increasing levels of physical exertion. The use of ICDs is increasing to control life-threatening ventricular arrhythmias. Before patients with ICDs start an exercise program, an exercise test is recommended to determine the safety of exercise and rule out any other underlying diagnoses. When prescribing exercise, the major concern with patients with ICDs is to avoid reaching the threshold heart rate that will cause the device to shock. Training heart rate should stay 20 beats $\cdot$ min^{-1} below the preset heart rate that produces a shock. Otherwise, no specific limitations govern the prescription of exercise in this select population. The use of CRT to restore the coordinated pumping action of the ventricles is becoming more widespread in patients with chronic heart failure and delayed conduction. This type of device allows patients to be more active and have a better quality of life. Further research is needed to examine the long-term benefit of this therapy and to identify patients who are most likely to benefit from it.

Case Study

Medical History

Mrs. JD is a 64 yr old female African American referred to cardiac rehabilitation 6 wk following implantation of a DDDR pacemaker with a vibration sensor. The indication for the pacemaker was marked sinus bradycardia and chronotropic incompetence. Her primary complaints were episodes of shortness of breath, undue fatigue, light-headedness, and weakness at rest and during exertion. She had no other significant cardiac history except for mild hypertension. Mrs. JD also has mild arthritis in her hands and knees. Her body mass index is 29. Currently, her symptoms at rest are resolved, but she complains of early exertional fatigue and shortness of breath while doing housework and taking short walks in her neighborhood. Her resting heart rate is paced at 60 beats $\cdot$ min^{-1}.

Initial Exercise Prescription

The pacemaker settings relevant to her exercise prescription were a lower rate limit of 60 beats $\cdot$ min^{-1}, a programmed **maximum tracking rate** to 110 beats $\cdot$ min^{-1}, and a programmed sensor rate to 110 beats $\cdot$ min^{-1}. Mrs. JD was given a standard exercise prescription consisting primarily of stationary cycling for 15 min and walking on a treadmill for 15 min. Exercises with handheld weights and flexibility exercises were prescribed for warm-up. Because she did not have an exercise stress test before starting the cardiac exercise program, her exercise intensity for aerobic exercise was set at 12 to 14 on the Borg RPE scale.

Response to Initial Exercise-Training Session

On the stationary cycle, Mrs. JD complained of shortness of breath and early fatigue, stopping at 4 min. Her heart rate peaked at 75 beats $\cdot$ min^{-1}. On the treadmill, she was able to walk for 10 min at 1.5 mph (2.4 kph), reaching a peak heart rate of 98 beats $\cdot$ min^{-1}. Her main complaint was shortness of breath.

Testing of the Pacemaker and Diagnosis

Because of limited exercise tolerance, Mrs. JD was referred to her cardiologist for a pacemaker evaluation. Her physician administered an office-based walking protocol during which Mrs. JD complained of shortness of breath after 3 min of "brisk" walking. Her heart rate reached a peak of 102 beats $\cdot$ min^{-1}. The diagnosis was that her increase in heart rate was inadequate for the amount of work being performed. As a result, the pacemaker was programmed to a maximum tracking rate of 110 beats $\cdot$ min^{-1} and a maximum sensor rate of 130 beats $\cdot$ min^{-1}, and the threshold of the sensor was lowered to be more sensitive to body movements.

Adjustment of the Exercise Prescription

In cardiac rehabilitation, the exercise prescription was changed to two bouts of 15 min each of treadmill walking with 2 min rest between bouts. The warm-up was unchanged. Mrs. JD tolerated 2.0 mph (3.2 kph) at 0% grade on the treadmill as prescribed, reporting only mild leg fatigue and shortness of breath. Her peak heart rate reached 122 beats $\cdot$ min^{-1}.

Discussion Questions

1. Why was the maximum tracking rate left unchanged when the pacemaker settings were adjusted?

2. Why was stationary cycling dropped as an exercise modality for this patient?

3. What would be the initial choice for progressing the exercise intensity on the treadmill as the patient improves her fitness—an increase in speed, an increase in grade, or both?

PART V

Diseases of the Respiratory System

Pulmonary diseases and conditions can play havoc with a person's ability to be physically active. The clinical exercise physiologist can play a role in diagnosis, functional assessment, and the recommendation of exercise training. Often these patients present in a rehabilitative setting in a pulmonary rehabilitation program. But given the level of association with other diseases (e.g., heart disease, cancer), these patients may also seek exercise therapy in cardiac rehabilitation programs and in general fitness center facilities. This section provides excellent chapters on three of the most common pulmonary conditions.

Chapter 21 examines chronic obstructive pulmonary disease, also known as COPD. Patients with COPD are most likely to seek out or be referred to pulmonary rehabilitation. Most are current or previous smokers, and they are often debilitated to the point of being unable to perform much physical activity without shortness of breath. The clinical exercise physiologist must learn the skills to educate and motivate these patients and implement appropriate exercise programming. This chapter is an excellent resource pertaining to these skills.

Chapter 22 explores asthma. Increasing numbers of people in the United States and the world are being diagnosed with asthma. This trend is likely a result of better recognition and diagnostic testing, although worsening environmental factors that trigger asthma attacks may also be contributing. Exercise-induced bronchospasm, a form of asthma, is also increasing in prevalence. People with asthma can benefit from exercise training and often are able to perform exercise independent of clinical supervision. As this chapter demonstrates, however, the clinical exercise physiologist can be a great asset to these patients in developing and implementing an exercise program to limit potential bouts of asthma and in helping with methods to treat an attack that is associated with exercise.

Chapter 23 deals with cystic fibrosis (CF). Patients with CF have multiple medical issues that affect not only the lungs but also the gastrointestinal tract, sinuses, and sweat glands. These parts of the body are primarily affected by inflammation and excess mucus production because of CF. The lungs are particularly affected, so a daily process of breaking up and expelling the mucus is a way of life for some people. Exercise is important for these individuals if they wish to maintain independence and functionality, but they must deal with potential respiratory difficulties and impaired thermoregulation. The clinical exercise physiologist can play an important role in helping these people deal with the effects of CF so that they can live more active and productive lives.

Chronic Obstructive Pulmonary Disease

Ann M. Swank, PhD

Michael J. Berry, PhD

C. Mark Woodard, MS

Chronic obstructive pulmonary disease (COPD) is defined by the American Thoracic Society as a disease characterized by the presence of airflow obstruction that is attributable to either **chronic bronchitis** or emphysema (15,16). Chronic bronchitis is a clinical diagnosis for patients who have chronic cough and sputum production. The American Thoracic Society defines it as the presence of a productive cough most days during 3 consecutive months in each of 2 successive years (15,137). The cough is a result of hypersecretion of mucus, which in turn is the result of an enlargement of the mucus-secreting glands. In contrast to the clinical diagnosis for chronic bronchitis, emphysema is a pathological or anatomical diagnosis marked by abnormal permanent enlargement of the respiratory bronchioles and the alveoli, the airspaces distal to the terminal bronchioles. Emphysema is accompanied by destruction of the lung parenchyma without obvious fibrosis (137,139,151). Most patients with COPD have both chronic bronchitis and emphysema, and the relative extent of each varies among patients (see figures 21.1 and 21.2 on p. 372). The World Health Organization's International Classification of Disease (ICD) codes used by nosologists to classify COPD are 490, 491, 492, 494, 495, and 496.

Patients with COPD experience acute exacerbations, or periods of worsening symptoms. The pathogenesis of an exacerbation is not well understood, and it may be difficult to define clinically. These exacerbations can lead to respiratory failure and are a major cause of hospitalizations in the United States. The major risk factors for the development of COPD include cigarette smoking, exposure to passive smoke, and air pollution.

At times, asthma (ICD code 493) has been subsumed under the rubric of COPD. Asthma is characterized by inflammation and hyperresponsiveness of the tracheobronchial tree to a variety of stimuli (16). Although asthma patients experience exacerbations or attacks, these exacerbations are interspersed with symptom-free periods when airway narrowing is completely or almost completely reversed. In contrast, most patients with chronic bronchitis do not exhibit significant reversibility of airway narrowing and present with residual symptoms between exacerbations. Although patients with COPD and asthma share similar clinical characteristics, the pathology of the two syndromes differs considerably, suggesting that they are different diseases (96). Accumulating evidence suggests that COPD is a disease of multiple organ systems in which ventilatory impairments and muscle

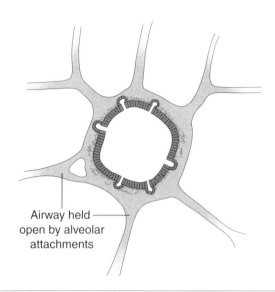

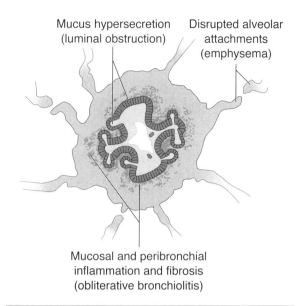

Mucus hypersecretion (luminal obstruction)

Disrupted alveolar attachments (emphysema)

Airway held open by alveolar attachments

Mucosal and peribronchial inflammation and fibrosis (obliterative bronchiolitis)

Figure 21.1 A normal airway that has little inflammation or mucus plugging and is being held open by parenchymal lung tissue.

Adapted from P.J. Barnes, "Chronic obstructive pulmonary disease," *The New England Journal of Medicine* 343: 269-280.

Figure 21.2 An obstructed airway that has significant inflammation and mucus plugging. Also shown is a loss of alveolar attachments, thus making airway collapse more likely.

Adapted from P.J. Barnes, "Chronic obstructive pulmonary disease," *The New England Journal of Medicine* 343: 269-280.

dysfunction contribute to the exercise intolerance seen in these patients (13). Because of the differences between COPD and asthma, they should be considered separately. Asthma will be considered in the next chapter.

SCOPE

COPD is the fourth-leading cause of death in the United States and is projected to be the third-leading cause of death for both males and females by 2020. Estimates are that 16 million people in the United States may be currently diagnosed with COPD and an additional 14 million may be in the early stages without symptoms. Additionally, among the top five causes of death, COPD was the only one that showed an increase (109). Besides being the fourth-leading cause of death in the United States, COPD is a major cause of morbidity and disability and a major healthcare cost. In 2000, COPD was given as the diagnosis at the time of discharge for 726,000 hospitalizations. The average length of stay for COPD hospitalizations was 6.3 d, and the mean cost of each of these hospitalizations was $10,684 (5,109). In 1995, 16,087,000 physician office visits were attributable to COPD. Respiratory tract infections in COPD patients have been shown to have a major effect on utilization of healthcare resources (67). Between 1990 and 1992,

800,000 Americans reported that emphysema caused them to limit their activity (109). Serres et al. (129) reported that COPD patients have lower levels of physical activity than age-matched controls.

PATHOPHYSIOLOGY

The anatomical, physiological, and pathological abnormalities associated with COPD often result in debilitation for the COPD patient. Because of the direct insult that cigarette smoke, the primary risk factor for the development of COPD, has on the lungs, it has long been thought that the lungs were the primary organs affected by COPD. Recent research suggests, however, that the disease process itself or certain aspects of the disease process adversely affect not only the lungs but also skeletal muscle.

Cigarette smoking affects the large airways (**bronchi**), the small airways (**bronchioles**), and the pulmonary parenchyma. The pathological conditions that develop are a result of the effect that cigarette smoke has on each of these structures. Additionally, the degree of airway reactivity of the individual patient will have an effect. Within the large airways, cigarette smoke causes the bronchial mucus glands to become enlarged and the gland ducts to become dilated. Excessive cough and sputum production are a result of these factors and are the characteristic symptoms of chronic bronchitis (150).

These alterations in the large airways have little effect on airflow or spirometry. The airflow obstruction that is characteristic of COPD does not occur until additional damage is incurred by the small airways and the lung parenchyma. The changes in the smaller airways include mucus plugging, inflammation, and an increase in the smooth muscle (see figure 21.2). These changes decrease the cross-sectional area of the airways and can have a profound effect on airflow and the COPD patient's spirometry (150). The spirometry of a COPD patient is characterized by reductions in expiratory flow rates including the **forced expiratory volume in 1 s** (FEV_1), the FEV_1/**forced vital capacity** (FVC) ratio, and the midexpiratory flow rate.

With **emphysema** destructive changes occur to the alveolar walls. The net effect of these changes is twofold. First, destruction of the alveolar walls results in a loss of the tethering or supportive effect that the alveoli have on the smaller airways. This tethering effect helps keep the airways open during expiration. Without this alveolar support, the smaller airways are likely to collapse during expiration, thus adding further to the airway obstruction (see figure 21.2). The second effect of destruction of the alveolar walls is to diminish the elastic recoil of the lungs, which in turn decreases the force that moves air out of the lungs. The combination of these two effects reduces airflow and increases the amount of work the respiratory muscles must perform to meet the ventilatory demands of the body.

The combined effects of airway obstruction and the reduced expiratory driving force increase the time needed for expiration. If inspiration occurs before the increased expiratory time requirement can be met, then the normal end-expiratory lung volume will not be reached, resulting in increased functional residual capacity and hyperinflation of the lungs. Furthermore, the diaphragm will assume a shorter more flattened position. Because the diaphragm is a skeletal muscle, it operates according to the length–tension relationship (103), whereby the tension developed by skeletal muscle is a function of its resting length. As a muscle is shortened or lengthened beyond its optimal length, the potential for tension development decreases (64). Because the diaphragm is shortened with hyperinflation, it has less force-generating potential (44,119,136). Some evidence suggests that the diaphragm adapts to these chronic changes by shortening the optimal length of its fibers (134). As such, each fiber would have the potential to generate its maximal force at its new length (57). Despite these adaptive changes, evidence still suggests

that COPD patients have decreased capacity to generate diaphragmatic pressure.

In normal healthy people, the end-expiratory lung volume decreases by approximately 200 to 400 ml with moderate exercise (73,131). In contrast, patients with COPD demonstrate an increase in the end-expiratory lung volume with exercise, leading to dynamic hyperinflation of the lungs (143). This resulting dynamic hyperinflation leads to further diaphragm weakness and may contribute to **dyspnea** and reduced exercise tolerance. Because the diaphragm is a skeletal muscle, it has been hypothesized that positive adaptations may result from training of the diaphragm muscle (90,125).

Besides the damaging effect that it has on the lungs, COPD may cause skeletal muscle dysfunction (36,141). This skeletal muscle dysfunction may contribute to the exercise intolerance seen in COPD patients. Several studies showed that COPD patients have diminished peripheral muscle strength (22,40,48,53,66,70). Patients with COPD have been found to have a 20% to 30% reduction in quadriceps strength compared with age-matched controls (48,66,70). These decreases in strength are accompanied by a reduction in muscle cross-sectional area (22,53). Other studies have reported reduced muscle mass in patients with COPD (52,127,165). An analysis of muscle biopsies from patients with moderate COPD showed decreases in type I fibers and an increase in the proportion of hybrid fibers compared with controls (65). Studies with advanced COPD patients have also reported a reduction in the proportion of type I fibers compared with control subjects (77,159). These cumulative findings are consistent with the report of reduced oxidative enzyme activities in these patients (99,100). Additionally, this reduction in type I fibers has been shown to be accompanied by a corresponding increase in type IIb fibers (159). Both chronic **hypoxemia** (74) and a lack of physical activity (159) may contribute to the changes in fiber types. Chronic steroid use has been suggested as a contributor to muscle weakness (47,48). Other possible contributors to the skeletal muscle abnormalities seen in COPD patients include chronic **hypercapnia**, inflammation, nutritional depletion, and comorbid conditions that may impair skeletal muscle function (6,13). Practical application 21.1 on p. 374 discusses malnutrition in chronic obstructive pulmonary disease.

One hallmark of patients with COPD is a reduction in airflow, which is most prominent during maximal efforts. This reduction is typically quantified by using the results of pulmonary function tests, with the FEV_1 being one of the standards used to assess disease severity and monitor disease history. Additionally, the FEV_1 has been shown to be a strong predictor of mortality rate from COPD (149).

Practical Application 21.1

Malnutrition in Chronic Obstructive Pulmonary Disease

Malnutrition is a problem for as many as 25% of all COPD patients (30,76). In addition, 50% of patients hospitalized for treatment of COPD demonstrate protein as well as calorie malnutrition (76). In a review of 90 COPD patients, researchers found that patients who required hospitalization and mechanical ventilation demonstrated the most severe nutritional decrements (58). Whereas the causes of malnutrition in COPD patients have not been clearly defined (161), the results of malnutrition have. Weight loss in COPD patients has been shown to be a predictor of mortality rate (163). Additionally, prolonged malnutrition results in deleterious changes to the diaphragm muscle such that its ability to generate force is decreased (81,92). This fact coupled with the fact that COPD patients exhibit dysfunction in other skeletal muscles suggests the need for nutritional support in these patients. A study demonstrated that when COPD patients are given sufficient calories in excess of their needs, they will gain weight and achieve significant improvement in ventilatory and peripheral muscle strength (162). Given the need for nutritional intervention and the positive outcomes that can result, the clinical exercise physiologist should consult with a nutritionist when working with underweight COPD patients.

In healthy nonsmokers the FEV_1 declines by 20 to 30 ml per year (82,138,147). In both men and women smokers, the rate of decline in FEV_1 increases (35,87,147). The rate of decline is both age and sex dependent, and the greatest rates occur in men between the ages of 50 and 70 (35). People who have characteristics compatible with an emphysematous form of COPD have a rate of decline in the FEV_1 of 70 ml per year (31).

In those who quit smoking, the decline of FEV_1 is less pronounced than the decline observed in those who continue to smoke (35,87). In fact, ex-smokers show rates of decline of FEV_1 similar to those of nonsmokers, and in younger ex-smokers the FEV_1 has been shown to increase following smoking cessation (35). Although complete cessation of smoking has beneficial effects on the decline of the FEV_1, the effect on FEV_1 decline of attempting to quit smoking and relapsing is equivocal. Sherrill et al. (132) reported that the rate of decline of the FEV_1 is steeper in ex-smokers who resume smoking compared with those who continue to smoke. More recently, Murray et al. (108) reported that attempts to quit smoking in patients with mild COPD slow the rate of FEV_1 decline compared with those who continue to smoke.

MEDICAL AND CLINICAL CONSIDERATIONS

This section discusses signs and symptoms, history and physical examination, diagnostic exercise testing, and treatment specifically related to COPD.

Signs and Symptoms

The acute respiratory illness is characterized by increased cough, purulent sputum production, wheezing, dyspnea, and occasional fever. With progression of the disease, the interval between these illnesses decreases (15). In the early disease stages, slowed expiration and wheezing are noted during the physical examination. Additionally, breath sounds are decreased, heart sounds may become distant, and course crackles may be heard at the base of the lungs (15).

History and Physical Exam

The diagnosis of COPD is made based on patient history, a physical examination, and the results of laboratory and radiographic studies. The diagnosis is suspected in patients who have a history of smoking and present with an acute respiratory illness or respiratory symptoms such as a productive cough. A smoking history of 70 or more **pack years** has been reported to be suggestive of COPD (18).

Diagnostic Testing

Results from pulmonary function tests are necessary for establishing a diagnosis of COPD and for determining the severity of the disease, but they cannot be used to distinguish between chronic bronchitis and emphysema. The FEV_1, FVC, FEV_1/FVC, and single-breath diffusing capacity are the primary pulmonary function tests recommended to aid in the diagnosis of COPD (14). In patients with COPD, the results from all these tests are less than what would be predicted for a person of similar age, sex, and stature. Other recommended tests include lung volume measurements and determination of arterial blood gas levels. Lung volume measurements often reveal an increase in total lung capacity, functional residual capacity, and residual volume. Arterial blood gases may reveal hypoxemia in the absence of hypercapnia in the early stages of the disease, with a worsening of hypoxemia and hypercapnia presenting in the later stages of the disease (15).

Because emphysema is defined in anatomic terms, the chest **roentgenogram** can sometimes be used to differentiate between emphysema and chronic bronchitis. Whereas patients with chronic bronchitis often have a normal chest roentgenogram, the roentgenogram of patients with advanced emphysema may reveal large lung volumes, **hyperinflation**, a flattened diaphragm, and vascular attenuation (15). Computed tomography has greater sensitivity and specificity than the chest roentgenogram and can be used for both qualitative and quantitative assessment of emphysema. Because the additional information gained from computed tomography will rarely alter therapy, it is infrequently used in the routine care of COPD patients. Although computed tomography is not recommended for routine use with COPD patients, it is the best way of recognizing emphysema and probably has a significant role in recognizing localized emphysema that is amenable to surgical treatment (151).

The American Thoracic Society proposes staging patients with COPD into distinct categories based on the degree of airflow obstruction. This organization suggests using the FEV_1 as the staging criterion. As such, patients with an FEV_1 greater than or equal to 50% of predicted are categorized as stage 1, or as having mild disease. Those with an FEV_1 between 35% and 49% of predicted are categorized as stage 2, or as having moderate disease. Finally, those with an FEV_1 less than 35% of predicted are categorized as stage 3, or as having severe disease (14,15). Most COPD patients are categorized as having mild disease.

Exercise Testing

Exercise testing is an integral component in the evaluation of patients with COPD. In patients with mild or moderate disease, symptoms generally do not present until increased demand is placed on the respiratory system, such as with exercise. In patients with severe disease, the functional capacity is reduced to such a level that even simple activities of daily living may impose a challenge

to the respiratory system. Most patients with moderate to severe COPD have reduced exercise capacity because of reduced ventilatory capacity in the face of an increased ventilatory demand (20,51,79). Because exercise places increased demand on the respiratory system, exercise testing provides an objective evaluation of the functional capacity of the COPD patient. Additionally, exercise testing can be used to detect COPD and cardiovascular disease, follow the course of the disease, detect exercise hypoxemia, determine the need for supplemental oxygen during exercise training, evaluate the response to treatment, and prescribe exercise (144).

Table 21.1 on p. 376 outlines procedures and guidelines for exercise testing of the COPD patient (11). During the exercise tests the minimum monitoring should include measurement of blood pressure, a 12-lead electrocardiogram (ECG), analysis of arterial oxygen saturation, and measurement of dyspnea. These should be measured before the start of the test, continuously throughout the test, and at the termination of the test. Blood pressure should be measured with the patient's arm relaxed and the manometer mounted at eye level. Automatic monitors for blood pressure measurement during exercise are available, and their use has been found acceptable (68). Placement for the 12-lead ECG is typically the Mason-Likar. The gold standard for the measurement of arterial oxygen saturation is co-oximetry using arterial blood. If this is unavailable, the use of a pulse oximeter is acceptable as long as the pulse oximeter has been validated during exercise. At oxygen saturations greater than 90%, these devices have a high degree of reliability (110). However, as oxygen saturation drops below 90%, their reliability worsens (110). Because of the problems with precisely defining the degree of hypoxemia with these instruments, these devices should be used to qualify whether desaturation is occurring and then to correct it with supplemental oxygen. The final variable that should be monitored during the exercise test in patients with COPD is dyspnea. A number of scales that have been validated and proved reliable are available for use (4,97,98). One particular dyspnea scale of interest to the exercise specialist is the Borg scale. This scale has been adapted for use during exercise testing and training with COPD patients (27).

If the equipment is available, gas exchange and ventilatory measurements should be obtained. These measures provide valuable information that can be used to prescribe exercise intensity more accurately, to evaluate the effectiveness of an exercise intervention, and to provide information regarding the extent of the lung disease (101,128). Of special concern when measuring gas exchange and ventilatory parameters in

Table 21.1 Exercise-Testing Recommendations, Procedures, and Guidelines for the COPD Patient

Test type	Mode	Protocol specifics	Clinical measures	Clinical implications	Special considerations
Cardiovascular	• Treadmill or cycle ergometer (preferred) • 6 min walk	• Duration of 8–12 min, small incremental increases in workload individualized to the patient • Treadmill—1 to 2 METs/stage • Cycle ergometer—ramped protocol 10, 15, or 20 W/min, stage protocol 25–30 W/3 min stage	• HR, 12-lead ECG (Mason-Likar placement) • BP • RPE, RPD • Oxygen saturation (pulse oximetry/arterial PaO_2) • Ventilation measures and gas exchange • Blood lactate • Distance	• Serious dysrhythmias, >2mm ST-segment depression or elevation, ischemic threshold, T-wave inversion with significant ST change • SBP >250 mmHg or DBP >115 mmHg • Maximum ventilations, $\dot{V}O_2$peak, lactate/ventilatory threshold • Note rest stop distance or time, dyspnea index, vitals	• No arm ergometer testing • Clients with COPD often have coexisting CAD • Breathing pattern may help analysis • Lactic acidosis may contribute to exercise limitation in some patients • Exercise testing in mid- to late afternoon is desirable • Useful for measurement of improvement throughout conditioning program
Strength	• Isokinetic/isotonic • Sit to stand, stair climbing, lifting	Time to 10 reps	• Peak torque • Maximum number of reps • 1RM		• Clients may become more dyspneic when lifting objects • Specific evaluation and training may be needed
Range of motion	• Sit-and-reach • Gait analysis • Balance	Hip, hamstring, and lower-back flexibility			Body mechanics, coordination, and work efficiency are often impaired

COPD patients is the patient who requires supplemental oxygen. These patients should be tested on an elevated fraction of inspired oxygen such that it equates with the flow rate established for the use of supplemental oxygen. Most commercially available gas exchange measurement systems have established procedures that allow for use of elevated fractions of inspired oxygen during exercise testing and conversions of oxygen flow rates to inspired oxygen fractions.

Exercise testing has been shown to be extremely safe, even in high-risk populations, with 1 or fewer deaths per 10,000 tests (61). To minimize risk to patients, recom-mended guidelines from the American College of Sports Medicine (12), the American Association of Cardiovascular and Pulmonary Rehabilitation (8), and the American Heart Association (60) should be closely followed. In general, the procedures for testing patients with COPD follow those for testing other at-risk populations. Before conducting the exercise test, the clinician should review information from a medical exam and history to identify contraindications to testing, as observed in the previously cited guidelines. An additional concern with COPD patients is the patient with accompanying pulmonary hypertension. Some experts advise caution with these individuals because of the risk of serious cardiac arrhythmias or even sudden death while testing (9).

The exercise mode, the test protocol, and the monitoring equipment are all fundamental considerations in exercise testing. The exercise mode should be one that will increase total-body oxygen demands by requiring the use of a large-muscle group. The most common exercise-testing modalities for the COPD patient are the treadmill and the bicycle ergometer. One exercise mode that should not be used routinely to test COPD patients is arm ergometry because patients with severe COPD often use the accessory muscles of inspiration for breathing at rest. As such, any additional burden placed on these muscles could result in significant symptoms and distress for the patient (38).

The testing protocol should start at a work rate that the patient can easily accomplish, have increments in the work rate that are progressively difficult, and last a total duration of 8 to 12 min. The initial stages should be of an intensity that allows the patient adequate time to warm up and become accustomed to the exercise bout. The work rate increments should be small and based on characteristics of the patient (e.g., sex, size, severity of disease, and previous level of physical activity). The incremental rate of work rate will affect the exercise responses of COPD patients (46). For example, the peak work rate achieved for a given level of oxygen consumption will be greater when the work rate is increased quickly. Unfortunately, because of severe deconditioning and extreme dyspnea that some patients with COPD experience when performing even mild physical activity, having a test that lasts the minimum recommended duration may not be possible.

The responses of the COPD patient to an exercise test will vary depending on the severity of the disease. In patients with mild disease, the results of the exercise test are consistent with those of normal individuals or may demonstrate abnormalities indicative of cardiovascular disease or deconditioning (54). Exercise responses in COPD patients with moderate and severe disease compared with age-matched healthy controls are shown in table 21.2. Peak oxygen uptake and peak work rates are usually reduced in COPD patients with moderate or severe disease (37,99,113,133,157). Concomitant with the lower peak oxygen consumption is a lower peak heart rate (HR_{peak}) and a greater heart rate reserve (predicted HR_{peak} minus measured HR_{peak}). Oxygen pulse is also low in COPD patients because they terminate exercise at a low work rate. In normal and deconditioned individuals and in cardiac patients, the ventilatory reserve (maximal voluntary ventilation minus the peak minute ventilation) is high at peak exercise. In contrast, the patient with COPD has a low ventilatory reserve, and, in some cases,

peak minute ventilation is equal to or even greater than the maximal voluntary ventilation (155). Additionally, the peak minute ventilation is lower than predicted (37). The partial pressure of oxygen in the arterial blood and the percentage saturation of hemoglobin in the arterial blood are often low at maximal exercise in the COPD patient with moderate or severe disease (15). Although it was traditionally thought that patients with COPD did not develop anaerobiosis and did not demonstrate a lactate threshold during incremental exercise, recent research suggests that these patients can develop a significant anaerobiosis and do demonstrate a lactate threshold, although these occur at relatively low work rates (37,145). Because of their ventilatory impairment, patients with moderate and severe COPD will not show a disproportionate increase in minute ventilation with the development of anaerobiosis (145). Thus, the detection of a ventilatory threshold may not be possible in these patients.

Exercise testing is an important tool in assessing the patient with COPD. But the test and equipment used must be designed to meet the needs of the patient and the clinician administering the test. Additionally, because of the abnormal responses of the COPD patient, care must be exercised when interpreting the results of these tests.

Treatment

After a diagnosis of COPD has been made, a multifaceted approach to the treatment and management of the patient should be adopted. Comprehensive treatment of the patient should include smoking cessation, oxygen therapy, pharmacological therapy, and pulmonary rehabilitation (including exercise training) (33). Because

Table 21.2 Exercise-Test Responses in Patients With Chronic Obstructive Pulmonary Disease Compared With Normal Healthy Subjects

Parameter	Finding
Peak work rate	Decreased
Peak oxygen consumption	Decreased
Peak heart rate	Decreased
Peak ventilation	Decreased
Heart rate reserve	Increased
Ventilatory reserve	Decreased
Arterial partial pressure of oxygen	Decreased
Arterial oxygen saturation	Decreased
Lactate threshold	Occurs at a lower work rate
Ventilatory threshold	Absent

smoking is a major cause of COPD, smoking cessation is a major therapy in the treatment of COPD patients. Smoking cessation is one of two interventions that have been shown to improve patient survival (15).

The second therapy that has been shown to improve survival in patients with COPD is long-term oxygen therapy. The British Medical Research Council study (106) and the National Heart, Lung and Blood Institute's Nocturnal Oxygen Therapy Trial (111) showed that patients who received long-term oxygen therapy experienced a significant reduction in mortality rates. Other benefits of long-term oxygen therapy include a reduction in **polycythemia** (91), decreased pulmonary artery pressure (1–3), and improved neuropsychiatric function (72).

Acute administration of supplemental oxygen has been shown to preserve exercise tolerance in hypoxemic patients. Whether patients with COPD undergo an acute bout of exercise with administration of supplemental oxygen (29,42,45,107,118,153) or are trained with supplemental oxygen (49,118,126,153,166), the benefits are significant. The goal of oxygen therapy is to reverse hypoxemia and prevent tissue hypoxia (15). For COPD patients to realize benefits from supplemental oxygen therapy during training, they must demonstrate hypoxemia during training. If oxygen is to be prescribed for COPD patients, the goal is to maintain the partial pressure of oxygen in arterial blood above 60 mmHg or the percentage saturation above 90. Therefore, the delivery method and the dosage of oxygen, or the flow rate, need to be considered. COPD patients who need supplemental oxygen during exercise often use a liquid oxygen supply. These systems, although more expensive, are lightweight and easily refilled from larger stationary sources. Oxygen concentrators cannot be used during exercise because of their weight and need for an electrical supply.

Pharmacological therapy (table 21.3) in patients with COPD is aimed at inducing bronchodilation, decreasing the inflammatory reaction, and managing and preventing respiratory infections (15). **Bronchodilator** therapy includes the use of α_2-agonists, anticholinergic agents, and theophylline. The use of α_2-agonists may result in tremors, anxiety, palpitations, and arrhythmias. Because

Table 21.3 Pharmacology

Medications	Class and primary effects	Exercise effects	Special considerations
Albuterol, bitolterol, epinephrine, formoterol, isoetharine, isoproterneol, levalbuterol, metaproterenol, perbuterol, salmeterol, terbutaline	Sympathomimetic bronchodilators	May increase or have no effect on HR, ECG; may increase, decrease, or have no effect on BP; increases exercise capacity by limiting bronchospasm	
Ipratropium	Anticholinergic bronchodilator	May increase or have no effect on HR, ECG; no change on BP	
Aminophylline, dyphylline, oxtriphylline, theophylline	Methylxanthine bronchodilators	May increase or have no effect on HR, ECG; no change on BP	May produce PVCs
Beclomethasone, budesonide, dexamethasone, flunisolide, fluticasone, triamcinolone	Corticosteroid anti-inflammatory antiasthmatic	No effect on HR, BP, ECG, or exercise capacity	
Cromolyn, nedocromil	Asthma prophylatics	No effect on HR, BP, ECG, or exercise capacity	
Montelukasat, zafirlukast, zileutan	Antileukotriene antiasthmatics		
Acetylcysteine, guaifenesin, potassium iodide	Mucolytics or expectorants		

of these problems, careful dosing and monitoring of patients with known cardiovascular disease are necessary (15). After a patient develops persistent symptoms, anticholinergic agents such as ipratropium bromide may be prescribed because their effect is more intense and of longer duration. Additionally, they may have less potentially deleterious side effects than α_2-agonists do.

Theophylline, one of the methylxanthines, is a third agent that may be used to induce bronchodilation. Besides its bronchodilator effects, theophylline increases cardiac output, decreases pulmonary vascular resistance, and may have anti-inflammatory effects (146,167). Despite the beneficial effects of theophylline, its popularity has declined because of its toxicity (75,130) and potential to have adverse interaction with other drugs (124). Inhaled corticosteroids are indicated if a significant bronchodilator response occurs or the patient has more severe disease with frequent **exacerbations**. Oral corticosteroids may

be necessary for acute exacerbations (33).

Pulmonary rehabilitation is defined as a multidimensional continuum of services directed at persons with pulmonary disease and their families. Pulmonary rehabilitation is usually delivered by an interdisciplinary team of specialists, with the goal of achieving and maintaining the individual's maximum level of independence and functioning in the community (59). These services typically include patient assessment, patient education, exercise training, psychosocial intervention, and patient follow-up (8,9). The various components of each of these services are listed in table 21.4. Practical application 21.2 describes client–clinician interaction for patients with COPD. The goals of pulmonary rehabilitation are to decrease airflow limitations, improve exercise capacity or physical function, prevent and treat secondary medical complications, decrease respiratory symptoms, and improve the patient's quality of life. As a result of

Table 21.4 Various Components of the Different Services Offered in Pulmonary Rehabilitation

Patient assessment	Patient training and education	Exercise training	Psychosocial interventions	Patient follow-up
Medical history	Anatomy and physiology	Mode, duration, frequency, and intensity	Identification of support systems	Outcome measurements of physical function and health-related quality of life
Pulmonary function tests	Pathophysiology of lung disease	Upper- and lower-extremity endurance training	Treatment of depression	Support groups
Symptom assessment	Description and interpretation of assessment tests	Upper- and lower-extremity strength training	Treatment of anxiety	Maintenance programs
Physical function assessment	Breathing retraining	Inspiratory muscle training	Anger management	
Nutritional assessment	Bronchial hygiene	Flexibility and posture	Sexuality issues	
Activities of daily living assessment	Medication information	Orthopedic limitations	Adaptive coping styles	
Educational assessment	Symptom management	Home exercise plans	Adherence to lifestyle modifications	
Psychosocial assessment	Activities of daily living and energy conservation Nutrition Psychosocial issues Smoking cessation		Relapse prevention	

Adapted from *Guidelines for pulmonary rehabilitation programs,* 2nd ed. American Association of Cardiovascular and Pulmonary Rehabilitation.

Client–Clinician Interaction

Often, the first interaction between the clinical exercise physiologist and the COPD patient occurs when the patient has been referred for pulmonary rehabilitation. Unfortunately, patients with COPD are often referred to pulmonary rehabilitation only after they have experienced an exacerbation of their disease or when their dyspnea has become so oppressive that they are severely disabled. As a result, these patients are often anxious, scared, frustrated, and depressed. The clinical exercise physiologist must be aware of these problems when working with the COPD patient and be able to present a positive, yet realistic, picture of the benefits of exercise training for the COPD patient.

Dyspnea, the primary symptom of COPD, often results in a vicious cycle of fear and anxiety followed by inactivity resulting in deconditioning, which results in further dyspnea. Unless this cycle can be broken, COPD patients are destined to lose their independence and become dependent on others to meet their most basic needs. COPD patients need to be made aware that they can learn to live with their dyspnea and that exercise can help reduce the intensity of dyspnea and the distress associated with dyspnea. These patients should be taught strategies that will help them to manage their dyspnea on a daily basis. These strategies include such things as monitoring the effects of various medications on dyspnea and avoiding factors that can result in dyspnea—such as stress.

The clinical exercise physiologist must also be aware that COPD patients often experience exacerbations or periods of worsening symptoms. As a result, these patients may not be able to exercise at their prescribed intensity or may miss exercise sessions completely. The patient should be encouraged to continue exercising, even if at a much lower intensity. If even extremely mild exercise is not possible, the patient should be encouraged to resume exercising after he or she has recovered from the exacerbation.

The successful clinical exercise physiologist can effectively interact with each person on a one-to-one basis. Such a professional is sensitive to the particular needs of the patient and is able to tailor each patient's program to meet his or her individual needs. As a result, the patient is able to take control of his or her disease with less fear and anxiety.

participating in a comprehensive pulmonary rehabilitation program, patients have demonstrated improvements in quality of life (63,84), sense of well-being (17), self-efficacy (123), and functional capacity (123). Additionally, functional status has been shown to be a strong predictor of survival in patients with advanced lung disease following pulmonary rehabilitation (28). Determining which of the specific components of pulmonary rehabilitation is responsible for these improvements is difficult, because they are all integrally related.

The American Thoracic Society recommends that patients with COPD be referred to a pulmonary rehabilitation program after they have been placed on optimal medical therapy and still demonstrate the following (15):

1. Severe symptoms

2. Several emergency room or hospital admissions within the previous year

3. Diminished functional status that limits their activities of daily living

4. Impairments in quality of life

Although patients who are referred to pulmonary rehabilitation typically have severe disease, research indicates that patients with mild and moderate disease will benefit from participation in the exercise component of a pulmonary rehabilitation program similarly to those with severe disease (24,156). Participation in a pulmonary rehabilitation program that includes at least 4 wk of exercise training can result in improvements that are clinically significant for COPD patients (85). The improvements realized by patients who participate in exercise rehabilitation are quality of life, specifically in the relief of dyspnea, and improvement in their perceptions of how well they can cope with their disease.

EXERCISE PRESCRIPTION

Practical application 21.3 on summarizes recommendations for exercise prescription for the COPD patient.

Exercise Prescription Recommendations for the COPD Patient

Training method	Mode	Intensity	Frequency	Duration	Progression	Goals	Special considerations and comments
Cardio-vascular	Large-muscle activities (walking, cycling, swimming)	RPE 11–13/20 (comfortable pace and endurance) Monitor dyspnea	One or two sessions, 3–7 d · wk^{-1}	30 min sessions	Emphasize progression of duration, increase intensity 2–3 mo to ensure compliance	Increase $\dot{V}O_2$peak, increase lactate threshold and ventilatory threshold, become less sensitive to dyspnea Develop more efficient breathing patterns, facilitate improvement in ADLs	Exercise compliance should be considered when determining exercise intensity Shorter intermittent sessions may be necessary initially
Strength	Free weights, isokinetic or isotonic machines	Low resistance, high reps (fatigue by 8–15 reps)	2–3 d · wk^{-1}		Resistance should be increased as strength increases 2–3 mo to goal	Increase maximal number of reps Increase isokinetic torque and work Increase lean body mass	Respiratory muscle weakness is common in pulmonary patients Upper-body exercise contributes to dyspnea Inspiratory muscles may require training
Range of motion	Stretching, tai chi		Daily			Improve gait, balance, breathing efficiency	
Neuro-muscular	Walking Balance exercises Breathing exercises		Daily			Improve gait, balance Improve breathing efficiency	

Because heart rate is not a reliable indicator of exercise tolerance for patients with COPD, intensity is monitored by dyspnea or ratings of perceived exertion. The ACSM recommends that the mode of exercise should be any aerobic exercise that involves large-muscle groups, such as walking or cycling. The minimal recommendation for frequency is three to five times per week with a minimal duration goal of 20 to 30 min of continuous activity (12).

EXERCISE TRAINING

Dyspnea and reduced exercise capacity are two of the most common complaints of COPD patients. In addition to the diaphragm muscle, the accessory muscles of inspiration (scalene, sternocleidomastoid, and serratus anterior) are activated during exercise. Even at low work rates, unsupported arm exercise results in greater levels of dyspnea compared with lower-extremity exercise in patients with COPD (38). Arm exercise requires the use of the accessory muscles of inspiration, thereby decreasing their participation in ventilation and increasing the work of the diaphragm. This observation may explain, in part, why patients with COPD complain of dyspnea when performing activities of daily living with their upper extremities (148). Thus, strategies aimed at

improving the function of these accessory muscles of inspiration, such as resistance training, could benefit COPD patients.

Skeletal muscle dysfunction contributes to the reduced exercise tolerance seen in COPD patients (36,141). Impaired muscle strength has been found to be a significant contributor to symptom intensity during exercise in COPD patients (70,141). Additionally, quadriceps muscle strength has been shown to be positively correlated with both the 6 min walk distance and maximal oxygen consumption in COPD patients (66,70). These observations suggest that resistance training may also prove beneficial for the rehabilitation of patients with COPD.

In general, four exercise-training strategies are recommended for improving respiratory and skeletal muscle dysfunctions. These include lower-extremity aerobic exercise training, ventilatory muscle training, upper-extremity resistance training, and whole-body resistance training. A brief discussion and supporting literature for these four exercise-training strategies follows.

Table 21.5 and practical application 21.4 review the literature about lower-extremity exercise and COPD. Cumulative results indicate strong evidence for the use of lower-extremity exercise as a therapeutic intervention for patients with COPD.

Table 21.5 Results of Controlled Randomized Clinical Trials Examining the Efficacy of Lower-Extremity Exercise Training

Author	Exercise capacity	Increased peak oxygen consumption	Improved quality of life
Ambrosino et al. (7)	Yes	Not measured	Not measured
Berry et al. (23)	Yes	No	Not measured
Booker (26)	No	Not measured	Yes
Busch and McClements (32)	Yes	Not measured	No
Cambach (34)	Yes	Not measured	Yes
Goldstein et al. (63)	Yes	Not measured	Yes
Jones et al. (78)	Yes	Not measured	Yes
Lake et al. (86)	Yes	Not measured	Yes
Larson et al. (88)	Yes	Yes	Yes
McGavin et al. (104)	Yes	No	Yes
Reardon et al. (121)	Yes	No	Not measured
Ries et al. (123)	Yes	Yes	No
Strijbos et al. (142)	Yes	Not measured	Yes
Toshima et al. (152)	Yes	Not measured	No
Weiner et al. (158)	Yes	Not measured	Not measured
Wijkstra et al. (160)	No	Not measured	Yes

Practical Application 21.4

Literature Review

The American College of Chest Physicians and the American Association of Cardiovascular and Pulmonary Rehabilitation released evidence-based guidelines for pulmonary rehabilitation (10). This document contains recommendations for pulmonary rehabilitation and reviews the supporting scientific evidence. Lower-extremity exercise training received a grade of A. This grade reflects the fact that strong scientific evidence supports the use of lower-extremity exercise training in COPD patients. This evidence is from the results of well-designed, well-conducted, controlled (both randomized and nonrandomized) trials with statistically significant results that support the use of lower-extremity exercise training. The results of controlled randomized clinical trials that have included lower-extremity exercise training as part of an intervention are shown in table 21.5. Shown in this table are the effects of lower-extremity exercise training on submaximal and maximal exercise capacity, peak oxygen consumption, and quality of life. In nearly all the studies, exercise capacity, as evaluated from time on the treadmill or a timed distance walk, was found to improve following lower-extremity exercise training. Whether improvements in exercise capacity will translate into improvements in domains such as physical function and activities of daily living has yet to be determined. As recently pointed out at a workshop convened by the National Institutes of Health to investigate the efficacy of pulmonary rehabilitation, limiting the evaluation of interventions to outcomes such as timed walks or physiological measures was myopic and provided incomplete measures of medical outcomes (59). The conclusion from this august group was that the success of therapeutic interventions should be based on a variety of medical outcomes such as health-related quality of life; respiratory symptoms; frequency of exacerbations; activities of daily living; cost–benefit relationships; use of healthcare resources; and mental, social, and emotional function (59).

Ventilatory muscle training is recommended for COPD patients to increase ventilatory muscle strength and endurance. The ultimate goal is to improve exercise capacity, relieve the symptoms of dyspnea, and improve health-related quality of life. Three strategies have been used to train the ventilatory muscles: (1) voluntary isocapnic **hyperpnea**, (2) inspiratory resistive loading, and (3) **inspiratory threshold loading**.

With voluntary isocapnic hyperpnea, the patient is instructed to breathe at as high a level of minute ventilation as possible for 10 to 15 min. With this technique, the patient is hyperventilating and, therefore, a rebreathing circuit must be used to maintain **isocapnia**. This rebreathing circuit is complex and not portable, and the patient requires constant monitoring to ensure isocapnia when using this device. Because of these problems, this type of training has not been used or studied extensively.

During inspiratory resistive loading, the patient breathes through inspiratory orifices of smaller and smaller diameter while attempting to maintain a normal breathing pattern. A potential problem with the use of this device is that the patient may slow his or her breathing frequency in an attempt to decrease the sensation of effort. Because of the change in the breathing pattern,

the load on the inspiratory muscles is reduced such that a training response may not occur.

With inspiratory threshold loading, the patient breathes through a device that permits air to flow through it only after a critical inspiratory pressure has been reached. These devices are small, do not require supervision, and avoid the problems associated with changing breathing patterns.

Results from studies that examine the efficacy of **ventilatory muscle training** in COPD patients are equivocal. Table 21.6 on p. 384 shows the results of randomized controlled clinical trials that examine the effects of inspiratory resistive loading and inspiratory threshold loading. Of the 18 studies presented, inspiratory muscle strength was found to increase in 9 of them and inspiratory muscle endurance was found to increase in 10. These results suggest that inspiratory muscle training does not add significantly to a program of general exercise conditioning in patients with COPD. In the evidence-based guidelines for pulmonary rehabilitation (10), inspiratory muscle training received a grade of B. This grade reflected the fact that the scientific evidence from both observational and controlled clinical trials provided inconsistent results. Because of this grade, it was recommended that inspiratory muscle training not be considered an essential

Table 21.6 Results of Randomized Clinical Trials Examining the Efficacy of Ventilatory Muscle Training

Reference	Type of training	Outcomes (compared with a control group)
Belman and Shadmehr (21)	Resistive loading	Improved inspiratory muscle strength and endurance
Berry et al. (23)	Threshold loading coupled with general exercise conditioning	No change in inspiratory muscle strength No change in 12 min walk distance No change in dyspnea ratings
Bjerre-Jepsen et al. (25)	Resistive loading	No change in inspiratory muscle endurance No change in exercise tolerance
Chen et al. (39)	Resistive loading coupled with standard pulmonary rehabilitation	Improved inspiratory muscle endurance Improved inspiratory muscle strength No change in maximal or constant load cycle exercise
Dekhuijzen et al. (50)	Resistive loading coupled with standard pulmonary rehabilitation	Improved inspiratory muscle strength No improvement in maximal work capacity Increased 12 min walk distance
Falk et al. (56)	Resistive loading	Decreased dyspnea Improved submaximal exercise time
Goldstein et al. (62)	Threshold loading coupled with standard pulmonary rehabilitation	Improved inspiratory muscle endurance No change in inspiratory muscle strength No change in exercise tolerance
Guyatt et al. (69)	Resistive loading	No improvement in inspiratory muscle strength or endurance No improvement in 6 min walk distance No improvement in health-related quality of life
Harver et al. (71)	Resistive loading	Improved inspiratory muscle strength Decreased dyspnea
Larson et al. (89)	Threshold loading	Improved inspiratory muscle strength Improved inspiratory muscle endurance Improved 12 min walk distance No improvement in health-related quality of life
Lisboa et al. (93)	Threshold loading	Improved inspiratory muscle strength and endurance Decreased dyspnea Increased 6 min walk distance
Lisboa et al. (94)	Threshold loading	Decreased dyspnea Improved 6 min walk distance
McKeon et al. (105)	Resistive loading	No change in inspiratory muscle strength Increased inspiratory muscle endurance No increase in maximal cycle exercise, 12 min walk distance, or treadmill walking
Noseda et al. (112)	Resistive loading	Increased inspiratory muscle endurance No change in maximal or constant load cycle exercise
Pardy et al. (117)	Resistive loading	Improved 12 min walk distance Improved submaximal exercise endurance

Reference	Type of training	Outcomes (compared with a control group)
Preusser et al. (120)	Threshold loading	Improved inspiratory muscle strength and endurance
		Improved 12 min walk distance
Wanke et al. (154)	Threshold loading coupled with general exercise conditioning	Improved inspiratory muscle strength and endurance
		Improved maximal exercise capacity
Weiner et al. (158)(165)	Threshold loading coupled with general exercise conditioning	Improved inspiratory muscle strength and endurance
		Improved 12 min walk distance
		Improved submaximal exercise time

component of pulmonary rehabilitation. But in patients who have decreased respiratory muscle strength and breathlessness and who remain symptomatic despite optimal therapy, inspiratory muscle training may be considered an adjunctive exercise therapy (41,114).

Specific recommendations regarding the intensity, frequency, or duration of training for inspiratory muscle training have not been developed. Most studies that have reported improvements in inspiratory muscle function have had patients perform inspiratory muscle training at a minimum of 30% of their maximal inspiratory pressure. The duration of this training has been at least 15 min and the frequency at least three times per week. These appear to be the minimal requisites of an exercise prescription if inspiratory muscle strength and endurance are to be improved.

Based on the results of preliminary studies that have evaluated upper-, lower-, and whole-body strength training in COPD patients, resistance training appears to offer distinct advantages over other forms of exercise training. As such, resistance training should be included in a comprehensive exercise rehabilitation program. Table 21.7 presents research regarding whole-body resistance training and COPD. Although studies are still limited in number, a reasonable recommendation of exercise dosage for demonstrating improvement in outcomes includes training 2 to 3 d per week with 8 to 10 repetitions and loads of 50 to 85% of one-repetition maximum (141).

Upper-extremity resistance training has been proposed as a training modality to help reduce dyspnea in COPD patients. Ventilatory muscle fatigue and dyspnea

Table 21.7 Review of Studies of Whole-Body Resistance Training for Individuals With COPD

Author	n	Age/FEV$_1$ % predicted	Training intensity: % 1RM	Strength gains	Outcomes
Bernard et al. (19)	45	65.5/42.2%	60–80%	8–20%	No difference between groups
Clark et al. (40)	43	49/77%	70%	7.6 kg increase in max lifts	Increased quality of life
Kaelin et al. (80)	50	68/39%	RPE 4–7	na	Increased physical function
Kongsgaard et al. (83)	18	65–80/46%	"Heavy RT"	14–18%	Increased health
Mador et al. (95)	32	74/44%	60%	17.5–26%	Increased health-related quality of life
Ortega et al. (115)	72	64/41%	70–85%	Significant increase in lifts	Fatigue and emotion
Panton et al. (116)	17	62/40%	32–64%	36%	Decreased effort with three of eight ADLs
Simpson et al. (135)	34	71.5/38 %	50–80%	73% max cycle ergometer test	Decreased shortness of breath with ADLs
Spruit et al. (140)	48	64/40%	70%	20–40%	Increased health-related quality of life
Wright et al. (164)	28	55.7/42	Maximal	18.7%	Increased quality of life

Note. RPE = rating of perceived exertion.

occur when COPD patients use their upper extremities to perform activities of daily living. Fatigue results because of the additional work that the accessory muscles of inspiration must perform in helping to support the arms during such activities (38). A summary of upper-extremity resistance training is shown in table 21.8. Cumulative results show that patients with COPD can tolerate and benefit from a training program consisting of upper-extremity strength exercise. But research has not conclusively demonstrated that upper-extremity resistance training alone will improve activities of daily living or physical function, whereas whole-body resistance training appears to be effective (table 21.7). Preliminary results, however, support the recommendation that upper-extremity resistance training be included in a comprehensive rehabilitation program (10).

These preliminary studies do not provide clear recommendations on the specific upper-body exercises that would benefit this population or on the resistance or number of repetitions that will provide the optimal

benefits. The exercises should probably involve the accessory muscles of inspiration. With respect to the amount of resistance used and number of sets and repetitions to be completed, the ACSM guidelines (12), the recommendations of Evans (55), and the recent recommendations of Storer et al. (141) should be followed.

Table 21.9 summarizes the benefits associated with exercise training on fitness components for the patient with COPD. Chronic obstructive pulmonary disease is a common condition that affects a large number of older people. The disease process spans several decades and eventually results in significant morbidity and mortality rates. Research suggests that exercise can be used as an effective therapeutic intervention in patients with COPD. The review presented here supports the notion that participation in an exercise program will decrease dyspnea and increase exercise capacity—two of the most common complaints of COPD patients. Despite the positive findings from previous research, a number of questions regarding the most effective exercise strategies

Table 21.8 Results of Trials Examining the Efficacy of Upper-Extremity Resistance Training

Reference	Type of training	Outcomes
Couser et al. (43)	Arm ergometry at 60% of maximal workload. Unsupported arm exercise that consisted of bilateral shoulder abduction and extension for 2 min. Weight was added as tolerated. Both groups performed leg cycle ergometry.	Following training, minute ventilation and oxygen consumption were lower during arm elevation. Respiratory muscle strength did not change following training. No differences were reported between the two groups.
Martinez et al. (102)	Arm ergometry at a workload that engendered an RPE of 12 to 14 and an RPD of 3. Unsupported arm exercise that consisted of five shoulder and upper-arm exercises for up to 3.5 min. Weight was added as tolerated. Both groups performed leg cycle ergometry at a workload that engendered an RPE of 12 to 14, an RPD of 3, and inspiratory muscle training.	No difference in improvements in 12 min walk, cycle ergometer test, or respiratory muscle function. Task-specific unsupported arm exercise tests improved in the group that performed unsupported arm exercise. Decreased oxygen consumption in unsupported arm exercise tests in the group that performed unsupported arm exercise.
Ries et al. (122)	Gravity resistance exercises that included five low-resistance, high-repetition exercises to improve arm and shoulder endurance. Proprioceptive neuromuscular facilitation that included lower frequency progressive resistance training with weights to improve arm and shoulder strength and endurance. Both groups participated in standard pulmonary rehabilitation that included walking.	Compared with a control group, both training groups improved on training test specific to the exercise modality. Patients reported subjective improvement in ability to perform activities of daily living using the upper extremities. No change in performance of cycle ergometry tests, simulated activities of daily living tests, or ventilatory muscle endurance tests.

Note. RPE = rating of perceived exertion; RPD = rating of perceived dyspnea.

Table 21.9 Benefits Associated with Exercise Training on Fitness Components

Cardiorespiratory endurance	Skeletal muscle strength	Skeletal muscle endurance	Flexibility	Body composition
• Cardiovascular reconditioning • Desensitization to dyspnea • Improved ventilatory efficiency	• Improved muscle strength • Better balance	• Improved muscle endurance	• Improved range of motion	• Improved body composition • Enhanced body image

for the COPD patient remain unanswered. We hope that future research will provide these answers.

CONCLUSION

This chapter presents background information regarding etiology, clinical history, signs and symptoms of COPD as well as exercise testing and prescription strategies for the exercise professional. Exercise testing and training any patient with chronic disease involves individualization of treatment based on all patient information available. Thus, it is crucial that the exercise professional be familiar with each patient's history prior to development of an exercise program and that the program is individualized for each patient.

Case Study

Medical History

Mr. DM is a 69 yr old white male who complains of shortness of breath on exertion and occasionally at rest. The patient does not report any symptoms suggestive of myocardial ischemia. Additional findings from the medical history include treatment for hypertension and prostate cancer diagnosed within the past 5 yr. The patient quit smoking cigarettes approximately 3 yr ago, and reports a 102 pack year smoking history (average of 2 packs per day for 51 yr). The patient was admitted to a local hospital for an exacerbation of respiratory symptoms approximately 4 mo before enrolling in the exercise program. The remainder of the medical history is unremarkable. The patient does not use supplemental oxygen at the time of enrollment into the exercise program; oxygen saturation at rest by pulse oximetry is 95%. The patient's score on the dyspnea subscale of the Chronic Respiratory Disease Questionnaire, a measure of health-related quality of life, is 5 (on a 1–7 scale), corresponding to "some shortness of breath" when performing activities of daily living. The patient also reports a sedentary lifestyle, rarely walking outside the home and not participating in any sport or recreational activities. Results of the preexercise medical exam reveal the following:

- Height and weight of 70 in. (178 cm) and 222 lb (101 kg) (body mass index = 31.8)
- Resting heart rate of 85 and blood pressure of 144/98
- Enlarged anteroposterior chest diameter and decreased breath sounds, prolonged expiration, and wheezes
- Regular pulse with no murmurs, gallops, or bruits noted
- Normal hearing and vision, absence of edema in lower extremities, and good mobility

Upon enrollment into the exercise program, the patient reports the following medications:

- Atrovent inhaler, eight puffs twice a day (anticholinergic bronchodilator)
- Doxapram HCL, 50 mg three times a day (respiratory stimulant)
- Furosemide, 40 mg four times a day (diuretic)

(continued)

- Hytrin, 2 mg four times a day (antihypertension drug 1-selective adrenoceptor-blocking agent)
- Prednisone, 5 mg four times a day (corticosteroid)
- Proventil, 0.5% twice a day (β-2-adrenergic bronchodilator)
- Serevent inhaler, two puffs twice a day (β-2-adrenergic bronchodilator)
- Theo-Dur, 300 mg twice a day (methylxanthine derivative)
- Ventolin inhaler, two puffs twice a day (β-2-adrenergic bronchodilator)

Pulmonary function testing reveals a forced vital capacity of 5.31 L (127% of predicted), an FEV_1 of 1.60 L (49% of predicted), an FEV_1/FVC ratio of 30%, and a maximal voluntary ventilation of 77 L (60% of predicted). After administration of 200 mg of albuterol by metered dose inhaler, the FVC improved by 80 ml and the FEV_1 improved by 20 ml. Blood gas analysis was not performed.

Diagnosis

- Stage 2 (moderate) obstructive lung disease with shortness of breath on exertion
- Obesity
- Hypertension
- Physical deconditioning

Exercise Test Results

The patient performs a graded exercise test on the treadmill with continuous 12-lead ECG monitoring and blood pressure assessments. Ratings of perceived dyspnea are assessed with the Borg scale (1–10), oxygen saturation is assessed by pulse oximetry, and respired gas analysis is performed with a metabolic cart. Resting data include heart rate of 88, blood pressure of 144/100, and oxygen saturation of 94%. The resting ECG is essentially normal. Mild nonspecific T-wave flattening is noted in the lateral chest leads. The patient is able to complete only the first stage of the graded exercise test using a modified Naughton protocol, walking for 2 min at 1.5 mph (2.4 kph) and 1.0% grade. Heart rate is 125 beats · min^{-1} (83% of age-predicted maximum), and blood pressure is 194/100 at maximal exercise. The patient reports a dyspnea rating of 5, corresponding to "strong shortness of breath" on the Borg scale. No ECG changes consistent with ischemia are noted, and the patient does not report chest tightness, pain, or pressure. Rare premature ventricular contractions (PVCs) are observed during exercise. Oxygen saturation at maximal exercise decreases to 85%, and the peak oxygen consumption is 14.7 ml · kg^{-1} · min^{-1}. The test is terminated because of shortness of breath and oxygen desaturation.

The patient also performs a 6 min walk for distance and a hands-over-head task for time before beginning the exercise program. The distance covered during the 6 min walking trial is 948 feet (289 m); oxygen saturation decreases to 85% and the patient reports a shortness of breath rating on the Borg dyspnea scale of 7 (severe shortness of breath or very hard breathing). The hands-over-head task is designed to assess upper-body strength and susceptibility to dyspnea when the patient uses the upper extremities. The task involves removing and replacing 10 lb (4.5 kg) weights along a row of six pegs positioned at shoulder height. The patient completes this task in 57.4 s with a dyspnea rating of 3 (moderate shortness of breath) and an oxygen saturation of 86%. The average time for subjects in our rehabilitation program to complete this task is 50.3 s.

Exercise Prescription

The primary consideration in prescribing exercise for this patient with chronic obstructive lung disease is his ability to maintain adequate oxygen saturation. The oxygen saturation values from the graded exercise test, the 6 min walk, and hands-over-head tasks indicate that the patient should be prescribed supplemental oxygen for use when exercising. In this case, the clinical exercise specialist can serve as patient advocate by providing the primary care physician or pulmonologist with documentation that supports the need for supplemental oxygen. During exercise, oxygen flow rate should be adjusted to maintain a minimum oxygen saturation of 90% or greater. The lack of ECG changes suggestive of myocardial ischemia during the treadmill test does not preclude the presence of coronary artery disease. Coronary artery disease is common in patients with COPD, and the diagnostic sensitivity of treadmill testing improves if the patient can attain a maximal or near-maximal level of exertion. In this case, the patient achieves a heart rate corresponding to approximately 83% of predicted. Signs and symptoms of myocardial ischemia should be carefully monitored during exercise training in this population. The exercise prescription for this patient includes the following components:

- Aerobic training through walking to improve functional capacity, perception of dyspnea, and ability to perform activities of daily living
- Upper-body strength training exercises with dumbbells (biceps curl, triceps extension, shoulder flexion, shoulder abduction, and shoulder shrugs) to increase muscular strength and lean body mass
- Stretching exercises three times weekly, performed after walking to improve joint range of motion and mobility
- Frequency of three times weekly in a supervised setting to maximize training effects, minimize fatigue and risk of injury, and maximize compliance
- Intensity of aerobic training exercise at a dyspnea rating of 3 to 5 on Borg dyspnea scale and intensity of strength training at two sets of each exercise with a weight that allows 12 to 15 repetitions of the movement to maximize training effects, minimize risk of untoward cardiovascular or pulmonary events, and maximize compliance
- Duration of 30 min per session (interval training may be required, especially early in the training program) to maximize training effects, minimize fatigue and risk of injury, and maximize compliance

Discussion Questions

1. How would the results of this patient's graded exercise test be expected to differ from those of a healthy age-matched nonsmoker?
2. What improvements can be expected in the graded exercise test and the other outcome measures because of this patient's participation in a program of exercise rehabilitation using the previously described exercise prescription?
3. How are the results of this patient's pulmonary function tests different from those of a healthy age-matched nonsmoker?
4. What physiological factors would account for these pulmonary function test differences?
5. Would involvement in an exercise program result in improvements in these pulmonary function tests? Why or why not?

22

Asthma

Brian W. Carlin, MD

Anil Singh, MD

Asthma represents a continuum of a disease process characterized by inflammation of the airway wall. An operational definition of asthma is a chronic inflammatory disorder of the airways with **airway hyperresponsiveness** that leads to recurrent episodes of wheezing, breathlessness, chest tightness, and coughing occurring particularly at night or in the early morning. These episodes are usually associated with widespread but variable airflow obstruction that is often reversible either spontaneously or with treatment (49).

SCOPE

Asthma is a worldwide problem that affects an estimated 300 million people (49). It affects at least 22 million Americans (up to 5% of the U.S. population) (14), and incidence rates of up to 3.9% per year are noted. Studies have shown that the prevalence of asthma in the world has been increasing over the last several decades by 5% per year (25,44,60). Most childhood asthma begins in infancy (before the age of 3), and viral infections are proposed to be a critical component in its development (67). The incidence is higher in some patient populations. Up to 23% of inner-city African Americans have asthma compared with 5% of Caucasians (38). Asthma morbidity (currently estimated at 4,000 people per year in the United States) varies depending not only on race but also in some instances on geographic locale (49). For example, asthma morbidity in Hispanics in the United States is highest in the Northeast (31).

Despite the availability of good medical therapy, the morbidity and mortality rates associated with asthma, both in the United States and worldwide, have increased, particularly within the African American population. The reasons for this increase in morbidity and mortality rates are unclear but include variability of the pathophysiology of the disease process between individuals, influence of environmental factors on the development and progression of the disease, inability of a patient to access healthcare effectively, and inability of the patient to comply with the recommended treatment regimen.

PATHOPHYSIOLOGY

Knowledge about the pathophysiology of asthma has increased significantly over the last decade. The many factors that influence the risk of asthma are divided into those that cause the development of asthma and those that trigger asthma symptoms. The former include host factors, and the latter include environmental factors (13) (table 22.1 on p. 392).

The mechanisms that influence the development and expression of asthma are complex and interactive. Genetic factors interact with environmental factors to determine asthma susceptibility (51). Although the clinical spectrum is highly variable, the presence of airway

Table 22.1 Factors Influencing the Development and Expression of Asthma

Host factors	Environmental factors
Genetic	Allergens
Obesity	Indoor (e.g., mites, domestic animals, cockroach allergen, fungi, molds)
Gender	Outdoor (e.g., pollens, molds, yeasts, fungi)
	Infections (e.g., viral)
	Tobacco smoke (passive and active)
	Air pollution (outdoor and indoor)
	Diet

inflammation is persistent even though the symptoms are episodic. This inflammation affects all airways but is most pronounced in the medium-sized bronchi. Over 100 cellular mediators (e.g., cytokines, cysteinyl leukotrienes, histamine, nitric oxide, and prostaglandin D2) are responsible for the development of the inflammation associated with asthma (8). The **CD4 lymphocyte** (Th2 subgroup) is currently believed to promote inflammation by the **eosinophils** and **mast cells** (29) with subsequent infiltration of these cells into the airway wall resulting in edema formation (37).

Besides this inflammatory response, structural changes occur in the airways (often described as airway remodeling) (30). Subepithelial fibrosis results from the deposition of collagen fibers under the basement membrane and is seen in all asthmatic patients, even before the onset of symptoms. Hypertrophy and hyperplasia of the airway smooth muscle occurs and contributes to the thickness of the airway wall. Proliferation of new blood vessels, because of greater expression of vascular endothelial growth factor, can contribute to airway wall edema. Finally, mucus hypersecretion, resulting from an increase in the number of goblet cells in the airway epithelium and increased size of submucosal glands, adds to the structural alterations of the airway (8,20,58).

Airway hyperresponsiveness, the characteristic functional abnormality of asthma, results in airway narrowing (46). It is linked to both inflammation and airway remodeling and is partially reversible with therapy. The mechanisms behind the hyperresponsiveness (including excessive contraction of the airway smooth muscle, thickening of the airway wall, and sensitization of the airway nerves leading to exaggerated bronchospasm), however, are not currently completely understood. Figure 22.1 shows airway remodeling.

MEDICAL AND CLINICAL CONSIDERATIONS

The medical and clinical considerations resulting from this pathophysiology lead to the development of increased airflow obstruction manifest as increased shortness of breath and wheezing. In most instances, these clinical manifestations can be treated successfully, but in some instances, the chronic changes associated with the airway remodeling can result in symptoms that are difficult to reverse.

Signs and Symptoms

Asthma is currently classified into various forms based on clinical and pulmonary function criteria (table 22.2) (41,49). Most patients with asthma develop symptoms on an intermittent basis, but some develop progressive disease. In some, escalation from a mild form of asthma to a severe exacerbation can occur at any time. Death from asthma can occur and may occur even in patients who have mild asthma (57).

A clinical diagnosis of asthma is prompted by symptoms such as episodic wheezing, breathlessness, cough, and chest tightness. Asthma symptoms may be intermittent, and both patients and physicians may overlook their significance, particularly in children. The symptoms may occur after an accidental exposure to allergens or may occur in relation to seasonal rhinitis. They may also be precipitated by exposure to dust mites, smoke, strong fumes, cold air, or exercise (13). Variable clinical presentations occur from one patient to the next. Up to 15% of patients with asthma fail to appreciate any type of discomfort following a 20% decrease in the forced expiratory volume in 1 s (FEV_1) (59). In some patients, the clinical symptoms may develop only after exercise (exercise-induced bronchospasm) and may be manifest only as a decrease in exercise tolerance without other symptoms.

History and Physical Examination

Several important questions must be considered in the diagnosis of asthma. Has the patient had an attack or recurrent attacks of wheezing? Does the patient have a cough at night? Does the patient wheeze or cough after exercise? Does the patient experience chest tightness, wheezing, or cough after exposure to airborne pollutants or allergens? Does the patient have colds that "go to the

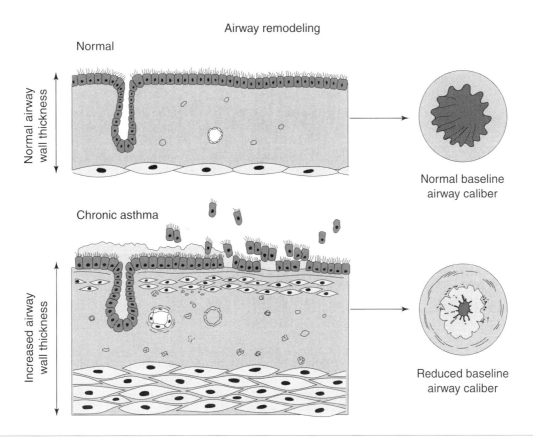

Figure 22.1   Normal tissue, swelling, and remodeling.

Table 22.2   Classification of Asthma Severity by Clinical Features Before Treatment

	Days with symptoms	Nights with symptoms	PEF or FEV$_1$
Intermittent	Symptoms less than once a week Brief exacerbations	Not more than twice a month	FEV$_1$ or PEF >80% predicted PEF or FEV$_1$ variability <20%
Mild persistent	Symptoms more than once a week but less than once a day Exacerbations may affect activity and sleep	Nocturnal symptoms more than twice a month	FEV$_1$ or PEF >80% predicted PEF or FEV$_1$ variability <20-30%
Moderate persistent	Symptoms daily Exacerbations may affect activity and sleep Daily use of inhaled short-acting β-agonists	Nocturnal symptoms more than once a week	FEV$_1$ or PEF 60–80% predicted PEF or FEV$_1$ variability >30%
Severe persistent	Symptoms daily Frequent exacerbations Limitation of physical activities	Frequent nocturnal asthma symptoms	FEV$_1$ or PEF <60% predicted PEF or FEV$_1$ variability >30%

chest" or take more than 2 wk to resolve? Does appropriate asthma treatment improve symptoms?

Because asthma symptoms are variable, the physical examination of the chest may in fact be normal. Wheezing confirms the presence of airflow limitation and is the most usual abnormal physical finding. In some, wheezing may be able to be reproduced only if the patient forcefully exhales. In a patient with severe asthma, wheezing may

Client–Clinician Interaction

To determine the level of exercise capability that a patient with asthma might be able to attain (as well as to follow during training sessions), the examiner should review the following features of the clinical assessment:

1. Presence or absence of symptoms (e.g., cough, wheezing, shortness of breath) while at rest or with exercise. A history of exercise limitation (attributable to the previously mentioned symptoms) in an otherwise asymptomatic patient should alert the examiner to the possibility of exercise-induced bronchospasm.

2. Use of medication before exercise (e.g., inhaled β-agonist, inhaled corticosteroid, oral leukotriene modifier).

3. Correct use of the medication (particularly when using a metered-dose inhaler).

4. Correct use of warm-up and cool-down periods during exercise.

A correct diagnosis of asthma (or exercise-induced bronchospasm) must be made and followed with appropriate use of medications before exercise to allow the patient to optimize his or her exercise capabilities.

be absent because of the extreme decrease in airflow and other symptoms such as cyanosis, drowsiness, difficulty in speaking, and tachycardia, and the use of the accessory muscle of respiratory may be present.

Diagnostic Testing

Although the diagnosis of asthma is often made based on the presence of the characteristic symptoms, pulmonary function tests assist in the diagnostic certainty of the presence of the disease. Patients often have poor recognition of their symptoms and insufficient perception of their disease severity. Lung function measurements provide an assessment of the severity of the airflow limitation, its variability, its reversibility, and confirmation of the diagnosis.

Spirometry is the recommended method of measuring airflow limitation and reversibility to help establish a diagnosis of asthma. Measurements of FEV_1 and forced vital capacity (FVC) are performed during a forced expiration by the patient. Airflow limitation is defined by a decrease in the FEV_1 to less than 80% of predicted and a decrease in the FEV_1/FVC to less than 65%. The degree of reversibility in FEV_1 is generally accepted as an increase in FEV_1 greater than 12% (or greater than 200 ml) following administration of a short-acting bronchodilator (e.g., albuterol sulfate) from the prebronchodilator value (54). Most asthma patients will not show reversibility at each assessment, particularly those on treatment. Flow-volume loops (also measured during spirometry) (figure 22.2) can be helpful to differentiate airway obstruction secondary to asthma (which will show

an improvement in flow rates following bronchodilator administration) or emphysema (which will not show an improvement in flow rates).

Peak expiratory flow measurements, made using a peak flow meter, can be helpful in both the diagnosis and the monitoring of asthma. These meters are inexpensive, portable, and ideal for patients to use in their home and everyday surroundings. Measurements of peak expiratory flow are not interchangeable with other measurements of lung function (such as FEV_1). Peak expiratory flow can underestimate the degree of airflow limitation, particularly as airflow limitation worsens. Values for peak expiratory flow obtained with different peak flow meters vary, and the range of predicted values is wide. Peak flow measures should be compared with the patient's own previous best measurement and can serve as a reference value for monitoring the effects of treatment. Because many patients have poor perception of symptoms, the use of peak flow measurements can help improve the control of asthma for a patient.

In the management of a patient with asthma, the terms *reversibility* and *variability* are often used. These terms refer to changes in symptoms accompanied by changes in airflow limitation that occur either spontaneously or as a result of treatment. The term *reversibility* is generally applied to rapid improvements in FEV_1 (or PEF) measured within minutes of inhalation of a short-acting bronchodilator. Variability refers to improvement or worsening of symptoms or lung function over time. Variability may occur over the course of a day, from day to day, or from week to week. Obtaining a history of variability is an important component of the diagnosis and ongoing control of asthma.

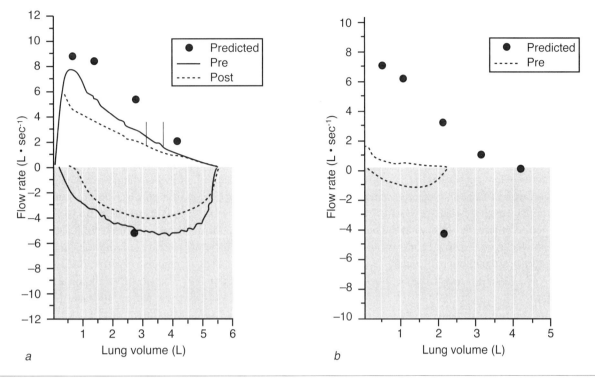

Figure 22.2 Flow-volume tracings of a patient with asthma and a patient with emphysema.

In some instances, normal spirometry may be present, and airflow limitation may only be shown following bronchial provocation testing or following exercise. Nonspecific airway irritants, such as methacholine or histamine, can be administered in aerosol form to determine whether any decline is present in the FEV_1 or FVC. A decline in FEV_1 greater than 20% following administration of the irritant is abnormal and indicates the likely presence of asthma. Exercise challenge may also be used to uncover airflow limitation occurring following exercise. Again, a decline in FEV_1 greater than 20% following exercise indicates the presence of exercise-induced bronchospasm (18).

Other studies may be helpful to substantiate the diagnosis of asthma. The chest roentgenogram may show hyperinflation of the lung (increase in the retrosternal airspace, diaphragm flattening). The chest roentgenogram may also be helpful to rule out other causes for the patient's symptoms (e.g., pneumonia, pneumothorax, congestive heart failure). The evaluation of airway inflammation associated with asthma may be done by examining sputum produced by the patient for eosinophilic or neutrophilic inflammation. In addition, levels of exhaled nitric oxide (FeNO) and carbon monoxide (FeCO) may be used as noninvasive markers of airway inflammation (32,34,55). Neither sputum eosinophilia nor FeNO has been prospectively evaluated, however, as

an aid in the diagnosis of asthma, but the latter is being studied to monitor the response to treatment (61).

Exercise Testing

In patients who have asthma diagnosed clinically and confirmed by spirometry, exercise testing is generally not performed unless a decline in exercise tolerance occurs that is out of proportion to the patient's symptoms or degree of airflow limitation. In some instances, for patients with such significant exercise limitation, standard exercise testing should be performed with a symptom-limited incremental test. Measurement of oxyhemoglobin saturation by pulse oximetry and monitoring of cardiac rhythm should be performed. Measurement of oxygen consumption, carbon dioxide production, and anaerobic threshold using a metabolic cart during progressive incremental exercise will provide a detailed assessment of an individual's response to maximal symptom-limited exercise and is helpful in making further decisions regarding an exercise prescription.

For most patients with asthma, there are few contraindications to exercise. Should a patient have acute bronchospasm, chest pain, or an increased level of shortness of breath above that usually experienced, the exercise testing should be withheld. If a patient has severe exercise deconditioning or other comorbid conditions (e.g., unstable angina, orthopedic limitations) exercise testing may not be able to be performed.

Exercise-induced bronchospasm (EIB) occurs in 50 to 100% of patients with asthma (19). The pathophysiology behind EIB is thought to relate to the consequences of heating and humidifying large volumes of air during exercise. Cooling and drying of the airways lead to inflammatory mediator release. The airways are narrowed by the bronchial smooth muscle and cellular abnormalities similar to those discussed previously (3,4). The symptoms of EIB are similar to those of asthma but are associated with short periods of intense physical activity. The typical response of a patient who has EIB involves a 10 min period of bronchodilation at the beginning of exercise followed by progressive bronchospasm peaking at 10 min following completion of exercise. Spontaneous resolution of EIB symptoms occurs over the ensuing 60 min (35,45).

It is important to confirm the presence of exercise-induced bronchospasm in patients who have symptoms primarily during or following exercise and have minimal, if any symptoms, while at rest. Exercise testing in this situation should be performed for patients who have the appropriate symptoms only during or after exercise and are suspected to have exercise-induced bronchospasm. The exercise test is performed to a symptom-limited maximum (on either a bicycle or a treadmill). The work intervals should be short (e.g., 2 min stages), and the increase in workload should be approximately 1 metabolic equivalent between stages. Ideally, a protocol should be chosen that will elicit a patient's maximal effort between 8 and 12 min. Immediately following exercise, spirometry should be performed and repeated at 15 and 30 min to determine whether airflow limitation has developed. Should the patient develop bronchospasm, an inhaled β-agonist (e.g., albuterol sulfate) should be administered.

Treatment

The general goals of asthma therapy include prevention of chronic asthma symptoms and asthma symptoms during the day and night, maintenance of normal activity levels, maintenance of normal (or near normal) lung function, satisfaction with asthma care, and minimal or no side effects while receiving optimal medications. These goals provide criteria that the clinician and patient should use to evaluate the patient's particular response to therapy. The exercise professional should determine the patient's personal goals, share the general goals of therapy with the patient, and then agree on the goals that will be the foundation for the treatment plan.

Previously, determination of asthma severity based on the level of symptoms, airflow limitation, and lung function variability into four categories (intermittent, mild persistent, moderate persistent, or severe persistent) was believed to be useful in determining the initial and ongoing treatment (table 22.2). Classification of severity is useful when making decisions to determine the initial treatment regimen but are no longer recommended as the basis for ongoing treatment (26). The classification has poor value in predicting what treatment will be required and what the patient's response to the individual treatment might be. Asthma control may be defined as a control of the manifestations of the disease. These include daytime symptoms, limitations of activities of daily living, nocturnal symptoms or awakenings, need for reliever or rescue treatment, and lung function abnormalities (49). Treatment should be aimed at complete control of all these clinical features of the disease. Routine, periodic assessment of this level of control is important in the ongoing management of the patient.

Asthma treatment can be administered in a variety of ways—inhaled, orally, or parenterally. Most therapy is administered by the inhaled route. In this instance the medication is delivered directly into the airways, producing higher local drug concentrations with much less risk of systemic side effects. Medications are classified as either controllers or relievers. Controllers are medications taken daily on a long-term basis to maintain asthma control. They act through an anti-inflammatory mechanism. They include inhaled and systemic glucocorticosteroids, leukotriene modifiers, long-acting inhaled b-agonists in combination with inhaled glucocorticosteroids, sustained-release theophylline, cromolyn, and anti-IgE therapy. The inhaled glucocorticosteroids are the most effective controller medications currently available (26).

Relievers are medications used on an as-needed basis. They act quickly to reverse bronchospasm and bronchoconstriction and thus relieve symptoms. They include rapid-acting b-2-agonists, inhaled anticholinergics, short-acting theophylline, and short-acting oral b-2-agonists.

A stepwise approach to the control of asthma should be undertaken—increasing medication dosages and types as needed and decreasing them whenever possible, based on the level of asthma control. For patients with persistent asthma, defined as daytime symptoms more than twice weekly or nighttime symptoms more than twice monthly, medications to control asthma and prevent exacerbations in the long term, as well as those to control acute symptoms, should be used.

For patients with EIB, prevention of the bronchospasm can be provided through a variety of means. An appropriate warm-up period before the actual exercise is an important nonpharmacological method that can be used. A warm-up period of 15 min of continuous exercise at 60% of maximal oxygen consumption can significantly decrease postexercise bronchoconstriction in moderately trained athletes. Interval warm-up may be used, but in some instances (based on eight 30 s runs at 100% maximal oxygen consumption with a 1.5 min rest between trials) this type of warm-up has not been found to reduce postexercise bronchoconstriction significantly. Thus at least 15 min of moderate-intensity exercise should precede significant exercise for active persons with asthma (19).

Pharmacological methods, which include inhaled b-agonists and leukotriene modifiers, are also used to prevent EIB. Although many patients used short-acting β-agonists to reduce EIB symptoms, long acting β-2-agonists(e.g., salmeterol xinafoate) and leukotriene modifiers (e.g., montelukast) have been shown to reduce EIB symptoms (21,40,50). Inhaled corticosteroids and cromolyn sodium have also been shown to be beneficial (28,39). Given the widely variable pathophysiological components of EIB, therapy must be individually tailored. A sample exercise prescription process is described on page 398.

EXERCISE PRESCRIPTION

Given the great variability of the pathophysiological processes among patients with asthma, the response to exercise varies widely as well (24,52,62). Some patients may be able to exercise at an Olympic level, whereas others may be unable to walk across the room without significant shortness of breath. The individual patient's exercise ability may vary from time to time depending on the current level of control of the disease, particularly during an exacerbation when the person's exercise ability may be extremely limited (42,52). The mode, frequency, time, intensity, and progression for health and fitness benefits should be similar to those used for a patient who does not have lung disease unless other comorbid illnesses, as described in practical application 22.2, are present.

EXERCISE TRAINING

A wide variety of physiological outcomes might be expected as a result of exercise training for patients with asthma. No effect, either adverse or beneficial, has been reported to occur in regard to static lung function measurements (spirometry) including bronchial hyperresponsiveness related to exercise training. But several physiological changes have been noted to occur following a training program. These include

increases in maximal oxygen uptake, oxygen pulse, and anaerobic threshold. Significant reductions in blood lactate level, carbon dioxide production, and minute ventilation at maximal exercise have also been shown to occur. Subjective responses also have been noted, particularly a reduction in perceived breathlessness at equivalent workloads, following exercise training. This latter response could be attributable to a central nervous system "desensitizing" effect, a decrease in minute ventilation at submaximal workloads, or an increase in the endorphin levels without a concomitant reduction in ventilatory chemosensitivity.

A variety of cardiopulmonary and metabolic responses to exercise in patients with asthma have been noted. From a cardiopulmonary perspective, treadmill exercise for patients with exercise-induced bronchospasm without prior treatment increases ventilation and perfusion inequality, physiological dead space, and arterial blood lactate levels (5). From a metabolic response perspective, a "blunted" sympathoadrenal response to exercise (7,63), an alteration in potassium homeostasis, and an excessive secretion of growth hormone have all been shown to occur. The role that each of these metabolic responses may play in the exercise limitation noted in patients with asthma is unknown. Again, given the variety of pathophysiological processes present in each patient who has asthma, a variety of cardiopulmonary responses to exercise can be expected.

One of the most confounding variables noted in patients with asthma who are attempting to exercise is the effect of dyspnea on exercise capability. The decision to exercise is often weighted against discontinuation of exercise because of the increasing levels of dyspnea experienced by the patient. In one study, "harmful anticipation" significantly increased the perception of visceral changes associated with exercise (47). A wide variability between the degree of airway obstruction, exercise tolerance, and the severity of breathlessness has been noted in several studies (43,53), but this only accounts for up to 63% of the variance in breathlessness that asthmatics rated during progressive incremental exercise. This complexity concerning the development of symptoms and exercise tolerance might be one reason that the diagnosis of exercise-induced bronchospasm is obscured. In one study that screened 503 children with asthma, an average of one child in each classroom had previously undiagnosed asthma (the symptoms of which were often absent or attributed to other causes) (49). Given these variable responses to exercise from a cardiopulmonary, metabolic, and symptom perspective, no unified conclusions can be made regarding such effects in patients with asthma as a whole. Individual

Exercise Prescription Summary

Mode	Frequency	Intensity	Duration	Special considerations
Treadmill (aerobic)	Five times per week	Just below anaerobic threshold	20-30 min per session	Optimize medication therapy before exercise
Walking/running (track, sidewalk)	Five times per week	Just below anaerobic threshold	20-30 min per session	Optimize medication therapy before exercise
Swimming	Five times per week	Just below anaerobic threshold	20–30 min per session	Optimize medication therapy before exercise

1. Assess patient's underlying respiratory status and goals for exercise.

2. Assess maximum level of exercise.

3. If maximum level of exercise has been determined by measurement of oxygen consumption and carbon dioxide production (cardiopulmonary exercise testing), begin exercise prescription at an initial intensity level just below the anaerobic threshold.

4. If such measurements are unavailable, begin exercise at a level of exercise at which the patient is comfortable performing for 5 min.

5. Instruct the patient to continue exercise for 20 to 60 min per session.

6. Have the patient perform sessions three to five times per week.

7. Increase exercise intensity by 5% with each session.

8. When maximal level of intensity is attained, increase exercise duration by 5%.

assessment of each patient is thus important when trying to determine the degree (and thus subsequent effects) of exercise intolerance present.

Training Schedules

A variety of training schedules have been used for patients with asthma. Various types of exercise including gym, games, distance running, swimming, cycling, altitude training, and treadmill running (to name just a few) have all been shown to improve exercise capability. The frequency of exercise training varied from study to study, ranging from once weekly to daily for periods ranging from 20 min up to 2 h per training session. Training periods of 6 to 8 wk were generally used. The intensity of exercise also varied from a gradual increase in exercise endurance to short, heavy increase in exercise endurance (15). In one study, 26 adults with mild to moderate asthma (FEV_1 63%) underwent a 10 wk supervised rehabilitation program with emphasis on individualized physical training. Daily exercise (swimming) for 2 wk was followed by twice weekly exercise. Exercise training intensity was measured by a target heart rate during the first 2 wk and then by perceived sense of exertion as measured by a Borg scale (1–10 scale) during the latter 8 wk. Each subject was encouraged to exercise to a Borg level of 7 to 8. All subjects were able to perform high-intensity exercise (80–90% of their maximum predicted heart rate), and improvements in cardiovascular conditioning and walk distance were observed after the program. Decrease in asthma symptoms and decrease in anxiety were noted following the training period (23).

Ongoing exercise following the initial training program has been shown to be effective for patients with asthma. Of 58 patients who had previously undergone a 10 wk rehabilitation program, 39 reported continuation of regular exercise. Cardiovascular conditioning (as measured by 12 min walk distance) and lung function values remained unchanged in all patients. But a significant decrease occurred in the number of emergency department visits over the 3 yr period compared with the year before entry into the rehabilitation program in these 39 patients. A decrease in asthma symptoms was noted only in a subgroup of patients (n = 26) who exercised one or two times per week. Continued exercise following a supervised rehabilitation program is helpful for patients with mild to moderate asthma (22).

Breathing exercises to strengthen the respiratory muscles have been used by some, but their overall effectiveness is controversial. Deep diaphragmatic breathing was used in 67 patients with asthma and significantly decreased the use of medical services and the intensity of asthma symptoms. But no significant change occurred in overall physical activity as measured by an inventory scale (25). Inspiratory muscle training that used a threshold inspiratory muscle-training device in a double-blind sham trial showed a significant increase in inspiratory muscle strength (as expressed by the maximum inspiratory pressure measured at residual volume) and respiratory muscle endurance. The training group also had a significant reduction in the amount of asthma symptoms, number of hospitalizations, and absence from work or school compared with the sham group (50).

In a recent Cochrane review, the effects of physical training in patients with asthma were evaluated. Thirteen studies involving 455 participants were reviewed. Physical training (at least 20–30 min of exercise two to three times per week for a minimum of 4 wk) had no effect on resting lung function or the number of days that a wheeze was present. Physical training improved cardiopulmonary fitness as measured by an increase in maximum oxygen uptake of 5.4 ml · kg^{-1} · min^{-1} (95% confidence interval 4.2 to 6.6) and maximum expiratory ventilation of 6.0 L · min^{-1} (95% confidence interval 1.5 to 10.4). Thus, in people with asthma, physical training can improve cardiopulmonary fitness without adverse effects on lung function or symptoms of wheeze (56).

Patients with exercise-induced bronchospasm should be encouraged to undergo exercise training as well. Instruction on preventive strategies that allow adequate control of airway inflammation and bronchoconstriction is important in the management of these patients. For each patient, the clinician should consider triggers for the development of asthma under such situations as being outside and exercising on a day with a high ozone concentration or high allergen counts in the atmosphere, or the development of symptoms following ingestion of certain foods within an hour or two before exercise (e.g., milk products, vegetables). As discussed previously, an adequate warm-up period and use of inhaled or oral medications before exercise should be stressed.

Comprehensive Rehabilitation

Comprehensive rehabilitation programs for patients with asthma include much more than just exercise training. Components of the initial patient assessment should include patient interview, medical history, diagnostic testing, symptoms and physical assessment, nutritional evaluation, activities of daily living assessment, educational and psychosocial history, and goal development. Actual program content should include education regarding the disease process, triggers of asthma, self-management of the disease (medication use, warning signs and symptoms associated with exacerbations, peak flow monitoring, metered-dose inhaler technique, importance of exercise warm-up and cool-down), activities of daily living, psychosocial intervention, and dietary intake and nutrition counseling. Follow-up and evaluation of outcomes are an important part of the rehabilitation process. Questionnaires used to assess the asthma patient's quality of life (measuring variables such as symptoms, emotions, exposure to environmental stimuli, and activity limitation) have been well validated and should be used as part of this assessment process (33). In addition, cost of medications and equipment, time lost from work or school, and utilization of healthcare resources (e.g., emergency department visits, calls to the patient's physician) can be assessed as part of the follow-up. Not all patients with asthma are candidates for such comprehensive rehabilitation programs.

The widespread acceptance, over many years, that patients with asthma cannot and should not exercise has led to many recommendations that these patients avoid exercise. Many parents have unnecessarily restricted their children who have asthma from exercise because of the fear that exercise may make the asthma worse. Attempts must be made to educate patients and their families about the importance of exercise and how the child who has asthma can safely perform such exercise. The use of β-agonist or leukotriene modifier therapy before exercise as well as the avoidance of conditions known to precipitate that person's asthma are important mainstays of the treatment and should be used aggressively when asthma plays a role in exercise intolerance.

Most patients with asthma can be managed effectively with a combination of medications and general exercise recommendations. In those who have moderate or severe persistent asthma, those who have failed medical therapy and have had a significant decline in their performance of the activities of daily living, or those in whom the disease process has had a drastic adverse effect on their lifestyle, comprehensive rehabilitation offers an effective means to improve overall quality of life.

Exercise Recommendations

Comprehensive supervised rehabilitation is helpful for patients with asthma. Appropriate control of the disease process is of primary importance. After the disease is under maximal medical control, improvements in aerobic capacity, muscle strength, and endurance can be maximized. The exercise prescription should be based on objectives measurement of exercise capabilities and individualized for each patient. A variety of training modalities can be used including treadmill, stationary bicycle, walking, and swimming. If weakness of a specific group of muscles is noted, exercises that address that particular muscle group should be offered.

For cardiorespiratory fitness, the exercise training should be 20 to 60 min in duration completed 3 to 5 d per week (1). The mode of exercise should take into consideration the patient's interests, past exercise experience, and availability of equipment while acknowledging the effects of the surrounding environment. Exposure to cold air, low humidity, or air pollutants should be minimized. Intermittent exercise or lower-intensity sports performed in the presence of warm, humid air are generally better tolerated. But no consensus has emerged about the optimal intensity level at which a patient with asthma should train (16,17).

The intensity prescription should be based on the clinical and exercise test data in conjunction with the patient's goals. If maximal oxygen consumption during exercise is obtained using a metabolic cart, training can be initiated at an intensity level of 50 to 85% of the heart rate reserve (maximal heart rate minus resting heart rate). Again, this intensity is below anaerobic threshold for most people (2). For patients with more limiting asthma, a target intensity based on perceived dyspnea (such as a Borg scale) may be more appropriate (10). General measures for resistance training and flexibility should be included as part of the exercise regimen for patients with asthma. Exercise training can be used in most populations including children as shown in practical application 22.3.

CONCLUSION

Asthma is a complex process that involves airway inflammation and bronchoconstriction. Environmental risk factors (such as indoor allergens, viral infections) or other triggers (such as exercise, cold air) can initiate an allergic response, resulting in airway inflammation and airway hyperresponsiveness. Airflow limitation occurs, and the patient develops symptoms such as chest tightness, wheezing, and shortness of breath. Exercise limitations and decreased levels of fitness are frequently noted in patients with asthma but in many instances are not considered important for some time following the initial development of symptoms. Exercise limitations and fitness levels can be improved in patients treated with an appropriate medication and exercise regimen.

Practical Application 22.3

Literature Review

In a recent study by Basaran and colleagues (9), the effects of regular submaximal exercise on quality of life, exercise capacity, and pulmonary function in children with asthma were studied. Sixty-two children with mild to moderate asthma (mean age 10.4 ± 2.1 yr) were randomly allocated into exercise and control groups. The exercise group underwent an intensive basketball-training program for 8 wk. Outcome measures included the Pediatric Asthma Quality of Life Questionnaire, exercise capacity, spirometry, and medication and symptom scores. The scores from the questionnaire improved in both groups but more so in the exercise-training group. The exercise group performed better in regard to exercise capacity and reduction of symptoms.

This study illustrates that children with mild to moderate asthma can safely perform regular, moderately intensive exercise. Improvements in exercise capacity and quality of life and reduction in symptoms can be obtained. Regular exercise should be encouraged in patients who have asthma who are otherwise under good medical control.

Medical History

Ms. JR is a 22 yr old Caucasian college senior. Throughout high school she was active in competitive sports including soccer, swimming, and field hockey. On occasion throughout high school she developed an increase in shortness of breath and a cough. Her primary care physician told her that she had bronchitis and that she should not worry about it. After starting college she continued with competitive soccer and swimming. At the end of a long run during soccer games she noted an increase in cough and a slight wheeze. She did not note any symptoms following swimming practice. She continued to exercise but noticed an increase in coughing and wheezing over the ensuing year.

Diagnosis

Her parents became concerned about her discomfort and tried to persuade her not to exercise because "it makes you feel much worse and could be dangerous." With such ongoing symptoms, she withdrew from soccer. She sought advice of the college physician, who told her that she might have asthma given the symptoms of wheezing. Spirometry was performed, which revealed an FEV_1 of 3.09 L (96% predicted), an FVC of 3.54 L (95% predicted), a **peak expiratory flow rate** (**PEFR**) of 6.97 L (95% predicted), and an FEV_1/FVC ratio of 87%. Given these results, showing normal pulmonary function, she was told that she did not have asthma but rather bronchitis and was told to continue her exercise (after a course of antibiotics). She continued to swim but noted that at the end of a training session she was slightly more short of breath than usual and had heaviness over her anterior chest region.

Exercise Test Results

She sought the advice of another physician, who ordered an exercise test with the measurement of expired gases during progressive incremental bike exercise. Spirometry was performed at 15, 30, and 60 min following the exercise test. Maximal oxygen consumption was 3.13 L · min^{-1} (52.2 ml · kg^{-1} · min^{-1}).

Flow rates were as follows:

- FEV_1 (L): preexercise = 3.09; 15 min postexercise = 2.87; 30 min postexercise = 2.20; 60 min postexercise = 2.24
- FVC (L): preexercise = 3.54; 15 min postexercise = 3.32; 30 min postexercise = 2.97; 60 min postexercise = 3.03
- PEFR (L · sec^{-1}): preexercise = 6.97; 15 min postexercise = 6.00; 30 min postexercise = 5.25; 60 min postexercise = 5.26

Exercise Prescription

As a result of these studies, a diagnosis of asthma (exercise induced) was made. The patient was started on a short-acting β-agonist (albuterol sulfate) administered 30 min before exercise. She was instructed to warm up for 15 min with low- to moderate-intensity exercise or swimming before starting a high-intensity swim practice. Exercise tolerance subsequently improved, and exercise-associated symptoms became rare (for the most part abated).

Discussion Questions

1. Why was the initial diagnosis of asthma not entertained?
2. How was the actual diagnosis of asthma (exercise induced) made? What tests should be useful in this determination?
3. How did the recommendations improve her exercise tolerance? Why was swimming initially better tolerated than soccer?
4. Discuss the intensity, frequency, and duration of exercise training for patients with asthma.
5. How would the development of asthma symptoms at the end of a 3 h practice session influence the choice of medication (e.g., short-acting versus long-acting β-agonist)?

Cystic Fibrosis

Michael J. Danduran, MS

Julie Biller, MD

Cystic fibrosis (CF) is a genetic disorder that affects the respiratory, digestive, and reproductive systems. Excessively viscid mucus causes obstruction of passageways including pancreatic and bile ducts, intestines, and bronchi. In addition, the sodium and chloride contents of sweat are increased.

SCOPE

CF is the most common life-shortening genetic disease in the Caucasian population. Currently, more than 23,000 patients have CF in the United States, and nearly 1,200 new patients are diagnosed each year (25). Sixty percent of all people with CF are younger than 18 yr of age. The remaining 40% make up the growing adult population. This growth in the adult-based CF population is further emphasized by the increased diagnosis of **adult variant CF** (64). The median survival age continues to improve, and it measured 37 yr of age in 2005. CF is inherited as an autosomal recessive disorder that affects approximately 1 in 3,000 live births in the Caucasian population and 1 in 17,000 in the African American population. Individuals with other ethnic backgrounds are also affected but in less frequency. The estimated total cost to treat CF in the United States continues to be noteworthy at greater than $900 million, representing a cost per CF patient of almost $40,000 per year (25).

PATHOPHYSIOLOGY

The gene for CF is located on chromosome 7 and results in the altered production of a protein called the **cystic fibrosis transmembrane conductance regulator** (CFTR), a protein that functions as a chloride channel regulated by cyclic adenosine triphosphate. More than 2,000 unique mutations of the CF gene have been identified, although one type, δF508, accounts for more than 70% of CF genes in the United States. The primary role of CFTR appears to be as a chloride channel, although other functions for CFTR have been proposed as well. The abnormal CFTR leads to abnormal sodium chloride and water movement across the cell membrane. When this occurs in the lungs, abnormal thick and dry mucus ensues, resulting in bronchial airway obstruction, bacterial infection, and inflammation. As this vicious cycle continues, lung tissue is progressively destroyed with eventual respiratory failure (figure 23.1 on p. 404). Lung disease accounts for more than 95% of the morbidity and mortality associated with CF. With aggressive intervention and early diagnosis, however, survival has been extended, with adults living well into their 30s and 40s. The average age at time of diagnosis is approximately 6 mo (25), but because many states are using newborn screening, the diagnosis can be made much earlier. Typically, one or more symptoms lead to

Therapy: Research

Gene therapy
Protein replacement
Diuretics (e.g., amiloride)

Therapy: Current

Mucolytics (e.g., DNase)
Mucus clearance techniques
 Chest physiotherapy
 Exercise
 Flutter® valve
 PEP valve
 ThAIRapy® vest
Antibiotics–oral, IV, Nebulized
Anti-inflammatory agents
Steroids
Ibuprofen
Lung transplant

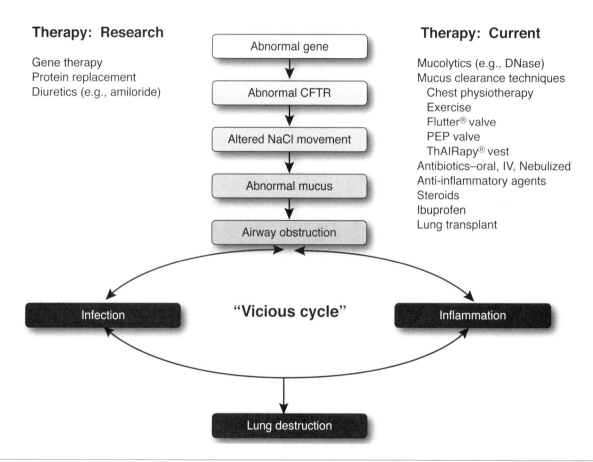

Figure 23.1 The vicious cycle of cystic fibrosis. CFTR = cystic fibrosis transmembrane conductance regulator.

diagnosis including the respiratory, gastrointestinal, sinus, and sweat gland systems. The underlying common theme to all these systems is the cellular abnormality of ion transport necessary for proper function of epithelial structures.

Respiratory System

At birth the lungs are normal on a histological basis. As the vicious cycle of infection, inflammation, and impaired mucus clearance ensues, the lungs become colonized with bacteria, and 80% of all patients with CF grow *Pseudomonas aeruginosa* later in life (25). Other bacteria such as *Staphylococcus aureus* occur in more than 40% of patients, and others occur less frequently. In infants who do not undergo newborn screening, a young child may present with acute signs of respiratory infection. Most children present with signs of chronic infection including cough, sputum production, wheeze, fever, and failure to thrive at the time of diagnosis. Many infants or young children with CF have been previously

misdiagnosed as having asthma, bronchitis, allergies, pneumonia, or bronchiolitis. Chest radiographs may indicate the presence of acute or chronic changes such as infiltrates, **bronchiectasis** (irreversibly irregular and dilated airways), or hyperlucency. When pulmonary function is assessed in older children (>5 yr) at the time of diagnosis, evidence of airways obstruction—reduced forced expiratory volume in 1 s (FEV_1) or forced expiratory flow between 25% and 75% of forced vital capacity (FEF_{25-75})—or hyperinflation (elevated residual volume and right ventricle/total lung capacity ratio) may exist. New technologies have allowed for infant assessment of pulmonary function, inviting therapeutic interventions to occur early on in patients more severely affected with CF. Exercise tolerance may become significantly compromised when compared with normative values. Ultimately, the progressive loss of lung tissue and airways obstruction lead to respiratory failure. The time course for this progression is variable. Some adults with CF experience little lung damage, and some children experience extensive lung disease.

Gastrointestinal and Nutritional Systems

In approximately 85% of individuals with CF, exocrine pancreatic insufficiency is present, resulting in malabsorption of important nutrients including fat and protein. Malabsorption can lead to frequent fatty stools (steatorrhea), malodorous stools, and abdominal pain. The combination of the need for increased caloric intake (attributable to increased resting energy expenditure, cough, and infection) and poor utilization of nutrients by malabsorption often leads to malnutrition or a constant struggle to maintain body weight. Maintaining a desirable body mass index (>22 women, >23 men) has been beneficial in long-term health of adults with CF. Additionally, other organs can be affected, resulting in liver disease, endocrine pancreatic insufficiency (CF-related diabetes mellitus), and gall bladder disease.

Sinuses

The development of pansinusitis and **nasal polyposis** is common for people with CF. For many, this finding may be inconsequential, although some individuals may find it difficult to breathe through the nose. Additionally, pansinusitis with associated bacterial colonization may contribute to the extent of lung disease. Some people require aggressive medical intervention (e.g., antibiotics, nasal irrigation, and endoscopic surgery).

Sweat Glands

Although all epithelial cells will demonstrate the chloride transport defect, the sweat glands are the organ on which the diagnostic test was based. The basis of the **sweat test** (i.e., pilocarpine iontophoresis analysis) rests on the presence of extremely high salt content in the sweat of individuals with cystic fibrosis. A sweat chloride concentration of greater than $60 \text{ mEq} \cdot dl^{-1}$ is highly suggestive for diagnosis of CF.

MEDICAL AND CLINICAL CONSIDERATIONS

The clinical manifestations of CF are variable with differing involvement of the pulmonary and gastrointestinal organ systems. Comprehensive evaluation that includes assessment of the signs and symptoms, diagnostic studies, and pulmonary function testing helps determine the severity of disease.

Signs and Symptoms

CF is usually diagnosed by the presence of classic signs and symptoms (table 23.1). Because of the expansive nature of the disease many systems are affected. Patient care must often be coordinated by a CF care team that may include pulmonologists, gastroenterologists, nurses, respiratory therapists, physical therapists or exercise clinicians, a social worker, a nutritionist, a psychologist, a genetic counselor, and pulmonary function technologists. Improved survival and a larger number of new patients diagnosed in adulthood has led many care teams to form adult care practices designed to address adult-specific needs. Nevertheless, respiratory and gastrointestinal support are the mainstay of therapy for patients with CF.

Medical Evaluation

A thorough medical history is necessary to identify potential risk factors that may limit exercise performance in individuals with CF. The history should focus on factors that may be present in the person with CF that can alter the pulmonary–cardiovascular–peripheral systems necessary for effective oxygen delivery and utilization during exercise. The most important consideration before testing a patient with CF is to determine the subject's level of pulmonary disease. Prior pulmonary function data can help predict which patient is likely to experience oxyhemoglobin desaturation with exercise testing. An **FEV$_1$** of less than 50% of predicted or a low resting **oxyhemoglobin**

Table 23.1 Clinical Signs and Symptoms of Cystic Fibrosis

System	Signs and symptoms
Respiratory	Chronic productive cough, pneumonia, wheezing, hyperinflation, exercise intolerance, *Pseudomonas aeruginosa* bronchitis
Gastrointestinal and nutrition	Steatorrhea, failure to thrive, biliary cirrhosis, intestinal obstruction, abdominal pain
Sinuses	Chronic sinusitis, nasal polyps
Sweat glands	Salty taste, recurrent dehydration, chronic metabolic acidosis
Other	Depression, infertility, pubertal delay, digital clubbing, family history

saturation places the person with CF at much greater risk of oxygen desaturation during exercise (36,44,55). A history of wheezing, chest tightness, or chest pain during exercise may indicate the presence of exercise-induced bronchoconstriction, which is seen in 22% to 55% of CF patients (50,84,89). Additional considerations, such as a history of pneumothorax or **hemoptysis** (coughing up blood) at rest or during exercise, should be reviewed. A history of nocturnal headaches or cyanosis may suggest advanced lung disease with associated **hypoxemia**. Because exercising at altitude may exaggerate hypoxemia, it should be determined whether the exercise testing or training program will occur at altitude (8,80). Few cardiovascular limitations exist for people with CF. A history of pulmonary hypertension or cor pulmonale requires consultation with a cardiologist before testing. Signs and symptoms of right-side heart failure should be sought (e.g., edema, venous congestion, hypoxemia). Peripheral factors such as **scoliosis**, **kyphosis**, and tight hamstrings are commonly present in individuals with CF and may reduce mechanical efficiency during exercise.

Other organ systems can be affected by CF and become an issue during acute exercise. Liver disease with associated ascites (abdominal distention) may interfere with respiratory muscle effectiveness, whereas liver-related bleeding disorders may be exacerbated with increased blood pressure during exercise. CF patients are also at increased risk of dehydration state, because exces-

sive salt loss with physical exertion is commonly seen in CF (69,70). Clinical signs of early dehydration should be discussed with those who plan to exercise. These signs include light-headedness, heat intolerance, flushed skin, decreased urine output, concentrated (dark yellow) urine production, nausea, headaches, and muscle cramps. Adequate hydration needs to be stressed for those who will exercise in warm, humid climates. Consumption of 4 oz (120 ml) of fluid every 20 min is a good general rule. For children who cannot readily quantify fluid amounts, eight gulps of fluid equals approximately 4 oz (120 ml).

The extent of malnutrition and body composition (e.g., lean muscle mass) should be noted, because these considerations may alter the mechanical load applied during exercise testing. Finally, the exercise professional can use validated physical activity questionnaires or diaries to help determine how physically active the individual is before developing a precise prescription. Options for assessing activity vary in format (e.g., recall questionnaires, activity diaries). Some tools may not be appropriate for younger individuals, but some are designed specifically for use in the pediatric population (e.g., previous day physical activity recall) (73). The sensitivity of these tools in this population or in children in general has been questioned, especially when recall is required. In adults, the ability to monitor activity may be less complicated. Refer to practical application 23.1 for a review of relevant literature.

Literature Review

Despite the pathophysiological manifestations of CF, a number of people with CF have performed at extremely high athletic levels, accomplishing many of the athletic endeavors attained by their non-CF counterparts. The short-term benefits of exercise for people with CF include the therapeutic aspects of enhanced mucus clearance, improved cardiopulmonary fitness, and positive psychological well-being. Furthermore, with the incidence of depression on the rise in both adolescents and adults with CF, the psychosocial benefits of improved self-esteem and greater sense of accomplishment associated with regular physical activity are important. The long-term benefits are more difficult to define, because the natural course of CF is complex and multifactorial. All aspects of fitness appear to show positive benefits in conjunction with an exercise-training program. Because many of the training programs used in people with CF have varied by duration, intensity, and modality, establishing causal relationships between training and disease progression is difficult. Exercise tolerance and long-term survival have been correlated with one another, but a causal relationship has not been established (67).

Cardiorespiratory Benefits

Some benefits in **static pulmonary function** have been seen after exercise training. In a group of individuals with advanced lung disease, increases in FEV_1 were seen in response to an inpatient bicycling program consisting of 20 min per session at 75% of prestudy maximal intensity (42). Additionally, improvements were observed in forced vital capacity (FVC), FEV_1, FEF_{25-75}, and peak expiratory flow following intensive exercise

(1 h swim, 2.5 km jog, several hours of hiking per day) during a 17 d elective stay in a pediatric rehabilitation hospital in the mountains of Austria (92).

Despite these findings, most exercise-training programs have not shown increases in spirometric indexes but rather have resulted in a slower deterioration of lung function when compared with the nonexercising group. Improvements in dynamic lung function and other parameters dependent on dynamic lung function following exercise training are well documented as evidenced by a lower resting heart rate, improved maximal oxygen consumption, increased maximal heart rate, increased physical work capacity, enhanced ventilatory threshold, and improved maximal minute ventilation. Exercise programs during a hospitalization for an infectious exacerbation can serve as an adjunct to traditional modalities including chest physiotherapy and bronchial drainage and have been associated with improvements in peak oxygen consumption and peak work capacity (1,20,76). Formal supervised training programs including running, cycling, and swimming programs, as well as structured camps, have helped to maximize compliance with exercise. Length of participation and intensity of training vary in these studies, and greater training effects are seen in the more intense programs (31,32,39,68).

Inefficiencies in the mechanism by which individuals with CF accomplish exercise are apparent. Recent investigations have suggested that much of the ventilatory compromise, especially in patients more severely affected with CF, may be attributed to low tidal volume and resultant hyperventilation (61,88). This marked increase in the respiratory rate decreases the air actively participating in gas exchange. Furthermore, a recent study found subtle slowing in the phase II oxygen kinetics in individuals with CF as deoxygenated blood returns from the periphery (41). Improvement in tidal volume through exercise training, in both an aerobic and an anaerobic fashion, may contribute to increased cardiorespiratory function and improve delayed oxygen kinetics.

Muscular Strength and Endurance

The benefits of exercise training on anaerobic function such as muscular strength and endurance has received increased attention in recent years. A program that included upper-body strength training demonstrated increased strength and physical work capacity as well as good compliance throughout, as more than 85% of the subjects completed the program (71). Both weight training and home cycling for 6 mo increased muscle strength in individuals with CF, although long-term adherence to these programs is of concern (39,87). Exercise programs focused on respiratory muscle training have also shown training adaptations as demonstrated by increased peak inspiratory pressures (4,81). The intensity of inspiratory muscle training has become better defined. High-intensity training (80% of maximum) was shown to improve muscle function significantly more than lower-intensity training did (33).

Body Composition and Nutrition

Children with CF who undergo regular exercise training (e.g., swimming, biking, running, weight lifting) are capable of increasing body mass despite increased caloric needs (6,20,43). Nutritional supplementation has been associated with improved aerobic exercise tolerance and respiratory muscle strength in some small case reports (22,85).

Psychological Well-Being

The long-term psychosocial benefits of exercise have been fairly well established in children with CF. The Quality of Well-Being Scale, designed to measure daily functioning, has been shown to correlate with exercise capacity in individuals with CF (48,72). Furthermore, improvements in self-concept and well-being have been shown to be associated with involvement in CF summer camps (49,92). The current rate of depression among adults with CF is now greater than 14%. As this number continues to grow, quality research in the area of well-being and quality of life becomes increasingly important.

Exercise Influence on CF Disease

The role of exercise in preventing the deterioration of lung function or occurrence of complications associated with CF continues to be unclear. Although studies have linked exercise tolerance with long-term prognosis, the preventive benefits of exercise were not established (63,67). Only one study to date has attempted to

(continued)

Practical Application 23.1 *(continued)*

examine the rate of exercise decline as a predictor of mortality. Measures of oxygen consumption were made annually over a 5 yr period. The rate of decline was calculated along with the subsequent 8 yr survival. On average, oxygen consumption fell by 2.1 ml $\cdot$ kg^{-1} $\cdot$ min^{-1} per year. Furthermore, those with an oxygen consumption of less than 32 ml $\cdot$ kg^{-1} $\cdot$ min^{-1} had an 8 yr mortality of almost 60%, and those above 45 ml $\cdot$ kg^{-1} $\cdot$ min^{-1} showed no mortality at all. This improved prognosis associated with exercise tolerance in CF has not been completely explained. The potential of an enhanced immune modulation is an attractive but speculative theory and requires much investigation. The ability of exercise to alter the immune system is now well recognized (65). This notion is consistent with the belief among the public and athletes that heavy exertional activities predispose an individual to illnesses, whereas moderate levels offer protective effects. Although some studies have shown a beneficial effect of chronic exercise on the immune system, this relationship has not been established in people with CF (11,12).

Diagnostic Testing

Many states have instituted a newborn screening for CF, allowing for earlier detection. In utero diagnosis has also become available. A positive sweat test or genetic mutation analysis can confirm the diagnosis of CF. Additional laboratory testing should be performed when the diagnosis of CF is considered:

- Sputum culture (positive for *P. aeruginosa* or other CF bacteria)
- Chest radiograph
- Static and dynamic lung assessment if age appropriate
- Blood sampling for complete cell count
- Liver function
- Nutritional parameters (e.g., total protein, albumin)
- Renal function (e.g., blood urea nitrogen, creatinine)
- Fat-soluble vitamins A, D, E, and K (prothrombin time/international normalized ratio PT/INR for vitamin K)
- Glucose

Assessment of static pulmonary function, as defined by the properties of the lung at rest or baseline, is essential in the acute and chronic management of individuals with CF. Simple spirometry, as well as assessment of lung volumes, diffusion capacity, and bronchodilator responsiveness, assists in the detection of an acute **pulmonary exacerbation**. Although some individuals with CF have little or mild lung disease, most people affected with CF will demonstrate varying degrees of airway obstruction with signs of hyperinflation. Additionally, almost 30% of people with CF will demonstrate signs of airway hyperreactivity when exposed to a bronchodilator. Declines in

FEV$_1$ or indexes of smaller airway function (e.g., FEF$_{25-75}$, FEF$_{50}$, FEF$_{75}$) over time may serve as warning signs of acute or chronic lung deterioration. In conjunction with static pulmonary function assessment, **dynamic pulmonary function**, as defined by lung function in response to changing physiological state (e.g., work, exercise, physiologic stress), also plays an important role as a diagnostic tool and in monitoring the patient's clinical condition. In fact, a single measure of aerobic exercise tolerance has been strongly correlated with long-term survival in patients with CF (2,63,67). Moreover, patients who have undergone repeated exercise evaluations over a 5 yr period were studied to look at mortality in years to follow. Those kids who maintained exercise tolerance had greater long-term survival when compared with those who showed a decline in exercise capacity (74). When healthcare workers assess the lungs under measurable stress (e.g., exercise), ventilatory limitations and impairment in other parameters such as oxygen saturations that depend on dynamic lung function may become apparent that were not noted when the patient was at rest. Regular assessment of exercise tolerance in patients with CF is an integral component of their medical care. The exercise clinician who treats and assesses individuals with CF should comprehensively understand both dynamic and static lung function measurements and the role that exercise testing serves in the management of this population.

Exercise Testing

The importance of performing a complete exercise evaluation before developing an exercise prescription for individuals with CF cannot be underestimated. This evaluation can help identify potential risk factors for this specific population. Although many people with CF have limitations in exercise performance, physical activity remains a vital part of the therapeutic management plan. An effective exercise program should help optimize all aspects of fitness including overall well-being, both physical and psychological. The exercise clinician plays an important role in conjunction with the patient, parents, and CF medical team in establishing realistic goals and developing an achievable exercise program to enhance the individual's quality of life (see practical application 23.2).

Contraindications

Although no absolute contraindications to exercise testing exist, special considerations need to be observed for the individual with a history of pulmonary hypertension, acute hemoptysis, pneumothorax, oxygen dependence, exercising at altitude, a bleeding disorder secondary to liver disease, and severe malnutrition. Monitoring during

Practical Application 23.2

Exercise Recommendations for Individuals With Advanced Lung Disease

Traditionally, people with CF and severe obstructive pulmonary disease have not received the attention of exercise clinicians because of an extremely conservative approach to their exercise participation. Fear of exercise-induced hypoxia leading to pulmonary hypertensive episodes, cardiac ischemia, and dyspnea as well as the perception of limited beneficial effects of training have all been cited as deterrents to regular physical activity. Although concerns of hypoxia exist for individuals with advanced CF, appropriate exercise prescriptions can be safely administered. The first consideration for a person with advanced lung disease who wishes to participate in regular physical activity is to determine whether exercise will induce oxyhemoglobin desaturation.

Table 23.2 Clinical Evaluation of the Individual With Cystic Fibrosis

Mode	Protocol specifics	Clinical measures	Clinical implications	Special considerations
Cardiorespiratory: treadmill, bicycle	Treadmill: Bruce, Balke Bike: James, Godfrey	Heart rate, blood pressure, EKG, oxygen uptake, SaO_2	Assesses endurance, PWC, risk of desaturation, $\dot{V}O_2$ max, serial tests	FEV_1 <50% = risk of desaturation, in severe patients reduced peak heart rate
Resistance: bicycle	Wingate, 1RM, grip strength, respiratory muscle function	Peak power, mean power, 1RM max. PI_{max}, PE_{max}	Assesses muscular strength and endurance, power, lung strength	Significantly affected by nutritional status
Flexibility: stretching	Sit-and-reach, assessment of range of motion	Hamstring and quadriceps flexibility	Thoracic kyphosis develops as disease severity progresses	Early detection can lead to stabilization of the abnormality
Nutritional: height, weight, BMI, skinfold	Skinfold assessed at the triceps (CF registry). Three-site assessment permitted	Stature, weight, BMI, % body fat	Desired BMI: males >23, females >22	Significant correlation between nutrition and performance on exercise tests

Note. BMI = body mass index; 1RM = one-repetition maximum (to assess strength); EKG = electrocardiogram (3 lead or 12 lead based on equipment); SaO_2 = oxygen saturation (measured in percent); PI_{max} = maximal inspiratory pressure; PE_{max} = maximal expiratory pressure; PWC = peak work capacity; $\dot{V}O_2$max = maximum oxygen consumption during 1 min of exercise; FEV_1 = forced expiratory volume in 1 s.

(continued)

Practical Application 23.2 *(continued)*

A baseline maximal aerobic exercise challenge should be administered with monitoring of pulse oximetry and electrocardiogram. If the subject completes the challenge without desaturation (defined as a value <90% on room air), then supplemental oxygen is not required. Although an individual may desaturate to less than 90% during the challenge, the point at which this occurs becomes critical. Because most aerobic exercise prescriptions use submaximal intensity levels (60-75% of maximal), the exercise clinician should determine whether the subject desaturated at this submaximal level. If so, supplemental oxygen needs to be administered for subsequent exercise training, whereas it is not required for those who do not desaturate. If supplemental oxygen is required, a repeat exercise challenge should be administered after an appropriate recovery period to document that the subject remains normoxic during exercise. The level of oxygen supplementation required should be recorded.

Special considerations for the person requiring oxygen for exercise should be made. The choice of activity may need to be modified to allow for the presence of oxygen tanks. Activities using stationary modalities (e.g., treadmills, bicycles) may be more appropriate for this group. Many health clubs are capable of accommodating people with these special needs.

Individuals with advanced lung disease are capable of gaining the beneficial effects of exercise training. Improved gas exchange, ventilation, aerobic tolerance, peripheral muscle adaptations, and sense of well-being have all been documented following exercise training (42,60). Appropriate exercise prescriptions can be developed for even the most debilitated person with CF. This population has not traditionally received the benefits of interacting with an exercise clinician. One could argue that individuals with severe CF have the most to gain from exercise that would help reestablish functional ability. By understanding the particular needs of this population, the exercise clinician can play a major role in achieving this goal.

Management of CF may require varying types and amounts of medications. A complete listing of the medications used is provided in chapter 3.

---

testing should include continuous pulse oximetry, electrocardiogram, and vital sign assessment. Supplemental oxygen and a short-acting bronchodilator (e.g., albuterol by inhaler) should be available to all patients with CF during or following testing as needed.

Recommendations

Following a complete history, the assessment of exercise capacity can be made. Typical cardiopulmonary responses in CF patients are listed in table 23.3. The precise protocol and testing location depend on several factors including the desired goals for training, the age of the person, the medical considerations of the disease, and the resources available. Finally, a clearly defined goal for testing should be determined so that individualized exercise programs can be designed. Such goals may include determining the heart rate at which oxyhemoglobin desaturation occurs, monitoring for improvement in response to medical therapy, or comparison to prior performance. Table 23.4 list clinical evaluation guidelines for individuals with CF.

Aerobic Testing

Numerous reproducible exercise protocols exist for testing maximal exercise performance in children and adults: the Godfrey, McMaster, and James protocols for

bicycle testing and the Bruce, modified Bruce, and Balke protocols for treadmill testing (62,77). Monitoring of pulse oximetry, electrocardiogram, and blood pressure response with exercise should be considered in all patients with CF, especially those with more advanced disease. Despite minimal risk associated with testing, standard practices for emergency management should be followed including access to a crash cart and supplemental oxygen as well as personnel trained in advanced life support and cardiopulmonary resuscitation.

The extremely comprehensive information provided by a maximal aerobic test allows the exercise clinician to develop an appropriate exercise prescription. But this form of testing requires sophisticated exercise equipment and highly trained technical staff and can result in a significant financial cost to the patient. A submaximal aerobic test may be useful in determining whether exercise desaturation or breathlessness takes place and can help verify the effectiveness of exercise prescriptions established from maximal tests. Submaximal assessment is used infrequently but may be easier to perform for the young child or adult with significant ventilatory limitations.

Gas exchange parameters are not routinely required for this test. A treadmill or bicycle similar to that used during a maximal test is the ideal equipment for the submaximal test. Traditional submaximal protocols require

Table 23.3 Cardiopulmonary Parameter Changes Among Individuals With Cystic Fibrosis

Parameter	Change
Static pulmonary function (rest or baseline)[a]	
Spirometry: FEV_1, FEF_{25-75}, tidal volume	Decreased
Lung volumes: RV, FRC, RV/TLC	Increased
Diffusion capacity: DL_{CO}	Decreased
Oxyhemoglobin saturation: SpO_2	Decreased
Dynamic cardiac and pulmonary function (in response to exercise or stress)[b]	
Aerobic capacity: PWC, $\dot{V}O_2$ max	Decreased
Breathing response: $\dot{V}_E$, $\dot{V}_E$/MVV, RR	Increased
Gas exchange: VCO_2, $EtCO_2$	Increased
Blood pressure response	Normal
Heart rate at rest	Increased
Heart rate during peak exercise	Decreased

Note. FEV_1 = Forced expiratory volume in 1 s; FEF_{25-75} = forced expiratory flow between 25% and 75% of the forced vital capacity; RV = residual volume; FRC = functional residual capacity; DL_{CO} = diffusion capacity of the lung by the carbon monoxide technique; SpO_2 = pulse oximetry; PWC = peak work capacity; $\dot{V}O_2$max = oxygen consumption; $\dot{V}_E$ = minute ventilation; $\dot{V}_E$/MVV = ratio of minute ventilation to maximal voluntary ventilation; RR = respiratory rate; VCO_2 = carbon dioxide production; $EtCO_2$ = end tidal carbon dioxide.
[a]Static pulmonary function is decreased and declines with advancing lung disease. [b]Abnormal parameters tend to follow extent of lung disease; aerobic performance weakly correlated with static lung function parameters.

Table 23.4 Muscle Endurance and Strength, Body Composition, and Flexibility Guidelines for Individuals With CF

Parameter	Change
Muscular endurance[a]	
WAnT mean power; isokinetic cycle ergometry power	Decreased
Muscle efficiency	Decreased
Muscular strength[b]	
Respiratory muscle strength (PI_{max}, PE_{max})	Decreased
Peripheral muscle strength	Decreased
Body composition[c]	
Body weight	Decreased
Lean muscle mass	Decreased
Body mass index	Decreased
Percent body fat (BIA, skinfold assessment)	Decreased
Flexibility[d]	
Peripheral muscle flexibility (hamstrings, quadriceps)	Decreased
Posture—extent of kyphosis	Increased

Note. WAnT = Wingate anaerobic test; PI_{max} = peak inspiratory pressure; PE_{max} = peak expiratory pressure; BIA = bioelectric impedance analysis.
[a]Muscle endurance decreases as lung disease progresses and may reflect impaired nutritional status and intrinsic cellular deficiencies.
[b]Decreases as disease progresses and may reflect nutritional status and loss of mechanical efficiency.
[c]Decreases in body composition reflect increased caloric expenditure with advanced lung disease, poor oral intake, and release of cachectic mediators.
[d]Reflects deconditioning associated with decreased activity or advanced disease.

workloads consistent with 75% of the age-predicted maximal heart rate. Individuals with CF, however, may not reach the theoretical age-predicted maximal heart rate secondary to ventilatory limitations. Exercise at 75% of age-predicted maximal heart rate for children with significant respiratory compromise may cause the child to approach or exceed his or her maximal capacity (66). Heart rate, blood pressure, and pulse oximetry should be monitored. The lowered technical demands, financial costs, and ease of repeat testing (e.g., tracking performance during treatment or over time) make the submaximal test an attractive alternative to the maximal test.

Lab-based exercise tests, whether maximal or submaximal, are not always convenient for individuals with CF. Over the past few years, several walking and running tests have been developed in an attempt to mimic real life more accurately and offer simple-to-administer testing protocols. Walk tests for 2, 6, and 12 min have been used for people with CF (38,40,79). These protocols allow the subject to walk over a set period at their own pace while heart rate and oxygen saturations are monitored. Total distance traveled, the development of exercise breathlessness, and oxygen desaturation are recorded and can be compared over time with prior tests. Outcome variables have been relatively well correlated to standardized maximal tests for individuals with mild to severe CF lung disease. Serial walk tests may provide simple yet valuable assessment of the value of supplemental oxygen or pulmonary rehab in the patient severely affected with CF. Shuttle tests have also been used for people with CF (16,17). In one version, the subject walks (or runs) at increasing speeds (set by an audio signal) over a set course until voluntary exhaustion occurs. Heart rate, pulse oximetry, and distance traveled are monitored during the test. Although both the walk tests and shuttle tests are relatively simple to perform, a limitation is that they depend on patient effort, which makes motivation by the test administrator essential. A 3 min step test has been developed as a modification of the master two-step exercise test used in adult cardiac testing (9). The subject steps on a single step set a standard height (6 in., or 15 cm) at a rate of 30 steps per minute for 3 min. The test is complete after 3 min or when the subject is unable to continue. The total number of steps can be tabulated along with change in heart rate, oxygen desaturation, and sensation of breathlessness (75). All these noninvasive tests are easy to perform and do not require sophisticated equipment. The utility of these tests for assessment of aerobic fitness in patients with milder CF remains unknown and they are generally reserved for patients with more severe disease.

Muscular Endurance

Although many tests have been proposed to assess muscular endurance in healthy individuals (90), relatively few protocols have been used for individuals with CF. The Wingate anaerobic test (WAnT) has been the most widely used test to assess both short-term mechanical power or strength (measured by peak power) and leg muscle endurance over a brief time (measured by mean power). The WAnT was designed to measure nonoxidative muscle function with peak power indicative of muscle power and mean power providing information about anaerobic muscle endurance. The test consists of a 30 s all-out sprint on a cycle ergometer against fixed resistance. Determination of resistance depends on lean muscle mass, but a standard starting point is 75 g of resistance per kilogram of body weight (46). The test is demanding, but patients with CF have been able to complete it (15,18). Although sophisticated equipment is available to perform the WAnT, a mechanical cycle ergometer (e.g., Monarck or Fleisch) can be adapted for the WAnT. Alternative protocols for testing muscle endurance in individuals with CF include use of an isokinetic cycle ergometer and cycling at supramaximal levels. Both of these protocols have been used in the research setting for testing children with CF (53,83) but are not readily accessible outside the academic exercise laboratory. Measurements of respiratory and peripheral muscle fatigue have been used as research tools to assess both respiratory and peripheral muscle function in individuals with CF (51,54). Many school-based fitness testing (push-ups, pull-ups, long jump, and high jump) measures can be performed by CF patients with mild to moderate disease because they generally do not have restrictions or limitations to any activities. These field tests may provide a fun and effective alternative to assessing both muscular strength and muscular endurance in young patients with CF.

Muscular Strength

Strength of both respiratory muscle and peripheral skeletal muscle groups has been assessed in people with CF. Peak inspiratory pressure determination specifically measures the muscles used for inspiration and consists of the subject's inspiring a breath of air at residual volume against an occluded airway. The greatest inspiratory subatmospheric pressure that can be developed is recorded. Similarly, peak expiratory pressure measures the strength of the abdominal and accessory muscles of breathing and consists of the subject's exhaling forcefully against an occluded airway, usually at total lung capacity. Inspiratory muscle training when corrected for workload can be effective in improving inspiratory muscle function and work capacity (33). These maneuvers are relatively easy to perform (78). The equipment required is usually part of a standard body plethysmography system. Alternatively, handheld direct reading manometers or electronic pressure transducers and recorders can be used. Peripheral skeletal muscle has been shown to respond to training in individuals with CF. A recent study demonstrated that a home-based strength training program resulted in increased strength and physical work capacity (71). Standard techniques including use of dynamometers, cable tensiometers, isokinetic muscle testing, and free weights can be applied to test specific muscle groups. The age of the subject will often determine the choice of test. For very young children, a child's own body weight can be used as a resistance tool according to the testing criteria of the President's Council on Physical Fitness and Sport (e.g., push-ups, pull-ups) (30). Expected muscular endurance and strength responses in children with CF are listed in table 23.4.

Body Composition

Because individuals with CF tend to be lower in both body weight and height than those without CF, assessment of anthropometrics becomes important. From a clinical perspective, monitoring body mass index as well as body composition is an important part of nutritional assessment. Typical body composition responses in CF relative to age- and gender-matched peers are listed in table 23.4. Clinically, the desirable body mass index for males with CF is greater than 23 and for females with CF is greater than 22. Patients below these values tend to have decreased clinical standing and poor long-term prognosis when compared with those above the desired value (26). For the exercise clinician, documenting body composition is essential in determining workloads needed for testing that depend on the amount of muscle mass present as well as for scaling absolute exercise data per muscle mass. Additionally, monitoring other interventions (nutritional or exercise) depends on the distribution of muscle and fat mass. Many techniques exist for determining fat distribution in healthy adults. Some of the underlying assumptions of these techniques are in question for children, for individuals with chronic lung disease, and in conditions associated with electrolyte disturbances. Currently, single-site (triceps) or multiple-site skinfold assessment has been the most commonly used technique for monitoring body fat in children with CF. Skinfold calipers (e.g., Harpenden, Lange) provide an inexpensive means for determining body composition. Use of pediatric reference equations for a child is manda-

tory (57). An alternative technique for body composition assessment uses bioelectrical impedance analysis through commercially available systems. Both skinfold and bioelectrical impedance assessments are easy to perform, inexpensive, and reproducible in the hands of a trained technician. A common practice by some CF specialty centers is to use both techniques as a means of establishing internal reliability of measurements.

Flexibility

For most people with CF, flexibility is not a major limiting factor in exercise performance. As lung disease advances, thoracic kyphosis ensues, and associated mechanical inefficiencies are seen with exercise (47). Some of these postural changes are associated with tight hamstrings, leading to potential exercise limitations and injury (47). Early identification of these abnormalities through routine assessment of large-muscle-group range of motion, as part of an exercise assessment, can lead to establishment of stretching programs and stabilization of abnormal posture.

Anticipated Responses

Individuals with CF have impaired exercise tolerance as demonstrated by reduced maximal oxygen consumption and peak work capacity compared with healthy children (34,35,58,86). The ratio of minute ventilation to the **maximal voluntary ventilation**, a marker of ventilatory limitation, may exceed 100% (normal 70–80%) and worsen as CF lung disease progresses (19,24). People with CF demonstrate expiratory airflow limitation as evidenced by tidal loop analysis during exercise (5). End tidal carbon dioxide, another marker of ventilatory limitation, increases with exercise and is related to the severity of lung disease (21,24,59). Alveolar ventilation appears normal in patients with mild lung disease with a compensatory increase in the tidal volume (21). But as disease severity increases, alveolar hypoventilation becomes evident as the tidal volume approaches and is limited by the vital capacity (56). As this occurs, breathing frequency increases as a compensatory factor but does not provide the minute ventilation necessary for increased exercise intensity. Gas exchange can also be compromised as evidenced by the lack of increase in the diffusion capacity of the lungs following exercise (92). In adult patients with moderate to severe disease, dead space ventilation increases. This appears to be secondary to reduced tidal volume in conjunction with increased respiratory rates resulting in less gas exchange with each breath (88).

Finally, the phase II component of oxygen kinetics (increased ventilation secondary to a return of deoxygenated blood from muscles) appears to be slowed in patients with CF, resulting in peripheral adaptations (41).

The cardiovascular system is generally able to keep up with the oxygen demands of the exercising muscle and becomes compromised only with advanced disease. Heart rate and blood pressure responses to exercise appear normal, although a lower peak heart rate is seen as the disease progresses (34). Muscular efficiency in children with CF can be reduced by up to 25%. The reduced efficiency may reflect altered aerobic pathways at the mitochondrial level (28). Studies of muscular strength demonstrate that CF patients have reduced muscle strength when compared with healthy controls (27,37,45,71,82). But recent studies on exercise evaluation in patients with CF suggest that positive strength gains are associated with training programs (45,71).

Studies using the WAnT have demonstrated decreased anaerobic performance in individuals with CF that is related to muscle mass quantity (15). Overall oxygen cost of work appears elevated during exercise for people with CF (29). Additionally, it appears that energy metabolism during exercise is abnormal in children with CF (13,90).

Treatment

Current treatment for CF is complex. Specialized CF care centers that offer the multiple-specialty care necessary for these individuals have emerged. Both preventive and acute management are required to optimize health for people with CF. Respiratory and gastrointestinal support are the mainstay of therapy for children with CF. The complexity of care increases in adult patients and those with severe lung disease. Treatment of the pulmonary component can be best viewed in terms of addressing the vicious cycle of infection and inflammation (see figure 23.1 on p. 404).

Current strategies are designed to intervene in this process at multiple levels, thus minimizing the progressive loss of lung tissue. A combination of mucolytic agents and daily mucus clearance techniques can help maintain good pulmonary hygiene. Because bacterial colonization leads to a brisk inflammatory response (e.g., cough, sputum production, increased work of breathing), use of antibiotics becomes necessary. The choice of oral, nebulized, or intravenous antibiotics is determined by the severity of the acute exacerbation. Exercise as part of the therapeutic medical regimen is standard care in most specialized CF centers. As progressive lung destruction and deterioration occur, the final choice of therapy is lung transplantation. Currently more than 170 patients with

CF within the CF registry have undergone lung transplantation. The approximate 3 yr survival rate is 60%.

Besides considering the pulmonary aspects of CF, the exercise professional must give attention to the nutritional aspects of CF. For the 85% of individuals with CF who are pancreatic insufficient, the use of pancreatic enzymes can help in the utilization of nutrients. Fat-soluble vitamins are also given as supplements. The use of a high-fat, high-calorie diet to ensure the consumption of sufficient calories is standard care for most patients with CF. Fluid management during exercise poses unique challenges for the individual with CF. Persons with CF tend to lose more salt in their sweat per surface area than do their non-CF counterparts. Thus, the sodium and chloride levels in the bloodstream often decrease after exercise, whereas these levels are maintained in those without CF (69). In addition, individuals with CF tend to underestimate their fluid needs during exercise. In one study, patients with CF lost twice as much body weight as healthy subjects did when drinking fluid only when thirsty (10). Although children with CF have a tendency to lose salt while exercising, especially in extremely hot, humid weather, most children consume sufficient salt. Ready access to a saltshaker or salty snacks (e.g. pretzels, potato chips) along with liberal fluid intake usually suffices. Anthropometric data including height, weight, body mass index (BMI), and percent body fat are vital markers in the nutritional status of the patient with CF. Nutritional growth is highly correlated to prognosis in patients with CF. Desirable levels for BMI in both males and females have been established to assist the nutritional maintenance of these patients. Exercise is routinely recommended for all people with CF, regardless of pulmonary status. In conjunction with regular chest physiotherapy, exercise can enhance clearance of mucus from the bronchial tree (7). Exercise alone, however, does not appear to be as effective as standard chest physiotherapy. Nevertheless, exercise therapy with either unsupervised or supervised pulmonary rehabilitation should be recommended because its positive physiologic outcomes have been proved even in patients severely affected with CF. Furthermore, with 13.5% of adult patients suffering from depression, the psychological benefits of exercise in the person with CF also appears to be an important benefit of regular exercise (49).

EXERCISE PRESCRIPTION

As lung disease progresses, lung function may become significantly compromised for individuals with CF. After the high-risk individual is identified, specific precautions may be needed before initiation of an exercise prescription. Use of supplemental oxygen while exercising for those who are prone to oxyhemoglobin desaturation has been beneficial in allowing successful exercise training and recovery (23,60,83). Family and medical personnel can encourage those who use supplemental oxygen to participate in exercise training. Although the clinical guidelines for exercise testing offered by the American College of Sports Medicine discourage testing in individuals with FEV_1 less than 60% of predicted (94), with appropriate medical direction patients with significant lung disease associated with CF have safely undergone clinical evaluation using a 6 min walk test or alternate submaximal evaluation. Participation in a formal exercise program or pulmonary rehabilitation can be established after the level of dyspnea and need for supplemental oxygen are established.

Additional concerns exist for the child who will be exercising at high altitude. Assessment may be warranted at sea level to determine the risk of desaturation at high altitude (8). Because individuals with CF are susceptible to dehydration, especially when exercising in warm weather, fluid intake should be carefully monitored and encouraged. Fluid intake every 20 to 30 min should suffice. Practical application 23.1 on p. 406 reviews the literature regarding exercise training for patients with CF.

Special Exercise Considerations

Conditions associated with CF such as CF-related diabetes, exercise-induced asthma, and liver disease may require special consideration. Case-specific guidelines from the CF physician should be made for these conditions before initiating an exercise program. Hypoglycemia and acute bronchospasm can be relatively easy to prevent. Although severe CF-related liver disease is uncommon, its presence can result in a bleeding tendency. People who have either enlarged visceral organs (e.g., liver, spleen) or a bleeding tendency should avoid contact sports. Prescribing exercise in CF patients with severe lung disease can be challenging. Specific application issues in individuals with advanced lung disease are discussed in practical application 23.2.

Exercise Recommendations

The goals of an exercise program for individuals with CF should include enhancing physical fitness, reducing the severity or recurrence of disease, and ensuring safe and enjoyable participation. To maximize compliance with any exercise program, activity should be care-

fully selected to enhance cardiopulmonary fitness and other exercise goals as described. Despite the increasing number of adults with CF, 60% of all individuals with CF are in the pediatric and adolescent age group. Exercise prescriptions should accommodate the special needs of these participants. Children, especially those under the age of 8, generally do not respond to a formal structured exercise program. Children respond well when an exercise program matches their muscular development, strength, and coordination with age-appropriate activities. Additionally, a gradual progression in the level of physical activity should allow for attainment of exercise goals while minimizing the risk of injury and noncompliance. This concept of gradual progression depends on the fitness parameter being addressed and is unique to different ages as well as disease severity. The use of "play" consisting of games and diversionary tactics may be most beneficial in meeting the preceding criteria while maintaining good compliance and teaching an active lifestyle. Reducing nonschool sedentary time by incorporating outside activities or tasks can be helpful. Older children (>8 yr) may be able to undergo a more structured program built on the mode, intensity, duration, and frequency of exercise. In adults with CF, many of the same compliance issues present in the general population exist. Time management issues related to career and family are compounded by daily treatment regimes and therapies designed to maintain disease stability. Exercise progression in this group should not only follow the guidelines of the American College of Sports Medicine but also emphasize a feeling of well-being and quality of life.

Recommendations for exercise prescription, presented in practical application 23.2, follow the guidelines established by the American College of Sports Medicine (94). These guidelines were developed to address each component of fitness. The guidelines have been adapted for adults with CF as well as for the pediatric population (26). When prescribing exercise in the younger age groups, including the parent in the development of an exercise program is vital to its success because many parents falsely perceive negative consequences of exercise for their child with CF (e.g., weight loss) (14). Furthermore, exercise compliance in children depends largely on the motivation and encouragement of the parent (see practical application 23.3 on p. 416). An exercise prescription for individuals with CF is presented in practical application 23.4 on p. 417.

Cardiorespiratory Training

The main objective of cardiorespiratory training is to improve aerobic capacity. Higher levels of aerobic fitness

Client–Clinician Interaction

Motivation may be the most powerful factor in determining whether an exercise program will succeed.

Motivation is unique for each person. Because 60% of the patients with CF are children or teenagers, motivational issues are especially important given adolescent issues of self-image, fitting in with peer groups, and establishing physical abilities. Adult patients with CF require motivation to maintain appropriate levels of fitness despite increasing time constraints associated with both career and life. The appropriate client–clinician interaction is important in ensuring successful exercise testing and program satisfaction that addresses both the physical and the emotional needs of the client.

Situational Motivational Tips for Clinical Testing

- Make the testing experience fun for younger children by creating a gamelike scenario, with cheering and enthusiasm throughout.
- Explain all procedures in detail. Forewarn the child what he or she will feel during testing (e.g., breathlessness, muscle fatigue, and cough).
- Listen to the child's questions and concerns.
- Pick an apparatus (treadmill, bike) that the person believes would allow him or her the greatest success.
- Introduce equipment to children in a way that is fun or easily understood. The pulse oximeter might be described as the "ET light" or as a secret spy decoder that introduces the patient by his or her fingerprint.
- Use positive motivational phases such as "You can do it," "You're almost there," "We are so proud of you," or "Only one more minute." Try to avoid using phrases that influence decisions such as "Do you need to stop?" or "Do you have to quit?"
- Adult patients should be motivated as well by ensuring that they understand that the exercise evaluation is a key component to the evaluation and treatment process.

Tips for Exercise Adherence

- Allow the child to play an active role in planning. Establish a partnership and develop the exercise program with the child.
- Address the child's or parents' concerns (e.g., increased weight loss, not being able to keep up with friends, poor body image) associated with exercise programs and facilities.
- Have the individual (adult or child) assist in setting exercise goals.
- In adults, acknowledge the concerns and barriers to adherence to the program and work to find alternative strategies.
- Understand that what is successful will be unique for each individual with CF (e.g., completing a marathon, being able to enjoy activities with family, being physically prepared for lung transplantation).
- In both children and adults, communicate frequently and address concerns before they lead to noncompliance.

Cystic fibrosis patients appear to benefit significantly from exercise programming. Pulmonary function tends to either improve or have a slower rate of deterioration following training. Muscle strength and endurance improvements are well documented. Additionally, body weight can be successfully maintained or increased during exercise intervention, and patient psychological well-being typically improves.

Exercise Prescription

	Disease severity: mild to moderate[a]			Disease severity: severe[b]		
	Aerobic	**Anaerobic**	**Flexibility**	**Aerobic**	**Anaerobic**	**Flexibility**
Type	Any enjoyable aerobic activity. Swim, bike, walk, jog, sports	Sprinting, push-ups, sit-ups, plyometrics, age-appropriate weight training	Stretching, yoga, Pilates can be used for gains as well as relaxation	Supplement O_2 if needed, which may limit choices; stationary ergometers work well	Increasing disease state increases risk. Light weights can be used to maintain tone	Enhanced chest wall mobility, body relaxation, and flexibility
Frequency	3–5 d per week	3–5 d per week	2–7 d per week	3–5 d per week	1–2 d per week	2–7 d per week
Intensity	70–80% of measured maximum	10–12 repetitions, low resistance. Inspiratory muscle training 80% of PI_{max}	Pain free 10–30 s for each stretch	Measuring a true max is vital because pulmonary limitations limit peak HR to 70–80%	Very light resistance; limit activities that would induce a Valsalva maneuver	Pain free, 10–30 s for each stretch
Duration	20–60 min	20–30 min	10 min	20–60 min	10–20 min	10 min
Progression	No more than 10% in any given 2 wk period	Work upper and lower body, progress when 10–12 reps are no longer challenging	Natural progression as flexibility improves	No more than 10% in any given 2 wk period	Minimal progression; maintaining aerobic activities will reduce muscle wasting	Natural progression as flexibility improves
Goals	Improve aerobic function, increase lung function	Increase respiratory strength, assist in increasing body mass	Reduce risk of injury, maintain or enhance chest wall mobility	May be formalized pulmonary rehabilitation program to move toward or prepare for transplant	Increase respiratory strength, improve performance of tasks of daily living	Maintain or enhance chest wall mobility in advancing disease
Comments	3–5 d per week in adults; daily activity in children. Retest yearly	Wingate, grip strength, shuttle run, 1RM	Flexibility issues are similar to those of general population	Formalized pulmonary rehabilitation may be best suited for clinical gains	Alternative measures of strength assessment are warranted	Exercise prescription should be an adjunct to regular CF treatments

Note. [a]Mild to moderate disease severity: minimal risk of desaturation, no sports or activity restrictions; forced expiratory volume in 1 s (FEV_1) >50%. [b]Severe disease: increased risk of desaturation, which may require supplemental oxygen; FEV_1 <50%.

have been associated with better quality of life and improved survival rates. The components of the exercise prescription are reviewed here to optimize the client's exercise-training program.

Mode

No specific activity has been identified as optimal for patients with CF and those with more severe lung dysfunction. Choice of modality depends on the subject's personal preference and need not be costly. Cardiopulmonary benefits have been seen with a multitude of activities, some of which require little or no equipment (e.g., walking, jogging). Treadmills, bicycles, or alternative aerobic modalities (e.g., elliptical trainers) can all be incorporated into a successful exercise program. For patients who experience desaturation, exercising in an environment where supplemental oxygen is available may limit some choices but should not prohibit participation in an exercise program.

Frequency

In general, physical activity should be encouraged daily in people with CF. But in a formalized exercise programs to ensure improvements in the cardiorespiratory conditioning of individuals with CF, 3 to 5 d of exercise per week appears optimal. Intense exercise beyond five times per week may lead to increased risk of injury. If more than five times per week is desired, the use of cross-training (e.g., strength training, stretching) is advised to allow adequate muscle recovery. Signs of increased fatigue and staleness may be a result of overtraining and should prompt a reduction in exercise frequency. For adults with CF, 5 d of exercise per week is optimal for enhanced fitness.

Intensity

Exercise intensity should range from 70% to 85% of the measured maximum heart rate, but lower intensities should be used for beginners. If a maximum cardiopulmonary test is not performed, then the general rule of using a maximum heart rate of 200 beats · min^{-1} for intense running and 195 for cycling can be applied for children and adolescents. After a child has completed puberty, the formula of 220 minus age can be applied as it would be for adults. Thus, a heart rate of 140 to 170 beats · min^{-1} is a reasonable estimate of the heart rate that should be attained for optimal cardiopulmonary benefit. The use of a steady-state protocol (e.g., treadmill or cycle ergometer) can ensure that the appropriate workload is established. This objective is easily met by choosing a submaximal workload and having the subject exercise for 10 min while measuring heart rate throughout. This appropriate intensity of exercise is necessary for obtaining gains in cardiorespiratory fitness.

Duration

Exercise sessions should last 20 to 60 min. Alternatively, two abbreviated sessions can provide similar benefits. Attention span may play a role in the child's ability to perform an activity for longer than 10 min. For the younger patient, varying the exercise sessions by interspersing different activities may minimize boredom and enhance compliance (e.g., 5 min of bike riding followed by 5 min of jumping rope followed by 5 min on the treadmill at various speeds and grades).

Progression

Because too rapid a progression in the exercise dose may cause the patient to lose enthusiasm for the activity, special attention should be given to advances in the exercise prescription. As in adults, no more than a 10% increase in activity duration should occur after any 2 wk period during the exercise program. Frequency can be gradually increased from 3 times per week for the beginner to the preferred 5 times per week over a 3 mo period. Finally, the progression of an individual's program should be based on the desired goals and the individual's particular needs.

Muscular Strength and Endurance Training

People with CF may benefit from resistance training through both a generalized increase in muscle strength and a decrease in residual air trapped in pulmonary dead space. Recent research has attempted to quantify the added benefits of resistance training in individuals with CF and special considerations are discussed next.

Mode

Strength-training programs using both supervised and home-based activities have been explored. Free weights, weight machines, and resistance against body weight can all be used to enhance peripheral muscle strength and endurance, provided that proper direction is given to the individual with CF. Anaerobic activities that mimic the way that children play such as plyometrics, sprinting, cycling, and other modalities that require high-intensity, short-burst duration can also develop muscle strength and endurance. Individuals with CF can perform plyometrics in an age-appropriate manner if proper technique and supervision are given. Other modalities that address multiple muscle groups are ideal for enhancing muscular strength and endurance. Respiratory muscle training should be done in consultation with a clinician trained in respiratory disorders.

Frequency

A frequency of three to five times a week is usually appropriate. Care should be taken to allow for adequate muscle recovery. Alternating major muscle groups when training minimizes muscle injury while maximizing training effects. Subtle signs of overuse injuries (e.g., muscle soreness, joint pain) should prompt a reduction in frequency.

Intensity

Muscle strength and endurance can be optimized through high-repetition, low-intensity resistance training or through other modalities described earlier. The American Academy of Pediatrics Committee on Sports Medicine does not recommend high-intensity resistance training for children because of the potential of musculoskeletal injury, epiphyseal fractures, ruptured intervertebral disks, and growth plate injury before a child reaches full maturation (Tanner stage 5) (3). But strength-training programs that use lower-intensity weights and modalities can be permitted if the planned program is appropriate for the child's stage of maturation (3). Whether adults with CF can safely participate in weight lifting is controversial. Lifting heavy weights can be associated with a Valsalva maneuver that results in increased thoracic pressures. In the susceptible patient with CF (e.g., history of **pneumothorax**, advanced lung disease), this increased thoracic pressure may result in a spontaneous pneumothorax. Consultation with a CF clinician before initiation of a weightlifting program is strongly encouraged. After an individual is cleared for participation, a resistance of 50% to 60% of one-repetition maximum is generally used. Three sets of 12 or more repetitions should produce strength gains.

In addition, inspiratory muscle training can significantly improve lung function in many CF patients. Suggested training intensities that produce significant improvements may be at approximately 80% of maximal inspiratory effort.

Duration

In children and adolescents with CF, the duration of each session depends on the number of muscle groups exercised. Generally, 10 to 30 min of properly performed activities can increase muscular strength and endurance. In adults with CF, 30 min of strength training should be sufficient when combined with inspiratory muscle training in addition to routine CF treatments.

Progression

In individuals with CF, the progression of a strength program should be slow. Repetitions or resistance should be increased only when the muscle has adapted to the current workload. The progression for increased resistance should occur when the individual is able to perform 8 to 12 repetitions without fatigue to the muscle. For non-weightlifting activities, activity should progress by no more than 10% during each 2 wk period.

Flexibility Training

Generalized flexibility exercises should be an adjunct to any exercise program. Adequate range of motion is essential for minimizing risk of skeletal injury and ensuring healthy aging. Increasing popularity of alternative exercise options such as yoga and Pilates has resulted in enhanced flexibility while increasing aerobic and anaerobic fitness. At low intensity these activities may also serve as "centering" or calming exercise options that allow the patient with anxiety or depression to relax or enhance mood. These activities can easily be done on a daily basis.

Mode

Because stretching can be performed with little or no equipment, it is one of the easiest aspects of fitness to address. Stretching may provide a protective mechanism against injury, and it can be a source of tension release and relaxation. Stretching exercises should focus on large-muscle groups and should be included before and after activity as part of an effective warm-up and cool-down.

Frequency

Stretching exercises should be considered a routine component of any exercise program. Stretching can occur before, during, and after an exercise session depending on the activity chosen. A stretching program of 2 or more days per week can yield positive results such as decreased tightening of the hamstrings and quadriceps and more efficient use of respiratory muscles during exercise. Stretching for relaxation or tension relief can be performed daily. Individuals with specific flexibility issues (e.g., posture abnormalities, tight hamstrings) may need a more comprehensive stretching program.

Intensity

Proper stretching technique will help ensure that an appropriate intensity is used for stretching. A proper stretch often feels like a gentle pull in the muscle. Stretching should not be forced. Proper breathing including exhaling before the stretch and inhaling afterward will help minimize injury. Finally, slowly releasing a stretch back into a neutral position will allow the muscle to recover.

Duration

The American College of Sports Medicine guidelines suggest that a stretch should last between 10 and 30 s (94). The use of progressive stretching that includes a 10 to 30 s stretch followed by an additional 10 to 30 s has been proposed as well.

Progression

The flexibility of a person will gradually improve as the muscle adapts to an increased stretch. Slow progression of a stretching program should occur over a 5 wk period. The length at which a stretch is held can be gradually progressed from an initial 10 to 30 s to 40 to 50 s by the end of 5 wk.

CONCLUSION

People with CF appear to benefit from exercise-training programs. Both static and dynamic pulmonary function either improve or have a slower rate of deterioration following training programs. Improved muscle strength and endurance following exercise programs are well documented. Body weight can be successfully maintained or increased during exercise intervention. The psychosocial implications of exercise are considerable, especially in the adult population, in which the rate of depression is greater than 14%. Improvements in quality of life and sense of well-being may outweigh the physiologic benefits of exercise in some cases. Finally, the potential effect of exercise on improving patient prognosis makes exercise an attractive therapeutic modality.

Case Study

Medical History

Mr. MD is a 10 yr old Caucasian male who was diagnosed with cystic fibrosis at 6 mo of age secondary to recurrent respiratory infections and failure to thrive. Mr. MD has done relatively well with intermittent respiratory infections that require antibiotic and hospital therapy. Because of his inability to consume adequate calories, a gastrostomy tube was placed to allow supplemental nocturnal nutrition. Mr. MD is comfortable with his gastrostomy tube and is not currently self-conscious about his appearance. His parents, however, are protective of his gastrostomy tube and deny him activities that may cause difficulties to this area.

Diagnosis

An exercise evaluation was performed as part of Mr. MD's medical care. An activity questionnaire revealed that Mr. MD enjoys most sports but believes that he is having increasing difficulty keeping up with other children, especially in prolonged aerobic activities. He owns a bicycle and a skateboard, and states that he uses them mostly in the summer when the weather is appropriate. His parents are active and set a good example for Mr. MD and his older sibling, who does not have CF. Before exercise testing, the following pulmonary function tests were obtained.

- Pulmonary function tests: FEV_1 = 65% of predicted; FEF_{25-75} = 48% of predicted
- Resting pulse oximetry = 96% on room air
- Residual volume = 195% of predicted
- Diffusing capacity of the lungs = 86% of predicted

Exercise Test Results

Maximal graded ergometry test (Godfrey protocol) results:

- Physical work capacity = 65 W (82% of predicted)
- $\dot{V}O_2$peak = 35.6 ml · kg^{-1} · min^{-1} (74% of predicted)
- Peak end tidal CO_2 = 38 mmHg
- Lowest exercise oxygen saturation = 92%
- Ratio of minute ventilation to the maximal voluntary ventilation = 96%

- Resting heart rate = 84 beats $\cdot$ min^{-1}
- Peak heart rate = 200 beats $\cdot$ min^{-1}
- Body composition: weight = 24.1 kg (5th percentile for age); height = 129 cm (10th percentile for age)
- Body mass index = 14.48 kg $\cdot$ m^{-2} (5th percentile for age)
- Percent body fat (bioelectrical impedance analysis) = 12%
- Flexibility: good posture; no muscle issues

Exercise Prescription

Overall, Mr. MD has moderate obstructive pulmonary disease with mild ventilatory limitations to exercise. Although oxygen saturations decrease with exercise, he can safely participate in aerobic activities without adverse effect. Overall, nutrition is only fair with lower than expected lean body mass. From a psychological perspective, Mr. MD appears to be a well-adjusted young man who is starting to recognize some of the limitations of having CF. His exercise tolerance is near normal. The biggest challenge will be to maintain or enhance his fitness as he ages. To that means, the following program was developed:

- Mr. MD was encouraged to select individualized activities based on enjoyment (e.g., bike riding, skateboarding, and team sports according to his interest), daily activities were encouraged, and intensity was set at self-limiting exertion (rating of perceived exertion: 6–7 on a 0–10 scale) because heart rate prescription can be difficult to control in young children.
- Incorporating a brief warm-up and cool-down into each exercise session was discussed.
- Parents were asked to help in maintaining motivation and were given strategies to enhance compliance with daily activity. In addition, parents were counseled regarding gastrostomy tubes and activity.
- Mr. MD will enroll in at least one winter sports activity to account for decreased activity associated with winter weather.
- The clinician will discuss the use of an appropriate reward system.
- The exercise clinician suggested formal reevaluation yearly to monitor progress, establish new goals, and offer motivation.

Discussion Questions

1. How does Mr. MD's gastrostomy tube pose a challenge in designing an exercise prescription?

2. What is the role of the exercise clinician in optimizing compliance with the exercise program?

3. Discuss the potential challenges in prescribing exercise to children in general.

4. What role do the parents play in the success or failure of an exercise program in a child? How is this role complicated by a child's having a chronic disease?

5. What anticipatory counseling should be offered before initiation of an exercise program?

6. What steps should the exercise clinician take should Mr. MD come back for his yearly evaluation and demonstrate a significant decline in exercise capacity?

Oncology and the Immune System

A chief compliant among patients with a chronic disease is loss of exercise tolerance, or fatigue. In patients with cancer and those who are HIV positive, this common symptom may be related to the disease itself or it may be brought on (or worsened) by the medications or other therapies used to treat the disorder. Part VI highlights the important role that regular exercise can have in helping to attenuate or reverse a patient's loss of exercise tolerance, regardless of the cause. And although an ever-increasing body of research describes the benefits of exercise in patients being treated for cancer or those who are HIV positive, immense opportunities await the current student who is interested becoming a clinical research scientist who can help address the many exercise-related treatment questions that surround disease-related fatigue and clinical outcomes.

Chapter 24 focuses on cancer. In the late 1960s and early 1970s the beneficial role of exercise testing and training in the diagnosis and treatment of patients with heart disease was becoming formalized. Today we stand at a similar threshold regarding the beneficial role of exercise in patients with cancer. Preliminary research evidence suggests that exercise helps reverse exercise intolerance and improves mood and quality of life. Still, many questions remain regarding safety, dose, rate of progression, clinical outcomes, and the optimal timing to interject exercise into a patient's treatment plan. This chapter summarizes what we know today about exercise in the prevention and treatment of cancer and causes us to think ahead, like the exercise pioneers did for cardiology three to four decades ago, about the great promise that awaits the use of exercise in yet another important health disorder.

Chapter 25 discusses human immunodeficiency virus. Over the last decade the medicines used to treat patients who are HIV positive have done much to attenuate viral load, maintain CD4+ count, and lessen disease-related morbidity and mortality. Unfortunately, however, like so many other health disorders that require complex medical therapies, the prescribed treatments often lead to other ailments that are themselves troublesome. Such is the case in patients who are HIV positive. Loss of skeletal muscle size, strength, and endurance; increased risk for cardiovascular disease; and fat redistribution are but a few of the common side effects that develop out of the treatment plans used in these patients. Fortunately, as pointed out in this chapter, many favorable exercise training–induced benefits can occur, even in patients who are receiving antiretroviral agents. All this suggests that the clinical exercise physiologist who works with patients who are HIV positive must understand not only principles of exercise testing and training but also the pathophysiology, evaluation, and therapies unique to the disorder.

Cancer

John R. Schairer, DO

Steven J. Keteyian, PhD, FACSM

The words *cancer* and *malignancy* are commonly used for the medical term **neoplasm**. Neoplasm is an abnormal growth of tissue that grows by cellular proliferation more rapidly than normal, shows partial or complete lack of structural organization, lacks functional coordination with the normal tissue, and usually forms a distinct mass of tissue that may be either benign or malignant (77). Malignant neoplasms are generally fast growing, have the ability to invade host tissue, are associated with large areas of necrosis because the tumor outgrows its blood supply, and can **metastasize** to other parts of the body. They are eventually fatal if untreated. Benign neoplasms, on the other hand, are generally slower growing, have well organized and well differentiated cells, and usually are not fatal. Cancer is a unique disease because it can originate in any organ system, can spread to other organ systems, and has multiple etiologies. Cancer affects all nationalities, races, and ages, as well as both men and women. The treatment varies with each cancer type and location and includes chemotherapy, radiation therapy, biotherapy, and surgery, individually or in combination.

The discussion about the role of exercise for patients with cancer is multifaceted. Physical inactivity is often cited as a risk factor for developing cancer. Exercise has been hypothesized to enhance immunity and alter function within the endocrine system, both of which could be beneficial in the primary and secondary prevention of cancer. Exercise is now increasingly used as an adjunc-

tive treatment for the exercise intolerance and other side effects that often occur as a result of cancer or its treatment. For the purpose of our discussion, we focus primarily on the four most common cancers: lung, breast, prostate, and colon or rectum.

SCOPE

Cancer is the major cause of morbidity and mortality throughout the world and accounts for 23% of all deaths annually in the United States (83). In the United States, cancer ranks as the second-leading cause of death, behind cardiovascular disease. About 1.3 million new cases of cancer are diagnosed in the United States annually. Forty percent of Americans will develop cancer during their lifetime, and 560,000 Americans die annually (83). Unlike heart disease, for which the age-adjusted death rate has declined since the 1950s, the age-adjusted cancer mortality rate increased until 1992, declined slightly until 1995, and then leveled off until now (2,46). With the decrease in deaths due to heart disease and the relatively stable incidence of deaths due to cancer, it has been postulated that cancer will become the leading cause of death in the United States within the next 25 yr.

The annual economic burden to care for the 100 million Americans who will develop cancer in their lifetime is estimated to be more than $205 billion: $74 billion for hospital costs and physician services, $17.5 billion in lost productivity, and $118 billion in lost productivity due to

425

premature death (2,8,46). But the story is not completely bleak. With continued improvement in diagnosis and treatment, cancer patients are living longer. The 5 yr survival rate for all cancers has increased to almost 65%. More than 10 million Americans who have survived cancer are alive today.

Cancer is found throughout the world, with about a threefold difference between countries having the highest frequency of cancer and those with the least. Geographic variation can be as much as 100-fold for specific cancers. For example, the death rate from upper gastrointestinal cancer is extremely high in South Africa, China, Japan, and Iran. This type of cancer is much less frequent in the United States, except in those areas that have a high incidence of alcoholism. The most common cancers in Western countries are lung, large bowel, and breast, whereas in southeast China, nasopharyngeal cancer is the most common malignancy.

Haenszel et al. (30) and Muir et al. (44) demonstrated that migrating populations tend to acquire the cancer incidence profile of their new country of residence, suggesting that genetics is less important in the genesis of cancer than environmental influences. Results from epidemiological studies show that more than two-thirds of cancer deaths might be prevented through lifestyle modification (19,46,31). One-third of cancer deaths are due to cigarette smoking, and another one-third are attributed to alcohol use, specific sexual practices, pollution, and dietary factors.

There are differences between men and women in the type and frequency of cancer and the likelihood of dying of cancer (2). For nearly all cancers, the incidence rates are higher in men than in women, with the exceptions of thyroid, gallbladder, and, of course, breast and uterine cancer. About 50% of men and 33% of women will develop cancer during their lifetime. Among men, the most common cancers in order of prevalence are prostate, lung, and colon or rectum. With women, the most common cancers are breast, colon or rectum, and lung. If death from cancer is considered, the order of the cancers changes. More men die of lung cancer followed by colon or rectum cancer and finally prostate cancer. Among women, lung cancer is now responsible for more deaths than breast cancer and colon or rectum cancer.

Race affects cancer incidence and cancer death rates. African American males have the highest cancer incidence and death rates, followed by white men. The incidence and death rates are similar between African American women and white women. Asian and Pacific Islanders, American Indians, and Hispanics have lower incidences and death rates than does the white population (2).

Age also affects the distribution and frequency of cancer. Among patients under the age of 15, the most common cancers are leukemia, brain, and endocrine. For patients over age 75, the most common cancers are lung, colon, and breast. Cancer incidence increases with age. Between birth and age 39 yr, 1 in 58 men and 1 in 52 women will develop cancer. The incidence of cancer in men and women between ages 60 to 79 years is 1 in 3 and 1 in 4, respectively (2).

PATHOPHYSIOLOGY

The final common pathway in virtually every instance is a cellular genetic mutation that converts a well-behaved cellular citizen of the body into a destructive renegade that is unresponsive to the ordinary checks and balances of the normal community of cells. (5)

The stem cell theory was developed to explain the sequential triggering mechanisms that cause cells to begin to specialize and develop new structures and discrete functions (figure 24.1). The stem cell is pluripotent, which means that it is an uncommitted cell with various developmental options still open. The process by which the **pluripotent stem cell** is able to develop special functions and structures within an organ system is called differentiation. Thus, some stem cells are triggered to differentiate and become hair cells, and some cells become cardiac myocytes. The pluripotent stem cell also has the capacity for self-renewal. But after a stem cell becomes committed to a cell line such as a hair cell or cardiomyocyte, it no longer has pluripotent and self-renewal properties and is destined to develop along its specialized pathway of differentiation. The best example is a pluripotent hematopoietic stem cell, with its capacity to form both red blood cells and white blood cells (neutrophils, basophils, etc.). After it commits to a specific cell line, however, it can no longer differentiate into other cell types or divide to form new cells.

The stem cell model for cancer proposes that tumors arise from carcinogenic-causing events occurring within the normal stem cells of a particular tissue. A cancer-causing insult is believed to produce a defect in the control of normal stem cell function, resulting in abnormalities in self-renewal, differentiation, and proliferation. In other words, the normal quality and quantity control for cell function and growth is lost.

The carcinogenic event for many cancers is unknown, but five broad categories have been identified:

1. Environment
2. Heredity

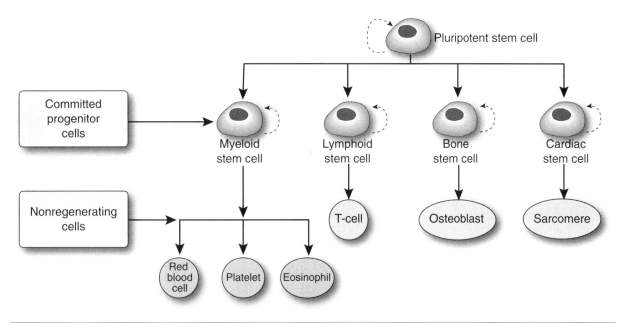

Figure 24.1 Stem cell sequence theory. This schematic presentation of the stem cell theory depicts how the uncommitted pluripotent stem cell develops or differentiates into several committed, nonregenerating cells with specialized functions.

3. Oncogenes

4. Hormones

5. Impaired immune system function

Cancers most likely arise when abnormalities are present in one or more of these categories (e.g., radiation exposure from the environment), causing cellular mutations that go unrecognized because of injury to the immune system as well.

Environmental factors are implicated in 60% to 90% of cancers, partly because of the known association between certain agents and the development of cancer (table 24.1 on p. 428). For example, lung cancer is more prevalent among miners and those who work with asbestos, chromate, or uranium. Exposure to certain solvents is associated with leukemia, and excessive exposure to sunlight is responsible for a higher incidence of skin cancer in farmers. Other factors, including lifestyle choices such as a diet high in fat, obesity, and a sedentary lifestyle, are associated with an increased likelihood of developing cancer. Cigarette smoking, another lifestyle choice, increases the risk for lung cancer because tobacco smoke is a carcinogen. Lifestyle choices and their possible role in causing cancer are discussed later in this chapter in the literature review in the practical application on p. 435. See as well "Lifestyle Habits to Reduce the Risk of Cancer."

Observations made in animal studies provide support for the proposition that hereditary or genetic factors

Lifestyle Habits to Reduce the Risk of Cancer

1. Don't smoke cigarettes.

2. Limit alcohol consumption to less than 1 oz (30 ml) per day.

3. Be physically active.

4. Eat five or more servings of fruit and vegetables, whole grains, and foods low in saturated and trans fat.

5. Practice safe sex.

6. Protect skin from excessive light.

7. Maintain healthy weight.

8. Have regular screenings.

cause cancer. Genes that act at the cellular level to cause uncontrolled proliferation of previously normal cells are called oncogenes. Possibly activated by heredity or the environment (e.g., radiation, chemicals, viruses), oncogenes produce a defect in the control of normal stem cell function. In humans, genetic factors have been implicated in colon (e.g., familial polyposis), breast, and stomach cancers. The presence of breast cancer in a

Table 24.1 Environmental Causes of Cancer

Agent	Cancer
Alcohol	Oropharynx, larynx, esophagus, liver
Arsenic and arsenic compounds	Lung, skin
Asbestos	Lung, gastrointestinal
Benzene	Leukemia, Hodgkin's
Beryllium and beryllium compounds	Lung, leukemia
Cadmium and cadmium compounds	Prostate
Chromium compounds	Lung
Ethylene oxide	Leukemia
Nickel compounds	Nasal, lung, leukemia
Radon	Lung
Strontium	Lung, leukemia
Tobacco	Mouth, pharynx, larynx, esophagus, lung, pancreas, bladder
Ultraviolet rays	Skin cancer (melanoma)
Uranium	Lung, leukemia
Vinyl chloride	Liver

female relative more than doubles a woman's likelihood of developing the disease.

Some hormones are thought to possess carcinogenic potential. For example, the ovary, breast, and uterus are hormonally sensitive; therefore, estrogen may play a role in the cancers of these organs in women. The role that hormones may play in breast cancer is discussed later in the literature review in the practical application on p. 435.

Finally, any discussion about the cause of cancer would be incomplete if it did not address the **immune system** of the body. The immune system is responsible for mediating the interaction between an individual's internal and external environment. The immune system is divided into two major categories: innate and acquired responses (table 24.2). The innate immune system response is nonspecific and immediate, beginning within minutes of an insult. The response occurs without "memory" for the eliciting stimulus. This process is called inflammation. The innate immune system is composed of a cellular component that includes the monocyte and macrophage system, neutrophils, and natural killer (NK) cells, as well as soluable factors such as C-reactive protein, interleukin-1, tumor necrosis factor, interferon, and oncastatin. The innate immune system represents our first line of defense against cancer. These cells are capable of lysing tumor cells without previous exposure.

The adaptive or acquired immune system is characterized by an antigen-specific response to a foreign

Table 24.2 Components of the Immune System

Immune system component	Description
Innate immune system	Nonspecific response
Monocyte	Mature to become macrophages
Macrophage	Nonspecific killing response of tumor cells by phagocytosis and cytolysis,
Neutrophils	a process by which cell-specific molecules are injected through
Natural killer cells	pores in the cell membrane and rupturing the cell membrane
Acquired or adaptive immune system	Antigen-specific response
Cytotoxic T-lymphocytes	Requires tumor antigens in association with class I major histocompatibility antigens

antigen or pathogen and generally takes several days or longer to activate. A key feature of acquired immunity is memory for the antigen, such that subsequent exposure leads to a more rapid and often more vigorous response. The primary cell type for this task is the cytotoxic T-lymphocyte.

The overall function of the immune system is to rid the body of foreign agents such as bacteria, viruses, and malignant cells. The immune system recognizes infectious agents and malignant cells because they contain abnormal antigens in their cell membranes. The immune

system can also inhibit subsequent formation of a tumor by countering factors responsible for its growth.

Cancer therapies such as chemotherapy, radiation therapy, and cancer surgery are known to be immune suppressive. The role that this suppression of the immune system plays on the cure and recurrence rates in the treatment of cancer is unknown. The important role that exercise plays in modulating the immune system has only recently been studied. Practical application 24.1 discusses the beneficial effects of acute and chronic exercise on immune function.

Practical Application 24.1

Acute and Chronic Exercise and Immunity

Several excellent review articles discuss the effect of exercise on immune system function (34,48,69,70,91). The working hypothesis in the field of exercise immunology is the inverted-J hypothesis (see figure below), which suggests that enhanced immune system function occurs with a chronic moderate exercise training program, but immune system function is depressed even below that of sedentary individuals after chronic exhaustive exercise. Clinically moderate to high levels of physical activity seem to be associated with decreased incidence or mortality rates for some cancers (68,76,38,58), and overtraining or intense competition may lead to immunosuppression as evidenced by an increased incidence of upper-respiratory infections in people who train intensely (51,65). To this point, no cancers have been shown to be associated with exhaustive exercise programs. Also, it is not known whether exercise plays its greatest role in preventing cancer by stimulating the immune system or through other mechanisms such as decreasing transit time through the bowel to prevent colon cancer or by decreasing the number of menstrual cycles to prevent breast cancer. Low to moderate exercise is defined as <60% $\dot{V}O_2$max and <60 min per bout, whereas exhaustive sessions are defined as >75% $\dot{V}O_2$max and >90 min in duration (10,47,49). The following is a discussion of each cell type, its response to an acute bout of exercise, and its response to an exercise-training program in the healthy population. At the end of this section we will include what is known about the influence of exercise on the immune system in the patient with cancer.

With an acute bout of exercise, macrophages, NK cells, and neutrophils appear to be the most responsive to the effects of exercise, increasing in both number and function (29,48,64,72) (see table 24.3 on

p. 430). Exercise, regardless of the intensity or duration, increases the number of monocytes in peripheral blood transiently (92). Both acute and moderately exhaustive exercise have been shown to enhance a variety of macrophage functions including chemotaxis (92,43), adherence (55,56), phagocytic (16,21), and cytotoxic (15,93) activity. With chronic training, macrophage activity is diminished when compared with an acute exercise bout (55,93,92,57), but the level of activity is still increased over sedentary control animals.

NK cells are extremely responsive to exercise, such that in recovery, cell activity and circulating numbers are transiently increased

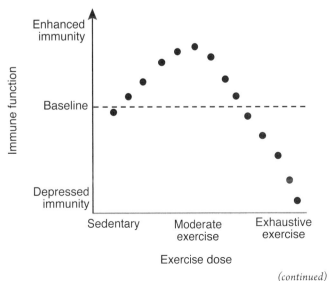

(continued)

Table 24.3 Effects of Exercise on the Immune System

Component	Effect of acute exercise	Effect of chronic exercise
Innate immune system		
NK cells	Immediate increase in cell concentration and cytoloytic activity Intense activity of prolonged duration (≥1 hr) decreases concentration and cytolytic activity 2 to 4 hr later	Increase in resting NK cell count and activity, in one study only
Macrophages	Immediate increase in monocyte and macrophage count Adherence unchanged Increased phagocytosis with moderate activity	Response is unclear. Resting monocyte count is unchanged May cause adaptations that alter exercise response
Neutrophils	Large and sustained increased concentration and activity with moderate exercise Most PMN functions decrease significantly after extreme exercise	Function is suppressed during periods of chronic high-intensity training
Acquired immune system		
T-lymphocytes	Moderate activity enhances cell concentration, with depressed levels 30 min postexercise Vigorous activity causes a transient decrease in proliferation	Unclear response

Note. NK = natural killer; PMN = polymorphonuclear leukocytes.

(29,50,51). Following high-intensity acute and chronic exercise, NK cell cytotoxic activity increases 40% to 100%. This effect has been observed in marathon runners, sedentary controls who have undergone exercise training (61), and elite cyclists during the summer (intense training period) versus the winter months (low training period) (81). Several studies using chronic moderate endurance activity over 8 to 15 wk reported no significant elevations in NK cell activity (4,52,53), suggesting that endurance activity must take place over a long period (i.e., years) before a response is observed.

Regardless of intensity, an acute exercise bout induces a profound leukocytosis that includes an increase in neutrophils. Although the number of neutrophils increases with moderate exercise during periods of high-intensity training, neutrophil function is reported to be suppressed. Pyne et al. (64,65) reported that elite swimmers who undertake intensive training have significantly lower neutrophil oxidative activity at rest than do controls and that cell function is further suppressed during strenuous exercise. Suppression of neutrophil function during periods of heavy training has been cited to partially explain the increased risk for upper-respiratory tract infections among some competitive athletes. These symptoms have never been linked to exercise-induced changes in immune function. High-intensity exercise is associated with increased numbers of cytotoxic T-lymphocytes on the order of 50% to 100% immediately after exercise. As with NK cells, this increase is transient and resolves in 30 min (47).

During recovery from high-intensity aerobic-type exercise, subjects experience a sustained **neutrophilia** and **lymphocytopenia** (23,74) and possible compromised host protection. NK cell activity and T-cell function are reduced as well. These changes are believed to be attributable to excessive elevations in cortisol occurring after prolonged exercise (45,49,54). Moderate exercise results in a lower cortisol response and is associated with a more favorable immune response. In summary, moderate exercise seems to enhance the body's immune system. But the significance of this finding in preventing and treating cancer is unknown.

The effects of chronic exercise are not well-documented. A cross-sectional study demonstrated elevation of NK cell activity in chronically trained athletes, while other members of the immune system remain unchanged in concentration (53,81). Several longitudinal studies using chronic, moderate-intensity endurance activity over 8 to 15 wk reported no significant responses in the immune system (4,52,53), suggesting that endurance activity must take place over a long period (i.e., years) before a response is observed.

The effect of exercise on the immune system of patients with cancer was reviewed by Fairly et al. (20). They identified six articles that examined the effect of exercise training on the immune system function in cancer survivors during and after cancer treatment. The studies reported favorable responses. But all six studies included small numbers of patients, used different exercise interventions, and included different patient populations. Four of the six studies demonstrated statistically significant improvements in the immune system function as a result of exercise, as demonstrated by improvements in NK cell cytolytic activity, monocyte function, proportion of circulating granulocytes, and duration of nutropenia. Two studies found no statistically significant improvement in immune function as a result of exercise. To date there is no evidence that exercise-induced changes to the immune system translate into reduced risk for developing cancer or preventing recurrence.

MEDICAL AND CLINICAL CONSIDERATIONS

There are four types of cancer: carcinoma, sarcoma, leukemia, and lymphoma. Carcinomas are cancers of epithelial tissues and include cancers of the skin, digestive tract, genitourinary tract, pulmonary system, and so on. Cancer of the breast, colon and rectal, lung, and prostate are examples of carcinoma. Carcinomas represent 90% of all cancers. Sarcomas are tumors of the connective tissue and bone and include cancer of the bone, muscles, and cartilage. Leukemias are cancers of the white blood cells. Lymphomas are cancers of the lymphatic system. Because of the diversity of organ system involvement, no single part or segment of the history and physical examination focuses only on cancer. As a result, the physician, nurse practitioner, or physician extender must perform a complete and accurate history and physical examination. Using this information, the clinical exercise physiologist needs to understand the historical and physical aspects of the specific cancer that the patient has in order to help assess progress and identify any exercise-related concerns.

Signs and Symptoms

Early on, the symptoms of cancer are usually nonspecific, such as weight loss, fatigue, nausea, and malaise. Only an astute clinician can make the diagnosis of cancer at this time, yet early detection is key to maximizing the patient's chance for survival.

Later on, the patient will develop symptoms specific for the involved organ, such as shortness of breath in lung cancer or jaundice attributable to biliary obstruction in pancreatic carcinoma. By this time, however, prognosis is much poorer. Social history is also important, revealing occupational exposure to carcinogens or habits such as smoking or ethanol ingestion. The family history may reveal familial predisposition to a cancer, one that requires closer surveillance in the future. The review of systems may reveal symptoms indicating that the cancer has already metastasized.

As mentioned earlier, detecting cancer during a physical examination is extremely difficult. Any such examination is usually orientated to examining for enlargement of an organ such as an enlarged lymph node in the case of lymphoma or a testicle in testicular cancer. Because most organs such as lung, pancreas, and kidney are deep within the body, the yield is low. Although examining for masses is important, a mass is often a later sign of cancer that often occurs after the cancer has metastasized.

Preexercise Evaluation

Besides the usual contraindications for exercise and exercise testing in patients free of cancer, additional considerations are unique to patients with cancer. The history should include both noncancer and cancer considerations. Noncancer considerations include other comorbid conditions such as age, diabetes, hypertension, fitness level, and orthopedic problems. Cancer-related issues of importance include type and stage of cancer, type of treatment, side effects of therapy, psychological status, and timing of tests and therapy. Other considerations include nutritional status, metabolic considerations such as electrolyte abnormalities, and fluid status. The physical examination should attempt to identify acute signs or symptoms that are cancer-type specific and would prevent participation in exercise. For example, women who have recently undergone axillary lymph node biopsy or breast surgery may not be able to participate in upper-body exercises. Another physical finding might include bone tenderness indicative of metastatic lesions to bones of the pelvis, back, or legs.

Other findings can include gait instability attributable to chemotherapy or central nervous system involvement, delayed wound healing, immune suppression, and bleeding that may occur as a result of bone marrow suppression or surgery. Finally, some complications may require modification of the exercise plan, such as nausea, vomiting, fatigue, and weakness. All these factors must be considered before an exercise test and when a patient enters an exercise program, and they must be continually reevaluated throughout the program. See "Contraindications to Exercise for Patients With Cancer."

Diagnostic Testing

Because the prevention of cancer is not always possible, the earliest detection of the disease is the next best strategy to reduce cancer mortality rates. To help accomplish this, the American Cancer Society recommends a series of screening procedures and evaluations, which are shown in table 24.4.

When cancer is suspected, the first diagnostic principle is that adequate tissue must be obtained to establish the diagnosis. Because the therapy used for each type and subtype of cancer is often unique, every effort must be made to obtain appropriate tissue samples, even if treatment is delayed for a short time. The process of obtaining a sample of tissue is called a **biopsy**.

The second diagnostic principle is to determine the extent of spread of the cancer, also known as **staging**. In leukemia, staging can be accomplished through routine history and physical examination, laboratory tests, chest

Contraindications to Exercise for Patients With Cancer

- Hemoglobin <10.0 g · dl^{-1}
- White blood cells <3,000 · μL^{-1}
- Neutrophil count <0.5 × 10^9 · ml^{-1}
- Platelet count <50 × 10^9 · ml^{-1}
- Fever >38 °C (100.4 °F)
- Unsteady gait (ataxia)
- Cachexia or loss of >35% of premorbid weight
- Limiting dyspnea with exertion
- Bone pain
- Severe nausea

X-ray, and bone marrow biopsy. With solid tumors, computed tomography (CT) and magnetic resonance imaging (MRI) in conjunction with a biopsy are often needed to determine the size of the tumor and the extent of its spread. The degree to which the cancer has spread is reflected in its stage, which guides the type of treatment most appropriate for the patient. An example of a simplified staging system is shown in table 24.5. Each cancer has a staging system unique to itself, one that takes into consideration pathogenic features, the modes of spread, and the curability of the disease.

Exercise Testing

Patients with cancer have the same incidence of coronary artery disease as the general population. Therefore, before beginning an exercise program, some patients with cancer may need to be screened for coronary artery disease as well as cancer-related issues. The American College of Sports Medicine (3) recommends that men over the age of 45, women over the age of 55, or individuals with two or more risk factors undergo exercise testing before beginning a vigorous exercise program. Practical application 24.2 on p. 434 describes client–clinician interaction for patients with cancer.

The ideal exercise test should last 8 to 12 min. Patients with cancer are frequently and markedly deconditioned because of both the disease process and the treatments. The average maximal work capacity in patients with malignancy is reduced to 3 to 6 metabolic equivalents or a peak $\dot{V}O_2$ of 11 to 21 ml · kg^{-1} · min^{-1} (71). Exercise is often limited by general or leg fatigue. Peak heart rate may be reduced as well. Protocols beginning at lower work rates and progressing more slowly should be considered (e.g., modified Bruce, Naughton-Balke, modified Balke). See table 24.6 on p. 434.

Treatment

There are four treatment options for cancer. The selection of which treatment option to use depends on the type, location, and stage of the cancer. Treatment options include the following:

- Surgery
- Radiation therapy
- Chemotherapy
- Biotherapy

Surgery is the oldest and most definitive treatment for cancer. The two types of surgery are curative and palliative. **Curative surgery** is the primary treatment

Table 24.4 American Cancer Society Recommendations for Early Detection

Test	Sex	Age	Frequency
Flexible sigmoidoscopy	M & F	50 and over	Every 3–5 yr
Fecal occult blood	M & F	50 and over	Every year
Digital rectum exam	M & F	40 and over	Every year
Prostate exam	M	50 and over	Every year
Prostatic specific antigen	M	50 and over	Every year
Papanicolaou (Pap) test	F	18 and over	Every year (after three normal exams, may be performed less often)
Pelvic exam	F	18–40	Every 1–3 yr
		Over 40	Every year
Endometrial tissue	F	Menopause	
Breast self-exam	F	20-40	Every month
Breast clinical exam	F	20 and over	Every 3 yr
		Over 40	Every year
Mammography	F	40–49	Every 1–2 yr
		Over 49	Every year
Chest X-ray			Not recommended
Sputum cytology			Not recommended

Note. The American Cancer Society recommends a series of a cancer-screening procedures. Although not all experts agree with all aspects of these recommendations, they serve as a useful guide to screen patients for cancer. M = males; F = females. High-risk males (African American males and males with a family history) should begin screening at age 45.

Table 24.5 Tissue-Nodal-Metastasis (TNM) Classification System for Breast Cancer

Tumor size (T)	Nodal involvement (N)	Metastastis (M)
Ts = insitu	NO = No nodal metastasis	MO = No distant metastasis
T1 = <2cm	N1 = Moveable axillary nodes	M1 = Distant metastasis
T2 = 2–5cm	N2 = Fixed axillary nodes	
T3 = >5cm	N3 = Internal mammary nodes	

These TNM categories are combined to give the stage (e.g., stage 1 = T1 NO MO).

for about one-third of cancers that are small and not yet metastasized. If the tumor is removed along with a small amount of surrounding normal tissue, the chance for a cure is good.

In **palliative surgery**, a large tumor mass is removed to make the patient more comfortable, to relieve obstruction of vital organs, and to reduce the tumor burden. For example, a colon cancer may be removed to prevent bowel obstruction or an ovarian cancer may be removed to prevent obstruction of a ureter. Decreasing tumor burden may also make the tumor more susceptible to radiation or chemotherapy. Palliative surgery does not usually change overall chances for survival.

Radiation therapy is thought to stop the growth of malignant cells by damaging the DNA within the cell. Radiation can be applied either from implanted internal sources (brachytherapy) or from external machines. Most radiation therapy is applied in small fractions, usually between 180 and 250 rads · day^{-1}. Doses above 4,000 rads over 4 wk increase the likelihood of developing radiation pericarditis. Radiation therapy can be used alone or in conjunction with surgery or chemotherapy. Radiation is good for a localized tumor that has not metastasized or tumors that are difficult to reach with surgery (e.g., within the brain).

Cells that grow the fastest are best treated by chemical agents that interfere with cell replication. Known as **chemotherapy**, these agents frequently result in a cure. Some cancer cells, however, may become resistant to chemotherapeutic agents. Using several drugs at the

Client–Clinician Interaction

Part of this individualized approach to using exercise in patients with cancer requires that you, as the clinical exercise professional, accomplish two things. First, become familiar with the type of therapy or therapies that your patient is undergoing. You may have to do some extra reading when you encounter a therapy that you are unfamiliar with. Also, talk with oncologists and surgeons about the agents or interventions used to treat cancer. Such discussions should include the clinical presentation of expected side effects and the natural history of the disease. Ultimately, you will improve your ability to interact with patients in a learned fashion.

Second, develop your ability to ask questions about how well your patient is tolerating therapy and his or her disease. Such ability usually comes from working with many patients, thus improving your skills in evaluating the interaction between exercise, cancer, and the cancer treatment. Because part of any exercise program includes regular follow-up, either in an exercise program or by telephone to the patient's home, you are in a unique position to establish patient confidence and assess clinical status. Obviously, you can take any concerns to the patient's doctors. You play an important role in long-term patient surveillance.

Clinical features that you should pay special attention to include sudden loss of exercise tolerance over several days to a week or so, increased shortness of breath with exertion, an inordinate increase in anxiety or depression as manifested by difficulty falling to sleep or lack of interest in social contact, and sudden changes in nutritional status. These are issues that you can work in to the routine follow-up phone calls or clinic visits that you might be using to evaluate exercise compliance and progress. At times the information that you gather may dictate that exercise be withheld for a time while a specific treatment protocol runs its course.

Because more evidence is needed to describe how exercise improves function and whether exercise lowers risk for cancer recurrence or future hospitalizations, the best approach may be to keep a patient's attention focused on how exercise can help him or her lessen fatigue, regain control, and improve quality of life. This means using exercise to help patients lead a more comfortable life, one that allows them to perform the activities that they enjoy without symptoms or limitations.

Table 24.6 Summary of Exercise Testing Relative to Patients With Cancer

Test type	Mode	Protocol specifics	Clinical measures	Clinical implications	Special consideration
Cardiovascular	Treadmill 6 min walk	Exercise to ≥85% of PMHR Choose a protocol that begins at lower work rates and progresses more slowly	METs Distance walked	Provides basis for determining starting point for exercise training	Cancer treatment may result in cardiomyopathy, pulmonary fibrosis, or neuropathy
Muscle strength and endurance	Machine weights	One-repetition maximum	kg	Same as above	Use machines rather than free weights to reduce risk of injury
Flexibility	Goniometry	Active stretching	degrees	Same as above	Avoid pain Assess upper extremity range of motion post mastectomy

same time (i.e., combination chemotherapy) is one way to minimize the resistance.

Cancer cells possess distinct surface protein antigens that are targets for antibody-directed or cell-mediated immunity. **Biotherapy** stimulates the immune response of the body to these protein antigens. Biotherapy also involves the production of antibodies outside the body, which are then administered to the cancer patient in an attempt to destroy the tumor. Biotherapy includes bone marrow transplantation and the use of cytokines such as a-interferon and interleukin-5.

EXERCISE PRESCRIPTION

The exercise program for cancer patients does not require electrocardiographic monitoring, although some supervision and instruction about heart rate monitoring and proper exercise technique are desirable. Initially, the exercise prescription should be reviewed with the patient, and the patient should be taught both how to take his or her pulse and the symptoms of an adverse response to exercise. The patient should also be instructed in the types of exercise best suited to his or her cancer. The actual exercise program can be performed at home, in a healthcare facility, or outdoors. If resistance training is being incorporated, the patient may benefit from temporarily enrolling in a medical fitness center. See practical application 24.3 for a review of the literature that addresses exercise training and primary and secondary prevention in apparently healthy people and patients with cancer. Practical application 24.4 on p. 438 summarizes the exercise prescription recommendations for patients with cancer.

Because cancer is a constellation of diseases, knowing which cancer the patient has and to what extent it has spread is important. The exercise professional working with the patient with cancer needs to assess the risk–benefit relationship case by case as well as the fluctuating clinical status of the individual patient within the program. The patient may experience setbacks such as side effects of chemotherapy and radiation therapy, additional surgical procedures, and progression of the cancer resulting in metastatic lesions with resultant pain and concern for pathologic fractures. If setbacks occur, exercise goals need to be modified accordingly. At some point it may even be necessary to suspend exercise if the risk–benefit relationship no longer justifies exercise as a component of the treatment. When possible, the patient's

---

Practical Application 24.3

Literature Review

The role of exercise in the treatment of disease is divided into primary prevention, secondary prevention, and adjunctive therapy. Primary prevention looks at whether exercise can prevent the disease. The role that exercise may play in primary prevention is promising. Secondary prevention refers to prevention of additional disease after the original disease has been treated. At this time, no evidence suggests that exercise helps prevent recurrence of cancer. Adjunctive therapy refers to treatment in addition to, or potentiating, usual treatment. Here again, exercise clearly plays a role.

Clinically, people who exercise generally have a healthier lifestyle. They tend to avoid carcinogens such as a diet high in fat, alcohol, and tobacco, and they tend to maintain their weight. The latter is important, because obesity is linked to increased risk for breast, colon, rectum, prostate, endometrial, and kidney cancers.

Data show that higher levels of physical activity lower overall cancer mortality (59,60,79). The American Cancer Society concluded in a 2006 report (2) that "there is the potential to reduce the overall incidence, morbidity, and mortality from certain kinds of cancer if successful physical activity and dietary interventions are implemented." In fact, for the nonsmoker, dietary and physical activity interventions are the most important modifiable determinants of cancer risk.

Determining a relationship between physical activity and cancer risk is hindered because it is not known in what time frame exercise needs to occur to decrease the cancer risk. Exercise may help block initiators of cancer, in which case exercise done consistently at a relatively young age may be most beneficial. This is the rationale behind studies that investigate whether participation in high school and college athletics reduces the risk of cancer during adulthood. Alternatively, exercise may counter the promoters of cancer cell replication, so that exercise during a later phase of the neoplastic process may be preferred to decrease the development of clinically significant disease (70,76). Therefore, the point at which exercise occurs during a person's lifespan may be an important factor relative to its effect on cancer development.

(continued)

One example is breast cancer. Large doses of estrogens, when given continuously, can cause breast carcinoma in susceptible strains of mice. Estrogen plays a role during four phases of a woman's life: menarche, first pregnancy, menopause, and postmenopause. Currently, it is not certain at which point or points in a woman's life cycle exercise exerts its greatest anticancer effect.

Several studies indicate that breast cancer risk is directly related to the cumulative number of ovulatory menstrual cycles (32,39,63,82,85). Intense exercise delays menarche, which can be thought of as favorable, because the risk of breast cancer increases twofold in women who experience menarche before age 12 versus at age 13 or older. Epidemiological data indicate that for every year menarche is delayed, breast cancer risk decreases by 5% to 15% percent (35,26). Moderate levels of activity have been shown to increase the risk of anovulatory cycles threefold. Delayed menarche and anovulatory cycles decrease the woman's exposure to estrogen and progesterone (9,28,85). The first full pregnancy induces differentiation of the breast and may change the sensitivity of the breast to both endogenous and exogenous risk factors (66,11). And women who experience natural or artificially induced menopause before the age of 45 have a markedly reduced risk of breast cancer compared with women whose menopause occurs after the age of 55 (78). Postmenopausal women generally have an increased body mass index, which is a significant risk factor for breast cancer (33,60). In postmenopausal women, fat tissue is the primary source of estrogen. Therefore, obesity increases the woman's exposure to estrogen.

The role of physical activity, either leisure time activity or occupational physical activity, in reducing overall cancer risk and site-specific cancer risk is gradually being defined (36,80,27). The relationship between colorectal cancer and physical activity has been studied the most. Forty-eight studies with 40,000 patients (80) demonstrated a 10% to 70% reduced risk for colon cancer in physically active individuals. Physical activity, however, offers no protective effect for rectal cancer. Decreased bowel transit time caused by physical activity may explain the observation that colon cancer frequency is reduced in physically active individuals while no change occurs in the frequency of rectal cancer.

Forty-one studies including 108,321 women (80) evaluated breast cancer and possible risk reduction with physical activity. Twenty-six studies demonstrated that both occupational and leisure time activity reduce breast cancer risk by about 30%. The results for breast cancer are less conclusive than those for colon cancer. Of the 12 studies in the literature regarding physical activity and the risk for developing endometrial cancer, 8 demonstrated a 20% to 80% reduction in risk. The data in the studies that evaluate the possible protective effect of exercise in preventing ovarian, prostate, and testicular cancer is quite favorable but inconsistent at this time. The association of physical activity with lung cancer is reported in 11 studies. Six of the studies support a protective effect of physical activity on lung cancer.

How much physical activity is needed to reduce the risk of cancer? A definitive answer is not currently available. Data from Blair et al. (6) indicate that the reduction in cancer risk occurred primarily between the very low fitness group and the moderately fit group, with no further decrease in risk among the more fit subjects. Paffenbarger et al. (59) reported a decrease in all-cause mortality for alumni who expend 1,500 or more kcal/week, but these authors did not break out cancer deaths in this group. One thousand kcal/week, 4 h of moderate activity per week, or 3 h of vigorous activity per week have a protective effect for colon cancer and breast cancer (73,84). Lee (40) also reported that exercising at moderate intensity, greater than 4 to 5 METs for 4 h per week, decreased the incidence of lung cancer.

The influence of exercise in patients with cancer can be divided into three areas: fatigue, psychosocial, and therapy-related side effects. Improving one of these areas can have a positive domino effect on other aspects of quality of life (22).

Fatigue

Having the energy to carry out activities of daily living as well as occupational, leisure, and social activities is inherent to many quality-of-life measures (42,62,75). Up to 70% of cancer patients recovering from surgery or receiving chemotherapy or radiotherapy report loss of energy. This important symptom persists in up to one-third of cancer survivors for years after cessation of treatment (24).

Among the many potential causes of fatigue in patients with cancer are preexisting conditions, the direct effect of the cancer, symptoms related to cancer, effects from the treatment used for the cancer, and the demands of dealing with cancer. One frequently underestimated factor contributing to fatigue is loss of physical fitness as a result of bed rest. When patients experience fatigue they are often told to "take it easy" and "get plenty of rest," further limiting their level of daily activities and perpetuating deconditioning and exercise intolerance. The exercise capacity of the cancer patient is markedly reduced, often with a $\dot{V}O_2$ peak of 11 to 25 ml $\cdot$ kg^{-1} $\cdot$ min^{-1}, or 3 to 6 METs. One-third or more of the decline in functional capacity is attributable to the consequences of physical inactivity (71). The rapid decline is seen in multiple systems but is most dramatic in the cardiopulmonary system. Additional consequences of physical inactivity include limited joint mobility, osteoporosis, impaired balance, and lowered pain threshold. Impaired physical fitness is an important contributing factor that lessens quality of life in patients with cancer.

Psychosocial

Exercise also has an immediate mood-elevating effect (68) and can help cancer victims psychologically, breaking the vicious cycle of physical inactivity and tissue loss. Because physical mobility is associated with health and well-being, rehabilitation efforts that incorporate exercise are associated with increased energy levels and decreased fatigue, which in turn improve comfort, concentration, appetite, and sleep habits.

Dependence on others for assistance with routine activities of daily living often intensifies the psychological responses to disease and treatment in patients with cancer. By fostering functional independence, exercise improves self-concept, self-esteem, confidence, self-image, sense of personal worth, control, and self-acceptance while at the same time providing a means for attaining a sense of control over one's life (1,41,90). Exercise also improves quality of life by decreasing feelings of depression, tension, anxiety, anger, hostility, helplessness, and pessimism (22). Winningham (87) reported that these changes in psychological state give patients hope by improving their feelings of well-being and their ability to cope and adapt to stress. With 10 million cancer survivors alive today, emphasis on quality-of-life issues is paramount.

Cancer Therapy Side Effects

Thirty-six studies (12) have examined the relationship between exercise and quality of life following cancer diagnosis. These studies have consistently demonstrated beneficial effects on a wide range of quality-of-life outcomes, regardless of exercise prescription, cancer rate, or cancer treatment.

Dimeo et al. (17,18) reported the effect of an exercise-training program on fatigue in 80 patients undergoing high-dose chemotherapy. The patients were randomly assigned to one of two groups. One group performed supine bicycling for 30 min per day during each day of their hospitalization for chemotherapy. A second group received chemotherapy only. After 7 wk, physical performance (+34%) and hemoglobin concentration were significantly greater for the training group. Heart rate and lactate concentration measured at a submaximal workload were significantly lower in the training group. The training group demonstrated reduced duration of neutropenia and thrombocytopenia, severity of diarrhea, severity of pain, and duration of hospitalization. Conversely, at the time of hospital discharge, fatigue and somatic complaints had increased significantly in the nonexercising group, but not in the training group. Finally, the training group had a significant positive change in measures of obsessive-compulsive traits, fear, interpersonal sensitivity, and phobic anxiety. These changes were not seen in the nonexercising group.

Knols et al. reviewed 27 randomized clinical trials and 7 controlled clinical trials that reported the benefits of exercise training in 1,844 cancer patients during and after treatment (37). The findings suggest that some cancer patients may benefit from exercise training both during and after cancer treatment. The results, however, may vary with the type and stage of cancer. Exercise may be effective in reducing symptoms such as fatigue, nausea, and difficulty sleeping and improving measures of physical function such as muscle strength, walking distance, and total energy expenditure. Exercise training may positively influence psychosocial variables such as psychological well-being, mood status, and quality of life. Because many of the studies are of moderate methodologic quality and small in sample size, positive results are reported in some but not all studies. Although NK cell activity and other measures of immune function increase in response to exercise training, there is no evidence at this time that exercise training reduces recurrence or affects longevity of cancer patients. Further research in the form of large randomized clinical trials is needed to define the benefits in patients who already have cancer.

Summary of Exercise Prescription for Patients with Cancer

Training method	Mode	Intensity	Frequency	Duration	Progression	Goals	Special considerations and comments
Aerobic	Walking or stationary bike. Perform exercises that use large-muscle groups	50 to 70% of heart rate reserve. Should be able to talk during exercise	Three to five times per week. Training program should also be flexible to accommodate any scheduled cancer treatments	30 to 40 min	30 sec to 2 min per day	Increase peak $\dot{V}O_2$ total work endurance	Patients that are unstable on their feet may benefit from use of a recumbent or stationary bike. Begin intensity at 50% of heart rate reserve and progress as long as perceived exertion is 11 to 13. If needed, divide exercise into two or three sessions per day and begin at 5 to 10 min
Resistance	Machines. Circuit training	50 to 60% of one-repetition maximum performed in an aerobic or circuit fashion	One or two times per week	One set of 12 to 15 reps		Increase muscle strength and endurance	Use machines as opposed to free weights to prevent injury from loss of control of the weights
Flexibility	Stretching. Yoga		Before and after exercise	5 to 10 min		Increase flexibility and range of motion	Stretch maximally but avoid pain especially in joints. 10 to 30 s for each major muscle group

care and exercise programming should be discussed with his or her physician.

Cancer patients face many factors that may require modification of their exercise programs. Because of their malignancy or possibly because of the therapy used to treat it, many patients will experience symptoms of extreme fatigue, shortness of breath, nausea, or vomiting. The exercise professional must consider these symptoms when planning an exercise program. Some patients may have cancers that involve the bones, in which case exercise of that extremity may cause pain or result in injury. Patients with metastatic disease to the pelvis or legs should avoid high-impact exercises and may benefit from exercise programs that allow them to sit down, such as stationary cycling or chair exercises. Stationary bicycle activities are also frequently used for patients who are recovering from breast cancer or thoracic surgery so that they can forgo upper-body exercise while maintaining lower-body conditioning. This mode also works well in patients with ataxia, central venous catheters, or lymphedema. Water aerobics may also reduce stress on the skeletal system and allow exercise to continue. Intravenous devices preclude some stretching exercises. Pool activities are acceptable in patients with indwelling central venous catheters, continent urinary devices, or colostomies but are contraindicated for patients with intravenous catheters, nephrostomy tubes, and urinary bladder catheters.

The type of cancer treatment and the treatment protocol are also important. For example, chemotherapy and radiation therapy both use protocols that dictate how and when the treatment is administered. Surgery has both an in-hospital and out-of-hospital recovery phase that must be considered. Finally, treatment-related side effects such as fatigue, depression, and nausea will likely influence, at the very least, intensity of exercise. Any exercise program needs to be flexible enough to work around these schedules and problems.

Resistance training to maintain or enhance muscle strength is important. Patients should perform strength training on machines rather than free weights to avoid any potential for bruising or bone fractures. Low-resistance, high-repetition workouts are recommended. High-intensity endurance or resistance exercise should be avoided during all cancer treatments. Patients who undergo blood tests should avoid resistance training for 36 h before blood is drawn.

As your patient's clinical exercise physiologist, you should be sure to remember several things:

- Many patients with cancer are inactive and experience mood disturbances such as anxiety and depression that are themselves associated with little interest for exercise. Be sure to increase exercise dose progressively in an individualized manner for every patient.

- For the first week or so of exercise training, patients may be a bit tired later in the day. To compensate for this, ask them to limit the amount of other, home-based activities for the first few days.

- A variety of treatment-related obligations and complications often interrupt exercise therapy for patients with cancer. Before patients even start an exercise regimen, inform them that setbacks and interruptions are not uncommon. Instead of not exercising at all or stopping the exercise program, patients should plan around interruptions as well as they can and continue to adhere to their programs whenever possible.

- Learn about the different types of treatments that your patients are receiving.

- Consider developing an exercise buddy system, which matches up cancer survivors who are already exercising with patients who are just starting out. Such group support from patients with similar medical problems may help improve short-term adherence.

- Ask patients daily how they are feeling and what barriers to exercise they may be experiencing.

- Stay in close contact with the patient's supervising physician.

EXERCISE TRAINING

Beginning exercise intensity should be modified in patients with cancer. Unsupervised patients should probably begin at lower intensities such as 50% of heart rate reserve, whereas supervised patients may be able to start at higher intensities such as 60 to 80% of heart rate reserve. With cancer patients the best approach is to err on the side of conservative exercise programming. Patients with cancer who present with severely impaired fitness may initially benefit from interrupted programs that incorporate several bouts of shorter-duration exercise. An excellent review by Maryl Winningham (87) discussed how to get a cancer patient started in an exercise program. As shown in table 24.7 on p. 440, patients with activity level categories of 0 and 1 can probably start with 15 to 20 min of continuous exercise. Patients in category 2 may need to start with periods of 5 to 10 min. When in doubt about where to start, the exercise professional should use the 50% rule: Ask the patient how far he or she can walk before becoming too tired and start at half that distance or time.

After the starting point has been determined, the exercise professional can build endurance in the patient by encouraging long slow distances as opposed to faster short distances. Exercise frequency is ideally set at three to five times per week, but this plan can vary based on the individual patient (14). Patients undergoing radiation therapy or chemotherapy have complex schedules and require flexibility relative to planning which days to exercise. The patient's exercise program should increase between 30 sec and 2 min each day, while the patient maintains a rating of perceived exertion between 11

Table 24.7 Categorizing Performance Levels

Activity level	Category	Exercise duration	Exercise frequency
Active, no limitations	0	15–20 min	Daily
Ambulatory, decrease leisure activity, can perform self-care	1	15–20 min	Daily
Ambulatory >50% of time, moderate fatigue, limited assistance with ADL	2	5–10 min	Two sessions daily
Ambulatory <50% of time, fatigue with mild exertion, requires assistance with ADL	3	5–10 min	Daily
Confined to bed	4	No exercise	

Note. ADL = activities of daily living.

and 14. Practical application 24.4 on p. 438 provides an exercise prescription summary for patients with cancer. Using these principles, Winningham developed the Winningham Aerobic Interval Training (WAIT) program, a supervised moderate-intensity 10 wk program for cancer patients (87).

Pain and fatigue are common symptoms for cancer patients. The exercise professional must be alert to distinguish the normal pain and fatigue of exercise from that caused by progression of disease, deconditioning, aging, or an exercise program that is too strenuous. Pain at the initiation of the exercise program is a sign to start at a lower level of intensity.

Finally, the exercise professional should keep in mind that one purpose of an exercise program is to improve fitness, regardless of whether the patient has cardiovascular disease, cancer, or is free of disease. The components of fitness include strength, endurance, and flexibility. Therefore, each exercise program should incorporate aerobic training, resistance training, and range-of-motion exercises to achieve maximum fitness.

Exercise training improves the ability of muscle cells to produce energy, whereas inactivity results in decreased muscle function and exercise tolerance and secondarily leads to fatigue. In patients with cancer, investigators have found muscle fiber atrophy and necrosis (86). These changes are partly responsible for the fatigue that patients with cancer experience when they exert themselves (41). Exercise can counter these effects and stimulate the body to maintain or improve its abilities to extract oxygen for energy production.

A recent study by Segal et al. (67) randomized 123 women with breast cancer to either 6 mo of regular exercise or a no-exercise control group. Those randomized to exercise trained five times per week. Although exercise capacity, as measured by $\dot{V}O_2$peak, was not different after training in the exercise group versus controls, physical

function as assessed by patient questionnaire improved. Winningham et al. (87,89) found a 40% increase in functional capacity among patients with cancer participating in an exercise group, whereas nonexercisers described twice the fatigue.

In 2000 and again in 2003, Courneya et al. (12,14) reviewed all exercise interventional trials conducted between 1988 and 2002. Forty-seven trials met their criteria for inclusion. The researchers divided the studies into two groups: those that studied patients with breast cancer and those that studied patients with other forms of cancer. They then further divided the studies into those done during cancer treatment and those done after cancer treatment. In all groups statistically significant beneficial effects of exercise were found for endurance, muscle strength and function (40 to 50%), body weight, flexibility, symptoms of fatigue and nausea, and overall quality of life. In general the studies on exercise performed during cancer treatment were of good quality, consisting of randomized control trials with appropriate controls. The studies on exercise performed following treatment were not as methodologically rigorous and involved smaller sample sizes.

CONCLUSION

Cancer is a constellation of diseases. It can begin in any organ and spread to other organ systems. Its treatment includes surgery, radiation therapy, chemotherapy, and, more recently, biotherapy. Quality-of-life issues are extremely important among patients with cancer. Exercise benefits these patients primarily through improving quality of life, such as by reducing fatigue, reducing dependence on others, and countering some of the side effects of cancer therapy. The role of exercise in preventing cancer or altering the natural history of the disease remains unknown.

Medical History

Mr. CB is a 68 yr old Caucasian male with metastatic renal cell carcinoma. He was a millwright for the city water department but is now retired. He has a medical history of dyslipidemia (total cholesterol = 178 mg · dl^{-1}, high-density lipoprotein cholesterol = 24 mg · dl^{-1}, triglyceride = 774 mg · dl^{-1}), hypertension, and myocardial infarction in August 2003. He also has a history of atrial fibrillation. In January 2005 his body mass was 83 kg, his blood pressure was 110/70 mmHg, and his heart rate was 72 beats per min. Current medications are metoprolol, isosorbide dinitrate, Solu-Cortef, fenofibrate, and aspirin.

Diagnosis

In 2000 during a follow-up visit for impotency in the Department of Urology, urine tests detected microscopic hematuria. The following year, still having difficulties of impotency, he complained of gross hematuria and right flank discomfort. An intravenous pyelogram suggested a renal mass. Results of a computed tomography scan showed a right renal mass that invaded the right kidney and the inferior vena cava. He underwent a right radical nephrectomy in November 2001 for stage IIIA renal cell carcinoma. Pathology showed invasion of the renal capsule, as well as the renal vein and inferior vena cava.

A follow-up computed tomography scan in April 2003 showed metastases in the lower lobe of the left lung. These nodules were considered too small for biopsy. In May 2003 he complained of radiating pain from the right buttock to the knee. A bone scan identified widespread bone metastases. Magnetic resonance imaging showed a mass in the left posterior lateral aspect of the lumbar vertebrae. He underwent 14 d of radiotherapy in June 2003, during which his back discomfort improved. As part of clinical trial, he also received two courses of interferon and chemotherapy.

In May 2004 Mr. CB underwent surgical decompression and excision of an L3 vertebrae tumor. He began immunotherapy with a-interferon in July 2004. Repeat magnetic resonance imaging and bone scans through January 2005 showed the disease to be stable.

Exercise Test Results

His most recent electrocardiogram showed normal sinus bradycardia with evidence of a previous inferior wall myocardial infarction. Resting heart rate was 53 beats per min. An exercise stress test completed in August 2003 was mildly suggestive of ischemia, with a V̇O$_2$peak of 17.6 ml · kg^{-1} · min^{-1}. Peak heart rate was 92 beats per min, and exercise was discontinued because of mild to moderate grade angina. His cardiac status is now stable with Canadian Cardiovascular Society grade 2 angina.

Exercise Prescription

During chemotherapy in 2003 Mr. CB initiated an exercise program. He exercised 3 to 4 d a week at a perceived exertion of 11 to 12 (Borg scale 6–20) and at a prescribed heart rate range of 72 to 82 beats per min. This heart rate range was free of any electrocardiographic evidence of ischemia. Exercise modalities included treadmill, dual-action bike, and rower. Exercise therapy was tolerated well, without complications or symptoms.

Discussion Questions

1. What effect does metoprolol have on heart rate response to exercise? Would you alter how you go about guiding exercise intensity for patients taking this drug?

2. Fatigue and mood disturbances are common complaints of patients with cancer, often attributable both to the disease itself and the treatments used to manage the disorder. Explain whether exercise should be used or withheld in and around those times when a patient is undergoing therapy. What clinical features and symptoms would lead you to consider (or not consider) exercise during this time? How might you quantify whether, in fact, a patient is or is not responding to an exercise regimen?

(continued)

3. This patient asks you to help him design and start a resistance-training program. Do his test results indicate that this type of training should be incorporated into his exercise regimen? Explain your answer and any concerns that you have with respect to resistance training in patients with cancer.

25

Human Immunodeficiency Virus

Barbara Smith, RN, PhD, FACSM

James Raper, CFNP, DNS

Michael Saag, MD

Human immunodeficiency virus (HIV) infection and **acquired immunodeficiency syndrome** (**AIDS**) are conditions that evolve from the initial infection to the time of death. Throughout this cycle, a person becomes increasingly debilitated, often to the degree that even low-level physical activity is difficult. Exercise training can help attenuate loss of physical functioning and may affect the progress of the disease cycle. This chapter reviews the disease and the recommendations for exercise in this population.

The case definition of AIDS has evolved over time, reflecting changes in technology and better understanding of the clinical manifestations of HIV infection. The first case definition was published in 1982 by the Centers for Disease Control and Prevention (CDC). The list of AIDS-defining-diseases at that time was drawn from the clinical experience of providers treating primarily homosexual men with AIDS.

First identified in 1983, the HIV-type 1 is a **retrovirus** that infects human cells bearing the CD4+ surface marker. The preferred CD4+ site is the **T-helper lymphocyte**. After infection is established, a complex array of pathogenic mechanisms occurs, typically over many years, and ultimately results in the depletion of CD4+ lymphocytes from the normal range of 500 to 1,400 cells · mm^{-3}. As the number of CD4+ declines, persons infected with HIV are predisposed to **opportunistic infections**, malignancies, **wasting**, and other complications of AIDS (5,26,49).

After HIV was established as the causative virus and the HIV antibody test became available in 1985, the case definition was expanded from the 1983 definition to include clinical conditions less closely associated with immune suppression. These included the following:

- Disseminated histoplasmosis, a type of fungus
- Chronic isosporiasis, an intestinal protozoa
- Certain non-Hodgkin's lymphomas, cancer of lymph tissue (14)

In 1987 the definition of AIDS was further expanded to include the following:

- Wasting (loss of 10% of body weight in 3 mo)
- HIV encephalopathy (inflammation of the brain)
- Extrapulmonary (outside the lung) tuberculosis (15,16)

In addition, these expanded definitions allowed for a presumptive diagnosis to accommodate changing diagnostic and treatment practices. In 1993 the CDC issued a revised classification system for defining AIDS using three CD4+ cell ranges and three clinical categories (12). This system formed a matrix of nine mutually exclusive categories (table 25.1). The most important changes were the inclusion of all patients with fewer than 200 CD4+ cells · mm^{-3} or a CD4+ count of less than 14% of total lymphocytes. These were the first laboratory criteria for defining AIDS. In addition, the following were added to the list of AIDS-defining clinical conditions:

- Recurrent bacterial pneumonia
- Invasive cervical cancer
- Pulmonary tuberculosis

Finally, in January 2000 the CDC made the most recent revision to the HIV case definition by combining reporting criteria for HIV infection and AIDS to a single case definition.

SCOPE

Since 1981, when the first case of AIDS was reported, this infection has presented a significant burden for the U.S. healthcare system. A total of 944,306 cases of AIDS and 529,113 AIDS-related deaths were reported in the United States up to the year 2004 (17). Estimates are that more than 1 million people in the United States are infected with HIV, or 1 person in about 300 (55), suggesting that many do not know that they are infected.

Identifying those at greatest risk of contracting HIV is important to determining whether screening needs to be performed (see "Indications for HIV-1 Testing"). Along with the traditional high-risk groups of men who have sex with men and intravenous drug users, other high-risk groups include recipients of blood transfusions before 1985 including those with **hemophilia**, children born to HIV-

infected women, and persons who engage in impulsive sexual behavior or with many partners. Increased HIV transmission by heterosexuals has resulted in a higher infection risk for sexually active adolescents and young adults, the mentally ill, and abusers of alcohol and other mood-altering drugs. Although other avenues of transmission (e.g., insect bites, saliva transfer, sweat) have been suggested, none is known to transmit the virus.

Indications for HIV-1 Testing

- Persons with other sexually transmitted diseases
- Persons in high-risk categories
- Injection drug users
- Men who have sex with men
- Hemophiliacs
- Sexual partners of persons with known HIV-1 infection or those at high risk
- Persons who consider themselves at risk or request the test
- Pregnant women
- Patients with tuberculosis
- Persons with an occupational exposure (e.g., deep penetrating wound, splashed with blood)
- Hospitalized persons age 15 to 24 in which seroprevalence is >1% or case rate is >1/1,000
- Persons who perform exposure-prone invasive procedures (surgeons)
- Blood, semen, and organ donors
- Persons with clinical or laboratory findings suggestive of HIV-1 infection

Table 25.1 1993 Centers for Disease Control Classification System or Matrix for HIV-1 Infection

CD4+ cells/mm3	Clinical category A	Clinical category B	Clinical category C
>500 cells · mm^{-3}	A1	B1	C1
200–499 cells · mm^{-3} (14–28%)	A2	B2	C2
<200 cells · mm^{-3} (<14%)	A3	B3	C3

Note. Shaded areas show the expanded AIDS surveillance case definition. These are reportable AIDS cases in the United States and its territories, effective January 1, 1993. Category A includes asymptomatic, or persistent, generalized lymphadenopathy, or acute HIV infection. Category B includes symptomatic, or not A or C. Category C includes AIDS-defining or indictator condition.

Reprinted from the CDC.

Since the early 1980s the demographic characteristics of those newly infected with HIV have changed. Data show a disproportionate effect on minority communities as reflected in AIDS incidence rates that are higher among African Americans and Hispanics than among Caucasians (17). Minority women account for 80% of all newly reported AIDS cases among women. For a more detailed description of the demographic characteristics of HIV–AIDS, access the CDC Web site at www.CDC.gov. Worldwide, the demographic characteristics of HIV–AIDS vary from country to country. The World Health Organization (WHO) estimated that more than 40 million people worldwide are living with HIV infection and approximately 20 million deaths from HIV–AIDS occurred from 1981 to 2003. Most of these have occurred in Africa. For a more detailed description of the worldwide demographic characteristics of HIV–AIDS or the demographic characteristics of a specific country, access the WHO Web site at www.who.org.

PATHOPHYSIOLOGY

The natural history of HIV infection from the time of transmission to death is 8 to 10 yr in an individual who does not receive antiretroviral therapy. HIV infection progresses through five stages (35):

1. Acute infection (seroconversion)
2. Early disease (CDC A1, B1, C1)
3. Middle disease (CDC A2, B2, or C2)
4. Late disease or AIDS (CDC A3, B3, or C3)
5. Advanced disease

Stage 1—Acute Infection

Primary HIV infection and seroconversion (the change from **HIV negative** to **HIV positive** antibody status) occur over a period of a few weeks to months. During this stage, there is a high concentration of HIV in plasma and a precipitous decline in the CD4+ cell count, indicating a decline in function of the immune system. Figure 25.1 on p. 446 depicts the process of HIV replication. CD4+ cell destruction may result from direct infection of the CD4+ cell by HIV or by the body's immune system response to the infection.

Eventually, the immune system gains control of viral replication, the HIV concentration in the blood (i.e., **viremia**) decreases, and viral levels gradually stabilize (44,47). Although some people may not experience any symptoms, most will have mild to moderate flulike symptoms during this period.

Stage 2—Early Disease

Stabilization of the level of plasma viral load signifies the beginning of clinical latency, or what is known as the early disease stage. At this time, plasma levels of CD4+ cells are >500 cells · mm^{-3}. During this period, the patient is clinically asymptomatic and generally has no findings on physical examination except for persistent generalized lymphadenopathy (enlargement of lymph nodes). At this time the lymph tissue serves as the major reservoir for HIV. This results in a relatively lower number of viral particles in peripheral blood than seen in stage 1.

This stage is the longest, often lasting many years. Chronic immune activation and persistent viral replication characterize this period, but the rates at which HIV cell death and replacement occur are nearly in balance. But over time the CD4+ cell count gradually declines at the rate of approximately 50 to 70 cells · mm^{-3} per year.

Stage 3—Middle Disease

This stage begins when CD4+ cells decrease to between 200 and 499 cells · mm^{-3} plasma. Although most patients remain asymptomatic during the middle stage of disease, many lesions or disorders of the skin and mucous membrane become evident. Patients with middle-stage HIV disease, when left untreated, have a 20% to 30% chance of developing an AIDS-defining illness (i.e., moving to stage 4) or dying within 18 to 24 mo (40).

Stage 4—Late Disease

Movement into this stage occurs when the plasma CD4+ cell count falls below 200 cells · mm^{-3}. At this point, the patient meets the CDC case definition of AIDS. During the late disease stage CD4+ cells are between 50 and 199 cells · mm^{-3}, and the risks associated with developing an AIDS-defining illness increase (12). HIV usually increases during this stage of disease and correlates with a sharp decrease in CD4+ cell count (57). In addition, patients with late-stage disease are at increased risk of developing opportunistic infections and malignancies that are included in the 1993 CDC list of AIDS-defining conditions.

Stage 5—Advanced Disease

When HIV infection is uncontrolled and CD4+ cells decrease to less than 50 cells · mm^{-3}, the final stage of the disease process is reached. This stage is characterized by a period of increasing opportunistic infection risk, certain cancers, wasting, and neurological complications.

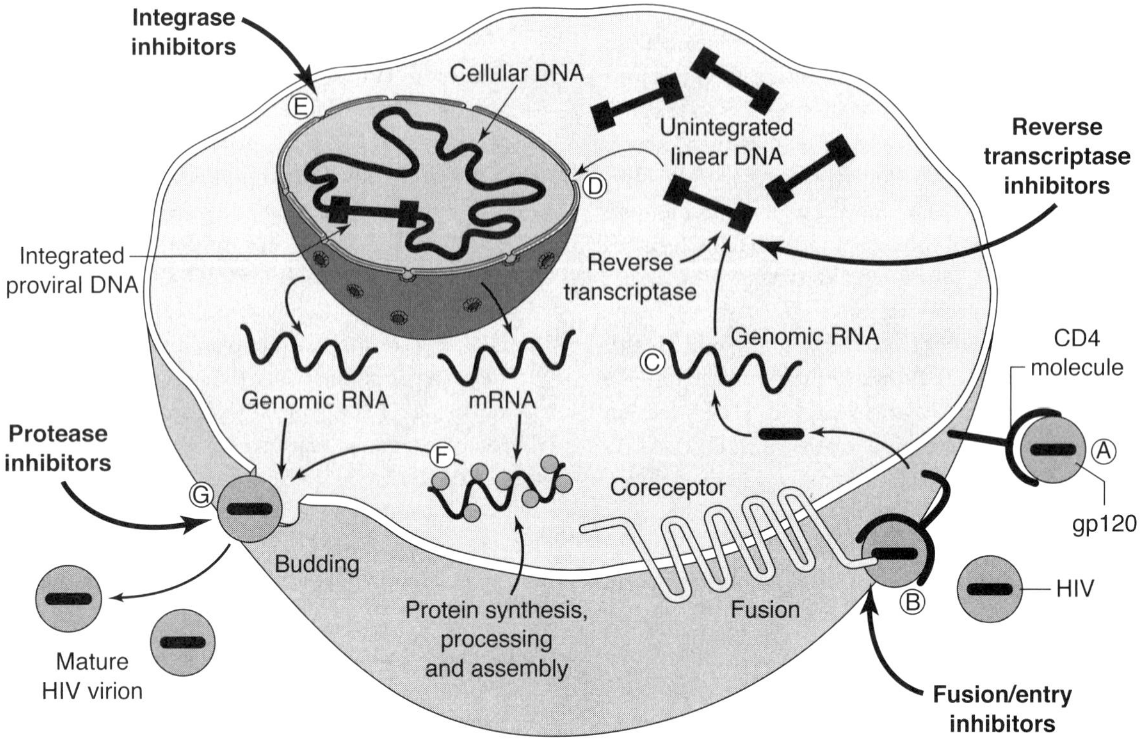

Figure 25.1   Replication of the HIV-1 virion. (*a*) HIV-1 virion with binding sites for CD4+ receptor. (*b*) Virion fuses (binds) with CD4+ receptors on cell surface and uncoating of virus core occurs. (*c*) After genetic material is in the cell, reverse transcription occurs. (*d*) Importation of viral genetic material into the cell nucleus occurs, and it is then incorporated into the genetic material of the cell. (*e*) Translation and transport occurs. (*f*) Assembly of new virions occurs. (*g*) New virus buds from the cell matures, and the process is repeated.

Each of these conditions is associated with severe immunosuppression, and often several opportunistic diseases coexist. Without aggressive antiretroviral treatment and aggressive treatment of the opportunistic infection, death is imminent.

MEDICAL AND CLINICAL CONSIDERATIONS

The diagnosis and treatment of HIV infection and AIDS depend on clinical evaluation and careful consideration of treatment options. This section reviews these topics.

Signs and Symptoms

Primary HIV infection (stage 1) is often unrecognized clinically. Many patients however, develop symptoms of viral illness (19). Following are the five most commonly described symptoms of primary HIV infection:

1. Fever
2. Sore throat
3. Fatigue
4. **Myalgia**
5. Average weight loss of 5 kg (68)

During the early disease stages (stages 1–2), up to 5% of patients will develop one or more of the following signs and symptoms:

1. Mild to moderate lymphadenopathy: increase in size of lymph nodes
2. Seborrheic dermatitis: dry, scaly skin
3. Psoriasis
4. Pruritic folliculitis: itchy red bumps
5. Fungal infection of the fingernails and toenails
6. Exaggerated responses to insect bites
7. Molluscum contagiosum: bumps under the skin containing a cheesy-looking substance
8. Taste distortion
9. Ulcers of the mouth

Muscle and joint pain, headache, and fatigue are commonly present during the middle disease stages (stages 2–4) and are related to higher levels of HIV virus in the blood. Other commonly reported middle-stage signs and symptoms include the following:

1. Recurrent fever blisters
2. Herpes simplex ulcerations
3. Shingles (i.e., varicella-zoster virus)
4. Removable white plaques on mucosal surfaces
5. Recurrent diarrhea
6. Skin problems
7. Intermittent fever
8. Wasting

During the late and advanced disease stages (stages 4–5), symptom progression often intensifies and includes multiple systems. AIDS-defining opportunistic conditions include the following:

1. Invasive cervical cancer
2. Disseminated histoplasmosis
3. Kaposi's sarcoma
4. *Mycobacterium avium* complex
5. *Pneumocystis carinii* pneumonia (PCP)
6. Tuberculosis
7. Wasting

Opportunistic infections are the primary cause of death in people with AIDS. *P. carinii* pneumonia is the primary diagnosis and affects up to 52% of patients with AIDS. Kaposi's sarcoma is the next most prevalent (26%). Cytomegalovirus retinitis is the leading cause of blindness in these patients and causes various other end-organ disease, resulting in painful swallowing, bloody diarrhea with abdominal pain, and weakness of the lower extremities. Various endocrine disorders and neurological symptoms are common in late- and advanced-stage disease (17).

History and Physical Exam

When an HIV-infected individual begins an exercise program, a medical evaluation must be conducted to detect conditions that might contraindicate exercise. A history of extreme fatigue, weakness or malaise, fever, chills, night sweats, or wasting indicates active infection and should be considered because exercise may worsen these conditions. Care should be taken to identify individuals who are at high risk for developing wasting,

such as those with uncontrolled viremia or those who are already actively wasting. Although wasting is not an absolute contraindication for performing an exercise test, the cause should be identified and treated before the person starts an aerobic exercise program. The clinical exercise professional should assess body weight at regular intervals for rapid or excessive weight loss. Increased caloric expenditure associated with exercise training could exacerbate weight loss.

Additionally, those with severe peripheral lipoatrophy (i.e., decrease or absence of peripheral fat, increased central obesity, and the development of a cervical fat pad) may need to be cautioned about the exacerbation of peripheral fat loss that may occur with aerobic training. Additional tips for working with HIV-infected people are provided in practical application 25.1 on p. 448. The physical examination should focus on temperature, resting heart rate, blood pressure, body weight, and indicators of cardiovascular disease (e.g., murmurs, arrhythmias, edema). Other findings indicative of disease progression include findings of psychomotor slowing, eye movement abnormalities, hyperreflexia, and peripheral neuropathy.

It is important to consider any recent history of changes in body weight (i.e., gain or loss, period of time, and distribution of fat) or other symptoms of metabolic disorders. Fat redistribution, characterized by peripheral lipoatrophy in HIV-infected men and women, is noted with specific HAART regimens (10,38,61,70,84). Metabolic disorders such as impaired glucose metabolism and reduced insulin sensitivity may occur in people on active antiretroviral therapy, both with and without the use of protease inhibitors (10,67,71,80,81).

Diagnostic Testing

Laboratory tests are required to make a conclusive diagnosis of HIV infection. Recent advances allow testing with a sample of either blood or saliva. Two techniques are used most commonly:

1. Enzyme immunoassay method
2. Western blot method

The enzyme immunoassay (EIA; also referred to as enzyme-linked immunosorbent assay, or ELISA) is the most commonly used test. It measures the presence of antibodies to HIV viral proteins. A single positive EIA is classified as HIV-reactive but not positive for HIV infection until it is confirmed by one or more subsequent tests. Patients must receive a thorough explanation of the need for the confirmatory subsequent test when they have a

Client–Clinician Interaction

"Healthy" patients who are HIV positive and choose to exercise often do so on their own or at a commercial health club. Therefore, the initial interaction between the clinical exercise professional and an HIV-infected patient is likely to occur at the patient's first visit to an exercise facility, often for treatment of established cardiac disease. These patients are often concerned about who is going to know about their diagnosis because of the stigma associated with their disease. The clinical exercise professional must be aware of this concern and assure the patient that only those staff members who need to know about their diagnosis will be told. The patient can also be assured that no staff member will discuss medical information with other patients. Additionally, the clinical exercise professional should not display any hesitancy or prejudice toward these patients.

An HIV-positive patient cannot be identified by simple observation. Additionally, one cannot rely on an individual to provide his or her HIV status because some are not willing to reveal this information and others do not know that they are HIV positive. Finally, other diseases are equally or more detrimental to health than HIV, such as hepatitis B and C. People with these and other conditions are more likely than a person with HIV to transmit their disease to someone else. Applying the universal precautions will greatly reduce risk of HIV and other disease transmission. The universal precautions listed below are suggested for working with any patient, including those with HIV, to reduce the risk of contracting an infectious disease.

Table 25.2 Universal Precautions for Working With Patients

Procedure	Wash hands	Gloves	Gown	Mask	Eyewear
Talking					
Adjusting intravenous fluid rate or other noninvasive equipment					
Examining patient without touching blood, body fluids, or mucous membranes (e.g., checking blood pressure)					
Examining patient including touching blood, body fluids, or mucous membranes (e.g., placing and removing mouth pieces, vigorous skin preparation for electrodes during exercise testing)	X				
Blood draw	X	X			
Inserting a venous catheter	X	X			
Handling soiled linen or other waste	X	X			
Intubating during cardiopulmonary resuscitation	X	X	X	X	X
Suctioning	X	X	X	X	X
Surgical procedures	X	X	X	X	X

Adapted from the University of Alabama at Birmingham (UAB), "Interdisciplinary Standard: Universal Precautions CDC Category: Standard Precautions." For more detail on this topic search the CDC Web site at www.CDC.gov.

reactive EIA test, because the test is not 100% accurate. The Western blot method is the most commonly used retest method. Various organizations have developed and published criteria for interpreting Western blot reactivity (13,20,82). Reactivity patterns that do not fit these criteria are classified as indeterminate. Indications for HIV testing include

1. adults in populations with an estimated prevalence greater than 1%,

2. pregnant women,

3. sexual assault victims,

4. persons with occupational exposure, and

5. anyone who requests the test (1).

Ongoing evaluation of disease severity is important to determine appropriate treatment strategies. Several studies demonstrate the value of using CD4+ cell count as the best predictive marker of relative risk for the development of HIV-related opportunistic disease (75). CD4+ count, however, provides only a surrogate marker for clinical progression (18,22). Rapid decay of plasma HIV RNA concentration (or viral load) occurs in patients treated with potent **antiretroviral medications** (29,54), and the degree of this effect is inversely related to disease progression (45,46,55). Decisions regarding the initiation of or changes in antiretroviral therapy are guided by plasma HIV RNA concentration, the CD4+ cell count, and the clinical condition of the patient. To establish the overall clinical condition and reveal existing comorbidities, the patient often undergoes the laboratory evaluations outlined in table 25.3.

Exercise Testing

Because a number of antiretroviral medications cause severe nausea and vomiting following ingestion, the graded exercise test should be scheduled at a time when the patient is not as likely to experience these symptoms. The exercise professional should discuss this concern with the patient to clarify the issue. See chapter 5 for general exercise-testing recommendations.

Depending on the individual's physical status, the exercise test protocol should be carefully selected. For patients with poor balance or coordination secondary to fatigue or neuromuscular effects, using a cycle ergometer may be better than treadmill walking. Standard test end points should be used along with standard electrocardiogram and blood pressure monitoring. As a substitute for maximal exercise testing, a submaximal exercise evaluation on a motor-driven treadmill or a cycle ergometer, step testing, or a field test such as the Cooper 12 min test can provide information regarding fitness status and improvement (2).

Many people who are in early, middle-, or late-stage disease can expect to have normal blood pressure, pulse rate, and respiratory responses to a graded exercise test. Some HIV medications known to cause peripheral neuropathy may also increase the risk of autonomic neuropathy, which in turn increases the risk of an abnormal blood pressure response, an elevated resting heart rate, and attenuated exercise heart rate response.

In general, exercise testing reveals that HIV-infected individuals have low peak $\dot{V}O_2$ values (32,73,76). These low peak $\dot{V}O_2$ values are seen even when there is no evidence of a pulmonary opportunistic infection (32). Possible reasons

Table 25.3 Laboratory Tests for Baseline Evaluation Following HIV-1 Diagnosis

Evaluation	Test rationale
CD4+ cells · mm^{-3}	Level of CD4+ cells and HIV disease stage
HIV RNA	Amount of virus and later to assess effectiveness of treatment
PPD skin test	Tuberculosis (comorbidity)
CBC	Presence of anemia, overall level of specific blood cells
Serology (VDRL/RPR)	Presence of syphilis, an STD
Serum chemistry	General health, status of kidney, and liver function
Hepatitis B and C	Assess for these viruses that are common comorbidities
G-6-PD	Test for presence or absence of this genetic disease, because it predisposes individuals to hemolytic anemia following exposure to certain drugs used in the treatment of HIV
Pap smear	Assess health of the cervix and test for other STDs
Chest X-ray	Evaluate the presence of PCP and other pulmonary complications

Note. PPD = purified protein derivative; CBC = complete blood count; VDRL = Venereal Disease Research Laboratories; RPR = rapid plasma reagin; STD = sexually transmitted disease; PCP = *Pneumocystis carinii* pneumonia.

for a reduced peak $\dot{V}O_2$ include effects of the infection and medications on skeletal muscle, and loss of cardiorespiratory endurance secondary to an increasingly sedentary lifestyle.

Treatment

The management of HIV infection has evolved rapidly. Therefore, the U.S. Department of Health and Human Services Panel of Clinical Practices for the Treatment of HIV recommends that the care of patients with HIV infection be supervised by an expert. Those working with HIV-infected individuals must have access to the latest available recommendations from the International AIDS Society–USA Panel or the HIV–AIDS Treatment Information Service (9). For best treatment results, all patients should be treated with the most up-to-date therapy as soon as possible after seroconversion. Often this does not occur until patients have reached stage 2 and demonstrate related symptoms. Recommendations to treat asymptomatic persons must be based on the person's willingness to accept therapy, the probability of adherence to the prescribed regimen, and the prognosis in terms of time to an AIDS-defining complication as predicted by plasma HIV RNA concentration and the CD4+ cell count (9). After the decision is made to initiate therapy, the goal is maximal viral suppression for as long as possible. Four categories of antiretroviral agents currently available are the mainstays of medical therapy. Each category is identified by its selective action in the life cycle of the virus.

1. Nucleoside analogs or nucleoside analog reverse transcriptase inhibitors (NRTIs)
2. Nonnucleoside reverse transcriptase inhibitors
3. **Protease inhibitors**
4. Entry (fusion) inhibitors

Chapter 3 (pharmacology) provides specific information about medications used to treat HIV infection. Treatment with these agents is complex. Patient education and involvement are important for optimal adherence to the treatment regimen. Therapeutic results are evaluated primarily by observing plasma HIV-RNA concentration. An effective or therapeutic response occurs when a 10-fold decrease occurs in HIV RNA copies at 8 wk and no virus is detectable 4 to 6 mo after initiation of treatment. Failure of therapy at 4 to 6 mo may be the result of nonadherence to the prescribed drug regimen, inadequate potency of medications, suboptimal levels of antiretroviral agents, medication resistance, and other factors that are poorly understood at this time. Patients whose therapy fails should be changed to new medications that are devoid of anticipated cross-resistance and for which clinical trial data support a high probability of suppressive viral response. The dramatic improvements in mortality and morbidity rates associated with HIV infection observed over the past decade are attributable, in part, to the widespread use of highly active antiretroviral therapy, or HAART (29,33,56). In addition, the development of guidelines for the use of prophylactic antimicrobial agents and vaccines to prevent opportunistic infections such as **cytomegalovirus**, *P. carinii* pneumonia, *M. avium* complex, and others have also contributed to improved outcome. The exercise professional working with these patients should understand that these patients often have many other comorbid conditions that require the use of other medications. These include chemotherapeutic agents for cancer, psychotropic drugs, medications for gastrointestinal distress, and other drugs to treat the side effects of the HIV medications. Some of these medications may have a direct or indirect effect on the ability to perform exercise.

With declining morbidity and mortality rates associated with HIV infection, there is much optimism about the future for HIV-infected patients. But healthcare providers must be aware of the following physiological effects of the HAART regimen that place patients at increased risk for developing coronary heart disease (28):

- Body fat redistribution or lipodystrophy
- Impaired glucose metabolism
- Insulin resistance
- Increased bleeding in patients with hemophilia
- Elevated blood lipids
- Osteopenia and osteoporosis, and osteonecrosis and avascular necrosis related to protease inhibitors or NRTIs

These effects have led some physicians to alter their approach to therapy in HIV-infected individuals. To date, however, little emphasis has been placed on the use of exercise to counteract these side effects.

EXERCISE PRESCRIPTION

Exercise training can counter some of the debilitating effects of HIV. By participating in a regular exercise program, patients can expect to improve their cardiorespiratory fitness, muscular strength and endurance, flexibility, and body composition. Additionally, exercise training may help improve psychological status, including reduction in depression and anxiety and improvement of mood and fatigue. Finally, exercise training does not appear to impair immune status in stable patients with HIV infection. In fact, exercise training may help retard the disease process and possibly improve immune status. When an HIV-positive individual requests guidance during exercise training, the clinical exercise professional should provide assistance for at least the first several exercise sessions. Goals include teaching the patient how to exercise at an appropriate intensity and with proper technique to optimize the benefits of the exercise and to reduce the likelihood of injury or overtraining syndrome. Early supervision may also increase the likelihood that a person will adhere to an exercise-training program. These patients have no special exercise facility requirements.

Special Exercise Considerations

The clinical exercise professional who conducts a graded exercise test or supervises exercise sessions for an HIV-infected person has little risk of infection. But before working with any clinical population, particularly those with blood-borne infections such as HIV, hepatitis B, hepatitis C, the exercise professional individual should become familiar with universal precautions to reduce the risk of contracting an infectious disease (see practical application 25.1). The CD4+ cell count should be evaluated within 3 mo after the initiation of an exercise-training program. This evaluation is particularly important for individuals who are not tested regularly. As stated previously, because HIV-positive patients are at risk of extreme fatigue, the exercise physiologist should not induce unnecessary fatigue as part of the exercise-training program. Each patient should be regularly evaluated for indications of fatigue, especially in the early stages of a training program. An easy method to evaluate fatigue is to use the rating of perceived exertion scale at a standardized work rate at least once a week.

Fatigue is inversely related to the CD4+ count and may be the result of the infection or treatment. The patient who experiences fatigue may become depressed and resist continued involvement in an exercise regimen. These patients should be encouraged to continue to exercise but at a reduced level. The goal of working at a reduced level is to prevent deconditioning and the subsequent cascade of increasing fatigue and reduced activity levels. An exception to encouraging continued exercise is if the patient is experiencing fever or active wasting. If that is the case, the patient should be immediately referred to his or her physician to evaluate and treat the underlying cause of the fever or wasting.

Nausea is a common side effect for HIV-infected patients who begin a new medication or have a change in the amount of medication that they take. If this occurs in association with exercise, patients may become discouraged with exercise because they feel sick. The clinical exercise professional should be prepared to deal with this situation by making the patient aware of this possibility and emphasizing that the medicine, not exercise, is likely causing the problem. The exercise professional can work with the patient and his or her physician to develop a schedule that takes into consideration the patient's eating, medication, and exercise schedules to prevent or reduce medication side effects during exercise training.

Osteonecrosis with avascular necrosis is another somewhat frequently encountered late complication in HIV disease that may be related to the HAART regimen (21,43). The most common site is the femoral head. Many patients have other risk factors for avascular necrosis including alcohol abuse, hyperlipidemia, lipid-lowering agents, testosterone therapy, corticosteriods use, and hypercoagulability. Screening of asymptomatic patients is not recommended. But the practitioner should maintain a high level of suspicion in patients who experience joint pain. A CT scan or MRI should be considered in patients with symptoms and risk. Bone density studies using dual-energy X-ray (DEXA) scanning show that osteopenia and osteoporosis are relatively common, although there is no clear association with specific HIV medications or lipoatrophy (7,21).

Exercise Recommendations

Recommendations for exercise prescription and training for HIV-infected patients are based on a combination of the American College of Sports Medicine's *Guidelines for Exercise Testing and Prescription* (2), studies of the effects of aerobic exercise on HIV-infected individuals, and general practical experience. These recommendations are concordant with recommendations from within *Physical Activity and Cardiovascular Health: NIH Consensus Statement*; *Physical Activity and Health: A Report of the Surgeon General*; and a number of the goals and objectives of the Healthy People 2010 program related to HIV, physical activity and fitness, nutrition and overweight, and heart disease and stroke (53,78,79). The following exercise prescription information is summarized in practical application 25.2 on p. 452.

Cardiovascular Training
As with any individual, to provoke continued adaptations of the cardiovascular system and skeletal muscles of HIV-infected individuals, an appropriate exercise prescription

must include an appropriate mode, frequency, duration, and intensity (2,3,6,36).

The exercise intensity must provide the highest level of total work without inducing excessive fatigue. Most people can exercise at a heart rate corresponding to between 40% and 85% of their peak $\dot{V}O_2$ (or heart rate reserve) achieved during an exercise evaluation. Those in the later stages of disease, or exhibiting indications of fatigue, should exercise at the lower end of this intensity range (e.g., 40–50%). If a heart rate range cannot be used, the rating of perceived exertion scale can be substituted. Training workload should be increased as tolerated by the individual.

The best approach is to increase workload in small, regular increments of duration initially until a desired duration (i.e., 30–60 min) is attained on most if not all days of the week as recommended by the U.S. surgeon general (2,79). Increases in pace or intensity can follow.

Because aerobic exercise training can increase the rate of weight loss, individuals should be weighed at least biweekly. If the patient loses more than 1 kg (2.2 lb) for 3 wk in a row, his or her healthcare provider should be consulted. Additionally, those with severe peripheral lipoatrophy may need to be cautioned about the exacerbation of peripheral fat loss that may occur in association with

Practical Application 25.2

Exercise Prescription Summary

Training method	Mode	Intensity	Frequency	Duration	Progression	Goals	Special considerations and comments
Aerobic	Walking, stationary cycling, jogging; most other exercise modes also acceptable	40–85% peak $\dot{V}O_2$	Most or all days	30–60 min	As tolerated	Maintain or increase cardio-respiratory endurance	Watch for excessive fatigue, which may signify disease exacerbation; consider a 3 mo CD4+ determination for those who do not have this regularly determined
Resistance	Free or machine weights; resistance cords	Begin at 50% of 1RM	2–4 d per week	Two to three sets per exercise, 8–12 repetitions per set	Advance progressively toward 80% of 1RM as tolerated	Focus on major muscle group strength and endurance	Do not allow patients to perform on consecutive days
Range of motion	Static stretching of all major muscle groups	Stretch just below point of pain	Daily	Hold each stretch for 30 s	Increase stretch as tolerated and avoid pain	Simple increase in range of motion	Avoid pain; may use towels, cords, and so on to increase "stretch"; relaxation exercises, such as yoga or tai chi–type programs may assist with increasing range of motion and reduce anxiety associated with the disease

aerobic training. Reductions in exercise-training volume may be necessary to reduce the rate of fat loss.

Resistance Training

Patients infected by HIV who perform resistance training demonstrate positive skeletal muscle adaptations that enhance strength and endurance. Patients report that activities of daily living become less difficult and less fatiguing. Additionally, increases in cross-sectional area are reported and may prevent wasting. Standard principles of training specificity and mode, frequency, duration, and overload (intensity) should be followed (3,6,36,83). See chapter 5 for general resistance-training recommendations. The following recommendations for resistance training are based on the American College of Sports Medicine's 1990 position statement, studies by Roubenoff et al., Yarasheski et al. (85), and general practice experience with resistance exercise in HIV-infected and other chronically ill populations (3,60,62). Strength-training exercises should include those that work both the upper and lower extremities, focusing on the large-muscle groups. Common exercises include leg press, squats, elbow flexion and extension, cable pull-

downs, bench and military presses, leg extension and curl, upright row, low-back extension, and bent knee curl-ups. Resistance training can be performed on 2 to 4 d of the week with a training bout every other day. People should abstain from consecutive days of resistance training to reduce the chance of excessive fatigue, immunosuppression, muscle damage, and opportunistic infection. Experienced exercise professionals can adapt these recommendations to meet the individual needs of their clients.

Range of Motion

As with any person performing a well-rounded exercise program, the HIV-positive patient should be instructed to perform range of motion routinely. General recommendations for range-of-motion exercise are presented in chapter 5.

EXERCISE TRAINING

Exercise training in patients with HIV has been studied. The following section reviews the literature for aerobic and resistance training. Additionally, practical application 25.3 summarizes the findings of several major studies.

Practical Application 25.3

Literature Review

This section summarizes the findings of several major HIV studies.

Aerobic Exercise

Research unequivocally shows that aerobic exercise improves physical and psychological endpoints in many chronic disease states including cardiovascular disease, hypertension, Parkinson's disease, pulmonary disease, renal dialysis, and cancer (11,23,25,27,31,41,59). Studies indicate that older men and women as well as women at risk for diabetes can increase strength and fat-free mass by resistance training (65,66,69). Despite the wealth of investigation on these and other clinical populations, only a few studies have documented the benefits of aerobic and resistive exercise in HIV-infected individuals (37,60,62,73,74,76). A review of exercise-training studies is provided in table 25.4 on p. 454.

Two investigations demonstrated that exercise-trained subjects could exercise longer on a cycle ergometer than non-trained HIV subjects could (37,60). Both studies, however, used relatively small samples of men and did not directly measure $\dot{V}O_2$. Stringer et al. (76) reported a modest improvement (0.2 L · min^{-1}) in directly measured $\dot{V}O_2$max after 6 wk of heavy training in HIV-positive patients. Smith et al. (73) reported a significant increase in time on a treadmill and a nonsignificant increase of peak $\dot{V}O_2$ (0.16 L · min^{-1}).

Lower peak $\dot{V}O_2$ values in some HIV-infected individuals may be related to the effects of medication or the HIV virus on the cardiovascular, pulmonary, or skeletal muscle systems or to comorbid conditions associated with HIV infection. Johnson et al. (32) reported that a group of HIV-infected active duty servicemen had a lower peak $\dot{V}O_2$ compared with an otherwise similar group of noninfected controls. The authors concluded that the lower peak $\dot{V}O_2$ was attributable to coronary artery disease.

Keyser et al. (34) concluded that the functional aerobic impairment that they observed in HIV+ adolescents resulted from HIV-related or pharmacologically mediated skeletal muscle dysfunction.

(continued)

Table 25.4 Review of Exercise-Training Studies

Author	HIV+ subject number	Type of exercise	Length of program	Primary exercise results
LaPerriere at al. (33)	10 HIV+ exercisers 6 HIV+ controls	Aerobic	10 wk	Increased estimated $\dot{V}O_2$max
Perna et al. (52)	28 HIV+	Aerobic	12 wk	Increased $\dot{V}O_2$peak
MacArthur et al. (35)	6 HIV+ exercisers completed	Exercise training	24 wk	Increased estimated $\dot{V}O_2$max
Rigsby et al. (54)	13 HIV+ exercisers 11 HIV+ controls	Aerobic/resistance	12 wk, 3 times per week	Increased total time at 150 W and increased strength
Roubenoff et al. (56)	24 HIV+ exercisers	Resistance	8 wk training + 8 wk ad libitum	Increased strength, increased lean mass
Roubenoff et al. (58)	10 HIV+ exercisers No controls	Aerobic and resistance	16 wk, 3 times per week	No aerobic measure, increased strength, increased lean mass
Smith et al. (67)	18 HIV+ exercisers 30 HIV+ controls	Aerobic	12 wk, 3 times per week	Increased time to exhaustion on treadmill
Stringer et al. (70)	HIV+ exercisers and HIV+ controls	Aerobic	6 wk	Increased measured $\dot{V}O_2$max
Baigis et al. (3)	52 HIV+ exercisers 47 HIV+ controls	Aerobic	15 wk	No improvement in $\dot{V}O_2$max
Terry et al. (77)	15 HIV+ exercisers 15 HIV+ controls	Aerobic	12 wk	Increased peak oxygen uptake
Yarasheski et al. (85)	18 HIV+ exercisers No controls	Resistive	16 wk	DEXA increase in whole-body lean mass, no decrease in fat mass

Sinnwell et al. (72) used magnetic resonance imaging and found a decreased muscle phosphocreatine level following a graded exercise test in individuals on AZT, compared with non-AZT-treated HIV-infected individuals. These authors suggested that the skeletal muscle cells in HIV-infected patients on AZT rely heavily on anaerobic pathways to generate adenosine triphosphate because of the effects of the drug. Mannix et al. (42), also using magnetic resonance imaging, determined that there was a decrease in oxidative, and an increase in anaerobic, sources of adenosine triphosphate (ATP) in skeletal muscle cells of stable AIDS patients compared with healthy controls. They concluded that the decrease in oxidative sources of ATP contributes to poor exercise tolerance and suggested that the decrease might be attributable to HIV-induced skeletal muscle mitochondrial damage.

In summary, the first few studies of aerobic exercise training in HIV-infected individuals are promising because they have demonstrated that aerobic exercise is safe and may be an effective health promotion strategy for people with HIV infection. But the mechanism for the reduced-exercise peak $\dot{V}O_2$ and its response to training requires further study.

Resistance Training

Only a few studies have described the use of resistance exercise in people infected with HIV. Roubenoff et al. (62) reported increased lean mass with the use of resistance training. Additionally, this study reported no decline in CD4+ cells or increase in HIV RNA. Roubenoff and colleagues (64) combined aerobic and resistance exercise during a 16 wk intervention using 10 HIV-infected men. The authors noted an increased one-repetition maximum with three of the four exercises tested (leg press, leg extension, and chest press). Body weight, lean mass, and bone mineral density, measured with the dual-energy X-ray absorptiometry method, did not change, but total body fat and abdominal fat decreased. Yarasheski and colleagues (85) reported increases in total body lean mass, as measured by DEXA, and reduced hypertriglyceridemia in HIV+ men on HAART.

These data suggest that resistance training is not harmful, and is potentially beneficial, for patients with HIV infection. A need remains for more investigation in this area.

Disease Prevention

Exercise may indirectly reduce HIV progression to AIDS by reducing psychological depression (50). Reductions in depression are associated with a decreased likelihood that a person will engage in behavior that may be detrimental to their health. Because the risk of HIV infection is related to behavior, those who exercise may indirectly reduce their risk of HIV infection, although direct evidence of the theory does not exist. A longitudinal observational study of a large HIV–AIDS clinic reported that HIV-infected patients who exercised three or more times per week had a lower risk of progressing to AIDS or death from AIDS at 1 yr follow-up (48). Regular exercise training of moderate intensity does not appear to activate an acute phase immune response. Also, it is reported that a single bout of exercise does not increase viral replication in HIV-infected adults (63). Additionally, exercise-training studies that use HIV-infected patients report either no change or improved CD4+ cell count. Thus, the use of moderate-intensity exercise that promotes general health, along with appropriate antiretroviral therapy and opportunistic infection prophylaxis, may reduce the severity of HIV infection and progression to AIDS.

HIV-infected people who may wish to engage in highly intense exercise can point to elite athletes such as Magic Johnson, Greg Louganis, and others who have participated in rigorous sporting events with no health consequences. Although no data indicate that intense exercise increases the morbidity and mortality rates associated with HIV infection, infected individuals may be wise to avoid prolonged, intense exercise. This recommendation is based on data from healthy individuals, which indicates that intense, prolonged exercise is associated with an increase in upper-respiratory infections and the suppression of specific immune factors (8,51,52). Each of these may be detrimental to the health of an HIV-infected person. These individuals should be encouraged to participate in moderate aerobic and resistance exercise training because of myriad potential physiological and psychological benefits. More longitudinal studies are needed to demonstrate the role of moderate exercise in slowing the progression of HIV infection and delaying the onset of opportunistic infections.

CONCLUSION

Although it may seem counterintuitive to recommend exercise training for HIV-infected patients because of the risk of fatigue, a properly designed and implemented exercise-training program can provide benefits for these patients. A clinical exercise professional can provide guidance for patients who wish to begin an exercise-training program. Future investigations should focus on the effects of exercise training on disease progression and mortality rates.

Case Study

Medical History

Mr. BP, a 39 yr old Caucasian, bisexual male was diagnosed with hypertension about 6 yr ago. He has had difficulty adjusting to antihypertensive medications, and thus his hypertension is not well controlled. He had no other significant medical history at that time.

(continued)

Diagnosis

He was first diagnosed with HIV infection about 2 1/2 yr ago following repeated bouts of upper-respiratory infection, persistent generalized lymphadenopathy, and an unexplained weight loss of 15 lb (7 kg) in 3 mo. Since his HIV diagnosis he has been diagnosed with clinical depression.

At the time of diagnosis his weight was 187 lb (85 kg), and his blood pressure was 166/90 mmHg. His laboratory work revealed the following:

- 206 CD4+ cells · mm^{-3}
- 98,650 HIV RNA copies · ml^{-1}
- Total cholesterol = 209 mg · dl^{-1}
- High-density lipoprotein cholesterol = 37 mg · dl^{-1}
- Triglycerides = 145 mg · dl^{-1}

He was placed on a HAART regime that included the protease inhibitor indinavir, plus two NRTIs, AZT, and ddI. He was also encouraged to see his internist regarding his blood pressure. Six months after beginning HAART his laboratory work was as follows:

- 389 CD4+ cells · mm^{-3}
- 1009 HIV RNA copies · ml^{-1}
- Total cholesterol = 215 mg · dl^{-1}
- High-density lipoprotein cholesterol = 39 mg · dl^{-1}
- Triglycerides = 160 mg · dl^{-1}

At 2 1/2 yr after diagnosis, Mr. BP continues on the same therapy. His weight has returned to an almost normal 199 lb (90 kg), and his most recent blood pressure was 144/86 mmHg. He reports that he is feeling better because his virus is "under control," and he is seeing a therapist to deal with his depression.

He also complains about getting a "real paunch" in his abdominal area. Laboratory work at his second-to-last clinic visit revealed the following:

- 347 CD4+ cells · mm^{-3}
- Less than 100 HIV RNA copies · ml^{-1}
- Total cholesterol = 238 mg · dl^{-1}
- High-density lipoprotein cholesterol = 37 mg · dl^{-1}
- Triglycerides = 678 mg · dl^{-1}

Laboratory work at his last visit revealed the following:

- 329 CD4+ cells · mm^{-3}
- Less than 100 HIV RNA copies · ml^{-1}
- Total cholesterol = 273 mg · dl^{-1}
- High-density lipoprotein cholesterol = 34 mg · dl^{-1}
- Triglycerides = 1,265 mg · dl^{-1}

At the second visit, his provider talked with him about increasing his exercise to help with his "paunch," to reduce his cholesterol and triglyceride levels, and to reduce his blood pressure. His physician also scheduled an exercise evaluation to determine his level of fitness and to attain some feedback from a clinical exercise physiologist about beginning an exercise program.

Exercise Test Results

At the time of the exercise evaluation Mr. BP's resting vitals were as follows:

- Heart rate = 72 beats $\cdot$ min^{-1}
- Blood pressure = 146/98 mmHg
- Respiratory rate = 22 $\cdot$ min^{-1}

The exercise evaluation resulted in the following maximal values:

- Heart rate = 184 beats $\cdot$ min^{-1}
- Blood pressure = 270/96 mmHg
- Respiratory rate = 44 $\cdot$ min^{-1}
- Peak oxygen consumption = 36 ml $\cdot$ min^{-1} $\cdot$ kg^{-1}

His resting electrocardiogram revealed normal sinus rhythm with nonspecific ST-segment changes. No ST-segment changes or arrhythmias were noted during exercise. Before beginning his exercise program, the patient was advised to revisit his internist and have his blood pressure and his antihypertensive medication reevaluated. The physician added a diuretic to the patient's medical regime. When he returned to the exercise facility the next week, his resting blood pressure was 132/90 mmHg.

Exercise Prescription

The clinical exercise professional who performed the exercise evaluation developed the following exercise prescription.

- Exercise mode: stationary cycling or walking or jogging on a treadmill or track for at least 20 min followed by 10 min of aerobic exercise on either a stair stepper, a cross-country ski machine, or an elliptical trainer.
- Exercise frequency: most if not all days each week.
- Exercise session duration: 30 min, not including warm-up and cool-down.
- Exercise intensity or workload: at a workload that will produce a heart rate between 139 and 167 (60–85% of heart rate reserve).
- Exercise progression: As cardiovascular and muscular adaptations occur in response to training, the workload should be adjusted to maintain the heart rate within the initially prescribed range. A few minutes can be added each week to reach a duration goal of 45 to 60 min.

The patient attended an average of four exercise sessions per week and over the next 12 wk increased his exercise session duration from 30 to 45 min. He lost approximately 7 lb (3 kg), and although he complained of still having a bit of a paunch, he was feeling considerably better about how he looked. His therapist had indicated that his depression was improving. His resting blood pressure was 132/84 mmHg, and his resting heart rate was 68. Follow-up laboratory work revealed the following:

- 317 CD4+ cells $\cdot$ mm^{-3}
- Less than 400 HIV RNA copies $\cdot$ ml^{-1}
- Total cholesterol = 249 mg $\cdot$ dl^{-1}
- High-density lipoprotein cholesterol = 39 mg $\cdot$ dl^{-1}
- Triglycerides = 746 mg $\cdot$ dl^{-1}

His follow-up exercise evaluation revealed a maximal heart rate of 188, maximal blood pressure of 220/84, and a maximal respiratory rate of 44. Peak oxygen consumption was measured at 39.8 ml $\cdot$ min^{-1} $\cdot$ kg^{-1}. His electrocardiogram revealed normal sinus rhythm with nonspecific ST-segment changes, and no changes were noted during exercise.

(continued)

Discussion Questions

1. How can skeletal muscle wasting be avoided in people with HIV by using exercise?

2. Develop a resistance-training exercise prescription for this case study patient.

3. What precautions must be taken to guarantee the patient's safety when performing an exercise evaluation on someone who has HIV infection and a CD4+ count of 125 cells $\cdot$ mm^{-3}?

4. What are the risks that a technician will contract HIV during an exercise evaluation? How can any risk be reduced?

5. Why was the patient sent back to his internist?

6. What is likely to limit exercise capacity in individuals with HIV infection?

7. What are the five stages of HIV infection, and what are the several implications for exercise testing and training?

8. What might explain the changes noted in the blood lipid profile of the case study patient?

PART VII

Disorders of the Bone and the Joints

Part VII contains three chapters specific to orthopedic and musculoskeletal conditions. These include a chapter on arthritis, a chapter on osteoporosis and osteopenia, and a chapter on low-back discomfort. These chapters were selected based on the large numbers of people living with these conditions. Certainly, the practicing clinical exercise physiologist is likely to deal with these conditions on an almost daily basis.

Chapter 26 focuses on arthritis. Arthritis currently afflicts up to 40 million people in the United States, predominantly older individuals but also a significant number of middle-aged people. This chapter provides a comprehensive review of the various types of arthritis and the importance of exercise for the treatment of this condition. Additionally, the chapter provides important information regarding techniques to use when designing and implementing an exercise program for the person with arthritis.

Chapter 27 examines osteoporosis and osteopenia. Although one typically thinks of women when consider-

ing osteoporosis, these bone conditions afflict an increasing number of older men. This chapter reviews these conditions and emphasizes the importance of exercise training in the long-term prevention of osteoporosis and osteopenia. The chapter also provides information regarding the development and implementation of exercise-training programs for individuals with established osteoporosis.

Chapter 28 focuses on low-back pain. Issues of low-back pain tend to affect people at younger ages than does either arthritis or osteoporosis. Low-back pain is also the most common reason for work disability in those under age 45. The clinical exercise physiologist should understand that exercise can play an important role in both the prevention and the long-term treatment of low-back pain. Issues such as improved cardiovascular conditioning and abdominal strength and their relationship to restoration of function and reduction in disability are presented.

26

Arthritis

Virginia B. Kraus, MD, PhD

Kim M. Huffman, MD, PhD

Arthritis is a condition that affects the **synovial joint**. It is characterized by inflammation, varying degrees of degeneration of joint structures, and pain. Finding the correct balance of rest and physical activity in the comprehensive management of arthritis is a challenge. Despite traditional beliefs that individuals with arthritis should avoid vigorous physical activity, regular physical activity reduces impairment and improves joint function without aggravating symptoms (130).

SCOPE

In the United States, arthritis is the leading cause of disability (32), and more than 15% (40 million) of Americans have some form of arthritis. The **prevalence** of arthritis increases with age, affecting 50% of persons age 65 yr or older (107), and it is the most prevalent condition in both women and men over 65, followed by high blood pressure and hearing impairment (21,199). Prevalence is higher for women (17%) than men (12%). In 1992 the total cost of arthritis and musculoskeletal conditions was equivalent to 2.5% of the gross national product, or $149 billion (203). By 2020 the prevalence of arthritis in the United States is estimated to reach 18%, owing primarily to the aging of the population (81,203). Encouraging exercise has been identified as one of the ways of reducing the disability and economic burden associated with musculoskeletal disease (24).

There are more than 100 different forms of arthritis. Osteoarthritis (OA) is the most common (12.1%, or 21 million Americans) and has the highest annual incidence. Rheumatoid arthritis (RA) occurs in 1% (2 million) of Americans. Other forms of arthritis include gout (1.6 million), **spondyloarthropathy** such as ankylosing spondylitis (AS) (0.4 million), and juvenile rheumatoid arthritis (30–50 thousand children under age 16) (108). Arthritis is also associated with connective tissue diseases including systemic lupus erythematosus, dermatomyositis, and scleroderma. Arthritis adversely affects psychosocial and physical function and is the leading cause of disability in later life (117). The effect of arthritis on social functioning is impressive: 25% of people with arthritis never leave their home or do so only with help, and 18% never participate in social activities (15). The energy cost of daily activity increases with increasing impairment (66,150) and contributes to prolonged inactivity. Inactivity negatively affects health by increasing the risk of cardiovascular disease, hypertension, diabetes, and obesity (136).

PATHOPHYSIOLOGY

There are genetic components to OA, RA, and AS (123). In addition, injury and joint microtrauma play a major role in OA (168). Each of these arthritides is distinct in presentation and clinical manifestations, yet they have common biological and physiological consequences. For instance, people with any type of arthritis have impaired exercise tolerance (62,134,160). The disease affects the following, each of which contributes to functional limitation in individuals with arthritis:

- Flexibility
- Biomechanical efficiency
- Muscle strength
- Endurance
- Speed
- Proprioception (16,92)

These impairments are usually more pronounced for women.

Muscle weakness is the longest recognized and best established correlate of lower-limb functional limitation in people with knee OA (16). In addition, joint motion is limited (83,95,151). The most common causes of disability are impairment of knee flexion, hip extension, and hip external rotation (186). Limited joint motion impairs joint nutrition (79,94). And, although the joint depends on dynamic loading for maintenance of joint function and joint nutrition, both chronic insufficient and excessive loads are deleterious to joints (27). Osteoarthritis is a dynamic disease process characterized by the uncoupling of the normal balance between degradation and repair of the components of the articular cartilage and subchondral bone (27,159).

Rheumatoid arthritis begins as an **autoimmune** inflammatory process of the synovial lining of the presenting joint and affects the hand joints, with possible involvement of feet, ankles, knees, elbows, and spine. Inflammation of the synovial lining results in erosion of articular cartilage and marginal bone with subsequent joint destruction.

Ankylosing spondylitis is a process of inflammation and erosion at the **entheses**. It involves the sacroiliac joint and the spine. Healing follows, during which new bone is formed, leading eventually to spinal fusion.

There are three levels of classifying arthritis disease stages:

1. Acute: reversible signs and symptoms in the joint related to **synovitis**
2. Chronic: stable but irreversible structural damage brought on by the disease process
3. Chronic with acute exacerbation of joint symptoms: increased pain and decreased range of motion and function, often related to overuse or superimposed injury

Each stage has disease-specific presentations, treatment considerations, and goals. Table 26.1 provides specifics about the stages of each type of these arthritic conditions.

MEDICAL AND CLINICAL CONSIDERATIONS

The various arthritides can be differentiated based on whether symptoms arise from the joint or from a periarticular location, the number of joints involved, their location, whether the distribution is symmetric or asymmetric, and the **chronicity** of disease (18,112). Pharmacological treatment of OA, RA, and AS varies dramatically, but all need to include exercise in their comprehensive management.

Table 26.1 Types of Arthritis, Stages, and Related Impairments

Type of arthritis	Disease stage	Related impairments
Osteoarthritis	Acute joint pain	Often insidious
Chronic radiographic joint disease	Chronic with exacerbation	Increased joint pain and swelling
Rheumatoid arthritis	Acute disease in multiple joints with limited range of motion	Joint stiffness, muscle weakness, and fatigue
Chronic irreversible joint deformity present	Chronic with exacerbation	Increased joint pain and swelling which may occur in one or more joints
Ankylosing spondylitis	Acute spinal pain and stiffness without significant decrease in mobility	
Chronic spinal ankylosis predominant with decreased spinal and thoracic mobility	Chronic with exacerbation	Increased pain and stiffness of the back or peripheral joints

Signs and Symptoms

In the evaluation of individuals with musculoskeletal complaints, the history and physical examination are the most informative elements. Restricted movement of a joint and tenderness to palpation along the axis of joint movement are indicative of arthritis. This contrasts with tenderness around the joint, which is more indicative of periarticular soft-tissue problems. The signs and symptoms of arthritis are as follows:

- Pain
- Stiffness
- **Joint effusion**
- Synovitis
- Deformity
- **Crepitus**

Joint pain may arise from pathological changes in the joint capsule and periarticular ligaments, **intraosseous** hypertension, muscle weakness, subchondral microfractures, **enthesopathy**, **bursitis**, and psychosocial factors including depression (23). Pain does not arise, however, from articular cartilage directly, because it is **aneural**. Stiffness from OA is intermittent and typically of short duration. Stiffness in those with RA and AS varies with the severity of inflammation. The prognosis of recent onset arthritis is aided by determining whether the duration of symptoms has exceeded 4 to 6 wk (91).

History and Physical Examination

The history is essential for determining the duration, location, extent, and severity of musculoskeletal symptoms. It is also useful for obtaining information on current level of functioning and any previous or ongoing efforts at an exercise intervention including any barriers or facilitators to exercise. The physical examination provides the majority of information for establishing the appropriate diagnosis and for recording specific information about any abnormalities of joint range of motion, alignment, or function. The physical examination is also used to assess joints for the four cardinal signs of inflammation: redness, swelling, pain, and heat. The combination of the history and physical examination provide the necessary information for establishing the stage of disease as described in table 26.1.

Diagnostic Testing

The American College of Rheumatology has developed diagnostic criteria for the classification of hip, knee, and hand OA, RA, and AS (4–6,11,69). Table 26.2 on p. 464 summarizes these criteria.

Joint imaging is often used to help confirm a particular arthritis diagnosis. The plain radiograph, which detects bony changes, is the traditional imaging modality. Typical features of OA include **osteophyte** formation, joint space narrowing, cysts, and **subchondral sclerosis**. RA joints develop erosions at joint margins from the invasion of synovial tissue at the intersection of cartilage and bone. Early AS is characterized radiographically as a squaring of the corners of the vertebrae. Later there is evidence of bone formation, ossification, or thin vertically oriented outgrowths that bridge the disc space and limit spinal motion (175). Currently, no specific diagnostic tests or markers of arthritis exist. But several useful serum and synovial fluid laboratory studies, in combination with joint imaging, are often used to support a particular arthritis diagnosis. These include rheumatoid factor, which is often negative early in RA but becomes positive in about 80% of people within the first 2 yr after disease onset (201). Recently, a blood test (antibodies to cyclic citrullinated protein or anti-CCP) became available with high specificity (95%) for RA and for predicting the future development of RA in both healthy individuals and patients with undifferentiated arthritis (13). Like rheumatoid factor, however, anti-CCP is not expressed by all people with RA.

Nonspecific measures of systemic inflammation, such as the sedimentation rate and C-reactive protein, are elevated in conjunction with active synovitis in RA and can be mildly elevated in active AS. C-reactive protein is also predictive of OA status (153,181), and especially high levels of the soluble receptors of TNF-α are associated with lower physical function, increased OA symptoms, and worse knee radiographic scores in older obese adults with knee OA (155). These findings underscore the fact that inflammation can be a manifestation of all three diseases. Synovial fluid analysis is helpful to differentiate the type of arthritis when joint swelling is present and fluid readily obtained. The leukocyte count increases from a normal value of 500 cells · mm^{-3} in a joint to 2,000 cells · mm^{-3} in OA and 5,000 to 15,000 cells · mm^{-3} in RA and AS. Macromolecules originating from joint structures and measurable in blood, synovial fluid, or urine reflect processes taking place locally in the joint (173).

Exercise Testing

Individuals with RA have an increased risk of cardiovascular disease, possibly related to the inflammatory disease process (14,45,93). Moreover, people with arthritis tend to be more deconditioned than are sedentary individuals without arthritis, increasing the risk of cardiovascular disease in general. A symptom-limited

Table 26.2 Distinguishing Characteristics and American College of Rheumatology (ACR) Diagnostic Criteria for Arthritis

Arthritis type	Distinguishing characteristics	Presentation	ACR criteria

OA	Joint pain
	Crepitus
	Gel phenomenon
		Affects hands, hips, knees, lumbar, and cervical spine
		Pain worsens throughout the day
		Affects any traumatized joint
			Knee clinical:
			  Knee pain and three of following:
			    a. Age >50 yr
			    b. Morning stiffness <30 min
			    c. Crepitus
			    d. Bony tenderness
			    e. Bony enlargement
			    f. No warmth
			
			Knee clinical and radiographic:
			  Knee pain and one of following:
			    a, b, or c clinical criteria (see above)
			    Osteophytes on knee X-ray
			
			Knee clinical and laboratory:
			  Knee pain and five of following:
			    a–f clinical criteria (see above)
			    ESR <44 mm · hour^{-1}
			    RF <1:40
			    Synovial fluid compatible with OA
			
			Hip combined clinical, laboratory, and radiographic:
			  Hip pain and one of following:
			    ESR <20 mm · hour^{-1}
			    Osteophytes on hip X-ray
			    Joint space narrowing on hip X-ray
			
			Hand clinical:
			  Hand pain or stiffness and three of following:
			    Bony enlargement of two or more DIPs
			    Bony enlargement of two or more of 2nd and 3rd DIPs, 2nd and 3rd PIPs, 1st CMC
			    Fewer than three swollen MCPs
			    Deformity of at least one of 2nd and 3rd DIPs, 2nd and 3rd PIPs, 1st CMC

RA	Hand pain
	Swelling
	Fatigue
	Prolonged morning stiffness
		Affects wrists, MCPs, and PIPs
		Symmetric
			At least four of following:
			  Morning stiffness >1 h*
			  Simultaneous swelling in three or more of PIPs, MCPs, wrists, elbows, knees, ankles, MTPs*
			  One swollen wrist, MCP, PIP*
			  Symmetric joint swelling*
			  Rheumatoid nodules
			  RF >1:40
			  Radiographic changes including erosions and periarticular osteopenia

Arthritis type	Distinguishing characteristics	Presentation	ACR criteria

AS	Low-back pain
	Low-back stiffness
		Early: discovertebral bone erosion
		
		Late: vertebral fusion and sacroiliac joint fusion (via bone formation, ossification)
			Clinical:
			  Inflammatory back pain (improves with activity and associated with stiffness and worsening with inactivity)
			  Limited lumbar spine mobility
			  Limited chest expansion
			At least one of preceding and at least one of following:
			  Radiographic
			  Grade 2 bilateral sacroiliitis
			  Grade 3-4 unilateral sacroiliitis

*Present for at least 6 wk.

Note. ACR criteria based on the following references: 4–6, 11, 169. OA = osteoarthritis; RA = rheumatoid arthritis; AS = ankylosing spondylitis; ESR = erythrocyte sedimentation rate; RF = rheumatoid factor; DIP = distal interphalangeal joint; PIP = proximal interphalangeal joint; CMC = carpal-metacarpal joint; MCP = metacarpal phalangeal joint.

exercise test should be considered to screen for coronary artery disease, when appropriate, in those wishing to exercise and to assist with the cardiovascular exercise prescription (26). An exercise test is also commonly used to assess cardiovascular status for surgical risk before joint replacement. Joint symptoms and fatigue, however, may adversely affect performance on an exercise test and may prohibit maximal testing (19,40).

Contraindications

Arthritis and musculoskeletal conditions are often listed as relative contraindications to graded exercise testing (26). But one study found that of those with severe end-stage hip or knee arthritis attributable to OA or RA, 95% were capable of performing a symptom-limited exercise test using cycle ergometry methods (158). The majority

Practical Application 26.1

Client–Clinician Interaction

The individual with arthritis should be screened for conditions that will guide the exercise prescription. Suggestions for questions that should be included as part of the exercise evaluation follow.

1. Individual's age, level of fitness, medications, personal goals, and lifestyle
2. Names of healthcare providers including primary care physician, rheumatologist, orthopedist, clinical exercise physiologist, and physical therapist
3. Musculoskeletal disease diagnosis
4. Pattern of joint involvement: (a) **symmetric** or **asymmetric**, (b) upper- or lower-extremity involvement, (c) joints affected
5. Severity of disease activity (acute, chronic, or chronic with acute exacerbation)
6. Comorbidities (other medical conditions, including pulmonary disease, fibromyalgia, Raynaud's phenomenon, Sjogren's syndrome, osteoporosis)
7. Surgical history including joint replacements
8. Previous treatment (whether successful or not)
9. Presence of fatigue
10. Adequacy of footwear

also achieved a respiratory exchange ratio greater than 1.0, indicating a metabolically maximal test. Approximately two-thirds of subjects were capable of completing tests by pedaling with their legs, and the remainder performed the same task using their arms. Standard contraindications for exercise testing should also be followed (chapter 5).

Recommendations and Anticipated Responses

Cycle ergometry is generally the preferred mode because of reduced baseline exercise capacity and the high frequency of lower-extremity impairment. Testing by arm ergometry may be necessary in those with severe lower-extremity joint pain or with significant deformity that might cause the extremity to strike the flywheel (such as a **varus** or valgus deformity of the knee). The treadmill may be used in those with minimal or no functional disability. Testing procedures for people with arthritis are similar to protocols recommended for elderly and deconditioned people (97). These tests should have small incremental changes in workload, for instance, increments of 10 to 15 $W \cdot min^{-1}$ on the cycle ergometer or the modified Naughton protocol when using a treadmill (22). Submaximal testing does not provide optimal diagnostic information about cardiovascular disease but can be used to predict aerobic capacity (135). Standard equations by the American College of Sports Medicine that estimate $\dot{V}O_2$ based on exercise testing are inaccurate in an elderly population with arthritis and tend to overestimate $\dot{V}O_2$peak. This in turn could lead to

overprescribing exercise intensity for training (22). The following equation was developed to predict $\dot{V}O_2$peak in seniors with knee OA or cardiovascular disease: $\dot{V}O_2$ ($ml \cdot min^{-1} \cdot kg^{-1}$) = 0.0698 × speed ($m \cdot min^{-1}$) + 0.8147 × grade (%) × speed ($m \cdot min^{-1}$) + 7.533 $ml \cdot min^{-1} \cdot kg^{-1}$. This equation, valid in both men and women, requires that participants use front handrails for support during the exercise test (22). This is an advantageous method of testing an arthritic individual with lower-extremity disability in whom standard nonhand support methods of treadmill testing may be hazardous. In general, cycle ergometry is the best method to use for testing because of its non-weight-bearing nature and reduced reliance on balance. But equations for estimating $\dot{V}O_2$ by cycle ergometry have not been validated for this population. Most of those with arthritis are able to achieve maximal cardiovascular effort using cycle ergometry (158). Recommendations for exercise testing are outlined in tables 26.3 and 26.4.

Treatment

A comprehensive treatment strategy for arthritis should strive to counteract inactivity and reduce disability and handicap. This approach may include use of the following:

- Pharmacological agents
- Braces and bandages
- Activity modification
- Exercise

Table 26.3   Cardiovascular Exercise Testing in Arthritis

Mode	Protocol specifics	Clinical measures	Clinical implications	Special considerations
Use a treadmill for those with minimal to mild joint impairment.	Use protocols with small increment increases (i.e., Modified Naughton) unless minimal disease activity and severity is present.	• Assess type of arthritis, degree of activity, and impairment. • Assess comorbidities, past surgical and medical history.	ACSM peak $\dot{V}O_2$ prediction equations may overestimate functional capacity.	With patient using handrails for support, use following equation to predict $\dot{V}O_2$max[a].
Use cycle ergometry for those with mild to moderate lower-extremity impairment.	Use protocols with small increment increases (i.e., 10–15 $W \cdot min^{-1}$) or ramping protocols.	• Assess type of arthritis, degree of activity and impairment. • Assess comorbidities, past surgical and medical history.	ACSM peak $\dot{V}O_2$ prediction equations may overestimate functional capacity.	Additional investigations are needed to improve prediction of peak $\dot{V}O_2$

Mode	Protocol specifics	Clinical measures	Clinical implications	Special considerations
Use arm ergometry for those with severe lower-extremity impairment.	Use arm ergometry specific protocols with small increment increases or ramping protocols.	Assess type of arthritis, degree of activity and impairment. Assess comorbidities, past surgical and medical history.	ACSM peak $\dot{V}O_2$ prediction equations may overestimate functional capacity.	Additional investigations are needed to improve prediction of peak $\dot{V}O_2$. Consider submaximal testing in those with severe impairment.

$^a\dot{V}O_2 \,(ml^{-1} \cdot min^{-1} \cdot kg) = 0.0698 \times speed \,(m \cdot min^{-1}) + 0.8147 \times grade \,(\%) \times speed \,(m \cdot min^{-1}) + 7.533 \,ml^{-1} \cdot min^{-1} \cdot kg \,(14).$

Table 26.4 Strength, Range of Motion, and Balance Testing in Arthritis

Test type	Mode	Protocol specifics	Clinical measures	Clinical implications*
Lower-extremity testing	Dynamometer Timed stands test	All testing in supine position except knee flexion and extension (while seated) Time to rise 10 times from standard chair	Reference ranges for 50–79 yr olds (9) Reference ranges (36)	Often (>50%) decreased in persons with arthritis
Upper-extremity and grip testing	Hydraulic dynamometer (Jamar) Electronic dynamometer (Grippit)	In seated position with unsupported arm flexed 90° at elbow Peak grip force Average sustained force	Reference ranges for position 2 (118) Reference ranges (118, 145)	Usually (>90%) decreased in persons with arthritis
Range of motion	Goniometer	Align device fulcrum with joint fulcrum	Reference ranges	Usually (>90%) decreased in persons with arthritis
Balance	Figure-eight walking* Berg balance scale*	Useful in those with limited or mild impairments* Track width = 150 mm Inner diameter = 1.5 m Outer diameter = 1.8 m Useful in those with moderate to severe impairments* Includes 14 single tasks beginning with sitting unsupported and progressing to standing on one leg	More than two oversteps during two circuits suggests decreased balance Median ranges for RA functional classes (147)	Often (>50%) decreased in persons with arthritis

*Based on investigations in persons with RA (53, 147).

Previously, the traditional standard of treatment for an arthritic joint was rest (28). Current practice has shifted toward joint mobilization because of substantial evidence of the beneficial effects and relative safety of exercise for individuals with arthritis (35,102,140,177,204). As depicted in figure 26.1, exercise interventions can interact at each stage in the progression from arthritis pathology to handicap and can help to mitigate the effects of the disease process including associated disability (98).

The implementation of exercise as part of a comprehensive therapeutic management strategy is a challenge. Physicians rarely recommended exercise to their patients with arthritis, and if doctors do recommend exercise, they provide little instruction (46). People with arthritis are typically poorly motivated to perform physical activity on their own. During leisure time, the period most amenable to efforts to increase physical activity, adults and children with arthritis are less active than healthy individuals (10,101,134). Persons with arthritis, especially those with multiple functional limitations, no access to exercise facilities, and needs for special equipment, are significantly less likely to engage in recommended amounts of moderate to vigorous exercise (179). Similarly, despite known benefits of community-based arthritis exercise programs such as the Arthritis Foundation YMCA Aquatics Program (AFYAP) and the

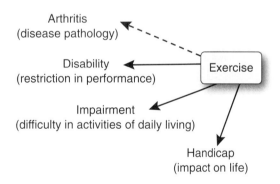

Figure 26.1 The World Health Organization classification of impairments, disabilities, and handicaps is presented to demonstrate the potential for a positive impact of exercise as an interactive and mitigating factor in this process. Strong evidence exists for benefits of exercise at the level of disability, impairment, and handicap with arthritis. Much remains to be learned about the effects of exercise on the level of disease pathology. Most studies show no worsening of arthritis with exercise, and a few suggest a beneficial effect on disease pathology itself.

Adapted from R. Shephard and P. Shek, 1997, "Autoimmune disorders, physical activity, and training, with particular reference to rheumatoid arthritis," *Exercise Immunology Review* 3:53-67.

land-based People with Arthritis Can Exercise (PACE), less than 1% of eligible persons with arthritis access these programs (90). Thus, individuals with arthritis require education and encouragement to increase and maintain appropriate physical activities. In addition to exercise, multiple nonpharmacological interventions are used in the rehabilitation of people with chronic arthritis (39):

- Education
- Physical and occupational therapy
- Braces and bandages
- Canes and other walking aids
- Shoe modification and orthotics
- Ice and heat modalities
- Weight reduction
- Avoidance of repetitive motion occupations
- Joint irrigation and joint surgery (in select circumstances)

The judicious use of rest is beneficial and generally recommended between days of dynamic exercise, particularly for those with systemic, multiple-joint disease. Joint replacement has markedly improved the quality of life for individuals with knee and hip arthritis. Surgery is usually reserved for unremitting pain or disability when conservative measures have failed. An estimated 652,000 joint (knee and hip) replacements are performed every year in the United States (103), and more than 70% of these are for OA (55).

Pharmacological therapies can be administered orally, topically, or directly into joints. Topical agents include capsaicin cream and topical nonsteroidal anti-inflammatory drugs (NSAIDs). Oral agents frequently used to treat arthritis include analgesics in the form of acetaminophen, opioids, and oral NSAIDs. Dosing of an NSAID 1 h before exercise may aid exercise compliance by minimizing pain and stiffness during activity. A variety of disease-modifying agents currently exist for RA and AS (methotrexate, leflunomide, sulfasalazine, hydroxychloroquine, and tumor necrosis factor inhibitors [etanercept, infliximab, and adalimumab]) but very few for OA. Glucosamine and chondroitin sulfate are naturally occurring constituents of **articular** cartilage. Although data regarding efficacy are conflicting (33,82), taken orally, these may reduce OA pain and may slow progression of the disease with minimal side effects (154,163,200). In persons with OA, being overweight is associated with increased pain and disability (7,41,54) and may worsen disease activity (49). The inclusion of dietary intervention in combination with exercise may be necessary when weight loss is a goal for the arthritic person (128). Other dietary considerations for OA management may include adequate (but not excess) intake of vitamin D and selenium (119,120), and for RA, n-3 fatty acids found in fish oil or cold-water fish. These may decrease inflammation (34). Exercise may have disease-modifying effects for arthritis because it decreases circulating CD4 (helper) lymphocyte counts, reduces the length of hospitalizations in RA patients (178), and reduces the number of inflamed joints in OA and RA (133). Exercise does not put a person at increased risk for OA. In fact, exercise is one of the mainstays of therapy for OA, RA, and AS recommended in national U.S. (2,87,88) and European (96,205,206) guidelines. In a few instances, however, exercise may promote arthritis. For instance, sports participation associated with an increased risk of injury may subsequently increase OA risk (105). In general, however, physical activity promotes joint health by promoting cartilage nutrition and weight loss, which decreases the risk of knee and hip OA.

EXERCISE PRESCRIPTION

Exercise benefits for the arthritic population include increased muscle strength, increased flexibility, improved sense of well-being, improved coping ability, weight control, enhanced quality of sleep, decreased blood pressure, and fewer heart attacks (140). The following are goals of exercise for the treatment of arthritis:

- Reduce inflammation and pain
- Prevent contractures and deformities
- Maintain or improve range of motion, muscle strength, and cardiovascular fitness

Immobilization and inactivity amplify the negative systemic and psychological manifestations that accompany arthritis (61). The effects of inactivity include rapid loss of strength (about 3–8% per week), loss of endurance, negative nitrogen balance reflecting a loss of muscle protein, and loss of cartilage matrix components (20,99,106). Because cartilage is **avascular**, it depends on normal repetitive loading of the joint for its nutrition and normal physiological function (89). Moreover, joints with **effusions** may develop synovial ischemia attributable to elevated intra-articular pressure. Walking and cycling increase synovial blood flow in inflamed knees (94). The intra-articular oxygen partial pressure increases during joint movement in both healthy joints and those affected by OA, but arthritic joints demonstrate a smaller increase than normal joints do (129). People with arthritis can achieve normal training heart rate ranges (125). Thus, standard heart rate range development techniques can be used (see chapter 5). Supervised exercise training for individuals with arthritis most often occurs in a group setting (versus personal training) and is well supported as a treatment modality (35,52,77,113,133,141,143,192). A person in an exercise program, whether supervised or unsupervised, requires education, skill acquisition and reinforcement, and regular monitoring by health professionals. A supervised group setting may be beneficial because of peer social support in a positive environment. Regular monitoring by clinical exercise professionals helps to ensure accuracy and safety of exercise performance. Compliance with exercise is associated with improved physical function (196). Access to a physical therapist or clinical exercise physiologist who has previously evaluated the individual may help decrease anxiety and improve compliance (182). Additionally, the clinical exercise physiologist may play a key role by encouraging regular attendance at sessions early in the treatment regimen, because session attendance in the initiation phase is the strongest predictor of session attendance in the later stages of the intervention (195).

Although in AS, improvements in spinal mobility appear superior with supervised programs, home-based exercise programs are clearly better than no exercise (37). For OA, supplementing home-based exercise with supervised classes may boost responses in pain and function (47,122,165). In addition, participants who do exercise partly at home during the maintenance stage of the intervention are more likely to adhere to the exercise program than peers who exercise only at a facility (195). Thus, a component of unsupervised and independent exercise is important to the long-term success of a person's exercise routine.

Many exercise options are available for unsupervised programs. These include cardiovascular and strengthening exercises through an independent water exercise program, including water walking and independent joint range-of-motion exercises. Videotapes of exercise programs suited to people with arthritis are available through the Arthritis Foundation. Other activities and exercises that can be performed throughout the day include chin tucks, corner pectoral stretches in a doorway, abdominal tightening, checking posture throughout the day in the mirror when in the bathroom, and extending walking time by taking the stairs or parking farther from a destination. For individuals with RA and severe morning stiffness, active range-of-motion exercises performed within 15 min of bedtime can decrease morning stiffness (30). Walking in climate-controlled buildings such as fitness centers or shopping malls is an excellent option for cardiovascular exercise. The ability to perform a variety of cardiovascular exercises may help maintain interest and compliance. In addition, cross-training may prevent overuse injuries. Range-of-motion exercises in a pool can be performed as a component of a supervised or unsupervised exercise program. Following the Arthritis Foundation's low-intensity, recreational aquatic program (AFYAP), available nationally in YMCAs and in private facilities, increases hip range of motion, isometric strength, and flexibility when performed two to three times a week over 6 to 8 wk (187,188). The AFYAP exercises, however, may not be of sufficient intensity to increase strength and ROM in joints not affected by arthritis (188) and can be supplemented by strength training and joint ROM activity on land.

Special Exercise Considerations

The principles of exercise training vary according to disease stage. In establishing an exercise prescription, the exercise professional should consider the success of various treatment modalities for particular joint impairments as well as the individual's affected areas, level of fitness, surgical history, comorbidities, medications, age, personal goals, and lifestyle. Inflammation and joint degeneration associated with the disease process cause a cycle of decreasing function and increasing impair-

ment. It is not yet clear whether therapeutic exercise alters the pathological process of arthritides. Performed judiciously, however, exercise therapy can prevent, retard, or correct the mechanical and functional limitations that may occur throughout the course of the disease.

Exercise therapy can induce pain, the primary complaint and disabling factor related to arthritis. This effect can make it difficult to motivate a person to maintain an exercise program. In an arthritic population that is generally older, is commonly sedentary, and may be using systemic steroids, avoidance of injury and pain is important to maintaining exercise compliance (17). The clinical exercise physiologist's role is to make recommendations that minimize these symptoms. Attention to affected joints is important because they have decreased range of motion, instability, reduced muscle strength and flexibility, poor joint proprioception, and increased pain. The clinical exercise physiologist should query the individual and the medical record to determine which joint sites are affected. This record may be refined with information from the individual's rheumatologist, primary care physician, or physical therapist. Impaired balance and increased fatigue are additional factors that must be considered in developing an exercise program. Site-specific exercise recommendations are listed in table 26.5.

Table 26.5 Arthritis Site-Specific Recommendations

Site	Condition	Presentation	Recommendation
Lower extremity			
Foot	Hallus valgus	Lateral deviation of the first digit of the foot leading to bursitis at the metatarsophalangeal joint (bunion)	Padding between first and second toe Prescription for footwear that decreases need for extension of the joint during walking If moderate to severe, avoid walking for exercise
Foot	Plantar fasciitis or heel spur	Sharp foot pain on weight bearing, especially in heel; worse with first steps upon wakening and with walking	Avoid weight-bearing activities that increase pain Wear shoes with supportive arch; may need soft or semirigid foot orthoses or heel cups Stretch calf muscles throughout the day Wear night splints to maintain stretch in dorsiflexion

Site	Condition	Presentation	Recommendation
	Lower extremity		
Leg	Musculotendinous inflammation of the anterior or posterior tibialis muscle (shin splints) or compartment syndrome	Aching pain in anterior lateral or posterial medial leg	Avoid painful activities (usually walking) for as much as 2 wk May need improved footwear or orthotics Evaluate training routine for sudden change that preceded onset Acute onset with excruciating pain and area hard to the touch requires immediate medical attention—send to emergency room
Hip or knee	Hip OA, knee OA	Gait deviation Pain with weight bearing or stair climbing	Patellar femoral OA, hip or groin pain without gait deviation or balance problems; rearward walking Patellofemoral OA; leg press; no leg extension Hip or groin pain: hip bridging, free-speed walking, stationary cycling If gait deviation is caused by pain or decreased joint ROM, may need to use cane or rolling walker in the hand opposing the painful limb
Hip	Bursitis	Lateral hip pain, may extend to lateral knee Cannot tolerate lying or single-leg stance on affected side	Control inflammation: ice lateral hip Should see doctor to rule out other problems and for medication to control inflammation Should see physical therapist
Knee	Valgus or varus deformity	Valgus deformity often called knock knees; varus deformity often called bow-legged	Avoid weight-bearing exercise Perform open-chain strengthening May need prescription for wedged insoles
	Upper extremity		
Shoulder	Shoulder pathology, possibly adhesive capsulitis, tendinitis, or bursitis; may be secondary to OA	Pain with overhead or end-range motion	Perform shoulder ROM in pain-free range in pool with UE submerged If pain disturbs function or sleep, should see doctor Would benefit from physical therapy
Hand	OA of carpal-metacarpal joint of the thumb	Pain in hand proximal to thumb	Avoid gripping activity during exercise Enlarge grips
Hand	Ulnar deviation in RA	Body of hand and fingers deviate to the small digit side of the hand	Avoid gripping activity during exercise Use large muscles and joints for functional activities

(continued)

Table 26.5 *(continued)*

Site	Condition	Presentation	Recommendation
Axial skeleton			
Lumbar spine	Spinal stenosis	Flexed low back when walking, standing, and sitting Symptoms increase with extension (standing, looking overhead) Often presents with claudication-type pain with walking	Tolerates flexion exercises, seated cardiovascular exercise (recumbent bicycle or stair stepper) Aquatic exercise Rolling walker for household or community ambulation
Cervical spine	Atlantoaxial subluxation in patients with RA	Facial sensory loss Vertigo with cervical extension Numbness or tingling of hands or feet Difficulty walking Loss of control of bowel or bladder Transient loss of consciousness when extending cervical spine May be asymptomatic	Avoid any passive or heavy resistive neck ROM Needs immobilization if unstable (many symptoms) Surgery if neurological signs progress Gentle active ROM, low repetitions, no extension
Cervical or lumbar spine	Nerve compression secondary to OA	Gradual, recurrent pain or pain after activity Numbness, tingling, or pain in the extremities, sometimes only with certain movements	Avoid activity that results in numbness, tingling, or pain in the extremities

Disease Staging

A primary consideration in developing an appropriate exercise prescription for a person with arthritis is the disease stage. The focus of exercise therapy for chronic stages of arthritis should be to maintain or improve function while minimizing or avoiding exacerbations. Most people referred for exercise therapy will be experiencing arthritis in a chronic stage. Table 26.6 lists the arthritis stages and their associated exercise-related particulars. Practical application 26.2 on p. 474 provides site-specific recommendations for the clinical exercise professional to follow for arthritic patients with disease-specific skeletal conditions.

Other Factors

Compliance with an exercise program can be a challenge for people with arthritis. Efforts on the part of the exercise professional to prevent musculoskeletal injury from exercise and to appreciate the specialized concerns of the individual with arthritis will facilitate overall enjoyment and compliance with the exercise program. Some special considerations for people with arthritis are described next.

Preventing musculoskeletal injury secondary to exercise

Because cardiovascular exercise involves high repetition of joint motion, risk of overuse injuries is present. Fortunately, injuries attributable to supervised exercise are infrequent. Estimates are that 2.2 minor injuries occur per 1,000 h of exercise, and 0.48 major injuries occur per 1,000 h of exercise that lead to reduction or discontinuation of exercise for at least 1 wk (35). Overuse of the soft tissue and bone may be minimized in this population by performing interval or cross-training during endurance exercise. Examples include the following:

1. Alternating cycle ergometry between 25% and 75% of the maximum work rate performed during a graded exercise test
2. Alternating between water walking and joint range-of-motion exercises in a pool setting
3. Walking and weight training
4. Walking and higher-intensity cardiovascular exercise such as recumbent stair stepping or cycling

Repeated loading of the lower limb during walking and other daily activity may contribute to the onset and

Table 26.6 Arthritis Stages, General Signs and Symptoms, and Exercise-Related Considerations

Stage	Signs and symptoms	Exercise considerations
Acute	Fatigue Joint pain Reduced joint tissue tensile strength attributable to inflammation Reduced joint nutrition	Teach energy conservation Perform non-weight-bearing and limited and slow ROM exercise Avoid stairs, carrying loads >10% body weight, fast walking, isotonic resistance training (40, 83) Perform isometrics at <40–50% of maximum voluntary contraction to limit blood flow reduction
Chronic	Permanent joint damage Pain at end of normal ROM Stiffness after rest Poor posture and ROM Joint deformities Pain with weight bearing Abnormal gait Weakness Contractures or adhesions Reduced aerobic endurance	Initiate walking and perform in water if necessary to reduce pain Perform low-back flexion and abdominal strengthening exercise Perform lower-body strengthening and ROM exercises (83) Avoid trunk extension (especially those with spinal stenosis) Use weights during resistance training that don't cause joint pain Maintain neutral spine position Avoid oral corticosteriods to reduce risk of osteoporosis and ligament laxity
Chronic with acute exacerbation	Inflammation and joint size greater than normal Joint tenderness, warmth, swelling Joint pain at rest and with motion Stiffness Functional limitations Hips and spine affected	Rest Normalize gait Same recommendations as for acute phase

Note. ROM = range of motion.

progression of knee OA. In contrast, the development of strong knee extensors with quadriceps-strengthening exercise decreases impulse loading of the lower limb during walking by slowing the deceleration phase occurring before heel strike. Therefore, in a normal joint, adequate quadriceps strength in deceleration may help prevent knee injury and knee OA. People may have laxity in the structures that support a joint because of the rheumatic process or because of the use of corticosteroids. Under these circumstances, the joint should be protected when the person performs exercise or normal activities. Protection is afforded by cautiously stretching to avoid extending beyond the functional range of motion. Vigorous stretching or manipulative techniques are contraindicated (100). In many cases, providing external support to a joint may be necessary. To protect smaller

joints during activities and exercises, larger muscles and joints should be used.

Fatigue

Fatigue is common in those with rheumatic disease and can profoundly affect quality of life. Fatigue beginning in the afternoon and lasting until evening, and morning stiffness lasting for 3 or 4 h, are common symptoms in people with RA, leaving only a few hours during midday when stiffness or fatigue are not a problem (201). Fatigue is a complex phenomenon related to exertion, deconditioning, depression, or a combination of these factors (20). A person who is fatigued cannot exercise for a long duration, can exercise only at lower intensity, cannot tolerate progression of exercise volume, tends to be less motivated, and may become frustrated. Appropriate

Exercise Prescription Based on Disease Staging

Training method	Mode	Intensity	Frequency	Duration	Goals	Special considerations and comments
Acute stage of arthritis						
Aerobic	Activities using large groups, minimizing involvement of affected joints	Low intensity	As prescribed by medical professional	Minimize cardiorespiratory endurance and function loss	Medical supervision required	
Strength training	Isometrics at functional joint angles	<50% maximum voluntary contraction	5–10 repetitions daily	6 s while exhaling	Minimize muscle atrophy	May increase blood pressure, intra-articular pressure, and joint contact forces
Range of motion	Passive or active-assisted ROM[a]	Within limits of pain	Twice daily	Several minutes	Prevent deformity	Consider supportive and assistive equipment and immobilizing splints for active, painful joints or those causing gait deviation
Chronic stage of arthritis						
Aerobic	Activities using large groups with repetitive motion: Walking Cycling Dancing Aquatics	60–80% MHR; 50–70% $\dot{V}O_2$ max; RPE 12–16 (Borg scale 6–20) or 3–6 (Borg scale (1–10) OA: 50–70% MHR (1) RA: 60–85% MHR (1)	Begin at 2–3 d per week and increase to 5–7 d per week; more frequently for lower-intensity activities	30–60 min	Improve cardiovascular fitness, function and quality of life and reduce pain	See text regarding the following: Injury Fatigue Joint replacement Time of day Aquatic therapy Footwear Pulmonary manifestations Corticosteroids Body composition

474

Training method	Mode	Intensity	Frequency	Duration	Goals	Special considerations and comments
<td colspan="7" align="center">**Chronic stage of arthritis** (continued)</td>
Strength training	Isotonic with isometric	Begin against gravity then progress to about 70% 1RM OA: 40–80% 1RM (3) RA: 50–80% 1RM (1)	Rest between sessions OA: 2 d per week (3) RA: 2–3 days per week (1)	OA: 8–10 exercises 4–6 repetitions One set (3) RA: 8–10 exercises 8–12 repetitions One or two sets (1)	To improve strength and function and reduce pain	Perform in pain-free range Use functional movement patterns Include all groups of both LEs, even with unilaterally distributed joint dysfunction
Range of motion	Static stretching (see chapter 5) Functional activities: sit to stand, stairs	To mild discomfort	Daily	10–30 s 3–5 repetitions	Improve range of motion and function, prevent deformity, and reduce pain	Target shortened muscle groups Perform active ROM within the normal range for all LE joints, even with unilaterally distributed joint dysfunction
<td colspan="7" align="center">**Chronic stage of arthritis with local exacerbation of symptoms**</td>
Aerobic	Follow recommendations for chronic stage	Improve cardiorespiratory endurance	<td colspan="4">Use uninvolved joints only</td>
Strength training	Exacerbated joints: follow recommendations for acute stage Nonexacerbated joints: follow recommendations for chronic stage	Minimize muscle atrophy				
Range of motion	Exacerbated joints: follow recommendations for acute stage Nonexacerbated joints: follow recommendations for chronic stage	Prevent deformity, reduce pain, improve range of motion and function				

Note. ROM = range of motion; MHR = age predicted maximum heart rate; RPE = rating of perceived exertion; OA = osteoarthritis; RA = rheumatoid arthritis; RM = repetition maximum; LE = lower extremity.

[a]Passive ROM involves no muscle work by the individual while an outside force (another person or passive motion machine) moves the body part through a range of motion. Active range of motion is movement of a body part by the individual performing the exercise without outside forces. Active assisted ROM involves partial assistance with motion by an outside force, whereby a portion of the motion of the limb may be provided by a mechanical device, another limb, or another person.

exercise training may decrease fatigue levels without worsening arthritis (141). As little as 15 min of exercise three times a week is sufficient to improve aerobic capacity in those with RA and fatigue (77).

Previous joint replacement

Total joint replacement surgery is common in people with arthritis. Decreased joint range of motion occurs after surgery but does not typically affect function. Individuals with lower-extremity joint replacement should avoid all high-impact activity. Those having had hip replacement should not flex their hip past 90° or adduct or internally rotate the hip past neutral. Many surgeons terminate these precautions after 2 to 6 mo. Without physician confirmation, a conservative recommendation is to maintain these precautions with all exercise activity.

Time of day

Morning stiffness is a problem for individuals with arthritis. The exercise professional must be sensitive to the daily variability of symptoms, the difficulty of arising and performing activities of daily living, and, as a result, the difficulty of early morning activity. Moreover, a change in ability to perform exercise during periods of inclement weather is frequently reported by individuals with rheumatologic conditions. A drop in barometric pressure along with an increase in humidity can increase pain (107). For those with inflammatory arthritis characterized by prolonged morning stiffness, exercise should be prescribed for the late morning or early afternoon (61).

Special considerations regarding aquatic therapy

RA is commonly associated with Raynaud's phenomenon and Sjogren's syndrome. Raynaud's phenomenon is a **vasospastic** problem presenting as blanching or cyanosis of the hands and feet when exposed to cold or emotional stress. It can result in pitting scars or gangrene. Individuals with Raynaud's phenomenon should avoid cool air and water and wear protective clothing including noncotton gloves, shirt, pants, and shoes. The choice of exercise modality should be dictated by the symptoms attributable to arthritis and Raynaud's phenomenon. Sjogren's syndrome is an autoimmune condition characterized by dry mouth and eyes. It is caused by the infiltration of salivary and lacrimal glands with lymphocytes. People with this condition may find chlorinated water and the air surrounding pools especially irritating to the eyes and should wear goggles when in a pool.

Footwear

Use of appropriate footwear can reduce the risk of injury related to poor lower-extremity mechanics and repeated shock. Lightweight commercial athletic shoes that include hind foot control, a supportive midsole of shock-absorbing materials, a continuous sole, and forefoot flexibility can improve shock attenuation and biomechanics. Individuals with OA, RA, and AS with biomechanical faults in the lower extremity may need custom-made rigid or semirigid orthotics from a podiatrist, orthotist, or physical therapist. People with lower-extremity arthritis should be advised to wear pool shoes to assist mobility and protect feet from injury during aquatic exercise.

Pulmonary manifestations of rheumatic disease

Pulmonary disease can be associated with arthritic conditions. Rheumatoid arthritis is associated with interstitial lung disease (8). Those with AS often have restrictive lung disease caused by impairment of chest expansion. In most cases, pulmonary manifestations of rheumatologic disease are not absolute contraindications to exercise. A vital capacity of ≤1 L should be a relative contraindication to participating in pool therapy because of the restrictive effects of water against the chest wall (60). Chapters 21 and 22 review exercise specifics for those with pulmonary limitations.

Ankylosing spondylitis

With AS, the bony fusion that occurs in the spine and sacroiliac joint cannot be prevented, but rehabilitative strategies can improve compromised range of motion and function (60). For this reason, therapy should focus on strengthening and stretching exercises to improve joint range of motion and on improving static and dynamic posture. Because individuals with AS tend to be younger and more active at diagnosis, exercise for AS can be performed at a higher intensity than for RA and OA (57,161). When peripheral joints are involved, disease pathology is similar to RA, and thus exercise recommendations specific for RA should apply. A phenomenon that can occur is called the last-joint syndrome. With this, bridging ossification between vertebral bodies occurs at every level except one (78). This sole mobile segment is exposed to considerable stresses during exercise and can present with localized pain and discitis. Rest, bracing, or surgical fusion may be necessary.

Corticosteroids

Systemic corticosteroids are a common treatment for RA and can lead to bone fragility and muscle atrophy. Muscle wasting from disuse attributable to pain and steroid-induced myopathy contribute to reductions in muscle strength (40). For those with RA who are using systemic steroids, exercise training should be modified so as not to affect disease activity adversely (113).

Body composition

Obesity is a modifiable factor that negatively affects various arthritides. In particular, obesity is a strong risk factor for OA **incidence**, progression, and disability (128). A high body mass index (BMI) is associated with OA of the knee, hands, and feet combined (56). The mechanism is related to an increase in mechanical stress on weight-bearing joints (56). A 5 kg weight loss decreases the risk by 50% of developing knee OA within 10 yr (56). Rheumatoid arthritis is often associated with cachexia and low BMI. Some people with RA, however, have high body mass indices along with obesity-related health risks. Weight loss is difficult in this population because loss of muscle mass must be avoided. A study of obese subjects with RA found that a program of moderate physical training, reduced dietary energy intake, and a high-protein, low-energy supplement was successful in achieving a significant weight loss without loss of lean tissue (51).

Exercise Recommendations

The sequencing of exercises for individuals with arthritis is similar to that of the general population, beginning with a warm-up and ending with a cool-down. A warm-up should be performed to increase tissue temperature throughout the body. As greater range and decreased stiffness evolve, the individual should be taught to judge whether increasing the range through which he or she is exercising is safe (176). Superficial heat has a symbiotic effect and may be used immediately before the warm-up. Skeletal muscle strengthening and cardiovascular conditioning exercises should be performed following warm-up and followed by a cool-down period. Flexibility exercises should be performed during the cool-down. Laliberte et al. provide extensive and useful examples and illustrations of various sequences of exercise appropriate for people with arthritis (104).

Mode

Isotonic exercise is preferred over isometric exercise for dynamic strength training during the chronic stage of arthritis (150). Isotonic exercise is advantageous because it closely corresponds to everyday activities and therefore promotes improved daily function. Low-intensity isometric exercise is preferred for muscle strengthening during the acute arthritic stage because it produces low articular pressures. Instructions should be to perform isometric contraction at no more than one-half of the individual's maximum contraction for 6 s, while exhaling. Isometric exercise should be targeted at one muscle group at a time. Aerobic training is ideally achieved by using a mode that minimizes the magnitude and rate of joint loading (189). The best types include walking, cycling, and pool exercise (52). Free-speed walking produces less hip joint contractile pressure than do isometric or standing dynamic hip exercises (189). Faster walking speeds, however, increase stress on knee joints (137). Poor lower-extremity biomechanics, joint instability, or poor proprioception may contribute to undesirable joint forces when walking speed is increased (131). Without therapy to improve these impairments, increasing gait speed may have deleterious effects. If walking is uncomfortable, if pain lasts more than 2 h after walking, or if the individual has complicated biomechanics of the lower extremity, an alternative cardiovascular exercise should be used such as cycle ergometry, recumbent stair stepping, upper-body ergometry (especially in the absence of osteoporosis), or water walking (35,102,133,193). Cycle ergometry should be conducted with the seat height and crank length adjusted to limit knee flexion and minimize pedal load, which decreases knee joint stress (115).

Water is a good medium for cardiovascular work, range-of-motion exercise, and low-level strengthening. The buoyant quality of water can help patients perform passive and active joint range of motion. Strengthening exercise may be performed in the water. Water can offer resistance to motion taking place against buoyancy when turbulence of the water is increased by attaching a float to an extremity. An increase in speed increases the resistance. Many people with arthritis are able to tolerate longer and more vigorous workouts in the water than on land. Compliance with aquatic exercise decreases with water temperatures colder than 84 °F (28.9 °C), and cardiovascular stress increases with temperatures above 98 °F (36.7 °C) (28). Contraindications to hydrotherapy include a history of uncontrolled seizures, incontinence of bowel or bladder, pressure sores or contagious skin rashes, and cognitive impairments that would jeopardize the patient's safety (97). If a great deal of assistance is needed with dressing, or changing clothes causes fatigue or joint pain, then pool therapy should not be used. The mode and sequencing of exercise are important considerations for people with AS. An exercise program for an individual with AS should begin as soon as possible after diagnosis of the condition, beginning with exercises to improve spinal and peripheral joint motion before initiating a strengthening program (60). Achieving functional range of motion of the hip joints should be emphasized, because a lack of such capability can be extremely disabling (31,60,149). The goal of muscle strengthening in this population is to maintain or approximate a neutral spine over the long term. A program to strengthen the back and hip extensors as well as general strengthening can be performed on land or in water. During strength training, the person should be initially supervised with a goal of independently maintaining proper posture during spinal extension. High-impact activities should be avoided, because they are stressful to the spinal and sacroiliac joints. Swimmers with limited spinal and neck motion should use a mask, snorkel, and fins to avoid trunk and neck rotation. Sports that encourage extension are preferred over activities that require flexion (60). Contact sports are contraindicated for those with cervical spine (60) or peripheral joint involvement.

Frequency

The key to exercise therapy for arthritic conditions is to manipulate the intensity and duration for aerobic exercise, or the number of repetitions for resistance exercise, to achieve a training effect without causing joint discomfort. Patients should begin conservatively and increase the frequency before increasing the intensity of exercise. People with arthritis should begin at about 3 d per week and work toward a goal of 5 to 7 d per week.

Intensity

Cardiovascular exercise intensity should be guided by heart rate or rating of perceived exertion (58,121). Standard training heart rate range development can be used (see chapter 5). A cardiovascular exercise intensity of 12 to 16 on the 6–20-point Borg scale or 3 to 6 on the 10-point Borg scale is recommended (see practical application 26.3 on p. 481).

Duration

Exercise duration and intensity are inversely related. Exercise intensity or the time or repetition variable of a specific exercise should be progressed when the exercise is not challenging and symptoms do not increase for two to three consecutive sessions. Conservative increments are recommended. In general, after 1 wk of consistent exercise without an increase in symptoms, the duration of aerobic and resistive exercise or repetitions during resistive exercise should be increased toward the maximum recommended. Exercises can be progressed each session by increasing the duration of exercise or the number of exercises performed or by increasing total time spent exercising by decreasing any rest periods. An increase in symptoms may require decreasing the intensity or duration or repetition of exercise for the affected joint. The "2 h pain rule" is a helpful maxim for regulating exercise intensity. A localized increase in pain after an activity that lasts more than 2 h suggests the need to decrease

the exercise intensity or duration or repetition of the next exercise session.

Outcomes

Finally, collecting outcomes data serially to monitor response to therapy objectively is useful. Table 26.4 lists a variety of physical function measures appropriate to arthritis patients that can be used to evaluate strength, range of motion, and balance. In addition, a number of standardized and validated instruments (questionnaires) are available for assessing arthritis pain, stiffness, and function as well as response to therapy for OA, RA, and AS (19,109,157,170,171).

EXERCISE TRAINING

The following section reviews the exercise-training literature that applies to those with OA, RA, and RS.

OA

The finding that the degree of severity of knee OA is associated with the level of cardiovascular deconditioning supports the concept that regular aerobic exercise should be performed by people with OA (164). Reviews of exercise for OA demonstrate positive effects on pain and disability as well as general safety (16,98,191,204). Seventeen randomized controlled trials of exercise for OA demonstrated small to moderate improvements in pain and physical function (59,191) and moderate to large improvements in patient global assessment (191). The first randomized study demonstrating that exercise prevented disability in individuals with knee OA used resistance training or walking (156). The largest controlled trial of exercise for knee OA compared education with home-based aerobic and resistance interventions with limited supervision for 15 mo following an initial 3 mo of supervised exercise (52). In this trial, both aerobic and resistance training improved disability by arresting functional decline. Similarly, reviews of randomized controlled trials showed that aerobic walking and quadriceps-strengthening programs provided comparable positive effects on pain and disability (166). Besides reducing pain and disability, therapeutic programs to strengthen knee extension and hip and ankle flexion in people with OA have resulted in decreased stiffness, improved strength, mobility, balance, gait, independence, and physical function (106,174). With respect to aerobic exercise, low-level exercise intensity (at 40% of heart rate reserve) may be as effective as high-intensity cycling (at 70% of heart rate reserve) in improving function, gait, and aerobic capacity, and decreasing pain in OA patients (115).

In addition to exercise interventions, weight loss has been shown to reduce knee-joint forces in OA (126), and a recent randomized controlled trial suggested that a combination of exercise and diet produced superior improvements in pain, function, and mobility than either exercise or diet interventions alone (127). Alternative modes of exercise such as yoga and tai chi also appear to impart improvements in pain, flexibility, and function for patients with osteoarthritis (162,180). The benefits of exercise training for those with arthritis is listed in table 26.7.

Table 26.7   Benefits of Exercise Training for Persons With Arthritis

Arthritis type	Cardiorespiratory endurance	Skeletal muscle strength and endurance	Flexibility	Body composition
OA	Aerobic exercise improves cardiorespiratory endurance, pain, depression, fatigue, function, health status, and gait (25,167).	Resistance training improves strength, function, health status, pain, and stiffness (25,167).	Dynamic exercise improves joint mobility, pain, and function (131). Aquatic exercise improves knee and hip range of motion, pain, and function (111). AFYAP and PACE program improve flexibility and isometric strength (187).	Combined diet, resistance and aerobic training produces weight loss, and improves function, mobility, and pain to a greater extent than diet or exercise alone (127).
RA	Aerobic exercise improves cardiorespiratory endurance, pain, function, and, with resistance training, mood. Aerobic training improves fitness without worsening disease activity (193).	Exercise that includes functional strengthening improves strength, pain, fitness, and mobility. Hand strengthening may improve dexterity and grip strength (152).	Joint mobility improves with dynamic exercise training (193). AFYAP and PACE program improve flexibility and isometric strength (187).	Combination of aerobic and resistance training decreases percent body fat without significant weight loss (68) and slows bone mineral density loss (42).
AS	Supervised physical therapy improves fitness, mobility, and function (85, 86); improvements in fitness and stiffness may mediate improvements in global health (84).	Few investigations are relevant.	Home or supervised exercise can improve spinal mobility, physical function, and patient global assessment (supervised only) (37). Some evidence suggests improvements in pain and depression (90). Relative to conventional physical therapy, flexibility and strengthening exercises that target specific muscle groups involved in AS improve axial mobility and function (110).	Few investigations are relevant.

Note. OA = osteoarthritis; RA = rheumatoid arthritis; AS = ankylosing spondylitis; AFYAP = Arthritis Foundation YMCA Aquatics Program; PACE = People With Arthritis Can Exercise (land-based community program).

RA

With RA, joint mobility, muscle strengthening, and aerobic conditioning are equally important. Patients with RA often have significant improvements, likely because of their severe baseline disability (204). Dynamic exercise has been shown to increase muscle strength and aerobic capacity without detrimental effects on disease activity, and in some cases disease activity may decrease (193). Also in persons with RA, functional strengthening and low- or high-intensity exercise produce improvements in pain and function; pain may be reduced more with low-intensity exercise than with high-intensity exercise (152).

Isometric exercises moderately increase quadriceps femoris muscle strength (80,114,192). Additionally, significant gains in function have been seen with programs that employ range-of-motion exercises (72). But when compared with static or isometric exercises for RA patients, greater benefits are observed with the use of dynamic aerobic or strengthening exercises (50,67,72,183). In addition, although combined sessions of intensive weight-bearing exercise and cycling may be more effective than lower-intensity exercise (192), this result may depend on the patient's level of disability. A trial of a combination of high-intensity aerobic and resistance training twice weekly for 2 yr demonstrated significant improvements in functional ability, aerobic fitness, muscle strength, and emotional health, and showed a decline in bone mineral density loss as compared with usual care (42,43). Remarkably, this program demonstrated a 78% adherence and satisfaction rate after 2 yr (138). Most investigations evaluate aerobic exercise at an intensity of 50 to 85% of maximum HR and strengthening beginning at 30 to 50% and progressing to 80% of maximum. These studies provide the basis for recommendations regarding exercise prescription in RA (185).

Although exercise training can improve strength and cardiorespiratory fitness, effects on disease activity and radiographic progression are less straightforward. In a number of investigations, exercise did not appear to worsen the appearance of RA on X-rays (44,71,76,144,184). Recent evidence suggests, however, that high-intensity weight-bearing exercise may promote radiographic progression in large joints (ankles and shoulders) with severe damage (139).

Nontraditional modalities such as dance and tai chi chuan may have beneficial effects on depression, anxiety, fatigue, tension, and lower-extremity range of motion (75,99,146,194). Water exercises, including seated immersion, have been shown to improve aerobic capacity and other physical and psychological measures in patients with RA (74,133). Notably, when performed at low speeds, exercise in water results in lower heart rates and $\dot{V}O_2$ than exercise of a comparable speed performed on land because of the buoyancy effect of water, but at higher speeds, the resistance of water results in higher heart rates and $\dot{V}O_2$ than land-based exercise (73).

Continuance of exercise after completing a supervised exercise program was associated with high physical activity levels and aerobic capacity at the start of the exercise program (132). Home exercise programs, however, have failed to establish significant improvements in physical impairments (38,182). Training effects, such as improved strength, can be maintained as long as exercise continues. Although these benefits wane rapidly on termination of exercise (69), self-directed training following a supervised program sustains many of the beneficial effects afforded by exercise (70).

AS

Investigations of exercise-training effects in patients with AS are relatively few, although exercise was among the 10 key recommendations for the management of AS recently developed using a combination of research-based evidence and expert consensus, as part of a collaboration between the ASsessment in AS (ASAS) International Working Group and the European League Against Rheumatism (EULAR) (206). The literature review from which these recommendations arose suggested that different types of exercise-based intervention could affect disease outcomes in AS (207). Given the goals of maintaining posture and functional ability in this disease where disability is mainly related to effects in the spine, most interventions use flexibility and muscle-strengthening programs. Daily exercise is considered vital to maintenance of spinal mobility, but long-term effects have not been studied (204).

Significant short-term improvement in spinal and hip range of motion of AS patients who are enrolled in intensive physical therapy has been demonstrated, and the performance of regular moderate (2–4 h per week) exercise is associated with functional improvement for patients with AS (172). A recent review of randomized controlled trials in AS evaluated six trials (37). In this review, as compared with no exercise, home-based programs demonstrated improvements in spinal mobility and physical function. Gains in patient global assessment and spinal mobility with home-based exercise can be enhanced with supervised exercise programs (37). Additionally, relative to weekly group physical therapy, improvements in pain, function, and patient global assessment were noted with combination spa-exercise therapy consisting of group exercises, walking, posture exercises, hydrotherapy, sports, and thermal treatment visits (i.e., sauna) for 3 wk followed by 37 additional weeks of combination group exercises, hydrotherapy, and sports (198). This combined spa-exercise therapy has been shown to be cost effective in the management of AS (197).

Practical Application 26.3

Fibromyalgia

Fibromyalgia is an increasingly recognized chronic pain syndrome possibly related to central neuromodulatory dysregulation (124). Fibromyalgia is not a form of arthritis but may easily be confused with arthritis because of its associated widespread musculoskeletal pain and so-called trigger or tender points, often with a periarticular location. It affects at least 2% of people, women more than men (64), and can be diagnosed using 1990 American College of Rheumatology criteria (202), namely widespread pain in combination with tenderness at 11 or more of the 18 specific tender point sites. These tender points occur in characteristic locations, and are associated with fatigue and exercise intolerance (64), but laboratory and radiographic joint studies are normal. Multiple investigations have demonstrated that muscle and joint inflammation or damage are not associated with fibromyalgia. But a lower pain threshold and altered pain-processing pathways provide clues to the underlying pathogenesis of this complicated syndrome (63). Fibromyalgia is often associated with depression and inadequate, nonrestorative sleep and commonly coexists with other rheumatologic conditions. Treatment of fibromyalgia requires a multidisciplinary approach employing exercise, education, and both pharmacologic and behavioral therapies for depression and sleep (64).

A multitude of evidence shows that aerobic exercise is beneficial for persons with fibromyalgia and can improve fitness, function, self-efficacy, and tender point pain threshold (29,65). But results have been inconsistent regarding the ability of exercise to improve self-reported pain, sleep, fatigue, or depression (29,65) Aquatic exercise may provide superior benefits in mood over land-based training (12). Strength training provides the best means of increasing muscular strength in persons with fibromyalgia, but few studies have evaluated the benefits of strength training on other aspects of fibromyalgia.

With respect to exercise testing, procedures outlined in chapter 5 can be followed in persons with fibromyalgia. One consideration is that at similar fitness levels, relative to age and sex-matched controls, persons with fibromyalgia report higher levels of perceived exertion during exercise testing (142). Additionally, persons with fibromyalgia are often deconditioned and may not reach maximal effort (148,190). In these cases, ventilatory threshold rather than peak $\dot{V}O_2$ is recommended as an indication of fitness levels (190). As they do for persons with arthritis, ACSM prediction equations appear to overestimate peak $\dot{V}O_2$ in persons with fibromyalgia (48). Although slight underestimations and overestimations, respectively, were noted with the FAST equation depicted in table 26.3 and the FOSTER equation (with handrail support: $\dot{V}O_2$ (ml·min^{-1}·kg^{-1}) = 0.694 × ACSM predicted + 3.33), these equations appear to provide clinically acceptable estimations of peak $\dot{V}O_2$ using a Duke–Wake Forest testing protocol (48).

For persons with fibromyalgia, exercise-training goals include improving fitness and function. Potential modes of exercise include walking, cycling, dancing, and water aerobics. Pool options, such as the Arthritis Foundation aquatics program for fibromyalgia or FIT (fibromyalgia interval training) may be ideal for individuals with significant complaints of pain with exercise or concerns that exercise will worsen pain symptoms (65). Some people with fibromyalgia may progress to high-intensity exercise and sustain important benefits from such training. For most individuals with fibromyalgia, however, a prudent approach is to begin training at low intensity, below the capacity of the individual, and progress slowly and gradually to moderate intensity (55–75% of maximum heart rate) for 20 to 30 min (65). Although exercise has not been shown to worsen self-reported pain in fibromyalgia, the clinical exercise physiologist should educate the person about transient increases in pain with exercise bouts and about methods of adjustment of exercise protocols. Such education will enhance adherence and long-term benefits of exercise training and should be an essential component of the exercise prescription for people with fibromyalgia (116). Recommendations for minimizing pain with strength training include limiting eccentric exercises, performing upper- and lower-extremity training on alternate days, and resting between repetitions (65).

Literature Review

The purpose of this investigation was to determine factors that promote or prevent physical activity in people with arthritis. Eligibility criteria included age of 18 yr or older and an arthritis diagnosis given by a physician. People were recruited by newspaper, radio, and posted advertisements. Using the physical activity module from the Behavioral Risk Factor Surveillance System of the Centers for Disease Control and Prevention, participants were characterized as exercisers (moderate activity ≥ 30 min $\cdot$ d^{-1} ≥ 3 d $\cdot$ wk^{-1}, vigorous activity ≥ 20 min $\cdot$ d^{-1} ≥ 3 d $\cdot$ wk^{-1}, or resistance training ≥ 20 min $\cdot$ d^{-1} ≥ 3 d $\cdot$ wk^{-1}) or nonexercisers (any exercise ≤ 1 d $\cdot$ wk^{-1} or any exercise ≤ 10 min $\cdot$ d^{-1} on 2 d $\cdot$ wk^{-1}). Composed of exercisers or nonexercisers, focus groups were conducted, and themes related to exercise barriers, benefits, and enablers (factors that would promote exercise) were identified in each. These were categorized into physical, psychological, social, and environmental themes.

For exercisers and nonexercisers, physical barriers to exercise included pain, fatigue, mobility, and comorbid conditions. Physical benefits and enablers relating to exercise included symptom management, increased mobility and function, improved strength and flexibility, and weight loss. Exercisers commonly described adapting or continuing exercise despite pain. In contrast, nonexercisers more often reported discontinuing exercise because of pain and expressed less confidence in the potential for benefits associated with exercise.

Psychological barriers segregated into themes of attitudes and beliefs, fears, and perceived negative outcomes. Exercisers commonly identified other life priorities as interfering with exercise. In contrast, nonexercisers emphasized postexercise pain and a fear of water as barriers to exercise. Psychological benefits and enablers included independence, improved attitudes and beliefs, emotional benefits, and enjoyment. Exercisers more commonly related that exercise promoted independence. Also, in general, exercisers reported psychological benefits related to recent exercise, but nonexercisers more often described benefits that occurred with exercise before their arthritis diagnosis.

Social barriers included lack of support from family, friends, and healthcare providers; lack of exercise partners; and competing role responsibilities such as family responsibilities. Social benefits and enablers included enjoying exercising with others, having others to exercise with, and encouragement regarding exercise from friends, family, and healthcare providers. Exercisers more often reported having these types of social supports, whereas nonexercisers reported that these supports were insufficient.

S. Wilcox, C. Der Ananian, J. Abbott, J. Vrazel, C. Ramsey, P.A. Sharpe, and T. Brady. "Perceived Exercise Barriers, Enablers, and Benefits Among Exercising and Nonexercising Adults with Arthritis: Results From a Qualitative Study." *Arthritis and Rheumatism* (*Arthritis Care and Research*) 2006; 55(4): 616–627.

CONCLUSION

The available data indicate that properly performed exercise is safe and effective for individuals with OA, RA, and AS (204). In the short term, exercise improves strength, enhances cardiovascular endurance, decreases stiffness, increases range of motion, decreases and prevents impairments, improves function, and prevents disability (191,192) (see table 26.7 on p. 479). These benefits are in addition to the well-accepted benefits of exercise to the general population. Precautions must be taken, however, to ensure that exercise is safe and comfortable for people with arthritis.

Medical History

Mrs. MZ is a 69 yr old African American woman with a 15 yr history of OA of the hands, cervical spine, right knee, and feet. She presented with a rotator cuff tear secondary to a fall the previous year, multiple sites of joint pain, up to 2 h of morning stiffness, and evening fatigue. She occasionally uses a cane for ambulation. Her chief complaints are pain in her knees and hips (right worse than left), feet, hands, right shoulder, and lower back.

Diagnosis

A rheumatologic evaluation revealed mild fibromyalgia superimposed on OA. She had no known symptoms or risk factors of cardiovascular disease.

Exercise Test Results

An exercise test using cycle ergometry was notable for a resting heart rate of 60 beats · min^{-1} and a peak metabolic equivalent (MET) of 7.7. She ambulated 1,985 feet (605 m) during a 12 min walk test with a perceived exertion of 15 on the Borg scale. Mrs. MZ underwent a 3 mo supervised exercise program. Pain, stiffness, and difficulty with activities of daily living improved dramatically by the end of the supervised exercise program. When she was ascending stairs, walking on a flat surface, shopping, getting in and out of a bath, and performing heavy domestic duties, pain with weight-bearing activity and morning stiffness improved from a preprogram difficulty of "very severe" to a postprogram difficulty of "mild." After completing the program, she stopped exercising for 3 mo both at home and at the Duke Center for Living because her husband experienced an illness. She resumed exercise 2 to 3 d per week and underwent follow-up exercise testing 10 mo after the initial test. The results were 68 beats · min^{-1} resting heart rate and a peak MET level of 5.4. Mrs. MZ ambulated 1,488 ft (454 m) in a 12 min walk test, a decrease of 25%. Gait changes were noted during the walk, and she complained of left foot pain on termination of the test. The discussion of these findings with her served as motivation to resume regular attendance with concentration on aerobic conditioning.

Exercise Prescription

Exercise commenced with the AFYAP I pool class for full-body range-of-motion exercises and low-level strengthening, lower-extremity muscle stretching, lower-extremity strengthening, and cardiovascular conditioning. The exercise program included physical therapy for her right shoulder. The first week of exercise consisted of lower-extremity muscle flexibility exercises and two strengthening exercises. Lower-extremity muscle flexibility exercises included seated hamstring stretch, calf stretch, and standing quadriceps muscle stretch. Isotonic leg presses and step-ups onto a 3 in (8 cm) rise were added but then discontinued because of knee pain. After 1 wk, Mrs. MZ tried walking to improve cardiovascular conditioning. She had an increase in foot and knee pain during and for approximately 24 h after walking. Cycle ergometry did not increase her joint symptoms and was performed at each session thereafter. Mrs. MZ was instructed to walk in the pool for cardiovascular conditioning for 15 to 20 min after the AFYAP I pool class. Lower-extremity strengthening through functional movements and isotonic machines was added each week. These exercises included standing hip abduction, isotonic seated hip abduction and adduction, and seated leg curl. Strengthening exercises were initiated at 25% of maximum voluntary contraction. Mrs. MZ was directed to report symptoms that occurred between sessions to the clinical exercise physiologist at the following session.

Strengthening exercises were progressed slowly because of an increase in knee pain with high repetitions and resistance. Mrs. MZ reported an increase in shoulder, knee, lateral thigh, medial elbow, lateral elbow, and foot pain intermittently throughout the program. Exercises associated with an increase in joint or muscle symptoms were revised or discontinued. Fibromyalgia pain may have played a role in elbow symptoms, because the pain was associated with the muscle–tendon junction, not the joint.

(continued)

Discussion Questions

1. How could Mrs. MZ's program have been changed to improve her cardiovascular status?

2. Which cardiovascular equipment might be best for Mrs. MZ?

3. Could Mrs. MZ's postprogram compliance have been improved? If so, how?

4. What initial exercise intensity should be prescribed when fibromyalgia accompanies arthritis

Osteoporosis

David L. Nichols, PhD

Chad D. Swank, MS

As middle age begins, people begin to lose small amounts of bone every year. In women, this bone loss accelerates during a 3 to 5 yr period after menopause. Because women typically have lower bone mass than men do and because of the accelerated decline following menopause, they tend to be more prone to osteoporosis. Osteoporosis has classically been defined, for older women and men, as a pathological condition associated with increased loss of bone mass, known as osteopenia, caused by increased bone resorption. As bone mineral density (BMD) declines, individuals are at risk for skeletal fractures. Osteoporosis is a major public health problem that results in significant morbidity, mortality, and economic burden. Fractures result in impairments that increase the risk of pneumonias and can accelerate other underlying diseases, such as coronary artery disease.

SCOPE

The risk of fragility fracture approximately doubles for each standard deviation decrease in the measured BMD. The level of BMD necessary to qualify as osteoporotic is still debated, primarily because of disagreement regarding what value should be used to identify the young adult mean BMD (the reference point at which BMD reaches its peak before age-related bone mineral loss begins, which occurs at the end of skeletal growth around age 18 to 20). The World Health Organization (WHO) has defined osteoporosis as a bone mineral density measure-

ment more than 2.5 standard deviations below the young adult mean. But the WHO classification should not be used with premenopausal women or young men (33). The consequences of poor bone health have continued to escalate in recent years despite heightened awareness, better understanding of prevention strategies, advances in technology for screening and diagnosis, and expanded treatment options. The major effect of osteoporosis is on older, postmenopausal, Caucasian women. Osteoporosis is thought of primarily as a disease of women, yet prevalence rates in men are as high as 12% (27).

In addition, low BMD is often found in young amenorrheic women. Osteoporosis now affects almost one out of every two women at some point in their lives. Medical costs are estimated at $12.2 to 17.9 billion per year (19). The mortality associated with fractures is substantial. Studies indicate that a one standard deviation decrease in BMD increases mortality from 10 to 40% (36).

PATHOPHYSIOLOGY

Changes in bone result from the continual process of bone resorption and **bone formation**—known as bone remodeling. Bone remodeling does the following:

- Maintains the architecture and strength of the bone
- Regulates calcium levels
- Prevents fatigue damage

Remodeling is also important during periods of growth, puberty, and adolescence, when the majority of adult bone mass is laid down (18,41). Bone modeling and remodeling often occur simultaneously, and distinctions between them are not always apparent, but in general, bone modeling refers to alterations in the shape of the bone such as changes in length. Bone modeling usually ceases around age 18 to 20, when the skeleton stops growing, whereas bone remodeling occurs throughout the lifespan. Bone remodeling is a complex process that has been more thoroughly described elsewhere (18,38). In brief, bone remodeling is performed by individual basic multicellular units (sometimes referred to as bone multicellular units (18) or bone modeling units) consisting of osteoclasts and osteoblasts (18,41). These basic multicellular units have two roles: osteoclasts erode old bone (resorption), whereas osteoblasts form new bone (formation). After growth has ceased, bone formation generally equals bone resorption; that is, the density of the bone remains unchanged (18,41). However, bone cells age, just like all other cells in the body. As a result, the amount of bone remodeling that occurs begins to decrease with increasing age. Osteoblasts seem to be more greatly affected by aging than osteoclasts, and thus bone formation decreases more than bone resorption, resulting in an age-related bone loss (18,41) that can lead to osteoporosis. The decrease in bone remodeling seen with aging may not be inevitable, however, and, as with other things, may be a result of the decreased physical activity and poorer nutrition of older people. Figure 27.1 illustrates the cycle of bone modeling and remodeling. On the other hand, the loss of bone mass seen at menopause, or in young amenorrheic women, is a direct result of the decrease in endogenous estrogen. This decrease in estrogen levels does not alter bone formation activity but does increase bone resorption, because bone resorption is inhibited by estrogen (41). Over her lifetime, a woman loses as much as 50% of her peak **trabecular bone** mass (46), and most of that loss is attributable to an estrogen deficiency (44). Although this loss occurs mostly after menopause, bone loss has been seen in the perimenopausal years (20,40) and is probably a result of decreased ovarian function (43).

The two most important factors in the development of osteoporosis are the amount of peak bone mass attained and the rate of bone loss (45). **Peak bone density** (or peak bone mass), for most purposes, is defined as the highest amount of bone mass attained during life. Although this definition is accurate, the term *peak bone mass* is often confused with the term *maximal bone mass*. Maximal bone mass should be defined as the highest bone mass that a person could possibly achieve. Maximal bone mass

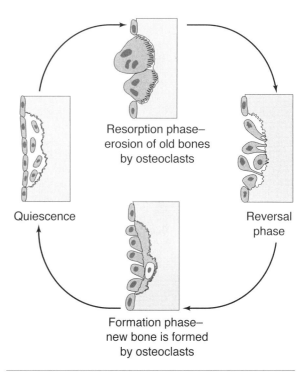

Figure 27.1 Bone modeling and remodeling.

would, theoretically, be controlled solely by genetic factors. Peak bone mass is also influenced to a certain extent by genetics, but the difference between maximal bone mass and peak bone mass is controlled by other factors such as physical activity, diet, and hormonal balance. It is unlikely that anyone ever reaches maximal bone mass.

MEDICAL AND CLINICAL CONSIDERATIONS

Osteoporosis may go undetected in many people until a fall or injury exposes the seriousness of the disease. A risk factor assessment can be conducted to determine the likelihood of the disease. Various sophisticated measurement techniques have evolved for assessing bone density that may be useful for disease diagnosis.

Signs and Symptoms

A variety of risk factors exist for osteoporosis (see table 27.1). A thorough medical history can help identify these risk factors. Although a variety of screening tools exist, they typically cannot accurately predict the presence of osteoporosis. Risk factors that may be easily measured have a much lower predictability than measures of BMD (6,39). But for a postmenopausal (or amenorrheic), Caucasian woman with low body weight who is not on

Table 27.1   Risk Factors for Osteoporosis or Decreased Bone Mineral Density

Inherited factors	Environmental factors
Caucasian or Asian	Below normal weight
Female	Loss of menstrual function
Osteoporotic fracture in first-degree relative	Low calcium intake
Height <67 in. (170 cm)	Inactivity
Weight <127 lb (58 kg)	Prolonged corticosteroid use
	Smoking
	Excessive alcohol intake
	Caffeine

hormone replacement therapy, the presence of any other risk factor would strongly suggest that this woman has low bone mass (58). Risk of osteoporotic fracture is related, for the most part, to a person's BMD, with fall frequency and **bone geometry** also playing a role. But because fragility fractures are linked to BMD, and BMD later in life is determined to a great extent by peak bone mass, striving to make peak bone mass equal maximal bone mass could be the best protection against osteoporosis. Unfortunately, a fracture is often the first sign that a person has osteoporosis. Loss of bone mass occurs without outward symptoms, and by the time a fracture occurs, a patient may have lost as much as 30% or more of peak bone density. Even the best available treatments can increase bone density by only about 12% (7,35). So after a person reaches the level of bone density that is considered osteoporotic, he or she is unlikely ever to return to a level of normal BMD. This circumstance is true even in young amenorrheic women, who, after having been amenorrheic for extended periods, never regain sufficient bone mass to return to normal, even with the resumption of regular menstrual periods (14,26).

History and Physical Exam

A thorough understanding of the bone status of each osteoporosis patient is crucial. History of fractures, age, and amount of BMD can help determine the level of caution that must be taken to minimize inappropriate mechanical stress on high-risk joints.

Risk factors for osteoporosis may be obtained from a patient's medical history during a physical examination. Women are at a greater risk than men are, especially women who are thin or have a small frame, as are those of advanced age. White or Asian women, especially those with a family history, have a greater risk of developing osteoporosis than other women. Postmenopausal women, including those who have had early or surgically induced menopause are at risk. Also younger women with abnormal or an absence of menstrual periods are at greater risk. Cigarette smoking, eating disorders such as anorexia nervosa or bulimia, low amounts of calcium in the diet, heavy alcohol consumption, inactive lifestyle, and use of certain medications, such as corticosteroids and anticonvulsants, are also risk factors. Nevertheless, although these risk factors may raise suspicions of the disease, diagnostic testing is necessary for confirmation.

Diagnostic Testing

The primary means of assessing bone health is to measure bone mineral density. Over the past 20 yr, rapid advances have been made in techniques used for measuring bone density (33). These advances have resulted in increased precision, less radiation exposure, and the ability to measure fracture-prone sites (22). BMD is most often measured in the spine, hip (femoral neck), and wrist, because these are the most common sites for fracture in osteoporosis. The accuracy of predicting BMD and fracture risk at one site based on measurement of BMD at another site is low (≤50%). Thus, a measurement at all three sites is preferable, especially the spine and hip, because the consequences of fracture are the highest at these two sites. Measurements of **bone mineral content** are usually expressed as the amount of bone mineral (primarily hydroxyapatite) per unit of area, which gives BMD. BMD is generally expressed as grams of bone mineral content per square centimeter of bone ($g \cdot cm^{-2}$) because most of the available technology can only measure area, not volume, of bone. **Quantitative computed tomography** is the only method currently available that provides an actual measurement of volumetric bone density. Other methods are available:

- **Single-photon absorptiometry**
- Dual-photon absorptiometry

- Single-energy X-ray absorptiometry
- **Dual-energy X-ray absorptiometry** (DXA)

All these methods express bone density as an area measurement (22). The most widely used densitometry technique is DXA. Because quantitative computed tomography has the advantage of providing a three-dimensional image, thus allowing a separation of trabecular and cortical bone, it would seem to be the measurement of choice. But quantitative computed tomography is a much more costly procedure and has a higher radiation exposure than DXA, and thus it is used less in clinical practice. DXA uses X-rays emitted at two different energy levels to distinguish bone tissue from the surrounding **soft tissue**. The information provided with a bone density measurement is not only the absolute value in terms of grams of bone mineral but also a comparison of that value to established normal values. A DXA image of a femoral neck scan, along with its accompanying printout, is presented in figure 27.2. The subject's BMD is compared with reference standards.

Another method for assessing bone health is **quantitative ultrasound**. Ultrasound does not measure bone density but rather measures two parameters called speed of sound and broadband ultrasound attenuation, which are related to the structural properties of bone. The advantage of ultrasound devices is that they are small, are portable, and use no ionizing radiation; thus, they may provide an attractive alternative to radiation-based densitometry. Biochemical markers of bone metabolism are also sometimes used to monitor therapy or progression of bone disease. Although these markers generally show poor correlation with actual measures of bone density ($r \leq .3$), they can be useful for assessing relative rates of bone formation or resorption. Another advantage of biochemical markers is that they can detect alterations in bone formation or resorption in a matter of weeks, whereas detecting changes in BMD may take several months. In addition, clinical trials have shown that a reduction in fracture risk occurs with antiresorptive therapy before detectable changes occur in BMD and thus monitoring of biochemical markers of bone turnover may increase the ability to predict future risk of fracture (9).

With the use of DXA technology and the aforementioned WHO guidelines on the level of bone density necessary to be considered osteoporotic, the diagnoses of osteoporosis might seem a simple matter. But a multitude of factors can affect not only the individual's bone density but also the accuracy of the bone density measurement (8). Several advances in tools to evaluate bone health have been emerging, and a thorough review of the diagnostic and evaluation procedures can be found elsewhere (8). The clinician should be familiar with the critical anatomical locations where fractures typically occur and should have fundamental understanding of the modalities used to evaluate BMD. After osteoporosis has been confirmed by low **bone mineral density** and frequent fractures, the appropriate treatment should follow.

Exercise Testing

The primary purposes of exercise testing are to aid in the diagnoses of coronary artery disease and to determine appropriate levels of exercise training. The value of exercise testing in a person with established osteoporosis should be carefully evaluated to make certain that any potential benefits outweigh the risks. Most people with osteoporosis will be women (29); the accuracy of using exercise testing for diagnoses is lower in women than in men (32). Exercise testing can be of significant benefit in generating an exercise prescription. In addition, most patients at risk for osteoporosis (postmenopausal women and elderly men) will be at a greater risk for heart disease, which could potentially be diagnosed with a stress test.

Contraindications

In addition, because fractures in osteoporotic patients often occur with little trauma, the impact associated with exercise testing could lead to fractures, even when walking or bicycle protocols are used. On the other hand, the American College of Sports Medicine does not specifically state that osteoporosis is an absolute contraindication to exercise testing (1).

Recommendations and Anticipated Responses

If an exercise stress test is to be used, one that uses a bicycle protocol is probably the best choice, because that

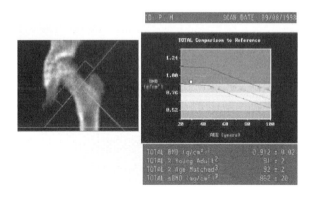

Figure 27.2 Femoral neck scan. Image of a proximal femur scan from a dual-energy x-ray absorptiometer.

would involve the least trauma and impact on the bones. But caution must still be taken when a bike protocol is used. The patient should maintain an upright posture at all times, because any sort of spinal flexion is contraindicated in people with osteoporosis. Treadmill protocols can be used if necessary, but a walking protocol should be used, and care should be taken to ensure that the patient does not trip or fall. No studies have specifically examined acute physiological responses to exercise testing in an osteoporotic population, but for those patients who can tolerate the exercise, there is no reason to believe that their responses will differ from those of individuals without osteoporosis. Osteoporosis, however, can sometimes mask the presence of coronary artery disease because it can prevent a person from achieving a heart rate and blood pressure necessary for accurate diagnoses. Pharmacological tests are available that can diagnose coronary artery disease without the use of exercise, and these types of tests might be more advisable in patients whose exercise capacity is clearly limited because of osteoporosis. A summary of exercise-testing guidelines is found in table 27.2.

Treatment

Exercise can be useful to help increase, or at least maintain, bone mass in patients with low BMD. A number of studies in postmenopausal women have shown that exercise can increase bone density or prevent further bone loss compared with nonexercising controls. But these studies have also pointed out that exercise without concomitant **estrogen replacement therapy** will usually result in further bone loss. The same situation can be seen in young amenorrheic athletes who continue to lose bone mass, despite their exercise training, as long as they continue without their menstrual cycles (14). Exercise should still be one of the first choices in the treatment of osteoporosis for both men and women, because it has the potential not only to increase bone mass but also to increase muscle strength and balance (which may help decrease falls). Several nonpharmacological and pharmacological agents are available to increase, or slow the loss of, bone mass:

- Calcium supplementation
- Vitamin D supplementation
- Estrogen (or hormone) replacement therapy
- **Selective estrogen receptor modulators** (SERMs)
- Bisphosphonates
- Parathyroid hormone (PTH)
- **Calcitonin**

The role of calcium intake as a risk factor for osteoporosis, and as a therapeutic intervention, is controversial (23,28). The uncertainty arises from conflicting data in the literature. The controversy may stem from the fact

Table 27.2 Exercise Testing

Test type	Mode	Protocol	Clinical measures	Clinical implications	Special considerations
Cardiovascular	Bicycle	Standard test	$\dot{V}O_2$, METS	Anticipated physiological responses likely to be similar to healthy population, although, no specific research has been done on osteoporotic populations regarding cardiovascular response to exercise	Minimize inappropriate mechanical stress on high-risk joints to avoid potential injury or fracture. Severe kyphosis can affect gait, balance, and center of gravity, and decrease vital capacity
Strength		Dynamometer			May be used to determine intensity of strength-training program, but evidence for its usefulness is lacking, and it may be contraindicated in some patients with osteoporosis
Range of motion		Sit-and-reach	Distance		Not suitable for patients with known or suspected osteopenia

that, although calcium is crucial to skeletal integrity, its role in the development of osteoporosis is more permissive than causal (23,28). The evidence suggests that calcium is necessary for bone structure, but its role is more passive depending on adequate hormonal regulation. The effect of calcium intake in postmenopausal women may depend on their stage in menopause. Despite the unclear role of calcium in osteoporosis prevention, a recent meta-analysis indicates that calcium supplementation has a small positive effect on bone density and also results in a significant reduction in vertebral fractures (53). Women are encouraged to take calcium supplements to prevent osteoporosis for several reasons.

- Calcium is necessary for skeletal integrity.
- Nutrition studies suggest that the average diet contains insufficient amounts of calcium.
- Calcium is not harmful.

Nevertheless, ensuring adequate dietary calcium intake is prudent because dietary consumption is low. Most of the current drugs with Food and Drug Administration (FDA) approval for osteoporosis are considered antiresorptive therapy (see table 27.3). They halt the loss of bone or even increase bone mass by inhibiting bone resorption, while having no effect on bone formation. Available drugs and their mechanism of action are discussed in chapter 3. An important point to note is that none of the current drug therapies or those under current investigation have been studied in the premenopausal woman with low bone mass, and they do not have FDA approval for use in this population. Thus, for the premenopausal woman with low bone mass, treatment should always focus on the underlying cause, which is most likely an estrogen deficiency. This person should be counseled on ways to regain her menstrual cycles (such as increasing energy intake or decreasing energy expen-

Table 27.3 Medical Therapies for the Treatment or Prevention of Osteoporosis

Drug class	Name of drug	Brand name
Estrogens[a]	Estrone sulfate	Ogen
	Conjugated estrogen	Premarin
	Transderman estrogen	Estraderm
	Estriopipate	Ortho-Est
	Esterified estrogen	Estratab
	Conjugated estrogen +	Premphase[b]
	Medroxyprogesterone acetate[b]	PremPro[b]
SERMs	Raloxifene[d]	Evista
	Tamoxifene[e]	Nolvadex
Bisphosphonates	Alendronate[d]	Fosamax
	Etidronate[e]	Didronel
	Risedronate[d]	Actonel
	Pamidronate[e]	Aredia
Calcitonin[c]	Synthetic salmon calcitonin	MiaCalcin
		Calcimar
Others	Calcitriol[f]	
	Sodium flouride[g]	
	Parathyroid hormone[h]	

Note. SERMS = selective estrogen receptor modulators.
[a]All estrogens have Food and Drug Administration (FDA) approval for prevention of osteoporosis, but only Premarin is approved for treatment. [b]Premphase and PremPro are estrogen and progesterone taken in combination, and both are FDA approved for treatment of osteoporosis. [c]Both calcitonins are approved for prevention, but only MiaCalcin is approved for treatment of osteoporosis. [d]FDA approved for both prevention and treatment of osteoporosis. [e]FDA approved but not with an osteoporosis indication. [f]Calcitriol is a vitamin D metabolite with FDA approval but not for osteoporosis. [g]Approval pending for an osteoporosis indication. [h]In clinical trials for treatment or prevention of osteoporosis.

diture). If this fails, or she refuses to comply, then some form of drug therapy may need to be considered to offset the inevitable bone loss seen with **amenorrhea**. The most reasonable choice would be some type of estrogen replacement, because it is the only osteoporosis therapy available with any data regarding treatment or safety efficacy in premenopausal women. Insufficient data are available, however, to make any specific treatment recommendations regarding dosing regimens in at-risk premenopausal women.

EXERCISE PRESCRIPTION

Although studies have shown that several forms of exercise training have the potential to increase BMD, the optimal training program for improving or maintaining skeletal integrity has yet to be defined. Recent studies have found significant correlations between BMD changes and exercise intensity, but the correlations are moderate at best, explaining less than 20% of the change in BMD (56). Current experimental knowledge indicates that an **osteogenic** exercise regime should include the following factors (10):

- Have load-bearing activities at high magnitude with few repetitions
- Create variable strain distributions throughout the bone structure (load the bone in directions to which it is unaccustomed)
- Be long term and progressive in nature

Resistance training (weightlifting) probably offers the best opportunity to meet these criteria on an individual basis, because it requires the least skill and has the added advantage of being highly adaptable to changes in both magnitude and strain distribution.

In addition, strength and muscle size increases have been demonstrated following resistance training, even in the elderly. Practical application 27.1 reviews the literature regarding swimming and bone health.

Exercise Recommendations

Resistance training combined with some sort of cardiovascular training (bicycling or walking) is probably the best exercise program for a patient with osteoporosis. Such a program will not only increase overall fitness and BMD but also greatly reduce the risk of falling, which is

Practical Application 27.1

Literature Review

Cross-sectional studies with athletes have indicated that swimming has little if any significant effect on BMD. Indeed, these studies have consistently found that swimmers had bone mineral densities that were significantly lower than those in other athletes and equal to, or in some cases lower than, than bone mineral densities in nonactive controls (25,96). Cross-sectional studies are subject to bias because of their design, but, unfortunately, longitudinal studies looking at the effects of swimming on bone density in humans have not been done, although longitudinal studies in rats have indicated that swim training can increase BMD (92,95). But a recent abstract presented at the American Society of Bone and Mineral Research meeting presented limited evidence that swimming may benefit skeletal integrity (106). The investigators examined the effects of 2 yr of swim training on BMD in postmenopausal women. Participants were divided into a swimming group (n = 22) and a control group (n = 19). Age, height, weight, calcium intake, and daily activity were similar between the groups. The swim program consisted of 1 h sessions, and the participants attended an average of 1.5 sessions per week. BMD of the lumbar spine and the proximal femur (femoral neck, Ward's area, and trochanter) was measured by DXA at baseline and at 1 and 2 yr. Leg extensor strength was also measured at the same time points. At the end of 2 yr, both groups showed a decrease in BMD at the lumbar spine with the rate of decline not different between the groups. At the sites on the proximal femur, however, the swim group had increases in BMD of 4.4%, 3.4%, and 5.7%, respectively, that were significantly higher than those of the control group (−0.2%, 1.4%, and 1.0%). Leg extensor strength also significantly increased in the swim group. Although only preliminary, these results suggest that swim training may have a role in exercise prescription for osteoporosis not only to help increase muscle strength, which would aid in fall prevention, but also to help prevent bone loss or even increase bone density.

Client—Clinician Interaction

Some special issues should be considered when developing an exercise program for a client with osteoporosis. Osteoporotic patients can fracture with little or no trauma, and thus high-impact activities such as running, jumping, or high-impact aerobics should be avoided in this population. Another activity that absolutely must not be done by people with osteoporosis is spinal flexion, which drastically increases the forces on the spine and increases the likelihood of a fracture; exercises to be avoided include sit-ups and toe touches. Other activities that should be avoided are those that may increase the chance of falling, such as trampolines, step aerobics, or exercising on slippery floors (see table 27.4).

Table 27.4 The Osteoporotic Patient and Exercise

Beneficial exercises	Exercises to avoid
Modified sit-ups in which only the head and shoulders come off the ground	Any type of abdominal crunches that involve the spine coming off the ground
Lying leg lifts: the person lies flat on a firm surface with legs straight out in front and then lifts legs 6 in. (15 cm) off the floor	Any movement involving spinal flexion such as toe touches or rowing machines
Back extension exercises	Jogging or running
Walking to help increase cardiovascular fitness and increase bone mineral density	High-impact aerobics or other activities that may jar the spine or hip
Strength-training exercises using either dumbbells or resistance-exercise machines; pay special attention to hips, thighs, back, and arms	Leg adduction or abduction or squat exercises with any significant resistance
Exercises to improve balance or agility such as standing on one foot (with assistance if needed) for ≥15 s	Exercises or activities that increase the chance of falling, such as using trampolines or exercising on slippery floors

one of the primary causes of fracture in osteoporosis. Although the physiological responses to exercise in an osteoporotic population have not been specifically investigated, they should not be substantially different than those of an age-matched individual without osteoporosis. Similarly, the goals of an exercise program for someone with or without osteoporosis should also be the same. For a person just beginning an exercise program, those goals should include an increase in cardiovascular fitness, increased muscular strength, and an increase (or at least no decrease) in BMD. Heart disease remains the number one killer of men and women by a wide margin. So the goal of both women and men should be to increase their physical activity to reduce the risk of heart disease. The 1996 surgeon general's report on health and physical activity (2) recommends approximately 30 min of moderate physical activity on most if not all days of the week. This goal would be worthwhile for all people,

including those with osteoporosis. But if the person with osteoporosis is just beginning an exercise program, the duration of exercise might need to be shortened initially to allow time for adjustment to the exercise to reduce the likelihood of musculoskeletal injury. As the person's fitness level increases, the amount of time exercising can be increased. For an osteoporotic person, or for that matter any elderly person, just beginning an exercise program, a simple walking program should provide the needed benefits along with safety. Practical application 27.3 contains exercise prescription guidelines for people with osteoporosis.

Cardiovascular Training

No known studies have specifically examined cardiovascular adaptations in osteoporotic patients, but older adults can increase their fitness levels 10% to 30% with prolonged endurance training (2,3). Because endurance training can decrease cardiovascular disease risk factors (i.e., high blood pressure and cholesterol), it should be incorporated as part of the exercise regimen of osteoporotic women and men. Weight-bearing endurance activities (e.g., walking) may be most beneficial for retaining bone density.

Resistance Training

Resistance training offers the most benefits for increases in muscular strength and bone density. Current recommendations suggest a single set of 15 repetitions of 8 to 10 exercises performed at least 2 d per week. This is a worthwhile goal for the person with osteoporosis, but

a less strenuous program may be needed initially. Care should be taken to avoid the exercises previously mentioned that are dangerous for people with osteoporosis. Table 27.5 summarizes exercise training guidelines.

Balance and Agility Training

Certain exercises are particularly beneficial for the osteoporotic patient. These include exercises designed to help with balance and agility to reduce falls. For instance, exercises that strengthen the quadriceps (i.e., knee extension) are helpful because poor strength in this muscle group has been linked to risk of falling. Squats

Table 27.5 Exercise Training

Cardiorespiratory endurance	Skeletal muscle strength	Skeletal muscle endurance	Flexibility	Body composition
Weight-bearing activities may be most beneficial for retaining bone density Endurance training can decrease cardiovascular disease risk factors	Increases in muscle strength and bone density with resistance training	Benefits seen in healthy populations are similar to what are expected in the osteoporotic patient	Benefits seen in healthy populations are similar to what are expected in the osteoporotic patient	Benefits seen in healthy populations are similar to what are expected in the osteoporotic patient

Practical Application 27.3

Exercise Prescription Guidelines

In designing an exercise program for the osteoporotic patient, the clinical exercise physiologist should bear in mind two primary goals:

1. Increase overall fitness (cardiovascular, muscular strength, balance, and flexibility)

2. Increase or at least maintain BMD

These goals must be weighed against the need to create an exercise program that is safe for the person with osteoporosis (see table 27.4 for exercises to avoid). Most people with osteoporosis are postmenopausal women or elderly men in whom the risk of cardiovascular disease is significant. Thus, a walking program designed to increase cardiovascular fitness is surely the best choice, because this has the potential of improving both cardiovascular fitness and bone density (bicycle training is unlikely to improve BMD because it is not weight bearing). If the patient has no recent history of exercise, the walking program should start slowly and progress gradually over time. The ultimate goal would be to walk five times a week for 30 to 45 min at a speed of at least 3 mph (4.8 kph) or faster if possible (but never reaching a jogging speed). To reach that goal, the person may need to start at a slower pace and shorter distance, but the number of times per week can start at five. The clinical exercise physiologist should just be sure that the patient can comfortably handle the speed and distance used for the first 2 wk. The speed and distance can then be increased gradually. The best choice is to increase only speed on a couple of days a week and increase the distance walked on the other days of the week.

A resistance-training program can help increase both bone density and muscular strength in elderly people. Although increases in bone density are generally only modest, strength gains can be much greater and will help reduce the risk

(continued)

Practical Application 27.3   *(continued)*

of falling, which is a major cause of hip fracture. The resistance-training program should begin slowly, starting out at 2 d per week and moving up to 3 or 4. One-repetition maximum testing (1RM) is not appropriate in this population. A 10RM test can be used but is not necessary. Initially, the resistance should be set low enough (even if it is the minimum weight) that the person can easily complete 15 repetitions of an exercise without undue strain. Exercises that target all the major muscle groups should be chosen, especially for the legs and back. But the client must avoid any exercise that involves spinal flexion (table 27.4). After the first 2 wk, during which the client does one set of 15 repetitions for each exercise, a second set for each exercise can be added. Following a couple of weeks of two sets for each exercise, resistance can be added to increase the intensity of the program. The resistance should be increased so that the client can complete no more than 10 to 12 repetitions on the second set. After that, the weight should be increased gradually as needed to maintain the 10- to 12-repetition limit on the second set. Exercises should always be performed with slow, controlled movements. Exercise prescription guidelines are listed in the table below.

Training method	Mode	Intensity	Frequency	Duration	Progression	Goals	Special considerations and comments
Aerobic	Walking or bicycling	Moderate pace	5 d · wk^{-1}	30 min	Increase speed and distance gradually after initial 2 wk	30–45 min at ≥3.0 mph (4.8 kph)	No jogging. Avoid activities that increase risk of falling (i.e., step aerobics). Patients with severe kyphosis may be limited to stationary cycling
Strength	Resistance training	15 reps of 8–10 exercises (may require less strenuous program initially)	2 d · wk^{-1}	One or two sets up to 30–60 min	Add a set after initial 2 wk	3–4 d · wk^{-1} at 10–12 reps · set^{-1}	Avoid spinal flexion; use slow and controlled movements. Target legs and back
Range of motion	Stretching		5–7 d · wk^{-1}	15–20 min		Increase or maintain range of motion	Stretching exercises involving spinal flexion should be avoided

with free weights should be avoided, however, because of the excess load applied to the spine as well as the potential for spinal flexion during the squat lift. A specific exercise that helps build hip and low-back strength as well as improve balance is standing on one foot for 5 to 15 s. Initially, patients should place their hands on a counter for support until they develop the strength and balance needed to perform the exercise without danger of falling. The osteoporotic patient should also be encouraged to do spine extension (but not spinal flexion) exercises. Spine extension exercises can be performed in a chair. These exercises can strengthen the back muscles, which should help reduce the development of a dowager's hump and possibly reduce the risk of vertebral fracture. Patients with osteoporosis should perform these and all exercises with slow and controlled movements; they should avoid jerky, rapid movement. More complete information on these and other exercises for the osteoporotic patient can be found elsewhere.

Exercise may be useful both for increasing bone density to help prevent osteoporosis and as a therapeutic modality for patients in whom osteoporosis is already present. As a means of preventing osteoporosis, individuals need to engage in weight-bearing activities on a regular basis. These activities may be most beneficial among young adults, when bone modeling is at its highest.

Nevertheless, excess exercise can sometimes be detrimental to bone health, especially if it leads to amenorrhea, as is often seen in female athletes. When a clinician is caring for an osteoporosis patient, however, caution must be observed in the type of exercise program to be used and the specific exercises done. Patients with severe osteoporosis who are just beginning an exercise program should probably be supervised until it is determined that they can properly perform the exercises without danger to themselves.

CONCLUSION

Osteoporosis is an increasing health problem and should be a concern of everyone, although the disease primar-

ily affects women. As the population ages, the cost and problems associated with osteoporosis will continue to increase. But osteoporosis should not be considered an inevitable consequence of aging. It is a preventable disease, but prevention must begin at an early age, perhaps even before puberty. Osteoporosis or **osteopenia** can be diagnosed with the use of DXA technology, and bone density measurements should be seriously considered in anyone with existing risk factors for **osteoporosis**. Drug and exercise therapies are available to treat osteoporosis and its related problems, but when bone density is low enough to be considered osteoporotic, most of that lost density cannot be regained. Therefore, prevention should be the primary focus.

Case Study

Medical History

Ms. RF, a 60 yr old Caucasian woman, is referred for the development of an exercise program. She is 9 yr past menopause and has never been on hormone replacement therapy. She smoked for 35 yr but quit 1 yr ago after having been diagnosed with atherosclerosis. She has recently developed high blood pressure and would like to begin an exercise program to help in that regard. She is 5 ft 2 in. (157 cm) tall and of normal weight for her height, but she has never exercised regularly.

Diagnosis

Based on the medical history, although there is currently no confirmation, low bone density should be suspected in Ms. RF. She is 9 yr postmenopausal, has never been on hormonal replacement therapy, and is advancing in age. Other risk factors include cigarette smoking, being Caucasian, and being petite in stature. A measure of BMD would likely confirm those suspicions.

Exercise Test Results

She recently underwent an exercise stress test that was normal, but no other physiological testing was done other than regular blood chemistries.

Exercise Prescription

A walking program is surely the best choice, because it has the potential of improving not only her cardiovascular fitness but also her bone density (bicycle training is unlikely to improve BMD because it is not weight bearing). Although her stress test was normal, given her age and lack of any exercise history, the walking program should start slow and gradually progress over time. The ultimate goal would be for her to walk five times a week for 30 to 45 min at a speed of at least 3 mph (4.8 kph) or faster if possible (but never reaching a jogging speed). To reach that goal, Ms. RF may need to start at a slower pace and shorter distance, but the number of times per week can start at five. After it is determined that the patient can comfortably handle the speed and distance used for the first 2 wk, the speed and distance can gradually be increased. The best choice is to increase only speed on a couple of days a week and increase the distance walked on the other days of the week. A resistance-training program should also be implemented for Ms. RF. Again, it should begin slowly, starting out at 2 d per week and moving up to 3 or 4. Initially, resistance should be low enough that she can easily complete 15 repetitions of an exercise without undue strain. Exercises that target all the major muscle groups should be chosen, especially for the legs and back. But Ms. RF should not perform any exercise that involves spinal flexion. She can do leg lifts or modified sit-ups, in which only the head is lifted off the floor, to strengthen the abdominal muscles.

(continued)

After the first 2 wk, during which Ms. RF does one set of 15 repetitions each exercise, she can do a second set for each exercise. Following a couple of weeks of two sets for each exercise, resistance can be added to increase the intensity of the program. The resistance should be increased so that she can complete no more than 10 to 12 repetitions on the second set. All exercises should be performed with slow, controlled movements. Based on Ms. RF's case history, such an exercise program should be well tolerated and provide optimal health benefits. Individual cases vary, however, and for a woman who has suffered fragility fractures as a result of osteoporosis, even the minimal exercise described here may not be tolerated initially. For such a woman, a more gradual, less intense program will be needed.

Discussion Questions

1. What other information should be obtained from Ms. RF?

2. What other physiological tests might be recommended?

3. Given the case history for Ms. RF presented here, and the fact that resistance training will offer little in the way of cardiovascular benefits beyond what she will achieve with the walking program, what will be the value of the resistance-training program? Will the risks outweigh the benefits?

28

Nonspecific Low-Back Pain

Jan Perkins, PhD

J. Tim Zipple, DScPT, OCS

Most people in modern society experience nonspecific low-back pain (NSLBP). This complicated and poorly understood phenomenon involves the interaction of a wide range of physical, social, and psychological factors. For an individual, the effect may be devastating; for society, the costs are enormous. Although exercise is widely used in the management of NSLBP, experts do not agree on the optimum type or dose of exercise. Furthermore, exercise prescription is complicated by differences between acute, subacute, and chronic NSLBP and by substantial variability in recommended management strategies.

NSLBP is an umbrella term that includes pain in the lumbosacral area caused by a variety of somatic (musculoskeletal) dysfunctions. A specific origin of pathology, such as herniated disk, vertebral fracture, congenital malformation, or general medical condition, has not been identified. Theories about causation appear to be as plentiful as the number of structures in the back. Indeed, "the actual origin of the pain is more of a philosophic statement of the training of the practitioner than hard, scientific fact" (4, p. 63). In addition, NSLBP is not a single entity but rather a syndrome that has pain and disability as the most important symptoms (39). The best definition of NSLBP may be simply pain experienced in the lumbosacral region in the absence of major identifiable pathology. The pain is typically diffuse and located in a region that includes the areas of the back below the ribs and above the distal fold of the buttocks (39).

This chapter focuses primarily on management of subacute, recurrent, or chronic NSLBP. Management of the acute phase of an episode of NSLBP mainly involves medical screening to exclude serious conditions, reassurance as to the benign nature of the problem, and simple symptom-based treatment (39,75). The minority of back pain caused by serious pathologies or requiring urgent medical or surgical intervention is not covered, and postsurgical rehabilitation routines are also excluded. For information on typical postsurgical rehabilitation, recovery paths, and reviews of surgical outcomes, refer to other resources (e.g., see references 18,21,62), which should be used in consultation with the healthcare workers caring for individual clients. Therapeutic exercise has not been shown to be any better than other treatments in promoting the resolution of an acute episode of NSLBP (39,77). The mainstay of prescribed treatment is to avoid prolonged bed rest and stay as physically active (without aggravating the condition) as possible (25,26,72). As pain improves and the condition becomes subacute, exercise can be more helpful (39,77). Exercise treatment may possibly help prevent recurrences, but more research is needed on this topic (12).

SCOPE

Most people, in all societal groups, will experience at least one episode of NSLBP during their lifetime (16). Back problems are "the most expensive musculoskeletal affliction and industrial injury and the most common cause of disability for Americans under the age of 45" (7, p. 192). Reported rates of back pain vary widely. A recent systematic review found annual rates between 14% and 93% in individuals with a history of back pain (28), whereas a population-based cohort study in North America found an annual **incidence** of 18.6%, with most episodes being mild and with a recurrence rate of 28.7% within 6 mo (10). Another study found an annual incidence of back pain of 49.1% in working-age adults (13). The variability likely reflects different locations, definitions, populations, and survey techniques. But all experts agree that NSLBP is a frequent aspect of life. Many people experience distress and some degree of temporary or mild disability, but only a small percentage become seriously disabled or go on to experience chronic back pain (39,75).

Up to a point, NSLBP increases with age. A cohort study that examined individuals at age 14 and used survey follow-up at age 38 found that **prevalence** of NSLBP increased as the participants aged (with lifetime prevalence at 70% by age 38) and that rates for men and women were similar (24). Many others have noted an increase in NSLBP with age, with a peak between age 45 and 60, followed by a decline in reported pain (25,34,44). This finding may be attributable to the inherent stability of the spine associated with the loss of elasticity and increased stiffness seen with aging. Older people may also be more cautious with lifting. Younger people are more likely than older people to experience complete resolution of an episode of pain, whereas those older people who continue to have pain have more frequent and prolonged pain (10). But back pain in childhood and adolescence tracks into adulthood, suggesting that we need to pay more attention to first-time episodes of NSLBP in young people to determine whether **prevention** at that stage can decrease later, more severe problems (27). The relatively high rates of NSLBP noted in young adults and adolescents in cohort studies and the association with sports in boys (12,27) matches our clinical experience, in which many adults describe a first episode of NSLBP in high school, frequently precipitated by athletic competition, heavy lifting, or trauma. The cost of managing the minority of NSLBP patients whose acute episode becomes chronic is enormous. NSLBP costs may exceed $50 billion per year in the United States. About 75% of this cost arises from expenses related to the minority who become disabled (20,72,75).

The average adult can expect between one and three episodes of NSLBP in a year. In most cases, a single episode will resolve spontaneously without producing major disability (10,39). In a small percentage of cases, acute episodes of NSLBP will lead to chronic NSLBP (10,39).

PATHOPHYSIOLOGY

NSLBP is a syndrome with many causes and consequently many pathologies. Up to 85% of low-back pain has no known cause (69). In this chapter, the focus is on this type of NSLBP, for which precise pathologic descriptions are impossible. Opinions as to the structure causing pain have varied over time and may indeed be more reflective of fashions in healthcare and contemporary technologies than actual pathologies.

An impressive list of structures with pain receptors could cause NSLBP, including the anterior and posterior ligaments, interspinous ligament, yellow ligament, posterior annular fibers of the disk, intervertebral joint capsules, vertebral fascia, blood vessel walls, and paravertebral muscles (79,82). Mechanical stimulation during microsurgery has been used to test tissues for pain sensitivity. Fascia, supraspinous and interspinous ligaments, spinous process, muscle, lamina, ligamentum flavum, facet capsule, facet synovium, annulus fibrosis, and nucleus pulposis were tested (30). Pain was "always" or "usually" caused with nerve root compression and with stimulation of the outer rim of the annulus fibrosis, vertebral endplate, anterior dura, and posterior longitudinal ligament (30). All other structures tested rarely or never produced pain consistently (30).

Many experts consider the intervertebral disk to be the most frequently implicated structure in nontraumatic NSLBP, and popular treatments are often based on theories relating to disk pathologies (4,73). Others disagree and consider the importance of the disk as the source of NSLBP to have been overrated (79). It makes intuitive sense to suppose that NSLBP may be caused by multiple anatomic structures that may all respond similarly to the appropriate regimen. An approach that focuses on risk factors and prevention rather than pathologies is best.

MEDICAL AND CLINICAL CONSIDERATIONS

NSLBP is common in our society. If an individual's medical history includes NSLBP, a prudent approach is to determine whether it is possibly associated with a more serious pathology that would require further physician evaluation before exercise testing and training. Medical screening is important for those with new-onset NSLBP for the identification of the minority of people with potentially serious pathology (38,39,72).

A number of risk factors are associated with NSLBP. But the strongest predictors of recurring NSLBP are the length of time between episodes of NSLBP (10,34,65) and a history of back pain (27,39). Numerous medical diagnostic tests can be used to assess NSLBP. These tests generally are not beneficial, however, unless it is thought that the NSLBP is associated with a serious pathology, such as **cauda equina syndrome**. After serious pathology has been ruled out, noninvasive treatment rather than surgery is strongly suggested in most cases. An exception would be a situation in which precise diagnosis is possible, such as a disk **herniation** with progressive neuromuscular deficits. In this case, surgical treatment has a high rate of success.

As a patient goes through treatment for acute to subacute NSLBP, consideration should be given to **secondary** prevention to decrease recurrence rates. In addition, because of the high incidence of NSLBP in our society, **primary** prevention is strongly encouraged, especially in occupational, household, and recreational pursuits with a high incidence of NSLBP. Many recent studies suggest that physicians and other healthcare providers should help patients change their perception of NSLBP as a serious disorder that results in permanent disability. They suggest shifting the emphasis back onto the patient by convincing her or him that self-care produces comparable effects to traditional medical care (12,58,75). Only 5% of people who experience an acute episode of NSLBP will go on to chronic pain and disability (39).

Signs and Symptoms

The person presenting with NSLBP may complain of localized or generalized lumbosacral region pain of variable intensity, duration, and frequency. Radiating pain with a specific distribution of sensory changes, numbness, or lower-extremity weakness can be associated with more serious pathology and can indicate specific tissue involvement. The client may have NSLBP with weight-bearing activities and complain of increasing pain with certain lumbar motions and postures. NSLBP may cause nocturnal discomfort that awakens the client when changing positions in bed. Symptoms are usually decreased with rest and anti-inflammatory medication. Any client presenting with **red flags** or other symptoms such as those listed in table 28.1 on p. 500, or with new undiagnosed symptoms, chest pain, heart palpitations, shortness of breath, hernia, or unremitting spinal pain that is not relieved by rest, should consult with a physician before initiating an exercise program. Clinicians should also be alert to signs and symptoms that may indicate an inflammatory arthritic disorder such as **ankylosing spondylitis** because a disorder of that type will require further medical evaluation and treatment.

A number of established risk factors pertain to NSLBP and disability. Prevalence increases with age into midlife. NSLBP is lower in those with greater endurance in back extensor muscles, and people who have had previous episodes of NSLBP are more likely to experience additional back pain. This risk increases as the interval since the last episode shortens. People with recurrent or persistent back pain have decreased flexibility of hamstring and back extensor muscles and lower trunk muscle strength. The best predictor of an episode of back pain (and the only one sufficiently discriminative to be valuable in job selection) is a history of previous episodes, with risk of recurrence being higher the more recent the previous episode (10,34,65). Other risk factors for back pain include obesity (40), smoking (22), and whole-body vibration such as with prolonged motor vehicle driving (56,74). Sedentary occupations are also a possible risk factor (56,74,79). Heavy lifting, or lifting with twisting regardless of the weight, are also risk factors for back pain (56,74). Research is now beginning to clarify the influence of genetics on degenerative disk disease (50).

Note that the associations do not necessarily imply causation. For example, although a modest correlation exists between back pain and obesity, it is not found in all studies and some experts believe that the association is unlikely to be causal (39,41). Instead, obesity may play a part in simple NSLBP becoming chronic or recurrent.

Psychosocial and work environment factors are more predictive of back pain and disability than are physical examination findings or mechanical stress at work (70,74,75,77). Psychological distress, dissatisfaction with employment, low levels of physical activity, poor self-rated health, and smoking status are, along with poor spinal movement, associated with persistent back pain after an acute episode (70,75,76). For many years, calls have been made for a shift from treatment of pain to management of disability (75,77). This proposal may be the key to cost control, because disability costs seem to be linked more to psychosocial determinants than to physical attributes. Failure to address these issues will lead to inadequate management of individual cases and continued high costs to society (20,75,77).

In attempting to predict long-term disability in the small percentage (1%) of people who recover from NSLBP, Waddell and Burton (77) describe "yellow flags" for risk of chronicity. These yellow flags are primarily associated with several categories of the biopsychosocial model and represent obstacles to recovery as is elaborated in table 28.2 on p. 500.

Several recent studies suggest that psychosocial factors may play a large role in determining successful outcomes

Table 28.1 Indications of Possible Severe Spinal or General Pathology

Red flags	• Age under 20 yr or over 55 yr
	• Significant medical history, such as the following:
	• Carcinoma
	• Systemic steroids
	• Human immunodeficiency virus (HIV)
	• Other major medical pathology
	• High-impact trauma such as a fall from a height or a motor vehicle accident (or low-impact trauma in a person with known or suspected osteoporosis)
	• Constant or progressive nonmechanical pain (not related to particular times or actions)
	• Spontaneous or persistent pain at night or with lying supine
	• Undiagnosed thoracic pain in addition to the low-back pain
	• Systemically unwell or with constitutional signs such as the following:
	• Unexplained weight loss
	• Fever
	• Nausea
	• Vomiting
	• Current or recent infection
	• Widespread neurologic signs or symptoms such as major or progressive motor weakness or reflex changes
	• Anesthesia around anus, perineum, or genitals or difficulties with bladder or bowel function (fecal incontinence, urinary retention or incontinence)
	• Severe and lasting restriction of flexion
Inflammatory disorders such as ankylosing spondylitis	• Insidious onset in young adulthood
	• Prolonged morning stiffness
	• Limitation of spinal movement in all directions
	• Peripheral joint inflammation
	• Family history of ankylosing spondylitis or related disorders
	• Psoriasis
	• Iritis

Based on S. Bigos et al., 1994, "Acute low back problems in adults," *Clinical Practice Guideline* No. 14. AHCPR Publication No. 95-0642 (Rockville, MD: Agency for Health Care Policy and Research, Public Health Service, U.S. Department of Health and Human Services), 7; S.J. Bigos and G.E. Davis, 1996, "Scientific application of sports medicine principles for acute low back pain," *Journal of Orthopedic and Sports Physical Therapy* 24: 192-207; B.W. Koes et al., 2001, "Clinical guidelines for the management of low back pain in primary care: An international comparison," *Spine* 26: 2504-2513; B.W. Koes, M.W. van Tulder, and S. Thomas, 2006, "Diagnosis and treatment of low back pain," *British Medical Journal* 332: 1430-1434; M. van Tulder, 2006, "European guidelines for the management of acute nonspecific low back pain in primary care," *European Spine Journal* 15(Suppl. 2): S169-S191; G. Waddell, 2004, *The back pain revolution*, 2nd ed. (Edinburgh: Chruchill Livingstone).

Table 28.2 Yellow Flags Indicating a Risk of Developing Chronic Pain

Personal and psychosocial obstacles	Work preparedness obstacles	Environmental and social obstacles
Dysfunctional attitudes, beliefs, and expectations about pain and disability	Physical or mental demands of work	Inappropriate medical information and advice about work
Inappropriate attitudes, beliefs, and expectation about healthcare	Occupational "stress"	Medical leave practices that sometimes reinforce illness behaviors
Uncertainty, anxiety, fear avoidance	Job dissatisfaction	Lack of occupational health provision

Personal and psychosocial obstacles	Work preparedness obstacles	Environmental and social obstacles
Depression, distress, low mood, negative emotions	Lack of social support at work, relationships with coworkers and employer	Employers' lack of understanding of common health problems and their modern management; assuming that they automatically mean sickness absence
Passive or negative coping strategies (e.g., catastrophizing)	Attribution of health condition to work-related activities	Coworkers' unhelpful attitudes and behaviors
Lack of motivation and readiness to change, failure to take personal responsibility for rehabilitation, awaiting a "fix," lack of effort	Beliefs that work is harmful and that return to work will cause further damage or be unsafe	Belief by many employers that symptoms must be "cured" before they can risk permitting return to work, for fear of reinjury and liability
Illness behavior	Self-perceptions of current and future "work-ability"	Loss of contact and lack of communication between worker, employer, and health professionals
	Beliefs about being too sick or disabled to contemplate return to work	Lack of suitable policies and practice for sickness absence, return to work, and so forth
	Beliefs that one cannot or should not become fully active or return to work until the health condition is completely "cured"	Rigidity of rules of employment, duties, and sick pay
	Expectation of increased pain or fatigue if work is resumed	Lack of modified work
	Low self-efficacy	Organizational size and structure; poor organizational culture
	Low expectations about return to work	Impending downsizing, termination of employment
	Beliefs and expectations about (early) retirement	Detachment and distance from the labor market

Adapted, by permission, from G. Waddell and A.K. Burton, 2005, "Concepts of rehabilitation for the management of low back pain," *Best Practice & Research Clinical Rheumatology* 19: 655-670.

with conservative management strategies. Studies found that strong encouragement by the practitioner and selection of an enjoyable general exercise program are equally effective and potentially more cost-effective than a management strategy that employs specific back exercises. Most recent literature recommends general aerobic exercise (swimming, walking, cycling), staying active, and general back school education with sound self-management strategies (16,25,26,29,38,39,58,64,66).

Conventional wisdom has long argued that the natural history of a single episode of acute NSLBP is fairly well defined, and 75% to 90% of people recover within a few weeks (75). Unfortunately, recent epidemiological studies have found that although improvement is usual, complete resolution is less typical and recurrences are common (28,37,75). The risk of additional episodes following a brief acute episode of back pain is 60% to 75% within a year (28,71). In line with the traditional perception that most new NSLBP will improve on its own, aggressive intervention is often reserved for the minority of people who do not improve in 2 mo. The increasing awareness of the recurrent nature of the problem has led some experts to reevaluate their approach. Instead they now suggest exercise therapy and self-management strategies to decrease the risks of recurrence and persistence of pain. Clients who are engaged and committed to self-management strategies are less likely to seek medical care with these recurrences of NSLBP.

HISTORY

Medical evaluation should have cleared the individual referred with NSLBP for major pathologies, but a clinical exercise physiologist should be alert to the possibilities that serious pathologies have been missed. Table 28.1 lists key red flag findings that indicate when medical evaluation is required.

The history of an individual with NSLBP should first focus on this screening for possible serious pathology. Questions should cover all the areas indicated in table 28.1. Following this, the interview should focus on the mechanisms of injury, both initial and for recurrences, because this information may guide the practitioner in selecting management strategies. For example, recurrences may be associated with specific situations such as spinal flexion with lifting, prolonged postures such as with driving, or particular sporting activities. Careful questioning and analysis of such patterns can guide physical examinations, suggest postures and activities to avoid or use in early rehabilitation, and indicate what education is needed during the rehabilitation process. For example, a patient with a history of pain recurrences following long periods of driving with the spine in flexion or neutral may benefit from exercises that avoid these postures, from modifying motor vehicle seating, and from education on incorporating breaks for stretching and brief walking into any long drives in the future. A history of usual vocational and recreational activities can also offer valuable hints for education on self-management strategies after resolution of the current episode. Asking about the benefit of treatments tried for previous episodes is also important.

PHYSICAL EXAMINATION

Again, the physical examination in NSLBP is primarily one of exclusion. An examiner should do a quick scan of posture and general range of motion. A neurologic screening should check for any abnormalities of sensation, motor function, or reflexes that may indicate serious pathologies. Assuming that these scans are clear, the physical examination should include range-of-motion and flexibility testing that specifically looks at common deficits seen in NSLBP. Hamstring and hip flexor tightness are common in people with NSLBP, and spinal flexion and particularly extension are frequently limited. As a minimum these should be tested. Palpation of paraspinal muscles may reveal muscle spasm and tenderness.

Ideally, the examination would incorporate aerobic testing and spinal and abdominal muscle strength test-ing. Back extensor and abdominal muscle weakness is common in individuals with NSLBP, and those who have a long history of recurrences may be considerably deconditioned even when compared with their sedentary peers without back pain. Unfortunately, if the client is being seen early after a flare-up of pain, strength and aerobic testing may be difficult and the results may be invalid because of pain limitations. In this case, formal testing of these functional capacities may have to be deferred. The examination should instead clear the patient for general safety for exercise using ACSM guidelines.

Diagnostic Testing

Diagnosis of specific low-back pathology is fraught with difficulty. Despite the use of highly sophisticated tests, experts are often unable to give definitive diagnoses, and it is argued that the diagnosis given relates more to the specialist who is consulted than to the patient's back (11,76). Radiographic imaging will identify loss of disk space, malalignment (e.g., **scoliosis, spondylolisthesis**), and osteoarthritic changes such as **osteophytes** or stenosis. Often these do not correlate with the severity of signs and symptoms. Abnormal findings on imaging studies are common and do not predict back pain recurrence (32,33). Some have suggested that imaging is currently overused and should be reserved for cases in which serious pathology is suspected or possibly when conservative management has failed (14,32,33,75). **Magnetic resonance imaging** (MRI) is better at soft-tissue examination than plain X-ray is, but neither technology can identify the pain receptors responsible for the reported pain. Use of diagnostic injections or neurolytic injections has shown some promise at identifying pain-sensitive structures and a specific cause of LBP.

Most people who see physicians for back pain do not have serious pathology. In primary care, approximately 1% of patients will have back pain from serious spinal pathology (75). Careful screening for red flag symptoms that indicate serious pathology is an essential part of primary healthcare management (6,7,38,39,72,75) (see table 28.1). For most cases in which an underlying systemic disease or clearly definable injury is not present, diagnosis may be less important than management.

In determining appropriate examination tests and measures used by clinical exercise physiologists, please consult chapter 4 of this text. Besides being able to measure and record vital signs, the exercise physiologist is capable of performing baseline postural observations and basic gait deviations related to velocity, cadence, and base of support. Exercise physiologists can perform palpation on common superficial anatomical landmarks to determine soft-tissue irritability and help locate the source of pathology. For a more thorough neuromusculoskeletal examination, a physician or physical therapist may be the appropriate referral.

Exercise Testing

NSLBP is a diagnostic category that does not in itself indicate a need for a graded exercise test (GXT). Maximal or submaximal testing, although useful for prescription, is not required in general clinical practice unless history indicates possible coronary artery disease or other medical conditions that would normally require a GXT before formulating an exercise prescription (61). American College of Sports Medicine (ACSM) guidelines suggest that older adults (men ≥45 or women ≥55) should have medical clearance including exercise testing and follow-up if vigorous exercise is planned (1). ACSM suggests similar precautions for those at increased risk for cardiovascular events and for those with known pulmonary, cardiac, or metabolic disease (1). People with NSLBP should be screened routinely to identify individuals in any of these categories. For formal GXT, any submaximal protocol that does not exacerbate their pain may be used. The selection may be made based on the client's history. For example, those who find the spinal-loading stress of walking painful may do better with a cycle ergometer test. Submaximal testing is suggested because pain may

prevent maximal testing. For muscle strength testing, submaximal tests are suggested. Education for the client is important in light of the emphasis on learning lifelong self-management and prevention strategies. Practical application 28.1 describes client–clinician interaction for patients with NSLBP.

Contraindications

A brief period of rest is commonly used for the first few days of an acute flare-up of back pain. During this period, GXT is inadvisable. Moreover, practical concerns relating to the patient's pain pattern and presentation can affect GXT administration and evaluation. Because both upper-extremity and lower-extremity ergometers and treadmills require coordinated trunk mobility and stability, pain may interfere with the individual's ability to perform the test or reach maximal exercise levels. Selection of the means of testing should be tailored to the individual. For some, the seated position could be most painful and ambulation may be preferred, whereas for others a stationary bicycle could provoke less pain than treadmill testing does. In any case, pain may prevent the person from reaching maximal exercise levels, and aggressive testing protocols could produce considerable posttesting soreness. A submaximal testing protocol using the exercise modality least likely to cause a flare-up in that person's pain may be most appropriate for individuals who report pain that is easily aggravated by activity.

Practical Application 28.1

Client–Clinician Interaction

Many people with NSLBP have not had the condition well explained to them. Consequently, they may have fears of exercise that prevent them from participating fully in exercise programs. They also have often been given testing results that mislead them into believing that serious pathology is present. After evaluation has ensured that the condition is NSLBP, a clinician can use the following points as the core of client education, both at the initial evaluation and as the client goes through the slow and often frustrating process of learning to manage NSLBP.

- Radiographic evidence of pathology does not correlate with level of spinal pain or disability.
- Studies provide evidence that adherence to a specific, progressive exercise program reduces the incidence and frequency of episodic NSLBP and improves function.
- Ergonomic adaptations and postural awareness reduce the incidence of episodic NSLBP.
- Minor lifestyle changes will permit compliance with self-management strategies for episodic NSLBP.
- Improvement in pain will not be instantaneous. Muscle strength can take 4 to 6 wk to improve, and loss of pain may not be noted until the spine is stable and has time to adjust to muscle and ligament changes.
- After the initial rehabilitation stage, continued general exercise is suggested.

Recommendations and Anticipated Responses

No specific GXT protocols have been developed for individuals with NSLBP. Any of the standard testing procedures recommended by the ACSM (1) and in chapter 5 may be suitable. As discussed previously, however, if a person is currently having an acute episode of NSLBP, GXT is not recommended because the person will be unable to give an appropriate effort. When a person's medical history reveals previous NSLBP, the mode of exercise testing should be carefully determined to avoid further exacerbation. An individual who has experienced multiple episodes of NSLBP and who reports interference with recreation or vocational activities may be deconditioned beyond that usually seen in sedentary individuals, and this circumstance should be taken into consideration in planning testing. High-impact test protocols, protocols that place the individual in sustained spine flexion, or maximal testing may provoke pain and limit validity of the results.

Treatment

Many medical management strategies are used for NSLBP. Common medical management includes a wide range of medications; prescription of exercise or passive modalities such as heat, massage, or **spinal traction**; facet joint injections; and surgeries such as **spinal discectomy**, **spinal decompression**, and **spinal fusion**. Current emphasis in primary care management of NSLBP is early return to activity, avoidance of needless surgery or use of unnecessary diagnostic tests, and ultimately cost-effective medical and self-management of back pain.

Surgical rates and types of surgery vary widely across geographic areas. Outcomes show no corresponding variation, and agreement is lacking on indications for surgery or on successful outcomes. Surgical rates have increased dramatically, but no decrease in disability has occurred (75). Acceptable clinical trials provide little evidence that the expensive option of surgery offers any benefit over nonsurgical management in most cases of back pain (75). Unequivocal disk herniation is an important exception to this general statement. When strict criteria are adhered to in selecting surgical patients, the success rate can be high (78). Unfortunately, the wide variation in surgical rates nationally and globally attests to the lack of strict application of criteria by all practitioners. The actual surgical management of disk pathology varies from partial excision of the offending protruding disk material, to **laminectomy**, to

significant disk excision with a stabilizing fusion using bone fragments from other parts of the body or metal spacers between the vertebrae.

The management of an isolated episode of acute back pain is less controversial than that of subacute, recurrent, or chronic pain. Current medical practice advocates conservative care for acute back pain in the absence of any red flag findings. Bed rest, if used at all, is limited to 24 to 48 h; passive treatment modalities (such as hot packs, **transcutaneous electrical nerve stimulation**, **ultrasound**, and traction) are used sparingly; and early resumption of normal activities with or without additional exercise training is advocated (6,7,12,15,72,75,78). Table 28.3 summarizes evidence-based management strategies for acute and chronic NSLBP.

Many patients with NSLBP will need pharmacologic intervention only briefly, with over-the-counter medications being most used. Generally, acetaminophen is the safest medication for low-back pain. Nonsteroidal anti-inflammatory drugs (NSAIDs), including aspirin and ibuprofen, are also effective at decreasing pain but tend to cause gastrointestinal irritation and ulcers. They also may cause renal pathology, blood thinning, or an allergic reaction. Either acetaminophen or NSAIDs, or a combination, is the typical first-line choice for NSLBP.

Muscle relaxants may also be prescribed to people with NSLBP. Although they may reduce pain, a number of side effects limit their use. They carry the adverse effects of drowsiness, impaired motor function, and increased reaction times, creating potential risks with operation of motor vehicles and power tools or machinery. Although some physicians still prescribe narcotic analgesics or opioids for relief of musculoskeletal pain, they also appear to have no greater benefit than safer analgesics, and are generally considered a poor choice for NSLBP. Narcotics can be addicting, and patients tend to become dependent on these medications during long-term use. As with muscle relaxants, a greater risk of injury occurs with the use of machinery or with driving under the influence of narcotics. Narcotics should never be taken with alcoholic beverages, which can compound the effects of the narcotics. Clients should be warned of the potential side effects listed in table 28.4 on p. 506.

The limited studies that look at primary and secondary prevention (discussed subsequently) make it possible to suggest that some intervention, preferably in a community or worksite setting, may be beneficial in avoiding recurrence of back pain after an initial nonspecific injury. Low-stress aerobic activity is usually considered safe within 2 wk, and "conditioning," especially of back extensors, is begun thereafter (6). Aerobic exercise, with fast walking being particularly suitable for people recovering

Table 28.3 Treatments for Acute and Chronic Low-Back Pain

Effectiveness	Acute LBP	Chronic LBP
Beneficial	Advice to stay active NSAIDs	Exercise therapy Intensive multidisciplinary treatment programs
Tradeoff*	Muscle relaxants	Muscle relaxants
Likely to be beneficial	Spinal manipulation Behavior therapy Multidisciplinary treatment programs	Analgesics Acupuncture Antidepressants Back schools Behavior therapy NSAIDs Spinal manipulation
Unknown	Analgesics Acupuncture Back schools Epidural steroid injections Lumbar supports Massage Multidisciplinary treatment TENS Traction Thermal modalities EMG biofeedback	Epidural steroid injections EMG biofeedback Lumbar supports Massage TENS Traction Local injections
Unlikely to be beneficial	Specific back exercises	
Ineffective or harmful	Prolonged bed rest	Facet joint injections

*Tradeoff refers to the balance between beneficial effects and undesirable side effects.

Reprinted, by permission, from B.W. Koes, M.W. van Tulder, and S. Thomas, 2006, "Diagnosis and treatment of low back pain," *British Medical Journal* 332: 1430-1434.

from a NSLBP episode (51), should continue as long as there is evidence that the patient is making functional progress (17). Few patients will have insurance coverage for this type of program in a rehabilitation center, because the current focus of the healthcare system is on short-term secondary prevention and tertiary prevention. Instead, people may need to continue exercise programs under the guidance of fitness professionals at health clubs and fitness or wellness centers. Many community wellness centers will have an important role in both secondary and primary prevention of nonspecific back pain. The healthcare specialist providing rehabilitation for the client should work collaboratively with the community wellness trainer to create a reasonable long-term maintenance program for the client.

Primary prevention is avoidance of a condition for which a person is at risk. Intervening before an injury or disease develops and preventing it from appearing would be the optimal form of preventive healthcare. For traumatic back pain, frequently caused by vehicle accidents or falls, primary prevention includes standard safety precautions in household, recreational, and occupational situations (4). Safety equipment such as seatbelts should be used in all appropriate situations, and watchful caution should be practiced during activities that place people at risk of falling or encountering moving vehicles. For NSLBP in general, leisure time physical activity for at least 3 h per week is associated with decreased incidence of back pain (23). In addition, a number of "best practices" in work and leisure activities can reduce the risk of sustaining a back injury (51).

Table 28.4 Pharmacological Agents Commonly Used With NSLBP

Drug categories and common generic names	Trade names	Side effects	Contraindications
Nonnarcotic analgesic • Aspirin • Acetaminophen • Ibuprofen • Naproxen • Piroxicam • Diflunisal	• Bufferin, Empirin • Tylenol, Anacin, Valadol, Datril • Motrin, Rufen, Nuprin • Naprosyn, Anaprox • Feldene • Dolobid	Gastrointestinal (GI) distress, GI ulcerations, allergic reactions, renal dysfunction	Patients with history of allergic reactions or gastrointestinal distress with these medications
Narcotic (opiate) analgesic • Meperidine • Hydromorphone • Methadone • Codeine • Morphine • Oxycodone	• Demerol, Mepergan • Dilaudid, Palladone • Dolophine • Codeine • Avinza, Roxicodone, Duramorph • Oxycontin, Oxyir, Tylox, Percodan, Percocet	Poor tolerance, gastrointestinal disturbances, sleep disturbances, drowsiness, increased reaction time, clouded judgment, misuse or abuse and dependence issues	Patients with history of poor tolerance, allergic reactions, or dependent personality types
Muscle relaxants • Cyclobenzaprine hydrochloride • Carisoprodol • Metaxalone	• Flexeril • Robaxin • Skelaxin • Soma • Parafon Forte • Relaxazone • Vanadom	CNS depressant, drowsiness, tachycardia, hives, mental depression, SOB, skin rash, itching	Allergies, blood disease caused by an allergy or reaction to any other medicine, drug abuse or dependence, kidney or liver disease, porphyria, epilepsy

A considerable body of literature exists regarding risk factors for NSLBP, but to summarize, primary prevention strategies include the following:

- Use safety equipment in work and leisure activities.
- Address risk factors of smoking, poor general fitness, obesity, stress, and poor seating.
- Perform balanced exercise programs that include both spinal flexion and extension.
- Avoid long periods in one position—take breaks to move the spine out of fixed positions and balance postures of flexion and extension.
- Avoid lifting with twisting and preserve the neutral spine curve during lifting.
- Avoid lifting activities immediately after rising or prolonged spinal flexion positions without breaks to move into extension.
- Be physically active and keep fit.

- Vary postures during prolonged sitting.
- Balance activity in flexion with activity in extension.

The more common nontraumatic low-back injuries are thought to result from a series of cumulative events in a person with risk factors such that a relatively minor incident precipitates symptoms (4). The best options for primary prevention of nontraumatic back pain come from addressing risk factors such as smoking, obesity, psychosocial stress, and poor seating for those exposed to occupational risk factors such as truck driving (4). Any awareness of early signs of dysfunction such as joint stiffness, minor aches and pains, or difficulties straightening the spine can minimize progression. Physical activity that promotes both cardiovascular and musculoskeletal fitness is beneficial, as is a balance of spinal flexion and extension activities and avoidance of prolonged loading of the spine in either flexion or extension. The spine is not meant for static positioning but for movement, and sustained postures or repetitive movements in any direction should be balanced with movement in the opposite direction.

In secondary prevention, treatment is used early in the course of a condition to cure that condition or prevent or slow its progression. In back pain, the aim is to prevent recurrences and to prevent acute and subacute back pain from becoming chronic.

Secondary prevention strategies for back pain include the following:

- Catch problems early before an injury becomes disabling.
- Avoid or minimize bed rest.
- Encourage early return to activity, even in the presence of some pain.
- Avoid aggressive spinal-loading exercises during early rehabilitation.
- Avoid loading the spine throughout the full range of motion, particularly in the general (nonathlete) population.
- Keep use of passive modalities to a minimum or use them to assist with a more active program.
- Be willing to adapt a program to individual needs.
- Use behavioral strategies to encourage participation.

After the resolution of the NSLBP episode continuing to follow the common-sense principles of primary prevention is important. Emphasizing early return to activity even with some continuing pain; weaning from pain medications; avoiding dependence on passive modalities such as heat, rest, transcutaneous electrical nerve stimulation, or massage; and strongly emphasizing self-management are the preferred management strategies (4,6,58,66,65). After recovery, a program that incorporates the measures suggested for primary prevention with extra consideration of the individual's particular at-risk activities will enhance outcomes.

EXERCISE PRESCRIPTION

Most programs of exercise for people with back pain include a combination of several forms of exercise and educational advice regarding lifestyle factors and general back care. With a single episode of back pain, it is not clear whether exercise will have any effect on the anticipated natural history of that episode. Exercise intervention after back pain has occurred is aimed at reducing risk factors and minimizing recurrences. With chronic and recurrent back pain, the evidence for more aggressive intervention is stronger. Increasing function and decreasing severity and frequency of back pain episodes should be the goal.

The clinical exercise physiologist may need to emphasize endurance over strength in the early training, adjusting repetitions and resistance to reflect this emphasis. Exercise will usually start with low levels of exercise done frequently and will progress through a system of exercising to quota. The focus should be on promoting function improvements rather than reducing pain. A **periodization** schedule based on estimated maximums from low-intensity testing and using ACSM guidelines for testing and progression (1,2,3) may be appropriate. In the absence of specific restrictions imposed by pathology, the exercise program should be designed to correct specific impairments found in initial comprehensive evaluation (62).

Adherence to exercise is a problem for many people with recurrent or chronic NSLBP. Although function can improve quickly, individuals with chronic back pain may need more than 2 mo of training to experience significant pain relief (49). Discontinuing exercise after the completion of a program is as common among individuals with back pain (17,18) as it is in the general population. Support, encouragement, and easy availability of follow-up programs in the community may be especially important until an exercise habit is well established and whenever a program has to be restarted after a lapse for any reason.

As a general guideline for exercise in people with NSLBP, watch for the following complaints:

- Pain that is severe enough to halt exercise
- Pain that persists for more than 3 h after cessation of exercise
- Pain that results in several days of disability or sleep disturbances

Exercise should be generally tolerable during the activity and should result only in mild musculoskeletal discomfort associated with delayed muscle soreness post exercise. Pain similar to the examples cited earlier or which results in referred or radiating pain into the posterior thigh or down into the lower extremity indicates that the exercise is too aggressive or irritating to sensitive neuromuscular structures in the lumbar spine. Radiating pain into the lower extremity, in itself, is cause for concern and usually indicates irritation of lumbar nerve roots.

Clients coming for treatment of NSLBP may have a long history of flare-ups of back pain and experience with a wide number of traditional and alternative approaches to manual therapy and exercise. Background information on several of these is given in practical application 28.2 on p. 508-510. Gradually theory-based research is developing improved guidelines for management and practical clinical protocols.

Literature Review

Exercises have been used to manage back pain for more than 100 yr, yet well-controlled trials on the benefits of exercise are lacking. "Some form of exercise is probably the most commonly prescribed therapy for patients recovering from low-back pain" (78, p. 363). Types of exercise favored have varied widely (50). Lumbar flexion exercises, usually based on the Williams flexion routine (80,81), have had their vogue, as have hyperextension exercises focusing on increasing paravertebral muscle strength and endurance. At present, a balanced approach that involves strengthening most spinal musculature and lower-extremity muscles is arguably the most popular strategy, usually in combination with stretching exercises and recommendations for some form of aerobic conditioning.

When prescribed based on an individual evaluation, exercises are hypothesized to have multiple beneficial effects, including reduction of disk herniation, improved joint mobility, strengthening of weakened muscles, and stretching of adaptively shortened ligaments, capsules, and muscles. Increasingly, behavioral approaches to management are also included to address aspects of the back pain syndrome other than the simple physical limitation.

Popular approaches in rehabilitation have included variations on the Swedish Back School approach (29), which mainly uses education on back care; functional restoration and work-hardening using individually prescribed exercise combined with aggressive disability management (59,64); and a wide range of manual therapy approaches.

Although these manual therapies are based on widely different theories of causation, many use surprisingly similar positions and techniques. The McKenzie approach, which uses treatments based on symptom response to movement and a theory of disk pathology (19,54), has become one of the more common manual therapy approaches in rehabilitation. Healthcare and wellness practitioners will also encounter people who have used a wide range of alternative approaches to therapeutic exercise and body work. The table below describes selected manual therapy and therapeutic exercise approaches that are frequently encountered.

Manual Therapy and Therapeutic Exercise Approaches

	Founder and history	Treatment philosophies
Manual therapy philsophies		
Chiropractic approach	Founder: Daniel David Palmer treated misalignments (subluxations) of the spine with manipulations (adjustments). Chiropractors began treatment of the spine in the late 1800s	Many chiropractors also use a variety of gentle techniques and joint oscillations to realign spinal segments that are disrupting normal neurological function
Osteopathic approach	Founder: Andrew Taylor Still, a medical doctor in the United States in 1874	Treatment of somatic dysfunctions with a variety of gentle muscle contractions and joint mobilizations, as well as high-velocity, low-amplitude thrust techniques (HVLA)
Australian approach	Accredited to Geoffrey Maitland, a physiotherapist from Australia	Treatment emphasis on subjective complaints and quality of pain, as well as behavior of pain and response to positions and mobilization techniques. Advocates use of comparable signs (pre–post treatment assessment) but avoids "diagnostic titles"
New Zealand approach	Accredited to Robin McKenzie and Stanley Paris, two physiotherapists from New Zealand	Approach emphasizes self-management strategies. Three predisposing factors to low-back pain: sitting postures, limited lumbar extension, and predominance of flexion activities in society. Divides back problems into postural, dysfunction, and derangement syndromes

	Founder and history	Treatment philosophies
Manual therapy philsophies (*continued*)		
Norwegian approach	Accredited to Oddvar Holten, a physiotherapist in Oslo, Norway. Introduced in the United States by Freddy Kaltenborn and Olaf Eventh	Uses a biomechanical approach to assess mobility of spinal and extremity joints. Uses Maitland and Kaltenborn oscillatory mobilization grading system to normalize passive joint mobility and improve function
Craniosacral therapy	Early work of William Garner Sutherland in early 1900s, popularized in the 1970s by John Upledger, an osteopathic physician in the United States	Treatment involves manipulation of the cranial sutures and sacrum to adjust the flow of cerebral spinal fluid, which oscillates caudally and cranially (craniosacral rhythm). Disruptions in the rhythm lead to a variety of autonomic and musculoskeletal dysfunctions
Variety of soft-tissue approaches	Myofascial release: Fred Mitchell, Sr., DO. Rolfing: Ida P. Rolf, PhD. Hellerwork and soma: Joseph Heller and Bill Williams, PhD. Bindegewemassage: Elisabeth Dicke, Swedish remedial techniques	Soft-tissue approaches run the gamut of gentle myofascial stretching to vigorous mobilization of deep tissues and internal cavities. Realignment and restructuring of tissues lead to normalization of posture and muscle function
Therapeutic exercise approaches		
Pilates	Joseph Pilates, developed while working as a nurse in World War I	Use of specialized equipment to promote balance, strength, proper posture, and agility. Currently in vogue as an exercise approach
Tai chi	Passed on from many Chinese generations including Wu style, Yang style, Ch'en style, and Chuan style	Use of slow, balanced movements of extremities around stable trunk through a series of 108 postures. Uses concept of power centering during standing postures, balancing ying and yang. Slow form used for strengthening and meditation; fast form used as defense
Feldenkrais	Developed by Moshe Feldenkrais, a physicist, judo expert, and athlete	Observed unnatural movement patterns and created a system of movement awareness exercises. Recognition of mental and emotional activity that perturb all aspects of human performance. Uses visual, auditory, and kinesthetic cues to alter movement
Alexander technique	Frederick M. Alexander, developed during late 1800s	Use of specific movements of the body to promote bodily awareness and health. Self-awareness of harmful movement patterns allows subject to correct and move in harmony with breathing pattern
Trager approach	Milton Trager, MD, who was a former boxer and acrobat. He developed the techniques for 50 yr and began teaching in the 1970s and 1980s	Concentration, repetition, and refinement of movement, targeted through unconscious mind. Use of active effortless movements called Mentastics
Yoga	Origins in Egypt and India 5,000 yr ago. Popular in the 1960s and 1970s, it has made a resurgence in the last 10 yr	Uses a series of stretching positions that liberate the natural flow of energy in the body, with slow meditative assumptions of the poses and strong emphasis on deep relaxation. Improves flexibility, posture, and body awareness. Improves circulation

(*continued*)

Practical Application 28.2 *(continued)*

One approach to using exercise in a clinical setting focuses on initially establishing normal movement patterns and then moving on to stretching tight structures, strengthening with early progression through increasing repetitions, and adding weighted training exercises (17). Great emphasis is placed on form and maintenance of good muscle control throughout the entire range of movement. This approach is compatible with the functional restoration philosophy of treatment and, in our opinion, is a logical dynamic approach to back care.

Some excellent attempts have been made to validate exercise selection and link experimental research and clinical practice. One interesting technique combines **electromyography** and computer modeling to estimate tissue loads during activities (35,51–53). In text specifically written for clinicians, McGill (51) synthesizes research evidence and makes specific suggestions for exercise and prevention. Although emphasizing that any program must be tailored to the specifics of the individual patient, he suggests that the ideal early exercise program for a patient with back pain would avoid loading the spine through range of motion yet would provide sufficient challenge to allow a muscle-conditioning effect. Whereas a healthy athlete is able to load the spine through full range of motion, McGill's approach is safer for the less trained individual. Other treatments that are currently in vogue unload the spine during early rehabilitation and then add progressive loading and resistance

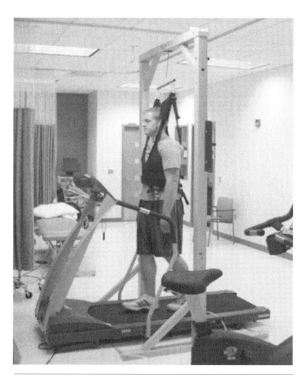

Figure 28.1   Typical unloading system.
Photo courtesy of J. Tim Zipple.

training as recovery occurs. Examples include some programs that incorporate aquatic therapy and use unloading harnesses for treadmill ambulation (65). Many vendors in the United States offer unloading equipment, but the basic premise with the use of the equipment is that reducing compressive forces on the lumbosacral region allows pain-free ambulation on a treadmill or during the performance of functional activities for the low back such as partial squats, heel raises, weight shifting, and step-ups onto a platform or step. This approach, however, is not considered traction for the spine, a common misconception among the uninformed. Figure 28.1 shows a typical unloading system found in an increasing number of U.S. clinics.

Supervision is valuable, particularly in the early stages, to ensure use of correct form and encourage exercise adherence. In one study, chronic back pain patients were randomly assigned to either supervised or unsupervised training for flexibility, muscle strength, and aerobic conditioning according to a periodization schedule (62). After 6 mo, marked differences were seen in the experimental groups. The group given independent exercise averaged only 31.95 of a planned 96 sessions. The other group, assigned a certified strength and conditioning specialist who supervised them, completed an average of 90.75 sessions. Because long-term follow-up was not done, it is not known whether benefits and exercise participation continued, but this work supports the common-sense idea that supervision enhances adherence. Other

studies have noted the benefits of supervised exercise as part of a comprehensive management model, particularly when pain has become chronic (34,39,47). The client is held accountable and experiences the positive benefits of social interaction with a healthcare provider, who acts as a cheerleader for the patient.

Special Exercise Considerations

After a person has been cleared by a medical professional, few contraindications to an exercise program exist. But many patients with recurrent or chronic pain may be so deconditioned and fearful of exercise or movement that progression should be slow and initial exercise levels low.

Special exercise considerations for patients with NSLBP include the following:

- The patient should obtain medical clearance for exercise.
- Patients experience a fear-avoidance behavior pattern when they believe that movement and activity will further damage or injure the spine.
- The exercise specialist should monitor for red flag findings.
- Deconditioning may be greater in these patients than in sedentary healthy individuals. This deconditioning may be more severe in smaller phasic muscle groups than in larger tonic muscle groups.
- The exercise specialist should use caution if loading through the spine and should consider unloading in some cases for pain management.
- Progression should be slow and initial exercise levels low.
- Smokers may need a slower progression.

One caution that should apply is to avoid overstressing or overtraining. With overtraining, large-muscle groups are allowed to compensate or substitute for smaller, deconditioned muscle groups. An emphasis on form and evaluation of postexercise response is important in determining exercise level. The goal is to provide adequate stress for a training effect but to avoid stressing any tissue beyond tolerable levels and to do this without using abnormal movement patterns (31). The clinical exercise specialist must remember that a person with NSLBP may have been extremely inactive for a considerable time and so will need to start with a lower-intensity program than would a sedentary but healthy individual. Persons with chronic NSLBP have been shown to have selective atrophy of type 2 muscle fibers in the back muscles. This condition has consequences for fatigue resistance and may help explain the poor back muscle endurance seen in people with back pain (57).

In general, the best approach is to avoid exercises that provide compression loading of the spine in injured individuals until late in rehabilitation. Early exercise focus should be on safe performance of rhythmic exercises with low resistance and an emphasis on correct form that avoids substitution patterns. Initially, exercises such as squats or calf raises may be done with less than full body weight. Progression can then occur to full weight bearing and finally to additional resistance. For most people, the use of handheld weights may be more appropriate

for adding resistance than a bar behind the neck and across the shoulders for exercises like lunges, squats, and calf raises. Similarly, leg presses that rely on adding weight through the shoulders and spine should usually be avoided in early treatment. Early resistance training is more easily tolerated with the use of pulley-assisted equipment (performing squats with a pulley assist) or reduced body-weight exercises (e.g., Total Gym squats or heel raises). A system such as a Total Gym machine allows the resistance progression to be objectified and provides a stabilized position for the spine during resistance exercises for the upper and lower extremities.

Unless they can be performed with supervision, dead lifts should be avoided or used with extreme caution with patients with a history of back pain. Some clinicians, however, use these in their rehabilitation programs for patients with back pain (17). The argument in favor of dead lifts states that they are an important means of strengthening the trunk extensor muscles. Correct training in technique can help avoid future exacerbations and permit retraining for stressful activities. This argument has validity, but the exercise remains controversial because of the high potential for reinjury. Unless an exercise advisor is confident of his or her ability to instruct people in this lift and is certain that the client will continue to use correct form, the dead lift is best left out of an exercise program.

Given the association between smoking and back pain, caution should be used in exercise prescription with smokers. Although most published work takes a judgmental approach to the issue of smoking or merely documents the association, a better approach may be to acknowledge the fact that some people will not be able to quit and to use a slower exercise progression for smokers than for nonsmokers (17). Becoming judgmental with patients who smoke often leads to feelings of resentment that impair the client–provider relationship. Clinical experience shows that only those who are committed to quitting smoking will benefit from active cessation programs.

Exercise Recommendations

Most people with NSLBP should finish rehabilitation with a general exercise prescription that matches ASCM guidelines for healthy adults (1–3). Adoption of these guidelines is appropriate in the early stages of exercise training to allow for common problems such as significantly decreased exercise tolerance, lack of flexibility, poor neuromuscular coordination, and pain with movement or loading through the spine. Practical application 28.3 includes general guidelines for prescribing exercise and anticipated benefits of exercise for the population with NSLBP. Although there is some suggestion that patient-specific programs based on expert biomechanical evaluation may be better than general exercise, this form of evaluation and treatment needs more research and access to specialized practitioners (19). For individuals with persistent pain and disability, a referral to such practitioners for initial treatment may be useful.

Cardiovascular Training

Aerobic exercise is often used in rehabilitation of NSLBP, although evidence is not conclusive (9). Despite this, most experts still recommend aerobic exercise with subacute and chronic back pain because of its known benefits in other areas of health, hypothesized psychological benefits, or the theoretical rationale with back pain. General fitness is considered desirable for many reasons and has several hypothesized benefits in general pain management.

Given known benefits of aerobic exercise for general health and mental well-being, including a graduated exercise program with an individualized exercise prescription is reasonable. Maximal or submaximal testing, although

Practical Application 28.3

Exercise Prescription Summary

Exercise Prescription

Mode	Intensity	Frequency	Duration	Progression	Goals	Special considerations and comments
Aerobic: • Brisk walking with arm movement ideal • Cycling (preferably recumbent) • Swimming • Elliptical training	• Moderate to high • 40–85% peak $\dot{V}O_2$ • Start at lower range, particularly if deconditioned • Borg 12–16	• 3–5 d a week • Preferably most or all days in early stages	• Build up to 20–60 min • May do by multiple 10 min bouts throughout day • Begin with several 2–5 min bouts if very deconditioned • Anticipate 4–6 wk program for substantial improvement	• Increase bouts (if used) to at least 10 min until goal is reached • ACSM-recommended progression for sedentary low-risk individuals unless comorbidities increase restrictions	• 30 min or more (20 if high intensity)	• Low impact best initially
Resistance: • Free weights • Machines • Resistance tubing or bands	• Submaximal suggested • 12–13 on Borg at start • Progress to 15–16	• 2–3 non-consecutive days a week	• Sessions of less than 1 h	• Follow ACSM guidelines for healthy sedentary individuals	• Avoid low repetitions and high intensity; use more repetitions and lower resistance to maximize endurance; back extensor and latissimus dorsi endurance increases	• Avoid unstable surfaces (e.g., exercise balls) early in training • Later unstable surfaces may enhance training • Emphasize good form • Limit spine loading through full range

Mode	Intensity	Frequency	Duration	Progression	Goals	Special considerations and comments
Range of motion: • Static stretching exercises	• Comfortable tightness without pain	• 2–7 d a week with higher number best	• Each stretch held 15–30 s	• Dictated by intensity	• Return ROM to normal	• Balance flexion and extension stretches of spine • Ensure hamstrings are stretched • Avoid flexion stretches soon after rising • Avoid standing toe touch

Exercise Testing

Test type	Mode	Protocol specifics	Clinical measures	Clinical implications	Special considerations
Cardiovascular	• Treadmill • Cycle ergometer • Stepping	• Ramp • Incremental	• BP • HR • Others as indicated by comorbidities or other ACSM guidelines • Subjective ratings of intensity and pain	• May be deconditioned beyond healthy sedentary level	• Difficulties with prolonged flexion may make recumbent bike more suitable than upright • Testing soon after rising not recommended
Strength	• Free weight or machines	• Submaximal testing	• BP • HR • Others as indicated by comorbidities or other ACSM guidelines • Subjective ratings of intensity and pain	• Many experience some discomfort, especially early in testing and training	• Ensure that spinal support and good posture is used at all times
Range of motion	• Standard flexibility testing	• Ensure hamstring flexibility tested		• Many experience some discomfort, especially early in testing and training • Avoid forcing into uncomfortable range	• Emphasize neutral spine while testing extremity flexibility • Avoid spinal flexion in first 1–2 h after rising

(continued)

Practical Application 28.3 *(continued)*

Benefits of Exercise Training

Cardiorespiratory endurance	Skeletal muscle strength	Skeletal muscle endurance	Flexibility	Body composition
• Improved performance • Enhances the effect of back exercise program • Benefits general health	• Increased tolerance of activities of daily living (ADLs), recreational activities, and occupational performance	• Back extensor and latissimus dorsi improvements may protect against recurrences	• Return to normal values (with associated strength gains) may decrease risk of recurrence	• Improved muscle mass and lowered BMI

useful for prescription, is not required in general clinical practice unless history indicates possible coronary artery disease (61) or other risk factors according to ACSM guidelines that make exercise testing advisable (1). Any aerobic activity that interests the client may be used, but in general, high-impact activities such as jogging or exercise requiring sustained spinal flexion (as in some bicycling) are considered poor choices. But the mode of activity that will aggravate NSLBP differs for each person. People with NSLBP who start an aerobic exercise program should start slowly and be sensitive to activities that precipitate NSLBP. Otherwise, aerobic exercise training should follow the ACSM guidelines for promoting health and fitness. Goals should be to return clients to exercise routines suggested for the general adult population. People with NSLBP may need to initiate training with short training bouts at lower than usual intensity until they achieve low-level exercise tolerance. Evidence of progressive low-back pain during and after exercise, residual postexercise muscle soreness lasting greater than 48 h, impairment in tolerance for activities of daily living, or evidence of neuropathy requires reevaluation by a qualified healthcare practitioner before exercise is resumed. Modifications in the parameters of resistance or fitness training will be prescribed by the appropriate healthcare provider.

Resistance Training

Available evidence suggests that the more important component of resistance training, at least for the typical NSLBP and for prevention, is endurance (5,46,53). **Dynamic endurance** may be more critical than **static endurance** (55). A well-designed program will consider both strength and endurance, although literature and clinical expertise support a somewhat heavier emphasis on endurance. Many people with back pain have dif-

ficulty tolerating positions for more than a brief period. Poor endurance can predispose them to injuries at relatively low loads. If sustained positions are required for vocational or avocational activities, the clinical exercise physiologist should consider training for these specific positions (17).

Most programs encourage strengthening of back and abdominal muscles. To strengthen all muscles involved, a variety of exercises are required. The following specific suggestions are from the series of articles by McGill and colleagues and the text by McGill (35,51–53) that focus on safety through minimizing spinal loading while providing adequate stimulus for muscle training. Based on **electromyography** studies for iliopsoas muscle activity and intervertebral disk pressure, no evidence exists to support the use of bent-knee sit-ups over straight-leg sit-ups. Instead of sit-ups, various types of curl-ups or crunches are suggested to strengthen rectus abdominis muscles, including curl-ups with one leg straight and hands used to maintain a neutral lumbar curve (figure 28.2a). Others believe that the people who perform crunches should independently stabilize the pelvis to promote lumbopelvic neuromuscular control. This neutral curve may be most important in early recovery.

For strengthening the lateral oblique muscles, the horizontal side support exercise shown in figure 28.2, c and d, is suggested (35,51–53). This exercise activates the lateral oblique muscles without causing high lumbar compressive loading. The exercise has the additional advantage of training quadratus lumborum, which is considered important in lumbar stabilization.

Back extensions are suggested for strengthening the erector spinae muscles (17). Hyperextensions (of back with fixed legs or legs with fixed spine) are an example of a once popular exercise that is not indicated and that has

Figure 28.2 Strengthening exercises commonly used in low-back rehabilitation.

not been shown to benefit patients. Evaluating individual responses to end-range exercise may help minimize problems and increase responsiveness to treatment. It is important to avoid loading the spine through range of motion in clients who are beginning a program (51–53).

Extension of both back and legs simultaneously in prone lying (figure 28.3 on p. 516) is often suggested (7), but high lumbar spine compression is produced (51–53,79). Back extensors can be strengthened in other ways. Alternatives include resisted limited-range back extension with fixed pelvis or less demanding exercises involving extension of one arm, one leg, or both in kneeling (figure 28.2, *e* and *f*). Equipment that isolates the lumbar spine from the pelvis and legs may produce greater

strength improvement than more traditional Roman chair exercises (61). But most programs do without such sophisticated equipment, and lack of access to equipment should not be a reason for avoiding lumbar extension training. A simple exercise to start a program that does not need special equipment and that can be used to teach form is the "good morning exercise" done with the pelvis blocked by a bench or table (figure 28.2*g*).

Upper-extremity and lower-extremity muscle strengthening should be part of any general prevention program. In the upper extremities, latissimus dorsi exercises are important because this muscle is involved in back protection and some movement initiation (17). Anatomical connections link lumbodorsal fascia with

Figure 28.3 Lumbar hyperextension exercise.

gluteals, hamstrings, and latissimus dorsi (8), connecting them to overall back function. Exercises such as the lunge and squat (with handheld weights providing resistance) provide training for trunk stabilization as well as desirable lower-extremity strengthening (17).

We concur with the suggestions from the literature given previously but would also consider specific rotator muscle strengthening. As suggested previously, a potential source of back pain may be the selective loss of strength and endurance of the smaller muscles of the low back such as the rotatores and multifidi. Specific exercises for these muscles may be helpful. Figure 28.2*f* shows an exercise that will encourage activation of the thoracolumbar rotators, but this exercise may allow substitution of the large-muscle groups. A simple seated exercise (later progressed to standing to promote lumbopelvic coordination) uses a pulley, rubber band, or tubing as light resistance (figure 28.4).

Having made the previous suggestions for specific back-strengthening exercises, we are obliged to indicate that many of the current studies that recommend therapeutic exercise for patients with NSLBP suggest a more generalized approach to exercise. General aerobic exercises such as walking, swimming, or light sport-related activities that the patient self-selects tend to be comparable or even more effective at positive outcomes than a specific back-training exercise program. A program that is enjoyable may be more likely

Figure 28.4 Seated resisted trunk rotation exercise.

to be followed long term, an important consideration for people with NSLBP. This recommendation seems to be related to the shift in philosophy in management strategies for patients with NSLBP to a more active approach, with simple back-school education and reliance on a self-care program.

Range of Motion

Providing general recommendations for flexibility exercises to someone with low-back pain is difficult (51–53). A stretching program must be developed based on the patient's history and physical examination findings. The limited data available do not support the view that greater flexibility of the spine prevents injury. Maintaining adequate flexibility at hips and knees for lifting is important, and if stretching of the spine is done, it is probably best done in an unloaded position (52). Deficits in lower-extremity flexibility should be addressed individually. Spinal flexibility should not be emphasized until the person has the muscular strength and endurance to control a mobile spine (53). We recommend starting with low-resistance, high-repetition strengthening to reeducate neuromuscular control of increased spinal movement that occurs with stretching. An important caveat is that flexion stretching after sleep or prolonged sitting is not recommended (51,67).

Individuals with a history of back pain have reduced back and hamstring flexibility (5). This finding is associated with recurrence or persistence of NSLBP. On the other hand, those with very mobile spines are at increased risk of experiencing NSLBP during the following year (5). Apart from this, there is no demonstrated link between spinal flexibility and back pain.

Most programs incorporate flexibility exercises into their exercise routines. In the absence of clear evidence, it seems reasonable to continue to do so unless hypermobility or instability is a particular problem. When in doubt, the best approach is to have the person evaluated for abnormal flexibility before advocating spinal stretching exercises. Nonspinal stretches, usually calf muscles and hamstrings stretches (17), are part of most back programs. Other areas and muscles that are frequently tight include the iliotibial band, hip adductors, hip flexors, and quadriceps. Specific muscle length tests should be administered by qualified individuals to establish an individualized stretching program.

A reasonable approach would be to evaluate for tightness and to design an exercise program that includes stretches to correct any impairments noted. A few commonly used stretches for the low back and lower extremities are shown in figure 28.5. Many variations are equally suitable. Any exercise that involves repeated

Figure 28.5 Stretches commonly used in low-back rehabilitation.

flexion should be balanced with stretches into extension. Caution should be used with people who have known disk pathology, because repeated flexion exercises can shift disk nuclear material posteriorly, potentially aggravating symptoms. Passive hyperextension stretches (such as pushing up from prone by using the arms with the hips staying in the floor) may be a better option than active extension stretches. A balance in stretching movements and inclusion of extension as well as flexion stretches is important. Patients can be instructed to move to a comfortable range of movement and hold a prolonged (15–30 s) stretch in the end portion of the range where tissue tension is developed. Sometimes, when working with clients with NSLBP, the addition of strategically timed

breathing can allow relaxation of the spinal muscles, enhancing the stretching effect.

Frequency

Only continued adherence to an exercise program is likely to offer lasting benefit (49). In a study with intensive training of people with chronic back pain, all subgroups of studied patients improved with the 3 mo training program, but a year later only those who continued training at least once a week had remained significantly improved (49). Training at least twice a week is usually suggested (51,61).

Given the population targeted in this chapter, two to three sessions a week is appropriate for a maintenance program. In early rehabilitation, daily exercise is often suggested. Frequency can decrease as intensity increases and as exercise tolerance approaches population norms. Clients who are empowered during their rehabilitation phase of recovery are more likely to adhere to a maintenance program after their subacute or chronic pain has dissipated. Having patients select exercises that they enjoy will provide motivation to continue compliance in self-care of the back after they have completed rehabilitation with a healthcare provider.

Intensity

Few guidelines are specific to the management of NSLBP regarding intensity selection. Those available usually suggest that exercise needs to be more intense than that normally provided for patients. Sometimes a very thin line separates exercise intensity that is sufficient to be effective from exercise that is too intense and leads to worsening of symptoms. Self-selected intensity or exercise to pain tolerance often leads to inadequate exercise levels. Although pain may not improve for several months in many people, an intensive program will result in greater functional and psychological benefit than a less aggressive approach. Some advocate a quota approach to prescribing exercise intensity to prevent underexercising and suggest using operant conditioning behavioral tactics (42,62).

In setting exercise intensity, we suggest a trial-and-error approach, such as the DeLorme method of determining maximal resistance, to select resistance weights that allow 20 to 30 repetitions with good neuromuscular control in a pain-free or minimal range of motion. This initial program will promote endurance and control of movement. As the person progresses, resistance should increase while the number of repetitions decreases to 8 to 12, compatible with ACSM training guidelines (2,3).

Duration

Weight training may require 30 to 60 min per session to complete a prescribed program. More than this may lead to decreased adherence. The total resistance-training program includes the intermittent rest periods between intensive sets to recuperate.

CONCLUSION

NSLBP is one of society's most common problems. Details of suitable exercise programs are limited, but research shows that most people with nonspecific back pain can use exercise to restore function and decrease disability. Programs should be initiated with modifications to accommodate individual impairments and should progress to maintenance programs that are as close as possible to those suggested for general fitness with healthy populations.

Although people with back pain may need extra encouragement to begin and continue a general exercise program, those who are able to do so can hope for considerable improvement in function and quality of life. Reassurance, education, and encouragement with self-management are important components of care. Individuals too rarely receive this information, and open communication with people who have back pain can be a key aspect of management. For most people, back pain is nonspecific and can be best managed with conservative yet active treatment strategies.

Case Study

Medical History

Mrs. AB is a 28 yr old Caucasian bank worker. She has had recurrent back pain since the age of 16. At that time she had an awkward fall while playing softball. She was taken to her local emergency department, where she was given the diagnosis of muscle strain and treated with painkillers and muscle relaxants. Although her back improved quickly, she believes that the pain never completely resolved. As she continued through high school she noticed that although she had continuing periods of pain, they seemed to be less severe when she was physically active.

At age 21 she experienced another acute episode of back pain that began suddenly when she sat down on a couch. This time she did not seek medical help but treated herself with over-the-counter pain medication. The pain slowly resolved over several months.

Diagnosis

Two years ago on July 4th weekend, she was camping with her husband and daughter. On the morning when they were packing up to go home, she bent over to help her husband pick up the tent and had a sudden onset of severe back pain. Her pain was in her back and right buttock. She saw both her physician and chiropractor for treatment. Her X-ray results were normal, but she was told that her MRI showed a bulging disk between the fourth and fifth lumbar vertebrae. All other medical screening was clear, and she was told that she did not have any serious medical problems. She was given a short course of muscle relaxants and stayed off work for 3 d. Her condition slowly improved but continued to bother her for the next several months.

Exercise Test Results

In October her chiropractor suggested that she start fitness exercise. The previous January she had joined her town's wellness center, but her attendance had been sporadic. She had completely stopped working out before she injured her back while camping, and the manager of the facility had put her membership on hold while she was receiving treatment for her back pain.

Following her chiropractor's advice, Mrs. AB returned to her fitness facility for a reevaluation and treatment program. Her physician had cleared her to begin an exercise program, and the fitness facility cleared her based on PAR-Q responses. The facility manager attempted a graded cycle ergometer test of aerobic capacity, but Mrs. AB was unable to complete it because of pain. Similarly, muscle strength testing of back extensors and abdominals was not possible. She was able to extend her hips only to neutral, and an attempt to test hamstring flexibility by reaching for her toes in sitting demonstrated a 13 in. (33 cm) distance from her fingertips to her toes.

She was able to do five repetitions of a trunk extension exercise from mid-flexion range to neutral spine position using a resistance of 30 lb (14 kg). She attempted latissimus pull-downs with 20 lb (9 kg) of resistance and a leg press that loaded through her shoulders, but she found the exercises too painful to continue to a formal strength evaluation. She was able to walk for 15 min on a treadmill without an incline at 3 mph (4.8 kph). She found this moderately painful.

Exercise Prescription

Under the guidance of the fitness center manager, she began a program of treadmill walking to tolerance and resumed a modified version of the program of resistance exercise she had started when she joined the facility. Her walking started with a 15 min walk at 3 mph (4.8 kph), and her goal was to increase to 4 mph (6.4 kph) and a distance of 2 mi (3.2 km). She did curl-ups using an exercise ball for abdominals, beginning with one set of 10 repetitions. She also started with one set of 10 repetitions of the following exercises: chest press (two varieties), seated rowing, leg curls, triceps pull-down, and knee extension. All were done on weight equipment rather than with free weights, and initial resistance was determined by selecting a weight that did not increase her back pain and allowed her to do the required number of repetitions without loss of form. On the back extension machine she started with one set of 5 repetitions with 30 lb (14 kg) of resistance in a very limited range.

The center is adjacent to the bank where she works, and Mrs. AB now goes there after work several days a week. Her employer is a major corporate sponsor of the wellness center, so she is able to take advantage of a reduced membership rate. Currently, she experiences some back pain at least weekly. She does not take time off work for pain but may modify her activities slightly. On days when pain bothers her, she takes over-the-counter pain medication. With the severe episodes described previously, she returned to work despite considerable pain within 3 to 5 d of the episode. Her supervisor insisted on buying an ergonomic chair for her use at work, and Mrs. AB finds the chair helpful. Her job allows her to change position frequently, and she is never required to either sit or stand for long periods. She has noticed that any prolonged posture aggravates her back pain for several days.

Similarly, she has found that beginning any new sporting activity also increases pain. Last summer she coached her daughter's softball team and found that the frequent squatting activities caused her back to flare up considerably. Depending on the activity, days or weeks may pass before the pain settles down to the usual level. She continues to see her chiropractor once a month.

Mrs. AB normally exercises three to four times a week at the wellness center. She has found that her episodes of back pain worsen and the frequent low levels of pain increase in severity whenever she fails to exercise regularly. She believes that her back is still improving but extremely slowly. Her goal continues to be complete elimination of pain.

In the first 6 mo after her injury Mrs. AB increased her exercise routine to include the following exercises: chest press (two varieties), seated rowing, leg curls, triceps pull-down, and knee extension. She does two sets

(continued)

of 30 abdominal crunches. For the previous exercises she does three sets of 12 to 15 repetitions, all on weight equipment. For the chest presses and rowing she now uses 40 lb (18 kg). The triceps pull-down is 20 lb (9 kg), the leg curls are 30 lb (14 kg), and the knee extension is 35 lb (16 kg). Her back extension exercises have increased to three sets of 20 reps with a resistance of 80 lb (36 kg), with extension only to neutral. She now walks 2 mi (3.2 km) on the treadmill at a speed of 4 mph (6.4 kph) and does light stretching of arms and legs before her workout. The flexibility of her spine has improved somewhat, but she is cautious of allowing extension of her back much beyond a straightened position with any exercise. Before the episode she had been using a leg press machine that loaded through the shoulders and a lat pull-down. Both were extremely painful after the injury, and she dropped them from her program. She has not attempted them since. Although she has continued to work out regularly, the weights have been the same on the machines for about a year. She uses the center treadmill for a walking program. She still has tight hamstrings, fingertips 6 in. (15 cm) from toes, and reduced active and passive back extension.

Discussion Questions

1. How well does Mrs. AB fit the profile of the typical person with nonspecific back pain?

2. In her presentation are there any red or yellow flags?

3. What characteristics in her work and recreational situation have helped her deal with her pain?

4. Do you think her current back program will produce her desired outcome (pain-free spinal mobility and physical tolerance of all work-related tasks)?

5. Do you believe, based on the recurrent nature of her symptoms, that present management strategies are adequate?

6. What specific recommendations for changes in her exercise program could you make for her based on the provided information?

7. List general principles of exercise that she should follow to decrease her risk of recurrence.

8. What advice would you give her for a suitable aerobic conditioning program?

9. Write an outline of advice that you would give her for those days when her job requires a lot of sitting activity.

10. Would you reintroduce the partial squats and lat pull-downs in her program? If so, what modifications could help her resume those exercises?

11. She wants to continue coaching her daughter's softball team. What strategies should she use to prevent flare-ups of her back pain from this activity?

12. How can she safely stretch into extension to regain some of the lost mobility in her spine?

PART VIII

Selected Neuromuscular Disorders

Part VIII contains four chapters that cover the major neuromuscular conditions likely to be encountered by the practicing clinical exercise physiologist. Although most of these conditions require some type of specialized care or rehabilitation, many persons who are minimally afflicted or significantly recovered will seek exercise-training advice and programs. The clinical exercise physiologist should be aware of these conditions and know how to design and implement exercise programs that address patients' needs.

Chapter 29 centers on spinal cord injury. This chapter addresses the degrees of spinal cord injury, the special care needed immediately after injury, and the barriers that these people face for the rest of their lives. Those who recover well and maintain a degree of function that allows them to perform regular exercise training may seek services at facilities staffed by clinical exercise physiologists. This chapter discusses the needs of these patients and their specific exercise issues.

Chapter 30 investigates multiple sclerosis (MS), a disease that typically affects grown individuals. This chapter reviews the many presentations of MS and the degree of functioning along the course of the disease. Lesser affected, higher functioning people may not have much need to alter a regular exercise-training routine, whereas individuals with greater disability may need significant adaptations to their exercise program.

Chapter 31 explores cerebral palsy. Cerebral palsy (CP) afflicts people at birth or in early childhood. Degrees of deficit are certainly present in these individuals, and this chapter provides a summary of specific issues that they face. Because aging can compound these deficits, the chapter provides important information with respect to designing, implementing, and adjusting exercise-training programs for these individuals.

Chapter 32 focuses on stroke. Although one might argue that stroke should be placed in the cardiovascular section because it is a vascular issue, the long-term effects are often related to neuromuscular function. For instance, a stroke that affects the functioning of limbs becomes an important issue in exercise training. This chapter discusses the important concerns for these patients that the clinical exercise physiologist may encounter.

Spinal Cord Injury

David R. Gater, Jr., MD, PhD

The spinal cord serves as the major conduit for motor, sensory, and autonomic neural information transmission between the brain and the body. A **spinal cord injury** (SCI) affects conduction of neural signals across the site of the injury or lesion. An SCI is classified by the lowest segment of the spinal cord with normal sensory and motor function on both sides of the body and may be defined as complete (without sensory in the lowest sacral segment) or incomplete (partial preservation of sensory or motor function below the neurological level, including the lowest sacral segment). Tetraplegia (the preferred term over quadriplegia) refers to "the impairment or loss of motor and/or sensory function in the cervical segments of the spinal cord due to damage of neural elements within the spinal canal" (65). Tetraplegia is distinguished from paraplegia in that it includes dysfunction of the arms, whereas both tetraplegia and paraplegia involve impairment of function of the trunk, legs, and pelvic organs. Table 29.1 on p. 524 lists the American Spinal Injury Association's definitions of five degrees (completeness) of SCI (65). Table 29.2 on p. 524 lists specific clinical syndromes as they relate to injuries to specific spinal cord locations.

SCOPE

Approximately 250,000 to 400,000 U.S. citizens have an SCI (75). The incidence of SCI approaches 52 new injuries per million per year. Almost 40% die before reaching a hospital. Hence, 8,000 to 10,000 Americans will survive a new SCI each year, and more than 90% will return to a private residence following rehabilitation. Men are affected four times as frequently as women, and 77% of SCI injuries occur in those between 16 and 45 yr old (54% for 16 to 30 yr olds). Most of these injuries occur in young men involved in motor vehicle collisions. Acts of violence, falls (mostly elderly), and sports injuries (mostly diving) are other primary causes of SCI. Pediatric SCI is often congenital, presenting at birth with incomplete formation and closure of neural and skeletal bony elements referred to as spina bifida or **myelomeningocele**. Recent advances in acute care have significantly reduced the number of complete SCIs, particularly in cervical injuries. Since 1994 30% of SCIs have resulted in **incomplete tetraplegia**, whereas only 23% are considered complete tetraplegia at the time of discharge from a rehabilitation facility. In the same period, 20% of SCIs resulted in **incomplete paraplegia**, whereas 26% still had **complete paraplegia** at the time of discharge. A small percentage of SCIs (<1%) completely resolve by the time of hospital discharge, without neurological deficits. The medical costs associated with SCI are staggering. Average lifetime medical costs exceed $1.1 million per person, and average lifetime foregone earnings and fringe benefits per case may exceed $2.1 million (45). Even with the advent of managed care, the initial hospitalization and rehabilitation length of stay will exceed 60 d and $80,000 for tetraplegia and 40 d and $50,000 for thoracic paraplegia. As SCI care and technology have improved, life expectancy has likewise improved. A 20 yr old man with newly acquired tetraplegia now has an additional

Table 29.1 American Spinal Injury Association Impairment Scale

Degree of impairment	Conditions of impairment
A	Complete: No motor or sensory function is preserved in the sacral segments S4–S5.
B	Incomplete: Sensory but not motor function is preserved below the neurological level and includes the sacral segments S4–S5.
C	Incomplete: Motor function is preserved below the neurological level, and more than half of key muscles below the neurological level have a muscle grade of <3.
D	Incomplete: Motor function is preserved below the neurological level, and at least half of key muscles below the neurological level have a muscle grade of ≥3.
E	Normal: Motor and sensory function are normal.

Adapted, by permission, from R. J. Marino et al., 2000, *International standards for neurological classification of spinal cord injury* (Chicago, IL: American Spinal Injury Association), 18-19.

Table 29.2 Clinical Syndrome Resulting in Spinal Cord Injury

Syndrome	Cause	Physiological deficits
Central cord syndrome	Incomplete cervical spinal cord injury with cord damage	Weakness, sensory deficits in upper extremities (less in lower)
Brown-Sequard syndrome	Unilateral cord lesion	Ipsilateral proprioceptive and motor deficit; contralateral pain impairment and temperature deficit below level of injury
Anterior cord syndrome	Anterior cord ischemia (T10–L2)	Below level of injury, pain impairment and temperature deficit
Conus medullaris syndrome	Upper and lower motor neuron damage	Bowel, bladder, lower-extremity areflexia, and flaccidity; preserved or facilitated reflexes
Cauda equia syndrome	Lumbsacral nerve root injury	Bowel, bladder, and lower-extremity areflexia and severe dysesthetic pain

life expectancy of 33 yr or more, whereas the 20 yr old with paraplegia may expect to live an additional 44 yr or more (28).

PATHOPHYSIOLOGY

The spinal cord, a portion of the central nervous system, links the conscious and subconscious functions of the brain with the peripheral and **autonomic nervous systems**. It extends through and is protected by the spinal column, a flexible segment of interdigitating bones and disks arranged to maximize mobility and reduce risk of injury. There are 33 total vertebrae (7 cervical, 12 thoracic, 5 lumbar, 5 sacral, and 3 to 5 coccygeal). A pair of nerves arises from each vertebral segment. Figure 29.1 depicts these segments and nerves. The spinal cord is about 25% shorter than the spinal column, and thus subsequent spinal nerves exit from the cord as the **cauda equina**. The vascular supply to the cord comes from an anterior and two posterior spinal arteries at each vertebral level, supplying the anterior two-thirds and posterior one-third of the cord, respectively. Neural tracts of the central nervous system and cell bodies of the peripheral nervous system are susceptible to primary and secondary injury. Primary injury can damage neural tracts, cell bodies, and vascular structures that supply the cord. Secondary injury occurs because of hemorrhage and local edema within the cord, which compromise vascular supply, resulting in local ischemia. Infarction of the gray matter will occur within 4 to 8 h after injury if blood flow cessation remains. Inevitably, necrosis, or cell death, occurs and can enlarge over the one or two vertebral levels above and below the area of trauma. Gliosis, astrocyte, and syringomyelia may form during the next several months. Formation of fibrous and glial scarring is the final phase of the injury process. Injury to the spinal cord obstructs the transmission of neural messages through the cord and results in the loss of somatic and autonomic control over trunk, limbs, and viscera below the site of the lesion. Systemic responses to exercise seen

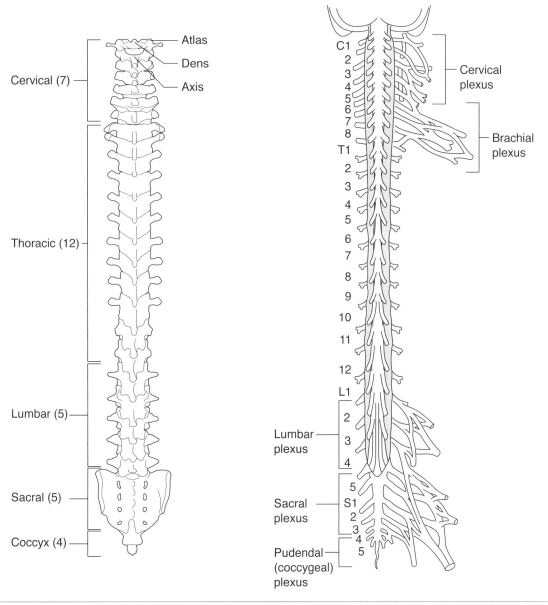

Figure 29.1   Anatomy of the spinal cord.

Reprinted from R. Behnke, 2001, *Kinetic Anatomy,* 1st ed. (Champaign, IL: Human Kinetics), 130, 171.

in nondisabled individuals are blunted in persons with SCI. This diminishes their ability to perform physical activity and exercise, both at the conscious (somatic) and subconscious (autonomic) level.

Somatic Nervous System Disruption

The somatic nervous system consists of motor (efferent) and sensory (afferent) neural pathways connecting the brain to the body by the spinal cord. Complete SCI interrupts transmission of these signals, and voluntary

movement and sensory perception are absent below the lesion. Neurological classification of SCI is standardized as seen in figure 29.2 on p. 526.

Autonomic Nervous System Disruption

The autonomic nervous system coordinates automatic life-sustaining processes and organizes visceral responses to somatic reactions. It is composed of sympathetic and parasympathetic divisions, which regulate the action of smooth muscle and glands. Essential functions of the

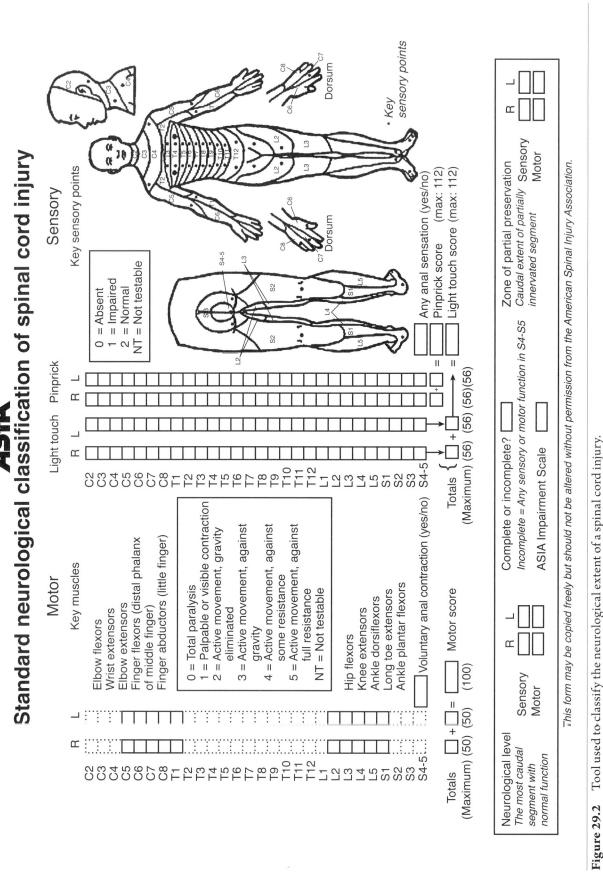

Figure 29.2 Tool used to classify the neurological extent of a spinal cord injury.

American Spinal Injury Association: International Standards for Neurological Classification of Spinal Cord Injury, revised 2002, Chicago, IL.

Table 29.3 Effects of Level of Injury on Somatic and Autonomic Function

Level of injury	Somatic: movements preserved include all above each level affected	Autonomic: functions preserved include all above each level affected
C1–3	Chin	Cranial level parasympathetic nerves are not affected by SCI
C4	Shoulders shrug and head turn	
C5	Shoulder movement and elbow flexion	
C6	Wrist extension	
C7	Elbow, wrist, finger extension	
C8	Lowest level of tetraplegia, finger flexion	
T1–10	Begins level of paraplegia; variable sensory deficits; intercostals control	Sympathetic nerves are from T1-L5; SCI above this level affects smooth muscles, organs, and glandular sympathetic mediated effects (e.g., heart rate, stroke volume, ventilation, sweating, splanchnic vasoconstriction, and skeletal muscle vasodilation)
T6–12	Trunk stabilizers	
T1–S1	Paraspinal muscles	
L2	Hip flexion	
L3	Knee extension	
L4	Ankle dorsiflexion	
L5	Toe extension; hip abduction	
S1	Ankle plantar flexors; hip extensors	Affects lower digestive structures, gall bladder, and bladder function
S2–S5	Bowel, bladder sphincter control	

American Spinal Injury Association: International Standards for Neurological Classification of Spinal Cord Injury, revised 2002, Chicago, IL.

autonomic nervous system during exercise include modulating heart rate, stroke volume, blood pressure, blood flow, ventilation, thermoregulation, and metabolism. The **sympathetic nervous system** and **parasympathetic nervous system** make up the autonomic nervous system. Function affected by level of SCI is provided in table 29.3.

Systemic Adaptation

The following sections review the systemic effects and adaptations to SCI.

• **Cardiovascular:** In response to acute cervical and upper thoracic SCI, bradycardia is common and attributable to loss of sympathetic nervous system influences, with no effect on the parasympathetic nervous system (57). This situation usually resolves in 2 to 6 wk (30,72). Reduced sympathetic influence on peripheral and splanchnic vascular beds reduces peripheral vascular resistance, which enhances venous pooling and **orthostatic hypotension**.

• **Autonomic dysreflexia:** For persons with SCI above T6, an uncontrolled outflow of sympathetic activity in response to stimuli below the SCI (e.g., distended bowel or bladder, lacerations, fractures, pressure sores, sunburn) may occur, resulting in life-threatening paroxysmal hypertension (16,23,70). Reflex bradycardia and vasodilation, manifested as flushing, headache, hyperhydrosis, piloerection, pupillary dilation, and blurred vision, may occur (16). Such symptoms suggest an autonomic crisis that can lead to intracerebral hemorrhages, seizures, arrhythmias, and death if not immediately and appropriately treated.

• **Pulmonary:** Ventilation is impaired in most SCI patients because of paralysis of rib cage and abdominal musculature, reduced pulmonary compliance, and reduced diaphragmatic excursion. Thus, some patients may require ventilator assistance. Tetraplegia below C5 typically spares voluntary control of the diaphragm, although inspiration remains impaired. Ventilation will worsen as the level of disability increases (21), although aggressive spirometry and exercise training may improve it (22).

- **Bowel and bladder function:** Cervical, thoracic, and lumbar spinal cord lesions increase the risk of gastric and duodenal ulcers, increase bowel motility and hyperreflexia, and eliminate voluntary control of defecation (94). Hyperreflexia of the bladder wall and sphincter muscles results in greater risk of vesicoureteral reflux, hydronephrosis, and acute renal failure if not appropriately managed. Bladder spasms and loss of voluntary sphincter control may lead to urinary incontinence. Urinary tract infections, as well as renal or bladder stones, are more frequent in SCI patients than in the nondisabled population.

- **Hyperreflexia: Spasticity**, seen in up to 60% of individuals with SCI (66), is attributable to central inhibition of spinal reflex arcs (100). Some people with SCI are able to use their spasticity to assist them in performing mobility and **activities of daily living** (ADL) tasks, but many find that the spasticity is painful, disrupts sleep, interferes with function, causes muscle contractures, and can lead to shear- or pressure-induced skin breakdown.

- **Thermoregulation:** The interruption of autonomic pathways in tetraplegia results in partially **poikilothermic** responses to thermal stress and reduced ability to dissipate environmental and internally generated heat by sweating (85). This occurs because most of the sweat response is limited to regions above the level of SCI.

- **Endocrine:** Autonomic dysfunction and somatic paralysis in SCI may also affect metabolic and hormonal function, including the sympathomedullary response and the adrenocortical–pituitary axis, resulting in flattened circadian rhythms and poorly regulated corticosteroid responses (97). Glucose intolerance often occurs and is often accompanied by hyperinsulinemia (1). Although thyroid function may be acutely altered in SCI, thyroid function tests are generally normal in healthy SCI adults (10). Conversely, testosterone and free testosterone levels in men with SCI are often reduced (93), whereas growth hormone release is blunted and chronically depressed (9). Following the acute phase of SCI, ovulatory menstrual cycles are fairly well preserved (80).

- **Osteopenia:** Neurogenic **osteopenia** in complete tetraplegia results from the withdrawal of stress and strain on bone (99). Bone mineral loss is rapid during the first 4 mo of SCI (39). Homeostasis at 67% of original bone mass is achieved at about 16 mo postinjury. Thus, the risk of fractures increases.

MEDICAL AND CLINICAL CONSIDERATIONS

The management of SCI is clinically complex and beyond the scope of the clinical exercise professional who is working with these patients on long-term fitness management. See "Management Issues During SCI Acute Hospitalization and Rehabilitation." Comorbidities that may affect the long-term rehabilitation of these patients include brain injury, thoracic contusions and fractures, intra-abdominal trauma, upper- and lower-extremity fractures, plexopathies, and peripheral neuropathies. Typically, medical management issues take precedence over fitness concerns during the first 3 mo postinjury, and the altered physiology must constantly be considered as the person with SCI is reintegrated into community and recreational pursuits.

Signs and Symptoms

The common signs and symptoms of chronic SCI are listed in table 29.4 on p. 530. Each of these poses potential issues during the daily life of the patient with SCI. The clinical exercise professional working with these patients should be aware of these and able to assist when needed.

History and Physical Exam

Comorbid medical issues common among patients with SCI that should be considered for exercise testing and training include respiratory complications, coronary artery disease, peripheral arterial disease, **circulatory hypokinesis** leading to hypotensive responses, obesity, type 2 diabetes mellitus, pressure sores, joint contractures, and osteopenia. A high incidence of smoking (35%) and upper-extremity overuse problems are present. Finally, medications should be listed and assessed for their effect on exercise tolerance. Commonly prescribed medications used in SCI, their indications, and common side effects are provided in table 29.5 on p. 530. Heart disease is the second leading cause of death in patients with SCI, accounting for approximately 22% of all deaths (45). Unfortunately, silent myocardial ischemia caused by disrupted visceral afferent fibers in higher levels of SCI may prevent an individual from recognizing symptoms such as angina (8). Amputations for peripheral arterial disease are sometimes required because of diminished healing and the development of ulcers. Obesity is a problem attributable to reduced (12–54%) basal energy expenditure. As a result, the risk for glucose intolerance and type 2 diabetes increases (41). Pressure ulcers occur

Management Issues
During SCI Acute Hospitalization and Rehabilitation

Acute Hospitalization Management Issues

- Spine management: imaging of cervical, thoracic, lumbar, sacral spine
- Surgical or orthotic stabilization of unstable spinal column injuries and spinal cord decompression
- Range-of-motion limitations to allow complete bony and soft-tissue healing of spinal elements

Respiratory Management Issues

- Assisted ventilation often required for high cervical injuries
- Secretion management essential because of impaired cough and increased parasympathetic nervous system influence on pulmonary secretions
- Assisted cough required for SCI above T6 attributable to intercostals and abdominal muscle paralysis

Cardiovascular Management Issues

- Relative bradycardia attributable to impaired sympathetic nervous system in SCI above T6; occasionally requires pacemaker placement
- Hypotension attributable to systemic vasodilation resulting from impaired sympathetic drive
- Venous stasis can result in deep venous thrombosis or pulmonary embolism

Functional Mobility Issues

- Upper-extremity range of motion, strengthening, and endurance within limitations of orthotics and medical management
- Bed mobility (including side to side, supine to prone to supine, supine to sit)
- Wheelchair mobility (including forward and backward propulsion, turning, uneven terrain, curbs, ramps, hills)
- Transfers (including bed to wheelchair to bed, wheelchair to toilet to wheelchair, wheelchair to bath to wheelchair, wheelchair to floor to wheelchair, wheelchair to car to wheelchair)
- Activities of daily living including feeding, grooming, dressing, bathing, toileting
- Bladder management training, typically with intermittent catheterization or alternative
- Bowel management training, typically with intermittent catheterization or alternative
- Skin management training with monitoring and pressure relief techniques
- Equipment evaluation for personal care, mobility, and public accessibility
- Home and vehicle evaluation for accessibility
- Psychological and social adjustment to spinal cord injury
- Introduction to vocational and recreational opportunities for persons with SCI

as the result of shear, friction, and unrelieved pressure, usually over a bony prominence, which damages underlying tissue. This problem is exacerbated in persons with tetraplegia because they are unable to feel the sensations of tingling, discomfort, and pain. Appropriately prescribed wheelchair-seating systems with scheduled pressure relief every 15 to 30 min should reduce the risk

of pressure ulcers in the exercising person with SCI (33). Phantom or neuropathic pain (e.g., burning, tingling, electrical sensations) may occur from a region below the SCI and can adversely affect exercise ability (17).

Upper-extremity overuse injuries in wheelchair-reliant individuals occur most frequently at the shoulder (24,91). Hip, knee, and plantar flexion contractures may result

Table 29.4 Signs and Symptoms of Spinal Cord Injury

Signs	Symptoms
Motor paralysis (BLOI)	Impaired or absent voluntary motor function
Sensory loss (BLOI)	Impaired or absent sensation
Hyperreflexia (UMN lesion)	Brisk DTRs, spasticity, spasms, clonus
Flaccidity (LMN lesion)	Flaccid paralysis with absent DTRs
Hypotension	Dizziness or loss of consciousness
Pulmonary dysfunction	Require accessory muscles of respiration
Neurogenic bladder	Urinary incontinence, urinary tract infection
Neurogenic bowel	Fecal incontinence, constipation

Adapted from A.H. Ropper, 1994, Trauma of the head and spine. In *Harrison's principles of internal medicine*, edited by K.J. Isselbacher et al. (St. Louis: McGraw Hill).

Table 29.5 Medications Commonly Used in Spinal Cord Injury

Medication (generic)	Brand name	Indication	Common side effects
Amitriptyline	Elavil	Neuropathic pain	Dry mouth, constipation, arrhythmias, hypotension
Carbamazepine	Tegretol	Neuropathic pain	Drowsiness, dizziness, aplastic anemia, agranulocytosis
Diazepam	Valium	Spasticity	Drowsiness, dry mouth, dizziness, hypotension
Enoxaparen	Lovenox	DVT or PE	Bleeding, bruising
Gabbapentin	Neurontin	Neuropathic pain	Drowsiness, dry mouth, dizziness, hypotension
Imipramine	Tofranil	Bladder incontinence	Dry mouth, constipation, arrhythmias, hypotension
Lioresol	Baclofen	Spasticity	Drowsiness, hallucinations, seizures
Nortripyline	Pamelar	Neuropathic pain	Dry mouth, constipation, arrhythmias, hypotension
Oxybutynin	Ditropan	Bladder spasms	Drowsiness, dry mouth, dizziness, hypotension
Phenoxybenzamine	Dibenzyline	Autonomic dysreflexia	Dizziness, hypotension
Prazosin	Mininpress	Autonomic dysreflexia	Dizziness, hypotension
Terazosin	Hytrin	Autonomic dysreflexia	Dizziness, hypotension
Tizanidine	Zanaflex	Spasticity	Drowsiness, dizziness, dry mouth, hypotension
Tolteridine	Detrol	Bladder spasms	Drowsiness, dizziness, dry mouth, hypotension
Warfarin	Coumadin	DVT or PE	Bleeding, bruising

from unbalanced muscle forces in wheelchair-reliant individuals over time, but these do not usually affect the person's function unless he or she desires to stand or walk (incomplete SCI). Heterotopic ossification (bony overgrowth within the joint space) can occasionally limit range of motion to such an extent that the affected individual loses the ability to transfer and perform certain ADLs. Management is with gentle range-of-motion exercises, nonsteroidal anti-inflammatory agents, bisphosphonates, and occasionally surgical resection (96).

Spasticity can affect exercise ability. Treatment includes the removal of any stimuli inducing increased tone, daily prolonged stretch of affected muscle groups, and pharmacological or surgical management (55). Understanding the level of SCI and associated medical problems can help the clinical exercise professional understand the physical limitations of each patient. Table 29.6 lists ADLs and functional ability by level of SCI.

Men 40 or more years of age and women 50 or more years of age who have SCI should be considered at moderate risk for untoward events during exercise, independent of traditional CAD risk. Persons with SCI should obtain medical clearance from a physician knowledgeable in SCI care before performing regular exercise.

Table 29.6 Functional Ability by Level of Spinal Cord Injury

Activity	C3–C4	C5	C6	C7	C8–T1	T2–T6	T7–T10	T11–L2	L3–S3
Feeding	D	Modified I	I	I	I	I	I	I	I
Grooming	D	Modified I	I	I	I	I	I	I	I
Upper-extremity dressing	D	Min A	Modified I	Modified I	I	I	I	I	I
Lower-extremity dressing	D	Modified I	Min A	Modified I	I	I	I	I	I
Bathing	D	D	Modified I	Modified I	I	I	I	I	I
Pulmonary hygiene	D	Assisted cough	Assisted cough	Assisted cough	Assisted cough	Assisted cough	I	I	I
Bowel management	D	D	Modified I	Modified I	I	I	I	I	I
Bladder management	D	D	Mod A	Modified I	I	I	I	I	I
Bed mobility	D	Mod A	Modified I	I	I	I	I	I	I
Wheelchair propulsion	D	Modified I	Modified I	I	I	I	I	I	I
Wheelchair transfers	D	Max A	Modified I	I	I	I	I	I	I
Pressure relief	D	D	I	I	I	I	I	I	I
Driving	D	Modified I	Modified I	Modified I	Modified I	Modified I	Modified I	Modified I	Modified I
Ambulation	N/A	N/A	N/A	N/A	Exercise only	KAFO + Loft	KAFO + Loft	KAFO + Loft	AFO

Note. D = totally dependent; Min A = minimal (25%) assistance required from another person; Mod A = moderate (50%) assistance required from another person; Max A = maximal (75%) assistance required from another person; Modified I = independent with modified equipment; I = independent; N/A = not applicable; KAFO = knee–ankle–foot orthosis (bracing) required; AFO = ankle-foot orthosis (bracing) required; Loft = Lofstrand (forearm) crutches required.

Diagnostic Testing

The diagnosis of SCI is largely based on the physical examination according to the American Spinal Injury Association's criteria listed in table 29.2 on p. 524. Motor function sensory levels (pinprick and light touch) are assessed, and the injury is listed as complete if no motor or sensory function is spared at the S4 to S5 level; any preservation of function spared at these sacral levels denotes an incomplete SCI. The diagnosis of SCI is assisted with the use of electrodiagnostic studies, most notably **somatosensory evoked potentials** (18). When assessing the exercise literature, the exercise professional should note the motor and impairment levels for SCI subjects, because the level and completeness of the injury significantly affect the degree to which the somatic and autonomic nervous systems contribute to exercise responses. The International Stoke Mandeville Wheelchair Sports Federation classification system may help the clinical exercise professional determine the degree of impairment as it relates to exercise (table 29.7 on p. 532).

The spine is considered unstable when two or more of the spinal columns are damaged (involving soft tissue or bony elements) and surgical stabilization is warranted (figure 29.3 on p. 533) (27). Note, however, that spinal cord damage may occur in the presence of apparent spine stability and that neurological dysfunction does not always occur in the presence of an unstable spine. Several examples of spinous fractures and dislocations are provided in table 29.8 on p. 533.

Body composition analysis is important in this population because BMI is not representative and obesity is a

Table 29.7 The International Stoke Mandeville Wheelchair Sports Federation Classification System for Spinal Cord Injury Competition in Sport

Level	Criteria
IA	All participants with cervical lesions with complete or incomplete quadriplegia who have involvement of both hands, weakness of triceps (up to and including grade 3 on manual muscle-testing scale[a]), and severe weakness of the trunk and lower extremities interfering with trunk balance and the ability to walk
IB	All participants with cervical lesions with complete or incomplete quadriplegia who have involvement of upper extremities but less than IA with preservation of normal or good triceps (4 or 5 on testing scale) and with a generalized weakness of the trunk and lower extremities interfering significantly with trunk balance and the ability to walk
IC	All participants with cervical lesions with complete or incomplete quadriplegia who have involvement of upper extremities but less than IB with preservation of normal or good triceps (4 or 5 on testing scale) and normal or good finger flexion and extension (grasp and release) but without intrinsic hand function and with a generalized weakness of the trunk and lower extremities interfering significantly with trunk balance and the ability to walk
IIA	Participants with complete or incomplete paraplegia below T1 down to and including T5 or comparable disability with total abdominal paralysis or poor abdominal muscle strength (0–2 on testing scale) and no useful trunk sitting balance
IIB	Participants with complete or incomplete paraplegia or comparable disability below T5 down to and including T10 with upper abdominal and spinal extensor musculature sufficient to provide some element of trunk sitting balance but not normal
III	Participants with complete or incomplete paraplegia or comparable below T10 down to and including L2 without quadriceps or very weak quadriceps with a value up to and including 2 on the testing scale and gluteal paralysis
IV	Participants with complete or incomplete paraplegia or comparable disability below L2 with quadriceps in grades 3–5
V	Participants with minimal muscle deficit

[a]Manual muscle testing scale: (1) muscle flicker without joint movement, (2) full joint range of motion with gravity eliminated, (3) full joint range of motion against gravity, (4) greater than antigravity but less than normal motor strength, (5) age- and sex-matched normal strength.

Reprinted, by permission, from the International Stroke Mandeville Wheelchair Sports Federation, 2002.

common problem. Although recently touted as the gold standard for determining body composition in SCI (89), dual-energy X-ray absorptiometry (DEXA) introduces significant error because it does not account for hydration status. Four-compartment body composition modeling should be used to determine percent body fat in persons with SCI until regression-based equations specific to DEXA in SCI are established (47). Additional testing appropriate for the person with SCI before exercise training may include pulmonary function testing, quantified strength and flexibility measures, DEXA to determine bone mineral density, radiographs of paralyzed extremities to exclude asymptomatic fractures, lipid profiles, and HbA1c to rule out glucose intolerance and diabetes.

Exercise Testing

Graded exercise testing can be used to assess aerobic fitness or training effects in asymptomatic or athletic populations, screen individuals at risk for heart disease, determine progress in rehabilitation, demonstrate maximal strength and power capacities, and assist with the exercise prescription (87). Additional benefits to persons with SCI include the opportunity to establish a relationship between fitness and posttraumatic return to gainful employment and to determine how the fitness level of a person with SCI changes over time (26).

Field testing is the easiest, least expensive, and most mobility-specific method of evaluation in selected wheelchair users (38). Recent reports, however, have failed to show significant correlations of field testing with actual $\dot{V}O_2$peak, possibly because of variability in terrain, wind speed, temperature, and humidity (96). Arm crank ergometry is the most often used test mode with SCI (83). Wheelchair ergometry is mobility specific for most. Several systems have been developed and tested, including wheelchairs mounted on a motorized treadmill (95), low-friction rollers (64), and specialized devices to simulate overground propulsion (20,60). When compared with arm crank ergometry, wheelchair ergometry results in similar or greater $\dot{V}O_2$peak responses with lower peak power output (44), indicating reduced mechanical efficiency.

Several devices are available to assess all-extremity oxygen consumption, which may be appropriate for persons with incomplete SCI and for monitoring aerobic fitness of those using combined upper-extremity and lower-extremity **functional electrical stimulation** (FES). An improved venous return, by inclusion of the lower-body muscle pump, increases stroke volume and cardiac output (63,69,77).

Protocols should employ incremental graded advances in resistance or power output requirement with periodic discontinuance for blood pressure, heart rate, and electrocardiogram determination (43,60). A typical wheelchair ergometry protocol employs an initial resistance of 25 W, with 5 to 10 W increases every 2 to 3 min to symptom-limited fatigue. People with SCI commonly achieve only 40 to 100 W at peak exercise. Population-specific prediction equations should be used when estimating $\dot{V}O_2$peak (51). Standard test termination rationale should be used (see chapter 5).

For people undergoing testing to rule out ischemic heart disease, postexercise echocardiography (59) or nuclear imaging studies (8) may improve the sensitivity of the exercise stress test. In a study using standard exercise testing, only 5 of 13 subjects with known myocardial ischemia had ST-segment changes indicative of ischemia (8). The arm crank or wheelchair ergometer should be adjusted appropriately to allow optimal efficiency and reduce musculoskeletal injuries at the shoulder, elbow, and wrist. Straps applied to the torso improve trunk stability. Wheelchair gloves or flexion mitts with Velcro straps can prevent blisters, lacerations, and abrasions, especially for those with tetraplegia whose hands and fingers are **insensate** or unable to grasp sufficiently.

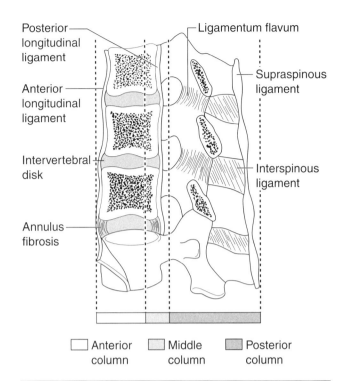

Figure 29.3 Denis three-column model.

Adapted, by permission, from F. Denis, 1983, "The three column spine and its significance in the classification of acute thoracolumbar spinal injuries," *Spine* 8(8): 817-831.

Velcro straps and cuffed weights are commonly used for resistance-training equipment modifications. Abdominal binders and leg wraps may improve pulmonary dynamics and venous return, which reduce the risk of hypotension.

Because of sympathetic impairment, peak heart rate rarely exceeds 120 beats · min^{-1} in those with complete tetraplegia and T1–T3 **paraplegia**. Although variable

Table 29.8 Types and Causes or Results of Vertebral Fractures and Dislocations

Type of injury	Common cause and result of fracture
Atlantoaxial dislocation	Rheumatoid arthritis; can result in respiratory failure
Atlanto-occipital dislocation	Most common in children; typically causes death
Jefferson's fracture	Burst ring of atlas from descending force on vertex of scull (e.g., diving accident)
Hangman's fracture	Hyperextension and longitudinal distraction of upper cervical spin (e.g., chin striking steering wheel in vehicular accident)
Teardrop fracture	Vertebral body compression with anterior bony fragment
Compression or burst fracture	Retropulsed bony fragment into spinal cord
Hyperextension of cervical spine	With cervical stenosis, central cord syndrome
Thoracolumbar fracture	High-impact spinal cord injury associated with rib fracture

Adapted from A.H. Ropper, 1994, Trauma of the head and spine. In *Harrison's principles of internal medicine*, edited by K.J. Isselbacher et al., (St. Louis: McGraw Hill).

responses occur in T4–T6 paraplegia, most persons with SCI below T7 are able to reach their age-adjusted peak heart rate. Similar trends are reported for blood pressure responses. In general, $\dot{V}O_2$peak and peak power output are significantly diminished in patients with SCI (6,13,53,62,84,86). But the lower the injury, the less the impairment. $\dot{V}O_2$peak values range from 12 ml $\cdot$ kg^{-1} $\cdot$ min^{-1} for tetraplegic patients to more than 30 ml $\cdot$ kg^{-1} $\cdot$ min^{-1} in low-level injury paraplegic patients. In these same groups, peak power output ranges from less than 40 W to more than 100 W, respectively.

Persons with SCI and a resting systolic blood pressure below 100 mmHg should be closely monitored during exercise. As such people approach peak exercise, the risk of a hypotensive response increases despite the use of leg wraps and abdominal binders. Should symptomatic hypotension occur, exercise testing should be halted and the person should be tilted back in his or her wheelchair to elevate the lower extremities above the level of the heart, promoting venous return.

Treatment

The care and management of the person with SCI have significantly advanced, extending longevity with improved quality of life but also unmasking the problems associated with physical inactivity, blunted autonomic and hormonal responses, and upper-extremity overuse syndromes common to chronic survivors of SCI. Clinical practice guidelines developed by the Consortium for Spinal Cord Medicine and published by the Paralyzed Veterans of America (www.pva.org) describe expected outcomes for SCI and discuss the prevention and management of autonomic dysreflexia, thromboembolism, bowel management, depression, and pressure ulcers in SCI. Future clinical practice guidelines must address the application and prescription of exercise including a 12-lead electrocardiogram and risk profile assessment, before performing an exercise test. Standard contraindications for exercise testing noted in chapter 5 should be applied to the SCI population. The following are common contraindications in patients with SCI. Without adequate assessment and treatment, these should be considered absolute reasons not to perform an exercise test:

- Autonomic dysreflexia resulting from recent fracture may precipitate spasms or increase the risk of fatty emboli, hypertensive crisis, or cerebrovascular events
- Orthostatic hypotension, with the risk of syncope
- Recent deep vein thrombosis or pulmonary embolism

- Pressure ulcers, which increase the risk of autonomic dysreflexia during exercise

Common relative contraindications include the following:

- Active tendinitis (e.g., rotator cuff, elbow flexors, wrist flexors/extensors)
- Chronic heterotopic ossification
- Peripheral neuropathy
- Pressure ulcers of grade 2 or less
- Spasticity

Bladder and bowel evacuation should be implemented immediately before the graded exercise test to minimize the risk of exertional incontinence or autonomic dysreflexia. Manual stimulation or **disimpaction** is required to maintain fecal continence, and, in some cases, colostomy is warranted for bowel management. Bladder management is most often performed by intermittent catheterization, although indwelling catheters and bladder diversion techniques are sometimes warranted. Environmental consideration must also be made because of the SCI patient's difficulty with regulation of body temperature.

EXERCISE PRESCRIPTION

The exercise prescription for a person with SCI should focus on the typical parameters (see chapter 5). Because of potential complexities, however, the prescription should ideally be developed by a team composed of an exercise physiologist (physiological responses), a physical therapist (orthopedic limitations), and a physician (medical concerns and oversight). The best approach is for the person with SCI to begin an exercise-training program under the supervision of either a physical therapist or a clinical exercise physiologist.

Special Exercise Considerations

Because of their reduced body temperature regulatory ability, persons with **tetraplegia** should exercise only in a mildly temperate climate (outside during appropriate weather or inside in a climate-controlled area) to avoid the risk of hypo- and hyperthermia. Appropriate seating and positioning are necessary to reduce the risk of pressure sores, autonomic dysreflexia, spasticity, and musculoskeletal trauma.

Few commercial fitness centers are able to accommodate the needs of the person with SCI, including

Client–Clinician Interaction

Patients with a SCI must overcome tremendous physical, emotional, spiritual, and intellectual obstacles to succeed in life. They must wake up early to bathe, dress, groom, and perform bowel and bladder care. Seated pressure relief must be performed every 15 to 20 min. They must manage bodily functions according to a timed regimen. Obstacles are everyday occurrences. As an exercise professional, you will find often that the person with SCI who consults you is one of the most disciplined, motivated, and enthusiastic clients you will have. Prepare to be surprised! You should understand the goals and obstacles of the person with SCI. To do this, you must have direct interaction with the patient. Whenever possible, speak at eye level with the client. Gain an understanding of the client's daily routine, environmental and transportation barriers that must be overcome, and concerns that the person may have about an exercise routine. Provide empathetic listening. Be prepared to discuss the application of exercise benefits to the client's functional abilities. Become familiar with accessible facilities in the community. You might consider spending a 24 h day without standing. During this time, perform community mobility and seated activities of daily living (including car, tub, and toilet transfers) from a wheelchair, to gain greater appreciation for your SCI client's perspective and needs. This experience can be eye opening, allowing you a glimpse into the world of the SCI patient. This activity may be helpful when considering the daily life of SCI patients and how they might incorporate an exercise or physical activity regimen.

wheelchair access. A recent survey of physical fitness facilities in a major city demonstrated that none of the 34 facilities reviewed met all of the 1990 Americans with Disabilities Act (ADA) requirements for accessibility (35). Other potential barriers include individual constraints attributable to transportation availability and required assistance. Finances may also be a concern for many patients. These issues have a direct effect on compliance.

Adapted or adaptable equipment, some of which is outlined in the exercise-testing recommendations section, is required for proper and safe exercise. Other recommendations include the use of upper armbands or Co-Ban tape to prevent abrasions at the medial upper arm with wheelchair propulsion. Abdominal binders and leg wraps may be used to facilitate improved pulmonary dynamics and greater venous return. FES–LCE (functional electrical stimulation–leg cycling exercise) and hybrid systems are now commercially available but remain expensive to purchase and maintain relative to arm crank ergometry. Velcro straps and cuffed weights are commonly used for resistance training to improve or create a grip.

Exercise Recommendations

Individuals with complete lesions at or above C4 are limited to using FES-LCD or hybrid exercise equipment because of problems with arm function. For instance,

the person with C5 tetraplegia will require exercises that do not entail active wrist extension, elbow extension, or grasp. Conversely, those with C6 lesions have active wrist extension but little or no elbow extension or grasp. Table 29.3 can be used to determine the effects of the level of SCI on the ability to exercise. Additionally, persons with incomplete SCI who have relative sparing of sensation in the lower extremities may not be able to tolerate FES–LCE or hybrid exercise modes because of pain.

The initial stages of an aerobic exercise program should focus more on developing the habit of exercise rather than the intensity and duration of exercise, because exercise adherence is poor in this population. Succinct, precise, and quantifiable goals will optimize chances of a successful outcome. Having the patient make the informed choice of exercise mode will positively affect patient compliance. Also, individuals should be told to expect delayed-onset muscle soreness (DOMS). The onset and duration of DOMS are similar to those for nondisabled people.

Cardiovascular Training

The most appropriate method to monitor and prescribe intensity of aerobic exercise for the individual with SCI is controversial. Heart rate responses in people with tetraplegia and high-level paraplegia are lower than in nondisabled people because of a poor sympathetic response (67). Although variable, 30% to 80% of heart rate reserve

Exercise Prescription Summary

Training method	Mode	Intensity	Frequency	Duration	Progression	Goals	Special considerations and comments
Aerobic	ACE Wheelchair ergometry Arm crank cycling Community wheeling Seated aerobics Aquatics Wheelchair recreation FES and leg cycling ergometry ACE and FES leg ergometry	RPE 11–14 50%–85% $\dot{V}O_2$ peak or peak power output 30–80% HR reserve 60–90% HR peak Talk test	3–7 d per week	20–60 min, continuous or interval	Slow (<5% per week)	Improved functional capacity Reduction in activity-affected cardio-vascular disease risk factors	Avoid exertional hypotension May initially require multiple sets for 5–10 min duration Monitor for autonomic dysreflexia Avoid thermal stress Include warm-up and cool-down
Resistance	Elastic bands Wrist weights Body weight Dumbbells Free weights Wheelchair accessible machines FES isokinetic	8–12 reps at 60–75% 1RM	One to three sets on 2–3 d per week	30–60 min	Increased resistance when reps achieved are ≥12	Improved strength Improved ability to ambulate using arms	Avoid Valsalva maneuver Provide spotter Use seat belt and chest strap for balance Use adaptive grip and mitts Do not exceed stress limits of wheelchair
Range of motion	Active assisted (anterior shoulder, pectoral, rotator cuff) Passive assisted (hip flexors, knee flexors, plantar flexors)	As tolerated	7 d per week	5–15 min		Reduction in contractures and spasticity Improved joint ROM for affected and nonaffected joints	Stretch to strain, no pain Avoid Valsalva maneuver Don't overstress insensate joints Provide midshaft support for long osteopenic limbs

Note. FES = functional electrical stimulation; ACE = arm crank ergometry; RPE = rating of perceived exertion; RM = repetition maximum; HR = heart rate; ROM = range of motion.

appears to correspond to 50% to 85% of $\dot{V}O_2$peak in those with high-level paraplegia and tetraplegia (54). Using a percentage of peak power output is also recommended (67). This method is cumbersome, however, because continual reevaluation is required to maintain optimal training intensity because peak power output increases with training. Rating of perceived exertion (RPE) may also be used to guide exercise-training intensity. This method is used successfully in cardiac transplant patients who, like patients with SCI, have reduced peak heart rates secondary to cardiac denervation (56). Although not investigated in patients with SCI, RPE values of 11 to 14 may be used during exercise training because this level corresponds to approximately 60% to 90% of $\dot{V}O_2$peak in nondisabled individuals. Another possible method is the talk test. Nondisabled people able to speak during exercise without feeling short of breath are typically at appropriate exercise intensity. Again, however, this method has not been validated in the SCI population.

The duration of a single aerobic exercise bout will vary depending on fitness level, but a goal is to follow the recommendations for the general population (2). Exercise progression should occur as tolerated to a maximum of 60 min duration at the prescribed intensity range.

Aerobic exercise should be performed no fewer than 3 d a week to maintain fitness but may occur up to 5 d a week for optimal gains without negative consequences, particularly when performed at relatively low intensity levels (37). Persons with SCI may require increased exercise frequency to optimize caloric expenditure for weight management but should be judiciously monitored to reduce the incidence of upper-extremity overuse syndromes. Persons with SCI who have $\dot{V}O_2$peak less than 10.5 ml $\cdot$ kg^{-1} $\cdot$ min^{-1} (<3 metabolic equivalents) may require multiple bouts of exercise daily, each lasting 5 to 15 min, until they are able to tolerate 20 to 30 min sessions.

Resistance Training

Minimally, scapular stabilization and rotator cuff resistance exercises should be used in all SCI patients capable of voluntary control of these muscles. Initial intervention should include two sets of 10 repetitions, with 6 s isometric contractions for shoulder protractors, retractors, elevators, and depressors, as well as for internal and external shoulder rotators. As the patient tolerates, progression to dynamic exercise should occur. Resistance band exercises are useful initially, but a plateau in gains can be expected because of the limitation in resistance of this device. Although dumbbells and free weights may be used under close supervision, paralyzed lower extremities and truncal musculature significantly reduce a person's ability to balance even small objects when lying supine or when seated without significant truncal support. When the person is using free weights or isotonic or isokinetic machines, wheelchair brakes should be set before lifting, and care should be taken not to exceed the weight and stress limitations of the wheelchair as provided by the manufacturer. Standard recommendations for intensity and progression should be followed as discussed in chapter 5.

Range of Motion

Range-of-motion exercise should be performed daily and should focus on all major joints, especially those with contracture and spasticity. Both active and passive assisted methods of static stretching can be used. During passive stretching, carefully working through the range of motion in joints lacking sensation is important, because the individual cannot determine when the maximal range of the joint has been reached. Supporting the midshaft of long bones in the patient who has osteopenia may be important to reduce the chance of fracture.

Practical Application 29.3

Exercise-Training Literature Review

Tasks such as feeding, grooming, hygiene, dressing, bathing, transfers, and toileting are referred to as activities of daily living (ADLs), whereas tasks of community mobility include traversing sidewalks, stairs or ramps, paths, and environmental barriers (e.g., curbs, speed bumps). Patients with chronic (>20 yr) SCI will require greater physical assistance as they age. The percentage of maximal heart rate while performing ADLs and community mobility tasks is higher in people with tetraplegia than in those with paraplegia, and an inverse relationship exists between physical capacity and physical strain (25,52). In one study, only 29% of SCI subjects with $\dot{V}O_2$peak less than 15 ml $\cdot$ kg^{-1} $\cdot$ min^{-1} were able to perform independent ADLs (49).

(continued)

Upper-Extremity Aerobic Training

Critical review of upper-extremity aerobic conditioning studies demonstrates variability in exercise prescription and results, partially attributable to the level of SCI. For instance, Gass et al. (40) trained seven subjects (four with C5–C6 lesions, three with T1–T4 lesions) five times weekly to exhaustion on a graded exercise test protocol by using a wheelchair on a treadmill ergometer. After 7 wk of training with this ergometry system, the mean $\dot{V}O_2$peak had increased from 9.5 to 12.7 ml $\cdot$ kg^{-1} $\cdot$ min^{-1} and endurance time had increased by 4.4 min, suggesting considerable change in functional capacity (40). Knutsson et al. (58) evaluated 20 SCI patients with complete and incomplete SCI between C5 and L1. For the 10 persons assigned to the training group, a 40% increase in peak work rate (40–57 W) was reported, although no significant change was noted in the three subjects with SCI above T6. Taylor et al. (92) assessed the effects of ACE performed 30 min per day, 5 d per week for 8 wk in individuals with paraplegia. The trained group significantly improved $\dot{V}O_2$peak from 22.8 to 26.3 ml $\cdot$ kg^{-1} $\cdot$ min^{-1} without significant changes in maximal heart rate, postexercise lactate, or body fat.

More recently, McLean and Skinner (67) matched 14 tetraplegic subjects, by peak power output, to either a supine or seated exercise-training regimen to assess for changes in postural position on stroke volume, cardiac output, and exercise capacity. Their subjects performed arm crank ergometry exercise in either a sitting or a supine position at 60% of their $\dot{V}O_2$peak three times a week for 10 wk with progressive increments in either duration or resistance; no significant differences were found in stroke volume or cardiac output, although absolute $\dot{V}O_2$peak increased from 720 to 780 ml $\cdot$ min^{-1}, suggesting peripheral adaptations.

Locomotor Training

Recent research indicates that locomotor training can greatly improve human walking ability after spinal cord injury (3,4,11,12,15,29). When therapists provide partial body weight support and manual assistance to spinal cord injury subjects during treadmill stepping, task-specific motor learning occurs and improves gait control and mechanics. Studies on patients with chronic incomplete spinal cord injury indicated that approximately 80% of wheelchair-reliant individuals became independent walkers after locomotor training (98). The National Institutes of Health recently funded a multimillion dollar clinical trial at several sites across North America for subjects with acute incomplete spinal cord injury to validate the effectiveness of locomotor training in clinical practice (31). Although improvements were noted, partial weight-bearing treadmill training (PWBTT) was not significantly better than "usual" rehabilitation in the acute setting, likely because of neural recovery often seen in the acute phase. A few trials are still ongoing in the chronic phase of SCI, and preliminary data are promising. Despite its probable benefits, only a handful of clinics in the United States currently offer manually assisted locomotor training with partial body weight support for gait rehabilitation after spinal cord injury, presumably because of the high cost of therapist labor and equipment for locomotor training. Reports on overall health parameters including aerobic fitness, lipid profiles, body composition changes, insulin sensitivity and glucose tolerance, and lower-extremity bone mineral density are promising but sparse to date (48,73,78).

Resistance Training

Because of the large number of shoulder and upper-extremity musculoskeletal problems encountered by persons with SCI, a prophylactic, structured, and progressively resistive strengthening program that focuses on scapular, rotator cuff, and pectoral muscles is likely to increase strength and reduce the risk for overuse injury. Such a program will probably improve the ability of these people to perform functional tasks in the community. Little information is available, however, to determine the effects of resistance training on strength, power, muscle mass, or functional abilities in persons with SCI. Nilsson et al. (74) reported increases in dynamic strength (16%) and endurance (80%) when comparing bench press before and after a 7 wk combined arm crank ergometry and resistance-training program in adults with paraplegia. Chawla et al. (19) reported that a resistance-training program including bench press, incline press, lateral raises, incline curls, lat pulls, and triceps stretch improved ADL function in 10 patients with SCI. Unfortunately, specific ADLs and quantitative measures of strength were not reported. From these few studies, it appears promising that specific strength training will benefit patients with SCI.

Functional Electrical Stimulation

In 1987 medical guidelines were developed for patient participation in FES rehabilitation, including medical criteria for inclusion and exclusion. Computerized FES is a neuromuscular aid used to restore purposeful movement of limbs paralyzed by upper motor neuron lesions. Numerous reviews are devoted to FES and its potential to stimulate beneficial exercise adaptations in patients with SCI (42,76,77). Briefly, FES of the lower extremities can be used to do the following:

Stimulate skeletal muscle strength (32,79,81) and endurance (32,50,68)

Improve energy expenditure and increase stroke volume (36,50)

Increase total body peak power, $\dot{V}O_2$peak, and ventilatory rate (5,50,68)

Reverse myocardial disuse atrophy (71)

Increase high-density lipoprotein levels and improve body composition (7)

Improve self-perception (88)

Increase lower-extremity bone mineral density (14,46)

Despite these encouraging findings, functional gains in upper-extremity strength, aerobic capacity, **community mobility**, and ADLs were not demonstrated in response to FES lower-extremity training. After 8 wk of inactivity that followed a 12 wk training program, the 45% improvement in $\dot{V}O_2$peak attained by FES plus leg ergometry training was reduced by approximately 50%, whereas power output returned to pretraining levels. Submaximal heart rate, peak heart rate, and peak ventilatory volume did not change at any time point.

A logical and intuitive progression in the development of FES lower-extremity exercise training has been the combined use of concurrent arm crank ergometry and FES leg cycle ergometry (LCE), termed **hybrid exercise**. As expected from the combination of upper- and lower-extremity exercise peak power, $\dot{V}O_2$peak, stroke volume, and cardiac output ($\dot{Q}$) significantly increase during hybrid exercise bouts with SCI subjects (34,69) and during combined upper-extremity rowing plus lower-extremity FES (61).

CONCLUSION

The patient with SCI presents with unique obstacles and considerations that the clinical exercise professional must be familiar with to provide safe and optimal exercise testing and training oversight. People with SCI tend to be sedentary and are excellent candidates for regular exercise training. Because of the many potential medical and possible exercise-related problems, a team approach to evaluation, exercise prescription development, and exercise-training guidance is recommended.

Case Study

Medical History

Ms. BF is a 28 yr old white female with thoracic paraplegia since age 19 caused by a motor vehicle accident. She played volleyball in high school and was athletic before the accident. Since her rehabilitation, she has been wheelchair reliant for community mobility but is otherwise independent for activities of daily living, including bowel and bladder management and driving with hand controls. She received an MBA this year, has a full-time job as an accountant for a law firm, but is otherwise quite sedentary. Her body weight has remained stable at 150 lb (68 kg) for the past 4 yr. Her body mass index (BMI) is 31. Other exam findings include blood pressure (BP) of 100/60; heart rate (HR) of 85; electrocardiogram (ECG) with normal sinus rhythm; normal heart, lung, and bowel sounds; intact skin; mildly reduced range of motion at the shoulder (extension, internal and external rotation); 5° hip, knee, and plantar flexion contractures bilaterally; 5/5 upper extremity with 0/5 lower-extremity motor strength; absent pinprick and light touch sensation everywhere below the xyphoid process; and 3+ (brisk)

(continued)

reflexes at both the knees and ankles. She reports poor sleep quality and excessive fatigue and intermittently notes pounding headaches and sweating associated with an overfull bladder. Her medications include Detrol for bladder spasms and Lioresol for lower-extremity spasms and spasticity.

Diagnosis

Ms. BF has T6 American Spinal Injury Association A paraplegia with neurogenic bowel and bladder, spasticity, and occasional autonomic dysreflexia. She likely has impaired pulmonary function attributable to lower intercostal, paraspinal, and abdominal muscle paralysis. Although she is fully functional with her upper extremities, she has poor to fair sitting balance because of paraspinal, abdominal, and lower-extremity paralysis. Her elevated BMI is probably the result of her sedentary lifestyle. The fact that she has completed a graduate degree and holds a full-time job suggests that she has adjusted well to her SCI.

Exercise Test Results

Ms. BF will most likely want to know how exercise can affect her current lifestyle and health, and whether the benefits will be worth investing time already committed to a busy and productive life. Encouraging her to share short- and long-term life goals will provide an opportunity to educate her about the acute and chronic benefits of exercise, particularly as they can improve her community mobility and modify her elevated risks for chronic diseases, including coronary artery disease, obesity, diabetes, osteoporosis, and upper-extremity overuse syndromes. As an initial step to increasing Ms. BF's physical activity level, an exercise evaluation is suggested. Ms. BF's exercise evaluation is performed on an upper-body ergometer by using a discontinuous protocol with an initial workload of 24 W at a constant 50 rev · min^{-1} and 6 W increments every 3 min. Her peak exercise performance yielded the following results: peak HR = 172, BP = 130/50, $\dot{V}O_2$peak = 16.4 ml · kg^{-1} · min^{-1}, peak power output = 42 W. ECG revealed isolated premature ventricular contractions (PVCs) with no evidence of ST-segment depression. Ms. BF stopped the test herself because of shoulder fatigue and dizziness, associated with a 10 mmHg decrease in systolic blood pressure. She experienced some exertional dyspnea but no other symptoms. In the past, Ms. BF's physician has asked her to try FES. She was initially evaluated and thought to be a good candidate but never seriously considered FES.

Exercise Prescription

Ms. BF is interested in improved community mobility, losing weight, and maintaining good health. She is motivated to begin a regular physical activity program. She will begin under the guidance of an outpatient physical therapist. Her goal is to perform her routine at a medical fitness center under the guidance of a clinical exercise physiologist. Her maximal heart rate response on the exercise test will be used to set a target heart rate zone based on her heart rate reserve, and she will be introduced to a wheelchair ergometer as well as the upper-body ergometer. Her goal is to perform between 20 and 60 min of aerobic exercise 3 to 5 d each week, although her initial workouts may be divided into two or three 10 min sessions, with 10 to 15 min recovery between sessions to reduce her risk of hypotension. Ms. BF will also begin a resistance-training program that emphasizes shoulder stabilization with resistance bands 2 to 3 d per week, progressing to machines or free weights as accessible. She will perform range-of-motion exercises daily to improve sitting posture and shoulder biomechanics as well as to prepare for an eventual trial with FES exercise.

Discussion Questions

1. What is your interpretation of the medical history?

2. What might you discuss with this patient regarding her examination findings and lifestyle, and how would you do this?

3. Discuss the exercise test with respect to mode, protocol, and her physiological responses. Considering these results, what are your recommendations for exercise?

4. Do you think that FES might benefit Ms. BF? Why or why not?

5. At what heart rate should she perform aerobic exercise?

6. What specific precautions should be considered for people with SCI who perform exercise?

7. What types of strength and range-of-motion exercises should she perform?

8. How might you keep Ms. BF motivated?

30

Multiple Sclerosis

Chad C. Carroll, PhD

Charles P. Lambert, PhD

Multiple sclerosis (MS) is a demyelinating disorder of the central nervous system that causes patches of sclerosis (plaques) in the white matter of the brain and spinal cord (4). These plaques can develop into permanent scars (106) that impair nerve transmission (40,114,115) and lead to an array of symptoms, such as muscle weakness, fatigue, and motor function difficulties (60). Sclerosis (hardening or scarring) results from the **demyelination** process, and multiple sclerosis refers to the multiple areas of demyelinated tissue throughout the central nervous system (4).

SCOPE

Multiple sclerosis is most common in Caucasians of northern European descent. Approximately 250,00 to 350,000 people are diagnosed in the United States (3). Women appear to be affected at nearly twice the rate as men and make up more than 60% of those diagnosed with the disease (24,42). Susceptibility to MS appears to be complex (44), possibly resulting from an interaction of several genetic and environmental factors. Several genes (HLA class II, apoE, IL-1ra, IL-1β, and TGFβ1) are reported to be associated with the course and severity of MS (35,44,98). The HLA-DR2 allele is most commonly associated with progression and onset of MS (44,56). But no single identifiable gene is associated with all aspects of MS. A genetic component of MS is supported by studies of twins, which report concordance rates of about 26%

in identical twins in contrast with only about 2% in other siblings (93). Additionally, approximately 20% of people with MS have a close relative with the disease (94). But a common environmental component that influences susceptibility to MS is suggested because the incidence of MS is greater in high-latitude areas (30) such as northern Europe, the northern United States, Canada, and southern Australia and New Zealand (95).

The initial diagnosis of MS is most common in people between the ages of 15 and 59 (95). Prevalence rates in the United States average about 58 cases per 100,000 people (42) but may be as high as 173 cases per 100,000 depending on geographic location (95). Incidence rates are about 3 cases per 100,000 live births (42). Multiple sclerosis can be personally devastating, as demonstrated by the fact that within 10 yr of diagnosis more than 50% of individuals with MS will become unemployed (41). Lifetime cost of MS healthcare can exceed $1,000,000 per afflicted person (1), and the cost increases with the severity of neurological dysfunction (11).

PATHOPHYSIOLOGY

Multiple sclerosis is believed to be an inflammatory autoimmune disease that affects the oligodendrocytes and the myelin sheath (26,62). As the myelin sheath deteriorates, plaques, or sclerosis, form, potentially leading to axonal destruction (105), and inhibition of normal nerve conduction can lead to an array of symptoms that affect the MS patients' ability to perform normal activities of daily living.

Autoreactive T-cells are thought to initiate an immune response against myelin, resulting in the CNS demyelination associated with MS. T-cells are activated by an unknown mechanism and infiltrate the blood–brain barrier (88). These activated T-cells enter the CNS, proliferate, and secrete lymphokines or cytokines, which recruit microglia, macrophages, and other immune cells to participate in oligodendrocyte death and myelin destruction (26,54,88). But it is not clear whether macrophages or T-cells are the primary cells that mediate demyelination (91). The autoimmune response is further enhanced by plasma cells, which produce antibodies against myelin and oligodendrocytes (88). Therefore, it seems likely that MS has an autoimmune basis, but the exact mechanisms responsible for the process of demyelination are still undetermined.

MEDICAL AND CLINICAL CONSIDERATIONS

The following section reviews the medical and clinical issues that patients with multiple sclerosis face and the generals methods used to evaluate these patients.

Signs and Symptoms

The initial symptoms of MS are associated with the areas of the central nervous system that are demyelinated to the greatest degree (table 30.1). Common initial symptoms include muscle weakness, **optic neuritis**, and somatosensory complaints (60,61,64). General weakness is present in most patients (60,83,103,104), and advanced MS often presents with muscle weakness and spasticity, markedly affecting voluntary movement.

Systemic fatigue is common and often rated by patients as the worst symptom of MS (2,8,28). Fatigue is typically noted after routine work such as house cleaning. Cerebellar symptoms occur but are difficult to distinguish from spasticity, weakness, **vertigo**, and sensory loss. Nystagmus is also a common sign of MS (60).

Psychological effects such as an impaired cognitive ability and memory loss can occur early after disease onset (84). A strong correlation exists between the level of physical disability and a defect of cognition (63). Depression is also associated with MS (3) and can lead to reduced participation in work, social activities, and other endeavors (85).

Urinary urgency, **incontinence**, and sexual dysfunction in both men and women are often present (60). An abnormal sweating response (21) and impaired thermoregulation are also common. Abnormal cardiovascular function during exercise (e.g., low peak heart rate) can be caused by autonomic cardiovascular reflex dysfunction and may occur in up to 50% of patients with MS (76,101,109).

History and Physical Exam

Multiple **sclerosis** can follow at least four courses of clinical progression (table 30.2) (58). A fifth course, relapsing progressive, has been defined, but its use overlaps other defined clinical courses and some recommend that this terminology not be used (58). Many people have occasional attacks but frequently recover and never incur any permanent disability (3). In other cases individuals may have frequent attacks, and although they do not completely recover they may never become "disabled."

Common early indicators of MS include visual impairments, motor function difficulties, and **paresthesia** (3,64). Muscle **spasticity** is usually not present until the disease reaches the advanced stages (64). Some people do not demonstrate clinical signs of MS during life, and MS is diagnosed only during autopsy. For many patients, MS progression is slow, and deterioration occurs over 10 to 25 yr. Ambulation can become increasingly difficult (15), although proper treatment can help delay or diminish the debilitation of the patient. Multiple sclerosis is a disease with a multitude of signs and symptoms (table 30.1), many of which are difficult for untrained individuals to recognize. Therefore, the clinical exercise professional should consult with a patient's physician before exercise testing and training. Some patients exhibit problems with vision, and issues related to sight during exercise should be addressed in these

Table 30.1 Common Signs and Symptoms of Multiple Sclerosis

Symptoms	Signs
Muscle weakness	Optic neuritis
Fatigue	Paresthesia
Spasticity	Nystagmus
Memory impairment	
Vertigo	
Sensory loss	
Balance problems	
Bladder or bowel incontinence	
Speech or vision disturbances	

Table 30.2 Clinical Courses of Multiple Sclerosis

Type	Characteristic
Relapsing-remitting	Characterized by disease relapses with either a full recovery or a deficit after recovery; no progression of disease symptoms in recovery stage
Primary progressive	Disease progression from onset with infrequent plateaus and temporary small improvements; clinical status is continuously worsening with no distinctive relapses
Secondary progressive	Begins as relapsing-remitting but continually progresses either with or without infrequent relapses, plateaus, and remissions
Progressive-relapsing	Progressive from onset with short definite relapses with or without full recovery

cases. Those with cognitive deficits may have impaired ability to comprehend the purpose of an exercise test, provide informed consent, provide feedback, and follow an exercise prescription. In such cases, a person responsible for the patient should accompany him or her. Loss of muscle strength, numbness, or poor flexibility can affect the patient's ability to perform certain types of exercise. These conditions must be considered on an individual basis. Additionally, poor coordination, balance, and gait abnormalities are common problems associated with MS and should be considered before exercise. Classification of the level of MS-related disability based on the medical evaluation can be made using the Kurtzke functional systems (table 30.3) and the Kurtzke expanded disability status scale (table 30.4 on p. 546) (51). These tables are used together to find an objective rating of a patient's level of functional disability and can provide important information regarding a patient's ability to perform certain types of exercise.

Table 30.3 Kurtzke Functional Systems

Category	Rating scale range
Pyramidal	0–6
Bowel and bladder	0–6
Cerebellar	0–5
Visual (optic)	0–6
Brainstem	0–5
Cerebral (mental)	0–5
Sensory	0–6
Other[a]	0–3

Note. Ratings range from normal function (0), to signs only without disability (1), to increasing levels of disability (2–6).
[a] Any other neurological findings attributed to multiple sclerosis (i.e., fatigue).
Adapted from J.A. Ponichtera-Mulcare et al., 1994, "Maximal aerobic exercise in ambulatory and semiambulatory patients with multiple sclerosis," *Medicine and Science in Sports and Exercise* 26:S29.

Review of the medical history should be used to determine any prescribed or over-the-counter medications that an individual is using. Medications to treat muscle spasticity often result in fatigue, which may limit maximal exercise capacity. Skeletal muscle fatigue may occur before cardiorespiratory fatigue occurs, which may limit the ability to diagnose coronary artery disease (i.e., reduce the predictive value of the test) using a standard exercise test. If a diagnostic test is desired, it may be best to use an imaging-type test.

Diagnostic Testing

From 1965 through the 1980s, the diagnostic criteria for MS were limited (64,100). See "Criteria for Diagnosing Multiple Sclerosis From 1965 to the Mid-1980s" on p. 547. When five of the six criteria were met, and included number 6, the clinical diagnosis of MS was satisfied (64). In the 1970s and 1980s, advances in imaging (e.g., computed tomography, or CT; **magnetic resonance imaging**, or MRI) (7,27) and the development of **evoked response testing** (16) allowed for the visualization and evaluation of sclerotic lesions. In addition, obtaining cerebrospinal fluid samples to detect specific immune factors has advanced diagnostic ability (64). In 1983 new diagnostic criteria that incorporated these diagnostic tools and information obtained from immunological analyses were developed (82). Even today, however, diagnosis initially focuses on the medical history and neurological examination. Subsequently, the decision to perform an MRI, a CT scan, or other testing should be made. Figure 30.1 shows an MRI of the brain of a MS patient with sclerotic lesions. Practical application 30.1 on p. 548 provides more information about imaging techniques.

Exercise Testing

Exercise testing is useful in patients with MS to determine the safety and effectiveness of exercise training. Optimally, each patient should be evaluated before testing.

Table 30.4 Kurtzke Expanded Disability Status Scale

Rating	Disability	Functional limitations	Comparison to Kurtzke FS rating
0	Normal neurological exam	None	0 in all FS categories
1.0	None	None	Grade 1 in one FS category
1.5	None	None	Grade 1 in more than one FS category
2.0	Minimal	Affects one FS	Grade 2 in one FS category
2.5	Minimal	Affects two FS	Grade 2 in two FS categories
3.0	Moderate	Affects three to four FS	Grade 2 in three to four FS categories
3.5	Moderate, fully ambulatory	Affects three to four FS	Grade 3 in one FS category and 1–2 in two categories
4.0	Mildly severe disability: fully ambulatory without aid; self-sufficient up to 12 h per day	Affects one or more FS	Grade 4 in one FS category with 0–1 in others
4.5	Moderately severe disability: same as 4.0 with some limitation of ADLs or needs minimal assistance	Affects one or more FS	Grade 4 in one FS category with 0–1 in others
5.0	Severe: walks only 200 m without rest or aid; impaired ADLs	Affects one or more FS	Grade 5 in one FS category with 0–1 in others
5.5	Severe: walks only 100 m without rest or aid; impaired ADLs	Affects one or more FS	Grade 5 in one FS category with 0–1 in others
6.0	Severe: intermittent or unilateral aid to walk 100 m	Affects two or more FS	Grade 3 or more in two or more FS categories
6.5	Very severe: constant aid (cane, crutches, braces) to ambulate 20 m	Affects two or more FS	Grade 3 or more in two or more FS categories
7.0	Extremely severe: unable to walk 5 m; needs aid 12 or more hours per day; can transfer self from chair	Affects more than one FS	Grade 4 or more in more than one FS category
7.5	Extremely severe: only takes a few steps; cannot wheel self; full day in wheelchair	Affects more than one FS	Grade 4 or more in more than one FS category
8.0	No ambulation; restricted to bed or chair; retains self-care and can use arm	Affects many FS	Grade 4 or more in many FS categories
8.5	Bedridden; minimal arm use; some self-care	Affects many FS	Grade 4 or more in many FS categories
9.0	Bedridden; no self-care; can talk and eat	Affects all FS	Mostly grade 4 or more for all FS categories
9.5	Bedridden; cannot communicate, eat, or swallow	Affects all FS	Grade 4 or more for all FS categories
10.0	Death attributable to multiple sclerosis		

Note. FS = functional systems; ADL = activities of daily living.

Resulting information from a test can be used to develop the exercise prescription. Standard graded exercise testing contraindications reviewed in chapter 5 should be followed for patients with MS (45). No specific contraindications are common among patients with MS.

The exercise professional should carefully consider the mode of exercise used because each has specific advantages. A cycle ergometer is most common because it requires little balance and those with ambulatory impairment can use it. For individuals who have **ataxia** or who walk using an assistive device, a cycle ergometer is best. Additionally, the resistance of a cycle ergometer can be adjusted to a level that is within most patients' capabilities. People with little ambulatory impairment could use a treadmill, although this determination should be made on an individual basis. The risk of falling should

Criteria for Diagnosing Multiple Sclerosis From 1965 to the Mid-1980s

1. Examination of objective neurological indicators
2. Onset of symptoms between the ages of 10 and 50 yr
3. Neurological signs and symptoms of **central nervous system** white matter disease
4. Dissemination of disease activity in time: two or more attacks that lasted at least 24 h and were separated by 1 mo or more, or 6 mo of progression of signs and symptoms
5. Two or more nonconnected anatomical areas affected
6. No alternative clinical explanation

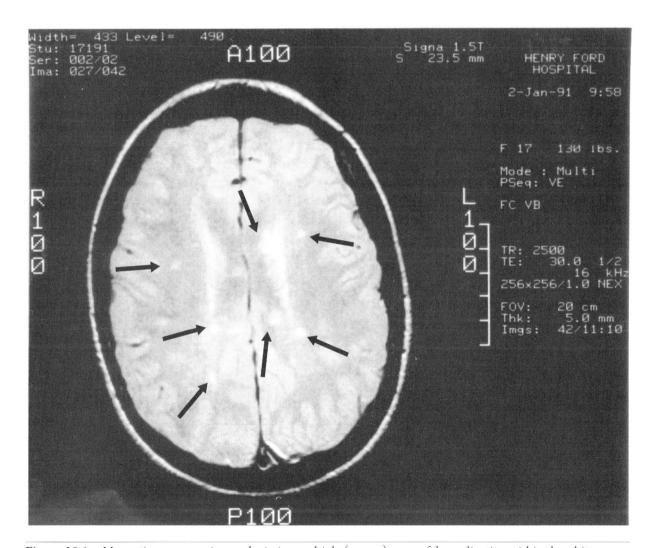

Figure 30.1 Magnetic resonance image depicting multiple (arrows) areas of demyelination within the white matter of the brain.

Reprinted, by permission, from C.A. Brawner and J.R. Schairer, 2000, "Multiple sclerosis: Case report from Henry Ford Hospital," *Clinical Exercise Physiology* 2(1): 17.

Imaging Techniques

Computed tomography (CT) is a method of scanning capable of producing high-resolution, high-contrast images (110). A complete CT picture is a composite of several sliced images (25). During this procedure, an X-ray beam and detector move through a 360° arc around the patient, who is on a motorized couch. A highly focused X-ray beam irradiates the subject. The image produced corresponds to the amount of X-ray beam that passes through each section and is inversely proportional to the density of the tissue (25). The scan takes only a few minutes to complete.

The detector system converts the X-rays to an electron stream. The stream is then digitized and quantified into a system of measurement known as the Hounsfield unit. These units are used to differentiate between tissue densities. Dense structures, such as bone, absorb a large quantity of X-ray and appear white. Air shows up as very dark or black. Water is typically the standard to which other tissues are compared and is given a Hounsfield unit of zero (110). CT scans of the brain may demonstrate ventricular enlargement, low-density periventricular abnormalities, or focal regions of enhancement attributable to contrast agent injection in patients with definite MS. In addition to its use in patients with MS, CT is also useful for detecting calcification of the coronary arteries.

Magnetic resonance imaging (MRI) uses magnets and radiofrequency waves to produce high-quality images of anatomy in several planes (7,25). MRI does not require the use of ionizing radiation. Instead, radiofrequency pulses pass through a patient who is placed in a large magnetic field. The computer assembles the image from 1 to 10 mm scanned slices or sections.

Tissues with high hydrogen proton numbers, such as fat, emit a strong signal. A computer program converts the signal to an intense or bright image. When less hydrogen is present, such as in bone, a signal with a low intensity will be produced, showing up as dark or black. The resulting image has excellent contrast resolution.

More than 90% of patients with MS will have a positive MRI or CT scan with increased brightness or whiteness at the affected areas of the brain. Because of its excellent resolution (greater ability to detect CNS lesions), MRI is the most commonly used imaging technique available for MS-affected tissue.

be considered; the choice can be quickly evaluated during the walk from the waiting room to the testing room.

A low-level protocol beginning with a 1 to 2 min warm-up period may be best for most patients (68). Each 2 to 3 min stage should increase by no more than 2 metabolic equivalents (METs) for the treadmill, 10 to 25 W for leg ergometry (81,89,99), and 8 to 12 W for arm ergometry. Some younger patients with little disability may tolerate a higher-level protocol. Table 30.5 reviews the physiological responses to exercise in patients with MS. Table 30.6 provides a summary of exercise-testing specifics.

Individuals with MS may have an attenuated systolic blood pressure response during exercise (77,101). This response may be related to autonomic dysfunction resulting from sclerotic **plaques** in paraventricular and cardiovascular autonomic nuclei (i.e., cardiovascular control center) region of the brain (109). Other evidence suggests that an attenuated pressor response may be attributable to a reduced skeletal muscle metabolic response that affects the skeletal muscle chemoreflex (69). Individuals

with MS may also have abnormal temperature regulation and sweating response (21), increasing the risk of hyperthermia during exercise. Use of electric fans to improve evaporative and convective cooling, attention to fluid replacement, and control of climate (i.e., room temperature 72–76 °F [22.2 °C–24.4 °C], low humidity) are recommended to reduce the risk of heat-related illness. Skeletal muscle fatigue may occur before patients reach peak cardiovascular levels. This circumstance may be caused by inability to recruit additional skeletal muscle or by impaired skeletal muscle metabolism (47,48).

Treatment

Common skeletal muscle problems associated with MS include spasticity, tremors, and pain (60). Spasticity reduces a person's ability to coordinate muscle contractions for movements such as walking. Tremors can also affect the ability to coordinate skeletal muscle contraction during physical activity. The primary treatment for these

Table 30.5 Physiological Responses During Exercise of Individuals With MS Relative to Normal Healthy Individuals

Physiological response	Response of MS patients relative to the response of normal healthy individuals
Submaximal oxygen consumption during treadmill walking	Increased (74)
Submaximal and maximal arterial blood pressure	Decreased (76)
Temperature regulation	Increased (14,72)
Skeletal muscle fatigue	Earlier (47,48)

Table 30.6 Exercise-Testing Summary

Test type	Mode	Protocol specifics	Clinical measures	Clinical implications	Special considerations
Cardiovascular	Cycle ergometry	2–5 min of light warm-up; 2 min stages with 10–25 W increases per stage	BP, HR, oxygen consumption	MS patients may have increased CV risk; respiratory dysfunction; very low fitness levels	Attenuated BP response; blunted HR response; impaired thermoregulation; excessive muscle weakness and fatigue
Strength	Machine weights or manual muscle testing may be used in more disabled patients	1RM testing when appropriate; 10 or 15RM for muscle endurance	Overhead lifting, stand from sitting position	Muscle weakness may be an underlying cause of reduced mobility and functional impairment	Poor flexibility; spasticity; muscle fatigue
Range of motion	Standard tests	Follow standard protocols; adaptations may be necessary for wheelchair-bound or bedridden patients	Goniometry can be used to quantify changes in ROM	Poor flexibility may contribute to poor ambulation and performance of ADLs	Contractures; spasticity

Note. BP = blood pressure; HR = heart rate; CV = cardiovascular; RM = repetition maximum; ADL = activities of daily living.

conditions in patients with MS is drug therapy. Several categories of medications are used to treat MS, including immunosuppressants, corticosteroids, immunomodulators, and antispastics (117). Specifics about these types of medications and others presented in this section can be found in chapter 3. General musculoskeletal pain can be treated with analgesics (acetaminophen, ibuprofen, aspirin), although these are not always effective (55). If pain is attributed to plaques within the nervous system, tricyclic antidepressants are commonly used and are effective (18).

Fatigue is common (2,8,28) and can either be systemic or result in chronic persistent tiredness (2). Fatigue can occur following physical activity or increased body temperature (3). Amantadine and pemoline (Cylert) have been used to reduce systemic fatigue in MS patients (50,116), but amantadine appears to be more effective (50).

Depression is also common in patients with MS (3). The use of tricyclic antidepressants or monoamine oxidase inhibitors is useful (97).

An increase in urinary frequency may be the result of an irritable bladder wall. This condition appears to be a manifestation of the demyelination associated with the disease but may also be secondary to bladder infection caused by inability to control bladder function. Imipramine (Tofranil), which is an antidepressant, and isopropamide (Darbid), oxybutynin (Ditropan), propantheline (Pro Banthine), and dicyclomine (Bentyl), which are all anticholinergics, can be used to reduce the irritative effects.

Immunosuppressive drugs are used in an attempt to suppress autoimmune effects on myelin. These medications may be best for patients with MS who have very slow disease progression. Immunomodulators are agents that alter the function of the immune system with a goal of delaying MS progression and short-term **exacerbation** (92). These drugs are effective for relapsing-remitting MS, and the reduction in the rate of **relapse** may be as much as 29% (43,57,92). Additionally, immunomodulator treatment reduces the number of new or enlarged lesions measured by MRI (59,65,75,107,118).

The use of corticosteroids does not appear to affect MS progression or exacerbation, but these drugs are effective at reducing the recovery time from an exacerbative event (19). But the clinical exercise professional must understand the side effects associated with long-term use, including bone softening, high blood pressure, and transient weight gain (19). Short-term side effects include acne, psychosis, infection, and ulcers (19). These side effects are further addressed in chapter 3.

A variety of interventions have been used in the treatment of MS. See "Interventional Treatments for Multiple Sclerosis." Of these, only **thalamotomy** and thalamic stimulation have demonstrated potential benefits (10,122). The evidence is equivocal, however, and these therapies should be used with caution. These costly therapies are not currently approved by the Food and Drug Administration and are thus considered investigational. Recently the use of neurol stem cell therapy has been proposed as a means to reestablish myelin in the CNS. This technique has proved effective at decreasing demyelination and axonal loss and improving function in an animal model of **multiple sclerosis** (79), but human trials are still in their early stages (39).

Natural killer and nonspecific suppressor cell function is impaired in MS patients (5,9). This circumstance could result in inability to suppress an acute autoimmune response, leading to disease exacerbation. Exercise improves natural killer and suppressor cell function in individuals free of disease (53,71), which may be beneficial to MS patients. More recent evidence suggests that exercise training can also increase serum levels of neurotropic factors in MS patients (34). An increase in these factors may possibly stimulate neural repair in

Interventional Treatments for Multiple Sclerosis

- Acupuncture
- Dorsal column stimulation
- Hyperbaric oxygen
- Transcutaneous nerve stimulation
- Thalamotomy
- Thalamic stimulation
- Sympathectomy
- Ganglionectomy
- Surgical spinal cord relaxation
- Vertebral artery surgery
- Ultrasound
- Magnetotherapy
- Dental occlusal therapy
- Replacement of mercury amalgam fillings
- Hysterectomy
- Surgical implantation of pig brain into the abdominal wall (i.e., cellular therapy)
- Stem therapy or transplantation

Adapted from W.A. Sibley, 1996, *Therapeutic claims in multiple sclerosis: A guide to treatments* (New York: Demos Vermande).

MS patients. Exercise training may also provide immunomodulatory or anti-inflammatory effects that could benefit MS patients (37,38). Additionally, exercise may reduce some secondary effects of MS including impaired bladder and bowel function (77). Finally, exercise may positively affect psychological health and quality of life in these patients (31,66,77,99), reduce **muscle weakness**, increase flexibility, and reduce **fatigue**.

EXERCISE PRESCRIPTION

When deciding about appropriateness of exercise and level of supervision, the clinical exercise physiologist can use the functional disability system presented in tables 30.3 and 30.4. Common disease-associated problems include fatigue, imbalance, incoordination, numbness or loss of sensation, and loss of strength. Supervision should be considered in these individuals, and the level of supervision should be appropriate to the level of disability. Those with known impairment of heart rate or blood pressure responses during exercise may require monitoring.

Special Exercise Considerations

Because people with MS have problems with thermoregulation, cooling with an electric fan and controlling room temperature are important. Although time consuming, precooling by body immersion in cool water has been successful in improving exercise tolerance and may be useful in some cases (119). In addition, fluid should be replaced during and after exercise at the rate at which fluid is lost from the body. Because of the potential for heat-related illness, patients with MS should be educated about this potential problem and the steps needed to avoid problems during exercise training.

Cardiovascular Training

With cardiovascular training, patients with MS attain the same type of improvement (e.g., increased peak $\dot{V}O_2$, improved psychological variables, decreased muscle

fatigue, and increased skeletal muscle strength and oxidative capacity) as healthy individuals do and at a similar rate (23,31,36,66,73,77,87,90,99,111,113,121). The risk of disease exacerbations is small (78). Practical application 30.3 on p. 552 provides a review of the exercise prescription.

To improve cardiovascular function, individuals with MS should perform standard aerobic exercise for 30 min at 50% to 70% of $\dot{V}O_2$peak for at least three sessions per week. Some patients, especially those with severe fatigue, disease exacerbations, or poor fitness, may benefit from lower initial exercise intensity (e.g., 40–50% of $\dot{V}O_2$peak). For those patients whose autonomic nervous system is affected and who have an attenuated exercise heart rate response, guiding exercise intensity using either the rating of perceived exertion scale, percent peak METs, or a work rate equivalent to the desired percent $\dot{V}O_2$peak is appropriate. Cycling is the most common method used to improve aerobic fitness in patients with MS and may be the modality of choice for those with ambulatory limitations. Swimming or walking can also be considered. Progression should be individualized and based on a patient's ability to tolerate exercise, focusing primarily on avoiding excessive fatigue and disease exacerbation.

Practical Application 30.2

Client–Clinician Interaction

The systemic fatigue noted by individuals with MS varies from day to day. The clinical exercise professional must anticipate the need to reduce exercise-training volume (i.e., intensity, duration) if a person is fatigued. Doing this requires daily assessment of an individual for indicators of fatigue. Verbal communication is the best means of assessment. Appropriate questions include "How do you feel today?" and "Are you tired today?"

The clinical exercise professional must also be prepared to recognize and discuss indicators of depression. But most patients are unlikely to divulge this information, so the clinician often must ask questions to discern the degree of depression. Indicators of stress may be related to depression. These signs include poor sleep habits, noncompliance with lifestyle change, and elevated scores on standardized questionnaires. The exercise clinician should share the positive effects of exercise training on various psychological variables (e.g., mood, depression, anxiety). Patients with excessive depression should be referred to a mental health professional.

Because people with MS may have cognitive problems, the exercise clinician may have to repeat directions or clarify explanations. Many individuals with MS die of heart disease and stroke, and this may occur at a younger age if the disease has forced them into a sedentary lifestyle. The exercise clinician should counsel these patients about the risk factors associated with coronary artery and cardiovascular disease in those with MS.

The exercise clinician should provide constant motivation to individuals with MS to exercise regularly. Common reasons for nonadherence to exercise training are fatigue and muscle soreness. The clinical exercise professional must stress that although exercise training may be fatiguing and cause soreness in the short term, the soreness will likely cease and fatigue will decline over the long term. Additionally, these patients should understand that resistance exercise will strengthen their skeletal muscles and allow them to perform activities of daily living with less fatigue.

Practical Application 30.3

Specific Exercise Prescription Recommendations

Training method	Mode	Intensity	Frequency	Duration	Progression	Goals	Special considerations and comments
Aerobic	Cycle ergometry	50–70% of $\dot{V}O_2$peak*; 60–75% of HR peak; 50–85% of HR reserve	Three times per week	About 30 min	Careful progression based on the individual's responses to the exercise; avoid excessive fatigue	Improving cardio-respiratory function and exercise capacity; increasing quality of life; reducing fatigue	Balance on some equipment; consider seated modes or handrail use on treadmill; temperature regulation; excessive fatigue; attenuated HR or BP response
Resistance	Machine weights; elastic bands; aquatic training	60–80% of 1RM*, one or two sets, 8–15 repetitions; 30–50% of 1RM*, one to three sets, 15 or more repetitions	3 times per week	About 30 min	Careful progression based on the individual's abilities otherwise follow standard PRT principles	Increasing muscle strength and decreasing muscle fatigue; increasing ability to perform ADLs	Unilateral weakness; balance; muscle spasticity; excessive fatigue; work to correct imbalances; rest at least 1 d between sessions; avoid free weights unless spotter is used
Range of motion	Static stretching; may be assisted by a partner or using a device such as a towel	NA	Five to seven times per week	About 15 min	Follow recommendations for healthy individuals	Increasing ROM, reducing or avoiding contractures	Balance; avoid excessive joint range and muscle spasticity aggravation; perform while seated or on the floor; perform after either warm-up or cool-down

Note. HR = heart rate; 1RM = one-repetition maximum; NA = not applicable; BP = blood pressure; PRT = progressive resistance training; ADL = activities of daily living.

*Largely dependent on patient's ability; assess on an individual basis.

Resistance Training

Strength training should follow standard training practices for clinical populations including recommendations for intensity, sets, repetitions, and frequency of exercise (see chapter 5). The training routine should focus on the major muscle groups using standard types of resis-

tive movements. If a patient is weakened, low resistance should be used. Avoiding free weights will eliminate problems with balance and coordination. For patients with MS, the best approach may be to perform strength training on noncardiovascular conditioning days to avoid excessive fatigue. But patients who are stable and have only minor side effects from the disease may be able to perform resistance training following an aerobic-training session. Modifications, such as training muscle groups unilaterally because of differences in strength between the limbs or because of range-of-motion limitation, may be required. These limitations need to be considered on a patient-by-patient basis. The clinical exercise professional can also use other resistance-training options such as band or tubing exercises, especially in subjects who are weak or confined to a wheelchair. Additionally, pool exercises should be considered. Muscle endurance training can be performed after strength training and should consist of one set to muscular fatigue. Intensity should be approximately 40% to 50% of the one-repetition maximum weight.

Range of Motion

Because patients with MS may have reduced range of motion of specific affected joints, a general flexibility program, designed to improve range of motion, is recommended. This should be performed 5 to 7 d per week and should follow a strength- or cardiovascular-training session. A static stretching routine focusing on all the major muscle groups and joints, with special emphasis on areas with spastic flare-ups, is recommended. Patients should be encouraged to stretch even when other exercise training is not performed on that day. Chapter 5 provides general information about range-of-motion training.

EXERCISE TRAINING

Differences in physiological variables in individuals with MS compared with non-MS subjects provide a basis for using exercise training in this population. These negatively affected variables include the following:

- Muscle strength (6,13,52,80)
- Muscle endurance (48,102)
- Motor unit firing rates (86)
- Muscle activation (22,86)
- Cardiorespiratory fitness (81)
- Respiratory muscle function (17,29,96)
- Muscle oxidative capacity (46)
- Walking speed (46)
- Habitual physical activity (67,70)

The following paragraphs provide a brief review of the effects of these physiological abnormalities and of the exercise-training studies performed to date. For more details refer to several recent reviews (12,38,120).

Cardiovascular Training

Peak oxygen consumption is reduced in patients with MS (89), as is their level of physical activity (70), suggesting that they may be at increased risk of diseases related to sedentary lifestyle. In contrast, patients with increased activity levels have smaller waist circumference, lower blood triglycerides, and lower blood glucose (108), indicating that increasing aerobic activity in these patients may have multiple benefits. Several studies (49,66,73,77,87,99,111,113) have evaluated the efficacy of aerobic exercise training in individuals with MS and report many positive benefits. These include improvements in $\dot{V}O_2$peak (77,87), decreased serum triglyceride levels (77), increased fatigue tolerance (73,111), improved health perception and quality of life (49,66,77,99), and increased walking speed and endurance (99,113). These changes suggest that muscle endurance training is beneficial for patients with MS.

No randomized controlled trials have assessed the effects of flexibility training in the MS patient population. Some treatment methods, however, have been shown to increase range of motion in these patients (20). Table 30.7 on p. 554 briefly reviews the anticipated responses to exercise training in patients with MS.

Muscle Fatigue

Skeletal muscle fatigue is more pronounced in MS than in non-MS subjects (102). Kent-Braun et al. (47) reported that the half-time of skeletal muscle phosphocreatine (PCr) resynthesis after exercise-induced depletion was slower in people with MS compared with non-MS controls. This impaired PCr resynthesis rate appears to be the result of a reduced concentration of oxidative enzymes in those with MS (46) and, thus, decreased skeletal muscle oxidative capacity. The skeletal muscle of patients with MS may also contain a greater proportion of type II and smaller proportion of type I muscle fibers (32,46). Greater reliance on these more "fatigable" type II muscle fibers could explain a portion of the muscle fatigue reported in MS patients. The reduced skeletal muscle oxidative capacity and fiber-type changes in patients with MS are thought to be consistent with their reduced physical activity levels (67,70). Therefore, exercise training may be a therapy to decrease fatigue in these patients by increasing muscle oxidative capacity and altering muscle fiber type. During voluntary exercise, however, skeletal muscle fatigue may result more from activation failure than from metabolic factors (48).

Muscle Weakness

Skeletal muscle fiber (32,46) and whole-muscle (46) cross-sectional area (CSA) are reduced in patients with MS in some but not all studies (13). Reduced fiber size may be because of the decreased physical activity (67,70) of these patients and may partially account for their reduced strength (80). Force, normalized to muscle fiber CSA, may also be reduced in MS patients (32). Thus, reductions in muscle size may not completely account for the reduced muscle strength in these patients. In healthy subjects, chronic resistance training increases muscle size and strength. In contrast, whole-muscle strength, but not size, is increased in patients with MS who complete 8 wk of resistance training (121). The lack of an increase in muscle mass could be because of the low intensity and volume of the training sessions. No other reports exist about the effects of resistance training on muscle size in individuals with MS.

Several additional studies have evaluated the effects of strength training in patients with MS (23,36,90,112, 121,123). Svensson et al. (112) reported increased muscle endurance after 4 to 8 wk of knee flexion training. Additionally, Gehlsen et al. (33) demonstrated improvement in strength, total work, and fatigue in both upper and lower extremities following a 10 wk aquatics exercise program. More recently, several studies using home-based or monitored resistance-training programs of 8 wk to 6 mo in duration report increased muscle strength (23,36,90,121,123). Other benefits include increased total walk time (89,123), decreased fatigue (36), and improved gait (36). The benefits are described in table 30.7.

CONCLUSION

It is best for patients with MS to begin an exercise routine as early in the disease process as possible. There is ample evidence that exercise, when performed properly, can limit the degree of physical disability experienced by an individual. While specific exercise-training recommendations do not yet exist for the MS population, it is prudent to maintain lower levels of exercise intensity and duration, especially in the initial stages of beginning an exercise routine and during periods of MS exacerbation. Those who tolerate exercise well may be able to increase their exercise volume up to normal, age-associated levels.

Table 30.7 Benefits Associated With Exercise Training in Individuals With MS

Cardiorespiratory endurance	Skeletal muscle strength	Skeletal muscle endurance	Flexibility	Body composition
Increased (77)	Increased (23,33,121)	Increased (112)	Likely increased to similar degree as healthy individuals	Benefits likely similar to healthy individuals

Case Study

Medical History

Mrs. NB, a 29 yr old Caucasian female, was diagnosed with MS 6 yr ago. At that time she had problems with ataxia and **diplopia**. She has had one to two exacerbations of MS per year since that time. An increase in disease stability was noted after she started recombinant interferon β-1b (Betaseron) 5 yr ago. She stopped the interferon therapy briefly during a pregnancy 4 yr ago, and after the pregnancy she had three exacerbations during the following year despite being back on Betaseron. With each exacerbation, her symptoms of ataxia, vertigo, and diplopia worsened. These symptoms involve primarily the left side of her body. These extremities also cramp on occasion. Overall, she functions reasonably well. She takes care of her 4 yr old daughter. She rests daily while her daughter goes to preschool. She had to quit working within a year after her diagnosis because of problems with ataxia and fatigue.

An MRI of her brain demonstrated multiple **white matter** lesions consistent with MS. A spinal fluid examination demonstrated elevated immunoglobulin G synthesis rate and oligoclonal bands consistent with the diagnosis of MS.

She is taking no other medication on a regular basis and has not been involved in a regular exercise program. She occasionally gets a urinary tract or upper-respiratory infection that necessitates antibiotics. Cramping in her extremities has not been bad enough to warrant a muscle relaxant on a regular basis. She notes that when she walks more than a couple of blocks, she feels weakness in her left leg and often needs to rest for a few minutes before walking farther. She also thinks that at times her left leg may give out.

Examination shows mild incoordination and diffuse hyperreflexia, which is more pronounced in the left side extremities. She also has a few beats of **nystagmus**.

Diagnosis

Her diagnosis is relapsing-remitting multiple sclerosis, grade 4.0 on the Kurtzke expanded disability status scale (tables 30.3 and 30.4). Grade 4.0 means that the individual is fully ambulatory without aid, is self-sufficient, is up and about some 12 h a day despite relatively severe disability, is able to walk at least 500 m without aid or rest, and that the disease affects one or more functional systems.

Exercise Test Results

An exercise test was performed to assess functional ability and to rule out cardiac origin of her arm weakness while walking. The patient performed a bicycle protocol of 3 min stages at 25, 50, and 75 W. No sign of electrocardiographic abnormalities was present. Exercise was discontinued because of volitional fatigue.

Resting values
Heart rate = 92 beats $\cdot$ min^{-1}
Blood pressure = 100/60 mmHg
Electrocardiogram: normal sinus rhythm and within normal limits

Peak exercise values
Heart rate = 188 beats $\cdot$ min^{-1}
Blood pressure = 150/50 mmHg
Rating of perceived exertion = 19 (6-20 scale)
Peak $\dot{V}O_2$ = 17.5 ml $\cdot$ kg$^{-1} \cdot$ min^{-1}
Peak METs = 5.0

Exercise Prescription

Because the patient is just beginning an exercise program and because of the ataxia and fatigue that she experiences, she should start out at low intensity and duration for both aerobic and strength training. Limb strength is reduced on the left side because of increased MS-related symptoms.

Cardiovascular training: She should begin aerobic training on a cycle ergometer at 60% of her heart rate reserve (about 150 beats $\cdot$ min^{-1}) for 30 min three times per week. The intensity should be increased gradually, as tolerated, over a few weeks to a heart rate that is 85% of heart rate reserve (about 172 beats $\cdot$ min^{-1}).

Strength training: She should perform resistance training two times per week at 40% of one-repetition maximum (1RM) on her left side and 60% of 1RM on her right side. This training will require using hand weights, dumbbells, or resistance-training machines. Initially, she should perform only one set of eight repetitions, focusing on the major muscle groups. Exercises such as leg extensions, leg curls, dumbbell chest press, and seated rowing are appropriate. The volume should be increased gradually over several weeks to two sets of eight repetitions and expand to other skeletal muscle groups.

Muscular endurance training: An endurance-training program for skeletal muscles can be administered on the days that resistance exercise is performed. A suggestion is two times per week, at 40% to 50% of 1RM, performing one set to muscular failure for the gastrocnemius, quadriceps, hamstring, biceps and triceps brachii, chest, and abdominal muscles.

Flexibility training: She should perform stretching exercise before or after each exercise session, focusing on the ankles, knees, hips, lower back, shoulders, wrists, and neck. Static stretches should be held for 10 to 20 s and performed at least twice for each muscle group.

(continued)

Discussion Questions

1. What can you say about the ability of the graded exercise test to detect myocardial ischemia in this patient versus others who might be more limited by MS?

2. How should the prescribed exercise be modified for a person with a greater level of disability?

3. How should the prescribed exercise be modified for a person with less disability?

4. What mode of aerobic and resistance exercise should this individual use?

5. What is an advantage of using non-weight-bearing exercise in a person with MS with a high degree of disability?

6. What physiological variables need to be monitored closely in this individual? Why?

7. Do you think people with MS should exercise during an exacerbation of disease activity? Why?

8. What are the potential benefits in terms of functional capacity of resistance training for people with MS?

31

Cerebral Palsy

Amy E. Rauworth, MS
James H. Rimmer, PhD

Cerebral palsy is defined as "a generic term for various types of non-progressive motor dysfunction present at birth or beginning in early childhood. Risk factors are both hereditary and acquired; depending on cause, classified as intrauterine, natal, and early postnatal; motor disturbances include **diplegia, hemiplegia, quadriplegia, choreo-athetosis,** and ataxia"(82).

Dr. William Little, an English surgeon, first documented the symptoms of what is now known as spastic diplegia in the early 1860s. He described infants that had difficulty grasping, walking, or crawling. Other symptoms included greater involvement of the legs than arms and no progression of the initial degree of involvement. For many years, this form of cerebral palsy was referred to as Little's disease and was linked to difficult births associated with asphyxiation. Later, Dr. Sigmund Freud suggested that the etiology of this disease might stem from problems with the development process of the brain during gestation (32). Despite this conclusion, complication during birth was the primary identifiable cause of cerebral palsy for many years.

In 1958 a definition was presented at the First International Study Group on Child Neurology and Cerebral Palsy (50). The study group defined cerebral palsy as "a persisting qualitative motor disorder due to non-progressive interference with development of the brain occurring before the growth of the central nervous system is complete." This description led to a consensus definition: "Cerebral palsy is a persistent but not changing disorder of movement and posture, appearing in the early stages of life and due to a non-progressive disorder of the brain, the result of interference during its development" (50). Bax annotated this definition in 1964 to read "a disorder of movement and posture due to a defect or lesion of the immature brain" (14).

More recently, advances in the understanding of the developmental process of infants with early brain damage led to the definition of cerebral palsy as "an umbrella term covering a group of non-progressive, but often changing motor impairment syndromes secondary to lesions or anomalies of the brain arising in the early stages of its development" (57). The heterogeneous nature of cerebral palsy evokes controversy even today in defining the condition. A proposed definition and classification was presented in 2005 (13): "Cerebral palsy describes a group of developmental disorders of movement and posture, causing activity restriction or disability that is attributed to disturbances occurring in the fetal or infant brain. The motor impairment may be accompanied by a seizure disorder and by impairment of sensation, cognition, communication and/or behaviour." This definition was established to broaden the classic definition and to include the common secondary conditions associated with cerebral palsy.

SCOPE

In the United States, cerebral palsy occurs in 1.5 to 2.5 per 1,000 live births (3). There are 500,000 to 764,000 persons in the United States with cerebral palsy. The incidence of cerebral palsy has remained stable or may have increased slightly because of the survival of premature infants with extremely low birth weight (64). The prevalence of cerebral palsy worldwide varies from 1.6 per 1,000 children under the age of 7 in China (46) to 2.0 to 2.5 per 1,000 live births in Australia (67). Cerebral palsy and low birth weight are also more prevalent in populations of lower socioeconomic status (60).

According to the Centers for Disease Control and Prevention, the lifetime cost of persons born in the year 2000 with cerebral palsy is estimated to reach $11.5 billion (20). This figure emphasizes the importance of primary and secondary prevention in managing cerebral palsy.

PATHOPHYSIOLOGY

The most common risk factors associated with the development of cerebral palsy are premature birth and low birth weight (25). Twin and higher-order multiple-birth infants are at increased risk for developing cerebral palsy (36,58). Although twinning, prematurity, and low birth weight are interrelated, these factors have been found to be independent risk factors of cerebral palsy (97). Stanely et al. (81) have suggested a "causal pathways" concept of interdependent events such as twinning, prematurity, and low birth weight that could result in cerebral palsy. Maternal or fetal abnormalities in coagulation such as thrombophilia and the presence of factor V Leiden mutation increase the risk for complications during pregnancy and birth that can lead to cerebral palsy (10,44,86). Maternal infection is also highly correlated with the development of cerebral palsy and has been directly related to brain lesions in low birth weight infants (98). Maternal fever

during term labor has been shown to increase the risk of cerebral palsy ninefold.

Other prenatal risk factors associated with increased risk of cerebral palsy include maternal abuse of drugs and cigarettes, congenital brain malformations, and genetic conditions (25). Complications during birth that result in anoxia or hemorrhage with direct brain damage and kernicterus result in increased risk for cerebral palsy. Estimates are that approximately 75% of all cases of cerebral palsy are a result of prenatal events (67). Postnatal events that occur before the age of 2 yr such as viral or bacterial meningitis; traumatic head injury from vehicular accidents or abuse; anoxia as a result of near drowning, cerebral vascular accidents, tumors, or surgery; and toxins that cause heavy metal encephalopathy are risk factors for cerebral palsy (25). In developing countries, common causes of cerebral palsy include septicemia, meningitis, and other conditions such as malaria (67).

MEDICAL AND CLINICAL CONSIDERATIONS

Patients with cerebral palsy face a host of medically related issues. The exercise physiologist should have a general familiarization with these issues to improve basic assessment skills when working with these patients. This section reviews some of the basic medical and clinical issues for patients with cerebral palsy.

Signs and Symptoms

The hallmark sign of the group of disorders identified as cerebral palsy is an impairment of voluntary motor

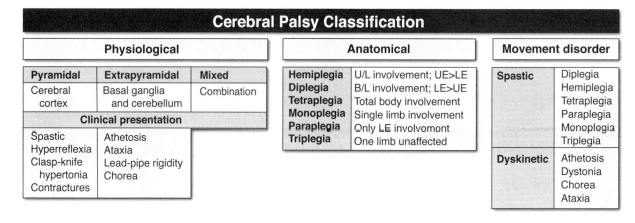

Figure 31.1 Cerebral palsy classification. U = upper; L = lower; E = extremity.

Adapted from C. Andersson and E. Mattsson, 2001, "Adults with cerebral palsy: A survey describing problems, needs, and resources, with special emphasis on locomotion," *Developmental Medicine and Child Neurology* 43: 76-82.

control. Cerebral palsy frequently involves one or more limbs and the trunk musculature. It is classified by the type of motor abnormality (spasticity, athetosis, ataxia), the area of the body that is affected (**monoplegia**, diplegia, hemiplegia, **triplegia, tetraplegia**), and the location of the lesion in the brain (**pyramidal, extrapyramidal,** mixed) (25). Figure 31.1 provides a detailed description of the classification of cerebral palsy.

Classification and Clinical Manifestation

Classification by movement disorder is often difficult because of the appearance of characteristics that represent multiple movement disorders. Abnormalities in movement include spasticity, athetosis, chorea, dystonia, and ataxia (see table 31.1 for clinical symptoms of muscle impairment associated with spastic paresis). The clinical manifestations of cerebral palsy can be related to the location of the lesions within the brain. Injury to the cortical system (pyramidal type) results in spasticity. Spasticity is present in approximately 75% of people with cerebral palsy (56). Injuries of the extrapyramidal type refer to lesions within the **basal ganglia** and **cerebellum**. This form of injury produces **dyskinetic** cerebral palsy, which includes movement disorders such as athetosis, **chorea**, ataxia, and **rigidity**. Having clinical

Table 31.1 Summary of Clinical Symptoms of Impaired Muscle Function in Spastic Paresis

Impairment of muscle activation	
Deficit symptoms (signs of reduction or loss of normal voluntary muscle activation)	
Paresis	Inadequate force
Loss of selective motor control	Impaired ability to activate and control selective or isolated movements across specific joints
Loss of dexterity of movement	Impaired ability to coordinate temporal and spatial activation of many muscles
Enhanced fatigability	Inadequate sustained force
Excess symptoms (signs of abnormal involuntary muscle activation)	
PASSIVE MOVEMENT	
Hypertonia (TSR)	Non-velocity-dependent resistance to passive movement experienced by the examiner as increased muscle tone, caused by continuous muscle activation (tonic stretch reflex activity)
Spasticity	Velocity-dependent resistance to passive movement experienced by the examiner as increased muscle tone caused by an increase in stretch reflex activity
Clonus	Involuntary rhythmic muscle contraction
ACTIVE MOVEMENT	
Mirror movements	Involuntary, simultaneous, contralateral movement of a muscle caused by voluntary movement of the same muscle on the ipsilateral side of the body
Associated abnormal postures	Involuntary abnormal muscle activity related to posture or the performance of any task
Cocontraction	Simultaneous involuntary contraction of the antagonist at voluntary contraction of the agonist
Abnormal reflexes	
Hyperreflexia of tendon jerks	Raised reaction to tendon tap, extension of the reflexogenic zone, spread of reflexes
Abnormal nociceptive flexion reflexes	Abnormal pathological reflexes to painful stimuli (i.e., Babinski reflex)
Abnormal musculocutaneous reflexes	Abnormal pathological reflexes to cutaneous stimuli (i.e., adductor reflex)
Changes in biomechanical properties of muscle and other connective tissues	
Contractures	Fixed shortening of muscle and tendon, resulting in a reduced range of motion
Hypertonia	Non-velocity-dependent resistance to passive stretch experienced by examiner as increased muscle tone, caused by biomechanical changes within muscle itself

Reprinted, with permission, from V.A. Scholtes et al., 2006, "Clinical assessment of spasticity in children with cerebral palsy: A critical review of available instruments," *Developmental Medicine and Child Neurology* 48: 64-73.

signs and symptoms of mixed forms of cerebral palsy is also common.

Cerebral palsy can also be classified by systems developed specifically for athletes with cerebral palsy. Disabled Sports and the Cerebral Palsy–International Sports and Recreation Association (CP–ISRA) created eight classes to determine movement quality of people with cerebral palsy (2). Class 1 delineates severe involvement, whereas class 8 describes minimal involvement. Classes 1 through 4 are identified by those who use wheelchairs for mobility, and classes 5 through 8 depict individuals who are ambulatory. See table 31.2 for a complete description of all classes. Additional classification systems have been developed, such as the Quantitative Sports and Functional Classification (QSFC) (40). This classification system determines potential functional ability and allows evaluation of improvement, which may be useful to the clinical exercise physiologist. This assessment consists of 26 items and involves skills such as transfers, touching behind the head, combination movement, kicking or thrusting with the foot, control of the wheelchair foot rest, and leg abduction.

Associated Conditions

Besides the neuromuscular impairments present in cerebral palsy, associated conditions such as impaired cognition, sensory deficits, seizure disorders, feeding problems, behavioral dysfunction, and emotional problems have been identified (28). (See table 31.3 for a list of associated conditions and general exercise guidelines.) The preva-

lence of intellectual disabilities varies with the type and severity of cerebral palsy and increases with the presence of **epilepsy** (60). Seizures occur in approximately half of all children with cerebral palsy. When seizures occur without a direct trigger, the condition is labeled epilepsy. Visual disturbances frequently occur in cerebral palsy. The most common type is **strabismus**, a condition in which a discrepancy is present between the right and left eye muscles. Individuals with hemiparesis often have an impairment or loss of sight in the normal field of vision of one eye, referred to as **hemianopia**. This condition can also occur in both eyes and affect the same area of the visual field. Approximately 90% of people with hemiplegic cerebral palsy display significant bilateral sensory deficits (60). Speech impairments such as **dysarthria** and **aphasia** occur most commonly in individuals with dyskinetic and tetraplegic cerebral palsy (30). Dental problems such as malocclusion are prevalent in children with cerebral palsy, resulting in a significant increased over-jet (an acute angle of the mandible) (31). **Scoliosis** is commonly seen in preteens, teens, and young adults. Scoliosis occurs most frequently in people with spastic cerebral palsy (73). Surgical management is sometimes necessary to prevent further progression of the spinal curvature. To view detailed information on the prevalence of **associated conditions**, see Odding et al. (60).

Secondary Conditions

The clinical symptoms and **secondary conditions** that result from cerebral palsy lead to a functional decline and

Table 31.2 Cerebral Palsy International Sport and Recreation Association Classes (CP-ISRA)

Classification	Description
Wheelchair users	
CP1	Athletes use power wheelchairs or assistance for mobility. They are unable to propel a wheelchair.
CP2	Athletes are able to propel a wheelchair but have very poor useful strength in their arms, legs, and trunk.
CP3	Athletes show fair trunk movement when pushing wheelchairs, but forward trunk movement is limited during forceful pushing.
CP4	Athletes have minimal limitations or control problems in their arms and trunk while pushing wheelchairs.
Self-ambulatory	
CP5	Athletes may need assistive devices for walking but not in standing or throwing. Athletes may have sufficient function to run but demonstrate poor balance.
CP6	Athletes do not have the capacity to remain still, and they show involuntary movements with all four limbs affected. They usually walk without assistive devices.
CP7	Athletes have movement and coordination problems on one-half of the body. They have good ability in the dominant side of the body.
CP8	Athletes have minimal involvement in one limb only.

Reprinted, by permission, from Cerebral Palsy International Sports and Recreation Association (CP-ISRA). *Classification and sports rule manual*, 9th ed. By permission of Taylor & Francis Ltd. Http://tandf.co.uk/journals.

Table 31.3 Associated Conditions Observed in Persons With Cerebral Palsy (33,45)

Associated conditions	Causative and related factors	Exercise guidelines
Bowel and bladder problems	• Incontinence • Neurogenic bladder	• Empty bladder before exercise session
Cognitive and behavioral problems	• Intellectual disabilities • About one-third of children who have cerebral palsy have mild intellectual disabilities, one-third have moderate or severe impairments, and the remaining third are intellectually normal. Mental impairment is even more common among children with spastic quadriplegia • Learning disabilities • ADD (attention deficit disorder) • ADHD (attention deficit hyperactivity disorder) • Sensory integration disorder	• Ask questions that require brief answers • Allow time for accommodation of a new task; repetition is important • Use precise language and simple words • Limit the number of instructions given at one time
Musculoskeletal problems	• Bone loss • Fractures • Hip dislocation • Hypertonia • Leg length discrepancies • Pelvic obliquity • Spasticity	• Incorporate weight-bearing exercise when appropriate • Assess balance to choose appropriate exercise modality to prevent falls • Be cautious when stretching or strengthening the hip musculature if history of dislocations is present • Decrease cadence of exercise activity to avoid the increase of spasticity
Oral motor dysfunction	• Dysphagia • Gastroesophageal reflux disease, or GERD • Oral hypersensitivity • Dysphagia, difficulty in swallowing • Drooling • Dysarthia • Poor motor control of the mouth • Feeding problems	• Provide a towel for possible drooling • Make adaptations to the mouthpiece during exercise testing or use a face mask
Respiratory problems	• Blocked or obstructed airways • Airway inflammation • Increased mucus production • Chemical or bacterial pneumonia • Lung damage • Diminished ability to clear secretions • Weakened pulmonary defenses	• Use breathing exercises (pursed-lip breathing, diaphragmatic breathing) to increase or maintain lung function
Seizures or epilepsy	A seizure occurs when the normal, orderly pattern of electrical activity in the brain is disrupted by uncontrolled bursts of electricity. When seizures recur without a direct trigger, such as fever, the condition is called epilepsy	• Consider the effects of medication on exercise and patient's endurance level

(continued)

Table 31.3 *(continued)*

Associated conditions	Causative and related factors	Exercise guidelines
Sensory and perception impairments	• Impaired perception to pain and pressure • Sensory integration dysfunction • Stereognosia • Visuo-auditory and visuo-spatial processing impairments	• Break down instructions and cues
Vision deficits	• Strabismus • Hemianopia	• Speak to someone in the appropriate field of vision • Provide an obstacle-free environment to prevent falls

are often exacerbated by the aging process (9,33). (See table 31.4 for a complete list of secondary conditions.) Secondary conditions have been identified as "changeable aspects of health status for people with disabilities" (51). They often begin in late childhood and progress throughout adulthood (62). The decline in health over time can be related to the effects of clinical signs and symptoms such as abnormal movement patterns that stress the structural stability of the orthopedic and musculoskeletal systems; **contractures** resulting in decreased range of motion and changes in posture such as spinal kyphosis; prolonged use of medication with undesired side effects such as osteoporosis and fractures; and poor nutrition as a result of gastrointestinal complications (9,33,41,55).

Deconditioning is an important secondary condition observed in persons with cerebral palsy. Ando and Ueda (9,33,41,55) reported that 35% of adults with cerebral palsy have reduced ability to perform activities of daily living over time. The time sequence for this decline could be as short as a period of 5 yr. Strauss, et al. (83) identified a decline in ambulatory function after the age of 60 with a subsequent increase in mortality. Secondary conditions such as poor joint alignment and overuse syndromes may also contribute to the loss of independent walking (55). Fatigue as a limiting factor for ambulation has been reported in several studies (37,95). Murphy, et al. reported that approximately 75% of their cohort discontinued ambulation by the age of 25 because of fatigue and inefficient movement (55). This cohort also reported pain in weight-bearing joints that subsequently decreased ambulation around age 45. Pain is commonly reported among people with cerebral palsy. Turk et al. reported that 84% of the adult women with cerebral palsy identified pain as a deterrent to participation in certain activities of daily living (91). The occurrence of pain is most commonly identified in the lower extremities, back, shoulders, and neck (55,70,91). In relation to pain in these areas, limited range of motion is commonly reported in the ankle, hip, and shoulder joints (75). Sandstrom et al.

Table 31.4 Secondary Conditions in Cerebral Palsy (33,45,92)

Secondary condition	Causative or related factors	Exercise prescription
Cardiovascular and cardiopulmonary conditions	• Inactivity • Weakness • Postural instability	Aerobic exercises: at least 3 d · wk⁻¹; start with low duration, gradually increasing to 20–60 min as tolerated. Examples: ergometers, chair aerobics, dancing
Communication disorders	• Learning disability • Perception issues • Speech difficulties (dysphasia) • Intellectual disability	Ask yes and no questions, or provide a pad and paper to assist in the communication process
Depression	• Communication difficulties • Low self-esteem • Social isolation	• Provide motivational coaching techniques • Use short-term goals to illustrate progress • Provide opportunity for group activity

Secondary condition	Causative or related factors	Exercise prescription
Fatigue and decreased endurance	• Overuse syndromes • Nerve entrapments	Begin with intermittent exercise sessions with frequent rest intervals (three 10 min sessions) and slowly progress exercise duration Increase rest intervals between sets and between exercises to avoid muscle fatigue (2–3 min after each set)
Mobility impairments	• Balance and coordination problems • Contractures • Hypertonicity	Incorporate balance, stability, and fall prevention exercises
	• Muscle weakness and atrophy	High volume (resistance × reps) and low intensity. Do large muscles before smaller muscles. Use multi- and single-joint exercises. Concentrate on eccentric muscle actions. Move muscle in a slow, controlled manner Start slowly with resistance-training program. Examples: chair aerobics, elastic resistance bands
	Contractures and musculoskeletal deformities (scoliosis, hip dislocation, pelvic obliquity, foot and ankle deformities such as clubfoot and flat foot)	Improve flexibility and ROM Modified interval training (changes in intensity as well as direction) Static technique for stretching, a minimum of $2-3 \text{ d} \cdot \text{wk}^{-1}$ to a position of mild discomfort; hold for 15–30 s; repeat three or four times for each stretch Use PNF and active assistive stretching to help achieve a full range of motion and help relax muscles in contraction. Examples: yoga, tai chi, ergometers
	Decreased bone density and strength	Use weight-bearing exercise Example: dancing
Obesity	• Food used as a reward • Decreased caloric expenditure • May lead to joint pain	Incorporate aerobic exercises Encourage client to increase general physical activity levels beyond structured exercise Pedometer may be useful to motivate those who ambulate With appropriate adaptations, pedometer can be utilized by wheelchair users
Pain	• Arthritis • Contractures • Musculoskeletal deformities • Nerve entrapments • Overuse syndromes • Osteoarthritis • Osteoporosis • Postural instability • Pressure ulcers	Incorporate non-weight-bearing activities for training (swimming, hand cycling) Use elastic bands, Swiss balls, and hydrotherapy Record pain scale in SOAP notes before and after exercise session

(continued)

Table 31.4   *(continued)*

Secondary condition	Causative or related factors	Exercise prescription
Poor dental health and hygiene	• Cavities • Gingival and periodontal disease • Occlusal disorders • Hyperactive gag reflex • Hypoplasia (incomplete calcification) of the primary teeth, reflecting an early disturbance in the development of enamel and dentin • Temporomadibular contractures may be present • Can cause hypoxemia	Make adaptations to mouth piece for $\dot{V}O_2$ testing; make sure fit is secure and seal is tight Use facemask for $\dot{V}O_2$ testing
Pressure sores or poor skin integrity	• Due to impaired sensory perception • Chronic infections • Skin disorders	Take precautions with skincare, positioning, and duration of activity If in a wheelchair, wheelchair push-ups for $20–30 \ min \cdot d^{-1}$ in addition to other forms of resistance training to avoid pressure sores Check placement of orthotics before and after exercise to ensure skin integrity

(2004) reported a high incidence of pain in the lower extremities of persons who can walk with or without assistance. The life span of people with cerebral palsy without significant comorbidities approaches that of the general population (83).

History and Physical Exam

The clinical exercise physiologist must consider all associated and secondary conditions related to cerebral palsy when developing an exercise program. Because of the heterogeneous nature of cerebral palsy, the exercise program should be tailored to the needs of the individual. When necessary, the clinical exercise physiologist should consult with a team of rehabilitative specialists, which may include physical therapists, occupational therapists, speech therapists, neurologists, or physiatrists, to develop an appropriate program. Regular reporting of progress to the team of specialists who work with the client is important. Accurate and detailed SOAP (subjective, objective, assessment, plan) notes are critical to facilitate quality communication among all healthcare providers who serve the client. The SOAP notes will also assist the client in tracking the occurrence of secondary conditions and may provide motivation and self-empowerment for the person's overall health pro-

motion program. The detailed SOAP notes will allow the clinical exercise physiologist to determine the most effective exercise progression. Pain levels, fatigue, and any medication changes should be noted before execution of the exercise session. See figure 31.2 for a sample SOAP note format.

Diagnostic Testing

The diagnosis of cerebral palsy is often a result of the failure of a child to reach developmental milestones, the presence of abnormal **muscle tone**, and persistence of primitive reflexes or qualitative differences in movement patterns (72). A physical examination, complete history, and specialized tests are required to diagnose cerebral palsy accurately (43). Tests such as the Movement Assessment of Infants (MAI), Bayley Motor Scale, and the Test of Infant Motor Performance (TIMP) are used to determine the diagnosis (19). Specialized technology such as cranial ultransonography, MRI, and CT can assist in determining the extent of the insult to the central nervous system (43). Experimental technology such as functional MRI and transcranial magnetic stimulation are currently used only for research purposes, but these modalities show promise in the diagnosis of cerebral palsy. According to Samson et al., an accurate diagnosis

---

PARTICIPANT S.O.A.P. NOTE

Participant name: _____ ID #: _____

Interviewer: _____ Date: _____

	Preexercise	Postexercise
Heart rate (BPM)		
Blood pressure (mmHg)		
Blood glucose if diabetic (mg/dl)		
Activities pain index if indicated (0-10)		

QUESTIONS (Yes/No):

_____ Any illnesses or changes in health? _____

_____ Eat breakfast/lunch today?

_____ Take medications today?

_____ Any changes or adjustments to medications? _____

Subjective statements made by participant: _____

_____

Objective modalities or activities: _____

_____

Assessment of participant progress and compliance: _____

_____

Plan for next visit: _____

Additional comments, including complications or significant events: _____

_____

_____

Figure 31.2 Sample SOAP note format.

can be made at 6 mo of age except in the mildest forms of cerebral palsy (74). The suggestion has been made that because of the appearance of the progression of signs and symptoms in the early stages of an infant's life, the extent of the condition may not be identified until 2 or 3 yr of age or later (24).

The exercise professional should understand the functional levels of the client and any other factors that may influence his or her health behavior, such as measures of quality of life. See "Functional Instruments and Health-Related Quality-of-Life Measures" on page 566 for a list of tests to determine those factors.

When working with an adult with cerebral palsy, it is important to have a comprehensive understanding of the factors that influence exercise testing and prescription, such as gait, **spasticity,** and balance.

Gait and Posture Assessment

The ability to ambulate and the extent to which spasticity, contractures, and deformities affect gait vary among individuals with cerebral palsy. During ambulation and activity, abnormalities in movement can be clearly identified. If the clinical exercise physiologist is knowledgeable of the types of deformities, contractures, and gait deviations associated with cerebral palsy, a precise exercise prescription can be developed to accommodate these issues and assist in the prevention of secondary conditions.

Functional Instruments and Health-Related Quality-of-Life Measures

Global function

Pediatric evaluation of disability inventory

Functional independence measure for children

Arm functional assessment tests and scales

Assessment of quality of movement for unilateral upper-limb function (Melbourne)

Quality of upper-extremity skills

House classification

Leg functional assessment instruments

Gross motor performance measure

Gross motor functional measure

Gross motor functional classification system

Physician rating scale (lower-extremity rating scale)

Spasticity assessment instruments

Ashworth scale

Modified Ashworth scale

Tardieu scale

Burke-Rahn-Marsden scale of dystonia

Barry-Albright dystonia scale

Health-related quality of life in cerebral palsy: generic instruments

Child health questionnaire

Pediatric musculoskeletal-functional health questionnaire

Peds QL 4.0

Health-related quality of life in cerebral palsy: disease-specific instruments

Caregiver questionnaire

Utility measures

Health utilities index, mark 3

Adapted from Koman, L.A., B.P. Smith, and J.S. Shilt, "Cerebral Palsy." *Lancet* 363: 1619–1631.

Practical Application 31.1

Use of the International Classification of Functioning, Disability, and Health (ICF) to Prepare Individualized Exercise Prescriptions for Persons With CP

The International Classification of Functioning, Disability, and Health (ICF) provides a comprehensive framework to assess the relationship between an individual's function, activities, and participation while also considering environmental and personal factors that influence the individual's overall health. According to the World Health Organization, the overall aim of the ICF classification is to provide a unified and standard language and framework for the description of health and health-related states. The classification defines components of health and some health-related components of well-being (such as education and labor). The domains contained in the ICF are referred to as health domains and health-related domains. These domains are described from the perspective of the body, the individual, and society in two basic categories: (1) body functions and structures (system level); and (2) activities and participation (person level and person-environment interaction). The ICF can be used as a clinical tool in exercise physiology to conduct a needs assessment or as an outcome evaluation. It allows the clinical exercise physiologist to identify the barriers and facilitators that affect the health of the client with cerebral palsy (1) (see figure 31.3) (62). For a detailed description of the application of the ICF model to cardiovascular and resistance training, see figure 31.4.

ICF Structure

Body functions are the physiological functions of body systems (including psychological functions).

Body structures are anatomical parts of the body such as organs, limbs and their components.

Impairments are problems in body function or structure such as a significant deviation or loss.

Activity is the execution of a task or action by an individual.

Participation is involvement in a life situation.

Activity limitations are difficulties an individual may have in executing activities.

Participation restrictions are problems an individual may experience in involvement in life situations.

Environmental factors make up the physical, social and attitudinal environment in which people live and conduct their lives.

Components

Part I. Functioning and Disability

   1. Body functions and structures

   2. Activities and participation

Part II. Contextual Factors

   1. Environmental factors

   2. Personal factors*

*Not classified in ICF, but included here to show contribution and impact

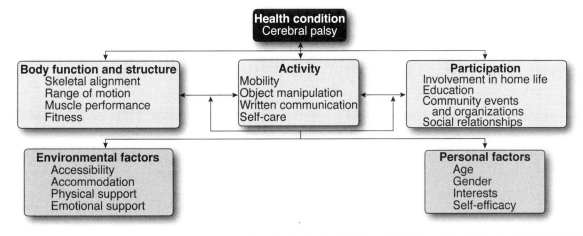

Figure 31.3   Relationships among components of health and contextual factors in children with CP. The conceptual framework is the ICF.

Reprinted from *Seminars in Pediatric Neurology,* Vol. II, R.J. Palisano, L.M. Snider and M.N. Orlin, "Recent advances in physical and occupational therapy for children with cerebral palsy from seminars in pediatric neurology," pg. 67, copyright 2004, with permission from Elsevier.

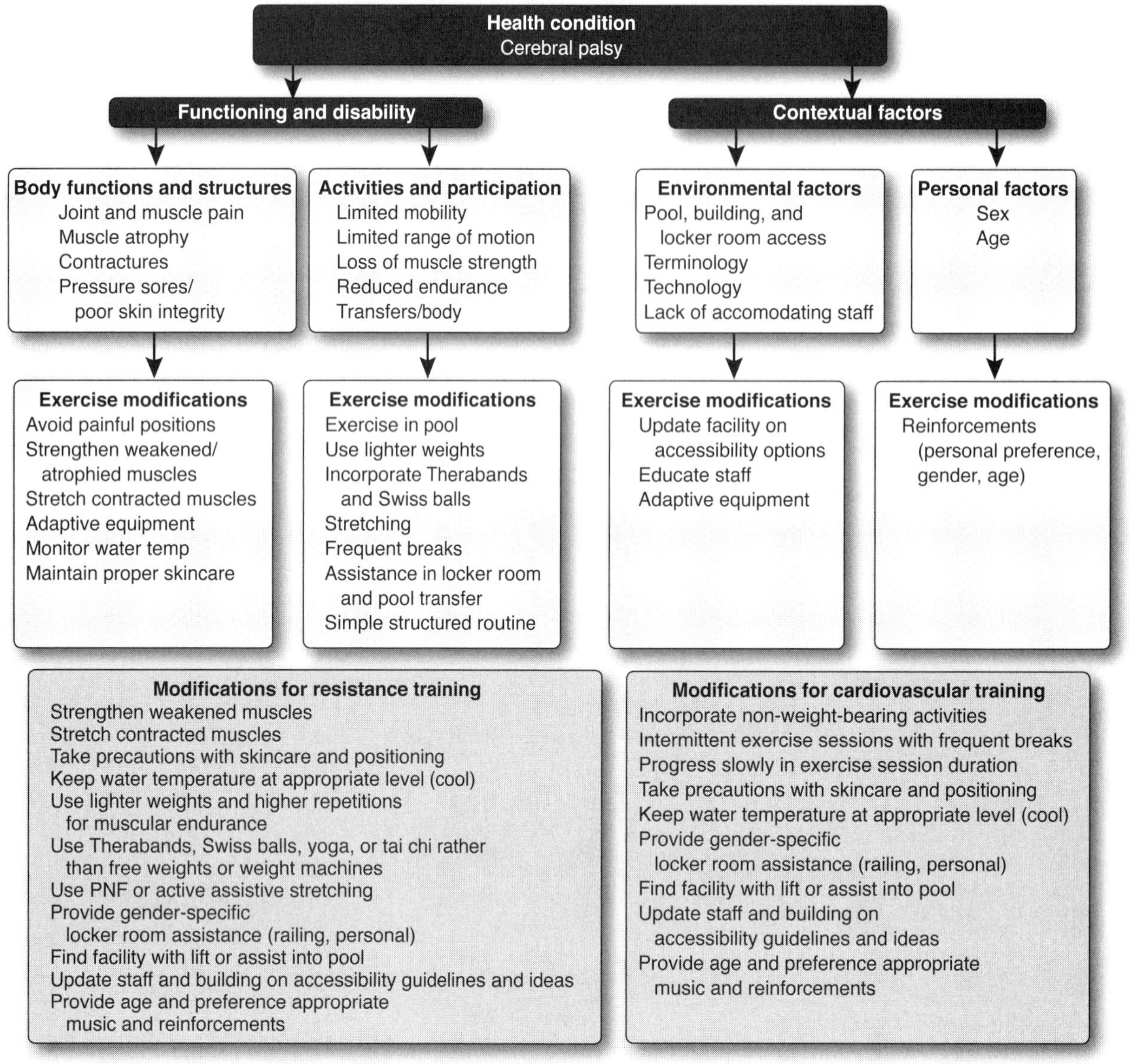
Health condition
Cerebral palsy
Functioning and disability
Contextual factors
Body functions and structures
Joint and muscle pain
Muscle atrophy
Contractures
Pressure sores/ poor skin integrity
Activities and participation
Limited mobility
Limited range of motion
Loss of muscle strength
Reduced endurance
Transfers/body
Environmental factors
Pool, building, and locker room access
Terminology
Technology
Lack of accomodating staff
Personal factors
Sex
Age
Exercise modifications
Avoid painful positions
Strengthen weakened/ atrophied muscles
Stretch contracted muscles
Adaptive equipment
Monitor water temp
Maintain proper skincare
Exercise modifications
Exercise in pool
Use lighter weights
Incorporate Therabands and Swiss balls
Stretching
Frequent breaks
Assistance in locker room and pool transfer
Simple structured routine
Exercise modifications
Update facility on accessibility options
Educate staff
Adaptive equipment
Exercise modifications
Reinforcements (personal preference, gender, age)
Modifications for resistance training
Strengthen weakened muscles
Stretch contracted muscles
Take precautions with skincare and positioning
Keep water temperature at appropriate level (cool)
Use lighter weights and higher repetitions for muscular endurance
Use Therabands, Swiss balls, yoga, or tai chi rather than free weights or weight machines
Use PNF or active assistive stretching
Provide gender-specific locker room assistance (railing, personal)
Find facility with lift or assist into pool
Update staff and building on accessibility guidelines and ideas
Provide age and preference appropriate music and reinforcements
Modifications for cardiovascular training
Incorporate non-weight-bearing activities
Intermittent exercise sessions with frequent breaks
Progress slowly in exercise session duration
Take precautions with skincare and positioning
Keep water temperature at appropriate level (cool)
Provide gender-specific locker room assistance (railing, personal)
Find facility with lift or assist into pool
Update staff and building on accessibility guidelines and ideas
Provide age and preference appropriate music and reinforcements

Figure 31.4 ICF model and cardiovascular and resistance training for cerebral palsy.

Practical Application 31.2

Client–Clinician Interaction

June Kailes, author of *Health, Wellness and Aging With Disability*, is a well-known national disability policy consultant. She is in her 50s and was diagnosed with ataxic cerebral palsy. June uses a scooter to ambulate and to manage musculoskeletal pain. Her job requires long hours on the computer, on the phone, on airplanes, and in meetings. She makes physical activity a priority because her job and lifestyle are sedentary. She works out regularly and can provide insight to the clinical exercise physiologist from both a personal and a professional context.

June's primary exercise goals include (1) pain reduction, (2) prevention of secondary conditions, (3) maintenance of function, and (4) weight control. June works with a personal trainer once a week on average and exercises independently two to four times per week. June's exercise program includes walking on the treadmill for 40 min followed by a comprehensive set of flexibility exercises. June needs a treadmill that can operate at 1.3 mph (2.1 kph). She uses a

Swiss ball to perform strength-training exercises that also challenge her balance. Using the treadmill for support, she performs standing exercises to improve her balance. In the evenings, June does floor exercises that focus on strength and flexibility for her back and core muscles.

When asked what motivates her most to continue her exercise program, June suggested that concerns about losing function and preventing increased pain keeps her exercise regimen as a high priority in her lifestyle. June said, "In the past we didn't age, we just died! We are the first generation to live this long, so the question is not, 'Will we live?' But 'How well will we live?'" She also stated, "Many of us will probably live longer than we think, so we have to think about what we can influence or change in terms of the quality of our years as we age." June advises the clinical exercise physiologist not to minimize the symptoms that she is experiencing and simply relate them to the aging process. "What's coming into sharp focus for many of us is that the changes brought about by regular aging (yet to be clearly defined) can play havoc with a person's ability to function." Like many of her fellow baby boomers, she states she expects more, not less, with age! When asked what she enjoys the most about exercise, June stated, "When it's over . . . a sense of accomplishment!"

She also encourages the exercise physiologist to be specific about the type and amount of exercise recommended. She believes that the more she knows about why she is doing an exercise and what benefits it will bring, the more likely she will be to include it in her activities. She advises the exercise physiologist to speak in nonclinical terms, to be creative in exercise design, to ask for the patient's suggestions or thoughts, and to challenge people with cerebral palsy. She stresses the importance of understanding the secondary and associated conditions of cerebral palsy and the importance of tailoring the program to her wants and needs. June often travels and sometimes encounters barriers such as the lack of accessible fitness centers in hotels or equipment that is not accommodating. When fitness centers lack accessible options, June walks the hotel halls and does flexibility and strength exercises in her room. June expects that exercise will always be part of her life. She sees the benefits and believes that the investment of time is more important for her than for people without disabilities.

Commonly observed postures include (1) shoulder internal rotation, elbow flexion, forearm pronation, wrist flexion, finger flexion, and thumb-in-palm; and (2) hip flexion, hip abduction, knee flexion, ankle equinus, **hindfoot valgus**, and toe flexion (43). Common gait abnormalities include toe walking, crouched gait, jump gait, and scissoring. Tables 31.5 and 31.6 describe the common contractures and the resulting deformities, gait deviations, and orthotics. Tables 31.7 and 31.8 present the type of contracture and discuss muscles that should be stretched and strengthened. Motion analysis is a tool used to understand gait abnormalities. Markers are placed on select body points to define body segments. Multiple cameras capture the motion, which is then analyzed following specific kinematic parameters. Motion analysis is used to determine the appropriate surgical procedure and to determine the effect of interventions that involve treatments such as orthotics, botulinum, physical therapy, and selected dorsal rhizotomy (54).

Spasticity Assessment

Spasticity is an involuntary velocity-dependent increase in muscle tone. Spasticity is often measured by tests such as the Ashworth Scale, Modified Ashworth Scale, Tardieu Scale, and Modified Tardieu Scale. These tests differ in how they measure spasticity. The Ashworth Scale measures muscle tone by scoring the resistance encoun-

tered in a specific muscle group by passively moving the limb through its range of motion on a 5-point scale (the Modified Ashworth Scale uses a 6-point scale) (78). The Tardieu Scale measure spasticity by passive movement of the joints at three specified velocities with the intensity and duration of the muscle reaction to the stretch rated on a 5-point scale with the angle notated. The Modified Tardieu Scale is less time consuming and requires the clinician to define the moment of the "catch" (the first occurrence or "showing" of resistance or spasticity when the muscle is stretched) that is experienced in the range of motion at a fast, passive stretch (78). Although the clinical exercise physiologist may not use these tests, they are important clinical assessments that should be understood. The clinical exercise physiologist must know the level of spasticity that a client experiences to allow detection of any increases in spasticity that may be exercise related. Spasticity levels should be monitored following each exercise bout. The exercise physiologist should observe the client's range of motion during functional activities such as reaching tasks before and after exercise. Inability to complete a task or decreased range of motion may indicate an increase in spasticity.

Balance Assessment

People with cerebral palsy often have poor balance because of the initial insult to the cerebellum which

Table 31.5 Common Contractures of the Lower Extremities

Joints	Common types of contractures	Type of deformities	Gait deviations	Orthotics
Hip	Flexion	Coxa vara	Scissoring gait	Hip knee ankle foot orthosis (HKAFO)
	Adduction			
	Internal rotation			
	Flexion	Coxa valga	Abducted gait	Hip knee ankle foot orthosis (HKAFO)
	Abduction		Wide-based gait	
	External rotation			
Knee	Flexion	Genu valgum	Crouched gait	Knee ankle foot orthosis (KAFO): medial T-strap
		Genu varum		Knee ankle foot orthosis (KAFO): lateral T-strap
Ankle	Plantar flexion	Pes equinus		Ankle foot orthosis (AFO)
	Pronation			Medial wedge
	Eversion	Hindfoot valgus		
	Plantarflexion	Pes equinus	Equinus gait	Ankle foot orthosis (AFO)
	Supination	Clubfoot		Lateral wedge
	Inversion	Hindfoot varus		

Adapted from L.A. Koman, B.P. Smith, and J.S. Shilt, 2004, "Cerebral palsy," *Lancet* 363: 1619-1631.

Table 31.6 Common Contractures of the Upper Extremities

Joints	Common types of contractures	Orthotics
Shoulder	Flexion	Orthotics not commonly used
	Adduction	
	Internal rotation	
Elbow	Flexion	Orthotics not commonly used
	Pronation	
Wrist, MCP, IP, and CMC (thumb)	Flexion	Resting splints: Wrist in 30°extension MCP: 60° flexion IP: neutral Thumb: opposition Functional splints: cock up splint Wrist: 30° extension Thumb: abduction

Note. MCP = metacarpophalangeal; IP = interphalangeal; CMC = carpometacarpal.

produces **ataxia,** abnormal involuntary movements, altered postural alignments, and gait abnormalities. It is important to assess the client's seated and standing balance along with static and dynamic balance control. Balance can be assessed by utilizing tests such as the Functional Reach Test or by observing the client stand from a seated position (27). If a client's balance is poor,

alterations to exercise testing and prescription should be made to ensure safety. Other tests used to determine balance include Clinical Test of Sensory Interaction and Balance (CTSIB), the Tinetti Balance Test of the Performance-Oriented Assessment of Mobility Problems, timed get up and go test, Berg Balance Scale, and the Physical Performance Test (16,21,52,66,68,88).

Table 31.7 Stretching and Strengthening of Muscles Associated With Common Contractures in the Lower Extremities

Joint	Types of contracture	Direction of stretch*	Muscle strengthening
Hip	Flexion, adduction, internal rotation	Extension, abduction, external rotation	Extensors: gluteus maximus, biceps femoris, semimembranosus, semitendinosus Abductors: gluteus maximus, gluteus medius, gluteus minimus, TFL, sartorius External rotators: gluteus maximus, sartorius
	Flexion, abduction, external rotation	Extension, adduction, internal rotation	Extensors: gluteus maximus, biceps femoris, semimembranosus, semitendinosus Adductors: adductor longus, adductor magnus, adductor brevis, gracilis, pectineus Internal rotators: TFL, adductor longus, adductor magnus, adductor brevis, pectineus
Knee	Flexion	Extension	Extensors: quadriceps femoris
Ankle and foot	Plantar flexion and eversion	Dorsiflexion and inversion	Dorsiflexors: tibialis anterior, extensor digitorum longus, extensor hallucis longus Invertors: tibialis posterior, flexor digitorum longus, flexor hallucis longus, popliteus
	Plantar flexion and inversion	Dorsiflexion and eversion	Dorsiflexors: tibialis anterior, extensor digitorum longus, extensor hallucis longus Evertors: peroneus longus, peroneus brevis

*Stretch the antagonist muscles.

Table 31.8 Stretching and Strengthening of Muscles Associated With Common Contractures in the Upper Extremities

Joint	Types of contracture	Direction of stretch*	Muscle strengthening
Shoulder	Flexion, adduction, internal rotation	Extension, abduction, external rotation	Extensors: pectoralis major, latissimus dorsi, posterior deltoid Abductors: middle deltoid, supraspinatus External rotators: infraspinatus, teres minor
Elbow	Flexion	Extension	Extensors: triceps, anconeus
Wrist	Flexion, ulnar deviation	Extension, radial deviation	Extensors: extensor carpi radialis longus and brevis, extensor digitorum, extensor carpi ulnaris Radial deviators: flexor carpi radialis, extensor carpi radialis longus and brevis
MCP, IP, CMC	Flexion at MCP, IP Adduction at CMC	Extension at MCP, IP Abduction at CMC	MCP extensors: extensor indices IP extensors: extensor indices, lumbricals CMC abductors: abductor pollicis longus and brevis

Note. MCP = metacarpophalangeal; IP = interphalangeal; CMC = carpometacarpal.

*Stretch the antagonist muscles.

EXERCISE TESTING

Exercise testing for those with cerebral palsy may be required if they are at risk for cardiovascular disease or for exercise prescription purposes. Because of the disorder and its complications, some modifications to the typical stress test may be required. These are reviewed in the next several sections. For general information regarding exercise testing see chapter 5.

Cardiovascular

Cardiovascular exercise testing can help the clinical exercise physiologist determine cardiovascular fitness and create an appropriate exercise prescription. The principle of specificity is important when choosing a modality to perform the graded exercise test. Bhambhani et al. tested the validity and reliability of maximal aerobic power in wheelchair athletes and determined that those who used wheelchairs for their primary mode of ambulation had higher $\dot{V}O_2$ max values during wheelchair ergometry. The individuals who used canes or no aids for ambulation had higher $\dot{V}O_2$ max values during leg cycle ergometry.

Wheelchair ergometry involves the use of the individual's own wheelchair and a wheelchair roller, which locks the chair in place and provides a stationary means of propelling the wheels. Calculating or controlling workload using wheelchair ergometry with a mechanism to measure power is often difficult. An arm ergometer allows control of the workload and still provides an accurate measure of oxygen consumption. Testing individuals with cerebral palsy can be difficult because the test may increase spasticity as a result of the increase in the speed of movement and temporarily affect coordination. The recommended protocol for arm ergometry begins with a starting power output of 0 to 15 W at 30 to 50 rpm and increasing 5 to 10 W every 1 to 2 min until volitional fatigue is reached (6). The exercise physiologist must make sure the client is in a stable position during testing and that the arm ergometer is secured into a fixed position. If the client has hemiplegia (a contracture or weakness in the hands) straps or an adaptive glove may be needed to secure the hand to the handlebar of the ergometer (6,63). Sufficient range of motion in the hand and wrist must be identified before attaching to the ergometer with any assistive devices. If the person lacks adequate of range of motion to perform this movement, an injury can result.

Leg cycle ergometry has been used effectively in this population and provides a stable testing method if balance is an issue. The protocol for cycle ergometry varies from 25 to 50 W at 50 to 60 rpm with an increase of 15 to 25 W for each 2 min increment until volitional fatigue (4,6). As a person with cerebral palsy reaches muscular fatigue, he or she may have difficulty continuing to pedal on a cycle ergometer because foot control and knee alignment may prevent performance of full revolutions (63). Other testing devices such as the Nu-Step Recumbent Stepper, Schwinn Air-Dyne, or treadmill can be used as testing modalities for individuals with cerebral palsy, depending on their functional ability. An advantage of the Nu-Step is the greater recruitment of musculature as compared with cycling and the ability to increase workload without increasing cadence. The Schwinn Air-Dyne cannot increase workload without increasing cadence (rpm), which may increase spasticity and affect coordination. Studies in the nondisabled population show that peak $\dot{V}O_2$ values achieved during treadmill walking correlate strongly with all-extremity exercise-testing values and do not differ significantly from nonwalking modalities (47). The treadmill is a useful testing device but may be limited in some persons with cerebral palsy because of limitations in balance and coordination. If a treadmill is used, the exercise professional must protect against increased spasticity and use a spotter. The client should be allowed to practice the testing methods to allow accommodation and therefore produce a more accurate test (38). During $\dot{V}O_2$ tests, it may be necessary to adjust the mouthpiece or use a facemask because of the oral motor dysfunctions and occlusal disorders. Table 31.9 reviews specifics for cardiovascular exercise testing.

Resistance Training

According to the American College of Sports Medicine (ACSM), the best format to determine dynamic strength testing (involving movement of the body or an external load) is the one-repetition maximum (1RM) (5). In some cases, a 1RM test is contraindicated, so a 6RM or 10RM test is used. But extrapolating those data and estimating a person's true 1RM is often problematic because of the marked variation in the number of repetitions that can be performed at a fixed percentage of a 1RM for different muscle groups (i.e., leg press versus bench press). Use of the Holton curve takes the 10RM of an individual and adjusts the score to estimate the 1RM. According to ACSM, an 8RM or 25RM can be used to test muscular strength and endurance of individuals with cerebral palsy (4,6). An alternative method of testing includes determining the number of repetitions that can be completed in 1 min. Performance level is graded by individual progression from one testing period to the next.

Table 31.9 Cardiovascular Exercise Testing

Mode	Protocol specifics	Clinical measures	Clinical implications	Special considerations
Wheelchair ergometry using rollers		ECG, HR, BP, consider $\dot{V}O_2$	Difficult to calculate or control workload	Allow practice time before testing, but do not trigger fatigue by allowing excessive practice. Make sure that client is in a stable position. Be cautious if using straps to attach the client's hand to the handle. Be sure that client has adequate range of motion to perform the exercise.
Arm ergometry	Begins with a starting power output of 0–15 W at 30–50 rpm and increasing 5–10 W every 1–2 min until volitional fatigue is reached	ECG, HR, BP, consider $\dot{V}O_2$	Easier to control workload	Allow practice time before testing, but do not trigger fatigue by allowing excessive practice. Make sure that client is in a stable position. Be cautious if using straps to attach the client's hand to the handle. Be sure that client has adequate range of motion to perform the exercise.
Cycle ergometry; Air-Dyne type cycle; Nu-Step Recumbent Stepper	Varies from 25–50 W at 50–60 rpm with an increase in 15–25 W for each 2 min increment until volitional fatigue is reached	ECG, HR, BP, consider $\dot{V}O_2$	As a person with cerebral palsy reaches muscular fatigue, he or she may have difficulty continuing to pedal on a cycle ergometer because foot control and knee alignment may prevent performance of full revolutions. Greater recruitment of musculature; can increase workload without increasing cadence. Easier to control workload	Allow practice time before testing, but do not trigger fatigue by allowing excessive practice. Make sure that client is in a stable position. Be cautious if using straps to attach the client's hand to the handle. Be sure that client has adequate range of motion to perform the exercise.
Treadmill	RPE (6–20) $\dot{V}O_2$max; METs 12-lead ECG	ECG, HR, BP, consider $\dot{V}O_2$	Cannot increase workload without increasing cadence (rpm) which may increase spasticity and impair coordination	Allow practice time before testing, but do not trigger fatigue by allowing excessive practice. May be limited in persons with CP because of limitations in balance and coordination. Protect against increased spasticity and use a spotter at all times.
Metabolic data (i.e., peak $\dot{V}O_2$)				For metabolic data, mouthpiece may be a problem; use mask but be sure to maintain a tight seal.

*Some tests may increase spasticity and affect coordination.

Simple field tests can be used to evaluate the muscular endurance of certain muscle groups. For example, the number of abdominal crunches that can be performed without any rest provides information regarding the endurance of the abdominal muscles. Certain resistance-training equipment can be adapted to measure muscular endurance by choosing a submaximal level of resistance and then measuring the number of reps that the person can perform before fatigue sets in. As an example, the YMCA bench press involves performing standardized reps at a rate of 30 lifts per minute. Women are tested using a 35 lb (16 kg) barbell, and men are tested with an 80 lb (36 kg) barbell. Individuals are scored based on the number of successful repetitions. The weight suggested in the YMCA bench press may not be appropriate for all individuals and should be determined by individual strength levels.

Other modalities that can be used to test strength include handheld dynamometers and isokinetic dynamometers such as the Cybex II (11,84).

When testing persons with cerebral palsy, the exercise professional should remember several specific issues that can affect the safety and effectiveness of the testing (4,6):

1. Co-contraction may offset strength in tested muscle groups (agonists).
2. Measure ROM in tested muscle groups to determine the safety of exercise testing and specific exercise prescription.
3. Test muscle groups unilaterally (may be more spasticity on one side or a significant strength difference).
4. Focus on stability, coordination, ROM, and timing.
5. Adaptations: wide benches, low seats, and trunk and pelvic strapping.
6. Machines are safer than free weights and provide greater fluidity to the movement.
7. Use a metronome to ensure appropriate fluidity. Be sure to use a slow cadence to decrease spasticity.
8. Use nonslip handgrips and gloves if necessary.
9. Always provide adequate practice before testing.

Table 31.10 presents an overview of muscular strength and endurance testing.

Table 31.10   Strength Exercise Testing

Mode	Protocol specifics	Clinical measures	Clinical implications	Special considerations (may apply to all strength-testing modes)
Free weights Weight machines	1RM = the heaviest weight that can be lifted or moved only once using proper exercise form	8 or 25RM Alternatively, determine the number of reps in 1 min		Machines are safer than free weights and provide greater "fluidity" to the movement Cocontraction may offset strength in tested muscle groups (agonists) Measure ROM in tested muscle groups Test muscle groups unilaterally (may be more spasticity on one side) Focus on stability, coordination, ROM, and timing Adaptations: wide benches, low seats, and trunk and pelvic strapping Use a metronome to ensure appropriate fluidity and be sure to use a slow cadence to decrease spasticity Use nonslip handgrips and gloves (if necessary) Always provide adequate practice before testing

Mode	Protocol specifics	Clinical measures	Clinical implications	Special considerations (may apply to all strength-testing modes)
Field tests: (i.e., abdominal crunches performed without rest)	Simple field tests can be used to evaluate muscular endurance of certain muscle groups	Determine the number of reps in 1 min		
Hand-held dynamometers and isokinetic dynamometers (i.e., Cybex II)		The dynamometer measures the amount of grip strength used to depress the handle in pounds or kilograms		
Hydraulic resistance				

Flexibility (Range of Motion)

As stated in previous sections, measuring the range of motion for each joint is crucial to designing an effective and safe exercise program for a person with cerebral palsy. Maintaining flexibility of joints facilitates movement; tissue damage can easily occur if an activity moves the structures of a joint beyond its shortened range of motion. Because flexibility is joint specific the clinical exercise physiologist must be aware of any contractures that are present to prevent injury. Tables 31.7 and 31.8 show what muscles should be stretched and what muscles should be strengthened. Common devices to measure flexibility include goniometers, electrogoniometers, and tape measures. The range of each joint can be measured by the use of a goniometer. This is most commonly done through passive motion but can be done through active motion as well. The measurement is done throughout one plane of motion at a time. Placing the joint in the zero position is important. The goniometer must be aligned with the bony prominence to measure the range of motion accurately. Other forms of flexibility tests include the sit-and-reach test (which measures low-back and hip-joint flexibility), and Apley's Scratch Test. Some of these tests may not be appropriate for all individuals with cerebral palsy. Table 31.11 on p. 576 presents a review of range-of-motion assessment.

Treatment

The medical management of cerebral palsy includes non-pharmacological, pharmacological, and surgical inter-ventions. See table 31.12 on p. 576 for a list of treatment options along with the characteristic function or outcome.

Nonpharmacological Treatments

The following sections provide an overview of several nondrug treatment regimens for patients with cerebral palsy.

Neuromuscular electrical stimulation

Neuromuscular electrical stimulation can be used functionally to cause a muscular contraction. If this method is effective, an increase in strength can be seen in 1 to 2 wk (43). As children with cerebral palsy enter their teen years, a progressive crouched gait and loss of ambulation occurs, which has been linked to gastrocnemius insufficiency (34). The use of serial casting and pharmacological interventions such as botulinum toxin A has been shown to increase ankle dorsiflexion and improve functional gait (17).

Assistive technology devices (ATDs)

Orthotics are used commonly in the management of clinical signs and symptoms of cerebral palsy. The main goal of an orthotic is to prevent deformity of the joints and musculature and increase function (12). The solid ankle foot orthosis (AFO) is frequently used to hold the foot in a neutral position and reduce equinus caused by a spastic gastrocsoleus (25). If the orthotic is removed during strengthening or stretching exercises, it must be replaced correctly. Improper positioning can cause skin degradation.

Table 31.11   Flexibility or Range-of-Motion (ROM) Testing

Mode	Protocol specifics	Clinical measures	Clinical implications	Special considerations
Goniometry, electrogonimeters, and tape measures	Align goniometer with the bony prominence to measure available range of motion. Measure during passive or active range of motion throughout available ROM.	Joint angles at full flexion or extension	None	ROM may be limited because of spasticity, athetosis, or contractures.
Sit-and-reach test Apley's Scratch Test				Tests may not be appropriate for all individuals.

Table 31.12   Treatment Options for the Management of Cerebral Palsy

Treatment options		Characteristics and functions
Nonpharmacologic treatments		
Therapeutic intervention	Physical therapy	Strength training
	Occupational therapy	Endurance training
	Speech therapy	Neuro developmental techniques (NDT)
		Patterning
		Conductive education
		Hippotherapy
		Aquatics
		Vojta techniques
		Pediatric massage
		Adeli, or Polish, suit program
		Constraint-induced movement therapy (CIMT)
		Inhibit spastic agonist muscles
		Improve motor development
Therapeutic modalities	Electrotherapeutic modalities neuromuscular electrical stimulation: Therapeutic NME Functional NMES	Maintain or improve ROM
		Delay or prevent deformity
		Provide joint stability
	Splinting	Maintains resting musculotendinous length
	Serial casting	Strengthen weak muscles
	Orthotics	Improve functionality by using ATDs
	Assistive technology and adaptive devices	Facilitate community ambulation and integration
	Spinal bracing	
Hyperbaric oxygen		Experimental

Treatment options		Characteristics and functions
Pharmacological		

(See chapter 3 for details of pharmacologic agents used to treat cerebral palsy)	Oral admin:
	  GABA agonists
	  α2—adrenergic agonists
	  Muscle relaxants
	  Seizure medications
	Parenteral admin:
	  Intrathecal baclofen
	  Baclofen pump
	  Botulinum toxin
	  Alcohol
	  Phenol
		Reduction of spasticity

Neuromuscular blockade or chemodenervation	Botulinum toxin A	
	Phenol	
	Alcohol	
Surgical		
Surgical techniques to reduce spasticity	Selective dorsal rhizotomy	Decreases spasticity by balancing spinal cord mediated facilitatory and inhibitory control
Orthopedic surgery	Neurectomy	
	Tenotomy	
	Arthrodesis	
	Osteotomy	
	Ostectomy	
	Tendon transfer	
	Tendon lengthening	
	Fractional myotendinous lengthening	
	Multisegmental spinal fusion	

Note. NMES = neuromuscular electrical stimulation.

People with cerebral palsy commonly use posterior walkers to facilitate ambulation. This device allows the individual to have better stability and balance while ambulating. The clinical exercise physiologist should be cognizant that ambulation with assistive devices requires greater oxygen demands independent of the presence or absence of a physical disability (53). The increased energy expenditure in ambulation should be considered when managing symptoms such as fatigue.

Wheelchair use may be necessary to facilitate community participation. Using a wheelchair results in a decrease in energy expenditure that can allow the person with cerebral palsy to participate fully and maximize functional outcomes. A manual wheelchair should be encouraged for individuals with adequate upper-body strength and optimal endurance levels. Power wheelchairs or scooters can be used for people with secondary conditions related to musculoskeletal and cardiovascular deconditioning.

Pharmacological Treatments

The management of spasticity and seizures is a primary goal for most of the pharmacological treatments used with clients who have cerebral palsy. Chapter 3 reviews the medications commonly used to treat cerebral palsy. Many of the medications cause dizziness, weakness, and fatigue. The exercise physiologist should know whether the client has taken the appropriate medications before exercise and should monitor any changes in medication that may cause undesired side effects during exercise.

Surgical Treatments

Surgical procedures are used to treat spasticity, to lengthen muscle, or to decrease the effects of contractures. Selective dorsal rhizotomy interrupts the afferent limb of the reflex arc, which reduces activity in the circuits that facilitate spasticity (64). Because of this procedure, transient muscle weakness is observed and requires orthotics and extensive physical therapy (43). Orthopedic surgical interventions are symptom dependent and consider age, disease severity, comorbidities, and overall well-being (43).

EXERCISE PRESCRIPTION

The frequency, intensity, and duration of physical activity used in an exercise prescription for the able-bodied population are well known; refer to chapter 5 for details. That guideline serves as the basis for exercise training. But the specifics for frequency, intensity, and duration of physical activity necessary for improving fitness and functional health outcomes in persons with cerebral palsy are less clear. Therefore, certain modifications need to be made to the training program while considering both associated and secondary conditions (69).

Cardiovascular Exercise

To minimize joint and muscle pain, the use of non-weight-bearing activities is recommended. These activities reduce pressure on joints while still enabling the individual to receive the same benefits from the exercise routine. Good examples include swimming or other water activity and hand cycling rather than running or walking on a hard surface. Other modalities include recumbent stepping and dual-action cycling. Individuals with hemiparesis or those who use a wheelchair can use recumbent steppers. Adaptive straps may be needed to secure the feet in place.

Interval training entails working for intense bouts of short duration and alternating them with relief periods. For a person with cerebral palsy, interval training can be used as well, but modifications may have to be made. For example, the level of intensity for the short bouts of duration should not be so great that the entire workout session has to be lessened because of secondary conditions (reduced endurance and joint and muscle pain). Training on an arm ergometer should include directional changes that

work both agonist and antagonist upper-body muscles (i.e., 5 min forward on the arm ergometer followed by 5 min backward on the arm ergometer). The speed of training may be changed throughout (slow 5 min, fast 3 min, slow 5 min, moderate 7 min, etc.). The person should maintain a cadence that does not increase spasticity.

Because of the reduced endurance levels of individuals with cerebral palsy, exercise may have to be broken down into intermittent exercise rather than performed in one continuous session. Instead of one 30 min session of cardiovascular exercise, shorter bouts of exercise with frequent rest intervals (3 to 10 min bouts) may be necessary. These exercise sessions can be performed throughout the day and should progress slowly in duration.

To avoid pressure sores, precautions must be taken with prolonged sitting periods during exercise. Proper positioning is important as well. Trunk stability may be inadequate to provide proper support and ensure safety. Therefore, additional support may be required. If aquatic training is used, maintaining proper skin care is important to ensure that the integrity of the skin is not deteriorating, which could lead to increased risk of pressure sore development.

Resistance Exercise

People with cerebral palsy can benefit by incorporating resistance training using devices such as elastic bands, Swiss balls, and hydrotherapy. Elastic bands and Swiss balls provide a variety of resistance exercise options throughout all multijoint movements (i.e., bicep curls, knee extensions, trunk rotations) without the added risk of injury seen in using free weights. Elastic bands offer many levels of resistance, and the individual should progress accordingly as with any other training program (least to most resistant: yellow, red, green, blue, black, and silver). If **athetosis** is present, elastic bands may be inappropriate because of the involuntary movement and decrease in stability of the exercise band. The most appropriate exercises for people with slow, writhing movements are cuffed weights or exercise machines that can provide fluidity of movement. Active-assistive exercise may be needed to perform the motion smoothly. The use of aquatic therapy enables the individual to exercise in a non-weight-bearing environment while still attaining exercise benefits. This modality is extremely useful in decreasing the effect of musculoskeletal pain.

According to ACSM, many program variables are needed to produce appropriate progression of resistance exercise for the individual's fitness level. With the secondary conditions in mind, these variables need to be adjusted to meet each individual's needs and capabilities. In general, a high volume (resistance × reps) should be used, using lower resistance and increasing the number of repetitions to account for muscle atrophy. The exercise professional must be aware of fatigue that may occur rapidly because of reduced muscular endurance. Large-muscle groups should be completed before small-muscle groups (i.e., quadriceps before calf muscles). The use of both multi- and single-joint exercises is not contraindicated (i.e., squat versus knee extension). Each muscle should move in a slow, controlled manner for each exercise. To ensure safety, the client should start slowly with the resistance-training program and progress at a comfortable pace. Resistance-training programs should also focus on enhancing or maintaining good range of motion in the affected limbs.

A common type of cerebral palsy that results in weakness or **paralysis** to the right or left side of the body is spastic hemiplegia. This condition often requires greater attention to developing strength on the weaker side of the body. The amount of improvement that can be made to the hemiplegic side depends on the amount of damage that was sustained to that part of the central nervous system. If the person has complete paralysis on one side of the body, resistance training should be substituted by flexibility training.

The strong pull of the hip adductors experienced by many persons with cerebral palsy will require a resistance-training program that places greater emphasis on strengthening the hip abductors. This suggestion does not necessarily mean that the hip adductors do not need to be strengthened. Although the adductor muscles are often extremely tight because of spasticity, they may also be very weak. Therefore, both sets of muscle must be strengthened, although the abductors might have to receive a greater amount of work. Make sure that the client has not had a hip dislocation before you have him or her work these muscle groups. If the client has a history of hip dislocation, seek advice from his or her primary care physician or health provider to determine whether hip exercises can be conducted safely.

Because balance is often affected in ambulatory persons with cerebral palsy, the exercise physiologist must protect clients from injury by developing safe resistance-training programs that do not expose them to high risk of injury. Some clients will be able to work on strength exercises in a standing position with physical assistance from the instructor, whereas others will have to perform the exercise routines from a chair. Before developing the resistance-training program, the clinical exercise physiologist should measure the client's static and dynamic balance to determine whether standing exercises are safe.

Range-of-Motion Exercise

Active and passive range of motion should be incorporated into a well-balanced flexibility program. When possible, functional activities that use all joints throughout their full range of motion can be used in conjunction with passive stretching. Each exercise session should begin and end with flexibility exercises. Active assistive stretching and proprioceptive neuromuscular facilitation (PNF) are effective stretching techniques that assist the client in achieving full range of motion. These methods also assist in the relaxation of muscles in contraction. Chapter 5 provides details about how to perform PNF stretching. Also, refer to tables 31.7 and 31.8 to determine the appropriate muscles to stretch when contractures are present. Before prescribing range-of-motion exercise, the exercise physiologist should know whether the client has leg length discrepancy, scoliosis, hip dislocations, or pelvis abnormalities. In some instances, flexibility exercises may exacerbate the problem. For example, a spastic muscle should not be overstretched. Tai chi and yoga are also good exercise choices to increase or maintain flexibility in individuals with cerebral palsy.

EXERCISE TRAINING

The heterogeneous nature of cerebral palsy creates a complex array of signs and symptoms that often vary between individuals. Understanding the primary, secondary, and associated conditions related to cerebral palsy is necessary to comprehend the beneficial effects of exercise for people with cerebral palsy.

Turk defines the primary condition as the fundamental source of disability (90). In the case of cerebral palsy the etiology is not always identifiable, but in most cases it is a result of an initial insult to the brain or disruption in normal development. The primary disability does not progress. The signs and symptoms, however, can be transient, and progression of secondary and associated conditions may occur. Associated conditions are aspects of the pathology of the primary condition (90). Associated conditions are not necessarily preventable but can often be managed through medical intervention. An example of an associated condition of cerebral palsy is spasticity. See table 31.3 for a list of associated conditions. Secondary conditions are defined as an injury, impairment, functional limitation, or disability that occurs because of the primary condition or pathology. Secondary conditions include physical problems, social concerns, and mental health difficulties. Secondary conditions also can develop when the primary disability interferes with the delivery of standard healthcare for the treatment or prevention of a health condition. See table 31.4 for a list of secondary conditions.

Persons with cerebral palsy report a decline in functional abilities with age (9,42,75). The decline with age may include symptoms such as fatigue (37,61), pain (9,33,55,61), progressive musculoskeletal deformity and dysfunction (33,55,61), and a decrease in functional ambulation (9,55,95).

Exercise can play an important role in the prevention of many secondary conditions and in the delay or prevention of the functional decline associated with aging. Most exercise research has focused on children and adolescents with cerebral palsy. The specific intensity, frequency, and duration that are most appropriate and effective for adults with cerebral palsy are still unknown (69). Nevertheless, the clinical exercise physiologist can utilize the successful studies to establish a basis for developing an appropriate exercise prescription.

Cardiovascular Exercise

The physical work capacity of adults with cerebral palsy is significantly lower than that of the general population (89). Studies suggest that persons with cerebral palsy require more muscle activity to perform tasks when compared with people without disabilities (48,49). Higher oxygen uptake, oxygen pulse, and RER occur at a given workload for individuals with cerebral palsy (70,93). Studies show that the $\dot{V}O_2$ max for people with cerebral palsy is similar to, or slightly lower than, that of people without disabilities (49,70). In conjunction with a greater level of energy expenditure, children with cerebral palsy often are smaller, have a higher body fat percentage, and are less physically active than children without disabilities (94). Other research reports poor fitness levels in children and adults with cerebral palsy (71,29).

Cardiovascular exercise can be beneficial for individuals with cerebral palsy. Aerobic exercise increases lower-extremity strength, decreases energy required to ambulate, and improves gross motor function for adolescents with spastic cerebral palsy (77). An 8 wk training program using a bicycle ergometer at 40% to 70% of peak $\dot{V}O_2$ twice weekly produced a 12% increase in cardiorespiratory fitness (65). This training program elicited an increase in physical work capacity as evaluated by a Schwinn Air-Dyne ergometer (65). Other studies that use cycle ergometry report significant increases in maximal heart rate and aerobic capacity in children with cerebral palsy (15,80).

Thorpe and Reilly used water walking in conjunction with lower-extremity resistance exercise and found an improved gross motor function for an adult male with cerebral palsy (87). This study was conducted over a 10 wk period with an exercise frequency of three times weekly. A review of aquatic exercise for children with cerebral palsy suggests that aquatic exercise may be beneficial for children with cerebral palsy, but more research is needed to determine the effect on fitness (39).

Resistance Training

Muscle weakness is common in people with cerebral palsy and directly affects functional ability (96). Several studies have elicited enhanced ambulation through resistance training. Andersson et al. found that individuals with cerebral palsy who were physically active reported improved ability to walk (8). Further research suggests that a 10 wk progressive strength-training program improves muscle strength and walking ability without increasing spasticity (7). Besides improving walking mobility, a progressive resistance-training program that used a 6RM for upper-body exercises three times a week increased muscular strength and wheelchair performance in children with cerebral palsy (59). A clinically feasible home-based training program that focused on lower-limb strength and physical activity in people with spastic diplegia showed that a short 6 wk program can lead to lasting changes in the strength of key lower-limb muscles that affect daily function of young people with cerebral palsy (26). Exercises that used virtual reality to practice isometric exercise that focus on ankle dorsiflexion are successful in children (18). The range of motion and hold times increased with the virtual reality exercise, increasing compliance over the conventional home exercise program. Community-based exercise is also effective in increasing strength and motor activity in adults over the age of 40 with cerebral palsy (85). Strength increases ranged from 17% to 22% over the 10 wk program. An added benefit was social interaction within the group. Resistance training in children with spastic cerebral palsy has demonstrated positive functional outcomes in individuals with a wide array of involvement (22). This study elicited strength increases in the hemiplegic side and an increase in walking pace. Dodd et al. completed a systematic review of the strength-training literature for individuals with cerebral palsy. The authors concluded that most training programs were conducted three times a week for a period of 6 to 10 wk with a progressive adjustment in resistance to provide sufficient intensity as suggested by the American College of Sports Medicine (5). Additional reviews report a general increase in muscle force production without an increase in spasticity (23).

CONCLUSION

Because individuals with disabilities and chronic health conditions are spending significantly less time in inpatient and outpatient rehabilitation facilities than they did 10 yr ago, clinical exercise physiologists will play a greater role in responding to the health and exercise needs of this population. As a result, the exercise physiologist will assist in the prevention of many secondary conditions that will improve the quality of life for people with cerebral palsy and other disabilities.

Case Study

Diagnosis

Ms. AT is a 22 yr old female with a history of premature birth (34 wk), low birth weight (4 lb 10 oz [2.1 kg]), an **APGAR** score of 4, and postnatal meningitis. At age 1 she was diagnosed with spastic diplegia.

Medical History

Ms. AT underwent a partial hamstring muscle release at age 10 and an Achilles tendon lengthening at age 12 for her gait abnormalities because of knee flexion contracture and **pes equinus** deformity. Postsurgery, Ms. AT has attended several physical therapy sessions with a focus on stretching and flexibility exercises. As a student within an inclusive school setting she was an active participant in sports events and other recreational activities, and she was mostly independent in day-to-day activities, requiring minimal to no assistance. At age 18 she underwent a surgical release for her hip adductor muscles to correct her scissoring gait. But for the past 2 yr she requires moderate assistance in ADLs and has trouble with activities such as walking and climbing stairs. Activity is usually followed by shortness of breath and fatigue. She has also gained 25 lb (11 kg) over the last 2 to 3 yr. She complains about excessive postural sway and loss of balance while walking and therefore uses her father's cane. She has no history of smoking or alcohol consumption and no history of hypertension and diabetes. But she has a familial history of diabetes and a history of bronchial asthma. She is currently on the following medications: Advair Diskus (BID) (fluticasone and salmeterol).

Exercise Test Results

Because of the fact that Ms. AT is ambulatory but has increased postural sway along with balance difficulties, leg cycle ergometry was chosen as the method of aerobic exercise testing. Her resting blood pressure and heart rate in a seated position were 134/80 mmHg and 82 beats · min^{-1}, respectively. A small face mask was used to provide an adequate seal. Toe clips were used to maintain proper foot positioning on the pedals. Ms. AT performed a 2 min warm-up without resistance. The leg cycle ergometry test began at 25 W at a cadence of 50 to 60 rpm. The protocol consisted of 2 min stages with an increase of 15 W per stage. The following data were collected during the exercise test.

The exercise test was terminated at 5:15 because of increased hip adductor spasticity, inability to maintain a speed of 50 rpm, and by request of the participant because of shortness of breath. No signs of EKG abnormalities were present during the test. Her $\dot{V}O_2$ at 5:00 was 17 ml · kg^{-1} · min^{-1}. The R <1.00 and the HR and $\dot{V}O_2$ failed to reach peak. This was not a valid peak $\dot{V}O_2$ test.

Because of the termination of the clinical leg cycle ergometry test, a more functional approach was chosen. The timed get up and go test was used as a baseline measure. This test used the following protocol.

Test Instructions

1. Starting position: Sit upright against the chair.
2. Stand up from the chair (time start).
3. Move quickly along the line toward the opposite end.
4. Touch the end and pivot back toward the chair.
5. Move quickly all the way back to the chair.

(continued)

6. Pivot.

7. Sit back down (time end).

The total distance used was 10 ft (3.05 m). The test was performed three times. Ms. AT's scores were 42 s, 38 s, and 40 s. (A score over 30 s suggests a higher risk of falls). Ms. AT used her assistive device during all tests.

Exercise Prescription

Using the information presented in the medical history and exercise-testing sections and in tables 31.9–11, develop an exercise prescription that encompasses aerobic, strength, flexibility, and neuromuscular training. Consider the different modes for each of the exercise-training domains and the frequency, duration, intensity, and progression of each.

Discussion Questions

1. How would the clinical exercise physiologist best determine the appropriate resistance levels that should be prescribed for this person?

2. What factors represented in the ICF model should the clinical exercise physiologist take into consideration to increase the adherence level and success of this client?

3. What fall prevention strategies should be incorporated into the exercise prescription for an individual who is at increased risk for falls?

4. Is the assistive device described in this case study appropriate for use by Ms. AT?

5. How should the clinical exercise physiologist incorporate weight-loss strategies into the goals for this client?

32

Stroke

Christopher J. Womack, PhD, FACSM

A stroke is the loss of blood flow to a region of the brain. This loss can occur because of a manifestation of cardiovascular disease, characterized by the buildup of atherosclerotic plaque in cerebrovascular arteries, ultimately resulting in what is known as an ischemic stroke. In most ischemic strokes, a blood clot ultimately seals off the narrowing artery. Strokes can also occur because of excessive bleeding in a cerebral artery, also known as a hemorrhagic stroke. The excessive bleeding and swelling in the brain prevents blood from flowing to brain cells downstream of the hemorrhage. Most strokes are ischemic, accounting for approximately 88% of all strokes (37). When a stroke occurs, neurons in the brain die, and the accompanying brain damage is the main cause of subsequent disability in stroke survivors. The resulting brain damage can impair voluntary muscle movement, speech, vision, and judgment. Strokes can also occur in the form of silent cerebral infarctions. These strokes are not coupled with clinically apparent acute symptoms.

SCOPE

Cardiovascular disease is the number one cause of mortality in U.S. adults (37). Of the causes of death from cardiovascular disease, stroke is the second-leading cause, encompassing 17% of all deaths attributable to cardiovascular disease (37). This figure amounts to 56.2 deaths per 100,000 people, making stroke the third-leading cause of death in the United States, behind coronary artery disease and cancer (37). The incidence of stroke

is such that one occurs every 45 s in the United States. As a result, approximately 5.5 million American adults have had a stroke (37). Each year, approximately 700,000 Americans will experience a stroke, and the vast majority of these will be new stroke events (37). Clearly, there are many survivors of stroke. As evidence of this, the number of noninstitutionalized stroke survivors increased from 1.5 to 2.4 million from the early 1970s to the early 1990s (28). Although this is a positive trend, in that stroke is taking fewer lives, a larger number of people now have a stroke-related disability. The effect that stroke has on quality of life is large, because 20% of stroke survivors require institutional care 3 mo after onset and 15% to 30% of stroke survivors are permanently disabled (37). In a study evaluating patients after onset of stroke, 26% of individuals were institutionalized and 30% were unable to walk without some assistance 6 mo after stroke onset (18). In total, over 1 million American adults reported functional limitations because of stroke (2). Because of the high rate of mortality and disability associated with this condition, indirect and direct cost for stroke have recently been estimated at $57.9 billion (37).

Gender

Overall, stroke incidence in men is 1.25 times that of women, but these differences disappear over the age of 80 (37). In fact, approximately 46,000 more women than men have strokes every year (37). The larger number for women is mainly due to the longer lifespan in women, but it also points out that stroke is an important health concern for women as well as men. The increased risk of stroke is likely due to the increased cardiovascular disease risk associated with menopause. Furthermore, estrogen replacement therapy can increase risk of cardiovascular disease and stroke. Estrogen alone increased risk of stroke by 39% in women posthysterectomy (4). Furthermore, estrogen plus progestin has been shown to increase risk of ischemic stroke in postmenopausal females (39).

Although menopause represents the primary state that places females at increased risk for stroke, pregnancy also temporarily increases risk. Pregnant women are at nearly a ninefold increased risk for stroke during the 6 wk postpartum period (19). This risk is even greater in African American women (15). In a study on postpartum stroke, it was observed that 4.1% of women with postpartum stroke died in the hospital, whereas 22% of survivors died at home after discharge (15).

Ethnicity

Table 32.1 summarizes the distinct racial influences on the incidence and mortality rate from stroke. From recent data, 3.6% of adult American Indians and Alaska Natives have had a stroke, contrasted with 3.3% of African American, 2.2% of Caucasians, and 2.0% of Asians (37). The mortality from stroke is correspondingly higher in African American males and females compared with Caucasians (78.8% for black males, 69.1% for black females, 51.9% for white males, 50.5 % for white females) (37). Furthermore, African Americans experience a higher degree of functional limitations following stroke than Caucasians do (1), although the exact cause of these differences is unknown. Residents of southeastern states have a higher prevalence of stroke (1), which likely reflects cultural differences regarding diet and physical activity. The additive effect of these geographical and racial influences combine to make African American males in southeastern states the population at greatest risk for stroke in the United States (1).

PATHOPHYSIOLOGY

The atherosclerotic process that causes cerebrovascular disease, and ultimately an ischemic stroke, proceeds in the same fashion as plaque progression in coronary artery disease (CAD). For a more detailed explanation of this process, see chapter 16. For this reason, the same traditional and nontraditional risk factors that are related to the development and progression of CAD and peripheral arterial disease (PAD) are also associated with the development of ischemic cerebrovascular disease. Isch-

Table 32.1 Stroke Incidence and Mortality for Selected Ethnic Groups (37)

	Prevalence (% of population)	Mortality rate
Caucasian		
Males	2.3	51.9
Females	2.6	50.5
African American		
Males	4.0	78.8
Females	3.9	69.1
Hispanic		
Males	*	44.3
Females	*	38.6
Total	2.2	
American Indian and Alaskan Native		
Males	*	37.1
Females	*	38.0
Total	3.1	
Asian		
Males	*	50.8
Females	*	45.4
Total	1.8	

*Data not available; mortality rate is referenced per 100,000 deaths within a specified group.

Based on T. Thom et al., 2006, "Heart disease and stroke statistics—2006 update: A report from the American Heart Association Statistics Committee and Stroke Statistics Subcommittee," *Circulation* 113: 85–151.

emic strokes can be further categorized as thrombotic, embolic, and hemodynamic. In the case of thrombotic strokes, in which an occlusive thrombus develops in or outside an ulcerated plaque, hypercoagulable states due to increased coagulation potential or decreased fibrinolytic potential are particularly important risk factors. Emboli that cause embolic strokes are typically from the carotid or other arteries. In these cases, the thrombus is not large enough to occlude the large vessel, but the embolus that breaks off ultimately lodges in a smaller cerebral artery or arteriole. Often, major strokes are preceded by transient ischemic attacks (TIAs), which are considered a major predictor of impending stroke (16). In a study of patients who reported to an emergency room with a TIA, approximately 10% of subjects experienced a stroke within 90 d. Perhaps even more compelling is the fact that 5% experienced a stroke within 2 d (16). All older populations should be screened for possible prior

TIA, because about 50% of patients who experience a TIA do not report it to a clinician (16).

Hypertension is the major risk factor for hemorrhagic stroke, which comprises approximately 10% of strokes (35). Hemorrhagic strokes can also be caused by aneurysm, drug use, brain tumor, congenital arteriovenous malformations, and anticoagulant medication. Hemorrhagic strokes are classified as either intracerebral, which refers to bleeding inside the brain, or subarachnoid, which refers to bleeding in and around the spaces that surround the brain (35). Unfortunately, there is usually little warning for a hemorrhagic stroke. Acute signs or symptoms include altered consciousness, headache, vomiting, and large elevations in blood pressure. Additionally, patients with subarachnoid hemorrhage may develop neck stiffness (35).

MEDICAL AND CLINICAL CONSIDERATIONS

Medical and clinical considerations for stroke include signs and symptoms that will be specific to the whether damage occurs on the right or left side of the brain. The clinician needs to be aware of the acute signs of the stroke and the clinical manifestation that occur after the event. Finally, the method of determining the definitive diagnosis of stroke is another important clinical decision.

Signs and Symptoms

Memory loss and paralysis are two of the more consistent symptoms of stroke. In the case of paralysis, the brain damage will cause paralysis on the opposite side of the body (i.e., right-brain damage causes left-side paralysis). Furthermore, right-brain damage can result in vision problems and awkward or inappropriate behavior, whereas left-brain damage causes speech and language problems and slow or cautious behavior. A patient suffering from acute stroke can have one of the following symptoms: (1) numbness or weakness of the face, arm, or leg; (2) confusion, speech problems, and cognitive defects; (3) impaired bilateral or unilateral vision; (4) impaired coordination and walking; and (5) headache.

History and Physical Exam

One of the main changes for the clinical exercise physiologist to be aware of is the hemiplegic gait of stroke patients. Concomitant risk factors for cardiovascular disease such as hypertension and diabetes will frequently be present. Furthermore, underlying CAD is often present (26), so

resting EKG changes and symptoms for ischemia should be evaluated. Most stroke patients will develop mental depression during the poststroke period (40), so a psychological referral may be necessary.

Diagnostic Testing

Ultrasound, MRI, and angiography are the main diagnostic tests that will be used to assess impending occlusions that could cause an ischemic stroke. The major diagnostic tool to determine hemorrhagic stroke is computerized tomography (CT) of the head (35). If testing occurs within 24 h, this method can reliably determine bleeding location unless it occurs in the region of the posterior fossa. In these instances, magnetic resonance imaging will be used (35).

Exercise Testing

Contraindications for exercise testing are the same as those for all patient populations. Because hypertension is the major risk factor for hemorrhagic stroke, particular attention should be paid to make sure that preexercise resting blood pressure is below the contraindicative values of systolic pressure of 200 mmHg and diastolic pressure of 110 mmHg. Furthermore, because ischemic stroke is highly associated with CAD, screening should ensure that symptoms of CAD such as unstable angina are not present.

Considerations for exercise testing in stroke patients are summarized in table 32.2 on p. 586. Not all stroke patients are capable of traditional treadmill graded exercise tests to determine functional capacity. Some have even suggested that submaximal measurements such as lactate or ventilatory threshold or oxygen pulse ($\dot{V}O_2$/HR) should be the criterion measures of cardiorespiratory fitness in stroke patients because these markers are easier to obtain and put the patients at a lower risk (22). Recent research has shown, however, that if patients achieve a self-selected walking speed of at least 0.5 mph (0.8 kph) during 30 ft (9 m) of floor walking, they are capable of performing a graded treadmill exercise test with handrail support as needed (26). Out of 30 patients with hemiparetic stroke, 29 were able to meet this criterion and achieve an average of 84% of age-predicted maximal heart rate. This intensity was sufficient to detect asymptomatic myocardial ischemia in 29% of patients without previously determined CAD. The treadmill test was at a self-selected walking velocity with grade increases of 2% every 2 min (26).

If a patient is unable to perform treadmill or cycle ergometry work, protocols are available that do not

Table 32.2 Exercise-Testing Considerations for Stroke Patients

Test type	Mode	Protocol specifics	Clinical measures	Clinical implications	Special considerations
Cardiovascular	Treadmill, cycle ergometry	Self-selected speed, 2% increase in grade every 2 min. Begin at 20 W and increase 10 W $\cdot$ min^{-1}	EKG, $\dot{V}O_2$/HR, ventilation	Myocardial ischemia can be detected	Patients unable to achieve at least 0.5 mph (0.8 kph) during floor walking can use cycle ergometry, arm ergometry, basic activities protocol, or bridging protocol
Strength	Free weights, exercise machines, dynamometry	10RM testing, handgrip	NA	NA	Assessment should be given before testing on functional ability to perform motion on paretic side
Range of motion	Goniometer, sit-and-reach	Range of motion in affected joints	NA	NA	NA

Based on T. Thom et al., 2006, "Heart disease and stroke statistics—2006 update: A report from the American Heart Association Statistics Committee and Stroke Statistics Subcommittee," *Circulation* 113: 85–151.

require a large amount of leg muscle mass. Tsuji et al. (38) developed an incremental protocol that uses an increasing rate of "bridging," which involved elevating the pelvis to a point of maximal hip extension. The protocol uses 4 min stages and increases the rate of bridges per minute from a starting point of 3, to 6, with 6 bridge per minute increases thereafter up to 24. About 89% of patients were able to complete this protocol, and the test–retest intraclass coefficients were over 0.90 for HR, over 0.70 for $\dot{V}O_2$, and 0.98 for the oxygen pulse. But the peak oxygen consumption does not appear to be higher than 2 METs for this protocol (38), suggesting that it should be used only for extremely deconditioned patients. Although 6 min walk distance is commonly used to assess cardiorespiratory fitness in several clinical populations, it was recently observed (29) that $\dot{V}O_2$peak obtained during cycle ergometry and 6 min walk distance are not associated in stroke patients. Rather, 6 min walk performance is primarily dictated by balance, knee extensor strength, and degree of muscle spasticity (29). Therefore, the 6 min walk may be a good addition to the battery of tests in that it is a functional outcome of these impairments, but it should not be viewed as a measure of cardiorespiratory fitness per se.

Treatment

For advanced atherosclerosis (70–99% occlusion) in the carotid arteries, carotid endarterectomy is a common surgery (35). In this procedure, most of the plaque is physically removed from the artery wall by the surgeon. Percutaneous transluminal carotid angioplasty has also

been used with some success to open cerebrovascular occlusions (35).

Pharmacological treatment depends on the type of stroke. As shown in table 32.3, this would include anticoagulants, antiplatelet medications, and antihypertensives, the latter being especially important for patients with hemorrhagic stroke. Antihypertensives such as labetolol, enalapril, and clonidine are the most common drugs for subarachnoid hemorrhage (35). Patients with ischemic stroke are often on medication, which could potentially include statins and diabetes medications, to control associated risk factors. Nitroglycerin or other vasodilators may be prescribed following an ischemic, hemodynamic stroke to lessen the chance of vasospasm. Patients with subarachnoid hemorrhage are typically given nimodipine to reduce vasospasm every 4 h until symptoms subside (35).

Standard rehabilitation for stroke patients usually includes physical therapy. The primary aim of this treatment is to restore balance, movement, and coordination. As such, basic strengthening exercises, passive movements to increase range of motion, assisted and unassisted walking, and functional movements such as chair stands and transferring from bed to chair are commonly prescribed. Because of the hemiparesis, patients must often relearn daily activities such as dressing and bathing, particularly if they are often able to perform these tasks with one hand. Therapy will also include instruction and practice with using assistive walking devices such as walkers and canes. Speech therapy is commonly necessary because of the effects of the stroke on speech control. Nutritional consultations with a registered

Table 32.3 Common Medications for Stroke Patients and Their Respective Effects

Medications	Class and primary effects	Exercise effects	Other effects	Special considerations
Warfarin	Anticoagulant	None		Patients should avoid activities with high risk for trauma or injury
Ticlopidine, clopidogrel, aspirin	Antiplatelet	None		
Ramipril, enalapril	ACE inhibitor	None	Decrease in resting and exercise BP	
Nimodipine	Ca-channel blocker	Increased exercise capacity in patients with angina	Possible increase in resting and exercise HR; decrease in resting and exercise BP	
Hydrochlorothiazide	Diuretics	None	Decrease in resting and exercise BP	

Based on T. Thom et al., 2006, "Heart disease and stroke statistics—2006 update: a report from the American Heart Association Statistics Committee and Stroke Statistics Subcommittee," *Circulation* 113: 85–151.

dietitian may be necessary if the patient is overweight or obese. Additionally, because depression is common in patients who have had a stroke, psychological referrals may be employed.

EXERCISE PRESCRIPTION

The average functional capacity of a stroke patient is approximately 14.4 ml $\cdot$ kg^{-1} $\cdot$ min^{-1} (25), which is of great concern because 20 ml $\cdot$ kg^{-1} $\cdot$ min^{-1} has been suggested as the minimum necessary for independent living (8). Therefore, a major goal of stroke rehabilitation should be to increase functional capacity. Standard therapy, however, does not provide enough of an aerobic stimulus to engender an increase in cardiorespiratory fitness (24). In support of this, Duncan et al. observed that only 12% of patients receiving physical therapy and 2% of patients receiving occupational therapy received therapy directed to improve this parameter (11). Authors also noted a distinct lack of progression in the rehabilitation of patients in usual care. Therefore, supplemental aerobic exercise should be prescribed for stroke patients. The clinical exercise physiologist should be aggressive about implementing this therapy in the patient's rehabilitation regimen. Patient education is important, because many patients may believe that their standard physical therapy is a sufficient amount of exercise. Besides the decrease in functional capacity, profound decreases occur in muscular strength and endurance, and these should be addressed. A particular challenge in terms of exercise prescription

for a stroke patient that may have led a sedentary lifestyle is the perceived volume of work per week necessary to engender appropriate changes. One suggested approach is that of Rimmer et al. (33), who observed significant improvements in body fat percentage, 10RM strength, $\dot{V}O_2$peak, and hamstring and lower-back flexibility in predominantly African American patients poststroke. The program consisted of 3 d per week for 60 min each session (30 min of cardiovascular training using a variety of modalities, 20 min of strength training on commercial strength-training machines, and 10 min of flexibility training). Therefore, impressive gains can be realized with a total of 3 h per week of training, a time commitment that should not place an excessive burden on patients. Considerations for all aspects of exercise prescription are shown in practical application 32.1 on p. 588.

Cardiorespiratory Endurance

Most studies that have observed improvements in patients with stroke employ a frequency, intensity, and duration that are similar to ACSM guidelines for healthy populations. Modes of training are listed in practical application 32.1 and can include walking (on ground or on a treadmill), water exercise, and cycle ergometry. A mixture of these modalities is warranted because some have specific supplementary benefits. Employing some weight-bearing exercise is highly recommended, because these activities have been shown to maintain bone mineral

Exercise Prescription Considerations for Stroke Patients

Training method	Mode	Intensity	Frequency	Duration	Progression	Goals	Special considerations and comments
Aerobic	Floor and treadmill walking, cycle ergometry, Nu-Step, water exercise	40–80% HRR	3–5 d · wk^{-1}	15–30 min	Progress from low to high intensity and duration. Because of biomechanical limitations, heart rate recommendations should be superseded by perceived effort	To increase functional capacity above 20 ml · kg^{-1} · min^{-1}	
Resistance	Elastic resistance bands, body-weight exercises, sandbags, active motion, water exercise, commercial strength-training equipment	As tolerated, up to 80% of 1RM	3–5 d · wk^{-1}	30–45 min	As tolerated	To improve gait-related parameters and activities of daily living	
Range of motion	Passive movement, proprioceptive neuromuscular facilitation (PNF)	Below point of discomfort	3–5 d · wk^{-1}	10–20 min	As tolerated		Emphasis on stretching muscles on the paretic side, particularly in muscle groups experiencing spasticity

Based on T. Thom et al., 2006, "Heart disease and stroke statistics—2006 update: A report from the American Heart Association Statistics Committee and Stroke Statistics Subcommittee," *Circulation* 113: 85–151.

density in patients poststroke (30). Treadmill exercise not only increases cardiorespiratory endurance but also may aid in improving balance, coordination, and gait abnormalities (20,31). Water-based exercise training has been shown to increase V̇O$_2$peak by approximately 22% and cause concomitant increases in isokinetic strength (7). Water exercises can also be a relatively safe way for these patients to train because risk of falls is avoided. Cycle ergometry training results in improved in stair-climbing

ability (17), suggesting that cycle ergometry training can lead to nonspecific improvements in performance related to ambulation. If either the patient or the clinician is concerned about falls during treadmill exercise, supported treadmill exercise, as discussed in chapter 28, can be employed (12). By using supported treadmill exercise, patients can train at the highest velocity possible without stumbling and with minimal risk. Stroke patients who perform 10 s intervals at these higher speeds experience changes in 10 m walking speed, stride length, and functional ambulation that are superior to those produced by standard treadmill training (31).

Resistance Training and Flexibility

Because of the neuromuscular compromise caused by stroke, resistance training is an important part of the rehabilitative exercise prescription. In this regard, ACSM recommendations for strength training could be easily followed with possible modification in the mode of training. Because of lowered muscular strength and endurance in these patients, strength-training exercise can consist of low-resistance modalities such as elastic bands, body-weight exercises, and sandbags. For patients who are extremely weak, exercises against gravity (e.g., shoulder and leg abductions), either with or without assistance, may be necessary until the patient is capable of using any of the other recommended modalities. Rimmer et al. (33) successfully employed commercial resistance-training equipment in poststroke patients, suggesting that these devices can be used, provided that the resistance can be decreased low enough to accommodate the specific demands for this population. Whatever mode is selected, both the concentric and eccentric movements should be included, because concentric-only training does not increase isokinetic strength in stroke patients (13). A portion of the exercise prescription should focus on exercises designed to improve activities of daily living. Examples include chair stands, stair climbing, ball kicking, balance

beam step-ups, and walking through obstacles. These activities require combinations of strength, endurance, and balance. Traditional flexibility exercises can be performed as tolerated for these patients. A raised platform with a stretching mat can be a useful implement because it greatly assists the patients with lying down and returning to a standing position following the exercises. The focus of the flexibility program should be on the paretic limbs, particularly muscle groups in which the patient is experiencing a large degree of muscle spasticity.

EXERCISE TRAINING

Functional capacity can be dramatically improved in stroke patients. The improved $\dot{V}O_2$peak, combined with the lowered oxygen cost of ambulation because of improved economy, results in a profound decrease in the relative intensity of normal ambulation. As an example, Macko et al. (27) observed that 6 mo of treadmill aerobic exercise resulted in a 20% reduction in the metabolic cost of ambulating at a constant, submaximal intensity because of both increased $\dot{V}O_2$peak and decreased $\dot{V}O_2$ during submaximal walking. This study also suggested that the improvements in exercise economy are maximized within a 3 mo period, while $\dot{V}O_2$peak continues to improve throughout the course of training (27). Exercise rehabilitation also improves functional ambulation, as evidenced by consistent increases in 6 min walk distances found throughout the literature (3,9,12,34). As stated earlier, 6 min walk performance is largely influenced by balance, strength, and spasticity

Table 32.4 Adaptations to Exercise Rehabilitation in Stroke Patients

Cardiorespiratory endurance	Skeletal muscle strength	Skeletal muscle endurance	Flexibility	Body composition
• Increased functional capacity • Improved walking economy • Increased oxygen pulse • Decreased submaximal blood pressure • Improved 6 min walk time • Increased voluntary physical activity	• Increased isokinetic strength • Improved upper-extremity function	• Improved timed up-and-go • Increased number of chair stands • Improved upper-extremity function • Increased number of steps in 15 s	• Improved range of motion • Decreased spasticity in muscles on the paretic side	• Minimal decrease in body fat percentage • Large decreases in overweight or obese patients who receive nutritional counseling in addition to exercise

Based on T. Thom et al., 2006, "Heart disease and stroke statistics—2006 update: A report from the American Heart Association Statistics Committee and Stroke Statistics Subcommittee," *Circulation* 113: 85–151.

of the lower paretic limb in stroke patients (29,30). Therefore, the degree of improvement in the 6 min walk is likely due to improvement in these factors either in place of or in conjunction with improved cardiorespiratory endurance. Table 32.4 on p. 589 presents an overview of the anticipated physiologic adaptations to exercise training in these patients.

Because of the aforementioned relationship between functional capacity and the ability to live independently (8), these adaptations are extremely meaningful to the quality of life of these patients. A direct relationship also exists between improvements in sensorimotor function because of exercise training and the magnitude of improvement in $\dot{V}O_2$peak, which again points to the importance of improving functional capacity. Because blood pressure is an important risk factor for both ischemic and hemorrhagic strokes, another important clinical adaptation is a decrease in blood pressure in response to acute exercise (32).

Because most studies provide a holistic intervention that includes aerobic and strength training, discerning the exact contribution that resistance training makes to the outcome variables is difficult. Nonetheless, rehabilitation programs that incorporate resistance training will typically yield impressive changes in both laboratory and functional evaluations. Isokinetic quadriceps and ankle plantar flexion strength can increase in the paretic limb without increases in muscle spasticity (36). Similar to the aerobic adaptations, strength improvements appear to increase performance in functional activities. Short-distance (5, 10, and 30 m) walking speed (9,34,36), timed up and go (9), number of steps performed in 15 s (9), and force production in the paretic limb when going from a seated to standing position (9) all increase with programs that incorporate resistance exercises, especially those designed to increase functional performance. Combining aerobic and resistance training can drastically improve quality of life and activities of daily living (36). The latter is an especially important finding given that physical activity is associated with risk for stroke (21).

CONCLUSION

Stroke remains a significant cause of death and disability. Exercise testing and training are important in both the prevention and treatment of those post-stroke. The clinical exercise physiologist will likely become an increasingly demanded allied health professional in the care of patients at-risk for, or poststroke. For the poststroke patient this is especially true for those with only minimal related loss of motor function.

Practical Application 32.2

Client–Clinician Interaction

Dealing with stroke patients presents some unique challenges that can be best addressed by emphasizing good communication with the patients and associated family members or caregivers. Aforementioned mental consequences include depression, inappropriate behavior, or slow or cautious behavior. The clinician should be sensitive to these considerations, establish appropriate boundaries for interaction with the patient, and, if necessary, make appropriate referrals to mental health professionals. Communication with family members or caregivers can help determine whether depression or other psychological conditions may be a concern. Furthermore, because many of the desired outcomes of rehabilitation are improved activities of daily living, some of the outcome data may be qualitative information received from family members or caregivers about the level of independence that the patient exhibits when completing daily tasks.

Literature Review

Pang et al. (29) determined (1) the relationship between 6 min walk distance and $\dot{V}O_2$peak achieved during a maximal cycle ergometry test and (2) the physical impairments due to stroke that were the most important determinants of 6 min walk performance. Because of the equipment, skill, and time necessary to perform graded exercise testing, a 6 min walk test is often used as a surrogate for direct measurement of functional capacity. Furthermore, 6 min walk performance is associated with direct determination of $\dot{V}O_2$peak in patients with cardiorespiratory disease (5,6,10,23) and obese females (14). But because stroke patients have limitations in strength, balance, and motor control that may influence ambulatory ability more than cardiorespiratory endurance, this test may be less predictive of $\dot{V}O_2$peak in this population. Sixty-three stroke survivors (average time since stroke of 5.5 yr) performed a maximal exercise test on the cycle ergometer to determine $\dot{V}O_2$peak and a 6 min walk test. Subjects were also evaluated for balance, muscle spasticity, knee extension strength, and body fat percentage. A significant but mild correlation was found between 6 min walk distance and $\dot{V}O_2$peak (r = .40). When a multiple regression analysis was performed, a combination of balance, muscle spasticity, and knee extension strength accounted for approximately 73% of the variance in 6 min walk distance. Of these factors, balance was the most important, accounting for approximately 66% of the variance. Results of this study suggest that 6 min walk is not an appropriate surrogate for determining $\dot{V}O_2$peak in poststroke patients. Additionally, this study illuminates the relative importance of these impairments with respect to a functional test that may reflect daily challenges. In light of this, perhaps the most significant clinical application of this study is that it underscores the need for improved balance and strength in this population in conjunction with increased functional capacity.

Case Study

Medical History

The patient is a 69 yr old male with a history of obesity, type 2 diabetes, and hypertension. He is 3 mo poststroke and has gone through basic physical therapy. He has right-side hemiplegia, speech impediment, and depression. His resting BP is 150/90.

Medications

Ticlopidine, HCTZ, and glucophage.

Diagnosis

The patient reported to his physician with numbness, dizziness, motor function impairment on the right side of his body, and vision problems. The patient was referred, and stroke was confirmed and diagnosed by MRI.

Exercise Test Results

The patient achieved 6 METs on a progressive cycle ergometer test to exhaustion with no significant EKG changes. Peak HR was 140, and peak BP was 210/100.

Exercise Prescription

The patient was prescribed aerobic exercise 3 d per week progressing up to 5 d per week and resistance exercises 3 d per week. The initial intensity for aerobic exercise was set at 40 to 50% HRR and an RPE of 12 to 15 on the Borg scale, progressing every 3 wk as tolerated. The resistance exercises performed were prescribed as tolerated for 10 to 15 repetitions.

(continued)

Initial duration for the aerobic exercise was 15 min, progressing every 3 wk as tolerated up to 1 h per session. Resistance exercises were to start with one set of exercises for 3 wk, progressing up to three sets. The initial mode of training was a mix of water-based aerobic exercises and supported treadmill exercises. Support on the latter is preferably in the form of handrails, but a support device can be used if it enhances patient comfort. These modes will serve to increase cardiorespiratory fitness and other factors related to walking performance such as balance and muscular strength. As frequency and duration progress, cycle ergometry exercise can be added. The patient's ability to perform resistance exercises with elastic bands and available strength-training equipment should be evaluated. Initial exercises may need to include several movements against gravity and assisted movements.

Because of comorbid conditions (diabetes, obesity, depression), the lower end of the ACSM recommended range for intensity and duration should be used initially. But because caloric expenditure and improved insulin sensitivity are secondary goals, frequency should progress to the upper end of the ACSM recommended range. A variety of modes should be employed to enhance patient interest, an important consideration because of mental depression. Tests for $\dot{V}O_2$ peak should be repeated, and the testing battery should include at least two functional tests (40 m walk speed, 6 min walk, timed up and go, chair stands) so that the patient and the clinician can progress as it relates to ambulation. To assess progress for body fat percentage, weight and an estimate of body fat percentage (skinfolds, girths, or impedance) can be used. Because this patient exhibits characteristics of metabolic syndrome, waist-to-hip ratio should be determined regularly.

Discussion Questions

1. What are the main factors that determine the mode of training in this patient population?

2. What common associated conditions (e.g., obesity, diabetes, hypertension) may alter the exercise prescription for stroke patients?

3. What will be the determining factors for the type of assessments performed?

GLOSSARY

α_1-antitrypsin (AAT)—Protein produced in the liver and found in the lungs that inhibits neutrophil elastase.

α_1-antitrypsin deficiency—A genetic disorder characterized by abnormally low levels of α_1-antitrypsin, thereby predisposing an individual to emphysema.

abdominal obesity—Characterized by more fat on the trunk and also known as android obesity. Increased fat in the abdominal region increases the risk for development of hypertension, type 2 diabetes, dyslipidemia, coronary artery disease, and premature death compared to gynoid obesity (increased fat in the hip and thigh).

absolute contraindications for test termination—Conditions that occur during a stress test that require the test to be terminated.

absolute oxygen uptake—Oxygen uptake expressed in liters per minute ($L \cdot min^{-1}$).

acquired immunodeficiency syndrome (AIDS)—A disease caused by HIV. See the text for the CDC case definition.

activities of daily living (ADLs)—Bathing, dressing, grooming, toileting, feeding, and transferring.

acute MI—The initial stages of an evolving MI.

adult variant cystic fibrosis—Adult diagnosis of cystic fibrosis secondary to presentation of respiratory symptoms.

afterload—The resistance against which the pumping chamber or ventricle of the heart works.

agonist—Denoting a muscle in a state of contraction, with reference to its opposing muscle, or antagonist.

airway hyperresponsiveness—The ability of the airway wall to be sensitive to various inhalants.

akinetic—Loss of movement of a left ventricular wall during the normal cardiac cycle.

amenorrhea—Absence of normal menses; for most studies, a woman is considered amenorrheic if she has fewer than three menses per year.

anabolic steroids—Testosterone derivatives or steroid hormones resembling testosterone that stimulate the building up of body tissues.

androgenic—Masculinizing effects; stimulation of male sex characteristics and male hair characteristics.

anemia—A decrease in the red blood cells that carry hemoglobin, resulting in reduced oxygen-carrying capacity.

aneural—Absence of nerve fibers.

aneurysm—Dilation of an artery that is connected with the lumen of the artery or cardiac chamber. Usually occurs because of a congenital or acquired (e.g., myocardial infarction) weakness in the wall of the artery or chamber. Forms of aneurysms include true, dissecting, and false.

angina pectoris—Constricting chest pain, often radiating to the left shoulder or arm, back, or neck and jaw regions, caused by ischemia of the heart muscle.

angina threshold—Point at which the supply of oxygen is less than the demand, leading to ischemia and producing symptoms of angina pectoris. Generally observed during physical or mental exertion in patients with significant coronary artery disease.

angioplasty—Reconstitution or recanalization of a blood vessel; may involve balloon dilation, mechanical stripping of intima, forceful injection of fibrinolytics, or placement of a stent.

angiotensin-converting enzyme inhibitor—Medication that prevents the conversion of angiotensin 1 to angiotensin 2, which ultimately decreases blood vessel vasoconstriction.

ankylosing spondylitis—A chronic rheumatic disease that causes inflammation, stiffness, and pain in the spine, sacroiliac joints (often an early indicator), and in some cases, the neck, hips, jaw, and rib cage. This disease may be accompanied by fever, loss of appetite, and heart and lung problems. It may cause spinal deformities and eventually causes the spinal segments to fuse (ankylose), causing the back to assume a rigid fixed posture.

anorexia nervosa—Loss of appetite associated with intense fear of becoming obese.

anovulatory—Not accompanied with the discharge of an ovum.

antagonist—Something opposing or resisting the action of another; certain structures, agents, diseases, or physiological processes that tend to neutralize or impede the action or effect of others.

anticoagulation therapy—Pharmacological delaying or preventing of blood coagulation (clotting).

antioxidant—An agent that inhibits oxidation and thus prevents the deterioration of other materials through oxidative processes.

antiresorptive therapy—A term used to describe the various drug therapies currently available for treatment of osteoporosis. The term originates from the fact that the therapies halt the loss of bone by inhibiting bone resorption.

antiretroviral medications—Medications demonstrated to be effective against the HIV virus, which is a retrovirus.

anuria—Suppression or arrest of urinary output, resulting from impairment of renal function or from obstruction in the urinary tract.

anxiety—A complex psychophysiological response to an environmental stressor, disaster, or trauma; more often manifested in people who are genetically vulnerable to the disorder.

APGAR—A noninvasive clinical test designed by Dr. Virginia Apgar (1953) carried out immediately on a newborn. The name is also an acronym for activity (muscle tone), pulse, grimace (reflex irritability), appearance (skin color), and respiration. A score is given for each sign at 1 min and 5 min after birth.

aphasia—Partial or total loss of the ability to articulate ideas or comprehend spoken or written language, resulting from damage to the brain caused by injury or disease.

apolipoprotein—The protein component of lipoprotein complexes that is a normal constituent of plasma chylomicrons, high-density lipoprotein, low-density lipoprotein, and very low-density lipoprotein in humans.

arterial intimal layer—The innermost layer of an artery, composed of endothelial cells.

arthrodesis—The surgical immobilization of a joint intended to result in bone fusion.

articular—Relating to a joint.

associated conditions—Conditions that accompany a primary disability but are not necessarily preventable. They can, however, be controlled with medication, surgery, or medical devices.

asthma—A continuum of disease processes characterized by inflammation of the airway wall.

astrocyte—Star-shaped neural cell that provides nutrients, support, and insulation for neurons of the central nervous system.

asymmetric—Denoting lack of symmetry between two or more like parts.

ataxia—Inability to coordinate voluntary muscle movements; unsteady movements and staggering gait or loss of ability to coordinate muscular movement most often caused by disorders of the cerebellum or the posterior columns of the spinal cord; may involve the limbs, head, or trunk.

atherectomy—Procedure used for revascularization of an obstructed coronary artery consisting of a catheter tipped with either a metal burr that grinds a calcified atheroma (rotational atherectomy) or a rotating cup-shaped blade housed in a windowed cylinder that cuts or shaves the atheroma (directional atherectomy).

atherogenic—Having the capacity to initiate, increase, or accelerate the process of the atherosclerosis.

atheroma—A mass of plaque of degenerated, thickened arterial intima occurring in atherosclerosis.

atherosclerosis—An extremely common form of arteriosclerosis in which deposits of yellowish plaques (atheroma) containing cholesterol, lipid material, and lipophages are formed within the intima and inner media of large and medium-sized arteries.

atherosclerotic—Relating to or characterized by atherosclerosis.

athetosis—A constant succession of slow, writhing, involuntary movements of flexion, extension, pronation, and supination of fingers and hands, and sometimes of toes and feet.

auscultation—Listening to the sounds made by various body structures as a diagnostic method.

autoimmune—Referring to cells or antibodies that arise from and are directed against the person's own tissues, as in autoimmune disease.

autonomic dysreflexia—Sudden, exaggerated reflex increase in blood pressure in persons with SCI above T6, sometimes accompanied by bradycardia, in response to a noxious stimulus originating below the level of SCI.

autonomic nervous system—Components of the nervous system responsible for coordinating life-sustaining processes and organization of visceral responses to somatic reactions.

A-V synchrony—The sequence and timing of the atria and ventricles during systole.

avascular—Absence of blood vessels.

balance—The ability to make adjustments to maintain body equilibrium.

ballistic flexibility (stretching)—Stretching using active muscle movement with a bouncing action.

bariatric surgery—A general description of several methods of surgical intervention designed to assist morbidly obese people lose weight.

basal ganglia—The caudate and lentiform nuclei of the brain and the cell groups.

behavior therapy—Strategies, based on learning principles, that provide tools for overcoming barriers to compliance.

β-agonist—A drug or hormone capable of combining with β-receptors to initiate drug actions.

β-blocker—Medication used to block β-receptors in the myocardium, which decreases myocardial work by decreasing HR and myocardial contractility.

β-receptor—A cell receptor that is activated by a β-agonist such as epinephrine, norepinephrine, or dopamine.

bifurcation—Point at which an artery branches to form two arteries.

biopsy—A surgical procedure whereby a sample of tissue is obtained.

biotherapy—Stimulation of the body's immune response system to cancer-specific protein antigens.

blood pressure—The force of circulating blood on the walls of the blood vessels as it circulates throughout the body. Chronic elevated blood pressures (systolic blood pressure ≥140 mmHg or diastolic blood pressure ≥90 mmHg) are associated with increased risk for cardiovascular disease.

body mass index (BMI)—Relative weight for height with a value of weight in kilograms divided by height in meters squared ($kg \cdot m^{-2}$).

bone formation (modeling)—The process by which new bone is formed and deposited within the existing bone matrix. Bone formation is accomplished primarily by bone cells called osteoblasts.

bone geometry—Refers to the overall cross-sectional area of the bone, cross-sectional area of the outer cortex, the number of cross-links in the trabecular bone, and other related factors. When dealing with the femoral neck, bone geometry also refers to the angle that the neck of the femur makes with the shaft of the femur.

bone mineral content—A measurement of the total amount of hydroxyapatite (calcium phosphate crystal) of bone and expressed as $g \cdot cm^{-2}$. It is synonymous with bone mass.

bone mineral density (BMD)—Relative amount of bone mineral per measured bone width. Values are expressed as $g \cdot cm^{-2}$.

bone modeling—Alterations in the shape of the bone such as changes in length.

bone remodeling—A constant state of formation and resorption.

bone resorption—The process of eroding old bone from the existing bone matrix so that new bone can be formed in its place. Resorption is accomplished primarily by bone cells called osteoclasts. Bone resorption is greatly increased in estrogen-depleted women.

botulinum toxin—A neurotoxin that blocks the release of acetylcholine from the motor endplates of the lower motor neuron at the myoneural junction, thereby preventing muscle contraction.

bradycardia—A heart rate of less than 60 beats $\cdot min^{-1}$.

bromocriptine—Ergot derivative that suppresses secretion of prolactin. Used to stimulate ovulation in the galactorrhea-amenorrhea syndrome.

bronchi—Large airways of the lungs.

bronchiectasis—Chronic dilatation of a bronchus usually associated with secondary bacterial infection.

bronchioles—Small airways of the lungs.

bronchodilator—A drug that relaxes the smooth muscles surrounding the bronchi and bronchioles.

bronchoprovocation—A type of pulmonary function testing in which a particular medication (e.g., methacholine) is aerosolized to induce bronchospasm.

bronchospasm—Spasmodic contraction of the smooth muscle of the bronchi, as occurs in asthma.

bruit—An acquired sound of venous or arterial origin caused by turbulent blood flow; heard by auscultation.

bulimia—Disorder that includes recurrent episodes of binge eating, self-induced vomiting and diarrhea, excessive exercise, strict diet, and exaggerated concern about body shape.

bursitis—Inflammation of one of the fluid-filled sacs located at sites of friction surrounding the joint.

calcitonin—Hormone responsible for calcium regulation and inhibits bone resorption.

cardiogenic shock—Lack of cardiac and systemic oxygen supply resulting from a decline in cardiac output secondary to serious heart disease; typically follows an MI.

cardiorespiratory fitness—Also known as aerobic capacity, this describes the ability of the body to perform high-intensity activity for a prolonged period without undue physical stress or fatigue.

cardiovascular autonomic neuropathy—Neural damage to the autonomic nerves of the cardiovascular system, which can result in a high resting and low peak exercise heart rate and severe orthostatic hypotension.

cardiovascular disease—A term used to describe disease conditions that affect heart, arteries, and or veins of the circulatory system. Cardiovascular disease most often refers to the atherosclerotic process occurring in the coronary arteries of the heart.

catecholamine—Chemicals released in the body that are major elements in the response to stress and exercise. Two catecholamines of interest are epinephrine and norepinephrine. Both exert, among other properties, a positive inotropic and chronotropic effect on cardiac function.

cauda equina—Lumbosacral spinal nerve roots forming a cluster at the terminal region of the spinal cord that resembles a horse's tail.

cauda equina syndrome—Severe compression of the cauda equina, resulting in loss of bowel or bladder function, loss of sensation in the buttocks and groin, and weakness in the legs. Cauda equina syndrome requires an emergency surgical intervention to prevent permanent damage.

CD4 lymphocyte—A type of white blood cell that is part of the immunological system of the body.

CD4+ cells—A membrane receptor found on T-helper lymphocytes (or T4 cells), it is the preferred target of HIV.

central nervous system—The brain and the spinal cord.

cerebellum—The trilobed structure of the brain, lying posterior to the pons and medulla oblongata and inferior to the occipital lobes of the cerebral hemispheres, that is responsible for the regulation and coordination of complex voluntary muscular movement as well as maintenance of posture and balance.

cerebral cortex—The thin, convoluted surface layer of gray matter of the cerebral hemispheres that consists of the frontal, parietal, temporal, and occipital lobes.

chemotherapy—Use of chemical agents to kill rapidly growing cancer cells.

cholesterol—A fatlike, waxy substance found throughout the body. High levels of cholesterol in the blood (>200 mg · dl^{-1}) increases the risk of developing cardiovascular disease.

chorea—State of excessive, spontaneous movements, irregularly timed; nonrepetitive and abrupt and unable to maintain voluntary muscle contraction.

choreo-athetosis—A type of athetosis marked by extreme range of motion, jerky involuntary movements that are more proximal than distal, and muscle tone fluctuating from hypotonia to hypertonia.

chronic bronchitis—Disease characterized by the presence of a productive cough most days during 3 consecutive months in each of 2 successive years.

chronicity—The state of being chronic.

chronic MI—The latest phase of an MI when the heart is stable.

chronic obstructive pulmonary disease (COPD)—Presence of airflow obstruction attributable to either chronic bronchitis or emphysema.

chronotropic assessment exercise protocol—Treadmill protocol used to determine whether heart rate response is appropriate throughout the length of the exercise test.

chronotropic incompetence—Lack of an appropriate increase in heart rate with physical exertion. Considered an abnormal response if peak heart rate does not reach within two standard deviations below a person's age-predicted maximum heart rate, assuming that the patient was highly motivated and not on medications to blunt heart rate response (i.e., β-blockers, calcium channel blockers).

circulatory hypokinesis—Insufficient vascular tone resulting in hypotension during exercise.

clasp-knife hypertonia—Increased muscle resistance to passive movement of a joint followed by a sudden release of the muscle; commonly seen in individuals with spasticity.

claudication—Limping, lameness, and pain that occur in individuals who have an ischemia response in the muscles of the legs, which is brought on with physical activity (e.g., walking). A scale can be used to determine the severity of claudication.

clonus—An abnormality in neuromuscular activity characterized by rapidly alternating muscular contraction and relaxation. A form of movement marked by contractions and relaxations of a muscle, occurring in rapid succession, after forcible extension or flexion of a part. Also called clonospasm.

club foot—Also known as talipes equinovarus, a congenital deformity affecting one or both feet characterized by the heel pointing downward and the forefoot turning inward. The heel cord (Achilles tendon) is tight, causing the heel to be drawn up toward the leg.

cognitive-behavioral therapy—For patients with major depressive disorder, treatment that focuses on modifying maladaptive thoughts and addresses deficits in behavior (e.g., unassertiveness, isolating oneself from others).

community mobility—Locomotion and transportation of an individual through his or her community.

complete paraplegia—Motor and sensory dysfunction of the trunk, legs, and pelvic organs resulting from SCI, without motor or sensory sparing below the level of the injury.

complete tetraplegia—Motor and sensory dysfunction of the arms, trunk, legs, and pelvic organs resulting from SCI, without motor or sensory function spared below the level of the injury.

computed tomography (CT)—Imaging technique that involves an X-ray, which gives a film with a detailed cross-section of the structure of the tissue.

congestive or chronic heart failure (CHF)—The symptom complex associated with shortness of breath, edema, and exercise intolerance attributable to abnormal left ventricular function resulting in congestion of blood in other bodily organs.

contracture—An abnormal, often permanent shortening, as of muscle or scar tissue, that results in distortion or deformity, especially of a joint of the body. Shortening of a muscle group and tendon usually observed in persons with spasticity.

coronary dissection—Separation of tissue within the lining of a coronary artery.

cortical bone—One of the two main types of bone tissue, it is hard, compact bone found mainly in the shafts of long bones. The other type is trabecular.

coxa valga—A hip deformity produced when the angle of the head of the femur with the shaft exceeds 120°. The greater the degree of coxa valga, the longer the resulting limb length.

coxa vara—A hip deformity produced when the angle made by the head of the femur with the shaft is below 120°. In coxa vara it may be 80° to 90°. Coxa vara occurs in rickets or may result from bone injury. The affected leg appears shortened, resulting in a limp.

C-reactive protein (CRP)—A β-globulin found in the serum of various persons with certain inflammatory, degenerative, and neoplastic diseases. CRP levels are often detectable in the blood of individuals with metabolic syndrome, suggesting chronic inflammation.

creatinine—End product of creatine metabolism excreted in the urine at a constant rate—a blood marker of renal function.

creatinine clearance—An index of the glomerular filtration rate, calculated by multiplying the concentration of creatinine in a timed volume of excreted urine by the milliliters of urine produced per minute and dividing the product by the plasma creatinine value.

crepitus—Crackling from the joint palpated on examination.

cross-training—The concept of training in another mode that allows for the development of physiology that will have a carryover effect to another mode; for example, resistance training is often performed to develop sport-specific strength.

CRT (cardiac resynchronization therapy)—Refers to the use of a biventricular pacemaker to restore the coordinated (or synchronized) pumping action of the ventricles when electrical conduction is delayed by bundle branch block, a common feature of chronic heart failure.

culprit lesion—The primary obstruction responsible for decreased blood flow through a coronary artery.

curative surgery—Surgery aimed at complete removal of tumor along with a small amount of surrounding normal tissue.

cystic fibrosis transmembrane conductance regulator (CFTR)—A protein responsible for moving salt across cell membranes.

cysts—Abnormal sacs containing gas, fluid, or a semisolid material, with a membranous lining.

cytokine—A generic term for nonantibody proteins released by one cell population on contact with a specific antigen, which act as intercellular mediators.

cytomegalovirus—One of a group of highly host-specific herpes viruses.

dementia—A progressive decline in mental function, memory, and acquired intellectual skills.

demyelination—The loss of the myelin covering that insulates the nerve tissue.

diabetic ketoacidosis—A type of metabolic acidosis caused by accumulation of ketone bodies in diabetes mellitus.

dialysate—The fluid that passes through the membrane in dialysis and contains the substances of greater diffusibility in solution.

diastolic—The pressure remaining in the arteries after cardiac contraction.

differential diagnosis—The determination of which of two or more diseases with similar symptoms is the one from which the patient is suffering, by a systematic comparison and contrasting of the clinical findings.

digoxin—A cardioactive steroid glycoside used to increase myocardial contractility.

diplegia—Paralysis of corresponding parts on both sides of the body. Fine motor function in the upper extremities may be affected and the trunk may be slightly affected, but the legs will be primarily affected.

diplopia—Double vision caused by a disorder of the nerves that innervate the extraocular muscles or by impaired function of the muscles themselves.

disability—Loss of physical function.

disimpaction—Manual removal of fecal material from rectal vault.

disordered eating—Inappropriate eating behaviors leading to insufficient energy intake.

distal—Away from the origin or centerline, as opposed to proximal.

Doppler ultrasonography—Application of the Doppler effect in ultrasound to detect movement of scatterers (usually red blood cells) by the analysis of the change in frequency of the returning echoes.

dorsal rhizotomy—A surgical procedure used to treat spasticity, particularly in young children, usually between 2 and 8 yr old with cerebral palsy, often referred to as selective dorsal rhizotomy. This surgical procedure permanently reduces spasticity by selectively cutting the abnormal sensory nerve rootlets.

double product—The value obtained by multiplying the heart rate and systolic blood pressure; an estimate of myocardial oxygen demand that is reproducible, such as at the ischemic threshold; used to determine confidence of results of a diagnostic exercise evaluation (i.e., value should be >24,000 for highest predictive confidence).

drug-eluting stent—Stent that slowly releases a drug (e.g., sirolimus), resulting in a reduction of restenosis rates.

dual-chamber pacemaker—Pulse generator that can pace or sense in the atrium or ventricle.

dual-energy X-ray absorptiometry (DXA)—A method for measuring bone mineral density and bone mineral content. It is based on the amount of radiation absorption, or attenuation, of the different body tissues. When bone mass is measured, the higher the attenuation of radiation by the bone, the greater the mass. Radiation exposure is minimal

(<5 mR) compared with a chest X-ray (100 mR) or lumbar X-ray (600 mR).

dual-photon absorptiometry—A method similar to DXA for measuring bone density but one that relies on a radionuclide source as opposed to X-ray. The photon intensity is not as great as with DXA, and precision is therefore reduced.

Duke nomogram—Five-step tool to estimate a person's prognosis (5 yr survival or average annual mortality rate) following completion of a maximum GXT.

dynamic endurance—Classification of exercise in which concentric–eccentric shifting occurs until muscular fatigue is induced. An example of dynamic endurance is performing biceps curls until fatigue occurs and the subject is unable to continue full motion against resistance.

dynamic flexibility (stretching)—Slow and constant stretch held for a period of time.

dynamic pulmonary function—Lung function in response to changing physiological conditions.

dysarthria—Difficulty speaking because of impairment of the tongue and other muscles essential for speech.

dyskinetic (pertaining to cardiac)—Denoting an outward or bulging movement of the myocardium during systole; often associated with aneurysm.

dyskinetic (pertaining to cerebral palsy)—Characterized by an abnormal amount and type of involuntary motion with varying amounts of tension, normal reflexes, and asymmetric involvement.

dyslipidemia—When blood lipid and lipoprotein concentrations are abnormally altered as a result of a combination of genetic, environmental, and pathological conditions.

dysmenorrhea—Pain in association with menstruation.

dyspareunia—Pain in the labial, vaginal, or pelvic areas during or after sexual intercourse.

dyspepsia—Stomach discomfort, including symptoms such as heartburn, gas, and acid reflux.

dysphagia—Difficulty in swallowing.

dyspnea—Shortness of breath, or labored or difficult breathing that is perceived by an individual at rest or with exertion (a.k.a., dyspnea on exercise, DOE). A scale can be used to determine the severity of dyspnea.

dystonia—Sustained muscle contractions that result in twisting and repetitive movements or abnormal posture.

eccentric lesion—A blockage that is equal distance away from the center of the artery—around the lining of the artery.

echocardiogram—An investigation of the heart and great vessels with ultrasound technology as a means to diagnose cardiovascular abnormalities.

echocardiography—Use of ultrasound images to evaluate the heart and great vessels.

economy—The rate of oxygen uptake necessary to perform a given activity.

ectopic pregnancy—Implantation of the fertilized ovum outside of the uterine cavity.

edema—A condition in which body tissue contains an excessive amount of fluid.

effective insulin—Insulin available for use by body tissues.

effusion—Excess synovial fluid within a joint.

ejection fraction (EF)—Percentage of blood that is ejected from the left ventricle per beat (normal 55–60%), EF = $[(EDV - ESV)/EDV] \times 100$, where EDV = end diastolic volume and ESV = end systolic volume. Normal is >55%. Decreases are noted with systolic heart failure to values below 35-40%.

elastin—Structural protein found in the walls of the alveoli.

electromyography—A diagnostic neurological test to study the potential (electrically measured activity) of a muscle at rest, the reaction of muscle to contraction, and the response to muscle insertion of a needle. The test is an aid in ascertaining whether a patient's illness is directly affecting the spinal cord, muscles, or peripheral nerves.

embolism—Obstruction of a blood vessel by foreign substances or a blood clot.

emphysema—Disease characterized by abnormal permanent enlargement of the respiratory bronchioles and the alveoli.

endocrine—Referring to glands that secrete hormones into the bloodstream.

endothelial cell—One of the squamous cells forming the lining of serous cavities, blood, and lymph vessels and the inner layer of the endocardium.

endothelial-derived relaxing factors—Diffusible substances produced by endothelial cells that cause vascular smooth muscle relaxation; nitric oxide is one such substance.

endothelium—A thin layer of cells that line the inner surface of blood vessels.

entheses—Sites where ligaments, tendons, or joint capsules are attached to bone.

enthesopathy—Inflammation at entheses.

environmental factors—Physical and social factors that can influence participation in physical activity (e.g., vehicular traffic, inclement weather, and unsafe neighborhoods).

eosinophils—A granular leukocyte that contains vasoactive amines.

epilepsy—The paroxysmal transient disturbances of brain function that may be manifested as episodic impairment or loss of consciousness, abnormal motor phenomena, psychic or sensory disturbances, or perturbation of the autonomic nervous system.

epistaxis—Bleeding from the nose.

erythrocyte sedimentation—The sinking of red blood cells in a volume of drawn blood.

erythropoiesis—Stimulation of red blood cell production.

estrogen replacement therapy—Therapy useful for protecting bone loss in postmenopausal women.

etiology—Cause.

evidence-based—Using the best available clinical research to guide treatment.

evoked response testing—Test in which brain electrical signals are recorded as they are elicited by specific stimuli of the somatosensory, auditory, and visual pathways.

evolving MI—Period of time after the acute onset of an MI when the myocardial tissue is transforming from ischemic to necrotic tissue.

exacerbation—A period of worsening symptoms.

excess body weight—Occurs when too few calories are expended and too many consumed for individual metabolic requirements. Overweight (>25) and obese (≥30) as defined by body mass index (kg · m^{-2}).

exertional ischemia—Myocardial ischemic response produced by exerting oneself physically.

extrapyramidal—The area of the brain that includes the basal ganglia and the cerebellum.

fatigue—That state, following a period of mental or bodily activity, characterized by a lessened capacity for work and reduced efficiency of accomplishment, usually accompanied by a feeling of weariness, sleepiness, or irritability; may also supervene when, from any cause, energy expenditure outstrips restorative processes and may be confined to a single organ.

fatty liver disease—Also know as steatorrhoeic hepatosis. This condition is characterized by the accumulation of triglycerides in the cells of the liver. Fatty liver disease is most commonly associated with excessive alcohol intake or obesity (nonalcoholic fatty liver disease).

fibrinolytic—Causing fibrinolysis, which is the breakdown of fibrin, a blood-coagulating protein.

foam cell—Smooth muscle cells that take up intimal lipid that accumulates in the cytoplasm and develops a bubbly appearance when observed microscopically.

follicle-stimulating hormone—Hormone produced by the anterior pituitary to stimulate the growth of the follicle in the ovary and spermatogenesis in the testes.

forced expiratory volume in 1 s (FEV$_1$)—Marker of airway obstruction; the maximum amount of air that can be exhaled in 1 s; may be expressed as an absolute value, a percentage of the forced vital capacity, or a percentage of a predicted value.

forced vital capacity (FVC)—The maximum amount of air that can be exhaled forcefully after a maximal inspiration.

frail—Having delicate health.

functional aerobic impairment (FAI)—Percentage of an individual's observed functional capacity that is below that expected for the individual's sex, age, and conditioning level. %FAI = (predicted $\dot{V}O_2$ – observed $\dot{V}O_2$)/(predicted $\dot{V}O_2$) × 100.

functional capacity—A person's maximum level of oxygen consumption. Can be measured at maximal effort with the use of a metabolic cart or predicted based on the maximum workload achieved.

functional electrical stimulation—Externally applied electrical stimulation of neuromuscular elements to activate paralyzed muscles in precise sequence and intensity to restore muscular function.

galactorrhea—Continuation of flow of milk after cessation of nursing.

gangrene—Necrosis of body tissues caused by obstruction, loss, or diminution of blood supply.

gastroepiploic—An artery with its origin in the stomach region used for coronary revascularization surgery.

gel phenomenon—The sensation of difficulty moving a joint after a period of joint rest or immobility.

genotype—The resultant expression of specific genes.

genu valgum—More commonly referred to as knock-knee deformity; a condition in which the knees angle in and touch when the legs are straightened.

genu varum—More commonly referred to as bowleg deformity; a condition in which the medial angulation of the leg in relation to the thigh results in an outward bowing of the legs.

geriatrics—A branch of medicine that deals with the problems and diseases associated with elderly people (>65 yr) and the aging process.

gestational diabetes—Carbohydrate intolerance of variable severity with onset or first recognition during pregnancy.

ghrelin—A potent appetite-increasing gut hormone.

gliosis—Excess of astroglia in damaged areas of the central nervous system.

glomerular filtration rate—The rate of filtrate formation at the glomerulus.

glomerulonephritis—An acute, subacute, or chronic, usually bilateral, diffuse inflammatory kidney disease that primarily affects the glomeruli.

glucagon—A hormone produced by the pancreas that stimulates the liver to release glucose, causing an increase in blood glucose levels, thus opposing the action of insulin.

glucose intolerance—A term used to describe a transitional state from normoglycemia to diabetes. Diagnosed when fasting blood glucose levels are ≥100 mg · dl^{-1} but <126 mg · dl^{-1} or when glucose levels are between 140 to 199 mg · dl^{-1} 2 h after a 75 g oral glucose tolerance test.

GLUT 4—Insulin-regulated glucose transporter responsible for the removal of glucose from blood and delivery to the inner cell membrane.

glycemic goals—A goal range for blood glucose concentration.

Golgi tendon organ—A proprioceptive sensory nerve ending embedded among the fibers of a tendon, often near the musculotendinous junction; it is compressed and activated by any increase of the tendon's tension, caused either by active contraction or passive stretch of the corresponding muscle.

gonadotropin-releasing hormone—Releasing hormone produced in the hypothalamus that acts on the pituitary and causes the release of gonadotropic substances, luteinizing hormone, and follicle-stimulating hormone.

graded exercise testing—Gradual increase in exercise workload to a predetermined point or until volitional fatigue, unless symptoms occur before this point. Generally completed on a treadmill or bicycle ergometer.

growth factor—A category of hormones responsible for stimulating the process of tissue growth.

hazard ratio—Multiplicative measure of association. Exposure to a certain risk factor or certain characteristic is associated with a fixed instantaneous risk, compared with the hazard in the unexposed.

HbA_{1C}—Refers to glycosylated hemoglobin. This form of hemoglobin is primarily used to identify the plasma glucose concentration. A very high HbA_{1C} (i.e., >7%) represents poor glucose control.

HDL cholesterol—high-density lipoprotein cholesterol. Binds to cholesterol and carries it back to the liver for extraction as bile. Low levels of HDL cholesterol (<40 mg · dl^{-1}) increase the risk of cardiovascular disease.

health belief model—Theory that proposes that only psychological variables influence health behaviors.

heart and lung bypass—Device for maintaining the functions of the heart and lungs while either or both are unable to continue to function adequately.

heart failure—The pathophysiological state in which an abnormality of cardiac function is responsible for failure of the heart to pump blood at a rate commensurate with the requirements of metabolizing tissues.

heart rate reserve (HRR)—The difference between a person's resting heart rate and maximal heart rate.

hematocrit—The percentage by volume of packed red blood cells in a sample of blood.

hematuria—Red blood cells in the urine.

hemianopia—Loss of vision for one-half of the visual field of one or both eyes.

hemiparesis—Paralysis affecting only one side of the body.

hemiplegia—Paralysis that affects only one side of the body.

hemolysis—Alteration or destruction of red blood cells.

hemophilia—A hereditary hemorrhagic diathesis caused by deficiency of coagulation factor VIII. Characterized by spontaneous or traumatic subcutaneous and intramuscular hemorrhages.

hemoptysis—Expectoration of blood arising from the respiratory system; for people with CF, this occurrence reflects further infection or advancing disease.

hemorrhagic—Of or relating to excessive bleeding.

herniation—Development of an abnormal protrusion or projection of an intervertebral disk.

high tone—Often referred to as spasticity or hypertonicity; excess tone in a muscle group.

hindfoot valgus—Excessive lateral deviation of the talocalcaneal complex relative to the trochanteric knee ankle (TKA) line.

hindfoot varus—Excessive medial deviation of the talocalcaneal complex relative to the trochanteric knee ankle (TKA) line.

histoplasmosis—Infection resulting from inhalation or, infrequently, ingestion of fungal spores. May cause pneumonia.

HIV negative—Describes an individual without antibodies to HIV viral proteins. An individual who has been recently infected with HIV may not have had an opportunity to have developed antibodies to the virus.

HIV positive—Describes an individual with antibodies to HIV viral proteins. Many times this is used to describe individuals who are infected but who have not yet developed an AIDS-defining condition or whose CD4+ cells are greater than 200 cells · mm^{-3}.

homocysteine—A homolog of cysteine.

hybrid—Combined use of concurrent upper-extremity exercise and lower-extremity functional electrical stimulation ergometry.

hydronephrosis—Kidney with dilated renal pelvis and collecting system attributable to ureteral obstruction or backflow (reflux) from bladder.

hypercapnia—An increased arterial carbon dioxide content.

hyperemia—Increased amount of blood in a part or organ.

hyperglycemia—An abnormally high concentration of glucose in the circulating blood, seen especially in people with diabetes mellitus.

hyperinflation—An overinflated lung resulting in a greater functional residual capacity and total lung capacity.

hyperkalemia—Excess concentrations of potassium in the bloodstream.

hyperlipidemia—Elevated lipid levels in the blood.

hyperparathyroidism—A state produced by the increased function of the parathyroid glands; results in dysregulation of calcium.

hyperpnea—More rapid and deeper breathing than normal.

hyperreflexia—A condition in which the deep tendon reflexes are exaggerated and are defined by overactive or overresponsive reflexes, which may include twitching and spastic tendencies.

hypertension—A condition in which blood pressure is chronically elevated. Diagnosed when resting systolic blood pressure ≥140 mmHg and or resting diastolic blood pressure ≥90 mmHg on two separate occasions.

hypertonia—Increased rigidity, tension, and spasticity of the muscles.

hypertrophy—An increase in cell size.

hyperuricemia—Presence of high levels of uric acid in the blood. Diagnosed when uric acid levels ≥7 mg · dl^{-1} in men and ≥6 mg · dl^{-1} in women.

hypoestrogenic—Decreased plasma estrogen levels.

hypoglycemia—Symptoms resulting from low blood glucose (normal glucose range 60–100 mg · dl^{-1}, or 3.3–5.6 mmol · L^{-1}) that are either autonomic or neuroglycopenic.

hypokalemia—Extreme potassium depletion in the circulating blood.

hypotension—Abnormally low blood pressure; typically associated with symptoms.

hypotonic—Having a lesser degree of tension.

hypovolemia—Diminished blood volume.

hypoxemia—Insufficient oxygenation of the blood; assessed by arterial blood gas or pulse oximetry.

hypoxia—A state of oxygen deficiency.

idiopathic—Denoting a disease of unknown cause.

immune system—System that mediates the body's interaction between internal and external environments. Helps rid the body of infectious agents and malignant cells.

immunosuppression—Suppression of immune responses produced primarily by a variety of immunosuppressive agents.

incidence—The frequency of occurrence of any event or condition over a period of time and in relation to the population in which it occurs.

incomplete paraplegia—Incomplete motor and sensory dysfunction of the trunk, legs, and pelvic organs resulting from SCI.

incomplete tetraplegia—Incomplete motor and sensory dysfunction of the arms, trunk, legs, and pelvic organs resulting from SCI.

incontinence—Lack of control of urination or defecation.

insensate—Lacking sensation.

inspiratory resistive loading—The act of inspiring air against a resistance greater than normal.

inspiratory threshold loading—The act of inspiring after attaining and proceeding at a predetermined inspiratory pressure (threshold point).

insulin resistance—A condition in which normal amounts of insulin secreted by the pancreas are inadequate to produce a normal insulin response in the muscle or liver. As a result, the pancreas secretes additional insulin thereby elevating insulin levels in the plasma. High levels of insulin in the plasma often lead to the development of type 2 diabetes or metabolic syndrome.

intermittent claudication—Recurrent cramping symptoms, usually of the calf muscles, at regular intervals when walking.

internal mammary—An artery with its origin in the chest region used for coronary revascularization surgery.

intima—The inner layer of blood vessels containing endothelial cells.

intraosseous—Within bone.

iritis—Inflammation of the iris of the eye.

ischemia—Deficiency of blood flow with respect to demand, attributable to functional constriction or actual obstruction of a blood vessel.

ischemic—A sustained deficiency in oxygen delivery.

ischemic heart disease—A pathological condition in which blood flow to the myocardium is reduced below the demand, resulting in a lack of oxygen delivery to cardiac tissue (i.e., coronary atherosclerosis).

isocapnia—Normal arterial carbon dioxide levels.

isokinetic—Denoting the condition in which muscle fibers shorten at a constant speed in such a manner that the tension developed may be maximal over the full range of joint motion.

isometric—Denoting the condition in which the ends of a contracting muscle are held fixed so that contraction produces increased tension at a constant overall length.

isotonic—Denoting the condition in which muscle fibers shorten with varying tension as the result of a constant load.

joint contractures—Reduced passive range of motion at a joint caused by shortened tendons, typically associated with unbalanced spasticity.

joint effusion—Increased fluid in synovial cavity of a joint.

joule (J)—A unit of energy; the heat generated, or energy expended, by an ampere flowing through an ohm for 1 s; equal to 107 erg and to a newton meter. The joule is an approved multiple of the SI fundamental unit of energy, the erg, and is intended to replace the calorie (4.184 J).

Kaposi's sarcoma—Firm, subcutaneous, brown-black or purple lesions usually observed on the face, chest, genitals, oral mucosa, or viscera.

ketones—A substance with the carbonyl group linking two carbon atoms.

kilocalorie (kcal)—A unit of heat content or energy. The amount of heat necessary to raise 1 g of water from 14.5 °C to 15.5 °C, times 1,000.

kyphosis—Excessive angulation of the spine resulting in increased anteroposterior diameter of the chest cavity; humpback; may reflect chronic pulmonary disease.

laminectomy—Removal of the bony arch on the posterior part of the vertebra that surrounds and protects the spinal cord, thereby creating more space for the protruding disk material.

lead-pipe rigidity—Diffuse muscular rigidity resembling the resistance to bending of a thin-walled metal pipe. Lead-pipe rigidity persists throughout the range of movement of a particular joint, as distinct from clasp-knife rigidity, which varies.

left ventricular dysfunction—Abnormal function of the left ventricle (i.e., poor wall motion).

leptin—A protein messenger from adipose tissue to the satiety center in the hypothalamus involved in regulating appetite.

lifestyle-based physical activity—Home- or community-based participation in many forms of activity that include much of a person's daily routine (e.g., transport, home repair and maintenance, yard maintenance).

low-calorie diet—A hypocaloric diet of 800 to 1,500 kcal per day.

low energy availability—Also known as energy deficit, energy drain, or negative energy balance. Results from low dietary energy intake and high energy expenditure.

low tone—Often referred to as flaccidity or hypotonia; decreased amount of tone in a muscle group.

lower rate limit—The rate at which the pulse generator begins pacing in the absence of intrinsic activity.

luteinizing hormone—Hormone secreted by the anterior lobe of the pituitary to stimulate the development of the corpus luteum.

LV diastolic dysfunction—Clinically, diagnosis is less exact than systolic dysfunction. Diagnosis is often made when the clinical syndrome of congestive HF (fatigue, dyspnea, and orthopnea) requires hospitalization in the presence of a relatively normal ejection fraction.

LV systolic dysfunction—Ejection fraction reduced below 45% (<30% considered severe) as measured by echocardiogram or another quantitative measure.

lymphocytes—Any of the mononuclear, nonphagocytic leukocytes found in the blood, lymph, or lymphoid tissues that are the body's immunologically competent cells.

lymphocytopenia—A reduction in the number of lymphocytes in the circulating blood.

macrovascular disease—Atherosclerosis that affects large vessels such as the aorta, femoral artery, and carotid artery.

magnetic resonance imaging (MRI)—Medical imaging that uses nuclear magnetic resonance as its energy source and allows higher resolution and no radiation compared with computed tomography. A detailed cross-section of the structure of the tissue can be obtained using MRI.

major depressive disorder—A diagnosis requiring endorsement of at least five symptoms, one of which must be either depressed mood or diminished interest or pleasure, that have been present during the same 2 wk period.

maladaptation—Adaptation to a progressive stimulus (e.g., exercise) that results in an overload to the system to the degree that performance is reduced and the risk of injury is increased.

mast cells—Connective tissue cells that are important in the defense mechanisms of the body needed during injury or infection.

maximal bone mass—The highest bone mass that a person could possibly achieve.

maximal oxygen consumption—The maximum amount of oxygen consumed (or used) by the body, usually measured under conditions of maximal exercise.

maximal voluntary ventilation—Amount of air maximally breathed in expressed as liters per minute.

maximum sensor rate—The maximum rate for a rate-responsive pacemaker that can be achieved under sensor control.

maximum tracking rate—The maximum rate at which the pulse generator will respond to atrial events.

medical nutrition therapy—The use of nutrition as a treatment for a clinical condition or disease.

menarche—The beginning of menstrual function.

mentation—The process of reasoning and thinking.

metabolic equivalent (MET)—An expression of the rate of energy expenditure at rest. 1 MET = 1 kcal $\cdot$ kg^{-1} $\cdot$ hr^{-1} = 3.5 ml $\cdot$ kg^{-1} $\cdot$ min^{-1}.

metabolic syndrome—A constellation of insulin resistance characterized by central obesity, elevated triglycerides, suppressed HDL-cholesterol, hypertension, or prediabetes.

metastases—The spread of a disease process from one part of the body to another, as in the appearance of neoplasms in parts of the body remote from the site of the primary tumor.

microalbuminuria—A condition in which the kidneys leak a small amount of albumin into the urine. Diagnosed with 24 h urine collections ($\geq$20 μg $\cdot$ min^{-1}) or when levels are $\geq$30 μg $\cdot$ min^{-1} on two separate occasions.

microvascular disease—Atherosclerosis that affects small blood vessels such as those of the kidney, eye, heart, and brain.

mode switching—A programmed feature of dual-chamber pacemakers that prevents tracking or matching every atrial impulse with a ventricular pacing pulse to prevent tracking of rapid atrial rates to the ventricle.

moderate-intensity physical activity—Activities that cause small increases in breathing and heart rate; ~50–70% $\dot{V}O_2$peak.

monoplegia—Paralysis of a single limb, muscle, or muscle group.

monounsaturated fats—Dietary fatty acid that contains one double bond along the main carbon chain.

morbidity—Manifestations of disease other than death.

morphology—Configuration or shape (e.g., shape of the ST segment: downsloping, upsloping, or horizontal).

mortality—Death.

motion artifact—Incidental activity that is picked up on an ECG during body movement.

multiple sclerosis—A debilitating disease characterized by multiple areas of scar tissue replacing myelin around axons in the central nervous system.

muscle spindle receptors—A fusiform end organ in skeletal muscle in which afferent and a few efferent nerve fibers terminate; this sensory end organ is particularly sensitive to passive stretch of the muscle in which it is enclosed.

muscle tone—Amount of tension in a muscle or muscle group at rest.

muscle weakness—Refers to skeletal muscle that lacks strength and power-generating capability.

muscular strength and endurance—The ability of skeletal muscles to perform hard or prolonged work.

musculoskeletal flexibility—The range of motion in a joint or sequence of joints.

myalgia—Pain in a muscle or muscles.

***Mycobacterium avium* complex**—Complex that consists of two predominant species, *M. avium* and *Mycobacterium intracellulare*. More than 95% of infections in patients with AIDS are caused by *M. avium*, whereas 40% of infections in immunocompetent patients are caused by *M. intracellulare*.

myelomeningocele—Congenital open neural tube defect with disruption of skin, bone, and neural elements; usually involves spinal cord dysfunction despite surgical closure.

myocardium—The heart muscle.

myoglobinuria—Excretion of myoglobin in the urine resulting from muscle degeneration.

nasal polyposis—Growths of tissue in the nose that may block the air passage through the nostril; not life threatening.

neoplasm—Abnormal tissue that grows by cellular proliferation more rapidly than normal and continues to grow after the stimuli that initiated the growth ceases. Structural organization and function of neoplastic tissue are partially or completely different from the normal tissue.

neurectomy—Partial or total excision or resection of a nerve.

neuroglycopenic hypoglycemia—Symptoms of hypoglycemia that include feelings of dizziness, confusion, tiredness, difficulty speaking, headache, and inability to concentrate.

neuromuscular—Of, relating to, or affecting both nerves and muscles.

neuropathy—A disease involving the cranial nerves or the peripheral or autonomic nervous system.

neuropeptide Y (NPY)—Central nervous system appetite stimulant.

neutral spine—The position in which the trunk and neck, and therefore the joints of the spine, are neither in flexion nor extension.

neutrophilia—An increase in neutrophilic leukocytes in blood or tissue.

nitroglycerine—Medication that is used to promote vasodilation in patients with angina pectoris.

nonischemic cardiomyopathy—Disease process involving cardiac muscle that is not related to ischemic heart disease; may be attributable to viral infection or alcohol abuse.

nuclear perfusion—Radioactive isotope that has the ability to perfuse through tissue, so that select organs can be imaged.

nucleoside reverse transcriptase inhibitor (NRTI)—A specific type of antiretroviral medication.

nystagmus—Rhythmic, involuntary movements of the eyes.

obesity—A condition of having an abnormally high proportion of body fat. Defined as having a body mass index greater than 30.

occipital—The posterior part of the skull.

old age—Between 65 and 74 yr of age.

oldest old—Older than 85 yr of age.

oligomenorrhea—Scanty or infrequent menstrual flow.

oliguria—A diminution in the quantity of urine excreted: specifically, less than 400 ml in a 24 h period.

omega-3 fatty acids—Long-chain polyunsaturated fatty acids that contain a double bond in the n-3 position.

opportunistic infections—Infections that are most commonly seen in people who are immunocompromised, such as individuals with late or advanced HIV-1 disease, cancer, or other immunocompromising conditions.

optic neuritis—Inflammation of the optic nerve.

orthopnea—Labored or difficult breathing while lying flat or supine.

orthostatic—Relating to upright or erect posture.

orthostatic hypotension—Decrease of at least 20 mmHg in systolic blood pressure when an individual moves from a supine position to a standing position.

ostectomy—Surgical excision of a bone or a portion of one.

osteoarthritis (OA)—Erosion of articular or joint cartilage that leads to pain and loss of function.

osteogenic—Increasing bone mass.

osteopenia—Reduced bone mineral density, defined as being between 1 and 2.5 standard deviations below the young adult mean.

osteophyte—A bony excrescence or outgrowth, usually branched in shape.

osteoporosis—A pathological condition associated with increased susceptibility to fracture and decreased bone mineral density more than 2.5 standard deviations below the young adult mean.

osteoporotic fracture—Broken bone caused by a reduction in the mass of the bone per unit of volume.

osteotomy—Operation for cutting through a bone to improve alignment or correct deformities.

overweight—A body mass index of 25.0 to 29.9.

oxygen uptake (consumption)—A measure of a person's ability to take in and use oxygen.

oxyhemoglobin saturation—Percentage of hemoglobin bound to oxygen; assessed noninvasively by pulse oximeter or invasively by arterial blood gas sampling.

pacemaker sensor—Sensor incorporated into the pulse generator that detects a physiological stimulus to control the heart rate to match physiological demands.

pack years—Number of packs of cigarettes smoked per day multiplied by the number of years smoked; for example, if a person smoked two packs a day for 20 yr, she or he would have a 40 pack year history.

palliative surgery—Surgery aimed at removal of tumor to make patient more comfortable, relieve organ obstruction, or reduce tumor burden.

pancreas—A gland lying behind the stomach that secretes pancreatic enzymes into the duodenum and that secretes insulin, glucagon, and somatostatin into the bloodstream.

pancreatic insufficiency—Inadequate exocrine function of the pancreas resulting in little or no production of pancreatic enzymes needed for digestion (i.e., lipase, amylase, protease); results in nutrient malabsorption.

pansinusitis—Chronic inflammation and infection involving all sinuses; commonly seen in individuals with CF.

papilledema—Swelling of the optic disk in the eye caused by severe hypertension.

paralysis—Loss or impairment of the ability to move a body part, usually as a result of damage to its nerve supply, loss of sensation over a region of the body, or complete loss of motor control.

paraplegia—Motor and sensory dysfunction of the trunk, legs, and pelvic organs resulting from SCI.

parasympathetic nervous system—Craniosacral portion of the central nervous system that promotes anabolic activity and energy conservation.

parathyroid hormone—A peptide hormone formed by the parathyroid glands, it raises the serum calcium when administered parenterally by causing bone resorption.

parenchyma—The essential or primary tissue of the lungs.

paresis—Refers to slight or partial paralysis, or partial weakness to one or more limbs.

paresthesia—A subjective feeling such as numbness, "pins and needles," or tingling.

paroxysmal nocturnal dyspnea—Sudden awakening caused by labored or difficult breathing.

peak bone density—The highest amount of bone mass achieved by a person during his or her lifetime. Peak bone mass is assumed to be achieved in the second or third decade of life. The age at which peak bone mass is achieved will also vary based on which bone site is being measured.

peak expiratory flow rate (PEFR)—The highest flow rate (exhalation of gas from the lung) that a person can generate during a forceful expiration.

pericardial effusion—Increased amounts of fluid within the pericardial sac, usually attributable to inflammation.

perineum—The area between the scrotum and anus in males and between the vulva and anus in females.

periodization—A system of fractioning larger periods of muscle training into smaller phases or cycles. Intensity, frequency, sets, repetitions, and rest periods are altered to reduce the risk of overtraining and minimize uncomfortable responses.

peripheral nervous system—Sensory and motor components of the nervous system that have extensions outside the brain and spinal cord.

peripheral neuropathy—Damage to the nerves of the legs or arms resulting in a loss of sensation (e.g., touch, temperature).

peripheral vascular disease—Disease of the vascular system that can be found in the periphery (i.e., commonly observed in the legs, which leads to claudication with physical exertion).

peritoneal dialysis—Dialysis performed by introducing fluid into the peritoneal cavity. Dialysis fluid can be cycled through the peritoneal cavity by a machine over a 10 to 12 h period daily (intermittent peritoneal dialysis) or exchanged every 4 h with the fluid staying in the peritoneal cavity between exchanges (continuous ambulatory peritoneal dialysis). The fluid is introduced through a catheter (tube) that is placed in the abdomen.

pes equinus—A condition marked by walking without touching the heel to the ground.

physical activity—Any bodily movement produced by skeletal muscles that results in caloric expenditure.

plaque—A yellow area or swelling on the intimal surface of an artery, produced by the atherosclerotic process of lipid deposition.

plaques—Scarring of axons in the central nervous system attributable to demyelination.

plasticity—The extent to which normal maturation of maximal aerobic power can be altered by changes in physical activity; also, the ability of tissue to change shape due to a stimulus.

platelet aggregation—The congregation of platelets, which are disk-shaped fragments found in the peripheral blood and involved in the clotting process.

plegia—Greater involvement of one or more limbs than paresis; often associated with paralysis.

pluripotent stem cell—Uncommitted cell with various developmental options pending.

***Pneumocystis carinii* pneumonia (PCP)**—An AIDS-defining condition caused by the parasite *P. carinii*.

pneumothorax—An acute collection of air in the pleural space; results in collapse of the affected lung; common in advancing cystic fibrosis lung disease.

poikilothermic—Having body temperature that varies with the environment.

point of maximal cardiac impulse (PMI)—Point identified by palpation and inspection of the chest wall during physical examination as the most prominent location for cardiac impulse.

polycystic kidney disease—Hereditary bilateral cysts distributed throughout the renal parenchyma resulting in markedly enlarged kidneys and progressive renal failure.

polycystic ovarian syndrome—Endocrine disturbance associated with primary anovulation and polycystic ovaries.

polycythemia—An abnormally elevated level of red cells in the blood.

polydipsia—Excessive thirst that is relatively prolonged.

polymorphonuclear leukocytosis—An elevation in neutrophilic leukocyte (white blood cell) count.

polyunsaturated fats—Dietary fatty acids that contain two or more double bonds along the main carbon chain.

polyuria—Excessive excretion of urine.

postprandial—1 to 2 h following a meal.

prevalence—The number of cases of a disease present in a specified population at a given time. This number may be given at one identified time (point prevalence) or in a specified period, such as 2 wk or a year (period prevalence).

prevention—Three general categories of intervention strategies to limit the effect of potential or established disease in the population.

primary risk reduction—Intervention geared toward removing or reducing the risk factors of disease.

primary amenorrhea—Delay of menarche beyond age 18.

protease inhibitor—A specific type of antiretroviral medication.

proteinuria—The presence of abnormal amounts of protein in the urine.

prothrombotic—Condition or agent that increases the risk of formation or presence of a thrombus.

pulmonary exacerbation—An episode of worsening lung disease caused by increased infection and inflammation.

pulsatile—Characterized by a rhythmical pulsation.

pyelonephritis—The disease process from the immediate and late effects of bacterial and other infections of the parenchyma and the pelvis of the kidney.

pyramidal—An area of the brain that consists of the cortical system, which controls all voluntary movements.

Q angle—Acute angle formed by a line from the anterior superior iliac spine of the pelvis through the center of the patella and a line from tibial tubercle through the patella.

quadriplegia—Paralysis of all four limbs; also called tetraplegia.

quality of life—Perception of life satisfaction.

quantitative computed tomography—The only method currently available that provides an actual measurement of volumetric bone density.

quantitative ultrasound—A device that measures structural properties of bone with sound waves. It uses no ionizing radiation like densitometry devices.

radiation therapy—Therapy meant to stop growth of malignant cells by damaging RNA within the cells.

radionuclide agent—Isotope (natural or artificial) that exhibits radioactivity. Used in nuclear cardiology medicine to image the myocardium for potential ischemia.

radionuclide imaging—A type of cardiac imaging that can detect ischemia and wall motion; uses an injected radioisotope (i.e., thallium 201 or technitium-99m sestamibi) that is scanned using X-ray.

ramping treadmill protocol—Continuous gradual increase in workload (speed and grade) over a select period of time.

range of motion—The total degrees of movement that a joint can move through.

rate pressure product (RPP)—Indirect indication of how hard the heart is working. RPP = systolic BP × HR the same as double-product.

rate-responsive pacing—Refers to a pacemaker that changes the rate by sensing a physiological stimulus. Other terms to describe this type of pacemaker are rate modulated, adaptive rate, or sensor driven.

rating of perceived exertion (RPE)—A person's perception of how hard he or she is working physically. There are currently two commonly used scales to assess RPE.

red flags—Signs or symptoms that act as warning signs of severe local or systemic pathology.

refractory—Resilient or resistant to treatment.

rejection—Immune response to foreign tissue (transplanted organ).

relapse—Reversion to an active disease process in MS after a remission.

relapse prevention model—A model used to help new exercisers anticipate problems with adherence. Factors that contribute to relapse include negative emotional or physiological states, limited coping skills, social pressure, interpersonal conflict, limited social support, low motivation, high-risk situations, and stress.

relative contraindications for test termination—Conditions that occur during a stress test that require strong clinical judgment concerning the safety of continuing the exercise test.

relative oxygen uptake—Oxygen uptake expressed in milliliters of oxygen per kilogram of body weight per minute $(ml \cdot kg^{-1} \cdot min^{-1})$.

remission—Recovery period from the active disease process.

renal replacement therapy—Medical technologies that serve as substitutes for renal function include hemodialysis, peritoneal dialysis, and transplantation. Without this therapy, the patient with no renal function would die.

reocclusion—To close again; reclosure.

reperfusion—The process of reinstituting blood flow to an area of tissue previously deprived of normal blood flow.

repetition maximum (RM)—The number of times that a weight can be lifted; 1RM is the maximal amount of weight that can be lifted one time only.

resolving MI—The phase of an MI in which necrotic tissue forms a scar.

respiratory failure—Failure of the respiratory system to keep gas exchange at an acceptable level.

restenosis—To become narrow or restricted again.

retrovirus—Viruses containing both RNA-dependent and DNA-synthesizing material.

revascularization—Restoration of blood flow to a body part.

rhabdomyolysis—Acute, potentially fatal disease of the skeletal muscle that entails destruction of the muscle as evidenced by myoglobinuria.

rheology—The study of the deformation and flow of liquids and semisolids.

rheumatoid arthritis (RA)—Inflammation of the joints attributable to autoimmune attack that leads to pain and loss of function.

rigidity—Stiffness or inflexibility that may result in the inability to bend or be bent.

roentgenogram—A photograph made with X-rays.

sclerosis—Tissue hardening that occurs because scar tissue replaces lost myelin around axons in the central nervous system.

scoliosis—Abnormal lateral curvature (side-bending and rotational components) of the spine that may be congenital or acquired by extremely poor posture, disease, or muscular weakness. Usually the curvature consists of two curves, the original abnormal curve and a second compensatory curve in the opposite direction (also referred to as an S-curve).

secondary—Intervention that promotes early detection and treatment of disease with the goals of preventing disease recurrences or progression, promoting recovery, and avoiding complications.

secondary amenorrhea—Cessation of menses in a woman who has previously menstruated.

secondary conditions—An injury, impairment, functional limitation, or disability that occurs as a result of the primary condition or pathology. Secondary conditions include physical problems, social concerns, and mental health difficulties. Secondary conditions also can develop when the primary disability interferes with the delivery of standard healthcare for the treatment or prevention of a health condition.

selective estrogen receptor modulators (SERMs)—Antiresorptive agents that have fewer side effects than estrogen replacement therapy and may be a good alternative to ERT for the woman with a history of breast cancer.

selective serotonin reuptake inhibitors—A class of antidepressant medication commonly used to treat depression.

self-efficacy—A person's belief in his or her capability to perform the behavior and perceived incentive to do so.

sensitivity—The proportion of affected people who give a positive test result for the disease that the test is intended to reveal.

set-point theory—A metabolic theory that postulates stability in weight in both overfeeding and calorie restriction situations.

sick sinus syndrome—Syndrome in which the sinus node is not functioning at an appropriate rate, leading to sinus bradycardia, pauses, arrest, or exit block. Syncopal episodes can be caused by this abnormality.

significant Q wave—Depicts a prior MI on an ECG. For Q waves to be considered significant, they must be either greater than or equal to 0.04 sec wide and one-third the height of the associated R wave.

single-chamber pacemaker—Pulse generator that can pace or sense in the atrium or ventricle.

single-photon absorptiometry—A method for determining bone mineral content by measuring the absorption by bone of a monoenergetic photon beam.

social cognitive theory—Theory that behavior change is affected by environmental influences, personal factors, and attributes of the behavior itself.

social support—Support and encouragement that a person receives from others to maintain behavior change.

soft tissue—The total amount of tissue in the body minus bone mass as determined by DXA.

somatic nervous system—Neural elements over which a person has conscious awareness and control.

somatosensory evoked potentials—Physiological or electrical stimulation of afferent peripheral nerve fibers with subsequent monitoring of (evoked) electrical activity in the central nervous system, usually at the levels of the spinal cord, midbrain, and cortex.

spasticity—Of, relating to, or characterized by spasms, an involuntary increase in muscle tone.

spastic-paresis—Characterized by a posture- and movement-dependent tone regulation disorder. The clinical symptoms are the loss or absence of tone in lying and an increase in tone in sitting, standing, walking, or running, depending on the degree of involvement.

specificity—The proportion of people with negative test results for the disease that the test is intended to reveal.

spina bifida—Congenital neural tube defect with varying degrees of skin, bone, and neural element involvement.

spinal cord injury (SCI)—Damage involving the spinal cord.

spinal decompression—Surgical intervention to excise bony or soft-tissue structures that exert pressure on neural tissues in the spine.

spinal discectomy—Surgical intervention to excise the portion of the herniated disk that is causing compression on neural tissue. The extent of the tissues removed is based on the extent of the intervertebral disk herniation.

spinal fusion—Surgical intervention to fixate unstable hypermobile vertebral segments by the use of metal plates, screws, wires, and autologous bony transplants.

spinal traction—Use of specialized harness systems and electronic winch or manually applied distractive forces on the spine in a variety of spinal and bodily positions. The purpose of this modality is to separate vertebrae and stretch the associated soft tissues, thus decompressing nerve roots and relieving symptoms.

spirometry—The measurement of the breathing capacity of the lungs.

spondyloarthropathy—A type of inflammatory arthritis involving ligament or tendon insertion sites (enthuses), leading to spinal and peripheral joint arthritis, usually in human lymphocyte antigen-B27-positive individuals.

spondylolisthesis—Forward subluxation (malalignment) of superior lumbar vertebra on an inferior vertebra, leading to traction or compression of nerve roots and intervertebral supportive soft tissues and causing associated irritation and nociceptive input. This condition may be benign, depending on the amount of the slippage.

staging—A system used to classify the extent and spread of cancer.

static endurance—Classification of exercise in which isometric contractions of muscle groups lead to anaerobic exhaustion. An example of static endurance is a prolonged trunk extension position until fatigue occurs and the subject is unable to hold the position.

static pulmonary function—Properties of the lung measured at rest or baseline.

stenosis—Constriction or narrowing of a passage or orifice. In spinal stenosis, congenital or degenerative narrowing of the intervertebral or vertebral foramen (opening) is present, leading to compressive forces on the nerve roots that travel through these openings.

stent—A stainless steel bridge, expanded by a balloon-tipped catheter, designed to hold open an area of stenosis within an artery.

sternotomy—The operation of cutting through the sternum.

strabismus—A deviation of the eye that the individual cannot overcome. The visual axes assume a position relative to each other different from that required by the physiological conditions. The various forms of strabismus are spoken of as tropias. Their direction is indicated by the appropriate prefix, as cyclotropia, esotropia, exotropia, hypertropia, and hypotropia. Also called cast, heterotropia, manifest deviation, and squint.

strategies to promote adherence—Techniques commonly used among exercise professionals to improve initiation and compliance to a structured exercise regimen.

stratify—The process of separating individuals or samples into subcategories based on variables of interest (e.g., sex, age, number of risk factors, symptoms).

strength—Maximal voluntary contractile force of a given muscle group or groups.

stress echocardiogram—Combining an exercise test with an echocardiogram. Resting and postexercise echocardiogram images are compared for wall motion abnormalities that can suggest an ischemia response.

subarachnoid—Space in the brain under the arachnoid membrane containing cerebrospinal fluid.

subchondral sclerosis—Thickening of the bone beneath the cartilage layer of an arthritic joint.

subendocardial—Referring to the endocardial surface of the heart.

subendocardial ischemia—Myocardial ischemic response beneath the endocardium.

subepithelial fibrosis—The structural changes noted beneath the epithelial layer of the bronchus resulting in scar tissue formation in this area.

sweat test—Diagnostic test used for cystic fibrosis; usually involves stimulation of the skin's sweat glands by chemical (i.e., pilocarpine) and electrical means (i.e., iontophoresis); an elevation of greater than 60 mEq/dl is highly suggestive of CF.

symmetric—Equality or correspondence in the form of parts on the opposite sides of any body.

sympathetic nervous system—Lumbosacral portion of the central nervous system that promotes the classic fight-or-flight response to a given stimuli.

syncope—Loss of consciousness caused by diminished cerebral blood flow.

synovial joint—A joint in which the opposing bony surfaces are covered with a layer of hyaline cartilage or fibrocartilage and is nourished and lubricated by synovial tissue.

synovitis—Swelling within a joint attributable to inflammation of the synovial lining.

syringomyelia—Chronic syndrome characterized pathologically by cavitation and gliosis of the spinal cord (usually cervical or thoracic), medulla, or both.

systemic—The arterial system supplying the body.

systolic—The pressure generated in the arteries by contraction of the heart muscle.

tachycardia—Heart rate greater than 100 beats · min^{-1}.

technetium 99m sestamibi—A radioisotope, introduced into the bloodstream by a catheter, that tags red blood cells and when imaged using a gamma camera can provide a measure of ventricular volume, ejection fraction, and regional ventricular wall motion at rest and during exercise. Used to depict myocardial ischemia.

tenotomy—Irreversible surgical section of severely contracted tendons attached to muscles that do not respond to any other type of spasticity control and are causing intractable pain and skin complications related to lack of physical movement.

tertiary—Intervention designed to reduce the functional effect of an illness or disability.

tetraplegia—Motor and sensory dysfunction of the arms, trunk, legs, and pelvic organs resulting from spinal cord injury.

thalamotomy—Destruction of a selected portion of the thalamus for the relief of pain and involuntary movements.

thallium 201—A white metallic substance with radioactivity, introduced into the bloodstream by a catheter, that is perfused into the myocardium; used in conjunction with stress testing (exercise and pharmacological) to image the myocardium to detect transient ischemia and tissue necrosis.

T-helper lymphocyte—Lymphocytes whose secretions and other activities coordinate the cellular and humoral immune responses.

theory of planned behavior—Theory that adds to the theory of reasoned action with the concept of perceived control over the opportunities, resources, and skills necessary to perform a behavior.

theory of reasoned action—Theory that performance of a given behavior (e.g., exercise) is primarily determined by the person's attitude toward the behavior and the influence of the person's social environment or subjective norm (i.e., beliefs about what other people think the person should do as well as the person's motivation to comply with the opinions of others).

thrombolytic—Agents that degrade fibrin clots by activating plasminogen, a naturally occurring modulator of hemostatic and thrombotic processes.

thrombosis—The formation, development, or existence of a clot or thrombus within the vascular system.

thromboxane—Vasoconstrictor and platelet activation substance.

thrombus—An aggregation of blood factors, primarily platelets and fibrin with entrapment of cellular elements, which frequently causes vascular obstruction at the point of formation.

trabecular bone—One of the two main types of bone tissue, also known as cancellous or spongy bone. Trabecular bone is made up of interlacing plates of bone tissue and is found mainly at the ends of long bones and within the vertebrae.

tracking—The concept that risk factors or other conditions that are expressed in childhood will persist and be expressed in adulthood.

transcutaneous electrical nerve stimulation—Use of small battery-operated or plugged-in devices for delivery of electrical current across the skin to provide patients with pain relief, artificial contraction of muscles, fatigue of spastic muscles, and pulsations to decrease swelling in a joint. The stimulation is given through electrode pads placed directly on the skin over the muscles selected to be stimulated or inhibited from nociceptive input or in areas determined by nerve supply or to acupuncture points. The underlying theories are based on the gate theory of pain control, in which sensory stimulation inhibits pain transmission at the spinal cord level, or stimulation of Aδ and C fibers, to cause the release of endogenous opiates.

transmural—Referring to effects on all tissue layers of the heart.

transmural ischemia—Myocardial ischemic response that occurs throughout the myocardial wall.

transtheoretical model of behavior change—Model wherein behavior change is conceptualized as a five-stage process or continuum related to a person's readiness to change: precontemplation, contemplation, preparation, action, and maintenance. People are thought to progress throughout the stages.

tremor—Repetitive, often regular, oscillatory movements caused by alternate, or synchronous, but irregular contraction of opposing muscle groups.

triglycerides—Chemical storage form of fat in the body. Hypertriglyceridemia ($\geq$150 mg · dl^{-1}) increases the risk of cardiovascular disease.

triplegia—A form of cerebral palsy that affects three limbs. The most common pattern is for both legs and one arm to be affected. Triplegia is sometimes thought of as hemiplegia overlapping with diplegia because the primary motor difficulty is with the legs.

true maximal heart rate—The highest heart rate as measured at, or near, the end of an exercise test; as opposed to peak heart rate estimated by the equation 220 – age.

type 2 diabetes mellitus—The most common form of diabetes affecting approximately 90% to 95% of all those with diabetes. With type 2 diabetes, abnormal blood glucose regulation is often a result of insulin resistance of the peripheral tissues and defective insulin secretion.

ultrasound—Use of sonic wave energy, created by a vibrating quartz crystal, used to deliver heat or medication to healing musculoskeletal structures. A variety of machines deliver the ultrasonic waves through a transducer rubbed directly over the skin using gel or water as transmitting medium.

upper rate limit—The highest rate at which ventricular pacing will track 1:1 each sensed atrial event.

uremia—A complex biochemical abnormality occurring in kidney failure, characterized by azotemia, chronic acidosis, anemia, and a variety of systemic and neurologic symptoms and signs.

valgus—Bowlegged deformity.

Valsalva maneuver—Forced exhalation with the glottis, nose, and mouth closed, resulting in increased intrathoracic pressure, slowing of the heart rate, and decreased return of blood to the heart.

variant angina—Angina pectoris occurring during rest; not necessarily preceded by exercise or an increase in heart rate.

varus—Knock-kneed deformity.

vascular pathologies—Manifestations of disease in blood vessels.

vasospastic—Contraction or spasm of the muscular coats of the blood vessels.

ventilatory derived lactate threshold (V-LT)—The point where a nonlinear increase in blood lactate occurs during exercise; when determined with ventilatory parameters is sometimes referred to as ventilatory threshold.

ventilatory muscle training—Specific exercises that are used to increase respiratory muscular strength.

ventricular hypertrophy—Muscle thickening in a pumping chamber of the heart

vertigo—A sensation of spinning or whirling motion.

very old age—Between 75 and 84 yr of age.

vesicoureteral reflux—Backflow of urine from the bladder into the upper urinary tracts.

vigorous-intensity physical activity—Activities that result in large increases in breathing and heart rate; >70% $\dot{V}O_2$peak.

viremia—Viral particles in the blood.

virion—A single, encapsulated piece of viral genetic material.

visceral fat—One of three compartments of abdominal fat. Others are retroperitoneal and subcutaneous.

wall motion—Relating to movement of the left ventricular segments of the heart; used to describe normal or abnormal movement during contraction and to calculate ejection fraction during two-dimensional echocardiography or some types of nuclear imaging.

wasting—Involuntary loss of more than 10% of body weight.

white matter—Regions of the brain and spinal cord that are largely or entirely composed of nerve fibers and contain few or no neuronal cell bodies or dendrites.

REFERENCES

CHAPTER 1

1. American Association of Cardiovascular and Pulmonary Rehabilitation: www.aacvpr.org
2. American Council on Exercise: www.acefitness.org
3. American College of Sports Medicine: www.acsm.org
4. American Society for Exercise Physiologists: www.asep.org
5. Canadian Society of Exercise Physiology: www.csep.ca
6. Cooper KH. Aerobics. New York: Evans; 1968.
7. Saltin B, et al. Response to submaximal and maximal exercise after bedrest and training. Circulation 1968; 38(Suppl 7).
8. U.S. Department of Health and Human Services. Physical activity and health: A report of the surgeon general. Atlanta: U.S. Department of Health and Human Services, Centers for Disease Control and Prevention, National Center for Chronic Disease Prevention and Health Promotion; 1996.
9. Wilmore JH, Costill DL, Kenney L. Physiology of sport and exercise. 4th ed. Champaign, IL: Human Kinetics; 2007.

CHAPTER 2

1. U.S. Department of Health and Human Services. Physiologic responses and long-term adaptations to exercise. In: Physical activity and health: a report of the surgeon general. Atlanta: U.S. Department of Health and Human Services, Centers for Disease Control and Prevention, National Center for Chronic Disease Prevention and Health Promotion; 1996. p 61–73.
2. Sidney KH, Shephard RJ, Harrison JE. Endurance training and body composition of the elderly. Am J Clin Nutr 1977;30:326–333.
3. Smith DM, Khairi MRA, Norton J, et al. Age and activity effects on rate of bone mineral loss. J Clin Invest 1976;58:716–721.
4. Paffenbarger RS, Hyde RT, Wing AL. Physical activity as an index of heart attack risk in college alumni. Am J Epidemiol 1978;108:161–175.
5. Tipton CH. Exercise, training, and hypertension: an update. Exerc Sport Sci Rev 1991;19:447–505.
6. Lee IM, Paffenbarger RS, Hsieh CC. Physical activity and risk of developing colorectal cancer among college alumni. J Natl Cancer Inst 1991;83:1324–1329.
7. Paffenbarger RS Jr, Hyde RT, Wing AL, Hsieh CC. Physical activity, all-cause mortality, and longevity of college alumni. N Engl J Med 1986;314:605–613.
8. Ntoumanis N, Biddle SJ. A review of motivational climate in physical activity. J Sports Sci 1999;17:643–665.
9. Shephard RJ. Exercise and aging: extending independence in older adults. Geriatrics 1993;48:61–64.
10. Caspersen CJ, Powell KE, Christenson GM. Physical activity, exercise, and physical fitness: definitions and distinctions for health-related research. Public Health Rep 1985;100:126–131.
11. U.S. Department of Health and Human Services. Healthy people 2010. 2nd ed. Understanding and improving health and objectives for improving health. 2 vols. Washington, DC: U.S. Government Printing Office; November 2000.
12. Pate RR, Pratt M, Blair SN, et al. Physical activity and public health: a recommendation from the Centers for Disease Control and Prevention and the American College of Sports Medicine. JAMA 1995;273:402–407.
13. Haskell WL, Lee I-M, Pate RR, Powell KE, Blair SN, Franklin BA, Macera CA, Heath GW, Thompson PD, Bauman A. Physical activity and public health. Updated recommendation for adults from the American College of Sports Medicine and the American Heart Association. Circulation 2007;116.
14. American College of Sports Medicine. Position stand on the recommended quantity and quality of exercise for developing and maintaining cardiorespiratory and muscular fitness in healthy adults. Med Sci Sports Exerc 1990;22:265–274.
15. Braith RW, Graves JE, Pollock JL, Leggett SL, Carpenter DM, Colvin AB. Comparison of 2 vs. 3 days/week of variable resistance training during 10- and 18-week programs. Int J Sports Med 1989;10:450–454.
16. Barry HC, Eathorne SW. Exercise and aging: issues for the practitioner. Med Clin North Am 1994;78:357–376.
17. Cady LD, Bischoff DP, O'Connell ER, Thomas PC, Allan JH. Strength and fitness and subsequent back injuries in firefighters. J Occup Med 1979;21:269–272.
18. West Virginia University, Department of Safety and Health Studies and Department of Sports and Exercise Studies. Physical fitness and the aging driver: phase I. Washington, DC: AAA Foundation of Traffic Safety; 1988.
19. Passmore R. The regulation of body weight in man. Proc Nutr Soc 1971;30:122–127.
20. Wood PD, Stefanick ML, Dreon DM, Frey-Hewitt D, Garay BC, Williams PT, Superko HR, Fortmann SP, Albers JJ, Vranizan KM, Ellworth NM, Terry RB, Haskell WL. Changes in plasma lipids and lipoproteins in overweight men during weight loss through dieting as compared with exercise. N Engl J Med 1988;319:1173–1179.
21. Sallis JF, Hovell MF. Determinants of exercise behavior. Exerc Sports Sci Rev 1990;18:307–330.
22. King AC. Intervention strategies and determinants of physical activity and exercise behavior in adult and older adult men and women. World Rev Nutr Diet 1997;82:148–158.
23. Pollock ML. Prescribing exercise for fitness and adherence. In: Dishman RK, editor. Exercise adherence. Champaign, IL: Human Kinetics; 1988. p 259–277.
24. Sallis JF, Hovell MF, Hofstetter CR. Predictors of adoption and maintenance of vigorous physical activity in men and women. Prev Med 1992;21:237–251.
25. Sallis JF, Hovell MF, Hofstetter CR, et al. A multivariate study of determinants of vigorous exercise in a community sample. Prev Med 1989;18:20–34.
26. King AC, Haskell WL, Taylor CB, Kraemer HC, DeBusk RF. Group- vs. home-based exercise training in healthy older men and women. JAMA 1991;266:1535–1542.
27. Kriska AM, Bayles C, Cauley JA, et al. Randomized exercise trial in older women: increased activity over two years and the factors associated with compliance. Med Sci Sports Exerc 1986;18:557–562.
28. Young DR, Haskell WL, Taylor CB, Fortmann SP. Effect of community health education on physical activity knowledge, attitudes, and behavior: the Stanford Five-City Project. Am J Epidemiol 1996;144:264–274.
29. Hassman PR, Ceci R, Backman L. Exercise for older women: a training method and its influences on physical and cognitive performance. Eur J Appl Physiol 1992;64:460–466.
30. King AC, Haskell WL, Young DR, Oka RK, Stefanick ML. Long-term effects of varying intensities and formats of physical activity

on participation rates, fitness, and lipoproteins in men and women aged 50 to 65 years. Circulation 1995;91:2596–2604.

31. American College of Sports Medicine. Guidelines for exercise testing and prescription. 4th ed. Philadelphia: Lea & Febiger; 1991.

32. Dishman RK. Increasing and maintaining exercise and physical activity. Behav Ther 1991;22:345–378.

33. U.S. Department of Health and Human Services. Understanding and promoting physical activity. In physical activity and health: a report of the surgeon general. Atlanta: U.S. Department of Health and Human Services, Centers for Disease Control and Prevention, National Center for Chronic Disease Prevention and Health Promotion; 1996. p 209–259.

34. Rosenstock IM. The health belief model: explaining health behavior through expectancies. In: Glanz K, Lewis FM, Rimer BK, editors. Health behavior and health education. Theory, research, and practice. San Francisco: Jossey-Bass; 1990. p 39–62.

35. Marlatt GA, George WH. Relapse prevention and the maintenance of optimal health. In: Shumaker SA, Schron EB, Ockene J, editors. The handbook of health behavior change. New York: Springer; 1990. p 44–63

36. Marcus BH, Stanton AL. Evaluation of relapse prevention and reinforcement interventions to promote exercise adherence in sedentary females. Res Q Exerc Sport 1993;64:447–452.

37. Ajzen I, Fishbein M. Understanding attitudes and predicting social behavior. Englewood Cliffs, NJ: Prentice Hall; 1980.

38 Ajzen I. Attitudes, personality, and behavior. Chicago: Dorsey Press; 1988.

39. Bandura A. Self-efficacy: toward a unifying theory of behavioral change. Psychol Rev 1977;84:191–215.

40. Bandura A. Social learning theory. Englewood Cliffs, NJ: Prentice Hall; 1977.

41. Bandura A. Social foundations of thought and action: a social–cognitive theory. Englewood Cliffs, NJ: Prentice Hall; 1986.

42. Prochaska JO, DiClemente CC. The transtheoretical approach: crossing traditional boundaries of change. Homewood, IL: Dorsey Press; 1984.

43. Long BJ, Calfas KJ, Patrick K, Sallis JF, Wooten WJ, Goldstein M, Marcus B, Schwenck T, Carter R, Torez T, Palinkas L, Heath G. Acceptability, usability, and practicality of physician counseling for physical activity promotion: project PACE. Am J Prev Med 1996;12:73–81.

44. Calfas KJ, Long BJ, Sallis JF, Wooten WJ, Pratt M, Patrick K. A controlled trial of physician counseling to promote the adoption of physical activity. Prev Med 1996;25:225–233.

45. Marcus BH, Banspach SW, Leffebvre RC, Rossi JS, Carleton RA, Abrams DB. Using the stages of change model to increase adoption of physical activity among community participants. Am J Health Promot 1992;6:424–429.

46. Wankel LM, Yardley JK, Graham J. The effects of motivational interventions upon the exercise adherence of high and low self-motivated adults. Can J Appl Sport Sci 1985;10:147–156.

47. Martin JE, Dubbert PM, Katell AD, et al. The behavioral control of exercise in sedentary adults: studies 1 through 6. J Consult Clin Psychol 1984;52:795–811.

48. McAuley E, Courneya DS, Rudolph DL, Lox CL. Enhancing exercise adherence in middle-aged males and females. Prev Med 1994;23:498–506.

49. Taylor CB, Miller NH, Smith PM, DeBusk RF. The effect of home-based, case-managed, multifactorial risk reduction program on reducing psychological distress in patients with cardiovascular disease. J Cardiopulm Rehabil 1997;17:157–162.

50. Dishman RK, Buckworth J. Increasing physical activity: a quantitative synthesis. Med Sci Sports Exerc 1996;28:706–719.

51. Dunn AL, Marcus BH, Kampert JB, et al. Comparison of lifestyle and structured interventions to increase physical activity and cardiorespiratory fitness: a randomized trial. JAMA 1999;281:327–334.

52. Winett R. A framework for health promotion and disease prevention programs. Am Psychol 1995;50:341–350.

53. King AC. Community and public health approaches to the promotion of physical activity. Med Sci Sports Exerc 1994;26:1405–1412.

54. Calfas KJ, Sallis JF, Zabinski BA, Wilfley BE, Rupp R, Prochaska JJ, Thompson T, Pratt M, Patrick K. Preliminary evaluation of a multicomponent program for nutrition and physical activity change in primary care: PACE+ for adults. Prev Med 2002;34:153–161.

55. Stokols D. Establishing and maintaining healthy environments: toward a social ecology of health promotion. Am Psychol 1992;47:6–22.

56. Baranowski T, Lin LS, Wetter DW, Resnicow K, Hearn MD. Theory as mediating variables: why aren't community interventions working as desired? Ann Epidemiol 1997;S7:S89–S95.

57. Sallis JF, Bauman A, Pratt M. Environmental and policy interventions to promote physical activity. Am J Prev Med 1998;15:379–396.

58. New South Wales Physical Activity Task Force. Simply active every day: a discussion document from the NSW Physical Activity Task Force on proposals to promote physical activity in NSW, 1997–2002. Summary report. Sydney, Australia: New South Wales Health Department; 1997.

59. McLeroy KR, Bibeau D, Steckler A, Glanz K. An ecological perspective on health promotion programs. Health Educ Q 1988;15:351–377.

60. Linenger JM, Chesson CV 2nd, Nice DS. Physical fitness gains following simple environmental change. Am J Prev Med 1991;7:298–310.

61. Epstein LH, Wing RR. Aerobic exercise and weight. Addict Behav 1980;5:371–388.

62. Blamey A, Mutrie N, Aitchison T. Health promotion by encouraged use of stairs. Br Med J 1995;311:289–290.

63. Brownell K, Stunkard AJ, Albaum J. Evaluation and modification of exercise patterns in the natural environment. Am J Psychiatry 1980;137:1540–1545.

64. Kahn EB, Ramsey LT, Brownson RC, Heath GW, Howze EH, Powell KE, Stone EJ, Rajab MW, Corso P. The effectiveness of interventions to increase physical activity: a systematic review. Am J Prev Med 2002;22(4S):73–107.

65. Heath GW, Brownson RC, Kruger J, Miles R, Powell KE, Ramsey LT. The effectiveness of urban design and land use and transport policies and practices to increase physical activity: a systematic review. J Phys Act Health 2006;1:S55–S71.

CHAPTER 3

1. United States Anti-Doping Agency. 2007 guide to prohibited substances and prohibited methods of doping. 7th ed. Colorado Springs, CO: USADA; 2006.

2. National Collegiate Athletic Association. NCAA banned-drug classes 2007–2008. [Online]. www1.ncaa.org/membership/ed_outreach/health-safety/drug_testing/banned_drug_classes.pdf

3. Sproat TT. Anemias. In: Dipiro JT, editor. Pharmacotherapy: a pathophysiologic approach. 4th ed. Stamford, CT: Appleton & Lange; 1999. p 1531–1548.

4. Balaban EP. Sports anemia. Clin Sports Med 1992;11(2):313–325.

5. Ineck B, Mason BJ, Thompson EG. Anemias. In: Dipiro JT, Talbert RL, Yee GC, et al., editors. Pharmacotherapy: a pathophysiologic approach. 6th ed. New York: McGraw-Hill; 2005. p 1805–1831.

6. Nielsen P, Nachtigall D. Iron supplementation in athletes. Current recommendations. Sports Med 1998;26(4):207–216.

7. Macdougall IC. An overview of the efficacy and safety of novel erythropoiesis stimulating protein (NESP). Nephrol Dial Transplant 2001;16(Suppl 3):14–21.

8. Aranesp (darbepoetin alfa) package insert. Amgem, Inc. March 2006.

9. Ekblom BT. Blood boosting and sport. Best Pract Res Clin Endocrinol Metab 2000;14(1):89–98.

10. Triplitt CL, Reasner CA, Isley WL. Diabetes mellitus. In: Dipiro JT, Talbert RL, Yee GC, et al., editors. Pharmacotherapy: a pathophysiologic approach. 6th ed. New York: McGraw-Hill; 2005. p 1333–1367.

11. Actos (pioglitazone) package insert. Eli Lily and Company. August 2004.

12. Avandia (rosigliatazone) package insert. GlaxoSmithKline. May 2006.

13. Mooradian AD, Thurman JE. Drug therapy of postprandial hyperglycemia. Drugs 1999;57(1):19–29.

14. Ristic S, Bates PC. Vildagliptin: a novel DPP-4 inhibitor with pancreatic islet enhancement activity for treatment of patients with type 2 diabetes. Drugs of Today 2006;42(8):519–531.

15. Ahren B, Landin-Olsson M, Jansson P, et al. Inhibition of dipeptidyl peptidase-4 reduces glycemia, sustains insulin levels, and reduces glucagons levels in type 2 diabetes. J Clin Endocrinol Metab 2004;89(5):2078–2084.

16. Triplitt C, Wright A, Chiquette E. Incretin mimetics and dipeptidyl peptidase-IV inhibitors: potential new therapies for type 2 diabetes mellitus. Pharmacotherapy 2006;26(3):360–374.

17. Symlin package insert. Amylin Pharmaceuticals. June 2005.

18. Byetta (exenatide) package insert. Amylin Pharmaceuticals. February 2007.

19. Lean M, Finer N. ABC of obesity management: part II-Drugs. BMJ 2006;333(7572):794–797.

20. St. Peter JV, Khan MA. Obesity. In: Dipiro JT, Talbert RL, Yee GC, et al., editors. Pharmacotherapy: a pathophysiologic approach. 6th ed. New York: McGraw-Hill; 2005. p 2659–2676.

21. Bray GA, Greenway FL. Current and potential drugs for treatment of obesity. Endocrine Rev 1999;20(6):805–875.

22. American Heart Association. Unpublished data. Dallas: American Heart Association; 2001.

23. Talbert RL. Ischemic heart disease. In: Dipiro JT, editor. Pharmacotherapy: a pathophysiologic approach. 4th ed. Stamford, CT: Appleton & Lange; 1999. p 182–210.

24. Goldstein S. Beta-blocking drugs and coronary heart disease. Cardiovasc Drugs Ther 1997;11:219–225.

25. Opie LH. Calcium channel antagonists in the treatment of coronary artery disease: fundamental pharmacological properties relevant to clinical use. Prog Cardiovasc Dis 1996;38:273–290.

26. Ferrari R. Major differences among the three classes of calcium antagonists. Eur Heart J 1997;18(Suppl A):A56–A70.

27. Bauman JL, Schoen MD. Arrhythmias. In: Dipiro JT, editor. Pharmacotherapy: a pathophysiologic approach. 4th ed. Stamford, CT: Appleton & Lange; 1999. p 232–264.

28. Chobanian AV, Bakris GL, Black HR, et al. The seventh report of the Joint National Committee on Prevention, Detection, Evaluation, and Treatment of High Blood Pressure (JNC-7). Hypertension 2003;42:1206–1252.

29. Valvo E, D'Angelo G. Diuretics in hypertension. Kidney Int 1997;59(Suppl):S36–S38.

30. Dormans TP, van Meyel JJM, Gerlag PGG, et al. Diuretic efficacy of high dose furosemide in severe heart failure: bolus injection versus continuous infusion. J Am Coll Cardiol 1996;28:376–382.

31. Kaplan NM. Calcium entry blockers in the treatment of hypertension. JAMA 1989;262:817–823.

32. Nadelmann J, Frishman WH. Clinical use of beta adreno receptor blockade in systemic hypertension. Drugs 1990;39:862–876.

33. Heidenreich PA, Lee TT, Massie BM. Effect of beta blockade on mortality in patients with heart failure: a meta-analysis of randomized clinical trials. J Am Coll Cardiol 1997;30:27–34.

34. Krumholz HM, Anderson JL, Brooks NH, et al. ACC/AHA clinical performance measures for adults with ST-elevation and non-ST-elevation myocardial infarction. A report of the American College of Cardiology/American Heart Association Task Force on Performance Measures (Writing Committee to Develop Performance Measures on ST-Elevation and Non-ST-Elevation Myocardial Infarction). Circulation 2006;113:732-761.

35. Perry HM. Central and peripheral sympatholytics. In: Izzlo JL, Black HR, editors. Hypertension primer. Dallas: American Heart Association; 1993. p 306–308.

36. Packer M, Fowler MB, Roecker EB, et al. Effect of carvedilol on the morbidity of patients with severe chronic heart failure: results of the Carvedilol Prospective Randomized Cumulative Survival (COPERNICUS) Study. Circulation 2002;106:2194–2199.

37. Henriksen T. Hypertension in pregnancy: use of antihypertensive drugs. Acta Obstet Gynecol Scand 1997;76:96–106.

38. Venkata C, Ram S, Featherston WE. Vasodilators. In: Izzlo JL, Black HR, editors. Hypertension primer. Dallas: American Heart Association; 1993. p 314–316.

39. Ferrario CM. Importance of the rennin–angiotensin–aldosterone system (RAS) in the physiology and pathology of hypertension. Drugs 1990;39(Suppl 2):1–8.

40. American College of Cardiology/American Heart Association Task Force on Practice Guidelines. Guideline update for the diagnosis and management of chronic heart failure in the adult—summary article. J Am Coll Cardiol 2005;46:1116–1143.

41. Antman EM, Anbe DT, Armstrong PW, et al. ACC/AHA guidelines for the management of patients with ST-elevation myocardial infarction—executive summary. Circulation 2004;110:588–636.

42. Messerli FH, Weber MA, Brunner HR. Angiotensin II receptor inhibition. A new therapeutic principle. Arch Intern Med 1996;156:1957–1965.

43. Black HR. Evolving role of aldosterone blockers alone and in combination with angiotensin-converting enzyme inhibitors or angiotensin II receptor blockers in hypertension management: a review of mechanistic and clinical data. Am Heart J 2004;147:564–572.

44. Kelly RA, Smith TW. Pharmacologic treatment of heart failure. In: Hardman JG, Limbird LE, Molinoff PB, et al., editors. Goodman and Gilman's pharmacological basis of therapeutics. 9th ed. New York: McGraw Hill; 1996. p 875–898.

45. Ansell JE, Buttaro ML, Thomas OV, et al. Consensus guidelines for coordinated outpatient oral anticoagulation therapy management. Ann Pharmacother 1997;31:604–615.

46. Ansell J, Hirsh J, Poller L, et al. The pharmacology and management of the vitamin K antagonists. Chest 2004;126(Suppl):204S–233S.

47. Braunwald E, Antman EM, Beasley JW, et al. ACC/AHA 2002 guideline update for the management of patients with unstable angina and non-ST-segment-elevation myocardial infarction—summary article. A report of the American College of Cardiology/American Heart Association Task Force on Practice Guidelines (Committee on the Management of Patients with Unstable Angina). J Am Coll Cardiol 2002;40(7):1366–1374.

48. Spinler SA, de Denus S. Acute coronary syndromes. In: Dipiro JT, Talbert RL, Yee GC, et al., editors. Pharmacotherapy: a pathophysiologic approach. 6th ed. New York, McGraw-Hill; 2005. p 291–319.

49. Smith SC Jr, Dove JT, Jacobs AK, et al. ACC/AHA guidelines for percutaneous coronary intervention (revision of the 1993 PTCA guidelines)—executive summary. J Am Coll Cardiol 2001;37(8):2215–2238.

50. Talbert RL. Hyperlipidemia. In: Dipiro JT, editor. Pharmacotherapy: a pathophysiologic approach. 4th ed. Stamford, CT: Appleton & Lange; 1999. p 350–373.

51. Anderson GL, Limacher M, Assaf AR, et al. Effects of conjugated equine estrogen in postmenopausal women with hysterectomy: the Women's Health Initiative randomized control trial. JAMA 2004;291:1701–1712.

52. Rossouw JE, Anderson GL, Prentice RL, et al. Risks and benefits of estrogen plus progestin in healthy postmenopausal women: principal results from the Women's Health Initiative randomized controlled trial. JAMA 2002;288:321–333.

53. Dailey JH, Gray DR, Bradberry JC, et al. Lipid-modifying drugs. In: McKenney JM, Hawkins D, editors. Handbook on the management of lipid disorders. 2nd ed. St. Louis: National Pharmacy Cardiovascular Council; 2001. p 124–166.

54. Mahley RW, Weisgraber KH, Farese RV. Disorders of lipid metabolism. In: Larsen PR, Kronenberg HM, Melmed S, et al., editors. Williams textbook of endocrinology. Orlando, FL: Saunders; 1998. p 1099–1154.

55. Ast M, Frishman WH. Bile acid sequestrants. J Clin Pharmacol 1990;30:99–106.

56. Belalcazar LM, Ballantyne CM. Defining specific goals of therapy in treating dyslipidemia in the patient with low high-density lipoprotein cholesterol. Prog Cardiovas Dis 1998;41:151–174.

57. Farmer JA, Gotto AM. Dyslipidemia and other risk factors for coronary artery disease. In: Brauwnwald E, editor. Heart disease: a textbook of cardiovascular medicine. Orlando, FL: Saunders; 1997. p 1126–1160.

58. Guidelines for the diagnosis and management of asthma. Expert Panel Report 2. Clinical Practice Guidelines. National Institutes of Health, National Heart, Lung, and Blood Institute. NIH Publication No. 97-4051, July 1997. p 1–153.

59. Milavetz G, Smith JJ. Cystic fibrosis. In: Dipiro JT, Talbert RL, Yee GC, et al., editors. Pharmacotherapy: a pathophysiologic approach. 6th ed. New York: McGraw-Hill; 2005. p 591–603.

60. McManus Balmer C, Wells Valley A, Iannucci A. Cancer treatment and chemotherapy. In: Dipero JT, Talbert RL, Yee GC, et al., eds. Pharmacotherapy: a pathophysiologic approach. 6th ed. New York: McGraw-Hill; 2005. p 2279–2328.

61. American Cancer Society, www.cancer.org

62. Calabresi P, Chabner BA. Chemotherapy of neoplastic diseases. In: Hardman JG, Limbird LE, Molinoff PB, et al., editors. Goodman & Gilman's the pharmacological basis of therapeutics. 10th ed. New York: McGraw-Hill; 2001. p 1381–1388.

63. Fletcher CV, Kakuda TN. Human immunodeficiency virus infection. In: Dipero JT, Talbert RL, Yee GC, et al., editors. Pharmacotherapy: a pathophysiologic approach. 6th ed. New York: McGraw-Hill; 2005. p 2255–2277.

64. DHHS Panel on Antiretroviral Guidelines for Adults and Adolescents—A Working Group of the Office of AIDS Research Advisory Council (OARAC). Guidelines for the use of antiretroviral agents in HIV-1-infected adults and adolescents; 2006 (www.AIDSinfo.nih.gov).

65. Kohl NE, Emini EA, Schleif WA, et al. Active human immunodeficiency virus protease is required for viral infectivity. Proc Natl Acad Sci USA 1988;85:4686–4690.

66. Gilbert DN, Moellering RC, Eliopoulos GM, Sande MA, editors. The Sanford guide to antimicrobial therapy 2006. 36th ed. Sperryville, VA: Antimicrobial Therapy; 2006. p 117–123.

67. National Osteoporosis Foundation. Physician's guide to prevention and treatment of osteoporosis; 2003. www.nof.org/physguide/table_of_contents.htm

68. O'Connell MB, Seaton TL. Osteoporosis and osteomalacia. In: Dipero JT, Talbert RL, Yee GC, et al., editors. Pharmacotherapy: a pathophysiologic approach. 6th ed. New York: McGraw-Hill; 2005. p. 1645–1669.

69. Rodan GA. Bone mass homeostasis and bisphosphonate action. Bone 1997;20:1–4.

70. Fogelman I, Ribot C, Smith R, et al. Risedronate reverses bone loss in postmenopausal women with low bone mass: results from a multinational, double-blind, placebo-controlled trial. BMD-MN study group. J Clin Endocrinol Metab 2000;85:1895–1900.

71. Rosen CJ, Hochberg MC, Bonnick SL, et al. Treatment with once weekly alendronate 70 mg compared with once weekly risedronate 35 mg in women with postmenopausal osteoporosis: a randomized double-blind study. J Bone Miner Res 2005;20:141–151.

72. Uchida S, Taniguchi T, Shimizu T, et al. Therapeutic effects of alendronate 35 mg once weekly and 5 mg once daily in Japanese patients with osteoporosis: a double-blind, randomized study. J Bone Miner Metab 2005;23:383–388.

73. Reid DM, Hughes RA, Laan RF, et al. Efficacy and safety of daily risedronate in the treatment of corticosteroid-induced osteoporosis in men and women: a randomized trial. European Corticosteroid-Induced Osteoporosis Treatment Study. J Bone Miner Res 2000;15:1006–1013.

74. Saag KG, Emkey R, Schnitzer TJ, et al. Alendronate for the prevention and treatment of glucocorticoid-induced osteoporosis. Glucocorticoid-Induced Osteoporosis Intervention Study Group. NEJM 1998;339:292–299.

75. Rizzoli R, Greenspan SL, Bone G, et al. Two-year results of once-weekly administration of alendronate 70 mg for the treatment of postmenopausal osteoporosis. J Bone Miner Res 2002;17:1988–1996.

76. Schnitzer T, Bone HG, Crepaldi G, et al. Therapeutic equivalence of alendronate 70 mg once-weekly and alendronate 10 mg daily in the treatment of osteoporosis. Alendronate Once-Weekly Study Group. Aging (Milano) 2000;12:1–12.

77. Emkey RD, Ettinger M. Improving compliance and persistence with bisphophonate therapy for osteoporosis. Am J Med 2006;119:S18–24.

78. Delmas PD, Bjarnason NH, Mitlak BH, et al. Effects of raloxifene on bone mineral density, serum cholesterol concentrations and uterine endometrium in postmenopausal women. NEJM 1997;l337:1641–1647.

79. Lufkin EG, Whitaker MD, Nickelsen T, et al. Treatment of established postmenopausal osteoporosis with raloxifene: a randomized trial. J Bone Miner Res 1998;13:1747–1754.

80. Walsh BW, Kuller LH, Wild RA, et al. Effects of raloxifene on serum lipids and coagulation factors in healthy postmenopausal women. JAMA 1998;279:1445–1451.

81. Cauley JA, Norton L, Lippman ME, et al. Continued breast cancer risk reduction in postmenopausal women treated with raloxifene: 4-year results from the MORE trial. Multiple outcomes of raloxifene evaluation. Breast Cancer Res Treat 2001;65:125–134.

82. Chesnut CH, Silverman S, Andriano K, et al. A randomized trial of nasal spray salmon calcitonin in postmenopausal women with established osteoporosis: the prevent recurrence of osteoporotic fractures study. Am J Med 2000;109:267–276.

83. Ishida Y, Kawai S. Comparative efficacy of hormone replacement therapy, etidronate, calcitonin, alfacalcidol, and vitamin K in postmenopausal women with osteoporosis: the Yamaguchi Osteoporosis Prevention Study. Am J Med 2004;117:549–555.

84. Karachalios T, Lyritis GP, Kaloudis J, et al. The effects of calcitonin on acute bone loss after pertrochanteric fractures. A prospective, randomized trial. J Bone Joint Surg Br 2004;86:350–358.

85. Garnero P, Hausherr E, Chapuy MC, et al. Markers of bone resorption predict hip fracture in elderly women: the EPIDOS Prospective Study. J Bone Miner Res 1996;11:1531–1538.

86. Body JJ, Gaich GA, Scheele WH, et al. A randomized, double-blind trial to compare the efficacy of teriparatide [recombinant human parathyroid hormone (1-34)] with alendronate in postmenopausal women with osteoporosis. J Clin Endocrinol Metab 2002;87:4528–4535.

87. Neer RM, Arnaud CD, Zanchetta JR, et al. Effect of parathyroid hormone (1-34) on fractures and bone mineral density in postmenopausal women with osteoporosis. NEJM 2001;344:1434–1441.

88. Ettinger BF, Genant HK, Cann CE. Long-term estrogen replacement therapy prevents bone loss and fractures. Ann Intern Med 1985;102:319–324.

89. Hillard TC, Whitcroft SJ, Marsh MS, et al. Long-term effects of transdermal and oral hormone replacement therapy on postmenopausal bone loss. Osteoporosis Int 1994;4:341–348.

90. Kohrt WM, Birge SJJ. Differential effects of estrogen treatment on bone mineral density of the spine, hip, wrist, and total body in late postmenopausal women. Osteoporos Int 1995;5:150–155.

91. Solomon CG, Dluhy RG. Rethinking postmenopausal hormone therapy. NEJM 2003;348:579–580.

92. Schuna AA. Rheumatoid arthritis. In: Dipero JT, Talbert RL, Yee GC, et al., editors. Pharmacotherapy: a pathophysiologic approach. 6th ed. New York: McGraw-Hill; 2005. p 1671–1683.

93. Trappe TA, White F, Lambert CP, et al. Effect of ibuprofen and acetaminophen on postexercise muscle protein synthesis. Am J Physiol Endrocrinol Metab 2002;282(3):E551–556.

94. Farquhar WB, Morgan AL, Zambraski EJ, et al. Effects of acetaminophen and ibuprofen on renal function in the stressed kidney. J Appl Physiol 1999;86(2):598–604.

95. Sanders LR. Exercise-induced acute renal failure associated with ibuprofen, hydrochlorothiazide, and triamterene. J Am Soc Nephrol 1995;5(12):2020–2023.

96. Carter JR, Sauder CL, Ray CA. Effect of morphine on sympathetic nerve activity in humans. J Appl Physiol 2002;93(5):1764–1769.

97. Cook DB, O'Connor PJ, Ray CA. Muscle pain perception and sympathetic nerve activity to exercise during opiod modulation. Am J Physiol Regul Integr Comp Physiol 2000;279(5):R1565–1573.

98. Santiago TV, Johnson J, Riley DJ, et al. Effects of morphine on ventilatory response to exercise. J Appl Physiol 1979;47(1):112–118.

99. Marquet P, Lac G, Chassain AP, et al. Dexamethasone in resting and exercising men. I. Effects on bioenergetics, minerals, and related hormones. J Appl Physiol 1999;87:175–182.

100. Price SR. Increased transcription of ubiquitin-proteasome system components: molecular responses associated with muscle atrophy. In J Biochem Cell Bio 2003;35(5):617–628.

101. Simon LS. Biologic effects of nonsteroidal antiinflammatory drugs. Curr Opin Rheumatol 1997;9:178–182.

102. Allison MC, Howatson AG, Torrance CJ, et al. Gastrointestinal damage associated with the use of nonsteroidal anti-inflammatory drugs. N Engl J Med 1992;327:749–754.

103. Bennett WM, Henrich WL, Stoff JS. The renal effects of nonsteroidal anti-inflammatory drugs: summary and recommendations. Am J Kidney Dis 1996;28(Suppl 1):S56–S62.

104. Lane JM. Anti-inflammatory medications: selective COX-2 inhibitors. J Am Acad Orthop Surg 2002;10(2):75–78.

105. Mukherjee D, Nissen SE, Topol EJ. Risk of cardiovascular events associated with selective COX-2 inhibitors. JAMA 2001;286:954–959.

106. Koman LA, Paterson Smith B, Shilt JS. Cerebral palsy. Lancet 2004;363:1619–1631.

107. Albright, AL. Baclofen in the treatment of cerebral palsy. J Child Neurol 1996;11:77–83.

108. Kirkwood CK, Melton ST. Anxiety disorders I: generalized anxiety, panic, and social anxiety disorders. In: Dipero JT, Talbert RL, Yee GC, et al., editors. Pharmacotherapy: a pathophysiologic approach. 6th ed. New York: McGraw-Hill; 2005. p 1285–1305.

109. Kando JC, Wells BG, Hayes PE. Depressive disorders. In: Dipero JT, Talbert RL, Yee GC, et al., editors. Pharmacotherapy: a pathophysiologic approach. 6th ed. New York: McGraw-Hill; 2005. p 1235–1255.

110. Doering PL. Substance-related disorders: overview of depressants, stimulants and hallucinogens. In: Dipero JT, editor. Pharmacotherapy: a pathophysiologic approach. 4th ed. Stamford, CT: Appleton & Lange; 1999. p 1083–1103.

111. Doering PL. Substance-related disorders: alcohol, nicotine, and caffeine. In: Dipero JT, editor. Pharmacotherapy: a pathophysi-

ologic approach. 4th ed. Stamford, CT: Appleton & Lange; 1999. p 1104–1117.

112. Parssinen M, Seppala T. Steroid use and long-term health risks in former athletes. Sports Med 2002;32(2):83–94.

113. Giorgi A, Weatherby RP, Murphy PW. Muscular strength, body composition and health responses to the use of testosterone enanthate: a double blind study. J Sci Med Sport 1999;2(4):341–355.

114. Demling RH, DeSanti L. The rate of restoration of bodyweight after burn injury, using the anabolic agent oxandrolone, is not age dependent. Burns 2001;27(1):46–51.

115. Wolf SE, Edelman LS, Kemalyan N, et al. Effects of oxandrolone on outcome measures in severely burned: a multicenter prospective randomized double-blind trial. J Burn Care Res 2006;27(2):131–139.

116. Volkow ND. Anabolic steroid abuse. National Institute on Drug Abuse Research Report Series. NIH publication number 06-3721, National Institutes of Health, August 2006.

117. Bond AJ, Choi PY, Pope HG. Assessment of attentional bias and mood in users and non-users of anabolic androgenic steroids. Drug Alcohol Depend 1995;37(3):241–245.

118. Pope HG Jr, Kouri EM, Hudson JI. Effects of supraphysiologic doses of testosterone on mood and aggression in normal men: a randomized controlled trial. Arch Gen Psychiatry 2000;57(2):133–140.

119. Heck AM, Calis KA, Yanovski JA. Pituitary gland disorders. In: Dipiro JT, editor. Pharmacotherapy: a pathophysiologic approach. 4th ed. Stamford, CT: Appleton & Lange; 1999. p 1281–1297.

120. Demling RH. Comparison of the anabolic effects and complications of human growth hormone and the testosterone analog, oxandrolone, after severe burn injury. Burns 1999;25(3):215–221.

CHAPTER 4

1. Ades PA, Savage PD, Brawner CA, Lyon CE, Ehrman JK, Bunn JY, Keteyian SJ. Aerobic capacity in patients entering cardiac rehabilitation. Circulation 2006;113:2706–2712.

2. Campeau L. Grading of angina pectoris. Circulation 1976;54:522–523.

3. Carhart RL, Ades PA. Gender differences in cardiac rehabilitation. Cardiol Clin 1998;16:37–43.

4. Centers for Disease Control and Prevention. Diabetes prevalence among American Indians, Alaskan Natives, and the overall population—United States. Morb Mortal Wkly Rep 2003;52:702.

5. Diamond GA, Forrester JS. Analysis of probability as an aid in the clinical diagnosis of coronary-artery disease. N Engl J Med 1979; 300:1350–1358.

6. Hirsch HT, Haskal ZJ, Hertzer NR, Bakal CW, Creager MA, Halperin JL, Hiratzka LF, et al. ACC/AHA 2005 practice guidelines for the management of patients with peripheral arterial disease (lower extremity, renal, mesenteric and abdominal aortic): executive summary: a collaborative report from the American Association for Vascular Surgery/Society for Vascular Surgery, Society for Cardiovascular Angiography and Interventions, Society for Vascular Medicine and Biology, Society of Interventional Radiology, and the ACC/AHA Task Force on Practice Guidelines (Writing Committee to Develop Guidelines for the Management of Patients with Peripheral Arterial Disease). Circulation 2006;113:1474–1547.

7. Kaplan, NM. The deadly quartet. Upper-body obesity, glucose intolerance, hypertriglyceridemia, and hypertension. Arch Int Med 1989;149:1514–1520.

8. Kloner, RA. The guide to cardiology. Greenwich, CT: Le Jacq Communications; 1995.

9. Lloyd-Jones DM, Larson MG, Beiser A, Levy D. Lifetime risk of developing coronary heart disease. Lancet 1999;353:89–92.

10. Périard D, Telenti A, Sudre P, Cheseaux J, Halfon P, Reymond MJ, Marcovina SM, et al. Atherogenic dyslipidemia in HIV-infected individuals treated with protease inhibitors. Circulation 1999;100:700–705.

11. Smith DW. Changing causes of death of elderly people in the United States, 1950–1990. Gerontology 1998;44:331–335.

12. Swartz MH. Textbook of physical diagnosis: history and examination. Philadelphia: Saunders; 1998.

13. Thom T, Haase N, Rosamond W, Howard VJ, Rumsfeld J, Manolio T, Zheng ZJ, et al. Heart disease and stroke statistics—2006 update: a report from the American Heart Association Statistics Committee and Stroke Statistics Subcommittee. Circulation 2006;113:e85–151.

CHAPTER 5

1. Adams GM. Exercise physiology: laboratory manual. 3rd ed. Boston: WCB McGraw-Hill; 1998.

2. American College of Sports Medicine. ACSM's guidelines for exercise testing and prescription. 7th ed. Baltimore: Lippincott Williams & Wilkins; 2005.

3. American College of Sports Medicine. Exercise for patients with coronary artery disease. Med Sci Sports Exerc 1994;26(3):i–v.

4. American College of Sports Medicine. Position stand. Exercise and physical activity for older adults. Med Sci Sports Exerc 1998;30(6):992–1008.

5. American College of Sports Medicine. Position stand. Exercise and type 2 diabetes. Med Sci Sports Exerc 2000;32:1345–1362.

6. American College of Sports Medicine. Position stand. Physical activity, physical fitness, and hypertension. Med Sci Sports Exerc 2004;36(3):533–553.

7. American College of Sports Medicine. Position stand. The recommended quantity and quality of exercise for developing and maintaining cardiorespiratory and muscular fitness, and flexibility in healthy adults. Med Sci Sports Exerc 1998;30(6):975–991.

8. Atterhog JH, Jonsson B, Samuelsson R. Exercise testing: a prospective study of complication rates. Am Heart J 1979;98:572–579.

9. Bartels R, Billings CE, Fox EL, Mathews DK, O'Brien R, Tanz D, Webb W. AAHPER Con 1968:123.

10. Blair SN, Kohl HW III, Paffenbarger RS Jr, Clark DG, Cooper KH, Gibbons LW. Physical fitness and all-cause mortality: a prospective study of healthy men and women. JAMA 1989;262:2395–2401.

11. Brawner CA, Keteyian SJ, Ehrman JK. The relationship of heart rate reserve to $\dot{V}O_2$ reserve in patients with heart disease. Med Sci Sports Exerc 2002;34(3):418–422.

12. Brawner CA, Keteyian SJ, Ehrman JK. Predicting maximum heart rate in patients with heart disease: influence of beta-adrenergic blockade therapy. Med Sci Sports Exerc 2002;34(5):s269.

13. Brubaker PH, Rejeski WJ, Law HD, et al. Cardiac patients' perception of work intensity during graded exercise testing: do they generalize to field testing? J Cardiopulm Rehabil 1994;14:127–133.

14. Butler RM, Belerwalters WH, Rodger FJ. The cardiovascular response to circuit weight training in patients with cardiac disease. J Cardiac Rehabil 1987;7:402–409.

15. Chapman, CB. Edward Smith (? 1818–1874), physiologist, human ecologist, reformer. J Hist Med Allied Sci 1967;22:1–26.

16. Cornelius WL, Craft-Hamm K. Proprioceptive neuromuscular facilitation flexibility techniques: acute effects on arterial blood pressure. Physician Sportsmed 1988;16(4):152–161.

17. Coyle EF, Martin WH, Sinacore DR, Joyner MJ, Hagberg JM, Holloszy JO. Time course of loss of adaptations after stopping prolonged intense endurance training. J Appl Physiol 1984;57(6):1857–1864.

18. DeMichele PD, Pollock ML, Graves JE, et al. Isometric torso rotation strength: effect of training frequency on its development. Arch Physiol Med Rehabil 1997;78:64–69.

19. Etnyre BR, Abraham LD. Antagonist muscle activity during stretching: a paradox re-assessed. Med Sci Sports Exerc 1988;20:285–289.

20. Etnyre BR, Lee JA. Chronic and acute flexibility of men and women using three different stretching techniques. Res Q Exerc Sport 1988;59:222–228.

21. Feigenbaum MS, Pollock ML. Strength training: rationale for current guidelines for adult fitness programs. Physician Sportsmed 1997;25:44–64.

22. Fleck SJ, Kraemer WJ. Designing resistance training programs. 2nd ed. Champaign, IL: Human Kinetics; 1997.

23. Foss ML, Keteyian SJ, editors. Fox's physiological basis for exercise and sport. 6th ed. Boston: McGraw-Hill; 1998.

24. Franklin BA, Dressendorfer R, Bonzbeim K, et al. Safety of exercise testing by non-physician health care providers: eighteen year experience [abstract]. Circulation 1995;92(Suppl I):1–37.

25. Frontera WR, Meredith CN, O'Reilly KP, Evans WJ. Strength conditioning in older men: skeletal muscle hypertrophy and improved function. J Appl Physiol 1988;68:1038–1044.

26. Gettman LR, Pollock ML, Durstine JL, Ward A, Ayres J, Linnerud AC. Physiological responses of men to 1, 3, and 5 days per week training program. Res Q 1976;47:638–646.

27. Gettman LR, Pollock ML. Circuit weight training: a critical review of its physiological benefits. Physician Sportsmed 1981;9:44–60.

28. Ghilarducci LE, Holly RG, Amsterdam EA. Effects of high resistance training in coronary artery disease. Am J Cardiol 1989;64:866–870.

29. Gibbons RJ, Balady GJ, Bricker JT, et al. ACC/AHA 2002 guideline update for exercise testing: summary article: a report of the American College of Cardiology/American Heart Association Task Force on Practice Guidelines (Committee to Update the 1997 Exercise Testing Guidelines). Circulation 2002;106:1883–1892.

30. Gordon NF, Kohl HW. Exercise testing and sudden cardiac death. J Cardiopulm Rehabil 1993;13:381–386.

31. Graves JE, Pollock ML, Jones AE, et al. Specificity of limited range of motion variable resistance training. Med Sci Sports Exerc 1989;21:84–89.

32. Graves JE, Pollock ML, Leggett SH, Braith RW, Carpenter DM, Bishop LE. Effect of reduced training frequency on muscular strength. Int J Sports Med 1988;9:316–319.

33. Hermansen L, Saltin B. Oxygen uptake during maximal treadmill and bicycle exercise. J Appl Physiol 1969;26:31–37.

34. Hurley BF, Seals DR, Ehsani AA, Cartier LJ, Dalsky GP, Hagberg JM, Holloszy JO. Effects of high-intensity strength training on cardiovascular function. Med Sci Sports Exerc 1984;16(5):483–488.

35. Jackson J, Sharkey B, Johnston L. Cardiorespiratory adaptations to training at specified frequencies. Res Q 1968;39:295–300.

36. Judge JO, Lindsey C, Underwood M, Winsemius D. Balance improvements in older women: effects of exercise training. Phys Ther 1993;73:254–265.

37. Karvonen M, Kentala K, Mustala O. The effects of training heart rate: a longitudinal study. Ann Med Exp Biol Fenn 1957;35:307–315.

38. Kibler WB, Goldberg C, Chandler TJ. Functional biomechanical deficits in running athletes with plantar fasciitis. Am J Sports Med 1991;266:185–196.

39. Knight JA, Laubach CA, Butcher RJ, et al. Supervision of clinical exercise testing by exercise physiologist. Am J Cardiol 1995;75:390–391.

40. Leach RE, James S, Wasilewski S. Achilles tendinitis. Am J Sports Med 1981;9:93–98.

41. Lee IM, Hsieh C, Paffenbarger RS. Exercise intensity and longevity in men. JAMA 1995;273:1179–1184.

42. Lem V, Krivokapich J, Child JS. A nurse-supervised exercise stress testing laboratory. Heart Lung 1985;14:280–284.

43. Mark DB, Shaw L, Harrell FE, et al. Prognostic value of a treadmill exercise score in outpatients with suspected coronary artery disease. N Engl J Med 1991;325:849–853.

44. McArdle WD, Magel JR. Physical work capacity and maximum oxygen uptake in treadmill and bicycle exercise. Med Sci Sports 1970;2:118–123.

45. McKay GA, Banister EW. A comparison of maximum oxygen uptake determination by bicycle ergometry of various pedaling frequencies and by treadmill running at various speeds. Eur J Appl Physiol 1976;35:191–200.

46. McMurray RG, Ainsworth BE, Harrell JS, Griggs TR, Williams OD. Is physical activity or aerobic power more influential on reducing cardiovascular disease risk factors? Med Sci Sports Exerc 1998;30(10):1521–1529.

47. Miyamura M, Kitamura K, Yamads A, Matsui H. Cardiorespiratory responses to maximal treadmill and bicycle exercise in trained and untrained subjects. J Sports Med Phys Fitness 1978;18:25–32.

48. Morgan RE, Adamson GT. Circuit weight training. London: G Bell and Sons, 1961.

49. Morgan W, Borg G. Perception of effort in the prescription of physical activity. In: Nelson T, editor. Mental health and emotional aspects of sports. Chicago: American Medical Association; 1976. p 126–129.

50. Nostratian F, Froelicher VF. ST elevation during exercise testing: a review. Am J Cardiol 1989;63:986–988.

51. Paffenbarger RS, Hyde RT, Wing AL, Hsieh CC. Physical activity, all-cause mortality, and longevity of college alumni. N Engl J Med 1986;314:605–613.

52. Pate RR, Pratt M, Blair SN, et al. Physical activity and public health: a recommendation from the Centers for Disease Control and Prevention and the American College of Sports Medicine. JAMA 1995;273:402–407.

53. Pina IL, Balady GJ, Hanson P, et al. Guidelines for clinical exercise testing laboratories. A statement for healthcare professionals from the committee on exercise and cardiac rehabilitation, American Heart Association. Circulation 1995;91(3):912–921.

54. Pollock ML, Cureton TK, Greninger L. Effects of frequency of training on working capacity, cardiovascular function, and body composition of adult men. Med Sci Sports 1969;1:70–74.

55. Pollock ML, Leggett SH, Graves JE, Jones A, Fulton M, Cirulli J. Effect of resistance training on lumbar extension strength. Am J Sports Med 1989;17:624–629.

56. Rochmis P, Blackburn H. Exercise tests: a survey of procedures, safety, and litigation experience in approximately 170,000 tests. JAMA 1971;217:1061–1066.

57. Rodgers GP, Ayanian JZ, Balady G, et al. American College of Cardiology/American Heart Association clinical competence statement on stress testing. A report of the American College of Cardiology/American Heart Association/American College of Physicians/American Society of Internal Medicine Task Force on Clinical Competence. Circulation 2000;102:1726–1738.

58. Sady SP, Wortman M, Blanke D. Flexibility training: ballistic, static or proprioceptive neuromuscular facilitation? Arch Phys Med Rehabil 1982;63:261–263.

59. Saltin B, Blomqvist G, Mitchell JH, Johnson RL, Wildenthal K, Chapman CB. Response to exercise after bed rest and after training. Circulation 1968;38(Suppl. VII):VII-1–VII-55.

60. Scherer D, Kaltenbach M. Frequency of life-threatening complications associated with exercise testing. Dtsch Med Wochenschr 1979;33:1161–1165.

61. Shetler K, Marcus R, Froelicher VF, et al. Heart rate recovery: validation and methodologic issues. J Am Coll Cardiol 2001;38:1980–1987.

62. Sidney KH, et al. Training: scientific basis and application. Taylor AW, editor. Springfield, IL: Charles C Thomas; 1972.

63. Squires RW, Allison TG, Johnson BD, and Gau GT. Non-physician supervision of cardiopulmonary exercise testing in chronic heart failure: safety and results of a preliminary investigation. J Cardiopulm Rehabil 1999;19:249–253.

64. Squires RW, Muri AJ, Anderson LJ, et al. Weight training during phase II (early outpatient) cardiac rehabilitation: heart rate and blood pressure responses. J Cardiac Rehabil 1991;11:360–364.

65. Stuart RJ, Ellestad MH. Upsloping S-T segments in exercise stress testing: six year follow-up study of 438 patients and correlation with 248 angiograms. Am J Cardiol 1976;37:19–22.

66. Swain D, Leutholtz B, King M, Haas L, Branch J. Relationship between % heart rate reserve and % VO_2 reserve in treadmill exercise. Med Sci Sports Exerc 1998;30:318–321.

67. Swain D, Leutholtz B. Heart rate reserve is equivalent to %VO_2 reserve, not to %VO_2max. Med Sci Sports Exerc 1997;29:410–414.

68. U.S. Department of Health and Human Services. Physical activity and health: a report of the surgeon general. Atlanta: U.S. Department of Health and Human Services, Centers for Disease Control and Prevention, National Center for Chronic Disease Prevention and Health Promotion; 1996.

69. Verstappen FTJ, Huppertz RM, Snoeckx LHEH. Effect of training specificity on maximal treadmill and bicycle ergometer exercise. Int J Sports Med 1982;3:43–46.

70. Weiner DA, Ryan TJ, McCabe CH, et al. Prognostic importance of a clinical profile and exercise test in medically treated patients with coronary artery disease. J Am Coll Cardiol 1984;3(3):772–779.

71. Welsch MA, Pollock ML, Brechue WF, Graves JE. Using the exercise test to develop the exercise prescription in health and disease. Primary Care 1994;21(3):589–609.

72. Will PM, Walter JD. Exercise testing: improving performance with a ramped Bruce protocol. Am Heart J 1999;138:1033–1037.

73. Williams, MA. Exercise testing in cardiac rehabilitation: exercise prescription and beyond. Cardiol Clin 2001;19:415–431.

74. Wilmore J, Davis JA, O'Brien RS, Vodak PA, Wolder G, Amsterdam EA. Physiological alterations consequent to 20-week conditioning programs of bicycling, tennis, and jogging. Med Sci Sports Exerc 1980;12:1–8.

75. Wilson GJ, Elliott BC, Wood GA. Stretch shorten cycle performance enhancement through flexibility training. Med Sci Sports Exerc 1992;24:116–123.

76. Worrell TW. Factors associated with hamstring injuries: an approach to treatment and preventative measures. Sports Med 1994;17:335–345.

CHAPTER 6

1. Adams FH. The physical working capacity of normal children. Pediatrics 1961;28:55.

2. Adams GM. Exercise physiology—laboratory manual. Boston: McGraw-Hill; 1998. p 85.

3. Alpert BS, Wilmore JH. Physical activity and blood pressure in adolescents. Pediatr Exerc Sci 1994;6(4):361–380.

4. American Academy of Pediatrics: the fourth report on the diagnosis, evaluation, and treatment of high blood pressure in children and adolescents. Pediatrics 2004;114(2):555.

5. American Alliance for Health, Physical Education, Recreation and Dance. AAHPERD health related physical fitness test manual. Washington, DC: American Alliance for Health, Physical Education, Recreation and Dance; 1992.

6. American Alliance for Health, Physical Education, Recreation and Dance. Physical best. Reston, VA: American Alliance for Health, Physical Education, Recreation and Dance; 1988.

7. American Alliance for Health, Physical Education, Recreation and Dance. Physical best teacher's guide. Champaign, IL: Human Kinetics; 1999.

8. American College of Sports Medicine. ACSM's guidelines for exercise testing and prescription. 6th ed. Philadelphia: Lippincott, Williams and Wilkins; 2000.

9. American College of Sports Medicine. Opinion statement on physical fitness in children and youth. Med Sci Sport Exer 1988;204:422–423.

10. American College of Sports Medicine. ACSM's Certified News 1997;8(1)2.

11. American Heart Association. 1999 heart and stroke facts statistical update. Dallas: American Heart Association; 1998.

12. Armstrong N, Williams L, Balding J, Gentle P, Kirby B. Cardiopulmonary fitness, physical activity patterns, and selected coronary risk factor variables in 11 to 16 year olds. Pediatr Exerc Sci 1991;3:219–228.

13. Bailey DA, Martin AD. Physical activity and skeletal health in adolescents. Pediatr Exerc Sci 1994;6:330–347.

14. Bao W, Srinivasan SR, Valdez R, Greenlund KJ, Wattigney WA, Berenson GS. Longitudinal changes in cardiovascular risk from childhood to young adulthood in offspring of parents with coronary artery disease. JAMA 1997; 278:1749-1754.

15. Bar-Or O. Pediatric sports medicine for the practitioner. New York: Springer-Verlag; 1983.

16. Bartosh SM, Aronson AJ. Childhood hypertension—an update on etiology, diagnosis and treatment. Pediatr Cardiol 1999;46(2):235–250.

17. Berenson GS, Sprinivasn SR, Weihang B, Newman WP, Tracy RE, Wattigney WA. Association between multiple cardiovascular risk factors and atherosclerosis in children and young adults. New Engl J Med 1998;338:1650–1656.

18. Blair SN. Changes in physical fitness and all-cause mortality. JAMA 1995;273:1093–1098.

19. Blair SN, Kohl HW III, Paffenbarger RS, Clark DG, Cooper KH, Gibbons LW. Physical fitness and all-cause mortality: a prospective study of healthy men and women. JAMA 1989;262:2395–2401.

20. Blimkie CJR, Sale DG. Strength development and trainability during childhood. In: Praash EV, editor. Pediatric anaerobic performance. Champaign, IL: Human Kinetics; 1986. p 196.

21. Blimkie C. Resistance training during preadolescence. Sports Med 1993;15(6):389–407.

22. Boileau RA, Lohman TG, Slaughter MH, Horswill CA, Stillman RJ. Problems associated with determining body composition in maturing youngsters. In: Brown EW, Branta CF, editors. Competitive sports for children and youth. Champaign, IL: Human Kinetics; 1998. p 3–16.

23. Bono MJ, Roby JJ, Micale FG, Sallis JF, Shepard WE. Validity and reliability of predicting maximum oxygen uptake via field tests in children and adolescents. Pediatr Exerc Sci 1991;3:250–255.

24. Cahill BR, editor. Proceedings of the conference on strength training and the prepubescent. American Orthopedic Society for Sports Medicine, Chicago, 1988.

25. Calfas KJ, Taylor WC. Effects of physical activity on psychological variables in adolescents. Pediatr Exerc Sci 1994;6:406–425.

26. Cassell CS, Benedict M, Uetrect G, Ranz M, Specker B. Bone mineral density in young gymnasts and swimmers. Med Sci Sports Exerc 1993;25(Suppl):S49.

27. Centers for Disease Control. Vigorous physical activity among high school students. MMWR Morb Mortal Wkly Rep 1990;41:33–35.

28. Chestnut C. Theoretical overview: bone development, peak bone mass, bone loss, and fracture risk. Am J Med 1991;91(5B):25–45.

29. Clark HH. Joint and body range of movement. Physical Fitness Research Digest 1975;5:16–18.

30. Cook S, Weitzman M, Auinger P, Nguyen M, Dietz WH. Prevalence of a metabolic syndrome phenotype in adolescents. Arch Pediatr Adolesc Med. 2003;157:821–827.

31. Cooper KH. A means of assessing maximal O_2 uptake. JAMA 1968;203:201–204.

32. Corbin CB, Pangrazi RP. Physical activity for children: a statement of guidelines. Reston, VA: National Association for Sport and Physical Education; 1999.

33. Cureton KJ, Sloniger MA, O'Bannon JP, Block DM, McCormick WP. A generalized equation for prediction of VO_2 peak from one-mile run/walk performance. Med Sci Sports Exerc 1995;27(3):445–451.

34. Dotson CO, Ross JG. Relationships between activity patterns and fitness. J Physical Educ Recreation Dance 1985;56(1):86–90.

35. Drinkwater BL. Physical exercise and bone health. J Am Med Womens Assoc 1990;45(3):91–97.

36. Eisenmann JC, et al. Assessing body composition among 3- to 8-year-old children: anthropometry, BIA, and DXA. Obes Res 2004;12:1633.

37. Fitnessgram test administration manual. 2nd ed. Champaign, IL: Human Kinetics; 2001.

38. Gibbons RJ, Balady GJ, Beasley WJ. ACC/AHA guidelines for exercise testing. A report of the American College of Cardiology/American Heart Association Task Force on Practice Guidelines (Committee on Exercise Testing). J Am Coll Cardiol 1997;30:260–315.

39. Grodjinovsky A, Inbar O, Dotan R, Bar-Or O. Training effect on the anaerobic performance of children as measured by the Wingate anaerobic test. In: Berg K, Eriksson BO, editors. Children and exercise IX. Baltimore: University Park Press; 1979. p 139–145.

40. Hansen HS, Froberg K, Hyldebrandt N, Nielsen JR. A controlled study of eight months of physical training and reduction of blood pressure in children: the Odense schoolchild study. BMJ 1991;303:682–685.

41. Harrell JS, Gansky SA, McMurray RG, Bangdiwala SI, Frauman AC, Bradley CB. School-based interventions improve heart health in children with multiple cardiovascular disease risk factors. Pediatrics 1998;102:371-380.

42. Harro M, Riddoch C, Armstrong N, Physical activity. In: Armstrong N, Van Mechehen W, editors. Pediatric exercise science and medicine. Oxford, UK: Oxford University; 2000. p 77.

43. Houtkooper LB, Going SB, Lohman TG, Roche AF, Van Loan M. Bioelectrical impedance estimation of fat-free mass in children and youth: a cross-validation study. J Appl Physiol 1992:72:377–73.

44. Kowalski KC, Crocker PRE, Faulkner RA. Validation of the Physical Activity Questionnaire for older children. Pediatr Exerc Sci 1997;9:174–186.

45. Krahenbuhl GS, Skinner JS, Kohrt NM. Development of aerobic power in children. Exerc Sport Sci Rev 1985;13:503–538.

46. Krahenbuhl GS, Pangrazi RP, Petersen GW, Burkett LN, Schneider MJ. Field testing of cardiorespiratory fitness in primary school children. Med Sci Sports Exerc 1978;10:208–213.

47. Krahenbuhl GS, Pangrazi RP, Burkett LN, Schneider MJ, Petersen GW. Field estimation of VO_2 max in children eight years of age. Med Sci Sports Exerc 1977;9:37–40.

48. Kramer WJ, Fleck SJ. Strength training for young athletes. Champaign, IL: Human Kinetics; 1993.

49. Kuczmarski RJ, et al. Varying body mass index cutoff points to describe overweight prevalence among U.S. adults: NHANES III (1988–1994). Obes Res 1997;5:542.

50. Lauer RM, Lee J, Clarke WR. Factors affecting the relationship between childhood and adult cholesterol levels: the Muscatine study. Pediatrics 1988;82:309–318.

51. Leger LA, Mercier D, Gadoury C, Lambert J. The multistage 20 meter shuttle run test for aerobic fitness. J Sports Sci 1988;6:93–101.

52. Lohman TG. Assessment of body composition in children. Pediatr Exerc Sci 1989;1(1):19–30.

53. Lohman TG. The use of skinfold to estimate body fatness in children and youth. J Physical Educ Recreation Dance 1987;58:98–102.

54. Mahon AD. Exercise training. In: Armstrong N, Van Mechelen W, editors. Pediatric exercise science and medicine. Oxford, UK: Oxford University Press; 2000. p 201–219.

55. Malina RM, Bouchard C. Growth, maturation, and physical activity. Champaign, IL: Human Kinetics; 1991. p 87–150.

56. Massicotte DR, Gauther R, Markon RP. Prediction of VO_2 max from running performance in children aged 10–17 years. J Sports Med Phys Fitness 1985;25:10–17.

57. Matkovic V, Fonatana D, Tominac C, Goel P, Chestnut C. Factors which influence peak bone mass formation: a study of calcium balance and the inheritance of bone mass in adolescent females. Am J Clin Nutr 1990;52:878–888.

58. Mayo Medical Laboratories. Test catalog. Rochester, MN: Mayo Press; 2001.

59. McGill HC, McMahan A, Zieske AW, Walcott JV, Malcom GT, Tracy RE, Strong JP. Effects of nonlipid risk in youth with a favorable lipoprotein profile. Circulation 2001;103:1546–1550.

60. McMahan CA, Gidding SS, Fayad AA, Zieske AW, Malcom GT. Risk scores predict atherosclerotic lesions in young people. Arch Intern Med 2005;165:883-890.

61. Melton LJ. Osteoporosis. In: Berg R, Cassels J, editors. The second fifty years: promoting health and preventing disability. Washington, DC: National Academy Press; 1990. p 76.

62. Milne C, Seefeldt V, Reuschlein P. Relationship between grade, sex, race, and motor development in young children. Res Q 1976;47:726.

63. Morrow JR, Freedson PS. Relationship between habitual physical activity and aerobic fitness in adolescents. Pediatr Exerc Sci 1994;6(4):315–329.

64. Must A, Strauss RS. Risks and consequences of childhood and adolescent obesity. Int J Obes Relat Metab Disord 1999;23(Suppl 2):5211.

65. Must A, Jacques PF, Dallal GE, Bajema CJ, Dietz WH. Long-term morbidity and mortality of overweight adolescents. N Engl J Med 1992;327:1350–1355.

66. National Center for Health Statistics. Prevalence of overweight among children and adolescents. www.cdc.gov/nchs/products/pubs/pubd/hestats/over99fig1.htm

67. National Cholesterol Education Program-Report of the Expert Panel on Blood Cholesterol Levels in Children and Adolescents NIH Publication No. 91-2732 p 39-55, 1991.

68. National Institutes of Health. Report of the Second Task Force on Blood Pressure Control in Children. Pediatrics 1999;103:1175–1182.

69. Orchard TJ, Donahue RP, Kuller LH, Hodges PN, Drash AL. Cholesterol screening in childhood: does it predict adult hypercholesterolemia? The Beaver County experience. J Pediatr 1983;103:687–691.

70. Pfeiffer RD, Francis RS. Effects of strength training on muscle development in pre-pubescent, pubescent, and postpubescent males. Phys Sportsmed 1986;14:134–143.

71. Public Health Service. Healthy people 2000 review 1992 (DHHS publication no. PHS 93-12321). Washington, DC: U.S. Department of Health and Human Services; 1993.

72. Public Health Service. Healthy people 2000: national health promotion and disease prevention objectives (DHHS publication no. PHS 91-50212). Washington, DC: U.S. Department of Health and Human Services; 1990.

73. Risser WL. Musculoskeletal injuries caused by weight training. Clin Pediatr 1990;29(6):305–310.

74. Risser WL. Weight training injuries in children and adolescents. Am Fam Physician 1991;44:2104–2108.

75. Rivera-Brown A, Rivera MA, Fontera UR. Applicability of criteria for $\dot{V}O_2$max in active adolescents. Pediatr Exerc Sci 1992;4:331–339.

76. Roberts S. Exercise prescription recommendations for children. ACSM Certified News 1997;7(1):3.

77. Rotstein AR, Dotan R, Bar-Or O, Tenenbaum G. Effects of training on anaerobic threshold, maximal aerobic power and anaerobic performance of preadolescent boys. Int J Sports Med 1986;7:281–286.

78. Rowland TW. Developmental exercise physiology. Champaign, IL: Human Kinetics; 1996.

79. Rowland TW. Exercise and children's health. Champaign, IL: Human Kinetics; 1990. p 27–83.

80. Rowland TW, editor. Pediatric laboratory testing—clinical guidelines. Champaign IL: Human Kinetics; 1993.

81. Safrit MJ. Complete guide to youth fitness testing. Champaign, IL: Human Kinetics; 1995.

82. Sale DG. Strength training in children. In: Gisolfland CV, Lamb DR, editors. Perspectives in exercise science and sports medicine (vol. 2). Indianapolis: Benchmark Press; 1989. p 165–222.

83. Sallis JF, Patrick K. Physical activity guidelines for adolescents: consensus statement. Pediatr Exerc Sci 1994;6:302–314.

84. Sallis JF, Buono MJ, Roby JJ, Micate FG, Nelson JA. Seven-day recall and other physical activity self-reports in children and adolescents. Med Sci Sports Exerc 1993;25:99–108.

85. Saltarelli W. The effects of pace training on children's performance time and heart rate response during a one-mile run. Doctoral dissertation. University of Toledo; 1989.

86. Sargeant AJ, Dolan P, Thorne A. Effects of supplementary physical activity on body composition, aerobic, and anaerobic power in 13-year-old boys. In: Binkborst RA, Kemper HCG, Saris WH, editors. Children and exercise XI. Champaign, IL: Human Kinetics; 1985. p 140–150.

87. Schlicker SA, Borra ST, Regan C. Nutr Rev 1996;52:11–20.

88. Sewell L, Micheli LJ. Strength training for children. J Pediatr Orthop 1986;6:143–146.

89. Slaughter MH, Lohman TG, Boileau RA, Horswill CA, Stillman RH, Van Loan MD, Bemben DA. Skinfold equations for estimation of body fatness in children and youth. Hum Biol 1988, 60(5):709-723.

90. Strong JP, Malcom GT, McMahan CA, Tracy RE, Newman WP, Herderick EE, Cornhill JF. Prevalence and extent of atherosclerosis in adolescents and young adults. JAMA 1999;281:727–735.

91. Taylor CB, Sallis JF, Needle R. The relation of physical activity and exercise to mental health. Public Health Rep 1985;100:195–202.

92. Tipton CM. Exercise training and hypertension: an update. Exerc Sport Sci Rev 1991;19:447–505.

93. Tortolero SR, Taylor TW, Murry NG. Physical activity, physical fitness and social, psychological and emotional health. In: Armstrong N, Van Mechehen W, editors. Pediatric exercise science and medicine. Oxford University Press; 2000. p 273–291.

94. Twisk JW, Mechelen W, van Kemper HCG, Post GB. The relation between "long term exposure" to lifestyle during youth and young adulthood and risk factors for cardiovascular disease. J Adolesc Health 1997;20:309.

95. Vrijens J. Muscle strength development in pre and postpubescent age. Med Sport 1978;11:152–158.

96. Welk GJ, Wood K. Physical activity assessments in physical education—a practical review of instruments and their use in the curriculum. J Physical Educ Recreation Dance 2000;71:30–40.

97. Weltman A, Janney C, Rians CB, Strand K, Berg B, Tippett S, Wise J, Cahill BR, Katch FI. The effects of hydraulic resistance strength training in pre-pubescent males. Med Sci Sports Exerc 1986;18:629–638.

98. Weston AT, Petosa R, Pate RR. Validation of an instrument for measurement of physical activity in youth. Med Sci Sports Exerc 1997;29:138–143.

99. Zieske AW, Malcom GT, Strong JP. Natural history and risk factors of atherosclerosis in children and youth: the PDAY study. Pediatr Pathol Mol Med 2002;21:213-237.

CHAPTER 7

1. Administration on Aging. A profile of older Americans. Washington, DC: Author; 2005.

2. American College of Sports Medicine. ACSM's resource manual for guidelines for exercise testing and prescription. 5th ed. Baltimore: Williams & Wilkins; 2006.

3. American College of Sports Medicine. ACSM's guidelines for exercise testing and prescription. 7th ed. Baltimore: Lippincott Williams & Wilkins; 2006.

4. Anniansson A, Grimby G, Hedberg M. Compensatory muscle fiber hypertrophy in elderly men. J Appl Physiol 1992;73:812–816.

5. Barry HC, Eathorne SW. Exercise and aging. Issues for the practitioner. Med Clin North Am 1994;78:357–376.

6. Bayles C. Frailty. In: Durstine JJ, editor. American College of Sports Medicine's exercise management for persons with chronic diseases and disabilities. Champaign, IL: Human Kinetics; 1997. p 112–118.

7. Binder EF, Birge SJ, Spina R, Ehsani AA, Brown M, Sinacore DR, Kohrt WM. Peak aerobic power is an important component of physical performance in older women. J Gerontol 1999;54A:M353–356.

8. Birdt TA. Alzheimer's disease and other primary dementia. In: Fauci AS et al., editors. Harrison's principles of internal medicine. New York: McGraw-Hill; 1998. p 2348–2356.

9. Blackman DK, Kamimoto LA, Smith SM. Overview: surveillance for selected public health indicators affecting older adults United States. MMWR 1999;48(SS08):1–6.

10. Blair SN, Kampert, JB, Kohl HW, et al. Influences of cardiorespiratory fitness and other precursors on cardiovascular disease and all-cause mortality in men and women. JAMA 1996;276:205–210.

11. Blazer DG. Social support and mortality in an elderly community population. Am J Epidemiol 1982;115:684–694.

12. Bloomfield SA. Bone, ligament and tendon. In: Gisolfi CV, Lamb DR, Nadel E, editors. Perspectives in exercise science and sports medicine. Carmel, IN: Cooper; 1995. p 175–227.

13. Boult C, Kane RL, Loius TA, Boult L, McCaffrey D. Chronic conditions that lead to functional limitation in the elderly. J Gerontol 1994;49:M28–M36.

14. Brown AB, McCartney N, Sale DG. Positive adaptations to weightlifting training in the elderly. J Appl Physiol 1990;69:1725–1733.

15. Buchner DM, Cress ME, Wagner EH, de Lateur BJ, Price R, Abrass IB. The Seattle FICSIT/Move It study: the effect of exercise on gait and balance in older adults. J Am Geriatr Soc 1993;41:321–325.

16. Campbell WW, Crim MC, Dallal GE, Young VR, Evans WJ. Increased energy requirements and changes in body composition with resistance training in older adults. Am J Clin Nutr 1994;60:167–175.

17. Chodzko-Zajko WJ, Moore KA. Physical fitness and cognitive function in aging. Exerc Sport Sci Rev 1994;22:195–220.

18. Colantonio A, Kasl SV, Ostfeld AM, Berkman LF. Psychosocial predictors of stroke outcomes in an elderly population. J Geront Biol Sci 1993;48:S261–S268.

19. Cononie CC, Goldberg AP, Rogus E, Hagberg JM. Seven consecutive days of exercise lowers plasma insulin responses to an oral glucose challenge in sedentary elderly. J Am Geriatr Soc 1994;42:394–398.

20. Cox JH, Cortright RN, Dohm GL, Houmard JA. Effect of aging on response to exercise training in humans: skeletal muscle GLUT-4 and insulin sensitivity. J Appl Physiol 1999;86:2019–2025.

21. Cress EM, Buchner DM, Questad KA, Esselman PC, deLateur BJ, Schwartz RS. Exercise: effects on physical functional performance in independent older adults. J Gerontol 1999;54A:M242–M248.

22. Davis JW, Ross PD, Nevitt MC, Wasnich RD. Risk factors for falls and for serious injuries on falling among older Japanese women in Hawaii. J Am Geriatr Soc 1999;47:792–798.

23. Deschenes MR. Effects of aging on muscle fibre type and size. Sports Med 2004;34:809–824.

24. Desai MM, Zhang P, Hennessy CH. Surveillance for morbidity and mortality among older adults—United States, 1995–1996. MMWR 1999;48(SS08):7–25.

25. DiPietro L, Caspersen CJ, Ostfeld AM. A survey for assessing physical activity among older adults. Med Sci Sports Exerc 1995;25:628–642.

26. Doherty TJ, Vandervoort AA, Taylor AW, Brown WF. Effects of motor units losses on strength in older men and women. J Appl Physiol 1993;74:868–874.

27. Ehsani AA, Ogawa T, Miller TR, Spina RJ, Jilka SM. Exercise training improves left ventricular systolic function in older men. Circulation 1991;83:96–103.

28. Engels H-J, Drouin J, Zhu W, Kazmierski JF. Effects of low-impact, moderate-intensity exercise training with and without wrist weights on functional capacities and mood status on older adults. Gerontology 1998;44:239–244.

29. Erikssen G, Liestol K, Bjornholt J, Thaulow E, Erikssen J. Changes in physical fitness and changes in mortality. Lancet 1998;352:759–762.

30. Ernst JM, Cacioppo JT. Lonely hearts: psychological perspectives on loneliness. Appl Prev Psychol 1999;8:1–22.

31. Evans WJ. Exercise training guidelines for the elderly. Med Sci Sports Exerc 1999;31:12–17.

32. Ferketich AM, Kirby TE, Alway SE. Cardiovascular and muscular adaptations to combined endurance and strength training in elderly women. Acta Physiol Scand 1998;164:259–267.

33. Ferrara CM, Goldberg AP, Ortmeyer HK, Ryan AS. Effects of aerobic and resistive exercise training on glucose disposal and skeletal muscle metabolism in older men. J Gerontol 2006;61A:480–487.

34. Fiatarone MA, O'Neill EF, Ryan ND, Clements KM, Solares GR, Nelson ME, Roberts SB, Kehaias JJ, Lipsizt LA, Evans WJ. Exercise training and nutritional supplementation for physical frailty in very elderly people. New Engl J Med 1994;330:1769–1775.

35. Fiatarone MA, Marks EC, Ryan ND, Meredith CN, Lipsitz LA, Evans WJ. High-intensity strength training in nonagenarians. Effects on skeletal muscle. JAMA 990;263:3029–3034.

36. Fleg JL, O'Connor F, Gerstenblith G, Becker LC, Clulow J, Schulman SP, Lakatta EG. Impact of age on the cardiovascular response to dynamic upright exercise in healthy men and women. J Appl Physiol 1995;78:890–900.

37. Fleg JL, Lakatta EG. Role of muscle loss in the age associated reduction in $\dot{V}O_2$max. J Appl Physiol 1988;65:1147–1151.

38. Frontera WR, Meredith CN, O'Reilly KP, Knuttgen HG, Evans WJ. Strength conditioning in older men: skeletal muscle hypertrophy and improved function. J Appl Physiol 1988;64:1038–1044.

39. Galganski ME, Fuglevand AJ, Enoka RM. Reduced control of motor output in a human hand muscle of elderly subjects during submaximal contractions. J Neurophysiol 1993;69:2108–2115.

40. Gallagher D, Ruts E, Visser M, Heshka S, Baumgartner RN, Wang J, Pierson RN, Pi-Sunyer FX, Heymsfield SB. Weight stability masks sarcopenia in elderly men and women. Am J Physiol Endocrinol Metab 2000;279:E366–E375.

41. Goran MI, Poehlman ET. Total energy expenditure and energy requirements in healthy elderly persons. Metabolism 1992;41:744–753.

42. Heath G. Exercise programming for older adults. In: American College of Sports Medicine's resource manual for guidelines for exercise testing and prescription. 3rd ed. Baltimore: Williams & Wilkins; 1998. p 516–520.

43. House JS, Landis KR, Umberson D. Social relationships and health. Science 1988;241:540–545.

44. Hoyert DL, Kochanek KD, Murphy SL. Deaths: final data for 1997. National vital statistics reports. Washington, DC: U.S. Department of Health and Human Services, Centers for Disease Control and Prevention, National Center for Health Statistics, National Vital Statistics Systems; 1999.

45. Hunter GR, Treuth MS, Weinsier RL, Kekes-Szabo T, Kell SH, Roth DL, Nicholson C. The effects of strength conditioning on older women to perform daily tasks. J Am Geriatr Soc 1995;43:756–760.

46. Kahn SE, Larson VG, Beard JC, Cain KC, Fellingham GW, Schwarts RS, Veith RC, Stratton JR, Cerqueira MD, Abrass IB. Effect of exercise on insulin action, glucose tolerance and insulin secretion in aging. Am J Physiol 1990;258:E937–E943.

47. Kamimoto LA, Easton AN, Maurice E, Husten CG, Macera CA. Surveillance for five health risks among older adults—United States, 1993–1997. MMWR 1999;48(SS08):89–130.

48. Kenney WL. Thermoregulation at rest and during exercise in healthy older adults. Exerc Sport Sci Rev 1997;25:41–76.

49. Kenney WL. The older athlete: exercise in hot environments. Sport Science Exchange 1993;6:44.

50. King AC, Martin JE. Physical activity promotion: adoption and maintenance. In: American College of Sports Medicine's resource manual for guidelines for exercise testing and prescription. 3rd ed. Baltimore: Williams & Wilkins; 1998. p 564–569.

51. King AC, Haskell WL, Taylor CB, Kraemer HC, DeBusk RF. Group vs home-based exercise training in healthy older men and women. A community-based clinical trial. JAMA 1991;266:1535–1542.

52. Klitgaard H, Mantoni M, Schiaffino S, Ausoni S, Gorza L, Laurent-Winter C, Schnohr P, Saltin B. Function, morphology and protein expression of aging skeletal muscle: a cross-sectional study of elderly men with different training backgrounds. Acta Physiol Scand 1990;140:41–54.

53. Koch M, Gottschalk M, Baker DI, Palumbo S, Tinetti ME. An impairment and disability assessment and treatment protocol for community-living elderly persons. Phys Ther 1994;74:286–294.

54. Kohrt WM, Malley MT, Coggan AR, Spina RJ, Agawa T, Ehsani AA, Bourey RE, Martin III WH, Holloszy JO. Effects of gender, age, and fitness level on response of $\dot{V}O_2$max to training in 60–71 year olds. J Appl Physiol 1991;71:2004–2011.

55. Kuczmarski RJ, Flegal KM, Campbell SM, Johnson CL. Increasing prevalence of overweight among U.S. adults. JAMA 1994;272:205–211.

56. Lakatta EG. Cardiovascular aging research: the next horizons. J Am Geriatr Soc 1999;47:613–625.

57. Lan C, Lai J, Chen S, Wong M. 12-month t'ai chi training in the elderly: its effect on health fitness. Med Sci Sports Exerc 1998;30:345–351.

58. Lemmer JT, Hurlbut DE, Martel GF, Tracy BL, Ivey FM, Metter EJ, Fozard JL, Fleg JL, Hurley BF. Age and gender responses to strength training and detraining. Med Sci Sports Exerc 2000;32:1505–1512.

59. Manton KG, Corder LS, Stallard E. Estimates of change in chronic disability and institutional incidence and prevalence rates in the U.S. elderly population from the 1982, 1984, and 1989 national long term care survey. J Gerontol 1993;48:S153–S166.

60. Mazzeo RS, Cavanagh P, Evans WJ, Fiatarone M, Hagberg J, McAuley E, Starzell J. Exercise and physical activity for older adults. Position stand. Med Sci Sports Exerc 1998;30:992–1008.

61. McMurdo MET, Rennie L. A controlled trial of exercise by residents of older people's home. Age Ageing 1993;22:11–15.

62. McMurdo MET, Burnett L. Randomized controlled trial of exercise in the elderly. Gerontology 1992;38:292–298.

63. Menshikova EV, Ritov VB, Fairfull L, Ferrell RE, Kelley DE, Goodpaster BH. Effects of exercise on mitochondrial content and function in aging skeletal muscle. J Gerontol 2006;61A:534–540.

64. Meredith CN, Frontera WR, Fisher WC, Hughes VA, Herland JC, Edwards J, Evans WJ. Peripheral effects of endurance training in young and old subjects. J Appl Physiol 1989;66:2844–2849.

65. Meijer EP, Westerterp KR, Verstappen FTJ. Effect of exercise training on total daily physical activity in elderly humans. Eur J Appl Physiol 1999;80:16–21.

66. Miller JP, Pratley RE, Goldberg AP, Gordon P, Rubin M, Treuth MS, Ryan AS, Hurley BF. Strength training increases insulin action in healthy 50- to 65-yr-old men. J Appl Physiol 1994;77:1122–1127.

67. Mills EM. The effect of low-intensity aerobic exercise on muscle strength, flexibility, and balance among sedentary elderly persons. Nurs Res 1994;43:207–211.

68. Morey MC, Schenkman M, Studenski SA, Chandler JM, Crowley JM, Sullivan Jr. RJ, Pieper CF, Doyle ME, Higginbotham MB, Horner RD, MacAller H, Puglisi CM, Morris KG, Weinberger M. Spinal-flexibility-plus aerobic versus aerobic-only training: effects of a randomized clinical trial on function in at-risk older adults. J Gerontol 1999;54A:M335–M342.

69. Morey MC, Cowper PA, Feussner JR, DiPasquale R, Crowley JM, Sullivan RJ. Two-year trends in physical performance following supervised exercise among community-dwelling older veterans. J Am Geriatr Soc 1991;39:549–554.

70. Morey MC, Cowper PA, Feussner JR, DiPasquale RC, Crowley JM, Kitzman DW, Sullivan RJ. Evaluation of a supervised exercise program in a geriatric population. J Am Geriatr Soc 1989;37:348–354.

71. Nelson ME, Fiatarone MA, Morganti CM, Trice I, Greenberg RA, Evans WJ. Effects of high-intensity strength training on multiple risk factors for osteoporotic fractures. JAMA 1994;272:1909–1914.

72. Nelson ME, Rejeski WJ, Blair SN, Duncan PW, Judge JO, King AC, Macera CA, Castandasceppa C. Physical activity and public health in older adults: recommendation from the American College of Sports Medicine and the American Heart Association. Med Sci Sports Exerc 2007;39:435–1445.

73. Nevitt MC, Cummings SR, Kidd S, Black D. Risk factors for recurrent non-syncopal falls. JAMA 1989;261:2663–2668.

74. Nieman DC. Fitness and sports medicine. Palo Alto, CA: Bull; 1990.

75. Ogawa T, Spina RJ, Martin III WH, Kohrt WM, Schechtman K, Holloszy JO, Ehsani AA. Effects of aging, sex and physical training on cardiovascular responses to exercise. Circulation 1992;86:494–503.

76. Orth-Gomer K, Rosengren A, Wilhelmsen L. Lack of social support and incidence of coronary heart-disease in middle-aged Swedish men. Psychol Med 1993;55:37–43.

77. Orth-Gomer K, Johnson JV. Social network interaction and mortality. A six-year follow-up study of a random sample of Swedish population. J Chron Dis 1987;40:949–957.

78. Pate RR, Pratt M, Blair SN, Haskell WL, Macera CA, Bouchard C, Buchner D, Ettinger W, Heath GW, King AC, Kriska A, Leon AS, Marcus BH, Morris J, Paffenbarger Jr. RS, Patrick K, Pollock ML, Rippe JM, Sallis JF, Wilmore JH. Physical activity and public health. A recommendation from the Centers for Disease Control and Prevention and the American College of Sports Medicine. JAMA 1995;273:402–407.

79. Piers LS, Soares MJ, McCormack LM, O'Dea K. Is there evidence for an age-related reduction in metabolic rate? J Appl Physiol 1998;85:2196–2204.

80. Poehlman ET, Arciero PJ, Goran MI. Endurance exercise in aging humans: effects on energy metabolism. Exerc Sport Sci Rev 1994;22:251–284.

81. Podsiadlo D, Richardson S. The timed "Up & Go": a test of basic functional mobility for frail elderly persons. J Am Geriatr Soc 1991;39:142–148.

82. Pollock ML, Franklin BA, Balady GJ, Chaitman BL, Fleg JL, Fletcher B, Limacher M, Pina IL, Stein RA, Williams M, Bazzarre T. Resistance exercise in individuals with and without cardiovascular disease. Benefits, rationale, safety and prescription. An advisory from the committee on exercise, rehabilitation and prevention, council on clinical cardiology, American Heart Association. Circulation 2000;101:828–833.

83. Prioux J, Ramonatxo M, Hayoit M, Mucci P, Prefaut C. Effect of aging on the ventilatory response and lactate kinetics during incremental exercise in man. Eur J Appl Physiol 2000;81:100–107.

84. Proctor DN, Balagopal P, Nair KS. Age-related sarcopenia in humans is associated with reduced synthetic rates of specific muscle proteins. J Nutr 1998;128:351S–355S.

85. Raab DM, Agre JC, McAdam M, Smith EL. Light resistance and stretching exercise in elderly women: effect upon flexibility. Arch Phys Med Rehabil 1988;69:268–272.

86. Resnick NM. Geriatric medicine. In: Fauci AS et al., editors. Harrison's principles of internal medicine. 14th ed. New York: McGraw-Hill; 1998. p 37–46.

87. Rohm Young D, Apple LJ, Jee S, Miller ER. The effects of aerobic exercise and t'ai chi on blood pressure in older people: results of a randomized trial. J Am Geriatr Soc 1999;47:277–284.

88. Rowell LB. Human cardiovascular control. 1993. Oxford, UK: Oxford University Press.

89. Ryan AS, Craig LD, Finn SC. Nutrient intakes and dietary patterns of older Americans: a national study. J Gerontol 1992;47:M145–M150.

90. Seeman TE, Kaplan GA, Knudsen L, Cohen R, Guralnik J. Social network ties and mortality among the elderly in the Alameda county study. Am J Epidemiol 1978;126:714–723.

91. Shephard RJ. The scientific basis of exercise prescribing for the very old. J Am Geriatr Soc 1990;38:62–70.

92. Skinner JS. Importance of aging for exercise testing and exercise prescription. In: Skinner JS, editor. Exercise testing and exercise prescription for special cases. 2nd ed. Philadelphia: Lea & Febiger; 1993. p 75–86.

93. Spina RJ. Cardiovascular adaptations to endurance exercise training in older men and women. Exerc Sport Sci Rev 1999;22:317–332.

94. Spina RJ, Miller TR, Bogenhagen WH, Schechtman KB, Ehsani AA. Gender-related differences in left ventricular filling dynamics in older subjects after endurance exercise training. J Gerontol 1996;51A:B232–B237.

95. Spina RJ, Ogawa T, Miller TR, Kohrt WM, Ehsani AA. Effect of exercise training on left ventricular performance in older women free of cardiopulmonary disease. Am J Cardiol 1993;71:99–104.

96. Spina RJ, Meyer TE, Peterson LR, Villareal DT, Rinder MR, Ehsani AA. Absence of left ventricular and arterial adaptations to exercise in octogenarians. J Appl Physiol 2004;97:1654–1659.

97. Spirduso WW. Physical dimensions of aging. Champaign, IL: Human Kinetics; 1995.

98. Stanton MW. The high concentration of U.S. health care expenditures. Research in Action, issue 19. AHRQ publication no. 06-0060, June 2006. Agency for Healthcare Research and Quality, Rockville, MD. www.ahrq.gov/research/ria19/expendria.htm

99. Steinbach U. Social networks, institutionalization, and mortality among elderly people in the United States. J Gerontol 1992;47:S183–S190.

100. Stevens JA, Harbraouck LM, Durant TM, Dellinger AM, Batabyal PK, Crosby AE, Valluru BR, Kresnow M, Huerro JL. Surveillance for injuries and violence among older adults. MMWR 1999;48(SS8):27–50.

101. Stratton JR, Levy WC, Cerqueria MD, Schwartz RS, Abrass IB. Cardiovascular responses to exercise. Effects of aging and exercise training in healthy men. Circulation 1994;89:1648–1655.

102. Sugisawa H, Liang J, Liu X. Social networks, social support, and mortality among older people in Japan. J Gerontol 1994;49:S3–S13.

103. Tanaka H, DeSouza CA, Jones PP, Stevenson ET, Davy KP, Seals DR. Greater rate of decline in maximal aerobic capacity with age in physically active vs. sedentary healthy women. J Appl Physiol 1997;83:1947–1953.

104. Tinetti ME, Baker DI, McAvay G, Claus EB, Garret P, Gottschalk M, Koch ML, Trainor K, Horwitz RI. A multifactorial intervention to reduce the risk of falling among elderly people living in the community. New Engl J Med 1994;331:821–827.

105. Tracy BL, Ivey FM, Hurlbut D, Martel GF, Lemmer JT, Siegel EL, Metter EJ, Fozard JL, Fleg JL, Hurley BF. Muscle quality II: effects of strength training in 65- to 75-yr-old men and women. J Appl Physiol 1999;86:195–201.

106. U.S. Department of Health and Human Services. 1996. Physical activity and health: a report of the surgeon general. Atlanta: U.S. Department of Health and Human Services, Centers of Disease Control and Prevention, National Center for Chronic Disease Prevention and Health Promotion.

107. Weiss EP, Spina RJ, Holloszy JO, Ehsani AA. Gender difference in the decline in aerobic capacity and its physiological determinants during later decades of life. J Appl Physiol 2006;101:938–944.

108. Withers RT, Smith DA, Tucker RC, Brinkman M, Clark DG. Energy metabolism in sedentary and active 49- to 70-yr-old women. J Appl Physiol 1998;84:1333–1340.

109. White TP. Skeletal muscle structure and function in older mammals. In: Gisolfi CV, Lamb DR, Nadel E, editors. Perspectives in exercise science and sports medicine. Carmel, IN: Cooper; 1995. p 115–174.

CHAPTER 8

1. American College of Sports Medicine. Position stand on the female athlete triad. Med Sci Sports Exerc 2007;39(10):1867–1882.

2. American College of Sports Medicine. Position stand on the female athlete triad. Med Sci Sports Exerc 1997;29(5):i–ix.

3. American College of Sports Medicine. ACSM's guidelines for exercise testing and prescription. 6th ed. Philadelphia: Lippincott Williams & Wilkins; 2000.

4. Agostini R, Drinkwater BL, Johnson MD. The female athlete triad [video]. ACSM's hot topics and fundamentals of sports medicine series: a physician's guide; 1996 (LDN: 476). Available from ACSM, PO Box 1440, Indianapolis, IN 46206-1440.

5. Bahrke MS, Yesalis CE, Brower KJ. Anabolic androgenic steroid abuse and performance-enhancing drugs among adolescents. Child Adolesc Psychiatr Clin N Am 1998;7(4):821–838.

6. Bale P. Body composition and menstrual irregularities of female athletes: are they precursors of anorexia? Sports Med 1994;17:347–352.

7. Banikarim C, Chacko MR, Kelder SH. Prevalence and impact of dysmenorrhea on Hispanic female adolescents. Arch Pediatr Adolesc Med 2000;154:1226–1229.

8. Bar-Or O. Thermoregulations in females from a life span perspective. In: Bar-Or O, Lamb DR, Clarkson PM, editors. Exercise and the female: a life span approach. Volume 9. Carmel, IN: Cooper; 1996. p 249–288.

9. Baum A. Eating disorders in the male athlete. Sports Med 2006;36:1–6.

10. Beals KA, Hill AK. The prevalence of disordered eating, menstrual dysfunction, and low bone mineral density among US collegiate athletes. Int J Sport Nutr Exerc Metab 2006;16(1):1–23.

11. Bechner-Melman R, Zohar AH, Ebstein RP, Elizur Y, Constantini N. How anorexic-like are the symptom and personality profiles of aesthetic athletes? Med Sci Sports Exerc 2006;38:628–636.

12. Beckvid-Henriksson G, Schnell C, Linden-Hirschberg A. Women endurance runners with menstrual dysfunction have prolonged interruption of training due to injury. Gynecol Obstet Invest 2000;49(1):41–46.

13. Bell R, Palma S. Antenatal exercise and birthweight. Aust NZ J Obstet Gynaecol 2000;40(1):70–73.

14. Bemben DA, Buchanan TD, Bemben MG, Knehans AW. Influence of type of mechanical loading, menstrual status, and training season on bone density in young women athletes. J Strength and Cond Res 2004;18:220–226.

15. Bungum TJ, Peaslee DL, Jackson AW, Perez MA. Exercise during pregnancy and type of delivery in nulliparae. J Obstet Gynecol Neonatal Nurs 2000;29(3):258–264.

16. Burt CW, Overpeck MD. Emergency visits for sport-related injuries. Annals of Emergency Medicine 2001;37:301–308.

17. Cann, CE, Martin MC, Genant HK, Jaffe RB. Decreased spinal mineral content in amenorrheic women. JAMA 1984;251:623–626.

18. Charkoudian N, Johnson JM. Modification of active cutaneous vasodilation by oral contraceptive hormones. J Appl Physiol 1997;83:2012–2018.

19. Charkoudian N, Johnson JM. Female reproductive hormones and thermoregulatory control of skin blood flow. Exerc Sport Sci Rev 2000;28:108–112.

20. Charkoudian N, Joyner MJ. Physiologic considerations for exercise performance in women. Clin Chest Med 2004;25:247–255.

21. Cheung CC, Thornton JE, Kuijper JL, Weigle DS, Clifton DK, Steiner RA. Leptin is a metabolic gate for the onset of puberty in the female rat. Endocrinology 1997;138:855–858.

22. Ciullo JV. Lower extremity injuries. In: Pearl AJ, editor. The athletic female. Champaign, IL: Human Kinetics; 1993. p 267–298.

23. Clapp JF. Exercise during pregnancy. In: Bar-Or O, Lamb DR, Clarkson PM, editors. Exercise and the female: a life span approach. Volume 9. Carmel, IN: Cooper; 1996. p 413–451.

24. Clapp JF. Exercise during pregnancy. A clinical update. Clin Sports Med 2000;19(2):273–286.

25. Clark N. Eating disorders among athletic females. In: Pearl AJ, editor. The athletic female. Champaign, IL: Human Kinetics; 1993. p 141–157.

26. Conn JM, Annest JL, Gilchrist J. Sports and recreation related injury episodes in the US population, 1997–99. Injury Prevention 2003;9:117–123.

27. Dale E, Gerlach DH, Wilhite AL. Menstrual dysfunction in distance runners. Obstet Gynecol 1979;54:47–53.

28. Dalsky GP. Effect of exercise on bone: permissive influence of estrogen and calcium. Med Sci Sports Exerc 1990;24:281–285.

29. Deitch JR, Starkey C, Walters SL, Moseley JB. Injury risk in professional basketball players: a comparison of Women's National Basketball Association and National Basketball Association athletes. Am J Sports Med 2006;34:1077–1083.

30. DiPietro L, Stachenfeld NS. Th e myth of the female athlete triad. Br J Sports Med 2006;40:490–493.

31. Drinkwater BL, Nilson K, Chesnut CH, Bremner J, Shainholtz S, Sothworth MB. Bone mineral content of amenorrheic and eumenorrheic athletes. N Engl J Med 1984;311:277–281.

32. Drinkwater BL. Amenorrhea, body weight, and osteoporosis. In: Brownell KD, Rodin J, Wilmore JH, editors. Eating, body weight and performance in athletes. Philadelphia: Lea & Febiger; 1992, p 235–247.

33. Drinkwater BL. The female athlete triad. Presented at the 13th Annual Conference on Exercise Sciences and Sports Medicine. San Juan, Puerto Rico, March 1999.

34. Duncombe D, Skouteris H, Wertheim EH, Kelly L, Fraser V, Paxton SJ. Vigorous exercise and birth outcomes in a sample of recreational exercisers: a prospective study across pregnancy. Aust NZ J Obstet Gynaecol 2006;46:288–292.

35. Dusek T. Influence of high intensity training on menstrual cycle disorders in athletes. Croat Med J 2001;42:79–82.

36. Erdelyi GJ. Gynecological survey of female athletes. J Sports Med Phys Fitness 1962;2:174–179.

37. Frisch RE, McArthur JW. Menstrual cycles: fatness as a determinant of minimum weight for height necessary for their maintenance or onset. Science 1974;185:949–951.

38. Garcia AM, Lacerda MG, Fonseca IA, Reis FM, Rodriguez LO, Silami-Garcia E. Luteal phase of the menstrual cycle increases sweating rate during exercise. Braz J Med Biol Res 2006;39:1255–1261.

39. Gorman C. Girls on steroids. Among young athletes, these dangerous drugs are all the rage. Is your daughter using them? Time Magazine 1998;152(6):93.

40. Gray J, Taunton JE, McKenzie DC, Clement DB, McConkey JP, Davidson RG. A survey of injuries to the anterior cruciate ligament of the knee in female basketball players. Int J Sports Med 1985;6(6):314–316.

41. Green GA, Uryasz FD, Petr TA, Bray CD. NCAA study of substance use and abuse habits of college student-athletes. Clin J Sport Med 2001;11:51–56.

42. Gruber AJ, Pope HG. Psychiatric and medical effects of anabolic-androgenic steroid use in women. Psychother Psychosom 2000;69(1):19–26.

43. Guyton AC. Textbook of medical physiology. 8th ed. Philadelphia: Saunders; 1991. p 899–914.

44. Harber VJ. Menstrual dysfunction in athletes: an energetic challenge. Exerc Sport Sci Rev 2000;28(1):19–23.

45. Heffernan AE. Exercise and pregnancy in primary care. Nurse Pract 2000;25(3):42,49,53–56.

46. Hewett TE. Neuromuscular and hormonal factors associated with knee injuries in female athletes. Strategies for intervention. Sports Med 2000;29(5):313–327.

47. Hind K, Truscott JG, Evans JA. Low lumbar spine bone mineral density in both male and female endurance runners. Bone 2006;39(4):880–885.

48. Hutchinson MR, Ireland ML. Knee injuries in female athletes. Sports Med 1995;19(4):288–302.

49. Inoue Y, Tanaka Y, Omori K, Kiwahara T, Ogura Y, Ueda H. Sex- and menstrual cycle-related differences in sweating and cutaneous blood flow in response to passive heat exposure. Eur J Appl Physiol 2005;94:323–332.

50. International Olympic Committee Medical Commission Working Group on Women in Sport. Position stand on the female athlete triad. Available from http://multimedia.olympic.org/pdf/en_report_917.pdf

51. Izzo A, Labriola D. Dysmenorrhea and sports activities in adolescents. Clin Exp Obstet Gynecol 1991;18:109–116.

52. Johnson C, Powers PS, Dick R. Athletes and eating disorders: the National Collegiate Athletic Association study. Int J Eat Disord 1999;26(2):179–188.

53. Johnson MD. Disordered eating. In: Agostini R, editor. Medical and orthopedic issues of active and athletic women. Philadelphia: Hanley & Belfus; 1994. p 141–151.

54. Klungland Torstveit M, Sundgot-Borgen J. The female athlete triad: are elite athletes at increased risk? Med Sci Sports Exerc 2005;37:184–193.

55. Klungland Torstveit M, Sundgot-Borgen J. The female athlete triad exists in both elite athletes and controls. Med Sci Sports Exerc 2005;37:1449–1459.

56. Kuwahara T, Inoue Y, Taniguchi M, Ogura Y, Ueda H, Kondo N. Effects of physical training on heat loss responses of young women to passive heating in relation to menstrual cycle. Eur J Appl Physiol 2005;94:376–386.

57. Latthe P, Latthe M, Say L, Gulmezoglu M, Khan K. WHO systematic review of prevalence of chronic pelvic pain: a neglected reproductive health morbidity. BMC Public Health 2006;6:177–183.

58. Lemcke, DP. Osteoporosis and menopause. In: Agostini R, editor. Medical and orthopedic issues of active and athletic women. Philadelphia: Hanley & Belfus; 1994. p 175–182.

59. Lloyd T, Triantafyllou SJ, Baker ER, Houts PS, Whiteside JA, Kalenak A, Stumpf PG. Women athletes with menstrual irregularity have increased musculoskeletal injuries. Med Sci Sports Exerc 1986;18(4):374–379.

60. Loucks AB. Effects of exercise training on the menstrual cycle: existence and mechanisms. Med Sci Sports Exerc 1990;22:275–280.

61. Loucks AB, Laughlin GA, Mortola JF, Girton L, Nelson JC, Yen SSC. Hypothalamic-pituitary-thyroid function in eumenorrheic and amenorrheic athletes. J Clin Endocrinol Metab 1992;75:514–518.

62. Loucks AB, Callister R. Induction and prevention of low-T3 syndrome in exercising women. Am J Physiol 1993;264:R924–R930.

63. Loucks AB, Heath EM. Induction of low-T3 syndrome in exercising women occurs at a threshold of energy availability. Am J Physiol 1994;266:R817–R823.

64. Loucks AB. The reproductive system. In: Bar-Or O, Lamb DR, Clarkson PM, editors. Exercise and the female: a life span approach. Volume 9. Carmel, IN: Cooper; 1996. p 41–70.

65. Loucks AB, Verdun M, Heath EM. Low energy availability, not stress of exercise, alters LH pulsatility in exercising women. J Appl Physiol 1998;84:37–46.

66. Loucks AB, Verdun M. Slow restoration of LH pulsatility by refeeding in energetically disrupted women. Am J Physiol 1998;275:R1218–R1226.

67. Lund-Hanssen H, Gannon J, Engebretsen L, Holen K, Hammer S. Isokinetic muscle performance in healthy female handball players and players with a unilateral anterior cruciate ligament reconstruction. Scand J Med Sci Sports 1996;6(3):172–175.

68. Malarkey WB, Strauss RH, Leizman DJ, Liggett M, Demers LM. Endocrine effects in female weight lifters who self-administer testosterone and anabolic steroids. Am J Obstet Gynecol 1991;165:1385–1390.

69. Marsh SA, Jenkins DG. Physiological responses to the menstrual cycle: implications for the development of heat illness in female athletes. Sports Med 2002;32:601–614.

70. Marshall, LA. Clinical evaluation of amenorrhea. In: Agostini R, editor. Medical and orthopedic issues of active and athletic women. Philadelphia: Hanley & Belfus; 1994. p 152–163.

71. Mendelsohn ME. Protective effects of estrogen on the cardiovascular system. Am J Cardiol 2002;89:12E–17E; discussion 17E–18E.

72. Mesaki N, Sasaki J, Shoji M, Iwasaki H, Eda M. Menstrual characteristics in college athletes. Nippon Sanka Fujinka Gakkai Zasshi 1984;36:247–254.

73. Millard-Stafford M, Sparling PB, Rosskopf LB, Snow TK, DiCarlo LJ, Hinson BT. Fluid intake in male and female runners during a 40-km field run in the heat. J Sports Sci 1995;13:257–263.

74. Moller-Nielsen J, Hammar M. Women's soccer injuries in relation to the menstrual cycle and oral contraceptive use. Med Sci Sports Exerc 1989;21(2):126–129.

75. Moquin A, Mazzeo RS. Effect of mild dehydration on the lactate threshold in women. Med Sci Sports Exerc 2000;32(2):396–402.

76. Murray B. Fluid replacement: the American College of Sports Medicine position stand. GSSI 1996;9(4):i–vii.

77. Myerson M, Gutin B, Warren MP, May MT, Contento I, Lee M, Pi-Sunyer FX, Pierson RN, Brooks-Gunn J. Resting metabolic rate and energy balance in amenorrheic and eumenorrheic runners. Med Sci Sports Exerc 1991;23:15–22.

78. Myklebust G, Maehlum S, Holm I, Bahr R. A prospective cohort study of anterior cruciate ligament injuries in elite Norwegian team handball. Scand J Med Sci Sports 1998;8(3):149–153.

79. Okonofua FE, Blogun JA, Ayangade SO, Fawole JO. Exercise and menstrual function in Nigerian university women. Afr J Med Med Sci 1990;19:185–190.

80. Otis CL. Exercise-associated amenorrhea. Clin Sports Med 1992;11(2):351–362.

81. Pomeroy C, Mitchell JE. Medical issues in the eating disorders. In: Brownell KD, Rodin J, Wilmore JH, editors. Eating, body weight and performance in athletes. Philadelphia: Lea & Febiger; 1992. p 202–221.

82. Practice Committee of the American Society for Reproductive Medicine. Current evaluation of amenorrhea. Fertil Steril 2004;82:266.

83. Reinking MF, Alexander LE. Prevalence of disordered-eating behaviors in undergraduate female collegiate athletes and nonathletes. J Athl Train 2005;40:47–51.

84. Rencken ML, Chesnut CH, Drinkwater BL. Bone density at multiple skeletal sites in amenorrheic athletes. JAMA 1996;276:238–240.

85. Rhea DJ. Eating disorder behavior of ethnically diverse urban female adolescent athletes and non-athletes. J Adolesc 1999;22(3):379–388.

86. Rickenlund A, Erikson MJ, Schenck-Gustafsson K, Hirschberg AL. Amenorrhea in female athletes is associated with endothelial dysfunction and unfavorable lipid profile. J Clin Endocrinol Metab 2005;90:1354–1359.

87. Rivera MA, Matos RM, Volquez B. Age of menarche and menstruation characteristics of Puerto Rican women athletes. PR Health Sci J 1990;9:179–183.

88. Robinson TL, Snow-Harter C, Taaffe DR, Gillis D, Shaw J, Marcus R. Gymnasts exhibit higher bone mass than runners despite similar prevalence of amenorrhea and oligomenorrhea. J Bone Miner Res 1995;10:26–35.

89. Roos H, Ornell M, Gardsell P, Lohmander LS, Lindstrand A. Soccer after anterior cruciate ligament injury: an incompatible combination? A national survey of incidence and risk factors and a 7-year follow-up of 310 players. Acta Orthop Scand 1995;66(2):107–112.

90. Root MPP. Recovery and relapse in former bulimics. Psychotherapy 1990;27:397–403.

91. Round table. Eating disorders in young athletes. Phys Sportsmed 1985;13:89–106.

92. Rumball JS, Legrun CM. Use of the preparticipation physical examination form to screen for the female athlete triad in Canadian interuniversity sport universities. Clin J Sport Med 2005;15:320–325.

93. Sanborn CF, Horea M, Siemers BJ, Dieringer KI. Disordered eating and the female athlete triad. Clin Sports Med 2000;19(2):199–213.

94. Sanborn CF, Wagner WW. The female athlete and menstrual irregularity. In: Puhl J, Brown CH, Voy RO, editors. Sport science perspectives for women. Champaign, IL: Human Kinetics; 1988. p 111–130.

95. Shangold MM. Sports and menstrual function. Phys Sportsmed 1980;8:66–69.

96. Shangold MM, Rebar RW, Wentz AC, Schiff I. Evaluation and management of menstrual dysfunction in athletes. JAMA 1990;263:1665–1669.

97. Slavin JL. Eating disorders in women athletes. In: Puhl J, Brown CH, Voy RO, editors. Sport science perspectives for women. Champaign, IL: Human Kinetics; 1988. p 189–197.

98. Slavin JL, Lutter JM, Cushman S, Lee V. Pregnancy and exercise. In: Puhl J, Brown CH, Voy RO, editors. Sport science perspectives for women. Champaign, IL: Human Kinetics; 1988. p 151–160.

99. Schneider S, Seither B, Tonges S, Schmitt H. Sports injuries: population based representative data on incidence, diagnosis, sequelae, and high risk groups. Br J Sports Med 2006;40:334–339.

100. Stachenfeld NS, DiPietro L, Palter SF, Nadel ER. Estrogen influences osmotic secretion of AVP and body water balance in postmenopausal women. Am J Physiol 1998;274:R187–195.

101. Stachenfeld NS, DiPietro L, Kokoszka CA, Silva C, Keefe DL, Nadel ER. Physiological variability of fluid regulation hormones in young women. J Appl Physiol 1999;86:1092–1096.

102. Stephenson LA, Kolka MA. Thermoregulation in women. Exerc Sport Sci Rev 1993;21:231–262.

103. Stephenson LA, Kolka MA. Esophageal temperature threshold for sweating decreases before ovulation in premenopausal women. J Appl Physiol 1999;86:22–28.

104. Sundgot-Borgen J, Torstveit MK. The female football player, disordered eating, menstrual function and bone health. Br J Sports Med 2007;41(Suppl I):i68–i72.

105. Trattner Sherman R, Thompson RA. Practical use of the International Olympic Committee Medical Commission position stand on the female athlete triad: a case example. Int J Eat Disord 2006;39(3):193–201.

106. Vereeke-West R. The female athlete: the triad of disordered eating, amenorrhea and osteoporosis. Sports Med 1998;26:63–71.

107. Warren MP, Brooks-Gunn J, Fox RP, Lancelot C, Newman D, Hamilton WG. Lack of bone accretion and amenorrhea: evidence for a relative osteopenia in weight-bearing bones. J Clin Endocrinol Metab 1991;72:847–853.

108. Warren MP, Shangold MM. Sports gynecology: problems and care of the athletic female. Cambridge, UK: Blackwell Science; 1997. p 26–30,113–136.

109. Warren MP, Perlroth NE. The effects of intense exercise on the female reproductive system. J Endocrinol 2001;170:3–11.

110. Weimann E. Gender-related differences in elite gymnasts: the female athlete triad. J Appl Physiol 2002;92:2146–2152.

111. Yeo S, Steele NM, Chang MC, Leclaire SM, Ronis DL, Hayashi R. Effect of exercise on blood pressure in pregnant women with a high risk of gestational hypertensive disorders. J Reprod Med 2000;45(4):293–298.

112. Yesalis CE, Barsukiewicz CK, Kopstein AN, Bahrke MS. Trends in anabolic-androgenic steroid use among adolescents. Arch Pediatr Adolesc Med 1997;151(12):1197–1206.

113. Zanker CL, Osborne C, Cooke CB, Oldroyd B, Truskott JG. Bone density, body composition and menstrual history of sedentary female former gymnasts, aged 20–32 years. Osteoporos Int 2004;15:145–154.

114. Zanker CL, Cooke CB. Energy balance, bone turnover, and skeletal health in physically active individuals. Med Sci Sports Exerc 2004;36:1372–1381.

CHAPTER 9

1. American Psychiatric Association. Diagnostic and statistical manual of mental disorders. 4th ed. Revised. Washington, DC: American Psychiatric Association; 2000.
2. Aneshensel CS, Huba GJ. Depression, alcohol use, and smoking over one year: a four-wave longitudinal causal model. J Abnormal Psychol 1983;92:134–150.
3. Babyak M, Blumenthal JA, Herman S, Khatri P, Doraiswamy M, Moore K, Craighead WE, Baldewicz TT, Krishnan KR. Exercise training for major depression: maintenance of therapeutic benefit at 10 months. Psychosom Med 2000;62:633–638.
4. Baldessarini RJ. Current status of antidepressants: clinical pharmacology and therapy. J Clin Psychiatry 1989;50:117–126.
5. Beck AT, Rush AJ, Shaw BF, Emery G. Cognitive therapy of depression. New York: Guilford Press; 1979.
6. Beck AT, Steer RA, Brown GK. Beck Depression Inventory. 2nd ed. San Antonio: Psychological Corporation; 1996.
7. Beck JS. Cognitive therapy: basics and beyond. New York: Guilford Press; 1995.
8. Blumenthal JA, Babyak M, Moore K, Craighead WE, Herman S, Khatri P, Waugh R, Napolitano MA, Forman LM, Appelbaum M, Doraiswamy M, Krishnan KR. Effects of exercise training on older adults with major depression. Arch Intern Med 1999;159:2349–2356.
9. Bonnet F, Irving K, Terra JL, Nony P, Berthezene F, Moulin P. Anxiety and depression are associated with unhealthy lifestyle in patients at risk of cardiovascular disease. Atherosclerosis 2005;178:339–344.
10. Brosse AL, Sheets ES, Lett HS, Blumenthal JA. Exercise and the treatment of clinical depression in adults: recent findings and future directions. Sports Med 2002;32:741–60.
11. Burns DD. Feeling good. The new mood therapy, revised and updated. New York: Avon Books; 1999.
12. Cassano P, Fava M. Depression and public health: an overview. J Psychosom Res 2002;53:849–857.
13. De Groot M, Anderson R, Freedland KE, Clouse RE, Lustman PJ. Association of depression and diabetes complications: a meta-analysis. Psychosom Med 2001;63:619–630.
14. Demyttenaere K, Enzlin P, Dewe W, Boulanger B, De Bie J, De Troyer W. Compliance with antidepressants in a primary care setting: beyond lack of efficacy and adverse events. J Clin Psychiatry 2001;62:30–33.
15. DiMatteo MR, Lepper HS, Croghan TW. Depression is a risk factor for noncompliance with medical treatment: meta-analysis of the effects of anxiety and depression on patient adherence. Arch Inter Med 2000;160:2101–2107.
16. Dunn AL, Trivedi MH, Kampert JB, Clark CG, Chambliss HO. Exercise treatment for depression: efficacy and dose response. Am J Prev Med 2005;28:1–8.
17. Eaton WW, Anthony JC, Gallo J, Cai G, Tien A, Romanoski A, et al. Natural history of Diagnostic Interview Schedule/DSM–IV major depression. The Baltimore Epidemiologic Catchment Area follow-up. Arch Gen Psychiat 1997;54:993–999.
18. Elkin I, Shea MT, Watkins JT, Imber SD, Sotsky SM, Collins JF, Glass DR, Pilkonis PA, Leber WR, Docherty JP, Fiester SJ, Parloff MB. NIMH treatment of depression collaborative research program: general effectiveness of treatments. Arch Gen Psychiatry 1989;46:971–982.
19. Gehi A, Haas D, Pipkin S, Whooley MA. Depression and medication adherence in outpatients with coronary heart disease: findings from the heart and soul study. Arch Intern Med 2005;165:2508–2513.
20. Glazer KM, Emery CF, Frid DJ, Banyasz RE. Psychological predictors of adherence and outcomes among patients in cardiac rehabilitation. J Cardiopulm Rehabil 2002;22:40–46.
21. Goodnick PJ, Hernandez M. Treatment of depression in comorbid medical illness. Expert Opin Pharmacother 2000;1:1367–1384.
22. Greenberg PE, Stiglin LE, Finkelstein SN, Berndt ER. The economic burden of depression in 1990. J Clin Psychiat 1993;54:405–418.
23. Herman S, Blumenthal JA, Babyak M, Khatri P, Craighead WE, Krishnan KR, Doraiswamy PM. Exercise therapy for depression in middle-aged and older adults: predictors of early dropout and treatment failure. Health Psychol 2002;21:553–563.
24. Hirschfeld RMA, Keller MB, Panico S, Arons BS, Barlow D, Davidoff F, Endicott J, Froom J, Goldstein M, Gorman JM, Guthrie D, Marek RG, Maurer TA, Meyer R, Phillips K, Ross J, Schwenk TL, Sharfstein SS, Thase ME, Wyatt RJ. The National Depressive and Manic-Depressive Association consensus statement on the undertreatment of depression. JAMA 1997;277:333–340.
25. Johnson J, Weissman MM, Klerman GL. Service utilization and social morbidity associated with depressive symptoms in the community. JAMA 1992;267:1478–1483.
26. Katon WJ, Rutter C, Simon G, Lin EHB, Ludman E, Ciechanowski P, Kinder L, Young B, Von Korff M. The association of comorbid depression with mortality in patients with type 2 diabetes. Diabetes Care 2005;28:2668–2672.
27. Kessler RC, Berglund P, Demler O, Jin R, Walters EE. Lifetime prevalence and age-of-onset distributions of DSM-IV disorders in the national comorbidity survey replication. Arch Gen Psychiat 2005;62:593–602.
28. Kessler RC, Chiu WT, Demler O, Walters EE. Prevalence, severity, and comorbidity of 12-month DSM-IV disorders in the national comorbidity survey replication. Arch Gen Psychiat 2005;62:617–627.
29. Lavoie KL, Fleet RP, Lesperance F, Arsenault A, Laurin C, Frasure-Smith N, Bacon SL. Are exercise stress tests appropriate for assessing myocardial ischemia in patients with major depressive disorder? Am Heart J 2004;148:621–627.
30. Lawlor DA, Hopker SW. The effectiveness of exercise as an intervention in the management of depression: systematic review and meta-regression analysis of randomised controlled trials. Brit Med J 2001;322:1–8.
31. Lehmann HE. Clinical evaluation and natural course of depression. J Clin Psychiat 1983;44:5–10.
32. Lett HS, Blumenthal JA, Babyak MA, Sherwood A, Strauman T, Robins C, Newman MF. Depression as a risk factor for coronary artery disease: evidence, mechanisms, and treatment. Psychosom Med 2004;66:305–315.
33. Lin EH, Katon W, Von Korff M, Rutter C, Simon GE, Oliver M, Ciechanowski P, Ludman EJ, Bush T, Young B. Relationship of depression and diabetes self-care, medication adherence, and preventive care. Diabetes Care 2004;27:2154–2160.
34. Mather AS, Rodriguez C, Guthrie MF, McHarg AM, Reid IC, McMurdo MET. Effects of exercise on depressive symptoms in older adults with poorly responsive depressive disorder. Br J Psychiat 2002;180:411–415.
35. McNeil JK, LeBlanc EM, Joyner M. The effect of exercise on depressive symptoms in the moderately depressed elderly. Psychol Aging 1991;6:487–488.
36. Melfi CA, Chawla AJ, Croghan TW, Hanna MP, Kennedy S, Sredl K. The effects of adherence to antidepressant treatment guidelines on relapse and recurrence of depression. Arch Gen Psychiat 1998;55:1128–1132.
37. Olfson M, Marcus SC, Druss B, Elinson L, Tanielian T, Pincus HA. National trends in the outpatient treatment of depression. JAMA 2002;287:203–209.
38. O'Neal HA, Dunn AL, Martinsen EW. Depression and exercise. Int J Sport Psychol 2000;31:110–135.
39. Pirl WF, Roth AJ. Diagnosis and treatment of depression in cancer patients. Oncology 1999;13:1293–1301.

40. Plant EA, Sachs-Ericsson N. Racial and ethnic differences in depression: the roles of social support and meeting basic needs. J Consult Clin Psych 2004;72:41–52.

41. Radloff LS. The CES-D Scale: a self-report depression scale for research in the general population. Appl Psych Meas 1977;1:385–401.

42. Robinson WD, Geske JA, Prest LA, Barnacle R. Depression treatment in primary care. J Am Board Fam Pract 2005;18:79–86.

43. Ruo B, Rumsfeld JS, Pipkin S, Whooley MA. Relation between depressive symptoms and treadmill exercise capacity in the heart and soul study. Am J Cardiol 2004;94:96–99.

44. Singh NA, Clements KM, Singh MAF. The efficacy of exercise as a long-term antidepressant in elderly subjects: a randomized controlled trial. J Gerontol 2001;56A:M497–M504.

45. Spitzer RL, Williams JBW, Kroenke K, et al. Utility of a new procedure for diagnosing mental disorders in primary care: the PRIME-MD 1000 study. JAMA 1994;272:1749–1756.

46. Thase ME, Kupfer DJ. Recent developments in the pharmacotherapy of mood disorders. J Consult Clin Psych 1996;64:646–659.

47. U.S. Preventive Services Task Force. Screening for depression: recommendations and rationale. Ann Intern Med 2002;136:760–776.

48. Ustun TB, Ayuso-Mateos JL, Chatterji S, Mathers C, Murray CL. Global burden of depressive disorders in the year 2000. British J Psychiat 2004;184:386–392.

49. Wang PS, Lane M, Olfson M, Pincus HA, Wells KB, Kessler RC. Twelve-month use of mental health services in the United States. Arch Gen Psychiat 2005;62:629–640.

50. Weissman MM, Bland RC, Canino GJ, Faravelli C, Greenwald S, Hwu H, Joyce PR, Karam EG, Lee C, Lellouch J, Lepine J, Newman SC, Rubio-Stipec M, Wells JE, Wickramaratne PJ, Wittchen H, Yeh E. Cross-national epidemiology of major depression and bipolar disorder. JAMA 1996;276:293–299.

51. Wells KB, Stewart A, Hays RD, Burnam MA, Rogers W, Daniels M, et al. The functioning and well-being of depressed patients. Results from the Medical Outcomes Study. JAMA 1989;262:914–919.

52. Whooley MA, Avins AL, Miranda J, Browner WS. Case-finding instruments for depression. J Gen Intern Med 1997;12:439–445.

53. Wittchen HU, Holsboer F, Jacobi F. Met and unmet needs in the management of depressive disorder in the community and primary care: the size and breadth of the problem. J Clin Psychiat 2001;62 Suppl 26:23–28.

54. Zigmond A, Snaith R. The hospital anxiety and depression scale. Acta Psychiat Scand 1983;67:361–370.

CHAPTER 10

1. Third report of the National Cholesterol Education Program (NCEP) Expert Panel on Detection, Evaluation, and Treatment of High Blood Cholesterol in Adults (Adult Treatment Panel III) final report. Circulation 2002;106:3143–3421.

2. ACSM's guidelines for exercise testing and prescription; 2006.

3. International Diabetes Federation. The IDF consensus worldwide definition of the metabolic syndrome. September 28, 2006. Internet communication.

4. Alberti KG, Zimmet PZ. Definition, diagnosis and classification of diabetes mellitus and its complications. Part 1: diagnosis and classification of diabetes mellitus provisional report of a WHO consultation. Diabet Med 1998;15:539–553.

5. Balkau B, Charles MA. Comment on the provisional report from the WHO consultation. European Group for the Study of Insulin Resistance (EGIR). Diabet Med 1999;16:442-443.

6. Ballantyne CM, Olsson AG, Cook TJ, Mercuri MF, Pedersen TR, Kjekshus J. Influence of low high-density lipoprotein cholesterol and elevated triglyceride on coronary heart disease events and response to simvastatin therapy in 4S. Circulation 2001;104:3046–3051.

7. Bray GA, Champagne CM. Obesity and the metabolic syndrome: implications for dietetics practitioners. J Am Diet Assoc 2004;104:86–89.

8. Denke MA, Pasternak RC. Defining and treating the metabolic syndrome: a primer from the Adult Treatment Panel III. Curr Treat Options Cardiovasc Med 2001;3:251–253.

9. Eckel RH, Wassef M, Chait A, Sobel B, Barrett E, King G, Lopes-Virella M, Reusch J, Ruderman N, Steiner G, Vlassara H. Prevention Conference VI: Diabetes and Cardiovascular Disease: Writing Group II: pathogenesis of atherosclerosis in diabetes. Circulation 2002;105:e138–e143.

10. Einhorn D, Reaven GM, Cobin RH, Ford E, Ganda OP, Handelsman Y, Hellman R, Jellinger PS, Kendall D, Krauss RM, Neufeld ND, Petak SM, Rodbard HW, Seibel JA, Smith DA, Wilson PW. American College of Endocrinology position statement on the insulin resistance syndrome. Endocr Pract 2003;9:237–252.

11. Ford ES, Giles WH, Dietz WH. Prevalence of the metabolic syndrome among US adults: findings from the third National Health and Nutrition Examination Survey. JAMA 2002;287:356–359.

12. Ford ES, Giles WH, Mokdad AH. Increasing prevalence of the metabolic syndrome among U.S. adults. Diabetes Care 2004;27:2444–2449.

13. Foreyt JP. Need for lifestyle intervention: how to begin. Am J Cardiol 2005;96:11E–14E.

14. Gimeno RE, Klaman LD. Adipose tissue as an active endocrine organ: recent advances. Curr Opin Pharmacol 2005;5:122–128.

15. Greenfield JR, Campbell LV. Insulin resistance and obesity. Clin Dermatol 2004;22:289–295.

16. Grundy SM, Brewer HB, Jr., Cleeman JI, Smith SC, Jr., Lenfant C. Definition of metabolic syndrome: report of the National Heart, Lung, and Blood Institute/American Heart Association conference on scientific issues related to definition. Circulation 2004;109:433–438.

17. Grundy SM, Cleeman JI, Daniels SR, Donato KA, Eckel RH, Franklin BA, Gordon DJ, Krauss RM, Savage PJ, Smith Jr SC, Spertus JA, Costa F. Diagnosis and management of the metabolic syndrome. An American Heart Association/National Heart, Lung, and Blood Institute Scientific Statement. Executive summary. Cardiol Rev 2005;13:322–327.

18. Hutley L, Prins JB. Fat as an endocrine organ: relationship to the metabolic syndrome. Am J Med Sci 2005;330:280–289.

19. Katzmarzyk PT, Church TS, Blair SN. Cardiorespiratory fitness attenuates the effects of the metabolic syndrome on all-cause and cardiovascular disease mortality in men. Arch Intern Med 2004;164:1092–1097.

20. Katzmarzyk PT, Leon AS, Wilmore JH, Skinner JS, Rao DC, Rankinen T, Bouchard C. Targeting the metabolic syndrome with exercise: evidence from the HERITAGE Family Study. Med Sci Sports Exerc 2003;35:1703–1709.

21. Liberopoulos EN, Mikhailidis DP, Elisaf MS. Diagnosis and management of the metabolic syndrome in obesity. Obes Rev 2005;6:283–296.

22. Malik S, Wong ND, Franklin SS, Kamath TV, L'Italien GJ, Pio JR, Williams GR. Impact of the metabolic syndrome on mortality from coronary heart disease, cardiovascular disease, and all causes in United States adults. Circulation 2004;110:1245–1250.

23. Masuzaki H, Flier JS. Tissue-specific glucocorticoid reactivating enzyme, 11 beta-hydroxysteroid dehydrogenase type 1 (11 beta-HSD1)—a promising drug target for the treatment of metabolic syndrome. Curr Drug Targets Immune Endocr Metabol Disord 2003;3:255–262.

24. Matsuda M, Shimomura I. Adipocytokines and metabolic syndrome—molecular mechanism and clinical implication. Nippon Rinsho 2004;62:1085–1090.

25. Matsuzawa Y, Funahashi T, Kihara S, Shimomura I. Adiponectin and metabolic syndrome. Arterioscler Thromb Vasc Biol 2004;24:29–33.

26. Park YW, Zhu S, Palaniappan L, Heshka S, Carnethon MR, Heymsfield SB. The metabolic syndrome: prevalence and associated risk factor findings in the US population from the Third National Health and Nutrition Examination Survey, 1988–1994. Arch Intern Med 2003;163:427–436.

27. Pitsavos C, Panagiotakos DB, Chrysohoou C, Kavouras S, Stefanadis C. The associations between physical activity, inflammation, and coagulation markers, in people with metabolic syndrome: the ATTICA study. Eur J Cardiovasc Prev Rehabil 2005;12:151–158.

28. Reaven GM. The metabolic syndrome: requiescat in pace. Clin Chem 2005;51:931–938.

29. Rennie KL, McCarthy N, Yazdgerdi S, Marmot M, Brunner E. Association of the metabolic syndrome with both vigorous and moderate physical activity. Int J Epidemiol 2003;32:600–606.

30. Reynolds K, He J. Epidemiology of the metabolic syndrome. Am J Med Sci 2005;330:273–279.

31. Rubins HB. Triglycerides and coronary heart disease: implications of recent clinical trials. J Cardiovasc Risk 2000;7:339–345.

32. Tangalos EG, Cota D, Fujioka K. Complex cardiometabolic risk factors: impact, assessment, and emerging therapies. J Am Med Dir Assoc 2006;7:1–10.

CHAPTER 11

1. Albright AL. Exercise precautions and recommendations for patients with autonomic neuropathy. Diabetes Spectrum 1998;11:231–237.

2. Albright A, Franz M, Hornsby G, Kriska A, Marrero D, Ullrich I, Verity, L. Exercise and type 2 diabetes. Med Sci Sports Exerc 2000;32:1345–1362.

3. Schwarz J, Moline K, Urban M, editors. A core curriculum for diabetes education. 2nd edition. Chicago: American Association of Diabetes Educators; 1993. p 203.

4. American College of Sports Medicine. Physical activity, physical fitness, and hypertension (position stand). Med Sci Sports Exerc 1993;25:i–x.

5. American College of Sports Medicine. ACSM's guidelines for exercise testing and prescription. 7th edition. Franklin B, Whaley M, Howley E, editors. Baltimore: Williams & Wilkins; 2005.

6. American College of Sports Medicine. ACSM's exercise management for persons with chronic diseases and disabilities. Champaign, IL: Human Kinetics; 1997.

7. American Diabetes Association. Handbook of exercise in diabetes. Ruderman N, Devlin J, Schneider S, editors. Alexandria, VA: American Diabetes Association; 2002. p 268–288.

8. American Diabetes Association. Economic costs of diabetes mellitus in the U.S. in 2007. Diabetes Care 2008;31:596-615.

9. American Diabetes Association. Therapy for diabetes mellitus and related disorders. 4th edition. Alexandria, VA: American Diabetes Association; 2004.

10. American Diabetes Association. Diagnosis and classification of diabetes mellitus. Diabetes Care 2008;31:S55–S60.

11. American Diabetes Association. Nutrition recommendations and interventions for diabetes. Diabetes Care 2008;31:S61–S78.

12. American Diabetes Association. Physical activity/exercise and diabetes. Diabetes Care 2004;27:S58–S62.

13. American Diabetes Association. National standards for diabetes self-management education. Diabetes Care 2008;31:S97–S104.

14. American Diabetes Association. Standards of medical care in diabetes—2008. Diabetes Care 2008;31:S12–S54.

15. Andersson DKG, Svaardsudd K. Long-term glycemic control related to mortality in type II diabetes. Diabetes Care 1995;18:1534–1543.

16. Andressen EM, Lee JA, Pecoraro RE, Koepsell TD, Hallstrom AP, Siscovick DS. Under reporting of diabetes on death certificates, King County, Washington. Am J Public Health 1993;83:1021–1024.

17. Bandura A. Social foundations of thought and action. Englewood Cliffs, NJ: Prentice Hall; 1986.

18. Bantle JP, Swanson JE, Thomas W, Laine DC. Metabolic effects of dietary sucrose in type II diabetic subjects. Diabetes Care 1996;19:1249–1256.

19. Barnard RJ, Lattimore L, Holly RG, Cherny S, Pritikin N. Response of non-insulin-dependent diabetic patients to an intensive program of diet and exercise. Diabetes Care 1982;5:370–374.

20. Berger M, Berchtold P, Cuppers HJ, et al. Metabolic and hormonal effects of muscular exercise in juvenile type diabetics. Diabetologia 1977;13:355–365.

21. Björntorp P. Portal adipose tissue as a generator of risk factors for cardiovascular disease and diabetes. Arteriosclerosis 1990;10:493–496.

22. Blair SN, Kohl HW, Paffenbarger RS, Clark DG, Cooper KH, Gibbons LW. Physical fitness and all-cause mortality. A prospective study of healthy men and women. JAMA 1989;262:2395–2401.

23. Blake GA, Levin SR, Koyal SN. Exercise induced hypertension in normotensive patients with NIDDM. Diabetes Care 1990;13:799–801.

24. Bogardus C, Ravussin E, Robins DC, Wolfe RR, Horton ES, Sims EAH. Effects of physical training with diet therapy on carbohydrate metabolism in patients with glucose intolerance and non-insulin-dependent diabetes mellitus. Diabetes 1984;33:311–318.

25. Bogardus C, Lillioja S, Mott DM, Hollenbeck C, Reaven G. Relationship between degree of obesity and in vivo insulin action in man. Am J Physiol 1985;248:E286–E291.

26. Brosseau JD. Occurrence of diabetes among decedents in North Dakota. Diabetes Care 1987;10:542–543.

27. Centers for Disease Control and Prevention. Trends in the prevalence and incidence of self-reported diabetes mellitus United States, 1980–1994. Morb Mortal Wkly Rep 1997;46:1027–1028.

28. Centers for Disease Control and Prevention. National diabetes fact sheet: general information and national estimates on diabetes in the United States, 2007. Atlanta: U.S. Department of Health and Human Services, Centers for Disease Control and Prevention; 2008.

29. Cowie CC, Eberhardt MS. Sociodemographic characteristics of persons with diabetes. In: Harris MI, Cowie CC, Stern MP, et al., editors. Diabetes in America (NIH publication 95-1468). 2nd edition. Bethesda, MD: National Institutes of Health, National Institute of Diabetes and Digestive and Kidney Diseases; 1995. p 85–101.

30. Davidson MB. Diabetes mellitus diagnosis and treatment. 4th edition. Philadelphia: Saunders, 1998.

31. DeFronzo R, Deibert D, Hendler R, Felig P. Insulin sensitivity and insulin binding to monocytes in maturity-onset diabetes. J Clin Invest 1979;63:939–946.

32. Diabetes Control and Complications Trial Research Group. The effect of intensive treatment of diabetes on the development and progression of long-term complications in insulin-dependent diabetes mellitus. New Engl J Med 1993;329:977–986.

33. Diabetes Prevention Program Research Group. The Diabetes Prevention Program. Design and methods for a clinical trial in the prevention of type 2 diabetes. Diabetes Care 1999;22:623–634.

34. Diabetes Research Working Group. Conquering diabetes: a strategic plan for the 21st century (NIH publication 99-4398). Bethesda, MD: National Institutes of Health; 1999.

35. Dowse GK, Zimmet PZ, Gareeboo H, Alberti KGMM, Tuomilehto J, Finch CF, Chitson P, Tulsidas H. Abdominal obesity and physical inactivity are risk factors for NIDDM and impaired glucose tolerance in Indian, Creole, and Chinese Mauritians. Diabetes Care 1991;14:271–282.

36. Duncan JJ, Gordon NF, Scott CB. Women walking for health and fitness: how much is enough? JAMA 1991;266:3295–3299.

37. Eriksson J, Taimela S, Eriksson K, Parvianen S, Peltonen J, Kujala U. Resistance training in the treatment of non-insulin-dependent diabetes mellitus. Int J Sports Med 1997;18:242–246.

38. Estacio RO, Regensteiner JG, Wolfel EE, Jeffers B, Dickenson M, Schrier RW. The association between diabetic complications and exercise capacity in NIDDM patients. Diabetes Care 1998;21:291–295.

39. Ewing DJ, Boland O, Neilson JM, Cho CG, Clarke B. Autonomic neuropathy, QT interval lengthening, and unexpected deaths in diabetic autonomic neuropathy. J Clin Endocrinol Metab 1991;54:751–754.

40. Feskens EJ, Loeber JG, Kromhout D. Diet and physical activity as determinants of hyperinsulinemia: the Zutphen elderly study. Am J Epidemiol 1994;140:350–360.

41. Fletcher GF, Blair SN, Blumenthal J, Caspersen C, Chaitman B, Epstein S. Statement on exercise: benefits and recommendations for physical activity programs for all Americans. Circulation 1992;86(1):340–344.

42. Fluckey JD, Hickey MS, Brambrink JK, Hart KK, Alexander K, Craig BW. Effects of resistance exercise on glucose tolerance in normal and glucose-intolerant subjects. J Appl Physiol 1994;77:1087–1092.

43. Franz MJ, Etzwiler DD, Joynes JO, Hollander PM. Learning to live well with diabetes. Minneapolis: DCI; 1991.

44. Franz M. Medical nutrition therapy. In: Diabetes mellitus: diagnosis and treatment. Philadelphia: Saunders; 1998. p 45–79.

45. Frisch RE, Wyshak G, Albright TE, Albright NL, Schiff I. Lower prevalence of diabetes in female former college athletes compared with nonathletes. Diabetes 1986;35:1101–1105.

46. Fujimoto WY, Leonetti DL, Kinyoun JL, Shuman WP, Stolov WC, Wahl PW. Prevalence of complications among second-generation Japanese-American men with diabetes, impaired glucose tolerance or normal glucose tolerance. Diabetes 1987;36:730–739.

47. Gordon ND, Kohl HW, Blair SN. Lifestyle exercise: a new strategy to promote physical activity for adults. J Cardiopulm Rehabil 1993;13:161–163.

48. Graham C, Lasko-Mccarthey P. Exercise options for persons with diabetic complications. Diabetes Educator 1990;16:212–220.

49. Harris MI, Couric CC, Reiber G, Boyko E, Stern M, Bennett P, editors. Diabetes in America (NIH publication 95-1468). 2nd edition. Washington, DC: U.S. Government Printing Office; 1995.

50. Helmrish SP, Ragland DR, Leung RW, Paffenbarger RW. Physical activity and reduced occurrence of non-insulin-dependent diabetes mellitus. New Engl J Med 1991;325:147–152.

51. Herman WH, Eastman RC, Songer TJ, Dasbach EJ. The cost-effectiveness of intensive therapy for diabetes mellitus. Endocrinol Metab Clin North Am 1997;26:679–695.

52. Holloszy JO, Schultz J, Kusnierkiewicz J, Hagberg JM, Ehsani AA. Effects of exercise on glucose tolerance and insulin resistance: brief review and some preliminary results. Acta Med Scand 1987;711(Suppl.):55–65.

53. Horton ES. Exercise and physical training: effects on insulin sensitivity and glucose metabolism. Diabetes Metab Rev 1986;2:1–17.

54. Horton ES. Role and management of exercise in diabetes mellitus. Diabetes Care 1988;11:201–211.

55. Hubinger A, Franzen A, Gries A. Hormonal and metabolic response to physical exercise in hyperinsulinemic and non-hyperinsulinemic type 2 diabetics. Diabetes Res 1987;4:57–61.

56. Hurley BF, Seals DR, Ehsani AA, Carter LJ, Dalsky GP, Hagberg JM, Holloszy JO. Effects of high-intensity strength training on cardiovascular function. Med Sci Sports Exerc 1984;16:483–488.

57. Kahn JK, Zola B, Juni J, Vinik A. Decreased exercise heart rate and blood pressure response in diabetic subjects with cardiac autonomic neuropathy. Diabetes Care 1986;9:389–394.

58. Kahn JK, Sisson JC, Vinik AI. Prediction of sudden cardiac death in diabetic autonomic neuropathy. J Nucl Med 1988;29:1605–1606.

59. Kelley S, Seraganian P. Physical fitness level and autonomic reactivity to psychosocial stress. J Psychosom Res 1984;28:279–287.

60. Kenny SJ, Aubert RE, Geiss LS. Prevalence and incidence of non-insulin-dependent diabetes. In: Harris MI, Cowie CC, Stern MP, et al., editors. Diabetes in America (NIH publication 95-1468). 2nd edition. Bethesda, MD: National Institutes of Health, National Institute of Diabetes and Digestive and Kidney Diseases; 1995. p 46–67.

61. Kissebah AH, Vydelingum N, Murray R, Evans DF, Hartz AJ, Kalkhoff RK, Adams PW. Relationship of body fat distribution to metabolic complications of obesity. J Clin Endocrinol Metab 1982;54:254–260.

62. Kjaer M, Hollenbeck CB, Frey-Hewitt B, Galbo H, Haskell W, Reaven GM. Glucoregulation and hormonal responses to maximal exercise in non-insulin-dependent diabetes. J Appl Physiol 1990;68:2067–2074.

63. Klem ML, Wing RR, Mcguire MT, Seagle HM, Hill JO. A descriptive study of individuals successful at longterm maintenance of substantial weight loss. Am J Clin Nutr 1997;66:239–246.

64. Kobberling J, Tillil H. Empiric risk figures for first degree relatives of non-insulin-dependent diabetics. In: Kobberling J, Tattersall R, editors. The genetics of diabetes mellitus (Serono Symposium No. 47). New York: Academic Press; 1982. p 201–209.

65. Koivisto VA, Defronzo RA. Exercise in the treatment of type 2 diabetes. Acta Endocrinol 1984;262:107–111.

66. Koivisto VA, Yki-Jarvinen H, Defronzo RA. Physical training and insulin sensitivity. Diabetes Metab Rev 1986;1:445–481.

67. Kolterman OG, Gray RS, Griffin J, Burstein P, Insel J, Scarlett JA, Olefsky JM. Receptor and postreceptor defects contribute to the insulin resistance in non-insulin-dependent diabetes mellitus. J Clin Invest 1981;68:957–969.

68. Krentz AJ, Ferner RE, Bailey CJ. Comparative tolerability profiles of oral antidiabetic agents. Drug Safety 1994;11:223–241.

69. Krotkiewski M, Lonnroth P, Mandroukas K, Wroblewski Z, Rebuffe-Scrive M, Holm G, et al. The effects of physical training on insulin secretion and effectiveness and on glucose metabolism in obesity and type 2 (non-insulin-dependent) diabetes mellitus. Diabetologia 1985;28:881–890.

70. Lampman RM, Schteingart DE. Effects of exercise training on glucose control, lipid metabolism, and insulin sensitivity in hypertriglyceridemia and non-insulin-dependent diabetes mellitus. Med Sci Sports Exerc 1991;23:703–712.

71. Lawrence RH. The effects of exercise on insulin action in diabetes. Better Med J 1926;1:648–652.

72. Laws A, Reaven GM. Physical activity, glucose tolerance, and diabetes in older adults. Ann Behav Med 1991;13:125–131.

73. Lindgarde F, Saltin B. Daily physical activity, work capacity and glucose tolerance in lean and obese normoglycaemic middle-aged men. Diabetologia 1981;20:134–138.

74. Loghmani E, Rickard K, Washburne L, Vandagriff J, Fineberg N, Golden M. Glycemic response to sucrose-containing mixed meals in diets of children with insulin-dependent diabetes mellitus. J Pediatr 1991;119:531–537.

75. Manson JE, Rimm EB, Stampfer MJ, Colditz GA, Willett WC, Krolewski AS, Rosner B, Hennekens CH, Speizer FE. Physical activity and incidence of non-insulin-dependent diabetes mellitus in women. Lancet 1991;338:774–778.

76. Manson JE, Nathan DM, Krolewski AS, Stampfer MJ, Willett WC, Hennekens CH. A prospective study of exercise and incidence of diabetes among US male physicians. JAMA 1992;268:63–67.

77. Marrero DG, Sizemore JM. Motivating patients with diabetes to exercise. In: Anderson BJ, Ruben RR, editors. Practical psychology for diabetes physicians: how to deal with key behavioral issues faced by health care teams. Alexandria, VA: American Diabetes Association; 1996.

78. Mayer-Davis EJ, D'agostino R, Karta AJ, Haffner SM, Rewers MJ, Saad M, Bergman RN. Intensity and amount of physical activity in relation to insulin sensitivity. JAMA 1998;279:669–674.

79. Metzget BE, Cho NH, Roston SM, Radvany R. Prepregnancy weight and antepartum insulin secretion predict glucose tolerance five years

after gestational diabetes mellitus. Diabetes Care 1993;16:1598–1605.

80. Miller WJ, Sherman WM, Ivy JL. Effect of strength training on glucose tolerance and post-glucose insulin response. Med Sci Sports Exerc 1984;16:539–543.

81. Minuk HL, Vranic M, Marliss EB, Hanna AK, Albisser AM, Zinman B. Glucoregulatory and metabolic response to exercise in obese non-insulin-dependent diabetes. Am J Physiol 1981;240:E458–E464.

82. Moss SE, Klein R, Klein BEK, Meuer MS. The association of glycemia and cause-specific mortality in a diabetic population. Arch Intern Med 1984;154:2473–2479.

83. Mourier A, Gautier J-F, Dekerviler E, Biagard AX, Villette J-M, Garnier JP, Duvallet A, Guezennec CY, Cathelineau G. Mobilization of visceral adipose tissue related to the improvement in insulin sensitivity in response to physical training in NIDDM. Diabetes Care 1997;20:385–392.

84. Olefsky JM, Kolterman OG, Scarlett JA. Insulin action and resistance in obesity and non-insulin-dependent type II diabetes mellitus. Am J Physiol 1982;243:E15–E30.

85. Pan X-P, Li G-W, Hu Y-H, Wang J, Yang W, Hu Z-X, Lin J, Xiao J-Z, Cao H-B, Liu P, Jiang X-G, Jiang Y-Y, Wang J-P, Zheng H, Zhang H, Bennet PH, Howard BV. Effects of diet and exercise in preventing NIDDM in people with impaired glucose tolerance. Diabetes Care 1997;20:537–544.

86. Paternostro-Bayles M, Wing RR, Robertson RJ. Effect of life-style activity of varying duration on glycemic control in type 2 diabetic women. Diabetes Care 1989;12:34–37.

87. Pereira M, Kriska A, Joswiak M, Dowse G, Collins V, Zimmet P, Gareeboo H, Chitson P, Hemraj F, Purran A, Fareed D. Physical inactivity and glucose intolerance in the multi-ethnic island of Mauritius. Med Sci Sports Exerc 1995;27:1626–1634.

88. Perry I, Wannamethee S, Walker M, et al. Prospective study of risk factors for development of non-insulin-dependent diabetes in middle aged British men. Br Med J 1995;310:560–564.

89. Portuese E, Orchard T. Mortality in insulin-dependent diabetes. In: Harris MI, Cowie CC, Stern MP, et al., editors. Diabetes in America (NIH publication 95-1468). 2nd edition. Bethesda, MD: National Institutes of Health, National Institute of Diabetes and Digestive and Kidney Diseases; 1995. p 221–232.

90. Reaven GM, Bernstein R, Davis B, Olefsky JM. Nonketotic diabetes mellitus: insulin deficiency or insulin resistance? Am J Med 1976;60:80–88.

91. Regensteiner JG, Shetterly SM, Mayer EJ, Eckel RH, Haskell WL, Baxter J, Hamman RF. Relationship between habitual physical activity and insulin area among individuals with impaired glucose tolerance. Diabetes Care 1995;18:490–497.

92. Reitman JS, Vasquez B, Klimes I, Nagulusparan M. Improvement of glucose homeostasis after exercise training in non-insulin-dependent diabetes. Diabetes Care 1984;7:434–441.

93. Rice B, Janssen I, Hudson R, Ross R. Effects of aerobic exercise and/or diet on glucose tolerance and plasma levels in obese men. Diabetes Care 1999;22:684–691.

94. Rogers MA, Yamamoto C, King DS, Hagberg JM, Ehsani AA, Holloszy JO. Improvement in glucose tolerance after 1 wk of exercise in patients with mild NIDDM. Diabetes Care 1988;11:613–618.

95. Ruderman NB, Ganda OP, Johansen K. The effect of physical training on glucose tolerance and plasma lipids in maturity-onset diabetes. Diabetes 1979;28(Suppl. 1):89–92.

96. Ruderman N, Chisholm D, Pi-Sunyer X, Schneider S. The metabolically obese, normal weight individual revisited. Diabetes 1998;47:699–713.

97. Ryan AS, Pratley RE, Goldberg AP, Elahi D. Resistive training increases insulin action in postmenopausal women. J Gerontol Biol Sci Med 1996;51:M199–M205.

98. Schneider SH, Khachadurian AK, Amorosa LF, Clemow L, Ruderman NB. Ten-year experience with an exercise-based outpatient lifestyle modification program in the treatment of diabetes mellitus. Diabetes Care 1992;15(Suppl. 4):1800–1810.

99. Schneider SH, Amorosa LF, Khachadurian AK, Ruderman NB. Studies on the mechanism of improved glucose control during regular exercise in type 2 (non-insulin-dependent) diabetes. Diabetologia 1984;26:325–360.

100. SEARCH for Diabetes in Youth Study Group. The burden of diabetes mellitus among US youth: Prevalence estimates from the SEARCH for diabetes in youth study. Pediatrics 2006;118:1510-1518.

101. Sherman WM, Ferrara C, Schneider B. Nutritional strategies to optimize athletic performance. In: Ruderman N, Devlin J, editors. The health professional's guide to diabetes and exercise. Alexandria, VA: American Diabetes Association; 1995. p 71–158.

102. Sonstroem RJ, Morgan WP. Exercise and self-esteem: rationale and model. Med Sci Sports Exerc 1989;21:329–337.

103. Sigal RJ, Kenny GP, Wassserman, DH, Castaneda-Sceppa C. Physical activity/exercise and type 2 diabetes. Diabetes Care 2004;27:2518–2539.

104. Sothmann MS, Horn TS, Hart BA, Gustafson AB. Comparison of discrete cardiovascular fitness groups on plasma catecholamine and selected behavioral responses to psychological stress. Psychophysiology 1987;24:47–54.

105. Trovati M, Carta Q, Cavalot F, Vitali S, Banaudi C, Lucchina PG, et al. Influence of physical training on blood glucose control, glucose tolerance, insulin secretion, and insulin action in non-insulin-dependent diabetic patients. Diabetes Care 1984;7:416–420.

106. Turner RC, Holman RR, Matthews D, Hockaday TDR, Peto J. Insulin deficiency and insulin resistance interaction in diabetes: estimation of their interaction in diabetes: estimation of their relative contribution by feedback analysis from basal plasma insulin and glucose concentrations. Metabolism 1979;28:1086–1096.

107. UKPDS Group. UK Prospective Diabetes Study 33: intensive blood-glucose control with sulphonylureas or insulin compared with conventional treatment and risk of complications in patients with type 2 diabetes. Lancet 1998;352:837–853.

108. Uusitupaa MIJ, Niskanen LK, Siitonen O, Voutilainen E, Pyorala K. Ten year cardiovascular mortality in relation to risk factors and abnormalities in lipoprotein composition in type 2 (non-insulin-dependent) diabetic and non-diabetic subjects. Diabetologia 1993;18:1534–1543.

109. Vasterling JJ, Sementilli ME, Burish TG. The role of aerobic exercise in reducing stress in diabetic patients. The Diabetes Educator 1988;14(3):197–201.

110. Vinik AI. Neuropathy. In: Ruderman N, Devlin J, editors. The health professional's guide to diabetes and exercise. Alexandria, VA: American Diabetes Association; 1995. p 183–197.

111. Vranic M, Wasserman D. Exercise, fitness, and diabetes. In: Bouchard C, Shephard RJ, Stephens T, Sutton J, McPherson B, editors. Exercise, fitness and health. Champaign, IL: Human Kinetics; 1990. p 467–490.

112. Wang JT, Ho LT, Tang KT, Wang LM, Chen YDI, Reaven GM. Effect of habitual physical activity on age-related glucose intolerance. J Am Geriatr Soc 1989;37:203–209.

113. Watts NB, Spanheimer RG, Digirolamo A, Gebhart SSP, Musey VC, Siddiq K, Phillips LS. Prediction of glucose response to weight loss in patients with non-insulin-dependent diabetes mellitus. Arch Intern Med 1990;150:803–806.

114. West KM. Epidemiology of diabetes and its vascular lesions. New York: Elsevier; 1978.

115. Wing RR, Koeske R, Epstein LH, Nowalk MP, Gooding W, Becker D. Long-term effects of modest weight loss in type II diabetic patients. Arch Intern Med 1987;147:1749–1753.

116. Wing RR. Behavioral strategies for weight reduction in obese type 2 diabetic patients. Diabetes Care 1989;12:139–144.

117. Wing RR, Epstein LH, Nowalk MP, Koeske R, Hagg S. Behavior change, weight loss, and physiological improvements in type II diabetic patients. J Consult Clin Psychol 1985;53:111–122.

118. Yki-Jarvinen H. Glucose toxicity. Endocrinol Rev 1992;13:415–431.

119. Zamboni M, Armellini F, Turcato E, Todesco T, Bissoli L, Bergamo-Andreis IA, Bosello O. Effect of weight loss on regional body fat distribution in premenopausal women. Am J Clin Nutr 1993;58:29–34.

120. Zimmet PZ. Kelly West Lecture 1991: challenges in diabetes epidemiology: from west to the rest. Diabetes Care 1992;15:232–252.

121. Zimmet PZ, Tuomi T, Mackay R, Rowley MJ, Knowles W, Cohen M, Lang DA. Latent autoimmune diabetes mellitus in adults (LADA): the role of antibodies to glutamic acid decarboxylase in diagnosis and prediction of insulin dependency. Diabetic Med 1994;11:299–303.

CHAPTER 12

1. ACSM. ACSM's guideline for exercise testing and prescription. 8th ed. American College of Sports Medicine. Baltimore: Lippincott Williams and Wilkins; 2009.

2. ACSM. ACSM position stand on the appropriate intervention strategies for weight loss and prevention of weight regain for adults. Med Sci Sports Exerc 2001;33(12):2145–2156.

3. Adams T, et al. Long-term mortality after gastric bypass surgery. N Engl J Med 2007;357:753–761.

4. Allison DB, Fontaine KR, Manson JE, Stevens J, VanItallie TB. Annual deaths attributable to obesity in the United States. JAMA 1999;282(16):1530–1538.

5. Anderson JW, Greenway FL, Fujioka K, et al. Bupropion SR enhances weight loss: a 48-week double-blind, placebo-controlled trial. Obes Res 2002;10:633–641.

6. Anderson JW, Vichitbandra S, Oian W, Kryscio RJ. Long-term weight maintenance after an intensive weight-loss program. J Am Coll Nutr 1999;18:620–627.

7. Anderson RE, Wadden TA, Bartlett ST, Vogt RA, Weinstock RS. Relation of weight loss in serum lipids and lipoproteins in obese women. Am J Clin Nutr 1995;62:350–357.

8. Andreyeva T, Sturm R, Ringel JS. Moderate and severe obesity have large differences in health care costs. Obes Res 2004;12:1936–1943.

9. Angulo P. Nonalcoholic fatty liver disease. N Engl J Med 2002;346:1221–1231.

10. Ballor DL, Katch VL, Becque MD, Marks CR. Resistance weight training during caloric restriction enhances lean body weight maintenance. Am J Clin Nutr 1988;47:19–25.

11. Behme MT. Leptin: product of the obese gene. Nutr Today 1996;31:138–141.

12. Blackburn GL, Dwyer J, Flanders WD, Hill JO, Keller CH, Ii-Sunyer FX, St Jeart ST, Willett WC. Report of the American Institute of Nutrition (AIN) Steering Committee on Healthy Weight. J Nurt 1994;124:2240–2243.

13. Blair SN, Kohl HW, Barlow CE, et al. Changes in physical fitness and all-cause mortality: a prospective study of healthy and unhealthy men. JAMA 1995;273:1093–1098.

14. Bouchard C, Despres JP, Mauriege P. Genetic and nongenetic determinants of regional fat distribution. Endocr Rev 1993;14(1):72–93.

15. Bouchard C, Tremblay A, Despres JP, Nadeau A, Lupien PJ, Theriault G, Dussault J, Moorjani S, Pinault S, Fournier G. The response to long-term overfeeding in identical twins. N Engl J Med 1990;322(21):1477–1482.

16. Bray GA, Hollander P, Klein S, Kushner R, Levy B, Fitchet M, Perry BH. A 6-month randomized, placebo-controlled, dose-ranging trial of topiramate for weight loss in obesity. Obes Res 2003;11:722–733.

17. Bray GA. Obesity: a time bomb to be defused. Lancet 1998;352 (9123):160–161.

18. Bray GA, York DA. Clinical review 90: leptin and clinical medicine: a new piece in the puzzle of obesity. J Clin Endocrinol Metab 1997;82:2771–2776.

19. Brownell, KD. The LEARN program for weight control. Dallas: American Health; 2002.

20. Calle EE, Thun MJ, Petrelli JM, Rodriguez C, Heath CW Jr. Body-mass index and mortality in a prospective cohort of U.S. adults. N Engl J Med 1999;341(15):1097–1105.

21. Cassidy SB, Schwartz S. Prader-Willi and Angelman syndromes. Disorders of genomic imprinting. Medicine (Baltimore) 1998;77:140.

22. Christou NV, Sampalis JS, Liberman M, et al. Surgery decreases long-term mortality, morbidity, and health care use in morbidly obese patients. Ann Surg 2004;240:416–423.

23. Church TS, Earnest CP, Skinner JS, Blair SN. Effects of different doses of physical activity on cardiorespiratory fitness among sedentary, overweight or obese postmenopausal women with elevated blood pressure: a randomized controlled trial. JAMA 2007;297(19):2081–2091.

24. Clinical guidelines on the identification, evaluation, and treatment of overweight and obesity in adults—the evidence report. National Institutes of Health. Obes Res 1998;6 Suppl 2:51S.

25. Colditz GA, Willett WC, Rotnitzky A, Manson JE. Weight gain as a risk factor for clinical diabetes mellitus in women. Ann Intern Med 1995;122(7):481–486.

26. Cummings DE, Weigle DS, Frayo RS, Breen PA, Ma MK, Dellinger EP, Purnell JQ. Plasma ghrelin levels after diet-induced weight loss or gastric bypass surgery. N Engl J Med 2002;346(21):1623–1630.

27. Curtis JP, Selter JG, Wang Y, et al. The obesity paradox body mass index and outcomes in patients with heart failure. Arch Intern Med 2005;165:55–56.

28. Davidson MH, Hauptman J, DiGirolamo M, et al. Weight control and risk factor reduction in obese subjects treated for 2 years with orlistat. A randomized controlled trial. JAMA 1999;281:235–242.

29. De Fillipis E, Cusi K, Ocampo G, et al. Exercise-induced improvement in vasodilatory function accompanies increased insulin sensitivity in obesity and type 2 diabetes mellitus. J Clin Endocrinol Metab 2006;91:4903–4910.

30. Department of Agriculture, Department of Health and Human Services. Nutrition and your health: dietary guidelines for Americans. 5th ed. Washington, DC: Government Printing Office; 2000.

31. Despres JP, Golay A, Sjostrom L. Effects of rimonabant on metabolic risk factors in overweight patients with dyslipidemia. N Engl J Med 2005;353:2121–2134.

32. Dixon JB, O'Brien PE. Changes in comorbidities and improvements in quality of life after LAP-BAND placement. Am J Surg 2002;184:51S.

33. Duncan GE, Perri MG, Theriaque DW, et al. Exercise training, without weight loss, increases insulin sensitivity and postheparin plasma lipase activity in previously sedentary adults. Diabetes Care 2003;26(3):944–945.

34. Dunn CL, Hannan PJ, Jeffery RW, Sherwood NE, Pronk NP, Boyle R. The comparative and cumulative effects of a dietary restriction and exercise on weight loss. Int J Obesity 2006;30:112–121.

35. Fagard RH. Physical activity in the prevention and treatment of hypertension in the obese. Med Sci Sports Exerc 1999;31(suppl):s–24–s630.

36. Finkelstein EA, Fiebelkorn IC, Wang G. National medical spending attributable to overweight and obesity: how much, and who's paying? Health Affair 2003;W3:219–226.

37. Flegal KM, Graubard BI, Williamson DF, Gail MH. Excess deaths associated with underweight, overweight, and obesity. JAMA 2005;293(15):1861–1867.

38. Fletcher GF, Balady GJ, Amsterdam EA, et al. Exercise standards for testing and training: a statement for healthcare professionals from the American Heart Association 2001;104(14):1694–1740.

39. Fogelholm M, Kukkonin-Harjula K, Nenonen A, Pasanen M. Effects of walking training on weight maintenance after a very low energy diet in premenopausal obese women: a randomized controlled trial. Arch Intern Med 2000;160(14):2177–2184.

40. Folsom AR, Rasmussen ML, Chambless LE, et al. Prospective associations of fasting insulin, body fat distribution, and diabetes with risk of ischemic stroke. Diabetes Care 1999;22:1077–1083.

41. Fontaine KR, Redden DT, Wang C, et al. Years of life lost due to obesity. JAMA 2003;289(2):187–193.

42. Fontbonne A, Charles MA, Juhan-Vague I, et al. The effect of metformin on the metabolic abnormalities associated with upper-body fat distribution. BIGPRO Study Group. Diabetes Care 1996;19:920–926.

43. Foreyt JP, et al. Diet, genetics and obesity. Food Technology 1997;51:70–73.

44. Foster GD, Wadden TA, Vogt RA, Brewer G. What is a reasonable weight loss? Patients' expectations and evaluations of obesity treatment outcomes. J Consult Clin Psychol 1997;65:79–85.

45. Franz MJ, VanWormer JJ, Crain AL, Boucher JL, et al. Weight-loss outcomes: a systematic review and meta-analysis of weight-loss clinical trials with a minimum 1-year follow-up. J Am Diet Assoc 2007;107:1755–1767.

46. Gallagher D, Heymsfield SB, Heo M, et al. Healthy percentage body fat ranges: an approach for developing guidelines based on body mass index. Am J Clin Nutr 2000;72:694–701.

47. Goldstein DJ, Rampey AHJ, Enas GG, et al. Fluoxetine: a randomized clinical trial in the treatment of obesity. Int J Obes Relat Metab Disord 1994;18:129–135.

48. Gortmaker SL, Must A, Perrin JM, et al. Social and economic consequences of overweight in adolescence and young adulthood. N Engl J Med 1993;329:1008–1012.

49. Gregg EW, Cheng YJ, Cadwell BL, Imperatore G, Williams DE, Flegal KM, Narayan KM, Williamson DF. Secular trends in cardiovascular disease risk factors according to body mass index in US adults. JAMA 2005;293(15):1868–1874.

50. Grundy SM, Cleeman JI, Daniels SR, et al. Diagnosis and management of the metabolic syndrome: an American Heart Association/National Heart, Lung, and Blood Institute scientific statement. Circulation 2005;112:2735–2752.

51. Gulati M, Pandey DK, Arnsdorf MF, et al. Exercise capacity and the risk of death in women: the St James Women Take Heart Project. Circulation 2003;108:1554–1559.

52. Gurm HS, Whitlow PL, Kip KE. The impact of body mass index on short- and long-term outcomes in patients undergoing coronary revascularization: insights from the bypass angioplasty revascularization investigation (BARI); J Am Coll Cardiol 2002;39:834–840.

53. Haslam D, Sattar N, Lean M. ABC of obesity. Obesity—time to wake up. BMJ 2006;333:640–642.

54. Jakicic JM, Marcus BH, Gallagher KI, Napolitano M, Lang W. Effect of exercise duration and intensity on weight loss in overweight, sedentary women. JAMA 2003;290:1323–1330.

55. James WP, Astrup A, Finer N, et al. Effect of sibutramine on weight maintenance after weight loss: a randomised trial. STORM study group. Sibutramine trial of obesity reduction and maintenance. Lancet 2000;356:2119–2125.

56. Kahn R, Buse J, Ferrannini E, Stern M. The metabolic syndrome: time for a critical appraisal: joint statement from the American Diabetes Association and the European Association for the Study of Diabetes. Diabetes Care 2005;28(9):2289–2304.

57. Keith SW, Redden DT, Katzmarzyk PT, et al. Putative contributors to the secular increase in obesity: exploring the roads less travels. Int J Obesity 2006;30:1585–1594.

58. Klem ML, Wing RR, McGuire MT, Seagle HM, Hill JO. A descriptive study of individuals successful at long term maintenance of substantial weight loss. Am J Clin Nutr 1997;66:239–246.

59. Knowler WC, Barrett-Connor E, Fowler SE, Hamman RF, Lachin JM, Walker EA, Nathan DM. Reduction in the incidence of type 2 diabetes with lifestyle intervention or metformin. N Engl J Med 2002;346(6):393–403.

60. Kuczmarski RJ, Flegal KM. Criteria for definition of overweight in transition: background and recommendations for the United States. Am J Clin Nutr 2000;72:1074–1081.

61. Leibel RL, Rosenbaum M, Hirsch J. Changes in energy expenditure resulting from altered body weight. N Engl J Med 1995;332(10):621–628.

62. Leibowitz SF, Hoebel BG. Behavioral neuroscience of obesity. In: Bray GA, Bouchard C, James WPT, editors. Handbook of obesity. New York: Marcel Dekker; 1997. p 313.

63. Li Z, Hong K, Wong E, Maxwell M, Heber D. Weight cycling in a very low-calorie diet programme has no effect on weight loss velocity, blood pressure and serum lipid profile. Diab Obes Metab 2007;9:379–385.

64. Li Z, Maglione M, Tu W, et al. Meta-analysis: pharmacologic treatment of obesity. Ann Intern Med 2005;142:532–546.

65. Look AHEAD Research Group. The Look AHEAD study: a description of the lifestyle intervention and the evidence supporting it. Obesity 2006;14(5):737–752.

66. MacLean LD, Rhode BM, Sampalis J, Forse RA. Results of the surgical treatment of obesity. Am J Surg 1993;165:155–160.

67. McCullough PA, Gallagher MJ, Dejong AT, Sandberg KR, Trivax JE, Alexander D, Kasturi G, Jafri SM, Krause KR, Chengelis DL, Moy J, Franklin BA. Cardiorespiratory fitness and short-term complications after bariatric surgery. Chest 2006;130(2):517–525.

68. Metropolitan Life Insurance Company. 1983 Metropolitan height and weight tables. Stat Bull Metropol Life Insur Co 1983;64:1–19.

69. Miller WCD, Koceja DM, Hamilton EJ. A meta-analysis of the past 25 years of weight loss research using diet, exercise, or diet plus exercise intervention. Int J Obes 1997;21:941–947.

70. Miller WR, Rollnick S. Motivational interviewing. New York: Guilford Press; 2002.

71. Mokdad AH, Marks JS, Stroup DF, Gerberding JL. Correction: actual causes of death in the United States, 2000. JAMA 2005;293:293–294.

72. Mun EC, Blackburn GL, Matthews JB. Current status of medical and surgical therapy for obesity. Gastroenterology 2001;120:669–681.

73. Mustajoki P, Pekkarinen T. Very low energy diets in the treatment of obesity Obesity Reviews 2001;2:61–72.

74. National Center for Health Statistics, United States, 2006. With chartbook on trends in the health of Americans. Hyattsville, MD; 2006.

75. National Institutes of Health. Clinical guidelines on the identification, evaluation, and treatment of overweight and obesity in adults. Bethesda, MD: National Institutes of Health; 1998.

76. NHLBI Obesity Education Initiative Expert Panel on the Identification, Evaluation, and Treatment of Overweight and Obesity in Adults. Clinical guidelines on the identification, evaluation, and treatment of overweight and obesity in adults—the evidence report. Obes Res 1998;6:51S–209S.

77. Niessner A, Richter B, Penka M, et al. Endurance training reduces circulating inflammatory markers in persons at risk of coronary events: impact on plaque stabilization? Atherosclerosis 2006;186:160–165.

78. Ogden CL, Carroll MD, Curtin LR, et al. Prevalence of overweight and obesity in the United States, 1999–2004. JAMA 2006;295:1549–1555.

79. Olshansky SJ, Passaro DJ, Hershow RC, Layden J, Carnes BA, Brody J, Hayflick L, Butler RN, Allison DB, Ludwig DS. A potential decline in life expectancy in the United States in the 21st century. N Engl J Med 2005;352(11):1138–1145.

80. Padilla J, Wallace JP, Park S. Accumulation of physical activity reduces blood pressure in pre- and hypertension. Med Sci Sports Exerc 2005;37:1264–1275.

81. Parr RB, Capozzi L. Counseling patients about physical activity. J Am Acad Phys Assist 1997;10:45–49.

82. Peeters A, Barendregt JJ, Willekens F, Mackenbach JP, Al Mamun A, Bonneux L. Obesity in adulthood and its consequences for life expectancy: a life-table analysis. Ann Intern Med 2003;138(1):24–32.

83. Pi-Sunyer FX, Aronne LJ, Heshmati HM, et al. Effect of rimonabant, a cannabinoid-1 receptor blocker, on weight and cardiometabolic risk factors in overweight or obese patients: RIO-North America: a randomized controlled trial. JAMA 2006;295:761–775.

84. Poirier P, Giles TD, Bray GA, Hong Y, et al. Obesity and cardio-vascular disease: pathophysiology, evaluation, and effect of weight loss. An update of the 1997 American Heart Association Scientific Statement on Obesity and Heart Disease from the Obesity Committee of the Council on Nutrition, Physical Activity, and Metabolism. Circulation 2006;113:898–918.

85. Potteiger JA, Jacobsen DJ, Donnelly JE, Hill JO. Glucose and insulin responses following 16 months of exercise training in overweight adults; the Midwest Exercise Trial. Metabolism 2003;52(90):1175–1181.

86. Ray CS, Sue DY, Bray JE, et al. Effects of obesity on respiratory function. Am Rev Respir Dis 1983;128:501–506.

87. Rexrode KM, Carey VJ, Hennekens CH, et al. Abdominal adiposity and coronary heart disease in women. JAMA 1998;280:1843–1848.

88. Rimm EB, Stamfer MJ, Giovannucci E, et al. Body size and fat distribution as predictors of coronary heart disease among middleaged and older US men. Am J Epidemiol 1995;141:1117–1127.

89. Ross R, Dagnone D, Jones PJ, et al. Reduction in obesity and related comorbid conditions after diet-induced weight loss or exercise-induced weight loss in men: a randomized, controlled trial. Ann Intern Med 2000;133:92–103.

90. Sjostrom L, Lindroos AK, Peltonen M, et al. Lifestyle, diabetes, and cardiovascular risk factors 10 years after bariatric surgery. N Engl J Med 2004;351:2683–2693.

91. Sjostrom L. Impact of body weight, body composition, and adipose tissue distribution of morbidity and mortality. In: Stunkard AJ, Wadden TA, editors. Obesity: theory and therapy. New York: Raven Press; 1993. p 13–41.

92. Sjostrom L, et al. Effects of bariatric surgery on mortality in Swedish obese subjects. N Engl J Med 2007;357:741–752.

93. Skelton NK, Skelton WP. Medical implications of obesity. Postgrad Med 1992;92:151–152.

94. Snow V, Barry P, Fitterman N, et al. Clinical Efficacy Assessment Subcommittee of the American College of Physicians. Pharmacologic and surgical management of obesity in primary care: a clinical practice guideline from the American College of Physicians. Ann Intern Med 2005;142:525–531.

95. Stafford RS, Farhat JH, Misra B, Schoenfeld DA. National patterns of physician activities related to obesity management. Arch Fam Med 2000;9:631–638.

96. Stampfer MJ, Maclure KM, Colditz GA, et al. Risk of symptomatic gallstones in women with severe obesity. Am J Clin Nutr 1992;55:652–658.

97. Stunkard AJ, Harris JR, Pedersen NL, McClearn GE. The body-mass index of twins who have been reared apart. N Engl J Med 1990;322(21):1483–1487.

98. Tataranni PA, Young JB, Bogardus C, Ravussin E. A low sympathoadrenal activity is associated with body weight gain and development of central adiposity in Pima Indian men. Obes Res 1997;5(4):341–347.

99. Thorpe KE, Florence CS, Howard DH, Joski P. The impact of obesity on rising medical spending. Health Affair 2004; Suppl Web Exclusives:W4–480–6.

100. Tsai AG, Wadden TA. An evaluation of major commercial weight loss programs in the United States. Ann Intern Med 2005;142:55–66.

101. U.S. Department of Health and Human Services, U.S. Department of Agriculture. Dietary guidelines for Americans, 2005. 6th ed. Washington, DC: U.S. Government Printing Office, January 2005. [Online]. www.healthierus.gov/dietaryguidelines

102. U.S. Department of Health and Human Services. Healthy people 2010: National Health Promotion and Disease Prevention objectives (DHHS publication PHS91-50212). Washington, DC: U.S. Department of Health and Human Services; 2000.

103. U.S. Department of Health and Human Services. Physical activity and health: a report of the surgeon general. Atlanta: U.S. Department of Health and Human Services, Center for Disease Control and Prevention, National Center for Chronic Disease Prevention and Health Promotion; 1996.

104. Wadden TA, Berkowitz RI, Womble LG, et al. Randomized trial of lifestyle modification and pharmacotherapy for obesity. N Engl J Med 2005;353:2111–2120.

105. Wadden TA, Vogt, RA, Anderson RE, et al. Exercise in the treatment of obesity: effects of four interventions on body composition, resting energy expenditure, appetite, and mood. J of Consulting and Clin Psychol 1997;65(2):269–277.

106. Wadden TA. Treatment of obesity by moderate and severe caloric restriction. Results of clinical research trials. Ann Intern Med 1993;119(7 Pt 2):688 693.

107. Wang JS. Exercise prescription and thrombogenesis. J Biomed Sci 2006;13:753–761.

108. Wang TJ, Parise H, Levy D, et al. Obesity and the risk of new-onset atrial fibrillation. JAMA 2004;292:2471–2477.

109. Wei M, Kampert JB, Barlow CE, et al. Relationship between low cardiorespiratory fitness and mortality in normal-weight, overweight, and obese men. JAMA 1999;282(16):1547–1553.

110. Weiss EP, Racette SB, Villareal DT, et al. Lower extremity muscle size and strength and aerobic capacity decrease with caloric restriction but not with exercise-induced weight loss. J Appl Physiol 2007;102:634–640.

111. Whitaker RC, Wright JA, Pepe MS, Seidel KD, Dietz WH. Predicting obesity in young adulthood from childhood and parental obesity. N Engl J Med 1997;337:869–873.

112. Willett WC, Manson JE, Stampfer MJ, et al. Weight, weight change, and coronary heart disease in women. Risk within the "normal" weight range. JAMA 1995;273:461–465.

113. Wing R, Tate D, Gorin A, et al. A self-regulation program for maintenance of weight loss. N Engl J Med 2006;355:1563–1571.

114. Wing RR, Phelan S. Long-term weight loss maintenance. Am J Clin Nutr 2005;82(suppl):222S–225S.

115. Wing RR. Physical activity in the treatment of the adulthood overweight and obesity: current evidence and research issues. Med Sci Sports Exerc 1999;31,Suppl 11:s547–s552.

116. Wilson PW, D'Agostino RB, Sullivan L, Parise H, Kannel WB. Overweight and obesity as determinants of cardiovascular risk: the Framingham experience. Arch Intern Med 2002;162(16):1867–1872.

117. World Health Organization. Obesity: preventing and managing the global epidemic. Report of a WHO consultation. World Health Organ Tech Rep Ser 2000;894:i.

118. Yusuf S, Hawken S, Ounpuu S, et al. Obesity and the risk of myocardial infarction in 27,000 participants from 52 countries: a case-control study. Lancet 2005;366:1640–1649.

CHAPTER 13

1. Allison TG, Cordeiro MAS, Miller TD, Daida HD, Squires RW, Gau GT. Prognostic significance of exercise induced systemic hypertension in healthy subjects. Am J Cardiol 1999;83:371–375.

2. American College of Sports Medicine position stand. Exercise and hypertension. Med Sci Sports Exerc 2004;36:533–553.

3. American College of Sports Medicine. ACSM position stand. The recommended quantity and quality of exercise for developing and maintaining cardiorespiratory and muscular fitness, and flexibility in healthy adults. Med Sci Sports Exerc 1998;30:975–991.

4. Barnard RJ, Wen SJ. Exercise and diet in the prevention and control of the metabolic syndrome. Sports Med 1994;18:218–228.

5. Benbassat J, Froom P. Blood pressure response to exercise as a predictor of hypertension. Arch Intern Med 1986;146:2053–2055.

6. Blair SN, Kohl HW III, Barlow CE, Gibbons LW. Physical fitness and all-cause mortality in hypertensive men. Ann Med 1991;23:307–312.

7. Braith R, Stewart KJ. Resistance exercise training. Its role in the prevention of cardiovascular disease. Circulation 2006;113:2642–2650.

8. Brown MD, Moore GM, Korytkowski MT, McCole SD, Hagberg JM. Improvement of insulin sensitivity by short-term exercise training in hypertensive African-American women. Hypertension 1997;30:1549–1553.

9. Chobanian AV, Bakris GL, Black HR, Cushman WC, Green LA, Izzo JL Jr, Jones DW, Materson BJ, Oparil S, Wright JT Jr, Roccella EJ. The seventh report of the Joint National Committee on Prevention, Detection, Evaluation, and Treatment of High Blood Pressure: the JNC 7 report. JAMA 2003;289:2560–2572.

10. Cornelissen VA, Fagard RH. Effect of resistance training on resting blood pressure: a meta-analysis of randomized controlled trials. J Hypertens 2005;23:251–259.

11. Cortez-Cooper MY, DeVan AE, Anton MM, Farrar RP, Beckwith KA, Todd JS, Tanaka H. Effects of high intensity resistance training on arterial stiffness and wave reflection in women. Am J Hypertens 2005;18:930–934.

12. Dengel DR, Hagberg JM, Pratley RE, Rogus EM, Goldberg AP. Improvements in blood pressure, glucose metabolism, and lipoprotein lipids after aerobic exercise plus weight loss in obese, hypertensive middle-aged men. Metabolism 1998;47:1075–1082.

13. Dickinson HO, Mason JM, Nicolson DJ, Campbell F, Beyer FR, Cook JV, Williams B, Ford GA. Lifestyle interventions to reduce raised blood pressure: a systematic review of randomized controlled trials. J Hypertens 2006;24:215–233.

14. Ebrahim S, Smith GD. Lowering blood pressure: a systematic review of sustained effects of non-pharmacological interventions. J Public Health Med 1998;20:441–448.

15. Fagard R, Staessen J, Amery A. Exercise blood pressure and target organ damage in essential hypertension. J Human Hypertens 1991;5:69–75.

16. Fields LE, Burt VL, Cutler JA, Hughes J, Roccella EJ, Sorlie P. The burden of adult hypertension in the United States, 1999–2000: a rising tide. Hypertension 2004;44:398–404.

17. Filipovsky J, Ducimetiere P, Safar ME. Prognostic significance of exercise blood pressure and heart rate in middle-aged men. Hypertension 1992;20:333–339.

18. Franz IW. Blood pressure measurement during ergometric stress testing. Z Kardiol 1996;85(Suppl 3):71–75.

19. Gibbons RJ, Balady GJ, Beasley JW, Bricker JT, Duvernoy WFC, Froelicher VF, Mark DB, Marwick TH, McCallister BD, Thompson PD, Winters WL Jr, Yanowitz FG. ACC/AHA guidelines for exercise testing: a report of the American College of Cardiology/American Heart Association Task Force on Practice Guidelines (Committee on Exercise Testing). J Am Coll Cardiol 1997;30:260–315.

20. Gordon NF. Hypertension. In: Durstine JL, Moore GE, editors. ACSM's exercise management for persons with chronic diseases and disabilities. Champaign, IL: Human Kinetics; 2003. p 76–80.

21. Guidry MA, Blanchard BE, Thompson PD, Maresh CM, Seip RL, Taylor AL, Pescatello LS. Am Heart J 2006;151:1322.e5–12.

22. Hagberg JM, Ferrell RE, Dengel DR, Wilund KR. Exercise training-induced blood pressure and plasma lipid improvements in hypertensives may be genotype dependent. Hypertension 1999;34:18–23.

23. Harris KA, Holly RG. Physiological response to circuit weight training in borderline hypertensive subjects. Med Sci Sports Exerc 1987;19:246–252.

24. Kaplan NM. Systemic hypertension: mechanisms and diagnosis. In: Braunwald E, editor. Heart disease: a textbook of cardiovascular medicine. Philadelphia: Saunders; 1997.

25. Kaplan NM. The deadly quartet. Arch Intern Med 1989;149:1514–1520.

26. Kelley GA, Kelley KS. Progressive resistance exercise and resting blood pressure: a meta-analysis of randomized controlled trials. Hypertension 2000;35:838–843.

27. LaFontaine T. Resistance training for patients with hypertension. Strength Condition 1997;19:5–9.

28. Matthews CE, Pate RR, Jackson KL, Ward DS, Macera CA, Kohl HW, Blair SN. Exaggerated blood pressure response to dynamic exercise and risk of future hypertension. J Clin Epidemiol 1998;51:29–35.

29. Miyachi M, Kawano H, Sugawara J, Takahashi K, Hayashi K, Yamazaki K, Tabata I, Tanaka H. Unfavorable effects of resistance training on central arterial compliance: a randomized intervention study. Circulation 2004;110:2858–2863.

30. Mundal R, Kjeldsen SE, Sandvik L, Erikssen G, Thaulow E, Erikssen J. Exercise blood pressure predicts cardiovascular mortality in middle-aged men. Hypertension 1994;24:56–62.

31. Pollock M, Franklin B, Balady G, Chaitman B, Fleg J, Fletcher B, Limacher M, Pina I, Stein R, Williams M, Bazzare T. AHA science advisory. Resistance exercise in individuals with and without cardiovascular disease. Circulation 2000;101:828–833.

32. Rakobowchuk M, McGowan CL, de Groot PC, Bruinsma D, Hartman JW, Phillips SM, MacDonald MJ. Effect of whole body resistance training on arterial compliance in young men. Exp Physiol 2005;90:645–651.

33. Reaven GM. Role of insulin resistance in human disease. Diabetes 1988;37:1595–1607.

34. Reisin E. Nonpharmacologic approaches to hypertension. Weight, sodium, alcohol, exercise, and tobacco considerations. Med Clin North Am 1997;81:1289–1303.

35. Sacks FM, Svetkey LP, Vollmer WM, Appel LJ, Bray GA, Harsha D, et al. Effects on blood pressure of reduced dietary sodium and the Dietary Approaches to Stop Hypertension (DASH) diet. DASH-Sodium Collaborative Research Group. N Engl J Med 2001;344:3–10.

36. Singh JP, Larson MG, Manolio TA, O'Donnell CJ, Lauer M, Evans JC, Levy D. Blood pressure response during treadmill testing as a risk factor for new-onset hypertension. The Framingham Heart Study. Circulation 1999;99(14):1831–1836.

37. Stedman's medical dictionary. 27th ed. Baltimore: Lippincott Williams & Wilkins; 2000. p 1784.

38. Tanaka H, Reiling MJ, Seals DR. Regular walking increases peak limb vasodilatory capacity of older hypertensive humans: implications for arterial structure. J Hypertens 1998;16:423–428.

39. American Heart Association. Heart disease and stroke statistics—2006 update: a report from the American Heart Association Statistics Committee and Stroke Statistics Subcommittee. Circulation 2006;113:e85–e151.

40. Vasan RS, Beiser A, Seshadri S, Larson MG, Kannel WB, D'Agostino RB, Levy D. Residual lifetime risk for developing hypertension in middle-aged women and men: the Framingham Heart Study. JAMA 2002;287:1003–1010.

41. Appel LJ, Brands MW, Daniels SR, Karanja N, Elmer PJ, Sacks FM. Dietary approaches to prevent and treat hypertension: a scientific statement from the American Heart Association. Circulation 2006;47:296–308.

42. Whaley M. American College of Sports Medicine's guidelines for exercise testing and prescription. 7th ed. Baltimore: Lippincott Williams & Wilkins; 2006.

43. Williams GH. Approach to the patient with hypertension. In: Braunwald E, Fauci A, Hauser S, Jameson J, Kasper D, Longo D, editors. Harrison's principles of internal medicine. New York: McGraw-Hill; 1998.

44. Xin X, He J, Frontini MG, Ogden LG, Motsamai OI, Whelton PK. Effects of alcohol reduction on blood pressure: a meta-analysis of randomized controlled trials. Hypertension 2001;38:1112–1117.

CHAPTER 14

1. Alhassan S, Reese KA, Mahurin J, Plaisance EP, Hilson BD, Garner JC, Wee SO, Grandjean PW. Blood lipid responses to plant stanol ester supplementation and aerobic exercise training. Metabolism 2006;55:541–549.

2. Allison TG, Squires RW, Johnson BD, Gau GT. Achieving National Cholesterol Education Program goals for low-density lipoprotein cholesterol in cardiac patients: importance of diet, exercise, weight control, and drug therapy. Mayo Clin Proc 1999;74:466–473.

3. American College of Sports Medicine. ACSM's guidelines for exercise testing and prescription. 7th ed. Philadelphia: Lippincott Williams & Wilkins; 2006

4. Ascherio A, Katan MB, Zock PL, Stampfer MJ, Willett WC. Trans fatty acids and coronary heart disease. N Engl J Med 340:1994–1998, 1999.

5. Asikainen TM, Kukkonen-Harjula K, Miilunpalo S. Exercise for health for early postmenopausal women: a systematic review of randomised controlled trials. Sports Med 2004;34:753–778.

6. Ast M, Frishman WH. Bile acid sequestrants. J Clin Pharmacol 1990;30:99–106.

7. Ballantyne CM. Low-density lipoproteins and risk for coronary artery disease. Am J Cardiol 1998;82:3Q–12Q.

8. Belalcazar LM, Ballantyne CM. Defining specific goals of therapy in treating dyslipidemia in the patient with low high-density lipoprotein cholesterol. Prog Cardiovasc Dis 1998;41:151–174.

9. Carew TE, Schwenke DC, Steinberg D. Antiatherogenic effect of probucol unrelated to its hypocholesterolemic effect: evidence that antioxidants in vivo can selectively inhibit low density lipoprotein degradation in macrophage-rich fatty streaks and slow the progression of atherosclerosis in the Watanabe heritable hyperlipidemic rabbit. Proc Natl Acad Sci USA 1987;84:7725–7729.

10. Carroll MD, Lacher DA, Sorlie PD, Cleeman JI, Gordon DJ, Wolz M, Grundy SM, Johnson CL. Trends in serum lipids and lipoproteins of adults, 1960–2002. JAMA 2005;294:1773–1781.

11. Center for Disease Control and Prevention. Trends in cholesterol screening and awareness of high blood cholesterol—United States, 1991–2003. MMWR 2005;54:865–870.

12. Crouse SF, O'Brien BC, Grandjean PW, Lowe RC, Rohack JJ, Green JS. Effects of training and a single session of exercise on lipids and apolipoproteins in hypercholesterolemic men. J Appl Physiol 1997;83:2019–2028.

13. Cybulsky MI, Gimbrone MA Jr. Endothelial expression of a mononuclear leukocyte adhesion molecule during atherogenesis. Science 1991;251:788–791.

14. Denke MA, Grundy SM. Individual responses to a cholesterol-lowering diet in 50 men with moderate hypercholesterolemia. Arch Intern Med 1994;154:317–325.

15. Duncan JJ, Gordon NF, Scott CB. Women walking for health and fitness. How much is enough? JAMA 1991;266:3295–3299.

16. Durstine JL, Davis PG, Ferguson MA, Alderson NL, Trost SG. Effects of short-duration and long-duration exercise on lipoprotein(a). Med Sci Sports Exerc 2001;33:1511–1516.

17. Durstine JL, Grandjean PW, Cox CA, Thompson PD. Lipids, lipoproteins, and exercise. J Cardiopulm Rehabil 2002;22:385–398.

18. Durstine JL, Haskell WL. Effects of exercise training on plasma lipids and lipoproteins. Exerc Sport Sci Rev 1994;22:477–521.

19. Ebisu T. Splitting the distance of endurance running: on cardiovascular endurance and blood lipids. Jpn J Physical Ed 1995;30:37–43.

20. Eisenmann JC. Blood lipids and lipoproteins in child and adolescent athletes. Sports Med 2002;32:297–307.

21. Farmer JA, Gotto AM. Dyslipidemia and other risk factors for coronary artery disease. In: Braunwald E, editor. Heart disease: a textbook of cardiovascular medicine. Orlando, FL: Saunders; 1997. p 1126–1160.

22. Frolkis JP, Zyzanski SJ, Schwartz JM, Suhan PS. Physician noncompliance with the 1993 National Cholesterol Education Program (NCEP-ATPII) guidelines. Circulation 1998;98:851–855.

23. Gordon PM, Goss FL, Visich PS, Warty V, Denys BJ, Metz KF, Robertson RJ. The acute effects of exercise intensity on HDL-C metabolism. Med Sci Sports Exerc 1994;26:671–677.

24. Gordon PM, Visich PS, Goss FL, Fowler S, Warty V, Denys BJ, Metz KF, Robertson J. Comparison of exercise and normal variability on HDL cholesterol concentrations and lipolytic activity. Int J Sports Med 1996;17:332–337.

25. Grandjean PW, Crouse SF, Rohack JJ. Influence of cholesterol status on blood lipid and lipoprotein enzyme responses to aerobic exercise. J Appl Physiol 2000;89:472–480.

26. Grundy S, Bilheimer D, Chait A, Clark L, Denke M, Havel R, Hazzard W, Hulley S, Hunninghake DB, Kreisberg R, Kris-Etherton P, McKenney J, Newman M, Schaefer E, Sobel B, Somelofski C, Weinstein M. National Cholesterol Education Program. Second report of the Expert Panel on Detection, Evaluation, and Treatment of High Blood Cholesterol in Adults (Adult Treatment Panel II). Circulation 1994;89:1333–1445.

27. Grundy SM, Cleeman JI, Merz CN, Brewer HB Jr, Clark LT, Hunninghake DB, Pasternak RC, Stone NJ, for the Coordinating Committee of the National Cholesterol Education Program, Endorsed by the National Heart, Lung, and Blood Institute, American College of Cardiology Foundation, and American Heart Association. Implications of recent clinical trials for the National Cholesterol Education Program Adult Treatment Panel III guidelines. Circulation 2004;110:227–239.

28. Grundy SM, Denke MA. Dietary influences on serum lipids and lipoproteins. J Lipid Res 1990;31:1149–1172.

29. Gupta AK, Ross EA, Myers JN, Kashyap ML. Increased reverse cholesterol transport in athletes. Metabolism 1993;42:684–690.

30. Harper CR, Jacobson TA. New perspectives on the management of low levels of high-density lipoprotein cholesterol. Arch Intern Med 1999;159:1049–1057.

31. Herrington DM, Reboussin DM, Brosnihan KB, Sharp PC, Shumaker SA, Snyder TE, Furberg CD, Kowalchuk GJ, Stuckey TD, Rogers WJ, Givens DH, Waters D. Effects of estrogen replacement on the progression of coronary-artery atherosclerosis. N Engl J Med 2000;343:522–529.

32. Imamura H, Teshima K, Miyamoto N, Shirota T. Cigarette smoking, high-density lipoprotein cholesterol subfractions, and lecithin: cholesterol acyltransferase in young women. Metabolism 2002;51:1313–1316.

33. Jenkins DJ, Wolever TM, Rao AV, Hegele RA, Mitchell SJ, Ransom TP, Boctor DL, Spadafora PJ, Jenkins AL, Mehling C. Effect on blood lipids of very high intakes of fiber in diets low in saturated fat and cholesterol. N Engl J Med 1993;329:21–26.

34. Katsanos CS, Moffatt RJ. Acute effects of premeal versus postmeal exercise on postprandial hypertriglyceridemia. Clin J Sport Med 2004;14:33–39.

35. Kelley GA, Kelley KS, Tran ZV. Aerobic exercise and lipids and lipoproteins in women: a meta-analysis of randomized controlled trials. J Women's Health 2004;13:1148–1164.

36. Kelley GA, Kelley KS, Vu Tran Z. Aerobic exercise, lipids and lipoproteins in overweight and obese adults: a meta-analysis of randomized controlled trials. Int J Obesity 2005;29:881–893.

37. Kromhout D, Bosschieter EB, de Lezenne Coulander C. The inverse relation between fish consumption and 20-year mortality from coronary heart disease. N Engl J Med 1985;312:1205–1209.

38. Kwiterovich PO Jr. The antiatherogenic role of high-density lipoprotein cholesterol. Am J Cardiol 1998;82:13Q–21Q.

39. Maeda K, Noguchi Y, Fukui T. The effects of cessation from cigarette smoking on the lipid and lipoprotein profiles: a meta-analysis. Prev Med 2003;37:283–290.

40. Mahley RW, Weisgraber KH, Farese RV. Disorders of lipid metabolism. In: Larsen PR, Kronenberg HM, Melmed S, Pololnsky KS, editors. Williams textbook of endocrinology. Orlando, FL: Saunders, 1998. p 1099–1154.

41. Mann GV, Teel K, Hayes O, McNally A, Bruno D. Exercise in the disposition of dietary calories; regulation of serum lipoprotein and cholesterol levels in human subjects. N Engl J Med 1955;253:349–355.

42. Marcovina SM, Koschinsky ML. Lipoprotein(a) as a risk factor for coronary artery disease. Am J Cardiol 1998;82:57U–86U.

43. Mensink RP, Katan MB. Effect of dietary trans fatty acids on high-density and low-density lipoprotein cholesterol levels in healthy subjects. N Engl J Med 1990;323:439–445.

44. Miyashita M, Burns SF, Stensel DJ. Exercise and postprandial lipemia: effect of continuous compared with intermittent activity patterns. Am J Clin Nutr 2006;83:24–29.

45. Moffatt RJ. Effects of cessation of smoking on serum lipids and high density lipoprotein-cholesterol. Atherosclerosis 1988;74:85–89.

46. Moffatt RJ, Biggerstaff KD, Stamford BA. Effects of the transdermal nicotine patch on normalization of HDL-C and its subfractions. Prev Med 2000;31:148–152.

47. Moffatt RJ, Stamford BA, Biggerstaff KD. Influence of worksite environmental tobacco smoke on serum lipoprotein profiles of female nonsmokers. Metabolism 1995;44:1536–1539.

48. National Heart Lung Blood Institute. Third report of the National Cholesterol Education Program Expert Panel on the Detection, Evaluation and Treatment of High Blood Cholesterol in Adults. Washington, DC: National Heart, Lung, Blood Institute of the National Institutes of Health, Report No. 01-3670; May 2001. p 1–40.

49. Oram JF, Vaughan AM. ABCA1-mediated transport of cellular cholesterol and phospholipids to HDL apolipoproteins. Curr Opinion Lipid 2000;11:253–260.

50. Ornish D, Brown SE, Scherwitz LW, Billings JH, Armstrong WT, Ports TA, McLanahan SM, Kirkeeide RL, Brand RJ, Gould KL. Can lifestyle changes reverse coronary heart disease? The Lifestyle Heart Trial. Lancet 1990;336:129–133.

51. Pajukanta P. Do DNA sequence variants in ABCA1 contribute to HDL cholesterol levels in the general population? J Clin Invest 2004;114:1244–1247.

52. Richard F, Marécaux N, Dallongeville J, Devienne M, Tiem N, Fruchart JC, Fantino M, Zylberberg G, Amouyel P. Effect of smoking cessation on lipoprotein A-I and lipoprotein A-I:A-II levels. Metabolism 1997;46:711–715.

53. Ross R. Atherosclerosis—an inflammatory disease. N Engl J Med 1999;340:115–126.

54. Sabaté J, Fraser GE, Burke K, Knutsen SF, Bennett H, Lindsted KD. Effects of walnuts on serum lipid levels and blood pressure in normal men. N Engl J Med 1993;328:603–607.

55. Seip RL, Moulin P, Cocke T, Tall A, Kohrt WM, Mankowitz K, Semenkovich CF, Ostlund R, Schonfeld G. Exercise training decreases plasma cholesteryl ester transfer protein. Arterioscler Thromb 1993;13:1359–1367.

56. Seip RL, Semenkovich CF. Skeletal muscle lipoprotein lipase: molecular regulation and physiological effects in relation to exercise. Exerc Sports Sci Rev 1998;26:191–218.

57. St-Pierre AC, Ruel IL, Cantin B, Dagenais GR, Bernard PM, Després JP, Lamarche B. Comparison of various electrophoretic characteristics of LDL particles and their relationship to the risk of ischemic heart disease. Circulation 2001;104:2295–2299.

58. Stamford BA, Matter S, Fell RD, Sady S, Papanek P, Cresanta M. Cigarette smoking, exercise and high density lipoprotein cholesterol. Atherosclerosis 1984;52:73–83.

59. Stefanick ML, Mackey S, Sheehan M, Ellsworth N, Haskell WL, Wood PD. Effects of diet and exercise in men and postmenopausal women with low levels of HDL cholesterol and high levels of LDL cholesterol. N Engl J Med 1998;339:12–20.

60. Stocker R, Keaney JF Jr. Role of oxidative modifications in atherosclerosis. Physiol Rev 2004;84:1381–1478.

61. Tall AR, Wang N, Mucksavage P. Is it time to modify the reverse cholesterol transport model? J Clin Invest 2001;108:1273–1275.

62. Thom T, Haase N, Rosamond W, Howard VJ, Rumsfeld J, Manolio T, Zheng ZJ, Flegal K, O'Donnell C, Kittner S, Lloyd-Jones D, Goff DC Jr, Hong Y, Adams R, Friday G, Furie K, Gorelick P, Kissela B, Marler J, Meigs J, Roger V, Sidney S, Sorlie P, Steinberger J, Wasserthiel-Smoller S, Wilson M, Wolf P. Heart disease and stroke statistics—2006 update: a report from the American Heart Association Statistics Committee and Stroke Statistics Subcommittee. Circulation 2006;113:e85–151.

63. Thompson PD, Crouse SF, Goodpaster B, Kelley D, Moyna N, Pescatello L. The acute versus the chronic response to exercise. Med Sci Sports Exerc 2001;33:S438–445.

64. Thompson PD, Cullinane E, Henderson LO, Herbert PN. Acute effects of prolonged exercise on serum lipids. Metabolism 1980;29:662–665.

65. Thompson PD, Cullinane EM, Sady SP, Flynn MM, Bernier DN, Kantor MA, Saritelli AL, Herbert PN. Modest changes in high-density lipoprotein concentration and metabolism with prolonged exercise training. Circulation 1988;78:25–34.

66. U.S. Department of Health and Human Services. Physical activity and health: a report of the surgeon general. Atlanta: U.S. Department of Health and Human Services, Centers for Disease Control and Prevention, National Center for Chronic Disease Prevention and Health Promotion; 1996.

67. Vasankari TJ, Kujala UM, Vasankari TM, Ahotupa M. Reduced oxidized LDL levels after a 10-month exercise program. Med Sci Sports Exerc 1998;30:1496–1501.

68. Warshafsky S, Kamer RS, Sivak SL. Effect of garlic on total serum cholesterol. A meta-analysis. Ann Intern Med 1993;119:599–605.

69. Williams PT. High-density lipoprotein cholesterol and other risk factors for coronary heart disease in female runners. N Engl J Med 1996;334:1298–1303.

70. Williams PT, Dreon DM, Blanche PJ, Krauss RM. Variability of plasma HDL subclass concentrations in men and women over time. Arterioscler Thromb Vasc Biol 1997;17:702–706.

71. Williams PT, Wood PD, Haskell WL, Vranizan K. The effects of running mileage and duration on plasma lipoprotein levels. JAMA 1982;247:2674–2679.

72. Wood PD, Stefanick ML, Williams PT, Haskell WL. The effects on plasma lipoproteins of a prudent weight-reducing diet, with or without exercise, in overweight men and women. N Engl J Med 1991;325:461–466.

73. Zhang JQ, Ji LL, Nunez G, Feathers S, Hart CL, Yao WX. Effect of exercise timing on postprandial lipemia in hypertriglyceridemic men. Can J Appl Physiol 2004;29:590–603.

74. Ziogas GG, Thomas TR, Harris WS. Exercise training, postprandial hypertriglyceridemia, and LDL subfraction distribution. Med Sci Sports Exerc 1997;29:986–991.

75. Zmuda JM, Yurgalevitch SM, Flynn MM, Bausserman LL, Saratelli A, Spannaus-Martin DJ, Herbert PN, Thompson PD. Exercise training has little effect on HDL levels and metabolism in men with initially low HDL cholesterol. Atherosclerosis 1998;137:215–221.

CHAPTER 15

1. United States Renal Data System. USRDS 1999 annual data report. Bethesda, MD: National Institutes of Health.

2. Oberly E. Renal rehabilitation: bridging the barriers. Madison, WI: Medical Education Institute; 1994.

3. Brenner BM, Stein JH. Chronic renal failure. New York: Churchill Livin`gstone; 1981.

4. Nissenson AR, Fine RN. Dialysis therapy. 2nd ed. Philadelphia: Hanley & Belfus; 1993.

5. Brenner P. Quality of life: a phenomenological perspective on explanation, prediction and understanding in nursing science. Adv Nurs Sci 1985;8:14.

6. United Network for Organ Sharing. UNOS update 1996. Washington, DC: Organ Procurement and Transplantation Network and Scientific Registry for Organ Transplantation; 1996.

7. Barnea N, Drory Y, Iaina A, et al. Exercise tolerance in patients on chronic hemodialysis. Isr J Med Sci 1980;16:1721.

8. Beasley RW, Smith A, Neale J. Exercise capacity in chronic renal failure patients managed by continuous ambulatory peritoneal dialysis. Aust N Z J Med 1986;16:5–10.

9. Goldberg AP, Geltman EM, Hagberg JM, Delmez JA, Haynes ME, Harter HR. Therapeutic benefits of exercise training for hemodialysis patients. Kidney Int 1983;S16:S303–S309.

10. Moore GE, Brinker KR, Stray-Gundersen J, Mitchell JH. Determinants of $\dot{V}O_2$ peak in patients with end-stage renal disease: on and off dialysis. Med Sci Sports Exerc 1993;25:18–23.

11. Moore GE, Parsons DB, Painter PL, Stray-Gundersen J, Mitchell J. Uremic myopathy limits aerobic capacity in hemodialysis patients. Am J Kid Dis 1993;22:277–287.

12. Painter PL, Nelson-Worel JN, Hill MM, et al. Effects of exercise training during hemodialysis. Nephron 1986;43:8792.

13. Painter PL, Messer-Rehak D, Hanson P, Zimmerman S, Glass NR. Exercise capacity in hemodialysis, CAPD and renal transplant patients. Nephron 1986;42:47–51.

14. Painter P. Exercise in end stage renal disease. Exerc Sport Sci Rev 1988;16:305–339.

15. Painter P. The importance of exercise training in rehabilitation of patients with end stage renal disease. Am J Kidney Dis 1994;24 (Suppl 1):S2–S9.

16. Ross DL, Grabeau GM, Smith S, et al. Efficacy of exercise for end-stage renal disease patients immediately following high-efficiency hemodialysis: a pilot study. Am J Nephrol 1989;9:376–383.

17. Shalom R, Blumenthal JA, Williams RS. Feasibility and benefits of exercise training in patients on maintenance dialysis. Kidney Int 1984;25:958–963.

18. Zabetakis PM, Gleim GW, Pasternak FL, Saraniti A, Nicholas JA, Michelis MF. Long-duration submaximal exercise conditioning in hemodialysis patients. Clin Nephrol 1982;18:17–22.

19. Johansen KL. Physical functioning and exercise capacity in patients on dialysis. Adv Ren Replace Ther 1999;6:141–148.

20. Copley JB, Lindberg JS. The risks of exercise. Adv Ren Replace Ther 1999;6:165–172.

21. Painter PL, Stewart AL, Carey S. Physical functioning: definitions, measurement, and expectations. Adv Ren Replace Ther 1999;6:110–124.

22. Painter PL, Carlson L, Carey S, Paul SM, Myll J. Physical functioning and health related quality of life changes with exercise training in hemodialysis patients. Am J Kidney Dis 2000;3:482–492.

23. DeOreo PB. Hemodialysis patient-assessed functional health status predicts continued survival, hospitalization and dialysis-attendance compliance. Am J Kidney Dis 1997;30:204–212.

24. Painter P, Hanson P, Messer-Rehak D, Zimmerman SW, Glass NR. Exercise tolerance changes following renal transplantation. Am J Kidney Dis 1987;10:452–456.

25. Painter PL, Luetkemeier MJ, Dibble S, et al. Health related fitness and quality of life in organ transplant recipients. Transplantation 1997;64:1795–1800.

26. Painter PL, Stewart AL, Carey S. Physical functioning: definitions, measurement, and expectations. Adv Ren Replace Ther 1999;6:110–124.

27. Miller TD, Squires RW, Gau GT, Frohnert PP, Sterioff S. Graded exercise testing and training after renal transplantation: a preliminary study. Mayo Clin Proc 1987;62:773–777.

28. Kempeneers G, Myburgh KH, Wiggins T, Adams B, van Zyl-Smith R, Noakes TD. Skeletal muscle factors limiting exercise tolerance of renal transplant patients: effects of a graded exercise training program. Am J Kidney Dis 1990;14:57–65.

29. Painter P, Blagg C, Moore GE. Exercise for the dialysis patient: a comprehensive program. Madison, WI: Medical Education Institute; 1995.

30. Carey S, Painter P. An exercise program for CAPD patients. Nephrol News Issues 1997;June:15–18.

31. Carlson L, Carey S. Staff responsibility to exercise. Adv Ren Replace Ther 1999;4:172–181.

32. Deligianis A, Kouidid E, Tassoulas E, Gigis P, Tourkantonis A, Coats A. Cardiac response to physical training in hemodialysis patients: an echocardiographic study at rest and during exercise. Int J Cardiol 1999;70:253–266.

33. Mercer T, Naish PF, Gleeson NP, Wilcock JE, Crawford C. Development of a walking test for the assessment of functional capacity in non-anemic maintenance dialysis patients. Nephrol Dial Transplant 1998;13(8):2023–2026.

34. Lo C, Li L, Lo WK. Benefits of exercise training in patients on continuous ambulatory peritoneal dialysis. Am J Kidney Dis 1998;32(6):1011–1018.

35. Kouidi E, Albani M, Natsis K, Magalopoulos A, Gigis P, Guiba-Tziampiri O, Tourkantonis A, Deligiannis A. The effects of exercise training on muscle atrophy in hemodialysis patients. Nephrol Dial Transplant 1998;13:685–699.

36. Deligiannis A, Kouidi E, Tourkantonis A. Effects of physical training on heart rate variability in patients on hemodialysis. Am J Cardiol 1999;84:197–202.

37. American College of Sports Medicine. Guidelines for exercise testing and prescription. 5th ed. Philadelphia: Williams & Wilkins; 1995.

38. Moore GE, Painter PL, Brinker KR, Stray-Gundersen J, Mitchell JH. Cardiovascular response to submaximal stationary cycling during hemodialysis. Am J Kidney Dis 1998;31:631–637.

39. Csuka M, McCarty DJ. Simple method for measurement of lower extremity muscle strength. Am J Med 1985;78:77–81.

40. Bohannon RW. Comfortable and maximum walking speed of adults aged 20–79 years: reference values and determinants. Age Ageing 1997;26:15–19.

41. Ware J. SF-36 health survey: manual and interpretation guide. Boston: The Health Institute; 1993.

42. Ware JE, Kosinski M, Keller SD. SF-36 physical and mental health summary scales: a user's manual. 2nd ed. Boston: The Health Institute; 1994.

CHAPTER 16

1. Abboud L, Hir J, Eisen I, Cohen A, Markiewicz W. The current value of exercise testing soon after acute myocardial infarction. Isr J Med Sci 1992;28:694–699.

2. Acute Infarction Ramipril Efficacy (AIRE) Study Investigators. Effect of ramipril on mortality and morbidity of survivors of AMI with clinical evidence of heart failure. Lancet 1993;342:821–828.

3. Acute myocardial infarction (chapter 202). In: Harrison's principles of internal medicine. 13th ed. New York McGraw-Hill; 1994.

4. Adachi H, Koike A, Obayashi T, Umezawa S, Aonuma K, Inada M, Korenaga M, Niwa A, Marumo F, Hiroe M. Does appropriate endur-

ance exercise training improve cardiac function in patients with prior myocardial infarction? Eur Heart J 1996;17:1511–1521.

5. Adams GM. Exercise physiology laboratory manual. 3rd ed. Boston: WCB McGraw-Hill; 1998.

6. Ades PA, Savage PD, Cress E, Brochu M, Lee M, Poehlman ET. Resistance training on physical performance in disabled older female cardiac patients. Med Sci Sports Exerc 2003;35(8):1265–1270.

7. Ades PA. Cardiac rehabilitation and secondary prevention of coronary heart disease. N Engl J Med 2001;345:892–902.

8. Alpert JS, Thygesen K. A call for universal definitions in cardiovascular disease. Circulation 2006;114:757–758.

9. Alpert JS, Thygesen K. Myocardial infarction redefined—a consensus document of the Joint European Society of Cardiology/American College of Cardiology Committee for the Redefinition of Myocardial Infarction. Eur Heart J 2000;21:1502–1513.

10. Ambrosio G, Betocchi S, Pace L, Losi MA, Perrone P, Soricelli A, et al. Prolonged impairment of regional contractile function after resolution of exercise-induced angina. Evidence of myocardial stunning in patients with coronary artery disease. Circulation 1996;94:2455–2464.

11. American Association of Cardiovascular and Pulmonary Rehabilitation. Guidelines for cardiac rehabilitation and secondary prevention programs. 4th ed. Champaign, IL: Human Kinetics; 2004.

12. American Heart Association. 2002 Heart and stroke statistical update. Dallas: American Heart Association; 2002.

13. Antman EF, Van de Werf F. Pharmacoinvasive therapy: the future of treatment for ST-elevation myocardial infarction. Circulation 2004;109:2480–2486.

14. Antman EM, Anbe DT, Armstrong PW, et al. ACC/AHA guidelines for the management of patients with ST-elevation myocardial infarction—executive summary: a report of the American College of Cardiology/American Heart Association Task force on practice guidelines (writing committee to revise the 1999 guidelines for the management of patients with acute myocardial infarction). Circulation 2004;110:588–636.

15. Armstrong L, Balady GJ, Berry MJ, et al. ACSM's guidelines for exercise testing and prescription. 7th ed. Philadelphia: Lippincott Williams & Wilkins; 2006.

16. Aronow WS. Management of older persons after myocardial infarction. J Am Geriatr Soc 1998;46(11):1459–1468.

17. Beard CM, Barnard RJ, Robbins DC, Ordovas JM, Schaefer EJ. Effects of diet and exercise on qualitative and quantitative measures of LDL and its susceptibility to oxidation. Arterioscler Thromb Vasc Biol 1996;16:201–207.

18. Beller GA. Coronary heart disease in the first 30 years of the 21st century: challenges and opportunities. Circulation 2001;103:2428–2435.

19. Beta-Blocker Heart Attack Trial Research Group. A randomized trial of propranolol in patients with acute myocardial infarction. I. Mortality results. JAMA 1982;247:1707–1714.

20. Blair SN, Kohl HW III, Paffenbarger RS Jr, Clark DG, Cooper KH, Gibbons LW. Physical fitness and all-cause mortality: a prospective study of healthy men and women. JAMA 1989;262:2395–2401.

21. Bodi B, Monmeneu JV, Marin F. Acute cardiac rupture complicating pre-discharge exercise testing. A case report with complete echocardiographic follow-up. Int J Cardiol 1999;68(3):333–335.

22. Bolli R. Basic and clinical aspects of myocardial stunning. Prog Cardiovasc Dis 1998;40:477–516.

23. Brawner CA, Vanzant MA, Ehrman JK, Foster C, Porcari JP, Kelso AJ, Keteyian SJ. Guiding exercise using the talk test among patients with coronary artery disease. J Cardiopulm Rehab 2006;26:72–75.

24. Buch P, Rasmussen S, Abildstrom SZ, Køber L, Carlsen J, Torp-Pedersen C. The long-term impact of the angiotensin-converting enzyme inhibitor trandolapril on mortality and hospital admissions in patients with left ventricular dysfunction after a myocardial infarction: follow-up to 12 years. Eur Heart J 2005;26:145–152.

25. Buxton AE, Lee KL, Fisher JD, Josephson ME, Prystowsky EN, Hafley G. A randomized study of the prevention of sudden death in patients with coronary artery disease. N Engl J Med 1999;341:1882–1890.

26. Carson P, Phillips R, Lloyd M, Tucker H, Neophytou M, Buch NJ, Gelson A, Lawton A, Simpson T. Exercise after myocardial infarction: a controlled trial. J R Coll Physicians Lond 1982;16:147–151.

27. Cassidy AE, Bielak LF, Zhou Y, Sheedy PF, Turner ST, Breen JF, Araoz PA, Kullo IJ, Lin X, Peyser PA. Progression of subclinical coronary atherosclerosis: does obesity make a difference? Circulation 2005;111:1877–1882.

28. Castelli WP. Categorical issues in therapy for coronary heart disease. Cardiology in practice 1985; Jan/Feb:267–273.

29. Chait A, Han CY, Oram JF, Heinecke JW. Lipoprotein-associated inflammatory proteins: markers or mediators of cardiovascular disease? J Lipid Res 2005;46:389–403.

30. Chobanian AV, Bakris GL, Black HR, et al. Seventh report of the Joint National Committee on prevention, detection, evaluation, and treatment of high blood pressure. Hypertension. 2003;42:1206–1252.

31. Clark AM, Hartling L, Vandermeer B, McAlister FA. Meta-analysis: secondary prevention programs for patients with coronary artery disease. Ann Intern Med 2005;143:659–672.

32. Clearfield MB. C-reactive protein: a new risk assessment tool for cardiovascular disease. JAOA 2005;105(9):409–416.

33. Couillard C, Després JP, Lamarche B, Bergeron J, Gagnon J, Leon AS, Rao DC, Skinner JS, Wilmore JH, Bouchard C. Effects of endurance exercise training on plasma HDL cholesterol levels depend on levels of triglycerides: evidence from men of the Health, Risk Factors, Exercise Training and Genetics (HERITAGE) Family Study. Arterioscler Thromb Vasc Biol 2001;21:1226–1232.

34. Cruikshank JM. Beta-blockers continue to surprise us. Eur Heart J 2000;21(5):354–364.

35. Daub WD, Knapik GP, Black WR. Strength training early after myocardial infarction. J Cardiopulm Rehabil 1996;16:100–108.

36. Davies JR, Rudd JH, Weissberg PL. Molecular and metabolic imaging of atherosclerosis. J Nucl Med 2004;45:1898–1907.

37. DeBusk RF, Convertino VA, Hung J, Goldwater D. Exercise conditioning in middle-aged men after 10 days of bed rest. Circulation 1983;68:245–250.

38. DeBusk RF, Haskell WL, Miller NH, Berra K, Taylor CB, Berger WE III, Lew H. Medically directed at-home rehabilitation soon after uncomplicated acute myocardial infarction: a new model for patient care. Am J Cardiol 1985;55:251–257.

39. DeBusk RF, Houston Miller N, Superko HR, Dennis CA, Thomas RJ, Lew HT, Berger WE III, Heller RS, Rompf J, Gee D, et al. A case-management system for coronary risk factor modification after acute myocardial infarction. Ann Intern Med 1994;120:721–729.

40. DeBusk RF, Houston N, Haskell W, Fry F, Parker M. Exercise training soon after myocardial infarction. Am J Cardiol 1979;44:1223–1229.

41. DeGroot DW, Quinn TJ, Kertzer R, Vroman NB, Olney WB. Lactic acid accumulation in cardiac patients performing circuit weight training: implications for exercise prescription. Arch Phys Med Rehabil 1998;79:838–841.

42. Doi M, Itoh H, Niwa A, Taniguchi K, Hiore M, Marumo F. Effect of training program based on anaerobic threshold in the early phase after acute myocardial infarction. Cardiologia 1997;42(10):1077–1082.

43. Ehsani AA, Biello DR, Schultz J, Sobel BE, Holloszy JO. Improvement of left ventricular contractile function by exercise training in patients with coronary artery disease. Circulation 1986;74(2):350–358.

44. Ehsani AA, Heath GW, Hagberg JM, Sobel BE, Holloszy JO. Effects of 12 months of intense exercise training on ischemic ST-segment

depression in patients with coronary artery disease. Circulation 1981;64(6):1116–1124.

45. Ehsani AA, Miller TR, Miller TA, Ballard EA, Schechtman KB. Comparison of adaptations to a 12-month exercise program and late outcome in patients with healed myocardial infarction and ejection fraction <45% and >50%. Am J Cardiol 1997;79:1258–1260.

46. Endmann DH, Schiffrin EL. Endothelial dysfunction. J Am Soc Nephrol 2004;15:1983–1992.

47. Falk E, Shah PK, Fuster V. Coronary plaque disruption. Circulation 1995;92:657–671.

48. Falk E. Pathogenesis of atherosclerosis. J Am Coll Cardiol 2006;47:C7–12.

49. Fletcher GF, Balady GJ, Amsterdam EA, et al. Exercise standards for testing and training: a statement for healthcare professionals from the American Heart Association. Circulation 2001;104:1694–1740.

50. Fonarow GC. In-hospital initiation of statin therapy in acute coronary syndromes: maximizing the early and long-term benefits. Chest 2005;128:3641–3651.

51. Foster C, Gal RA, Murphy P, Port SC, Schmidt DH. Left ventricular function during exercise testing and training. Med Sci Sports Exerc 1997;29:297–305.

52. Foster C, Kalsdoffir AE, Porcari JP, Meyer K. Myocardial infarction. In: LeMura LM, Von Du Villard SP, editors. Clinical exercise physiology. application and physiological principles. Philadelphia: Lippincott Williams & Wilkins; 2004.

53. Franklin BA. Myocardial infarction. In: Moore GE, Roberts SO, editors. ACSM's exercise management for persons with chronic diseases and disabilities. Champaign, IL: Human Kinetics; 1997.

54. Giannuzzi P, Tavazzi L, Temporelli PL, Corra U, Imparato A, Gattone M, Giordano A, Sala L, Schweigher C, Malinverni C. Long-term physical training and left ventricular remodeling after anterior myocardial infarction: results of the exercise in anterior myocardial infarction (EAMI) trial. J Am Coll Cardiol 1993;22:1821–1829.

55. Gibbons RJ, Balady GJ, Bricker JT, et al. ACC/AHA 2002 guideline update for exercise testing: a report of the American College of Cardiology/American Heart Association Task Force on Practice Guidelines (committee on exercise testing). 2002. American College of Cardiology Web site. www.acc.org/clinical/guidelines/exercise/dirIndex.htm

56. Giri S, Thompson PD, Kiernan FJ, Clive J, Fram DB, Mitchel JF, Hirst JA, McKay RG, Waters DD. Clinical and angiographic characteristics of exertion-related acute myocardial infarction. JAMA 1999;282(18):1731–1736.

57. Golshahi J, Rajabi P, Golshahi F. Frequency of atherosclerotic lesions in coronary arteries of autopsy specimens in Isfahan Forensic Medicine Center. JRMS 2005;1:16–19.

58. Gordon NF, Hoh HW. Exercise testing and sudden cardiac death. J Cardiopulm Rehabil 1993;13:381–386.

59. Grace SL, Abbey SE, Shnek ZM, Irvine J, Franche R, Stewart DE. Cardiac rehabilitation II: referral and participation. Gen Hosp Psychiat 2002;24:127–134.

60. Green DJ, Walsh JH, Maiorana A, Burke V, Taylor RR, O'Driscoll GJ. Comparison of resistance and conduit vessel nitric oxide-mediated vascular function in vivo: effects of exercise training. J Appl Physiol 2004;97:749–755.

61. Grodzinski E, Jette M, Blumchen G, Borer JS. Effects of a four-week training program on left ventricular function as assessed by radionuclide ventriculography. J Cardiopulm Rehabil 1987;7:518–524.

62. Grundy SM, Becker D, Clark LT, et al. Third report of the National Cholesterol Education Program (NCEP) expert panel on detection, evaluation, and treatment of high blood cholesterol in adults (Adult Treatment Panel III) final report. Circulation 2002;106:3143–3421.

63. Hamalainen H, Luurila OJ, Kallio V, Arstila M, Hakkila J. Long-term reduction in sudden deaths after a multifactorial intervention programme in patients with myocardial infarction: 10-year results of a controlled investigation. Eur Heart J 1989;10:55–62.

64. Harrington RA, Becker RC, Ezekowitz M, Meade TW, O'Connor CM, Vorchheimer DA, Guyatt GH. Antithrombotic therapy for coronary artery disease: the seventh ACCP conference on antithrombotic and thrombolytic therapy. Chest 2004;126:513–548.

65. Haskell WL, Savin W, Oldridge N, DeBusk R. Factors influencing estimated oxygen uptake during exercise testing soon after myocardial infarction. Am J Cardiol 1982;50:299–304.

66. Hjalmarson A, Elmfeldt D, Herlitz J, et al. Effect on mortality of metoprolol in acute myocardial infarction. A double-blind randomised trial. Lancet 1981;8251:823–827.

67. Horne R, James D, Petrie K, Weinman J, Vincent R. Patients' interpretation of symptoms as a cause of delay in reaching hospital during acute myocardial infarction. Heart 2000;83:388–393.

68. Hung J, Gordon EP, Houston N, Haskell WL, Goris ML, DeBusk RF. Changes in rest and exercise myocardial perfusion and left ventricular function 3 to 26 weeks after clinically uncomplicated acute myocardial infarction: effects of exercise training. Am J Cardiol 1984;54:943–950.

69. Jain A, Myers GH, Sapin PM, O'Rourke RA. Comparison of symptom-limited and low level exercise tolerance tests early after myocardial infarction. J Am Coll Cardiol, 1993;22:1816–1820.

70. Jensen-Urstad K, Samad BA, Bouvier F, Hulting J, Hojer J, Ruiz H, Jensen-Urstad M. Prognostic value of symptom limited versus low level exercise stress test before discharge in patients with myocardial infarction treated with thrombolytics. Heart 1999;82:199–203.

71. Joao I, Cotrim C, Duarte JA, do Rosario L, Freire G, Pereira H, Oliveira LM, Catarino C, Carrageta M. Cardiac rupture during exercise stress echocardiography: a case report. J Am Soc Echocardiogr 2000;13(8):785–787.

72. Jolly K, Lip GYH, Sandercock J, Greenfield SM, Raftery JP, Mant J, Taylor R, Lane D, Wai Lee K, Stevens AJ. Home-based versus hospital-based cardiac rehabilitation after myocardial infarction or revascularisation: design and rationale of the Birmingham Rehabilitation Uptake Maximisation Study (BRUM): a randomised controlled trial. BMC Cardiovascular Disorders 2003;3:10–21.

73. Juneau M, Colles P, Theroux P, de Guise P, Pelletier G, Lam J. Symptom-limited versus low level exercise testing before hospital discharge after myocardial infarction. J Am Coll Cardiol 1992;20(4):927–933.

74. Kadish AH, Buxton AE, Kennedy HL, Hirshfeld JW, Lorell BH, Rodgers GP, Tracy CM, Weitz HH. ACC/AHA clinical competence statement on electrocardiography and ambulatory electrocardiography. A report of the ACC/AHA/ACP-ASIM task force on clinical competence (ACC/AHA committee to develop a clinical competence statement on electrocardiography and ambulatory electrocardiography). J Am Coll Cardiol 2001;38(7):2091–2100.

75. Kannel WB, Sorlie BP, Castelli WP, McGee D. Blood pressure and survival after myocardial infarction: the Framingham Study. Am J Cardiol 1980;45:326–330.

76. Kay IP, Kittelson J, Stewart RAH. Collateral recruitment and "warm-up" after first exercise in ischemic heart disease. Am Heart J 2000;140:121–125.

77. Kazemi MBS, Eshraghian K, Omrani GR, Lankarani KB, Hosseini E. Homocysteine level and coronary artery disease. Angiology 2006;57:9–14.

78. Kemp M, Donovan J, Higham H, Hooper J. Biochemical markers of myocardial injury. British J Anaesthesia 2004;93(1):63–73.

79. Kiechl S, Werner P, Egger G, Oberhollenzer F, Mayr M, Xu Q, Poewe W, Willeit J. Active and passive smoking, chronic infections, and the risk of carotid atherosclerosis: prospective results from the Bruneck Study. Stroke 2002;33:2170–2176.

80. Kristensen SD, Andersen HR, Thuesen L, Krusell LR, Bøtker HE, Lassen JF, Nielsen TT. Should patients with acute ST elevation MI be transferred for primary PCI? Heart 2004;90:1358–1363.

81. Krivokapich J, Child JS, Gerber RS, Lem V, Moser D. Prognostic usefulness of positive or negative exercise stress echocardiography for predicting coronary events in ensuing twelve months. Am J Cardiol 1993;71:646–651.

82. Lapin ES, Murray JA, Bruce RA, Winterschield L. Changes in maximal exercise performance in evaluation of saphenous vein bypass surgery. Circulation 1973;47:1164–1173.

83. Lavie CJ, Milani R. Benefits of cardiac rehabilitation in the elderly. Chest 2004;126:1010–1012.

84. Lee I, Sesso HD, Paffenbarger Jr RS. Physical activity and coronary heart disease risk in men: does the duration of exercise episodes predict risk? Circulation 2006;102:981–986.

85. Lee KWJ, Hill JS, Walley KR, Frohlich JJ. Relative value of multiple plasma biomarkers as risk factors for coronary artery disease and death in an angiography cohort. CMAJ 2006;174(4):461–466.

86. Leon AS, Franklin BA, Costa F, et al. Cardiac rehabilitation and secondary prevention of coronary heart disease: an American Heart Association scientific statement from the council on clinical cardiology (subcommittee on exercise, cardiac rehabilitation, and prevention) and the council on nutrition, physical activity, and metabolism (subcommittee on physical activity), in collaboration with the American Association of Cardiovascular and Pulmonary Rehabilitation. Circulation 2005;111:369–376.

87. Levine MJ, Leonard BM, Burke JA, Nash ID, Safian RD, Diver DJ, Baim DS. Clinical and angiographic results of balloon-expandable intracoronary stents in right coronary artery stenoses. J Am Coll Cardiol 1990;16:332–339.

88. Lichtlen P, Nikutta P, Jost S, et al., and the INTACT study group. Anatomical progression of coronary artery disease in humans as seen by prospective, repeated, quantitated coronary angiography: relation to clinical events and risk factors. Circulation 1992;86:828–838.

89. Mahmarian JJ, Moye LA, Chinoy DA, Sequeira RF, Habib GB, Hanry WJ, et al. Transdermal nitroglycerin patch therapy improves left ventricular function and prevents remodeling after acute myocardial infarction: results of a multicenter prospective randomized, double-blind, placebo-controlled trial. Circulation 1998;97(20):2017–2024.

90. Malfatto G, Facchini M, Sala L, Branzi G, Bragato R, Leonetti G. Effects of cardiac rehabilitation and betablocker therapy on heart rate variability after first acute myocardial infarction. Am J Cardiol 1998;81:834–840.

91. Marchionni N, Fattirolli F, Fumagalli S, Oldridge N, Del Lungo F, Morosi L, Burgisser C, Masotti G. Improved exercise tolerance and quality of life with cardiac rehabilitation of older patients after myocardial infarction: results of a randomized, controlled trial. Circulation 2003;107:2201–2206.

92. Marra S, Paolillo V, Spadaccini F, Angeleno PF. Long-term follow-up after a controlled randomized post-myocardial infarction rehabilitation programme: effects on morbidity and mortality. Eur Heart J 1985;6:656–663.

93. Mayou R, Sleight P, MacMahon D, Florencio MJ. Early rehabilitation after myocardial infarction. Lancet 1981;2(8260–61):1399–1401.

94. McAlister FA, Lawson FM, Teo KK, Armstrong PW. Randomised trials of secondary prevention programmes in coronary heart disease: systematic review. Br Med J 2001:957–962.

95. McClelland RL, Chung H, Detrano R, Post W, Kronmal RA. Distribution of coronary artery calcium by race, gender, and age results from the Multi-Ethnic Study of Atherosclerosis (MESA). Circulation 2006;113:30–37.

96. McConnell TR, Klinger TA, Gardner JK, Laubach CA Jr, Herman CE, Hauck CA. Cardiac rehabilitation without exercise tests for post-myocardial infarction and post-bypass surgery patients. J Cardiopulm Rehabil 1998;18:458–463.

97. McNeilly RH, Pemberton J. Duration of last attack in 998 fatal cases of coronary artery disease and its relation to possible cardiac resuscitation. Br Med J 1968;3:139–142.

98. Medicare. www.medicare.gov/coverage/Home.asp

99. Mehta J, Mehta P, Horalek C. The significance of platelet-vessel wall prostaglandin equilibrium during exercise-induced stress. Am Heart J 1983;105:895–904.

100. Mellett PA, Keteyian SJ, Davenport MJ, Fedel FJ, Stein PD. Existing criteria for electrocardiographic monitoring during cardiac rehabilitation versus observed events. Med Sci Sports Exerc 1994;26:S47.

101. Milani RV, Lavie CJ, Cassidy MM. Effects of cardiac rehabilitation and exercise training programs on depression in patients after major coronary events. Am Heart J 1996;132:726–732.

102. Miller NH, Haskell WL, Berra K, DeBusk RF. Home versus group exercise training for increasing functional capacity after myocardial infarction. Circulation 1984;70(4):645–649.

103. Mineo K, Takizawa A, Shimamoto M, Yamazaki F, Kimura A, Chino N, Izumi S. Graded exercise in three cases of heart rupture after acute myocardial infarction. Am J Phys Med Rehabil 1995;74:453–457.

104. Mittleman MA, Maclure M, Tofler GH, et al. Triggering of acute myocardial infarction by heavy physical exertion: protection against triggering by regular exertion. N Engl J Med 1993;329:1677–1683.

105. Moon JCC, De Arenaza DP, Elkington AG, et al. The pathologic basis of Q-wave and non–Q-wave myocardial infarction: a cardiovascular magnetic resonance study. J Am Coll Cardiol 2004;44:554–560.

106. Morris GC, Reul GJ, Howell JF. Follow-up results of distal coronary artery bypass for ischemic heart disease. Am J Cardiol 1972;29:180.

107. Mukherjee D, Fang J, Chetcuti S, Moscucci M, Kline-Rogers E, Eagle KA. Impact of combination evidence-based medical therapy on mortality in patients with acute coronary syndromes. Circulation 2004;109:745–749.

108. Myers J, Dziekan G, Goebbels U, Dubach P. Influence of high-intensity exercise training on the ventilatory response to exercise in patients with reduced ventricular function. Med Sci Sports Exerc 1999;31(7):929–937.

109. Myers J, Goebbels U, Dzeikan G, Froelicher V, Bremerich J, Mueller P, Buser P, Dubach P. Exercise training and myocardial remodeling in patients with reduced ventricular function: one-year follow-up with magnetic resonance imaging. Am Heart J 2000;139(2 Pt 1):252–261.

110. Myers L, Coughlin SS, Webber LS, Srinivasan SR, Berenson GS. Prediction of adult cardiovascular multifactorial risk status from childhood risk factor levels. The Bogalusa Heart Study. Am J Epidemiol 1995;142(9):918–924.

111. Newby DE, McLeod AL, Uren NG, Flint L, Ludlam CA, Webb DJ, Fox KAA, Boon NA. Impaired coronary tissue plasminogen activator release is associated with coronary atherosclerosis and cigarette smoking: direct link between endothelial dysfunction and atherothrombosis. Circulation 2001;103:1936–1941.

112. Norwegian Multicenter Study Group. Timolol-induced reduction in patients surviving acute myocardial infarction. N Engl J Med 1981;304:801–807.

113. Nowak TJ, Handford AG. Essentials of pathophysiology: concepts and applications for health care professionals. 2nd ed. St. Louis: WCB McGraw-Hill; 1999.

114. O'Connor GT, Buring JE, Yusuf S, Goldhaber SZ, Olmstead EM, Paffenbarger RS, Hennekens CH. An overview of randomized trials of rehabilitation with exercise after myocardial infarction. Circulation 1989;80(2):234–244.

115. Oldridge NB, Guyatt G, Jones N, Crowe J, Singer J, Feeny D, McElvie R, Runions J, Streiner D, Torrance G. Effects on quality of life with comprehensive rehabilitation after acute myocardial infarction. Am J Cardiol 1991;67:1084–1089.

116. Oldridge NB, Guyatt GH, Fischer ME, Rimm AA. Cardiac rehabilitation after myocardial infarction: combined experience of randomized clinical trials. JAMA 1988;260:945–950.

117. Oldridge NB. Cardiac rehabilitation and risk factor management after myocardial infarction. Clinical and economic evaluation. Wien Klin Wochenschr 1997;109(Suppl 2):6–16.

118. Paffenbarger RS, Hyde RT, Wing AL, Hsieh CC. Physical activity, all-cause mortality, and longevity of college alumni. N Engl J Med 1986;314:605–613.

119. Paul AK, Hasegawa S, Yoshioka J, Tsujimura E, Yamaguchi H, Tokita N, Maruyama A, Xiuli M, Nishimura T. Exercise-induced stunning continues for at least one hour: evaluation with quantitative gated single-photon emission tomography. Eur J Nucl Med 1999;26:410–415.

120. Pedersen TR, for the Scandinavian Simvastatin Survival Study group. Randomised trial of cholesterol lowering in 4444 patients with coronary heart disease: the Scandinavian Simvastatin Survival Study (4S). Lancet 1994;344:1383–1389.

121. Perk J, Veress G. Cardiac rehabilitation: applying exercise physiology in clinical practice. Eur J Appl Physiol 2000;83:457–462.

122. Perry K, Petrie KJ, Ellis CJ, Horne R, Moss-Morris R. Symptom expectations and delay in acute myocardial infarction patients. Heart 2001;86:91–93.

123. Plowman SA, Smith DL. Exercise physiology for health, fitness, and performance. 2nd ed. San Francisco: Benjamin Cummings; 2003.

124. Poirier P, Giles TD, Bray GA, Hong Y, Stern JS, Pi-Sunyer X, Eckel RH. Obesity and cardiovascular disease: pathophysiology, evaluation and effect of weight loss. Circulation 2006;113:898–918.

125. Pollock ML, Franklin BA, Balady GJ, et al. Resistance exercise in individuals with and without cardiovascular disease: benefits, rationale, safety, and prescription. An advisory from the committee on exercise, rehabilitation, and prevention, council on clinical cardiology, American Heart Association. Circulation 2000;101:828–833.

126. Pollock ML, Gettman LR, Milesis CA, Bah MD, Durstine JL, Johnson RB. Effects of frequency and duration of training on attrition and incidence of injury. Med Sci Sports 1977;9:31–36.

127. Pollock ML. How much exercise is enough? Physician Sportsmed 1978;8:50–64.

128. Probstfield JL. How cost-effective are new preventive strategies for cardiovascular disease? Am J Cardiol 2003;91:22G–27G.

129. Raggi P, Callister TQ, Shaw LJ. Progression of coronary artery calcium and risk of first myocardial infarction in patients receiving cholesterol-lowering therapy. Arterioscler Thromb Vasc Biol 2004;24:1272–1277.

130. Rauramaa R, Li G, Vaisanen SB. Dose-response and coagulation and hemostatic factors. Med. Sci. Sports Exerc 2001;33(6):S516–S520.

131. Rea TD, Heckbert SR, Kaplan RC, Smith NL, Lemaitre RN, Psaty BM. Smoking status and risk for recurrent coronary events after myocardial infarction. Ann Intern Med 2002;137:494–500.

132. Roger VL, Killian JM, Weston SA, Jaffe AS, Kors J, Santrach PJ, Tunstall-Pedoe H, Jacobsen SJ. Redefinition of myocardial infarction: prospective evaluation in the community. Circulation 2006;114:790–797.

133. Roman O, Gutierrez M, Luksic I, Chavez E, Camuzzi AL, Villalon E, Klenner C, Cumsille F. Cardiac rehabilitation after acute myocardial infarction: 9-year controlled follow-up study. Cardiology 1983;70:223–231.

134. Rossouw JE, Lewis B, Rifkind BM. The value of lowering cholesterol after myocardial infarction. N Engl J Med 1990;323:1112–1119.

135. Russell MW, Huse DM, Drowns S, Hamel EC, Hartz SC. Direct medical costs of coronary artery disease in the United States. Am J Cardiol 1998;81(9):1110–1115.

136. Ryan TJ, Anderson JL, Antman EM, Braniff BA, Brooks NH, Califf RM, et al. ACC/AHA guidelines for the management of patients with acute myocardial infarction. J Am Coll Cardiol 1996;28(5):1328–1428.

137. Ryan TJ, Antman EM, Brooks NH, et al. 1999 update: ACC/AHA guidelines for the management of patients with acute myocardial infarction: executive summary and recommendations: a report of the American College of Cardiology/American Heart Association task force on practice guidelines (committee on management of acute myocardial infarction). Circulation 1999;100:1016–1030.

138. Sacks FM. Do statins play a role in the early management of the acute coronary syndrome? Eur Heart J Supplements 2004;6(Supplement A):A32–A36.

139. Sakamoto H, Aikawa M, Hill CC, Weiss D, Taylor WR, Libby P, Lee RT. Biomechanical strain induces class A scavenger receptor expression in human monocyte/macrophages and THP-1 cells. Circulation 2001;104:109–114.

140. Schroeder EB, Chambless LE, Liao D, Prineas RJ, Evans GW, Rosamond WD, Heiss G. Diabetes, glucose, insulin, and heart rate variability. The Atherosclerosis Risk in Communities (ARIC) study. Diabetes Care 2005;28:668–674.

141. Shah PK. Mechanisms of plaque vulnerability and rupture. J Am Coll Cardiol 2003;41:15S–22S.

142. Shaw LW, for the project staff. Effects of a prescribed supervised exercise program on mortality and cardiovascular morbidity in patients after a myocardial infarction: the national exercise and heart disease project. Am J Cardiol 1981;48:39–46.

143. Silvet H, Spencer F, Yarzebski, J, Lessard, D, Gore JM, Goldberg RJ. Community wide trends in the use and outcomes associated with beta-blockers in patients with acute myocardial infarction. The Worcester Heart Attack Study. Arch Intern Med 2003;163:2175–2183.

144. Sinclair AJ, Conroy SP, Davies M, Bayer AJ. Post-discharge home-based support for older cardiac patients: a randomised controlled trial. Age and Ageing 2005;34:338–343.

145. Singh VN, Schocken DD, Williams K, Stamey R. Cardiac rehabilitation. eMedicine 2006; www.emedicine.com/pmr/topic180.htm

146. Sivarajan ES, Bruce RA, Lindskog BD, Almes MJ, et al. Treadmill test responses to an early exercise program after myocardial infarction: a randomized study. Circulation 1982;65(7):1420–1428.

147. Smith SC, Allen J, Blair SN, et al. AHA/ACC guidelines for secondary prevention for patients with coronary and other atherosclerotic vascular disease: 2006 update. Circulation 2006;113:2363–2372.

148. Solomon SD, Zelenkofske S, McMurry JJV, et al. Sudden death in patients with myocardial infarction and left ventricular dysfunction, heart failure, or both. N Engl J Med 2005;352(52):2581–2588.

149. Soriano JB, Hoes AW, Meems L, Grobbee DE. Increased survival with beta-blockers: importance of ancillary properties. Prog Cardiovasc Dis 1997;39(5):445–456.

150. Sparrow D, Dawber TR, Colton T. The influence of cigarette smoking on prognosis after a first myocardial infarction. J Chronic Dis 1978;31:425–432.

151. Stary HC, Chandler AB, Dinsmore RE, et al. A definition of advanced types of atherosclerotic lesions and a histological classification of atherosclerosis. Circulation 1995;92:1355–1374.

152. Steffen-Batey L, Nichaman NZ, Goff DC, Frankowski RF, Hanis CL, Ramsey DJ, Labarthe DR. Change in level of physical activity and risk of all-cause mortality or reinfarction: the Corpus Christi Heart Project. Circulation 2000;102:2204–2209.

153. Steg PG, Bonnefoy E, Chabaud S, Lapostolle F, Dubien P, Cristofini P, Leizorovicz A, Touboul P. Impact of time to treatment on mortality after prehospital fibrinolysis or primary angioplasty data from the CAPTIM randomized clinical trial. Circulation 2003;108:2851–2856.

154. Stenestrand U, Wallentin L. Early statin treatment following acute myocardial infarction and 1-year survival. JAMA 2001;285:430–436.

155. Stern MJ, Gorman PA, Kaslow P. The group counseling vs exercise therapy: a controlled intervention with subjects following myocardial infarction. Arch Intern Med 1983;143:1719–1725.

156. Stewart KJ, McFarland LD, Weinhofer JJ, Cottrell E, Brown CS, Shapiro EP. Safety and efficacy of weight training soon after acute myocardial infarction. J Cardiopulm Rehabil 1998;18:37–44.

157. Taylor CB, Houston-Miller N, Ahn DK, Haskell W, DeBusk RF. The effects of exercise training programs on psychosocial improvement in uncomplicated postmyocardial infarction patients. J Psychosom Res 1986;30:581–587.

158. Taylor RS, Brown A, Ebrahim S, et al. Exercise-based rehabilitation for patients with coronary heart disease: systematic review and meta-analysis of randomized controlled trials. Am J Med 2004;116:982–692.

159. The Israeli Sprint Study Group. Secondary Prevention Reinfarction Israeli Nifedipine Trial (SPRINT). A randomized intervention trial of nifedipine in patients with acute myocardial infarction. Eur Heart J 1988;9(4):354–364.

160. The MIAMI Trial Research Group. Metoprolol in Acute Myocardial Infarction (MIAMI). A randomised placebocontrolled international trial. Eur Heart J 1985;6:199–226.

161. Thom T, Haase N, Rosamond W, et al. Heart disease and stroke statistics—2006 update: a report from the American Heart Association Statistics Committee and Stoke Statistics Committee. Circulation 2006;113:e85–e151.

162. Thompson PD, Buchner D, Pina IL, et al. Exercise and physical activity in the prevention and treatment of atherosclerotic cardiovascular disease: a statement from the council on clinical cardiology (subcommittee on exercise, rehabilitation, and prevention) and the council on nutrition, physical activity, and metabolism (subcommittee on physical activity). Circulation 2003;107:3109–3116.

163. Thompson PD. Exercise prescription and proscription for patients with coronary artery disease. Circulation 2005;112:2354–2363.

164. Thompson PD. The cardiovascular complications of vigorous physical activity. Arch Intern Med 1996;156:2297–2302.

165. Thompson, PD. Exercise rehabilitation for cardiac patients. Physician Sportsmed Online 2001;1(29).

166. Thun JJ, Hoefsten DE, Lindholm MG, Mortensen LS, Andersen HR, Nielsen TT, Kober L, Kelbaek H. Simple risk stratification at admission to identify patients with reduced mortality from primary angioplasty. Circulation 2005;112:2017–2021.

167. Tran ZV, Brammell HL. Effects of exercise training on serum lipid and lipoprotein levels in post-MI patients. A meta-analysis. J Cardiopulmonary Rehabil 1989;9:250–255.

168. Travel M. Stress testing in cardiac evaluation: current concepts with emphasis on the ECG. Chest 2000;119:907–925.

169. Tsimikas S, Brilakis ES, Miller ER, McConnell JP, Lennon RJ, Kornman KS, Witztum JL, Berger PB. Oxidized phospholipids, Lp(a) lipoprotein, and coronary artery disease. N Engl J Med 2005;353:46–57.

170. Tzoulaki I, Murray GD, Lee AJ, Rumley A, Lowe GDO, Fowkes FGR. C-reactive protein, interleukin-6, and soluble adhesion molecules as predictors of progressive peripheral atherosclerosis in the general population: Edinburgh Artery Study. Circulation 2005;112:976–983.

171. Van Camp SP. Exercise-related sudden death: risks and causes. Physician Sportsmed 1988;16(5):97–112.

172. Van de Werf F, Ardissino D, Betriu A, et al. Management of acute myocardial infarction in patients presenting with ST-segment elevation. Eur Heart J 2003;24:28–66.

173. Vonder Muhll I, Daub B, Black B, Warburton D, Haykowsky M. Benefits of cardiac rehabilitation in the ninth decade of life in patients with coronary heart disease. Am J Cardiol 2002;90:645–648.

174. Warburton DER, Nicol CW, Bredin SSD. Health benefits of physical activity: the evidence. CMAJ 2006;174(6):801–809.

175. Waters D, Lesperance J, Galdstone P, et al. for the CCAIT study group. Effects of cigarette smoking on the angiographic evolution of coronary atherosclerosis—a Canadian coronary atherosclerosis intervention substudy. Circulation 1996;94:614–621.

176. Weissberg PL. Coronary artery disease. Atherogenesis: current understanding of the causes of atheroma. Heart 2000;83:247–252.

177. Westendorp RGJ. What is healthy aging in the 21st century? Am J Clin Nutr 2006;83(suppl):404S–409S.

178. Wilhelmsen L, Sanne H, Elmfeldt D, Grimby G, et al. A controlled trial of physical training after myocardial infarction: effects on risk factors, nonfatal reinfarction, and death. Prev Med 1975;4:491–508.

179. Willich SN, Lewis M, Lowel H, Arntz HR, Shubert F, Schroder R. Physical exertion as a trigger of acute myocardial infarction. N Engl J Med 1993;329:1684–1690.

180. Wilson K, Gibson N, Willan A, Cook D. Effect of smoking on mortality after myocardial infarction. Arch Intern Med 2000;160:939–944.

181. Wilt TJ, Bloomfield HE, MacDonald R, Nelson D, Rutks I, Ho M, Larsen G, McCall A, Pineros S, Sales A. Effectiveness of statin therapy in adults with coronary heart disease. Arch Intern Med 2004;164:1427–1436.

182. Wolff B, Grabe HJ, Völzke H, Lüdemann J, Kessler C, Dahm JB, Freyberger HJ, John U, Felix SB. Relation between psychological strain and carotid atherosclerosis in a general population. Heart 2005;91:460–464.

183. Wong ND, Wilson PWF, Kannel WB. Serum cholesterol as a prognostic factor after myocardial infarction: the Framingham Study. Ann Intern Med 1991;115:687–693.

184. Wu AHB, Apple FS, Gibler WB, Jesse RL, Warshaw MM, Valdes Jr R. National Academy of Clinical Biochemistry Standards of Laboratory Practice: recommendations for the use of cardiac markers in coronary artery diseases. Clinical Chemistry 1999;45(7):1104–1121.

185. Yu C, Lau C, Chau J, McGee S, Kong S, Man-Yung Cheung B, Sheung-Wai Li L. A short course of cardiac rehabilitation program is highly cost effective in improving long-term quality of life on patients with recent myocardial infarction or percutaneous coronary intervention. Arch Phys Med Rehabil 2004;85:1915–1922.

186. Zeiher AM, Drexler H, Wollschlager H, Just H. Modulation of coronary vasomotor tone in humans progressive endothelial dysfunction with different early stages of coronary atherosclerosis. Circulation 1991;83:391–401.

CHAPTER 17

1. ACC/AHA guidelines for the management of patients with acute myocardial infarction. A report of the American College of Cardiology/American Heart Association task force on practice guidelines. Writing committee to revise the 1999 guidelines for the management of patients with acute myocardial infarction. J Am Coll Cardiol 2004;44:671–719.

2. ACC/AHA guidelines for coronary artery bypass graft surgery: executive summary and recommendations. A report of the American College of Cardiology/American Heart Association task force on practice guidelines (committee to revise the 1991 guidelines for coronary artery bypass graft surgery). Circulation 1999;100:1464–1480.

3. ACC/AHA/SCAI 2005 guidelines update for percutaneous coronary intervention. A report of the American College of Cardiology/American Heart Association task force on practice guidelines (ACC/AHA/SCAI writing committee to update the 2001 guidelines for percutaneous coronary intervention). Circulation 2006;113;156–175.

4. AHA Scientific Statement. Cardiac rehabilitation and secondary prevention of coronary heart disease. Circulation 2005;111:369–376.

5. American Association of Cardiovascular and Pulmonary Rehabilitation. Guidelines for cardiac rehabilitation and secondary prevention programs. 4th ed. Champaign, IL: Human Kinetics; 2004.

6. American College of Sports Medicine. ACSM's guidelines for exercise testing and prescription. 7th ed. Baltimore: Lippincott Williams & Wilkins; 2007.

7. Carlier M, Meier B, Finci L, Karpuz H, Nukta E, Righett A. Early stress tests after successful coronary angioplasty. Cardiology 1993;83:339–344.

8. Dash H. Delayed coronary occlusion after successful percutaneous transluminal coronary angioplasty: association with exercise testing. Am J Cardiol 1982;52:1143–1144.

9. Dolansk MA, Moore SM. Effects of cardiac rehabilitation on the recovery outcomes of older adults after coronary artery bypass surgery. J Cardiopulmonary Rehabil 2004;24:236–244.

10. Dendale P, Berger J, Hansen D, Vaes J, Benit E, Weymans W. Cardiac rehabilitation reduces the rate of major cardiac events after percutaneous coronary intervention. Euro J Cardiovascular Nurs 2005;4:113–116.

11. Faris J, Stotts N. The effect of percutaneous transluminal coronary angioplasty on quality of life. Prog Cardiovasc Nurs 1990;5:132–140.

12. Foley JB, Chisholm RJ, Common AA, Langer A, Armstrong PW. Aggressive clinical pattern of angina at restenosis following coronary angioplasty in unstable angina. Am Heart J 1992;124:1174–1180.

13. Franklin BA, Berra K. The case for cardiac rehabilitation after coronary revascularization: achieving realistic outcome assessments. J Cardiopulmonary Rehabil 2002;22:418–420.

14. Gardner JK, McConnell TR, Klinger TA, Herman CP, Hauck CA, Laubach CA Jr. Quality of life and self-efficacy. Gender and diagnosis considerations for management during cardiac rehabilitation. J Cardiopulmonary Rehabil 2003;23:299–306.

15. Gaw B. Motivation to change lifestyle following percutaneous transluminal coronary angioplasty. Dimens Crit Care Nurs 1992;11:68–74.

16. Gaw-Ens B, Laing GP. Risk factor reduction behaviors in coronary angioplasty and myocardial infarction patients. Can J Cardiovasc Nurs 1994;5:4–12.

17. Gillis C. The family dimension of cardiovascular care. Can J Cardiovasc Nurs 1991;2:3–7.

18. Goodman JM, Pallandi DV, Reading JR, Plyley ML, Liu PP, Kavanagh T. Central and peripheral adaptations after 12 weeks of exercise training in postcoronary artery bypass surgery patients. J Cardiopulm Rehabil 1999;19:144–150.

19. Hambrecht R, Walther C, Mobius-Winkler S, Geilen S, Linke A, Conradi K, Erbs S, Kluge R, Kendziorra K, Sabri O, Sick P, Schuler G. Percutaneous coronary angioplasty compared with exercise training in patients with stable coronary artery disease. A randomized trial. Circulation 2004;109:1371–1378.

20. Hannan EL, Racz MJ, Arani DT, McCallister BD, Walford G, Ryan TJ. A comparison of short- and long-term outcomes for balloon angioplasty and coronary stent placement. J Am Coll Cardiol 2000;36:395–403.

21. Jensen K, Banwart L, Venhaus R, Popkess-Vawter S, Perkins S. Advanced rehabilitation nursing care of coronary angioplasty patients using self-efficacy theory. J Adv Nurs 1993;18:926–931.

22. Kelemen, MH Resistive training safety and assessment guidelines for cardiac and coronary prone patients. Med Sci Sports Exerc 1989;21:675–677.

23. Kestin, AS, Ellis PA, Barnard MR, Errichetti A, Rosner BA, Michelson AD. Effects of strenuous exercise on platelet activation state and reactivity. Circulation 1993;88:1502–1511.

24. King SB III. Interventions in cardiology: what does and does not work. Am J Cardiol 2000;86:3H–5H.

25. Kligfield P, McCormick A, Chai A, Jaccobson A, Feuerstadt P, Hao S. Effect of age and gender on heart rate recovery after submaximal exercise during cardiac rehabilitation in patients with angina pectoris, recent acute myocardial infarction, or coronary bypass surgery. Am J Cardiol 2003;92:601–603.

26. Kubo H, Hirai H, Machii K. Exercise training and the prevention of restenosis after percutaneous transluminal coronary angioplasty (PTCA). Ann Acad Med 1992;21:42–46.

27. Lan C, Chen SY, Hsu CJ, Chiu SF, Lai JS. Improvement of cardio-respiratory function after percutaneous transluminal coronary angioplasty or coronary artery bypass grafting. Am J Phys Med Rehabil 2002;81:336–341.

28. Lan C, Chen S, Chiu S, Hsu C, Lai J, Kuan P. Poor functional recovery may indicate restenosis in patients after coronary angioplasty. Arch Phys Med Rehabil 2003;84:1023–1027.

29. Levine GN, Kern MJ, Berger PB, Brown DL, Klein LW, Kereiakes DJ, Sanborn TA, Jacobs AK, for the American Heart Association Diagnostic and Interventional Catheterization Committeee and Council on Clinical Cardiology. Management of patients undergoing percutaneous coronary revascularization. Ann Int Med 2003;139:123–136.

30. Lissper J, Sudin O, Ohman A, Hofman-Bang C, Fydan L, Nygren A. Long-term effects of lifestyle behavior change in coronary artery disease: effects on recurrent coronary events after percutaneous coronary intervention. Health Psychology 2005;24:41–48.

31. McConnell TR, Klinger TA, Gardner JK, Laubach CA Jr, Herman CE, Hauck CA. Cardiac rehabilitation without exercise tests for post-myocardial infarction and post-bypass surgery patients. J Cardiopulm Rehabil 1998;18:458–463.

32. Milani RV, Lavie CJ, Cassidy MM. Effects of cardiac rehabilitation and exercise training programs on depression in patients after major coronary events. Am Heart J 1996;132:726–732.

33. Moser DK, Dracup KA, Marsden C. Needs of recovering cardiac patients and their spouses compared views. Int J Nurs Stud 1993;30:105–114.

34. Newton K, Sivarajan E, Clarke J. Patient perceptions of risk factor changes and cardiac rehabilitation outcomes after myocardial infarction. J Cardiac Rehabil 1985;5:159–168.

35. Ong ATL, van Domburg RT, Aoki J, Sonnenschein K, Lemos PA, Serruys PW. Sirolimus-eluting stents remain superior to bare-metal stents at two years: medium-term results from the Rapamycin-Eluting Stent Evaluated at Rotterdam Cardiology Hospital (RESEARCH) Registry. J Am Coll Cardio 2006;47:1356–1360.

36. Oxford JL, Selwyn AP, Ganz P, Popma JJ, Rogers C. The comparative pathobiology of atherosclerosis and restenosis. Am J Cardiol 2000;86(Suppl):6H–11H.

37. Pashkow FJ. Issues in contemporary cardiac rehabilitation. J Am Coll Cardiol 1993;21:822–834.

38. Patterson JM. 1989. Illness beliefs as a factor in patient spouse adaptation to treatment for coronary artery disease. Fam Sys Med 1989;7:428–443.

39. Pierson LM, Norton HJ, Herbert WG, Pierson ME, Ramp WK, Kiebzak GM, Fedor JM, Cook JW. Recovery of self-reported functional capacity after coronary artery bypass surgery. Chest 2003;123:1367–1374.

40. Przybojewski JZ, Welch HFH. Acute coronary thrombus formation after stress testing following percutaneous transluminal coronary angioplasty: a case report. S Afr Med J 1985;67:378–382.

41. Robertson D, Keller C. Relationships among health beliefs, self-efficacy, and exercise adherence in patients with coronary artery disease. Heart Lung 1992;21:56–63.

42. Samuels B, Schumann J, Kiat H, Friedman J, Berman DS. Acute stent thrombosis associated with exercise testing after successful percutaneous transluminal angioplasty. Am Heart J 1995;130:1210–1222.

43. Shaw DK, Deutsch DT, Schall PM, Bowling RJ. Physical activity and lean body mass loss following coronary artery bypass graft surgery. J Sports Med Phys Fitness 1991;31:67–74.

44. Sigwart U. Drug-eluting stents are safe and effective: right or wrong? J Am Coll Cardiol 2006;47:1361–1362.

45. Stewart K, Badenhop D, Brubaker PH, Keteyian SJ, King M. Cardiac rehabilitation following percutaneous revascularization, heart transplant, heart valve surgery, and for chronic heart failure. Chest 2003;123:2104–2111.

46. Weisz G, Leon MB, Holmes DR Jr, Kereiakes DJ, Clark MR, Cohen BM, Ellis SG, Coleman P, Hill C, Shi C, Cutlip DE, Kuntz RE, and Moses JW. Two-year outcomes after sirolimus-eluting stent implantation: results from the Sirolimus-Eluting Stent in de Novo Native Coronary Lesions (SIRIUS) Trial. J Am Coll Cardiol 2006;47:1350–1355.

47. Yu C, Lau C, Chau J, McGhee S, Kong S, Cheung BM. A short course of cardiac rehabilitation program is highly cost effective in improving long-term quality of life in patients with recent myocardial infarction or percutaneous coronary intervention. Arch Phys Med Rehabil 2004;85:1915–1922.

CHAPTER 18

1. American Association of Cardiovascular and Pulmonary Rehabilitation. Guidelines for cardiac rehabilitation and secondary prevention programs. Champaign, IL: Human Kinetics; 1999.

2. Afzal A, Brawner CA, Keteyian SJ. Exercise training in heart failure. Prog Cardiovasc Dis 1998;41:175–190.

3. Arvan S. Exercise performance of the high risk acute myocardial infarction patient after cardiac rehabilitation. Am J Cardiol 1988;62:197–201.

4. Barnard KL, Adams KJ, Swank AM, Kaelin M, Kushnik MR, Denny DM. Combined high intensity and aerobic training in congestive heart failure patients. J Strength Conditioning Res 2000;14:383–388.

5. Beaver WL, Wasserman K, Whipp BJ. A new method for detecting anaerobic threshold by gas exchange. J Appl Physiol 1986;60:2020–2027.

6. Belardinelli R, Georgiou D, Cianci G, Purcaro A. Randomized, controlled trial of long-term moderate exercise training in chronic heart failure. Circulation 1999;99:1173–1182.

7. Belardinelli R, Georgiou D, Scoccoo V, et al. Low intensity exercise training in patients with heart failure. J Am Coll Cardiol 1995;26:975–982.

8. Braith RW, Welsch MA, Feigenbaum MS, et al. Neuroendocrine activation in heart failure is modified by endurance exercise training. J Am Coll Cardiol 1999;34:1170–1175.

9. Braith RW, Mills RM, Welsch MA, Keller JW, Pollock ML. Resistance training restores bone mineral density in heart transplant recipients. J Am Coll Cardiol 1996;28:1471–1477.

10. Brubaker PH, Kee-Chan J, Stewart KP, Fray B, Moore B, Kitzman DW. Chronotropic incompetence and its contribution to exercise intolerance in older heart failure patients. J Cardiopulmonary Rehabil 2006;26:86–89.

11. Coats AJS, Adamopoulos S, Radaelli A. Controlled trial of physical training in chronic heart failure. Circulation 1992;85:2119–2131.

12. Cohen-Solal A, Chabernaud JM, Gourgon R. Comparison of oxygen uptake during bicycle exercise in patients with chronic heart failure and in normal subjects. J Am Coll Cardiol 1990;16:80–85.

13. Colucci WS, Braunwald E. Pathophysiology of heart failure. In: Braunwald E, Zipes DP, Libby P, editors. A textbook of cardiovascular medicine. Philadelphia: Saunders; 2001. p 503.

14. Colucci WS, Ribeiro JP, Rocco MB. Impaired chronotropic response to exercise in patients with congestive heart failure. Circulation 1989;80:314–323.

15. Conn RH, Williams RS, Wallace AG. Exercise responses before and after physical conditioning in patients with severely depressed left ventricular function. Am J Cardiol 1982;49:296–300.

16. Currnier D, Galinier M, Pathak A, Fourcade J, Bousquet M, Senard J, Fauvel J, Bounhoure J, Montastruc J. Rehabilitation of patients with congestive heart failure with or without β-blockade therapy. J Cardiac Failure 2001;7:241–248.

17. Drexler H, Hayoz D, Munzel T, et al. Endothelial function in chronic congestive heart failure. Am J Cardiol 1992;69:1596–1601.

18. Dubach P, Myers J, Dziekan G, et al. Effect of exercise training on myocardial remodeling in patients with reduced left ventricular function after myocardial infarction. Circulation 1997;95:2060–2067.

19. Dubach P, Myers J, Dziekan G, et al. Effect of high intensity exercise training on central hemodynamic responses to exercise in men with reduced left ventricular function. J Am Coll Cardiol 1997;29:1591–1598.

20. Duscha BD, Kraus WE, Keteyian SJ, et al. Capillary density of skeletal muscle. J Am Coll Cardiol 1999;33:1956–1963.

21. Ehrman JK, Keteyian SJ, Shepard R, et al. Cardiovascular responses of heart transplant recipients to graded exercise testing. J Appl Physiol 1992;73:260–264.

22. Fossier JF, Vernochet P, Bertrand P, Charbonnier B, Monpere C. Influence of carvedilol on the benefits of physical training in patients with moderate chronic heart failure. European J Heart Failure 2001;3:335–342.

23. Foss ML, Keteyian SJ. Fox's physiological basis for exercise and sport. 6th ed. Boston: McGraw-Hill; 1998.

24. Giannuzzi P, Tavazzi L, Temporelli PL, et al. Long-term physical training and left ventricular remodelling after anterior myocardial infarction: results of the exercise in anterior myocardial infarction (EAMI) trial. J Am Coll Cardiol 1993;22:1821–1829.

25. Giannuzzi P, Temporelli PL, Corra U, Tavazzi L, for the ELVD-CHF Study Group. Antiremodeling effect of long term exercise training in patients with stable chronic heart failure. Circulation 2003;108:554–559.

26. Gielen S, Adams V, Mobius-Winkler S, Linke A, S Erbs, Yu J, Kempf W, Schubert A, Schuler G, Hambrecht R. Anti-inflammatory effects of exercise training in the skeletal muscle of patients with chronic heart failure. J Am Coll Cardiol 2003;42:861–868.

27. Gielen S, Adams V, Linke A, Erbs S, Mobius-Winkler S, Schubert A, Schuler G, Hambrecht R. Exercise training in chronic heart failure: correlation between reduced local inflammation and improved oxidative capacity in the skeletal muscle. European Soc Cardiol 2005;12:393–400.

28. Hambrecht R, Fiehn E, Weigl C, et al. Regular physical exercise corrects endothelial dysfunction and improves exercise capacity in patients with chronic heart failure. Circulation 1998;98:2709–2715.

29. Hambrecht R, Fiehn E, Yu J, et al. Effects of endurance training on mitochondrial ultrastructure and fiber type distribution in skeletal muscle of patients with stable chronic heart failure. J Am Coll Cardiol 1997;29:1067–1073.

30. Hambrecht R, Niebauer J, Fiehn E, et al. Effects on cardiorespiratory fitness and ultrastructural abnormalities of leg muscles. J Am Coll Cardiol 1995;25:1239–1249.

31. Hambrecht R, Gielen S, Linke A, et al. Effects of exercise training on left ventricular function and peripheral resistance in patients with chronic heart failure. JAMA 2000;283:3095–3101.

32. Higginbotham MB, Morris KG, Conn EH, et al. Determinations of variable exercise performance among patients with severe left ventricular dysfunction. Am J Cardiol 1983;51:52–60.

33. Hunt, SA, Abraham WT, Chin MH, Feldman AM, Francis GS, Gantias TG, Jessup M, Konstam MA, Mancini DM, Michl K, Oats JA, Rahko PS, Silver MA, Stevenson LW, Yancy CW. ACC/AHA 2005 guideline update for the diagnosis and management of chronic heart failure in the adult: a report of the American College of Cardiology/American Heart Association task force on practice guidelines (writing committee to update the 2001 guidelines for the evaluation and management of heart failure). American College of Cardiology Web site. www.acc.org/clinical/guidelines/failure//index.pdf

34. Kataoka T, Keteyian SJ, Marks CRC, et al. Exercise training in a patient with congestive heart failure on continuous dobutamine. Med Sci Sports Exerc 1994;26:678–681.

35. Keteyian SJ, Levine AB, Brawner CA, et al. Exercise training in patients with heart failure. A randomized, controlled trial. Ann Intern Med 1996;124:1051–1057.

36. Keteyian SJ, Brawner CA, Schairer JR, et al. Effects of exercise training on chronotropic incompetence in patients with heart failure. Am Heart J 1999;138:233–240.

37. Keteyian SJ, Ehrman J, Fedel F, Rhoads K. Heart rate-perceived exertion relationship during exercise in orthotopic heart transplant patients. J Cardiopulm Rehabil 1990;10:287–293.

38. Keteyian SJ, Duscha BD, Brawner CA, Green HJ, Marks CRC, Schachat FH, Annex BH, Kraus WE. Differential effects of exercise training in men and women with chronic heart failure. Am Heart J 2003;145:912–918.

39. Kiilavuori K, Sovijarvi A, Naveri H, et al. Effect of physical training on exercise capacity and gas exchange in patients with chronic heart failure. Chest 1996;110:985–991.

40. Kiilavuori K, Toivonen L, Naveri H, Leinonen H. Reversal of autonomic derangements by physical training in chronic heart failure assessed by heart rate variability. Eur Heart J 1995;16:490–496.

41. Kitzman K, Little WC, Brubaker PH, Anderson RT, Hundley WG, Marburger CT, Brosnihan B, Morgan TM, Stewart KP. Pathophysiological characterization of isolated diastolic heart failure in comparison to systolic heart failure. JAMA 2002;288:2144–2150.

42. Kinugawa T, Ogino K, Kitamura H, et al. Response of sympathetic nervous system activity to exercise in patients with congestive heart failure. Eur J Clin Invest 1991;221:542–546.

43. Kobashigawa JA, Leaf DA, Lee N, et al. A controlled trial of exercise rehabilitation after heart transplantation. N Engl J Med 1999;340:72–277.

44. Kubo SH, Rector TS, Bank AJ, et al. Endothelial-dependent vasodilatation is attenuated in patients with heart failure. Circulation 1991;84:1586–1596.

45. Lee AP, Ice RP, Blessey R, et al. Long term effects of physical training in coronary patients with impaired ventricular function. Circulation 1979;60:1519–1526.

46. Linke A, Schoene N, Gielen S, Hofer J, Erbs S, Schuler G, Hambrecht R. Endothelial dysfunction in patients with chronic heart failure: systemic effects of lower-limb exercise training. J Am Coll Cardiol 2001;37:392–397.

47. Mancini DM, Walter G, Reichek N, et al. Contribution of skeletal muscle atrophy to exercise intolerance and altered muscle metabolism in heart failure. Circulation 85;1992:1364–1373.

48. McKelvie RS, Teo KK, Roberts R, McCartney N, Humen D, Montague T, Hendrican K, Yusuf S. Effects of exercise training in patients with heart failure: the exercise rehabilitation trial (EXERT). Am Heart J 2002;144:23–30.

49. Meyer K, Samek L, Schwaibold M, et al. Interval training in patients with severe chronic heart failure: analysis and recommendations for exercise of procedures. Med Sci Sports Exerc 1997;29:306–312.

50. Meyer K, Schwaibold M, Westbrook S, et al. Effects of short-term exercise training and activity restriction on functional capacity in patients with severe chronic congestive heart failure. Am J Cardiol 1996;78:1017–1022.

51. Minotti JR, Christoph I, Oka R, Weiner MW, Wells L, Massie BM. Impaired skeletal muscle function in patients with congestive heart failure. J Clin Invest 1991;88:2077–2082.

52. Myers J, Gullestad L, Vagelos R, et al. Cardiopulmonary exercise testing and prognosis in severe heart failure: 14 ml/kg/min revisited. Am Heart J 2000;139:78–84.

53. Myers J, Buchanan N, Walsh D, et al. Comparison of the ramp versus standard exercise protocols. J Am Coll Cardiol 1991;17:1334–1342.

54. Normandin EA, Camaione DN, Clark BA III, et al. A comparison of conventional versus anaerobic threshold exercise prescription methods in subjects with left ventricular dysfunction. J Cardiopulm Rehabil 1993;13:110–116.

55. Page E, Cohen-Solal A, Jondeau G, et al. Comparison of treadmill and bicycle exercise in patients with chronic heart failure. Chest 1994;106:1002–1006.

56. Passino C, Severino S, Poletti R, Piepoli MF, Mammini C, Clerico A, Gabutti A, Nassi G, Emdin M. Aerobic training decreases B-type natriuretic peptide expression and adrenergic activation in patients with heart failure. J Am Coll Cardiol 2006;47:1835–1839.

57. Robbins M, Francis G, Pashkow FJ, et al. Ventilatory and heart rate responses to exercise. Circulation 1999;100:2411–2417.

58. Roveda F, Middlekauff HR, Rondon MUPB, Reis SF, Souza M, Nastari L, Barretto ACP, Krieger EM, Negrao CE. The effects of exercise training on sympathetic neural activation in advanced heart failure. J Am Coll Cardiol 2003;42:854–860.

59. Squires RW, Lavie CJ, Brandt TR, et al. Cardiac rehabilitation in patients with severe ischemic left ventricular dysfunction. Mayo Clin Proc 1987;62:997–1002.

60. Sullivan MJ, Knight DJ, Higginbotham MB, Cobb FR. Relation between central and peripheral hemodynamics during exercise in patients with chronic heart failure. Circulation 1989;80:769–781.

61. Sullivan MJ, Green HJ, Cobb FR. Skeletal muscle biochemistry and histology in ambulatory patients with long-term heart failure. Circulation 1991;81:518–527.

62. Sullivan MJ, Higginbotham MB, Cobb FR. Exercise training in patients with severe left ventricular dysfunction. Circulation 1988;78:506–515.

63. Taylor D, Edwards L, Boucek M, Trulock E, Deng M, Keck B, Hertz M. Registry of the International Society for Heart and Lung Transplantation: twenty-second official adult heart transplant report—2005. J Heart Lung Transplantation 2005;24:945–955.

64. Thom T, Hasse N, Rosamond W, Howard VJ, Rumsfeld J, Manolio T, Zheng ZJ, Flegal K, O'Connell C, Kittner S, Lloyd-Jones D, Goff DC, Hong Y, Adams R, Friday G, Furie K, Gorelick P, Kissela B, Marler J, Meigs J, Roger V, Sidney S, Sorlie P, Steinberger J, Wasserthiel-Smoller S, Wilson M, Wolf P. Heart disease and stroke statistics—2006 update: a report from the American Heart Association statistics committee and stroke statistics subcommittee. Circulation 2006;113:e85–151.

65. Toepher M, Meyer K, Maier P, et al. Influence of exercise training and restriction of activity on autonomic balance in patients with severe congestive heart failure. Clin Sci 1996;91(Suppl):116.

66. Toth MJ, Gottlieb SS, Fisher ML, et al. Skeletal muscle atrophy and peak oxygen consumption in heart failure. Am J Cardiol 1997;79:1267–1269.

67. Volterrani M, Clark AL, Ludman PF, et al. Predictors of exercise capacity in chronic heart failure. Eur Heart J 1994;15:801–809.

68. Willams MA, Haskell WL, Ades PA, et al. Resistance exercise in individuals with and without cardiovascular disease: 2007 update. Circulation 2007;116:572–584.

69. Wilson JR, Martin JL, Schwartz D, et al. Exercise intolerance in patients with chronic heart failure: role of impaired nutritive flow to skeletal muscle. Circulation 1984;69:1079–1087.

70. Zhang YY, Wasserman K, Sietsmea KE, et al. O_2 uptake kinetics in response to exercise. Chest 1993;103:735–741.

CHAPTER 19

1. Aboyans V, Criqui MH, Denenberg JO, Knoke JD, Ridker PM, Fronek A. Risk factors for progression of peripheral arterial disease in large and small vessels. Circulation 2006;113(22):2623–0. [Epub 2006 May 30.]

2. Al Zahrani HA, Al Bar HM, Bahnassi A, Abdulaal AA. The distribution of peripheral arterial disease in a defined population of elderly high-risk Saudi patients. Int Angiol 1997;16(2):123–128.

3. Alpert J, Larson O, Lassen N. Exercise and intermittent claudication. Blood flow in the calf muscle during walking studied by the zenon-133 clearance methods. Circulation 1969;39:353–359.

4. American College of Sports Medicine. ACSM's guidelines for exercise testing and prescription. 7th ed. Media, PA: Lippincott Williams & Wilkins; 2005.

5. Amirhamzeh MM, Chant HJ, Rees JL, Hands LJ, Powell RJ, Campbell WB. A comparative study of treadmill tests and heel raising exercise for peripheral arterial disease. Eur J Vasc Endovasc 1997;13(3):301–305.

6. Arfvidsson B, Karlsson J, Dahllof AG, Lundholm K, Sullivan M. The impact of intermittent claudication on quality of life evaluated by the Sickness Impact Profile technique. Eur J Clin Invest 1993;23(11):741–745.

7. Ashworth NL, Chad KE, Harrison EL, Reeder BA, Marshall SC. Home versus center based physical activity programs in older adults. Cochrane Db Syst Rev 2005;(1):CD004017.

8. Bagger JP, Helligsoe P, Randsbaek F, Kimose HH, Jensen BS. Effect of Verapamil in intermittent claudication: a randomized, double-blind, placebo controlled, cross-over study after individual dose-response assessment. Circulation 1997;95:411–414.

9. Baker JD, Dix D. Variability of Doppler ankle pressure with arterial occlusive disease: an evaluation of ankle index and brachial-ankle pressure gradient. Surgery 1981;89:134–137.

10. Barbadimos AN, Zohman LR. Intravenous dipyridamole thallium imaging V combined arm-leg cycle stress testing of patients unable to exercise on the treadmill. Am J Phys Med Rehab 1999;78(2):111–116.

11. Barletta G, Perna S, Sabba C, Catalano A, O'Boyle C, Brevetti G. Quality of life in patients with intermittent claudication: relationship with laboratory exercise performance. Vasc Med 1996;1(1):3–7.

12. Bartels C, Bechtel JF, Hossmann V, Horsch S. Cardiac risk stratification for high-risk vascular surgery. Circulation 1997;95(11):2473–2475.

13. Bendermacher BL, Willigendael EM, Teijink JA, Prins MH. Supervised exercise therapy versus non-supervised exercise therapy for intermittent claudication. Cochrane Db Syst Rev 2006;(2):CD005263.

14. Brandsma JW, Robeer BG, van den Heuvel S, Smit B, Wittens CH, Oostendorp RA. The effect of exercises on walking distance of patients with intermittent claudication: a study of randomized clinical trials. Phys Ther 1998;78(3):278–286;discussion 286–288. [Published erratum appears in Phys Ther 1998;78(5):547.]

15. Buchwald H, Bourdages HR, Campos CT, Nguyen P, Williams SE, Boen JR. Impact of cholesterol reduction on peripheral arterial disease in the Program on the Surgical Control of the Hyperlipidemias (POSCH). Surgery 1996;120(4):672–679.

16. CAPRIE Steering Committee. A randomized, blinded, trial of clopidogrel versus aspirin in patients at risk of ischemic events. Lancet 1996;348:1329–1339.

17. Coffman JD, Mannick JA. Failure of vasodilator drugs in arteriosclerosis obliterans. Ann Intern Med 1972;76(1):35–39.

18. Coffman JD. Pathophysiology of obstructive arterial disease. Herz 1988,13(6):343–350.

19. Coffman JD. Intermittent claudication—be conservative. New Engl J Med 1991;325:577–578.

20. Criqui MH, Denenberg JO, Bird CE, Fronek A, Klauber MR, Langer RD. The correlation between symptoms and non-invasive test results in patients referred for peripheral arterial disease testing. Vasc Med 1996;1(1):65–71.

21. Criqui MH, Denenberg JO, Langer RD, Fronek A. The epidemiology of peripheral arterial disease: importance of identifying the population at risk. Vasc Med 1997;2(3):221–226.

22. Dahllof AG, Bjorntorp P, Holm J, Schersten T. Metabolic activity of skeletal muscle in patients with peripheral arterial insufficiency. Eur J Clin Invest 1974;4(1):9–15.

23. Dahllof AG, Holm J, Schersten T, Sivertsson R. Peripheral arterial insufficiency, effect of physical training on walking tolerance, calf blood flow, and blood flow resistance. Scand J Rehabil Med 1976;8(1):19–26.

24. Davidoff GN, Lampman RM, Westbury L, Deron J, Finestone HM, Islam S. Exercise testing and training of dysvascular amputation: safety and efficacy of arm ergometry. Arch Phys Med Rehab 1992;73:334–338.

25. de Vries SO, Fidler V, Kuipers WD, Hunink MG. Fitting multistate transition models with autoregressive logistic regression: supervised exercise in intermittent claudication. Med Decis Making 1998;18(1):52–60.

26. Depairon M, Zicot M. The quantitation of blood flow/metabolism coupling at rest and after exercise in peripheral arterial insufficiency, using PET and 15-0 labeled tracers. Angiology 1996;47(10):991–999.

27. Desvaux B, Abraham P, Colin D, Leftheriotis G, Saumet JL. Ankle to arm index following maximal exercise in normal subjects and athletes. Med Sci Sport Exer 1996;28(7):836–839.

28. Dormandy J, Mahir M, Ascady G, Balsano F, De Leeuw P, Blombery P, Bousser MG, Clement D, Coffman J, Deutshinoff A, et al. Fate of the patient with chronic leg ischaemia. A review article. J Cardiovasc Surg 1989;30(1):50–57.

29. Dormandy JA, Murray GD. The fate of the claudicant: a prospective study of 1969 claudicants. Eur J Vascular Surg 1991;5(2):131–133.

30. Eames RA, Lange LS. Clinical and pathological study of ischaemic neuropathy. J Neurol Neurosur Ps 1967;30(2):215–226.

31. Edwards AT, Blann AD, Suarez-Mendez VJ, Lardi AM, McCollum CN. Systemic response in patients with intermittent claudication after treadmill exercise. Brit J Surg 1994;81(12):1738–1741.

32. Ekroth R, Dahllof AG, Gundevall B, Holm J, Schersten T. Physical training with intermittent claudication: indications, methods, and results. Surgery 1978;84(5):640–643.

33. England JD, Regensteiner JG, Ringel SP, Carry MR, Hiatt WR. Muscle denervation in peripheral arterial disease. Neurology 1992;42(5):994–999.

34. Ericsson B, Haeger K, Lindell SE. Effect of physical training on intermittent claudication. Angiology 1970;21(3):188–192.

35. Ernst E. Exercise: the best therapy for intermittent claudication? Brit J Hosp Med 1992;48(6):303–304,307.

36. Ernst E, Fialka V. A review of the clinical effectiveness of exercise therapy for intermittent claudication. Arch Int Med 1993;153(20):2357–2360.

37. Ernst EE, Matrai A. Intermittent claudication, exercise, and blood rheology. Circulation 1987;76(5):1110–1114.

38. Federman DG, Trent JT, Froelich CW, Demirovic J, Kirsner RS. Epidemiology of peripheral vascular disease: a predictor of systemic vascular disease. Ostomy Wound Manage 1998;44(5):58–62,64,66 passim.

39. Feinberg RL, Gregory RT, Wheeler JR, Snyder SO Jr, Gayle RG, Parent FN III, Patterson RB. The ischemic window: a method for the objective quantitation of the training effect in exercise therapy for intermittent claudication. J Vasc Surg 1992;16(2):244–250.

40. Fletcher GF, Balady G, Froelicher VF, Hartley LH, Haskell WL, Pollock ML. Exercise standards: a statement for healthcare professionals from the American Heart Association Writing Group. Circulation 1995;91(2):580–615.

41. Fowkes FG, Housley E, Riemersma RA, Macintyre CC, Cawood EH, Prescott RJ, Ruckley CV. Smoking, lipids, glucose intolerance, and blood pressure as risk factors for peripheral atherosclerosis compared with ischemic heart disease in the Edinburgh Artery Study. Am J Epidemiol 1992;135(4):331–340.

42. Gallasch D, Diehm C, Dofer C, Schmitt T, Stage A, Morl H. Effect of physical training on the blood flow properties in patients with intermittent claudication. Klin Wochenschr 1985;63(12):554–559.

43. Gardner AW. Claudication pain and hemodynamic responses to exercise in younger and older peripheral arterial disease patients. J Gerontol 1993;48(5):M231–M236.

44. Gardner AW. Dissipation of claudication pain after walking: implications for endurance training. Med Sci Sport Exer 1993;25(8):904–910.

45. Gardner AW, Poehlman ET. Exercise rehabilitation programs for the treatment of claudication pain. A meta-analysis. JAMA 1995;274(12):975–980.

46. Gardner AW, Katzel LI, Sorkin JD, Killewich LA, Ryan A, Flinn WR, Goldberg AP. Improved functional outcomes following exercise rehabilitation in patients with intermittent claudication. J Gerontol A-Biol 2000;55(10):M570–M577.

47. Gardner AW, Skinner JS, Bryant CX, Smith LK. Stair climbing elicits a lower cardiovascular demand than walking in claudication patients. J Cardiopulm Rehabil 1995;15(2):34–142.

48. Gardner AW, Skinner JS, Vaughan NR, Bryant CX, Smith LK. Comparison of three progressive exercise protocols in peripheral vascular occlusive disease. Angiology 1992;43(8):661–671.

49. Gardner AW, Montgomery PS, Flinn WR, Katzel LI. The effect of exercise intensity on the response to exercise rehabilitation in patients with intermittent claudication. J Vasc Surg 2005;42(4):702–709.

50. Garg PK, Tian L, Criqui MH, Liu K, Ferrucci L, Guralnik JM, Tian J, McDermott MM. Physical activity during daily life and mortality in patients with peripheral arterial disease. Circulation 2006;114:242–248.

51. Gerhard M, Baum P, Raby KE. Peripheral arterial-vascular disease in women: prevalence, prognosis, and treatment. Cardiology 1995;86(4):349–355.

52. Gey DC, Lesho EP, Manngold J. Management of peripheral arterial disease. Am Fam Physician 2004;69(3):525–532.

53. Hall JA, Barnard J. The effects of an intensive 26-day program of diet and exercise on patients with peripheral vascular disease. J Card Rehabil 1982;2:569–574.

54. Hankey GJ, Norman PE, Eikelboom JW. Medical treatment of peripheral arterial disease. JAMA 2006;295(5):547–553.

55. Harrison DG, Ohara Y. Physiologic consequences of increased vascular oxidant stresses in hypercholesterolemia and atherosclerosis: implications for impaired vasomotion. Am J Cardiol 1995;75(6):75B–81B.

56. Hertzer NR. The natural history of peripheral vascular disease. Implications for its management. Circulation 1991;83(2 Suppl.):I-2–I19.

57. Hiatt WR, Regensteiner JG, Wolfel EE. Special populations in cardiovascular rehabilitation. Peripheral arterial disease, non-insulin-dependent diabetes mellitus, and heart failure. Cardiol Clin 1993;11(2):309–321.

58. Hiatt WR, Wolfel EE, Meier RH, Regensteiner JG. Superiority of treadmill walking exercise versus strength training for patients with peripheral arterial disease. Implications for the mechanism of the training response. Circulation 1994;90(4):1866–1874.

59. Hiatt WR, Regensteiner JG, Wolfel EE, Carry MR, Brass EP. Effect of exercise training on skeletal muscle histology and metabolism in peripheral arterial disease. J Appl Physiol, 1996;81(2):780–788.

60. Hiatt WR. Current and future drug therapies for claudication. Vasc Med 1997;2(3):257–262.

61. Hiatt WR, Hoag S, Hamman RF. Effect of diagnostic criteria on the prevalence of peripheral arterial disease. The San Luis Valley Diabetes Study. Circulation 1995;91(5):1472–1479.

62. Hiatt WR, Regensteiner JG, Hargarten ME, Wolfel EE, Brass EP. Benefit of exercise conditioning for patients with peripheral arterial disease. Circulation 1990;81(2):602–609.

63. Hiatt WR, Hirsch AT, Regensteiner JG, Brass EP. Clinical trials for claudication. Assessment of exercise performance, functional status, and clinical end points. Vascular clinical trialists. Circulation 1995;92(3):614–621.

64. Hiatt WR. Medical treatment of peripheral arterial disease and claudication. New Engl J Med 2001;344:1608–1621.

65. Higgins D, Santamore WP, Walinsky P, Nemir P Jr. Hemodynamics of human arterial stenoses. Int J Cardiol 1985;8:177–192.

66. Hilleman DE. Management of peripheral arterial disease. Am J Health-Syst Ph 1998;55(19 Suppl. 1):S21–S27.

67. Hirsch AT, Haskal ZJ, Hertzer NR, Bakal CW, Creager MA, Halperin JL, Hiratzka LF, Murphy WRC, Olin JW, Puschett JB, Rosenfield KA, Sacks D, Stanley JC, Taylor LM Jr, White CJ, White J, White RA, Antman EM, Smith SC Jr, Adams CD, Anderson JL, Faxon DP, Fuster V, Gibbons RJ, Halperin JL, Hiratzka LF, Hunt SA, Jacobs AK, Nishimura R, Ornato JP, Page RL, Riegel B. 2005 ACC/AHA guidelines for the management of patients with peripheral arterial disease (lower extremity, renal, mesenteric, and abdominal aortic): executive summary. A collaborative report from the American Association for Vascular Surgery/Society for Vascular Surgery, Society for Cardiovascular Angiography and Interventions, Society for Vascular Medicine and Biology, Society of Interventional Radiology, and the ACC/AHA Task Force on Practice Guidelines (Writing Committee to Develop Guidelines for the Management of Patients With Peripheral Arterial Disease): endorsed by the American Association of Cardiovascular and Pulmonary Rehabilitation; National Heart, Lung, and Blood Institute; Society for Vascular Nursing; Trans-Atlantic Inter-Society Consensus; and Vascular Disease Foundation. Circulation 2006;113(11):e463–654.

68. Hodges LD, Sandercock GR, Das SK, Brodie DA. Cardiac pumping capability in patients with peripheral vascular disease. Clin Physiol Funct I 2006;26(3):185–190.

69. Holm J, Dahllof AG, Bjorntorp P, Schersten T. Enzyme studies in muscles of patients with intermittent claudication. Effect of training. Scand J Clin Lab Inv 1973;31(Suppl. 128):201–205.

70. Holm J, Dahllof AG, Schersten T. Metabolic activity of skeletal muscle in patients with peripheral arterial insufficiency. Effect of arterial reconstructive surgery. Scand J Clin Lab Inv 1975;35(1):81–86.

71. Johnson EC, Voyles WF, Atterbom HA, Pathak D, Sutton MF, Greene ER. Effects of exercise training on common femoral artery blood flow in patients with intermittent claudication. Circulation 1989;80(5 Pt 2):III59–III72.

72. Jonason T, Ringqvist I, Oman-Rydberg A. Home-training of patients with intermittent claudication. Scand J Rehabil Med 1981;13(4):137–141.

73. Jones PP, Skinner JS, Smith LK, John FM, Bryant CX. Functional improvements following StairMaster vs. treadmill exercise training for patients with intermittent claudication. J Cardiopulm Rehabil 1996;16(1):47–55.

74. Kang SS, Wong PW, Malinow MR. Hyperhomocysteinaemia as a risk factor for occlusive vascular disease. Annu Rev Nutr 1992;12:279–298.

75. Kannel WB. The demographics of claudication and the aging of the American population. Vasc Med 1996;1(1):60–64.

76. Kannel WB, D'Agostino RB, Belanger AJ. Update on fibrinogen as a cardiovascular risk factor. Ann Epidemiol 1992;2:457–466.

77. Kannel WB, McGee DL. Update on some epidemiologic features of intermittent claudication: the Framingham Study. J Am Geriatr Soc 1985;33(1):13–18.

78. Khaira HS, Hanger R, Shearman CP. Quality of life in patients with intermittent claudication. Eur J Vasc Endovasc 1996;11(1):65–69.

79. Kirkpatrick UJ, Mossa M, Blann AD, McCollum CN. Repeated exercise induces release of soluble P-selectin in patients with intermittent claudication. Thromb Haemostasis 1997;78(5):1338–1342.

80. Kokesh J, Kazmers A, Zierler RE. Pentoxifylline in the nonoperative management of intermittent claudication. Ann Vascular Surg 1991;5(1):66–70.

81. Labs KH, Dormandy JA, Jaeger KA, Stuerzebecher CS, Hiatt WR. Basel PAD Clinical Methodology Group. Transatlantic conference on clinical trial guidelines in peripheral arterial disease: clinical trial methodology. Circulation 1999;100:e75–e81.

82. Lampman RM. Exercise prescription for chronically ill patients. Am Fam Physician 1997;55(6):2185–2192.

83. Larsen OA, Lassen NA. Effect of daily muscular exercise in patients with intermittent claudication. Lancet 1966;2(7473):1093–1096.

84. Laursen JB, Rajagopalan S, Galis Z, Tarpey M, Freeman BA, Harrison DG. Role of superoxide in angiotensin II-induced but not catecholamine induced hypertension. Circulation 1997;95(3):588–593.

85. Leng GC, Fowler B, Ernst E. Exercise for intermittent claudication. Cochrane Db Syst Rev 2000;2:CD000990 (Medline).

86. Lewis DR, Day A, Jeremy JY, Newcombe PV, Brookes ST, Baird R, Smith FC, Lamont PM. Vascular surgical society of Great Britain and Ireland: systemic effects of exercise in claudicants are associated with neutrophil activation. Brit J Surg 1999;86(5):699–700.

87. Lundgren F, Dahllof AG, Lundholm K, Schersten T, Volkmann R. Intermittent claudication, surgical reconstruction or physical training? A prospective randomized trial of treatment efficiency. Ann Surg 1989;209(3):346–355.

88. MacGregor AS, Price JF, Hau CM, Lee AJ, Carson MN, Fowkes FG. Role of systolic blood pressure and plasma triglycerides in diabetic peripheral arterial disease. The Edinburgh Artery Study. Diabetes Care 1999;22(3):453–458.

89. Mannarino E, Pasqualini L, Innocente S, Scricciolo V, Rignanese A, Ciuffetti G. Physical training and antiplatelet treatment in stage II peripheral arterial occlusive disease: alone or combined? Angiology 1991;42(7):513–521.

90. Mannarino E, Pasqualini L, Menna M, Maragoni G, Orlandi U. Effects of physical training on peripheral vascular disease: a controlled study. Angiology 1989;40(1):5–10.

91. McDermott MM, Liu K, Ferrucci L, Criqui MH, Greenland P, Guralnik JM, Tian L, Schneider JR, Pearce WH, Tan J, Martin GJ. Physical performance in peripheral arterial disease: a slower rate of decline in patients who walk more. Ann Intern Med 2006;144(1):10–20.

92. Meijer WT, Hoes AW, Rutgers D, Bots ML, Hofman A, Grobbee DE. Peripheral arterial disease in the elderly: the Rotterdam study. Arterioscl Throm Vas 1998;18(2):185–192.

93. Money SR, Herd JA, Isaacsohn JL, Davidson M, Cutler B, Heckman J, Forbes WP. Effect of cilostazol on walking distances in patients with intermittent claudication caused by peripheral vascular disease. J Vasc Surg 1998;27(2):267–275.

94. Montgomery PS, Gardner AW. The clinical utility of a six-minute walk test in peripheral arterial occlusive disease patients. J Am Geriatr Soc 1998;46(6):706–711.

95. Morris JN, Hardman AE. Walking to health. Sports Med 1997;23(5):306–332. [Published erratum appears in Sports Med 1997;24(2):96.]

96. Murabito JM, D'Agostino RB, Silbershatz H, Wilson WF. Intermittent claudication. A risk profile from the Framingham Heart Study. Circulation 1997;96(1):44–49.

97. Murphy TP. Medical outcomes studies in peripheral vascular disease. J Vasc Interv Radiol 1998;9(6):879–889.

98. Murphy TP, Khwaja AA, Webb MS. Aortoiliac stent placement in patients treated for intermittent claudication. J Vasc Interv Radiol 1998;9(3):421–428.

99. Mustonen P, Lepantalo M, Lassila R. Physical exertion induces thrombin formation and fibrin degradation in patients with peripheral atherosclerosis. Arterioscl Throm Vas 1998;18(2):244–249.

100. Myhre K, Sorlie DG. Physical activity and peripheral atherosclerosis. Scand J Soc Med 1982;(Suppl.)29:195–201.

101. Navab M, Berliner JA, Watson AD, Hama SY, Territo MC, Lusis AJ, Shih DM, Van Lenten BJ, Frank JS, Demer LL, Edwards PA, Fogelman AM. The yin and yang of oxidation in the development of the fatty streak. A review based on the 1994 George Lyman Duff Memorial Lecture. Arterioscl Throm Vas 1996;16(7):831–842.

102. Nestares T, Lopez-Jurado M, Urbano G, Seiquer I, Ramirez-Tortosa MC, Ros E, Mataix J, Gil A. Effects of lifestyle modification and lipid intake variations on patients with peripheral vascular disease. Int J of Vitam Nutr Res 2003;Oct;73(5):389–398.

103. Newman AB, Simonsick EM, Naydeck BL, Boudreau RM, Kritchevsky SB, Nevitt MC, Pahor M, Satterfield S, Brach JS, Studenski SA, Harris TB. Association of long-distance corridor walk performance with mortality, cardiovascular disease mobility limitation, and disability. JAMA 2006;May 3;295(17):2018–2026.

104. Ohara Y, Peterson TE, Harrison DG. Hypercholesterolemia increases endothelial superoxide anion production. J Clin Invest 1993;91(6):2546–2551.

105. Ohara Y, Peterson TE, Sayegh HS, Subramanian RR, Wilcox JN, Harrison DG. Dietary correction of hypercholesterolemia in the rabbit normalizes endothelial superoxide anion production. Circulation 1995;92(4):898–903.

106. Ohta T, Sugimoto I, Takeuchi N, Hosaka M, Ishibashi H. Indications for and limitations of exercise training in patients with intermittent claudication. Vasa 2002;21:23–27.

107. Oka RK, Altman M, Giacomini JC, Szuba A, Cooke JP. Abnormal cardiovascular response to exercise in patients with peripheral arterial disease: implications for management. J Vasc Nurs 2005;23(4):130–136.

108. Ouriel K. Peripheral arterial disease. Lancet 2001;358:1257–1264.

109. Patterson RB, Pinto B, Marcus B, Colucci A, Braun T, Roberts M. Value of a supervised exercise program for the therapy of arterial claudication. J Vasc Surg 1997;25(2):312–319.

110. Pedersen TR, Kjekshus J, Pyorala K, Olsson AG, Cook TJ, Musliner TA, Tobert JA, Haghfelt T. Effect of simvastatin on ischemic signs and symptoms in the Scandinavian simvastatin survival study (4S). Am J Cardiol 1998;81(3):333–335.

111. Pernow B, Saltin B, Wahren R, Cronestrand R, Ekestroom S. Leg blood flow and muscle metabolism in occlusive arterial disease of the leg before and after reconstructive surgery. Clin Sci Mol Med 1975;49(3):265–275.

112. Poredos P, Zizek B. Plasma viscosity increase with progression of peripheral arterial atherosclerotic disease. Angiology 1996;47(3):253–259.

113. Price JF, Lee AJ, Fowkes FG. Hyperinsulinaemia: a risk factor for peripheral arterial disease in the non-diabetic general population. J Cardiovasc Risk 1996;3(6):501–505.

114. Priebe M, Davidoff G, Lampman RM. Exercise testing and training in patients with peripheral vascular disease and lower extremity amputation. Western J Med 1991;154:598–601.

115. Radack K, Wyderski RJ. Conservative management of intermittent claudication. Ann Intern Med 1990;113:135–146.

116. Reaven GM. Banting lecture: role of insulin resistance in human disease. Diabetes 1988;37:1595–1607.

117. Regensteiner JG, Steiner JF, Hiatt WR. Exercise training improves functional status in patients with peripheral arterial disease. J Vasc Surg 1996;23(1):104–115.

118. Regensteiner JG, Gardner A, Hiatt WR. Exercise testing and exercise rehabilitation for patients with peripheral arterial disease: status in 1997. Vasc Med 1997;2(2):147–155.

119. Regensteiner JG, Hiatt WR. Exercise rehabilitation for patients with peripheral arterial disease. Exerc Sport Sci Rev 1995;23:1–24.

120. Regensteiner JG, Meyer TJ, Krupski WC, Cranford LS, Hiatt WR. Hospital vs home-based exercise rehabilitation for patients with peripheral arterial occlusive disease. Angiology 1997;48(4):291–300.

121. Regensteiner JG, Hiatt WR. Current medical therapies for patients with peripheral arterial disease: a critical review. Am J Med 2002;112:49–57.

122. Ridker PM, Cushman M, Stampfer MJ, Tracy RP, Hennekens CH. Plasma concentration of C-reactive protein and risk of developing peripheral vascular disease. Circulation 1998;97(5):425–428.

123. Robeer GG, Brandsma JW, van den Heuvel SP, Smit B, Oostendorp RA, Wittens CH. Exercise therapy for intermittent claudication: a

review of the quality of randomized clinical trials and evaluation of predictive factors. Eur J Vasc Endovasc 1998;15(1):36–43.

124. Rosfors S, Bygdeman S, Arnetz BB, Lahnborg G, Skoldo L, Eneroth P, Kallner A. Long-term neuroendocrine and metabolic effects of physical training in intermittent claudication. Scand J Rehabil Med 1989;21(1):7–11.

125. Rosfors S, Arnetz BB, Bygdeman S, Skoldo L, Lahnborg G, Eneroth P. Important predictors of the outcome of physical training in patients with intermittent claudication. Scand J Rehabil Med 1990;21(3):135–137.

126. Ruell PA, Imperial ES, Bonar FJ, Thursby PF, Gass GC. Intermittent claudication. The effect of physical training on walking tolerance and venous lactate concentration. Eur J Appl Physiol O 1984;52(4):420–425.

127. Rutherford RB, Baker JD, Ernst C, Johnston KW, Porter JM, Ahn S, Jones DN. Recommended standards for reports dealing with lower extremity ischemia: revised version. J Vasc Surg 1997;26(3):517–538.

128. Sakurai T, Masushita M, Nishikimi N, Nimura Y. Effect of walking distance on the change in ankle-brachial pressure index in patients with intermittent claudication. Eur J Vasc Endovasc 1997;13(5):486–490.

129. Salmasi AM, Nicolaides A, Al-Katoubi A, Sonecha TN, Taylor PR, Serenkuma S, Eastcott HH. Intermittent claudication as a manifestation of silent myocardial ischemia: a pilot study. J Vasc Surg 1991;14(1):76–86.

130. Saltin B. Physical training in patients with intermittent claudication. In: Cohen LS, Mock MB, Ringqvist I, editors. Physical conditioning and cardiovascular rehabilitation. New York: Wiley; 1981. p 181–196.

131. Sanne H, Sivertsson R. The effect of exercise on the development of collateral circulation after experimental occlusion of the femoral artery in the cat. Acta Physiol Scand 1968;73(3):257–263.

132. Schoop W. Mechanism of beneficial action of daily walking training of patients with intermittent claudication. Scand J Clin Lab Inv 1973;(Suppl. 128)31:197–199.

133. Skinner JS, Strandness DE. Exercise and intermittent claudication II: effect of physical training. Circulation 1967;36:23–29.

134. Sorlie D, Myhre K. Effects of physical training in intermittent claudication. Scand J Clin Lab Inv 1978;38(3):217–222.

135. Stewart AL, Greenfield S, Hays RD, Rogers WH, Berry SD, McGlynn EA, Ware JE Jr. Functional status and well-being of patients with chronic conditions. Results from the Medical Outcomes Study. JAMA 1989;262(7):907–913. [Published erratum appears in JAMA 1989;262(18):2542.]

136. Stewart KJ, Hiatt WR, Regensteiner JG, Hirsch AT. Exercise training for claudication. New Engl J Med 2002;347:1941–1951.

137. Tan KH, de Cossart L, Edwards PR. Exercise training and peripheral vascular disease. Brit J Surg 2000;87(5):553–562.

138. Tesfamariam B. Free radicals in diabetic endothelial cell dysfunction. Free Radical Bio Med 1994;16(3):383–391.

139. Tsang GM, Sanghera K, Gosling P, Smith FC, Paterson IS, Simms MH, Shearman CP. Pharmacological reduction of the systemically damaging effects of local ischaemia. Eur J Vascular Surg 1994;8(2):205–208.

140. Turton EP, Spark JI, Mercer KG, Berridge DC, Kent PJ, Kester RC, Scott DJ. Exercise-induced neutrophil activation in claudicants: a physiological or pathological response to exhaustive exercise. Eur J Vascular Surg 1998;16(3):192–196.

141. Ubbink DT, Spincemaille GH, Reneman RS, Jacobs MJ. Prediction of imminent amputation in patients with non-reconstructible leg ischemia by means of microcirculatory investigations. J Vasc Surg 1999;30(1):114–121.

142. Verhaeghe R. Epidemiology and prognosis of peripheral obliterative arteriopathy. Drugs 1998;56(Suppl. 3):1–10.

143. Whyman MR, Fowkes FG, Kerracher EM, Gillespie IN, Lee AJ, Housley E, Ruckley CV. Is intermittent claudication improved by percutaneous transluminal angioplasty? A randomized controlled trial. J Vasc Surg 1997;26:551–557.

144. Williams LR, Ekers MA, Collins PS, Lee JF. Vascular rehabilitation: benefits of a structured exercise/risk modification program. J Vasc Surg 1991;14(3):320–326.

145. Womack CJ, Sieminski DJ, Katzel LI, Yataco A, Gardner AW. Improved walking economy in patients with peripheral arterial occlusive disease. Med Sci Sport Exer 1997;29(10):1286–1290.

146. Woodburn KR, Rumley A, Murtagh A, Lowe GD. Acute exercise and markers of endothelial injury in peripheral arterial disease. Eur J Vasc Endovasc 1997;14(2):140–142.

147. Wyatt MG, Scott PM, Scott DJ, Poskitt K, Baird RN, Horrocks M. Effect of weight on claudication distance. Brit J Surg 1991;78(11):1386–1388.

148. Young DF, Cholvin NR, Kirkeeide RL, Roth AC. Hemodynamics of arterial stenoses at elevated flow rates. Circ Res 1977;41(1):99–107.

149. Zetterquist S. The effect of active training on the nutritive blood flow in exercising ischemic legs. Scand J Clin Lab Inv 1970;25:101–111.

CHAPTER 20

1. Alexander T, Friedman DB, Levine BD, Pawelczyk JA, Mitchell JH. Cardiovascular responses during static exercise. Studies in patients with complete heart block and dual chamber pacemakers. Circulation 1994;89:1643–1647.

2. Alt E, Combs W, Willhaus R, Condie C, Bambl E, Fotuhi P, Pache J, Schömig A. A comparative study of activity and dual sensor: activity and minute ventilation pacing responses to ascending and descending stairs. Pacing Clin Electrophysiol 1998;21:1862–1868.

3. Alt E, Matula M. Comparison of two activity-controlled rate-adaptive pacing principles: acceleration versus vibration. Cardiol Clin 1992;10:635–658.

4. Bacharach DW, Hilden TS, Millerhagen JO, Westrum BL, Kelly JM. Activity-based pacing: comparison of a device using an accelerometer versus a piezoelectric crystal. Pacing Clin Electrophysiol 1992;15:188–196.

5. Barold SS, Barold HS. Optimal cardiac pacing in patients with coronary artery disease. Pacing Clin Electrophysiol 1998;21:456–461.

6. Bodenhamer RM, Grantham RN. Mode selection: the therapeutic challenge: adaptive-rate pacing. St. Paul: Cardiac Pacemakers; 1993. p 19–52.

7. Carmouche DG, Bubien RS, Kay GN. The effect of maximum heart rate on oxygen kinetics and exercise performance at low and high workloads. Pacing Clin Electrophysiol 1998;21:679–686.

8. Erdelitsch-Reiser E, Langenfeld H, Millerhagen J, Kochsiek K. New concept in activity-controlled pacemakers: clinical results with an accelerometer-based rate adaptive pacing system. Pacing Clin Electrophysiol 1992;15:2245–2249.

9. Greco EM, Guardini S, Citelli L. Cardiac rehabilitation in patients with rate responsive pacemakers. Pacing Clin Electrophysiol 1998;21:568–575.

10. Gregoratos G, Abrams J, Epstein AE, Freedman RA, Hayes DL, Hlatky MA, Kerber RE, Naccarelli GV, Schoenfeld MH, Silka MJ, Winters SL, Gibbons RI, Antman EM, Alpert JS, Hiratzka LF, Faxon DP, Jacobs AK, Fuster V, Smith SC Jr. ACC/AHA/NASPE 2002 guideline update for implantation of cardiac pacemakers and antiarrhythmia devices: summary article. A report of the American College of Cardiology/American Heart Association Task Force on Practice Guidelines (ACC/AHA/NASPE Committee to Update the 1998 Pacemaker Guidelines). J Cardiovasc Electrophysiol 2002;13(11):1183–99. J Am Coll Cardiol 1998;31:1175–1209.

11. Harper GR, Pina IL, Kutalek SP. Intrinsic conduction maximizes cardiopulmonary performance in patients with dual chamber pacemakers. Pacing Clin Electrophysiol 1991;14:1787–1791.

12. Hayes DL, Von Feldt L, Higano ST. Standardized informal exercise testing for programming rate adaptive pacemakers. Pacing Clin Electrophysiol 1991;14:1772–1776.

13. Holmes DR. Hemodynamics of cardiac pacing. In: Furman S, Hayes DL, Holmes DR, editors. A practice of cardiac pacing. 2d ed. Mount Kisco, NY: Futura; 1989. p 167–191.

14. Joglar JA, Hamdan MH, Welch PJ, Page RL. Interaction of a commercial heart rate monitor with implanted pacemakers. Am J Cardiol 1999;83:790–792,A10.

15. Kruse I, Arnman K, Conradson TB, Rydén L. A comparison of the acute and long-term hemodynamic effects of ventricular inhibited and atrial synchronous ventricular inhibited pacing. Circulation 1982;65:846–855.

16. Leung SK, Lau CP, Tang MO, Leung Z, Yakimow K. An integrated dual sensor system automatically optimized by target rate histogram. Pacing Clin Electrophysiol 1998;21:1559–1566.

17. Medtronic: therapies for medical conditions. Minneapolis: Medtronic; 1999.

18. Nazarian S, Roguin A, Zviman MM, Lardo AC, Dickfeld TL, Calkins H, Weiss RG, Berger RD, Bluemke DA, Halperin HR. Clinical utility and safety of a protocol for noncardiac and cardiac magnetic resonance imaging of patients with permanent pacemakers and implantable-cardioverter defibrillators at 1.5 tesla. Circulation 2006;114(12):1277–1284.

19. Philippon F. Cardiac resynchronization therapy: device-based medicine for heart failure. J Card Surg 2004;19(3):270–274.

20. Seidl K, Rameken M, Vater M, Senges J. Cardiac resynchronization therapy in patients with chronic heart failure: pathophysiology and current experience. Am J Cardiovasc Drugs 2002;2(4):219–226.

21. Sharp CT, Busse EF, Burgess JJ, Haennel RG. Exercise prescription for patients with pacemakers. J Cardiopulm Rehabil 1998;18:421–431.

22. Shukla HH, Flaker GC, Hellkamp AS, James EA, Lee KL, Goldman L, Orav EJ, Lamas GA. Clinical and quality of life comparison of accelerometer, piezoelectric crystal, and blended sensors in DDDR-paced patients with sinus node dysfunction in the mode selection trial (MOST). Pacing Clin Electrophysiol 2005;28(8):762–770.

23. Sparks PB. Cardiac resynchronisation therapy. Heart Lung Circ 2004;13(Suppl 3):S56–59.

24. Steendijk P, Tulner SA, Bax JJ, Oemrawsingh PV, Bleeker GB, van Erven L, Putter H, Verwey HF, van der Wall EE, Schalij MJ. Hemodynamic effects of long-term cardiac resynchronization therapy: analysis by pressure-volume loops. Circulation 2006;113(10):1295–1304.

25. Sulke N, Dritsas A, Chambers J, Sowton E. Is accurate rate response programming necessary? Pacing Clin Electrophysiol 1990;13:1031–1044.

26. Wilkoff B, Corey J, Blackburn G. A mathematical model of the chronotropic response to exercise. J Electrophysiol 1989;3:176–180.

27. Wilkoff BL, Miller RE. Exercise testing for chronotropic assessment. Cardiol Clin 1992;10:705–717.

28. Wood MA, Stambler BS, Ellenbogen KA. Patient management: optimal programming of adaptive-rate pacemakers: adaptive-rate pacing. St. Paul: Cardiac Pacemakers, 1993; p 86–110.

CHAPTER 21

1. Abraham AS, Cole RB, Bishop JM. Reversal of pulmonary hypertension by prolonged oxygen administration to patients with chronic bronchitis. Circ Res 1968;23:147–157.

2. Abraham AS, Cole RB, Green ID, Hedworth-Whitty RB, Clarke SW, Bishop JM. Factors contributing to the reversible pulmonary hypertension of patients with acute respiratory failure studies by serial observations during recovery. Circ Res 1969;24:51–60.

3. Abraham AS, Hedworth-Whitty RB, Bishop JM. Effects of acute hypoxia and hypervolaemia singly and together, upon the pulmonary circulation in patients with chronic bronchitis. Clin Sci 1967;33:371–380.

4. Adams L, Chronos N, Lane R, Guz A. The measurement of breathlessness induced in normal subjects: validity of two scaling techniques. Clin Sci 1985;69:7–16.

5. Agency for Health Care Policy and Research. Clinical classifications for health policy research: hospital inpatient statistics, 1995. HCUP-3 Research Note 1999;16–17.

6. Agusti AGN, Noguera A, Sauleda J, Sala E, Pons J, Busquets X. Systemic effects of chronic obstructive pulmonary disease. Eur Respir J 2003;21:347–360.

7. Ambrosino N, Paggiaro PL, Macchi M, Filieri M, Toma G, Lombardi FA, Del Cesta F, Parlanti A, Loi AM, Baschieri L. A study of short-term effect of rehabilitative therapy in chronic obstructive pulmonary disease. Respiration 1981;41:40–44.

8. American Association of Cardiovascular and Pulmonary Rehabilitation. Guidelines for cardiac rehabilitation and secondary prevention programs. 4th ed. Champaign, IL: Human Kinetics; 2004.

9. American Association of Cardiovascular and Pulmonary Rehabilitation. Guidelines for pulmonary rehabilitation programs. 2nd ed. Champaign, IL: Human Kinetics; 1998.

10. American College of Chest Physicians/American Association of Cardiovascular and Pulmonary Rehabilitation Pulmonary Rehabilitation Guidelines Panel. Pulmonary rehabilitation: joint ACCP/AACVPR evidence-based guidelines. Chest 1997;112:1363–1396.

11. American College of Sports Medicine. ACSM's exercise management for persons with chronic diseases and disabilities. 2nd ed. Champaign, IL: Human Kinetics; 2002.

12. American College of Sports Medicine. ACSM's guidelines for exercise testing and prescription. 7th ed. Baltimore: Williams & Wilkins; 2005.

13. American Thoracic Society and European Respiratory Society. Skeletal muscle dysfunction in chronic obstructive pulmonary disease. Am J Respir Crit Care Med 1999;159:S1–40.

14. American Thoracic Society. Evaluation of impairment/disability secondary to respiratory disorders. Am Rev Respir Dis 1986;133:1205–1209.

15. American Thoracic Society. Standards for the diagnosis and care of patients with chronic obstructive pulmonary disease. Am J Respir Crit Care Med 1995;152:S77–S152.

16. American Thoracic Society. Standards for the diagnosis and care of patients with chronic obstructive pulmonary disease (COPD) and asthma. Am Rev Respir Dis 1987;136:225–244.

17. Atkins CJ, Kaplan RM, Timms RM, Reinsch S, Lofback K. Behavioral exercise programs in the management of chronic obstructive pulmonary disease. J Consult Clin Psychol 1984;52:591–603.

18. Badgett RG, Tanaka DJ, Hunt DK, Jelley MJ, Feinberg LE, Steiner JF, Petty TL. Can moderate chronic obstructive pulmonary disease be diagnosed by historical and physical findings alone? Am J Med 1993;94:188–196.

19. Barnard KL, Adams KJ, Swank AM, Mann E, Denny DM. Injuries and muscle soreness during the one repetition maximum assessment in a cardiac rehabilitation population. J Cardiopulm Rehabil 1999;19:52–58.

20. Belman MJ. Exercise in chronic obstructive pulmonary disease. Clin Chest Med 1986;7:585–597.

21. Belman MJ, Shadmehr R. Targeted resistive ventilatory muscle training in chronic obstructive pulmonary disease. J Appl Physiol 1988;65:2726–2735.

22. Bernard S, LeBlanc P, Whittom F, Carrier G, Jobin J, Belleau R, Maltais F. Peripheral muscle weakness in patients with chronic obstructive pulmonary disease. Am J Respir Crit Care Med 1998;158:629–634.

23. Berry MJ, Adair NE, Sevensky KS, Quinby A, Lever HM. Inspiratory muscle training and whole-body reconditioning in chronic obstructive pulmonary disease. Am J Respir Crit Care Med 1996;153:1812–1816.

24. Berry MJ, Rejeski WJ, Adair NE, Zaccaro D. Exercise rehabilitation and chronic obstructive pulmonary disease stage. Am J Respir Crit Care Med 1999;160:1248–1253.

25. Bjerre-Jepsen K, Secher NH, Kok-Jensen A. Inspiratory resistance training in severe chronic obstructive pulmonary disease. Eur J Respir Dis 1981;62:405–411.

26. Booker HA. Exercise training and breathing control in patients with chronic airflow limitation. Physiotherapy 1984;70:258–260.

27. Borg GA. Psychophysical bases of perceived exertion. Med Sci Sports Exerc 1982;14:377–381.

28. Bowen JB, Votto JJ, Thrall RS, Haggerty MC, Stockdale-Woolley R, Bandyopadhyay T, ZuWallack RL. Functional status and survival following pulmonary rehabilitation. Chest 2000;118:697–703.

29. Bradley BL, Garner AE, Billiu D, Mestas JM, Forman J. Oxygen-assisted exercise in chronic obstructive lung disease. The effect on exercise capacity and arterial blood gas tensions. Am Rev Respir Dis 1978;118:239–243.

30. Braun SR, Keim NL, Dixon RM, Clagnaz P, Anderegg A, Shrago ES. The prevalence and determinants of nutritional changes in chronic obstructive pulmonary disease. Chest 1984;86:558–563.

31. Burrows B, Bloom JW, Traver GA, Cline MG. The course and prognosis of different forms of chronic airways obstruction in a sample from the general population. N Engl J Med 1987;317:1309–1314.

32. Busch AJ, McClements JD. Effects of a supervised home exercise program on patients with severe chronic obstructive pulmonary disease. Phys Ther 1988;68:469–474.

33. Calverley PM. Modern treatment of chronic obstructive pulmonary disease. Eur Respir J Suppl 2001;34:60S–66S.

34. Cambach W, Chadwick-Straver RV, Wagenaar RC, van Keimpema AR, Kemper HC. The effects of a community-based pulmonary rehabilitation programme on exercise tolerance and quality of life: a randomized controlled trial. Eur Respir J 1997;10:104–113.

35. Camilli AE, Burrows B, Knudson RJ, Lyle SK, Lebowitz MD. Longitudinal changes in forced expiratory volume in one second in adults. Effects of smoking and smoking cessation. Am Rev Respir Dis 1987;135:794–799.

36. Casaburi R. Skeletal muscle function in COPD. Chest 2000;117:267S–271S.

37. Casaburi R, Patessio A, Ioli F, Zanaboni S, Donner CF, Wasserman K. Reductions in exercise lactic acidosis and ventilation as a result of exercise training in patients with obstructive lung disease. [See comment]. Am Rev Respir Dis 1991;143:9–18.

38. Celli BR, Rassulo J, Make BJ. Dyssynchronous breathing during arm but not leg exercise in patients with chronic airflow obstruction. N Engl J Med 1986;314:1485–1490.

39. Chen H, Dukes R, Martin BJ. Inspiratory muscle training in patients with chronic obstructive pulmonary disease. Am Rev Respir Dis 1985;131:251–255.

40. Clark CJ, Cochrane LM, Mackay E, Paton B. Skeletal muscle strength and endurance in patients with mild COPD and the effects of weight training. Eur Respir J 2000;15:92–97.

41. Collins EG, Langbein WE, Fehr L, Maloney C. Breathing pattern retraining and exercise in persons with chronic obstructive pulmonary disease. AACN Clin Issues 2001;12:202–209.

42. Cotes JE, Gilson JC. Effect of oxygen on exercise ability in chronic respiratory insufficiency. Lancet 1956;1:872–876.

43. Couser JI Jr, Martinez FJ, Celli BR. Pulmonary rehabilitation that includes arm exercise reduces metabolic and ventilatory requirements for simple arm elevation. Chest 1993;103:37–41.

44. Danon J, Druz WS, Goldberg NB, Sharp JT. Function of the isolated paced diaphragm and the cervical accessory muscles in C1 quadriplegics. Am Rev Respir Dis 1979;119:909–919.

45. Dean NC, Brown JK, Himelman RB, Doherty JJ, Gold WM, Stulbarg MS. Oxygen may improve dyspnea and endurance in patients with chronic obstructive pulmonary disease and only mild hypoxemia. Am Rev Respir Dis 1992;146:941–945.

46. Debigare R, Maltais F, Mallet M, Casaburi R, LeBlanc P. Influence of work rate incremental rate on the exercise responses in patients with COPD. Med Sci Sports Exerc 2000;32:1365–1368.

47. Decramer M, de Bock V, Dom R. Functional and histologic picture of steroid-induced myopathy in chronic obstructive pulmonary disease. Am J Respir Crit Care Med 1996;153:1958–1964.

48. Decramer M, Lacquet LM, Fagard R, Rogiers P. Corticosteroids contribute to muscle weakness in chronic airflow obstruction. Am J Respir Crit Care Med 1994;150:11–16.

49. Degre S, Sergysels R, Messin R, Vandermoten P, Salhadin P, Denolin H, De Coster A. Hemodynamic responses to physical training in patients with chronic lung disease. Am Rev Respir Dis 1974;110:395–402.

50. Dekhuijzen PN, Folgering HT, van Herwaarden CL. Target-flow inspiratory muscle training during pulmonary rehabilitation in patients with COPD. Chest 1991;99:128–133.

51. Dillard TA. Ventilatory limitation of exercise. Prediction in COPD. Chest 1987;92:195–196.

52. Engelen MP, Schols AM, Baken WC, Wesseling GJ, EF Wouters. Nutritional depletion in relation to respiratory and peripheral skeletal muscle function in out-patients with COPD. Eur Respir J 1994;7:1793–1797.

53. Engelen MP, Schols AM, Does JD, Wouters EF. Skeletal muscle weakness is associated with wasting of extremity fat-free mass but not with airflow obstruction in patients with chronic obstructive pulmonary disease. Am J Clin Nutr 2000;71:733–738.

54. Epstein SK, Celli BR. Cardiopulmonary exercise testing in patients with chronic obstructive pulmonary disease. Cleve Clin J Med 1993;60:119–128.

55. Evans WJ. Exercise training guidelines for the elderly. Med Sci Sports Exerc 1999;31:12–17.

56. Falk P, Eriksen AM, Kolliker K, Andersen JB. Relieving dyspnea with an inexpensive and simple method in patients with severe chronic airflow limitation. Eur J Respir Dis 1985;66:181–186.

57. Farkas GA, Roussos C. Adaptability of the hamster diaphragm to exercise and/or emphysema. J Appl Physiol 1982;53:1263–1272.

58. Fiaccadori E, Del Canale S, Coffrini E, Vitali P, Antonucci C, Cacciani G, Mazzola I, Guariglia A. Hypercapnic-hypoxemic chronic obstructive pulmonary disease (COPD): influence of severity of COPD on nutritional status. Am J Clin Nutr 1988;48:680–685.

59. Fishman, AP. Pulmonary rehabilitation research. Am J Respir Crit Care Med 1994;149:825–833.

60. Fletcher GF, Balady G, Froelicher VF, Hartley LH, Haskell WL, Pollock ML. Exercise standards. A statement for healthcare professionals from the American Heart Association. Circulation 1995;91:580–615.

61. Gibbons LW, Mitchell TL, Gonzalez V. The safety of exercise testing. Prim Care 1994;21:611–629.

62. Goldstein R, De Rosie J, Long S, Dolmage T, Avendano MA. Applicability of a threshold loading device for inspiratory muscle testing and training in patients with COPD. Chest 1989;96:564–571.

63. Goldstein RS, Gort EH, Stubbing D, Avendano MA, Guyatt GH. Randomised controlled trial of respiratory rehabilitation. Lancet 1994;344:1394–1397.

64. Gordon AM, Huxley AF, Julian FJ. The variation in isometric tension with sarcomere length in vertebrate muscle fibres. J Physiol (Lond) 1966;184:170–192.

65. Gosker HR, van Mameren H, van Dijk PJ, Engelen MP, van der Vusse GJ, Wouters EF, Schols AM. Skeletal muscle fibre-type shifting and metabolic profile in patients with chronic obstructive pulmonary disease. Eur Respir J 2002;19:617–625.

66. Gosselink R, Troosters T, Decramer M. Peripheral muscle weakness contributes to exercise limitation in COPD. Am J Respir Crit Care Med 1996;153:976–980.

67. Greenberg SB, Allen M, Wilson J, Atmar RL. Respiratory viral infections in adults with and without chronic obstructive pulmonary disease. Am J Respir Crit Care Med 2000;162:167–173.

68. Griffin SE, Robergs RA, Heyward VH. Blood pressure measurement during exercise: a review. Med Sci Sports Exerc 1997;29:149–159.

69. Guyatt G, Keller J, Singer J, Halcrow S, Newhouse M. Controlled trial of respiratory muscle training in chronic airflow limitation. Thorax 1992;47:598–602.

70. Hamilton AL, Killian KJ, Summers E, Jones NL. Muscle strength, symptom intensity, and exercise capacity in patients with cardio-respiratory disorders. Am J Respir Crit Care Med 1995;152:2021–2031.

71. Harver A, Mahler DA, Daubenspeck JA. Targeted inspiratory muscle training improves respiratory muscle function and reduces dyspnea in patients with chronic obstructive pulmonary disease. Ann Intern Med 1989;111:117–124.

72. Heaton RK, Grant I, McSweeny AJ, Adams KM, Petty TL. Psychologic effects of continuous and nocturnal oxygen therapy in hypoxemic chronic obstructive pulmonary disease. Arch Intern Med 1983;143:1941–1947.

73. Henke KG, Sharratt M, Pegelow D, Dempsey JA. Regulation of end-expiratory lung volume during exercise. J Appl Physiol 1988;64:135–146.

74. Hildebrand IL, Sylven C, Esbjornsson M, Hellstrom K, Jansson E. Does chronic hypoxaemia induce transformations of fibre types? Acta Physiol Scand 1991;141:435–439.

75. Holford N, Black P, Couch R, Kennedy J, Briant R. Theophylline target concentration in severe airways obstruction—10 or 20 mg/L? A randomised concentration-controlled trial. Clin Pharmacokinet 1993;25:495–505.

76. Hunter AM, Carey MA, Larsh HW. The nutritional status of patients with chronic obstructive pulmonary disease. Am Rev Respir Dis 1981;124:376–381.

77. Jakobsson P, Jorfeldt L, Brundin A. Skeletal muscle metabolites and fibre types in patients with advanced chronic obstructive pulmonary disease (COPD), with and without chronic respiratory failure. Eur Respir J 1990;3:192–196.

78. Jones DT, Thomson RJ, Sears MR. Physical exercise and resistive breathing training in severe chronic airways obstruction: are they effective? Eur J Respir Dis 1985;67:159–166.

79. Jones NL, Jones G, Edwards RH. Exercise tolerance in chronic airway obstruction. Am Rev Respir Dis 1971;103:477–491.

80. Kaelin ME, Swank AM, Adams KJ, Barnard KL, Berning JM, Green A. Cardiopulmonary responses, muscle soreness, and injury during the one repetition maximum assessment in pulmonary rehabilitation patients. J Cardiopulm Rehabil 1999;19:366–372.

81. Kelsen SG, Ference M, Kapoor S. Effects of prolonged undernutrition on structure and function of the diaphragm. J Appl Physiol 1985;58:1354–1359.

82. Knudson RJ, Slatin RC, Lebowitz MD, Burrows B. The maximal expiratory flow-volume curve. Normal standards, variability, and effects of age. Am Rev Respir Dis 1976;113:587–600.

83. Kongsgaard M, Backer V, Jorgensen K, Kjaer M, Beyer N. Heavy resistance training increases muscle size, strength and physical function in elderly male COPD patients: a pilot study. Respir Med 2004;98:1000–1007.

84. Lacasse Y, Guyatt GH, Goldstein RS. The components of a respiratory rehabilitation program: a systematic overview. Chest 1997;111:1077–1088.

85. Lacasse Y, Wong E, Guyatt GH, King D, Cook DJ, Goldstein RS. Meta-analysis of respiratory rehabilitation in chronic obstructive pulmonary disease. Lancet 1996;348:1115–1119.

86. Lake FR, Henderson K, Briffa T, Openshaw J, Musk AW. Upper-limb and lower-limb exercise training in patients with chronic airflow obstruction. Chest 1990;97:1077–1082.

87. Lange P, Groth S, Nyboe GJ, Mortensen J, Appleyard M, Jensen G, Schnohr P. Effects of smoking and changes in smoking habits on the decline of FEV_1. Eur Respir J 1989;2:811–816.

88. Larson JL, Covey MK, Wirtz SE, Berry JK, Alex CG, Langbein WE, Edwards L. Cycle ergometer and inspiratory muscle training in chronic obstructive pulmonary disease. Am J Respir Crit Care Med 1999;160:500–507.

89. Larson JL, Kim MJ, Sharp JT, Larson DA. Inspiratory muscle training with a pressure threshold breathing device in patients with chronic obstructive pulmonary disease. Am Rev Respir Dis 1988;138:689–696.

90. Leith DE, Bradley M. Ventilatory muscle strength and endurance training. J Appl Physiol 1976;41:508–516.

91. Levine BE, Bigelow DB, Hamstra RD, Beckwitt HJ, Mitchell RS, Nett LM, Stephen TA, Petty TL. The role of long-term continuous oxygen administration in patients with chronic airway obstruction with hypoxemia. Ann Intern Med 1967;66:639–650.

92. Lewis MI, Sieck GC, Fournier M, Belman MJ. Effect of nutritional deprivation on diaphragm contractility and muscle fiber size. J Appl Physiol 1986;60:596–603.

93. Lisboa C, Munoz V, Beroiza T, Leiva A, Cruz E. Inspiratory muscle training in chronic airflow limitation: comparison of two different training loads with a threshold device. Eur Respir J 1994;7:1266–1274.

94. Lisboa C, Villafranca C, Leiva A, Cruz E, Pertuze J, Borzone G. Inspiratory muscle training in chronic airflow limitation: effect on exercise performance. Eur Respir J 1997;10:537–542.

95. Mador MJ, Bozkanat E, Aggarwal A, Shaffer M, Kufel TJ. Endurance and strength training in patients with COPD. Chest 2004;125:2036–2045.

96. Magnussen H, Richter K, Taube C. Are chronic obstructive pulmonary disease (COPD) and asthma different diseases? Clin Exp Allergy 1998;28(Suppl 5):187–194.

97. Mahler DA, Rosiello RA, Harver A, Lentine T, McGovern JF, Daubenspeck JA. Comparison of clinical dyspnea ratings and psychophysical measurements of respiratory sensation in obstructive airway disease. Am Rev Respir Dis 1987;135:1229–1233.

98. Mahler DA, Weinberg DH, Wells CK, Feinstein AR. The measurement of dyspnea. Contents, interobserver agreement, and physiologic correlates of two new clinical indexes. Chest 1984;85:751–758.

99. Maltais F, LeBlanc P, Whittom F, Simard C, Marquis K, Belanger M, Breton MJ, Jobin J. Oxidative enzyme activities of the vastus lateralis muscle and the functional status in patients with COPD. Thorax 2000;55:848–853.

100. Maltais F, Simard AA, Simard C, Jobin J, Desgagnes P, LeBlanc P. Oxidative capacity of the skeletal muscle and lactic acid kinetics during exercise in normal subjects and in patients with COPD. Am J Respir Crit Care Med 1996;153:288–293.

101. Marciniuk DD, Gallagher CG. Clinical exercise testing in chronic airflow limitation. Med Clin North Am 1996;80:565–587.

102. Martinez FJ, Vogel PD, Dupont DN, Stanopoulos I, Gray A, Beamis JF. Supported arm exercise vs unsupported arm exercise in the rehabilitation of patients with severe chronic airflow obstruction. Chest 1993;103:1397–1402.

103. McCully KK, Faulkner JA. Length-tension relationship of mammalian diaphragm muscles. J Appl Physiol 1983;54:1681–1686.

104. McGavin CR, Gupta SP, Lloyd EL, McHardy GJ. Physical rehabilitation for the chronic bronchitic: results of a controlled trial of exercises in the home. Thorax 1977;32:307–311.

105. McKeon JL, Turner J, Kelly C, Dent A, Zimmerman PV. The effect of inspiratory resistive training on exercise capacity in optimally treated patients with severe chronic airflow limitation. Aust N Z J Med 1986;16:648–652.

106. Medical Research Council Working Party. Long term domiciliary oxygen therapy in chronic hypoxic cor pulmonale complicating chronic bronchitis and emphysema. Lancet 1981;1:681–686.

107. Miller WF, Taylor HF. Exercise training in the rehabilitation of patients with severe respiratory insufficiency due to pulmonary emphysema. South Med J 1962;55:1216–1221.

108. Murray RP, Anthonisen NR, Connett JE, Wise RA, Lindgren PG, Greene PG, Nides MA. Effects of multiple attempts to quit smoking and relapses to smoking on pulmonary function. J Clin Epidemiol 1998;51:1317–1326.

109. National Heart, Lung, and Blood Institute. Morbidity and mortality: 1998 chartbook on cardiovascular, lung and blood diseases. Bethesda, MD: National Institutes of Health; 1998. p 1–128.

110. Nickerson BG, Sarkisian C, Tremper K, Nickerson BG, Sarkisian C, Tremper K. Bias and precision of pulse oximeters and arterial oximeters. Chest 1988;93:515–517.

111. Nocturnal Oxygen Therapy Trial Group. Continuous or nocturnal oxygen therapy in hypoxemic chronic obstructive lung disease: a clinical trial. Ann Intern Med 1980;93:391–398.

112. Noseda A, Carpiaux JP, Vandeput W, Prigogine T, Schmerber J, Noseda A, Carpiaux JP, Vandeput W, Prigogine T, Schmerber J. Resistive inspiratory muscle training and exercise performance in COPD patients. A comparative study with conventional breathing retraining. Bull Eur Physiopathol Respir 1987;23:457–463.

113. Oelberg DA, Medoff BD, Markowitz DH, Pappagianopoulos PP, Ginns LC, Systrom DM, Oelberg DA, Medoff BD, Markowitz DH, Pappagianopoulos PP, Ginns LC, Systrom DM. Systemic oxygen extraction during incremental exercise in patients with severe chronic obstructive pulmonary disease. Eur J Appl Physiol 1998;78:201–207.

114. Orozco-Levi M. Structure and function of the respiratory muscles in patients with COPD: impairment or adaptation? Eur Respir J Suppl 2003;46:41S–51S.

115. Ortega F, Toral J, Cejudo P, Villagomez R, Sanchez H, Castillo J, Montemayor T. Comparison of effects of strength and endurance training in patients with chronic obstructive pulmonary disease. Am J Respir Crit Care Med 2002;166:669–674.

116. Panton LB, Golden J, Broeder CE, Browder KD, Cestaro-Seifer DJ, Seifer FD. The effects of resistance training on functional outcomes in patients with chronic obstructive pulmonary disease. Eur J Appl Physiol 2004;91:443–449.

117. Pardy RL, Rivington RN, Despas PJ, Macklem PT. Inspiratory muscle training compared with physiotherapy in patients with chronic airflow limitation. Am Rev Respir Dis 1981;123:421–425.

118. Pierce AK, Paez PN, Miller WF. Exercise training with the aid of a portable oxygen supply in patients with emphysema. Am Rev Respir Dis 1965;91:653–659.

119. Polkey MI, Kyroussis D, Hamnegard CH, Mills GH, Green M, Moxham J. Diaphragm strength in chronic obstructive pulmonary disease. Am J Respir Crit Care Med 1996;154:1310–1317.

120. Preusser BA, Winningham ML, Clanton TL. High- vs low-intensity inspiratory muscle interval training in patients with COPD. Chest 1994;106:110–117.

121. Reardon J, Awad E, Normandin E, Vale F, Clark B, ZuWallack RL. The effect of comprehensive outpatient pulmonary rehabilitation on dyspnea. Chest 1994;105:1046–1052.

122. Ries AL, Ellis B, Hawkins RW. Upper extremity exercise training in chronic obstructive pulmonary disease. Chest 1988;93:688–692.

123. Ries AL, Kaplan RM, Limberg TM, Prewitt LM. Effects of pulmonary rehabilitation on physiologic and psychosocial outcomes in patients with chronic obstructive pulmonary disease. Ann Intern Med 1995;122:823–832.

124. Rodrigo C, Rodrigo G. Treatment of acute asthma. Lack of therapeutic benefit and increase of the toxicity from aminophylline given in addition to high doses of salbutamol delivered by metered-dose inhaler with a spacer. Chest 1994;106:1071–1076.

125. Rollier H, Bisschop A, Gayan-Ramirez G, Gosselink R, Decramer M. Low load inspiratory muscle training increases diaphragmatic fiber dimensions in rats. Am J Respir Crit Care Med 1998;157:833–839.

126. Rooyackers JM, Dekhuijzen PN, Van Herwaarden CL, Folgering HT. Training with supplemental oxygen in patients with COPD and hypoxaemia at peak exercise. Eur Respir J 1997;10:1278–1284.

127. Schols AM, Soeters PB, Dingemans AM, Mostert R, Frantzen PJ, Wouters EF. Prevalence and characteristics of nutritional depletion in patients with stable COPD eligible for pulmonary rehabilitation. Am Rev Respir Dis 1993;147:1151–1156.

128. Schwaiblmair M, Beinert T, Seemann M, Behr J, Reiser M, Vogelmeier C. Relations between cardiopulmonary exercise testing and quantitative high-resolution computed tomography associated in patients with alpha-1-antitrypsin deficiency. Eur J Med Res 1998;3:527–532.

129. Serres I, Gautier V, Varray A, Prefaut C. Impaired skeletal muscle endurance related to physical inactivity and altered lung function in COPD patients. Chest 1998;113:900–905.

130. Shannon M. Predictors of major toxicity after theophylline overdose. Ann Intern Med 1993;119:1161–1167.

131. Sharratt MT, Henke KG, Aaron EA, Pegelow DF, Dempsey JA. Exercise-induced changes in functional residual capacity. Respir Physiol 1987;70:313–326.

132. Sherrill DL, Enright P, Cline M, Burrows B, Lebowitz MD. Rates of decline in lung function among subjects who restart cigarette smoking. Chest 1996;109:1001–1005.

133. Shuey CB Jr, Pierce AK, Johnson RL Jr. An evaluation of exercise tests in chronic obstructive lung disease. J Appl Physiol 1969,27.256–261.

134. Similowski T, Yan S, Gauthier AP, Macklem PT, Bellemare F. Contractile properties of the human diaphragm during chronic hyperinflation. N Engl J Med 1991;325:917–923.

135. Simpson K, Killian K, McCartney N, Stubbing DG, Jones NL. Randomised controlled trial of weightlifting exercise in patients with chronic airflow limitation. Thorax 1992;47:70–75.

136. Smith J, Bellemare F. Effect of lung volume on in vivo contraction characteristics of human diaphragm. J Appl Physiol 1987;62:1893–1900.

137. Snider GL. Chronic obstructive pulmonary disease: a definition and implications of structural determinants of airflow obstruction for epidemiology. Am Rev Respir Dis 1989;140:S3–S8.

138. Snider GL, Faling LJ, Rennard SI. Chronic bronchitis and emphysema. In: Murray JF, Nadel JA, editors. Textbook of respiratory medicine. Philadelphia: Saunders; 1994. p 1342.

139. Snider GL, Kleinerman J, Thurlbeck WM, Bengali ZK. The definition of emphysema. Report of a National Heart, Lung, and Blood Institute, Division of Lung Diseases workshop. Am Rev Respir Dis 1985;132:182–185.

140. Spruit MA, Gosselink R, Troosters T, De Paepe K, Decramer M. Resistance versus endurance training in patients with COPD and peripheral muscle weakness. Eur Respir J 2002;19:1072–1078.

141. Storer TW. Exercise in chronic pulmonary disease: resistance exercise prescription. Med Sci Sports Exerc 2001;33:S680–S692.

142. Strijbos JH, Postma DS, van Altena R, Gimeno F, Koeter GH. A comparison between an outpatient hospital-based pulmonary rehabilitation program and a home-care pulmonary rehabilitation program in patients with COPD. A follow-up of 18 months. Chest 1996;109:366–372.

143. Stubbing DG, Pengelly LD, Morse JL, Jones NL. Pulmonary mechanics during exercise in subjects with chronic airflow obstruction. J Appl Physiol 1980;49:511–515.

144. Sue DY. Exercise testing in the evaluation of impairment and disability. Clin Chest Med 1994;15:369–387.

145. Sue DY, Wasserman K, Moricca RB, Casaburi R. Metabolic acidosis during exercise in patients with chronic obstructive pulmonary disease. Use of the V-slope method for anaerobic threshold determination. Chest 1988;94:931–938.

146. Sullivan P, Bekir S, Jaffar Z, Page C, Jeffery P, Costello J. Anti-inflammatory effects of low-dose oral theophylline in atopic asthma. [Erratum appears in Lancet 1994;343:1512.] Lancet 1994;343:1006–1008.

147. Tager IB, Segal MR, Speizer FE, Weiss ST. The natural history of forced expiratory volumes. Effect of cigarette smoking and respiratory symptoms. Am Rev Respir Dis 1988;138:837–849.

148. Tangri S, Woolf CR. The breathing pattern in chronic obstructive lung disease during the performance of some common daily activities. Chest 1973;63:26–127.

149. Thomason MJ, Strachan DP. Which spirometric indices best predict subsequent death from chronic obstructive pulmonary disease? Thorax 2000;55:785–788.

150. Thurlbeck WM. Pathophysiology of chronic obstructive pulmonary disease. Clin Chest Med 1990;11:389–403.

151. Thurlbeck WM, Muller NL. Emphysema: definition, imaging, and quantification. Am J Roentgenol 1994;163:1017–1025.

152. Toshima MT, Kaplan RM, Ries AL. Experimental evaluation of rehabilitation in chronic obstructive pulmonary disease: short-term effects on exercise endurance and health status. Health Psychol 1990;9:237–252.

153. Wadell K, Henriksson-Larsen K, Lundgren R. Physical training with and without oxygen in patients with chronic obstructive pulmonary disease and exercise-induced hypoxaemia. J Rehabil Med 2001;33:200–205.

154. Wanke T, Formanek D, Lahrmann H, Brath H, Wild M, Wagner C, Zwick H. Effects of combined inspiratory muscle and cycle ergometer training on exercise performance in patients with COPD. Eur Respir J 1994;7:2205–2211.

155. Wasserman K, Hansen JE, Sue DY, Whipp BJ. Principles of exercise testing and interpretation. Philadelphia: Lea & Febiger; 1986. p 1–274.

156. Wedzicha JA, Bestall JC, Garrod R, Garnham R, Paul EA, Jones PW. Randomized controlled trial of pulmonary rehabilitation in severe chronic obstructive pulmonary disease patients, stratified with the MRC dyspnoea scale. Eur Respir J 1998;12:363–369.

157. Wehr KL, Johnson RL Jr. Maximal oxygen consumption in patients with lung disease. J Clin Invest 1976;58:880–890.

158. Weiner P, Azgad Y, Ganam R. Inspiratory muscle training combined with general exercise reconditioning in patients with COPD. Chest 1992;102:1351–1356.

159. Whittom F, Jobin J, Simard PM, Leblanc P, Simard C, Bernard S, Belleau R, Maltais F. Histochemical and morphological characteristics of the vastus lateralis muscle in patients with chronic obstructive pulmonary disease. Med Sci Sports Exerc 1998;30:1467–1474.

160. Wijkstra PJ, Ten Vergert EM, van Altena R, Otten V, Kraan J, Postma DS, Koeter GH. Long term benefits of rehabilitation at home on quality of life and exercise tolerance in patients with chronic obstructive pulmonary disease. Thorax 1995;50:824–828.

161. Wilson DO, Rogers RM, Openbrier D. Nutritional aspects of chronic obstructive pulmonary disease. Clin Chest Med 1986;7:643–656.

162. Wilson DO, Rogers RM, Sanders MH, Pennock BE, Reilly JJ. Nutritional intervention in malnourished patients with emphysema. Am Rev Respir Dis 1986;134:672–677.

163. Wilson DO, Rogers RM, Wright EC, Anthonisen NR. Body weight in chronic obstructive pulmonary disease. The National Institutes of Health Intermittent Positive-Pressure Breathing Trial. Am Rev Respir Dis 1989;139:1435–1438.

164. Wright PR, Heck H, Langenkamp H, Franz KH, Weber U. Influence of a resistance training on pulmonary function and performance measures of patients with COPD. Pneumologie 2002;56:413–417.

165. Wuyam B, Payen JF, Levy P, Bensaidane H, Reutenauer H, Le Bas JF, Benabid AL. Metabolism and aerobic capacity of skeletal muscle in chronic respiratory failure related to chronic obstructive pulmonary disease. Eur Respir J 1992;5:157–162.

166. Zack MB, Palange AV. Oxygen supplemented exercise of ventilatory and nonventilatory muscles in pulmonary rehabilitation. Chest 1985;88:669–675.

167. Ziment I. Pharmacologic therapy of obstructive airway disease. Clin Chest Med 1990;11:461–486.

CHAPTER 22

1. American College of Sports Medicine. ACSM's guidelines for exercise testing and prescription. 7th ed. Baltimore: Lippincott Williams & Wilkins; 2005.

2. American College of Sports Medicine. Position stand. The recommended quantity and quality of exercise for developing and maintaining cardiorespiratory and muscular fitness, and flexibility in healthy adults. Med Sci Sports Exerc 1998;30:975–991.

3. Anderson SD, Daviskas E. The mechanism of exercise-induced asthma is . . . . J Allergy Clin Immunol 2000;106:453–459.

4. Anderson SD, Holzer K. Exercise induced asthma: is it the right diagnosis in elite athletes? J Allergy Clin Immunol 2000;106:419–428.

5. Anderson SD, Silverman M, Walker SR. Metabolic and ventilatory changes in asthmatic patients during and after exercise. Thorax 1972;27:718–725.

6. Banzett RB, Dempsey JA, O'Donnell DE, Wambolt MZ. NHLBI workshop summary. Symptom perception and respiratory sensation in asthma. Am J Respir Crit Care Med 2000;162:1178–1182.

7. Barnes PJ, Brown MJ, Silverman M, et al. Circulating catecholamines in exercise and hyperventilation induced asthma. Thorax 1981;36:435–440.

8. Barnes PJ, Chung KF, Page CP. Inflammatory mediators of asthma: an update. Pharmacol Rev 1998;50:515–596.

9. Basaran S, Guler-Uysal F, Ergen N, Seydaoglu G, et al. Effect of physical exercise on quality of life, exercise capacity, and pulmonary function in children with asthma. J Rehabil Med 2006;38:130–135.

10. Borg G. Perceived exertion as an indicator of somatic stress. Scand J Rehabil Med 1970;2:92–98.

11. Boulet LP, Boulet V, Milot J. How should we quantify asthma control? A proposal. Chest 2002;122:2217–2223.

12. Busse WW. Inflammation in asthma: the cornerstone of the disease and target therapy. J Allergy Clin Immunol 1998;102:S17–22.

13. Busse WW, Lemanske RF Jr. Asthma. N Engl J Med 2001;344:350–362

14. Centers for Disease Control and Prevention. Forecasted state-specific estimates of self-reported asthma prevalence—United States. MMWR 1998;47:1022–1025.

15. Clark CJ, Cochrane LM. Physical activity and asthma. Curr Opin Pulmon Med 1999;5:68–75.

16. Clark CJ, Cochrane LM. Assessment of work performance in asthma for determination of cardiorespiratory fitness and training capacity. Thorax 1988;43:745–798.

17. Cochrane LM, Clark LJ. Benefits and problems of a physical training programme for asthmatic patients. Thorax 1990;45:345–351.

18. Cockroft DW. Bronchoprovocation methods: direct challenges. Clin Rev Allergy Immunol 2003;24:19–26.

19. Cypcar D, Lemanske DF. Asthma and exercise. Clin Chest Med 1994;15:351–368.

20. Ebina M, Yaegashi H, Chiba R, et al. Hyperreactive site in the airway tree of asthmatic patients recorded by thickening of bronchial muscles: a morphometric study. Am Rev Respir Dis 1990;141:1327–1332.

21. Edelman JM, Turpin JA, Bronsky EA, et al. Oral montelukast compared with inhaled salmeterol to prevent exercise-induced bronchoconstriction. Ann Intern Med 2000;132:97–104.

22. Emtner M, Finne M, Stalenheim G. A three year followup of asthmatic patients participating in a 10-week rehabilitation program with emphasis on physical training. Arch Phys Med Rehabil 1998;78:539–544.

23. Emtner M, Herela M, Stalenheim G. High intensity physical training in adults with asthma. Chest 1996;109:323–330.

24. Garfinkel SK, Kesten S, Chapman KR, et al. Physiologic and non-physiologic determinants of aerobic fitness in mild to moderate asthma. Am Rev Respir Dis 1992;145:741–745.

25. Girodo M, Ekstrand KA, Metivier GJ. Deep diaphragmatic breathing: rehabilitation exercise for the asthmatic patient. Arch Phys Med Rehabil 1992;73:717–720.

26. Global strategy for asthma management and prevention; 2006. www.ginaasthma.com

27. Helenius I, Haahtela T. Allergy and asthma in elite summer sport athletes. J Allergy Clin Immunol 2000;106:444–452.

28. Henriksen JM, Dahl R. Effect of inhaled budesonide alone and in combination with low-dose terbutaline in children with exercise induced asthma. Am Rev Respir Dis 1983;128:993–997.

29. Holgate ST. The cellular and mediator basis of asthma in relation to natural history. Lancet 1997;350(Suppl II):5–9.

30. Holgate ST, Davies DE, Lackie PM, Wilson SJ, Puddicombe SM, Lordan JL. Epithelial-mesenchymal interactions in the pathogenesis of asthma. J Allergy Clin Immunol 2000;105:193–204.

31. Homa DM, Mannino DM, Lara M. Asthma mortality in U.S. Hispanics of Mexican, Puerto Rican, and Cuban Heritage, 1990–1995. Am J Respir Crit Care Med 2000;161:504–509.

32. Horvath I, Barnes PJ. Exhaled monoxides in asymptomatic atopic subjects. Clin Exp Allergy 1999;29:1276–1280.

33. Juniper EF, Guyatt GH, Ferrie PJ, et al. Measuring quality of life in asthma. Am Rev Respir Dis 1993;147:832–838.

34. Kharitonov S, Alving K, Barnes PJ. Exhaled and nasal nitric oxide measurement: recommendations. The European Respiratory Society Task Force. Eur Respir J 1997;10:1683–1693.

35. Kowabori I, Pierson WE, Loveday LC, et al. Incidence of exercise induced asthma in children. J Allergy Clin Immunol 1996;58:447–455.

36. Laitinen A, Laitinen LA. Airway morphology: endothelium/basement membrane. Am J Respir Crit Care Med 1994;150:514–517.

37. Laitinen LA, Laitinen A, Haahtela T. Airway mucosal inflammation even in patients with newly diagnosed asthma. Am Rev Respir Dis 1993;147:697–704.

38. Lang DM, Polansky M. Patterns of asthma mortality in Philadelphia 1969 to 1991. N Engl J Med 1994;331:1542–1546.

39. Lee TM, Brown MJ, Nagy L, et al. Exercise induced release of histamine and neutrophil chemotactic factor in atopic asthmatics. J Allergy Clin Immunol 1982;70:73–81.

40. Leff JA, Busse WW, Pearlman D, et al. Montelukast, a leukotriene-receptor antagonist, for the treatment of mild asthma and exercise induced bronchospasm. N Engl J Med 1998;339:147–152.

41. Levy ML, Fletcher M, Price PB, Hausen T, Halbert RJ, Yawn BP. International Primary Care Respiratory Group guidelines: diagnosis of respiratory diseases in primary care. Prim Care Respir J 2006;15:20–34.

42. Ludwick SK, Jones JW, Jones TK, et al. Normalization of cardio-pulmonary endurance in severely asthmatic children after bicycle ergometry therapy. J Pediatr 1986;109:446–451.

43. Mahler DA, Faryniarz K, Lentine T, et al. Measurement of breathlessness during exercise in asthmatics: predictor variables, reliability, and responsiveness. Am Rev Respir Dis 1991;144:9–44.

44. Mannino DM, Homa DM, Pertowski CA, et al. Surveillance for asthma—United States, 1960–1995. MMWR 1998;47:1–27.

45. McFadden ER Jr, Gilbert FA. Exercise induced asthma. N Engl J Med 1994;330:1362–1367.

46. McParland BE, Macklem PT, Pare PD. Airway wall remodeling: friend or foe? J Appl Physiol 2003;95:426–434

47. Meyer R, Froner-Herwig B, Sporkel H. The effect of exercise and induced expectations on visceral perception in asthmatic patients. J Psychosom Res 1990;34:454–460.

48. Mosli M, Fabian D, Holt S, Beasley R. The global burden of asthma: executive summary of the GINA Dissemination Committee report. Allergy 2004;59:469–478.

49. National Heart, Lung, and Blood Institute. National Asthma Education Program. Expert Panel Report 3: guidelines for the diagnosis and management of asthma. Full report 2007. (NIH Publications 2007:08-4051). Bethesda, MD: U.S. Department of Health and Human Services; July 2007. www.nhlbi.nih.gov/guidelines/asthma/asthgdln.pdf

50. Nelson JA, Strauss L, Skowronski M, et al. Effect of long term salmeterol treatment on exercise induced asthma. N Engl J Med 1998;339:141–146.

51. Ober C. Perspectives on the past decade of asthma genetics. J Allergy Clin Immunol 2005;116:274–278.

52. Orenstein DM, Reed ME, Grogan FT, et al. Exercise conditioning in children with asthma. J Pediatr 1985;106:556–560.

53. Peel ET, Soutar CA, Seaton A. Assessment of variability of exercise tolerance limited by breathlessness. Thorax 1988;43:960–964.

54. Pelligrino R, Viegi G, Brusasco B, Crapo RO, Burgos F, Casaburi R, et al. Interpretative strategies for lung function tests. Eur Respir J 2005;26:948–968

55. Pizzichini MM, Popov TA, Efthimiadis A, Hussack P, Evans S, Pizzichini E, et al. Spontaneous and induced sputum to measure indices of airway inflammation in asthma. Am J Respir Crit Care Med 1996;154(4Pt1):866–869.

56. Ram FS, Robinson SM, Black PN, Picol J. Physical training for asthma. Cochrane Db Syst Rev 2005 Oct 19;(4):CD001116.

57. Robertson CF, Rubinfeld AR, Bowes G. Pediatric asthma deaths in Victoria: the mild at risk. Pediatr Pulmonol 1992;13:95–100.

58. Roche WR, Beasley R, Williams JH, Holgate ST. Subepithelial fibrosis in the bronchi of asthmatics. Lancet 1989;8637:520–524.

59. Rubenfeld A, Pain M. Perception of asthma. Lancet 1976;1(7965):882–884.

60. Senthilselvan A. Prevalence of physician diagnosed asthma in Saskatchewan, 1981–1990. Chest 1999;114:388–392.

61. Smith AD, Cowan JO, Brassett KP, Herbison GP, Taylor DR. Use of exhaled nitric oxide measurements to guide treatment in chronic asthma. N Engl J Med 2005;352:2163–2173.

62. Strunk RC, Rubin D, Kelly L, et al. Determination of fitness in children with asthma: use of standardized tests for functional endurance, body fat composition, flexibility, and abdominal strength. Am J Dis Child 1988;142:940–944.

63. Tsanakas JN, Milner RD, Bannister OM, et al. Free running asthma screening test. Arch Dis Child 1988;63:261–265.

64. Wang L, McParland BE, Pare PD. The functional consequences of structural changes in the airways: implications for airway hyper-responsiveness in asthma. Chest 2003;123 (3 Suppl):356S–652S.

65. Warren JB, Keynes RJ, Brown MJ, et al. Blunted sympathoadrenal response to exercise in asthmatic subjects. Br J Dis Chest 1982;76:147–150.

66. Weiner P, Axgad R, Ganam R, Weiner R. Inspiratory muscle training in patients with bronchial asthma. Chest 1992;102:1357–1361.

67. Zeiger FS, Dawson C, Weiss S. Relationships between duration of asthma and asthma severity among children in the childhood asthma management program (CAMP). J Allergy Clin Immunol 1999;103:376–387.

CHAPTER 23

1. Alison JA, Donnelly PM, Lennon M, Parker S, Torzillo P, Mellis C, Bye PTP. The effect of a comprehensive, intensive inpatient treatment program on lung function and exercise capacity in patients with cystic fibrosis. Phys Ther 1994;74:583–593.

2. Alison J, Duong B, Robinson M, Regnis J, Donnelly P, Bye PTP. Level of aerobic fitness and survival in adults with cystic fibrosis. Am J Respir Crit Care Med 1997;155(Part 2):A642.

3. American Academy of Pediatrics: Committee on Sports Medicine Policy Statement. Strength training, weight and power lifting, and body building by children and adolescents. Pediatrics 1990;86:801–803.

4. Asher MI, Pardy RL, Coates AL, Thomas E, Macklem PT. The effects of inspiratory muscle training in patients with CF. Am Rev Respir Dis 1982;126:855–859.

5. Babb TG. Mechanical ventilatory constraints in aging, lung disease, and obesity: perspectives and brief review. Med Sci Sports Exerc 1999;31(Suppl 1):S12–S22.

6. Bakker W. Nutritional state and lung disease in cystic fibrosis. Neth J Med 1992;41:130–136.

7. Baldwin DR, Hill AL, Peckham DG, Knox AJ. Effect of addition of exercise to chest physiotherapy on sputum expectoration and lung function in adults with cystic fibrosis. Respir Med 1994;88:49–53.

8. Balfour-Lynn IM, Carr SB, Madge S, Prasad A, MacAlister L, Laverty A, Dinwiddie R. Effect of altitude on exercise testing in children with cystic fibrosis. Proceedings of the 20th European Working Group on Cystic Fibrosis. Brussels; 1995.

9. Balfour-Lynn IM, Prasad SA, Laverty A, Whitehead BF, Dinwiddie R. A step in the right direction: assessing exercise tolerance in children with cystic fibrosis. Pediatr Pulmonol 1998;25:278–284.

10. Bar-Or O, Blimkie CJ, Hay JA, MacDougall JD, Ward DS, Wilson WM. Voluntary dehydration and heat intolerance in cystic fibrosis. Lancet 1992;339:696–699.

11. Boas SR, Danduran MJ, McBride AL, McColley SA, O'Gorman MRG. Post exercise immune correlates in children with and without cystic fibrosis. Med Sci Sports Exerc 2000;32:1997–2004.

12. Boas SR, Danduran MJ, McColley SA, Beeman K, O'Gorman MRG. Immune modulation following aerobic exercise in children with cystic fibrosis. Int J Sports Med 2000;21:294–301.

13. Boas SR, Danduran MJ, McColley SA. Energy metabolism during anaerobic exercise in children with cystic fibrosis and asthma. Med Sci Sports Exerc 1999;31:1242–1249.

14. Boas SR, Danduran MJ, McColley SA. Parental attitudes about exercise in cystic fibrosis. Int J Sports Med 1999;20:334–338.

15. Boas SR, Joswiak ML, Nixon PA, Fulton JA, Orenstein DM. Factors limiting anaerobic performance in adolescent males with cystic fibrosis. Med Sci Sports Exerc 1996;28:291–298.

16. Bradley J, Howard J, Wallace E, Elborn S. Validity of a modified shuttle test in adult cystic fibrosis. Thorax 1999;54:437–439.

17. Bradley J, Howard J, Wallace E, Elborn S. Reliability, repeatability, and sensitivity of the modified shuttle test in adult cystic fibrosis. Chest 2000;117:1666–1671.

18. Cabrera ME, Lough MD, Doershuk CF, DeRivera GA. Anaerobic performance—assessed by the Wingate test—in patients with cystic fibrosis. Pediatr Exerc Sci 1993;5:78–87.

19. Canny GJ. Ventilatory response to exercise in cystic fibrosis. Acta Paediatr Scand 1985;74:451–452.

20. Cerny FJ, Cropp GJA, Bye MR. Hospital therapy improves exercise tolerance and lung function in cystic fibrosis. Am J Dis Child 1984;138:261–265.

21. Cerny FJ, Pullano TP, Cropp GJA. Cardiorespiratory adaptations to exercise in cystic fibrosis. Am Rev Respir Dis 1982;126:217–220.

22. Charge TD, Drury D, Pianosi P, Kopelman H, Coates AL. Nutritional rehabilitation and changes in respiratory strength, function, and maximal exercise capacity in cystic fibrosis. Am Rev Respir Dis 1991;143(Suppl):A300.

23. Coates AL. Oxygen therapy, exercise, and cystic fibrosis. Chest 1992;101:2–4.

24. Cropp GJ, Pullano TP, Cerny FJ, Nathanson IT. Exercise tolerance and cardiorespiratory adjustments at peak work capacity in cystic fibrosis. Am Rev Respir Dis 1982;126:211–216.

25. Cystic Fibrosis Foundation. Patient registry, 2004. Annual data report. Bethesda, MD; October 2005.

26. Danduran MJ, Boas SR. Fibrosi cistica ed esercizio fisico. Il Fisioterapista 2000;6:36–41.

27. Darbee J, Watkins M. Isokinetic evaluation of muscle performance in individuals with cystic fibrosis. Pediatr Pulmonol 1987;3(Suppl):140–141.

28. deMeer K, Jeneson JAL, Gulmans VAM, van der Laag J, Berger R. Efficiency of oxidative work performance of skeletal muscle in patients with cystic fibrosis. Thorax 1995;50:980–983.

29. deMeer K, Gulmans VAM, van der Laag J. Peripheral muscle weakness and exercise capacity in children with cystic fibrosis. Am J Respir Crit Care Med 1999;159:748–754.

30. Docherty D. Field tests and test batteries. In: Docherty D, editor. Measurement in pediatric exercise science. Champaign, IL: Human Kinetics; 1996. p 285–327.

31. Dunlevy CL, Douce FH, Hill E, Baez S, Clutter J. Physiological and psychological effects of low-impact aerobic exercise on young adults with cystic fibrosis. J Cardiopulm Rehabil 1994;14:47–51.

32. Edlund LD, French RW, Herbst JJ, Ruttenberg HD, Ruhling RO, Adams TD. Effects of a swimming program on children with cystic fibrosis. Am J Dis Child 1986;140:80–83.

33. Enright S, Chatham K, Ionesco AA, Unnithan VB, Shale DJ. Inspiratory muscle training improves lung function and exercise capacity in adults with cystic fibrosis. Chest 2004;126:405–411.

34. Freeman W, Stableforth DE, Cayton RM, Morgan MDL. Endurance exercise capacity in adults with cystic fibrosis. Respir Med 1993;87:541–549.

35. Godfey S, Mearns M. Pulmonary function and response to exercise in cystic fibrosis. Arch Dis Child 1971;46:144–151.

36. Goldring RM, Fishman AP, Turino GM, Cohen HI, Denning CR, Andersen DH. Pulmonary hypertension and cor pulmonale in cystic fibrosis of the pancreas. J Pediatr 1964;65:501–524.

37. Griffiths DM, Miller L, Flack E, Connett GJ. Reduced grip strength in children with cystic fibrosis. Neth J Med 1999;54:S62.

38. Guillen MAJ, Posadas AS, Asensi JRV, Moreno RMG, Rodriguez MAN, Gonzalez AS. Reproducibility of the walking test in patients with cystic fibrosis. An Esp Pediatr 1999;51:475–478.

39. Gulmans VAM, deMeer K, Brackel HJL, Faber JAJ, Berger R, Helders PJM. Outpatient exercise training in children with cystic fibrosis: physiological effects, perceived competence, and acceptability. Pediatr Pulmonol 1999;28:39–46.

40. Gulmans VAM, van Veldhoven NHMJ, de Meer K, Helders PJM. The six-minute walking test in children with cystic fibrosis: reliability and validity. Pediatr Pulmonol 1996;22:85–89.

41. Hebestreit H, Hebestreit A, Trusen A, Hughson RL. Oxygen uptake kinetics are slowed in cystic fibrosis. Arch Dis Child 2004;89:928–933.

42. Heijerman HGM, Bakker W, Sterk PJ, Dijkman JH. Oxygen-assisted exercise training in adult cystic fibrosis patients with pulmonary limitation to exercise. Int J Rehabil Res 1991;14:101–115.

43. Heijerman HGM. Chronic obstructive lung disease and respiratory muscle function: the role of nutrition and exercise training in cystic fibrosis. Respir Med 1993;B:49–51.

44. Henke KG, Orenstein DM. Oxygen saturation during exercise in cystic fibrosis. Am Rev Respir Dis 1984;129:708–711.

45. Hussy J, Gormley J, Leen G, Greally P. Peripheral muscle strength in young males with cystic fibrosis. J Cyst Fibros 2002;1(3):116–121.

46. Inbar O, Bar-Or O, Skinner JS. The Wingate anaerobic test. Champaign, IL: Human Kinetics; 1996.

47. Jenesma M, Concannon D, Gallagher CG. An evaluation of spinal posture and hamstring muscle flexibility in cystic fibrosis adults. Pediatr Pulmonol 1998;17:350.

48. Kaplan RM, Anderson JP, Wu AW, Mathews WC, Kozin F, Orenstein DM. The quality of well-being scale. Application in AIDS, cystic fibrosis, and arthritis. Med Care 1989;27(Suppl):27–43.

49. Kaplan TA, McKey RM, Toraya N, Moccia G. Impact of CF summer camp. Clin Pediatr 1992;31:161–167.

50. Kaplan TA, Moccia G, McKey RM. Unique pattern of pulmonary function after exercise in patients with cystic fibrosis. Pediatr Exerc Sci 1994;6:275–286.

51. Keens TG, Krastius IRB, Wannemaker EM, Levison H, Crozier DN, Bryan AC. Ventilatory muscle endurance training in normal subjects and patients with cystic fibrosis. Am Rev Respir Dis 1977;116:853–860.

52. Kriemler S, Wilk B, Schurer W, Wilson WM, Bar-Or O. Preventing dehydration in children with cystic fibrosis who exercise in the heat. Med Sci Sports Exerc 1999;31:774–779.

53. Lands LC, Heigenhauser GJ, Jones NL. Analysis of factors limiting maximal exercise performance in cystic fibrosis. Clin Sci 1992;83:391–397.

54. Lands LC, Heigenhauser GJF, Jones NL. Respiratory and peripheral muscle function in cystic fibrosis. Am Rev Respir Dis 1993;147:865–869.

55. Lebecque P, Lapierre JG, Lamarre A, Coates AL. Diffusion capacity and oxygen desaturation effects on exercise in patients with cystic fibrosis. Chest 1987;91:693–697.

56. Levison H, Cherniack RM. Ventilatory cost of exercise in chronic obstructive pulmonary disease. J Appl Physiol 1968;25:21–27.

57. Lohman T. Advances in body composition assessment. Champaign, IL: Human Kinetics; 1992.

58. Loutzenhiser JK, Clark R. Physical activity and exercise in children with cystic fibrosis. J Pediatr Nurs 1993;8:112–119.

59. Marcotte JE, Grisdale RK, Levison H, Coates AL, Canny GJ. Multiple factors limit exercise capacity in cystic fibrosis. Pediatr Pulmonol 1986;2:274–281.

60. Marcus CL, Bader D, Stabile M, Wang CI, Osher AB, Keens TG. Supplemental oxygen and exercise performance in patients with severe cystic fibrosis. Chest 1992;101:52–57.

61. McKone EF, Barry SC, FitzGerald MX. The role of arterial hypoxemia and pulmonary mechanics in exercise limitation in adults with cystic fibrosis. J Appl Physiol 2005;99(3):1012–1018.

62. McKone EF, Barry SC, FitzGerald MX, Gallagher CG. Reproducibility of maximal exercise ergometer testing in patients with cystic fibrosis. Chest 1999;116:363–368.

63. Moorcroft AJ, Dodd ME, Webb AK. Exercise testing and prognosis in adult cystic fibrosis. Thorax 1997;52:291–293.

64. Nick J, Rodman D. Manifestations of cystic fibrosis diagnosed in adulthood. Curr Op Pulm Med 2005;11(6):513–518.

65. Nieman DC. The immune response to prolonged cardiorespiratory exercise. Am J Sports Med 1996;24:S98–S103.

66. Nixon PA, Orenstein DM. Exercise testing in children. Pediatr Pulmonol 1988;5:107–122.

67. Nixon PA, Orenstein DM, Kelsey SF, Doershuk CF. The prognostic value of exercise testing in patients with cystic fibrosis. N Engl J Med 1992;327:1785–1788.

68. Orenstein DM, Franklin BA, Doershuk CF, Hellerstein HK, Germann KJ, Horowitz JG, Stern RC. Exercise conditioning and cardiopulmonary fitness in cystic fibrosis: the effects of a three-month supervised running program. Chest 1981;80:392–398.

69. Orenstein DM, Henke KG, Costill DL, Doechuk CF, Lemon PJ, Stern RC. Exercise and heat stress in cystic fibrosis patients. Pediatr Res 1983;17:267–269.

70. Orenstein DM, Henke KG, Green CG. Heat acclimation in cystic fibrosis. J Appl Physiol 1984;57:408–412.

71. Orenstein DM, Hovell MF, Mulvhill M, Keating KK, Hofstetter CR, Kelsey S, Morris K, Nixon PA. Strength vs aerobic training in children with cystic fibrosis: a randomized controlled trial. Chest 2004;126:1204–1214.

72. Orenstein DM, Nixon PA, Ross EA, Kaplan RM. The quality of well-being in cystic fibrosis. Chest 1989;95:344–347.

73. Pate RR, Barabowski T, Dowda M, Trost S. Tracking of physical activity in young children. Med Sci Sports Exerc 1996;28:92–96.

74. Pianosi P, LeBlanc J, Almudevar A. Peak oxygen uptake and mortality in children with cystic fibrosis. Thorax 2005;60:50–54.

75. Prasad SA, Randall SD, Balfour-Lynn IM. Fifteen-count breathlessness score: an objective measure for children. Pediatr Pulmonol 2000;30:56–62.

76. Rachinsky SV, Kapranow NI, Tatochenko VK, Simonowa OI, Turina JE. Submaximal physical loads in cystic fibrosis. Acta Univ Carol 1990;36:198–200.

77. Rowland TM. Aerobic exercise testing protocols. In: Rowland TW, editor. Pediatric laboratory exercise testing: clinical guidelines. Champaign, IL: Human Kinetics; 1991. p 19–41.

78. Ruppel GL. Spirometry and related test. In: Ruppel GL, editor. Manual of pulmonary function testing. St. Louis: Mosby; 1998. p 52–53.

79. Ruter K, Staab D, Magdorf K, Kleinau I, Paul K, Hetzer R, Wahn U. The 12-minute walk test as assessment for lung transplantation in CF patients. Neth J Med 1999;54(Suppl):S58.

80. Ryujin DT, Samuelson WM, Marshall BC. Oxygen saturation in adult cystic fibrosis patients during exercise at 1500 meters above sea level. Pediatr Pulmonol 1998;17(Suppl):331.

81. Sawyer EH, Clanton TL. Improved pulmonary function and exercise tolerance with inspiratory muscle conditioning in children with cystic fibrosis. Chest 1993;104:1490–1497.

82. Selvadurai HC, Blimkie CJ, Cooper PJ, Mellis CM, Van Asperen PP. Gender differences in habitual activity in children with cystic fibrosis. Arch Dis Child 2004;89:928–933.

83. Shah AR, Gozal D, Keens TG. Determinants of aerobic and anaerobic exercise performance in cystic fibrosis. Am J Respir Crit Care Med 1998;157:1145–1150.

84. Silverman M, Hobbs FDR, Gordon IRS, Carswell F. Cystic fibrosis, atopy, and airways lability. Arch Dis Child 1978;53:873–877.

85. Skeie B, Askanazi J, Rothkopf MM, Rosenbaum SH, Kvetan V, Ross E. Improved exercise tolerance with long term parenteral nutrition in cystic fibrosis. Crit Care Med 1987;15:960–962.

86. Stanghelle JK, Hjeltnes N, Michalsen H, Bangstad HJ, Skyberg D. Pulmonary function and oxygen uptake during exercise in 11 year old patients with cystic fibrosis. Acta Paediatr Scand 1986;75:657–661.

87. Strauss GD, Osher A, Wang C, Goodrich E, Gold F, Colman W, Stabile M, Dobrenchuk A, Keens TG. Variable weight training in cystic fibrosis. Chest 1987;92:273–276.

88. Thin AG, Dodd JD, Gallagher CG, Fitzgerald MX, Mcloughlin P. Effect of respiratory rate on airway deadspace ventilation during exercise in cystic fibrosis. Respir Med 2004;98:1063–1070.

89. van Haren EHJ, Lammers JWJ, Festen J, van Herwaarden CL. Bronchial vagal tone and responsiveness to histamine, exercise and bronchodilators in adult patients with cystic fibrosis. Eur Respir J 1992;5:1083–1088.

90. Van Praagh E, Franca NM. Measuring maximal short term power output during growth. In: Van Praagh E, editor. Pediatric anaerobic performance. Champaign, IL: Human Kinetics; 1998. p 151–190.

91. Ward SA, Tomezsko JL, Holsclaw DS, Paolone AM. Energy expenditure and substrate utilization in adults with cystic fibrosis and diabetes mellitus. Am J Clin Nutr 1999;69:913–919.

92. Zach M, Oberwaldner B, Hausler F. Cystic fibrosis: physical exercise versus chest physiotherapy. Arch Dis Child 1982;57:587–589.

93. Zelkowitz PS, Giammona ST. Effects of gravity and exercise on the pulmonary diffusing capacity in children with cystic fibrosis. J Pediatr 1969;74:393–398.

94. Zwiren LD, Manos TM. Exercise testing and prescription considerations throughout childhood. In: Rottman JL, editor. ACSM's resource manual for guidelines for exercise testing and prescription. Baltimore: Williams & Wilkins; 1998. p 507–515.

CHAPTER 24

1. Airstars J. Fatigue in the cancer patient: a conceptual approach to a clinical problem. Oncol Nurs Forum 1987;14:25–34.

2. American Cancer Society. Cancer facts and fiction—2006. Atlanta: American Cancer Society; 2006.

3. American College of Sports Medicine. Guidelines for exercise testing and prescription. 7th ed. Philadelphia: Lippincott Williams & Wilkins; 2006.

4. Baslund B, Lyngberg K, Andersen V, Halkjaer-Kristensen J, Hansen M, Klokker M, Pedersen BK. Effect of 8 wk of bicycle training on the immune system of patients with rheumatoid arthritis. J Appl Physiol 1993;75:1691–1695.

5. Bennett JC, Plum F, Cecil F. Textbook of medicine. 20th ed. Philadelphia: Saunders; 1996.

6. Blair SN, Kohl HW, Paffenbarger RS, et al. Physical fitness and all-cause mortality. A prospective study of healthy men and women. JAMA 1989;262:2395–2401.

7. Blesch KS, Paice JA, Wickham R, Harte N, Schnoor DK, Purl S, Rehwalt K, Kopp PL, Manson S, Coveny SB, McHale M, Cahill M. Correlates of fatigue in people with breast or lung cancer. Oncol Nurs Forum 1991;18:81–87.

8. Brown J. The national economic burden of cancer: an update. J Natl Cancer Inst 1990;82:1811–1814.

9. Burnstein L, Ross RK, Lobo RA, et al. The effects of moderate physical activity on menstrual cycle patterns in adolescence: implications for breast cancer prevention. Br J Cancer 1987;55:681–685.

10. Butterworth E, Nehlsen-Cannarella SL. The effects of high- versus moderate-intensity exercise on natural killer cell cytotoxic activity. Med Sci Sports Exerc 1993;25:1126–1134.

11. Colditz GA, Frazier AL. Models of breast cancer show that risk is set by events of early life: prevention efforts must shift focus. Cancer Epidemiol Biomarkers Prev 1995;4:567–571.

12. Courneya KS, Mackey JR, Jones LW. Coping with cancer—can exercise help? Phys Sportsmed 2000;28:49–73.

13. Courneya KS. Exercise in cancer survivors: an overview of research. Med Sci Sports Exerc 2003;35:1846–1852.

14. Courneya KS, Mackey JR, Mckenzie DC. Exercise for breast cancer survivors. Research evidence and clinical guidelines. Phys Sportsmed 2002;30:33–42.

15. Davis JM, Kohut MI, Jackson DA, Hertler-Colbert LM, Mayer EP, Ghafar A. Exercise effects on lung tumor metastases and in vitro alveolar macrophage anti-tumor cytotoxicity. Am J Physiol 1998;274:R1454–R1459.

16. De La Fuente M, Martin MI, Ortega E. Changes in the phagocytic function of peritoneal macrophages from old mice after strenuous physical exercise. Comp Immunol Microbiol Infect Dis 1990;13:189–198.

17. Dimeo FC, Fetscher S, Lange W, Mertelsmann R, Keul J. Effects of aerobic exercise on the physical performance and incidence of treatment-related complications after high-dose chemotherapy. Blood 1997;90:3390–3394.

18. Dimeo FC, Stieglitz RD, Novelli-Fischer U, Fetscher S, Keul J. Effects of physical activity on the fatigue and psychologic status of cancer patients during chemotherapy. Cancer 1999;85:2273–2277.

19. Doll R, Peto R. The causes of cancer: quantitative estimates of avoidable risks of cancer in the United States today. J Natl Cancer Inst 1981;66:1191–1308.

20. Fairly AS, Courneya KS, Field CJ, Mackey JR. Physical exercise and immune system function in cancer survivors. Cancer 2002;94:539–551.

21. Fehr H, Lotzerich G, Michna H. Human macrophage function and physical exercise: phagocytic and histochemical studies. Eur J Appl Physiol 1989;58:613–617.

22. Ferrell B, Grant M, Funk B, Garcia N, Otis-Green S, Schaffner M. Quality of life in breast cancer. Cancer Patient 1996;4:331–340.

23. Fields CJ, Gougeon R, Marliss EB. Circulating mononuclear cell numbers and function during intense exercise and recovery. J Appl Physiol 1991;71:1089–1097.

24. Fobair PR, Hoppe T, Bloom J, Cox R, Varghese A, Spiegel D. Psychosocial problems among survivors of Hodgkin's disease. J Clin Oncol 1986;4:805–814.

25. Forner MA, Collazos ME, Barriga C, De La Foente M, Rodriguez AB, Ortega E. Effect of age on adherence and chemotaxis capacities of peritoneal macrophages: influence of physical activity stress. Mech Ageing Dev 1994;75:179–189.

26. Francis K. Physical activity: breast and reproductive cancer. Compr Ther 1996;22:94–99.

27. Friedenreich CM. Physical activity and cancer prevention: from observational to interventional research. Cancer Epidemiol Biomarkers Prev 2001;10:287–301.

28. Frisch R, Wyshak G, Albright N, Albright T. Lower prevalence of breast cancer and cancers of the reproductive system among former college athletes compared to non-athletes. Br J Cancer 1985;2:885–891.

29. Gabriel H, Schwartz L, Born P, Kindermann W. Differential mobilization of leukocyte and lymphocyte subpopulations into the circulation during endurance exercise. Eur J Appl Physiol 1992;65:529–534.

30. Haenszel W. Cancer mortality among the foreign born in the United States. J Natl Cancer Inst 1961;26:37–132.

31. Harvard report on cancer prevention. Volume 1: causes of human cancer. Cancer Cause Control 1996;7:1.

32. Henderson B, Ross R, Judd H. Do regular ovulatory cycles increase breast cancer risk? Cancer 1985;56:1206–1208.

33. Hershcopf R, Bradlow H. Obesity, diet, endogenous estrogens, and the risk of hormone-sensitive cancer. Am J Clin Nutr 1987;45:283–289.

34. Hoffman-Goetz L. Exercise, natural immunity, and tumor metastasis. Med Sci Sports Exerc 1994;26:157–163.

35. Hsieh C, Trichopoulos D, Katsouyanni K. Age at menarche, age at menopause, height and obesity as risk factors for breast cancer: association and interactions in an international case control study. Int J Cancer 1990;46:796–800.

36. Kliningham RB. Physical activity and the primary prevention of cancer. Primary Care: Clinics in Office Practice 1998;25:515–536.

37. Knols R, Aaronson NK, Ubelhart D, et al. Physical exercise in cancer patients during and after medical treatment: a systematic review of randomized and controlled trials. J Clin Oncol 2005;23:3830–3842.

38. Kohl H, LaPorte R, Blair S. Physical activity and cancer. Sports Med 1988;6:222–237.

39. Vecchia C, Decarli A, Di Pietro S, Franceschi S, Negri E, Parazzini F. Menstrual cycle patterns and the risk of breast disease. Eur J Cancer Clin Oncol 1985;21:417–422.

40. Lee I-M, Sesso HD, Paffenbarger RS. Physical activity and risk of lung cancer. Int J Epidemiol 1999;28:620–625.

41. MacVicar MG, Winningham ML, Nickel JL. Effects of aerobic interval training on cancer patient's functional capacity. Nurs Res 1989;38:348–351.

42. McSweeney AJ, Grant I, Heaton RK, Adams KM, Timms RM. Life quality of patients with chronic obstructive pulmonary disease. Arch Intern Med 1982;142:473–478.

43. Michna H. The human macrophage system: activity and functional morphology. Bibl Anat 1988;31:1–38.

44. Muir C, Waterhouse J, editors. Cancer incidence in five continents. Vol. 5. Lyon, France: IARC; 1987.

45. Munck A, Guyre PM, Holbrook NJ. Physiological functions of glucocorticoids in stress and their relation to pharmacological actions. Endocr Rev 1984;5:25–44.

46. National Cancer Institute. Cancer trends progress report—2005 update. U.S. National Institutes of Health. www.cancer.gov

47. Nehlsen-Cannarella SL, Nieman DC, Jessen J, Chang L, Gusewitch G, Blix GG, Ashley E. The effects of acute moderate exercise on lymphocyte function and serum immunoglobulin levels. Int J Sports Med 1991;12:391–398.

48. Nieman DC, Nehlsen-Cannarella SL. The immune response to exercise. Sem Hematol 1994;31:166–179.

49. Nieman, DC, Simandle S, Henson DA, et al. Lymphocyte proliferation response to 2.5 hours of running. Natl J Sports Med 1995;16:404–408.

50. Nieman DC, Henson DA, Gusewitch G, Warren BJ, Dotson RC, Butterworth DE, Nehlsen-Cannarella SL. Physical activity and immune function in elderly women. Med Sci Sports Exerc 1993;25:823–831.

51. Nieman DC. Exercise, upper respiratory infection, and the immune system. Med Sci Sports Exerc 1994;26:128–139.

52. Nieman DC, Berk LS, Simpson-Westerberg M, Arabatzis K, Youngberg W, Tan SA, Eby WC. Effects of long endurance running on immune system parameters and lymphocyte function in experienced marathoners. Int J Sports Med 1989;10:317–323.

53. Nieman DC, Buckley KS, Henson DA, et al. Immune function in marathon runners versus sedentary controls. Med Sci Sports Exerc 1995;27:986–992.

54. Nieman DC, Ahle JC, Henson DA, et al. Indomethacin does not alter natural killer cell response to 2.5 hours of running. J Appl Physiol 1995;79:748–755.

55. Ortega E, Forner MA, Barriga C, De La Fuente M. Stimulation of phagocytic function in guinea pig peritoneal macrophages by physical activity stress. Euro J Appl Physiol 1992;64:323–327.

56. Ortega E, Forner MA, Barriga C. Exercise-induced stimulation of murine macrophage chemotaxis: role of corticosterone and prolactin as mediators. J Physiol (Lond) 1997;498:729–734.

57. Ortega E, Forner MA, Barriga C, De La Fuente M. Effect of age and of swimming-induced stress on the phagocytic capacity of peritoneal macrophages from mice. Mech Ageing Dev 1993;70:53–63.

58. Paffenbarger R, Hyde R, Wing A, Hsieh C. Physical activity and incidence of cancer in diverse populations: a preliminary report. Am J Nutr 1987;45:312–317.

59. Paffenbarger R, Hyde R, Wing A, Hsieh C. Physical activity, all-cause mortality and longevity of college athletes. N Engl J Med 1986;314:605–613.

60. Paffenbarger R, Kampert J, Chang H. Characteristics that predict risk of breast cancer before and after menopause. Am J Epidemiol 1980;112:258–267.

61. Pedersen BK, Ullum H. NK cell response to physical activity. Possible mechanism of action. Med Sci Sports Exerc 1994;26:140–146.

62. Pickard-Holley S. Fatigue in cancer patients. Cancer Nurs 1991;14:13–19.

63. Pike M, Krailo M, Henderson B. Hormonal risk factors, breast tissue age and the age incidence of breast cancer. Nature 1983;302:767–770.

64. Pyne DB. Regulation of neutrophil function during exercise. Sports Med 1994;17:245–258.

65. Pyne DB, Baker MS, Fricker PA, McDonald WA, Nelson WJ. Effects of an intensive 12-wk training program by elite swimmers on neutrophil oxidative activity. Med Sci Sports Exerc 1995;27:536–542.

66. Russo J, Russo IH. Role of differentiation in the pathogenesis and prevention of breast cancer. Endocr Relat Cancer 1997;4:7–21.

67. Segal R, Johnson D, Smith J, Colletta S, Guyton J, et al. Structured exercise improves physical functioning in women with stages I and II breast cancer: results of a randomized controlled trial. J Clin Oncol 2001;19:657–665.

68. Shephard RJ. Physical activity and the healthy mind. Can Med Assoc J 1983;128:525–530.

69. Shephard RJ, Shek NP. Cancer, immune function, and physical activity. Can J Appl Physiol 1995;20:1–25.

70. Shephard RJ, Shek NP. Associations between physical activity and susceptibility to cancer. Sports Med 1998;26:193–315.

71. Shephard RJ. Exercise in the prevention and treatment of cancer: an update. Sports Med 1993;15:258–280.

72. Shinkai S, Shore S, Shek PN, Shepard RJ. Acute exercise and immune function: relationship between lymphocyte activity and changes insubset counts. Int J Sports Med 1992;13:452–461.

73. Slatery ML, Potter J, Caan B, et al. Energy balance and colon cancer—beyond physical activity. Cancer Res 1997;57:75–80.

74. Smith JA, Telford RD, Mason JB, Weidman, MJ. Exercise training and neutrophil microbial activity. Int J Sports Med 1990;11:179–187.

75. Spitzer WO. State of science of 1987: quality of life and functional status as target variables for research. J Chronic Dis 1987;40:465–471.

76. Sternfeld B. Cancer and the protective effect of physical activity: the epidemiological evidence. Med Sci Sports Exerc 1992;24:1195–1209.

77. The American heritage Stedman's medical dictionary. Boston: Houghton Mifflin; 2002.

78. Thompson, HJ. Effects of exercise intensity and duration on the induction of mammary carcinogenesis. Cancer Res Suppl 1994;54:1608–1635.

79. Thompson W. Exercise and health: fact or hype? South Med J 1994;87:567–574.

80. Thune I, Furberg AS. Physical activity and cancer risk: dose response and cancer, all sites and site specific. Med Sci Sports Exerc 2001;33:S530–S550.

81. Tvede N, Steensberg J, Baslund B, Halkjaer-Kristensen J, Pedersen BK. Cellular immunity in highly-trained elite racing cyclists and controls during periods of training with high and low intensity. Scand J Sports Med 1991;1:163–166.

82. Trichopoulos D, MacMahon B, Cole P. The menopause and breast cancer risk. J Natl Cancer Inst 1972;48:605–613.

83. U.S. Mortality Public Use Data Tape 2003. National Center for Health Statistics. Centers for Disease Control and Prevention; 2006.

84. Verloop J, Rookus MA, Van Der Koy, K, Leuwen FE. Physical activity and breast cancer risk in women. J Natl Cancer Inst 2000;92:128–135.

85. Vihko R, Apter D. Endocrine characteristics of adolescent menstrual cycles: impact of early menarche. J Steroid Biochem Mol Biol 1984;20:231–236.

86. Walmolts J, Peter K, Lewis R, Engel WK. Type 2 muscle fiber atrophy: an early systemic effect of cancer [abstract]. Neurology 1975;25:374.

87. Winningham ML. Walking program for people with cancer: getting started. Cancer Nurs 1991;14:270–276.

88. Winningham ML, MacVicar MG. The effect of aerobic exercise on patient reports of nausea. Oncol Nurs Forum 1988;15:447–450.

89. Winningham ML, Nail LM, Burke MB, et al. Fatigue and the cancer experience: the state of the knowledge. Oncol Nurs Forum 1994;21:23–36.

90. Winningham ML, MacVicar MG, Bondoc M, Anderson JI, Minton JP. Effect of aerobic exercise on body weight and composition in patients with breast cancer on adjuvant chemotherapy. Oncol Nurs Forum 1989;16:683–689.

91. Woods JA, Davis JM, Smith JA, Nieman DC. Exercise and cellular innate immune function. Med Sci Sports Exerc 1999;31:57–66.

92. Woods JA, Davis JM. Exercise, macrophages and cancer defense. Med Sci Sports Exerc 1994;26:147–156.

93. Woods JA, Davis JM, Mayer EP, Ghaffar A, Pate RR. Exercise increases inflammatory macrophage anti-tumor cytotoxicity. J Appl Physiol 1993;75:879–886.

CHAPTER 25

1. Aberg JA, Gallant JE, Anderson J, Oleske JM, Libman H, Currier JS, et al. Primary care guidelines for the management of persons infected with human immunodeficiency virus: recommendations of the HIV Medicine Association of the Infectious Diseases Society of America. Clin Infect Dis 2004;39:609-629.

2. American College of Sports Medicine. ACSM's guidelines for exercise testing and prescription. 7th ed. Philadelphia: Lippincott Williams & Wilkins; 2006.

3. American College of Sports Medicine. The recommended quantity and quality of exercise for developing and maintaining cardiorespiratory and muscular fitness in healthy adults: a position stand. Med Sci Sports Exerc 1990;22(2):265–274.

4. Baigis J, Korniewicz D, Chase G, Butz A, Jacobson D, Wu A. Effectiveness of a home-based exercise intervention for HIV-infected adults: a randomized trial. J Assoc of Nurses AIDS Care 2002;13(2):33–45.

5. Barre-Sinoussi F, Chermann JC, Rey F. Isolation of a T-lymphotrophic retrovirus from a patient at risk for acquired immune deficiency syndrome (AIDS). Science 1983;220:868–871.

6. Brooks GA, Fahey TD, White TP. Exercise physiology: human bioenergetics and its applications. 2nd ed. Mountain View, CA: Mayfield; 1996.

7. Brown TT, McComsey GA. Osteopenia and osteoporosis in patients with HIV: a review of current concepts. Curr Infect Dis Rep 2006;Mar;8(2):162–170.

8. Cannon JG. Exercise and resistance to infection. J Appl Physiol 1983;74:973–981.

9. Carpenter CCJ, Fischl MA, Hammer SM, et al. Antiretroviral therapy for HIV infection in 1998: updated recommendations of the International AIDS Society—USA Panel. JAMA 1998;280(1):78-88.

10. Carr AS, Samaras K, Burton S, et al. A syndrome of peripheral lipodystrophy, hyperlipidaemia, and insulin resistance in patients receiving HIV protease inhibitors [abstract]. AIDS 1998;12(7):F51–58.

11. Casaburi RA, Patessio A, Ioli F, Zanaboni S, Donner CF, Wasserman K. Reductions in exercise in lactic acidosis and ventilation as a result of exercise training in patients with obstructive lung disease. Am Rev Respir Dis 1991;143:9–18.

12. Centers for Disease Control and Prevention. 1993 revised classification system for HIV infection and expanded surveillance case definition for AIDS among adolescents and adults. MMWR 1992;41:RR–17.

13. Centers for Disease Control and Prevention. Interpretation and use of Western blot assay for serodiagnosis of HIV infection. MMWR 1989;S7:1–7.

14. Centers for Disease Control and Prevention. Revision of the case definition of acquired immunodeficiency syndrome for national reporting—United States. MMWR 1985;34:373–375.

15. Centers for Disease Control and Prevention. Revision of the CDC surveillance case definition for acquired immunodeficiency syndrome. MMWR CDC Surveillance Summary 1987;36(1S):1S–5S.

16. Centers for Disease Control and Prevention. Update: acquired immunodeficiency syndrome—United States. MMWR 1987;36(31):522–526.

17. Centers for Disease Control and Prevention. HIV/AIDS Surveillance Report. Volume 16. Atlanta: U.S. Public Health Service; 2006. p 1–42. www.cdc.gov/hiv/topics/surveillance/resources/reports/2004report/table7.htm

18. Choi S, Lagakos S, Schooley R, Volberding P. CD4 lymphocytes are an incomplete surrogate marker for clinical progression in persons with asymptomatic HIV infection taking zidovudine. Ann Intern Med 1993;118(9):674–680.

19. Cooper DA, Gold JA, MacLean P. Acute retrovirus infection: definition of a clinical illness associated with seroconversion. Lancet 1985;1:137–140.

20. CRS. Serologic diagnosis of human immunodeficiency virus infection by Western blot testing. JAMA 1988;260:674–679.

21. Curtis JR, Smith B, Weaver M, Landers K, Lopez-Ben R, Raper J, Saag M, Venkataraman, R, Saag K. Ethnic variations in the prevalence of metabolic bone disease among HIV positive patients with lipodystrophy. AIDS Res Hum Retroviruses 2006;22(2):125–131.

22. De Gruttola V, Wulfson M, Fischl M, Tsiatis A. Modeling the relationship between survival and CD4 lymphocytes in patients with AIDS and AIDS-related complex. J Acquir Immune Defic Syndr Hum Retrovirol 1993;6(4):359–365.

23. Dimeo F, Rumberger BG, Kaul J. Aerobic exercise as therapy for cancer fatigue. Med Sci Sports Exerc 1998;30(4):475–478.

24. Dolan SE, Frontera W, Librizzi J, Ljungquist K, Juan S, Dorman R, Cole ME, Kanter JR, Grinspoon S. Effects of a supervised home-based aerobic and progressive resistance training regimen in women infected with human immunodeficiency virus: a randomized trial. Arch Intern Med. 2006 Jun 12;166(11):1225-1231.

25. Fish AF, Smith BA, Frid DJ, Christmas SK, Post D, Montalto NJ. Step treadmill exercise training and blood pressure reduction in women with mild hypertension. Prog Cardiovasc Nurs 1997;12(1):28–35.

26. Gallo RC, Salahuddin SZ, Popovic M. Frequent detection and isolation of cytopathic retroviruses (HTLVIII) from patients with AIDS and at risk for AIDS. Science 1984;224:500–503.

27. Goldberg AP, Hagberg J, Delmez JA, et al. The metabolic and psychological effects of exercise training in hemodialysis patients. Am J Clin Nutr 1980;33:1620–1628.

28. Henry K, Melroe H, Huebsch J, et al. Severe premature coronary artery disease with protease inhibitors. Lancet 1998;351:1328.

29. Hogg RS, Heath KV, Yip B, et al. Improved survival among HIV-infected individuals following initiation of antiretroviral therapy. JAMA 1998;279:450–454.

30. Hughes MD, Johnson VA, Hirsch MS. Monitoring plasma HIV RNA levels in addition to CD4+ lymphocyte count improves assessment of antiretroviral therapeutic response. Ann Intern Med 1997;126:929–938.

31. Hurwitz A. The benefit of a home exercise regimen for ambulatory Parkinson's disease patients. J Neurosci Nurs 1989;21(3):180–184.

32. Johnson JE, Anders GT, Blanton HM, et al. Exercise dysfunction in patients seropositive for the human immunodeficiency virus. Am Rev Respir Dis 1990;141:618–622.

33. Katlama C, Valantin MA, Calvez V, et al. ALTIS PLUS: long term d 4T-3TC with and without ritonavir. In: Program and Abstracts of the 5th Conference on Retroviruses and Opportunistic Infections; February 1–5, 1998, Chicago, Abstract 376, 1998.

34. Keyser RE, Peralta L, Cade T, Miller S, Anixt J. Functional aerobic impairment in adolescents seropositive for HIV: a quasiexperimental analysis. Arch Phys Med Rehabil 2000;81:1479–1484.

35. Kilby JM, Saag MS, editors. Natural history of HIV disease. 2nd edition. Baltimore: Williams & Wilkins; 1999. p 49–58.

36. Lamb DR. Physiology of exercise. New York: Macmillan; 1984.

37. LaPerriere A, Fletcher MA, Antoni MH, Klimas NG, Ironson G, Schneiderman N. Aerobic exercise training in an AIDS risk group. Int J Sports Med 1991;12(Suppl. 1):S53–S57.

38. Lui A, Karter D, Turett G. Another case of breast hypertrophy in a patient treated with indinavir. Clin Infect Dis 1998;26:1482.

39. MacArthur RD, Levine SD, Birk TJ. Supervised exercise training improves cardiopulmonary fitness in HIV-infected persons. Med Sci Sports Exerc 1993;25(6):684–688.

40. MacDonell KB, Chimiel JS, Poggensee L. Predicting progression to AIDS: combined usefulness of CD4 lymphocyte counts and p24 antigenemia. Am J Med 1990;89:706–712.

41. MacVicar MG, Winningham ML, Nickel J. Effects of aerobic interval training on cancer patients' functional capacity. Nurs Res 1989;38(6):348–351.

42. Mannix ET, Boska MD, Ryder KD, Newcomer B, Manfredi F, Farber MO. Impaired oxidative energy metabolism in AIDS: a 31P MRS study. Official publication of the Federation of American Societies for Experimental Biology 1996;10(3):A375, Abstract 2172.

43. Mary-Krause M, Billaud E, Poizot-Martin I, Simon A, Dhiver C, Dupont C, Salmon D, Roudiere L, Costagliola D. Clinical Epidemiology Group of the French Hospital Database. AIDS 2006;20(12):1627–1635.

44. Mellors JW, et al. Quantitation of HIV RNA in plasma predicts outcome after seroconversion. Ann Intern Med 1995;122(8):573–579.

45. Mellors JW, Kingsley LA, Rinaldo CR. Quantitation of HIV RNA in plasma predicts outcome after seroconversion. Ann Intern Med 1995;122:573–579.

46. Mellors JW, Rinaldo CR, Gupta P. Prognosis in HIV infection predicted by the quantity of virus in plasma. Science 1996;272:1167–1170.

47. Mellors JW, Rinaldo CR, Jr, Gupta P, White RM, Todd JA, Kingsley LA. Prognosis in HIV infection predicted by the quantity of virus in plasma. Science 1996;272:1167–1170.

48. Mustafa T, Sy FS, Macera CA, et al. Association between exercise and HIV disease progression in a cohort of homosexual men. Ann Epidemiol 1999;9(2):127–131.

49. Neumann M, Harrison J, Saltareli M. Splicing variability in HIV type 1 revealed by quantitative RNA polymerase chain reaction. AIDS Res Hum Retroviruses 1994;10:1531–1532.

50. Neidig J, Smith BA, Brashers D, Para M, Fass B. Aerobic exercise training: affects on psychological parameters. J Nurses AIDS Care 2003;14(2):30–40.

51. Nieman DC. Immune response to heavy exertion. J Appl Physiol 1997;82:1385–1394.

52. Nieman DC, Johanssen LM, Lee JW, Cermak J, Arabatzis K. Infectious episodes in runners before and after the Los Angeles marathon. J Sports Med Phys Fitness 1990;30:316–328.

53. NIH. Physical activity and cardiovascular health: NIH consensus statement. 1995. p 1–33.

54. O'Brien TR, Blattner WA, Waters D. Serum HIV RNA levels and time to development of AIDS in the multicenter hemophilia cohort study. JAMA 1996;276:105–110.

55. O'Brien WA, Hartigan PA, Daar ES. Changes in plasma HIV RNA levels and CD4+ lymphocyte counts predict both response to antiretroviral therapy and therapeutic failure. Ann Intern Med 1996;126:939–945.

56. Palella FJ, Delaney KM, Mooreman AC, et al. Declining morbidity and mortality among patients with advanced human immunodeficiency virus infection. N Engl J Med 1998;338:853–860.

57. Pantaleo G, Fauci AS. New concepts in the immunopathogenesis of HIV infection. Annu Rev Microbiol 1996;50:825–854.

58. Perna FM, LaPerriere A, Klimas N, Ironson G, Perry A, Pavone J, et al. Cardiopulmonary and CD4 changes in response to exercise training in early symptomatic HIV infection. Med Sci Sports Exerc 1999;31(7):973–979.

59. Preusser BA, Winningham ML, Clanon TL. High- vs low-intensity inspiratory muscle interval training in patients with COPD. Chest 1994;106(1):110–117.

60. Rigsby LW, Dishman RK, Jackson AW, MaClean GS, Raven PB. Effects of exercise training on men seropositive for the human immunodeficiency virus-1. Med Sci Sports Exerc 1992;24(1):6–12.

61. Roth VR, Kravcik S, Angel JB. Development of cervical fat pads following therapy with human immunodeficiency virus type 1 protease inhibitors. Clin Infect Dis 1998;27:65–67.

62. Roubenoff R, McDermott A, Wood M, Suri J. Feasibility of increasing lean body mass in HIV-infected adults using progressive resistance training. Presented at the 12th International AIDS Conference, Geneva, Switzerland, 1999.

63. Roubenoff R, Skolnik PR, Shevitz A, et al. Effect of a single bout of acute exercise on plasma human immunodeficiency virus RNA levels. J Appl Physiol 1999;86(4):1197–1201.

64. Roubenoff R, Weiss L, McDermott A, et al. A pilot study of exercise training to reduce trunk fat in adults with HIV-associated fat redistribution. AIDS 1999;13:1373–1375.

65. Ryan AS, Pratley RE, Elahi D, Goldberg AP. Resistive training increases fat-free mass and maintains RMR despite weight loss in postmenopausal women. J Appl Physiol 1995;79(3):818–823.

66. Ryan AS, Treuth MS, Rubin MA, et al. Effects of strength training on bone mineral density: hormonal and bone turnover relationships. J Appl Physiol 1994;77(4):1678–1684.

67. Saint-Marc T, Touraine JL. Effects of metformin on insulin resistance and central adiposity in patients receiving effective protease inhibitor therapy. AIDS 1999;13(8):1000–1002.

68. Schacker T, Collier AC, Hughes J. Clinical and epidemiologic features of primary HIV infection. Ann Intern Med 1996;125(4):125–257.

69. Schneider BA, Roehrig KL, Smith BA, Sherman WM. Resistive exercise: strength, body composition, glucose tolerance and insulin action in African American women. Unpublished manuscript.

70. Shaw AJ, McLean KA, Evans B. Disorders of fat distribution in HIV infection. Int J STD AIDS 1998;9:595–599.

71. Shikuma CM, Waslien C, McKeague J, et al. Fasting hyperinsulinemia and increased waist-to-hip ratios in non-wasting individuals with AIDS. AIDS 1999;13:1359–1365.

72. Sinnwell TM, Kumaraaswamy S, Soueidan S, et al. Metabolic abnormalities in skeletal muscle of patients receiving zidovudine therapy observed by 31P in vivo magnetic resonance spectroscopy. J Clin Invest 1995;96:126–131.

73. Smith B, Neidig JL, Nickel J, Mitchell GL, Para MD, Fass RJ. Aerobic exercise training: effects on physiological fatigue, dyspnea, increased weight and central fat in HIV-infected adults. AIDS, in press.

74. Spence DW, Galantino LA, Mossberg KA, Zimmerman SO. Progressive resistance exercise: effect on muscle function and anthropometry of a select AIDS population. Arch Phys Med Rehabil 1990;71:645–649.

75. Stein DS, Korvick JA, Vermund SH. CD4+ lymphocyte cell enumeration for prediction of clinical course of human immunodeficiency virus disease. J Infect Dis 1992;165:352–363.

76. Stringer WW, Berezovskaya M, O'Brien WA, Beck CK, Casaburi R. The effect of exercise training on aerobic fitness, immune indices, and quality of life in HIV+ patients. Med Sci Sports Exerc 1998;30(1):11–16.

77. Terry L, Sprinz E, Stein R, Medeiros N, Oliveira J, Ribeiro J. Exercise training in HIV-1-infected individuals with dyslipidemia and lipodystrophy. Med Sci Sports Exerc 2006;38(3):411–417.

78. U.S. Department of Health and Human Services. Objectives for improving health. 2 volumes. Washington, DC: U.S. Government Printing Office; 2000.

79. U.S. Department of Health and Human Services. Physical activity and health: a report of the surgeon general. Atlanta: Center for Disease Control and Prevention, National Center for Chronic Disease Prevention and Health Promotion; 1996.

80. Visnegarwala F, Krause KL, Musher DM. Severe diabetes associated with protease inhibitor therapy. Ann Intern Med 1997;127(11):947.

81. Walli R, Herfort O, Michl GM, et al. Treatment with protease inhibitors associated with peripheral insulin resistance and impaired oral glucose tolerance in HIV infected patients. AIDS 1998;12(15):F167–F173.

82. World Health Organization. Proposed WHO criteria for interpreting results from Western blot assays for HIV, HIV-2, and HTLV-I/HTLV II. Wkly Epidemiol Rec 1990;37:281–283.

83. Wilmore JH, Costill DL. Physiology of sport and exercise. Champaign, IL: Human Kinetics; 1994.

84. Wurtz R. Abnormal fat distribution and use of protease inhibitors. Lancet 1998;351:1735–1736.

85. Yarasheski KE, Tebas P, Stanerson B, Claxton S, Marin D, Bae K, Kennedy M, Tantisiriwat W, Powderly W. Resistance exercise training reduces hypertriglyceridemia in HIV-infected men treated with antiviral therapy. J Appl Physiol 2001;90:133–138.

CHAPTER 26

1. Work group recommendations: 2002 Exercise and Physical Activity Conference, St. Louis, Missouri. Session V: evidence of benefit of exercise and physical activity in arthritis. Arthritis Rheum 2003;49:453–454.

2. American College of Rheumatology Subcommittee on Rheumatoid Arthritis guidelines. Guidelines for the management of rheumatoid arthritis: 2002 update. Arthritis Rheum 2002;46:328–346.

3. American Geriatrics Society Panel on Exercise and Osteoarthritis. Exercise prescription for older adults with osteoarthritis pain: consensus practice recommendations. A supplement to the AGS clinical practice guidelines on the management of chronic pain in older adults. J Am Geriatr Soc 2001;49:808–823.

4. Altman R, Alarcon G, Appelrouth D, Bloch D, Borenstein D, Brandt K, Brown C, Cooke TD, Daniel W, Feldman D, et al. The American College of Rheumatology criteria for the classification and reporting of osteoarthritis of the hip. Arthritis Rheum 1991;34:505–514.

5. Altman R, Alarcon G, Appelrouth D, Bloch D, Borenstein D, Brandt K, Brown C, Cooke TD, Daniel W, Gray R, et al. The American College of Rheumatology criteria for the classification and reporting of osteoarthritis of the hand. Arthritis Rheum 1990;33:1601–1610.

6. Altman R, Asch E, Bloch D, Bole G, Borenstein D, Brandt K, Christy W, Cooke TD, Greenwald R, Hochberg M, et al. Development of criteria for the classification and reporting of osteoarthritis. Classification of osteoarthritis of the knee. Diagnostic and Therapeutic Criteria Committee of the American Rheumatism Association. Arthritis Rheum 1986;29:1039–1049.

7. Anderson JJ, Felson DT. Factors associated with osteoarthritis of the knee in the first national Health and Nutrition Examination Survey (HANES I). Evidence for an association with overweight, race, and physical demands of work. Am J Epidemiol 1988;128:179–189.

8. Anderson R. Rheumatoid arthritis. In: Schumacher HR Jr, editor. Primer on rheumatic diseases. Atlanta: Arthritis Foundation; 1993.

9. Andrews AW, Thomas MW, Bohannon RW. Normative values for isometric muscle force measurements obtained with hand-held dynamometers. Phys Ther 1996;76:248–259.

10. Anonymous. Prevalence of leisure-time physical activity among persons with arthritis and other rheumatic conditions—United States, 1990–1991. MMWR Morb Mortal Wkly Rep 1997;46:389–393.

11. Arnett FC, Edworthy SM, Bloch DA, McShane DJ, Fries JF, Cooper NS, Healey LA, Kaplan SR, Liang MH, Luthra HS, et al. The American Rheumatism Association 1987 revised criteria for the classification of rheumatoid arthritis. Arthritis Rheum 1988;31:315–324.

12. Assis MR, Silva LE, Alves AM, Pessanha AP, Valim V, Feldman D, Neto TL, Natour J. A randomized controlled trial of deep water running: clinical effectiveness of aquatic exercise to treat fibromyalgia. Arthritis Rheum 2006;55:57–65.

13. Avouac J, Gossec L, Dougados M. Diagnostic and predictive value of anti-CCP (cyclic citrullinated protein) antibodies in rheumatoid arthritis: a systematic literature review. Ann Rheum Dis 2006;65:845–851.

14. Bacon PA, Townend JN. Nails in the coffin: increasing evidence for the role of rheumatic disease in the cardiovascular mortality of rheumatoid arthritis. Arthritis Rheum 2001;44:2707–2710.

15. Badley EM. The impact of disabling arthritis. Arthritis Care Res 1995;8:221–228.

16. Baker K, McAlindon T. Exercise for knee osteoarthritis. Curr Opin Rheumatol 2000;12:456–463.

17. Barry H. Activity for older persons and mature athletes. In: Safran M, McKeag D, Van Camp S, editors. Manual of sports medicine. New York: Lippincott-Raven; 1998. p 184–189.

18. Barth WF. Office evaluation of the patient with musculoskeletal complaints. Am J Med 1997;102:3S–10S.

19. Bellamy N, Buchanan WW, Goldsmith CH, Campbell J, Stitt LW. Validation study of WOMAC: a health status instrument for measuring clinically important patient relevant outcomes to antirheumatic drug therapy in patients with osteoarthritis of the hip or knee. J Rheumatol 1988;15:1833–1840.

20. Belza B. The impact of fatigue on exercise performance. Arthritis Care Res 1994;7:176–180.

21. Berman A, Studenski S. Musculoskeletal rehabilitation. Clin Geriatr Med 1998;14:641–659.

22. Berry MJ, Brubaker PH, O'Toole ML, Rejeski WJ, Soberman J, Ribisl PM, Miller HS, Afable RF, Applegate W, Ettinger WH. Estimation of $\dot{V}O_2$ in older individuals with osteoarthritis of the knee and cardiovascular disease. Med Sci Sports Exerc 1996;28:808–814.

23. Brandt K, Heilman D, Slemenda C, Mazzuca S, Braunstein E. Knee pain in elderly community subjects. Differences in lower extremity muscles strength, body weight, and depression scores among those with and without radiographic evidence of osteoarthritis. Tran Orthop Res 1999;24:222.

24. Brooks PM. The burden of musculoskeletal disease—a global perspective. Clin Rheumatol 2006;25:778–781.

25. Brosseau L, MacLeay L, Robinson V, Wells G, Tugwell P. Intensity of exercise for the treatment of osteoarthritis. Cochrane Db Syst Rev 2003;CD004259.

26. Bryant C, Mahler D, Froelicher V, Miller N. ACSM's guideline for exercise testing and prescription. 5th ed. Baltimore: Williams & Wilkins; 1995. p 371.

27. Bullough P, Cawston T. Pathology and biochemistry of osteoarthritis. In: Doherty M, editor. Osteoarthritis. London: Wolfe; 1994. p 29–58.

28. Bunning RD, Materson RS. A rational program of exercise for patients with osteoarthritis. Semin Arthritis Rheum 1991;21:33–43.

29. Busch A, Schachter CL, Peloso PM, Bombardier C. Exercise for treating fibromyalgia syndrome. Cochrane Db Syst Rev 2002;CD003786.

30. Byers PH. Effect of exercise on morning stiffness and mobility in patients with rheumatoid arthritis. Res Nurs Health 1985;8:275–281.

31. Calin A, Elswood J, Rigg S, Skevington SM. Ankylosing spondylitis—an analytical review of 1500 patients: the changing pattern of disease. J Rheumatol 1988;15:1234–1238.

32. CDC. Centers for Disease Control and Prevention. Prevalence of disabilities and associated health conditions among adults—United States. Morb Mortal Wkly Rep 2001;50:120–125.

33. Clegg DO, Reda DJ, Harris CL, Klein MA, O'Dell JR, Hooper MM, Bradley JD, Bingham CO 3rd, Weisman MH, Jackson CG, Lane NE, Cush JJ, Moreland LW, Schumacher HR Jr, Oddis CV, Wolfe F, Molitor JA, Yocum DE, Schnitzer TJ, Furst DE, Sawitzke AD, Shi H, Brandt KD, Moskowitz RW, Williams HJ. Glucosamine, chondroitin sulfate, and the two in combination for painful knee osteoarthritis. N Engl J Med 2006;354:795–808.

34. Cleland LG, Hill CL, James MJ. Diet and arthritis. Baillieres Clin Rheumatol 1995;9:771–785.

35. Coleman EA, Buchner DM, Cress ME, Chan BK, de Lateur BJ. The relationship of joint symptoms with exercise performance in older adults. J Am Geriatr Soc 1996;44:14–21.

36. Csuka M, McCarty DJ. Simple method for measurement of lower extremity muscle strength. Am J Med 1985;78:77–81.

37. Dagfinrud H, Kvien TK, Hagen KB. Physiotherapy interventions for ankylosing spondylitis. Cochrane Db Syst Rev 2004;CD002822.

38. Daltroy LH, Robb-Nicholson C, Iversen MD, Wright EA, Liang MH. Effectiveness of minimally supervised home aerobic training in patients with systemic rheumatic disease. Br J Rheumatol 1995;34:1064–1069.

39. Daly MP, Berman BM. Rehabilitation of the elderly patient with arthritis. Clin Geriatr Med 1993;9:783–801.

40. Danneskiold-Samsoe B, Grimby G. The relationship between the leg muscle strength and physical capacity in patients with rheumatoid arthritis, with reference to the influence of corticosteroids. Clin Rheumatol 1986;5:468–474.

41. Davis MA, Ettinger WH, Neuhaus JM. Obesity and osteoarthritis of the knee: evidence from the National Health and Nutrition Examination Survey (NHANES 1). Semin Arthritis Rheum 1990;20:34–41.

42. de Jong Z, Munneke M, Lems WF, Zwinderman AH, Kroon HM, Pauwels EK, Jansen A, Ronday KH, Dijkmans BA, Breedveld FC, Vliet Vlieland TP, Hazes JM. Slowing of bone loss in patients with rheumatoid arthritis by long-term high-intensity exercise: results of a randomized, controlled trial. Arthritis Rheum 2004;50:1066–1076.

43. de Jong Z, Munneke M, Zwinderman AH, Kroon HM, Jansen A, Ronday KH, van Schaardenburg D, Dijkmans BA, Van den Ende CH, Breedveld FC, Vliet Vlieland TP, Hazes JM. Is a long-term high-intensity exercise program effective and safe in patients with rheumatoid arthritis? Results of a randomized controlled trial. Arthritis Rheum 2003;48:2415–2424.

44. de Jong Z, Munneke M, Zwinderman AH, Kroon HM, Ronday KH, Lems WF, Dijkmans BA, Breedveld FC, Vliet Vlieland TP, Hazes JM, Huizinga TW. Long term high intensity exercise and damage of small joints in rheumatoid arthritis. Ann Rheum Dis 2004;63:1399–1405.

45. del Rincon ID, Williams K, Stern MP, Freeman GL, Escalante A. High incidence of cardiovascular events in a rheumatoid arthritis cohort not explained by traditional cardiac risk factors. Arthritis Rheum 2001;44:2737–2745.

46. Dexter PA. Joint exercises in elderly persons with symptomatic osteoarthritis of the hip or knee. Performance patterns, medical support patterns, and the relationship between exercising and medical care. Arthritis Care Res 1992;5:36–41.

47. Deyle GD, Allison SC, Matekel RL, Ryder MG, Stang JM, Gohdes DD, Hutton JP, Henderson NE, Garber MB. Physical therapy treatment effectiveness for osteoarthritis of the knee: a randomized comparison of supervised clinical exercise and manual therapy procedures versus a home exercise program. Phys Ther 2005;85:1301–1317.

48. Dominick, KL, Gullette EC, Babyak MA, Mallow KL, Sherwood A, Waugh R, Chilikuri M, Keefe FJ, Blumenthal JA. Predicting peak oxygen uptake among older patients with chronic illness. J Cardiopulm Rehabil 1999;19:81–89.

49. Dougados M, Gueguen A, Nguyen M, Thiesce A, Listrat V, Jacob L, Nakache JP, Gabriel KR, Lequesne M, Amor B. Longitudinal radiologic evaluation of osteoarthritis of the knee. J Rheumatol 1992;19:378–384.

50. Ekdahl C. Muscle function in rheumatoid arthritis. Assessment and training. Scand J Rheumatol Suppl 1990;86:9–61.

51. Engelhart M, Kondrup J, Hoie LH, Andersen V, Kristensen JH, Heitmann BL. Weight reduction in obese patients with rheumatoid arthritis, with preservation of body cell mass and improvement of physical fitness. Clin Exp Rheumatol 1996;14:289–293.

52. Ettinger WH Jr, Burns R, Messier SP, Applegate W, Rejeski WJ, Morgan T, Shumaker S, Berry MJ, O'Toole M, Monu J, Craven T. A randomized trial comparing aerobic exercise and resistance exercise with a health education program in older adults with knee osteoarthritis. The Fitness Arthritis and Seniors Trial (FAST). JAMA 1997;277:25–31.

53. Eurenius E, Stenstrom CH. Physical activity, physical fitness, and general health perception among individuals with rheumatoid arthritis. Arthritis Rheum 2005;53:48–55.

54. Felson DT. The epidemiology of knee osteoarthritis: results from the Framingham Osteoarthritis Study. Semin Arthritis Rheum 1990;20:42–50.

55. Felson DT. Weight and osteoarthritis. Am J Clin Nutr 1996;63:430S–432S.

56. Felson DT, Anderson JJ, Naimark A, Walker AM, Meenan RF. Obesity and knee osteoarthritis. The Framingham Study. Ann Intern Med 1988;109:18–24.

57. Fisher LR, Cawley MI, Holgate ST. Relation between chest expansion, pulmonary function, and exercise tolerance in patients with ankylosing spondylitis. Ann Rheum Dis 1990;49:921–925.

58. Frangolia DD, Rhodes EC. Metabolic responses and mechanisms during water immersion running and exercise. Sports Med 1996;22:38–53.

59. Fransen M, McConnell S, Bell M. Exercise for osteoarthritis of the hip or knee. Cochrane Db Syst Rev 2003;CD004286.

60. Gall V. Exercise in the spondyloarthropathies. Arthritis Care Res 1994;7:215–220.

61. Galloway MT, Jokl P. The role of exercise in the treatment of inflammatory arthritis. Bull Rheum Dis 1993;42:1–4.

62. Gecht MR, Connell KJ, Sinacore JM, Prohaska TR. A survey of exercise beliefs and exercise habits among people with arthritis. Arthritis Care Res 1996;9:82–88.

63. Giesecke T, Gracely RH, Grant MA, Nachemson A, Petzke F, Williams DA, Clauw DJ. Evidence of augmented central pain processing in idiopathic chronic low back pain. Arthritis Rheuma 2004;50:613–623.

64. Goldenberg DL, Burckhardt C, Crofford L. Management of fibromyalgia syndrome. JAMA 2004;292:2388–2395.

65. Gowans SE, deHueck A. Effectiveness of exercise in management of fibromyalgia. Curr Opin Rheumatol 2004;16:138–142.

66. Gussoni M, Margonato V, Ventura R, Veicsteinas A. Energy cost of walking with hip joint impairment. Phys Ther 1990;70:295–301.

67. Hakkinen A, Hakkinen K, Hannonen P. Effects of strength training on neuromuscular function and disease activity in patients with recent-onset inflammatory arthritis. Scand J Rheumatol 1994;23:237–242.

68. Hakkinen A, Hannonen P, Nyman K, Lyyski T, Hakkinen K. Effects of concurrent strength and endurance training in women with early or longstanding rheumatoid arthritis: comparison with healthy subjects. Arthritis Rheum 2003;49:789–797.

69. Hakkinen A, Malkia E, Hakkinen K, Jappinen I, Laitinen L, Hannonen P. Effects of detraining subsequent to strength training on neuromuscular function in patients with inflammatory arthritis. Br J Rheumatol 1997;36:1075–1081.

70. Hakkinen A, Sokka T, Kautiainen H, Kotaniemi A, Hannonen P. Sustained maintenance of exercise induced muscle strength gains and normal bone mineral density in patients with early rheumatoid arthritis: a 5 year follow up. Ann Rheum Dis 2004;63:910–916.

71. Hakkinen A, Sokka T, Kotaniemi A, Hannonen P. A randomized two-year study of the effects of dynamic strength training on muscle strength, disease activity, functional capacity, and bone mineral density in early rheumatoid arthritis. Arthritis Rheum 2001;44:515–522.

72. Hakkinen A, Sokka T, Lietsalmi AM, Kautiainen H, Hannonen P. Effects of dynamic strength training on physical function, Valpar 9 work sample test, and working capacity in patients with recent-onset rheumatoid arthritis. Arthritis Rheum 2003;49:71–77.

73. Hall J, Grant J, Blake D, Taylor G, Garbutt G. Cardiorespiratory responses to aquatic treadmill walking in patients with rheumatoid arthritis. Physiother Res Int 2004;9:59–73.

74. Hall J, Skevington SM, Maddison PJ, Chapman K. A randomized and controlled trial of hydrotherapy in rheumatoid arthritis. Arthritis Care Res 1996;9:206–215.

75. Han A, Robinson V, Judd M, Taixiang W, Wells G, Tugwell P. Tai chi for treating rheumatoid arthritis. Cochrane Db Syst Rev 2004;CD004849.

76. Hansen TM, Hansen G, Langgaard AM, Rasmussen JO. Longterm physical training in rheumatoid arthritis. A randomized trial with different training programs and blinded observers. Scand J Rheumatol 1993;22:107–112.

77. Harkcom TM, Lampman RM, Banwell BF, Castor CW. Therapeutic value of graded aerobic exercise training in rheumatoid arthritis. Arthritis Rheum 1985;28:32–39.

78. Haslock I. Ankylosing spondylitis: management. In: Klippel J, Dieppe P, editors. Rheumatology. St Louis: Mosby; 1994. p 3.29.21–23.29.10.

79. Hasselbacher P. Joint physiology. In: Klippel JH, Dieppe PA, editors. Rheumatology. London: Mosby; 1994. p 1.3.1–1.3.6

80. Hazes JM, van den Ende CH. How vigorously should we exercise our rheumatoid arthritis patients? Ann Rheum Dis 1996;55:861–862.

81. Helmick CG, Lawrence RC, Pollard RA, Lloyd E, Heyse SP. Arthritis and other rheumatic conditions: who is affected now, who will be affected later? National Arthritis Data Workgroup. Arthritis Care Res 1995;8:203–211.

82. Herrero-Beaumont G, Roman J, Trabado MC, Blanco FJ, Benito P, Martin-Mola E, Paulino J, Marenco J, Porto A, Laffon A, Araujo D, Figueroa M, Branco J. Effects of glucosamine sulfate on 6-month control of knee osteoarthritis symptoms versus placebo and acetaminophen: results from the Glucosamine Unum in Die Efficacy (GUIDE) trial. Arthritis Rheum 2005;52:240.

83. Hicks JE. Exercise in patients with inflammatory arthritis and connective tissue disease. Rheum Dis Clin North Am 1990;16:845–870.

84. Hidding A, van der Linden S. Factors related to change in global health after group physical therapy in ankylosing spondylitis. Clin Rheumatol 1995;14:347–351.

85. Hidding A, van der Linden S, de Witte L. Therapeutic effects of individual physical therapy in ankylosing spondylitis related to duration of disease. Clin Rheumatol 1993;12:334–340.

86. Hidding A, van der Linden S, Gielen X, de Witte L, Dijkmans B, Moolenburgh D. Continuation of group physical therapy is necessary in ankylosing spondylitis: results of a randomized controlled trial. Arthritis Care Res 1994;7:90–96.

87. Hochberg MC, Altman RD, Brandt KD, Clark BM, Dieppe PA, Griffin MR, Moskowitz RW, Schnitzer TJ. Guidelines for the medical management of osteoarthritis. Part I. Osteoarthritis of the hip. American College of Rheumatology. Arthritis Rheum 1995;38:1535–1540.

88. Hochberg MC, Altman RD, Brandt KD, Clark BM, Dieppe PA, Griffin MR, Moskowitz RW, Schnitzer TJ. Guidelines for the medical management of osteoarthritis. Part II. Osteoarthritis of the knee. American College of Rheumatology [see comments]. Arthritis Rheum 1995;38:1541–1546.

89. Hoffman DF. Arthritis and exercise. Prim Care 1993;20:895–910.

90. Hootman JM, Helmick CG. Projections of US prevalence of arthritis and associated activity limitations. Arthritis Rheum 2006;54:226–229.

91. Hubscher O. Pattern recognition in arthritis. In: Klippel J, Dieppe P, editors. Rheumatology. Philadelphia: Mosby; 1998. p 2.3.1–2.3.6.

92. Hurley MV. The role of muscle weakness in the pathogenesis of osteoarthritis. Rheum Dis Clin North Am 1999;25:283–298,vi.

93. Hurt-Camejo E, Paredes S, Masana L, Camejo G, Sartipy P, Rosengren B, Pedreno J, Vallve JC, Benito P, Wiklund O. Elevated levels of small, low-density lipoprotein with high affinity for arterial matrix components in patients with rheumatoid arthritis: possible contribution of phospholipase A2 to this atherogenic profile. Arthritis Rheum 2001;44:2761–2767.

94. James MJ, Cleland LG, Gaffney RD, Proudman SM, Chatterton BE. Effect of exercise on 99mTc-DTPA clearance from knees with effusions. J Rheumatol 1994;21:501–504.

95. Jokl P. Prevention of disuse muscle atrophy in chronic arthritides. Rheum Dis Clin North Am 1990;16:837–844.

96. Jordan KM, Arden NK, Doherty M, Bannwarth B, Bijlsma JW, Dieppe P, Gunther K, Hauselmann H, Herrero-Beaumont G, Kaklamanis P, Lohmander S, Leeb B, Lequesne M, Mazieres B, Martin-Mola E, Pavelka K, Pendleton A, Punzi L, Serni U, Swoboda B, Verbruggen G, Zimmerman-Gorska I, Dougados M. EULAR Recommendations 2003: an evidence based approach to the management of knee osteoarthritis: report of a task force of the Standing Committee for International Clinical Studies Including Therapeutic Trials (ESCISIT). Ann Rheum Dis 2003;62:1145–1155.

97. Kenney W. ACSM's guidelines for exercise testing and prescription. 5th ed. Baltimore: Williams & Wilkins; 1995. p 373.

98. Keysor JJ. Does late-life physical activity or exercise prevent or minimize disablement? A critical review of the scientific evidence. Am J Prev Med 2003;25:129–136.

99. Kirsteins AE, Dietz F, Hwang SM. Evaluating the safety and potential use of a weight-bearing exercise, Tai-Chi Chuan, for rheumatoid arthritis patients. Am J Phys Med Rehabil 1991;70:136–141.

100. Kisner C, Colby L. Therapeutic exercise: foundations and techniques. 2nd ed. Philadelphia: Davis; 1990. p 713.

101. Klepper SE. Effects of an eight-week physical conditioning program on disease signs and symptoms in children with chronic arthritis. Arthritis Care Res 1999;12:52–60.

102. Kovar PA, Allegrante JP, MacKenzie CR, Peterson MG, Gutin B, Charlson ME. Supervised fitness walking in patients with osteoarthritis of the knee. A randomized, controlled trial. Ann Intern Med 1992;116:529–534.

103. Kurtz S, Mowat F, Ong K, Chan N, Lau E, Halpern M. Prevalence of primary and revision total hip and knee arthroplasty in the United States from 1990 through 2002. J Bone Joint Surg Am 2005;87:1487–1497.

104. Laliberte R, Kraus V, Rooks D. The everyday arthritis solution. Pleasantville, NY: Reader's Digest; 2003. p 239.

105. Lane NE. Exercise: a cause of osteoarthritis. J Rheumatol Suppl 1995;43:3–6.

106. Lane NE, Buckwalter JA. Exercise: a cause of osteoarthritis? Rheum Dis Clin North Am 1993;19:617–633.

107. Lawrence J. The influence of climate on rheumatoic complaints. In: Rheumatism in populations. London: William Heinemann Medical Books; 1977. p 505–517.

108. Lawrence RC, Helmick CG, Arnett FC, Deyo RA, Felson DT, Giannini EH, Heyse SP, Hirsch R, Hochberg MC, Hunder GG, Liang MH, Pillemer SR, Steen VD, Wolfe F. Estimates of the prevalence of arthritis and selected musculoskeletal disorders in the United States. Arthritis Rheum 1998;41:778–799.

109. Lequesne MG, Mery C, Samson M, Gerard P. Indexes of severity for osteoarthritis of the hip and knee. Validation—value in comparison with other assessment tests. Scand J Rheumatol Suppl 1987;65:85–89.

110. Lim HJ, Moon YI, Lee MS. Effects of home-based daily exercise therapy on joint mobility, daily activity, pain, and depression in patients with ankylosing spondylitis. Rheumatol Int 2005;25:225–229.

111. Lin SY, Davey RC, Cochrane T. Community rehabilitation for older adults with osteoarthritis of the lower limb: a controlled clinical trial. Clin Rehabil 2004;18:92–101.

112. Loeser RF Jr. Evaluation of musculoskeletal complaints in the older adult. Clin Geriatr Med 1998;14:401–415.

113. Lyngberg KK, Harreby M, Bentzen H, Frost B, Danneskiold-Samsoe B. Elderly rheumatoid arthritis patients on steroid treatment tolerate physical training without an increase in disease activity. Arch Phys Med Rehabil 1994;75:1189–1195.

114. Machover S, Sapecky AJ. Effect of isometric exercise on the quadriceps muscle in patients with rheumatoid arthritis. Arch Phys Med Rehabil 1966;47:737–741.

115. Mangione KK, McCully K, Gloviak A, Lefebvre I, Hofmann M, Craik R. The effects of high-intensity and low-intensity cycle ergometry in older adults with knee osteoarthritis. J Gerontol A Biol Sci Med Sci 1999;54:M184–190,

116. Mannerkorpi K. Exercise in fibromyalgia. Curr Opin Rheumatol 2005;17:190–194.

117. March LM, Bachmeier CJ. Economics of osteoarthritis: a global perspective. Baillieres Clin Rheumatol 1997;11:817–834.

118. Massy-Westropp N, Rankin W, Ahern M, Krishnan J, Hearn TC. Measuring grip strength in normal adults: reference ranges and a comparison of electronic and hydraulic instruments. J Hand Surg [Am] 2004;29:514–519.

119. McAlindon TE, Felson DT, Zhang Y, Hannan MT, Aliabadi P, Weissman B, Rush D, Wilson PW, Jacques P. Relation of dietary intake and serum levels of vitamin D to progression of osteoarthritis of the knee among participants in the Framingham Study. Ann Intern Med 1996;125:353–359.

120. McAlindon TE, Jacques P, Zhang Y, Hannan MT, Aliabadi P, Weissman B, Rush D, Levy D, Felson DT. Do antioxidant micronutrients protect against the development and progression of knee osteoarthritis? Arthritis Rheum 1996;39:648–656.

121. McCardle W, Katch F, Katch V. Exercise physiology: energy, nutrition, and human performance. 2nd ed. Philadelphia: Lea & Febiger; 1986. p 696.

122. McCarthy CJ, Mills PM, Pullen R, Roberts C, Silman A, Oldham JA. Supplementing a home exercise programme with a class-based exercise programme is more effective than home exercise alone in the treatment of knee osteoarthritis. Rheumatology (Oxford) 2004;43:880–886.

123. McCurdy D. Genetic susceptibility to the connective tissue diseases. Curr Opin Rheumatol 1999;11:399–407.

124. Mease PJ, Clauw DJ, Arnold LM, Goldenberg DL, Witter J, Williams DA, Simon LS, Strand CV, Bramson C, Martin S, Wright TM, Littman B, Wernicke JF, Gendreau RM, Crofford LJ. Fibromyalgia syndrome. J Rheumatol 2005;32:2270–2277.

125. Melton-Rogers S, Hunter G, Walter J, Harrison P. Cardiorespiratory responses of patients with rheumatoid arthritis during bicycle riding and running in water. Phys Ther 1996;76:1058–1065.

126. Messier SP, Gutekunst DJ, Davis C, DeVita P. Weight loss reduces knee-joint loads in overweight and obese older adults with knee osteoarthritis. Arthritis Rheum 2005;52:2026–2032.

127. Messier SP, Loeser RF, Miller GD, Morgan TM, Rejeski WJ, Sevick MA, Ettinger WH Jr, Pahor M, Williamson JD. Exercise and dietary weight loss in overweight and obese older adults with knee osteoarthritis: the Arthritis, Diet, and Activity Promotion Trial. Arthritis Rheum 2004;50:1501–1510.

128. Messier SP, Loeser RF, Mitchell MN, Valle G, Morgan TP, Rejeski WJ, Ettinger WH. Exercise and weight loss in obese older adults with knee osteoarthritis: a preliminary study. J Am Geriatr Soc 2000;48:1062–1072.

129. Miltner O, Schneider U, Graf J, Niethard FU. Influence of isokinetic and ergometric exercises on oxygen partial pressure measurement in the human knee joint. Adv Exp Med Biol 1997;411:183–189.

130. Minor M. Rest and exercise. In: Wegener S, Belza B, Gall E, editors. Clinical care in the rheumatic diseases. Atlanta: American College of Rheumatology; 1997. p 73–78.

131. Minor MA. Exercise in the treatment of osteoarthritis. Rheum Dis Clin North Am 1999;25:397–415,viii.

132. Minor MA, Brown JD. Exercise maintenance of persons with arthritis after participation in a class experience. Health Educ Q 1993;20:83–95.

133. Minor MA, Hewett JE, Webel RR, Anderson SK, Kay DR. Efficacy of physical conditioning exercise in patients with rheumatoid arthritis and osteoarthritis. Arthritis Rheum 1989;32:1396–1405.

134. Minor MA, Hewett JE, Webel RR, Dreisinger TE, Kay DR. Exercise tolerance and disease related measures in patients with rheumatoid arthritis and osteoarthritis. J Rheumatol 1988;15:905–911.

135. Minor MA, Johnson JC. Reliability and validity of a submaximal treadmill test to estimate aerobic capacity in women with rheumatic disease. J Rheumatol 1996;23:1517–1523.

136. Minor MA, Lane NE. Recreational exercise in arthritis. Rheum Dis Clin North Am 1996;22:563–577.

137. Minor MA, Sanford MK. The role of physical therapy and physical modalities in pain management. Rheum Dis Clin North Am 1999;25:233–248,viii.

138. Munneke M, de Jong Z, Zwinderman AH, Jansen A, Ronday HK, Peter WF, Boonman DC, van den Ende CH, Vliet Vlieland TP, Hazes JM. Adherence and satisfaction of rheumatoid arthritis patients with a long-term intensive dynamic exercise program (RAPIT program). Arthritis Rheum 2003;49:665–672.

139. Munneke M, de Jong Z, Zwinderman AH, Ronday HK, van Schaardenburg D, Dijkmans BA, Kroon HM, Vliet Vlieland TP, Hazes JM. Effect of a high-intensity weight-bearing exercise program on radiologic damage progression of the large joints in subgroups of patients with rheumatoid arthritis. Arthritis Rheum 2005;53:410–417.

140. Neuberger GB, Kasal S, Smith KV, Hassanein R, DeViney S. Determinants of exercise and aerobic fitness in outpatients with arthritis. Nurs Res 1994;43:11–17.

141. Neuberger GB, Press AN, Lindsley HB, Hinton R, Cagle PE, Carlson K, Scott S, Dahl J, Kramer B. Effects of exercise on fatigue, aerobic fitness, and disease activity measures in persons with rheumatoid arthritis. Res Nurs Health 1997;20:195–204.

142. Nielens H, Boisset V, Masquelier E. Fitness and perceived exertion in patients with fibromyalgia syndrome. Clin J Pain 2000;16:209–213.

143. Nordemar R. Physical training in rheumatoid arthritis: a controlled long-term study. II. Functional capacity and general attitudes. Scand J Rheumatol 1981;10:25–30.

144. Nordemar R, Ekblom B, Zachrisson L, Lundqvist K. Physical training in rheumatoid arthritis: a controlled long-term study. I. Scand J Rheumatol 1981;10:17–23.

145. Nordenskiold UM, Grimby G. Grip force in patients with rheumatoid arthritis and fibromyalgia and in healthy subjects. A study with the Grippit instrument. Scand J Rheumatol 1993;22:14–19.

146. Noreau L, Moffet H, Drolet M, Parent E. Dance based exercise program in rheumatoid arthritis. Feasibility in individuals with American College of Rheumatology functional class III disease. Am J Phys Med Rehabil 1997;76:109–113.

147. Noren AM, Bogren U, Bolin J, Stenstrom C. Balance assessment in patients with peripheral arthritis: applicability and reliability of some clinical assessments. Physiother Res Int 2001;6:193–204.

148. Norregaard J, Bulow PM, Lykkegaard JJ, Mehlsen J, Danneskiold-Samsooe B. Muscle strength, working capacity and effort in patients with fibromyalgia. Scand J Rehabil Med 1997;29:97–102.

149. O'Driscoll SL, Jayson MI, Baddeley H. Neck movements in ankylosing spondylitis and their responses to physiotherapy. Ann Rheum Dis 1978;37:64–66.

150. O'Grady M, Fletcher J, Ortiz S. Therapeutic and physical fitness exercise prescription for older adults with joint disease: an evidence-based approach. Rheum Dis Clin North Am 2000;26:617–646.

151. O'Reilly S, Jones A, Doherty M. Muscle weakness in osteoarthritis. Curr Opin Rheumatol 1997;9:259–262.

152. Ottawa Panel. Ottawa Panel evidence-based clinical practice guidelines for therapeutic exercises in the management of rheumatoid arthritis in adults. Phys Ther 2004;84:934–972.

153. Otterness IG, Zimmerer RO, Swindell AC, Poole AR, Saxne T, Heinegard D, Ionescu M, Weiner E. An examination of some molecular markers in blood and urine for discriminating patients with osteoarthritis from healthy individuals. Acta Orthop Scand 1995;66(Suppl 266):148–150.

154. Pavelka K, Gatterova J, Olejarova M, Machacek S, Giacovelli G, Rovati LC. Glucosamine sulfate use and delay of progression of knee osteoarthritis: a 3-year, randomized, placebo-controlled, double-blind study. Arch Intern Med 2002;162:2113–2123.

155. Penninx BW, Abbas H, Ambrosius W, Nicklas BJ, Davis C, Messier SP, Pahor M. Inflammatory markers and physical function among older adults with knee osteoarthritis. J Rheumatol 2004;31:2027–2031.

156. Penninx BW, Messier SP, Rejeski WJ, Williamson JD, DiBari M, Cavazzini C, Applegate WB, Pahor M. Physical exercise and the prevention of disability in activities of daily living in older persons with osteoarthritis. Arch Intern Med 2001;161:2309–2316.

157. Pham T, van der Heijde D, Altman RD, Anderson JJ, Bellamy N, Hochberg M, Simon L, Strand V, Woodworth T, Dougados M. OMERACT-OARSI initiative: Osteoarthritis Research Society International set of responder criteria for osteoarthritis clinical trials revisited. Osteoarthritis Cartilage 2004;12:389–399.

158. Philbin EF, Ries MD, French TS. Feasibility of maximal cardiopulmonary exercise testing in patients with end-stage arthritis of the hip and knee prior to total joint arthroplasty. Chest 1995;108:174–181.

159. Poole AR, Rizkalla G, Ionescu M, Reiner A, Brooks E, Rorabeck C, Bourne R, Bogoch E. Osteoarthritis in the human knee: a dynamic process of cartilage matrix degradation, synthesis and reorganization. Agents Actions Suppl 1993;39:3–13.

160. Rall LC, Roubenoff R. Body composition, metabolism, and resistance exercise in patients with rheumatoid arthritis. Arthritis Care Res 1996;9:151–156.

161. Rasmussen JO, Hansen TM. Physical training for patients with ankylosing spondylitis. Arthritis Care Res 1989;2:25–27.

162. Raub JA. Psychophysiologic effects of Hatha Yoga on musculoskeletal and cardiopulmonary function: a literature review. J Altern Complement Med 2002;8:797–812.

163. Reginster JY, Deroisy R, Rovati LC, Lee RL, Lejeune E, Bruyere O, Giacovelli G, Henrotin Y, Dacre JE, Gossett C. Long-term effects of glucosamine sulphate on osteoarthritis progression: a randomised, placebo-controlled clinical trial. Lancet 2001;357:251–256.

164. Ries MD, Philbin EF, Groff GD. Relationship between severity of gonarthrosis and cardiovascular fitness. Clin Orthop Relat Res1995;313:169–176.

165. Roddy E, Doherty M. Changing life-styles and osteoarthritis: what is the evidence? Best Pract Res Clin Rheumatol 2006;20:81–97.

166. Roddy E, Zhang W, Doherty M. Aerobic walking or strengthening exercise for osteoarthritis of the knee? A systematic review. Ann Rheum Dis 2005;64:544–548.

167. Roddy E, Zhang W, Doherty M, Arden NK, Barlow J, Birrell F, Carr A, Chakravarty K, Dickson J, Hay E, Hosie G, Hurley M, Jordan KM, McCarthy C, McMurdo M, Mockett S, O'Reilly S, Peat G, Pendleton A, Richards S. Evidence-based recommendations for the role of exercise in the management of osteoarthritis of the hip or knee—the MOVE consensus. Rheumatology (Oxford) 2005;44:67–73.

168. Roos EM. Joint injury causes knee osteoarthritis in young adults. Curr Opin Rheumatol 2005;17:195–200.

169. Rudwaleit M, Khan MA, Sieper J. The challenge of diagnosis and classification in early ankylosing spondylitis: do we need new criteria? Arthritis Rheum 2005;52:1000–1008.

170. Ruof J, Sangha O, Stucki G. Comparative responsiveness of 3 functional indices in ankylosing spondylitis. J Rheumatol 1999;26:1959–1963.

171. Salaffi F, Stancati A, Neri R, Grassi W, Bombardieri S. Measuring functional disability in early rheumatoid arthritis: the validity, reliability and responsiveness of the Recent-Onset Arthritis Disability (ROAD) index. Clin Exp Rheumatol 2005;23:S31–42.

172. Santos H, Brophy S, Calin A. Exercise in ankylosing spondylitis: how much is optimum? J Rheumatol 1998;25:2156–2160.

173. Saxne T. Differential release of molecular markers in joint disease. Acta Orthop Scand 1995;66(Suppl 266):80–83.

174. Schilke JM, Johnson GO, Housh TJ, O'Dell JR. Effects of muscle-strength training on the functional status of patients with osteoarthritis of the knee joint. Nurs Res 1996;45:68–72.

175. Schweitzer M, Resnick D. Enthesopathy. In: Klippel J, Dieppe P, editors. Rheumatology. St Louis: Mosby; 1994. p 3.27.21–23.27.26.

176. Scully R, Barnes M, editors. Physical therapy. Philadelphia: Lippincott; 1989.

177. Semble EL, Loeser RF, Wise CM. Therapeutic exercise for rheumatoid arthritis and osteoarthritis. Semin Arthritis Rheum 1990;20:32–40.

178. Shephard RJ, Shek PN. Autoimmune disorders, physical activity, and training, with particular reference to rheumatoid arthritis. Exerc Immunol Rev 1997;3:53–67.

179. Shih M, Hootman JM, Kruger J, Helmick CG. Physical activity in men and women with arthritis national health interview survey, 2002. Am J Prev Med 2006;30:385–393.

180. Song R, Lee EO, Lam P, Bae SC. Effects of tai chi exercise on pain, balance, muscle strength, and perceived difficulties in physical functioning in older women with osteoarthritis: a randomized clinical trial. J Rheumatol 2003;30:2039–2044.

181. Spector TD, Hart DJ, Nandra D, Doyle DV, Mackillop N, Gallimore JR, Pepys MB. Low-level increases in serum C-reactive protein are present in early osteoarthritis of the knee and predict progressive disease. Arthritis Rheum 1997;40:723–727.

182. Stenstrom CH. Home exercise in rheumatoid arthritis functional class II: goal setting versus pain attention. J Rheumatol 1994;21:627–634.

183. Stenstrom CH. Therapeutic exercise in rheumatoid arthritis. Arthritis Care Res 1994;7:190–197.

184. Stenstrom CH, Lindell B, Swanberg E, Swanberg P, Harms-Ringdahl K, Nordemar R. Intensive dynamic training in water for rheumatoid arthritis functional class II—a long-term study of effects. Scand J Rheumatol 1991;20:358–365.

185. Stenstrom CH, Minor MA. Evidence for the benefit of aerobic and strengthening exercise in rheumatoid arthritis. Arthritis Rheum 2003;49:428–434.

186. Steultjens MP, Dekker J, van Baar ME, Oostendorp RA, Bijlsma JW. Range of joint motion and disability in patients with osteoarthritis of the knee or hip. Rheumatology (Oxford) 2000;39:955–961.

187. Suomi R, Collier D. Effects of arthritis exercise programs on functional fitness and perceived activities of daily living measures in older adults with arthritis. Arch Phys Med Rehabil 2003;84:1589–1594.

188. Suomi R, Lindauer S. Effectiveness of Arthritis Foundation Aquatic Program on strength and range of motion in women with arthritis. Journal of Aging and Physical Activity 1997;5:341–351.

189. Tackson SJ, Krebs DE, Harris BA. Acetabular pressures during hip arthritis exercises. Arthritis Care Res 1997;10:308–319.

190. Valim V, Oliveira LM, Suda AL, Silva LE, Faro M, Neto TL, Feldman D, Natour J. Peak oxygen uptake and ventilatory anaerobic threshold in fibromyalgia. J Rheumatol 2002;29:353–357.

191. van Baar ME, Assendelft WJ, Dekker J, Oostendorp RA, Bijlsma JW. Effectiveness of exercise therapy in patients with osteoarthritis of the hip or knee: a systematic review of randomized clinical trials. Arthritis Rheum 1999;42:1361–1369.

192. van den Ende CH, Hazes JM, le Cessie S, Mulder WJ, Belfor DG, Breedveld FC, Dijkmans BA. Comparison of high and low intensity training in well controlled rheumatoid arthritis. Results of a randomised clinical trial. Ann Rheum Dis 1996;55:798–805.

193. Van den Ende CH, Vliet Vlieland TP, Munneke M, Hazes JM. Dynamic exercise therapy in rheumatoid arthritis: a systematic review. Br J Rheumatol 1998;37:677–687.

194. Van Deusen J, Harlowe D. The efficacy of the ROM Dance Program for adults with rheumatoid arthritis. Am J Occup Ther 1987;41:90–95.

195. van Gool CH, Penninx BW, Kempen GI, Miller GD, van Eijk JT, Pahor M, Messier SP. Determinants of high and low attendance to diet and exercise interventions among overweight and obese older adults. Results from the arthritis, diet, and activity promotion trial. Contemp Clin Trials 2005.

196. van Gool CH, Penninx BW, Kempen GI, Rejeski WJ, Miller GD, van Eijk JT, Pahor M, Messier SP. Effects of exercise adherence on physical function among overweight older adults with knee osteoarthritis. Arthritis Rheum 2005;53:24–32.

197. Van Tubergen A, Boonen A, Landewe R, Rutten-Van Molken M, Van Der Heijde D, Hidding A, Van Der Linden S. Cost effectiveness of

combined spa-exercise therapy in ankylosing spondylitis: a randomized controlled trial. Arthritis Rheum 2002;47:459–467.

198. van Tubergen, A, Landewe R, van der Heijde D, Hidding A, Wolter N, Asscher M, Falkenbach A, Genth E, The HG, van der Linden S. Combined spa-exercise therapy is effective in patients with ankylosing spondylitis: a randomized controlled trial. Arthritis Rheum 2001;45:430–438.

199. Verbrugge LM, Patrick DL. Seven chronic conditions: their impact on US adults' activity levels and use of medical services. Am J Public Health 1995;85:173–182.

200. Verbruggen G. Chondroprotective drugs in degenerative joint diseases. Rheumatology (Oxford) 2006;45:129–138.

201. Weiss TE, Gum OB, Biundo JJ Jr. Rheumatic diseases. 1. Differential diagnosis. Postgrad Med 1976;60:141–150.

202. Wolfe F, Smythe HA, Yunus MB, Bennett RM, Bombardier C, Goldenberg DL, Tugwell P, Campbell SM, Abeles M, Clark P, et al. The American College of Rheumatology 1990 criteria for the classification of fibromyalgia. Report of the Multicenter Criteria Committee. Arthritis Rheum 1990;33:160–172.

203. Yelin E, Callahan LF. The economic cost and social and psychological impact of musculoskeletal conditions. National Arthritis Data Work Groups. Arthritis Rheum 1995;38:1351–1362.

204. Ytterberg SR, Mahowald ML, Krug HE. Exercise for arthritis. Baillieres Clin Rheumatol 1994;8:161–189.

205. Zhang W, Doherty M, Arden N, Bannwarth B, Bijlsma J, Gunther KP, Hauselmann HJ, Herrero-Beaumont G, Jordan K, Kaklamanis P, Leeb B, Lequesne M, Lohmander S, Mazieres B, Martin-Mola E, Pavelka K, Pendleton A, Punzi L, Swoboda B, Varatojo R, Verbruggen G, Zimmermann-Gorska I, Dougados M. EULAR evidence based recommendations for the management of hip osteoarthritis: report of a task force of the EULAR Standing Committee for International Clinical Studies Including Therapeutics (ESCISIT). Ann Rheum Dis 2005;64:669–681.

206. Zochling J, van der Heijde D, Burgos-Vargas R, Collantes E, Davis JC Jr, Dijkmans B, Dougados M, Geher P, Inman RD, Khan MA, Kvien TK, Leirisalo-Repo M, Olivieri I, Pavelka K, Sieper J, Stucki G, Sturrock RD, van der Linden S, Wendling D, Bohm H, van Royen BJ, Braun J. ASAS/EULAR recommendations for the management of ankylosing spondylitis. Ann Rheum Dis 2006;65:442–452.

207. Zochling J, van der Heijde D, Dougados M, Braun J. Current evidence for the management of ankylosing spondylitis: a systematic literature review for the ASAS/EULAR management recommendations in ankylosing spondylitis. Ann Rheum Dis 2006;65:423–432.

CHAPTER 27

1. American College of Sports Medicine position stand. Osteoporosis and exercise. Med Sci Sports Exerc 1995;27:i–vii.

2. Surgeon general's report on physical activity and health. From the Centers for Disease Control and Prevention. JAMA 1996;276:522.

3. American College of Sports Medicine position stand. Exercise and physical activity for older adults. Med Sci Sports Exerc 1998;30:992–1008.

4. Anderson GL, Limacher M, Assaf AR, Bassford T, Beresford SA, Black H, Bonds D, Brunner R, Brzyski R, Caan B, Chlebowski R, Curb D, Gass M, Hays J, Heiss G, Hendrix S, Howard BV, Hsia J, Hubbell A, Jackson R, Johnson KC, Judd H, Kotchen JM, Kuller L, LaCroix AZ, Lane D, Langer RD, Lasser N, Lewis CE, Manson J, Margolis K, Ockene J, O'Sullivan MJ, Phillips L, Prentice RL, Ritenbaugh C, Robbins J, Rossouw JE, Sarto G, Stefanick ML, Van Horn L, Wactawski-Wende J, Wallace R, Wassertheil-Smoller S. Effects of conjugated equine estrogen in postmenopausal women with hysterectomy: the Women's Health Initiative randomized controlled trial. JAMA 2004;291:1701–1712.

5. Baran D. Osteoporosis. Efficacy and safety of a bisphosphonate dosed once weekly. Geriatrics 2001;56:28–32.

6. Bensen R, Adachi JD, Papaioannou A, Ioannidis G, Olszynski WP, Sebaldt RJ, Murray TM, Josse RG, Brown JP, Hanley DA, Petrie A, Puglia M, Goldsmith CH, Bensen W. Evaluation of easily measured risk factors in the prediction of osteoporotic fractures. BMC Musculoskelet Disord 2005;6:47.

7. Body JJ, Gaich GA, Scheele WH, Kulkarni PM, Miller PD, Peretz A, Dore RK, Correa-Rotter R, Papaioannou A, Cumming DC, Hodsman AB. A randomized double-blind trial to compare the efficacy of teriparatide [recombinant human parathyroid hormone (1-34)] with alendronate in postmenopausal women with osteoporosis. J Clin Endocrinol Metab 2002;87:4528–4535.

8. Bonnick SL. Bone densitometry in clinical practice. Application and interpretation. Totowa, NJ: Humana Press; 1998.

9. Bonnick SL, Shulman L. Monitoring osteoporosis therapy: bone mineral density, bone turnover markers, or both? Am J Med 2006;119:S25–S31.

10. Borer KT. Physical activity in the prevention and amelioration of osteoporosis in women: interaction of mechanical, hormonal and dietary factors. Sports Med 2005;35:779–830.

11. Cauley JA, Norton L, Lippman ME, Eckert S, Krueger KA, Purdie DW, Farrerons J, Karasik A, Mellstrom D, Ng KW, Stepan JJ, Powles TJ, Morrow M, Costa A, Silfen SL, Walls EL, Schmitt H, Muchmore DB, Jordan VC, Ste-Marie LG. Continued breast cancer risk reduction in postmenopausal women treated with raloxifene: 4-year results from the MORE trial. Multiple outcomes of raloxifene evaluation. Breast Cancer Res Treat 2001;65:125–134.

12. Chesnut CH, Silverman S, Andriano K, Genant H, Gimona A, Harris S, Kiel D, LeBoff M, Maricic M, Miller P, Moniz C, Peacock M, Richardson P, Watts N, Baylink D. A randomized trial of nasal spray salmon calcitonin in postmenopausal women with established osteoporosis: the prevent recurrence of osteoporotic fractures study. Am J Med 2000;109:267–276.

13. Delmas PD, Bjarnason NH, Mitlak BH, Ravoux AC, Shah AS, Huster WJ, Draper M, Christiansen C. Effects of raloxifene on bone mineral density, serum cholesterol concentrations, and uterine endometrium in postmenopausal women. N Engl J Med 1997;337:1641–1647.

14. Drinkwater BL, Nilson K, Chesnut CH III, Bremner WJ, Shainholtz S, Southworth MB. Bone mineral content of amenorrheic and eumenorrheic athletes. N Engl J Med 1984;311:277–281.

15. Emkey RD, Ettinger M. Improving compliance and persistence with bisphosphonate therapy for osteoporosis. Am J Med 2006;119:S18–S24.

16. Ettinger BF, Genant HK, Cann CE. Long-term estrogen replacement therapy prevents bone loss and fractures. Ann Intern Med 1985;102:319–324.

17. Fogelman I, Ribot C, Smith R, Ethgen D, Sod E, Reginster JY. Risedronate reverses bone loss in postmenopausal women with low bone mass: results from a multinational, double-blind, placebo-controlled trial. BMD-MN Study Group. J Clin Endocrinol Metab 2000;85:1895–1900.

18. Frost HM. Some ABC's of skeletal pathophysiology. 6. The growth/modeling/remodeling distinction. Calcif Tissue Int 1991;49:301–302.

19. Gabriel SE, Tosteson AN, Leibson CL, Crowson CS, Pond GR, Hammond CS, Melton LJ III. Direct medical costs attributable to osteoporotic fractures. Osteoporos Int 2002;13:323–330.

20. Gambacciani M, Spinetti A, Taponeco F, Cappagli B, Maffei S, Manetti P, Piaggesi L, Fioretti P. Bone loss in perimenopausal women: a longitudinal study. Maturitas 1994;18:191–197.

21. Garnero P, Hausherr E, Chapuy MC, Marcelli C, Grandjean H, Muller C, Cormier C, Breart G, Meunier PJ, Delmas PD. Markers of bone resorption predict hip fracture in elderly women: the EPIDOS Prospective Study. J Bone Miner Res 1996;11:1531–1538.

22. Genant HK, Engelke K, Fuerst T, Gluer C, Grampp S, Harris ST, Jergas M, Lang T, Lu Y, Majumdar S, Mathur A, Takada M. Nonin-

vasive assessment of bone mineral and structure: state of the art. J Bone Miner Res 1996;11:707–730.

23. Heaney RP, Weaver CM. Newer perspectives on calcium nutrition and bone quality. J Am Coll Nutr 2005;24:574S–581S.

24. Hillard TC, Whitcroft SJ, Marsh MS, Ellerington MC, Lees B, Whitehead MI, Stevenson JC. Long-term effects of transdermal and oral hormone replacement therapy on postmenopausal bone loss. Osteoporos Int 1994;4:341–348.

25. Ishida Y, Kawai S. Comparative efficacy of hormone replacement therapy, etidronate, calcitonin, alfacalcidol, and vitamin K in postmenopausal women with osteoporosis: the Yamaguchi Osteoporosis Prevention Study. Am J Med 2004;117:549–555.

26. Jonnavithula S, Warren MP, Fox RP, Lazaro MI. Bone density is compromised in amenorrheic women despite return of menses: a 2-year study. Obstet Gynecol 1993;81:669–674.

27. Kamel HK. Male osteoporosis: new trends in diagnosis and therapy. Drugs Aging 2005;22:741–748.

28. Kanis JA. The use of calcium in the management of osteoporosis. Bone 1999;24:279–290.

29. Kanis JA, Melton LJ, Christiansen C, Johnston CC, Khaltaev N. The diagnosis of osteoporosis. J Bone Miner Res 1994;9:1137–1141.

30. Karachalios T, Lyritis GP, Kaloudis J, Roidis N, Katsiri M. The effects of calcitonin on acute bone loss after pertrochanteric fractures. A prospective, randomised trial. J Bone Joint Surg Br 2004;86:350–358.

31. Kohrt WM, Birge SJJ. Differential effects of estrogen treatment on bone mineral density of the spine, hip, wrist and total body in late postmenopausal women. Osteoporos Int 1995;5:150–155.

32. Kwok Y, Kim C, Grady D, Segal M, Redberg R. Meta-analysis of exercise testing to detect coronary artery disease in women. Am J Cardiol 1999;83:660–666.

33. Lewiecki EM, Watts NB, McClung MR, Petak SM, Bachrach LK, Shepherd JA, Downs RW Jr. Official positions of the international society for clinical densitometry. J Clin Endocrinol Metab 2004;89:3651–3655.

34. Lufkin EG, Whitaker MD, Nickelsen T, Argueta R, Caplan RH, Knickerbocker RK, Riggs BL. Treatment of established postmenopausal osteoporosis with raloxifene: a randomized trial. J Bone Miner Res 1998;13:1747–1754.

35. Marcus R, Wong M, Heath H III, Stock JL. Antiresorptive treatment of postmenopausal osteoporosis: comparison of study designs and outcomes in large clinical trials with fracture as an endpoint. Endocr Rev 2002;23:16–37.

36. Mussolino ME, Madans JH, Gillum RF. Bone mineral density and mortality in women and men: the NHANES I epidemiologic follow-up study. Ann Epidemiol 2003;13:692–697.

37. Neer RM, Arnaud CD, Zanchetta JR, Prince R, Gaich GA, Reginster JY, Hodsman AB, Eriksen EF, Ish-Shalom S, Genant HK, Wang O, Mitlak BH. Effect of parathyroid hormone (1-34) on fractures and bone mineral density in postmenopausal women with osteoporosis. N Engl J Med 2001;344:1434–1441.

38. Parfitt AM. The cellular basis of bone remodeling: the quantum concept reexamined in light of recent advances in the cell biology of bone. Calcif Tissue Int 1984;36 Suppl 1:S37–45.

39. Pongchaiyakul C, Nguyen ND, Eisman JA, Nguyen TV. Clinical risk indices, prediction of osteoporosis, and prevention of fractures: diagnostic consequences and costs. Osteoporos Int 2005;16:1444–1450.

40. Pouilles JM, Tremollieres F, Ribot C. Vertebral bone loss in perimenopause. Results of a 7-year longitudinal study. Presse Med 1996;25:277–280.

41. Raisz LG. Pathogenesis of osteoporosis: concepts, conflicts, and prospects. J Clin Invest 2005;115:3318–3325.

42. Reid DM, Hughes RA, Laan RF, Sacco-Gibson NA, Wenderoth DH, Adami S, Eusebio RA, Devogelaer JP. Efficacy and safety of daily risedronate in the treatment of corticosteroid-induced osteoporosis in men and women: a randomized trial. European Corticosteroid-Induced Osteoporosis Treatment Study. J Bone Miner Res 2000;15:1006–1013.

43. Richardson SJ, Nelson JF. Follicular depletion during the menopausal transition. Annals of the New York Academy of Science 1990;592:13–20.

44. Richelson LS, Wahner HW, Melton LJ, Riggs BL. Relative contributions of aging and estrogen deficiency to postmenopausal bone loss. N Engl J Med 1984;311:1273–1275.

45. Riggs BL, Melton LJ. Involutional osteoporosis. N Engl J Med 1986;311:1676–1686.

46. Riggs BL, Wahner HW, Dunn WL, Mazess RB, Offord KP, Melton LJ. Differential changes in bone mineral density of the appendicular and axial skeleton with aging: relationship to spinal osteoporosis. J Clin Invest 1981;67:328–335.

47. Rizzoli R, Greenspan SL, Bone G III, Schnitzer TJ, Watts NB, Adami S, Foldes AJ, Roux C, Levine MA, Uebelhart B, Santora AC, Kaur A, Peverly CA, Orloff JJ. Two-year results of once-weekly administration of alendronate 70 mg for the treatment of postmenopausal osteoporosis. J Bone Miner Res 2002;17:1988–1996.

48. Rodan GA. Bone mass homeostasis and bisphosphonate action. Bone 1997;20:1–4.

49. Rosen CJ, Hochberg MC, Bonnick SL, McClung M, Miller P, Broy S, Kagan R, Chen E, Petruschke RA, Thompson DE, de Papp AE. Treatment with once-weekly alendronate 70 mg compared with once-weekly risedronate 35 mg in women with postmenopausal osteoporosis: a randomized double-blind study. J Bone Miner Res 2005;20:141–151.

50. Rossouw JE, Anderson GL, Prentice RL, LaCroix AZ, Kooperberg C, Stefanick ML, Jackson RD, Beresford SA, Howard BV, Johnson KC, Kotchen JM, Ockene J. Risks and benefits of estrogen plus progestin in healthy postmenopausal women: principal results From the Women's Health Initiative randomized controlled trial. JAMA 2002;288:321–333.

51. Saag KG, Emkey R, Schnitzer TJ, Brown JP, Hawkins F, Goemaere S, Thamsborg G, Liberman UA, Delmas PD, Malice MP, Czachur M, Daifotis AG. Alendronate for the prevention and treatment of glucocorticoid-induced osteoporosis. Glucocorticoid-Induced Osteoporosis Intervention Study Group. N Engl J Med 1998;339:292–299.

52. Schnitzer T, Bone HG, Crepaldi G, Adami S, McClung M, Kiel D, Felsenberg D, Recker RR, Tonino RP, Roux C, Pinchera A, Foldes AJ, Greenspan SL, Levine MA, Emkey R, Santora AC, Kaur A, Thompson DE, Yates J, Orloff JJ. Therapeutic equivalence of alendronate 70 mg once-weekly and alendronate 10 mg daily in the treatment of osteoporosis. Alendronate Once-Weekly Study Group. Aging (Milano) 2000;12:1–12.

53. Shea B, Wells G, Cranney A, Zytaruk N, Robinson V, Griffith L, Hamel C, Ortiz Z, Peterson J, Adachi J, Tugwell P, Guyatt G. Calcium supplementation on bone loss in postmenopausal women. Cochrane Db Syst Rev 2004;CD004526.

54. Solomon CG, Dluhy RG. Rethinking postmenopausal hormone therapy. N Engl J Med 2003;348:579–580.

55. Uchida S, Taniguchi T, Shimizu T, Kakikawa T, Okuyama K, Okaniwa M, Arizono H, Nagata K, Santora AC, Shiraki M, Fukunaga M, Tomomitsu T, Ohashi Y, Nakamura T. Therapeutic effects of alendronate 35 mg once weekly and 5 mg once daily in Japanese patients with osteoporosis: a double-blind, randomized study. J Bone Miner Metab 2005;23:382–388.

56. Vainionpaa A, Korpelainen R, Vihriala E, Rinta-Paavola A, Leppaluoto J, Jamsa T. Intensity of exercise is associated with bone density change in premenopausal women. Osteoporos Int 2006;17:455–463.

57. Walsh BW, Kuller LH, Wild RA, Paul S, Farmer M, Lawrence JB, Shah AS, Anderson PW. Effects of raloxifene on serum lipids and

coagulation factors in healthy postmenopausal women. JAMA 1998;279:1445–1451.

58. Weinstein L, Ullery B, Bourguignon C. A simple system to determine who needs osteoporosis screening. Obstet Gynecol 1999;93:757–760.

CHAPTER 28

1. American College of Sports Medicine. ACSM's guidelines for exercise testing and prescription. 7th ed. Philadelphia: Lippincott Williams & Wilkins; 2006.

2. American College of Sports Medicine. ACSM position stand on progression models in resistance training for healthy adults. Med Sci Sport Exer 2002;34:364–380.

3. American College of Sports Medicine. ACSM position stand on the recommended quantity and quality of exercise for developing and maintaining cardiorespiratory and muscular fitness and flexibility in healthy adults. Med Sci Sport Exer 1998;30:975-991.

4. Barbis JM. Prevention and management of low back pain. In: Rothman J, Levine RE, editors. Prevention practice: strategies for physical therapy and occupational therapy. Philadelphia: Saunders; 1992. p 63–72.

5. Biering-Sörensen F. Physical measurements as risk indicators for low-back trouble over a one year period. Spine 1984;9:106–119.

6. Bigos S, Bowyer O, Braen G, Brown K, Deyo R, Haldeman S, Hart J, Johnson E, Keller R, Kiddo D, Liang M, Nelson R, Nordin M, Owen B, Pope M, Schwartz R, Stewart D, Susman J, Triano J, Tripp L, Turk D, Watts C, Weinstein J. Acute low back problems in adults. Clinical Practice Guideline No. 14. AHCPR Publication No. 95-0642. Rockville, MD: Agency for Health Care Policy and Research, Public Health Service, U.S. Department of Health and Human Services; 1994.

7. Bigos SJ, Davis GE. Scientific application of sports medicine principles for acute low back pain. J Orthop Sport Phys 1996;24:192-207.

8. Bogduck N. Clinical anatomy of the lumbar spine and sacrum. 4th. ed. New York: Churchill Livingstone; 2005.

9. Cady LD, Bischoff DP, O'Connell ER, Thomas PC, Allan JH. Strength and fitness and subsequent back injuries in firefighters. JOM-J Occup Med 1979;21:269–272.

10. Cassidy JD, Côté P, Carroll LJ, Kristman V. Incidence and course of low back pain episodes in the general population. Spine 2005;30:2817–2823.

11. Cherkin DC, Deyo RA, Wheeler K, Ciol MA. Physician variation in diagnostic testing for low back pain. Who you see is what you get. Arthritis Rheum 1994;37:15–22.

12. Croft PR, Dunn KM, Raspe H. Course and prognosis of back pain in primary care: the epidemiological perspective. Pain 2006;122:1–3.

13. Croft PR, Macfarlane GJ, Papageorgiou AC, Silman AJ. Outcome of low back pain in general practice: a one year follow-up study. BMJ 1998;316:1356–1359.

14. Deyo RA. Diagnostic evaluation of LBP: reaching a specific diagnosis is often impossible. Arch Intern Med 2002;162:1444–1448.

15. Deyo RA, Diehl AK, Rosenthal M. How many days of bed rest for acute low back pain? A randomized clinical trial. New Engl J Medicine 1986;315:1064–1070.

16. Deyo RA, Weinstein JN. Low back pain. New Engl J Med 2001;344:363–370.

17. D'Orazio BP, editor. Low back pain handbook. Boston: Butterworth-Heinemann; 1999.

18. D'Orazio BP, Tritsch C, Vath SA, Rennie MA. Postoperative lumbar rehabilitation. In: D'Orazio BP, editor. Low back pain handbook. Boston: Butterworth-Heinemann; 1999. p 277–308.

19. Fritz JM, Delitto A, Erhard RE. Comparison of classification-based physical therapy with therapy based on clinical practice guidelines for patients with acute low back pain. Spine 2003;28:1363–1372.

20. Frymoyer JW, Cats-Baril WL. An overview of the incidences and costs of low back pain. Orthop Clin N Am 1991;22:263–271.

21. Gibson JNA, Waddell G. Surgery for degenerative lumbar spondylosis. Cochrane Db Syst Rev 2005;4:CD001352.

22. Goldberg MS, Scott SC, Mayo NE. A review of the association between cigarette smoking and the development of nonspecific back pain and related outcomes. Spine 2000;25:995–1014.

23. Harreby M, Hesselsøe G, Kjer J, Neergaard K. Low back pain and physical exercise in leisure time in 38-year-old men and women: a 25-year prospective cohort study of 640 school children. Eur Spine J 1997;6:181–186.

24. Harreby M, Kjer J, Hesselsøe G, Neergaard K. Epidemiological aspects and risk factors for low back pain in 38-year-old men and women: a 25-year prospective cohort study of 640 school children. Eur Spine J 1996;5:312–318.

25. Hayden JA, van Tulder MW, Malmivaara A, Koes BW. Exercise therapy for treatment of non-specific low back pain. Cochrane Db Syst Rev 2006;3:CD000335.

26. Hayden JA, van Tulder MW, Malmivaara A, Koes BW. Meta-analysis: exercise therapy for treatment of non-specific low back pain. Ann Intern Med 2005;142:765–775.

27. Hestbaek L, Leboeuf-Yde C, Kyvik KO, Manniche C. The course of low back pain from adolescence to adulthood: eight-year follow-up of 9600 twins. Spine 2006;31:468–472.

28. Hestbaek L, Leboeuf-Yde C, C Manniche. Low back pain: what is the long-term course? A review of studies of general patient populations. Eur Spine J 2005;12:149–165.

29. Heymans MW, van Tulder MW, Esmail R, Bombardier C, Koes BW. Back schools for non-specific low-back pain. Cochrane Db Syst Rev 2004;4:CD000261.

30. Hochschule, SH, Cotler HB, Guyer RD. The tissue origin of mechanical low back pain and sciatica as identified by spinal microsurgery. In: Rehabilitation of the spine. St. Louis: Mosby-Year Book; 1993. p 595–600.

31. Jackson R. Postural dynamics: functional causes of low back pain. In: D'Orazio BP, editor. Low back pain handbook. Boston: Butterworth-Heinemann; 1999. p 159–191.

32. Jarvik JG, Deyo RA. Diagnostic evaluation of low back pain with emphasis on imaging. Ann Intern Med 2002;37:586–597.

33. Jarvik JG, Hollingworth W, Heagerty PJ, Haynor DR, Boyko EJ, Deyo RA. Three-year incidence of low back pain in an initially asymptomatic cohort: clinical and imaging risk factors. Spine 2005;30:1541–1548.

34. Jayson MI. ABC of work related disorders: back pain. BMJ 1996;313:355–358.

35. Juker D, McGill SM, Kropf P, Steffen T. Quantitative intramuscular myoelectric activity of lumbar portions of psoas and the abdominal wall during a wide variety of tasks. Med Sci Sport Exer 1998;30:301–310.

36. Karjalainen K, Malmivaara A, van Tulder M, Roine R, Juhiainen M, Hurri H, Koes B. Multidisciplinary biopsychosocial rehabilitation for subacute low-back pain among working age adults. Cochrane Db Syst Rev 2003;2:CD002193.

37. Kent PM, Keating J. The epidemiology of low back pain in primary care. Chiropr Osteop 2005;13:13 [Online Journal] DOI:10.1186/1746-1340-13-13. www.chiroandosteo.com/content/13/1/13

38. Koes BW, van Tulder MW, Ostelo R, Burton AK, Waddell G. Clinical guidelines for the management of low back pain in primary care: an international comparison. Spine 2001;26:2504–2513.

39. Koes BW, van Tulder MW, Thomas S. Diagnosis and treatment of low back pain. BMJ 2006;332:1430–1434.

40. Leboeuf-Yde C. Body weight and low back pain. A systematic literature review of 56 journal articles reporting on 65 epidemiological studies. Spine 2000;25:226–237.

41. Leboeuf-Yde C, Kyvik KO, Bruun NH. Low back pain and lifestyle. Part II: obesity. Information from a population-based sample of 29,424 twin subjects. Spine 1999;24:779–784.

42. Linton SJ. Chronic back pain: activities training and physical therapy. Behav Med 1994;20:105–111.

43. Linton SJ. Chronic back pain: integrating psychological and physical therapy. Behav Med 1994;20:101–104.

44. Loney PL, Stratford PW. The prevalence of low back pain in adults: a methodological review of the literature. Phys Ther 1999;79:384–396.

45. Long A, Donelson R, Fung T. Does it matter which exercise? Spine 2004;29:2593–2602.

46. Luoto S, Heliovaara M, Hurri H, Alaranta M. Static back endurance and the risk of low back pain. Clin Biomech 1995;10:323–324.

47. Maher CG. Effective physical treatment for chronic low back pain. Orthop Clin N Am 2004;35:57–64.

48. Manniche C. Clinical benefit of intensive dynamic exercises for low back pain. Scand J of Med Sci Spor 1996;6:82–87.

49. Manniche C, Lundberg E, Christensen I, Bentzen L, Hesselsøe G. Intensive dynamic back exercises for chronic low back pain: a clinical trial. Pain 1991;47:53–63.

50. Matsui H, Kanamori M, Ishihara H, Yudoh K, Naruse Y, Tsuji H. Familial predisposition for lumbar degenerative disc disease. A case-control study. Spine 1998;23:1029–1034.

51. McGill S. Low back disorders. Evidence-based prevention and rehabilitation. Champaign, IL: Human Kinetics; 2002.

52. McGill SM. Low back exercises: evidence for improving exercise regimens. Phys Ther 1998;78:754–765.

53. McGill SM. Low back exercises: prescription for the healthy back and when recovering from injury. In: Roitman JL, editor. American College of Sports Medicine's resource manual for guidelines for exercise testing and prescription. 3rd ed. Baltimore: Williams & Wilkins; 1998. p 116–126.

54. McKenzie RA. The lumbar spine: mechanical diagnosis & therapy. 2nd ed. Volumes 1 and 2. Waikanae, New Zealand: Spinal Publications New Zealand; 2003.

55. Moffroid MT. Endurance of trunk muscles in persons with chronic low back pain: assessment, performance, training. J Rehabil Res Dev 1997;34:440–447.

56. National Research Council and Institute of Medicine. Musculoskeletal disorders and the workplace: low back and upper extremities. Washington, DC: National Academy Press; 2001.

57. Ng JK-F, Richardson CA, Kippers V, Parnianpour M. Relationship between muscle fiber composition and functional capacity of back muscles in healthy subjects and patients with back pain. J Orthop Sport Phys 1998;26:389–402.

58. Nordin M, Welser S, Campello MA, Pietrek M. Self-care techniques for acute episodes of low back pain. Best Pract Res Cl Rh 2002;16:89–104.

59. Oldridge NB, Stoll JE. Spinal disorders and low back pain. In: Skinner JS, editor. Exercise testing and exercise prescription for special cases. 2nd ed. Philadelphia: Lea & Febiger; 1997. p 139–152.

60. Ostelo RWJG, de Vet HCW, Waddell G, Kerckhoffs MR, Leffers P, van Tulder MW. Rehabilitation after lumbar disc surgery. Cochrane Db Syst Rev 2002:CD003007.

61. Pollock ML, Leggett SH, Graves JE, Jones A, Fulton M, Cirulli J. Effective resistance training of lumbar extensor strength. Am J Sport Med 1989;17:624–629.

62. Rainville J, Sobel J, Hartigan C, Monlux G, Bean J. Decreasing disability in chronic back pain through aggressive spine rehabilitation. J Rehabil Res Dev 1997;34:383–393.

63. Ritz S, Lorren T, Simpson S, Mondry T, Comer M. Rehabilitation of degenerative disease of the spine. In: Hochschuler SH, Cotler HB, Guyer RD, editors. Rehabilitation of the spine. Science and practice. St. Louis: Mosby; 1993. p 457–477.

64. Schonstein E, Kenny DT, Keating J, Koes BW. Work conditioning, work hardening and functional restoration for workers with back and neck pain. Cochrane Db Syst Rev 2003;3; CD001822.

65. Smedley J, Inskip H, Cooper C, Coggon D. Natural history of low back pain. A longitudinal study in nurses. Spine 1998;23:2422–2426.

66. Snook SH. Self-care guidelines for the management of nonspecific low back pain. J Occup Rehabil 2004;14:243–253.

67. Snook SH, Webster BS, McGorry RW. The reduction of chronic nonspecific low back pain through the control of early morning lumbar flexion: 3-year follow-up. J Occup Rehabil 2002;12:13–19.

68. Snook SH, Webster BS, McGorry RW, Fogleman MT, McCann KB. The reduction of chronic nonspecific low back pain through the control of early morning lumbar flexion: a randomized controlled trial. Spine 1998;23:2601–2607.

69. Spitzer WO, LeBlanc FE, Dupuis M, Abenhaim L, Belanger AY, Bloch R, Bombardier C, Cruess RL, Drouin G, Duval-Hesler N, Lafamme J, Lamoureux G, Nachemson A, Page JJ, Rossignol M, Salmi LR, Salois-Arsenault S, Suissa S, Wood-Dauphinee S. Scientific approach to the assessment and management of activity-related spinal disorders. A monograph for clinicians. Report of the Quebec Task Force on Spinal Disorders. Spine 1987;12(7S):S5–S59.

70. Thomas E, Silman AJ, Croft PR, Papageorgiou AC, Jayson MIV, Macfarlane GJ. Predicting who develops chronic low back pain in primary care: a prospective study. BMJ 1999;318:1662–1667.

71. Van den Hoogen HJM, Koes BW, van Eijk JTM, Bouter LM, Devillé W. On the course of low back pain in general practice: a one year follow up study. Ann Rheum Dis 1998;57:13–19.

72. Van Tulder M, Becker A, Bekkering T, Breen A, del Real MT, Hutchinson A, Koes B, Laerum E, Malmivaara A. European guidelines for the management of acute nonspecific low back pain in primary care. Eur Spine J 2006;15(Suppl 2):S169–S191.

73. Videman T, Batié MC. A critical review of the epidemiology of idiopathic back pain. In: Weinstein JN, Gordon SL, editors. Low back pain: a scientific and clinical overview. Rosemont, IL: American Academy of Orthopaedic Surgeons; 1996. p 317–332.

74. Vingård E, Nachemson A. Work-related influences on neck and low back pain. In: Nachemson A, Jonsson E, editors. Neck and back pain: the scientific evidence of causes, diagnosis, and treatment. Philadelphia: Lippincott Williams & Wilkins; 2000. p 97–126.

75. Waddell G. The back pain revolution. 2nd ed. Edinburgh: Churchill Livingstone; 2004.

76. Waddell G. Subgroups within "nonspecific" low back pain. J Rheum 2005;32:395–396.

77. Waddell G, Burton AK. Concepts of rehabilitation for the management of low back pain. Best Pract Res Cl Rh 2005;19:655–670.

78. Weisel SW, Feffer HL, Rothman RH. A lumbar spine algorithm. In: Weinstein JN, Wiesel SW, editors. The lumbar spine. Philadelphia: Saunders; 1990. p 358–368.

79. White AA III, Panjabi MK. The clinical biomechanics of spine pain. In: Clinical biomechanics of the spine. 2nd ed, Philadelphia: Lippincott; 1990. p 379–474.

80. Williams PC. Lesions of the lumbosacral spine, part I. J Bone Joint Surg 1937;19:343–363.

81. Williams PC. Lesions of the lumbosacral spine, part II. J Bone Joint Surg 1937;19:690–703.

82. Wyke B. The neurological basis of thoracic spine pain. Rheumatol Phys Med 1970;10:356–367.

CHAPTER 29

1. Aksnes AK, Hjeltnes N, Wahlstrom EO, Katz A, Zierath JR, Wallberg-Henriksson H. Intact glucose transport in morphologically altered denervated skeletal muscle from quadriplegic patients. Am J Physiol 1996;271(3 Pt 1):E593–600.

2. American College of Sports Medicine. Position stand: the recommended quantity and quality of exercise for developing and maintaining cardiorespiratory and muscular fitness in healthy adults. Med Sci Sport Exer 1998;30(6):975–991.

3. Barbeau H, Ladouceur M, Norman KE, Pepin A, Leroux A. Walking after spinal cord injury: evaluation, treatment, and functional recovery. Arch Phys Med Rehab 1999;80(2):225–235.

4. Barbeau H, Ladouceur M, Mirbagheri MM, Kearney RE. The effect of locomotor training combined with functional electrical stimulation in chronic spinal cord injured subjects: walking and reflex studies. Brain Res Rev 2002;40(1–3):274–291.

5. Barstow TJ, Scremin AM, Mutton DL, Kunkel CF, Cagle TG, Whipp BJ. Gas exchange kinetics during functional electrical stimulation in subjects with spinal cord injury. Med Sci Sport Exer 1995;27(9):1284–1291.

6. Barstow TJ, Scremin AM, Mutton DL, Kunkel CF, Cagle TG, Whipp BJ. Peak and kinetic cardiorespiratory responses during arm and leg exercise in patients with spinal cord injury. Spinal Cord 2000;38(6):340–345.

7. Bauman WA, Alexander LR, Zhong Y-G, Spungen AM. Stimulated leg ergometry training improves body composition and HDL-cholesterol values. J Am Paraplegia Soc 1994;17(4):201.

8. Bauman WA, Raza M, Chayes Z, Machac J. Tomographic thallium-201 myocardial perfusion imaging after intravenous dipyridamole in asymptomatic subjects with quadriplegia. Arch Phys Med Rehab 1993;74:740–744.

9. Bauman WA, Spungen AM. Disorders of carbohydrate and lipid metabolism in veterans with paraplegia or quadriplegia: a model of premature aging. Metabolism 1994;43(6):749–756.

10. Bauman WA, Spungen AM. Metabolic changes in persons after spinal cord injury. Phys Med Rehabil Clin N Am 2000;11(1):109–140.

11. Behmran AL, Harkema SJ. Locomotor training after human spinal cord injury: a series of case studies. Phys Ther 2000;80(7):688–700.

12. Behrman AL, Lawless-Dixon AR, Davis SB, et al. Locomotor training progression and outcomes after incomplete spinal cord injury. Phys Ther 2005;85(12):1356–1371.

13. Bernard PL, Mercier J, Varray A, Prefaut C. Influence of lesion level on the cardioventilatory adaptations in paraplegic wheelchair athletes during muscular exercise. Spinal Cord 2000;38(1):16–25.

14. Bloomfield SA, Mysiw WJ, Jackson RD. Bone mass and endocrine adaptations to training in spinal cord injured individuals. Bone 1996;19(1):61–68.

15. Bose P, William W, Telford R, et al. Neuroplasticity following spinal cord injury (SCI) and locomotor training. Paper presented at Society for Neuroscience, 2005; Washington, DC.

16. Braddom RL, Rocco JF. Autonomic dysreflexia: a survey of current treatment. Am J Phys Med Rehab 1991;70:234–241.

17. Bryce TN, Ragnarsson KT. Pain after spinal cord injury. Phys Med Rehabil Clin N Am 2000;11(1):157–168.

18. Bursell JP, Little JW, Stiens SA. Electrodiagnosis in spinal cord injured persons with new weakness or sensory loss: central and peripheral etiologies. Arch Phys Med Rehab 1999;80(8):904–909.

19. Chawla JC, Bar C, Creber I, Price J, Andrews B. Techniques for improving the strength and fitness of spinal cord injured patients. Paraplegia 1979–80;17:185–189.

20. Cooper RA. A force/energy optimization model for wheelchair athletics. IEEE T Syst Man Cyb 1990;20(2):444–449.

21. Cooper RA, Baldini FD, Langbein WE, Robertson RN, Bennett P, Monical S. Prediction of pulmonary function in wheelchair users. Paraplegia 1993;31(9):560–570.

22. Crane L, Klerk K, Ruhl A, Warner P, Ruhl C, Roach KE. The effect of exercise on pulmonary function in persons with quadriplegia. Paraplegia 1994;32:435–441.

23. Curt A, Nitsche B, Rodic B, Schurch B, Dietz V. Assessment of autonomic dysreflexia in patients with spinal cord injury. J Neurol Neurosurg Psychiatry 1997;62(5):473–477.

24. Curtis KA, Black K. Shoulder pain in female wheelchair basketball players. J Orthop Sport Phys 1999;29(4):225–231.

25. Dallmeijer AJ, Hopman MTE, van AS HHJ, van der Woude LHV. Physical capacity and physical strain in persons with tetraplegia, the role of sport activity. Spinal Cord 1996;34:729–735.

26. Davis GM. Exercise capacity of individuals with paraplegia. Med Sci Sport Exer 1993;25(4):423–432.

27. Denis F. The three column spine and its significance in the classification of acute thoracolumbar spinal injuries. Spine 1983;8(8):817–831.

28. DeVivo MJ, Krause JS, Lammertse DP. Recent trends in mortality and causes of death among persons with spinal cord injury. Arch Phys Med Rehab 1999;80(11):1411–1419.

29. Dietz V, Harkema SJ. Locomotor activity in spinal cord-injured persons. J Appl Physiol 2004;96(5):1954–1960.

30. Dixit S. Bradycardia associated with high cervical spinal cord injury. Surg Neurol 1995;43:514.

31. Dobkin B, Apple D, Barbeau H, et al. Weight-supported treadmill vs overground training for walking after acute incomplete SCI. Neurology 2006;66(4):484–492.

32. Faghri PD, Glaser RM, Figoni SF. Functional electrical stimulation leg cycle ergometer exercise: training effects on cardiorespiratory responses of spinal cord injured subjects at rest and during submaximal exercise. Arch Phys Med Rehab 1992;73(11):1085–1093.

33. Ferguson-Pell MW. Technical considerations: seat cushion selection. In: Todd SP, editor. Choosing a wheelchair system. Journal of Rehabilitation Research and Development Clinical Supplement #2. Baltimore: Department of Veterans Affairs, Veterans Health Administration, Rehabilitation Research and Development Service, Scientific and Technical Publications Section; 1992. p 47–73.

34. Figoni SF, Glaser RM, Collins SR. Peak physiologic responses of trained quadriplegics during arm, leg and hybrid exercise in two postures [Abstract]. Med Sci Sport Exer 1995;27(5):S83.

35. Figoni FS, McLain L, Bell AA, et al. Accessibility of physical fitness facilities in the Kansas City metropolitan area. Top Spinal Cord Inj Rehabil 1998;3(3):66–78.

36. Figoni SF, Rodgers MM, Glaser RM, Hooker SP, Feghri PD, Ezenwa BN, Mathews T, Suryaprasad AG, Gupta SC. Physiologic responses of paraplegics and quadriplegics to passive and active leg cycle ergometry. J Am Paraplegia Soc 1990;13(3):33–39.

37. Franklin BA, editor. ACSM's guidelines for exercise testing and prescription. 6th edition. American College of Sports Medicine. Philadelphia: Lippincott Williams & Wilkins; 2000.

38. Franklin BA, Swnatek KI, Grais SL, Johnstone KS, Gordon S, Timmis GC. Field test estimation of maximal oxygen consumption in wheelchair users. Arch Phys Med Rehab 1990;71:574–578.

39. Garland DE, Stewart CA, Adkins RH, Hu SS, Rosen C, Liotta FJ, Weinstein DA. Osteoporosis after spinal cord injury. J Orthop Res 1992;10(3):371–378.

40. Gass GC, Watson J, Camp EM, Court HJ, McPherson LM, Redhead P. The effects of physical training on high level spinal lesion patients. Scand J Rehabil Med 1980;12:61–65.

41. Gater DR, Yates JW, Clasey JL. Relationship between glucose intolerance and body composition in spinal cord injury. Med Sci Sport Exer 2000;32(5):S148.

42. Glaser RM. Functional neuromuscular stimulation. Exercise conditioning of spinal cord injured patients. Int J Sports Med 1994;15(3):142–148.

43. Glaser RM, Janssen TWJ, Suryaprasad AG, Gupta SC, Mathews T. The physiology of exercise. In: Apple DF, editor. Physical fitness: a guide for individuals with spinal cord injury. Baltimore: Department of Veterans Affairs, Veterans Health Administration, Rehabilitation Research & Development Service, Scientific and Technical Publications Section; 1996. p 1–24.

44. Glaser RM, Sawka MN, Brune MF, Wilde SW. Physiological responses to maximal effort wheelchair ergometry and arm crank ergometry. J Appl Physiol 1980;48:1060–1064.

45. Go BK, DeVivo MJ, Richards JS. The epidemiology of spinal cord injury. In: Stover SL, DeLisa JA, Whiteneck GG, editors. Spinal cord injury: clinical outcomes from the model systems. Rockville, MD: Aspen; 1995. p 21–51.

46. Hangartner TN, Rodgers MM, Glaser RM, Barre PS. Tibial bone density loss in spinal cord injured patients: effects of FES exercise. J Rehabil Res Dev 1994;31(1):50–61.

47. Heymsfeld SB, Lichtman S, Baumgartner RN, et al. Body composition of humans: comparison of two improved four-compartment models that differ in expense, technical complexity, and radiation exposure. Am J Clin Nutr 1990;52(1):52–58.

48. Hicks AL, Adams MM, Ginis KM, et al. Long-term body-weight-supported treadmill training and subsequent follow-up in persons with chronic SCI: effects on functional walking ability and measures of subjective well-being. Spinal Cord 2005;43(5):291–298.

49. Hjeltnes N, Jansen T. Physical endurance capacity, functional status and medical complications in spinal cord injured subjects with long-standing lesions. Paraplegia 1990;28:428–432.

50. Hooker SP, Figoni SF, Rodgers MM, Glaser RM, Mathews T, Suryaprasad AG, Gupta SC. Physiologic effects of electrical stimulation leg cycle exercise training in spinal cord injured persons. Arch Phys Med Rehab 1992;73(5):470–476.

51. Hooker SP, Greenwood JD, Hatae DT, et al. Oxygen uptake and heart rate relationship in persons with spinal cord injury. Med Sci Sport Exer 1993;25(10):1115–1119.

52. Janssen TWJ, van Oers CAJM, Rozendaal EP, Willemsen EM, Hollander AP, van der Woude LHV. Changes in physical strain and physical capacity in men with spinal cord injuries. Med Sci Sport Exer 1996;28(5):551–559.

53. Janssen TW, van Oers CA, van Kamp GJ, TenVoorde BJ, van der Woude LH, Hollander AP. Coronary heart disease risk indicators, aerobic power, and physical activity in men with spinal cord injuries. Arch Phys Med Rehab 1997;78(7):697–705.

54. Janssen TWJ, van Oers CAJM, van der Woude LHV, Hollander AP. Physical strain in daily life of wheelchair users with spinal cord injuries. Med Sci Sport Exer 1994;26(6):661–670.

55. Katz RT. Management of spasticity. Am J Phys Med Rehabil 1988;67:108–116.

56. Kavanagh T. Physical training in heart transplant recipients. J Cardiovasc Risk 1996;3(2):154–159.

57. Kawamoto M, Sakimura S, Takasaki M. Transient increase of parasympathetic tone in patients with cervical spinal cord trauma. Anaesth Intensive Care 1993;21:218–221.

58. Knutsson E, Lewenhaupt-Olsson E, Thorsen M. Physical work capacity and physical conditioning in paraplegic patients. Paraplegia 1973;11(3):205–216.

59. Langbein WE, Edwards SC, Louie EK, Hwang MH, Nemchausky BA. Wheelchair exercise and digital echocardiography for the detection of heart disease. Rehabil Res Dev Reports 1996;34:324–325.

60. Langbein WE, Maki KC, Edwards LC, Hwang MH, Sibley P, Fehr L. Initial clinical evaluation of a wheelchair ergometer for diagnostic exercise testing: a technical note. J Rehabil Res Dev 1994;31(4):317–325.

61. Laskin JJ, Ashley EA, Olenik LM, Burnham R, Cumming DC, Steadward RD, Wheeler GD. Electrical stimulation assisted rowing exercise in spinal cord injured people. A pilot study. Paraplegia 1993;31(8):534–541.

62. Lassau-Wray ER, Ward GR. Varying physiological response to arm-crank exercise in specific spinal injuries. J Physiol Anthropol Appl Human Sci 2000;19(1):5–12.

63. Loudon JK, Cagle PE, Figoni SF, Nau KL, Klein RM. A submaximal all-extremity exercise test to predict maximal oxygen consumption. Med Sci Sport Exer 1998;30(8):1299–1303.

64. Lundberg A. Wheelchair driving: evaluation of a new training outfit. Scand J Rehabil Med 1987;12:67–72.

65. Marino RJ, Ditunno JF, Donovan WH, et al. International standards for neurological and functional classification of spinal cord injury, revised 2000. Chicago: American Spinal Injury Association; 2000.

66. Maynard FM, Karunas RS, Adkins RH, Richards JS, Waring III WP. Management of the neuromusculoskeletal systems. In: Stover SL, DeLisa JA, Whiteneck GG, editors. Spinal cord injury: clinical outcomes from the model systems. Bethesda, MD: Aspen; 1995. p 145–169.

67. McLean KP, Skinner JS. Effect of body training position on outcomes of an aerobic training study on individuals with quadriplegia. Arch Phys Med Rehab 1995;76:139–150.

68. Mohr T, Podenphant J, Biering-Sorensen F, Galbo H, Thamsborg G, Kjaer M. Increased bone mineral density after prolonged electrically induced cycle training of paralyzed limbs in spinal cord injured man. Calcif Tissue Int 1997;61(1):22–25.

69. Mutton DL, Scremin AME, Barstow TJ, Scott MD, Kunkel CF, Cagle TG. Physiologic responses during functional electrical stimulation leg cycling and hybrid exercise in spinal cord injured subjects. Arch Phys Med Rehab 1997;78:712–718.

70. Naftchi NE. Mechanism of autonomic dysreflexia. Contributions of catecholamine and peptide neurotransmitters. Ann N Y Acad Sci 1990;579:133–148.

71. Nash MS, Bilsker MS, Kearney HM, Ramirez JN, Applegate B, Green BA. Effects of electrically-stimulated exercise and passive motion on echocardiographically derived wall motion and cardiodynamic function in tetraplegic persons. Paraplegia 1995;33(2):80–89.

72. Nash MS, Bilsker S, Marcillo AE, Isaac SM, Botelho LA, Klose J, Green BA, Rountree MT, Shea JD. Reversal of adaptive left ventricular atrophy following electrically stimulated exercise training in human tetraplegics. Paraplegia 1991;29:590–599.

73. Nash MS, Jacobs PL, Johnson BM, Field-Fote E. Metabolic and cardiac responses to robotic-assisted locomotion in motor-complete tetraplegia: a case report. J Spinal Cord Med 2004;27(1):78–82.

74. Nilsson S, Staff PH, Pruett ED. Physical work capacity and the effect of training on subjects with long-standing paraplegia. Scand J Rehabil Med 1975;7(2):51–56.

75. National Spinal Cord Injury Association. Spinal cord injury statistics. Available at http://www.spinalcord.org/news.php?dep=17&page=94&list=1623 (Last accessed 4/8/08).

76. Petrofsky JS. Thermoregulatory stress during rest and exercise in heat in patients with a spinal cord injury. Eur J Appl Physiol 1992;64(6):503–507.

77. Phillips W, Burkett LN, Munro R, Davis M, Pomeroy K. Relative changes in blood flow with functional electrical stimulation during exercise of the paralyzed lower limbs. Paraplegia 1995;33(2):90–93.

78. Phillips SM, Stewart BG, Mahoney DJ, et al. Body-weight-support treadmill training improves blood glucose regulation in persons with incomplete spinal cord injury. J Appl Physiol 2004;97(2):716–724.

79. Ragnarsson KT, Pollack S, O'Daniel W, Edgar R, Petrofsky J, Nash MS. Clinical evaluation of computerized functional electrical stimulation after spinal cord injury: a multicenter pilot study. Arch Phys Med Rehab 1988;69:672–677.

80. Reame NE. A prospective study of the menstrual cycle and spinal cord injury. Am J Phys Med Rehabil 1992;71:15–21.

81. Rodgers MM, Glaser RM, Figoni SF, Hooker SP, Ezenwa BN, Collins SR, Mathews T, Suryaprasad AG, Gupta SC. Musculoskeletal responses of spinal cord injured individuals to functional neuromuscular stimulation-induced knee extension exercise training. J Rehabil Res Dev 1991;28(4):19–26.

82. Ropper AH. Trauma of the head and spine. In: Isselbacher KJ, Braunwald E, Wilson JD, Martin JB, Fauci AS, Kasper DL, editors. Harrison's principles of internal medicine. 13th ed. St. Louis: McGraw-Hill; 1994.

83. Sawka MN, Foley ME, Pimental NA, Toner MM, Pandolf KB. Determination of maximal aerobic power during upper-body exercise. J Appl Physiol 1983;54:113–117.

84. Schmid A, Huonker M, Barturen JM, Stahl F, Schmidt-Trucksass A, Konig D, Grathwohl D, Lehmann M, Keul J. Catecholamines, heart rate, and oxygen uptake during exercise in persons with spinal cord injury. J Appl Physiol 1998;85(2):635–641.

85. Schmidt KD, Chan CW. Thermoregulation and fever in normal persons and in those with spinal cord injuries. Mayo Clin Proc 1992;67(5):469–475.

86. Schneider DA, Sedlock DA, Gass E, Gass G. $\dot{V}O_2$peak and the gas-exchange anaerobic threshold during incremental arm cranking in able-bodied and paraplegic men. Eur J Appl Physiol Occup Physiol 1999;80(4):292–297.

87. Sharkey BJ, Graetzer DG. Specificity of exercise, training, and testing. In: Durstine JL, King AC, Painter PL, Roitman JL, Zwiren LD, editors. ACSM's resource manual for guidelines for exercise testing and prescription. 2nd ed. Philadelphia: Lea & Febiger; 1993. p 82–92.

88. Sipski ML, Delisa JA, Schweer SA. Functional electrical stimulation bicycle ergometry: patient perceptions. Am J Phys Med Rehabil 1989;68(3):147–149.

89. Spungen AM, Bauman WA, Wang J, Pierson RN. Measurement of body fat in individuals with tetraplegia: a comparison of eight clinical methods. Paraplegia 1995;33(7):402–408.

90. Subarrao JV, Garrison SJ. Heterotopic ossification: diagnosis and management, current concepts and controversies. J Spinal Cord Med 1999;22(4):273–283.

91. Subarrao JV, Klopfstein J, Turpin R. Prevalence and impact of wrist and shoulder pain in patients with spinal cord injury. J Spinal Cord Med 1995;18(1):9–13.

92. Taylor AW, McDonell E, Brassard L. The effects of an arm ergometer training programme on wheelchair subjects. Paraplegia 1986;24:105–114.

93. Tsitouras PD, Zhong YG, Spungen AM, Bauman WA. Serum testosterone and growth hormone/insulin-like growth factor-I in adults with spinal cord injury. Horm Metab Res 1995;27(6):287–292.

94. Ugalde V, Litwiller SE, Gater DR. Physiatric anatomic principles, bladder and bowel anatomy for the physiatrist. Physical Medicine and Rehabilitation: State of the Art Reviews 1996;10:547–568.

95. Van Der Woude LHV, Veeger HEJ, Rozendal RH, Van Ingen Schenau GJ, Rooth F, Van Nierop P. Wheelchair racing: effects of rim diameter and speed on physiology and technique. Med Sci Sport Exer 1988;20(5):492–500.

96. Vinet A, Bernard PL, Poulain M, Varray A, Le Gallais D, Micallef JP. Validation of an incremental field test for the direct assessment of peak oxygen uptake in wheelchair-dependent athletes. Spinal Cord 1996;34(5):288–293.

97. Wang YH, Huang TS. Impaired adrenal reserve in men with spinal cord injury: results of low- and high-dose drenocorticotropin stimulation tests. Arch Phys Med Rehabil 1999;80(8):863–866.

98. Wernig A, Nanassy A, Muller S. Maintenance of locomotor abilities following Laufband (treadmill) therapy in para- and tetraplegic persons: follow-up studies. Spinal Cord 1998;36(11):744–749.

99. Wolff J. Das gesetz der transformation der knochen. Berlin: Ahirshwald; 1892.

100. Young RR. Spasticity: a review. Neurology 1994;44(Suppl 9):S12–S20.

CHAPTER 30

1. Burden of illness of multiple sclerosis: part I: cost of illness. The Canadian Burden of Illness Study Group. Can J Neurol Sci 1998;25:23–30.

2. Multiple Sclerosis Council for Clinical Practice Guidelines. (1998). Fatigue and multiple sclerosis: Evidence-based management strategies for fatigue in multiple sclerosis. Washington, DC: Paralyzed Veterans of America.

3. Multiple sclerosis: hope through research. National Institute of Neurological Disorders and Stroke. http://www.ninds.nih.gov/disorders/multiple_sclerosis/detail_multiple_sclerosis.htm (Last accessed 4/7/08).

4. Stedman's medical dictionary for the health professions and nursing. Baltimore: Lippincott Williams & Wilkins; 2005.

5. Antel J, Bania M, Noronha A, Neely S. Defective suppressor cell function mediated by T8+ cell lines from patients with progressive multiple sclerosis. J Immunol 1986;137:3436–3439.

6. Armstrong LE, Winant DM, Swasey PR, Seidle ME, Carter AL, Gehlsen G. Using isokinetic dynamometry to test ambulatory patients with multiple sclerosis. Phys Ther 1983;63:1274–1279.

7. Arnold DL, Matthews PM. MRI in the diagnosis and management of multiple sclerosis. Neurology 2002;58:S23–31.

8. Bakshi R. Fatigue associated with multiple sclerosis: diagnosis, impact and management. Mult Scler 2003;9:219–227.

9. Benczur M, Petranyi GG, Palffy G, Varga M, Talas M, Kotsy B, Foldes I, Hollan SR. Dysfunction of natural killer cells in multiple sclerosis: a possible pathogenetic factor. Clin Exp Immunol 1980;39:657–662.

10. Bittar RG, Hyam J, Nandi D, Wang S, Liu X, Joint C, Bain PG, Gregory R, Stein J, Aziz TZ. Thalamotomy versus thalamic stimulation for multiple sclerosis tremor. J Clin Neurosci 2005;12:638–642.

11. Bourdette DN, Prochazka AV, Mitchell W, Licari P, Burks J. Health care costs of veterans with multiple sclerosis: implications for the rehabilitation of MS. VA Multiple Sclerosis Rehabilitation Study Group. Arch Phys Med Rehabil 1993;74:26–31.

12. Brown TR, Kraft GH. Exercise and rehabilitation for individuals with multiple sclerosis. Phys Med Rehabil Clin N Am 2005;16:513–555.

13. Carroll CC, Gallagher PM, Seidle ME, Trappe SW. Skeletal muscle characteristics of people with multiple sclerosis. Arch Phys Med Rehabil 2005;86:224–229.

14. Cartlidge NE. Autonomic function in multiple sclerosis. Brain 1972;95:661–664.

15. Chan A, Heck CS. Mobility in multiple sclerosis: more than just a physical problem. Int J MS Care 2000;3:35–40.

16. Chiappa KH. Use of evoked potentials for diagnosis of multiple sclerosis. Neurol Clin 1988;6:861–880.

17. Chiara T, Martin AD, Davenport PW, Bolser DC. Expiratory muscle strength training in persons with multiple sclerosis having mild to moderate disability: effect on maximal expiratory pressure, pulmonary function, and maximal voluntary cough. Arch Phys Med Rehabil 2006;87:468–473.

18. Clifford DB, Trotter JL. Pain in multiple sclerosis. Arch Neurol 1984;41:1270–1272.

19. Compston A. Treatment and management of multiple sclerosis. In: Compston A, Ebers G, Lassmann H, McDonald I, Matthews B, Wekerle H, editors. McAlpine's multiple sclerosis. London: Churchill Livingstone; 1998. p 437–498.

20. Crippa A, Cardini R, Pellegatta D, Manzoni S, Cattaneo D, Marazzini F. Effects of sudden, passive muscle shortening according to Grimaldi's method on patients suffering from multiple sclerosis: a randomized controlled trial. Neurorehabil Neural Repair 2004;18:47–52.

21. Davis SL, Wilson TE, Vener JM, Crandall CG, Petajan JH, White AT. Pilocarpine-induced sweat gland function in individuals with multiple sclerosis. J Appl Physiol 2005;98:1740–1744.

22. de Haan A, de Ruiter CJ, van Der Woude LH, Jongen PJ. Contractile properties and fatigue of quadriceps muscles in multiple sclerosis. Muscle Nerve 2000;23:1534–1541.

23. DeBolt LS, McCubbin JA. The effects of home-based resistance exercise on balance, power, and mobility in adults with multiple sclerosis. Arch Phys Med Rehabil 2004;85:290–297.

24. Duquette P, Pleines J, Girard M, Charest L, Senecal-Quevillon M, Masse C. The increased susceptibility of women to multiple sclerosis. Can J Neurol Sci 1992;19:466–471.

25. Erkonen WE. Radiology 101: the basics and fundamentals of imaging. Philadelphia: Lippincott-Raven; 1998.

26. Ewing C, Bernard CC. Insights into the aetiology and pathogenesis of multiple sclerosis. Immunol Cell Biol 1998;76:47–54.

27. Filippi M, Dousset V, McFarland HF, Miller DH, Grossman RI. Role of magnetic resonance imaging in the diagnosis and monitoring of multiple sclerosis: consensus report of the White Matter Study Group. J Magn Reson Imaging 2002;15:499–504.

28. Fisk JD, Pontefract A, Ritvo PG, Archibald CJ, Murray TJ. The impact of fatigue on patients with multiple sclerosis. Can J Neurol Sci 1994;21:9–14.

29. Foglio K, Clini E, Facchetti D, Vitacca M, Marangoni S, Bonomelli M, Ambrosino N. Respiratory muscle function and exercise capacity in multiple sclerosis. Eur Respir J 1994 ;7:23–28.

30. Frankel DI. Multiple sclerosis. In: Umpherd DA, editor. Neurological rehabilitation. 2nd ed. St. Louis: Mosby; 1990. p 531–550.

31. Freeman J, Allison R. Group exercise classes in people with multiple sclerosis: a pilot study. Physiother Res Int 2004;9:104–107.

32. Garner DJ, Widrick JJ. Cross-bridge mechanisms of muscle weakness in multiple sclerosis. Muscle Nerve 2003;27:456–464.

33. Gehlsen GM, Grigsby SA, Winant DM. Effects of an aquatic fitness program on the muscular strength and endurance of patients with multiple sclerosis. Phys Ther 1984;64:53–657.

34. Gold SM, Schulz KH, Hartmann S, Mladek M, Lang UE, Hellweg R, Reer R, Braumann KM, Heesen C. Basal serum levels and reactivity of nerve growth factor and brain-derived neurotrophic factor to standardized acute exercise in multiple sclerosis and controls. J Neuroimmunol 2003;138:99–105.

35. Green AJ, Barcellos LF, Rimmler JB, Garcia ME, Caillier S, Lincoln RR, Bucher P, Pericak-Vance MA, Haines JL, Hauser SL, Oksenberg JR. Sequence variation in the transforming growth factor-beta1 (TGFB1) gene and multiple sclerosis susceptibility. J Neuroimmunol 2001;116:116–124.

36. Gutierrez GM, Chow JW, Tillman MD, McCoy SC, Castellano V, White LJ. Resistance training improves gait kinematics in persons with multiple sclerosis. Arch Phys Med Rehabil 2005;86:1824–1829.

37. Heesen C, Gold SM, Hartmann S, Mladek M, Reer R, Braumann KM, Wiedemann K, Schulz KH. Endocrine and cytokine responses to standardized physical stress in multiple sclerosis. Brain Behav Immun 2003;17:473–481.

38. Heesen C, Romberg A, Gold S, Schulz KH. Physical exercise in multiple sclerosis: supportive care or a putative disease-modifying treatment. Expert Rev Neurother 2006;6:347–355.

39. Hintzen RQ. Stem cell transplantation in multiple sclerosis: multiple choices and multiple challenges. Mult Scler 2002;8:155–160.

40. Ingram DA, Thompson AJ, Swash M. Central motor conduction in multiple sclerosis: evaluation of abnormalities revealed by transcutaneous magnetic stimulation of the brain. J Neurol Neurosurg Psychiatry 1988;51:487–494.

41. Inman RP. Disability indices, the economic costs of illness, and social insurance: the case of multiple sclerosis. Acta Neurol Scand Suppl 1984;101:46–55.

42. Jacobson DL, Gange SJ, Rose NR, Graham NM. Epidemiology and estimated population burden of selected autoimmune diseases in the United States. Clin Immunol Immunopathol 1997;84:223–243.

43. Johnson KP, Brooks BR, Cohen JA, Ford CC, Goldstein J, Lisak RP, Myers LW, Panitch HS, Rose JW, Schiffer RB, Vollmer T, Weiner LP, Wolinsky JS. Extended use of glatiramer acetate (Copaxone) is well tolerated and maintains its clinical effect on multiple sclerosis relapse rate and degree of disability. Copolymer 1 Multiple Sclerosis Study Group. Neurology 1998;50:701–708.

44. Kantarci OH, de Andrade M, Weinshenker BG. Identifying disease modifying genes in multiple sclerosis. J Neuroimmunol 2002;123:144–159.

45. Kenney WL, Humphrey RH, Bryant CX. ACSM's guidelines for exercise testing and prescription. Baltimore: Williams & Wilkins; 1995.

46. Kent-Braun JA, Ng AV, Castro M, Weiner MW, Gelinas D, Dudley GA, Miller RG. Strength, skeletal muscle composition, and enzyme activity in multiple sclerosis. J Appl Physiol 1997;83:1998–2004.

47. Kent-Braun JA, Sharma KR, Miller RG, Weiner MW. Postexercise phosphocreatine resynthesis is slowed in multiple sclerosis. Muscle Nerve 1994;17:835–841.

48. Kent-Braun JA, Sharma KR, Weiner MW, Miller RG. Effects of exercise on muscle activation and metabolism in multiple sclerosis. Muscle Nerve 1994;17:1162–1169.

49. Kileff J, Ashburn A. A pilot study of the effect of aerobic exercise on people with moderate disability multiple sclerosis. Clin Rehabil 2005;19:165–169.

50. Krupp LB, Coyle PK, Doscher C, Miller A, Cross AH, Jandorf L, Halper J, Johnson B, Morgante L, Grimson R. Fatigue therapy in multiple sclerosis: results of a double-blind, randomized, parallel trial of amantadine, pemoline, and placebo. Neurology 1995;45:1956–1961.

51. Kurtzke JF. Rating neurologic impairment in multiple sclerosis: an expanded disability status scale (EDSS). Neurology 1983;33:1444–1452.

52. Lambert CP, Archer RL, Evans WJ. Muscle strength and fatigue during isokinetic exercise in individuals with multiple sclerosis. Med Sci Sports Exerc 2001;33:1613–1619.

53. Lambert CP, Flynn MG, Braun WA, Mylona E. Influence of acute submaximal exercise on T-lymphocyte suppressor cell function in healthy young men. Eur J Appl Physiol 2000;82:151–154.

54. Lassmann H. Neuropathology in multiple sclerosis: new concepts. Mult Scler 1998;4:93–98.

55. Leuschen MP, Filipi M, Healey K. A randomized open label study of pain medications (naproxen, acetaminophen and ibuprofen) for controlling side effects during initiation of IFN beta-1a therapy and during its ongoing use for relapsing-remitting multiple sclerosis. Mult Scler 2004;10:636–642.

56. Lincoln MR, Montpetit A, Cader MZ, Saarela J, Dyment DA, Tiislar M, Ferretti V, Tienari PJ, Sadovnick AD, Peltonen L, Ebers GC, Hudson TJ. A predominant role for the HLA class II region in the association of the MHC region with multiple sclerosis. Nat Genet 2005;37:1108–1112.

57. Liu C, Blumhardt LD. Randomized, double-blind, placebo-controlled study of subcutaneous interferon beta-1a in relapsing-remitting multiple sclerosis: a categorical disability trend analysis. Mult Scler 2002;8:10–14.

58. Lublin FD, Reingold SC. Defining the clinical course of multiple sclerosis: results of an international survey. National Multiple Sclerosis Society (USA) Advisory Committee on Clinical Trials of New Agents in Multiple Sclerosis. Neurology 1996;46:907–911.

59. Mancardi GL, Sardanelli F, Parodi RC, Melani E, Capello E, Inglese M, Ferrari A, Sormani MP, Ottonello C, Levrero F, Uccelli A, Bruzzi P. Effect of copolymer-1 on serial gadolinium-enhanced MRI in relapsing remitting multiple sclerosis. Neurology 1998;50:1127–1133.

60. Matthews B. Symptoms and signs of multiple sclerosis. In: Compston A, Ebers G, Lassmann H, McDonald I, Matthews B, Wekerle H, editors. McAlpine's multiple sclerosis. 3rd edition. London: Churchill Livingstone; 1998. p 145–190.

61. McAlpine D. Multiple sclerosis: a reappraisal. Edinburgh: Churchill Livingstone; 1972.

62. McDonald I. Pathophysiology of multiple sclerosis. In: Compston A, Ebers G, Lassmann H, McDonald I, Matthews B, Wekerle H. McAlpine's multiple sclerosis. London: Churchill Livingstone; 1998. p 359–378.

63. McIntosh-Michaelis SA, Roberts MH, Wilkinson SM, Diamond ID, McLellan DL, Martin JP, Spackman AJ. The prevalence of cognitive impairment in a community survey of multiple sclerosis. Br J Clin Psychol 1991;30(Pt 4):333–334.

64. Miller A. Diagnosis of multiple sclerosis. Semin Neurol 1998;18:309–316.

65. Monassier L, Brandt CM, Bousquet P. Effects of centrally-acting glutamatergic modulators on cardiovascular responses to stress in humans. J Cardiol 2001;37(Suppl 1):77–84.

66. Mostert S, Kesselring J. Effects of a short-term exercise training program on aerobic fitness, fatigue, health perception and activity level of subjects with multiple sclerosis. Mult Scler 2002;8:161–168.

67. Motl RW, McAuley E, Snook EM. Physical activity and multiple sclerosis: a meta-analysis. Mult Scler 2005;11:459–463.

68. Mulcare JA. Multiple Sclerosis. In: Durstine JL, editor. Exercise management for persons with chronic diseases and disabilities. 1st ed. Champaign, IL: Human Kinetics; 1997. p 189–193.

69. Ng AV, Dao HT, Miller RG, Gelinas DF, Kent-Braun JA. Blunted pressor and intramuscular metabolic responses to voluntary isometric exercise in multiple sclerosis. J Appl Physiol 2000;88:871–880.

70. Ng AV, Kent-Braun JA. Quantitation of lower physical activity in persons with multiple sclerosis. Med Sci Sports Exerc 1997;29:517–523.

71. Nieman DC, Henson DA, Gusewitch G, Warren BJ, Dotson RC, Butterworth DE, Nehlsen-Cannarella SL. Physical activity and immune function in elderly women. Med Sci Sports Exerc 1993;25:823–831.

72. Noronha MJ, Vas CJ, Aziz H. Autonomic dysfunction (sweating responses) in multiple sclerosis. J Neurol Neurosurg Psychiatry 1968;31:19–22.

73. Oken BS, Kishiyama S, Zajdel D, Bourdette D, Carlsen J, Haas M, Hugos C, Kraemer DF, Lawrence J, Mass M. Randomized controlled trial of yoga and exercise in multiple sclerosis. Neurology 2004;62:2058–2064.

74. Olgiati R, Burgunder JM, Mumenthaler M. Increased energy cost of walking in multiple sclerosis: effect of spasticity, ataxia, and weakness. Arch Phys Med Rehabil 1988;69:846–849.

75. Paty DW, Li DK. Interferon beta-1b is effective in relapsing-remitting multiple sclerosis. II. MRI analysis results of a multicenter, randomized, double-blind, placebo-controlled trial. UBC MS/MRI Study Group and the IFNB Multiple Sclerosis Study Group. Neurology 1993;43:662–667.

76. Pepin EB, Hicks RW, Spencer MK, Tran ZV, Jackson CG. Pressor response to isometric exercise in patients with multiple sclerosis. Med Sci Sports Exerc 1996;28:656–660.

77. Petajan JH, Gappmaier E, White AT, Spencer MK, Mino L, Hicks RW. Impact of aerobic training on fitness and quality of life in multiple sclerosis. Ann Neurol 1996;39:432–441.

78. Petajan JH, White AT. Recommendations for physical activity in patients with multiple sclerosis. Sports Med 1999;27:179–191.

79. Pluchino S, Quattrini A, Brambilla E, Gritti A, Salani G, Dina G, Galli R, Del Carro U, Amadio S, Bergami A, Furlan R, Comi G, Vescovi AL, Martino G. Injection of adult neurospheres induces recovery in a chronic model of multiple sclerosis. Nature 2003;422:688–694.

80. Ponichtera JA, Rodgers MA, Glaser RM, Gupta SC. Concentric and eccentric isokinetic lower extremity strength in multiple sclerosis and able bodied. J Orthop Sports Phys Ther 1992;16:114–122.

81. Ponichtera-Mulcare JA, Mathews T, Glaser RM, Gupta SC. Maximal aerobic exercise in ambulatory and semiambulatory patients with multiple sclerosis. Med Sci Sports Exerc 1994;26:S29.

82. Poser CM, Paty DW, Scheinberg L, McDonald WI, Davis FA, Ebers GC, Johnson KP, Sibley WA, Silberberg DH, Tourtellotte WW. New diagnostic criteria for multiple sclerosis: guidelines for research protocols. Ann Neurol 1983;13:227–231.

83. Poser S, Wikstrom J, Bauer HJ. Clinical data and the identification of special forms of multiple sclerosis in 1271 cases studied with a standardized documentation system. J Neurol Sci 1979;40:159–168.

84. Rao SM, Leo GJ, Bernardin L, Unverzagt F. Cognitive dysfunction in multiple sclerosis. I. Frequency, patterns, and prediction. Neurology 1991;41:685–691.

85. Rao SM, Leo GJ, Ellington L, Nauertz T, Bernardin L, Unverzagt F. Cognitive dysfunction in multiple sclerosis. II. Impact on employment and social functioning. Neurology 1991;41:692–696.

86. Rice CL, Vollmer TL, Bigland-Ritchie B. Neuromuscular responses of patients with multiple sclerosis. Muscle Nerve 1992;15:1123–1132.

87. Rodgers MM, Mulcare JA, King DL, Mathews T, Gupta SC, Glaser RM. Gait characteristics of individuals with multiple sclerosis before and after a 6-month aerobic training program. J Rehabil Res Dev 1999;36:183–188.

88. Rodriguez M. Multiple sclerosis: basic concepts and hypothesis. Mayo Clin Proc 1989;64:570–576.

89. Romberg A, Virtanen A, Aunola S, Karppi SL, Karanko H, Ruutiainen J. Exercise capacity, disability and leisure physical activity of subjects with multiple sclerosis. Mult Scler 2004;10:212–218.

90. Romberg A, Virtanen A, Ruutiainen J, Aunola S, Karppi SL, Vaara M, Surakka J, Pohjolainen T, Seppanen A. Effects of a 6-month exercise program on patients with multiple sclerosis: a randomized study. Neurology 2004;63:2034–2038.

91. Rosenblum D, Saffir M. The natural history of multiple sclerosis and its diagnosis. Phys Med Rehabil Clin N Am 1998;9:537–549,v.

92. Rudick RA, Goodkin DE, Jacobs LD, Cookfair DL, Herndon RM, Richert JR, Salazar AM, Fischer JS, Granger CV, Simon JH, Alam JJ, Simonian NA, Campion MK, Bartoszak DM, Bourdette DN, Braiman J, Brownscheidle CM, Coats ME, Cohan SL, Dougherty DS, Kinkel RP, Mass MK, Munschauer FE, Priore RL, Whitham RH, et al. Impact of interferon beta-1a on neurologic disability in relapsing multiple sclerosis. The Multiple Sclerosis Collaborative Research Group (MSCRG). Neurology 1997;49:358–363.

93. Sadovnick AD. Genetic epidemiology of multiple sclerosis: a survey. Ann Neurol 1994;36 (Suppl 2):S194–203.

94. Sadovnick AD, Baird PA, Ward RH. Multiple sclerosis: updated risks for relatives. Am J Med Genet 1988;29:533–541.

95. Sadovnick AD, Ebers GC. Epidemiology of multiple sclerosis: a critical overview. Can J Neurol Sci 1993;20:17–29.

96. Savci S, Inal-Ince D, Arikan H, Guclu-Gunduz A, Cetisli-Korkmaz N, Armutlu K, Karabudak R. Six-minute walk distance as a measure of functional exercise capacity in multiple sclerosis. Disabil Rehabil 2005;27:1365–1371.

97. Schiffer RB. Wineman NM. Antidepressant pharmacotherapy of depression associated with multiple sclerosis. Am J Psychiatry 1990;147:1493–1497.

98. Schmidt S, Barcellos LF, DeSombre K, Rimmler JB, Lincoln RR, Bucher P, Saunders AM, Lai E, Martin ER, Vance JM, Oksenberg JR, Hauser SL, Pericak-Vance MA, Haines JL. Association of polymorphisms in the apolipoprotein E region with susceptibility to and progression of multiple sclerosis. Am J Hum Genet 2002;70:708–717.

99. Schulz KH, Gold SM, Witte J, Bartsch K, Lang UE, Hellweg R, Reer R, Braumann KM, Heesen C. Impact of aerobic training on immune-endocrine parameters, neurotrophic factors, quality of life and coordinative function in multiple sclerosis. J Neurol Sci 2004;225:11–18.

100. Schumacker GA, Beebe G, Kibler RF, Kurland LT, Kurtzke JF, McDowell F, Nagler B, Sibler WA, Tourtellotte WW, Willmon TL. Problems of experimental trials of therapy in multiple sclerosis: report by the panel on the evaluation of experiment trials of therapy in multiple sclerosis. Ann NY Acad Sci 1965;122:552–568.

101. Senaratne MP, Carroll D, Warren KG, Kappagoda T. Evidence for cardiovascular autonomic nerve dysfunction in multiple sclerosis. J Neurol Neurosurg Psychiatry 1984;47:947–952.

102. Sharma KR, Kent-Braun J, Mynhier MA, Weiner MW, Miller RG. Evidence of an abnormal intramuscular component of fatigue in multiple sclerosis. Muscle Nerve 1995;18:1403–1411.

103. Shepherd DI. Clinical features of multiple sclerosis in north-east Scotland. Acta Neurol Scand 1979;60:218–230.

104. Shibasaki H, McDonald WI, Kuroiwa Y. Racial modification of clinical picture of multiple sclerosis: comparison between British and Japanese patients. J Neurol Sci 1981;49:253–271.

105. Silber E, Sharief MK. Axonal degeneration in the pathogenesis of multiple sclerosis. J Neurol Sci 1999;170:11–18.

106. Silby WA. Therapeutic claims in multiple sclerosis: a guide to treatments. New York: Demos Vermande; 1996.

107. Simon JH, Jacobs LD, Campion M, Wende K, Simonian N, Cookfair DL, Rudick RA, Herndon RM, Richert JR, Salazar AM, Alam JJ, Fischer JS, Goodkin DE, Granger CV, Lajaunie M, Martens-Davidson AL, Meyer M, Sheeder J, Choi K, Scherzinger AL, Bartoszak DM, Bourdette DN, Braiman J, Brownscheidle CM, Whitham RH, et al. Magnetic resonance studies of intramuscular interferon beta-1a for relapsing multiple sclerosis. The Multiple Sclerosis Collaborative Research Group. Ann Neurol 1998;43:79–87.

108. Slawta JN, McCubbin JA, Wilcox AR, Fox SD, Nalle DJ, Anderson G. Coronary heart disease risk between active and inactive women with multiple sclerosis. Med Sci Sports Exerc 2002;34:905–912.

109. Sterman AB, Coyle PK, Panasci DJ, Grimson R. Disseminated abnormalities of cardiovascular autonomic functions in multiple sclerosis. Neurology 1985;35:1665–1668.

110. Stimac GK. Introduction to diagnostic imaging. Philadelphia: Saunders; 1992.

111. Surakka J, Romberg A, Ruutiainen J, Aunola S, Virtanen A, Karppi SL, Maentaka K. Effects of aerobic and strength exercise on motor fatigue in men and women with multiple sclerosis: a randomized controlled trial. Clin Rehabil 2004;18:737–746.

112. Svensson B, Gerdle B, Elert J. Endurance training in patients with multiple sclerosis: five case studies. Phys Ther 1994;74:1017–1026.

113. van den Berg M, Dawes H, Wade DT, Newman M, Burridge J, Izadi H, Sackley CM. Treadmill training for individuals with multiple sclerosis: a pilot randomised trial. J Neurol Neurosurg Psychiatry 2006;77:531–533.

114. van der Kamp W, Maertens de Noordhout A, Thompson PD, Rothwell JC, Day BL, Marsden CD. Correlation of phasic muscle strength and corticomotoneuron conduction time in multiple sclerosis. Ann Neurol 1991;29:6–12.

115. Waxman SG. Conduction in myelinated, unmyelinated, and demyelinated fibers. Arch Neurol 1977;34:585–589.

116. Weinshenker BG, Penman M, Bass B, Ebers GC, Rice GP. A double-blind, randomized, crossover trial of pemoline in fatigue associated with multiple sclerosis. Neurology 1992;42:1468–1471.

117. Weinstock-Guttman B, Jacobs LD. What is new in the treatment of multiple sclerosis? Drugs 2000;59:401–410.

118. White AT, Petajan JH. Physiological measures of therapeutic response to interferon beta-1a treatment in remitting-relapsing MS. Clin Neurophysiol 2004;115:2364–2371.

119. White AT, Wilson TE, Davis SL, Petajan JH. Effect of precooling on physical performance in multiple sclerosis. Mult Scler 2000;6:176–180.

120. White LJ, Dressendorfer RH. Exercise and multiple sclerosis. Sports Med 2004;34:1077–1100.

121. White LJ, McCoy SC, Castellano V, Gutierrez G, Stevens JE, Walter GA, Vandenborne K. Resistance training improves strength and functional capacity in persons with multiple sclerosis. Mult Scler 2004;10:668–674.

122. Wishart HA, Roberts DW, Roth RM, McDonald BC, Coffey DJ, Mamourian AC, Hartley C, Flashman LA, Fadul CE, Saykin AJ.

Chronic deep brain stimulation for the treatment of tremor in multiple sclerosis: review and case reports. J Neurol Neurosurg Psychiatry 2003;74:1392–1397.

123. Yates HA, Vardy TC, Kuchera ML, Ripley BD, Johnson JC. Effects of osteopathic manipulative treatment and concentric and eccentric maximal-effort exercise on women with multiple sclerosis: a pilot study. J Am Osteopath Assoc 2002;102:267–275.

CHAPTER 31

1. ICF: international classification of functioning, disability and health. Geneva: World Health Organization; 2001.

2. Cerebral Palsy International Sports and Recreation Association (CP-ISRA). Classification and sports rule manual. 9th ed. www.cpisra.org

3. Albright AL. Spasticity and movement disorders in cerebral palsy. J Child Neurol 1996;11(Suppl 1):S1–4.

4. American College of Sports Medicine. ACSM's exercise management for persons with chronic diseases and disabilities. 2nd ed. Champaign, IL: Human Kinetics; 2003.

5. American College of Sports Medicine. ACSM's guidelines for exercise testing and prescription. 7th ed. Philadelphia: Lippincott Williams & Wilkins; 2006.

6. American College of Sports Medicine. ACSM's resources for clinical exercise physiology: musculoskeletal, neuromuscular, neoplastic, immunologic and hematologic conditions. Philadelphia: Lippincott Williams & Wilkins; 2002. p 304.

7. Andersson C, Grooten W, Hellsten M, Kaping K, Mattsson E. Adults with cerebral palsy: walking ability after progressive strength training. Dev Med Child Neurol 2003;45:220–228.

8. Andersson C, Mattsson E. Adults with cerebral palsy: a survey describing problems, needs, and resources, with special emphasis on locomotion. Dev Med Child Neurol 2001;43:76–82.

9. Ando N, Ueda S. Functional deterioration in adults with cerebral palsy. Clin Rehabil 2000;14:300–306.

10. Arias F, Romero R, Joist H, Kraus FT. Thrombophilia: a mechanism of disease in women with adverse pregnancy outcome and thrombotic lesions in the placenta. J Matern Fetal Med 1998;7:277–286.

11. Ayalon M, Ben-Sira D, Hutzler Y, Gilad T. Reliability of isokinetic strength measurements of the knee in children with cerebral palsy. Dev Med Child Neurol 2000;42:398–402.

12. Barry, MJ. Physical therapy interventions for patients with movement disorders due to cerebral palsy. J Child Neurol 1996;11(Suppl 1):S51–60.

13. Bax DA, Siersema PD, Van Vliet AH, Kuipers EJ, Kusters JG. Molecular alterations during development of esophageal adenocarcinoma. J Surg Oncol 2005;92:89–98, discussion 99.

14. Bax M. Terminology and classification of cerebral palsy. Dev Med Child Neurol 1964;6:295–307.

15. Berg K. Effect of physical activation and of improved nutrition on the body composition of school children with cerebral palsy. Acta Paediatrica Scandinavica Suppl 1970;204.

16. Berg KO, Wood-Dauphinee SL, Williams JI, Maki B. Measuring balance in the elderly: validation of an instrument. Can J Public Health 1992;83(Suppl 2):S7–11.

17. Booth MY, Yates CC, Edgar TS, Bandy WD. Serial casting vs combined intervention with botulinum toxin A and serial casting in the treatment of spastic equinus in children. Pediatr Phys Ther 2003;15:216–220.

18. Bryanton C, Bosse J, Brien M, McLean J, McCormick A, Sveistrup H. Feasibility, motivation, and selective motor control: virtual reality compared to conventional home exercise in children with cerebral palsy. Cyberpsychol Behav 2006;9:123–128.

19. Campbell SK, Vander Linden DW, Palisano RJ, editors. Physical therapy for children. 3rd ed. St. Louis: Saunders Elsevier; 2005.

20. Centers for Disease Control and Prevention. Economic costs associated with mental retardation, cerebral palsy, hearing loss, and vision impairment—United States, 2003. MMWR Morb Mortal Wkly Rep 2004;57–59.

21. Cohen H, CA Blatchly, LL Gombash. A study of the clinical test of sensory interaction and balance. Phys Ther 1993;73:346–351, discussion 351–344.

22. Damiano DL, Kelly LE, Vaughn CL. Effects of quadriceps femoris muscle strengthening on crouch gait in children with spastic diplegia. Phys Ther 1995;75:658–667, discussion 668–671.

23. Darrah J, Wessel J, Nearingburg P, O'Connor M. Evaluation of a community fitness program for adolescents with cerebral palsy. Pediatr Phys Ther 1999;11:18–23.

24. Davis, DW. Review of cerebral palsy, part II: identification and intervention. Neonatal Netw 1997;16:19–25; quiz 26–19.

25. DeLuca PA. The musculoskeletal management of children with cerebral palsy. Pediatr Clin North Am 1996;43:1135–1150.

26. Dodd KJ, Taylor NF, Graham HK. A randomized clinical trial of strength training in young people with cerebral palsy. Dev Med Child Neurol 2003;45:652–657.

27. Duncan PW, Weiner DK, Chandler J, Studenski S. Functional reach: a new clinical measure of balance. J Gerontol 1990;45:M192–197.

28. Eicher PS, Batshaw ML. Cerebral palsy. Pediatr Clin North Am 1993;40:537–551.

29. Fernandez JE, Pitetti KH, Betzen MT. Physiological capacities of individuals with cerebral palsy. Human Factors 1990;32:457–466.

30. Flett P, Saunders B. Ophthalmic assessment of physically disabled children attending a rehabilitation centre. J Paediatr Child Health 1993;29:132–135.

31. Franklin DL, Luther F, Curzon ME. The prevalence of malocclusion in children with cerebral palsy. Eur J Orthod 1996;18:637–643.

32. Freud S. Die infantile cerebrallahmung. Specielle Pathologie und Therapie. Vol. IX. Vienna: Alfred Holder; 1897. 327 p.

33. Gajdosik CG, Cicirello N. Secondary conditions of the musculoskeletal system in adolescents and adults with cerebral palsy. Phys Occup Ther Pediatr 2001;21:49–68.

34. Glanzman AM, Kim H, Swaminathan K, Beck T. Efficacy of botulinum toxin A, serial casting, and combined treatment for spastic equinus: a retrospective analysis. Dev Med Child Neurol 2004;46:807–811.

35. Grether JK, Nelson KB. Maternal infection and cerebral palsy in infants of normal birth weight. JAMA 1997;278:207–211.

36. Grether JK, Nelson KB, Cummins SK. Twinning and cerebral palsy: experience in four northern California counties, births 1983 through 1985. Pediatrics 1993;92:854–858.

37. Jahnsen R, Villien L, Stanghelle JK, Holm I. Fatigue in adults with cerebral palsy in Norway compared with the general population. Dev Med Child Neurol 2003;45:296–303.

38. Keefer DJ, Wayland T, Caputo JL, Apperson K, McGreal S, Morgan DW. Within- and between-day stability of treadmill walking VO_2 in children with hemiplegic cerebral palsy. Stability of walking VO_2 in children with CP. Gait Posture 2005;22:177–181.

39. Kelly M, Darrah J. Aquatic exercise for children with cerebral palsy. Dev Med Child Neurol 2005;47:838–842.

40. Khalili MA. Quantitative sports and functional classification (QSFC) for disabled people with spasticity. Br J Sports Med 2004;38:310–313.

41. King W, Levin R, Schmidt R, Oestreich A, Heubi JE. Prevalence of reduced bone mass in children and adults with spastic quadriplegia. Dev Med Child Neurol 2003;45:12–16.

42. Klingbeil H, Baer HR, Wilson PE. Aging with a disability. Arch Phys Med Rehabil 2004;85:S68–73, quiz S74–65.

43. Koman LA, Smith BP, Shilt JS. Cerebral palsy. Lancet 2004;363:1619–1631.

44. Kraus FT. Cerebral palsy and thrombi in placental vessels of the fetus: insights from litigation. Hum Pathol 1997;28:246–248.

45. Krigger KW. Cerebral palsy: an overview. Am Fam Physician 2006;73:91–100.

46. Liu JM, Li S, Lin Q, Li Z. Prevalence of cerebral palsy in China. Int J Epidemiol 1999;28:949–954.

47. Loudon JK, Cagle PE, Figoni SF, Nau KL, Klein RM. A submaximal all-extremity exercise test to predict maximal oxygen consumption. Med Sci Sports Exerc 1998;30:1299–1303.

48. Lundberg A. Maximal aerobic capacity of young people with spastic cerebral palsy. Dev Med Child Neurol 1978;20:205–210.

49. Lundberg A. Oxygen consumption in relation to work load in students with cerebral palsy. J Appl Physiol 1976;40:873–875.

50. MacKeith RC, MacKenzie ICK, Polani PE. Memorandum on terminology and classification of "cerebral palsy." Cereb Palsy Bull 1959;5:27–35.

51. Marge M. Health promotion for persons with disabilities: Moving beyond rehabilitation. Am J Health Promot 1988;2:29–35.

52. Mathias S, Nayak US, Isaacs B. Balance in elderly patients: the "get up and go" test. Arch Phys Med Rehabil 1986;67:387–389.

53. Mattsson E, Andersson C. Oxygen cost, walking speed, and perceived exertion in children with cerebral palsy when walking with anterior and posterior walkers. Dev Med Child Neurol 1997;39:671–676.

54. Morton R. New surgical interventions for cerebral palsy and the place of gait analysis. Dev Med Child Neurol 1999;41:424–428.

55. Murphy KP, Molnar GE, Lankasky K. Medical and functional status of adults with cerebral palsy. Dev Med Child Neurol 1995;37:1075–1084.

56. Murphy NA, Irwin MC, Hoff C. Intrathecal baclofen therapy in children with cerebral palsy: efficacy and complications. Arch Phys Med Rehabil 2002;83:1721–1725.

57. Mutch L, Alberman E, Hagberg B, Kodama K, Perat MV. Cerebral palsy epidemiology: where are we now and where are we going? Dev Med Child Neurol 1992;34:547–551.

58. Nelson KB, Ellenberg JH. Childhood neurological disorders in twins. Paediatr Perinat Epidemiol 1995;9:135–145.

59. O'Connell DG, Barnhart R. Improvement in wheelchair propulsion in pediatric wheelchair users through resistance training: a pilot study. Arch Phys Med Rehabil 1995;76:368–372.

60. Odding E, Roebroeck ME, Stam HJ. The epidemiology of cerebral palsy: incidence, impairments and risk factors. Disabil Rehabil 2006;28:183–191.

61. Overeynder JC, Turk MA. Cerebral palsy and aging: a framework for promoting the health of older persons with cerebral palsy. Top Geriatr Rehabil 1998;13:19–24.

62. Palisano RJ, Snider LM, Orlin MN. Recent advances in physical and occupational therapy for children with cerebral palsy. Semin Pediatr Neurol 2004;11:66–77.

63. Parker DF, Carriere L, Hebestreit H, Bar-Or O. Anaerobic endurance and peak muscle power in children with spastic cerebral palsy. Am J Dis Child 1992;146:1069–1073.

64. Petersen MC, Palmer FB. Advances in prevention and treatment of cerebral palsy. Ment Retard Dev Disabil Res Rev 2001;7:30–37.

65. Pitetti KH, Fernandez J, Lanciault M. Feasibility of an exercise program for adults with cerebral palsy: a pilot study. Adapt Phys Activ Q 1991;8:333–341.

66. Podsiadlo D, Richardson S. The timed "Up & Go": a test of basic functional mobility for frail elderly persons. J Am Geriatr Soc 1991;39:142–148.

67. Reddihough DS, Collins KJ. The epidemiology and causes of cerebral palsy. Aust J Physiother 2003;49:7–12.

68. Reuben DB, Siu AL. An objective measure of physical function of elderly outpatients. The Physical Performance Test. J Am Geriatr Soc 1990;38:1105–1112.

69. Rimmer JH. Physical fitness levels of persons with cerebral palsy. Dev Med Child Neurol 2001;43:208–212.

70. Rose J, Haskell WL, Gamble JG. A comparison of oxygen pulse and respiratory exchange ratio in cerebral palsied and nondisabled children. Arch Phys Med Rehabil 1993;74:702–705.

71. Rose J, Medeiros JM, Parker R. Energy cost index as an estimate of energy expenditure of cerebral-palsied children during assisted ambulation. Dev Med Child Neurol 1985;27:485–490.

72. Rosenbaum P. Cerebral palsy: what parents and doctors want to know. BMJ 2003;326:970–974.

73. Saito N, Ebara S, Ohotsuka K, Kumeta H, Takaoka K. Natural history of scoliosis in spastic cerebral palsy. Lancet 1998;351:1687–1692.

74. Samson JF, Sie LTL, de Groot L. Muscle power development in preterm infants with periventricular flaring or leukomalacia in relation to outcome at 18 months. Dev Med Child Neurol 2002;44:734–740.

75. Sandstrom K, Alinder J, Oberg B. Descriptions of functioning and health and relations to a gross motor classification in adults with cerebral palsy. Disabil Rehabil 2004;26:1023–1031.

76. Sanger TD, Delgado MR, Gaebler-Spira D, Hallett M, Mink JW. Classification and definition of disorders causing hypertonia in childhood. Pediatrics 2003;111:e89–97.

77. Schlough K, Nawoczenski D, Case LE, Nolan K, Wigglesworth JK. The effects of aerobic exercise on endurance, strength, function and self-perception in adolescents with spastic cerebral palsy: a report of three case studies. Pediatr Phys Ther 2005;17:234–250.

78. Scholtes VA, Becher JG, Beelen A, Lankhorst GJ. Clinical assessment of spasticity in children with cerebral palsy: a critical review of available instruments. Dev Med Child Neurol 2006;48:64–73.

79. Schwartz L, Engel JM, Jensen MP. Pain in persons with cerebral palsy. Arch Phys Med Rehabil 1999;80:1243–1246.

80. Shinohara TA, Suzuki N, Oba M, Kawasumi M, Kimizuka M, Mita K. Effect of exercise at the AT point for children with cerebral palsy. Bull Hosp Jt Dis 2002–2003;61:63–37.

81. Stanley F, Blair E, Alberman E. Cerebral palsies: epidemiology and causal pathways. London: Mac Keith Press; 1999.

82. Stedman TL. Stedman's medical dictionary. 28th ed. Philadelphia: Lippincott Williams & Wilkins; 2006

83. Strauss D, Shavelle R. Life expectancy of adults with cerebral palsy. Dev Med Child Neurol 1998;40:369–375.

84. Taylor NF, Dodd KJ, Graham HK. Test-retest reliability of hand-held dynamometric strength testing in young people with cerebral palsy. Arch Phys Med Rehabil 2004;85:77–80.

85. Taylor NF, Dodd KJ, Larkin H. Adults with cerebral palsy benefit from participating in a strength training programme at a community gymnasium. Disabil Rehabil 2004;26:1128–1134.

86. Thorarensen O, Ryan S, Hunter J, Younkin DP. Factor V Leiden mutation: an unrecognized cause of hemiplegic cerebral palsy, neonatal stroke, and placental thrombosis. Ann Neurol 1997;42:372–375.

87. Thorpe DE, Reilly MA. The effect of aquatic resistive exercise on lower extremity strength, energy expenditure, functional mobility, balance and self-perception in an adult with cerebral palsy: a retrospective case report. Aquatic Physical Therapy 2000;8:18–24.

88. Tinetti ME. Performance-oriented assessment of mobility problems in elderly patients. J Am Geriatr Soc 1986;34:119–126.

89. Tobimatsu Y, Nakamura R, Kusano S, Iwasaki Y. Cardiorespiratory endurance in people with cerebral palsy measured using an arm ergometer. Arch Phys Med Rehabil 1998;79:991–993.

90. Turk MA. Secondary conditions and disability. In: Field MJ, Jette AM, Martin L, editors. Workshop on disability in America. Washington, DC: The National Academies Press; 2006.

91. Turk MA, Geremski CA, Rosenbaum PF, Weber RJ. The health status of women with cerebral palsy. Arch Phys Med Rehabil 1997;78:S10–17.

92. United Cerebral Palsy. Exercise principles and guidelines for persons with cerebral palsy and neuromuscular disorders. www.ucp.org/ucp_channeldoc.cfm/1/15/11500/11500-11500/639

93. Unnithan VB, Dowling JJ, Frost G, Bar-Or O. Role of cocontraction in the O_2 cost of walking in children with cerebral palsy. Med Sci Sports Exerc 1996;28:1498–1504.

94. Van den Berg-Emons RJ, Van Baak MA, Speth L, Saris WH. Physical training of school children with spastic cerebral palsy: effects on daily activity, fat mass and fitness. Int J Rehabil Res 1998;21:179–194.

95. van der Dussen L, Nieuwstraten W, Roebroeck M, Stam HJ. Functional level of young adults with cerebral palsy. Clin Rehabil 2001;15:84–91.

96. Wiley ME, Damiano DL. Lower-extremity strength profiles in spastic cerebral palsy. Dev Med Child Neurol 1998;40:100–107.

97. Williams K, Hennessy E, Alberman E. Cerebral palsy: effects of twinning, birthweight, and gestational age. Arch Dis Child Fetal Neonatal Ed 1996;75:F178–182.

98. Yoon BH, Jun JK, Romero R, Park KH, Gomez R, Choi JH, Kim IO. Amniotic fluid inflammatory cytokines (interleukin-6, interleukin-1beta, and tumor necrosis factor-alpha), neonatal brain white matter lesions, and cerebral palsy. Am J Obstet Gynecol 1997;177:19–26.

CHAPTER 32

1. Health disparities experienced by black or African Americans—United States. MMWR Morb Mortal Wkly Rep 2005;54:1–3.

2. Prevalence of disabilities and associated health conditions among adults—United States, 1999. MMWR Morb Mortal Wkly Rep 2001;50:120–125.

3. Ada L, Dean CM, Hall JM, Bampton J, Crompton S. A treadmill and overground walking program improves walking in persons residing in the community after stroke: a placebo-controlled, randomized trial. Arch Phys Med Rehabil 2003;84:1486–1491.

4. Anderson GL, Limacher M, Assaf AR, Bassford T, Beresford SA, Black H, Bonds D, Brunner R, Brzyski R, Caan B, Chlebowski R, Curb D, Gass M, Hays J, Heiss G, Hendrix S, Howard BV, Hsia J, Hubbell A, Jackson R, Johnson KC, Judd H, Kotchen JM, Kuller L, LaCroix AZ, Lane D, Langer RD, Lasser N, Lewis CE, Manson J, Margolis K, Ockene J, O'Sullivan MJ, Phillips L, Prentice RL, Ritenbaugh C, Robbins J, Rossouw JE, Sarto G, Stefanick ML, Van Horn L, Wactawski-Wende J, Wallace R, Wassertheil-Smoller S. Effects of conjugated equine estrogen in postmenopausal women with hysterectomy: the Women's Health Initiative randomized controlled trial. JAMA 2004;291:1701–1712.

5. Cahalin L, Pappagianopoulos P, Prevost S, Wain J, Ginns L. The relationship of the 6-min walk test to maximal oxygen consumption in transplant candidates with end-stage lung disease. Chest 1995;108:452–459.

6. Cahalin LP, Mathier MA, Semigran MJ, Dec GW, DiSalvo TG. The six-minute walk test predicts peak oxygen uptake and survival in patients with advanced heart failure. Chest 110;1996:325–332.

7. Chu KS, Eng JJ, Dawson AS, Harris JE, Ozkaplan A, Gylfadottir S. Water-based exercise for cardiovascular fitness in people with chronic stroke: a randomized controlled trial. Arch Phys Med Rehabil 2004;85:870–874.

8. Cress ME, Meyer M. Maximal voluntary and functional performance levels needed for independence in adults aged 65 to 97 years. Phys Ther 2003;83:37–48.

9. Dean CM, Richards CL, Malouin F. Task-related circuit training improves performance of locomotor tasks in chronic stroke: a randomized, controlled pilot trial. Arch Phys Med Rehabil 2000;81:409–417.

10. Delahaye N, Cohen-Solal A, Faraggi M, Czitrom D, Foult JM, Daou D, Peker C, Gourgon R, Le Guludec D. Comparison of left ventricular responses to the six-minute walk test, stair climbing, and maximal upright bicycle exercise in patients with congestive heart failure due to idiopathic dilated cardiomyopathy. Am J Cardiol 1997;80:65–70.

11. Duncan PW. Stroke recovery and rehabilitation research. J Rehabil Res Dev 2002;39:ix–xi.

12. Eich HJ, Mach H, Werner C, Hesse S. Aerobic treadmill plus Bobath walking training improves walking in subacute stroke: a randomized controlled trial. Clin Rehabil 2004;18:640–651.

13. Engardt M, Knutsson E, Jonsson M, Sternhag M. Dynamic muscle strength training in stroke patients: effects on knee extension torque, electromyographic activity, and motor function. Arch Phys Med Rehabil 1995;76:419–425.

14. Hulens M, Vansant G, Claessens AL, Lysens R, Muls E. Predictors of 6-minute walk test results in lean, obese and morbidly obese women. Scand J Med Sci Sports 2003;13:98–105.

15. James AH, Bushnell CD, Jamison MG, Myers ER. Incidence and risk factors for stroke in pregnancy and the puerperium. Obstet Gynecol 2005;106:509–516.

16. Johnston SC, Gress DR, Browner WS, Sidney S. Short-term prognosis after emergency department diagnosis of TIA. JAMA 2000;284:2901–2906.

17. Katz-Leurer M, Shochina M, Carmeli E, Friedlander Y. The influence of early aerobic training on the functional capacity in patients with cerebrovascular accident at the subacute stage. Arch Phys Med Rehabil 2003;84:1609–1614.

18. Kelly-Hayes M, Beiser A, Kase CS, Scaramucci A, D'Agostino RB, Wolf PA. The influence of gender and age on disability following ischemic stroke: the Framingham Study. J Stroke Cerebrovasc Dis 2003;12:119–126.

19. Kittner SJ, Stern BJ, Feeser BR, Hebel R, Nagey DA, Buchholz DW, Earley CJ, Johnson CJ, Macko RF, Sloan MA, Wityk RJ, Wozniak MA. Pregnancy and the risk of stroke. N Engl J Med 1996;335:768–774.

20. Laufer Y, Dickstein R, Chefez Y, Marcovitz E. The effect of treadmill training on the ambulation of stroke survivors in the early stages of rehabilitation: a randomized study. J Rehabil Res Dev 2001;38:69–78.

21. Lee IM, Paffenbarger RS Jr. Physical activity and stroke incidence: the Harvard Alumni Health Study. Stroke 1998;29:2049–2054.

22. Liu M, Tsuji T, Hase K, Hara Y, Fujiwara T. Physical fitness in persons with hemiparetic stroke. Keio J Med 2003;52:211–219.

23. Lucas C, Stevenson LW, Johnson W, Hartley H, Hamilton MA, Walden J, Lem V, Eagen-Bengsten E. The 6-min walk and peak oxygen consumption in advanced heart failure: aerobic capacity and survival. Am Heart J 1999;138:618–624.

24. MacKay-Lyons MJ, Makrides L. Cardiovascular stress during a contemporary stroke rehabilitation program: is the intensity adequate to induce a training effect? Arch Phys Med Rehabil 2002;83:1378–1383.

25. Mackay-Lyons MJ, Makrides L. Exercise capacity early after stroke. Arch Phys Med Rehabil 2002;83:1697–1702.

26. Macko RF, Katzel LI, Yataco A, Tretter LD, DeSouza CA, Dengel DR, Smith GV, Silver KH. Low-velocity graded treadmill stress testing in hemiparetic stroke patients. Stroke 1997;28:988–992.

27. Macko RF, Smith GV, Dobrovolny CL, Sorkin JD, Goldberg AP, Silver KH. Treadmill training improves fitness reserve in chronic stroke patients. Arch Phys Med Rehabil 2001;82:879–884.

28. Muntner P, Garrett E, Klag MJ, Coresh J. Trends in stroke prevalence between 1973 and 1991 in the US population 25 to 74 years of age. Stroke 2002;33:1209–1213.

29. Pang MY, Eng JJ, Dawson AS. Relationship between ambulatory capacity and cardiorespiratory fitness in chronic stroke: influence of stroke-specific impairments. Chest 2005;127:495–501.

30. Pang MY, Eng JJ, Dawson AS, McKay HA, Harris JE. A community-based fitness and mobility exercise program for older adults with chronic stroke: a randomized, controlled trial. J Am Geriatr Soc 2005;53:1667–1674.

31. Pohl M, Mehrholz J, Ritschel C, Ruckriem S. Speed-dependent treadmill training in ambulatory hemiparetic stroke patients: a randomized controlled trial. Stroke 2002;33:553–558.

32. Potempa K, Lopez M, Braun LT, Szidon JP, Fogg L, Tincknell T. Physiological outcomes of aerobic exercise training in hemiparetic stroke patients. Stroke 1995;26:101–105.

33. Rimmer JH, Riley B, Creviston T, Nicola T. Exercise training in a predominantly African-American group of stroke survivors. Med Sci Sports Exerc 2000;32:1990–1996.

34. Salbach NM, Mayo NE, Wood-Dauphinee S, Hanley JA, Richards CL, Cote R. A task-orientated intervention enhances walking distance and speed in the first year post stroke: a randomized controlled trial. Clin Rehabil 2004;18:509–519.

35. Stewart DG. Stroke rehabilitation. 1. Epidemiologic aspects and acute management. Arch Phys Med Rehabil 1999;80:S4–7.

36. Teixeira-Salmela LF, Olney SJ, Nadeau S, Brouwer B. Muscle strengthening and physical conditioning to reduce impairment and disability in chronic stroke survivors. Arch Phys Med Rehabil 1999;80:1211–1218.

37. Thom T, Haase N, Rosamond W, Howard VJ, Rumsfeld J, Manolio T, Zheng ZJ, Flegal K, O'Donnell C, Kittner S, Lloyd-Jones D, Goff DC Jr, Hong Y, Adams R, Friday G, Furie K, Gorelick P, Kissela B, Marler J, Meigs J, Roger V, Sidney S, Sorlie P, Steinberger J, Wasserthiel-Smoller S, Wilson M, Wolf P. Heart disease and stroke statistics—2006 update: a report from the American Heart Association Statistics Committee and Stroke Statistics Subcommittee. Circulation 2006;113:e85–151.

38. Tsuji T, Liu M, Tsujiuchi K, Chino N. Bridging activity as a mode of stress testing for persons with hemiplegia. Arch Phys Med Rehabil 1999;80:1060–1064.

39. Wassertheil-Smoller S, Hendrix SL, Limacher M, Heiss G, Kooperberg C, Baird A, Kotchen T, Curb JD, Black H, Rossouw JE, Aragaki A, Safford M, Stein E, Laowattana S, Mysiw WJ. Effect of estrogen plus progestin on stroke in postmenopausal women: the Women's Health Initiative: a randomized trial. JAMA 2003;289:2673–2684.

40. Williams LS, Ghose SS, Swindle RW. Depression and other mental health diagnoses increase mortality risk after ischemic stroke. Am J Psychiatry 2004;161:1090–1095.

INDEX

A

abdominal obesity 181
abnormal eating behaviors 150
abnormal muscle tone 564
Aboyans, V. 335
absorptiometry 128, 487-488
ACE inhibitors 37t, 40, 240, 290, 322
acquired immunodeficiency syndrome. *See* AIDS
acromegaly 58
activities of daily living, with spinal cord injury 528
acute coronary syndromes, metabolic agents 38-41
acute treatment, myocardial infarction 289
Adams, G.M. 104
Ades, P.A. 62, 296
adiposity 113, 182-183
adjunctive therapy 435
adolescents. *See* children and adolescents
advanced lung disease, exercise recommendation 409-410
aerobic capacity 18, 116, 119-121, 121-122, 123t
aerobic exercise
    and blood lipid profile 259-260, 261
    and blood pressure 240-241
    cerebral palsy patients 580
    children and adolescents 112-113
    cystic fibrosis patient tolerance 408
    for elderly 145
    high-intensity negative effects 430
    HIV-infected patients 453-454
    nonspecific low-back pain 504-505, 512, 514
    peripheral arterial disease patients 343
    plasticity in children 128
    post-myocardial infarction patients 295
    spinal cord injury 535, 537
    stroke patients 587
aerobic testing, cystic fibrosis patients 410, 412
age
    arthritis prevalence 461
    associated physiological changes 137t
    basic to clinical knowledge 62
    cancer frequency 426
    effects on organ systems 136-137t
    nonspecific low-back pain 498, 499
    and obesity 214
    peripheral arterial disease 333, 345
    spinal cord injury 523
    transient ischemic attacks 584-585
agility training 493-494
AIDS
    about 443-444
    disease prevention 455
    medical and clinical considerations 446-447, 449-450
    metabolic agents 45-46
    progression from HIV 455
    scope 444-445
airways
    airway hyperresponsiveness 391, 392, 408
    airway with obstruction 372f
    inflammation with asthma 391
    normal airway 372f
    obstruction of 372-373
    remodeling 393f

alcohol consumption 238, 426, 431
aldosterone blockers 41
Alexander, Frederick M. 509t
Alexander, L.E. 151
alkylating agents 43, 45
allergies 64
Alpert, B.S. 113
alpha-adrenergic blocking agents 40
alpha/beta-adrenergic blocking agents 40
alpha-glucosidase inhibitors 34
alveolar hypoventilation 413
alveolar ventilation 413
alveolar walls 373
Alzheimer's disease 138, 141
amenorrhea
    about 160
    bone loss with 491
    evaluation of athletic amenorrhea 160
    in female athlete triad 149, 150
    in female-specific issues 151-155
    laboratory tests for diagnosis 161t
    low energy availability hypothesis 154-155
    physical and psychological stress hypothesis 154
    potential causes in athletes 154t
American Academy of Orthopedic Surgeons 12
American Association of Cardiovascular and Pulmonary Rehabilitation
    about 12
    cardiac patients and resistance training 296
    exercise testing guidelines 287, 376
    functional capacity assessment 80
    pulmonary rehabilitation 383
    revascularization patients exercise guidelines 307
American Cancer Society 12, 432, 433t, 435
American College of Cardiology
    about 12
    cardiorespiratory endurance testing 116
    exercise testing precautions 287
    "Guidelines for Implantation of Cardiac Pacemakers and Antiarrhythmia Devices" 364
    heart failure stages 319
    indications for graded exercise testing 79
    pacemakers 351-352
American College of Chest Physicians 383
American College of Rheumatology 463, 464t, 481
American College of Sports Medicine
    aerobic capacity testing 120
    aerobic exercise prescription 130
    blood pressure guidelines 273
    cardiac patients and resistance training 296
    certification 8, 10
    diabetes 197
    exercise physiologist definition 4
    exercise prescription 95-96, 99, 101, 511
    exercise progression guidelines 415
    exercise testing guidelines 287, 376, 415, 432, 488, 503, 507, 514, 587
    female athlete triad 149-150
    graded exercise testing 80
    Guidelines for Exercise Testing and Prescription 129, 259, 337, 451

health and fitness promotion 514
hypertension 237, 238, 241, 243
inaccurate $\dot{V}O_2$max measurements in elderly 466, 481
Medicine and Science in Sports and Exercise monthly journal 12
metabolic syndrome 183-184
obesity 215
one-repetition maximum 572
physical activity benefits 18
physical activity for elderly 141-142
physical examination of elderly 138
post-myocardial infarction patients 295, 296
quality of life emphasis 415
resistance training 103, 104, 453, 578-579, 580
revascularization patients exercise guidelines 307
safety for exercise 502
training guidelines 518
American Council on Exercise 4, 8
American Diabetes Association 12, 192, 200
American Heart Association
    about 12
    cardiac patients and resistance training 296
    cardiorespiratory endurance testing 116
    exercise testing guidelines 77-78, 253, 287
    graded exercise testing guidelines 79
    "Guidelines for Implantation of Cardiac Pacemakers and Antiarrhythmia Devices" 364
    heart failure stages 319
    myocardial infarction 281
    obesity 215
    pacemakers 351-352
    physical activity for elderly 141-142
    post-myocardial infarction patient risks 287
    resistance training guidelines 241
    revascularization patients exercise guidelines 307
    trans fatty acids 255
American Orthopedic Society for Sports Medicine 123
American Society of Bone and Mineral Research 491
American Society of Exercise Physiologists 4, 8
American Spinal Injury Association 523, 524t, 531
Americans with Disabilities Act 535
American Thoracic Society 371, 375, 380
amputation, and diabetes 208
anabolic agents 48
anabolic steroids 56-57, 157, 165
anaerobic activities, cystic fibrosis patients 418
anaerobic capacity, testing in children and adolescents 122-123
Andersson, C. 580
Ando, N. 562
Andreyeva, T. 214
androgenic effects 56
anemia 31-32, 73
aneurysm 93
angina 35, 38, 89t, 282, 289
angina pectoris 283
angina threshold 364

ABOUT THE EDITORS

Jonathan K. Ehrman

Paul M. Gordon

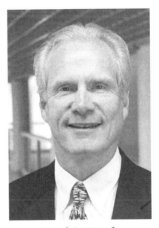

Paul S. Visich

Steven J. Keteyian

Jonathan K. Ehrman, PhD, FACSM, is associate program director of preventive cardiology and director of the weight management program at Henry Ford Hospital in Detroit. He has a 22-year background in clinical exercise physiology and is certified as an ACSM clinical exercise specialist and program director. He recently served for three years as the chair of the exercise specialist credentialing committee for ACSM and is currently a member at large of the credentialing board.

Dr. Ehrman is an author of over 100 manuscripts and abstracts, four books and chapters, and is senior editor of the sixth edition of *ACSM's Resource Manual for Guidelines for Graded Exercise Testing and Training.* He is a fellow of the American College of Sports Medicine and a member of the American Association of Cardiovascular and Pulmonary Rehabilitation, and the American Heart Association. Dr. Ehrman earned his PhD in clinical exercise physiology from The Ohio State University in Columbus.

Paul M. Gordon, PhD, MPH, FACSM, is an ACSM clinical exercise specialist and director of the Laboratory for Physical Activity and Health Related Research in the School of Medicine at the University of Michigan in Ann Arbor. He has 15 years of teaching experience in clinical exercise physiology curricula and has directed several cardiopulmonary rehabilitation programs. Dr. Gordon has served as an examiner and coordinator for the ACSM exercise specialist certification and is a contributing author for the *ACSM's Guidelines for Exercise Testing and Prescription.*

Dr. Gordon is an American College of Sports Medicine fellow, a fellow of the Centers for Disease Control Physical Activity Research Program, and a National Institutes for Health Study Section member. He earned his PhD in exercise physiology and an MPH in epidemiology from the University of Pittsburgh.

Paul S. Visich, PhD, MPH, has more than 14 years of experience in clinical exercise physiology and is the director of the Human Performance Laboratory in the College of Health Professions at Central Michigan University. He worked 12 years in a clinical setting that included cardiac and pulmonary rehabilitation and primary disease prevention. His research interests involve the assessment of cardiovascular disease risk factors in children, the influence of resistance training in elderly populations, and altitude physiology.

Dr. Visich is a member of the Registered Clinical Exercise Physiology Committee and previous chair for the Professional Education Committee for the American College of Sports Medicine. He is the author of more than 70 published scientific articles and abstracts. He earned a PhD in exercise physiology and an MPH in epidemiology from the University of Pittsburgh.

Steven J. Keteyian, PhD, FACSM, has more than 30 years of experience working as a clinical exercise physiologist. He is program director of preventive cardiology at the Henry Ford Hospital in Detroit. Over the course of his career, Dr. Keteyian has focused on exercise, physical activity, and health in both healthy individuals and those with chronic diseases. He is the author of more than 55 scientific articles and chapters in books and four textbooks.

Dr. Keteyian is a member of the American Association of Cardiovascular and Pulmonary Rehabilitation and the American Heart Association. He is a fellow of the American College of Sports Medicine. He earned his PhD from Wayne State University in Detroit.